Springer

Berlin
Heidelberg
New York
Barcelona
Hong Kong
London
Milan
Paris
Singapore
Tokyo

H.G. Baumgarten · M. Göthert (Editors)

Serotoninergic Neurons and 5-HT Receptors in the CNS

With 101 Figures and 35 Tables

 Springer

Professor Dr. med. H. G. Baumgarten, M.D.
Freie Universität Berlin
Institut für Anatomie
Königin-Luise-Str. 15
14195 Berlin
Germany

Professor Dr. med. M. Göthert
Rheinische Friedrich-Wilhelms-Universität Bonn
Institut für Pharmakologie und Toxikologie
Reuterstr. 2b
53113 Bonn
Germany

ISBN-13: 978-3-540-66715-5   e-ISBN: 978-3-642-60921-3

DOI: 10.1007/978-3-642-60921-3

Second Printing 1999 (Originally published in the Handbook of Experimental
Pharmacology: Volume 129, ISBN 3-540-62666-2, 1997)

Cataloging-in-Publication Data applied for

Serotoninergic neurons and 5-HT receptors in the CNS / Ed.: H. G. Baumgarten ; M. Göthert. – 2., print. –
Berlin ; Heidelberg ; New York ; Barcelona ; Hongkong ; London ; Mailand ; Paris ; Singapur ; Tokio :
Springer, 2000

© Springer-Verlag Berlin Heidelberg 2000

Softcover reprint of the hardcover 1st Edition 2000

Coverdesign: design & production, Heidelberg

SPIN: 10750275   52/3020 - 5 4 3 2 1 0 - Printed on acid-free paper

Preface

In 1966, Vittorio Erspamer, the editor of the first monograph within the Handbook of Experimental Pharmacology series that dealt with 5-hydroxy-tryptamine (Vol. XIX) wrote in the Preface: "In a decade of intense and dedicated work, an immense amount of experimental data has been collected, the significance of which . . . has spread into all fields of biology and medicine . . .". Three decades later, a dramatic further increase in our knowledge of all neuroscientific aspects of the serotoninergic system in the CNS has been achieved, and we are witnessing persisting interest in the biology of serotonin of scientists working in various basic biomedical and clinical disciplines. The scientific advance was made possible by significant improvements in analytical and morphological technologies of high sensitivity and cellular/subcellular resolution (e.g., patch clamp techniques, in vivo microdialysis, electrophysiological recording techniques, quantitative ligand autoradiography, immunohistochemistry, cytochemistry, catalytic enzyme histochemistry, and PET/SPECT techniques) and in molecular biology (e.g., in situ hybridization, PCR cloning, DNA transfection studies, and targeted gene disruption). Particular progress has been made in the anatomy, physiology, and pharmaco'ogy of serotoninergic neurons and their modulatory role in brain functions, in serotonin receptors and their transduction mechanisms, and in the potential role of serotonin in neuropsychiatric diseases, such as eating disorders, antisocial personality disorders, obsessive-compulsive disorder, seasonal affective disorder, and major depression.

Serotoninergic neurons and mechanisms deserve continuing attention because 5-HT neurons are obligatory architectural components of neuronal networks in all presently existing forms of vertebrates and invertebrates, and because serotoninergic neurons provide neuronal networks with modulatory capabilities. They are essential for basic properties of the nervous system, namely development and adult neuroplasticity, long-term control of gene expression for peptides in target neurons, control of electrical excitability, maintenance of coherency of network functioning in response to afferent input and gating of tonic and phasic motor activity and timing of motor coordination during waking. Further functions significantly influenced by serotoninergic neurons are neuroendocrine secretion and autonomic tone, adaptation of behavior in response to external and internal demand, and facilitation of social behavior by restraining of irritability and aggression.

The present monograph contains 28 contributions on basic biological and applied medical aspects of central serotoninergic neurons and their receptors in the vertebrate CNS. Chapters 1–3 deal with the development, adult plasticity, anatomy, projections, target cell relationship, and physiology of 5-HT neurons. Chapters 4–6 cover recent findings on the molecular biology and regulation of tryptophan hydroxylase and of serotonin-storing vesicles. Chapters 7–19 deal with the distribution, molecular biology, transductional, electrophysiological and pharmacological characteristics of 5-HT receptors and their functional role. Chapters 20–24 focus on the role of 5-HT receptors in hormone secretion, in the action of hallucinogens, in acute effects of alcohol and the development of alcoholism, in the pathophysiology of migraine, and in nociception. Chapters 25 and 26 review and discuss recent data on the molecular biology, and the distribution and function of monoamine oxidase and of the serotonin transporter protein. Chapters 27 and 28 are devoted to animal models of integrated serotoninergic functions and the psychiatric use of drugs which interact with the serotonin transporter and with serotonin receptor subtypes.

Berlin   H.G. BAUMGARTEN
Bonn   M. GÖTHERT

List of Contributors

AGHAJANIAN, G.K., Departments of Psychiatry and Pharmacology,
Yale University School of Medicine and the Abraham Ribicoff Research
Facilities, Connecticut Mental Health Center, 34 Park Street,
New Haven, CT 06508, USA

ANDRADE, R., Department of Psychiatry, Wayne State University School
of Medicine, 2309 Scott Hall, 540 E. Canfield, Detroit, MI 48201,
USA

AZMITIA, E.C., New York University, Department of Biology and Center
for Neural Science, 1009 Main Building, 100 Washington Square East,
New York, NY 10003-6688, USA, (email ecal@is2.NYU.edu)

BAUMGARTEN, H.G., Freie Universität Berlin, Universitätsklinikum
Benjamin Franklin, Institut für Anatomie,
Königin-Luise-Str. 15, D-14195 Berlin, Germany

BENJAMIN, D., Division of Neuropharmacology, Center of Alcohol Studies,
Rutgers University, Allison Rd. Bldg. 3570, Piscataway, N.J. 08855-0969,
USA

BLIER, P., Neurobiological Psychiatry Unit, Department of Psychiatry,
Research and Training Building, McGill University,
1033 avenue des Pins Ouest, Montreal, QC, Canada H3A 1A1

BOCKAERT, J., CNRS UPR 9023, "Mécanismes Moléculaires des
Communications Cellulaires", CCIPE, 141, Rue de la Cardonille,
F-34094 Montpellier Cedex 5, France

BRANCHEK, T.A., Department of Pharmacology, Synaptic Pharmaceutical
Corporation, 215 College Road, Paramus, NJ 07652-1410, USA

BRILEY, M., Institut de Recherche Pierre Fabre, Parc Industrial
de la Chartreuse, F-81100 Castres, France

CASTANON, N., Center for Neurobiology and Behavior, College of Physicians
& Surgeons of Columbia University, 722 West, 168th Street,
10032 New York, NY, USA

CHEN, K., Department of Molecular Pharmacology and Toxicology,
University of Southern California, School of Pharmacy, 1985 Zonal
Avenue, Los Angeles, CA 90033, USA

CHOPIN, P., Centre de Recherche Pierre Fabre, 17, Avenue Jean Moulin,
F-81100 Castres, France

COSTALL, B., Postgraduate Studies in Pharmacology, The School of Pharmacy,
University of Bradford, Bradford, West Yorkshire BD7 1DP,
United Kingdom

DE MONTIGNY, C., Neurobiological Psychiatry Unit,
Department of Psychiatry, Research and Training Building,
McGill University, 1033 avenue des Pins Ouest, Montreal, QC,
Canada H3A 1A1

DUMUIS, A., CNRS UPR 9023, "Mécanismes Moléculaires des
Communications Cellulaires", CCIPE, 141, Rue de la Cardonille,
F-34094 Montpellier Cedex 5, France

EDWARDS, R.H., Departments of Neurology and Physiology,
UCSF School of Medicine, Third and Parnassus Avenues, San Francisco,
CA 94143-0435, USA

FAGNI, L., CNRS UPR 9023, "Mécanismes Moléculaires des
Communications Cellulaires", CCIPE, 141, Rue de la Cardonille,
F-34094 Montpellier Cedex 5, France

FILLION, G., Institut Pasteur, Unité de Pharmacologie
Neuroimmunoendocrinienne, 28, rue du Dr. Roux,
F-75724 Paris Cedex 15, France

FILLION, M.P., Institut Pasteur, Unité de Pharmacologie
Neuroimmunoendocrinienne, 28, rue du Dr. Roux,
F-75724 Paris Cedex 15, France

FINN III, J.P., Departments of Neurology and Physiology,
UCSF School of Medicine, Third and Parnassus Avenues, San Francisco,
CA 94143-0435, USA

FORNAL, C.A., Princeton University, Program in Neuroscience, Green Hall,
Princeton, NJ 08544-1113, USA

GERSHON, M.D., Department of Anatomy and Cell Biology,
Columbia University, College of Physicians and Surgeons,
630 W. 168th Street, New York, NY 10032, USA

GÖTHERT, M., Rheinische Friedrich-Wilhelms-Universität Bonn,
Institut für Pharmakologie und Toxikologie, Reuterstr. 2b,
D-53113 Bonn, Germany

GRIMSBY, J., Department of Molecular Pharmacology and Toxicology, University of Southern California, School of Pharmacy, 1985 Zonal Avenue, Los Angeles, CA 90033, USA

GROZDANOVIC, Z., Freie Universität Berlin, Universitätsklinikum Benjamin Franklin, Institut für Anatomie, Königin-Luise-Str. 15, D-14195 Berlin, Germany

HAMON, M., INSERM U288, Neurogiologie Cellulaire et Fonctionelle, Faculté de Médecine Pitié-Salpêtriére, 91, Boulevard de l'Hopital, F-75634 Paris Cedex 13, France

HARTIG, P.R., DuPont Merck, CNS Diseases Research, Experimental Station, P.O. Box 80400, Wilmington, DE 19880-0440, USA

HEN, R., Center for Neurobiology and Behavior, College of Physicians & Surgeons of Columbia University, 722 West, 168th Street, 10032 New York, NY, USA

HOYER, D., Novartis Pharma Ltd., Nervous System Research, S-386/745, CH-4002 Basel, Switzerland

HYDE, E.G., Departments of Biochemistry and Psychiatry, Case Western Reserve University, School of Medicine, 11100 Euclid Avenue, Cleveland, OH 44106-5000, USA

JACOBS, B.L., Princeton University, Program in Neuroscience, Green Hall, Princeton, NJ 08544-1113, USA

JOH, T.H., Laboratory of Molecular Neurobiology, Cornell University Medical College at The Winifred Masterson Burke Medical Research Institute, Inc., 785 Mamaroneck Avenue, White Plains, NY 10605, USA

KEANE, P.E., Sanofi Recherche, 195, Route D'Espagne, B.P. 1169, F-31036 Toulouse, France

LÊ, A.D., Department of Pharmacology, University of Toronto and Addiction Research Foundation, 33 Russe Street, Toronto, Ontario, Canada M5S 2S1

LESCH, K.-P., Department of Psychiatry, University of Würzburg, Füchsleinstr. 15, D-97080 Würzburg, Germany

LIU, Y., Departments of Neurology and Physiology, UCSF School of Medicine, Third and Parnassus Avenues, San Francisco, CA 94143-0435, USA

MARIEN, M., Centre de Recherche Pierre Fabre, 17, Avenue Jean Moulin, F-81100 Castres, France

MASSOT, O., Institut Pasteur, Unité de Pharmacologie Neuroimmunoendocrinienne, 28, rue du Dr. Roux, F-75724 Paris Cedex 15, France

MENGOD, G., Department of Neurochemistry, IIBB, CSIC, Jordi Girona 18.26, E-08034 Barcelona, Spain

MERICKEL, A., Departments of Neurology and Physiology, UCSF School of Medicine, Third and Parnassus Avenues, San Francisco, CA 94143-0435, USA

MORET, C., Centre de Recherche Pierre Fabre, 17, Avenue Jean Moulin, F-81100 Castres, France

MOSKOWITZ, M.A., Stroke and Neurovascular Regulation, Massachusetts General Hospital, Harvard Medical School, 149 13th Street, Room 6403, Charlestown, MA 02129, USA

NAYLOR, R.J., Postgraduate Studies in Pharmacology, The School of Pharmacy, University of Bradford, Bradford, West Yorkshire BD7 1DP, United Kingdom

NICHOLS, D.E., Department of Medicinal Chemistry and Molecular Pharmacology, School of Pharmacy and Pharmacal Sciences, Purdue University, 1333 Robert E. Heine Pharmacy Building, West Lafayette, IN 47907-1333, USA

PALACIOS, J.M., Research Center, Laboratorios Almirall, E-08034 Barcelona, Spain

PETER, D., Rockefeller University, 1280 York Avenue, Box 304, New York, NY 10021, USA

PRUDHOMME, N., Institut Pasteur, Unité de Pharmacologie Neuroimmunoendocrinienne, 28, rue du Dr. Roux, F-75724 Paris Cedex 15, France

RAMBOZ, S., Center for Neurobiology and Behavior, College of Physicians & Surgeons of Columbia University, 722 West, 168th Street, 10032 New York, NY, USA

ROMACH, M.K., Department of Psychiatry and Centre for Research in Women's Health, University of Toronto, 76 Grenville Street, Toronto, Ontario, Canada M5S 1B2

ROTH, B.L., Departments of Biochemistry and Psychiatry, Case Western Reserve University, School of Medicine, 11100 Euclid Avenue, Cleveland, OH 44106-5000, USA

ROUSSELLE, J.C., Institut Pasteur, Unité de Pharmacologie
Neuroimmunoendocrinienne, 28, rue du Dr. Roux,
F-75724 Paris Cedex 15, France

SAUDOU, F., Institut de Génétique et de Biologie Moléculaire et Cellulaire
du CNRS-LGME, U184 de l'INSERM, Parc d'Innovation, B.P.-163,
F-67404 Illkirch Cedex, France

SAWYNOK, J., Department of Pharmacology, Dalhousie University,
Sir Charles Tupper Building, Halifax, NS, Canada B3H 4H7

SCHLICKER, E., Rheinische Friedrich-Wilhelms-Universität Bonn,
Institut für Pharmakologie und Toxikologie, Reuterstr. 2b,
D-53113 Bonn, Germany

SELLERS, E.M., Department of Pharmacology, Medicine and Psychiatry and
Centre for Research in Women's Health, University of Toronto,
76 Grenville Street, Toronto, Ontario, Canada M5S 1B2

SHIH, J.C., Department of Molecular Pharmacology and Toxicology,
University of Southern California, School of Pharmacy,
1985 Zonal Avenue, Los Angeles, CA 90033, USA

SOUBRIÉ, P., Sanofi Recherche, 371 rue du Professeur J. Blayac,
F-34184 Montpellier, France

TAMIR, H., Division of Neuroscience, New York State Psychiatric Institute,
722 West 168th Street, New York, NY 10032, USA

TOTH, M., Department of Pharmacology, Cornell University, Medical College,
1300 York Avenue, Room LC519, New York, NY 10021, USA

VAN DE KAR, L.D., Department of Pharmacology,
Loyola University of Chicago, Stritch School of Medicine,
2160 South First Avenue, Maywood, IL 60153, USA

WAEBER, C., Stroke and Neurovascular Regulation,
Massachusetts General Hospital, Harvard Medical School,
149 13th Street, Room 6403, Charlestown, MA 02129, USA

WIEDERHOLD, D., Novartis Pharma Ltd., Nervous System Research,
S-386/862, CH-4002 Basel, Switzerland

WHITAKER-AZMITIA, P.M., Department of Psychiatry,
State University of New York, Health Science Center, Stony Brook,
NY 11794, USA. (pwhitaker@epo.som.sunysb.edu)

ZGOMBICK, J.M., Department of Pharmacology,
Synaptic Pharmaceutical Corporation, 215 College Road, Paramus,
NJ 07652-1410, USA

Contents

CHAPTER 4

Tryptophan Hydroxylase: Molecular Biology and Regulation
T.H. JOH. With 4 Figures 117

CHAPTER 5

Molecular Analysis of Serotonin Packaging into Secretory Vesicles
D. PETER, J.P. FINN III, A. MERICKEL, Y. LIU, and R.H. EDWARDS ... 131

CHAPTER 6

**Regulation of the Environment of the Interior of
Serotonin-Storing Vesicles**

CHAPTER 7

**Molecular Biology and Transductional Characteristics of
5-HT Receptors**

CHAPTER 8

5-Hydroxytryptamine Receptor Histochemistry: Comparison of Receptor mRNA Distribution and Radioligand Autoradiography in the Brain
G. MENGOD, J.M. PALACIOS, K.H. WIEDERHOLD,
and D. HOYER. With 5 Figures

CHAPTER 9

The Main Features of Central 5-HT$_{1A}$ Receptor

CHAPTER 10

**Functional Neuropharmacology of Compounds Acting at
5-HT$_{1B/D}$ Receptors**
M. BRILEY, P. CHOPIN, M. MARIEN, and C. MORET. With 1 Figure 269

CHAPTER 11

**5-HT-Moduline, an Endogenous Peptide Modulating Serotoninergic
Activity via a Direct Interaction at 5-HT$_{1B/1D}$ Receptors**

CHAPTER 12

**Regulation of 5-HT Release in the CNS by Presynaptic 5-HT
Autoreceptors and by 5-HT Heteroreceptors**

CHAPTER 13

Behavioral Consequences of 5-HT$_{1B}$ Receptor Gene Deletion
N. Castanon, S. Ramboz, F. Saudou, and R. Hen. With 7 Figures ...   351

CHAPTER 14

Pharmacology of 5-HT$_2$ Receptors
B.L. Roth and E.G. Hyde. With 12 Figures   367

CHAPTER 15

Regulation of 5-HT$_{2A}$ Receptor at the Molecular Level
M. Toth and D. Benjamin   395

CHAPTER 16

Neuropharmacology of 5-HT$_3$ Receptor Ligands
B. Costall and R.J. Naylor   409

CHAPTER 17

5-HT$_4$ Receptors: An Update

CHAPTER 18

Molecular Biology and Potential Functional Role of 5-HT$_5$, 5-HT$_6$, and 5-HT$_7$ Receptors

T.A. BRANCHEK and J.M. ZGOMBICK. With 3 Figures 475

CHAPTER 19

Electrophysiology of 5-HT Receptors
G.K. AGHAJANIAN and R. ANDRADE. With 7 Figures . . . .

CHAPTER 20

**5-HT Receptors Involved in the Regulation of
Hormone Secretion**
L.D. VAN DE KAR. With 2 Figures . . . .

CHAPTER 21

Role of Serotoninergic Neurons and 5-HT Receptors in the Action of Hallucinogens

CHAPTER 22

The Interaction of the Serotoninergic and Other Neuroregulatory Systems in Alcohol Dependence

CHAPTER 27

Animal Models of Integrated Serotoninergic Functions:
Their Predictive Value for the Clinical Applicability of Drugs
Interfering with Serotoninergic Transmission
P.E. KEANE and P. SOUBRIÉ 707

CHAPTER 28

CHAPTER 1

Development and Adult Plasticity of Serotoninergic Neurons and Their Target Cells

E.C. AZMITIA and P.M. WHITAKER-AZMITIA

A. Introduction

Twenty-five years of work on the development and neuroplasticity of the serotonin system has revealed many specific parallels between these two states. The trophic factors involved in developmental processes appear to be similar to those involved in adult neuroplasticity and the maintenance of neural connections. We once spoke of the trophic role serotonin plays in brain development prior to the time it assumes its role as a neurotransmitter in the mature brain. Now we speak of the role serotonin plays throughout the life of the brain, as a factor regulating both the maintenance and plasticity of mature neuronal and glial phenotypes.

Serotonin (5-HT) functions as a major trophic factor at many developmental stages. It can influence embryogenesis in many organisms (starfish, amphibians, rat) by influencing maturation of the oocyte and regulating cleavage division (BUZNIKOV 1984; BUZNIKOV et al. 1993). Endocrine and mast cells within the lung, spleen, kidney, GI tract, reproductive organs, salivary glands, and heart contain serotonin (GARATTINI and VALZELLI 1965). In these areas, 5-HT serves as a maturation, growth and differentiating factor (DI SANT'AGNESE and CROCKETT 1994). 5-HT is known to stimulate endothelial cells that line the blood vessels (PAKALA et al. 1994), ependymal cells that line the ventricles (CHAN-PALAY 1976), and mesodermal cells that compose the craniofacial region (MOISEIWITSCH and LAUDER 1995).

In brain, the neurons that produce the neurotransmitter serotonin make up one of the most widely distributed neuronal systems in the mammalian brain. This neuronal network is also one of the earliest systems to develop. The turnover rate of serotonin is highest in the immature mammalian brain. High-affinity serotoninergic receptors peak during perinatal periods in both rodents and primates. This wide distribution, precocious development, and high level of activity are the keys to the trophic role of serotonin in the brain.

This review will first describe the development of the serotoninergic neurons and serotoninergic receptors (Sect. B.I). Next, we cover the plasticity of serotoninergic neurons in a damaged brain (Sect. B.II). In the second part of the review, we focus on the role of serotonin as a trophic factor. Serotonin's regulation of the maturation of its target cells occurs in the immature brain as well as the mature brain. Finally, we offer a concept of the dynamic instability of neurons. This new concept provides an explanation for serotonin's regulation of the fluctuations between mature and immature neuronal phenotypes.

B. Serotonin Growth

I. Immature Brain

1. Neurons

The anatomical and neurochemical development of the serotonin neuronal system has been most extensively studied in the rat (Olson and Seiger 1972; Lidov and Moliver 1982c; Lauder et al. 1982, 1985; Lauder 1990; Artigas et al. 1985), although the development has been described in several species, including man (Nobin and Bjorklund 1973), chick (Ahmad and Zamenhof 1978), leech (Stuart et al. 1987), sheep (Richards et al. 1990), *Xenopus* (Messenger and Warner 1989), fruitfly (Valles and White 1988), cat (Gu et al. 1990), and mouse (Ni and Jonakait 1989; Hohmann et al. 1988). There is a surprisingly similar developmental pattern across species. In virtually all cases, serotoninergic neurons are one of the earliest cell types to express a specific neurotransmitter. Moreover, all of these species show that peak amounts of serotonin occur early in development, and are normally not reached again in adulthood.

In rats, serotonin-containing cells are evident by gestational day (GD) 12 and their proliferation occurs up to GD 18 (Wallace and Lauder 1983). Through varying degrees of migration, these cells develop into the adult raphe nuclei by GD 18 (Lidov and Molliver 1982a). Although axonal projections are evident as early as GD 13, the subsequent development of the axons is very complex. In many cases, serotonin terminals overshoot their final innervation density, and are removed (D'Amato et al. 1987; Rhoades et al. 1990). In other regions, serotonin terminals are found during development in areas from which they are virtually absent in the adult (Leslie et al. 1992). Adult-like innervation of the cerebral cortex is extremely heterogeneous and may continue up to postnatal (PD) 21 (Lidov and Molliver 1982a). Not only do the serotonin terminal numbers change, but the percentage of terminals making classical synaptic connections also decreases (Crissman et al. 1993). As the serotoninergic neurons develop, the amount of serotonin produced, released, and metabolized also increases rapidly. The enzyme which is responsible for the synthesis of serotonin, tryptophan hydroxylase, is virtually saturated by the very high levels of tryptophan. During development, serotonin is transiently expressed in cells which may eventually not be serotonin neurons (Ni and Jonakait 1989; Becquet et al. 1990). During the whole period that serotonin terminals develop, they express high amounts of the growth-associated protein GAP-43 (Bendotti et al. 1991), a key element in development and plasticity.

Descending pathways also begin to develop almost immediately after the cells are differentiated, with the caudalmost regions of the spinal cord being reached by GD 17. However, as with the ascending projections, there is a period of overproduction of terminals, with the final adult pattern being reached at PD 21 (RAJAOFETRA et al. 1989).

In the human, serotoninergic neurons are first evident by 5 weeks of gestation (SUNDSTROM et al. 1993) and undergo a pronounced increase by 10 weeks of gestation (KONTUR et al. 1993; SHEN et al. 1989; LEVALLOIS et al. 1993). Serotonin levels peak in the early years of life, before 2 years of age, and then decline to adult levels by 5 years of age, when they are approximately one-sixth of those seen in the young child (HEDNER et al. 1986; TOTH and FEKETE 1986). The diurinal rhythm in serotonin is present immediately after birth (ATTANASIO et al. 1986).

2. Receptors

In order for immature serotoninergic neurons to show neurotrophic activity, there must be serotoninergic receptors in the developing brain. Although seven receptor families have been identified in the mature brain, their occurrence and function in the developing brain have not been well studied. The pharmacology and second messenger linkage of immature receptors may not be identical to those of the adult, and thus the receptors are difficult to characterize (WHITAKER-AZMITIA 1991). There may also be developmentally determined splice variants in the receptor. In spite of the difficulties, a clear picture of the role of various receptors is emerging. The principle receptors involved in brain development appear to be the 5-HT_{1A}, 5-HT_{1B}, and 5HT_2 receptors. These seven transmembrane, G-protein receptors will be discussed in the relevant sections below.

The 5-HT_{1A} receptor is intronless and is termed a "transiently expressed receptor" – that is at specific times in development very high amounts are expressed which then decrease as the animal ages. This peak in receptor number has been shown in rats (DAVAL et al. 1987), humans (BAR-PELED et al. 1991), and sheep (RICHARDS et al. 1990). In human fetal tissue, the peak is found between 16 and 22 weeks of gestation (BAR-PELED et al. 1991).

In rats, the 5-HT_{1A} receptor is first evident at GD 12 in brain stem and increases to a peak at GD 15 after which time it declines and is never as high again (HILLION et al. 1994). The peak in later developing target regions, such as cerebellum (DAVAL et al. 1987) and visual cortex (DYCK and CYNADER 1993), occurs later. In the serotonin cell line, RN46A, the 5-HT_{1A} receptor is 20-fold higher in the undifferentiated cell than in the differentiated cell. The authors suggest that the cell body 5-HT_{1A} receptors may mediate autoregulation of serotonin development (EATON et al. 1995; see below). Although the 5-HT_{1A} receptor is the principle neurotrophic receptor, it also has other developmentally specific behavioral roles. In the 1st week of neonatal life, this receptor can cause behavioral activation, decreased suckling (KIRSTEIN and SPEAR 1988), and decreased calling for the mother (WINSLOW and INSEL 1991). The 5-HT_{1A} receptor also regulates the respiratory rhythm, measurable as early as GD 18

(DiPASQUALE et al. 1994). This may have interesting implications for sudden infant death syndrome (SIDS), where serotonin deficits are reported (SPARKS and HUNSAKER 1991; MAURIZI 1985).

The 5-HT$_{1B}$ receptor has never been described in fetal tissue; however, it is present by early neonatal life (PRANZATELLI and GALVAN 1994). Thus, while a release-regulating autoreceptor appears to affect fetal serotonin development (see below), it is not identical to the autoreceptor of the adult brain, the 5-HT$_{1B}$ autoreceptor. The receptor, however, shows a transient expression in the neonatal period and may be important in the cortex for pattern and barrel field formation (LESLIE et al. 1992). Recent studies have shown the presence of the 5-HT$_{1B}$ receptor on transiently expressed serotonin terminals in rat somatosensory cortex and visual cortex. The fact that these receptors are in the regions where serotonin terminals hyperinnervate and are later removed suggests 5-HT autoregulation is present in the cortex postnatally. Behaviorally, 4-day-old rat pups respond to m-chlorophenylpiperazine (mCPP) by changes in mouthing and probing behaviors, as well as showing a general behavioral activation (KIRSTEIN and SPEAR 1988).

The predominant 5-HT$_2$ receptor in the neonatal period is the 5-HT$_{2C}$ receptor, changing to the 5-HT$_{2A}$ as the animal ages (IKE et al. 1995). The 5-HT$_2$ receptor-mediated production of inositol phosphates is approximately tenfold higher in the immature brain than in the adult brain (CLAUSTRE et al. 1988). This time period is too late to influence serotonin autoregulation of fiber outgrowth; however, the receptor may play a role in branching, terminal sprouting, synaptogenesis, mitogenesis, and glycogen breakdown (see Sect. E). In addition, the receptor may play a role in determining the number of glucocorticoid receptors (MEANEY et al. 1994). The 5-HT$_2$ receptor can be referred to as a programmable receptor – that is events during development may affect the number, affinity, or function of these receptors in the adult brain. For example, both prenatal and postnatal stress on the mother significantly increase the number of 5-HT$_2$ receptors in the offspring, even after they have become adults (PETERS 1988).

Other 5-HT receptors have been suggested to have trophic activity during development. The 5-HT$_3$ receptor serves as an ionophore for K^+/Ca^{2+}. It has two different splice variants expressed over the course of development (MIQUEL et al. 1995) and influences spinal cord development (BELL et al. 1992). The 5-HT$_{5B}$ receptor colocalizes with the gene producing the reeler mutation in mice (MATTHES et al. 1993). Future research could thus focus on these and other areas.

II. Mature Brain

This chapter focuses on the plasticity of serotoninergic neurons. In the adult, the developmental processes of serotonin outgrowth, overgrowth, and terminal elimination are still present and are referred to as plasticity. Plasticity of adult brain and spinal cord is a relatively new concept, for at one time the neural system was considered rigid and incapable of change. However, in the last 25 years firm morphological data have accumulated supporting the existence of 5-HT growth in the adult brain.

Serotonin sprouting is induced by a variety of conditions in the adult brain. Moreover, in a manner analogous to the overgrowth of developing serotonin terminals, serotonin sprouts often exceed normal levels. Serotonin sprouting can be of three types, depending on the means by which it is induced.

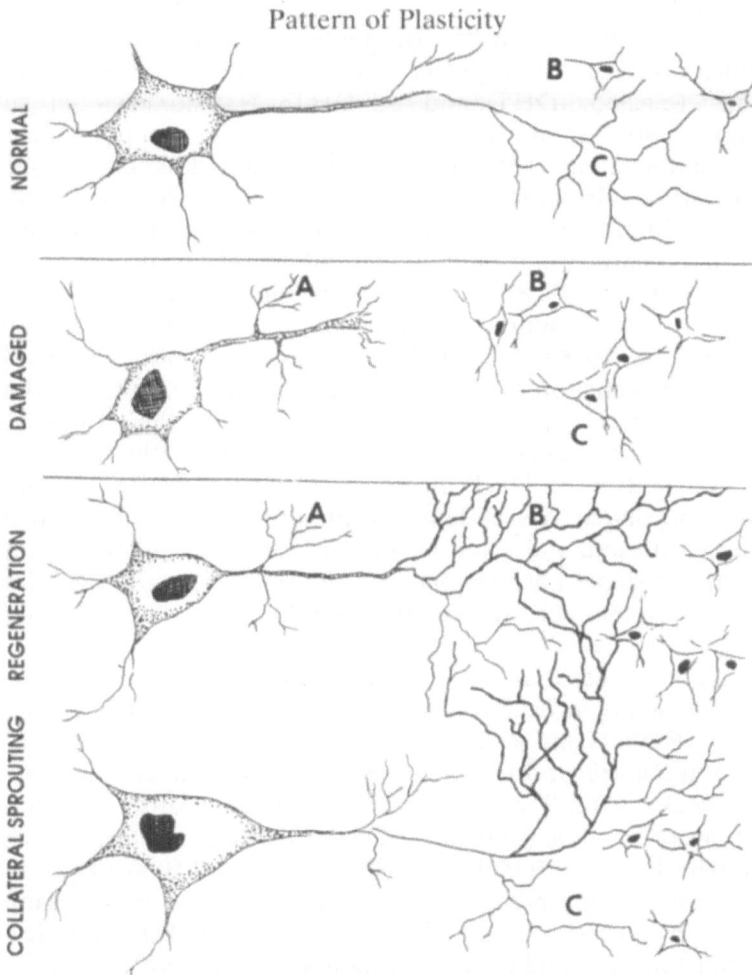

Pattern of Plasticity

Fig. 1. Three patterns of plasticity are shown: proximal sprouting (damming), regeneration, and collateral sprouting. *In the top panel,* a normal neuron grows into regions B and C. After damage to the axon, sprouts emerge proximal to the injury (region A). Regenerating axons hyperinnervate a more proximal region (B), while hypoinnervating a more distal target (C). A neighboring neuron of the same (homotypic) or different (heterotypic) chemical type sprouts into the region (C) vacated by the injured neuron

Regenerative sprouting is sprouting from a damaged serotonin neuron. When the damaged fibers do not reach their original target, neighboring serotonin terminals can sprout (homotypic collateral sprouting). If a neighboring neuron of a different transmitter specificity is damaged, the serotonin neurons can sprout in its place (heterotypic sprouting). These types of sprouting are discussed in the following sections and are shown in Fig. 1.

1. Regeneration

Serotonin fibers can be damaged by physical or chemical methods. There are a number of serotonin-specific chemical toxins, including dihydroxytryptamines (e.g., 5,6-DHT and 5,7-DHT; BAUMGARTEN et al. 1971, 1973; BJORKLUND et al. 1981) and substituted amphetamines [e.g., para-chloroamphetamine (PCA) and 3,4-methylenedioxymethamphetamine (MDMA); SANDERS-BUSH et al. 1975; MAMOUNAS and MOLLIVER 1988; FISCHER et al. 1995]. Within 24 h of a lesion, the proximal tips of the lesioned 5-HT fibers become enlarged and bulbous, resembling classical degenerating profiles (RAMON Y CAJAL 1928). The stumps produce numerous fine processes within several days (BJORKLUND et al. 1981; FRANKFURT and AZMITIA 1983; ZHOU and AZMITIA 1986). These enlarged stumps and processes have a morphological resemblance to the growth cones and filipodia of developing neurons. In many of these regenerating serotonin neurons, the growth-associated protein GAP-43 is expressed once again(ALONSO et al. 1995).

One month after 5,6-DHT injections into adult rats, 5-HT fluorescent fibers from the bulbar descending inferior group show innervation of the inferior olivary nucleus and several cranial motor nuclei and by 8 months grow down the entire spinal cord (BJORKLUND et al. 1981). The 5-HT fibers that reinnervate the spinal cord re-establish functional regulation (NYGREN et al. 1974).

The ascending serotoninergic groups can also show regeneration. In the rat hypothalamus, direct intracerebral injections of 5,7-DHT lead to a rapid loss of 5-HT-IR fibers and a corresponding increase in female sexual activity. By 1 month, the levels of 5-HT increase to 50% of control and reach control levels by 8 weeks. After 9 weeks there is no significant difference in the types of appositions or synaptic frequency made by the regenerating serotoninergic axons compared to the original fibers (FRANKFURT and BEAUDET 1987). These regenerating serotoninergic axons grow back to some of the same nuclei (dorsomedial hypothalamus and zona incerta), re-establish synaptic and neurotransmitter function, and restore hormonally dependent behavior.

Serotoninergic fibers are capable of both retraction and extension in the adult mammalian brain, but do not always re-establish their original distribution. In the rat, 5-HT fibers damaged by 5,6-DHT or 5,7-DHT can hyperinnervate the brain stem olivary nuclei by 7–8 months (WIKLUND and BJORKLUND 1980) and the hypothalamus by 9 weeks (FRANKFURT and AZMITIA 1983), but distant targets remain hypoinnervated. MDMA given to monkeys produces a substantial degeneration of fibers throughout the forebrain (FISCHER et al. 1995). The regeneration of the damaged sprouts hyper-innervates many targets proximal to the medial forebrain bundle (MFB), such as the amgydala and hypothalamus, but areas such as the frontal cortex are hypoinnervated even after 18 months. The overshoot over normal innervation levels is similar to the hyperinnervation produced by neonatal 5-HT fibers in the spinal cord and cortex. In the rat neonate and adult, this hyperinnervation is transient (AXT et al. 1994). There are morphological and functional consequences of this unbalanced and dynamic regeneration (see Sect. D).

2. Homotypic Collateral Sprouting

Undamaged 5-HT neurons can also sprout in the adult brain. In the case of homotypic sprouting, serotoninergic neurons expand their normal innervation area and substitute for the damaged 5-HT fibers (AZMITIA et al. 1978). In this type of plasticity, normal fiber density and function can be restored. This plastic response was first seen in the CNS of adult rats after 5,7-DHT selective destruction of the serotoninergic hippocampal afferents traveling in the cingu-lum bundle (CB) and monitoring the induced expansion of the undamaged serotoninergic fibers traveling in the fornix-fimbria (FF). After a 2-week delay, the 5-HT fibers in the FF begin to grow towards the neural targets left by the lesioned CB 5-HT fibers and within 42 days the normal innervation density and behavior are restored. Hyperinnervation is evident after 90 days (AZMITIA et al. 1978). When 5-HT fibers are specifically damaged, there is a period of time (2–3 weeks) before new 5-HT collateral sprouts from undamaged neu-rons are formed (AZMITIA and ZHOU 1986). A soluble factor from hippocam-pal extract produces increased trophic activity in culture 2 weeks after 5,7-DHT lesions (AZMITIA et al. 1991a). This is in marked contrast to the production of nerve growth factor (NGF), which increases within a few days and peaks in the hippocampus at 2 weeks post-lesion (CRUTCHER and COLLINS 1986). The slow increase in the level of neurotrophic activity is due, at least in part, to the expression of a soluble trophic factor by the target cells (the astroglial factor, S-100β) and the induction of the receptor signal for release (astroglial 5-HT$_{1A}$ receptors).

3. Heterotypic Collateral Sprouting

The phenomenon of heterotypic collateral sprouting (also referred to as reac-tive synaptogenesis) is faster than homotypic collateral sprouting, occurring within days and being essentially complete by 2 weeks (see AZMITIA and ZHOU 1986; COTMAN and LYNCH 1976). The chemical specificity of the lesioned fiber is critical for homotypic collateral sprouting but not for reactive synaptogenesis. In addition, the proximity of the fiber to the site of the lesion is critical in heteroypic sprouting, whereas in homotypic collateral sprouting the collateral sprouts can travel long distances (up to 1 cm).

Removal of serotoninergic fibers does not induce heterotypic sprouting from other nonserotoninergic neurons. However, serotoninergic fibers quickly expand into denervated sites formerly occupied by other types of neurons. The best-studied example is the serotonin sprouting after neonatal removal of dopamine fibers in the striatum (see Sect. C.I.2). Serotoninergic fibers can hyperinnervate a sensory target area after damage to the peripheral neurons. Both neonatal and adult eye removal increases 5-HT-IR fiber sprouting in the upper stratum griseum of the hamster superior colliculus (RHOADES et al. 1990). In these animals, the degree of sprouting is greater in the adult than in the neonate. In experimentally induced degranulation in the cerebellum, the 5-HT innervation is reorganized (SOTELO and BEAUDET 1979).

These studies indicate that 5-HT fibers can respond to removal of heterotypic fibers; however, the serotoninergic sprouts do not assume the function of the lost systems even when the sprouting occurs in the neonate (READER et al. 1995). Rather the fibers appear to invade the area targeting the astrocytes formed in response to the degenerating neurons, possibly releasing S-100β (see Sect. C.II.1). The heterotypic sprouting by serotoninergic neurons forms typical serotonin connections in both pattern and ultrastructure (LIDOV and MOLLIVER 1982b; DESCARRIES et al. 1992).

III. Transplantation: Mature and Immature Interactions

The most direct way to compare the trophic abilities of immature fetal cells and mature adult cells is by brain transplantation. The integration of cells of different maturational states uncovers neurotrophic aspects of both the immature and mature serotoninergic system. The first transplantation studies inserted fetal serotoninergic neurons into a cavity in the entorhinal cortex of the rat (BJORKLUND et al. 1976). The fetal 5-HT fibers grew into the hippocampal areas vacated by the degenerating cortical fibers. In later studies, the fetal 5-HT neurons were successfully microinjected into the hippocampus of adult and aged rats (for review see JACOBS and AZMITIA 1992). The 5-HT fetal neurons mature normally and show proper electrophysiological and pharmacological responses. In the adult hypothalamus, raphe fetal transplants can restore both the 5-HT fiber loss and the behavioral effects on female sexual activity produced by the 5,7-DHT lesion (LUINE et al. 1984). In the adult hippocampus, fetal raphe transplants reduce cognitive deficits induced by loss of 5-HT and cholinergic afferents (SEGAL and RICHTER-LEVIN 1989). Thus, neuronotrophic interactions occur between cells of different ages and these interactions are mutually beneficial to the morphology and function of both cell types. This is relevant to the aged brain, where the endogenous serotoninergic fibers are retracting and degenerating (see Sect. E.II.2).

Over time, the fetal neurons hyperinnervate nearby areas which they normally innervate. Hyperinnervation of target areas by fetal serotoninergic neurons has been demonstrated by morphological, neurochemical, and functional studies (see AZMITIA et al. 1990a, review). In fact, prior removal of endogenous 5-HT fibers greatly enhances the speed and magnitude of the hyperinnervation. Transplants made into 5,7-DHT-lesioned animals show a dramatic increase in high-affinity uptake (345% over control hippocampus) and 5-HT levels (225% over control hippocampus). The hyperinnervation by serotoninergic fibers is maintained for at least 5 months throughout the dorsal part of the hippocampus. The ability to hyperinnervate occurs during development, after damage to adult neurons and when immature neurons are introduced into an adult target area. Thus, the final innervation density produced by individual 5-HT raphe neurons and sustained by the target region is not rigidly set at a given point (programmed), but can be modified throughout life.

C. Trophic Factors

In the first section, we described serotonin development and plasticity in both the immature and mature brains and spinal cords. Now we present various molecules that can regulate serotoninergic growth. These range in size from small ions to proteins (reviewed by AzMITIA et al. 1988). However, knowing that a molecule has trophic activity does not provide a mechanism for how the molecule is regulated nor how it exerts its action. These questions are essential and provide insights into how individual sets of neurons can thrive or wither in the neural environment. The final level of analysis will be the study of the interactions between trophic factors since normally trophic factors will not act in isolation.

I. Neurotransmitters

1. Serotonin

Regulation of a cell's function by a substance that the cell produces is a common regulatory mechanism in biological systems. Many neurotransmitters depend on autoreceptors to limit their production, neuronal firing rate, or neurotransmitter release. The same principle is true for the development of serotonin neurons. As serotonin reaches a critical amount in the mammalian brain, it causes the inhibition of outgrowth of the serotonin neruons. We have termed this phenonemon "autoregulation of development."

Increasing the levels of serotonin in the extracellular media in tissue culture has effects on the growth and survival of serotoninergic neurons (WHITAKER-AzMITIA and AzMITIA 1986b). Increasing the amount of serotonin in the culture, or the addition of the nonspecific agonist 5-MT, inhibits the outgrowth of serotoninergic neurons. In invertebrates, serotonin has been shown to be present and direct application of high concentrations of 5-HT to growth cones produces a retraction (GOLDBERG and KATER 1989). In vertebrate studies, using prenatal treatment with serotoninergic agents, we have also found decreased serotonin growth in rats (SHEMER et al. 1991).

Furthermore, increasing tryptophan (the precursor for serotonin synthesis) during gestation causes delayed outgrowth of serotonin neurons, particularly in cortex (Huether et al. 1992). Finally, animals treated prenatally with inhibitors of serotonin breakdown, monoamine oxidase inhibitors, show significant loss of serotonin terminals in the adult offspring, which again is most apparent in cortex (Whitaker-Azmitia et al. 1994).

The cellular mechanism for serotoninergic autoregulation of growth is not known. It appears to require an intracellular increase and release of 5-HT but does not correlate directly with any known serotonin receptor. Thus, cocaine, MDMA, MDA, and PCA are inhibitory to developing and adult serotoninergic neurons. These drugs all stimulate cytoplasmic release of 5-HT and inhibit its uptake and breakdown.

2. Dopamine

Many studies have shown the importance of dopamine in the regulation of serotonin neuronal development (Berger et al. 1985; Towle et al. 1989). Removal of dopamine by selective lesioning of neonates has been shown to lead to overgrowth of serotonin terminals into the caudate nucleus. The overexpressed serotonin terminals have appropriate ultrastructural details for serotoninergic, not dopaminergic, terminals (Descarries et al. 1992) and release sufficient serotonin to achieve tissue levels three times control values (Jackson and Abercrombie 1992). In our studies with selective prenatal dopamine receptor agonists, we have shown dopamine inhibition of 5-HT growth can be mediated through the D_1 dopamine receptor (Whitaker-Azmitia et al. 1990a).

In the adult, dopamine effects on serotonin may involve different mechanisms. 6-Hydroxydopamine (6-OH-DA) injection into the striatum of an adult rat removes dopamine fibers but does not induce a sprouting of 5-HT axons (Snyder et al. 1986). However, if 6-OH-DA is injected into the substantia nigra (SN), serotoninergic fibers will sprout robustly (Zhou et al. 1991a). The principle difference between these two injection sites is that dopamine fibers afferent to the serotoninergic neurons are eliminated only in the SN injection. However, transplantations of dopaminergic neurons after neonatal 6-OH-DA do not result in a withdrawal of the new serotoninergic fibers although other reactive systems do normalize (Abrous et al. 1993).

II. Protein Growth Factors

Many protein growth factors have been cloned and tested on serotonin neurons, but few have been found to have trophic activity. The most active factor on serotoninergic growth identified to date is S-100β, derived from glial cells, which acts chiefly on neurite extension and stabilization. Brain-derived neurotrophic factor (BDNF) may promote survival and sprouting of serotoninergic neurons.

1. S-100β

In the first study attempting to identify a serotoninergic growth factor, we tested a number of recognized neurotrophic protein molecules. S-100β (a Ca^{2+}-binding protein), but not NGF, epidermal growth factor (EGF), or insulin, promotes the extension of serotoninergic fibers at very low concentrations (1 ng/ml) (AZMITIA et al. 1990a). Calmodulin, another soluble Ca^{2+}-binding protein, is without effect. S-100β increases sprouting and branching in the serotoninergic cultures without any effect on cell survival, and has no effect on the growth of dopaminergic neurons in cultures (LIU and LAUDER 1992). Insulin-like growth factor-II promotes the maturation of dopaminergic neurons but has no effect on the serotoninergic neurons. In the brain, S-100β is synthesized and stored in astrocytes, tanocytes, and radial glial cells. In certain brain stem cholinergic neurons, S-100-IR has been observed (YANG et al. 1995). In the developing brain stem, S-100β is transiently expressed lateral to the raphe nuclei when serotonin neurons are developing (VAN HARTESVELDT et al. 1986) and is present in hippocampus prior to the onset of target maturation (AKBARI et al. 1994).There is direct evidence for an active role for S-100β in serotoninergic sprouting. The C-6 cell line, a glioma line that secretes S-100β, transplanted into an adult rat hippocampus induces the sprouting of serotoninergic fibers (UEDA et al. 1995a). However, C-6 cells transfected with S-100β antisense gene but which do not synthesize S-100β (VAN ELDIK et al. 1992) do not induce 5 HT fiber growth when transplanted. Interestingly, the few 5-HT-IR fibers that are present in the glioma mass lacking S-100β are enlarged, irregular, and distorted, resembling the 5-HT fibers seen in the aged brain.

The serotoninergic neurons have reciprocal actions with S-100β. The release of S-100β from astrocytes is induced by 5-HT_{1A} agonists (WHITAKER-AZMITIA et al. 1990b). In the adult, removal of 5-HT induces a fall in S-100β (HARING et al. 1993) and an enhanced stimulation of this protein by 5-HT_{1A} agonists (AZMITIA et al. 1992, 1995). The trophic actions of S-100β do not require neuronal support. Injections of ibotenic acid into either the caudate or hippocampus induce rapid death of endogenous neurons by an excitatory amino acid mechanism (ZHOU et al. 1995). In this neuron-free inner zone, S-100β is increased. A peripherial outer zone forms which contains neuronal and reactive glial cells, but S-100β is not increased. 5-HT-IR fibers show a rapid and robust sprouting into the S-100β-enriched inner zone and no detectable increase in the neuronal and glial outer zone. The regulation of S-100β by serotoninergic neurons should produce a cascade of trophic effects. S-100β has profound trophic effects, causing neurite outgrowth in chick cortical cultures (KLIGMAN and MARSHAK 1985), glial proliferation (SELINFREUND et al. 1991), and increased survival of motoneurons (BHATTACHARYYA et al. 1992). In a mutant mouse lacking S-100β, the polydactyl Nagoya mouse, there is little development of cortex, and serotonin terminals are absent (UEDA et al. 1994). Treatment with antibodies raised against S-100β blocks cortical synaptogenesis. Infusion into the visual cortex blocks the rearrangement normally seen after eye enucleation in kittens (Muller et al. 1993). In adult animals, S-100β is involved in long-term potentiation (LTP), a model of learning (Lewis and Teyler 1986; Baido et al. 1992). This role in LTP implies a role for serotonin in adult synaptic plasticity.

2. Brain-Derived Neuronotrophic Factor

Brain-derived neuronotrophic factor (BDNF) has been shown by several laboratories to stimulate survival of dopaminergic, cholinergic, and GABAergic neurons in mescenphalic cultures (Roback et al. 1992; Hyman et al. 1991). In the immortalized serotoninergic cell line, RN46A under depolarizing conditions, the B_{max} of 5-HT$_{1A}$ receptor binding is increased 3.5-fold by BDNF treatment (Eaton et al. 1995). In human fetal ventral mesencephalic cultures, BDNF (10 ng/ml) increases dopamine neuronal survival 2.5– to 3.5-fold, and 5-HT levels 2-fold (Spenger et al. 1995). In primary midbrain raphe cultures, BDNF increases 5-HT content, somal size and number and the process length of 5-HT-IR neurons (Nishi et al. 1996b). In addition, the S-100β-IR is increased in glial cells.

In the intact animal, BDNF stimulates the sprouting of 5-HT fibers in the adult cortex of control rats (Mamounas et al. 1995). A more dramatic degree of sprouting is seen after prior destruction of 5-HT fibers with PCA. The mechanism of action may involve activation of the tyrosine receptor kinase-B (TRK-B) receptor, located on 5-HT neurons (Merlio et al. 1992). BDNF effects on 5-HT sprouting persist for long periods (months) after the infusion of BDNF is stopped (Mamounas et al. 1995). Thus it appears that BDNF can initiate the sprouting and recruit long-lasting, possibly astroglial, mechanisms to maintain the sprouts. BDNF has been shown to activate brain astrocytes (Roback et al. 1995). In recent studies, antibodies raised against S-100β blocked the neurite extension but not the increases in 5-HT-IR somal size and number (Nishi et al. 1996b). These cellular changes are sufficient to argue that glial cells are an active and long-acting participant in the actions of BDNF.

3. Other Factors

A number of novel proteins exist which have suspected trophic effects on serotonin neurons. Transgenic animals overexpressing transforming growth factor-alpha (TGF-α) show gender-specific changes in serotonin development (Hilakivi-Clarke et al. 1995). Interestingly, the aggressive behavior normally seen in these animals is reversed by treatment with serotonin uptake inhibitors (Hilakivi-Clarke and Goldberg 1993). There is also evidence suggesting that the basic fibroblast growth factors, FGF-2 (Chadi et al. 1993) and FGF-5 (Lindholm et al. 1994), are trophic to serotonin neurons.

III. Neuropeptides

Neuropeptides are derived from larger proteins and have been shown to have both neuroendocrine and neurotransmitter roles. A number of neuropeptides show growth regulatory effects on serotonin neurons. Adrenocorticotropic hormone (ACTH), ACTH 4–10, and Org 2766 at various concentrations stimulate the outgrowth of serotoninergic neurons in culture (AZMITIA and DE KLOET 1987). However, these peptides have no effect when serotoninergic neurons are in contact with their normal target cells. Thus, ACTH is stimulatory only during the period when 5-HT neurons are deprived of their normal target stimulation, as occurs during early development and after lesions. The serotoninergic cell line RN46A also responds to ACTH by developing a more mature serotoninergic phenotype (EATON et al. 1995). In the intact brain, ACTH has little effect once the serotoninergic neurons have reached their target (VAN LUIJTELAAR and TONNAER 1992). It may be pertinent that ACTH releases S-100β from astroglial cells (SUZUKI et al. 1987).

Neuropeptide Y (NPY) influences serotoninergic sprouting. 5-HT fetal neurons cotransplanted with fetal suprachiasmatic nucleus into the adult rat eye chamber produce very few 5-HT fibers (UEDA et al. 1994). However, if NPY neurons from the lateral geniculate nucleus are also transplanted into the eye chamber, the 5-HT fibers are stimulated to invade the suprachiasmatic nucleus. Thus, a peptidergic fiber can condition a target region to permit secondary serotoninergic ingrowth. This may be relevant to late terminal sprouting, which occurs after the terminals first reach the target area.

Leu-enkephalin has a potent inhibitory action on serotoninergic maturation (DAVILA-GARCIA and AZMITIA 1989, 1990). Daily dosing of leu-enkephalin (18 nM) for either 3 or 5 days in culture results in a 40% decrease in the 5-HT uptake without any effect on cell survival. Continuous blockade of the opiate receptor with naloxone, an opiate antagonist, has been shown to stimulate the 5-HT fiber growth in culture. The inhibitory effect of opiates is enhanced in the presence of target cells in co-culture experiments. It is interesting to note that morphine inhibits the regenerative capacity of a lesioned serotonin system (GORIO et al. 1993). Thus, the opiate system is inhibitory to serotoninergic growth in immature and mature neurons.

Substance P, which is colocalized with serotonin from GD 13 and throughout adulthood (NI and JONAKAIT 1989), has an influence on developing serotonin neurons. Serotonin also influences the expression of preprotachykinin (PPT), the substance P precursor. Both adults and developing animals show an increase in PPT following serotonin depletion with para-chlorop-henylalanine (PCPA) (WALKER et al. 1990). In cultures of serotoninergic neurons, substance P is stimulatory. However, this trophic influence appears to be only important in the rat brain if the endogenous serotonin influence is lacking (JONSSON and HALLMAN 1983). The trophic action of neuropeptides on serotoninergic growth appears to be markedly influenced by secondary factors that already exist in the target areas and are, in turn, influenced by serotoninergic neurons.

IV. Steroids

Studies on the involvement of glucocorticoids in serotonin metabolism, physiology, development, and plasticity span at least 30 years. The majority of studies indicate that increasing glucocorticoid levels results in a stimulation of serotoninergic growth, while eliminating glucocorticoids by adrenalectomy leads to a decline in serotoninergic morphology and function.

In developing animals, stress accelerates the appearance of serotonin-mediated behaviors and serotonin receptors (Peters 1988, 1990). These stimulatory changes can be long-lasting. Cold exposure at PD 1–6 causes increased serotonin content in the adult brain (Giulian et al. 1974). Injections of ACTH and nicotine, both of which increase glucocorticoids, result in a rise in 5-HT sprouting (King et al. 1991; Alves et al. 1993). ACTH exposure leads to hastened 5-HT development and hyperinnervation, but also results in premature retraction of the 5-HT sprouts (Alves et al. 1996). Adrenalectomy blocks the normal development of serotoninergic neurons (Sze 1980). In adult animals, adrenalectomy causes shrinkage of the cell body and dendrites of serotoninergic neurons in the raphe nuclei. This effect is reversed in rats given dexamethasone for 72 h (Azmitia et al. 1993). Homotypic collateral sprouting seen after 5,7-DHT lesions is blocked by prior adrenalectomy (Zhou and Azmitia 1985). Corticosterone injections re-instate the ability to sprout.

V. Attachment Factors

In order to move, there must be traction. In the adult brain, the amount of free surface for new growth is limited. In the adult brain, especially when it is uninjured, most suitable substrates are already occupied. Serotoninergic fibers will extend long distances but in a slow and random manner. This can be greatly improved by microinjectioning adhesion molecules, such as laminin, into the brain (Zhou and Azmitia 1988; Fig. 2) or by activation of astrocytes by removal of pre-existing neurons with excitochemicals (Zhou and Chiang 1995).

VI. Summary

Serotonin may be unique among neurotransmitter systems for its incredible capacity for regrowth. There are many factors which contribute to this unique capacity: (1) 5-HT fibers are largely unmyelinated and not limited to following a precise pathway. In fact 5-HT fibers can grow along myelinated axons, blood vessels, ependymal cells, and epithelial cells. (2) 5-HT fibers respond to the ubiquitous glial trophic factor S-100β. (3) 5-HT neurons respond trophically to a variety of neuronal and endocrine factors. (4) The adult 5-HT neurons contain abundant messages for GAP43 (a growth-associated protein normally seen in growth cones). (5) 5-HT neurons have TRK-B receptors and are responsive to BDNF.

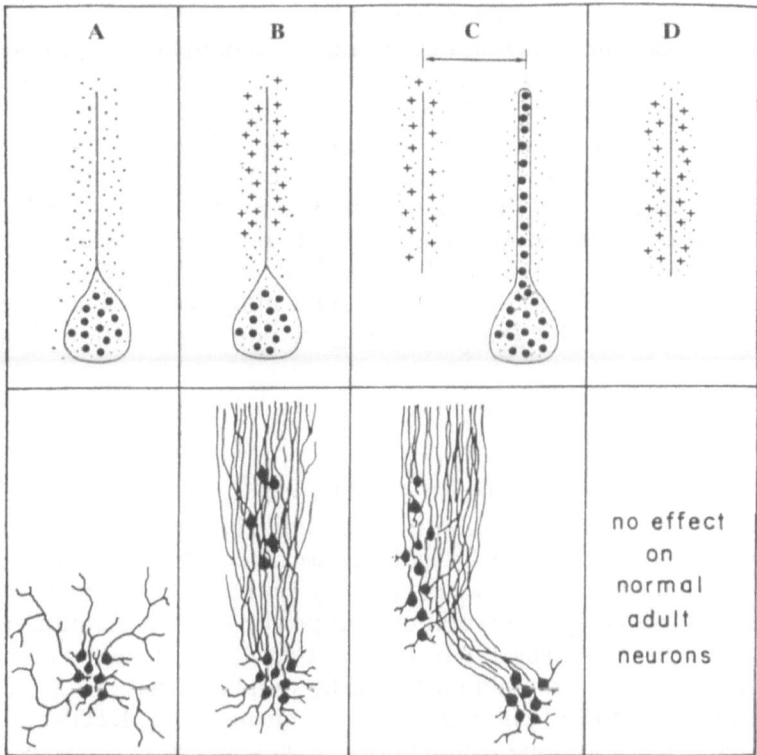

Fig. 2A–D. Diagrams showing the effect of laminin on developing and adult neurons: *upper panels*, experimental designs; *lower panels*, results. **A** Vehicle control: *upper panel*, dissociated brain stem raphe, locus ceruleus, and nigra cells (*solid circles*) were transplanted in an adult brain region, and Hank's solution vehicle (*dots*) was added dorsal to the graft site through a micropipette immediately after grafting. *Lower panel*, the fibers of the grafted neurons grew randomly with no recognizable pattern. Few neurons, if any, were observed on the graft tract. **B** Laminin group: *upper panel*, laminin (*crosses*) was infected dorsal to the transplant. *Lower panel*, the fibers of transplanted neurons grew preferentially along the laminin-treated tract. A number of neurons migrated up to the tract. **C** Parallel injection group: *upper panel*, raphe and locus ceruleus cells were transplanted in a brain region and a thin layer of cells was deposited in the pipette tract during retracting. Laminin solution was added in a tract 0.3–1 mm lateral and parallel to the grafted tract immediately after transplantation. *Lower panel*, fibers of transplanted neurons grew toward, and a number of neurons migrated to, the laminin-treated tract, when they were placed less than 0.5 mm away. **D** Laminin control: *upper panel*, laminin solution was injected in a brain region where no graft was made. *Lower panel*, the undamaged adult 5-HT, NE, and DA neurons did not respond to laminin treatment. *Solid circle*, transplanted cells; *dots*, Hank's medium (vehicle); *crosses*, laminin

D. Serotoninergic Influences on Target Tissue Maturation

I. Introduction

Serotonin plays a complex role in the maturation of diverse neuronal and glial cells, and dramatically regulates synapse density in both the immature and mature brain. Since serotonin is so widely distributed in brain (and in peripheral tissues), most regions of the brain and spinal cord (and body) are dependent on serotonin for reaching maturation. Thus, disturbances in serotonin during development would produce profound deficits in a variety of structures in the mature brain, as well as throughout the body. In the adult brain and spinal cord, 5-HT can continue to exert trophic activity on target cells. In fact a loss of the adult phenotype follows the removal of 5-HT. The loss of the mature morphology of adult neurons appears to be fully reversible when 5-HT is restored. The rapid fluctuation between immature and mature states indicates the mature neural morphology is not fixed after development and that the ability to change is consistent with normal brain function.

II. Immature Brain

The earliest studies on the effects of 5-HT on target development used the approach of removing serotonin, with PCPA, and observing the consequences. PCPA, an inhibitor of tryptophan hydroxylase, reduces 5-HT levels but does not destroy the terminals (Dewar et al. 1992). Prenatal depletion of serotonin with PCPA causes delayed maturation and prolonged proliferation of target regions (Lauder and Krebs 1978). In neonates, PCPA results in a reduced rate of growth for the forebrain measured by weight (Hole 1972a–c). Depletion of serotonin at PD 3 (with the selective neurotoxin 5,7-DHT) results in a decreased density of granule cell dendritic spines, which worsens as the animals grow into adolescence (Haring and Yan 1993). In electrophysiological studies, these animals show a decreased population excitatory postsynaptic potential (EPSP). In studies recently completed in our laboratories, we have found that depleting animals of serotonin with PCPA during peak synaptogenesis (PND 10–20) causes permanent (up to 180 days) loss of the dendrites (Fig. 3). Moreover, these animals have profound changes in learning and memory. Similar findings have been reported in developing chicks, where serotonin depletion causes a delay in hippocampal neuronal differentiation and a decreased number of synapses in the cerebral cortex and hippocampus (Okado et al. 1993).

The earliest direct studies of serotonin's influence on target development were done in tissue culture. Using hippocampal slice cultures, Chubakov and colleagues have shown that serotonin plays a role in neurite outgrowth in the target region, increases electrical interconnections, and promotes synaptogenesis (Chumasov et al. 1978; Chubakov et al. 1986a, b, 1993).

Much of the trophic effect of serotonin on target tissues is elicited through the 5-HT$_{1A}$ receptor. Very early in development, the 5-HT$_{1A}$ receptor plays a role in migration of cranial neural crest cells (Moiseiwitsch and Lauder 1995). In the visual cortex of dark-reared rats, there is a prolonged peak of the 5-HT$_{1A}$ receptor and a corresponding increased period of cortical synaptic plasticity (Mower 1991). Septal cholinergic neurons express 5-HT$_{1A}$ receptors early in development. Chronic application of various 5-HT$_{1A}$ agonists to cultures of septal cholinergic neurons promote choline acetyltransferase activity and neurite outgrowth and branching, but have no effect on cell survival (RIAD

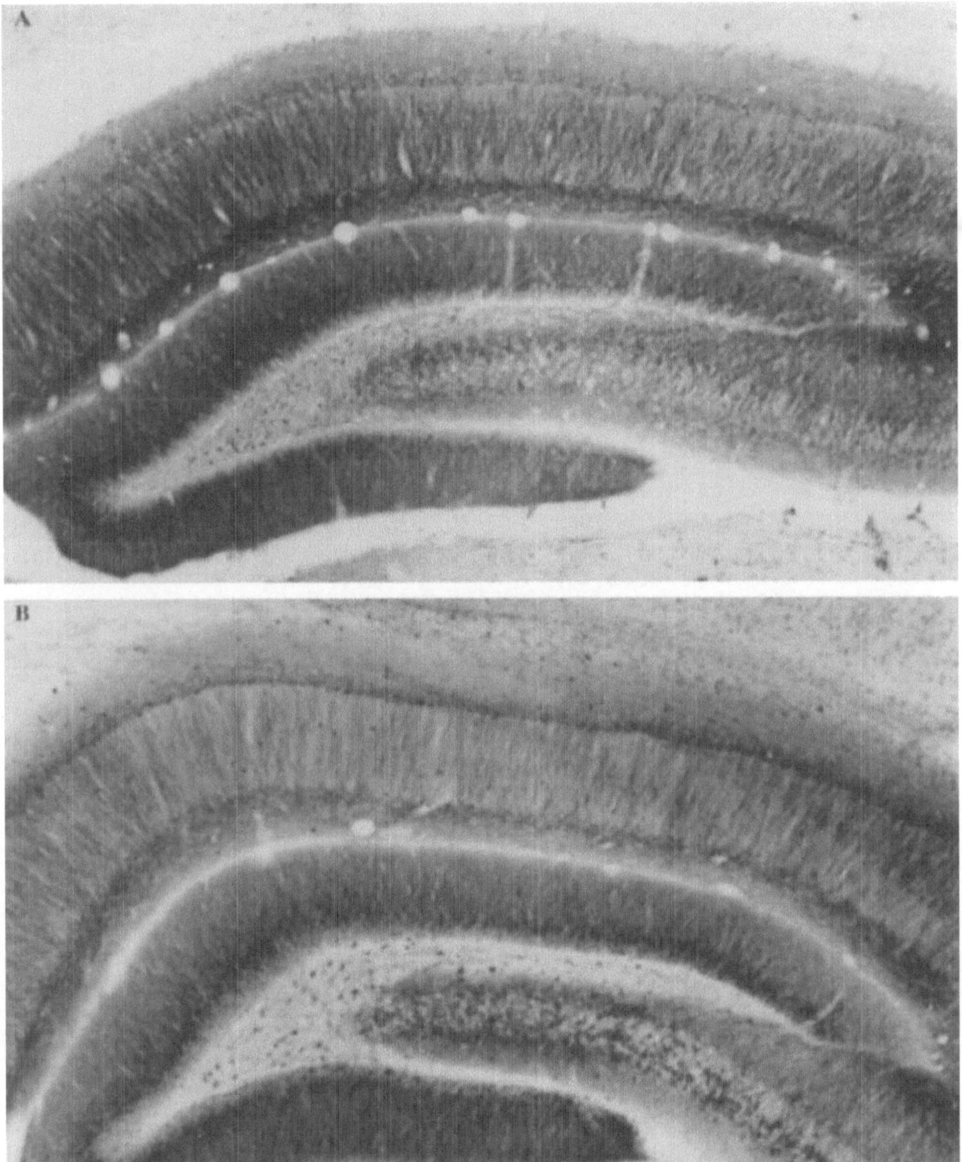

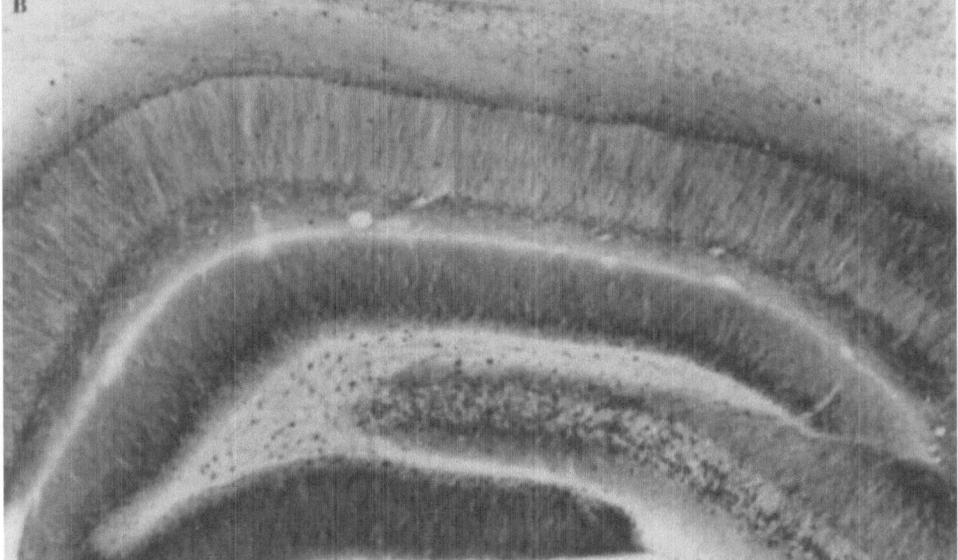

Fig. 3A,B. Effects of neonatal serotonin depletion on adult dendritic density. Rat pups were treated with PCPA from PD 10 to 20. On PN 190, the animals were perfused and immunocytochemically stained with the dendritic marker MAP-2. **A** Control hippocampus. **B** PCPA-treated hippocampus, showing a significant loss of dendrites

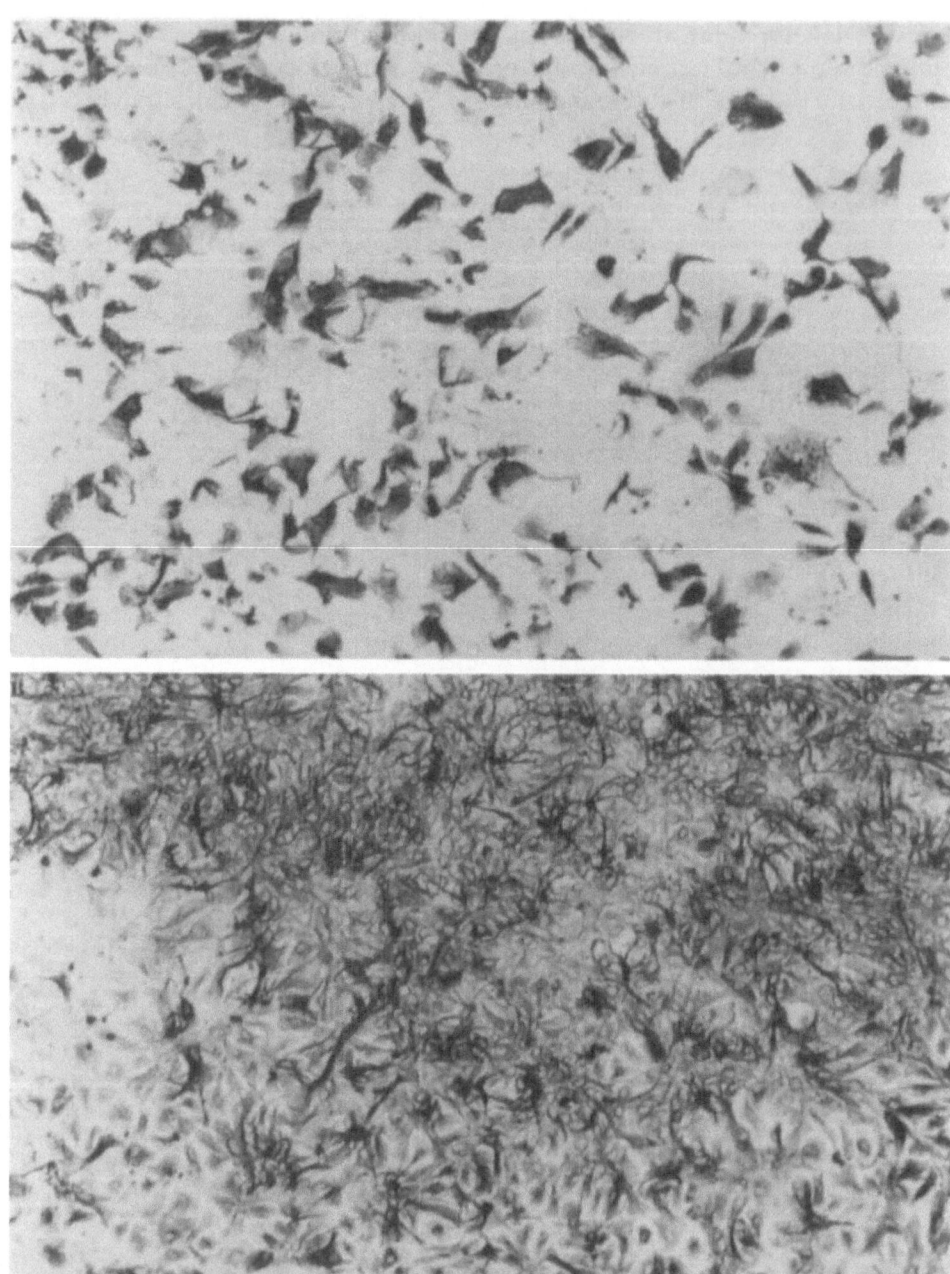

Fig. 4A,B. Effects of a 5-HT$_{1A}$ agonist on astroglia maturation, in culture. Cortical astroglial cultures were produced from 4-day-old rat pups. **A** Control culture. **B** Culture treated for 24 h with a 5-HT$_{1A}$ agonist. The cells are immunochemically stained for the astroglial protein glial fibrillary acidic protein (GFAP)

et al. 1994). These effects are blocked by a 5-HT$_{1A}$ antagonist. 5-HT$_{1A}$ agonists have also been reported to increase the number of synaptophysin-IR varicosites in hippocampal cell cultures (NISHI et al. 1996a). This effect was shared with S-100β and glucocorticoids. The effects of PCPA depletion in the neonate are reversed by the 5-HT$_{1A}$ agonist buspirone (HARING et al. 1995). Neonate rats treated with a 5-HT-depleting drug or 5-HT$_{1A}$ antagonist (NAN-190) show a loss of spines on hippocampal pyramidal neurons (YAN et al. 1995).

The 5-HT$_{1B}$ receptor may be involved in postnatal regulation of target development. Besides an autoreceptor type, there are also 5-HT$_{1B}$ receptors in these regions which are not on serotoninergic neurons. Interestingly, serotonin has been shown to be important for pattern formation in somatosensory cortex (i.e., physical representation of somatic body parts). Thus 5-HT$_{1B}$ receptors may be involved in this aspect of target tissue maturation (RHOADES et al. 1994; LESLIE et al. 1992).

Several researchers have also suggested a trophic role for the 5-HT$_{2A}$ receptor, including actions on synaptogenesis (NIITSU et al. 1995). Treatment from E11-F17 of chick embryos with the 5-HT$_{2A}$ receptor agonist (2,5-dimethoxy-4-iodoamphetamine, DOI) increased (20% above control) and an antagonist (ketanserin) decreased (30%–40% below control) the synaptic density in lateral motor column of the spinal cord.

Neurons are not the only target of serotoninergic neurons. 5-HT receptors are distributed on many cells types in the brain and body. We have focused on astrocytes. The actions of 5-HT have been studied in primary glial cultures and in glial-derived cell lines. Astroglial cells possess a number of different neurotransmitter receptors, including 5-HT$_{1A}$ and 5-HT$_{2A}$ receptors. When the 5-HT$_{1A}$ receptor is stimulated, the astroglial cell responds by releasing S-100β and attaining a mature morphology with a shift from a flattened morphology to a process-bearing morphology (WHITAKER-AZMITIA and AZMITIA 1990; Fig. 4). In this treatment there is an effective feedback inhibition since as the glial cell matures, it loses binding to the 5-HT$_1$ receptors (WHITAKER-AZMITIA and AZMITIA 1986a). The 5-HT$_2$ receptor on astroglial cells upregulates glycogenolysis (POBLETE and AZMITIA 1995). Thus, 5-HT, by stimulating both the 5-HT$_{1A}$ receptor and the 5-HT$_{2A/C}$ receptor, regulates the release of both S-100β and glucose, two factors necessary for neuronal sprouting. The pharmacology of these glial receptors has been discussed (WHITAKER-AZMITIA 1988).

III. Mature Brain

Similar to its function during development where serotonin induces maturation, in the adult serotonin is needed to maintain maturation. In adult rats, PCPA, given for 8 days, produces up to a 50% decrease in the number of nonmonoaminergic synapses in somatosensory cortex (CHENG et al. 1994). In an immunocytochemical study using the same injection schedule, a marked

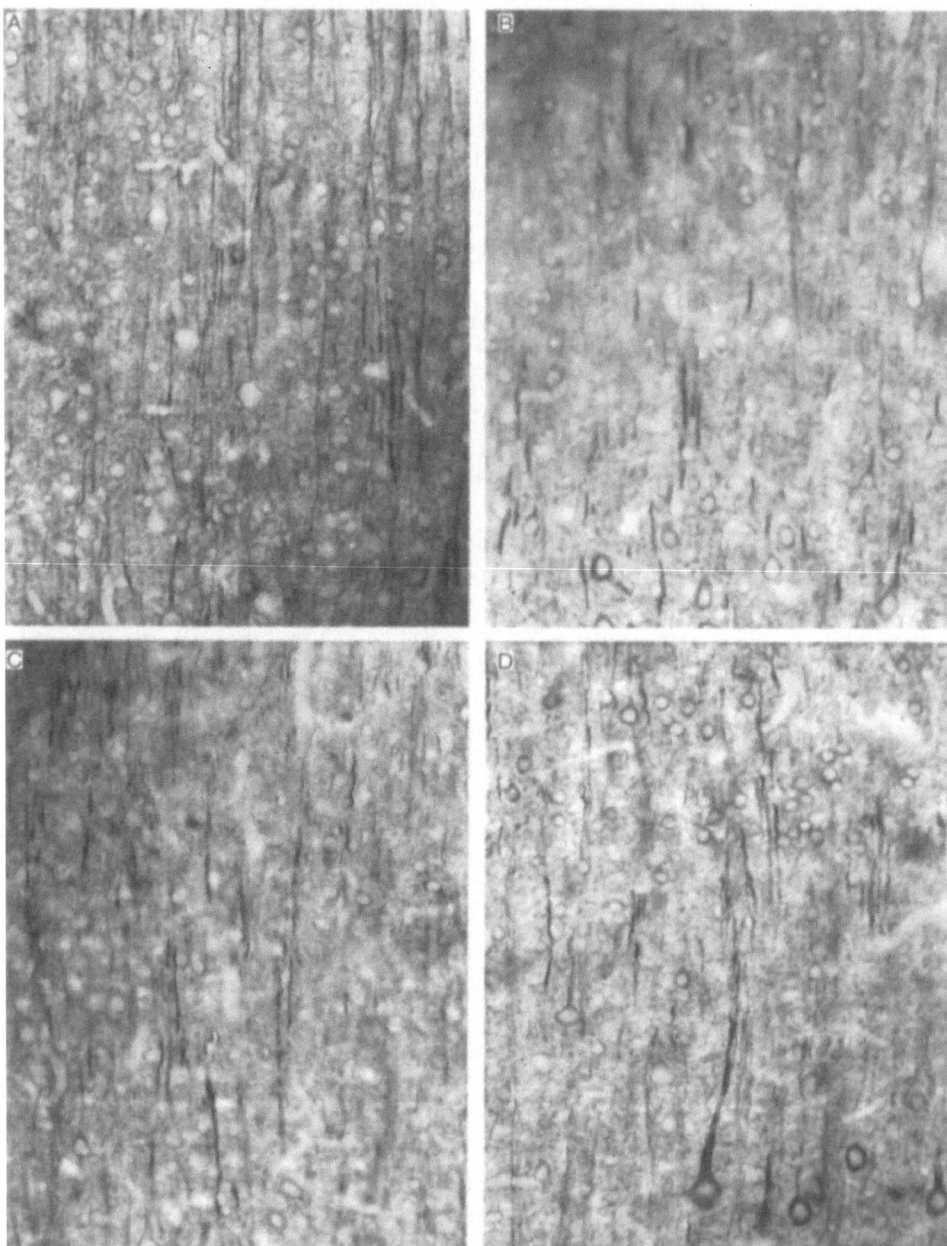

Fig. 5A-D. MAP-2 immunoreactivity (IR) in the parietal cortex of the rat. **A** Sham-injected control animal. The large pyramidal cell bodies are *at the lower end of the figure* and the dendrites extend towards the superficial surface (*top*). **B** MAP-2 IR 2 weeks after an injection of PCA (10mg/kg). There is a reduction in the dendritic staining but the soma labeling appears to be more intense. **C** Cortex after the rat received PCA injections as above but also received three daily injections of ipsapirone (1mg/kg) immediately before perfusion on day 14. There is an increase in dendritic staining. **D** Cortex treated as in **B** except that the animals were given dexamethasone (10μg/ml) the last 3 days before death. As in **C**, there is an apparent increase in the dendritic labeling (see AZMITIA et al. 1995)

reduction was seen in both MAP-2-IR and synaptophysin-IR in adult rat hippocampus (WHITAKER-AZMITIA et al. 1995).

PCA, which depletes serotoninergic axons, damages nonserotoninergic elements as evidenced by the massive increase in Fink-Heimer silver-stain for degenerating neurons seen throughout cortex (COMMINS et al. 1987). Two weeks after injection of PCA (2×10 mg/kg), a substantial reduction of MAP-2-IR and synaptophysin-IR also occurs throughout the brain (AZMITIA et al. 1995). The hippocampus shows the most dramatic loss but losses could be seen in most areas of the brain where 5-HT had been depleted. We can speculate that regions showing the most plasticity normally (hippocampus) would be most sensitive to the inhibitory effects of serotonin depletion.

As in the immature brain, much of serotonin's influence on target tissues of the mature brain appears to relate to actions of the 5-HT$_{1A}$ receptor. In adult rats, the decreases in MAP-2-IR and synaptophysin-IR, seen 2 weeks after injections of PCA (2×10 mg/kg, s.c.) in hippocampus, parietal cortex, temporal pole, and hypothalamus, are rapidly restored towards control levels after 3 days of injections with a 5-HT$_{1A}$ agonist (AZMITIA et al. 1995; Fig. 5). The degree of recovery is most apparent in hippocampus.

Long-term removal of glucocorticoids, by adrenalectomy, results in loss of the adult phenotype of granule neurons in the dentate gyrus (SLOVITER et al. 1989; LIAO et al. 1993) and a re-expression of the transitory peak in the 5-HT$_{1A}$ receptor (CHALMERS et al. 1993; MEIJER and DEKLOET 1994). The morphological changes induced by the loss of glucocorticoids can be rapidly (24–72 h) reversed by either steroid replacement or a 5-HT$_{1A}$ agonist.

In adult rats, after hemisection of the thoracic spinal cord, there is a fall in 5-HT levels and fibers (SARUHASI and YOUNG 1994). The fibers can regenerate within several weeks. After injections of mianserin, a 5-HT$_{2A/C}$ antagonist, there is a marked suppression of 5-HT regeneration which persists for as long as a week. Mianserin treatment after this time is no longer effective. The reason for the transient nature of this response is unknown. It is not known if the trophic effect takes place directly on neurons or is mediated through glial cells.

The loss of synapses, terminals, and dendrites can be considered to be a retraction of the mature state rather than evidence of neuropathological degeneration. This is consistent with the rapid reversibility of the morphology when 5-HT is replaced or the 5-HT$_{1A}$ receptor is activated. Serotonin plays an active role in maintaining the mature neuronal phenotype, probably by stabilization of the cytoskeleton of the cell. This subject is further discussed in the section on dynamic instability of neurons.

Serotoninergic effects on glial cells have also been seen in the adult. Depletion of 5-HT results in a depletion of S-100β levels and immunoreactivity (AZMITIA et al. 1992; 1995; HARING et al. 1993). Other astrocytic proteins are also affected by 5-HT. Glial fibrillary acidic protein (GFAP), an insoluble fibrillary marker protein, shows a small transient increase after 5-HT damage (WILSON and MOLIVER 1994; BENDOTTI et al. 1994). Microglial protein markers increase 3 weeks after treatment with PCA and return to normal levels by 9 weeks (see Wilson and Molliver 1994). The increase in microglial cells may have important implications for the actions of cytokines. Microglial cells release a number of cytokines, the most abundant being interleukin-1 (IL-1). IL-1β, but not IL-2 or IL-6, has been shown to stimulate the release and turnover of serotonin in a number of brain regions, including the hypothalamus, frontal cortex, and hippocampus (Zalcman et al. 1994). A central role for IL-1 in the activation of hypothalamic 5-HT systems and pituitary-adrenal axis was established using an IL-1 antagonist in conjunction with restraint stress (Shintani et al. 1995). In the hippocampus, IL-1β infusion into the ventricles increases 5-HT and corticosterone extracellular levels and inhibits behavioral activity (locomotion, grooming, eating, drinking) (Linthorst et al. 1995). Furthermore, glucocorticoids are able to modulate the release of 5-HT by IL-1β and also inhibit the synthesis of IL-1β, so preventing further stimulation of 5-HT (Gemma et al. 1994). The glucocorticoid receptor antagonist completely prevents IL-1β activation of the serotoninergic system. IL-1 can also stimulate the release of S-100β from astrocytes, and both IL-1 and S-100β are increased in brains with Alzheimer's disease (Griffin et al. 1989). Therefore, cytokines can be added to the list of glial bioactive substances which have feedback interactions on both steroids and serotonin.

E. Neuronal Dynamic Instability

I. Model

Neuroscientists commonly assume that the brain develops and ages linearly, progressing from immature to mature to aged and, finally, to death. In contrast, we are proposing a model based on the dynamic instability of neurons, which predicts brain morphology is in a constant flux. A schematic summary of this model appears in Fig. 6.

The main component of the cytoskeleton that gives neurons their shape is microtubules. These microtubules consist of long polymers of tubulin, which spontaneously depolymerize if they are not actively polymerizing (Mitchison and Kirschner 1984). Microtubule-associated proteins (MAPs), such as MAP-2 (primarily in dendrites) and tau (primarily in axons), stabilize microtubules. The phosphorylation by protein kinase-C (PKC) and PKA of MAPs disrupts their binding to microtubules (Jameson and Caplow 1981). In dendrites, phosphorylation of MAP-2 may permit the formation of dendritic spines (Aoki and Siekevitz 1988) and dendritic arborization (Diez-Guerra and Avila 1993). However, sustained phosphorylation of MAP-2 is associated with dendritic retraction (Matesic and Lin 1994) and may lead to cognitive loss (Jope and Johnson 1992).

GAP-43 is another key cytoplasmic component key to the formation of synaptic connections, in both the developing and the mature brain. GAP-43 plays a role in the outgrowth and guidance of the growth cone (Benowitz and

Routtenberg 1987; Aigner and Caroni 1995; Strittmatter et al. 1995). Like the MAPs, the phosphorylation state of GAP-43 is important in determining its function.

LINEAR MODEL

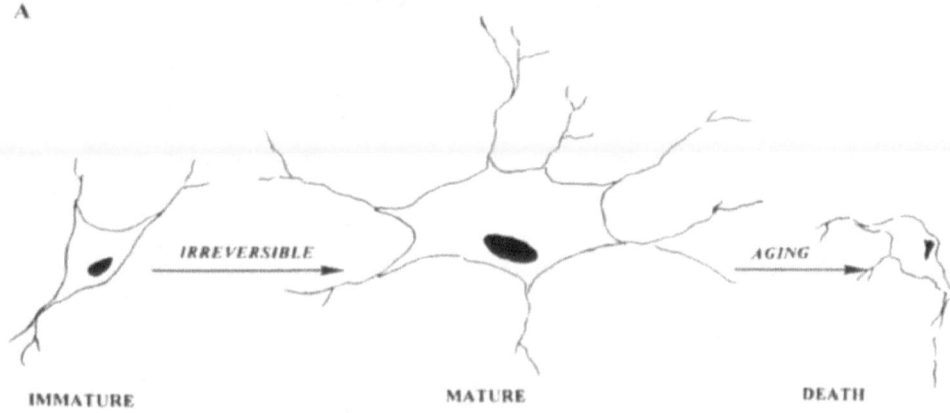

NEURONAL INSTABILITY MODEL

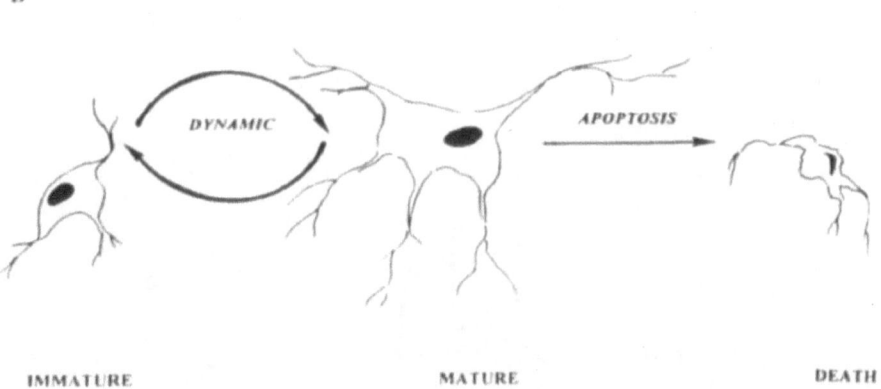

Fig. 6A,B. Schematic drawing illustrating the differences between a linear and a dynamic instability model. In **A** the linear progression of neurons proceeds from an immature to a mature to a dead neuron. The three stages are all essentially irreversible. Pathology is any decrease in size or loss of processes and synapses. In **B** the immature and mature stages are in dynamic equilibrium, and the balance is determined by the action of trophic factors such as steroids, serotonin, and S-100β. Decreases in the mature phenotype may represent a normal fluctuation in morphology. Death is a consequence of apoptosis, which may be common in hypermature neurons

We have established that much of serotonin's action on development and plasticity is through release of the trophic substance S-100β. The molecular mechanism for the trophic actions of S-100β could involve the inhibition of PKC-induced phosphorylation of GAP-43 (Sheu et al. 1994), MAP-2, and tau (Hesketh and Baudier 1986; Baudier et al. 1987). Thus, in the developing and mature brain, serotonin can regulate the cytoskeleton, and thus cell morphology, through the 5-HT$_{1A}$-receptor-linked release of S-100β.

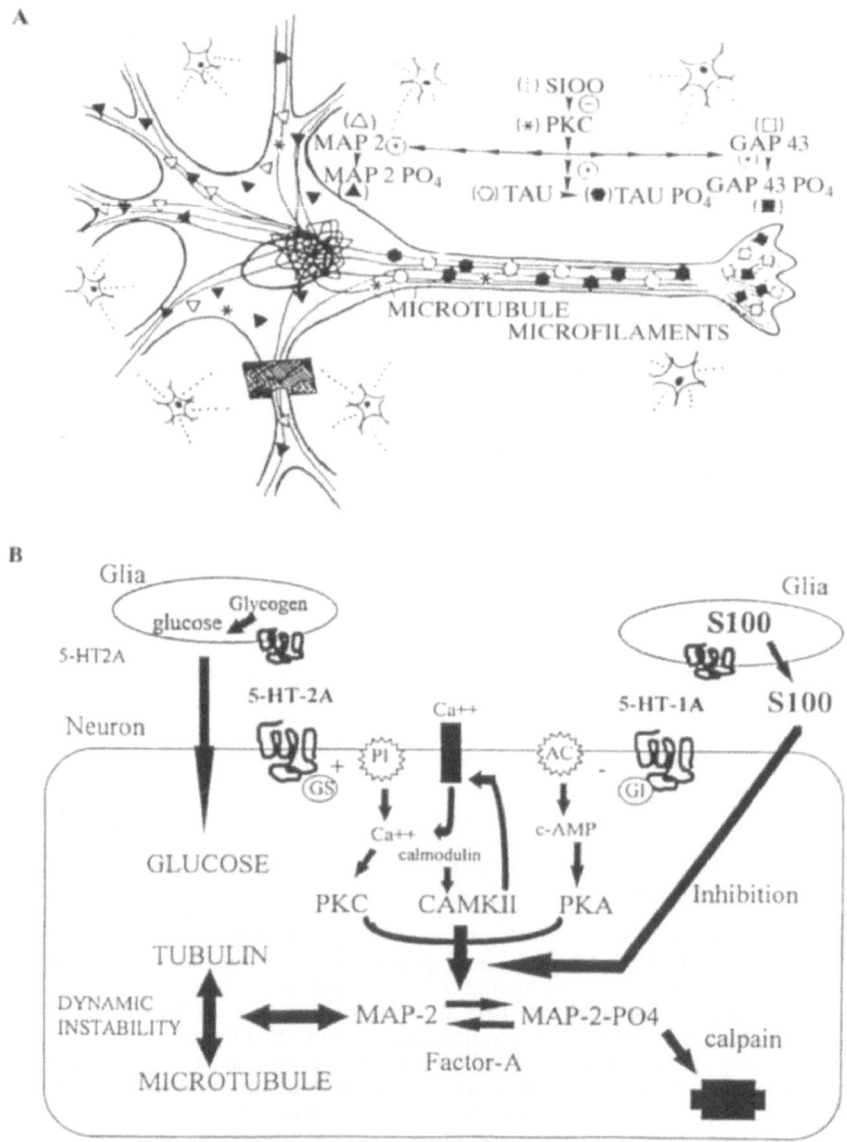

Serotonin may also have a more direct influence on the cytoskeleton, through regulation of protein kinases. 5-HT$_{1A}$ receptors may stabilize microtubules through inhibition of cAMP production (SHENKER et al. 1985), while 5-HT$_{2A}$ receptors may loosen the long polymers of tubulin by stimulating PKC-induced phosphorylation of MAPs.

These observations would predict that the loss of S-100β or 5-HT causes collapse of the mature neuronal cytoskeleton. In the most dramatic cases, a cell can lose Nissl staining without the cell being dead, but existing in a state of animated suspension (AZMITIA and LIAO 1994). The volume and metabolic activity may both be measurably decreased, although there is no actual loss of cells. However, increasing the serotonin or S-100β again would result in stabilization of the cytoskeleton and reappearance of the mature morphology. Thus, the daily and yearly fluctuations in serotonin and steroids are probably sufficient to induce rhythmic morphological changes. At the extreme, loss of serotonin or S-100β could cause mental disorders, by reducing the anatomical circuitry. However, increasing the serotonin again could restore normal function.

A schematic summary of the possible regulation by 5-HT on the neuronal cytoskeleton appears in Fig. 7.

II. Clinical Implications

Animal studies show loss of brain structure and volume, reversible by treatments which increase serotonin. For example, depletion of serotonin, with PCA or PCPA, produces a loss of dendrites and brain volume that is reversed

---

Fig. 7A,B. Effects of serotonin on cytoskeletal proteins and neuronal morphology. An overview of the sites is presented where S-100β influences the morphology of the neuron (A) and a detailed molecular mechanism to illustrate the dynamic instability of microtubules. **A** Distribution of microtubules and microtubule-associated proteins (MAP-2 and tau) in dendrites and axons, respectively. Growth cones show high levels of both microfilaments and growth-associated protein (GAP-43). Protein kinase-C (PKC) phosphorylates MAP-2, tau, and GAP-43. Phosphorylated MAP-2 and tau do not bind to microtubules and facilitate tubulin depolymerization. S-100β inhibits PKC phosphorylation of MAP-2, tau, and GAP-43. **B** The biochemical actions of the 5-HT$_{1A}$ and 5-HT$_{2A}$ receptors are shown. 5-HT$_{1A}$ receptor activation results in release of S-100β from glial cells and the inhibition of protein kinase-A (PKA) in neurons. In contrast, the 5-HT$_{2A}$ receptor elevates Ca^{2+} levels and PKC in neurons (KRAMER et al. 1995) and releases glucose from glial cells. PKC and PKA will desensitize the 5-HT$_{1A}$ receptor (RAYMOND and OLSEN 1994). Glucose can increase ATP levels and ATP-Mg^{2+}-dependent protein (factor FA). Factor A acts as either a phosphatase- or kinase-activating factor depending on whether the pH is basic or acidic, respectively (YANG et al. 1993). Under favorable (basic) conditions, branching and synaptogenesis occur (NIITSU et al. 1995). However, under acidic conditions, cytoskeletal collapse and neurite retraction occur (AZMITIA et al. 1990b). The influence of the high-affinity 5-HT receptor stabilizes microtubules while that of the lower affinity 5-HT receptor destabilizes the cytoskeleton

by a 5-HT$_{1A}$ agonist (Whitaker-Azmitia et al. 1995; Azmitia et al. 1995). In this final section we present a variety of human conditions wherein their etiology or treatment involves the serotonin system. Reversible changes in brain volume occur in several of these disorders.

1. Psychiatric Implications

There may also be some examples of serotonin regulation of brain morphology in human studies. An animal model of anorexia shows dramatic loss of 5-HT fibers (Son et al. 1994). In humans, serotonin uptake inhibitors are effective in treatment of anorexia nervosa (Anorexia Nervosa Study Group 1995). In addition, a reduced brain volume appears in MRI studies of anorexic children, and normalizes when the children are well (Hentschel et al. 1995; Shear et al. 1994). This observation, especially since recovery can be enhanced with 5-HT$_{1A}$ agonists, indicates pharmacologically induced changes in human brain volume are possible.

Although serotonin drugs are useful in treating alcoholics, the most direct link between alcohol intake and serotonin involves animal studies. Prenatal exposure to alcohol and cocaine interferes with serotonin's role in development, and thus may contribute to some of the damage seen in fetal alcohol syndrome or crack babies. In both of these models, 5-HT$_{1A}$ agonists reverse the damage seen in neonatal rats (Akbari et al. 1994; Tajuddidn and Druse 1993). Interestingly, rats bred for alcohol preference show marked disruption of serotoninergic projections (Zhou et al. 1991). In human alcoholics, transitory changes in brain volume correlate with the disease state (Jernigan et al. 1991).

Patients with post-traumatic stress disorder (PTSD) with an alcohol dependency respond well to sertraline, a specific serotonin reuptake inhibitor (SSRI) (Brady et al. 1995). Patients with this disorder also show a loss of brain volume as detected by MRI-related measurements (Bremner et al. 1995).

Certain types of depression result from a loss of brain 5-HT levels. Selected brain regions show reduced 5-HT fibers and increased 5-HT$_{1A}$ receptor binding (Arango et al. 1995). PCPA and a tryptophan-free drink induce a rapid onset of clinical depression (Delgado et al. 1990). The loss of neuronal morphology could explain many clinical symptoms of depression. These include withdrawal, lethargy, loss of rhythms, vegetative state, inability to feel pleasure and reward, loss of pleasant memories, cognitive deficits, and loss of appetite. Loss of 5-HT leads to neuronal de-differentiation (de-maturation) and then depression. The loss of cortical gray matter seen after severe depression may reflect the instability of a mature phenotype, and the emergence of immature cells.

2. Alzheimer's Disease

The majority of studies performed in rodents and humans indicate that the 5-HT system becomes dysfunctional during aging. Decreased levels of 5-HT (Venero et al. 1993) and human 5-HT$_{1A}$ receptor bindings (Dillon et al. 1991) occur during aging. Aberrant 5-HT-IR fibers appear in the caudate-putamen and the hippocampus in the aged rat (VAN LUIJTELAAR et al. 1988; VAN LUIJTELAAR and TONNAER 1992).

The time that the 5-HT system begins to degenerate appears to be variable. Injection of adult rats with PCA produces first a loss of 5-HT fibers in the cortex, followed by a dramatic hyperinnervation after several months (AXT et al. 1994). When the animals are examined over 1 year later, the age-related reduction of 5-HT cortical fibers is more dramatic than normally seen. Similarly, ACTH prenatal injections can lead to an enhanced innervation by 5-HT fibers; however, as the animals age there is a more precipitous fall of 5-HT than seen in control animals (ALVES et al. 1996). It would appear that hyperinnervation results in accelerated maturation followed by early age-related withdrawal of the fibers. This idea might explain the observations that reduced 5-HT levels from feeding rats low tryptophan or PCPA produced markedly prolonged growth and delayed aging (SEGALL and TIMIRAS 1975; SEGALL et al. 1978). The slower the rate of growth, the more extended the growth period.

The age-related loss of 5-HT may be attributed to the disruptions of 5-HT_{1A} receptors or changes in S-100β and IL-1 (reviewed in AZMITIA et al. 1992). S-100β is localized to chromosome 21 (ALLORE et al. 1988) within the obligate region for Down's syndrome. Nearly every Down's child develops Alzheimer's disease as young adults. Blood levels of S 100 correlate with the degree of mental retardation in Down's syndrome and with the amount of plaque formation in Alzheimer's disease (GRIFFIN et al. 1989). In addition, Alzheimer's disease appears to also compromise the 5-HT_{2A} receptors (MARCUSSON et al. 1984, 1987; CROSS et al. 1984). The dentate gyrus of the aged shows a decline in both serotoninergic fibers and S-100β immunoreactivity (UEDA et al. 1995). Whether the serotonin system or the S-100 system fails first remains to be determined.

The changes described are consistent with the dramatic loss of synapses and brain volume observed in Alzheimer's disease (MASLIAH and TERRY 1993; OBARA et al. 1994; BOBINSKI et al. 1995). The model of dynamic instability of neurons would predict that the neurons are not dying in Alzheimer's brains, but regressing to immature states induced by disruption of the serotonin system and glial S-100β.

Acknowledgements. The authors would like to thank Alice Borella for her excellent artwork and patience. Dr. Pamela Hou contributed to many aspects of organizing this chapter. The manuscript was critically read by Nancy Kheck and Jose Poblete. The research described in this chapter is mainly supported by the National Institute on Aging through a Program Project Grant AG10208 entitled "S-100β: a neuronal-glial link to Alzheimer's disease."

References

Abrous DN, Manier M, Mennicken F, Feuerstein C, Le Moal M, Herman JP (1993) Intrastriatal transplants of embryonic dopaminergic neurons counteract the increase of striatal enkephalin immunostaining but not serotoninergic sprouting elicited by a neonatal lesion of the nigrostriatal dopaminergic pathway. Eur J Neurosci 5:128–136

Ahmad G, Zamenhof S (1978) Serotonin as a growth factor for chick embryo brain. Life Sci 22:963–970

Aigner L, Caroni P (1995) Absence of persistent spreading, branching, and adhesion in GAP-43-depleted growth cones. J Cell Biol 128:647

Akbari HM, Whitaker-Azmitia PM, Azmitia EC (1994) In utero exposure to cocaine results in decreased availability of the trophic S-100b: a possible mechanism for cocaine-induced microcephaly. Neurosci Lett 170:141–144

Allore R, O'Hanlon D, Price R, Neilson K, Willard HF, Cox DR, Marks A, Dunn RJ (1988) Gene encoding the beta subunit of S-100 protein is on chromosome 21: implications for Down's syndrome. Science 239:1311–1313

Alonso G, Ridet JL, Oestreicher AB, Gispen WH, Privat A (1995) B-50 (GAP-43) immunoreactivity is rarely detected within intact catecholaminergic and serotonergic axons innervating the brain and spinal cord of the adult rat, but is associated with these axons following lesion. Exp Neurol 134:35–48

Alves SE, Akbari HM, Azmitia EC, Strand FL (1993) Neonatal ACTH and corticosterone alter hypothalamic monoamine innervation and reproductive parameters in the female rat. Peptides 14:379–384

Alves SE, Akbari HM, Azmitia EC, Strand FL (1996) Adult decrease in monoamines after neonatal ACTH injections. Peptides (in press)

Anorexia Nervosa Study Group (1995) Anorexia nervosa: directions for future research. Int J Eat Disord 17:235–241

Aoki C, Siekevitz P (1988) Plasticity in brain development. Sci Am 259:56–64

Arango V, Underwood M, Gubbi A, Mann J (1995) Localized alterations in pre- and postsynaptic serotonin binding sites in the ventrolateral prefrontal cortex of suicide victims. Brain Res 688:121–133

Artigas F, Sunol C, Tusell JM, Martinez E, Gelpi E (1985) Comparative ontogenesis of brain tryptamine, serotonin, and tryptophan, J Neurochem 44:31–37

Attanasio A, Rager K, Gupta D (1986) Ontogeny of circadian rhythmicity for melatonin, serotonin, and N-acetylserotonin in humans. J Pineal Res 3:251–256

Axt KJ, Mamounas LA, Molliver ME (1994) Structural features of amphetamine neurotoxicity in the brain. In: Cho AK, Segal DS (eds) Amphetamine and its analogs: psychopharmacology, toxicology and abuse. NIDA publication, Washington, DC, pp 315–367

Azmitia EC, de Kloet ER (1987) ACTH neuropeptide stimulation of serotoninergic neuronal maturation in tissue culture: modulation by hippocampal cells. In: de Kloet ER, Wiegant VM, de Wied D (eds) Progress in brain research, vol 72. Elsevier, Amsterdam, pp 311–317

Azmitia EC, Liao B (1994) Dexamethasone reverses adrenalectomy-induced dedifferentiation in midbrain raphe-axis. Ann N Y Acad Sci 746:180–193

Azmitia EC, Zhou FC (1986) Chemically induced homotypic collateral sprouting of hippocampal serotonergic afferents. Exp Brain Res 13:129–141

Azmitia EC, Buchan AM, Williams JH (1978) Structural and functional restoration by collateral sprouting of hippocampal 5-HT axons. Nature 274:374–377

Azmitia EC, Whitaker-Azmitia PM, Bartus R (1988) Use of tissue culture models to study neuronal regulatory trophic and toxic factors in the aged brain. Neurobiol Aging 9:743–758

Azmitia EC, Frankfurt M, Davila M, Whitaker-Azmita PM, Zhou FC (1990a) Plasticity of fetal and adult CNS serotonergic neurons: role of growth-regulatory factors. Neuropharmacol Serotonin 16:343–365

Azmitia EC, Dolan K, Whitaker-Azmitia PM (1990b) S-100b but not NGF, EGF, insulin or calmodulin is a CNS serotonergic growth factor. Brain Res 516:354–356

Azmitia EC, Lama P, Segal M, Whitaker-Azmitia PM, Murphy RB, Zhou FC (1991a) Activity of hippocampal extract on development of [^{3}H]5-HT high-affinity uptake in dissociated microcultures. Int J Dev Neurosci 9:251–258

Azmitia EC, Whitaker-Azmitia PM (1991b) Awakening the sleeping giant: anatomy and plasticity of the brain serotonergic system. J Clin Psychiatry 52:4–16

Azmitia EC, Griffin SW, Marshak DR, Van Eldik LT and Whitaker-Azmitia PM (1992) 5-HT$_{1A}$ and S-100β; a neuronal/glial link to Alzheimer's disease. In: Yu ACH, Hertz L, Norenberg MD, Sykova E, Waxman S (eds) Neuronal-astrocyte interactions: implications for normal and pathological CNS function. Prog Brain Res 94:459–473

Azmitia EC, Liao B, Chen Y (1993) Increase of tryptophan hydroxylase enzyme protein by dexamethasone in adrenalectomized rat midbrain. J Neurosci 13:5041–5055

Azmitia EC, Rubinstein VJ, Strafaci JA, Rios JC, Whitaker-Azmitia PM (1995) 5-HT$_{1A}$ agonist and dexamethasone reversal of para-chloroamphetamine induced loss of MAP-2 and synaptophysin immunoreactivity in adult rat brain. Mol Brain Res 677:181–192

Baido AI, Shiriaeva NV, Khichenko VI, Liuboslavskaia PN, Starostina MV (1992) Development of long-term post-tetanic potentiation and changes in S-100 protein contents in hippocampal slices of rats in different functional states (in Russian). Biull Eksp Biol Medit 113:645–646

Bar-Peled O, Gross-Isseroff R, Ben-Hur H, Hoskins I, Groner Y, Biegon A (1991) Fetal brain exhibits a prenatal peak in the number of 5-HT$_{1A}$ receptors. Neurosci Lett 127:173–176

Baudier J, Mochly-Rosen D, Newton A et al (1987) Comparison of S-100β protein with calmodulin: interactions with melittin and microtubule-associated τ proteins and inhibition of phosphorylation of τ proteins by protein kinase C. Biochem J 25:2886–2893

Baumgarten HG, Bjorklund A, Lachenmayer L, Nobin A, Stenevi U (1971) Long-lasting selective depletion of brain serotonin by 5,6-dihydroxytryptamine. Acta Physiol Scand [Suppl] 373:1–16

Baumgarten HG, Björklung A, Lachenmayer L, Nobin A (1973) Evaluation of the effects of 5,7-dihydroxatryptamine on serotonin and catecholamine neurons in the rat CNS. Acta Physiol Scand [Suppl] 391:1–19

Becquet D, Francious-Bellan AM, Boudouresque F, Faudon D, Hery F, Guillaume V, Hery M (1990) Serotonin synthesis from tryptophan by hypothalamic cells in serum-free medium culture. Dev Brain Res 54:142–146

Bell J III, Zhang XN Whitaker-Azmitia PM (1992) 5-HT$_3$ receptor-active drugs alter development of spinal serotonergic innervation: lack of effect of other serotonergic agents. Brain Res 571:293–297

Bendotti C, Servadio A, Samanin RJ (1991) Distribution of GAP-43 mRNA in the brain stem of adult rats as evidenced by an in situ hybridization: localization within monoaminergic neurons. J Neurosci 11:600–607

Bendotti C, Baldessari S, Pende M, Tarizzo G, Miari A, Presti ML, Mennini T, Samanin R (1994) Does GFAP mRNA and mitochondrial benzodiazepine receptor binding detect serotonergic neuronal degeneration in rat? Brain Res Bull 34:389–394

Benowitz LI, Routtenberg A (1987) A membrane phosphoprotein associated with neural development, axon regeneration, phospholipid metabolism and synaptic platisticity. TINS 10:527

Berger TW, Kaul S, Stricker EM, Zigmond MJ (1985) Hyperinnervation of the striatum by dorsal raphe afferents after dopamine-depleting brain lesions in neonatal rats. Brain Res 336:354–358

Bhattacharyya A, Oppenheim RW, Prevette D, Moore BW, Brackenbury R, Ratner N (1992) S100 is present in developing chicken neurons and Schwann cells and promotes motor neuron survival in vivo. J Neurobiol 23:451–466

Bjorklund A, Stenevi U, Svendgaard NA (1976) Growth of transplanted monoaminer-
    gic neurons into the adult hippocampus along the perforant path. Nature 262:787–
    790
Bjorklund A, Wiklund L, Descarries L (1981) Regeneration and plasticity of central
    serotonergic neurons: a review. J Physiol (Paris) 77:247–255
Bobinski M, Wegiel J, Wisniewski HM, Tarnawski M, Reisberg B, Mlodzik B, de Leon
    MJ, Miller DC (1995) Atrophy of hippocampal formation subdivisions correlate
    with stage and duration of Alzheimer's disease. Dementia 6:205–210
Brady KT, Sonne SC, Roberts JM (1995) Sertraline treatment of comorbid posttrau-
    matic stress disorder and alcohol dependency. J Clin Psychiatry 56:502–505
Bremner JD, Randall P, Scott TM et al (1995) MRI based measurement of hippocam-
    pal volume in patients with combat-related posttraumatic-stress disorder. Am J
    Psychiatry 152:973–981
Buznikov GA (1984) The action of neurotransmitters and related substances on early
    embryogenesis. Pharmacol Ther 25:23–59
Buznikov, GA, Nikitina LA, Galanov AY, Malchenko LA, Trubnikova OB (1993) The
    control of oocyte maturation in the starfish and amphibians by serotonin and its
    antagonist. Int J Dev Biol 37:363–364
Chadi G, Tinner B, Agnati LF, Fuxe K (1993) Basic fibroblast growth factor (bFGF,
    FGF-2) immunoreactivity exists in the noradrenaline, adrenaline and 5-HT nerve
    cells of the rat brain. Neurosci Lett 160:171–176
Chalmers DT, Kwak SP, Mansour A, Akil H, Watson SJ (1993) Corticosteroids regu-
    late brain hippocampal 5-HT$_{1A}$ receptor mRNA expression. J Neurosci 13:914–923
Chan-Palay V (1976) Serotonin axons in the supra- and subependymal plexuses and in
    the leptomeninges: their roles in local alterations of cerebrospinal fluid and vaso-
    motor activity. Brain Res 102:103–130
Cheng L, Hamaguchi K, Ogawa M, Hamada S, Okado N (1994) PCPA reduces both
    monoaminergic afferents and nonmonoaminergic synapses in the cerebral cortex.
    Neurosci Res 19:111–115
Chubakov AR, Gromova EA, Konovalov GV, Chumasov EI, Sarkisova EF (1986a)
    Effects of serotonin on the development of a rat cerebral cortex tissue culture.
    Neurosci Behav Physiol 16:490–497
Chubakov AR, Gromova EA, Konovalov GV, Sarkisova EF, Chumasov EI (1986b)
    The effects of serotonin on the morphofunctional development of rat cerebral
    neocortex in tissue culture. Brain Res 369:285–97
Chubakov AR, Tsyganova VG, Sarkisova EF (1993) The stimulating influence of the
    raphe nuclei on the morphofunctional development of the hippocampus during
    their combined cultivation. Neurosci Behav Physiol 23:271–276
Chumasov EI, Chubakov AR, Konovalov GV, Gromova EA (1978) Effect of seroto-
    nin on the growth and differentiation of the cellular elements of the hippocampus
    during cultivation (in Russian). Arkh Anat Gistol Embriol 74:98–106
Claustre Y, Rouquier L, Scatton B (1988) Pharmacological characterization of sero-
    tonin-stimulated phosphoinositide turnover in brain regions of the immature rat.
    J Pharmacol Exp Ther 244:1051–1056
Cotman CW, GS Lynch (1976) Reactive synaptogenesis in the adult nervous system:
    the effects of partial deafferentation on new synapse formation. In: Barondes SH
    (ed) Neuronal recognition. Plenum, New York, pp 69–108
Commins DL, Axt KJ, Vosmer G, Seiden LS (1987) Endogenously produced
    5,6-dihydroxytryptamine may mediate the neurotoxic effects of para-
    chloroamphetamine. Brain Res 419:253–261
Crissman RS, Arce EA, Bennett-Clarke CA, Mooney RD, Rhoades RW (1993) Re-
    duction in the percentage of serotonergic axons making synapses during the devel-
    opment of the superficial layers of the hampster's superior colliculus. Dev Brain
    Res 75:131–135
Cross AJ, Crow TJ, Johnson JA, Perry EK, Perry RH, Blessed G, Tomlinson
    BE (1984) Studies on neurotransmitter receptor systems in neocortex and

hippocampus in senile dementia of the Alzheimer-type. J Neurol Sci 64: 109–117

Crutcher KA, Collins F (1986) Entorhinal lesions result in increased nerve growth factor-like growth-promoting activity in medium conditioned by hippocampal slices. Brain Res 399:383–389

D'Amato RJ, Blue ME, Largent BL, Lynch DR, Ledbetter DJ, Molliver ME, Snyder SH (1987) Ontogeny of the serotonergic projection to rat neocortex: transient expression of a dense innervation to primary sensory areas. Proc Natl Acad Sci USA 84:4322–4326

Daval G, Verge D, Becerril A, Gozlan H, Spampinato U, Hamon M (1987) Transient expression of 5-HT$_{1A}$ receptor binding sites in some areas of the rat CNS during postnatal development. Int J Dev Neurosci 5:171–189

Davila-Garcia MI, Azmitia EC (1989) Effects of acute and chronic administration of leu-enkephalin on cultured serotonergic neurons. Dev Brain Res 49:97–103

Davila-Garcia MI, Azmitia EC (1990) Neuropeptides as positive or negative neuronal growth regulatory factors: effects of ACTH and leu-enkephalin on cultured serotonergic neurons. In: Lauder JM, Privat A, Giacobini E, Timiras P, Vernidakis A (eds) Molecular aspects of development and aging of the nervous system. Plenum, New York

Delgado PL, Charney DS, Price LH, Aghajanian GK, Landis H, Heninger GR (1990) Serotonin function and the mechanism of antidepressant action. Arch Gen Psychiatry 47:411–418

Descarries L, Soghomonian JJ, Garcia S, Doucet G, Bruno JP (1992) Ultrastructural analysis of the serotonin hyperinnervation in adult rat neostriatum following neonatal dopaminergic denervation with 6-hydroxydopamine. Brain Res 569:1–13

Dewar KM, Grondin L, Carli M, Reader TA (1992) [^{3}H]Paroxetine binding and serotonin content of rat cortical areas, hippocampus, neostratium, ventral mesencephalic tegmentum, and midbrain raphe nuclei region following p-chlorophenylalanine and p-chloroamphetamine treatment. J Neurochem 58:250–257

Diez-Guerra FJ, Avila J (1993) MAP2 phosphorylation parallels dendrite arborization in hippocampal neurons in culture. Neuroreport 4:419–22

Dillon KA, Gross-Isseroff R, Israeli M, Biegon A (1991) Autoradiographic analysis of serotonin 5-HT$_{1A}$ receptor binding in the human brain postmortem: effects of age and alcohol. Brain Res 554:56–64

Di Pasquale E, Monteau R, Hilaire G (1994) Endogenous serotonin modulates the fetal respiratory rhythm: an in vitro study in the rat. Brain Res Dev Brain Res 80:222–232

Di Sant'Agnese PA, Cockett AT (1994) The prostatic endocrine-paracrine (neuroendocrine) regulatory system and neuroendocrine differentiation in prostatic carcinoma: a review and future directions in basic research. J Urol 152:1927–1931

Dyck CH, Cynader MS (1993) Autoradiographic localization of serotonin subtypes in cat visual cortex: transient regional, laminar, and columnar distribution during postnatal development. J Neurochem 13:4316–4338

Eaton MJ, Staley JK, Globus MY, Whittemore SR (1995) Developmental regulation of early serotonergic neuronal differentiation: the role of brain-derived neurotrophic factor and membrane depolarization. Dev Biol 170:169–182

Fischer C, Hatzidimitriou G, Wlos J, Katz J, Ricaurte G (1995) Reorganization of ascending 5-HT axon projections in animals previously exposed to the recreational drug (±)3,4-methylenedioxymethamphetamine (MDMA, ecstasy). J Neurosci 15:5476–5485

Frankfurt M, Azmitia EC (1983) The effect of intracerebral injections of 5,7-dihydroxytryptamine and 6-hydroxydopamine on the serotonin immunoreactive cell bodies and fibers in the adult rat hypothalamus. Brain Res 261:91–99

Frankfurt M, Beaudet A (1987) Ultrastructural organization of regenerated serotonin axons in the dorsomedial hypothalamus of the adult rat. J Neurocytol 6:799–809

Garattini S, Valzelli L (1965) Serotonin. Elsevier, Amsterdam

Gemma C, De Luigi A, De Simoni MG (1994) Permissive role of glucocorticoids on interleukin-1 activation of the hypothalamic serotonergic system. Brain Res 651:169–173

Giulian D, McEwen BS, Pohorecky LA (1974) Altered development of the rat brain serotonergic system after disruptive neonatal experience. Proc Natl Acad Sci USA 71:4106–4110

Godridge H, Reynolds GP, Czudek C, Calcutt NA, Benton M (1987) Alzheimer-like neurotransmitter deficits in adult Down's syndrome brain tissue. J Neurol Neurosurg Psychiatry 50:775–778

Goldberg JI, Kater SB (1989) Expression and function of the neurotransmitter serotonin during development of Helisoma nervous system. Dev Biol 131:483–495

Gorio A, Di Giulio AM, Germani E, Bendot C, Bertelli A, Mantegazza P (1993) Perinatal morphine treatment inhibits pruning effect and regeneration of serotoninergic pathways following neonatal 5,7-HT lesions. J Neurosci Res 34:462–471

Griffin WST, Stanley LC, Ling C, White L, MacLeod V, Perrot LJ, White CL, Araoz C (1989) Brain interleukin 1 and S-100 immunoreactivity are elevated in Down's syndrome and Alzheimer's disease. Proc Natl Acad Sci USA 86:7611–7615

Gu Q, Patel B, Singer W (1990) The laminar distribution and postnatal development of serotonin-immunoreactive axons in the cat primary visual cortex. Exp Brain Res 81:257–266

Haring JH, Yan W (1993) Effects of neonatal serotonin depletion on the development of rat dentate granule cells. Soc Neurosci Abstr 19:673

Haring JH, Hagan A, Olson J, Rodgers B (1993) Hippocampal serotonin levels influence the expression of S100B detected by immunocytochemistry. Brain Res 631:119–123

Haring JH, Yan W, Wilson CC (1995) 5-HT$_{1A}$ receptors mediate the effects of 5-HT on developing dentate granule cells. Soc Neurosci Abst 21:862

Hedner J, Lundell KH, Breese GR, Mueller RA, Hedner T (1986) Developmental variations in CSF monoamine metabolites during childhood. Biol Neonate 49:190–197

Hentschel F, Schmidbauer M, Detzner U, Blanz B, Schmidt MH (1995) Reversible changes in brain volume in anorexia nervosa. Z Kinder Jugenpsychiatr 23:104–112

Hesketh J, Baudier J (1986) Evidence that S-100 proteins regulate microtubule assembly and stability in rat brain extracts. Int J Biochem 18:691–695

Hilakivi-Clarke LA, Goldberg R (1993) Effect of tryptophan and serotonin uptake inhibitors on behavior in male transgenic transforming growth factor alpha mice. Eur J Pharmacol 237:101–108

Hilakivi-Clarke LA, Corduban TD, Taira T, Hitri A, Deutsch S, Korpi ER, Goldberg R, Kellar KJ (1995) Alterations in brain monoamines and GABA A receptors in transgenic mice overexpressing TGF alpha. Pharmacol Biochem Behav 50:593–600

Hillion J, Catelon M, Riad M, Hamon M, deVitry F (1994) Neuronal localization of 5-HT$_{1A}$ receptor mRNA and protein in rat embryonic brain stem cultures. Dev Brain Res 79:195–202

Hohmann CF, Hamon R, Batshaw ML, Coyle JT (1988) Transient postnatal elevation of serotonin levels in mouse neocortex. Brain Res 471:163–166

Hole K (1972a) Behavior and brain growth in rats treated with p-chlorophenylalanine in the first weeks of life. Dev Psychobiol 5:157–173

Hole K (1972b) Reduced 5-hydroxyindole synthesis reduces postnatal brain growth in rats. Eur J Pharmacol 18:361–366

Hole K (1972c) The effects of cyproheptadine, methysergide, BC 105 and reserpine on brain 5-hydroxytryptamine and brain growth. Eur J Pharmacol 19:156–159

Huether G, Thomke F, Adler L (1992) Administration of tryptophan-enriched diets to pregnant rats retards the development of the serotonergic system in their offspring. Brain Res Dev Brain Res 68:175–181

Hyman C, Hofer M, Barde YA, Juhasz M, Yancopoulos GD, Squinto SP, Lindsay RM (1991) BDNF is a neurotrophic factor for dopaminergic neurons of the substantia nigra. Nature 350:230–232

Ike J, Canton H, Sanders-Busch E (1995) Developmental switch in the hippocampal serotonin receptor linked to phosphoinositide hydrolysis. Brain Res 678:49–54

Jackson D, Abercrombie ED (1992) In vivo neurochemical evaluation of striatal serotonergic innervation in striatum depleted of dopamine at infancy. J Neurochem 58:890–897

Jacobs BL, Azmitia EC (1992) Structure and function of the brain serotonin system. Physiol Rev 72:165–229

Jameson L, Caplow M (1981) Modification of microtubule steady-state dynamics by phosphorylation of the microtubule-associated proteins. Proc Natl Acad Sci USA 78:3413–3417

Jernigan TL, Butters N, DiTraglia G, Schafer K, Smith T, Irwin M, Grant I, Schuckit M, Cermak LS (1991) Reduced cerebral grey matter observed in alcoholics using magnetic resonance imaging. Alcohol Clin Exp Res 15:418–427

Jonsson G, Hallman H (1983) Effect of substance P on the 5,7-dihydroxytryptamine induced alteration of the postnatal development of central serotonin neurons. Med Biol 61:105–122

Jope RS, Johnson GV (1992) Neurotoxic effects of dietary aluminium. Ciba Found Symp 169:254–262; discussion 262–267

King JA, Davila-Garcia M, Azmitia EC, Strand FL (1991) Differential effects of prenatal and postnatal ACTH or nicotine exposure on 5-HT high affinity uptake in the neonatal rat brain. Int J Dev Neurosci 9:281–286

Kirstein CL, Spear LP (1988) 5-HT$_{1A}$, 5-HT$_{1B}$ and 5-HT$_2$ receptor agonists induce differential behavioral responses in neonatal rat pups. Eur J Pharmacol 150:339–345

Kligman D, Marshak DR (1985) Purification and characterization of a neurite extension factor from bovine brain. Proc Natl Acad Sci USA 82:7136–7139

Kontur PJ, Leranth C, Redmond DE Jr, Roth RH, Robbins RJ (1993) Tyrosine hydroxylase immunoreactivity and monoamine and metabolite levels in cryopreserved human fetal ventral mesencephalon. Exp Neurol 121:172–180

Kramer HK, Poblete JC, Azmitia EC (1995) 3,4-Methylenedioxymethamphetamine ("Ecstasy") promotes the translocation of protein kinase C (PKC): requirement of viable serotonin nerve terminals. Brain Res 680:1–8

Lauder JM (1990) Ontogeny of the serotonergic system in the rat: serotonin as a developmental signal. Ann N Y Acad Sci 600:297–314

Lauder JM, Krebs H (1978) Serotonin as a differentiation signal in early neurogenesis. Dev Neurosci 1:15–30

Lauder JM, Towle AC, Patrick K, Henderson P, Krebs H (1985) Decreased serotonin content of embryonic raphe neurons following maternal administration of p-chlorophenylalanine: a quantitative immunocytochemical study. Brain Res 352:107–114

Lauder JM, Wallace JA, Krebs H, Petrusz P, McCarthy K (1982) In vivo and in vitro development of serotonergic neurons. Brain Res Bull 9:605–625

Leslie MJ, Bennett-Clarke CA, Rhoades RW (1992) Serotonin 1B receptors form a transient vibrissa-related pattern in the primary somatosensory cortex of the developing rat. Brain Res 69:143–148

Levallois C, Calvet MC, Petite D (1993) Immunocytochemical demonstration of monoamine-containing cells in human fetal brain stem dissociated cultures. Brain Res 75:141–145

Lewis D, Teyler TJ (1986) Anti-S-100 serum blocks long-term potentiation in the hippocampal slice. Brain Res 383:159

Liao B, Miesak B, Azmitia EC (1993) Loss of 5-HT$_{1A}$ receptor mRNA in the dentate gyrus of long-term adrenalectomized rats and rapid reversal by dexamethasone. Mol Brain Res 19:328–332

Lidov HGW, Molliver ME (1982a) An immunohistochemical study of serotonin neuron development in the rat: ascending pathways and terminal fields. Brain Res Bull 8:389–430

Lidov HG, Molliver ME (1982b) The structure of cerebral cortex in the rat following prenatal administration of 6-hydroxydopamine. Brain Res 255:81–108

Lidov HGW, Molliver ME (1982c) Immunohistochemical study of the development of serotonergic neurons in the rat CNS. Brain Res Bull 9:559–604

Lindholm D, Harikka J, da Penha Berzaghi M, Castren E, Tzimagiorgis G, Hughes RA, Thoenen H (1994) Fibroblast growth factor-5 promotes differentiation of cultured rat septal cholinergic and raphe serotonergic neurons: comparison with the effects of neurotrophins. Eur J Neurosci 6:244–252

Linthorst AC, Flachskamm C, Muller-Preuss P, Holsboer F, Reul JM (1995) Effect of bacterial endotoxin and interleukin-1 beta on hippocampal serotonergic neurotransmission, behavioral activity, and free corticosterone levels: an in vivo microdialysis study. J Neurosci 15:2920–2934

Liu JP, Lauder JM (1992) S-100 beta and insulin-like growth factor-II differentially regulate growth of developing serotonin and dopamine neurons in vitro. J Neurosci Res 33:248–256

Luine VN, Renner KJ, Frankfurt M, Azmitia EC (1984) Facilitated sexual behavior reversed and serotonin restored by raphe nuclei transplanted into denervated hypothalamus. Science 1436–1439

Mamounas LA, Molliver ME (1988) Evidence for dual serotonergic projections to neocortex: axons from the dorsal and median raphe nuclei are differentially vulnerable to the neurotoxin p-chloroamphetamine (PCA). Exp Neurol V:23–36

Mamounas LA, Blue ME, Siuciak JA, Altar CA (1995) Brain-derived neuronotrophic factor promotes the survival and sprouting of serotonergic axons in rat brain. J Neurosci 15:7929–7939

Marcusson JO, Morgan DG, Winblad BF, Finch CE (1984) Serotonin-2 binding sites in human frontal cortex and hippocampus: selective loss of S-2A sites with age. Brain Res 311:51–56

Marcusson JO, Alafuzoff I, Backstrom IT, Ericson E, Gottfries CG, Winblad B (1987) 5-Hydroxytryptamine-sensitive [^{3}H]imipramine binding of protein nature in the human brain. II. Effect of normal aging and dementia disorders. Brain Res 425:137–145

Masliah E, Terry R (1993) The role of synaptic proteins in the pathogenesis of disorders of the central nervous sytem. Brain Pathol 3:77–85

Matesic DF, Lin RC (1994) Microtubule-associated protein 2 as an early indicator of ☞-induced neurodegeneration in the gerbil forebrain. J Neurochem 63:1012–1020

Matthes H, Boschert U, Amlaiky N, Grailhe R, Plassat JL, Muscatelli F, Mattei MG, Hen R (1993) Mouse 5-hydroxytryptamine 5A and 5-hydroxytryptamine 5B receptors define a new family of serotonin receptors: cloning, functional expression, and chromosomal localization. Mol Pharmacol 43:313–319

Maurizi CP (1985) Could supplementary dietary tryptophan prevent sudden infant death syndrome? Med Hypotheses 17:149–154

Meaney MJ, Diorio J, Francis D, LaRocque S, O'Donnell D, Smythe JW, Sharma S, Tannenbaum B (1994) Environmental regulation of the development of glucocorticoid receptor systems in the rat forebrain: the role of serotonin. Ann N Y Acad Sci 746:260–273

Meijer OC, deKloet ER (1994) Corticosterone suppresses the expression of 5-HT$_{1A}$ receptor mRNA in rat dentate gyrus. Eur J Pharmacol 266:255–261

Merlio JP, Ernforns P, Jaber M, Persson H (1992) Molecular cloning of rat trk C and distribution of cells expressing messenger RNA's for members of the trk family in the rat central nervous system. Neuroscience 51:513–532

Messenger NJ, Warner AE (1989) The appearance of neural and glial markers during early development of the nervous system in the amphibian embryo. Development 107:43–54

Micquel MC, Emerit MB, Gingrich JA, Nosjian A, Hamon M, el Mestikawy E (1995) Developmental changes in the differential expression of two serotonin 5-HT-3 receptor splice variants in the rat. J Neurochem 65:475–483

Mitchison TM, Kirschner MW (1984) Dynamic instability of microtubule growth. Nature 312:237–242

Moiseiwitsch JR, Lauder JM (1995) Serotonin regulates mouse cranial neural creast migration. PNAS 93:7182–7186

Mower GD (1991) Comparison of serotonin 5-HT$_1$ receptors and innervation in the visual cortex of normal and dark-reared cats. J Comp Neurol 312:223–230

Muller CM, Akhavan AC, Bette M (1993) Possible role of S-100 in glia-neuronal signalling involved in activity dependent plasticity in the developing mammalian cortex. J Chem Neuroanat 6:215

Ni L, Jonakait GM (1989) Ontogeny of substance P-containing neurons in relation to serotonin-containing neurons in the central nervous system of the mouse. Neuroscience 30:257–269

Niitsu Y, Hamada S, Hamaguchi K, Mikuni M, Okado N (1995) Regulation of synaptic density by 5-HT$_{2A}$ receptor agonist and antagonist in the spinal cord of chicken embryo. Neurosci Lett 195:159–162

Nishi M, Whitaker-Azmitia PM, Azmitia EC (1996a) Effects of 5-HT-1a, S100β and steroids on synaptophysin immunoreactivity in cultured hippocampal neurons. Synapse 23:1–9

Nishi M, Poblete JC, Whitaker-Azmitia PM, Azmitia EC (1996b) Brain derived neurotrophic factor and S100b; trophic interactions on cultured serotonergic neurons. *Neurosci. Net* 10003, www.neurosclence.com)

Nobin A, Bjorklund A (1973) Topography of the monoamine neuron systems in the human brain as revealed in fetuses. Acta Physiol Scand 388:1–40

Nygren LG, Fuxe K, Jonsson G, Olson L (1974) Functional regeneration of 5-hydroxytryptamine nerve terminals in the rat spinal cord following 5,6-dihydroxytryptamine induced degeneration. Brain Res 78:377–394

Obara K, Meyer JS, Mortel KF, Muramatsu K (1994) Cognitive decline correlates with decreased cortical volume and perfusion in dementia of Alzheimer type. J Neurol Sci 127:96–102

Okado N, Cheng L, Tanatsugu S, Hamaguchi K (1993) Synaptic loss following removal of serotonergic fibers in newly hatched and adult chickens. J Neurobiol 24:687–691

Olsen L, Seiger A (1972) Early prenatal ontogeny of central monoamine neurons in the rat: fluorescence histochemical observation. Z Anat Entwicklungsgesch 137:301–316

Pakala R, Willweson JT, Benedict CR (1994) Mitogenic effect of serotonin on vascular endothelial cells. Circulation 90:1919–1926

Peters DA (1988) Effects of maternal stress during different gestational periods on the serotonergic system in adult rat offspring. Pharmacol Biochem Behav 31:839–843

Peters DA (1990) Maternal stress increases fetal brain and neonatal cerebral cortex 5-hydroxytryptamine synthesis in rats: a possible mechanism by which stress influences brain development. Pharmacol Biochem Behav 35:943–947

Poblete JC, Azmitia EC (1995) Activation of glycogen phosphorylase by serotonin and 3,4-methylenemethamphetamine in astroglial-rich primary cultures: involvement of the 5-HT$_{2A}$ receptor. Brain Res 680:9–15

Pranzatelli MR, Galvan I (1994) Ontogeny of [^{125}I]iodocyanopindolol-labelled 5-hydroxytryptamine 1B-binding sites in the rat CNS. Neurosci Lett 167:166–170

Ramon y Cajal S (1928) Degeneration and regeneration in the nervous system. Oxford University Press, London

Rajaofetra N, Sandillon F, Geffard M, Privat A (1989) Pre- and post-natal ontogeny of serotonergic projections to the rat spinal cord. J Neurosci Res 22:305–321

Raymond JR, Olsen CL (1994) Protein kinase A induces phosphorylation of the human 5-HT$_{1A}$ receptor and augments its desensitization by protein kinase C in CHO-K1 cells. Biochemistry 33:11264–11269

Reader TA, Radja F, Dewar KM, Descarries L (1995) Denervation, hyperinnervation, and interactive regulation of dopamine and serotonin receptors. Ann N Y Acad Sci 757:293–310

Rhoades RW, Mooney RD, Chiaia NL, Bennett-Clarke CA (1990) Development and plasticity of the serotoninergic projection to the hamster's superior colliculus. J Comp Neurol 299:151–166

Rhoades RW, Bennett-Clarke CA, Shi MY, Mooney RD (1994) Effects of 5-HT on thalamocortical synaptic transmission in the developing rat. J Neurophysiol 72:2438–2450

Riad M, Emerit MB, Hamon MT (1994) Neurotrophic effects of ipsapirone and other 5-HT$_{1A}$ receptor agonists on septal cholinergic neurons in culture. Dev Brain Res 82:245–258

Richards GE, Gluckman PD, Ball K, Mannelli SC, Kalamaras JA (1990) Expanded ontogeny of neurotransmitters and their metabolites in the brains of fetal and newborn lambs. J Dev Physiol 14:331–336

Roback JD, Diede SJ, Downen M, Lee HJ, Kwon J, Large TH, Otten U, Wainer BH (1992) Expression of neurotrophins and the low-affinity NGF receptor in septal and hippocampal reaggregates cultures: local physiologic effects of NGF synthesized in the septal region. Dev Brain Res 70:123–133

Roback JD, Marsh HN, Downen M, Palfrey HC, Wainer BH (1995) BDNF-activated signal transduction in rat cortical glial cells. Eur J Neurosci 7:849–862

Sanders-Bush E, Bushing JA, Sulser F (1975) Long-term effects of p-chloroamphetamine and related drugs on central serotonergic mechanisms. J Pharmacol Exp Ther 192:33–41

Saruhashi Y, Young W (1994) Effect of mianserin on locomotory function after thoracic spinal cord hemisection in rats. Exp Neurol 129:207–216

Segal M, Richter-Levin G (1989) Raphe cells grafted into the hippocampus can ameliorate spatial memory deficits in rats with combined serotonergic/cholinergic deficiencies. Brain Res 478:184–186

Segall PE, Timiras PS (1975) Age-related changes in the thermoregulatory capacity of tryptophan-deficient rats. Fed Proc 34:83–85

Segall PE, Ooka H, Rose K, Timiras PS (1978) Neural and endocrine development after chronic tryptophan deficiency in rats: I. Brain monoamine and pituitary responses. Mech Aging Dev 7:1–17

Selinfreund RH, Barger SW, Pledger WJ, Van Eldik LJ (1991) Neurotrophic protein S-100 beta stimulates glial cell proliferation. PNAS 88:3554

Shear PK, Jernigan TL, Butters N (1994) Volumetric magnetic resonance imaging quantification of longitudinal brain changes in abstinent alcoholics. Alcohol Clin Exp Res 18:172–176

Shemer AV, Azmitia EC, Whitaker-Azmitia PM (1991) Dose-related effects of prenatal 5-methoxytryptamine (5-MT) on development of serotonin terminal density and behavior. Dev Brain Res 39:59–65

Shen WZ, Luo ZB, Zheng DR, Yew DT (1989) Immunohistochemical studies on the development of 5-HT (serotonin) neurons in the nuclei of the reticular formations of human fetuses. Pediatr Neurosci 15:291–295

Shenker A, Maayani S, Weinstein H, Green JP (1985) Two 5-HT receptors linked to adenylate cyclase in guinea pig hippocampus are discriminated by 5-carboxamidotryptamine and spiperone. Eur J Pharmacol 109:427–432

Sheu F-S, Azmitia EC, Marshak DR, Routtenberg R (1994) Glial-derived S-100b protein selectively inhibits recombinant β protein kinase C (PKC) phosphorylation of neuron-specific protein F1/GAP-43. Mol Brain Res 21:62–66

Shintani F, Nakaki T, Kanba S, Sato K, Yagi G, Shiozawa M, Aiso S, Kato R, Asai M (1995) Involvement of interleukin-1 in immobilization stress-induced increase in plasma adrenocorticotropic hormone and in release of hypothalamic monoamines in the rat. J Neurosci 15:1961–1970

Sloviter RS, Valiquette G, Abrams GM, Ronk EC, Sollas AL, Paul LA, Neubort S (1989) Selective loss of hippocampal granule cells in the mature rat brain after adrenalectomy. Science 243:535–538

Son JH, Baker H, Park DH, Joh TH (1994) Drastic and selective hyperinnervation of central serotonergic neurons in a lethal neurodevelopmental mouse mutant, Anorexia (anx). Brain Res Mol Brain Res 25:129–134

Sotelo C, Beaudet A (1979) Influence of experimentally induced agranularity on the synaptogenesis of serotonin nerve terminals in rat cerebellar cortex. Proc R Soc Lond B Biol Sci 206:133–138

Sparks DL, Hunsaker JC III (1991) Sudden infant death syndrome: altered aminergic-cholinergic synaptic markers in hypothalamus. J Child Neurol 6:335–339

Spenger C, Hyman C, Studer L, Egli M, Evtouchenko L, Jackson C, Dahl-Jorgensen A, Lindsay RM, Seiler RW (1995) Effects of BDNF on dopaminergic, serotonergic, and GABAergic neurons in cultures of human fetal ventral mesencephalon. Exp Neurol 133:50–63

Strittmatter SM, Fankhauser C, Huang PL, Masimo H, Fishman MC (1995) Neuronal pathfinding is abnormal in mice lacking the neuronal growth cone protein GAP-43. Cell 80:445

Stuart DK, Blair SS, Weisblat DA (1987) Cell lineage, cell death, and the developmental origin of identified serotonin- and dopamine-containing neurons in the leech. J Neurosci 7:1107–1122

Sundstrom E, Kolare S, Souverbie F, Samuelsson EB, Pschera H, Lunell NO, Seiger A (1993) Neurochemical differentiation of human bulbospinal monoaminergic neurons during the first trimester. Brain Res 75:1–12

Suzuki F, Kato K, Kato T, Ogasawara N (1987) S-100 protein in clonal astroglioma cells is released by adrenocorticotropic hormone and corticotropin-like intermediate-lobe peptide. J Neurochem 49:1557–1563

Sze PY (1980) Glucocorticoids as a regulatory factor for brain tryptophan hydroxylase during development. Dev Neurosci 3:217–223

Tajuddidn NF, Druse MJ (1993) Treatment of pregnant alcohol consuming rats with buspirone: effects on serotonin and 5-hydroxyindoleacetic acid content in offspring. Alcohol Clin Exp Res 17:110–114

Toth G, Fekete M (1986) 5-Hydroxyindole acetic excretion in newborns, infants and children. Acta Paediatr Hung 27:221–226

Towle AC, Criswell HE, Maynard EH, Lauder JM, Joh TH, Mueller RA, Breese GR (1989) Serotonergic innervation of the rat caudate following a neonatal 6-hydroxydopamine lesion: an anatomical, biochemical and pharmacological study. Pharmacol Biochem Behav 34:367–374

Ueda S, Gu XF, Whitaker-Azmitia PM, Naruse I, Azmitia EC (1994) Neuro-glial neurotrophic interaction in the S-100b retarded mutant mouse (Polydactylys nagoya). I. Immunocytochemical and neurochemical studies. Brain Res 633:275–283

Ueda S, Kokotos ET, Bell J III, Azmitia EC (1995a) Serotonergic sprouting into transplanted C-6 gliomas is blocked by S-100β antisense gene. Mol Brain Res 29:365–368

Ueda S, Takeuchi Y, Sawada T, Kawata M (1995b) Age-related decrease of serotonergic fibers and S100 beta immunoreactivity in the rat dentate gyrus. Neuroreport 10:1445–1448

Valles AM, White K (1988) Serotonin-containing neurons in Drosophila melanogaster: development and distribution. J Comp Neurol 268:414–428

Van Eldik LJ, Barger SW, Welsh MJ (1992) Antisense approaches to the function of glial cell proteins. Ann N Y Acad Sci 660:219–230

Van Hartesveldt C, Moore B, Hartman B (1986) Transient midline raphe glial structure in the developing rat. J Comp Neurol 253:175–184

Van Luijtelaar MG, Tonnaer JA, Steinbusch HW (1992) Serotonergic fibres degenerating in the aging rat brain or sprouting from grafted fetal neurons are not affected by the neurotrophic ACTH analogue Org 2766. J Chem Neuroanat 5: 315–25
Van Luijtelaar MGPA, Steinbusch HWM, Tonnaer JADM (1988) Aberrant morphology of serotonergic fibers in the forebrain of the aged rat. Neurosci Lett 95:93–96
Venero JL, de la Roza C, Machado A, Cano J (1993) Age-related changes on monoamine turnover in hippocampus of rats. Brain Res 631:89–965
Walker PD, Schotland S, Hart RP, Jonakait GM (1990) Tryptophan hydroxylase inhibition increases preprotachykinin mRNA in developing and adult medullary raphe nuclei. Brain Res 8:113–119
Wallace JA, Lauder JM (1983) Development of the serotonergic system in the rat embryo: an immunocytochemical study. Brain Res Bull 10:459–479
Whitaker-Azmitia PM (1988) Astroglial serotonin receptors. In: Kimelberg H (ed) Glial cell receptors. Raven, New York
Whitaker-Azmitia PM (1991) Role of serotonin and other neurotransmitter receptors in brain development: basis for developmental pharmacology. Pharmacol Rev 43:553–561
Whitaker-Azmitia PM, Azmitia EC (1986a) 5-Hydroxytryptamine binding to brain astroglial cells: differences between intact and homogenized preparations and mature and immature cultures. J Neurochem 46:1186–1189
Whitaker-Azmitia PM, Azmitia EC (1986b) Autoregulation of fetal serotonergic neuronal development: role of high affinity serotonin receptors. Neurosci Lett 67:307–312
Whitaker-Azmitia PM, Azmitia EC (1989) Stimulation of astroglial serotonin receptors produces media which regulates development of serotonergic neurons. Brain Res 497:80–85
Whitaker-Azmitia PM, Quartermain D, Shemer AV (1990a) Prenatal treatment with SKF 38393, a selective D-1 receptor agonist: long-term consequences on ^{3}H-paroxetine binding and on dopamine and serotonin receptor sensitivity. Dev Brain Res 57:181–187
Whitaker-Azmitia PM, Murphy R, Azmitia EC (1990b) Stimulation of astroglial 5-HT$_{1A}$ receptors releases the serotonergic growth factor, protein S-100, and alters astroglial morphology. Brain Res 528:155–158
Whitaker-Azmitia PM, Clarke C, Azmitia EC (1993) Localization of 5-HT$_{1A}$ receptors to astroglial cells in adult rats: implications for neuronal-glial interactions and psychoactive drug mechanism of action. Synapse 14:201–205
Whitaker-Azmitia PM, Zhang X, Clarke C (1994) Effects of gestational exposure to monoamine oxidase inhibitors in rats: preliminary behavioral and neurochemical studies. Neuropsychopharmacology 11:125–132
Whitaker-Azmitia PM, Borella A, Raio N (1995) Serotonin depletion in the adult rat causes loss of the dendritic marker MAP-2: a new animal model of schizophrenia? Neuropsychopharmacology 12:269–272
Wiklund L, Bjorklund A (1980) Mechanism of regrowth in the bulbospinal serotonin system following 5,6-dihydroxytryptamine-induced axotomy. II Fluorescence histochemical observations. Brain Res 191:109–127
Wilson MA, Molliver ME (1994) Microglial response to degeneration of serotonergic axon terminals Glia 11:18–34
Winslow JT, Insel TR (1991) Serotonergic modulation of the rat pup ultrasonic isolation call: studies with 5HT$_1$ and 5HT$_2$ subtype-selective agonists and antagonists. Psychopharmacology 105:513–520
Yan W, Wilson CC, Panneton WM, Haring JH (1995) Neonatal 5-HT depletion reduces dentate granule cell spine density. Soc Neurosci Abstr 21: 862

Yang Q, Hamberger A, Hyden H, Wang S, Stigbrand T, Haglid KG (1995) S-100β has a neuronal localisation in the rat hindbrain revealed by an antigen retrieval method. Brain Res 696:49–61

Yang SD, Song JS, Liu HW, Chan WH (1993) Cyclic modulation of cross-linking interactions of microtubule-associated protein-2 with actin and microtubules by protein kinase FA. J Protein Chem 12:393–402

Zalcman S, Green-Johnson JM, Murray L, Nance DM, Dyck D, Anisman H, Greenberg AH (1994) Cytokine-specific central monoamine alterations induced by interleukin-1, -2 and -6. Brain Res 643:40–49

Zhou FC, Azmitia EC (1985) Effect of adrenalectomy and corticosterone on the sprouting of serotonergic fibers in hippocampus. Neurosci Lett 54:111–116

Zhou FC, Azmitia EC (1986) Induced homotypic collateral sprouting of serotonergic fibers in the hippocampus. II. An immunocytochemistry study. Brain Res 373:337–348

Zhou FC, Azmitia EC (1988) Laminin facilitates and guides fiber growth of transplanted neurons in adult brain. J Chem Neuroanat 1:133–146

Zhou FC, Chiang YH (1995) Excitochemicals-induced trophic bridging directs axonal growth of transplanted neurons to distal target. Cell Transplant 4:103–125

Zhou FC, Bledsoe S, Murphy J (1991a) Serotonergic sprouting is induced by dopamine-lesion in substantia nigra of adult rat brain. Brain Res 556:108–116

Zhou FC, Bledsoe S, Lumeng Li T-K (1991b) Immunostained serotonergic fibers are decreased in selected brain regions of alcohol-preferring rats. Alcohol 8:425–431

Zhou FC, Azmitia EC, Bledsoe S (1995) Rapid serotonergic fiber sprouting in response to ibotenic acid lesion in the striatum and hippocampus. Brain Res Dev Brain Res 84:89–98

CHAPTER 2

Anatomy of Central Serotoninergic Projection Systems

H.G. BAUMGARTEN and Z. GROZDANOVIC

A. Introduction

Serotoninergic neurons are integral parts of central and/or peripheral nervous networks in diverse forms of invertebrates and vertebrates (PARENT 1981a,b), suggesting that these neurons provide animals – across phylogeny – with capacities essential for adapting to changing internal and/or external demands. Clues as to their functional role(s) may already be gained from their morphology: giant metacerebral serotoninergic neurons in molluscs have abundantly collateralized axons (COTTRELL 1977) as do some of the large multipolar, extensively ramifying serotoninergic neurons of the mammalian raphe nuclei/ extraraphe reticular 5-HT cell groups (WATERHOUSE et al. 1986; VERTES 1991; VAN BOECKSTAELE et al. 1993; VERTES and KOCSIS 1994; HOLMES et al. 1994), implying that they innervate multiple target networks along the neuraxis. This enables them to coordinate and harmonize activities (or response properties) in diverse networks with state-dependent determinants such as the prevailing level of central motor tone, (somato)-sensory processing and autonomic regulation across the sleep/wake cycle (JACOBS and FORNAL 1993). By way of abundantly collateralizing, a limited number of neurons are capable of modulating electrical activity and afferent input responsivity in multiple targets in a coordinate fashion. Therefore, certain serotoninergic neurons of the brainstem represent archetypical reticular-type multipolar neurons which resemble the Golgi type 1 neurons of the brainstem reticular core described by SCHEIBEL and SCHEIBEL (1958). The straightforward, primitive architecture of these serotoninergic neurons and their regular rhythmic firing patterns contrast with the diversity and selective distribution pattern of the numerous serotonin receptor subtypes which are responsible for the bewildering modulatory roles of serotonin in autonomic and endocrine, motor and sensory functions, and in emotional and complex (intelligent) adaptive behavior.

B. Historical Perspective

KOELLIKER (1891) and CAJAL and RAMON (1911) were the first to describe large multipolar cells in the center (i.e., midline or raphe) division of the vertebrate brainstem the projections of which remained enigmatic. In 1960, BRODAL and coworkers (1960a) noted, as a result of degeneration studies and nonspecific axon staining in cats, that the raphe nuclei entertain massive connections with the forebrain. The raphe nuclei represent the richest source of neuronal serotonin synthesized in the mammalian brain. Although the presence of serotonin in the central nervous system had been known since 1953 (Twarog and Page 1953; Amin et al. 1954), its cellular distribution remained elusive until the discovery of a specific method for the retention of serotonin in tissue (by freeze-drying) and its conversion into a fluorophore (by gaseous formaldehyde treatment: Falck et al. 1962), readily detected by fluorescence microscopy. The application of this method to the brainstem of the rat provided the first precise map of the distribution of serotoninergic neurons in the mammalian CNS (Dahlström and Fuxe 1964). Due to the rapid decomposition of the fluorophore by UV light, a comprehensive map of the projections and pattern and density of innervation of target areas by the delicate serotoninergic axons called for an improved method for visualizing of serotonin in tissue sections. In 1978, Steinbusch and associates developed a selective and sensitive immunohistochemical technique for serotonin by generating antibodies to serotonin linked to albumin by means of formaldehyde (Steinbusch et al. 1978). This technique, which replaced the Falck-Hillarp method, permits a precise analysis of serotoninergic axons and their target relationship by light and electron microscopy. The technological repertoire for demonstrating serotoninergic neurons in the mammalian brain has recently become enriched by the generation of antibodies towards aromatic amino acid hydroxylases (PH8; Haan et al. 1987; Cotton et al. 1988); these antibodies cross-react with tryptophan hydroxylase. PH8 immunohistochemistry has been particularly valuable in studies on the human brain (cf. Baker et al. 1990, 1991a,b; Halliday et al. 1992). The latest addition to the catalogue of methods for the identification of serotoninergic structures comprises antibodies directed against antidepressant-sensitive sites of the serotonin transporter protein (Quian et al. 1995).

An important methodological achievement which has assisted in clarifying of anatomical, physiological and pharmacological aspects of serotonin functions in the CNS was provided by the development of selective serotoninergic neurotoxins (5,6- and 5,7-dihydroxytryptamine; 5,6- and 5,7-DHT; Baumgarten et al. 1971, 1972, 1973a) and their application as tools for selectively ablating serotoninergic projections to targets (cf. reviews by Björklund et al. 1974; Baumgarten and Björklund 1976; Baumgarten et al. 1977, 1978b, 1982; Baumgarten and Zimmermann 1992). In 1972, Sanders-Bush and coworkers described the long-term actions of p-chloroamphetamine (p-CA) on forebrain 5-hydroxyindoles in the rat, a finding subsequently confirmed by many researchers (cf. Fuller 1992 for references). While 5,7-DHT has neurotoxic effects on all types of serotoninergic neurons in the mammalian brain, p-CA preferentially damages (ascending) axonal projections of the dorsal raphe nucleus; its selectivity, however, is not absolute and depends on the dose and dosing-schedule applied and on the composition of the different raphe nuclei by 5-HT neurons giving rise to fine versus beaded fibers

(MAMOUNAS and MOLLIVER 1988; MOLLIVER et al. 1990; SCHMIDT and KEHNE 1990; FULLER 1992; HALASY et al. 1992; WILSON et al. 1993; FULLER and HENDERSON 1994; AXT et al. 1994; HENSLER et al. 1994; McQUADE and SHARP 1995). p-CA and the ring-substituted amphetamine derivatives methylenedioxyamphetamine (MDA) and methylenedioxymethamphetamine (MDMA) have significantly contributed to our current understanding of the anatomy and differential drug-vulnerability of the serotoninergic projection systems (MAMOUNAS et al. 1991; MOLLIVER et al. 1990; WILSON et al. 1993; AXT et al. 1994). These methods have largely replaced the nonspecific lesioning techniques which yielded equivocal and controversial concepts on the physiological roles of CNS serotoninergic systems.

The application of 5,6-DHT and 5,7-DHT to the developing and adult rat brain has led to the discovery of an unprecedented, prolific, life-time growth potential of partially axotomized 5-HT neurons (BAUMGARTEN et al. 1973b, 1974, 1975; NOBIN et al. 1973; LYTLE et al. 1974; FRANKFURT and AZMITIA 1983; JONSSON and HALLMAN 1983; ZHOU and AZMITIA 1986). Under optimum circumstances, sprouting serotoninergic axons may regenerate to re-establish their original terminal innervation pattern following a transitory phase of cell body-close hyperinnervation (NOBIN et al. 1973; BJÖRKLUND and WIKLUND 1980; WIKLUND and BJÖRKLUND 1980; BJÖRKLUND et al. 1981; FRANKFURT and AZMITIA 1983; ZHOU and AZMITIA 1986). Successful regeneration may result in restoration of temporarily altered modulation of functions influenced by serotoninergic transmission (NYGREN et al. 1974; WUTTKE et al. 1977, 1978; BAUMGARTEN et al. 1978a,b). While regeneration and recovery of function have been shown to occur in young adult rats following discrete axonal injury by selective neurotoxins (NYGREN et al. 1974; WUTTKE et al. 1977, 1978; BJÖRKLUND et al. 1981; SCANZELLO et al. 1993; FISCHER et al. 1995), it has yet to be shown that regeneration of chemically injured 5-HT axons to distant target areas (e.g., the neocortex) occurs in nonhuman primates following similar lesion paradigms (RICAURTE et al. 1992; AXT et al. 1994; STEELE et al. 1994; FISCHER et al. 1995). This topic is discussed in more detail in Chap. 1, this volume.

C. Nuclei of Origin of Serotoninergic Projections

In mammals, most serotoninergic neurons reside within the confines of the raphe nuclei as defined by cytoarchitectonic parameters (TABER et al. 1960; BRAAK 1970; BAKER et al. 1990, 1991a,b; cf. Tables 1–4, Fig. 1). Nonetheless, significant numbers of cells (up to 20%) are displaced into components of the adjacent reticular formation of the brainstem (DAHLSTRÖM and FUXE 1964; STEINBUSCH et al. 1978; STEINBUSCH 1981, 1984; NIEUWENHUYS 1985). Lateralization of serotoninergic neurons appears to be associated with more advanced and complex levels of CNS organization in mammals such as represented in, e.g., cats, monkeys and man. Furthermore, there is a relation-

ship between the degree of neocortical, striatal and thalamic expansion in
primates and man (compared to nonprimates such as the cat and the rat) and
the evolvement of the rostral raphe complex and the nonraphe lateral
serotoninergic components, reflected by a relative and absolute increase in the
number of 5-HT neurons and the degree of subnuclear differentiation of
the dorsal and median (central) raphe nuclei (compare data in Tables 5–7).

Table 1. Cytoarchitectonic classification of the human oral raphe nuclei (mesopontine
raphe complex) [Ohm et al. (1989), combined lipofuscin pigment-Nissl stain according
to Braak (1970)]

Nucleus raphes dorsalis
   Supratrochlear part (wing-like lateral extensions dorsal to the trochlear nucleus)
   Intercalate part [central portion dorsal to the medial longitudinal fasciculus
     (MLF)]
   Interfascicular part (between MLF)
   Dorsofascicular part (suprajacent to MLF)
Nucleus raphes linearis (lamellar cell mass rostral to the caudal pole of the
    oculomotor nuclei)
Nucleus raphes centralis (sive medianus) (ventral to MLF)

Table 2. Cytoarchitectonic classification of the human dorsal raphe nucleus [Baker et
al. (1990), Nissl stain]

Dorsal raphe nucleus
   Interfascicular part (between MLF)
   Ventral part (dorsomedial to MLF)
   Ventrolateral part (dorsolateral to the trochlear nucleus)
   Dorsal part (dorsal to the ventrolateral subnucleus; two-thirds of the lateral wing)
   Caudal part (caudalmost portion of the nucleus, flanking the median sulcus of the
    aqueduct, suprajacent to the interfascicular subnucleus)

Table 3. Cyto- and chemoarchitectonic classification of the raphe nuclei in the human
brainstem [Törk and Hornung (1990), PH8 immunohistochemistry and cresyl violet]

Caudal linear nucleus (lamellar cell mass rostral to the caudal pole of the oculomotor
    nuclei)
Dorsal raphe nucleus
   Median (interfascicular part) (between MLF)
   Ventrolateral part (supratrochlear)
   Dorsal part (ventral to the floor of aqueduct)
   Lateral part (beyond central gray)
   Caudal part (dorsal to the MLF in subependymal gray of pons)
Median raphe nucleus (superior central raphe nucleus)
Raphe magnus nucleus (paramedian pontine cell cluster ventral to MLF)
Raphe obscurus nucleus (paramedian group of cells below IVth ventricle extending
    throughout the entire medulla oblongata)
Raphe pallidus nucleus (ventral paramedian cluster of cells in the rostral part of the
    medulla oblongata)

Depending on the methods applied (Nissl stain, pigment stain with or without Nissl stain, Nissl stain with serotonin – or PH8 – immunohistochemistry), authors have arrived at different concepts on the parcellation of the dorsal and median raphe nuclei into subnuclei (cf. Tables 1–7). In view of the fact that approximately two-thirds of the neuronal population within the dorsal raphe complex of the human, the nonhuman primate, the cat and the rat is composed of 5-HT-containing cells which are the main projection neurons of this nuclear complex, the most valid approach to functionally meaningful subgrouping

Table 4. 5-HT-containing neurons in the brainstem of mammals (except humans)[a] [B nomenclature for the rat brainstem according to DAHLSTÖM and FUXE (1964)]

Rostral group of cells
  Raphe nuclei
    Nucleus linearis caudalis (B8)
    Nucleus raphes dorsalis
      Principal part (B7)
        Medial portion: mediodorsal part (superior), medioventral part
          (interfascicular)
        Lateral portion (supratrochlear part)
      Caudal part (B6)
    Nucleus raphes medianus
      Principal (rostral) part (B8)
      Caudal part (B5)
    Nucleus raphes pontis (distinguished in some but not all species, considered as
      the caudal part of the median raphe nucleus)
  Cells outside the raphe nuclei
    Cells in the rostral part of the interpeduncular nucleus (B8)
    Cells in supralemniscal area (B9)
    Cells within nucleus (reticularis) pontis oralis (B8, B9)
    Cells in periventricular gray of the pons extending into the nucleus locus
      ceruleus (B6)
    Cells in subcoerulean area/periolivary field

Caudal group of cells
  Raphe nuclei
    Nucleus raphes magnus (B3)
    Nucleus raphes obscurus
      Principal part (B2)
      Dorsolateral part (B4)
    Nucleus raphes pallidus (B1, B4)
  Cells outside the raphe nuclei
    Cells in rostral ventrolateral medulla (B3)
    Cells in magnocellular field and paragigantocellular nucleus (B3)
    Cells in caudal ventrolateral medulla (within fibers of the medial lemniscus and
      among fascicles of the spinoreticular tract) (B1)
    Cells within area postrema

[a] Immunohistochemistry: STEINBUSCH (1981, 1984); LIDOV and MOLLIVER (1982); JACOBS et al. (1984); AZMITIA and GANNON (1986); HORNUNG and FRITSCHY (1988); NIEUWENHUYS (1985); TÖRK (1985, 1990); AZMITIA and WHITAKER-AZMITIA (1991); JACOBS and AZMITIA (1992); HALLIDAY et al. (1995).
The subdivisions listed here follow suggestions by TÖRK (1990) and JACOBS and AZMITIA (1992).

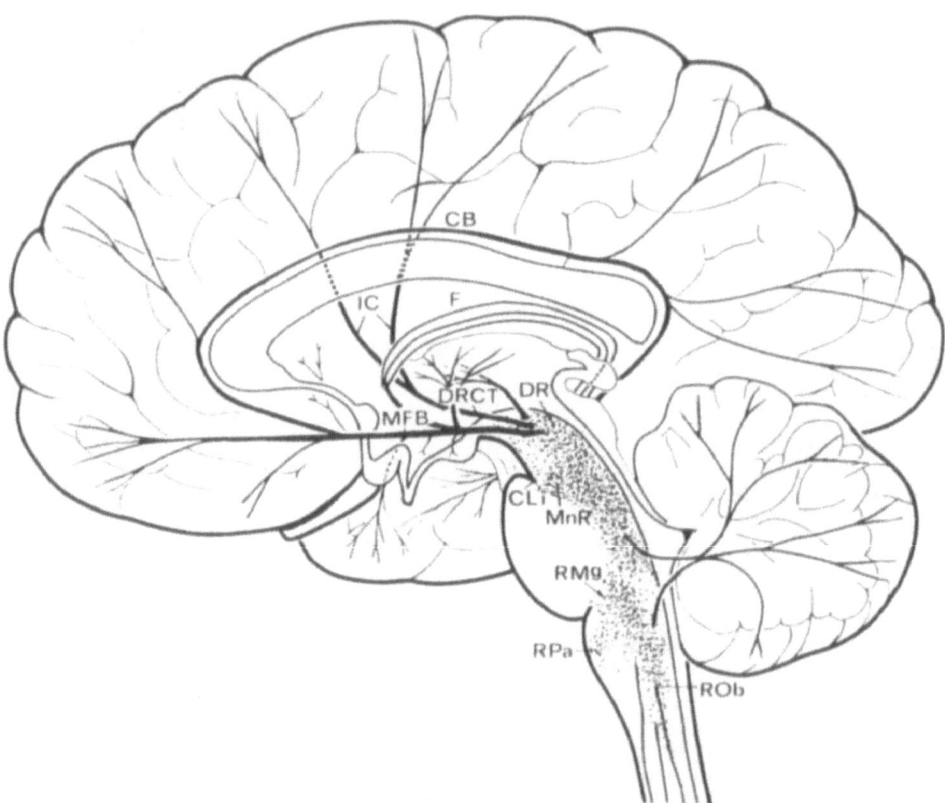

Fig. 1. Mediosagittal view of the human brain showing the distribution of the serotonin-containing nerve cells (*stippled area*) within the raphe nuclei and the major ascending fiber tracts. *CB*, cingulum bundle; *CLi*, caudal linear nucleus; *DR*, dorsal raphe
nucleus; *DRCT*, dorsal raphe cortical tract; *F*, fornix; *IC*, internal capsule; *MFB*, medial
forebrain bundle; *MnR*, median raphe nucleus; *RMg*, raphe magnus nucleus; *ROb*,
raphe obscurus nucleus; *RPa*, raphe pallidus nucleus

would consist of combining Nissl staining, serotonin, or PH8, immunohistochemistry and tracing of neurons projecting towards the same or to functionally related target fields. In case of the oral raphe complex, the closest to this
ideal is represented in retrograde tracing studies combined with 5-HT immunohistochemistry performed in the nonhuman primate (WILSON and
MOLLIVER 1991a). These studies show a rostrocaudal and mediolateral topographic gradient of serotoninergic neurons in the dorsal (but not the median)
raphe nucleus according to their preferential cortical projection targets (the
primary projection versus the association areas); this organization is not
strictly congruent with the subnuclear architecture revealed in Nissl-stained
material. Retrograde tracing studies performed in rats (O'HEARN and
MOLLIVER 1984; IMAI et al. 1986; BENNETT-CLARKE et al. 1994) support the idea
that there is only an approximative congruence of subnuclear architecture in

Table 5. Number of 5-HT neurons within the rostral and caudal raphe complex and in juxtaraphe position of the cat [WIKLUND et al. (1981), Falck-Hillarp method and cresyl violet stain]

	Number of 5-HT neurons	Percentage of cells in nucleus/area
Rostral raphe complex		
Nucleus linearis intermedius	2 100	25
Nucleus raphes dorsalis	24 500	70
Nucleus centralis superior	7 400	35
Caudal raphe complex		
Nucleus raphes magnus	2 400	15
Nucleus raphes obscurus	2 300	35
Nucleus raphes pallidus	8 000	50
Non-midline cells (total)	13 600	
Lateral extensions of ventral raphe in medulla oblongata	1 700	

Table 6. Number of 5-HT neurons in the medullary raphe nuclei and in medial reticular formation of the cat [HOLMES et al. (1994), serotonin immunohistochemistry]

Nucleus raphes magnus	4280
Nucleus raphes obscurus et pallidus	3780
Outside the raphe (mainly within magnocellular tegmental flield)	1540

the dorsal raphe nucleus and the spatial distribution of 5-HT neurons which share similar or functionally related projection targets. While the subnuclear organization of the dorsal raphe nucleus elaborated in Nissl material shows reasonable cross-species correspondence, the parcellation of the human dorsal raphe complex in pigment-stained material differs from that in Nissl-stained sections. Comparison of pigment architectonics (OHM et al. 1989; Table 1) with Nissl and PH8-immunoreactive mapping of dorsal raphe boundaries and sub-nuclear organization (Tables 2, 3) reveals that pigment architectonics do not precisely match with 5-HT neuron distribution patterns and Nissl-based nuclear subdivisions as pigment staining labels nonserotoninergic in addition to 5-HT-containing cells that are also located beyond the confines of the raphe zone and should therefore be considered as components of the midbrain reticular formation (see BAKER et al. 1991a for a detailed discussion of these discrepancies). The classification of the raphe nuclei and their subdivisions given in Table 4 may thus be considered a compromise, which is largely in agreement with the nomenclature proposed by a subcommittee of the Serotonin Club (cf. JACOBS and AZMITIA 1992); this table also lists the extraraphe serotonin neurons of the mammalian brain which appear to evolve in parallel with progressive complexity in vertebrate forebrain organization.

Table 7. Number of 5-HT neurons within the raphe nuclei and in the juxtaraphe position of the human brainstem [Baker et al. (1991), PH8 immunohistochemistry and Nissl stain]

	PH8-positive cells	Percentage of Nissl-stained cells
Dorsal raphe nucleus	165 000	70
Dorsal part (B7)	46 900	61
Ventral part (B7)	26 200	82
Ventrolateral part (B7)	29 900	62
Interfascicular part (B7)	26 400	107
Caudal part (B6)	35 600	66
Median raphe nucleus	64 305	43
Dorsal part	2 355	–
Midline part	39 825	81
Paramedian part	22 125	23
5-HT neurons in the oral pontine reticular nucleus	30 745	–
Dorsal subnucleus	11 610	–
Central subnucleus	19 125	–
5-HT neurons in supralemniscal positon (B9)	28 900	–

–, not distinguished in Nissl-stained material.

Table 8. Size and shape of PH8-immunoreactive neurons in the rostral raphe/extraraphe cell groups of the human brain [Data compiled from Baker and Halliday (1995) and Baker et al. (1991a,b)]

	Average size (μm)	Shape(s)
Dorsalis raphe		
Subnucleus caudalis	20 ± 5	Fusiform, ovoid and triangular; radiating and tufted dendrites (isodendritic)
Subnucleus dorsalis	24 ± 5	Large fusiform or ovoid; radiating dendrites (isodendritic)
Subnucleus ventrolateralis	26 ± 5	Large fusiform and triangular
Subnucleus ventralis	25 ± 5	Round, fusiform and triangular
Subnucleus interfascicularis	23 ± 5	Elongated and fusiform; dorsoventrally oriented dendrites
Medianus raphe		
Subnucleus dorsalis	25 ± 4	Large and round or small bipolar
Subnucleus medianus	26 ± 6	Small fusiform and bipolar
Subnucleus paramedianus	34 ± 3	Large ovoid with radiating dendrites (isodendritic)
Oral pontine reticular nucleus		
Subnucleus dorsalis	27 ± 5	Medium-sized, bipolar
Subnucleus centralis	26 ± 5	Medium-sized and small bipolar (isodendritic)
Supralemniscal nucleus	33 ± 6	Large bipolar

The 5-HT-containing neurons within the different raphe nuclei and their subdivisions differ in their mean perikaryal diameter, shape, polarity and dendritic architecture and domain (TÖRK and HORNUNG 1990; BAKER et al. 1990, 1991a,b; BAKER and HALLIDAY 1995; cf. Table 8), suggesting that they differ in their afferent and efferent connectivity pattern and degree of axonal collateralization. The finding of heterogeneous populations of 5-HT neurons within the raphe nuclei is evident not only in immunohistochemical pictures but also from cytoarchitectonic and morphometric studies using Nissl-stained and Golgi-impregnated material in nuclei rich in 5-HT cells such as the dorsal raphe nucleus and the medullary raphe nuclei (Fox et al. 1976; DANNER and PFISTER 1980; DIAZ-CINTRA et al. 1981; HÖLZEL and PFISTER 1981; BAKER et al. 1990; BAKER and HALLIDAY 1995). These have been described as large and multipolar (isodendritic), as large or medium-sized fusiform (isodendritic) and bipolar, as small bipolar or as triangular (Table 8). Based on the average diameter measured, even the largest multipolar 5-HT neurons do not reach the size of gigantocellular (glutamatergic) neurons such as represented in the medial reticular tegmental field.

A specific feature of 5-HT-containing neurons in the dorsal raphe nucleus is the abundance of dendro-dendritic, dendrosomatic and axodendritic contacts derived from recurrent axon collaterals (FELTEN and HARRIGAN 1980; CHAZAL and RALSTON 1987). This type of connectivity may be the substrate of the unique electrophysiology of these neurons: the rhythmicity of these spontaneously and regularly discharging neurons which is under inhibitory control of serotonin acting on somatodendritic 5-HT$_{1A}$-autoreceptors coupled to Gi protein.

D. Coexistence of Serotonin, Peptides, GABA and Nitric Oxide Synthase in 5-HT Neurons

The serotonin-containing neurons within the different rostral and caudal raphe nuclei and extraraphe cell groups are nonhomogeneous with respect to their pattern of peptide/amino acid/nitric oxide synthase (NOS) colocalization (HÖKFELT et al. 1987; CULLHEIM and ARVIDSSON 1995). Galanin (GAL) coexists with 5-HT in a significant number of serotoninergic neurons of almost all mesencephalic and medullary raphe nuclei in the rat (except the nucleus raphe magnus; MELANDER et al. 1986). GAL-immunoreactive bulbospinal serotoninergic axons in the cat and rat are concentrated in the ventral part of the spinal motor neuron column (ARVIDSSON et al. 1991b). Coexistence of serotonin and substance P (SP) is a prominent feature of the rat, cat and human lower brainstem raphe nuclei (B1–B3) and the extraraphe serotoninergic cell groups (CHAN-PALAY et al. 1978; HÖKFELT et al. 1978, 1987; JOHANSSON et al. 1981; PRIESTLEY and CUELLO 1982; LOVICK and HUNT 1983; BOWKER et al. 1983; MARSON and LOEWY 1985; HALLIDAY et al. 1988; BOWKER and ABBOTT 1990; KACHIDIAN et al. 1991; DEAN et al. 1993; JOHNSON et al. 1993; ARVIDSSON et al. 1994). As substance P and neurokinin A are products of a

common precursor, preprotachykinin, it is not surprising that neurokinin A
coexists with SP/5-HT-containing neurons confined to the lower medullary
raphe neurons in, e.g., the rat (Nevin et al. 1994; Halliday et al. 1995). A
significant proportion of the medullary 5-HT/SP neurons which project to the
lower brainstem and spinal cord somatomotor and to the spinal cord viscero-
motor nuclei synthesize thyrotropin-releasing hormone (TRH) in addition to
5-HT and SP (Sasek et al. 1990; Kachidian et al. 1991; Arvidsson et al. 1994).
Almost all 5-HT/SP-containing neurons of the lower brainstem 5-HT cell
groups in the rat exhibit glutamate/aspartate-like immunoreactivity following
colchicine treatment (Nicholas et al. 1992), suggesting that excitatory amino
acid transmitters coexist with serotonin and one (or more than one) peptide(s)
in a significant fraction of the bulbospinal pathways involved in the initiation/
modulation of tonic and rhythmic motor instructions. While tachykinins appear to be absent from serotonin-containing neurons of the rostral raphe
nuclei and extraraphe cell clusters in the rat (Halliday et al. 1995), a significant fraction of neurons in the human dorsal raphe complex (with the exception of the caudal subnucleus) store both SP and 5-HT (about 30% of the total
population of Nissl-positive neurons; Baker et al. 1991b). In addition to
tachykinins and excitatory amino acids, bulbospinal neurons of the lower
medullary raphe and extraraphe cell groups in the monkey have been shown
to express calcitonin gene-related peptide (CGRP) mRNA (Arvidsson et al.
1990); their preferential target seems to be ventral horn motoneurons
(Arvidsson et al. 1991a).

Leu- and Met-enkephalin (ENK) are colocalized in a fraction of 5-HT
neurons of the midline medullary raphe neurons (mainly of the nucleus raphe
pallidus and obscurus) in the cat and rat (Glazer et al. 1981; Hunt and Lovick
1982; Charnay et al. 1984; Leger et al. 1986; Arvidsson et al. 1992; Millhorn
et al. 1989; Sasek and Helke 1989; Kachidian et al. 1991; Tanaka et al. 1993).
In the cat and the monkey, the spinal cord segments receive a low degree of 5-
HT/ENK afferents which originate in the medullary raphe nuclei and in the
rostroventrolateral reticular nucleus (Arvidsson et al. 1992); 5-HT/ENK-
containing axon terminals are clustered in the motor nucleus of the ventral
horn (Tashiro et al. 1988, 1990; Arvidsson et al. 1992). In the monkey, axon
terminals ensheath somatodendritic portions of medium-sized ventral horn
neurons, whereas 5-HT/ENK fibers in the cat terminate in the neuropil surrounding motoneurons (Arvidsson et al. 1992). The potential coexistence of
5-HT and ENK in bulbospinal pathways innervating dorsal horn lamine (I, IIa
and V; Tashiro et al. 1988) has not been substantiated in lesion experiments
(cf. Cullheim and Arvidsson 1995). The majority of bulbospinal 5-HT neurons which project to the dorsal horn laminae (Skagerberg and Björklund
1985) do not coexpress any of the peptides listed above (GAL, tachykinins,
CGRP, ENK) under physiological conditions (Wessendorf and Elde 1987;
Arvidsson et al. 1990).

Although GABA has been claimed to coexist with serotonin in 5-HT
neurons of the dorsal raphe nucleus and of the medullary raphe nuclei

(NANOPOULOS et al. 1982; BELIN et al. 1983; MILLHORN et al. 1987; KACHIDIAN et al. 1991), more recent semiquantitative immunolabeling studies indicate that GABA/5-HT coexistence is a rare event in the nucleus raphe magnus and obscurus of the rat (3.6% and 1.5% of 5-HT neurons, respectively) whereas GABA/5-HT cells are virtually absent in the remaining raphe/extraraphe serotoninergic neurons of the rat (JONES et al. 1991; STAMP and SEMBA 1995). GABA and 5-HT appear to be contained in separate and distinct populations of neurons in the magnocellular tegmental field and in the nucleus raphe magnus and obscurus of the cat (HOLMES et al. 1994).

NOS, which catalyzes the formation of the free radical signaling molecule nitric oxide (NO), and its associated NADPH-diaphorase catalytic activity are expressed in a high percentage of serotoninergic neurons of the mesopontine raphe nuclei and extraraphe cell clusters (JOHNSON and MA 1993; DUN et al. 1994). The highest concentration of 5-HT/NOS neurons is found in the medial subnuclei of the dorsal raphe complex, which preferentially project to telencephalic cortical areas. In view of the fact that in the central as well as the peripheral nervous system NOS generally colocalizes with nonmonoaminergic neurons (for a review, see BLOTTNER et al. 1995), it is tempting to speculate that the colocalization of 5-HT and NOS in distinct dorsal raphe-cortical projection subsystems renders these systems exquisitely susceptible to the potentially neurotoxic effects of NO interacting with reactive oxygen species that are generated during monoamine oxidase catalyzed oxidation of 5-HT (BAUMGARTEN and ZIMMERMANN 1992). It has already been suggested that NO is involved in an ill-defined cascade of events leading to selective dorsal raphe-cortical terminal axon destruction following administration of substituted amphetamines such as p-CA and MDMA (BENMANSOUR and BRUNSWICK 1992; CLIKEMAN et al. 1992).

E. Projections to the Spinal Cord, Brainstem and Cerebellum

The majority of the serotoninergic projections to the spinal cord originate in the caudal midline medullary raphe nuclei (B1–B3) and in the extraraphe cells of the magnocellular tegmental field and ventrolateral medulla (HOLSTEGE and KUYPERS 1987; SKAGERBERG and BJÖRKLUND 1985; JACOBS and AZMITIA 1992; TÖRK 1985, 1990; HALLIDAY et al. 1995). 5-HT-containing axons which descend in the dorsolateral funiculus and terminate preferentially in the dorsal horn laminae I and II originate in the rostral magnocellular tegmental field and ventrolateral medulla and in the raphe magnus nucleus (KWIAT and BASBAUM 1992; LI et al. 1993b; SIDDALL et al. 1994). 5-HT-containing axons which descend in the ventrolateral funiculus and innervate target structures in the intermediate zone (intermediolateral nucleus) and ventral horn laminae (somatomotor neurons) originate mainly in the midline medullary raphe nuclei (raphe obscurus and pallidus) and the caudolateral tegmental cell groups.

The majority of the descending serotoninergic axons collateralize extensively to supply multiple spinal cord segments (Bowker and Abbott 1990; Kausz 1991). The cervical spinal cord segments receive additional input from descending collaterals of the rostral raphe complex and from its lateral reticular extensions (B6–B9) (Bowker et al. 1981a,b, 1982; Skagerberg and Björklund 1985; Zemlan et al. 1984). Spinally projecting medullary 5-HT neurons, many of which coexpress SP (Lovick and Hunt 1983; Marson and Loewy 1985; Dean et al. 1993; Arvidsson et al. 1994; Cullheim and Arvidsson 1995) and which terminate in the somatomotor column, may have brainstem collaterals that supply branchio- and somatomotor neurons of the pontomedullary cranial nerve nuclei (V, VII, XI, XII) and neurons of the medullary lateral parvocellular and medial magnocellular reticular tegmental fields in the rat, thus providing a substrate for concerted effects on skeletomotor, cardiovascular and respiratory activities (Tallaksen-Greene et al. 1993; Li et al. 1993a; Fort et al. 1990). Partial destruction of serotoninergic neurons in the magnocellular medullary reticular tegmental field and the adjacent raphe nuclei in the cat by the excitotoxin quisqualate results in degeneration of 5-HT neurons which have both local and distant 5-HT terminals in the medullary and pontine lateral, and in the gigantocellular and subcoerulear tegmental fields, as well as 5-HT axons in the caudal spinal trigeminal nucleus and the ventral horn pointing to rather extensive collateralization of individual medullary 5-HT neurons (Holmes et al. 1994).

Retrograde tracing studies indicate that individual serotoninergic neurons of the nucleus raphe magnus project towards the principal sensory and the spinal trigeminal nucleus as well as the dorsal horn laminae of the spinal cord, suggesting concerted modulatory actions on nociceptive primary afferent input into the spinal cord and brainstem and on secondary nocisensory processing (Li et al. 1993c). Efferent projections of the nucleus raphe magnus towards the periaqueductal gray, the parabrachial nucleus, the locus ceruleus and the noradrenergic cell groups A5 and A7 demonstrate the involvement of the medullary raphe in several supraspinal components of central pain modulatory systems, all of which entertain relationships to the dorsal horn laminae of the spinal cord (Sim and Joseph 1992). In addition, almost all of these supraspinal pain modulatory centers and the medullary raphe nuclei receive descending collateral projections from the dorsal raphe nucleus (Vertes and Kocsis 1994). The solitary complex of the brainstem receives serotoninergic afferences from multiple sources including the dorsal raphe nucleus and the nucleus raphe magnus (Thor and Helke 1987; Schaffar et al. 1984; Zemlan et al. 1984). These findings reveal that several brainstem structures receive overlapping serotoninergic input from more than one raphe/extraraphe 5-HT cell group, and that spinally and tegmentally projecting 5-HT neurons of the lower medullary raphe are subject to serotoninergic control by afferences from the rostral raphe complex and the lateral medullary tegmental cell groups (Beitz 1982; Lakos and Basbaum 1988; Vertes and Kocsis 1994; Thompson et al. 1995).

The cerebellar nuclei in the cat (the medial and lateral nucleus and the interposed nuclei) receive serotoninergic afferents from multiple sources, in particular from 5-HT neurons located in the locus ceruleus, in the dorsal tegmental nucleus and in the dorsal raphe nucleus (KITZMAN and BISHOP 1994). The medial nucleus receives additional projections from the lower medullary raphe nuclei, from the extraraphe 5-HT cells in the periolivary reticular formation and from the superior central nucleus. The latter nucleus is the source of additional projections to the lateral nucleus whereas the interposed nuclei receive additional input from the raphe magnus nucleus. Although the cerebellar nuclei entertain a topographic relationship with defined areas of the cerebellar cortex, the serotoninergic innervation to the cerebellar cortex does not arise from collaterals of neurons innervating the cerebellar nuclei but from separate populations of extraraphe 5-HT neurons located in the lateral tegmental field and paramedian reticular nucleus of the medulla oblongata (KERR and BISHOP 1991; KITZMAN and BISHOP 1994). By contrast, it appears likely that a fraction of the medullary raphe and extraraphe 5-HT neurons which project to the cerebellar nuclei and cortex in the rat also contribute to the indolaminergic innervation of the inferior olive. The source of the serotoninergic projections to the inferior olive of the rat has been investigated (BISHOP and HO 1986; COMPOINT and BUISSET-DELMAS 1988); the cells of origin are located in the medullary raphe nuclei (with the exception of the raphe magnus nucleus) and their lateral reticular extensions and overlap largely with those projecting to the cerebellum. The definite clarification of their identity as simultaneous contributors to the serotoninergic innervation of the inferior olive and the cerebellum would be interesting in view of the role of serotonin as an enhancer of the firing rate and synchronization of the oscillatory activity of groups of olivocerebellar projection neurons and of groups of Purkinje cells, resulting in improved timing of motor sequences in movement coordination (SUGIHARA et al. 1995).

Several upper brainstem structures receive afferents from the median raphe nucleus, among which are the pontine reticular formation, the midbrain central tegmental field and retrorubral area, the periaqueductal gray, the dorsal raphe nucleus, the interpeduncular nucleus and the dopaminergic cell clusters in the ventral tegmental area and medial portions of the substantia nigra, pars compacta (BOBILLIER et al. 1976; CONRAD et al. 1974; AZMITIA 1978; AZMITIA and SEGAL 1978; VERTES and MARTIN 1988). In addition, the dorsal midbrain tegmentum, the tectum and the pretectum are sparsely innervated by median raphe fibers. According to the detailed anterograde tracing data by VERTES and KOCSIS (1994), there is limited overlap by dorsal raphe projection fields in those brainstem areas which receive prominent input by median raphe neurons. Areas of overlap comprise the central gray, parts of the pontine and midbrain reticular formation and the dopaminergic midbrain nuclei.

The dorsal raphe nucleus, which accounts for as much as 40%–50% of the total number of 5-HT-containing neurons in the mammalian brainstem, maintains substantial connections with almost all of the remaining raphe nuclei, with the medial magnocellular mesencephalic, pontine and medullary reticular tegmental fields (known to contain a significant proportion of the extraraphe 5-HT neurons), and with the two major cholinergic brainstem nuclei (the laterodorsal and pedunculopontine nucleus). These dorsal raphe connections may be involved in inter-raphe activity coordination, in the facilitation of central motor tone for trunk and proximal limb muscles maintaining posture, and in the suppression of PGO-spiking during waking. The dorsal raphe complex, in particular its rostral component, projects to several brainstem areas involved in cardiovascular regulation, among which are the parabrachial nuclei, the rostroventrolateral medulla, magnocellular tegmental fields and the caudal raphe nuclei (STEININGER et al. 1992; PETROV et al. 1992a; VERTES and KOCSIS 1994).

The pontine noradrenergic nucleus locus ceruleus (LC) is a major site of termination of dorsal raphe efferents (PASQUIER et al. 1976; SAKAI et al. 1977; CEDARBAUM and AGHAJANIAN 1978; IMAI et al. 1986). Serotonin appears to be involved in restraining the range of firing of noradrenergic LC neurons in response to glutamate and to counteract LC hyperactivity during opiate withdrawal (ASTON-JONES et al. 1991; AKAOKA and ASTON-JONES 1993).

F. Projections to the Forebrain

The majority of the serotoninergic projections to the di- and telencephalon originate in the rostral mesopontine raphe nuclei (i.e., the nucleus raphe dorsalis, medianus and pontis B5, 6, 7, 8) and in the extraraphe tegmental 5-HT neurons located in the supralemniscal area B9 (cf. AZMITIA 1978, 1987; AZMITIA and SEGAL 1978; VAN DE KAR and LORENS 1979; PARENT et al. 1981; KÖHLER et al. 1982; O'HEARN and MOLLIVER 1984; TÖRK 1985, 1990; MOLLIVER 1987; VERTES and MARTIN 1988; WILSON and MOLLIVER 1991a,b; VERTES 1991; GONZALO-RUIZ et al. 1995; HALLIDAY et al. 1995). Many forebrain target areas receive parallel afferents from 5-HT neurons in the dorsal and the median raphe nucleus. The distinction between dorsal raphe and median raphe efferents which provide axon terminals to the same target area is facilitated by differences in the morphology and neurotoxin sensitivity of the axons and axon terminals of either source: corticopetal axons emanating from the dorsal raphe nucleus are generally very fine and have pleomorphic tiny varicosities of granular or fusiform shape, those originating in the median raphe nucleus being mainly (but not exclusively) characterized by large spherical or ovoid varicosities and by variations in axonal diameter ("beaded" or "coarse" axons; KOSOFSKY and MOLLIVER 1987; HORNUNG et al. 1990). An ill-defined percentage of axons arising from the median raphe nucleus 5-HT neurons are of the fine-fiber type (HALASY et al. 1992). The delicate, profusely branching axons of dorsal raphe origin (and their median raphe equivalents) are exquisitely sensitive to neurotoxic injury by certain substituted amphetamines such as p-CA, MDA and MDMA (MOLLIVER et al. 1990; MAMOUNAS et al. 1991; McQUADE and SHARP 1995). The sensitivity of dorsal raphe-derived axons towards these amphetamine neurotoxins appears to be relatively consistent across a variety of species including rats, guinea pigs and nonhuman primates (BATTAGLIA et al. 1987; COMMINS et al. 1987; SCHMIDT 1987; O'HEARN et al. 1988; MAMOUNAS and MOLLIVER 1988; RICAURTE et al. 1988; INSEL et al. 1989; WILSON et al. 1989; SCHMIDT and KEHNE 1990; MOLLIVER et al. 1990) and may also apply to humans (RICAURTE et al. 1990, 1992; STEELE et al. 1994; McCANN et al. 1994; AXT et al. 1994). The chemical ablation of the dorsal raphe afferents to given target areas thus permits a semiquantitative estimate of the relative contribution of either projection system to the density, pattern and topography of the regional serotoninergic terminal innervation. However, due to the presence of neurons in the median raphe nucleus which give rise to fine fibers that are susceptible to damage by p-CA (HARING et al. 1992; McQUADE and SHARP 1995), the dorsal raphe nucleus selectivity of p-CA is limited. Consequently, 5,7-DHT-induced selective lesions of the dorsal raphe nucleus do not exactly mirror the extent of regional 5-HT fiber denervations and transporter ligand-binding reductions caused by p-CA (HENSLER et al. 1994).

Using p-CA as a tool to lesion fine 5-HT axons, a specific and restricted pattern of distribution of beaded, amphetamine-resistant axon terminals has been described in certain forebrain areas of the rat which, according to retro- and anterograde tracing data, originate mainly in the median raphe nucleus (MAMOUNAS and MOLLIVER 1988; MOLLIVER and MAMOUNAS 1991; HARING et al. 1992). Median raphe forebrain serotoninergic projections provide beaded axon terminals to selective layers of the cerebral cortex, hippocampus and olfactory bulb, to the septum, amygdala and hypothalamus and to the ependymal lining of the lateral ventricles (MAMOUNAS et al. 1991; MOLLIVER et al. 1990; WILSON et al. 1993; AXT et al. 1994).

The ascending projections of the the dorsal and median raphe nuclei follow several distinct trajectories on their way towards their telodiencephalic targets. In subprimate species, the medial forebrain bundle (MFB) serves as the major pathway for ascending 5-HT axons of dorsal and median raphe origin (AZMITIA and SEGAL 1978). Two out of six fiber bundles converge into the MFB via tegmental radiations: the dorsal raphe and the median raphe forebrain tract. The dorsal raphe forebrain tract intermingles with ventrolateral components of the MFB and distributes fibers to lateral forebrain areas (basal ganglia, amygdala and entorhinal cortex). The median raphe forebrain tract joins ventromedial components of the MFB and distributes fibers to medial forebrain areas (septum, hippocampus and cingulate cortex). The dorsal raphe cortical tract encroaches on ventral aspects of the medial lemniscus in the posterior hypothalamus and enters the internal capsule to fan out into the basal ganglia and parietotemporal neocortex. This tract is the major pathway for corticopetal 5-HT axons in the nonhuman primate (AZMITIA and GANNON 1986). The dorsal raphe periventricular tract forms below the aqueduct and proceeds rostrally towards ventricle-close regions of the thalamus and hypothalamus. The dorsal raphe arcuate tract heads laterally to supply the geniculate body and the suprachiasmatic nuclei. The raphe medial tract receives fibers from both the median and the dorsal raphe nucleus and turns ventrally to distribute axons to the interpeduncular nucleus and midline mamillary body (AZMITIA 1978, 1987; AZMITIA and SEGAL 1978; AZMITIA and GANNON 1986). The serotoninergic nature of these anterogradely traced ascending pathways in the rat has been verified by significant reductions in the transport of radiolabel following prior destruction of 5-HT-containing axons by intradiencephalic microinjections of 5,7-DHT. Axons ascending from the midbrain into the forebrain via the different raphe trajectories either separate from the fiber bundles or ramify to give off collaterals which join classical myelinated forebrain tracts as links to their respective target fields.

The innervation territories of the dorsal and the median raphe projections in the rat have been mapped in detail by anterograde transport of *Phaseolus vulgaris* leucoagglutinin and tritiated leucin, respectively (VERTES 1991; VERTES and MARTIN 1988). Although no attempt has been made to verify the neurochemical identity of the ascending raphe projections, it seems justified to claim that the results of these tracing studies by and large reflect serotoninergic projections because 5-HT neurons account for the majority of projection neurons in these cranial raphe nuclei (60%–85% of retrogradely labeled raphe-cortical projection neurons in the dorsal and median raphe nucleus of the rat and monkey are 5-HT-containing; O'HEARN and MOLLIVER 1984; WILSON and MOLLIVER 1991a,b). Furthermore, the pattern and density of terminal innervation throughout the forebrain revealed by nonspecific anterograde labeling match closely with the results of 5-HT axon terminal immunolabeling patterns and densities (STEINBUSCH 1981, 1984). The vast number of efferent connections established by the dorsal raphe 5-HT neurons have stimulated speculation as to the functional roles of this expansive modulator system. The behavioral, cognitive, emotional and perception deficits in humans following intake of MDMA (MCCANN and RICAURTE 1991, 1994) suggest a role of dorsal raphe efferents in the regulation of affective state, anxiety, sensory perception, and mental and memory capacity. Animal experiments suggest general roles of forebrain 5-HT in the control of impulsive eating, sexual and social behavior, and anxiety, irritability and certain forms of aggressive behavior (BAUMGARTEN and LACHENMAYER 1972, 1985; SOUBRIE 1986, 1988; BEVAN et al. 1989; SAUDOU et al. 1994; TECOTT et al. 1995; BAUMGARTEN and GROZDANOVIC 1995).

According to the anterograde tracing data by VERTES (1991), the dorsal raphe nucleus is an important source of afferents to the midline and intralaminar nuclei of the thalamus, the central, lateral and basolateral nuclei of the amygdala, posteromedial sections of the striatum and ventral pallidum, the nucleus accumbens and olfactory tubercle, the diagonal band nuclei and the remaining basal forebrain nuclei rich in cholinergic neurons. The entire neocortex receives input from the dorsal raphe nucleus, in particular from its rostral part. The dorsal raphe projects heavily to the piriform, insular and frontal cortices, and moderately densely to the occipital, entorhinal, orbitofrontal, infralimbic and anterior cingulate cortices. Within the neocortex, 5-HT-containing axons of dorsal raphe origin are particularly enriched in layers which receive afferences from specific thalamic nuclei (IV and V). Neurons located in the caudal half of the dorsal raphe nucleus contribute to the innervation of the hippocampus, which is also a major target of median raphe afferents.

The dorsal raphe projections to primary versus association areas of the cerebral cortex in the rat and monkey exhibit a crude somatotopic arrangement with clustering of neurons in discrete spatial parts of the dorsal raphe nuclear complex and significant proportions of collateralized neurons which innervate functionally related cortical (and possibly also subcortical) target areas (O'HEARN and MOLLIVER 1984; WILSON and MOLLIVER 1991a; WATERHOUSE et al. 1986, VAN BOCKSTAELE et al. 1993). By combining retrograde horseradish peroxidase (HRP) tracing with serotonin-immunohistochemistry, GONZALO-RUIZ et al. (1995) have shown that clusters of 5-HT neurons within (ipsilateral) subnuclear compartments of the dorsal raphe complex of the rat maintain a somatotopic relationship with major subdivisions of the anterior thalamic nuclei. These findings support the involvement of the dorsal raphe in limbic aspects of information processing. The dorsal raphe projections to the substantia nigra, caudate putamen and amygdala in the rat similarly reveal a crude somatotopic arrangement (VAN DER KOOY and HATTORI 1980; STEINBUSCH et al. 1981; IMAI et al. 1986); a limited number of neurons which project in a collateralized fashion to both targets are clustered in the interfascicular part of the dorsal raphe nucleus. Dorsal raphe/periaqueductal 5-HT neurons have been found to project simultaneously to the caudal spinal or principal sensory trigeminal nucleus, the ventrolateral orbital cortex, nucleus accumbens and amygdala (LI et al. 1993c), suggesting coordinating modulatory roles in the processing of primary sensory and complex limbic aspects of pain-related neural systems. Other tracing studies show that serotoninergic neurons of the dorsal raphe collateralize to innervate cortical, subcortical and endocrine components of limbic circuits (prefrontal cortex, central nucleus of the amygdala, nucleus accumbens and paraventricular hypothalamic cell groups), allowing for concerted modulation of complex adaptive behavior, emotion and neuroendocrine secretion (VAN BOECKSTAELE et al. 1993; PETROV et al. 1994). Thus, the dorsal raphe nucleus in mammals harbors clusters of neurons which project to functionally interrelated cortical and subcortical target areas in part by collateralization of individual cells.

By contrast, no distinct rostrocaudal somatotopic arrangement of neurons projecting to functionally heterogeneous (though interconnected) neocortical areas (frontal, primary motor and somatosensory cortex, parietal association areas 5 and 7) has been found in the median raphe nucleus of the rat and the macaque (O'HEARN and MOLLIVER 1984; WILSON and MOLLIVER 1991a,b). This finding has been interpreted as suggestive of rather divergent and widespread connections of individual median raphe neurons and implies rather global modulatory effects on network excitability and/or afferent input responsivity. This assumption needs verification by single cell tracer injection, which would outline the total telodendritic domain of individual median raphe neurons.

According to nonspecific anterograde tracing data (Vertes and Martin 1988), median raphe neurons project heavily to several "limbic" forebrain structures including the medial mamillary nucleus, the lateral habenula, the medial preoptic area, diagonal band nuclei, the septum, nucleus accumbens, olfactory tubercle and the hippocampal formation. Additional diencephalic targets include the dorsal, dorsomedial and lateral hypothalamic nuclei and the anterior intralaminary complex and paraventricular nucleus of the thalamus. Within the neocortex, median raphe terminals are mainly confined to the outer layers (I-III). In the main olfactory bulb, the median raphe axon terminals are restricted to the glomerular layer (Mamounas et al. 1991; McLean and Shipley 1987). The median raphe nucleus provides narrow bands of dense axon terminals to sublayers of the stratum lacunosum moleculare of CA1-CA3 of the dorsal and ventral hippocampus and to the infragranular zone of the dentate gyrus (Vertes and Martin 1988).

The serotoninergic afferences to the neocortex are supplemented by projections from the supralemniscal nucleus (B9). Using retrograde dye labeling, O'Hearn and Molliver (1984) describe a crude somatotopic arrangement of neurons within the supralemniscal nucleus according to their cortical target: cells projecting to the occipital cortex are located more rostrally than those projecting to the parietal cortex. Cells sending their axons towards the frontal cortex have a wider rostrocaudal distribution across the nucleus B9. The frontal and parietal cortex receive bilaterally symmetric projections whereas the projection to the occipital cortex is predominantly ipsilateral. Only one-third of the retrogradely labeled neuron population in the supralemniscal area are serotonin containing, and the number of cortically projecting neurons in this extraraphe cell group is significantly smaller than the number of raphe-cortical projection neurons in the median and dorsal raphe nucleus (O'Hearn and Molliver 1984). Consequently, this extraraphe 5-HT cell group accounts for only a minor component of the serotoninergic afferents to the cortex in the rat.

G. Afferent Connections of the Raphe Nuclei

The mesopontine raphe nuclei have been shown to receive afferent input from limbic forebrain areas, e.g., the prefrontal and cingulate cortex, nucleus accumbens, the basal forebrain cholinergic nuclei, septal area, lateral habenula, preoptic and hypothalamic area (Pasquier et al. 1976; Aghajanian and Wang 1977; Sakai et al. 1977; Kawasaki and Sato 1981; Maciewicz et al. 1981; Kalen et al. 1985; Kalen 1988). A crude topographic organization of the septal projections to the raphe nuclei has been described (Kalen 1988). The medial septal nucleus projects to the caudal portion of the dorsal raphe and to the entire median raphe complex; the horizontal limbs of the diagonal band efferents terminate mainly in the rostral portion of the dorsal raphe and those originating in the vertical limb of the diagonal band distribute throughout the dorsal and median raphe nuclei. According to KALEN (1988), the distribution of afferent terminal labeling within the raphe overlaps with the distribution of 5-HT neurons across the dorsal raphe nucleus rather than its cytoarchitectonic boundaries or subnuclear divisions. Part of these limbic afferences use excitatory amino acids (EAA) as transmitters, e.g., those from the lateral habenula (KALEN et al. 1985, 1989). These excitatory afferents from the lateral habenula may terminate directly on serotoninergic neurons and increase their firing rate. EAA actions on 5-HT neurons may participate in stress-induced increases in hippocampal, striatal and cortical 5-HT release (KALEN 1988; KALEN et al. 1989; PEI et al. 1990; INOUE et al. 1993; KAWAHARA et al. 1993; VAHABZADEH and FILLENZ 1994; CADOGAN et al. 1994); "stress"-induced increases in forebrain 5-HT release may rather be a reflection of the level of behavioral activation/arousal caused by the alerting effects of "stress" (Chap. 3, this volume). Other limbic-mediated inputs may terminate on inhibitory GABAergic interneurons in the dorsal raphe, which in turn synapse with 5-HT neurons (WANG et al. 1992); such limbic-mediated inhibitory influences on the activity of the dorsal raphe may have a role in the modulation of anxiety, aversion and avoidance behavior.

Some of the neurochemically ill-defined hypothalamic and extrahypothalamic basal forebrain afferences to the rostral raphe may well conform to peptidergic systems [e.g., corticotropin-releasing hormone (CRH), arginine vasopressin (AVP) and adrenocorticotropic hormone (ACTH) projection systems], which transmit the effects of stress directly onto the brainstem serotoninergic and noradrenergic modulator neurons (SWANSON et al. 1983; ROMAGNANO and JOSEPH 1983; SAWCHENKO and SWANSON 1982).

The dorsal raphe nucleus has been considered a center for coordination of the general level of behavioral activation/central motor tone with appropriate modulation of cardiovascular and respiratory activity. In harmony with this concept, the raphe nuclei receive peptidergic afferences from the lateral and medial parabrachial nuclei (NT-, CGRP- and GAL-containing projections; PETROV et al. 1992b), catecholaminergic input from several brainstem noradrenergic (A1, 2, 5, 6) and adrenalinergic (C1, 3) cell groups (FUXE 1965; LOIZOU 1969; HÖKFELT et al. 1974, 1984; CHU and BLOOM 1974; SAKAI et al. 1977; LEVITT and MOORE 1979; LINDVALL and BJÖRKLUND 1974, 1978; WESTLUND and COULTER 1980; TANAKA et al. 1994; PEYRON et al. 1996) and neurochemically ill-defined afferences from pneumotaxic brainstem areas (medial parabrachial and Koelliker-Fuse complex, retrofacial nucleus and ventrolateral medulla; GANG et al. 1991; ZAGON 1993) and from viscerosensory components of the solitary tract nucleus (HERBERT 1992).

The work of LEVINE and JACOBS (1992) suggests that the mesopontine 5-HT neurons may receive collateral input from ascending auditory pathways which use EAAs as transmitters. Presentation of series of clicks to awake cats results in phasic increases of firing of dorsal raphe 5-HT neurons, which are reduced in magnitude by iontophoretic application of kynurenic acid. The EAA antagonist had no effect on spontaneous activity of the 5-HT neurons, i.e., the neurons are not dependent on external driving input for maintenance of their (intrinsic) activity. Similar relationships may hold for other sensory systems as well (e.g., for visual stimuli; Heym et al. 1982; Fornal and Jacobs 1988; Chap. 3, this volume). The spontaneous activity of dorsal raphe 5-HT neurons during waking is also independent of noradrenergic mechanisms (Heym et al. 1981; Chap. 3, this volume) despite the fact that the dorsal raphe nucleus receives substantial noradrenergic innervation from multiple brainstem sources (Westlund and Coulter 1980; Peyron et al. 1996).

The nucleus raphe magnus (NRM) is known to serve as a relay center for descending oligoneuronal pain modulatory pathways which originate in the mesencephalic periaqueductal gray (PAG) and the adjacent reticular nucleus cuneiformis (NCF). Pharmacological evidence suggests that raphe-spinal NRM neurons (many of which are serotonin-containing) receive both glutamatergic and cholinergic afferents from the NCF and PAG in addition to peptidergic (NT, SP, ENK) and serotoninergic input (Basbaum and Fields 1984; Behbehani and Zemlan 1986; Richter and Behbehani 1991).

Serotonin-containing neurons located in the medial medullary reticular formation (mainly in the magnocellular tegmental field and the raphe nuclei) of the cat are in contact with terminals of GABAergic (Golgi type II) inter-neurons, which represent a major population of neurons in this medullary reticular field (Holmes et al. 1994). According to nonspecific deafferentation studies, this medial raphe and reticular tegmental field of the brainstem receive input from ascending spinoreticular and from descending cor-ticoreticular, tectoreticular and raphereticular projections (Brodal 1957; Brodal et al. 1960b). These afferents may terminate on GABAergic inter-neurons. Much of the modulatory input into the caudal raphe and extraraphe 5-HT neurons would thus be indirectly transmitted via GABAergic interneu-rons. Transneuronal tracing studies will be required to verify such indirect afferent connections of the medullary 5-HT neurons.

H. Terminal Axon Innervation Patterns and Target Relationship

The topography of the 5-HT innervation of the CNS has been extensively studied using (a) fluorescence histochemistry; (b) biochemical analysis of the regional 5-HT or 5-hydroxyindoleacetic acid (5-HIAA) levels, [^{3}H]5-HT reuptake, or 5-HT synthesis rate; (c) various autoradiographic methods, such as injection of [^{3}H]amino acids, or incubation in [^{3}H]5-HT or [^{3}H]transporter ligands (e.g., cyano/nitroimipramine, paroxetine, citalopram); and (d) im-munohistochemistry using antibodies against synthesizing enzymes or serotonin–protein conjugates. 5-HT immunohistochemistry has proved particularly valuable in unveiling the extent of terminal arbors issued by 5-HT neurons. The terminal distribution of 5-HT fibers in various brain regions of several mammalian species has been reviewed in detail by AZMITIA (1978), STEINBUSCH (1984), AZMITIA and GANNON (1986), TAKEUCHI (1988) and JACOBS and AZMITIA (1992). The results of studies covered by these reviews as well as more recent investigations will be presented, devoting special attention to termination patterns of 5-HT axons in the forebrain, where interspecies differences are most conspicuous.

The current concepts on the regional distribution and laminar arrangement of the 5-HT-containing nerve fibers in the cerebral cortex began to evolve in the mid 1960s following FUXE's description of a diffuse innervation of the rat cortex by sparse 5-HT-containing terminal varicosities located mainly in the molecular layer using the fluorescence histochemical technique (FUXE 1965; FUXE et al. 1968). Subsequent autoradiographic work by DESCARRIES and coworkers, published in the mid 1970s, revealed a more extensive intracortical network of presumed 5-HT axon terminals than was delineated by the histofluorescence method (DESCARRIES et al. 1975; BEAUDET and DESCARRIES 1976, 1987). Following superfusion of the rat frontoparietal neocortex with [^{3}H]5-HT, aggregates of silver grains representing axonal varicosites labeled by 5-HT uptake were found unevenly distributed in the upper five cortical layers, gradually increasing in density towards the superficial layers. It was, however, not until the introduction of immunochemical staining methods for the demonstration of 5-HT neurons by Steinbusch and associates in the late 1970s (STEINBUSCH et al. 1978) that the 5-HT-containing axon terminals could be directly visualized and their morphological characteristics assessed.

The 5-HT innervation of the cerebral neocortex of the rat has been studied with immunohistochemical techniques by LIDOV et al. (1980), STEINBUSCH (1981), TAKEUCHI et al. (1983), KOSOFSKY and MOLLIVER (1987) and LÜTH and SEIDEL (1987). In the lateral neocortex, a dense network of 5-HT axons is spread relatively uniformly among different cytoarchitectonic areas and across cortical laminae. Most of the 5-HT terminals are present in layer I. Quantitative data are available on the density of 5-HT innervation in seven cytoarchitectonic areas of the anterior half of the rat cortex (AUDET et al. 1989). The mean regional density amounts to 5.8×10^6 varicosities/mm^3 (range 4.4–7.1 $\times$ 10^6). In Cg3 (cingulate cortex, area 3), AIDc (caudal agranular insular cortex, dorsal part), Pir (prepiriform cortex–primary olfactory cortex) and AIDr (rostral agranular insular cortex, dorsal part), significantly higher regional fiber densities exist than in Cg2 (cingulate cortex, area 2), Par1 (parietal cortex, area 1–primary somatosensory cortex) and Fr1 (frontal cortex, area 1–primary motor cortex). On the basis of these data, it is estimated that a cortical target neuron is innervated by approximately 145–230 5-HT varicosities; this figure equals $^1/_{200}$ of the total number of cortical terminals/mm^3.

Unlike the situation in the rodent, the pattern of the regional and laminar distribution of 5-HT terminals in the neocortex of nonhuman primates is more variable, supporting the idea that the increasing complexity and differentiation of the primate cortex is mirrored by the growing specialization of the cortical 5-HT afferents. The existence of a topographically distinct pattern of cortical 5-HT distribution was first suggested by BROWN and colleagues (1979) on the basis of measurements of 5-HT and 5-HIAA concentration in various cortical areas of rhesus monkey. The full extent of the interregional variation in the neocortical 5-HT innervation of several monkey species was clearly established in subsequent immunohistochemical studies (MORRISON et al. 1982; TAKEUCHI and SANO 1983, 1984; KOSOFSKY et al. 1984; MORRISON and FOOTE 1986; CAMPBELL et al. 1987; DEFELIPE and JONES 1988; DE LIMA et al. 1988; WILSON and MOLLIVER 1991b). In general, the density of 5-HT axon terminals is high in primary sensory areas, with the primary visual cortex being more heavily innervated than the primary auditory, primary somatosensory and secondary visual cortical areas, moderate in association areas and low in the primary motor area. With the exception of the densely innervated primary visual cortex, there are only minor variations in the laminar distribution of 5-HT fibers. In most accounts, a bilaminar pattern of 5-HT innervation has been disclosed, consisting of a superficial and a deep fiber plexus, the exact positions of which seem to differ in various cortical regions and according to different authors. As an example, in the primary somatosensory cortex, especially areas 3a and 3b, WILSON and MOLLIVER (1991b) observed an accentuated concentration of 5-HT fibers in layers II and outer III, a narrow strip of low fiber density in the deepest part of layer III and an increase in innervation density in layers IV–VI. By contrast, in the description of DEFELIPE and JONES (1988) the superficial plexus extends from the middle of layer III up to the pia mater, with the highest density of fibers in layer I, whereas the deep plexus resides in layers V–VI. In an earlier study of several neocortical areas, TAKEUCHI and SANO (1983) concluded that there was a strong preference for layer IV. Using autoradiography with [^{3}H]5-HT, AZMITIA and GANNON (1986) have found that in all regions examined (i.e., entorhinal, superior temporal, precentral, postcentral, frontal and calcarine cortex) 5-HT axons are enriched in layer I and layer IV. A detailed account of the predominant orientation of 5-HT fibers in different cortical regions and layers is presented by WILSON and MOLLIVER (1991b).

A striking laminar pattern of 5-HT innervation is seen in the primary visual cortex. In the squirrel monkey (*Saimi sciureus*), a New World monkey, 5-HT fibers are concentrated in layers IVA and IVC (MORRISON et al. 1982). Thus, 5-HT afferents are in a strategic position to modulate cortical processing of primary visual information conveyed by the geniculostriate projection. A more differential segregation of 5-HT axons is observed in cynomolgus monkey (*Macaca fascicularis*), an Old World monkey, in which layer IVC is further subdivided into IVCa and IVCb and layer V into VA and VB (LUND 1973). Clusters of 5-HT terminals are prominent in layers III–IVCa, reaching a peak in IVB, and layers VA–VI (KOSOFSKY et al. 1984; MORRISON and FOOTE 1986). These zones are separated by a narrow band of sparse innervation density occupying layer IVCb. This is corroborated by quantitative data showing that axonal varicosities are most numerous in sublayers IVB and IVCa (DE LIMA et al. 1988). A similar laminar organization of 5-HT axons is documented in the autoradiographic study of BERGER et al. (1988). These findings are challenged in part by TAKEUCHI and SANO (1983, 1984), who in a study of 5-HT innervation of *Macaca fuscata*, another Old World monkey, maintain that axonal varicosities are most abundant "in the upper portion of layer IVc of Brodmann's nomenclature, correponding to layer IVCb of Lund" (TAKEUCHI and SANO 1984). Obviously, differences in the interpretation of Lund's original work on sublaminar division of the monkey visual cortex are responsible for seemingly divergent findings. With these disparities in mind, it is worth noting that the sublaminar pattern of 5-HT axon distribution in the primary visual cortex of Old World monkeys points to a differential influence of the 5-HT system on visual processing in the terminal zones of the magnocellular (layer IVCa) versus parvocellular (layers IVA and IVCb) streams of the geniculostriate pathway.

A detailed analysis of the 5-HT innervation of the extrastriate cortical and subcortical visual regions in squirrel and cynomolgus monkeys has been published by MORRISON and FOOTE (1986). The density of 5-HT fibers in area 18 and the inferotemporal cortex is lower than that in area 17, and that in area 7 lower than in occipital and temporal areas. In both area 18 and the inferotemporal gyrus, the most pronounced concentration of 5-HT axons is observed in layer IV; no laminar preference of 5-HT termination is found in area 7. The visual thalamic nuclei, i.e., the lateral geniculate nucleus, the pulvinar complex and the lateral posterior nucleus, are characterized by a relatively uniform, moderate to high level of 5-HT innervation. A preponderance for the magnocellular layers of the lateral geniculate nucleus is evident in the cynomolgus but not squirrel monkey, probably reflecting the preferential 5-HT innervation of layer IVCa, the magnocellular recipient layer of the primary visual cortex of cynomolgus. In the mesencephalon, the superior colliculus is found to contain the densest accumulation of 5-HT of axon terminals in the intermediate layers.

Contrary to the situation in the neocortex, the limbic and to some extent also the paralimbic cortical areas in the rat exhibit a more heterogeneous topographic distribution and a distinct sublaminar organization of 5-HT terminals (LIDOV et al. 1980; KÖHLER et al. 1980, 1981; KÖHLER 1982; IHARA et al. 1988). The morphological data are supported by biochemical analyses (PAASONEN et al. 1957; SAAVEDRA et al. 1974a; BROWNSTEIN et al. 1975). The laminar pattern of the 5-HT axon distribution in the anterior cingulate cortex is similar to that of the lateral neocortex. By sharp contrast, a more restricted laminar pattern is found in the granular portion of the posterior cingulate cortex, characterized by concentrations of 5-HT axons in layers I, III and VI and a paucity of axons in layers II and V (LIDOV et al. 1980). Compared to the rat, the 5-HT innervation of the primate cingulate cortex is more dense. The anterior cingulate cortex is more heavily innervated than the posterior cingulate cortex, which lacks the striking laminar segregation of axons observed in the rat (Wilson and Moliver 1991b). In the hippocampal formation of the rat, dense accumulations of 5-HT axons are present in the stratum lacunosum-moleculare of CA1, the strata lacunosum-moleculare, oriens and radiatum of CA2 and CA3 as well as in the immediately infragranular portion of the polymorphic cell layer of the dentate gyrus (Köhler 1982; Ihara et al. 1988). The preferential 5-HT innervation of the stratum lacunosum-moleculare is also observed in other mammalian species, both nonprimates (chipmunk, hamster, cat and dog) and subhuman primates (Ihara et al. 1988). In the monkey, the most abundant accumulation of 5-HT fibers is found in the stratum lacunosum-moleculare (of CA2 and CA3). A moderate number of axons is noted in the strata moleculare, radiatum and oriens; the strata pyramidale and lucidum are only sparsely innervated by 5-HT terminals (Ihara et al. 1988; Wilson and Molliver 1991b). Within the dentate gyrus, the outer third of the molecular layer exhibits a very dense band of 5-HT axons. A moderate density of axons is observed in the dentate hilus. The granular cell layer has a very low density of 5-HT axons (Wilson and Molliver 1991b). Throughout the different subdivisions of the retro-hippocampal region of the rat, namely the entorhinal area, the parasubiculum, the retrosplenial area, the presubiculum and the subiculum, a rather diffuse pattern of 5-HT innervation exists (Köhler et al. 1980, 1981). As a rule, layer I contains the densest concentration of 5-HT fibers. A dense cluster of 5-HT terminals is found in layers III and IV of a spatially limited portion of the lateral entorhinal area, suggesting a preferential influence of the 5-HT system on a restricted segment of the lateral perforant path and, therefore, the hippocampal circuitry.

A quantitative analysis of the density of the hippocampal 5-HT innervation has been performed by Oleskevich and Descarries (1990). The mean innervation density is estimated at 2.7×10^6 varicosities/mm^3, being significantly higher in subiculum (3.6×10^6) and Ammon's horn (3.1×10^6) than in dentate gyrus (2.2×10^6). In accordance with immunohistochemical data, the innervation is particularly dense in the stratum moleculare of subiculum, CA1 and CA3 as well as in the stratum oriens and stratum radiatum of CA3. The highest innervation density occurs in the stratum oriens of CA3-a (5.2×10^6). From these figures, the average number of 5-HT varicosities per target neuron has been calculated to be about 130 in CA3 and 20–35 in dentate gyrus.

In the olfactory bulb of the rat, cat and monkey, 5-HT fibers are seen in all layers except the olfactory nerve layer. In the rat and cat, 5-HT fibers preferentially innervate the glomerular layer (Steinbusch 1981; Takeuchi et al. 1982; McLean and Shipley 1987). By contrast, in the monkey olfactory bulb 5-HT axons are only sparsely distributed in the glomerular layer; 5% of the glomeruli (or less) are occupied by 5-HT axons (Takeuchi et al. 1982). The internal plexiform layer receives the most profuse 5-HT innervation.

The terminal pattern of 5-HT axon ramification in the subcortical limbic structures of the telencephalon has been investigated in the rat. Distinct zones of the septum are found to contain differing levels of 5-HT innervation (STEINBUSCH 1981; KÖHLER et al. 1982; GALL and MOORE 1984). The richest terminal networks are seen in the ventral part of the lateral septum and an area bordering the medial edge of the islands of Calleja (KÖHLER et al. 1982). The diagonal band of Broca contains numerous fibers en passage (KÖHLER et al. 1982). According to STEINBUSCH (1981), the amygdala receives a well-developed 5-HT innervation. Very dense terminal arborizations are found in the rostral and medial part of the nucleus amygdaloideus basalis, the nucleus amygdaloideus medialis posterior, and the medial and lateral part of the nucleus amygdaloideus posterior.

The existence of a topographic organization of the 5-HT system in the basal ganglia has been established by both biochemical (FAHN et al. 1971; PALKOVITS et al. 1974; SAAVEDRA 1977; TERNAUX et al. 1977; MACKAY et al. 1978; BECOPOULOS et al. 1979; SHANNAK and HORNYKIEWICZ 1980; CROSS and JOSEPH 1981; WALSH et al. 1982) and morphological methods. The distribution of 5-HT axons in the basal ganglia components has been explored in several histofluorescence (FUXE 1965; AGHAJANIAN et al. 1973) and autoradiographic (CALAS et al. 1976; CHAN-PALAY 1977; ARLUISON and DE LA MANCHE 1980; PARENT et al. 1981) studies, but more recent immunohistochemical investigations have revealed the full extent of the 5-HT innervation in the basal ganglia (STEINBUSCH 1981; PASIK and PASIK 1982; PASIK et al. 1984a,b; MORI et al. 1985a,b, 1987; O'HEARN et al. 1988; SOGHOMONIAN et al. 1987, 1989; LAVOIE and PARENT 1990; CORVAJA et al. 1993). In a series of comparative studies, MORI and coworkers have examined the localization of 5-HT fibers and varicosities in the substantia nigra (MORI et al. 1987), the subthalamic nucleus (MORI et al. 1985a) and the striatum (MORI et al. 1985b) of the rat, cat and monkey (*Macaca fuscata*). In the substantia nigra of all species investigated, 5-HT fibers are more numerous in the pars reticulata than the pars compacta. In contrast to the generally uniform arrangement of 5-HT fibers in the pars reticulata of the rat and cat, a more varied distribution is observed in the monkey pars reticulata. The pars lateralis is more densely innervated in the cat and monkey than in the rat. Interspecies differences in the topographic variation of 5-HT fibers are also evident in other parts of the basal ganglia. In the subthalamic nucleus of the monkey, 5-HT fibers are more abundant and more unevenly distributed than in the cat and rat. The globus pallidus and the entopeduncular nucleus of the rat and cat as well as the internal pallidal segment of the monkey are more heavily and more diffusely innervated than the external pallidal segment in the monkey. In the striatum of all animals studied, 5-HT fibers are accumulated in the ventral, medial and caudal sectors. LAVOIE and PARENT (1990) studied the distribution of 5-HT axons and terminal varicosities in the basal ganglia of the squirrel monkey (*Saimi sciureus*). The results are similar to those obtained in the macaque monkey. The substantia nigra is the most densely innervated part of the basal ganglia. The fibers arborize more profusely in the pars reticulata and the pars lateralis than the pars compacta. The subthalamic nucleus contains a dense plexus of 5-HT fibers; only a few isolated varicose axons are concentrated in the ventrolateral part of the nucleus. In the globus pallidus, fibers intermingled with axonal varicosities are more numerous in the internal than the external segment. The density of fibers in the substantia innominata is somewhat less than in the adjacent pallidal complex. Regional differences are also observed in the striatum. 5-HT fibers and axon terminals are concentrated in the ventral striatum, including the nucleus accumbens and deep layers of the olfactory tubercle, the ventrolateral part of the putamen, and the ventromedial sector of the caudate nucleus. The bed nucleus of the stria terminalis is also very densely labeled. The finding of zones displaying a weak 5-HT-immunolabeling which are in register with similar zones of low tyrosine hydroxylase-immunoreactivity has led to the proposal that the 5-HT innervation in the striatum is predominantly directed to the extrastriosomal matrix compartment.

The intrastriatal variation in the density of the 5-HT innervation in the rat has been quantitatively evaluated by SOGHOMONIAN et al. (1987). The number of 5-HT varicosities averages $2.6 \times 10^6/mm^3$ of tissue, ranging from 1.5 to $4.8 \times 10^6/mm^3$. Values are significantly higher ventrally than dorsally, caudally than rostrally, and medially than laterally. In the nucleus accumbens, the density of 5-HT terminals is estimated at $3 \times 10^6/mm^3$. A very high innervation density is found in the globus pallidus ($4.5 \times 10^6/mm^3$). It has been calculated that between 100 and 200 5-HT axon terminals impinge upon a single neostriatal neuron.

The distributional pattern of 5-HT-immunoreactive fibers in the thalamus of the rat has been elucidated by STEINBUSCH (1981) and CROPPER et al. (1984). After pretreatment with L-tryptophan and pargyline, an inhibitor of monoamine oxidase, CROPPER et al. (1984) found the 5-HT innervation to be densest in nucleus ventralis corporis geniculati lateralis, certain midline nuclei, nucleus anterior ventralis, the intralaminar nuclei, and nucleus lateralis dorsalis. Moderate innervation was seen in nucleus reticularis, nucleus anterior dorsalis, nucleus ventralis medialis, nucleus lateralis pars posterior, the posterior complex, and nucleus dorsalis corporis geniculati lateralis. A rather sparse fiber network was encountered in nucleus corporis geniculati medialis, the ventral nuclei (with the exception of nucleus ventralis medialis), nucleus medialis dorsalis, nucleus anterior medialis, and some midline nuclei. It is clear from this description that 5-HT axons are differentially distributed to functionally diverse thalamic nuclei. Those nuclei involved in limbic circuits and transmission of sensory information are preferentially innervated.

A comparative study in rat, cat and monkey lateral geniculate nucleus was carried out by UEDA and SANO (1986). Invariably, the ventral part is more densely innervated than the dorsal part. In the rat, 5-HT fibers are most numerous in the magnocellular portion of the ventral lateral geniculate nucleus and the intrageniculate leaflet. In the cat and monkey, an even distribution of 5-HT fibrers is observed in the ventral lateral geniculate nucleus. By contrast, 5-HT fibers terminate diffusely in the dorsal lateral geniculate nucleus of the rat, but are clustered in the C-complex and the medial interlaminar nucleus of the cat and in the S layer and the interlaminar zones of the monkey.

It has long been known that the hypothalamus contains relatively high concentrations of 5-HT. The localization of 5-HT nerve fibers in the hypothalamus has been initially visualized by FUXE and coworkers using the fluorescence histochemical technique (FUXE 1965; FUXE et al. 1969). A more profuse distribution of 5-HT fibers has been suggested on the basis of biochemical measurements of the intrahypothalamic 5-HT content by SAAVEDRA et al. (1974b). Autoradiography after [^{3}H]5-HT uptake has then been used to detect a more elaborate 5-HT fiber network (DESCARRIES and BEAUDET 1978; PARENT et al. 1981). Subsequently, the successful application of immunohistochemical methods has produced comprehensive maps on the terminal arrangement of 5-HT fibers in the hypothalamus of several mammalian species, including rat (STEINBUSCH 1981; STEINBUSCH and NIEUWENHUYS 1982; STEINBUSCH et al. 1981), guinea pig (WAREMBOURG and POULAIN 1985), cat (UEDA et al. 1983b) and monkey (KAWATA et al. 1984). In all species examined, a widespread but heterogeneous distribution of 5-HT fibers is found. In the preoptic region of all species, certain nuclei receive a particularly dense 5-HT innervation, namely the nucleus praeopticus suprachiasmaticus in rats, the nucleus praeopticus periventricularis in guinea pigs, the nucleus praeopticus lateralis in cats and the nucleus praeopticus medialis in monkeys. In the anterior hypothalamus of rats, hamsters, guinea pigs and cats, the nucleus suprachiasmaticus is the most densely innervated structure (UEDA et al. 1983a), but it is almost devoid of 5-HT fibers in monkeys. The anterior hypothalamic area is densely innervated in rats but much less so in monkeys and, especially, cats. Both the nucleus paraventricularis and nucleus supraopticus display a low-to-moderate density of innervation, being higher in cats than in rats and monkeys. The terminal fields of 5-HT fibers in the paraventricular and supraoptic nuclei have been precisely delineated by SAWCHENKO et al. (1983). Fibers are more numerous in the parvocellular division of the paraventricular nucleus than in the magnocellular division, or in the supraoptic nucleus. Within the parvocellular division of the paraventricular nucleus, higher concentrations of fibers are seen in the periventricular parts, where neurons producing pituitary-releasing factors are located which project to the median eminence, and in the dorsal and ventral medial parts, which project to the autonomic centers of the lower brainstem and spinal cord. In the magnocellular division of the paraventricular nucleus and in the supraoptic nucleus, the oxytocin-producing neuroendocrine neurons are preferentially innervated. At the tuberal level, a constant feature in all species is a large contingent of 5-HT fibers in the nucleus ventromedialis. The nucleus dorsomedialis of cats is also richly supplied. Throughout the complex of mamillary nuclei in monkeys, 5-HT fibers are abundantly distributed. A more heterogeneous distribution is found in rats and cats.

The microscopic distribution of 5-HT in the pituitary of several mammals has been evaluated using autoradiography after incubation in [^{3}H]5-HT (Calas et al. 1974; Chan-Palay 1977; Descarries and Beaudet 1978) and immunohistochemistry (Steinbusch 1981; Steinbusch et al. 1981; Westlund and Childs 1982; Jennes et al. 1982; Sano et al. 1982; Leranth et al. 1983; Kawata et al. 1984; Warembourg and Poulain 1985; Saland et al. 1986). Most authors agree that 5-HT fibers are present in the infundibulum, both the external and the internal layer, and the posterior lobe of the neurohypophysis. Similarly, 5-HT fibers are constantly found in the intermediate lobe. By contrast, the anterior lobe has been shown to be innervated by 5-HT fibers in rats but not in monkeys.

The relationship of 5-HT axon terminals to target neurons has been extensively studied. Comprehensive reviews dealing with this topic have been published by Beaudet and Descarries (1987) and Soghomonian et al. (1988). The incidence of 5-HT terminals establishing synaptic contacts to targets in the cerebral cortex has been a matter of considerable debate (see Parnavelas and Papadopoulos 1989). In their initial autoradiographic study, Descarries and coworkers (1975) reported synaptic complexes in only 5% of 5-HT boutons in the frontoparietal cortex of rats. The synaptic incidence extrapolated stereologically for whole varicosities was later estimated to be 15%–20% (Beaudet and Descarries 1984). The existence of a substantially larger proportion of 5-HT varicosities engaged in junctional complexes in the rat cortex was communicated by Molliver et al. (1982), Parnavelas and McDonald 1983; Parnavelas et al. (1985) and Papadopoulos et al. (1987a,b). For example, Papadopoulos et al. (1987a,b), using electron microscopic immunohistochemistry and serial section analysis, found that approximately 90% of 5-HT profiles formed conventional synapses in the visual and the frontoparietal cortex of the rat. This issue was re-evaluated by Descarries' group using a similar method (Seguela et al. 1989). The incidence of synaptic boutons among the serially examined 5-HT terminals was calculated at 30%–40%. DeFelipe and Jones (1988), employing a comparable methodological approach in the sensorimotor cortex of the monkey, concluded that less than 3% of the 5-HT varicosities formed structurally differentiated synaptic contacts. The reason(s) for the observed discrepancies can only be guessed at (differences in the technique used or in the interpretation of the results, interspecies variabilty, etc). An intriguing explanation was offered by Törk (1990), who proposed that the relative density of the fine varicose and the beaded axon systems (vide supra) – which seem to have differential nuclear origin, axon morphology, areal and laminar distribution, susceptibility to the neurotoxic effects of certain substituted amphetamine analogs and, possibly, synaptic organization – in a given cortical (or subcortical) region may be responsible for the divergent findings obtained in studies on the synaptic incidence of 5-HT varicosities. In most electron microscopic studies, the synaptic 5-HT terminals are observed to make either asymmetrical or symmetrical contacts with dendritic branches or spines of cortical neurons (Descarries et al. 1975; Takeuchi and Sano 1984;

AZMITIA and GANNON 1986; PAPADOPOULOS et al. 1987a,b; DeFELIPE and JONES 1988; DE LIMA et al. 1988; SEGUELA et al. 1989). The cellular targets of 5-HT axons in the cerebral cortex are mostly obscure. TÖRK and collaborators have described elaborate basket-like arbors, which are issued by beaded varicose axons, around the somata and proximal dendrites of single cortical neurons in the supragranular layers of the cat (MULLIGAN and TÖRK 1987, 1988, 1993) and marmoset (HORNUNG et al. 1990) cerebral cortex. In preliminary ultrastructural studies, the beaded 5-HT boutons have been reported to form axosomatic and axodendritic synaptic contacts with cortical neurons (TÖRK et al. 1986; TÖRK 1990). Using double-labeling immunohistochemistry and antisera directed towards 5-HT and glutamate decarboxylase and/or GABA, the 5-HT-containing pericellular baskets have been shown to encircle a subpopulation of GABAergic neurons (TÖRK et al. 1988; HORNUNG and CELIO 1992). Using a comparable approach, DeFELIPE et al. (1991) have demonstrated an interaction of 5-HT boutons with GABAergic neurons in the primary auditory cortex of the cat. In the neocortex of the marmoset, the GABAergic interneurons belong to a subset of neurons containing the calcium-binding protein calbindin (HORNUNG and CELIO 1992). On the basis of these data it has been postulated that the raphe-cortical 5-HT projection pathway is comprised of two morphologically and functionally separate systems, namely the "diffuse" fine varicose axon subsystem influencing a multitude of diverse target neurons and the "selective" beaded axon system focussing onto a limited population of inhibitory GABAergic interneurons.

ANDERSON et al. (1986) first provided evidence for a distinct paucity of synaptic contacts in the dentate gyrus of the rat. In a detailed survey on the ultrastructural characteristics of 5-HT axon terminals innervating the hippocampus of the rat, a synaptic incidence of 20%–30% has been determined (OLESKEVICH et al. 1991). No significant differences between the CA3 region of Ammon's horn and the dentate gyrus have been found. The postsynaptic elements were mainly dendritic shafts, bearing predominently asymmetrical membrane specializations. FREUND and associates (reviewed by FREUND 1992), using a combination of anterograde tracing and pre- and postembedding immunohistochemical staining procedures, have conclusively shown that the median raphe afferents preferentially innervate GABAergic interneurons in the hippocampus and dentate gyrus of the rat. In the CA1 and CA3 sectors of the Ammon's horn, 5-HT afferents impinge selectively upon the calbindin D_{28K}-containing GABAergic interneurons in the strata radiatum and lacunosum-moleculare (FREUND et al. 1990). In the dentate gyrus, however, also GABAergic interneurons other than those containing calbindin D_{28K} are targeted by the median raphe axons (HALASY et al. 1992). Similarly, in the hippocampal formation of the marmoset, a close association between 5-HT baskets and calbindin-containing interneurons has been reported (HORNUNG and CELIO 1992).

In the only electron microscopic study on the 5-HT innervation of the olfactory bulb, performed in the rat, HALASZ et al. (1978), using autoradiography after local injection of [³H]5-HT, have presented circumstantial evidence that the presumable 5-HT terminals form axodendritic synapses with periglomerular cells in the glomerular layer and with granule cells in the external plexiform and granular layers.

Several reports deal with the synaptic and target relationships of 5-HT axons in the striatum of rats (ARLUISON and DE LA MACHE 1980; SOGHOMONIAN et al. 1989), cats (CALAS et al. 1976) and monkeys (PASIK et al. 1984a; PASIK and PASIK 1982). There is general agreement that a minority (less than 20%) of 5-HT terminals show junctional complexes. The synapsing 5-HT boutons are found to be engaged in asymmetrical axodendritic and axospinous synapses. The latter type of contact predominates in the striatum of the monkey (PASIK et al. 1984a; PASIK and PASIK 1982).

Information on the synaptic organization of 5-HT varicosities in the globus pallidus is limited. In an immunohistochemical study in monkeys, PASIK et al. (1984b) have found that 5-HT endings form asymmetrical synapses with the dendritic shafts and crests of pallidal neurons.

The fine structure of 5-HT axons in the substantia nigra has been investigated in rats (PARIZEK et al. 1971; MORI et al. 1987; CORVAJA et al. 1993), guinea pigs (NEDERGAARD et al. 1988) and monkeys (PASIK et al. 1984a). Morphologically distinguishable synaptic junctions are rarely seen between 5-HT terminals and nigral neurons. In the rat, less than 10% of the 5-HT varicosities exhibit synaptic membrane specializations (MORI et al. 1987). Symmetrical synapses of the "en passant" type are observed between 5-HT terminals and dendrites or, less often, somata of nigral neurons. However, most of the 5-HT varicosities are either in close apposition to the non-5-HT peridendritic axon terminals or exist freely in the neuropil. In another study performed in rats, CORVAJA et al. (1993) have found mainly asymmetrical synapses between 5-HT boutons and dendritic elements of neurons in the substantia nigra, pars reticulata. Clues as to the chemical nature of the postsynaptic targets were derived from experiments in which anterograde tracing from the dorsal raphe nucleus was combined with immunohistochemistry for tyrosine hydroxylase (TH). At the electron microscopic level, direct synaptic contacts of the anterogradely labeled raphe-nigral axons with TH-containing neurons could be revealed. These data strongly suggest that the 5-HT input in the substantia nigra is directed to the dendrites of dopaminergic neurons in the pars reticulata. The 5-HT innervation of dopaminergic neurons in the substantia nigra, pars reticulata of the guinea pig has been previously investigated by NEDERGAARD et al. (1988). Axonal endings immunoreactive for 5-HT have been found to make asymmetrical synaptic contacts with proximal and distal dendrites and spines of TH-containing neurons.

That dopaminergic neurons in the ventral tegmental area of the rat represent one of the synaptic targets of 5-HT axons in this brain region was suggested by HERVE et al. (1987) on the basis of double-labeling studies showing asymmetrical synaptic contacts between [³H]5-HT-labeled varicosities and TH-immunoreactive dendrites.

The ultrastructural relationships of 5-HT neurons in the hypothalamus of rats have been studied by BOSLER et al. (1984), KISS et al. (1984, 1988), BOSLER and BEAUDET (1985), KISS and HALASZ (1985, 1986) and LIPOSITS et al. (1987). The connections between 5-HT axon terminals and hypothalamic neurons are predominently of the nonjunctional type. The incidence of synaptic contacts varies among different hypothalamic nuclei, however. For example, in the very densely innervated suprachiasmatic nucleus the synaptic incidence is estimated at approximately 40% (BOSLER and BEAUDET 1985). By contrast, in the arcuate nucleus, which receives a sparse 5-HT innervation, very few junctional varicosities are found (KISS and HALASZ 1986). Direct synaptic connections between 5-HT terminals and neurons containing vasoactive intestinal peptide (suprachiasmatic nucleus; KISS et al. 1984; BOSLER and BEAUDET 1985), luteinizing hormone-releasing hormone (preoptic area; KISS and HALASZ 1985), dopamine (arcuate nucleus; KISS and HALASZ 1986), corticotropin-releasing factor (paraventricular nucleus; LIPOSITS et al. 1987) or somatostatin (anterior periventricular nucleus; KISS et al. 1988) have been demonstrated. These contacts are mostly seen on dendrites, less commonly so on somata of target neurons. Moreover, close appositions between 5-HT varicosities and GABAergic terminals are observed in the suprachiasmatic nucleus (BOSLER et al. 1984).

References

Aghajanian GK, Wang RY (1977) Habenular and other midbrain raphe afferents demonstrated by a modified retrograde tracing technique. Brain Res 122:229–242

Aghajanian GK, Kuhar MJ, Roth RH (1973) Serotonin-containing neuronal perikarya and terminals: differential effects of p-chlorophenylalanine. Brain Res 54:85–101

Akaoka H, Aston-Jones G (1993) Indirect serotonergic agonists attenuate neuronal opiate withdrawal. Neuroscience 54:561–565

Amin AH, Crawford TBB, Gaddum JH (1954) The distribution of substance P and 5-hydroxytryptamine in the central nervous system of the dog. J Physiol (Lond) 126:596–618

Anderson KJ, Holets VR, Mazur PC, Lasher RS, Cotman CW (1986) Immunocytochemical localization of serotonin in the rat dentate gyrus following raphe transplants. Brain Res 369:21–28

Arluison M, de la Manche (1980) High-resolution radioautographic study of the serotonin innervation of the rat corpus striatum after intraventricular administration of [³H]5-hydroxytryptamine. Neuroscience 5:229–240

Arvidsson U, Schalling M, Cullheim S, Ulfhake B, Terenius L, Verhofstad A, Hökfelt T (1990) Evidence for coexistence between calcitonin gene-related peptide and serotonin in the bulbospinal pathway in the monkey. Brain Res 532:47–57

Arvidsson U, Ulfhake B, Cullheim S, Terenius L, Hökfelt T (1991a) Calcitonin gene-related peptide in monkey spinal cord and medulla oblongata. Brain Res 558:330–334

Arvidsson U, Ulfhake B, Cullheim S, Hökfelt T, Theodorsson E (1991b) Distribution of ¹²⁵I-galanin binding sites, immunoreactive galanin and coexistence with 5-hydroxytryptamine in the cat spinal cord: biochemical, histochemical and experimental studies at the light and electron microscopic level. J Comp Neurol 308:115–138

Arvidsson U, Cullheim S, Ulfhake B, Ramirez V, Dagerlind Å, Luppi P-H, Kitahama K, Jouvet M, Terenius L, Åman K, Hökfelt T (1992) Distribution of enkephalin and its relation to serotonin in cat and monkey spinal cord and brain stem. Synapse 11:85–104

Arvidsson U, Cullheim S, Ulfhake B, Luppi P-H, Kitahama K, Jouvet M, Hökfelt T (1994) Quantitative and qualitative aspect on the distribution of 5-HT and its coexistence with substance P and TRH in cat ventral medullary neurons. J Chem Neuroanat 7:3–12

Aston-Jones G, Akaoka H, Charlety P, Chouvet G (1991) Serotonin selectively attenuates glutamate-evoked activation of noradrenergic locus coeruleus neurons. J Neurosci 11:760–769

Audet MA, Descarries L, Doucet G (1989) Quantified regional and laminar distribution of the serotonin innervation in the anterior half of adult rat cerebral cortex. J Chem Neuroanat 2:29–44

Axt KJ, Mamounas LA, Molliver ME (1994) Structural features of amphetamine neurotoxicity in the brain. In: Cho AK, Segal DS (eds) Amphetamine and its analogs. Academic, San Diego, pp 315–367

Azmitia EC (1978) The serotonin-producing neurons of the midbrain median and dorsal raphe nuclei. In: Iversen LL, Iversen SD, Snyder SH (eds) Handbook of psychopharmacology, vol 9: chemical pathways in the brain. Plenum, New York, pp 233–314

Azmitia EC (1987) The CNS serotonergic system: progression toward a collaborative organization. In: Meltzer HY (ed) Psychopharmacology. The third generation of progress. Raven, New York, pp 61–73

Azmitia EC, Gannon PJ (1986) The primate serotonergic system: a review of human and animal studies and a report on Macaca fascicularis. Adv Neurol 43:407–468

Azmitia EC, Segal M (1978) An autoradiographic analysis of differential ascending projections of the dorsal and median raphe nuclei in the rat. J Comp Neurol 179:641–668

Azmitia EC, Whitaker-Azmitia PM (1991) Awakening the sleeping giant: anatomy and plasticity of the brain serotonergic system. J Clin Psychiatry 52 [Suppl]:4–16

Baker KG, Halliday GM (1995) Ascending noradrenergic and serotonergic systems in the human brainstem. In: Tracey DJ, Paxinos G, Stone J (eds) Neurotransmitters in the human brain. Adv Behav Biol 43:155–171

Baker KG, Halliday GM, Törk I (1990) Cytoarchitecture of the human dorsal raphe nucleus. J Comp Neurol 301:147–161

Baker KG, Halliday GM, Halasz P, Hornung J-P, Geffen LB, Cotton RGH, Törk I (1991a) Cytoarchitecture of serotonin-synthesizing neurons in the pontine tegmentum of the human brain. Synapse 7:301–320

Baker KG, Halliday GM, Hornung J-P, Geffen LB, Cotton RGH, Törk I (1991b) Distribution, morphology and number of monoamine-synthesizing and substance P-containing neurons in the human dorsal raphe nucleus. Neuroscience 42:757–775

Basbaum AI, Fields HL (1984) Endogenous pain control system: brainstem spinal pathways and endorphin circuitry. Annu Rev Neurosci 7:309–338

Battaglia G, Yeh SY, O'Hearn E, Molliver ME, Kuhar M, De Souza EB (1987) 3,4-Methylenedioxymethamphetamine (MDMA) and 3,4-methylenedioxyamphetamine (MDA) preferentially destroy serotonin terminals in rat brain: quantification of neurodegeneration by measurement of ^{3}H-paroxetine-labeled serotonin uptake sites. J Pharmacol Exp Ther 242:911–916

Baumgarten HG, Björklund A (1976) Neurotoxic indoleamines and monoamine neurons. Annu Rev Pharmacol Toxicol 16:101–111

Baumgarten HG, Grozdanovic Z (1995) Psychopharmacology of central serotonergic systems. Pharmacopsychiatry 28 [Suppl II]:73–79

Baumgarten HG, Lachenmayer L (1972) 5,7-Dihydroxytryptamine: improvement in chemical lesioning of indoleamine neurons in the mammalian brain. Z Zellforsch 135:399–414

Baumgarten HG, Lachenmayer L (1985) Anatomical features and physiological properties of central serotonin neurons. Pharmacopsychiatry 18:180–187

Baumgarten HG, Zimmermann B (1992) Neurotoxic phenylalkylamines and indolealkylamines. In: Herken H, Hucho F (eds) Handbook of experimental pharmacology, vol 102. Springer, Berlin Heidelberg New York, pp 225–291

Baumgarten HG, Björklund A, Lachemayer L, Nobin A, Stenevi U (1971) Long-lasting selective depletion of brain serotonin by 5,6-dihydroxytryptamine. Acta Physiol Scand [Suppl] 373:1–15

Baumgarten HG, Evetts KD, Holman RB, Iversen LL, Vogt M, Wilson G (1972) Effects of 5,6-dihydroxytryptamine on monoaminergic neurones in the central nervous system of the rat. J Neurochem 19:1587–1597

Baumgarten HG, Björklund A, Lachenmayer L, Nobin A (1973a) Evaluation of the effects of 5,7-dihydroxytryptamine on serotonin and catecholamine neurons in the rat CNS. Acta Physiol Scand [Suppl] 391:1–19

Baumgarten HG, Lachenmayer L, Björklund A, Nobin A, Rosengren E (1973b) Long-term recovery of serotonin concentrations in the rat CNS following 5,6-dihydroxytryptamine. Life Sci 12:357–364

Baumgarten HG, Björklund A, Lachenmayer L, Rensch A, Rosengren E (1974) De- and regeneration of the bulbospinal serotonin neurons in the rat following 5,6- or 5,7-dihydroxytryptamine treatment. Cell Tissue Res 152:271–281

Baumgarten HG, Björklund A, Nobin A, Rosengren E, Schlossberger HG (1975) Neurotoxicity of hydroxylated tryptamines: structure-activity relationships. 1. Long-term effects on monoamine content and fluorescence morphology of central monoamine neurons. Acta Physiol Scand [Suppl] 429:1–27

Baumgarten HG, Lachenmayer L, Björklund A (1977) Chemical lesioning of indoleamine pathways. In: Myers AD (ed) Methods in psychobiology, vol III. Academic, London, pp 47–98

Baumgarten HG, Björklund A, Wuttke W (1978a) Neuronal control of pituitary LH, FSH and prolactin secretion: the role of serotonin. In: Scott D, Kozlowski GP, Weindl A (eds) Brain-endocrine interaction III. Neural hormones and reproduction. Karger, Basal, pp 327–343

Baumgarten HG, Klemm HP, Lachenmayer L, Björklund A, Lovenberg W, Schlossberger HG (1978b) Mode and mechanism of action of neurotoxic indoleamines: a review and progress report. Ann N Y Acad Sci 305:3–24

Baumgarten HG, Jenner S, Björklund A, Klemm HP, Schlossberger HG (1982) Serotonin neurotoxins. In: Osborne NN (ed) Biology of serotonergic transmission. Wiley, Chichester, pp 249–277

Beaudet A, Descarries L (1976) Quantitative data on serotonin nerve terminals in adult rat neocortex. Brain Res 111:301–309

Beaudet A, Descarries L (1984) Fine structure of monoamine terminals in cerebral cortex. In: Descarries L, Reader T, Jasper HH (eds) Monoamine innervation of cerebral cortex. Liss, New York, pp 77–93

Beaudet A, Descarries L (1987) Ultrastructural identification of serotonin neurons. In: Steinbusch HWM (ed) Methods in the neurosciences, monoaminergic neurons: light microscopy and ultrastructure. Wiley, Chichester, pp 265–313

Becopoulos NG, Radmond DE, Roth RH (1979) Serotonin and dopamine metabolites in brain regions and cerebral spinal fluid of a primate species, effects of ketamine and fluphenazine. J Neurochem 32:1215–1218

Behbehani MM, Zemlan FP (1986) Response of nucleus raphe magnus neurons to electrical stimulation of nucleus cuneiformis: role of acetylcholine. Brain Res 369:110–118

Beitz AJ (1982) The sites of origin of brain stem neurotensin and serotonin projections to the rodent nucleus raphe magnus. J Neurosci 2:829–842

Belin MF, Nanopoulos D, Didier M, Aguera M, Steinbusch H, Verhofstad A, Maitre M, Pujol JF (1983) Immunohistochemical evidence for the presence of gamma-aminobutyric acid and serotonin in one nerve cell. A study on the raphe nuclei of the rat using antibodies to glutamate decarboxylase and serotonin. Brain Res 275:329–339

Benmansour S, Brunswick DJ (1992) Protection of amphetamine-derivative-induced neurotoxicity to serotonin neurons by nitric oxide (NO) synthesis inhibition or N-methyl-D-aspartate (NMDA) receptor antagonism. Soc Neurosci Abstr 18:913

Bennett-Clarke CA, Hankin MH, Leslie MJ, Chiaia NL, Rhoades RW (1994) Patterning of the neocortical projections from the raphe nuclei in perinatal rats: investigation of potential organizational mechanisms. J Comp Neurol 348:277–290

Berger B, Trottier S, Verney C, Gaspar P, Avaraez C (1988) Regional and laminar distribution of the dopamine and serotonin innervation in the macaque cerebral cortex: a radioautographic study. J Comp Neurol 273:99–119

Bevan P, Olivier B, Schipper J, Mos J (1989) Serotoninergic function and aggression in animals. In: Mylecharane EJ, Angus JA, de la Lande IS, Humphrey PPA (eds) Serotonin – actions, receptors, pathophysiology. Macmillan, Houndmills, pp 101–108

Bishop GA, Ho RH (1986) Cell bodies of origin of serotonin-immunoreactive afferents to the inferior olivary complex of the rat. Brain Res 399:369–373

Björklund A, Wiklund L (1980) Mechanisms of regrowth of the bulbospinal serotonin system following 5,6-dihydroxytryptamine-induced axotomy. I. Biochemical correlates. Brain Res 191:109–127

Björklund A, Baumgarten HG, Nobin A (1974) Chemical lesioning of central monoamine axons by means of 5,6-dihydroxytryptamine and 5,7-dihydroxytryptamine. Adv Biochem Psychopharmacol 10:13–33

Björklund A, Wiklund L, Descarries L (1981) Regeneration and plasticity of central serotoninergic neurons: a review. J Physiol (Paris) 77:247–255

Blottner D, Grozdanovic Z, Gossrau R (1995) Histochemistry of nitric oxide synthase in the nervous system. Histochem J 27:785–811

Bobillier P, Seguin S, Petitjean F, Salvert D, Touret M, Jouvet M (1976) The raphe nuclei of the cat brain stem: a topographic atlas of their efferent projections as revealed by autoradiography. Brain Res 113:449–486

Bosler O, Beaudet A (1985) VIP neurons as prime synaptic targets for serotonin afferents in rat suprachiasmatic nucleus: a combined radioautographic and immunocytochemical study. J Neurocytol 14:749–763

Bosler O, Beaudet A, Tappaz M (1984) Morphological evidence for cellular interactions between serotoninergic terminals and VIP and GABAergic elements in rat suprachiasmatic nucleus. Excerpta Medica 7th Int Congr Endocrinol 652:486

Bowker RM, Abbott LC (1990) Quantitative re-evaluation of descending serotonergic and nonserotonergic projections from the medulla of the rodent: evidence for extensive co-existence of serotonin and peptides in the same spinally projecting neurons, but not from the nucleus raphe magnus. Brain Res 512:15–25

Bowker RM, Steinbusch HWM, Coulter JD (1981a) Serotonergic and peptidergic projections to the spinal cord demonstrated by a combined retrograde HRP histochemical and immunocytochemical staining method. Brain Res 211:412–416

Bowker RM, Westlund KN, Coulter JD (1981b) Origins of serotonergic projections to the spinal cord in rat: an immmunocytochemical retrograde transport study. Brain Res 226:187–199

Bowker RM, Westlund KN, Coulter JD (1982) Organization of descending serotonergic projections to the spinal cord. Prog Brain Res 57:239–265

Bowker RM, Westlund KN, Sullivan MC, Wilber JF, Coulter JD (1983) Descending serotonergic, peptidergic and cholinergic pathways from the raphe nuclei: a multiple transmitter complex. Brain Res 288:33–48

Braak H (1970) Über die Kerngebiete des menschlichen Hirnstammes. II. Die Raphekerne. Z Zellforsch 107:123–141

Brodal A (1957) The reticular formation of the brain stem: anatomical aspects and functional correlations. Oliver and Boyd, Edinburgh

Brodal A, Taber E, Walberg F (1960a) The raphe nuclei of the brain stem in the cat. II. Efferent connections. J Comp Neurol 114:239–260

Brodal A, Taber E, Walberg F (1960b) The raphe nuclei of the brain stem in the cat. III. Afferent connections. J Comp Neurol 114:261–282

Brown RM, Crane AM, Goldman PS (1979) Regional distribution of monoamines in the cerebral cortex and subcortical structures of the rhesus monkey: concentration and in vivo synthesis rates. Brain Res 168: 133–150

Brownstein MJ, Palkovits M, Saavedra JM, Kizer JS (1975) Tryptophan hydroxylase in the rat brain. Brain Res 97:163–166

Cadogan AK, Kendall DA, Fink H, Marsden CA (1994) Social interaction increases 5-HT release and cAMP efflux in the rat ventral hippocampus in vivo. Behav Pharmacol 5:299–305

Cajal S, Ramon Y (1911) Histologie du systeme nerveux de l'homme et des vertebres. Maloine, Paris

Calas A, Alonso G, Arnauld E, Vincent JD (1974) Demonstration of indolaminergic fibres in the median eminence of the duck, rat and monkey. Nature 250:241–243

Calas A, Besson MJ, Gaughy C, Alonso G, Glowinski J, Cheramy A (1976) Radioautographic study of in vivo incorporation of ^{3}H-monoamines in the cat caudate nucleus: identification of serotoninergic fibers. Brain Res 118:1–13

Campbell MJ, Lewis DA, Foote SL, Morrison JH (1987) Distribution of choline acetyltransferase-, serotonin-, dopamine-β-hydroxylase-, tyrosine hydroxylase-immunoreactive fibers in monkey primate auditory cortex. J Comp Neurol 261:209–220

Cedarbaum JM, Aghajanian GK (1978) Efferent projections to the rat locus coeruleus as determined by a retrograde tracing technique. J Comp Neurol 178:1–16

Chan-Palay V (1977) Indoleamine neurons and their processes in the normal rat brain and in chronic diet-induced thiamine deficiency demonstrated by uptake of ^{3}H-serotonin. J Comp Neurol 176:467–494

Chan-Palay V, Jonsson G, Palay SL (1978) Serotonin and substance P coexist in neurons of the rat's central nervous system. Proc Natl Acad Sci USA 75:1582–1586

Charnay Y, Paulin C, Dray F, Dubois PM (1984) Distribution of enkephalin in human fetus and infant spinal cord: an immunofluorescence study. J Comp Neurol 223:415–423

Chazal G, Ralston HJ III (1987) Serotonin-containing structures in the nucleus raphe dorsalis of the cat: an ultrastructural analysis of dendrites and axon terminals. J Comp Neurol 259:317–329

Chu NS, Bloom FE (1974) The catecholamine-containing neurons in the cat dorsolateral pontine tegmentum: distribution of the cell bodies and some axonal projections. Brain Res 66:1–21

Clikeman JA, Wei S, Turkanis SA, Finnegann KT (1992) Inhibitors of nitric oxide synthesis reduce the neurotoxic effects of methamphetamine in mice. Soc Neurosci Abstr 18:913

Commins DL, Axt KJ, Vosmer G, Seiden LS (1987) Endogeneously produced 5,6-dihydroxytryptamine may mediate the neurotoxic effects of parachloroamphetamine. Brain Res 419:253–261

Compoint C, Buisseret-Delmas C (1988) Origin, distribution and organization of the serotoninergic innervation in the inferior olivary complex of the rat. Arch Ital Biol 126:99–110

Conrad LCA, Leonard CM, Pfaff DW (1974) Connections of the median and dorsal raphe nuclei in the rat: an autoradiographic and degeneration study. J Comp Neurol 156:179–206

Corvaja N, Doucet G, Bolam JP (1993) Ultrastructure and synaptic targets of the raphe-nigral projection in the rat. Neuroscience 55:417–427

Cotton RGH, McAdam W, Jennings IG, Morgan FJ (1988) A monoclonal antibody to aromatic amino acid hydroxylases. Identification of the epitope. Biochem J 255:193–196

Cottrell GA (1977) Identified amine-containing neurons and their synaptic connexions. Neuroscience 2:1–18

Cropper EC, Eisenman JS, Azmitia EC (1984) An immunocytochemical study of the serotonergic innervation of the thalamus of the rat. J Comp Neurol 224:38–50

Cross AJ, Joseph MH (1981) The concurrent estimation of the major monoamine metabolites in human and nonhuman primate brain by HPLC with fluorescence and electrochemical detection. Life Sci 28:499–505

Cullheim S, Arvidsson U (1995) The peptidergic innervation of spinal motoneurons via the bulbospinal 5-hydroxytryptamine pathway. Prog Brain Res 104:21–40

Dahlström A, Fuxe K (1964) Evidence for the existence of monoamine-containing neurons in the central nervous system. I. Demonstration of monoamines in the cell bodies of brain stem neurons. Acta Physiol Scand 62 [Suppl 232]:1–55

Danner H, Pfister C (1980) Untersuchungen zur Zytoarchitektonik des Nucleus raphe dorsalis der Ratte. J Hirnforsch 21:655–664

Dean C, Marson L, Kampine JP (1993) Distribution and co-localization of 5-hydroxytryptamine, thyrotropin-releasing hormone and substance P in the cat medulla. Neuroscience 57:811–822

DeFelipe J, Jones EG (1988) A light and electron microscopic study of serotonin-immunoreactive fibers and terminals in the monkey sensory-motor cortex. Exp Brain Res 71:171–182

DeFelipe J, Hendry SHC, Hashikawa T, Jones EG (1991) Synaptic relationships of serotonin-immunoreactive terminal baskets on GABA neurons in the cat auditory cortex. Cereb Cortex 1:117–133

de Lima AD, Bloom FE, Morrison JH (1988) Synaptic organization of serotonin-immunoreactive fibers in primary visual cortex of the macaque monkey. J Comp Neurol 274:280–294

Descarries L, Beaudet A (1978) The serotonin innervation of adult rat hypothalamus. In: Vincent JD, Korolon C (eds) Cell biology of hypothalamic neurosecretion, vol 280. Coll Internat CNRS, Paris, pp 135–153

Descarries L, Beaudet A, Watkins KC (1975) Serotonin nerve terminals in adult rat neocortex. Brain Res 100:563–588

Diaz-Cintra S, Cintra L, Kemper T, Resnick O, Morgane PJ (1981) Nucleus raphe dorsalis: a morphometric Golgi study in rats of three age groups. Brain Res 207:1–16

Dun NJ, Dun SL, Förstermann U (1994) Nitric oxide synthase immunoreactivity in rat pontine medullary neurons. Neuroscience 59:429–445

Fahn S, Libsch LR, Cutler RW (1971) Monoamines in the human neostriatum: topographic distribution in normals and in Parkinson's disease and their role in akinesia, rigidity, chorea and tremor. J Neurol Sci 14:427–455

Falck B, Hillarp NA, Thieme G, Torp A (1962) Fluorescence of catecholamines and related compounds condensed with formaldehyde. J Histochem Cytochem 10:348–354

Felten DL, Harrigan P (1980) Dendrite bundles in nuclei raphe dorsalis and centralis superior of the rabbit: a possible substrate for local control of serotonergic neurons. Neurosci Lett 16:275–280

Fischer C, Hatzidimitriou G, Wlos J, Katz J, Ricaurte G (1995) Reorganization of ascending 5-HT axon projections in animals previously exposed to the recreational drug (±)3,4-methylenedioxymethamphetamine (MDMA, "ecstasy"). J Neurosci 15:5476–5485

Fornal CA, Jacobs BL (1988) Physiological and behavioral correlates of serotonergic single unit activity. In: Osborne NN, Hamon M (eds) Neuronal serotonin. Wiley, New York, pp 305–345

Fort P, Luppi PH, Sakai K, Salvert D, Jouvet M (1990) Nuclei of origin of monoaminergic, peptidergic and cholinergic afferents to the cat trigeminal motor nucleus: a double-labeling study with cholera-toxin as a retrograde tracer. J Comp Neurol 301:262–275

Fox GQ, Pappas GD, Purpura DP (1976) Morphology and fine structure of the feline neonatal medullary raphe nuclei. Brain Res 101:385–410

Frankfurt M, Azmitia EC (1983) The effect of intracerebral injections of 5,7-dihydroxytryptamine and 6-hydroxydopamine on the serotonin immunoreactive cell bodies and fibers in the adult rat hypothalamus. Brain Res 261:91–99

Freund TF (1992) GABAergic septal and serotonergic median raphe afferents preferentially innervate inhibitory interneurons in the hippocampus and dentate gyrus. In: Ribak CE, Gall CM, Mody I (eds) The dentate gyrus and its role in seizures. Elsevier, Amsterdam, pp 79–91

Freund TF, Gulyas AI, Acsady L, Görcs T, Toth K (1990) Serotonergic control of the hippocampus via local inhibitory interneurons. Proc Natl Acad Sci USA 87:8501–8505

Fuller RW (1992) Effects of p-chloramphetamine on brain serotonin neurons. Neurochem Res 17:449–456

Fuller RW, Henderson MG (1994) Neurochemistry of halogenated amphetamines. In: Cho AK, Segal DS (eds) Amphetamine and its analogs. Academic, San Diego, pp 209–242

Fuxe K (1965) Evidence for the existence of monoamine neurons in the central nervous system. IV. Distribution of monoamine nerve terminals in the central nervous system. Acta Physiol Scand [Suppl] 247:39–85

Fuxe K, Hökfelt T, Ungerstedt U (1968) Localization of indolealkylamines in CNS. Adv Pharmacol 6A:235–251

Fuxe K, Hökfelt T, Ungerstedt U (1969) Distribution of monoamines in the mammalian central nervous system by histochemical methods. In: Hooper G (ed) Metabolism of amines in the brain. Macmillan, London, pp 10–22

Gall C, Moore RY (1984) Distribution of enkephalin, substance P, tyrosine hydroxylase, and 5-hydroxytryptamine immunoreactivity in the septal region of the rat. J Comp Neurol 225:212–227

Gang S, Mizuguchi A, Aoki M (1991) Axonal projections from the pontine pneumotaxic region to the nucleus raphe magnus in cats. Respir Physiol 85:329–339

Glazer EJ, Steinbusch H, Verhofstad, Basbaum AI (1981) Serotonin neurons in nucleus raphe dorsalis and paragigantocellularis of the cat contain enkephalin. J Physiol (Paris) 77:241–245

Gonzalo-Ruiz A, Lieberman AR, Sanz-Anquela JM (1995) Organization of serotoninergic projections from the raphe nuclei to the anterior thalamic nuclei in the rat: a combined retrograde tracing and 5-HT immunohistochemical study. J Chem Neuroanat 8:103–115

Haan EA, Jennings IG, Cuello AC, Nakata H, Chow CW, Kushinsky R, Brittingham J, Cotton RGH (1987) A monoclonal antibody recognizing all three aromatic amino acid hydroxylases allows identification of serotonergic neurons in the human brain. Brain Res 426:19–27

Halasy K, Miettinen R, Szabat E, Freund TF (1992) GABAergic interneurons are the major postsynaptic targets of median raphe afferents in the rat dentate gyrus. Eur J Neurosci 4:144–153

Halasz N, Ljungdahl A, Hökfelt T (1978) Transmitter histochemistry of the rat olfactory bulb. II. Fluorescence histochemical, autoradiographic and electron microscopic localization of monoamines. Brain Res 154:253–271

Halliday GM, Li YW, Joh TH, Cotton RGH, Howe PRC, Geffen LB, Blessing WW (1988) Distribution of substance P-like immunoreactive neurons in the human medulla oblongata: co-localization with monoamine-synthesizing neurons. Synapse 2:353–370

Halliday GM, McCann HL, Pamphlett R, Brooks WS, Creasey H, McCusker E, Cotton
    RGH, Broe GA, Harper CG (1992) Brain stem serotonin-synthesizing neurons in
    Alzheimer's disease: a clinicopathological correlation. Acta Neuropathol (Berl)
    84:638–650
Halliday G, Harding A, Paxinos G (1995) Serotonin and tachykinin systems. In:
    Paxinos G (ed) The rat nervous system, 2nd edn. Academic Press, San Diego, pp
    929–974
Haring JH, Meyerson L, Hoffman TL (1992) Effects of parachloroamphetamine upon
    the serotonergic innervation of the rat hippocampus. Brain Res 577:253–260
Hensler JG, Ferry RC, Labow DM, Kovachich GB, Frazer A (1994) Quantitative
    autoradiography of the serotonin transporter to assess the distribution of seroton-
    ergic projections from the dorsal raphe nucleus. Synapse 17:1–15
Herbert H (1992) Evidence for projections from medullary nuclei onto serotonergic
    and dopaminergic neurons in the midbrain dorsal raphe nucleus of the rat. Cell
    Tissue Res 270:149–156
Herve D, Pickel VM, Joh TH, Beaudet A (1987) Serotonin axon terminals in the
    ventral tegmental area of the rat: fine structure and synaptic input to dopaminergic
    neurons. Brain Res 435:71–83
Heym J, Trulson ME, Jacobs BL (1981) Effects of adrenergic drugs on raphe unit
    activity in freely moving cats. Eur J Pharmacol 74:117–125
Heym J, Trulson ME, Jacobs BL (1982) Raphe unit activity in freely moving cats:
    effects of phasic auditory and visual stimuli. Brain Res 232:29–39
Hökfelt T, Fuxe K, Goldstein M, Johansson O (1974) Immunohistochemical evidence
    for the existence of adrenaline neurons in the rat brain. Brain Res 66:235–251
Hökfelt T, Ljungdahl Å, Steinbusch HWM, Verhofstad A, Nilsson G, Brodin E,
    Pernow B, Goldstein M (1978) Immunohistochemical evidence of substance P-like
    immunoreactivity in some 5-hydroxytryptamine-containing neurons in the rat cen-
    tral nervous system. Neuroscience 3:517–538
Hökfelt T, Johansson O, Goldstein M (1984) Central catecholamine neurons as re-
    vealed by immunohistochemistry with special reference to adrenaline neurons. In:
    Björklund A, Hökfelt T (eds) Handbook of chemical neuroanatomy, vol 2: Clas-
    sical transmitters in the CNS, part I. Elsevier, Amsterdam, pp 157–379
Hökfelt T, Johansson O, Holets V, Meister B, Melander T (1987) Distribution of
    neuropepetides with special reference to their coexistence with classical transmit-
    ters. In: Meltzer HY (ed) Psychopharmacology. The third generation of progress.
    Raven, New York, pp 401–416
Hölzel B, Pfister C (1981) Untersuchungen zur Topographie und Zytoarchitektonik
    der Raphe-Kerne der Ratte. J Hirnforsch 22:697–708
Holmes CJ, Mainville LS, Jones BE (1994) Distribution of cholinergic, GABAergic
    and serotonergic neurons in the medial medullary reticular formation and
    their projections studied by cytotoxic lesions in the cat. Neuroscience 62:1155–
    1178
Holstege JC, Kuypers HGJM (1987) Brainstem projections to spinal motoneurons: an
    update. Neuroscience 23:809–821
Hornung J-P, Celio MR (1992) The selective innervation by serotoninergic axons of
    calbindin-containing interneurons in the neocortex and hippocampus of the mar-
    moset. J Comp Neurol 320:457–467
Hornung J-P, Fritschy J-M (1988) Serotoninergic system in the brainstem of the mar-
    moset: a combined immunocytochemical and three-dimensional reconstruction
    study. J Comp Neurol 270:471–487
Hornung J-P, Fritschy J-M, Törk I (1990) Distribution of two morphologically distinct
    subsets of serotoninergic axons in the cerebral cortex of the marmoset. J Comp
    Neurol 297:165–181
Hunt SP, Lovick TA (1982) The distribution of serotonin, met-enkephalin and β-
    lipotropin-like immunoreactivity in neuronal perikarya of the cat brainstem.
    Neurosci Lett 30:139–145

Ihara N, Ueda S, Kawata M, Sano Y (1988) Immunohistochemical demonstration of serotonin-containing nerve fibers in the mammalian hippocampal formation. Acta Anat 132:335–346

Imai H, Steindler DA, Kitai ST (1986) The organization of divergent axonal projections from the midbrain raphe nuclei in the rat. J Comp Neurol 243:363–380

Inoue T, Koyama T, Yamashita I (1993) Effect of conditioned fear stress on serotonin metabolism in the rat brain. Pharmacol Biochem Behav 44:371–374

Insel TR, Battaglia G, Johanssen J, Marra S, DeSouza EB (1989) 3,4-Methylenedioxymethamphetamine ("ecstasy") selectively destroys brain serotonin nerve terminals in rhesus monkeys. J Pharmacol Exp Ther 249:713–720

Jacobs BL, Azmitia EC (1992) Structure and function of the brain serotonin system. Physiol Rev 72:165–229

Jacobs BL, Fornal CA (1993) 5-HT and motor control: a hypothesis. Trends Neurosci 16:346–352

Jacobs BL, Gannon PJ, Azmitia EC (1984) Atlas of serotonergic cell bodies in the cat brainstem: an immunocytochemical analysis. Brain Res Bull 13:1–31

Jennes L, Beckman WC, Stumpf WE, Grzanna R (1982) Anatomical relationships of serotoninergic and noradrenergic projections with the GnRH system in septum and hypothalamus. Exp Brain Res 46:331–338

Johansson O, Hökfelt T, Pernow B, Jeffcoate SL, White N, Steinbusch HMW, Verhofstad AAJ, Emson PC, Spindel E (1981) Immunohistochemical support for three putative transmitters in one neuron: coexistence of 5-hydroxytryptamine-, substance P- and thyrotropin releasing hormone-like immunoreactivity in medullary neurons projecting to the spinal cord. Neuroscience 6:1857–1881

Johnson MD, Ma PM (1993) Localization of NADPH diaphorase activity in monoaminergic neurons of the rat brain. J Comp Neurol 332:391–406

Johnson H, Ulhake B, Dagerlind Å, Bennett GW, Fone KCF, Hökfelt T (1993) The serotoninergic bulbospinal system and brainstem-spinal cord content of serotonin-, TRH-, and substance P-like immunoreactivity in the aged rat with special reference to the spinal cord motor nucleus. Synapse 15:63–89

Jones BE, Holmes CJ, Rodriguez-Veiga E, Mainville L (1991) GABA-synthesizing neurons in the medulla: their relationship to serotonin-containing and spinally projecting neurons in the rat. J Comp Neurol 313:349–367

Jonsson G, Hallman H (1983) Effect of substance P on the 5,7-dihydroxytryptamine-induced alteration of the postnatal development of central serotonin neurons. Med Biol 61:105–122

Kachidian P, Poulat P, Marlier L, Privat A (1991) Immunohistochemical evidence for the coexistence of substance P, thyrotropin-releasing hormone, GABA, methionin-enkephalin, and leucin-enkephalin in the serotonergic neurons of the caudal raphe nuclei: a dual labeling in the rat. J Neurosci Res 30:521–530

Kalen P (1988) Regulation of brain stem serotonergic and noradrenergic systems. Thesis, Department of Medical Cell Research University of Lund, Grahns Boktryckeri, pp 1–206

Kalen P, Karlson M, Wiklund L (1985) Possible excitatory amino acid afferents to the nucleus raphe dorsalis of the rat investigated with retrograde wheat germ agglutinin and D-[³H]aspartate tracing. Brain Res 360:285–297

Kalen P, Strecker RE, Rosengren E, Björklund A (1989) Regulation of striatal serotonin release by the lateral habenula-dorsal raphe pathway in the rat as demonstrated by in vivo microdialysis: role of excitatory amino acids and GABA. Brain Res 492:187–202

Kausz M (1991) Arrangement of neurons in the medullary reticular formation and raphe nuclei projecting to thoracic, lumbar and sacral segments of the spinal cord in the cat. Anat Embryol (Berl) 183:151–163

Kawahara H, Yoshida M, Yokoo H, Nishi M, Tanaka M (1993) Psychological stress increases serotonin release in the rat amygdala and prefrontal cortex assessed by in vivo microdialysis. Neurosci Lett 162:81–84

Kawasaki T, Sato Y (1981) Afferent projections to the caudal part of the dorsal nucleus of the raphe in cats. Brain Res 211:439–444

Kawata M, Takeuchi Y, Ueda S, Matsuura T, Sano Y (1984) Immunohistochemical demonstration of serotonin-containing nerve fibers in the hypothalamus of the monkey, Macaca fuscata. Cell Tissue Res 236:495–503

Kerr CWH, Bishop GA (1991) Topographical organization in the origin of serotoninergic projections to different regions of the cat cerebellar cortex. J Comp Neurol 304:502–515

Kiss J, Halasz B (1985) Demonstration of serotoninergic axons terminating on luteinizing hormone-releasing hormone in the preoptic area of the rat using a combination of immunocytochemistry and high resolution autoradiography. Neuroscience 14:69–78

Kiss J, Halasz B (1986) Synaptic connections between serotoninergic axon terminals and tyrosine hydroxylase-immunoreactive neurons in the arcuate nucleus of the rat hypothalamus. A combination of electron microscopic autoradiography and immunocytochemistry. Brain Res 364:284–294

Kiss J, Leranth C, Halasz B (1984) Serotoninergic endings on VIP-neurons in the suprachiasmatic nucleus and on ACTH-neurons in the arcuate nucleus of the rat hypothalamus. A combination of electron microscopic autoradiography and electron microscopic immunocytochemistry. Neurosci Lett 44:119–124

Kiss J, Csaky A, Halasz B (1988) Demonstration of serotoninergic axon terminals on somatostatin-immunoreactive neurons of the anterior periventricular nucleus of the rat hypothalamus. Brain Res 442:23–32

Kitzman PH, Bishop G (1994) The origin of serotoninergic afferents to the cat's cerebellar nuclei. J Comp Neurol 340:541–550

Koelliker A (1891) Der feinere Bau des verlängerten Markes. Anat Anz 6:427–431

Köhler C (1982) On the serotonergic innervation of the hippocampal region: an analysis imploying immunohistochemistry and retrograde fluorescent tracing in the rat brain. In: Palay SL, Chan-Palay V (eds) Cytochemical methods in neuroanatomy. Liss, New York, pp 387–405

Köhler C, Chan-Palay V, Haglund L, Steinbusch H (1980) Immunohistochemical localization of serotonin nerve terminals in the lateral entorhinal cortex of the rat: demonstration of two separate patterns of innervation from the midbrain raphe. Anat Embryol (Berl) 160:121–129

Köhler C, Chan-Palay V, Steinbusch H (1981) The distribution and orientation of serotonin fibers in the entorhinal and other retrohippocampal areas. Anat Embryol (Berl) 161:237–264

Köhler C, Chan-Palay V, Steinbusch H (1982) The distribution and origin of serotonin-containing fibers in the septal area: a combined immunohistochemical and fluorescent retrograde tracing study in the rat. J Comp Neurol 209:91–111

Kosofsky BE, Molliver ME (1987) The serotonergic innervation of cerebral cortex: different classes of axon terminals arise from dorsal and median raphe nuclei. Synapse 1:153–168

Kosofsky BE, Molliver ME, Morrison JH, Foote SL (1984) The serotonin and norepinephrine innervation of primary visual cortex in the cynomolgus monkey (Macaca fascicularis). J Comp Neurol 230:168–178

Kwiat GC, Basbaum AI (1992) The origin of brainstem noradrenergic and serotonergic projections to the spinal cord dorsal horn in the rat. Somatosens Mot Res 9:157–173

Lakos S, Basbaum AI (1988) An ultrastructural study of the projections from the midbrain periaqueductal gray to spinally projecting, serotonin-immunoreactive neurons of the medullary raphe magnus in the rat. Brain Res 443:383–388

Lavoie B, Parent A (1990) Immunohistochemical study of the serotoninergic innervation of the basal ganglia in the squirrel monkey. J Comp Neurol 299:1–16

Leger L, Charnay Y, Dubois PM, Jouvet M (1986) Distribution of enkephalin-immunoreactive cell bodies in relation to serotonin-containing neurons in the raphe nuclei of the cat: immunohistochemical evidence for the coexistence of enkephalins and serotonin in certain cells. Brain Res 362:63–73

Leranth CS, Palkovits M, Krieger DT (1983) Serotonin immunoreactive nerve fibers and terminals in the rat pituitary – light- and electron-microscopic studies. Neuroscience 9:289–296

Levine ES, Jacobs BL (1992) Neurochemical afferents controlling the activity of serotonergic neurons in the dorsal raphe nucleus: microiontophoretic studies in the awake cat. J Neurosci 12:4037–4044

Levitt P, Moore RY (1979) Origin and organization of brainstem catecholamine innervation in the rat. J Comp Neurol 186:505–528

Li Y-Q, Takada M, Mizuno N (1993a) The sites of origin of serotoninergic afferent fibers in the trigeminal motor, facial, and hypoglossal nuclei in the rat. Neurosci Res 17:307–313

Li Y-Q, Takada M, Shinonaga Y, Mizuno N (1993b) Collateral projections of single neurons in the nucleus raphe magnus to both the sensory trigeminal nuclei and spinal cord in the rat. Brain Res 602:331–335

Li Y-Q, Takada M, Shinonaga Y, Mizuno N (1993c) Direct projections from the midbrain periaqueductal gray and the dorsal raphe nucleus to the trigeminal sensory complex in the rat. Neuroscience 54:431–443

Lidov HGW, Molliver ME (1982) Immunohistochemical study of the development of serotonergic neurons in the rat CNS. Brain Res Bull 9:559–604

Lidov HGW, Grzanna R, Molliver ME (1980) The serotonin innervation of the cerebral cortex in the rat – an immunohistochemical analysis. Neuroscience 5:207–227

Lindvall O, Björklund A (1974) The organization of the ascending catecholamine neuron system in the rat brain as revealed by the glyoxylic acid fluorescence method. Acta Physiol Scand 412:1–48

Lindvall O, Björklund A (1978) Organization of catecholamine neurons in the rat central nervous system. In: Iversen LL, Iversen SD, Snyder SH (eds) Handbook of psychopharmacology, vol 9: chemical pathways in the brain. Plenum, New York, pp 139–231

Liposits Z, Phelix C, Paull WK (1987) Synaptic interaction of serotonergic axons and corticotropin releasing factor (CRF) synthesizing neurons in the hypothalamic paraventricular nucleus of the rat: a light and electron microscopic immunocytochemical study. Histochemistry 86:541–549

Loizou LA (1969) Projections of the nucleus locus coeruleus in the albino rat. Brain Res 15:563–566

Lovick TA, Hunt SP (1983) Substance P-immunoreactive and serotonin-containing neurones in the ventral brainstem of the cat. Neurosci Lett 36:223–228

Lüth H-J, Seidel I (1987) Immunhistochemische Charakterisierung serotoninerger Afferenzen im visuellen System der Ratte. J Hirnforsch 28:591–560

Lund JS (1973) Organization of neurons in the visual cortex, area 17, of the monkey (Macaca mulatta). J Comp Neurol 147:455–496

Lytle LD, Jacoby JH, Nelson M, Baumgarten HG (1974) Long-term effects of 5,7-dihydroxytryptamine administered at birth on the development of brain monoamines. Life Sci 15:1203–1217

Maciewicz R, Taber-Pierce E, Ronner S, Foote WE (1981) Afferents of the central superior raphe nucleus in the cat. Brain Res 216:414–421

Mackay AVP, Yates CM, Wright A, Hamilton P, Davies P (1978) Regional distribution of monoamines and their metabolism in the human brain. J Neurochem 30:841–848

Mamounas LA, Molliver ME (1988) Evidence for dual serotonergic projections to neocortex. Axons from the dorsal and median raphe nuclei are differentially vulnerable to the neurotoxin p-chloroamphetamine (PCA). Exp Neurol 102:23–36

Mamounas LA, Mullen CA, O'Hearn E, Molliver ME (1991) Dual serotoninergic projections to forebrain in the rat: morphological distinct 5-HT axon terminals

exhibit differential vulnerability to neurotoxic amphetamine derivatives. J Comp
    Neurol 314:558–586
Marson L, Loewy AD (1985) Topographic organization of substance P and monoam-
    ine cells in the ventral medulla of the cat. J Auton Nerv Syst 14:271–285
McCann UD, Ricaurte GA (1991) Lasting neuropsychiatric sequelae of (±)3,4-
    methylenedioxymethamphetamine ("ecstasy") in recreational users. J Clin
    Psychopharmacol 11:302–305
McCann UD, Ricaurte GA (1994) Use and abuse of ring-substituted amphetamine. In:
    Cho AK, Segal DS (eds) Amphetamine and its analogs. Academic, San Diego, pp
    371–385
McCann UD, Ridenour A, Shaham Y, Ricaurte GA (1994) Serotonin neurotoxicity
    after (±)3,4-methylenedioxymethamphetamine (MDMA; "ecstasy"): a controlled
    study in humans. Neuropsychopharmacology 10:129–138
McLean JH, Shipley MT (1987) Serotonergic afferents to the rat olfactory bulb: I.
    Origins and laminar specificity of serotonergic inputs in the adult rat. J Neurosci
    7:3016–3028
McQuade R, Sharp T (1995) Release of cerebral 5-hydroxytryptamine evoked by
    electrical stimulation of the dorsal and median raphe nuclei: effect of a neurotoxic
    amphetamine. Neuroscience 68:1079–1088
Melander T, Hökfelt T, Rökaeus Å, Cuello AC, Oertel WH, Verhofstad A, Goldstein
    M (1986) Coexistence of galanin-like immunoreactivity with catecholamines, 5-
    hydroxytryptamine, GABA, and neuropeptides in the rat CNS. J Neurosci 6:3640–
    3654
Millhorn DE, Hökfelt T, Seroogy K, Oertel W, Verhofstad AAJ, Wu J-Y (1987)
    Immunohistochemical evidence for colocalization of gamma-aminobutyric acid
    and serotonin in neurons of the ventral medulla oblongata projecting to the spinal
    cord. Brain Res 410:179–185
Millhorn DE, Hökfelt T, Verhofstad AAJ, Terenius L (1989) Individual cells in the
    raphe nuclei of the medulla oblongata of the rat that contain immunoreactivities
    for both serotonin and enkephalin project to the spinal cord. Exp Brain Res
    75:536–542
Molliver ME (1987) Serotonergic neuronal systems: what their anatomic organization
    tells us about function. J Clin Psychopharmacol 7 [Suppl]:3–23
Molliver ME, Mamounas LA (1991) Dual serotonergic projections to rat olfactory
    bulb: p-chloroamphetamine (PCA) selectively damages dorsal raphe axons while
    sparing median raphe axons. Soc Neurosci Abstr 17:1180
Molliver ME, Grzanna R, Lidov HGW, Morrison JH, Olschowka JA (1982) Monoam-
    ine systems in the cerebral cortex. In: Chan-Palay V, Palay SL (eds) Cytochemical
    methods in neuroanatomy. Liss, New York, pp 255–278
Molliver ME, Berger UV, Mamounas LA, Molliver DC, O'Hearn E, Wilson MA
    (1990) Neurotoxicity of MDMA and related compounds: anatomic studies. Ann N
    Y Acad Sci 600:640–664
Mori S, Takino T, Yamada H, Sano Y (1985a) Immunohistochemical demonstration of
    serotonin nerve fibers in the subthalamic nucleus of the rat, cat and monkey.
    Neurosci Lett 62:305–309
Mori S, Ueda S, Yamada H, Takino T, Sano Y (1985b) Immunohistochemical demon-
    stration of serotonin nerve fibers in the corpus striatum of the rat, cat and monkey.
    Anat Embryol (Berl) 173:1–5
Mori S, Matsuura T, Takino T, Sano Y (1987) Light and electron microscopic immuno-
    histochemical studies of serotonin nerve fibers in the substantia nigra of the rat, cat
    and monkey. Anat Embryol (Berl) 176:13–18
Morrison JH, Foote SL (1986) Noradrenergic and serotoninergic innervation of corti-
    cal, thalamic, and tectal visual structures in Old and New World monkeys. J Comp
    Neurol 243:117–138
Morrison JH, Foote SL, Molliver ME, Bloom FE, Lidov HGW (1982) Noradrenergic
    and serotonergic fibers innervate complementary layers in monkey primary

visual cortex: an immunohistochemical study. Proc Natl Acad Sci USA 79:2401–2405

Mulligan KA, Törk I (1987) Serotonergic axons form basket-like terminals in cerebral cortex. Neurosci Lett 81:7–12

Mulligan KA, Törk I (1988) Serotoninergic innervation of the cat cerebral cortex. J Comp Neurol 270:86–110

Mulligan KA, Törk I (1993) Serotoninergic innervation of area 17 in the cat. Cereb Cortex 3:108–121

Nanopoulos D, Belin MF, Maitre M, Vincendon G, Pujol JF (1982) Immunocytochemical evidence for the existence of GABAergic neurons in the nucleus raphe dorsalis. Possible existence of neurons containing serotonin and GABA. Brain Res 232:375–389

Nedergaard S, Bolam JP, Greenfield SA (1988) Facilitation of a dendritic calcium conductance by 5-hydroxytryptamine in the substantia nigra. Nature 333:174–177

Nevin K, Zhuo H, Helke CJ (1994) Neurokinin A coexists with substance P and serotonin in ventral medullary spinally projecting neurons of the rat. Peptides 15:1003–1011

Nicholas AP, Pieribone VA, Arvidsson U, Hökfelt T (1992) Serotonin-, substance P- and glutamate/aspartate-like immunoreactivities in medullo-spinal pathways of rat and primate. Neuroscience 48:545–559

Nieuwenhuys R (1985) Chemoarchitecture of the brain. Springer, Berlin Heidelberg New York

Nobin A, Baumgarten HG, Björklund A, Lachenmayer L, Stenevi U (1973) Axonal degeneration and regeneration of the bulbospinal indoleamine neuron systems after 5,6-dihydroxytryptamine treatment. Brain Res 56:1–24

Nygren LG, Fuxe K, Jonsson G, Olson L (1974) Functional regeneration of 5-hydroxytryptamine nerve terminals in the rat spinal cord following 5,6-dihydroxytryptamine-induced degeneration. Brain Res 78:377–394

O'Hearn E, Molliver ME (1984) Organization of raphe-cortical projections in rat: a quantitative retrograde study. Brain Res Bull 13:709–726

O'Hearn E, Battaglia G, De Souza EB, Kuhar MJ, Molliver ME (1988) Methylenedioxyamphetamine (MDA) and methylenedioxymethamphetamine (MDMA) cause selective ablation of serotoninergic axon terminals in forebrain: immunocytochemical evidence for neurotoxicity. J Neurosci 8:2788–2803

Ohm TG, Heilmann R, Braak H (1989) The human oral raphe system. Anat Embryol (Berl) 180:37–43

Oleskevich S, Descarries L (1990) Quantified distribution of the serotonin innervation in adult rat hippocampus. Neuroscience 34:19–33

Oleskevich S, Descarries L, Watkins KC, Seguela P, Daszuta A (1991) Ultrastructural features of the serotonin innervation in adult rat hippocampus: an immunocytochemical description in single and serial thin sections. Neuroscience 42:777–791

Paasonen MK, MacLean PD, Giarman NJ (1957) 5-Hydroxytryptamine (serotonin, enteramine) content of structures of the limbic system. J Neurochem 1:326–333

Palkovits M, Brownstein M, Saavedra JM (1974) Serotonin content of the brain stem nuclei in the rat. Brain Res 80:237–249

Papadopoulos GC, Parnavelas JG, Buijs R (1987a) Monoaminergic fibers form conventional synapses in the cerebral cortex. Neurosci Lett 76:275–279

Papadopoulos GC, Parnavelas JG, Buijs R (1987b) Light and electron microscopical immunocytochemical analysis of the serotonin innervation of the rat visual cortex. J Neurocytol 16:883–892

Parent A (1981a) Comparative anatomy of the serotoninergic systems. J Physiol (Paris) 77:147–156

Parent A (1981b) The anatomy of serotonin-containing neurons across phylogeny. In: Jacobs BL, Gelperin A (eds) Serotonin neurotransmission and behaviour. MIT Press, Cambridge, MA, pp 3–34

Parent A, Descarries L, Beaudet A (1981) Organization of ascending serotonin systems in the adult rat brain. A radioautographic study after intraventricular administration of [^{3}H]5-hydroxytryptamine. Neuroscience 6:115–138

Parizek J, Hassler R, Bak IJ (1971) Light and electron microscopic autoradiography of substantia nigra of rat after intraventricular administration of tritium labelled norepinephrine, dopamine, serotonin and the precursors. Z Zellforsch 115:137–148

Parnavelas JG, McDonald JK (1983) The cerebral cortex. In: Emson PC (ed) Chemical neuroanatomy. Raven, New York, pp 505–549

Parnavelas JG, Papdopoulos GC (1989) The monoaminergic innervation of the cerebral cortex is not diffuse and nonspecific. Trends Neurosci 12:315–319

Parnavelas JG, Moises HC, Speciale SG (1985) The monoaminergic innervation of the rat visual cortex. Proc R Soc Lond B 223:319–329

Pasik P, Pasik T (1982) Serotoninergic afferents in the monkey neostriatum. Acta Biol Acad Sci Hung 33:277–288

Pasik P, Pasik T, Holstein GR, Saavedra JP (1984a) Serotoninergic innervation of the monkey basal ganglia: an immunocytochemical light and electron microscopy study. In: McKenzie JS, Kemn RE, Wilcock LN (eds) The basal ganglia: structure and function. Plenum, New York, pp 115–129

Pasik P, Pasik T, Pecci-Saavedra J, Holstein GR, Yahr MD (1984b) Serotonin in pallidal neuronal circuits: an immunocytochemical study in monkeys. Adv Neurol 40:63–76

Pasquier DA, Anderson C, Forbes WB, Morgane PJ (1976) HRP tracing of the lateral habenula-midbrain raphe nuclei connections in the rat. Brain Res Bull 1:443–451

Pei Q, Zetterströme T, Fillenz M (1990) Tail pinch-induced changes in the turnover and release of dopamine and 5-hyroxytryptamine in different regions of the rat. Neuroscience 35:133–138

Petrov T, Krukoff TL, Jhamandas JH (1992a) The hypothalamic paraventricular and lateral parabrachial nuclei receive collaterals from raphe nucleus neurons: a combined double retrograde and immunocytochemical study. J Comp Neurol 318:18–26

Petrov T, Jhamandas JH, Krukoff TL (1992b) Characterization of peptidergic efferents from the lateral parabrachial nucleus to identified neurons in the rat dorsal raphe nucleus. J Chem Neuroanat 5:367–373

Petrov T, Krukoff TL, Jhamandas JH (1994) Chemically defined collateral projections from the pons to the central nucleus of the amygdala and hypothalamic paraventricular nucleus in the rat. Cell Tissue Res 277:289–295

Peyron C, Luppi P-H, Fort P, Rampon C, Jouvet M (1996) Lower brainstem catecholamine afferents to the rat dorsal raphe nucleus. J Comp Neurol 364:402–413

Priestley JV, Cuello AC (1982) Coexistence of neuroactive substances as revealed by immunohistochemistry with monoclonal antibodies. In: Cuello AC (ed) Cotransmission. Macmillan, London, pp 165–188

Quian Y, Melikian HE, Rye DB, Levey AI, Blakely RD (1995) Identification and characterization of antidepressant-sensitive serotonin transporter proteins using site-specific antibodies. J Neurosci 15:1261–1274

Ricaurte GA, Forno LS, Wilson MA, DeLÖanney LE, Irwin I, Molliver ME, Langston JW (1988) (±)3,4-Methylenedioxymethamphetamine selectively damages central serotonergic neurons in nonhuman primates. J Am Med Assoc 260:51–55

Ricaurte GA, Finnegan KT, Irwin I, Langston JW (1990) Aminergic metabolites in cerebrospinal fluid of humans previously exposed to MDMA: preliminary observations. Ann N Y Acad Sci 600:699–710

Ricaurte G, Martello AL, Katz JL, Martello MB (1992) Lasting effects of (±)3,4-methylenedioxymethamphetamine (MDMA) on central serotonergic neurons in nonhuman primates: neurochemical observations. J Pharmacol Exp Ther 261:616–622

Richter RC, Behbehani MM (1991) Evidence for glutamic acid as a possible neurotransmitter between the mesencephalic nucleus cuneiformis and the medullary nucleus raphe magnus in the lightly anesthetized rat. Brain Res 544:279–286

Romagnano MA, Joseph SA (1983) Immunocytochemical localization of $ACTH_{1-39}$ in the brainstem of the rat. Brain Res 276:1–16

Saavedra JM (1977) Distribution of serotonin and synthesizing enzymes in discrete areas of the brain. Fed Proc 36:2134–2144

Saavedra JM, Brownstein M, Palkovits M (1974a) Serotonin distribution in the limbic system of the rat. Brain Res 79:437–441

Saavedra JM, Palkovits M, Brownstein M, Axelrod J (1974b) Serotonin distribution in the nuclei of the rat hypothalamus and preoptic region. Brain Res 77:157–165

Sakai K, Salvert D, Touret M, Jouvet M (1977) Afferent connections of the nucleus raphe dorsalis in the cat as visualized by the horseradish peroxidase technique. Brain Res 137:11–35

Saland LC, Wallace JA, Comunas F (1986) Serotonin-immunoreactive nerve fibers of the rat pituitary: effects of anticatecholamine and antiserotonin drugs on staining patterns Brain Res 368:310–318

Sanders-Bush E, Bushing J, Sulser F (1972) Long-term effects of p-chloroamphetamine on tryptophan hydroxylase activity and on the levels of 5-hydroxytryptamine and 5-hydroxyindoleacetic acid in brain. Eur J Pharmacol 20:385–388

Sano Y, Takeuchi Y. Matsuura T, Kawata M, Yamada H (1982) Immunohistochemical demonstration of serotonin nerve fibers in the cat neurohypophysis. Histochemistry 75:293–299

Sasek CA, Helke CJ (1989) Enkephalin-immunoreactive neuronal projections from the medulla oblongata to the intermediolateral cell column: relationship to substance P-immunoreactive neurons. J Comp Neurol 287:484–494

Sasek CA, Wessendorf MW, Helke CJ (1990) Evidence for co-existence of tryrotropin-releasing hormone, substance P and serotonin in ventral medullary neurons that project to the intermediolateral cell column in the rat. Neuroscience 35:105–119

Saudou F, Amara DA, Dierich A, LeMeur M, Ramboz S, Segu L, Buhot MC, Hen R (1994) Enhanced aggressive behavior in mice lacking 5-HT_{1B} receptor. Science 265:1875–1878

Sawchenko PE, Swanson LW (1982) Immunohistochemical identification of neurons in the paraventricular nucleus of the hypothalamus that project to the medulla or to the spinal cord in the rat. J Comp Neurol 205:260–272

Sawchenko PE, Swanson LW, Steinbusch HWM, Verhofstad AAJ (1983) The distribution and cells of origin of serotonergic inputs to the paraventricular and supraoptic nuclei of the rat. Brain Res 277:355–360

Scanzello CR, Hatzidimitriou G, Martello AL, Katz JL, Ricaurte GA (1993) Serotonergic recovery after ($\pm$)3,4-(methylenedioxy)methamphetamine injury: observations in rats. J Pharmacol Exp Ther 264:1484–1491

Schaffar N, Jean A, Calas A (1984) Radioautographic study of serotoninergic axon terminals in the rat trigeminal motor nucleus. Neurosci Lett 44:31–36

Scheibel ME, Scheibel AB (1958) Structural substrates for integrative pattern in the brain stem reticular core. In: Jasper HH, Proctor LD, Knighton RS, Noshay WC (eds) Reticular formation of the brain. Little and Brown, Boston, pp 31–68

Schmidt CJ (1987) Neurotoxicity of the psychedelic amphetamine, methylenedioxymeth-amphetamine. J Pharmacol Exp Ther 240:1–7

Schmidt CJ, Kehne JH (1990) Neurotoxicity of MDMA: neurochemical effects. Ann N Y Acad Sci 600:665–681

Seguela P, Watkins KC, Descarries L (1989) Ultrastructural relationships of serotonin axon terminals in the cerebral cortex of the adult rat. J Comp Neurol 289:129–142

Shannak KS, Hornykiewicz O (1980) Brain monoamines in the rhesus monkey during long-term neuroleptic administration. Adv Biochem Psychopharmacol 24:315–323

Siddall BJ, Polson JW, Dempney RAL (1994) Descending antinociceptive pathway from the rostral ventrolateral medulla: a correlative anatomical and physiological study. Brain Res 645:61–68

Sim LJ, Joseph SA (1992) Efferent projections of the nucleus raphe magnus. Brain Res Bull 28:679–682

Skagerberg G, Björklund A (1985) Topographic principles in the spinal projections of serotonergic and non-serotonergic brainstem neurons in the rat. Neuroscience 15:445–480

Soghomonian J-J, Doucet G, Descarries L (1987) Serotonin innervation in adult rat neostriatum. I. Quantified regional distribution. Brain Res 425:85–100

Soghomonian J-J, Beaudet A, Descarries L (1988) Ultrastructural relationships of central serotonin neurons. In: Osborne NN, Hamon M (eds) Neuronal serotonin. Wiley, Chichester, pp 57–92

Soghomonian J-J, Descarries L, Watkins KC (1989) Serotonin innervation in adult rat neostriatum. II. Ultrastructural features: a radioautographic and immunocytochemical study. Brain Res 481:67–86

Soubrie P (1986) Reconciling the role of central serotonin neurons in human and animal behavior. Behav Brain Sci 9:319–164

Soubrie P (1988) Serotonin and behavior, with special regard to animal models of anxiety, depression and waiting ability. In: Osborne NN, Hamon M (eds) Neuronal serotonin. Wiley, Chichester, pp 255–270

Stamp JA, Semba K (1995) Extent of colocalization of serotonin and GABA in the neurons of the rat raphe nuclei. Brain Res 677:39–49

Steele TD, McCann UD, Ricaurte GA (1994) 3,4-Methylenedioxymethamphetamine (MDMA, "ecstasy"): pharmacology and toxicology in animals and humans. Addiction 89:539–551

Steinbusch HWM (1981) Distribution of serotonin-immunoreactivity in the central nervous system of the rat – cell bodies and terminals. Neuroscience 6:557–618

Steinbusch HWM (1984) Serotonin-immunoreactive neurons and their projections in the CNS. In: Björklund A, Hökfelt T, Kuhar MJ (eds) Handbook of chemical neuroanatomy, vol 3: classical transmitters and transmitter receptors in the CNS, part II, Elsevier, Amsterdam, pp 68–125

Steinbusch HWM, Nieuwenhuys R (1982) Localization of serotonin-like immunoreactivity in the central nervous system and pituitary of the rat, with special references to the innervation of the hypothalamus. In: Haber B, Gabay S, Issidorides MR, Alivisatos SGA (eds) Serotonin: current aspects of neurochemistry and function. Adv Exp Med Biol 133:7–36

Steinbusch HWM, Verhofstad AAJ, Joosten HWJ (1978) Localization of serotonin in the central nervous system by immunohistochemistry: description of a specific and sensitive technique and some applications. Neuroscience 3:811–819

Steinbusch HWM, Nieuwenhuys R, Verhofstad AAJ, van der Kooy D (1981) The nucleus raphe dorsalis of the rat and its projection upon the caudatoputamen. A combined cytoarchitectonic, immunohistochemical and retrograde transport study. J Physiol (Paris) 77:157–174

Steininger TL, Reye DB, Wainer BH (1992) Afferent projections to the cholinergic pedunculopontine tegmental nucleus and adjacent midbrain extrapyramidal area in the albino rat. I. Retrograde tracing studies. J Comp Neurol 321:515–543

Sugihara I, Lang EJ, Llinas R (1995) Serotonin modulation of inferior olivary oscillations and synchronicity: a multiple-electrode study in the rat cerebellum. Eur J Neurosci 7:521–534

Swanson LW, Sawchenko PE, River J, Vale WW (1983) Organization of ovine corticotropin-releasing factor immunoreactive cells and fibers in the rat brain: an immunohistochemical study. Neuroendocrinology 36:165–186

Taber E, Brodal A, Walberg F (1960) The raphe nuclei of the brain stem of the cat. I. Normal topography and cytoarchitecture and general discussion. J Comp Neurol 114:161–188

Takeuchi Y (1988) Distribution of serotonin neurons in the mammalian brain. In: Osborne NN, Hamon M (eds) Neuronal serotonin. Wiley, Chichester, pp 25–56

Takeuchi Y, Sano Y (1983) Immunohistochemical demonstration of serotonin nerve fibers in the neocortex of the monkey (Macaca fuscata). Anat Embryol (Berl) 166:155–168

Takeuchi Y, Sano Y (1984) Serotonin nerve fibers in the primary visual cortex of the monkey. Anat Embryol (Berl) 169:1–8

Takeuchi Y, Kimura H, Sano Y (1982) Immunohistochemical demonstration of serotonin nerve fibers in the olfactory bulb of the rat, cat and monkey. Histochemistry 75:461–471

Takeuchi Y, Kimura H, Matsuura T, Yonezawa T, Sano Y (1983) Distribution of serotonergic neurons in the central nervous system: a peroxidase-antiperoxidase study with anti-serotonin antibodies. J Histochem Cytochem 31:181–185

Tallaksen-Greene SJ, Elde R, Wessensorf MW (1993) Regional distribution of serotonin and substance P co-existing in nerve fibers and terminals in the brainstem of the rat. Neuroscience 53:1127–1142

Tanaka M, Okamura H, Yanaihara N, Tanaka Y, Ibata Y (1993) Differential expression of serotonin and met-enkephalin-Arg6-Gly7-Leu8 in neurons of the rat brain stem. Brain Res Bull 30:561–570

Tanaka M, Okamura H, Tamada Y, Nagatsu I, Tanaka Y, Ibata Y (1994) Catecholaminergic input to spinally projecting serotonin neurons in the rostral ventromedial medulla oblongata of the rat. Brain Res Bull 35:23–30

Tashiro T, Satoda T, Takahashi O, Matsushima R, Mizuno N (1988) Distribution of axons exhibiting both enkephalin- and serotonin-like immunoreactivities in the lumbar cord segments: an immunohistochemical study in the cat. Brain Res 440:357–362

Tashiro T, Satoda T, Matsushima R, Mizuno N (1990) Distribution of axons showing both enkephalin- and serotonin-like immunoreactivities in the lumbar cord segments of the Japanese monkey (Macaca fuscata). Brain Res 512:143–146

Tecott LH, Sun LM, Akana SF, Strack AM, Lowenstein DH, Dallman MF, Julius D (1995) Eating disorder and epilepsy in mice lacking 5-HT$_{2C}$ serotonin receptors. Nature 374:542–546

Ternaux JP, Hery F, Bourgoin S, Adrien J, Glowinski J, Hamon M (1977) The topographical distribution of serotoninergic terminals in the neostriatum of the rat and the caudate nucleus of the cat. Brain Res 121:311–326

Thompson AM, Moore KR, Thompson G (1995) Distribution and origin of serotoninergic afferents to guinea pig cochlear nucleus. J Comp Neurol 351:104–116

Thor KB, Helke CJ (1987) Serotonin- and substance P-containing projections to the nucleus tractus solitarii of the rat. J Comp Neurol 265:275–293

Törk I (1985) Raphe nuclei and serotonin containing systems. In: Paxinos G (ed) The rat nervous system. Academic, Sydney, pp 43–78

Törk I (1990) Anatomy of the serotonergic system. Ann N Y Acad Sci 600:9–35

Törk I, Hornung J-P (1990) Raphe nuclei and the serotonergic system. In: Paxinos G (ed) The human nervous system. Academic, San Diego, pp 1001–1022

Törk I, Hornung J-P, Mulligan KA, van der Loos H (1986) Synaptic connections of serotonergic axons in the molecular layer of the cat's neocortex. Neurosci Lett [Suppl] 26:S104

Törk I, Hornung J-P, Somogyi P (1988) Serotonergic innervation of GABAergic neurons in the cerebral cortex. Neurosci Lett [Suppl] 30:S131

Twarog BM, Page IH (1953) Serotonin content of some mammalian tissues and urine and a method for its determination. Am J Physiol 175:157–161

Ueda S, Sano Y (1986) Distributional pattern of serotonin-immunoreactive nerve fibers in the lateral geniculate nucleus of the rat, cat and monkey (Macaca fuscata). Cell Tissue Res 243:249–253

Ueda S, Kawata M, Sano Y (1983a) Identification of serotonin- and vasopressin immunoreactivities in the suprachiasmatic nucleus of four mammalian species. Cell Tissue Res 234:237–248

Ueda S, Kawata M, Takeuchi Y, Sano Y (1983b) Immunohistochemical demonstration of serotonin nerve fibers in the hypothalamus of the cat. Anat Embryol (Berl) 168:314–330

Vahabzadeh A, Fillenz M (1994) Comparison of stress-induced changes in noradrenergic and serotonergic neurons in the rat hippocampus using microdialysis. Eur J Neurosci 6:1205–1212

Van Boeckstaele EJ, Biswas A, Pickel VM (1993) Topography of serotonin neurons in the dorsal raphe nucleus that send axon collaterals to the rat prefrontal cortex and nucleus accumbens. Brain Res 624:188–198

Van de Kar LD, Lorens SA (1979) Differential innervation of individual hypothalamic nuclei and other forebrain regions by the dorsal and median midbrain raphe nuclei. Brain Res 162:45–54

Van der Kooy D, Hattori T (1980) Dorsal raphe cells with collateral projections to the caudate-putamen and substantia nigra: a fluorescent retrograde double labeling study in the rat. Brain Res 169:1–7

Vertes RP (1991) A PHA-L analysis of ascending projections of the dorsal raphe nucleus in the rat. J Comp Neurol 313:643–668

Vertes RP, Kocsis B (1994) Projections of the dorsal raphe nucleus to the brainstem: PHA-L analysis in the rat. J Comp Neurol 340:11–26

Vertes RP, Martin GF (1988) Autoradiographic analysis of ascending projections from the pontine and mesencephalic reticular formation and the median raphe nucleus in the rat. J Comp Neurol 275:511–541

Walsh FX, Bird ED, Stevens JT (1982) Monoamine transmitters and their metabolites in the basal ganglia of Huntington's disease and control postmortem brain. Adv Neurol 35:165–169

Wang Q-P, Ochiai H, Nakai Y (1992) GABAergic innervation of serotonergic neurons in the dorsal raphe nucleus of the rat studied by electron microscopy double immunostaining. Brain Res Bull 29:943–948

Warembourg M, Poulain P (1985) Localization of serotonin in the hypothalamus and the mesencephalon of the guinea-pig. Cell Tissue Res 240:711–721

Waterhouse BD, Mihailoff GA, Baack JC, Woodward DJ (1986) Topographical distribution of dorsal and median raphe neurons projecting to motor, sensorimotor, and visual cortical areas in the rat. J Comp Neurol 249:460–476

Wessendorf MW, Elde R (1987) The coexistence of serotonin- and substance P-like immunoreactivity in the spinal cord of the rat as shown by immunofluorescent double labeling. J Neurosci 7:2352–2363

Westlund KN, Childs GV (1982) Localization of serotonin fibers in the rat adenohypophysis. Endocrinology 111:1761–1763

Westlund KN, Coulter JD (1980) Descending projections of the locus coeruleus and subcoeruleus/medial parabrachial nuclei in monkey: axonal transport studies and dopamine-β-hydroxylase immunocytochemistry. Brain Res Rev 2:235–264

Wiklund L, Björklund A (1980) Mechanism of regrowth in the bulbospinal serotonin system following 5,6-dihydroxytryptamine-induced axotomy. II. Fluorescence histochemical observations. Brain Res 191:129–155

Wiklund L, Leger L, Persson M (1981) Monoamine cell distribution in the cat brainstem. A fluorescence histochemical study with quantification of indolaminergic and locus coeruleus cell groups. J Comp Neurol 203:613–647

Wilson MA, Molliver ME (1991a) The organization of serotonergic projections to cerebral cortex in primates: retrograde transport studies. Neuroscience 44:555–570

Wilson MA, Molliver ME (1991b) The organization of serotonergic projections to cerebral cortex in primates: regional distribution of axon terminals. Neuroscience 44:537–553

Wilson MA, Ricaurte GA, Molliver ME (1989) Distinct morphologic classes of serotonergic axons in primates exhibit differential vulnerability to the psychotropic drug 3,4-methylenedioxymethamphetamine. Neuroscience 28:121–137

Wilson MA, Mamounas LA, Fasman KH, Axt KJ, Molliver ME (1993) Reactions of 5-HT neurons to drugs of abuse: neurotoxicity and plasticity. NIDA Res Monogr 136:155–187

Wuttke W, Baumgarten HG, Björklund A, Fenske M, Klemm HP (1977) De- and regeneration of brain serotonin neurons following 5,7-dihydroxytryptamine treatment: effects on serum LH, FSH, and prolactin levels in male rats. Brain Res 134:317–331

Wuttke W, Hancke JL, Höhn KG, Baumgarten HG (1978) Effect of intraventricular injection of 5,7-dihydroxytryptamine on serum gonadotropins and prolactin. Ann N Y Acad Sci 305:423–436

Zagon A (1993) Innervation of serotonergic medullary raphe neurons from cells of the rostral ventrolateral medulla in rats. Neuroscience 55:849–867

Zemlan FP, Behbehani MM, Beckstead RM (1984) Ascending and descending projections from nucleus reticularis magnocellularis and nucleus reticularis gigantocellularis: an autoradiographic and horseradish peroxidase study in the rat. Brain Res 292:207–220

Zhou FC, Azmitia EC (1986) Induced homotypic collateral sprouting of serotonergic fibers in the hippocampus. II. An immunocytochemistry study. Brain Res 373:337–348

CHAPTER 3

Physiology and Pharmacology of Brain Serotoninergic Neurons

B.L. Jacobs and C.A. Fornal

A. General Introduction

Full appreciation of CNS drug actions is dependent on understanding the basic physiology of their targeted neurotransmitter system(s). For the past 20 years, we have been studying the factors that regulate the functional activity of the brain serotonin system in behaving animals. This chapter describes our research efforts in three subject areas:

1. Studies examining the effects of physiological, environmental, and behavioral manipulations on serotoninergic neuronal activity in behaving cats.
2. A complementary program to this electrophysiology, which employs similar manipulations, but uses in vivo microdialysis to measure changes in extracellular levels of serotonin in various "postsynaptic" brain areas in behaving rats or cats.
3. Experiments that examine drug effects on serotoninergic neuronal activity in behaving cats in order to elucidate the pharmacology and physiology of the somatodendritic 5-HT$_{1A}$ autoreceptor.

Overall, this research has provided us with a unique perspective for drug action on the CNS serotoninergic system and, more broadly, on the basic role of serotonin in physiology and behavior.

B. Anatomical Perspective

The functional significance of serotonin in the CNS is a reflection of its anatomical organization (Jacobs and Azmitia 1992). The serotoninergic system is primitive in two interrelated ways. First, the basic plan for the cell bodies of these neurons is highly conserved from the simplest to the most complex vertebrates. This implies that the function of this system must, in some way, be common to the physiology and behavior of fish, amphibians, and reptiles, as well as mammals, including infrahuman and human primates. Second, virtually all of the cell bodies are located in the brainstem (in a few species a small percentage are found in the rostral spinal cord or ventral diencephalon), especially on or near the midline. As with other brainstem-mediated functions, this implies an involvement in basic processes, especially those associated with axial functions, such as controlling the proximal limb and trunk muscles, and respiration.

In vertebrates, the cell bodies of the vast majority of brain serotoninergic neurons are localized in or near the raphe nuclei (Jacobs and Azmitia 1992). These midline clusters can be divided into two groups. The superior group, localized in the pons/mesencephalon, contains the two nuclei which supply most of the serotonin to the forebrain: the nucleus centralis superior (NCS) and dorsal raphe nucleus (DRN). The inferior group, localized in the medulla, contains the three nuclei which supply most of the serotonin to the spinal cord: the nucleus raphe magnus (NRM), nucleus raphe obscurus (NRO), and nucleus raphe pallidus (NRP).

C. Activity of Serotoninergic Neurons in Behaving Cats

I. Introduction

Over the past 15 years much of the research in our laboratory has been devoted to studying the single unit activity (extracellularly recorded action potentials) of brain serotoninergic neurons in conscious, freely moving cats. The method that we employ allows us to examine neuronal activity over long periods of time (typically for many hours, and sometimes for several days), despite gross bodily movements. In the following subsections we describe the activity of serotoninergic neurons in the various raphe nuclei, studied across a variety of experimental conditions.

II. Sleep-Wake-Arousal Cycle

With very few exceptions the activity of most neurons in the brain changes significantly across the sleep-wake-arousal cycle. The nature and magnitude of these changes provide important basic information about these cells and may even provide clues to their role in brain function. The activity of serotoninergic neurons in behaving animals has thus far been examined only in the domestic cat. It would of course be extremely valuable to have results from other species regarding the generality of these data.

The largest cluster of brain serotoninergic neurons is in the DRN. During the quiet waking state the activity of DRN serotoninergic neurons is slow and highly regular (Trulson and Jacobs 1979), just as it is when examined under anesthesia and even in vitro (Aghajanian et al. 1968; Mosko and Jacobs 1976). From a quiet waking rate of approximately 3 spikes/s, the activity of these neurons is typically increased by approximately 10%–30% in response to activating or arousing stimuli. Reciprocally, the activity of these neurons declines as the cat becomes drowsy, and becomes even slower upon entering slow-wave sleep. Finally, the culmination of this state-dependent decrease in single unit activity occurs as the cat enters rapid eye movement (REM) sleep, when the activity falls virtually silent.

In general, the pattern of activity across the sleep-wake-arousal cycle in DRN serotoninergic neurons is closely paralleled by serotoninergic neurons in the other major groups: NCS, NRM, and NRO/NRP (FORNAL and JACOBS 1988). There are, however, some differences which may be of functional significance. For example, NRM and NRO/NRP neurons generally display a higher spontaneous firing rate than DRN or NCS neurons during comparable behavioral states. The activity of NRO/NRP neurons, as well as a subset of NCS neurons, is not as strongly related to behavioral state as DRN, NRM, and the majority of NCS cells. The former did not show as steep a decline across the sleep-wake arousal cycle, and their activity, although significantly reduced, is not completely suppressed during REM sleep. Moreover, NRO/NRP neurons are the only ones whose activity is generally unresponsive to activating or arousing stimuli.

In conclusion, the activity of the brain serotoninergic system as a whole appears to display a direct relationship to the level of behavioral arousal. As will be seen below, this conclusion is also supported by neurochemical evidence.

Because brain serotonin metabolism is reported to be influenced by the light-dark cycle, it is of interest to examine whether the light-dark cycle can influence the activity of cat DRN serotoninergic neurons (TRULSON and JACOBS 1983). When state is held constant (i.e., when the activity of a given neuron during REM sleep or during quiet waking is studied at various times during both the light and dark phases of the daily cycle), the activity was found to be invariant. Thus, the activity of these cells is determined primarily by the behavioral state of the animal and is not significantly influenced by the light-dark cycle or, by inference, the circadian cycle. Therefore, the variation in serotonin metabolism observed across the light-dark cycle may simply be secondary to the distribution of sleep and waking across this cycle and the accompanying changes in neuronal activity. This latter hypothesis is consistent with neurochemical data described below.

Our most recent experiments on this general theme explored the effects of sleep deprivation on serotoninergic neurons (GARDNER et al. 1995; GARDNER 1996). Most, if not all, current pharmacological treatments for depression affect the serotoninergic system. One intervention that at least temporarily alleviates depressive symptoms in patients is one night of total sleep deprivation. To evaluate whether there might be a serotoninergic component involved in this treatment, we examined the effects of sleep deprivation on the discharge rate of serotoninergic neurons in the DRN of behaving cats. Cats were prevented from sleeping for a 24-h period. Activity of single neurons during quiet waking and active waking behavior was recorded at 3-h intervals for the duration of the 24-h deprivation period and during a 6-h recovery sleep period. The 5-HT$_{1A}$ autoreceptor agonist 8-hydroxy-2-(di-n-propylamino) tetralin (8-OH-DPAT), which reduces the firing rate of DRN serotoninergic cells, was administered systemically before and after the sleep deprivation period, in order to assess whether the sensitivity of this receptor changed over the course of the deprivation. Mean neuronal activity increased during the deprivation period by approximately 20% during active waking and by 10% during quiet waking. The magnitude of this effect reached its maximum on average at the 15-h time point. The firing rates of DRN serotoninergic neurons remained elevated above the pre-deprivation baseline for the entire 24-h period in all the test conditions, but after a 6-h recovery sleep period, neuronal activity returned to its pre-deprivation baseline level. The ability of 8-OH-DPAT to reduce the firing rate of serotoninergic cells was reduced by 10% after the 24-h period of sleep deprivation, suggesting that the negative-feedback, somatodendritic 5-HT$_{1A}$ autoreceptor was desensitized by the sleep deprivation. Overall, these data demonstrate that sleep deprivation can increase serotoninergic neurochemical system activity by causing an increase in the firing rate of individual neurons. These results therefore provide a plausible explanation for why, at least in part, sleep deprivation exerts an antidepressant effect.

III. Afferent Control of Neuronal Activity

Anatomical studies employing retrogradely transported tracers indicate that a number of brainstem and forebrain sites project to the DRN (Aghajanian and Wang 1977). However, little information is available concerning which of these inputs are activated (or inhibited), and under what conditions this occurs.

Studies of rat brain DRN serotoninergic neurons carried out in vitro indicate that they are autoactive, driven by an intrinsic pacemaker mechanism. The oscillating membrane properties of these neurons are what give rise to the slow and highly regular spontaneous activity which is characteristic of virtually all brain serotoninergic neurons (VanderMaelen and Aghajanian 1983). In addition to such intrinsic ionic mechanisms, there is evidence for a facilitatory noradrenergic modulation of this pacemaker activity (Baraban and Aghajanian 1980). The anatomical origin and functional role of this input remain obscure. However, it is possible that this noradenergic input is responsible for the approximately 10%–30% increase in DRN serotoninergic neuronal activity observed in response to activating or arousing stimuli.

As discussed above, it is well known that serotoninergic neuronal activity significantly slows during sleep, falling silent during REM sleep. What is responsible for this decreased activity? We utilized multibarrel microiontophoresis in head restrained, awake cats to address this question (Levine and Jacobs 1992). The iontophoretic application of the γ-aminobutyric acid (GABA) antagonist bicuculline during sleep restored the reduced activity of DRN serotoninergic neurons to the waking level (the cat, of course, remained asleep). However, if the same antagonist was applied during waking, there was no change in neuronal activity. These results indicate that a GABAergic input to brain serotoninergic neurons becomes activated during sleep and exerts a powerful inhibitory influence exclusively during that state. In another study in this series, we demonstrated that the excitatory response of serotoninergic neurons to phasic sensory inputs (such as a click) could be blocked by iontophoretic application of the excitatory amino acid antagonist kynurenic acid. This antagonist, however, exerted no effect on the spontaneous or basal activity of these neurons. In the final study in this series, we found that the iontophoretic application of an α_1-adrenergic agonist (phenylephrine) did not alter the spontaneous, waking activity of these neurons, suggesting that the aforementioned noradrenergic input to these neurons already exerts its maximal effect during waking.

IV. Response to Challenges/Stressors

One strategy for initially exploring the behavioral and physiological roles of a brain neurochemical system is to examine its activity across a wide variety of somewhat intense conditions, especially those that are biologically and/or ecologically relevant. It was assumed that this would reveal at least some characteristics of the stimuli to which brain serotoninergic neurons are programmed to respond. Thus, cats were exposed to the following strong environmental or physiological conditions while recording the activity of serotoninergic neurons in the DRN, NCS, or NRM: a heated environment or administration of a pyrogen; drug-induced increases or decreases in arterial blood pressure; insulin-induced hypoglycemia; phasic or tonic mildly painful stimuli; loud noise; physical restraint; or a natural enemy (dog) (AUERBACH et al. 1985; FORNAL et al. 1987, 1989, 1990; WILKINSON and JACOBS 1988). Despite the fact that all of these conditions evoked strong behavioral responses and/or physiological changes indicative of sympathetic activation, none of them significantly activated serotoninergic neuronal activity beyond the level normally seen during an undisturbed active waking state. The following two descriptions exemplify the results from this series of experiments.

The activity of DRN serotoninergic neurons was examined in response to both increased ambient temperature and pyrogen-induced fever, stimuli eliciting opposite thermoregulatory responses of heat loss and heat gain, respectively (FORNAL et al. 1987). Neuronal activity remained unaffected as ambient temperature was increased from 25°C to 43°C. Following prolonged heat exposure, cats displayed intense continuous panting, relaxation of posture, and a progressive rise in body/brain temperature (range 0.5°–2.0°C); yet no change in serotoninergic neuronal activity occurred. In a parallel study, a synthetic pyrogen (muramyl dipeptide) was administered, resulting in increased body/brain temperature within 30 min and lasting for approximately 6 h. The peak elevation of body temperature was typically 1.5°–2.5°C, yet, once again, no change in neuronal activity was observed. Consistent with these electrophysiological results, utilizing in vivo microdialysis, we have also found that pyrogen administration produces no change in extracellular levels of serotonin in the anterior hypothalamus, a primary thermoregulatory center of the cat (WILKINSON et al. 1991).

A large body of evidence implicates central serotonin in analgesia, especially those serotoninergic neurons localized in the NRM and projecting to the dorsal horn of the spinal cord. Accordingly, we examined the activity of NRM serotoninergic neurons in behaving cats exposed to a variety of phasic or tonic, mildly painful stimuli (Auerbach et al. 1985). No change in neuronal activity was produced by these stimuli relative to the discharge rate during an undisturbed active waking baseline. There was also no change in serotoninergic neuronal activity in response to the systemic administration of morphine, in a dose that produced analgesia. These results have recently been confirmed in a study reporting that identified NRM serotoninergic neurons in the rat were not activated by painful stimuli eliciting the withdrawal reflex (Potrebic et al. 1994).

In summary, these data indicate that mild to relatively strong stressors, drawn from a number of different environmental and physiological categories, do not significantly perturb the CNS serotoninergic system, when compared to data gathered under appropriate control conditions.

V. Relation to Motor Activity

A fundamental feature of REM sleep is the powerful inhibition of motoneurons controlling antigravity muscle tone, and the resulting paralysis. Because our previous work had shown that the activity of serotoninergic neurons is almost totally suppressed during REM sleep, we examined the possibility that there might be a relationship between these two phenomena. Lesions of the dorsomedial pons produce a condition which permits investigation of this issue. Cats with this lesion enter a stage of sleep which by all criteria appears to be REM sleep except that antigravity muscle tone is present, and the animals are thus capable of movement and even coordinated locomotion. In both waking and slow wave sleep, the activity of DRN serotoninergic neurons in these pontine-lesioned cats was similar to that of normal animals (Trulson et al. 1981). However, when these animals entered REM sleep, neuronal activity increased, instead of displaying the decrease typical of this state. Those animals displaying the greatest amount of muscle tone and overt behavior during REM sleep showed the highest levels of neuronal activity, with some of their serotoninergic neurons discharging at a level approximating that of the waking state.

Microinjection of carbachol, a cholinomimetic agent, into this same pontine area, produces a condition somewhat reciprocal to non-atonia REM sleep. These animals are awake, as demonstrated by their ability to track visual stimuli, but are otherwise paralyzed. However, unlike the normal waking state where serotoninergic neurons are tonically active, DRN serotoninergic neurons were inactive in these paralyzed animals (Steinfels et al. 1983). In the same study, we also found that a centrally acting muscle relaxant completely suppressed serotoninergic neuronal activity, but peripheral neuromuscular block had no effect on neuronal activity.

In sum, these data suggest that a strong positive relationship exists between tonic level of motor activity (muscle tone) and the firing rate of DRN serotoninergic neurons. To date, this issue has not been explored for neurons in any of the other serotoninergic nuclei.

In recent experiments we have observed much more specific relationships between serotoninergic neuronal activity and motor function. When cats engage in a variety of types of central pattern generator-mediated oral-buccal activities, such as chewing/biting, licking, or grooming, approximately one-fourth of DRN serotoninergic neurons increase their activity by as much as two- to fivefold (FORNAL et al. 1996a). In contrast, the rest of the serotoninergic neurons in this nucleus maintain their slow and rhythmic activity. These increases in neuronal activity often precede the initiation of movement by several seconds, but they invariably terminate coincident with the end of the behavioral sequence. Equally impressive is the fact that even brief (1–5 s) spontaneous pauses in these behaviors are accompanied by an immediate decrease in neuronal activity to baseline levels, or below. The increased neuronal activity during these central pattern generator-mediated behaviors is typically tonic, but is occasionally modulated in phase with a particular aspect of the repetitive behavior. During a variety of other nonrhythmic episodic or purposive movements, even those involving oral-buccal responses, such as yawning, no increase, or even a decrease, in neuronal activity is seen. In addition, there is often an inverse relationship between the activity of these serotoninergic neurons and the occurrence of eye movements. This could, of course, be a relationship to another variable such as attentional changes, rather than to eye movements per se. In fact, during dramatic attentional shifts, such as those occurring during orienting movements in response to novel or imperative stimuli, the activity of DRN serotoninergic neurons may fall silent for several seconds. This occurs in association with large eye movements, turning of the head toward the stimulus, and suppression of ongoing behaviors.

Somewhat surprisingly, most of these DRN neurons can also be activated by somatosensory stimuli applied to the head and neck region, while the same stimuli applied to the rest of the body surface are typically ineffective. The level of increased activity produced by this somatosensory stimulation often approaches that seen spontaneously during oral-buccal movements (i.e., two- to fivefold).

Serotoninergic neurons in the rostral pons (DRN and NCS) provide almost the entire serotoninergic innervation of the forebrain, whereas those in the caudal medulla (NRP and NRO) are the source of much of the serotoninergic innervation of the spinal cord (JACOBS and AZMITIA 1992). Therefore, it is interesting to compare the response properties of neurons in these two separate groups.

Contrary to pontine serotoninergic neurons, where only a subgroup of neurons are activated during central pattern generator-mediated behaviors, virtually all medullary serotoninergic neurons are activated under at least some of these conditions (Veasey et al. 1995). The degree of activation, however, is much less impressive, i.e., 50%–100% above baseline versus 100%–400% above baseline for the DRN. In this context, it may be important to note that the basal, quiet waking discharge rate of the medullary serotoninergic neurons is approximately twice that of the pontine serotoninergic neurons, 5–6 spikes/s versus 2–3 spikes/s. There also appears to be at least some degree of response specificity for these neurons. Thus, virtually all medullary serotoninergic neurons are activated during treadmill-induced locomotion, but only subgroups are activated during hyperpnea (induced by exposure to carbon dioxide), or during chewing/licking. Many of these individual neurons are activated in association with more than one of these motor activities. In most cases there is a strong positive correlation between magnitude of neuronal activation and speed of locomotion and/or depth of respiration. As with pontine serotoninergic neurons the increased activity of medullary serotoninergic neurons is sometimes phase-locked to the behavior (e.g., in association with the step cycle) and typically is tightly coupled to the onset and offset of the behavior. Unlike pontine serotoninergic neurons, medullary serotoninergic neurons are not activated by somatosensory stimuli applied to any region of the body surface. Nor is their activity significantly changed during orientation to strong or novel stimuli. Finally, when DRN neurons were examined under the identical conditions, none of them were activated during treadmill-induced locomotion, but some were activated during carbon dioxide-induced hyperpnea (Jacobs et al. 1994; Veasey et al., to be published).

VI. Conclusions

The activity of brain serotoninergic neurons, regardless of the nuclei in which they are found, changes dramatically across the sleep-wake-arousal cycle. At least in part, this appears to be correlated with changes in level of tonic motor activity that accompany these alterations in behavioral state. Somewhat surprisingly serotoninergic neurons are only minimally responsive to any of a broad range of environmental or physiological challenges/stressors. These results do not, however, preclude serotonin playing an important role in the organismic stress response. For example, this could be dependent on the presence of a tonic level of serotoninergic activity occurring conjointly with increased glucocorticoid release.

Beyond the level of neuronal activity attained during waking, many serotoninergic neurons display a further, often dramatic, elevation of neuronal activity during the expression of repetitive, central pattern generator-mediated motor outputs. The particular motor pattern that is associated with increased neuronal activity varies to some extent with the serotoninergic nucleus in which the neuron is localized.

We hypothesize that these data fit into the existing literature on serotonin in the following manner (Jacobs and Fornal 1993, 1995). Activation of brain serotoninergic neurons facilitates motor output and, simultaneously, suppresses sensory information processing. In addition, serotoninergic neurons play an auxiliary role in coordinating appropriate autonomic and neuroendocrine outputs to the ongoing tonic or repetitive motor activity. Finally, when serotoninergic neuronal activity is suppressed, for example during orientation, motor output is disfacilitated and sensory processing is disinhibited, permitting more focussed information processing to occur.

D. Release of Serotonin in Forebrain Sites

I. Introduction

Traditionally, functional activity within the CNS has been measured by means of neuronal activity. As described in the previous section, this is the primary manner in which we have evaluated the brain serotoninergic system. However, it is also possible, and perhaps even likely, that neuronal activity does not, under all conditions, directly and linearly reflect neurotransmitter release at postsynaptic target sites. There are a number of reasons, both theoretical and empirical, underlying this possibility. Therefore, we also employ in vivo brain microdialysis as a method to more directly assess neurotransmitter release in behaving animals.

II. Sleep-Wake-Arousal Cycle

In the study described above, in which we examined serotonin release in the hypothalamus in response to pyrogen administration, we also assessed changes across the sleep-wake-arousal cycle (WILKINSON et al. 1991). Extracellular serotonin in the anterior hypothalamic/preoptic area and caudate nucleus of the freely moving cat was measured using in vivo microdialysis. Behavioral state, from REM sleep through active waking, was quantified based on both behavioral and polygraphic criteria. In the first phase of this study we found that extracellular levels of serotonin in the hypothalamus and caudate nucleus were significantly increased in relation to increased levels of behavioral arousal. In a subsequent more detailed analysis, extracellular levels of serotonin in the hypothalamus displayed a strong positive correlation ($r = 0.82$) with behavioral state across the entire sleep-wake-arousal continuum. Finally, consistent with electrophysiological experiments, administration of 8-OH-DPAT, a specific $5-HT_{1A}$ autoreceptor agonist, which is known to decrease serotoninergic neuronal activity (see below), significantly reduced extracellular levels of serotonin in both the caudate nucleus and hypothalamus. In general, these data confirm the basic validity of using single unit activity as an index of the functional activity of the brain serotoninergic system.

In studies using rats we have examined changes in extracellular brain serotonin levels across the daily light-dark transition. These data support our electrophysiological data (described above) showing that behavioral state/

motor activity is a more important determinant of serotoninergic activity than the light-dark cycle. When we examined serotonin levels in the cerebellum across the light-dark transition we found that there was approximately a 40% increase in the dark (Mendlin et al. 1996). However, behavioral activity also increases dramatically in the dark. Thus, when we took this into account we found that there was a 0.97 correlation between serotonin levels and behavioral activity (percentage time spent in an active waking state).

We have also examined this issue in a number of forebrain sites in the rat (Rueter and Jacobs 1996). Serotonin levels increased significantly during the first $^{1}/_{2}$ h of the dark phase in the hippocampus, corpus striatum, amygdala, and prefrontal cortex. As in the cerebellum, serotonin levels covaried significantly with changes in time spent in active waking behaviors. When serotonin levels were compared during periods where either high levels of behavioral activity were equated in the light and dark, or where low levels of behavioral activity were equated in the light and dark, there was no significant difference. These data support our previous findings that serotonin release is importantly tied to behavioral state/behavioral activity, and is only indirectly related to the light-dark cycle insofar as it underlies the circadian distribution of behavioral activity.

III. Response to Challenges/Stressors

Employing in vivo microdialysis in rats, we have returned to the issue of stress and serotoninergic function that we had previously examined electrophysiologically (Rueter 1995; Rueter and Jacobs 1995). We addressed several different issues in these studies: (1) Do stressors produce a larger effect than an activating but nonstressful stimulus? (2) Do different stressors produce a different magnitude or pattern of change in levels of extracellular serotonin across four forebrain sites: amygdala, prefrontal cortex, hippocampus, and corpus striatum? (3) Is a given brain region differentially affected by different stressors? (4) Are any differences noted in (2) and (3) accounted for by brainstem site of origin of the serotoninergic input to these four forebrain sites? To answer this, microdialysis samples were taken every 30 min during the dark phase of the light-dark cycle while animals were exposed to various stressors. We found no evidence for a "stress response" in any of the forebrain sites that could not be accounted for by general activation of the organism. Thus, tail pinch, forced swimming, or exposure to a cat did not produce any larger increase (~30%–50%) in extracellular serotonin than that seen in these same sites during a non-stressful condition: spontaneous feeding. (Increases in plasma glucocorticoid levels confirmed the "stressful" nature of the stimuli.) There was also no clear-cut evidence that the different stressors, which clearly evoke different types of behavioral and physiological responses, produce a different pattern of serotonin release across the four target sites. Nor was there any compelling evidence that any one stressor had a stronger overall effect, independent of brain region, than any other one. Somewhat interesting, however, is the fact that serotonin levels in the corpus striatum were significantly less responsive overall (i.e., across all experimental manipulations) than serotonin levels in the other three sites. Thus, it is possible that serotonin which derives from the DRN exclusively (corpus striatum) is more refractory to change than serotonin which derives from the NCS alone (dorsal hippocampus) or from it and the DRN (amygdala and frontal cortex).

IV. Conclusions

In summary, these studies employing in vivo microdialysis to measure changes in brain levels of serotonin in rats are consistent with our previous single unit studies in cats. The serotoninergic system is not stress activatable per se. Instead, the system appears to be responsive to more general factors such as behavioral state/behavioral activity. More specifically, our single unit studies suggest that the functional activity of the serotonin system is directly related to levels of tonic and repetitive (central pattern generator-mediated) motor activity. This is an issue that we wish to continue to explore with in vivo microdialysis.

E. Pharmacological Studies of Serotonin Somatodendritic Autoreceptors

I. Introduction

It is becoming increasingly apparent that serotonin autoreceptors play an important role in the regulation of the activity of brain serotoninergic neurons. Furthermore, these receptors appear to play a critical role in the mechanism of action of several classes of psychotherapeutic drugs (e.g., non-benzodiazepine anxiolytics and certain antidepressants). In the following sections, we provide a brief summary of the electrophysiological responses of serotoninergic neurons to drugs acting on the 5-HT_{1A} autoreceptor, and describe recent studies from our laboratory indicating that these receptors exert a tonic inhibitory influence on the activity of serotoninergic neurons under physiological conditions. Finally, the role of somatodendritic serotonin autoreceptors in the mechanism of action of selective serotonin reuptake inhibitors (SSRIs) is discussed in light of recent studies which suggest that the therapeutic efficacy of these antidepressant drugs might be potentiated by agents which block the serotonin autoreceptor.

II. Effects of Agonist and Antagonist Drugs

In the early 1970s, AGHAJANIAN and coworkers (1972), using single cell recording in combination with microiontophoresis, demonstrated the presence of serotonin-sensitive receptors located on the somata and dendrites of serotoninergic neurons in the midbrain raphe. The activation of these "autoreceptors" resulted in an inhibition of neuronal activity. These receptors are thought to be part of an intrinsic negative feedback mechanism, whereby the local release of serotonin in the raphe region acts to dampen the impulse activity of central serotoninergic neurons (Mosko and Jacobs 1977; Aghajanian and Wang 1978; Aghajanian and VanderMaelen 1986). Thus, agents which increase the synaptic availability of brain serotonin (e.g., serotonin precursors, releasers, or reuptake inhibitors) depress the activity of central serotoninergic neurons by activating somatodendritic autoreceptors. This autoreceptor-mediated inhibition results from a hyperpolarization of the cell membrane through an increase in potassium ion conductance (Aghajanian and Lakoski 1984).

Historically, the first drugs reported to exert a preferential action at the serotonin autoreceptor were hallucinogenic compounds such as D-lysergic acid diethylamide (LSD) and 5-methoxy-N,N-dimethyltryptamine (5-MeODMT) (Haigler and Aghajanian 1974; De Montigny and Aghajanian 1977). These drugs were instrumental in the initial pharmacological characterization of the serotonin autoreceptor, although they do not discriminate between different serotonin receptor subtypes. In this respect, a new era in serotonin autoreceptor research began with the discovery of 8-OH-DPAT, a highly potent and selective 5-HT$_{1A}$ agonist. This compound displays high affinity ($K_i =$ ~2 nM) and selectivity (>400-fold) for the 5-HT$_{1A}$ site, relative to other 5-HT-binding sites (5-HT$_{1B,1C,1D}$, 5-HT$_2$, or 5-HT$_3$) (Van Wijngaarden et al. 1990). In intracellular studies, 8-OH-DPAT mimicked the hyperpolarizing action of serotonin on serotoninergic DRN neurons recorded in vitro, at concentrations approximately 1000 times lower than that of serotonin (Williams et al. 1988). 8-OH-DPAT also inhibits the firing of midbrain serotoninergic neurons recorded in vivo in anesthetized animals, whether administered systemically (Lum and Piercey 1988; Sinton and Fallon 1988; Blier et al. 1989) or applied directly onto these cells by microiontophoresis (Sprouse and Aghajanian 1986; Blier et al. 1990b). The inhibitory action of 8-OH-DPAT is blocked by several nonselective 5-HT$_{1A}$ antagonists, including spiperone (Lum and Piercey 1988; Williams et al. 1988; Blier et al. 1989, 1993a; Escandon et al. 1994), (−)-tertatolol (Jolas et al. 1993; Lejeune et al. 1993; Prisco et al. 1993), and (S)-UH 301 [(S)-5-fluoro-8-hydroxy-2-(di-n-propylamino) tetralin] (Arborelius et al. 1994). More recently, the selective 5-HT$_{1A}$ antagonists (S)-WAY 100135 [N-$tert$-butyl-3-(4-(2-methoxyphenyl)piperazin-1-yl)-2-phenylpropanamide] (Fletcher et al. 1993; Lejeune et al. 1993; Lanfumey et al. 1993; Mundey et al. 1994) and WAY 100635 {N-[2-[4-(2-methoxyphenyl)-1-piperazinyl]ethyl]-N-(2-pyridinyl) cyclohexanecarboxamide} (Forster et al. 1995; Mundey et al. 1996) have been shown to block the suppression of serotoninergic DRN neurons produced by 8-OH-DPAT. These electrophysiological data are consistent with the characterization of the serotonin autoreceptor as being of the 5-HT$_{1A}$ subtype. In agreement with this, immunohistochemical studies have shown that 5-HT$_{1A}$ receptors in the midbrain raphe nuclei are localized almost exclusively on the cell membrane of serotoninergic neurons (SOTELO et al. 1990).

Since the introduction of 8-OH-DPAT, a large number of chemically diverse compounds with 5-HT$_{1A}$ agonist properties have been synthesized and shown to suppress the firing of serotoninergic neurons (VANDERMAELEN et al. 1986; SPROUSE and AGHAJANIAN 1987; MILLAN et al. 1992, 1993; KIDD et al. 1993; MATHESON et al. 1994; FOREMAN et al. 1994). One chemical class of compounds which has attracted considerable attention is the pyrimidinyl-piperazines, in particular buspirone, ipsapirone, gepirone, and tandospirone. These compounds display moderate to high affinity for central 5-HT$_{1A}$ receptors and have anxiolytic and antidepressant properties in humans. The clinical efficacy of these drugs is largely attributed to their actions at 5-HT$_{1A}$ autoreceptors. Buspirone is currently marketed as a non-benzodiazepine anxiolytic agent.

LY293284 {(−)-4R-6-acetyl-4-(di-*n*-proplyamino)-1,3,4,5-tetrahydrobenz-[*c,d*]indole} is the most potent and selective 5-HT$_{1A}$ receptor agonist described to date. This compound has an extremely high affinity for the 5-HT$_{1A}$ site (K_i = 0.07 nM) and is reported to be approximately 45 times more potent than 8-OH-DPAT in suppressing the activity of serotoninergic DRN neurons in anesthetized rats (FOREMAN et al. 1994).

In behaving cats, intravenous administration of 8-OH-DPAT or the anxiolytic compounds buspirone and ipsapirone inhibited the spontaneous firing rate of serotoninergic neurons in a dose-dependent manner (FORNAL et al. 1994a). The onset of action of these compounds is rapid (less than 20 s), and complete suppression of neuronal activity is typically observed at low doses (i.e., 5–20 μg/kg). 8-OH-DPAT (ED$_{50}$ = 1.5 μg/kg) was approximately four times more potent than ipsapirone (ED$_{50}$ = 6.0 μg/kg) or buspirone (ED$_{50}$ = 6.8 μg/kg) in producing neuronal inhibition. The potency of these drugs in awake animals is similar to that reported in anesthetized rats (VANDERMAELEN et al. 1986; SPROUSE and AGHAJANIAN 1987; LUM and PIERCEY 1988; SINTON and FALLON 1988; BLIER et al. 1989). Furthermore, spiperone blocked the inhibitory action of 8-OH-DPAT, buspirone, and ipsapirone on serotoninergic neuronal activity in awake cats (FORNAL et al. 1994a). Recently, we have demonstrated the efficacy of (*S*)-WAY 100135 and WAY 100635 as 5-HT$_{1A}$ autoreceptor antagonists in the awake cat (FORNAL et al. 1996b).

Presynaptic (somatodendritic) 5-HT$_{1A}$ autoreceptors appear to be extremely sensitive to other drugs which have little or no apparent intrinsic activity at postsynaptic 5-HT$_{1A}$ receptors. Thus, compounds such as BMY 7378 {8-[2-[4-(2-methoxyphenyl)-1-piperazinyl]ethyl]-8-azaspiro[4,5]decane-7,9-dione} and NAN 190 {1-(2-methoxyphenyl)-4-[4-(2-phthalimido)butyl]piperazine}, which at one time were thought to display only antagonist properties at central 5-HT$_{1A}$ receptors, have since been shown to potently activate 5-HT$_{1A}$ autoreceptors and completely inhibit the firing of serotoninergic neurons in both anesthetized rats (CHAPUT and DE MONTIGNY 1988; COX et al. 1993; FLETCHER et al. 1993; MILLAN et al. 1993) and freely moving cats (Fornal et al. 1994b). The greater sensitivity of presynaptic receptors to the agonist properties of 5-HT$_{1A}$ ligands may be due to a large receptor reserve in the raphe region and/or to a more efficient receptor-effector coupling (Cox et al. 1993; Meller et al. 1990).

In addition to confirming the marked sensitivity of serotoninergic DRN neurons to 5-HT$_{1A}$ agonists in awake animals, our results indicate that these drugs are more effective in suppressing neuronal activity during low levels of behavioral arousal (e.g., drowsiness) than during high levels of behavioral arousal (e.g., active waking). Thus, following administration of a 5-HT$_{1A}$ agonist, neurons which were otherwise completely inhibited when cats were inactive (e.g., drowsiness), resumed firing, albeit at a lower rate, when animals spontaneously became more aroused (e.g., active waking). Apparently, during periods of increased behavioral arousal, an increase in excitatory input to serotoninergic neurons appears to counteract the direct inhibitory action of 5-HT$_{1A}$ agonists, particularly at low doses. These findings may be of clinical significance, since they suggest that the efficacy of "buspirone-like" anxiolytic drugs in modifying serotoninergic neuronal activity (and terminal release of serotonin) may vary over the course of the day, as the level of behavioral arousal changes.

In general, the magnitude of neuronal inhibition produced by 5-HT$_{1A}$ agonist drugs in awake animals was negatively correlated with the spontaneous baseline firing rate of serotoninergic DRN neurons (i.e., the faster the spontaneous discharge rate of a cell, the less responsive the cell was to each agonist). This relationship appears to be a general characteristic of serotoninergic neurons recorded in vivo in unanesthetized animals, since we previously reported a similar relationship for serotoninergic neurons in each of the four major raphe nuclei examined in response to systemic injections of 5-MeODMT and LSD (Jacobs et al. 1983; Fornal and Jacobs 1988). We have proposed that the sensitivity or density of somatodendritic autoreceptors on individual neurons may be an important factor in determining both their level of spontaneous activity and responsiveness to serotonin autoreceptor agonists (Jacobs et al. 1983). Thus, neurons with relatively few or with less sensitive autoreceptors would hypothetically display a faster tonic firing rate due to less feedback inhibition, and would show less of a response to serotonin agonist drugs.

There is pharmacological evidence that the sensitivity of serotonin somatodendritic autoreceptors may vary across different raphe cell groups. For example, previous studies in our laboratory conducted in behaving cats indicated that serotoninergic neurons in the caudal raphe nuclei (i.e., NRP and NRM) are substantially less responsive to serotonin autoreceptor agonists such as 5-MeODMT and LSD than serotoninergic neurons in the rostral raphé nuclei (i.e., DRN and NCS) (Heym et al. 1982; Jacobs et al. 1983; Fornal and Jacobs 1988). Recently, we have shown in awake cats that serotoninergic DRN neurons are more sensitive than NCS serotoninergic neurons to systemic administration of the selective 5-HT$_{1A}$ agonists 8-OH-DPAT and ipsapirone

(TADA et al. 1991). These findings are consistent with previous electrophysiological studies in anesthetized rats which utilized either systemic drug injections (SINTON and FALLON 1988) or the direct application of drugs onto serotoninergic neurons (BLIER et al. 1990b), and may explain why 5-HT$_{1A}$ agonists such as 8-OH-DPAT and ipsapirone are more potent in inhibiting serotonin synthesis in DRN-innervated regions (e.g., striatum) than in regions innervated by the median raphe (e.g., hippocampus) following systemic administration (GOBERT et al. 1995). Preliminary results in our laboratory also indicate that serotoninergic neurons in NRP/NRO are relatively insensitive to 8-OH-DPAT (VEASEY et al. 1995).

III. Role in the Regulation of Neuronal Activity

Although it is well established that the activity of central serotoninergic neurons is under negative feedback control, which is mediated by somatodendritic 5-HT$_{1A}$ autoreceptors (AGHAJANIAN and WANG 1978; AGHAJANIAN and VANDERMAELEN 1986; SPROUSE and AGHAJANIAN 1987), it is not known whether these receptors are tonically activated by endogenous serotonin under physiological conditions. Previous studies conducted in anesthetized animals have reported little or no change in the firing rate of serotoninergic DRN neurons following either pharmacological blockade (LUM and PIERCEY 1988; BLIER et al. 1989, 1993a; MILLAN et al. 1992) or inactivation of 5-HT$_{1A}$ autoreceptors (INNIS and AGHAJANIAN 1987; BLIER et al. 1993b). Thus, these studies do not support the hypothesis that 5-HT$_{1A}$ autoreceptor-mediated inhibition plays a role in the regulation of central serotoninergic neuronal activity. These negative results, however, might not extend beyond the anesthetized preparation.

Recently we have examined the effects of blockade of the 5-HT$_{1A}$ autoreceptor on the activity of central serotoninergic neurons recorded in behaving cats (FORNAL et al. 1994a). In direct contrast to studies in anesthetized animals, systemic administration of spiperone (0.25 and 1 mg/kg) produced a rapid, dose-dependent increase in the firing rate of serotoninergic DRN neurons, suggesting that under physiological conditions these neurons are controlled by tonic feedback inhibition. Furthermore, as expected of a negative feedback system, the autoreceptor-mediated mechanism appears to be engaged when neurons are firing at a relatively high rate and disengaged at lower firing rates. Thus, during periods when serotoninergic neurons are relatively inactive (e.g., sleep), autoreceptor blockade with spiperone has little, if any, effect on tonic spontaneous activity. However, during periods when serotoninergic neurons are activated (e.g., active waking), and the release of serotonin in the vicinity of 5-HT$_{1A}$ autoreceptors should have an inhibitory effect on neuronal activity, autoreceptor blockade with spiperone results in a significant increase in the tonic firing rate of these neurons (FORNAL et al. 1994a). Spiperone also increases the firing rate of serotoninergic neurons located in other raphe nuclei (e.g., NCS and NRP) in awake cats (TADA et al.

1991, unpublished observations). These findings suggest that serotoninergic neurons throughout the brainstem may be under tonic feedback inhibition. Thus, autoreceptor-mediated control of neuronal activity may constitute a general serotoninergic regulatory mechanism.

The effect of spiperone on serotoninergic neuronal activity appears to be due to a blockade of 5-HT_{1A} autoreceptors, since the increase in neuronal activity produced by spiperone was strongly correlated with its antagonist action at the 5-HT_{1A} autoreceptor, as assessed by the blockade of 8-OH-DPAT-induced neuronal suppression (FORNAL et al. 1994a). The effect of spiperone is not related to its dopaminergic D_2 or serotoninergic 5-HT_2 antagonist properties, inasmuch as haloperidol and ritanserin did not increase the firing rate of serotoninergic neurons under the same conditions. Because spiperone is not selective for 5-HT_{1A} receptors, more selective 5-HT_{1A} antagonists are needed in order to unequivocally demonstrate the physiological role of somatodendritic 5-HT_{1A} autoreceptors. It is only recently that such compounds have been developed.

Along these lines, we have examined the effects of two reportedly selective 5-HT_{1A} antagonists, (S)-WAY 100135 and its more potent analog WAY 100635 (FLETCHER et al. 1993; FORSTER et al. 1995), on the activity of serotoninergic DRN neurons in behaving cats (FORNAL et al. 1996b). Our results show that (S)-WAY 100135 (0.025–1.0 mg/kg, i.v.) moderately depressed neuronal activity at all doses tested, and was found to be a relatively weak autoreceptor antagonist, since it only partially blocked the inhibitory action of 8-OH-DPAT on neuronal activity. The suppression of serotoninergic DRN neuronal activity produced by WAY 100135 is consistent with previous results obtained in anesthetized rats (FLETCHER et al. 1993; LEJEUNE et al. 1993; HADDJERI and BLIER 1995) and anesthetized cats (ESCANDON et al. 1994). In contrast, WAY 100635, at doses as low as 0.025 mg/kg i.v., strongly blocked the action of 8-OH-DPAT and significantly increased neuronal activity. The stimulatory action of WAY 100635, like that of spiperone, was evident during wakefulness, when serotoninergic neurons typically display a relatively high level of activity, but not during slow-wave or REM sleep, when serotoninergic neurons display little or no spontaneous activity. The neurochemical basis for the state-dependent action of 5-HT_{1A} antagonists on serotoninergic unit activity appears to be a state-dependent decrease in raphe serotonin release (PORTAS and MCCARLEY 1994) and subsequent reduction in autoreceptor stimulation.

The antagonist activity of WAY 100635 at 5-HT_{1A} autoreceptors closely paralleled its ability to increase the activity of serotoninergic DRN neurons. Thus, WAY 100635 significantly increased neuronal activity at the same doses which blocked the action of 8-OH-DPAT, and the doses which produced maximal neuronal activation also produced the strongest autoreceptor blockade. Furthermore, the duration of the neuronal activation produced by WAY 100635 paralleled the duration of its antagonist action at 5-HT_{1A} autoreceptors. These results are consistent with the hypothesis that WAY

100635 increases serotoninergic neuronal activity by blocking 5-HT$_{1A}$ autoreceptors.

Both spiperone and WAY 100635 increased the activity of serotoninergic neurons to about the same maximum level at doses producing comparable autoreceptor blockade, although WAY 100635 was approximately ten times more potent than spiperone. Furthermore, the magnitude of the neuronal increase produced by both drugs was inversely related to the spontaneous baseline firing rate of serotoninergic neurons. Thus, slower-firing cells, in general, displayed a larger percentage increase in neuronal activity than faster-firing cells. This suggests that fast-firing cells may be under less feedback inhibition than slower-firing cells, which may explain why they fire faster. Interestingly, for each drug, a small subgroup of cells ($\sim$10%) was encountered which exhibited little or no increase in activity following systemic administration of spiperone or WAY 100635, although all of these cells showed the characteristic inhibitory response to 8-OH-DPAT. These results suggest that 5-HT$_{1A}$ autoreceptors on some cells may only be of pharmacological significance, because these cells may lack autoregulatory feedback inputs in situ. On the other hand, 5-HT$_{1A}$ autoreceptors on cells sensitive to 5-HT$_{1A}$ autoreceptor antagonists (e.g., spiperone and WAY 100635) may play an important physiological role in regulating neuronal activity. The weak or inconsistent stimulatory action of 5-HT$_{1A}$ antagonists observed in the anesthetized preparation may be due to the relatively low firing rate (typically 1 spike/s) of serotoninergic neurons under anesthesia and/or to the elimination of the influence of behavioral state (JOLAS et al. 1993; LEJEUNE et al. 1993; PRISCO et al. 1993; ARBORELIUS et al. 1994; FORSTER et al. 1995; GARTSIDE et al. 1995; MUNDEY et al. 1996). Thus, there may be too little serotonin released onto 5-HT$_{1A}$ autoreceptors to exert significant feedback inhibition.

Overall, our studies provide strong evidence that serotonin autoreceptors are activated under normal conditions by endogenous serotonin release, and that the level of autoinhibition varies as a function of the basal firing rate and the behavioral state of the animal. One function of these receptors may be to dampen or prevent excessive neuronal activation in response to increased tonic and phasic excitatory inputs, thereby maintaining a given steady-state concentration of synaptic serotonin in projection areas. Alterations in normal autoreceptor function may therefore lead to pronounced changes in central serotonin neurotransmission and may play a role in the etiology of various affective disorders (e.g., anxiety and depression) in which serotonin has been implicated.

In summary, while both (S)-WAY 100135 and WAY 100635 possess 5-HT$_{1A}$ antagonist properties, only WAY 100635 appears to act as a selective autoreceptor antagonist, since it increased, rather than depressed, neuronal activity in awake cats. The results obtained with WAY 100635 confirm our previous findings obtained with spiperone, and support further the hypothesis that 5-HT$_{1A}$ autoreceptor-mediated feedback inhibition operates under physiological conditions.

IV. Role in Therapeutic Action of Antidepressant Drugs

A major problem associated with current antidepressant drugs, including the recently developed class of compounds known as SSRIs, is the slow onset of their clinical action, typically 2–3 weeks. In addition, it has been estimated that approximately 30% of all patients treated with an antidepressant drug fail to demonstrate a significant improvement in mood, despite an adequate treatment regimen (HOLDEN 1991). Consequently, there has been considerable interest in the development of pharmacological approaches aimed at accelerating the clinical response to antidepressant drugs and increasing their therapeutic effectiveness.

SSRIs (e.g., fluoxetine) are presently the most widely prescribed class of antidepressant drugs and have proven effective in the treatment of major depression. The therapeutic action of these agents is generally believed to result from their ability to enhance central serotonin neurotransmission by increasing the synaptic availability of serotonin. Although SSRIs rapidly inhibit serotonin reuptake upon acute administration, several weeks of treatment are required to elicit a clinical response. A widely held hypothesis suggests that the delayed onset of action of these drugs may be due to the indirect activation of 5-HT_{1A} autoreceptors following reuptake inhibition (BLIER et al. 1990a; BLIER and DE MONTIGNY 1994).

The serotonin transporter, which inactivates serotonin released into the synapse, is present in both serotoninergic nerve terminal and cell body regions (i.e., raphe nuclei). Systemic administration of SSRIs preferentially increases the concentration of extracellular serotonin in the midbrain raphe (INVERNIZZI et al. 1992; ARTIGAS 1993; GARTSIDE et al. 1995), which leads to an inhibition of serotoninergic neurons through increased activation of 5-HT_{1A} autoreceptors. This decrease in serotoninergic neuronal activity following acute administration limits the ability of SSRIs to increase serotonin levels in the forebrain (e.g., frontal cortex), and may thus explain the failure of these drugs to produce an immediate therapeutic effect.

The long-term administration of these drugs is believed to desensitize the 5-HT_{1A} autoreceptor to the inhibitory feedback action of serotonin, thus allowing serotoninergic neurons to resume their normal firing activity in the presence of continued reuptake inhibition (CHAPUT et al. 1986; BLIER et al. 1990a). The resumption of serotoninergic neuronal activity (and hence neurotransmitter release) is thought to lead to an overall enhancement of serotonin neurotransmission which, in turn, may mediate the therapeutic effect. The time course of these changes is consistent with the delayed onset of action of these drugs in the clinic.

Theoretically, the same functional consequences seen with autoreceptor desensitization (i.e., decreased autoinhibition) might be achieved much more rapidly by administering drugs which selectively block 5-HT_{1A} autoreceptors and thus prevent the inhibitory feedback action of serotonin. It has been hypothesized that the combined administration of a highly potent and selective 5-HT$_{1A}$ autoreceptor antagonist and an SSRI might lead to a more rapid or more effective antidepressant response (BLIER et al. 1989; ARTIGAS 1993; BLIER and DE MONTIGNY 1994). Preliminary open-label clinical trials using (−)-pindolol, a β-blocker with 5-HT$_{1A}$ antagonist properties, support this idea (ARTIGAS et al. 1994; BLIER and BERGERON 1995). Although the precise mechanism underlying the therapeutic action of pindolol remains to be determined, neurochemical studies have shown that 5-HT$_{1A}$ antagonists in general potentiate the ability of SSRIs to increase extracellular serotonin levels in forebrain areas, presumably by blocking autoreceptor-mediated feedback inhibition of serotoninergic neuronal activity (HJORTH 1993; HJORTH and AUERBACH 1994; GARTSIDE et al. 1995).

Since the reestablishment of serotoninergic neuronal activity during sustained reuptake inhibition is thought to play a crucial role in the antidepressant action of SSRIs, we recently examined the ability of two novel 5-HT$_{1A}$ receptor antagonists {(S)-WAY 100135 and WAY 100635} to restore the activity of serotoninergic DRN neurons after acute treatment with the SSRI fluoxetine (FORNAL et al. 1995). In behaving cats, administration of fluoxetine (5 mg/kg, i.v.) decreased the firing rate of serotoninergic DRN neurons to approximately 5% of baseline levels. Subsequent administration of either (S)-WAY 100135 (0.5 mg/kg, i.v.) or WAY 100635 (0.1 mg/kg, i.v.) rapidly reversed the neuronal inhibition produced by fluoxetine, indicating that both compounds block the action of serotonin at 5-HT$_{1A}$ autoreceptors. WAY 100635 proved to be considerably more effective than (S)-WAY 100135, however, since it completely reversed the effect of fluoxetine at low doses (e.g., 0.025 mg/kg, i.v.), and further elevated the firing rate of these neurons above pre-fluoxetine baseline levels, while (S)-WAY 100135 only partially reversed the neuronal suppression produced by fluoxetine. After the peak drug effect was reached, neuronal activity steadily returned to the suppressed fluoxetine baseline level, as the antagonist activity of these compounds dissipated over time (i.e., washout). At a dose of 0.1 mg/kg i.v., WAY 100635 significantly antagonized the inhibitory action of fluoxetine on neuronal activity for approximately 2 h. In contrast, (S)-WAY 100135 (0.5 mg/kg, i.v.) antagonized the effect of fluoxetine for only 15–30 min. Thus, the antagonist action of (S)-WAY 100135 was relatively short lived compared to that of WAY 100635. The ability of these compounds to block the action of other SSRIs (i.e., citalopram and paroxetine) on DRN serotoninergic neuronal activity has recently been demonstrated in anesthetized rats (ARBORELIUS et al. 1995; GARTSIDE et al. 1995; HAJÓS et al. 1995). Interestingly, (S)-WAY 100135 and WAY 100635 appear to be substantially less effective in the anesthetized animal, since the inhibitory response to SSRIs was not reversed in every cell studied, and in those cases where reversal occurred, the magnitude of the effect was relatively small compared to that observed in awake animals with fluoxetine. The reason for this is not clear, but may be related to a diminished excitatory input to serotoninergic neurons under anesthesia.

In addition, we found that the ability of (S)-WAY 100135 and WAY 100635 to reverse the effect of fluoxetine on the activity of serotoninergic neurons was highly influenced by the behavioral state of the animal. These antagonists maximally elevated neuronal activity when cats were behaviorally active, and were less effective or completely ineffective in increasing neuronal activity after fluoxetine treatment during periods of behavioral quiescence (e.g., drowsiness or sleep). Since activation of an inhibitory GABAergic input during sleep is thought to mediate the state-dependent suppression of serotoninergic neuronal activity (LEVINE and JACOBS 1992), pharmacological blockade of 5-HT$_{1A}$ autoreceptors in fluoxetine-treated animals would not be expected to restore neuronal activity during sleep.

V. Conclusions

Considerable progress has been made in our understanding of the physiological and functional significance of the serotonin somatodendritic autoreceptor, largely due to the recent development of several highly selective 5-HT$_{1A}$ receptor ligands. Drugs which potently activate somatodendritic 5-HT$_{1A}$ autoreceptors, and thereby inhibit serotonergic cell firing and decrease serotonin release, have therapeutic efficacy in the treatment of anxiety. The clinical significance of drugs which *block* central 5-HT$_{1A}$ receptors, however, remains unclear. Our results suggest that 5-HT$_{1A}$ autoreceptor antagonists such as (S)-WAY 100135, and in particular WAY 100635, may be useful adjuncts to SSRI drug therapy, since the combined drug regimen should greatly enhance serotonin neurotransmission and produce a superior antidepressant response in the clinic. Furthermore, the apparent interaction between behavioral state and drug effect seen in unanesthetized animals suggests the potential utility of behavioral strategies (e.g., increasing general activity and/or the use of sleep deprivation) in further augmenting the effects of combining an SSRI and a 5-HT$_{1A}$ antagonist. The concomitant blockade of nerve terminal autoreceptors, which control the amount of serotonin released per nerve impulse, would also be expected to potentiate the effect of this drug combination on serotonin release. The ability of 5-HT$_{1A}$ antagonists such as WAY 100635 to increase the spontaneous firing rate of serotoninergic neurons, when administered alone, suggests that these drugs may also have beneficial effects in other disorders (e.g., obesity, obsessive compulsive behavior, chronic pain) which respond to treatments which increase serotonin neurotransmission. However, currently available 5-HT$_{1A}$ receptor antagonists also block postsynaptic 5-HT$_{1A}$ receptors, which are present in high concentrations in many brain regions (e.g., hippocampus, septum, and amygdala). This blockade could conceivably counteract the effects of increased serotonin release at postsynaptic 5-HT$_{1A}$ receptors, which may mediate some or all of the antidepressant effect. In this respect, drugs which specifically block 5-HT$_{1A}$ autoreceptors may prove to be more effective.

References

Aghajanian GK, Lakoski JM (1984) Hyperpolarization of serotonergic neurons by serotonin and LSD: studies in brain slices showing increased K^+ conductance. Brain Res 305:181–185

Aghajanian GK, VanderMaelen CP (1986) Specific systems of the reticular core: serotonin. In: Mountcastle VB, Bloom FE, Geiger SR (eds) Handbook of physiology. The nervous system. Intrinsic regulatory systems of the brain, sect 1, vol IV, chap 4. American Physiological Society, Bethesda, p 237

Aghajanian GK, Wang RY (1977) Habenular and other midbrain raphe afferents demonstrated by a modified retrograde tracing technique. Brain Res 122:229–242

Aghajanian GK, Wang RY (1978) Physiology and pharmacology of central serotonergic neurons. In: Lipton MA, DiMascio A, Killam KF (eds) Psychopharmacology: a generation of progress. Raven, New York, p 171

Aghajanian GK, Foote WE, Sheard MH (1968) Lysergic acid diethylamide: sensitive neuronal units in the midbrain raphe. Science 161:706–708

Aghajanian GK, Haigler HJ, Bloom FE (1972) Lysergic acid diethylamide and serotonin: direct actions on serotonin-containing neurons in rat brain. Life Sci 11: 615–622

Arborelius L, Backlund Höök B, Hacksell U, Svensson TH (1994) The 5-HT$_{1A}$ receptor antagonist (S)-UH-301 blocks the (R)-8-OH-DPAT-induced inhibition of serotonergic dorsal raphe cell firing in the rat. J Neural Transm 96:179–186

Arborelius L, Nomikos GG, Grillner P, Hertel P, Backlund Höök B, Hacksell U, Svensson TH (1995) 5-HT$_{1A}$ receptor antagonists increase the activity of serotonergic cells in the dorsal raphe nucleus in rats treated acutely or chronically with citalopram. Naunyn Schmiedebergs Arch Pharmacol 352:157–165

Artigas F (1993) 5-HT and antidepressants: new views from microdialysis studies. Trends Pharmacol Sci 14:262

Artigas F, Pere V, Alvarez E (1994) Pindolol induces a rapid improvement of depressed patients treated with serotonin reuptake inhibitors. Arch Gen Psychiatry 51:248–251

Auerbach S, Fornal C, Jacobs BL (1985) Response of serotonin-containing neurons in nucleus raphe magnus to morphine, noxious stimuli, and periaqueductal gray stimulation in freely moving cats. Exp Neurol 88:609–628

Baraban JM, Aghajanian GK (1980) Suppression of firing activity of 5-HT neurons in the dorsal raphe by alpha-adrenoceptor antagonists. Neuropharmacology 19:355–363

Blier P, Bergeron R (1995) Effectiveness of pindolol with selected antidepressant drugs in the treatment of major depression. J Clin Psychopharmacol 15:217–222

Blier P, de Montigny C (1994) Current advances and trends in the treatment of depression. Trends Pharmacol Sci 15:220–226

Blier P, Steinberg S, Chaput Y, de Montigny C (1989) Electrophysiological assessment of putative antagonists of 5-hydroxytryptamine receptors: a single-cell study in the rat dorsal raphe nucleus. Can J Physiol Pharmacol 67:98–105

Blier P, de Montigny C, Chaput Y (1990a) A role for the serotonin system in the mechanism of action of antidepressant treatments: preclinical evidence. J Clin Psychiatry 51 [Suppl 4]:14–20

Blier P, Serrano A, Scatton B (1990b) Differential responsiveness of the rat dorsal and median raphe 5-HT systems to 5-HT$_1$ receptor agonists and p-chloroamphetamine. Synapse 5:120–133

Blier P, Lista A, de Montigny C (1993a) Differential properties of pre- and postsynaptic 5-hydroxytryptamine$_{1A}$ receptors in the dorsal raphe and hippocampus. I. Effect of spiperone. J Pharmacol Exp Ther 265:7–15

Blier P, Lista A, de Montigny C (1993b) Differential properties of pre- and postsynaptic 5-hydroxytryptamine$_{1A}$ receptors in the dorsal raphe and hippocampus. II. Effect of pertussis and cholera toxins. J Pharmacol Exp Ther 265:16–23

Chaput Y, de Montigny C (1988) Effects of the 5-hydroxytryptamine$_1$ receptor antagonist, BMY 7378, on 5-hydroxytryptamine neurotransmission: electrophysiological studies in the rat central nervous system. J Pharmacol Exp Ther
246:359–370

Chaput Y, de Montigny C, Blier P (1986) Effects of a selective 5-HT reuptake blocker,
citalopram, on the sensitivity of 5-HT autoreceptors: electrophysiological studies
in the rat brain. Naunyn Schmiedebergs Arch Pharmacol 333:342–348

Cox RF, Meller E, Waszczak BL (1993) Electrophysiological evidence for a large
receptor reserve for inhibition of dorsal raphe neuronal firing by 5-HT$_{1A}$ agonists.
Synapse 14:297–304

de Montigny C, Aghajanian GK (1977) Preferential action of 5- methoxytryptamine
and 5-methoxydimethyltryptamine on presynaptic serotonin receptors: a comparative iontophoretic study with LSD and serotonin. Neuropharmacology
16:811–818

Escandon NA, Zimmermann DC, McCall RB (1994) Characterization of the
serotonin$_{1A}$ receptor antagonist activity of WAY-100135 and spiperone. J
Pharmacol Exp Ther 268:441–447

Fletcher A, Bill DJ, Bill SJ, Cliffe IA, Dover GM, Forster EA, Haskins JT, Jones D,
Mansell HL, Reilly Y (1993) WAY 100135: a novel, selective antagonist at presynaptic and postsynaptic 5-HT$_{1A}$ receptors. Eur J Pharmacol 237:283–291

Foreman MM, Fuller RW, Rasmussen K, Nelson DL, Calligaro DO, Zhang L, Barrett
JE, Booher RN, Paget CJ Jr, Flaugh ME (1994) Pharmacological characterization
of LY293284: a 5-HT$_{1A}$ receptor agonist with high potency and selectivity. J
Pharmacol Exp Ther 270:1270–1281

Fornal CA, Jacobs BL (1988) Physiological and behavioral correlates of serotonergic
single-unit activity. In: Osborne NN, Hamon M (eds) Neuronal serotonin. Wiley,
Chichester, p 305

Fornal CA, Litto WJ, Morilak DA, Jacobs BL (1987) Single-unit responses of serotonergic dorsal raphe nucleus neurons to environmental heating and pyrogen administration in freely moving cats. Exp Neurol 98:388–403

Fornal CA, Litto WJ, Morilak DA, Jacobs BL (1989) Single-unit responses of serotonergic neurons to glucose and insulin administration in behaving cats. Am J Physiol
257:R1345–R1353

Fornal CA, Litto WJ, Morilak DA, Jacobs BL (1990) Single-unit responses of serotonergic dorsal raphe neurons to vasoactive drug administration in freely moving cats.
Am J Physiol 259:R963–R972

Fornal CA, Litto WJ, Metzler CW, Marrosu F, Tada K, Jacobs BL (1994a) Single-unit
responses of serotonergic dorsal raphe neurons to 5-HT$_{1A}$ agonist and antagonist
drug administration in behaving cats. J Pharmacol Exp Ther 270:1345–1358

Fornal CA, Marrosu F, Metzler CW, Tada K, Jacobs BL (1994b) Effects of the putative
5-hydroxytryptamine$_{1A}$ antagonists BMY 7378, NAN 190 and (−)-propranolol on
serotonergic dorsal raphe unit activity in behaving cats. J Pharmacol Exp Ther
270:1359–1366

Fornal CA, Metzler CW, Martin FJ, Jacobs BL (1995) 5-HT$_{1A}$ autoreceptor antagonists
restore serotonergic neuronal activity after acute fluoxetine treatment in behaving
cats. Soc Neurosci Abstr 21:977

Fornal CA, Metzler CW, Marrosu F, Ribiero-do-Valle LE, Jacobs BL (1996a) A
subgroup of dorsal raphe serotonergic neurons in the cat is strongly activated
during oral-buccal movements, Brain Res 716:123–133

Fornal CA, Metzler CW, Gallegos RA, Veasey SC, McCreary AC, Jacobs BL (1996b)
WAY-100635, a potent and selective 5-HT$_{1A}$ antagonist, increases serotonergic
neuronal activity in behaving cats: comparison with (S)-WAY-100135. J
Pharmacol Exp Ther 278:752–762

Forster EA, Cliffe IA, Bill DJ, Dover GM, Jones D, Reilly Y, Fletcher A (1995) A
pharmacological profile of the selective 5-HT$_{1A}$ receptor antagonist, WAY-100635.
Eur J Pharmacol 281:81–88

Gardner JP (1996) The effects of sleep deprivation on serotonergic neuronal activity in the dorsal raphe nucleus of the cat. PhD dissertation, Princeton University

Gardner JP, Fornal CA, Jacobs BL (1995) The effects of sleep deprivation on serotonergic neuronal activity in the behaving cat. Soc Neurosci Abstr 21: 193

Gartside SE, Umbers V, Hajós M, Sharp T (1995) Interaction between a selective 5-HT_{1A} receptor antagonist and an SSRI in vivo: effects on 5-HT cell firing and extracellular 5-HT. Br J Pharmacol 115:1064–1070

Gobert A, Lejeune F, Rivet J-M, Audinot V, Newman-Tancredi A, Millan MJ (1995) Modulation of the activity of central serotonergic neurons by novel serotonin$_{1A}$ receptor agonists and antagonists: a comparison to adrenergic and dopaminergic neurons in rats. J Pharmacol Exp Ther 273:1032–1046

Haddjeri N, Blier P (1995) Pre- and post-synaptic effects of the 5-HT_3 agonist 2-methyl-5-HT on the 5-HT system in the rat brain. Synapse 20:54–67

Haigler HJ, Aghajanian GK (1974) Lysergic acid diethylamide and serotonin: a comparison of effects on serotonergic neurons and neurons receiving a serotonergic input. J Pharmacol Exp Ther 188:688–699

Hajós M, Gartside SE, Sharp T (1995) Inhibition of median and dorsal raphe neurones following administration of the selective serotonin reuptake inhibitor paroxetine. Naunyn Schmiedebergs Arch Pharmacol 351:624–629

Heym J, Steinfels GF, Jacobs BL (1982) Medullary serotonergic neurons are insensitive to 5-MeODMT and LSD. Eur J Pharmacol 81:677–680

Hjorth S (1993) Serotonin 5-HT_{1A} autoreceptor blockade potentiates the ability of the 5-HT reuptake inhibitor citalopram to increase nerve terminal output of 5-HT in vivo: a microdialysis study. J Neurochem 60:776–779

Hjorth S, Auerbach SB (1994) Further evidence for the importance of 5-HT_{1A} autoreceptors in the action of selective serotonin reuptake inhibitors. Eur J Pharmacol 260:251–255

Holden C (1991) Depression: the news isn't depressing. Science 254:1450–1452

Innis RB, Aghajanian GK (1987) Pertussis toxin blocks 5-HT_{1A} and $GABA_B$ receptor-mediated inhibition of serotonergic neurons. Eur J Pharmacol 143:195–204

Invernizzi R, Belli S, Samanin R (1992) Citalopram's ability to increase the extracellular concentrations of serotonin in the dorsal raphe prevents the drug's effect in the frontal cortex. Brain Res 584:322–324

Jacobs BL, Azmitia EC (1992) Structure and function of the brain serotonergic system. Physiol Rev 72:165–229

Jacobs BL, Fornal CA (1993) 5-HT and motor control: a hypothesis. Trends Neurosci 16:346–352

Jacobs BL, Fornal CA (1995) Activation of 5-HT neuronal activity during motor behavior. Semin Neurosci 7:401–408

Jacobs BL, Heym J, Rasmussen K (1983) Raphe neurons: firing rate correlates with size of drug response. Eur J Pharmacol 90:275–278

Jacobs BL, Metzler CW, Fornal CA, Veasey SC (1994) Response profiles of serotonergic neurons to motor challenges differ for pontine and medullary raphe neurons. Soc Neurosci Abstr 20:378

Jolas T, Haj-Dahmane S, Lanfumey L, Fattaccini CM, Kidd EJ, Adrien J, Gozlan H, Guardiola-Lemaitre B, Hamon M (1993) (–)Tertatolol is a potent antagonist at pre- and postsynaptic serotonin 5-HT_{1A} receptors in the rat brain. Naunyn Schmiedebergs Arch Pharmacol 347:453–463

Kidd EJ, Haj-Dahmane S, Jolas T, Lanfumey L, Fattaccini C-M, Guardiola-Lemaitre B, Gozlan H, Hamon M (1993) New methoxy-chroman derivatives, 4[N-(5-methoxy-chroman-3-yl)N-propylamino]butyl-8-azaspiro(4,5)-decane-7,9-dione [(+)-S 20244] and its enantiomers, (+)-S 20499 and (–)-S 20500, with potent agonist properties at central 5-hydroxytryptamine$_{1A}$ receptors. J Pharmacol Exp Ther 264:863–872

Lanfumey L, Haj-Dahmane S, Hamon M (1993) Further assessment of the antagonist properties of the novel and selective 5-HT$_{1A}$ receptor ligands (+)-WAY 100135 and SDZ 216–525. Eur J Pharmacol 249:25–35

Lejeune F, Rivet J-M, Gobert A, Canton H, Millan MJ (1993) WAY 100,135 and (–)-tertatolol act as antagonists at both 5-HT$_{1A}$ autoreceptors and postsynaptic 5-HT$_{1A}$ receptors in vivo. Eur J Pharmacol 240:307–310

Levine ES, Jacobs BL (1992) Neurochemical afferents controlling the activity of serotonergic neurons in the dorsal raphe nucleus: microiontophoretic studies in the awake cat. J Neurosci 12:4037–4044

Lum JT, Piercey MF (1988) Electrophysiological evidence that spiperone is an antagonist of 5-HT$_{1A}$ receptors in the dorsal raphe nucleus. Eur J Pharmacol 149:9–15

Matheson GK, Pfeifer DM, Weiberg MB, Michel C (1994) The effects of azapirones on serotonin$_{1A}$ neurons of the dorsal raphe. Gen Pharmacol 25:675–683

Meller E, Goldstein M, Bohmaker K (1990) Receptor reserve for 5-hydroxytryptamine$_{1A}$-mediated inhibition of serotonin synthesis: possible relationship to anxiolytic properties of 5-hydroxytryptamine$_{1A}$ agonists. Mol Pharmacol 37:231–237

Mendlin A, Martin FJ, Rueter LE, Jacobs BL (1996) Neuronal release of serotonin in the cerebellum of behaving rats: an in vivo microdialysis study. J Neurochem 67:617–622

Millan MJ, Rivet J-M, Canton H, Lejeune F, Bervoets K, Brocco M, Gobert A, Lefebvre De Ladonchamps B, Le Marouille-Girardon S, Verriele L, Laubie M, Lavielle G (1992) S 14671: a naphthylpiperazine 5-hydroxytryptamine$_{1A}$ agonist of exceptional potency and high efficacy possessing antagonist activity at 5-hydroxytryptamine$_{1C/2}$ receptors. J Pharmacol Exp Ther 262:451–463

Millan MJ, Rivet JM, Canton H, Lejeune F, Gobert A, Widdowson P, Bervoets K, Brocco M, Peglion J-L (1993) S 15535: a highly selective benzodioxopiperazine 5-HT$_{1A}$ receptor ligand which acts as an agonist and an antagonist at presynaptic and postsynaptic sites respectively. Eur J Pharmacol 230:99–102

Mosko SS, Jacobs BL (1976) Recording of dorsal raphe unit activity in vitro. Neurosci Lett 2:195–200

Mosko SS, Jacobs BL (1977) Electrophysiological evidence against negative neuronal feedback from the forebrain controlling midbrain raphe unit activity. Brain Res 119:291–303

Mundey MK, Fletcher A, Marsden CA (1994) Effect of the putative 5-HT$_{1A}$ antagonists WAY 100135 and SDZ 216–525 on 5-HT neuronal firing in the guinea-pig dorsal raphe nucleus. Neuropharmacology 33:61–66

Mundey MK, Fletcher A, Marsden CA (1996) Effects of 8-OH-DPAT and 5-HT$_{1A}$ antagonists WAY100135 and WAY100635 on guinea-pig behaviour and dorsal raphe 5-HT neurone firing. Br J Pharmacol 117:750–756

Portas CM, McCarley RW (1994) Behavioral state-related changes of extracellular serotonin concentration in the dorsal raphe nucleus: a microdialysis study in the freely moving cat. Brain Res 648:306–312

Potrebic SB, Field HL, Mason P (1994) Serotonin immunoreactivity is contained in one physiological cell class in the rat rostral ventromedial medulla. J Neurosci 14:1655–1665

Prisco S, Cagnotto A, Talone D, De Blasi A, Mennini T, Esposito E (1993) Tertatolol, a new β-blocker, is a serotonin (5-hydroxytryptamine$_{1A}$) receptor antagonist in rat brain. J Pharmacol Exp Ther 265:739–744

Rueter LE (1995) Serotonin release in multiple forebrain sites of the rat: a microdialysis study examining whether different experimental manipulations produce a differential magnitude and/or pattern of release. PhD dissertation, Princeton University

Rueter LE, Jacobs BL (1995) Serotonin release during stressful and non-stressful conditions: is the serotonergic system specifically stress-activatable? Soc Neurosci Abstr 21:1691

Rueter LE, Jacobs BL (1996) Changes in brain serotonin at the light-dark transition: correlation with behavior. Neuroreport 7:1107–1111

Sinton CM, Fallon SL (1988) Electrophysiological evidence for a functional differentiation between subtypes of the 5-HT$_1$ receptor. Eur J Pharmacol 157:173–181

Sotelo C, Cholley B, El Mestikawy S, Gozlan H, Hamon M (1990) Direct immunohistochemical evidence of the existence of 5-HT$_{1A}$ autoreceptors on serotonergic neurons in the midbrain raphe nuclei. Eur J Neurosci 2:1144–1154

Sprouse JS, Aghajanian GK (1986) (–)-Propranolol blocks the inhibition of serotonergic dorsal raphe cell firing by 5-HT$_{1A}$ selective agonists. Eur J Pharmacol 128:295–298

Sprouse JS, Aghajanian GK (1987) Electrophysiological responses of serotonergic dorsal raphe neurons to 5-HT$_{1A}$ and 5-HT$_{1B}$ agonists. Synapse 1:3–9

Steinfels GF, Heym J, Strecker RE, Jacobs BL (1983) Raphe unit activity in freely moving cats is altered by manipulations of central but not peripheral motor systems. Brain Res 279:77–84

Tada K, Fornal CA, Marrosu F, Metzler CW, Jacobs BL (1991) Single-unit responses of n. raphe dorsalis and centralis superior neurons to 5-HT$_{1A}$ drugs in behaving cats. Soc Neurosci Abstr 17:1437

Trulson ME, Jacobs BL (1979) Raphe unit activity in freely moving cats: correlation with level of behavioral arousal. Brain Res 163:135–150

Trulson ME, Jacobs BL (1983) Raphe unit activity in freely moving cats: lack of diurnal variation. Neurosci Lett 36:285–290

Trulson ME, Jacobs BL, Morrison AR (1981) Raphe unit activity during REM sleep in normal cats and in pontine lesioned cats displaying REM sleep without atonia. Brain Res 226:75–91

VanderMaelen CP, Aghajanian GK (1983) Electrophysiological and pharmacological characterization of serotonergic dorsal raphe neurons recorded extracellularly and intracellularly in rat brain slices. Brain Res 289:109–119

VanderMaelen CP, Matheson GK, Wilderman RC, Patterson, LA (1986) Inhibition of serotonergic dorsal raphe neurons by systemic and iontophoretic administration of buspirone, a non-benzodiazepine anxiolytic drug. Eur J Pharmacol 129:123–130

Van Wijngaarden I, Tulp MThM, Soudijn W (1990) The concept of selectivity in 5-HT receptor research. Eur J Pharmacol 188:301–312

Veasey SC, Fornal CA, Metzler CW, Jacobs BL (1995) Response of serotonergic caudal raphe neurons in relation to specific motor activities in freely moving cats. J Neurosci 15:5346–5359

Veasey SC, Fornal CA, Metzler CW, Jacobs BL (to be published) Single-unit responses of serotoninergic dorsal raphe neurons to specific motor challenges in freely moving cats. Neuroscience

Wilkinson LO, Jacobs BL (1988) Lack of response of serotonergic neurons in the dorsal raphe nucleus of freely moving cats to stressful stimuli. Exp Neurol 101:445–457

Wilkinson LO, Auerbach SB, Jacobs BL (1991) Extracellular serotonin levels change with behavioral state but not pyrogen-induced hyperthermia. J Neurosci 11:2732–2741

Williams JT, Colmers WF, Pan ZZ (1988) Voltage- and ligand activated inwardly rectifying currents in dorsal raphe neurons in vitro. J Neurosci 8:3499–3506

Note Added in Proof. Recently, we examined the action of pindolol on the activity of serotoninergic DRN neurons in the awake, freely-moving cat (FORNAL et al., unpublished observations). Contrary to current belief, our results indicate that pindolol acts as an agonist rather than an antagonist at the 5-HT$_{1A}$ autoreceptor. Systemic administration of pindolol produced a dose-dependent suppression of serotoninergic neuronal activity (ED_{50} = 0.25 mg/kg, i.v.). This effect was completely reversed by the subsequent administration of the 5-HT$_{1A}$ antagonist WAY-100635. In drug antagonism studies, pindolol (0.1–5 mg/kg, i.v.) failed to antagonize the effects of either a low dose of fluoxetine (0.5 mg/kg, i.v.), which mildly suppressed neuronal activity, or a high dose of fluoxetine (5 mg/kg, i.v.), which completely suppressed neuronal activity. Further more, in addition to lacking efficacy as an antagonist in these experiments, pindolol produced a further decrease in neuronal activity in animals treated with the low dose of fluoxetine. Pindolol also failed to antagonize the suppression of serotoninergic neuronal activity produced by the 5-HT$_{1A}$ agonist 8-OH-DPAT. Since pindolol apparently lacks presynaptic 5-HT$_{1A}$ antagonist activity, the clinical efficacy of the drug in augmenting the antidepressant response to SSRIs, such as fluoxetine, may be unrelated to a restoration of serotoninergic neuronal activity.

CHAPTER 4

Tryptophan Hydroxylase: Molecular Biology and Regulation

T.H. Joh

A. Introduction

Tryptophan hydroxylase [TPH, tryptophan 5-monooxygenase; L-tryptophan, tetrahydrobiopterin:oxygen oxidoreductase (5-hydroxylating), E.C. 1.14.16.4] catalyzes the first and rate-limiting step in the biosynthesis of serotonin (5-HT, 5'-hydroxytryptamine) (GRAHAME-SMITHE 1964; LOVENBERG et al. 1967; JEQUIRE et al. 1969) (Fig. 1). Within the brain of vertebrates, this enzyme is selectively expressed in the serotoninergic neurons of the brainstem raphe nuclei and adjacent reticular formations. The serotoninergic projection system is the most extensive monoaminergic system in the brain. Ill-defined abnormalities in serotoninergic as well as catecholaminergic neurotransmission are said to contribute to various affective brain disorders. Since regulation of 5-HT levels in the brain is directly related to TPH activity, understanding TPH regulation is an important step toward elucidating the functions of 5-HT in the central nervous system.

TPH also catalyzes the first step in melatonin biosynthesis in the pineal gland (Fig. 1). Although melatonin synthesis is mainly controlled by serotonin N-acetyltransferase (SNAT, EC 2.3.1.87) activity, which is regulated by the sympathetic impulse flow increasing during the night period of the light/dark cycle (see review article of KRAUSE and DUBOCOVICH 1990), pineal gland TPH activity is also enhanced in the dark cycle. This enhanced activity of both TPH and SNAT is due to increased levels of cAMP. In fact, the structure of the TPH gene has been recently characterized and its 5'-upstream region contains several possible *cis*-acting elements, including the cyclic AMP response element CRE (STOLL et al. 1990; BOULARAND et al. 1995a,b). Whether CRE plays an important role in TPH gene transcription is under dispute (BOULARAND et al. 1995a; REED et al. 1995).

In spite of the numerous studies done in the past on 5-HT, knowledge of its synthesizing enzyme TPH, such as its structure, characterization and regulation, have not progressed compared to that of other monoamine enzymes, such as tyrosine hydroxylase. This is because of its low abundance and difficulty of purification, and thus cloning of its gene. However, molecular biological studies on TPH are slowly but steadily progressing toward an understanding of its gene structure and transcriptional and translational regulation. In this chapter, a summary is given of the recent reports in the field

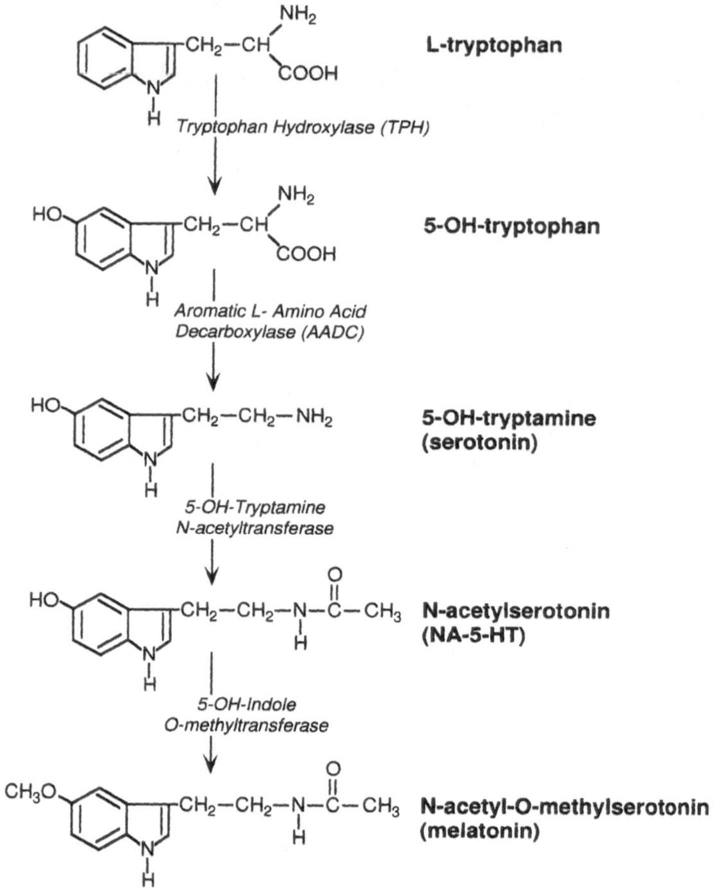

Fig. 1. Biosynthesis of serotonin and melatonin. Note that tryptophan hydroxylase is the first and rate-limiting enzyme in the serotonin biosynthetic pathway. For melatonin biosynthesis, serotonin *N*-acetyltransferase is believed to be the rate-limiting enzyme

which have been published mostly after 1990, including studies on the structural characterization and regulation of the TPH gene.

B. Structure

Recently, isolation and characterization of full-length complementary DNA for TPH led to the determination of TPH structure. Full-length cDNA for rabbit TPH was cloned from the rabbit pineal gland cDNA library, and its amino acid sequence is found to be highly homologous to those of tyrosine hydroxylase and phenylalanine hydroxylase (GRENETT et al. 1987). Two-thirds of the C-terminal of these proteins constituted the enzyme activity core, while the N-terminal third constituted domains for substrate specificity. Based on these results, three enzymes were identified as belonging to a gene family of aromatic L-amino acid hydroxylases (GRENETT et al. 1987). In *Drosophila*, only two aromatic L-amino acid hydroxylase genes exist: one encoding tyrosine hydroxylase, and the other encoding both TPH and phenylalanine hydroxylase (NECKAMEYER and WHITE 1992).

Rat TPH cDNA was cloned from the rat pineal cDNA library, from which a complete amino acid sequence was deduced (DARMON et al. 1988; DELORT et al. 1989). Cloning and characterization of cDNA encoding TPH from rat central serotoninergic neurons revealed the identical sequence to that of pineal TPH (KIM et al. 1991). Mouse TPH cDNA was cloned from P815 mouse mastocytoma cell cDNA library (STOLL et al. 1990). A complete coding sequence of human TPH was also identified (BOULARAND et al. 1990). The results of the human TPH cDNAs cloning and anchored polymerase chain reaction revealed diversity of the human TPH mRNA, which is restricted to their 5'-untranslated region. In this study, they identified four human TPH mRNA species which were transcribed from a single transcriptional initiation site. Their diversity resulted from differential splicing of three intron-like regions and of three exons located in the 5'-untranslated region (BOULARAND et al. 1995a). Diversity in rat TPH mRNA has also been reported: two rat TPH cDNA clones different at the 3'-untranslated regions (DARMON et al. 1988) and two others different in their 5'-untranslated regions (DELORT et al. 1989) were isolated from a rat pineal gland library. A further mRNA present in the rat pineal gland as well as the raphe nuclei has been identified whose expression is under differential control in the two tissues (DUMAS et al. 1989). Discrepancies in the ratios between TPH mRNA and TPH protein in these two tissues have been reported (DUMAS et al. 1989; KIM et al. 1991).

Human TPH was assigned to chromosome 11p (LEDLEY et al. 1987), and was further defined to 11p15.3–p14 by in situ hybridization (CRAIG et al. 1991). The mouse TPH locus was mapped by Southern blot analysis of somatic cell hybrids and by an interspecific backcross to a position in the proximal half of chromosome 7 (STOLL et al. 1990), and the homology between mouse chromosome 7 and human chromosome 11 was further defined (STOLL et al. 1990; STUBBS et al. 1994).

C. Neurochemistry

As mentioned above, TPH is a member of the aromatic L-amino acid hydroxylase gene family, which includes tyrosine hydroxylase and phenylalanine hydroxylase. TPH has proven difficult to purify due to its low abundance in the brain (TONG and KAUFMAN 1975; JOH et al. 1975), and the enzyme has remained the least characterized of all members of this enzyme gene family. Since full-length TPH cDNAs from various species were cloned and characterized (GRENETT et al. 1987; DARMON et al. 1988; BOULARAND et al. 1990; STOLL et al. 1990; HART et al. 1991; KIM et al. 1991), attempts have been made to express and purify it from *E. coli* (Tipper et al. 1994; Park et al. 1994a). For structural analysis and determination of enzyme active center regions, deletion and mutation of human TPH cDNA sequences and expression in *E. coli* were attempted (Yang and Kaufman 1994). The results showed that removal of the N-terminal 164 amino acids completely abolished TPH activity. Removal of the 91 amino acids from the N-terminal resulted in a sevenfold reduction in specific activity. While the removal of 36, 55, or 112 amino acids from the C-terminus abolished the activity, deletion of 19 residues decreased the specific activity by 11-fold (Yang and Kaufman 1994). It was concluded that TPH consists of an N-terminal regulatory domain, a catalytic core, and a small C-terminal region of uncertain but important function (Yang and Kaufman 1994).

Regulation of TPH activity by protein phosphorylation has been studied by various laboratories in the past. Activation of TPH by cAMP-dependent protein kinases was observed in brain slices and synaptosomal fractions, but not in soluble extracts (Kuhn et al. 1978; Boadle-Biber 1979; Kuhn and Lovenberg 1982; Sawada et al. 1985; Garber and Makman 1987; Ehret et al. 1989; Osborne and Barnett 1989; Ehret et al. 1991; Foguet et al. 1993). Recombinant rabbit TPH was found to be phosphorylated by cAMP-dependent protein kinase, suggesting TPH activation by phosphorylation (Vrana et al. 1994). In vitro, Ca^{2+}/calmodulin-dependent protein kinase incorporates phosphate into the TPH molecule, leading to increased TPH activity (Hamon et al. 1978; Kuhn et al. 1978; Yamaguchi et al. 1979; Ehret et al.

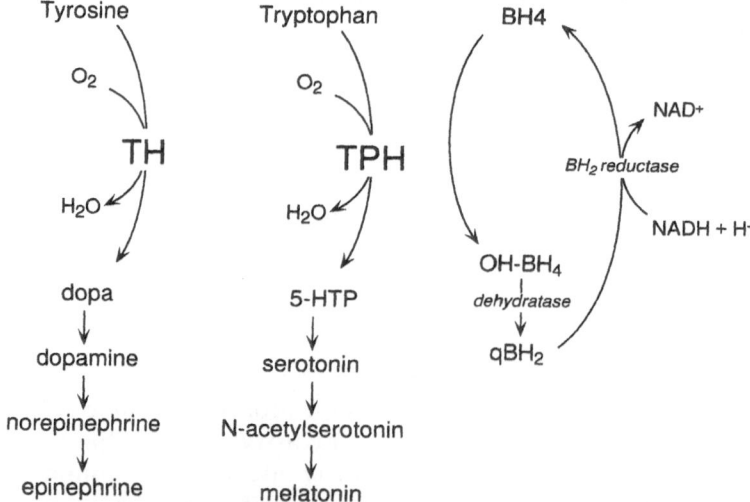

Fig. 2. Tetrahydrobiopterin requirement for tryptophan hydroxylase reaction. Tetrahydrobiopterin is required as an essential cofactor of all aromatic L-amino acid hydroxylases, which include tyrosine hydroxylase, phenylalanine hydroxylase, and tryptophan hydroxylase. *TH*, tyrosine hydroxylase; *TPH*, tryptophan hydroxylase; *BH₄*, tetrahydrobiopterin; *qBH₂*, quinoid form of biopterin

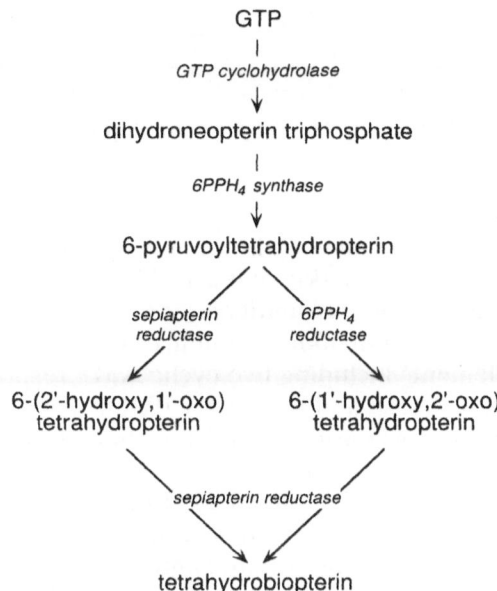

Fig. 3. Biosynthetic pathway of tetrahydrobiopterin. The first enzyme, GTP-cyclohydrolase, is the rate-limiting enzyme in this pathway

1989). However, no information is available with regard to the site(s) of TPH phosphorylation by these protein kinases.

These aromatic l-amino acid hydroxylases are tetrahydrobiopterin (BH4)-dependent enzymes, where BH_4 acts as an essential proton donor for the catalytic activity (Fig. 2). Thus, 5-HT biosynthesis in vivo requires BH_4 and its biosynthetic mechanisms in addition to catalytic activities of TPH and aromatic L-amino acid decarboxylase (AADC), the second enzyme in the biosynthetic pathway. BH_4, after giving two protons to the hydroxylation reaction, is converted to the quinonoid form of biopterin (qBH_2) via 4α-OH BH_4 (Fig. 3). qBH_2 returns to BH_4 by NADH-dependent catalytic reaction of dihydropteridin reductase (Fig. 2). Furthermore, recent results of immunohistochemistry (NAGATSU et al. 1996) and in situ hybridization (LENTZ and KAPATOS 1996; HWANG et al. 1996) of GTP-cyclohydrolase (GTPCH), the first enzyme in the biosynthetic pathway for BH_4 (Fig . 3), showed that it is specifically localized in monoaminergic neurons of the central nervous system and is heavily localized in central serotoninergic neurons.

D. Gene Expression and Regulation

Genomic DNA encoding TPH was screened from a BALB/c murine genomic library (STOLL and GOLDMAN 1991). Restriction mapping and sequence analysis of identified clones revealed that the gene contains 11 exons and covers a

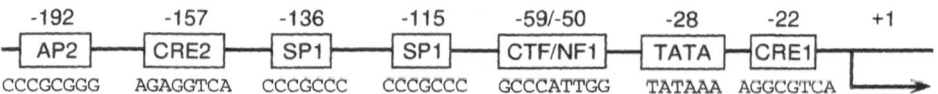

Fig. 4. *Cis*-acting DNA motifs in the 5′-upstream sequence of the rat tryptophan hydroxylase gene

region of DNA of approximately 21 kb. Total length of the TPH gene is suspected to be approximately 40 kb, and a 12-kb upstream and 14-kb downstream segment of the gene were identified (STOLL and GOLDMAN 1991). The nucleotide sequence of the 5′-flanking sequences of mouse TPH showed several important *cis*-elements including two cyclic AMP response elements and SP1 as shown in Fig. 4 (STOLL and GOLDMAN 1991).

The human TPH gene was also isolated and characterized (BOULARAND et al. 1995a). It spans a region of 29 kb containing at least 11 exons, and the variably spliced boundaries of the human TPH gene are very similar to those of human tyrosine hydroxylase and phenylalanine hydroxylase. The promoter region of the human TPH gene was isolated and characterized (BOULARAND et al. 1995b), and transient transfection assays using pinealocyte culture showed that the promoter region from −73 to +2 is sufficient to direct cAMP-dependent transcription, although it does not contain a motif exhibiting a significant identity to the CRE- or AP-2-binding site. Together with gel mobility shift analysis, it was suggested that an inverted CCAAT box, located between −73 and −51, is essential for cAMP inducibility of human TPH promoter (BOULARAND et al. 1995b). In similar analysis using P815-HTR mastocytoma cells, REED et al. (1995) demonstrated that sequences between nucleotides −343 and −21 were responsible for the mouse TPH transcriptional activation. DNase I footprint analysis using nuclear protein extracted from these cells revealed a protein-DNA interaction between nucleotides −77 and −46. It was further suggested that the transcription factor NF-Y binds to a GGCCAAT motif in the TPH proximal promoter and activates transcription (REED et al. 1995). However, preliminary reports indicate that one of the CRE (CRE1) in the mouse TPH gene upstream (Fig. 4) is responsible for the cAMP-dependent induction of the TPH gene (CHO et al. 1993).

Tryptophan hydroxylase is expressed in a tissue-specific manner in transgenic mice (HUH et al. 1994). Although the 6.1-kb 5′-upstream region of the mouse TPH gene directs expression of lacZ to major serotoninergic brain regions and pineal gland in transgenic mice (HUH et al. 1994), some ectopic expression of TPH using this gene construct was observed. While the 6.1-kb upstream region showed tissue-specific and high expression in the pineal gland (HUH et al. 1994; CHO et al. 1996), a smaller 5′-flanking sequence of 1.1 kb directed no detectable serotoninergic tissue-specificity. These results suggest firstly that DNA elements critical for serotoninergic tissue-specific expression reside between −6.1 kb and −1.1 kb of the 5′-flanking region of the mouse TPH gene, and secondly that this region confers a restricted tissue-specific expression (HUH et al. 1994).

E. Immortalized Serotoninergic Cell Lines

In order to study regulation of gene expression of TPH and to elucidate intracellular mechanisms underlying TPH regulation, immortalized serotonin-producing cell lines are essential. Although cloning of immortalized cell lines from naturally occurring cancer cells is a common method, recent technological advances allow us to produce immortalized cells through manipulation of retroviral oncogenes with dissociated embryonic neuronal cells. Two such studies in the field of serotonin research have been published (WHITE et al. 1994; SON et al. 1996).

A retrovirus encoding the temperature-sensitive mutant of SV40 T antigen was used to infect dissociated rat medullary raphe cells at embryonic day 13, and a neuronal cell line, RN46A, was cloned by serial dilution (WHITE et al. 1994). These cells divide at 33°C and cease dividing under differentiation conditions (39°C), taking on a neuronal morphology. At 39°C, they express low levels of TPH immunoreactivity and enhanced levels of neurospecific enolase as well as all molecular forms of neurofilament. The effects of various neurotrophic factors, KCl and forskolin on TPH immunoreactivity were examined (WHITE et al. 1994). This cell line was further characterized (EATON et al. 1995), and various neurotrophic factor receptors were identified. 5-HT synthesis in this cell line requires initial treatment with brain-derived neurotrophic factor (BDNF), followed by growth under partial membrane-depolarizing conditions. Several data with differentiation conditions or with BDNF under depolarizing conditions indicate that distinct aspects of 5-HT metabolism are differentially regulated during development. The authors suggested that 5-HT may function as a developmental signal in an autocrine loop during early serotoninergic differentiation (EATON et al. 1995).

SON et al. (1996) used transgenic technology to produce immortalized cell line from neuroendocrine pinealocytes which produce 5-HT and melatonin. They established transgenic mice carrying a construct consisting of a 6.1-kb 5'-upstream region of the mouse TPH gene, which is known to direct the restricted expression of lacZ reporter gene to the pineal gland (HUH et al. 1994) fused to the SV40 T-antigen. The animals developed highly invasive pineal tumors which were utilized to establish the immortalized pinealocyte-derived cell line. The cells expressed both TPH and serotonin N-acetyltransferase (SON et al. 1996), and thus this cell line may be a valuable in vitro model system for the study of the rhythmic nature of pineal function as well as TPH gene regulation.

F. TPH Regulation at mRNA Levels

As mentioned above, studies of TPH gene regulation at transcription and translation levels have not progressed as rapidly as those of other monoamine enzyme genes. Obvious reasons for this have been the low abundance of steady-state levels of TPH mRNA in the central nervous system for in vivo studies, as well as the lack of proper immortalized cell lines for in vitro studies.

Some pharmacological studies on drug effects on TPH activity are available which are relevant to the physiological significance of 5-HT. Therefore, in this section, some recent publications on TPH gene regulation at mRNA levels are summarized.

The steady-state levels of TPH mRNA are influenced in vivo by levels of 5-HT (BIGUET et al. 1986; FOSTER et al. 1991; WESSEL and JOH 1992; CORTES et al. 1993; LI et al. 1993; PARK et al. 1993a,b, 1994b), although high concentrations (up to $5 \times 10^{-4} M$) of 5-HT did not inhibit in vitro TPH activity (YOUDIM et al. 1975; LOVENBERG 1977; PARK et al. 1994b). p-Chlorophenylalanine (PCPA) is an irreversible inhibitor of TPH that potently and specifically depletes brain 5-HT (KOE and WEISSMAN 1966; GAL et al. 1970; LOVENBERG et al. 1973). A single dose of PCPA (300 mg/kg. i.p.) almost completely depletes brain 5-HT and TPH activity, which slowly returns to normal after 2 weeks (LOVENBERG et al. 1973; PARK et al. 1989; 1994b). Recent studies (CORTES et al. 1993; PARK et al. 1994b) suggest that TPH mRNA levels are at their highest when 5-HT and TPH protein levels are at their lowest. The data suggest that PCPA produces a decrease in 5-HT concentration, which then causes upregulation of TPH gene transcription. In contrast, when a monoamine oxidase inhibitor is administered, 5-HT levels increase in the 5-HT nerve terminals, and the conversion of ^{3}H-tryptophan to ^{3}H-5-HT markedly decreases, implying a decrease in TPH activity (MACON et al. 1971; CARLSSON and LINDQVIST 1972; HAMON et al. 1972). However, this study was done at the enzyme activity level and not at the transcriptional level. Since high concentrations of 5-HT do not inhibit in vitro TPH activity, these results strongly suggest that end-product inhibition may regulate TPH gene transcription, although nuclear run-on assay or transient transfection analysis in cell lines may be necessary to prove this hypothesis (CORTES et al. 1993; PARK et al. 1989; 1994b). However, BENDOTTI et al. (1993) demonstrated that repeated high doses of d-fenfluramine (10 mg/kg, i.p. twice daily for 4 days), which markedly reduce 5-HT concentrations in the hippocampus and striatum of rat brain up to 1 month, increased TPH mRNA levels in the nucleus raphe dorsalis. These data are also in agreement with the suggestion that end-product inhibition regulates TPH gene transcription. In this report, the authors also examined the effects of 5,7-dihydroxytryptamine (5,7-DHT) and demonstrated that intracerebroventricular injection of 5,7-DHT (150 μg/20 μl) induced a marked and long-lasting reduction of 5-HT and TPH in the hippocampus and striatum of rats (BENDOTTI et al. 1993). Thirty days after injection, 5,7-DHT markedly reduced the number of labeled neurons in the dorsal and ventral regions of the nucleus raphe dorsalis and raised the levels of TPH mRNA in the spared neurons at all times examined, suggesting that 5,7-DHT damages 5-HT nerve terminals and perikarya (BENDOTTI et al. 1993). Similar results have also been reported from use of in situ hybridization and immunocytochemical staining of neonatal animals after 5,7-DHT treatment. PARK et al. (1993) showed that intracisternal (i.c.) injection of 5,7-DHT in neonatal rats (24 μg free base) caused a marked reduction in TPH and aromatic L-amino acid decarboxylase (AADC) message levels as well as the number of 5-HT and AADC-immunoreactive cells within the dorsal raphe nucleus as early as 1 week after the injection. Even 15 weeks after drug administration, recovery did not occur. In 1975, BAUMGARTEN et al. demonstrated long-lasting reductions of tryptophan hydroxylase activity in various brain regions following i.c. administration of 5,7-DHT in newborn rats. These studies indicate that neonatal treatment with 5,7-DHT produces a marked and permanent reduction in the number of central serotoninergic neurons and that the neuronal loss is region-specific (PARK et al. 1993).

Central serotoninergic systems may mediate stress-induced stimulation of the hypothalamic-pituitary-adrenal (HPA) axis and, reciprocally, ACTH and glucocorticoids may affect central serotoninergic systems (CHAOULOFF 1993). However, the presently available data are controversial and do not clearly define the relationship between stress hormones and the central serotoninergic systems (AZMITIA and MCEWEN 1969; LOVENBERG et al. 1973; NECKERS and SZE 1975; AZMITIA and MCEWEN 1976; SZE et al. 1976; AZMITIA et al. 1993). To date, there have been no studies addressing the efficacy of glucocorticoid treatment and the role of ACTH in TPH mRNA regulation. Preliminary results in the author's laboratory indicate that TPH activity and mRNA in the rat dorsal raphe are significantly increased 10 days after hypophysectomy (PARK and JOH 1996, data not published). However, it is not known whether ACTH acts directly or indirectly through glucocorticoids on TPH activity. AZMITIA et al. (1993), using specific TPH antibodies, demonstrated that 49kDa immunoreactive protein was greater (50%–75%) in immunoblots of midbrain raphe samples from adrenalectomized rats given dexamethasone in their drinking water (10mg/l) for 12–96h. Quantification of immunostaining per cell soma indicated that amount of antibody immunoreactivity in neurons from both the lateral wing subdivision of the dorsal raphe nucleus and in the supralemniscal nucleus, B-9, was 80% higher in the adrenalectomized-dexamethasone-treated than in adrenalectomized animals (AZMITIA et al. 1993). They suggested that a part of the increase in TPH staining may be a consequence of cellular hypertrophy due to dexamethasone treatment of the adrenalectomized rats.

It is known that TPH activity in the pineal gland depends on the activity of innervating noradrenergic input from the superior cervical ganglia, which varies during the day/night cycle (KRAUSE and DIBOCOVICH 1990). Steady-state levels of TPH mRNA also vary with the light/dark cycle (JOH et al. 1996, preliminary results, data not published). Furthermore, any significant physiological activities on the superior cervical ganglia may increase TPH gene regulation in the pineal gland.

G. Conclusion

NIELSEN et al. (1994) reported that in some individuals a genetic variant of the TPH gene may influence 5-hydroxyindoleacetic acid (5-HIAA) concentration in the cerebrospinal fluid and predisposition of suicidal behavior. This report may represent the first step toward an understanding of the linkage between the TPH gene and abnormal behaviors controlled by 5-HT. More studies are required to substantiate the potential relationships between TPH polymorphism and abnormal behavior. We are now in an early stage of understanding the importance of the genetic control of 5-HT actions in animal brains. There are various disadvantages in studying the TPH system, such as the low abundance of TPH in the brain, the lack of proper immortalized cell lines, and the complexity of 5-HT action related to other monoamines. However, progress made within the past several years of TPH gene research is apparent, and the author believes that there will be a great advance in this field in the coming years.

The author acknowledges the incomplete compilation and citation of the relevant publications, especially those which deal with the various compounds affecting TPH activity.

References

Azmitia EC, McEwen BS (1969) Corticosterone regulation of tryptophan hydroxylase in midbrain of the rat. Science 166:1274–1276

Azmitia EC, McEwen BS (1976) Early response of rat brain tryptophan hydroxylase activity to cycloheximide, puromycin and corticosterone. J Neurochem 27:773–778

Azmitia EC, Liao B, Chen YS (1993) Increase of tryptophan hydroxylase enzyme protein by dexamethasone in adrenalectomized rat midbrain. J Neurosci 13:5041–5055

Baumgarten HG, Victor SJ, Lovenberg W (1975) A developmental study of the effects of 5,7-dihydroxytryptamine on regional tryptophan hydroxylase in rat brain. Psychopharmacol Commun 1:75–88

Bendotti C, Baldessari S, Ehret M, Tarizzo G, Samanin R (1993) Effect of d-fenfluramine and 5,7-dihydroxytryptamine on the levels of tryptophan hydroxylase and its mRNA in rat brain. Brain Res Mol Brain Res 19:257–261

Biguet NF, Buda M, Lamouroux A, Samolyck D, Mallet J (1986) Time course of the changes of TH mRNA in rat brain and adrenal medulla after single injection of reserpine. EMBO J 5:287–291

Boadle-Biber MC (1979) Activation of tryptophan hydroxylase from slices of rat brain stem incubated with N^6,O^2-dibutyryl adenosine-3′:5′-cyclic monophosphate. Biochem Pharmacol 29:669–672

Boularand S, Darmon MC, Yanem Y, Launay JM, Mallet J (1990) Complete coding sequence of human tryptophan hydroxylase. Nucleic Acids Res 18:4257

Boularand S, Darmon MC, Ravassard P, Mallet J (1995a) Characterization of the human tryptophan hydroxylase gene promoter. Transcriptional regulation by cAMP requires a new motif distinct from the cAMP-responsive element. J Biol Chem 270:3757–3764

Boularand S, Darmon MC, Mallet J (1995b) The human tryptophan hydroxylase gene. An unusual splicing complexity in the 5′-untranslated region. J Biol Chem 270:3748–3756

Carlsson A, Lindqvist M (1972) The effect of L-tryptophan and some psychotrophic drugs on the formation of 5-hydroxytryptophan in the mouse in vivo. J Neural Transm 33:23–43

Chaouloff F (1993) Physiopharmacological interactions between stress hormones and central serotonergic systems. Brain Res Rev 18:1–32

Cho SH, Huh SO, Park DH, Joh TH, Son JH (1993) Mouse tryptophan hydroxylase (TPH) promoter regions mediating basal and regulated expression of TPH gene in vitro. Neurosci Abst 19:35.5

Cho SH, Son JH, Park DH, Aoki C, Song X, Smith G, Joh TH (1996) Reduced sympathetic innervation after alteration of target cell neurotransmitter phenotype in transgenic mice. Proc Natl Acad Sci USA 93:2862–2866

Cortes R, Mengod G, Celada P, Artigas F (1993) p-Chlorophenylalanine increases tryptophan-5-hydroxylase mRNA levels in the rat dorsal raphe: a time course study using in situ hybridization. J Neurochem 60:661–764

Craig SP, Boularand S, Darmon MC, Mallet J, Craig IW (1991) Localization of human tryptophan hydroxylase (TPH) to chromosome 11p15.3–p14 by in situ hybridization. Cytogenet Cell Genet 56:157–159

Darmon MC, Guibert B, Leviel V, Ehret M, Maitre M, Mallet J (1988) Sequence of two mRNAs encoding active rat tryptophan hydroxylase. J Neurochem 51:312–316

Delort J, Dunas JR, Mallet J (1989) An efficient strategy for cloning 5' extremities of rare transcripts permits isolation of multiple 5'-untranslated regions of rat tryptophan hydroxylase mRNA. Nucleic Acid Res 17:6439–6448

Dumas S, Darmon MC, Delort J, Mallet J (1989) Differential control of tryptophan hydroxylase expression in raphe and in pineal gland: evidence for a role of translation efficiency. J Neurosci Res 24:537–547

Eaton MJ, Staley JK, Globus MY, Whittemore SR (1995) Developmental regulation of early serotonergic neuronal differentiation: the role of brain-derived neurotrophic factor and membrane depolarization. Dev Biol 170:169–182

Ehret M, Cash CD, Hamon M, Maitre M (1989) Formal demonstration of the phosphorylation of rat brain tryptophan hydroxylase by Ca^{2+}/calmodulin-dependent protein kinase. J Neurochem 52:1886–1891

Ehret M, Pevet P, Maitre M (1991) Tryptophan hydroxylase synthesis is induced by 3',5'-cyclic adenosine monophosphate during circadian rhythm in the rat pineal gland. J Neurochem 57:1516–1521

Foguet M, Harricker JA, Schmuck K, Leubbert H (1993) Long-term regulation of serotonergic activitiy in the rat brain via activation of protein kinase A. EMBO J 12:903–910

Foster OJF, Biswas S, Lightman SL (1991) Neuropeptide Y and tyrosine hydroxylase mRNA levels in the locus ceruleus show similar increases after reserpine treaament. Neuropeptides 18:137–141

Gal EM, Roggeveen AE, Millard SA (1970) D,L-[2-^{14}C] p-chlorophenylalanine as an inhibitor of tryptophan 5-hydroxylase. J Neurochem 17:1221–1235

Garber SL, Makman MH (1987) Regulation of tryptophan hydroxylase activity by a cyclic AMP-dependent mechanism in rat striatum. Mol Brain Res 3:1–10

Grahame-Smithe DG (1964) Tryptophan hydroxylation in brain. Biochem Biophys Res Comm 16:586–592

Grenett HE, Ledley FD, Reed LL, Woo SLC (1987) Full-length cDNA for rabbit tryptophan hydroxylase: functional domains and evolution of aromatic amino acid hydroxylases. Proc Natl Acad Sci USA 84:5530–5534

Hamon M, Bourgoin S, Morot-Yaudry Y, Glowinski J (1972) End product inhibition of serotonin synthesis in the rat striatum. Nature 237:184–187

Hamon M, Bourgoin S, Hery F, Simonnet G (1978) Activation of tryptophan hydroxylase by adenosine triphosphate, magnesium and calcium. Mol Pharmacol 14:99–110

Hart RP, Yang R, Riley LA, Green TL (1991) Post-transcriptional control of tryptophan hydroxylase gene expression in rat brain stem and pineal gland. Mol Cell Neurosci 2:71–77

Huh SO, Park DH, Cho JY, Joh TH, Son JH (1994) A 6.1 kb 5′ upstream region of the mouse tryptophan hydroxylase gene directs expression of E. coli lacZ to major serotonergic brain regions and pineal gland in transgenic mice. Mol Brain Res 24:145–152

Hwang O, Baker H, Gross S, Joh TH (1996) Localization of GTP cyclohydrolase in monoaminergic but not nitric oxide producing cells. (submitted for publication)

Jequire E, Robinson DS, Lovenberg W, Sjoerdsma A (1969) Further studies on tryptophan hydroxylase gene. Biochem Pharmacol 18:1071–1081

Joh TH, Shikimi T, Pickel VM, Reis DJ (1975) Brain tryptophan hydroxylase: purification of, production of antibodies to, and cellular and ultrastructural localization in serotonergic neurons of rat midbrain. Proc Natl Acad Sci USA 72:3575–3579

Kim KS, Wessel TC, Stone DM, Carver CH, Joh TH, Park DH (1991) Molecular cloning and characterization of cDNA encoding tryptophan hydroxylase from rat central serotonergic neurons. Mol Brain Res 9:277–283

Koe BK, Weissman A (1966) p-Chlorophenylalanine: a specific depletor of brain serotonin. J Pharmacol Exp Ther 154:499–516

Krause DN, Dubocovich ML (1990) Regulatory sites in the melatonin system of mammals. Trends Neurosci 13:464–470

Kuhn DM, Lovenberg W (1982) Role of calmodulin in the activation of tryptophan hydroxylase. Fed Proc 41:2258–2264

Kuhn DM, Vogel RL, Lovenberg W (1978) Calcium-dependent activation of tryptophan hydroxylase by ATP and magnesium. Biochem Biophys Res Commun 82:759–766

Ledley FD, van Tunen P, Ledbetter D, Gerhardt T, Jones C, Woo SLC (1987) Assignment of tryptophan hydroxylase to human chromosome 11p: duplication and rearrangement in aromatic amino acid hydroxylase evolution. Biochem 24:3389–3394

Lentz SI, Kapatos G (1996) Tetrahydrobiopterin biosynthesis in the rat brain: heterogeneity of GTP cyclohydrolase I mRNA expression in monoamine-containing neurons. Neurochem Intern 28:569–582

Li XM, Juorio AV, Boulton AA (1993) NSD-1015 alters the gene expression of aromatic L-amino acid decarboxylase in rat PC12 pheochromocytoma cells. Neurochem Res 18:915–919

Lovenberg W (1977) The enzymology of tryptophan hydroxylase. In: Usdin E, Weiner N, Yudim MBH (eds) Structure and function of monoamine enzyme. Marcel Dekker, New York, p 43

Lovenberg W, Jequier E, Sjoerdsma A (1967) Tryptophan hydroxylase: measurement in pineal gland, brainstem and carcinoid tumor. Science 155:217–219

Lovenberg W, Besselaar GH, Bensinger RE, Jackson RL (1973) Physiologic and drug-induced regulation of serotonin synthesis. In: Barchas J, Usdin E (eds) Serotonin and behavior. Academic, New York, pp 49–59

Macon JB, Sokoloff L, Glowinski J (1971) Feedback control of rat brain 5-hydroxytryptamine synthesis. J Neurochem 18:323–331

Nagatsu I, Ichinose H, Sakai M, Titani K, Suzuki M (1996) Immunocytochemical localization of GTP cyclohydrolase I in the brain, adrenal gland and liver of mice. J Neural Transm 102:175–188

Neckameyer WS, White K (1992) A single locus encodes both phenylalanine hydroxylase and tryptophan hydroxylase activities in Drosophila. J Biol Chem 267:4199–4206

Neckers L, Sze PY (1975) Regulation of 5-hydroxytryptamine metabolism in mouse brain by adrenal glucocorticoids. Brain Res 93:123–132

Nielsen DA, Goldman D, Virkkunen M, Tokola R, Rawlings R, Linnoila M (1994) Suicidality and 5-hydroxyindoleacetic acid concentration associated with a tryptophan hydroxylase polymorphism. Arch Gen Psychiatry 51:34–38

Osborne NN, Barnett NL (1989) Serotonin levels in the rabbit retina are elevated following intraocular injection of forskolin. J Neurochem 53:1955–1958

Park DH, Paivarinta H, Joh TH (1989) Tryptophan hydroxylase activity in hypothalamus and brainstem of neonatal and adult rats treated with hydrocortisone or parachlorophenylalanine. Neurosci Res 7:76–80

Park DH, Stone DM, Baker H, Wessel TC, Kim KS, Towle AC, Joh TH (1993a) Changes in activity and mRNA for rat tryptophan hydroxylase and aromatic L-amino acid decarboxylase of brain serotonergic cell bodies and terminals following neonatal 5,7-dihydroxytryptamine. Brain Res 609:59–66

Park DH, Wessel TC, Joh TH (1993b) Acute effects of reserpine on tryptophan hydroxylase activity and mRNA in rat brain. Brain Res 620:331–334

Park DH, Stone DM, Kim KS, Joh TH (1994a) Characterization of recombinant mouse tryptophan hydroxylase expressed in Escherichia coli. Mol Cell Neurosci 5:87–93

Park DH, Stone DM, Baker H, Kim KS, Joh TH (1994b) Early induction of rat brain tryptophan hydroxylase (TPH) mRNA following parachlorophenylalanine (PCPA) treatment. Mol Brain Res 22:20–28

Reed GE, Kirchner JE, Carr LG (1995) NF-Y activates mouse tryptophan hydroxylase transcription. Brain Res 682:1–12

Sawada M, Kanamori T, Hayakawa T, Nagatsu T (1985) Changes in tryptophan hydroxylase and cyclic AMP-dependent and calcium-calmodulin dependent protein kinases in raphe serotonergic neurons of 5,7-dihydroxytryptamine treated rats. Neurochem Intern 7:761–763

Son JH, Chung JH, Huh SO, Park DH, Peng C, Rosenblum MG, Chung YI, Joh TH (1996) Immortalization of neuroendocrine pinealocytes from transgenic mice by targeted tumorigenesis using the tryptophan hydroxylase promoter. Mol Brain Res 37:32–40

Stoll J, Goldman D (1991) Isolation and structural characterization of the murine tryptophan hydroxylase gene. J Neurosci 28:457–465

Stoll J, Kozak CA, Goldman D (1990) Characterization and chromosomal mapping of a cDNA encoding tryptophan hydroxylase from mouse mastocytoma cell line. Genomics 7:88–96

Stubbs L, Rinchik EM, Goldberg E, Rudy B, Handel MA, Johnson D (1994) Clustering of six human 11p15 gene homologs within a 500-kb interval of proximal mouse chromosome 7. Genomics 24:324–332

Sze PY, Neckers L, Towle AC (1976) Glucocorticoids as a regulatory factor for brain tryptophan hydroxylase. J Neurochem 26:169–173

Tipper JP, Citron BA, Ribeiro P, Kaufman S (1994) Cloning and expression of rabbit and human tryptophan hydroxylase cDNA in Escherichia coli. Arch Biochem Biophys 315:445–453

Tong JH, Kaufman S (1975) Tryptophan hydroxylase: purification and some properties of the enzyme from rabbit hindbrain. J Biol Chem 250:4152–4158

Vrana KE, Rucker PJ, Kumer SC (1994) Recombinant rabbit tryptophan hydroxylase is a substrate for cAMP-dependent protein kinase. Life Sci 55:1045–1052

Wessel TC, Joh TH (1992) Parallel upregulation of catecholamine-synthesizing enzymes in rat brain and adrenal gland: effects of reserpine and correlation with immediate early gene expression. Mol Brain Res 15:349–360

White LA, Eaton MJ, Castro MC, Klose KJ, Globus MY, Shaw G, Whittemore SR (1994) Distinct regulatory pathways control neurofilament expression and neurotransmitter synthesis in immortalized serotonergic neurons. J Neurosci 14:6744–6753

Yamaguchi T, Fujisawas H (1979) Regulation of rat brainstem tryptophan 5-monooxygenase. Calcium dependent reversible activation by ATP and magnesium. Arch Biochem Biophys 198:219–226

Yang XJ, Kaufman S (1994) High-level expression and deletion mutagenesis of human tryptophan hydroxylase. Proc Natl Acad Sci USA 91:6659–6663

Youdim MBH, Hamon M, Bourgoin S (1975) Properties of partially purified pig brain stem tryptophan hydroxylase. J Neurochem 25:407–414

CHAPTER 5

Molecular Analysis of Serotonin Packaging into Secretory Vesicles

D. Peter, J.P. Finn III, A. Merickel, Y. Liu, and R.H. Edwards

A. Introduction

The synthesis of serotonin by mast cells and platelets as well as neurons suggests a diverse role for this transmitter in inflammation and vasomotor control as well as neurotransmission. In all of these cases, serotonin is released by regulated exocytosis from vesicular stores. The transmitter, therefore, requires packaging into vesicles for regulated release. Indeed, an active transport mechanism functions to translocate both serotonin and the other monoamines from the site of synthesis in the cytoplasm into the lumen of secretory vesicles. Vesicles containing serotonin then undergo exocytosis in response to neural activity.

Vesicular monoamine transport differs in biological role, bioenergetics and pharmacology from the neurotransmitter transport activities found at the plasma membrane. Plasma membrane transport functions to terminate the action of many neurotransmitters and uses the cotransport of sodium to drive reuptake (Kanner and Schuldiner 1987). Further, distinct plasma membrane transport activities mediate the reuptake of dopamine, norepinephrine and serotonin (Amara and Kuhar 1993). In terms of pharmacology, cocaine and many antidepressants appear to act by inhibiting this class of transporters.

The vesicular transporters, in contrast to those at the plasma membrane, translocate transmitters from the cytoplasm into vesicles in preparation for regulated release by exocytosis. In addition, a single transport activity recognizes all of the monoamine transmitters with similar affinity, occurring in dopaminergic, noradrenergic and serotoninergic neurons (Johnson 1988). The vesicular transporter also uses a proton antiport mechanism to drive the transmitter into vesicles and, in the case of monoamines, shows inhibition by reserpine rather than cocaine or antidepressants (Kanner and Schuldiner 1987; Johnson 1988).

I. Bioenergetics

Classical studies have characterized the bioenergetic mechanism responsible for vesicular monoamine transport. Using secretory granules from the bovine adrenal medulla (chromaffin granules) as an abundant source of transport activity, these studies found that a distinct vesicular protein, a vacuolar-type

H⁺-ATPase, provides the energy necessary to drive monoamine accumulation within vesicles. In this respect, transmitter packaging resembles other processes such as nutrient uptake in bacteria and ATP production in mitochondria that also rely on a proton electrochemical gradient. In bacteria, nutrient uptake involves cotransport of protons. In mitochondria, the F_1F_0 H⁺-ATPase uses an H⁺ gradient derived from an electron transport chain to synthesize ATP. The vesicular H⁺-ATPase, in contrast, uses the energy derived from ATP hydrolysis to pump protons into the lumen of secretory vesicles. If an anion such as chloride accompanies the movement of protons, this leads primarily to a pH gradient with the inside of the vesicle acidic relative to the outside. If an anion does not accompany proton translocation, this leads primarily to an electrical gradient with the inside of the vesicle positive relative to the outside. The energy stored in this proton electrochemical gradient is then used to package the neurotransmitter. Specifically, studies of vesicular monoamine transport suggest that two protons from the lumen of the vesicle are transported out of the vesicle in exchange for one cytoplasmic amine (KNOTH et al. 1981). Although both the electrical and chemical gradients are potential sources of energy, classical studies also suggest that the vesicular monoamine transporter uses primarily the pH component (JOHNSON 1988).

II. Pharmacology

Two potent inhibitors of vesicular monoamine transport, reserpine and tetrabenazine, have helped to define the physiological roles of monoamines. These inhibitors also provide invaluable tools to characterize the mechanism and regulation of monoamine transport.

In vitro, reserpine potently and rapidly inhibits vesicular amine transport expressed by platelets (BRODIE et al. 1957), chromaffin granules (KIRSHNER 1962) and synaptic vesicles (ANGELIDES 1980). In vivo, reserpine depletes both peripheral and central monoamine stores (CARLSSON 1965). In terms of physiological effects, the drug lowers blood pressure and causes a syndrome resembling depression (FRIZE 1954). In addition, it can produce a parkinsonian state characterized by bradykinesia and rigidity. Reserpine also ameliorates hyperkinetic disorders presumably due to increased dopaminergic neurotransmission (JANKOVIC and ORMAN 1988).

Previous studies of transport and drug binding have characterized the interaction of reserpine with the monoamine transporter expressed in bovine chromaffin granules. Reserpine competitively inhibits vesicular monoamine transport and substrates compete with reserpine for binding at concentrations close to their K_m for transport, suggesting that reserpine binds near the site of amine recognition (WEAVER and DEUPREE 1982; SCHERMAN and HENRY 1984). Reserpic acid, a membrane impermeant derivative of reserpine, only inhibits monoamine uptake when applied to the outside of the vesicle, suggesting that reserpine also binds to the cytoplasmic face of the protein (CHAPLIN et al. 1985). In addition, the binding of reserpine is stimulated by the imposition of a proton electrochemical gradient across the vesicle membrane (RUDNICK et al.1990). In contrast to the two protons required for transport, reserpine binding appears to require a single proton (RUDNICK et al. 1990). Taken together, the results suggest that reserpine inhibits one part of the transport cycle and in particular the step where translocation of a proton induces the reorientation of the substrate recognition site to the cytoplasmic face of the membrane. Drug-binding studies have also indicated that reserpine dissociates extremely slowly from the transport protein ($T_{1/2} > 16$ h) (RUDNICK et al. 1990), enabling its use to label the transporter essentially irreversibly.

The synthetic benzoquinoline tetrabenazine also inhibits vesicular monoamine transport. Clinically, tetrabenazine has proven more useful than reserpine for the treatment of hyperkinetic disorders because it affects central monoamine stores more than peripheral (CARLSSON 1965) and does not produce the postural hypotension caused by reserpine (JANKOVIC and ORMAN 1988). In vitro, tetrabenazine inhibits the uptake of monoamines into chromaffin granules. However, in contrast to reserpine, competition studies demonstrate that amine substrates inhibit tetrabenazine binding only poorly (SCHERMAN and HENRY 1984). Thus, tetrabenazine is considered to bind at a site distinct from reserpine and the usual monoamine substrates. In addition, tetrabenazine binding is not accelerated by a pH gradient (SCHERMAN and HENRY 1984). Competition studies demonstrate, however, that a relationship exists between the sites for reserpine and tetrabenazine binding. Tetrabenazine inhibits reserpine binding to the transporter, and reserpine inhibits tetrabenazine binding, although at concentrations of reserpine much higher than those needed to inhibit transport (SCHERMAN and HENRY 1984). In addition, brief pretreatment with tetrabenazine in vivo can protect against the long-term effects of reserpine, presumably by preventing the essentially irreversible binding of reserpine to the transport protein (CARLSSON 1965).

These classic pharmacological observations suggest a model of the transport cycle in which the translocation of a proton out of the vesicle reorients the substrate recognition site to the cytoplasmic face of the membrane where it can bind either substrate or reserpine. Tetrabenazine has been considered to inhibit both transport and reserpine binding by preventing this conformational change (DARCHEN et al. 1989). The second proton not required for reserpine binding presumably translocates either simultaneously with substrate movement into the vesicle or shortly thereafter, promoting release of the substrate into the lumen of the vesicle.

In addition to reserpine and tetrabenazine, amphetamine and related psychostimulants interfere with the storage of monoamine transmitter. Amphetamines act by promoting the translocation of large amounts of monoamine into the synapse (DI CHIARA and IMPERATO 1988; CARBONI et al. 1989).However, rather than stimulating the fusion of secretory vesicles with the plasma membrane, amphetamines appear to induce release through nonexocytotic mechanisms. Inhibition of plasma membrane monoamine transport blocks the action of amphetamines (Fischer and Chu 1979; Liange and Rutledge 1982), indicating a role for this activity in amphetamine action and suggesting that reversal of flux by the transporter mediates the release of monoamines from the cytoplasm into the synapse (Rudnick and Wall 1992a–c; Sulzer et al. 1993). Indeed, the amphetamine derivatives 3,4-methylenedioxymethamphetamine (MDMA or ecstasy), p chloroamphetamine (PCA) and fenfluramine all interact with the plasma membrane monoamine transporters in vitro (Rudnick and Wall 1992a–c). Efflux of the entire cytoplasmic pool of monoamine into the synapse would, however, have little effect on the synaptic concentration unless the normally low cytoplasmic levels ($\sim 0.15\,mM$ for dopamine) were increased by efflux from storage vesicles that contain high concentrations of transmitter (0.3–$3\,M$). The mechanism by which amphetamines cause vesicular efflux remains unclear.

In one model for amphetamine action, the plasma membrane transporter catalyzes the exchange of extracellular amphetamine for cytoplasmic monoamine. The lipophilic nature of amphetamine then enables it to diffuse back out of the cell for another round of exchange, eventually redistributing transmitter from vesicular stores into the cytoplasm. In contrast to this exchange-diffusion model for amphetamine action, the drug may act directly on vesicular stores. In this model, amphetamines act as weak bases and alkalinize the interior of secretory vesicles. In the absence of a driving force for active transport, vesicular efflux of monoamines increases the cytoplasmic concentrations and thus enables plasma membrane transport to mediate efflux into the synapse (Sulzer and Rayport 1990). Indeed, classical studies using bovine chromaffin granules have shown that they retain preloaded transmitter for up to 1 h even after dilution into substrate-free medium that should promote redistribution out of the vesicle (Maron et al. 1983). Only dissipation of the pH gradient induces vesicular efflux. Nonetheless, the role of the vesicular transport proteins in flux reversal remains unclear. Reserpine and tetrabenazine apparently do not block efflux from chromaffin granules induced by amphetamines or by dissipation of ΔpH. Studies in primary culture also show that reserpine does not block the alkalinization of intracellular compartments by amphetamines in ventral tegmental neurons (Sulzer and Rayport 1990). Indeed, alkalinization of the vesicle interior could simply result in deprotonation of the monoamine transmitters and nonspecific leakage of the uncharged molecules across the lipid bilayer.

Amphetamine action may nonetheless involve vesicular amine transport and several observations support a direct interaction. At low micromolar concentrations, MDMA and PCA inhibit transport of monoamines into vesicles (Rudnick and Wall 1992a,b). In addition, the physiological effects of reserpine can be blocked in vivo by amphetamines (Carlsson 1965), suggesting that amphetamines may interact with the site of amine recognition. Further, MDMA and fenfluramine appear to inhibit uptake by bovine chromaffin granules by interacting directly with the transport proteins at the site of amine recognition, but they also reduce the transmembrane pH gradient and so may have indirect effects as well (SCHULDINER et al. 1993b). Indeed, PCA lacks a direct effect on the transport protein and apparently acts only through a reduction in the transmembrane pH gradient (SCHULDINER et al. 1993b).

Vesicular monoamine transport proteins may participate in the action of amphetamines through several distinct mechanisms. First, the transporter may exchange cytoplasmic amphetamine for intravesicular monoamine. Subsequent diffusion of the lipophilic drug back out of the vesicle then enables another round of exchange, leading eventually to dissipation of the pH gradient through many rounds of futile active transport. Alternatively, vesicular transport may accumulate lumenal amphetamine and, acting as a weak base, the drug may promote dissipation of the pH gradient. Monoamine efflux induced by dissipation of the pH gradient may then occur through the transporter even though reserpine and tetrabenazine have failed to inhibit it (SULZER and RAYPORT 1990). Failure of the drugs to inhibit efflux may simply result from dissipation of ΔpH, which clearly does inhibit reserpine binding. Thus, in contrast to the plasma membrane transporters which mediate the efflux of monoamines across the plasma membrane caused by amphetamine (GIROS et al. 1996), a role for the vesicular transport proteins remains less clear.

B. Biochemical Analysis and Molecular Cloning

To identify the proteins responsible for vesicular monoamine transport, investigators have previously relied on protein purification in conjunction with an assay for activity. Indeed, functional reconstitution of vesicular transport in artificial membrane vesicles has been achieved for monoamines and other classical transmitters (MAYCOX et al. 1988; CARLSON et al. 1989; HELL et al. 1990; STERN-BACH 1990; HELL et al. 1991). In crude membrane preparations from tissue, the addition of ATP suffices to activate the endogenous H^+-ATPase. In reconstituted detergent extracts, a diffusion potential or co-reconstitution with bacteriorhodopsin can supply the driving force needed for transport. However, the difficulties inherent in this assay have forced investigators to rely on other determinations such as drug binding to assay protein purification. Schuldiner and colleagues have in particular used the almost irreversible binding of ^{3}H-reserpine to follow the transport protein during its purification from bovine chromaffin granules (STERN-BACH et al. 1990). They then demonstrated the functional reconstitution of a highly purified fraction. Tetrabenazine and ketanserin have similarly been used to assay the fractions obtained during purification (ISAMBERT et al. 1992; VINCENT and NEAR 1991). However, the molecular cloning of a cDNA encoding vesicular amine transport did not depend on purification of the protein responsible.

I. Vesicular Monoamine Transport Suppresses MPTP+ Toxicity

The administration of N-methyl-1,2,3,6-tetrahydropyridine (MPTP) to humans and certain experimental animals causes the relatively selective degeneration of the same midbrain dopamine cell populations that degenerate in Parkinson's disease (PD) (Langston et al. 1983). Like PD, the MPTP syndrome also responds to treatment with L-dopa (Jenner et al. 1984). Since MPTP toxicity provides an excellent model for PD, understanding its pathogenesis may yield insight into the cause of the idiopathic disorder. After administration, MPTP readily crosses the blood-brain barrier due to its lipophilic character and is converted to the active metabolite N-methyl-4-phenylpyridinium (MPP$^+$) by monoamine oxidase B (Langston et al. 1984; Markey et al. 1984; Heikkila et al. 1984). MPP$^+$ is then recognized by plasma membrane amine transporters which accumulate the toxin. The selective expression of these transporters on monoamine cells presumably accounts for the selectivity of degeneration (Javitch et al. 1985). Once inside the cell, MPP$^+$ enters mitochondria and inhibits respiration (Krueger et al. 1990).

Several features of MPTP toxicity have proven relevant for idiopathic PD. Interestingly, the monoamine oxidase inhibitor deprenyl prevents the MPTP syndrome by blocking the formation of MPP$^+$ and also slows the rate of progression in PD (Parkinson Study Group 1989). Mitochondrial abnormalities predicted by the MPTP model have also been identified in multiple tissues from patients with PD (Schapira et al. 1990; Shoffner et al. 1991). However, an exogenous toxin responsible for idiopathic PD has not yet been found (Tanner and Langstrom 1990). In addition, plasma membrane transporters for norepinephrine and serotonin as well as dopamine mediate uptake of the toxin into multiple monoamine cell groups, leaving unexplained the relatively selective degeneration of dopamine neurons in the substantia nigra. In particular, adrenal chromaffin cells and postganglionic sympathetic neurons accumulate MPP$^+$ but do not generally die after exposure to the toxin or in idiopathic PD. Interestingly, it has been demonstrated that MPP$^+$ is a substrate for the vesicular amine transport activity expressed by bovine chromaffin granules (Daniels and Reinhard 1988; Scherman et al. 1988), suggesting one additional way to influence toxicity.

Consistent with the relative resistance of adrenal medullary cells to MPTP toxicity, the rat PC12 pheochromocytoma cell line also shows resistance to MPP$^+$. Even though it expresses a plasma membrane monoamine transport activity to accumulate the toxin and has been used as a model system to study MPTP toxicity, PC12 cells show more resistance to MPP$^+$ than Chinese hamster ovary (CHO) cells, a fibroblast cell line that lacks amine transporters. To characterize the mechanism responsible for resistance of PC12 cells to MPP$^+$, a PC12 cDNA library was transfected into the relatively MPP$^+$-sensitive CHO cells and transfectants were selected in $1\,mM$ MPP$^+$ (Liu et al. 1992a). This procedure led to the isolation of an MPP$^+$-resistant stable CHO transformant (Liu et al. 1992a). Importantly, reserpine inhibited the resistance of this clone to MPP$^+$, suggesting a role for vesicular amine transport. Further, the MPP$^+$-resistant CHO cells showed a particulate pattern of staining for dopamine after they were nonspecifically loaded with high concentrations of the transmitter and its distribution visualized by virtue of its intrinsic fluorescence. Wild-type CHO cells sensitive to MPP$^+$ and resistant cells treated with reserpine showed a more diffuse pattern of staining. These observations suggest that vesicular amine transport sequesters the toxin in secretory vesicles, away from its primary site of action in mitochondria, and so protects the cell.

II. Two Genes Encode Vesicular Amine Transport

Plasmid rescue of the cDNAs integrated into the genome of MPP$^+$-resistant stable CHO transformants yielded a single clone that alone confers resistance to MPP$^+$ (LIU et al. 1992b). Membrane vesicles prepared from the transfected cells also show robust uptake of ^{3}H-dopamine with the expected dependence on a pH gradient, appropriate affinity for substrates and pharmacology similar to that reported for bovine chromaffin granules (LIU et al. 1992b). In addition, membranes prepared from the MPP$^+$-resistant cells bind to reserpine with two distinct affinities (0.9 and 5 nM) and the imposition of a pH gradient accelerates drug binding (SCHULDINER et al. 1993a). Thus, a non-neural cell such as the CHO cell can support vesicular amine transport activity even though it lacks synaptic vesicles.

1. RNA Analysis

Northern analysis demonstrated expression of the cloned sequences in the adrenal gland but not the brain (LIU et al. 1992b). However, classical studies clearly indicate that central monoamine populations also have the capacity to store monoamines in regulated secretory vesicles. Further, these studies have suggested that a single activity mediates vesicular amine transport in both the adrenal gland and the brain. Since we could not detect expression of the original cloned sequences (initially referred to as the chromaffin granule amine transporter or CGAT, now referred to as vesicular monoamine transporter-1 or VMAT1) in the brain, we therefore considered that a distinct but closely related protein might occur in the central nervous system. Indeed, screening of a rat brainstem library with the VMAT1 cDNA as probe led to the isolation of a distinct but highly related cDNA clone (initially termed the synaptic vesicle amine transporter or SVAT, now known as VMAT2) (LIU et al. 1992b). In addition, Hoffman and colleagues isolated a cDNA identical to VMAT2 from a rat basophilic leukemia cell line while screening for plasma membrane serotonin uptake (ERICKSON et al. 1992).

In situ hybridization shows expression of VMAT2 mRNA by monoamine populations including dopaminergic cell groups of the substantia nigra and the ventral tegmental area, noradrenergic cells of the locus coeruleus, nucleus tractus solitarius and A2 and A5 cell groups as well as serotoninergic populations in the raphe (Liu et al. 1992b). These results strongly suggest that a single activity mediates vesicular amine transport in all of these central monoamine populations and that the protein responsible differs from the predominant form expressed by the adrenal gland.

Surprisingly, the sequence of a peptide derived from material purified from bovine chromaffin granules showed more similarity to rat VMAT2 than VMAT1 (Stern-Bach et al. 1992). Indeed, the sequence of the bovine cDNA also shows more similarity to VMAT2 (Howell et al. 1994). Subsequent work has demonstrated the existence of a VMAT1 homologue in the cow adrenal, indicating that both sequences occur in this tissue and suggesting that the relative proportion may vary in different species. Human cDNAs encoding VMAT2 have also been isolated (Surratt et al. 1993; Peter et al. 1993; Erickson et al. 1993) and show striking amino acid identity to the rat sequence (92.5%), with most of the divergence occurring within the large lumenal loop and the N- and C-termini. Consistent with the sequence similarity, the biochemical function of human VMAT2 resembles that of rat VMAT2 (Erickson et al. 1993).

2. Sequence Relationship to Bacterial Antibiotic Resistance Proteins

The amino acid sequences of rat VMAT1 and VMAT2 predict hydrophobic proteins with 12 transmembrane α-helical domains and a large lumenal loop between transmembrane domains (TMD) 1 and 2 (Liu et al. 1992b). The large lumenal loop contains several potential sites for N-linked glycosylation and the hydrophilic domains predicted to reside in the cytoplasm contain multiple sites for phosphorylation by a variety of kinases including cAMP-dependent protein kinase and protein kinase C. The two VMATs show a high degree of primary sequence homology, with 78% similarity and 62% identity. Most of the divergence occurs within the large lumenal loop and N- and C-termini. Interestingly, there is no significant homology between the vesicular amine transporters and any of the plasma membrane amine transporters that recognize the same transmitters as substrates and also contain 12 predicted TMDs. However, the N-terminal half of the VMATs shows remote homology to a class of bacterial transporters that includes tetracycline resistance proteins from a number of organisms, a bacterial multidrug resistance transporter and a protein conferring resistance to methylenomycin (Liu 1992b). Supporting the functional significance of the sequence similarity, the VMATs share several properties with the bacterial transporters (Liu et al. 1992b). They all act to remove toxins (MPP$^+$ or antibiotics) from the cytoplasm of the cell and they recognize a diverse group of compounds as substrate (Neyfakh et al. 1991). In addition, the VMATs and the bacterial transporters both use proton exchange to drive transport activity (Kaneko et al. 1985). Further, the drug reserpine that inhibits vesicular amine transport also inhibits the bacterial multidrug resistance transporter (Neyfakh et al. 1991). These functional as well as structural relationships suggest that the VMATs may have evolved from ancient bacterial detoxification systems, and support a role in detoxification as well as signaling.

3. Implications for Parkinson's Disease

Although no exogenous toxin such as MPTP has been identified in idiopathic PD, postmortem studies have shown evidence of oxidative stress. Free radicals such as the superoxide and hydroxyl radicals oxidize lipid and protein, damaging the cell. The source of the oxidative stress in PD is unknown, but may involve exogenous or endogenous toxins. One of the best candiates for an endogenous toxin is dopamine itself. Dopamine is highly toxic to neural cells in culture, more so than similar concentrations of MPP^+ (MICHEL and HEFTI 1990; ROSENBERG 1988). Catalase protects against dopamine toxicity, indicating that oxidation of dopamine is likely to play a role in this toxicity (ROSENBERG 1988). If dopamine itself plays a role in the oxidative stress seen in PD, then regulation of dopamine metabolism and distribution of dopamine in the cell are clearly of great importance. This indicates a possible role for the VMATs in control of the distribution of intracellular dopamine and therefore of the oxidative stress placed on dopaminergic neurons. Further, the locus coeruleus and dorsal raphe express higher levels of VMAT2 than dopamine cells in the midbrain (VANDER BORGHT et al. 1995). This may account for the relative sparing of noradrenergic and serotoninergic neurons in both MPTP toxicity and idiopathic PD.

4. Chromosomal Localization

A variety of observations have implicated monoamines in many behavioral phenomena, including sensory perception, motor control, motivation, the organization of thought, mood and consciousness as well as the control of heart rate, vascular tone and blood pressure. In particular, monoamines appear to have an important role in human psychiatric illness including depression and schizophrenia as well as neural degeneration (in Parkinson's disease). It is therefore of great interest to determine if the VMAT genes show linkage to any human neuropsychiatric disorders. Using mouse-human hybrid somatic cell lines as well as fluorescent in situ hybridization (FISH), the human gene encoding VMAT1 was localized to chromosome 8p21.3 and the human VMAT2 gene to chromosome 10q25 (PETER et al. 1993). Deletions in either of these regions are associated with rare disorders that suggest a possible involvement of the transporters, but no human disease gene has thus far been linked to these loci. More recently, direct sequencing of the VMAT2 cDNA from multiple patients with affective disorders has excluded a mutation in the protein-coding region as a cause for this condition (LESCH et al. 1994). Nonetheless, mutations in the VMATs may contribute to other human disorders.

C. Functional Properties

I. Transport

Although previous studies have suggested that all monoamine cells express a single transport activity, the identification of two distinct transport proteins suggested that they may differ in function. To address this possibility, we examined the biochemical properties of VMAT1 and VMAT2 in terms of substrate affinity and drug sensitivity, using vesicles prepared from stable CHO cell transformants resistant to MPP^+ and from monkey kidney COS cells transiently transfected with the two cDNAs (Peter et al. 1994). In both of these heterologous expression systems, the VMATs localize to endosomes that contain the vacuolar H^+-ATPase required to provide the driving force for transport (Liu et al. 1994), as do a number of other synaptic vesicle proteins when expressed in non-neural cells. Analysis of vesicular transport activity in these systems further demonstrates characteristics similar to those previously reported from bovine chromaffin granules with several unexpected differences. In terms of substrate specificity, both VMATs have the highest apparent affinity for serotonin but VMAT2 has a threefold higher affinity than VMAT1 for most monoamine substrates as well as MPP^+. In addition, VMAT2 has a strikingly higher (10- to 100-fold) affinity for histamine than VMAT1.

II. Drug Binding

In terms of pharmacology, transport catalyzed by both VMAT1 and VMAT2 shows equal sensitivity to inhibition by reserpine (Peter et al. 1994). Monoamines also inhibit the binding of ^{3}H-reserpine with a potency similar to their apparent affinity for transport, consistent with previous data using bovine chromaffin granules (Peter et al. 1994). In addition, reserpine binds to each of the VMATs with two distinct affinities and the imposition of a pH gradient accelerates binding (Schuldiner et al. 1993a). Using reserpine binding to determine the number of transporters, we have also calculated a turnover number of ~400/min for both VMATs (Peter et al. 1994). Although low relative to other transporters, this value greatly exceeds previous estimates from bovine chromaffin granules. In addition, the transport assay is performed at 29°C, so the rate at physiological temperatures is presumably much higher. In contrast to the similar potency of reserpine for both VMATs, tetrabenazine inhibits VMAT2 with tenfold greater potency than VMAT2 (Peter et al. 1994). Tetrabenazine also inhibits reserpine binding to VMAT2 and with much greater potency than to VMAT1. The inhibition of reserpine binding correlates with the inhibition of transport, suggesting that the inhibition derives from interference with substrate recognition. To determine whether the difference in sensitivity to tetrabenazine results from differences in binding or in the conformational change that results from drug binding, we measured the binding of ^{3}H-tetrabenazine and found no binding to VMAT1 despite strong binding to VMAT2. Therefore, the difference in sensitivity to tetrabenazine derives simply from a difference in drug binding. The differential sensitivity of VMAT1 and VMAT2 also helps to explain previous pharmacologic observations that whereas reserpine depletes both central and peripheral amine stores (presumably as a result of inhibiting both VMAT1 and VMAT2), tetrabenazine depletes primarily central stores (presumably as a result of selectivly inhibiting VMAT2) (CARLSSON 1965).

We have also examined the interaction of the VMATs with the atypical monoamine substrate histamine. Despite the apparent dependence on hydroxyl groups for substrate recognition, histamine lacks hydroxyl groups and still undergoes storage in the regulated secretory granules of neurons as well as mast cells. To determine if VMAT2 recognizes histamine as a substrate, we performed the transport assay using ^{3}H-histamine. Despite the absence of a hydroxyl group, histamine still undergoes specific transport by VMAT2, although at lower levels than serotonin. However, histamine does not compete with reserpine for binding to VMAT2, suggesting that histamine may interact with a distinct site (MERICKEL and Edwards 1995). Indeed, reserpine inhibits the transport of histamine only at very high concentrations if at all. Thus, histamine may interact with VMAT2 in a very different manner from most substrates, presumably because it lacks hydroxyl groups and so must rely on a different mechanism for recognition.

III. Flux Reversal

In addition to their role in regulated exocytotic release, secretory vesicles that contain neurotransmitter also exhibit efflux into the cytoplasm. Bovine chromaffin granules loaded with monoamines and then diluted into medium without transmitter retain their contents for up to 1 h despite a large concentration gradient favoring efflux (JOHNSON 1988). Since the core of aggregated protein in chromaffin granules may trap the transmitter and so prevent efflux, similar experiments have been performed with chromaffin ghosts. Chromaffin ghosts are prepared through the repeated hypo-osmotic lysis of chromaffin granules and thus lack a dense core. These ghosts also retain preloaded monoamines at prolonged intervals after dilution, demonstrating that the insoluble complex in the granule is not responsible for retention of the neurotransmitter (MARON et al. 1983). Indeed, efflux from chromaffin ghosts occurs only after disruption of the proton electrochemical gradient such as by the addition of proton ionophores, suggesting that the gradient acts as a kinetic barrier to the reversal of flux (MARON et al. 1983). However, neither reserpine nor tetrabenazine inhibit efflux from preloaded chromaffin ghosts, indicating either that the transporter does not mediate efflux or that the transporter adopts a conformation that cannot be recognized by the inhibitors. In support of these observations with chromaffin ghosts, synaptic vesicles purified from the brain also show net efflux of monoamines that is not sensitive to the two inhibitors (FLOOR et al. 1995). However, in this system, the drugs did inhibit an exchange reaction in

which preloaded vesicles are diluted into medium containing neurotransmitter. This reduces the concentration gradient favoring net efflux but enables
extravesicular monoamine to exchange with transmitter inside the vesicle. The
results thus support a role for the vesicular transport protein in exchange but
not net efflux, suggesting again either that the transporter mediates exchange
but not efflux or that reserpine and tetrabenazine cannot inhibit net efflux
catalyzed by the transporter. To determine if the conditions of the experiment
prevent inhibition by the drugs, we have examined the efflux of preloaded
neurotransmitter from secretory vesicles. Using a heterologous expression
system, we found that altering the conditions of the experiment enables
reserpine to inhibit net efflux from preloaded vesicles (D. PETER, unpublished
observations). This establishes a role for the VMATs in net efflux as well as
exchange reactions. In addition, we have found that, like synaptic vesicles
(FLOOR et al. 1995), considerable net efflux occurs after dilution despite the
presence of a proton electrochemical gradient. The origin of the discrepancy
between these results with synaptic vesicles and secretory vesicles, on the one
hand, and with chromaffin granules and ghosts on the other, remains unclear.
In summary, vesicular efflux may undergo regulation and has important physiological implications for the concentrations of transmitter inside the vesicle
and in the cytoplasm. Vesicular efflux also has relevance for the mechanism of
amphetamine action.

To increase synaptic concentrations of neurotransmitter through the reversal of plasma membrane monoamine transporters, the amphetamines must
first redistribute vesicular stores to the cytoplasm. Although reserpine and
tetrabenazine have previously failed to inhibit vesicular efflux in membrane
preparations, recent observations show that reserpine inhibits amphetamine-
induced reductions in quantal size as measured directly by amperometry
(SULZER et al. 1996), indicating that VMATs mediate the vesicular efflux. To
assess the interaction with methamphetamine using the cloned VMATs, we
have determined the effect of the drug on transport and found potent inhibition of VMAT2 at concentrations considerably below those at which it dissipates ΔpH (PETER et al. 1994). Further, similar concentrations of amphetamine
do not inhibit transport by VMAT1. These observations suggest a direct
interaction of amphetamines with VMAT2 that could promote efflux through
several distinct mechanisms: uptake of the weak base followed by alkalinization and monoamine efflux; exchange of amphetamine for lumenal monoamine followed by diffusion of the drug back to the cytoplasm for another round
of exchange and eventual redistribution of monoamine from the vesicle into
the cytoplasm. The precise contribution of these various mechanisms to amphetamine action and toxicity remains unknown.

D. Distribution

To understand the biological significance of differences in function between
VMAT1 and VMAT2, we have raised antipeptide antibodies to both rat

VMATs (PETER et al. 1995). Using these antibodies, we find VMAT1 immunoreactivity in adrenal chromaffin cells. Other studies show VMAT1 immunostaining in paracrine cells of the gastrointestinal tract and small intensely fluorescent (SIF) cells of sympathetic ganglia (WEIHE et al. 1994). Thus, non-neural cells express VMAT1.

With the exception of dopaminergic interneurons in the olfactory bulb, all central monoamine populations express VMAT2. In situ hybridization shows high levels of expression by monoamine cell groups and immunocytochemistry reveals staining of processes as well as cell bodies (PETER et al. 1995). However, both in situ hybridization and immunocytochemistry have failed to detect any expression in dopaminergic neurons of the olfactory bulb. Indeed, screening of an olfactory bulb cDNA library using the rat VMAT1 and VMAT2 as the probe did not yield any related cDNA clones (A. ROGHANI, personal communication), suggesting either that a remotely related transporter functions in amine transport in the olfactory bulb, that release of neurotransmitter occurs by another mechanism such as nonvesicular release or that the levels of expression are too low to detect. In the periphery, VMAT2 occurs in a small number of chromaffin cells, sympathetic ganglion cells, enteric neurons and enterochromaffin cells of the stomach (WEIHE et al. 1994; PETER et al. 1995). Thus, although adrenal chromaffin cells and a variety of other cells including sympathetic ganglion cells all derive from the neural crest, they express distinct VMATs.

The expression patterns observed for VMAT1 and VMAT2 correlate with previous observations. Clinically, tetrabenazine depletes central monoamine stores to a much greater extent than peripheral stores, producing much less hypotension than reserpine (CARLSON 1965; JANKOVIC and ORMAN 1988). Although a variety of explanations have been provided to account for this differential effect of tetrabenazine but not reserpine, our functional studies in vitro show that VMAT2 is tenfold more sensitive to tetrabenazine than VMAT1. The expression of VMAT2 in the brain and VMAT1 in the adrenal gland thus accounts for the preferential depletion of central monoamine stores by tetrabenazine. In addition, the functional studies show that VMAT2 has a 10- to 100-fold higher affinity for histamine than VMAT1. Correspondingly, histamine cells in both the brain and the gastrointestinal tract express VMAT2 rather than VMAT1.

E. Subcellular Localization

Regulated exocytosis of neurotransmitter occurs from two types of secretory vesicle. Large dense-core vesicles (LDCVs) store neural peptides (KELLY 1991) whereas small synaptic vesicles (SSVs) contain classical transmitters such as acetylcholine (ACh), γ-aminobutyric acid (GABA), glycine and glutamate (SUDHOF and JAHN 1991). Interestingly, monoamine storage appears different from that of other classical transmitters. Monoamines occur in both LDCVs (chromaffin granules of the adrenal medulla) (JOHNSON 1988) and SSVs in the brain (SMITH 1972; THURESON-KLEIN 1983).

The site of storage in LDCVs or SSVs is physiologically significant because it determines the role of the transmitter. LDCVs undergo exocytosis with a longer latency than SSVs and their release requires stimulation at higher frequency. In addition, exocytosis of the two vesicle types differs in its sensitivity to α-latrotoxin (MATTEOLI et al. 1988) and intracellular calcium (RANE et al. 1987; MARTIN and MAGISTRETTI 1989), indicating differences in the mechanism of release. Further, LDCVs and SSVs differ in their subcellular location. SSVs cluster at synaptic sites in the nerve terminal, supporting a role in fast synaptic transmission. In contrast, LDCVs occur at much lower levels at the nerve terminal but also occur in the cell body and dendrites, consistent with their release of neuromodulators such as peptides. Thus, the location of the VMATs in either LDCVs or SSVs will determine the sensitivity, mode, site and hence the physiological role of monoamine release. The existence of two VMATs further suggests that these proteins may differ in their location. Immunostaining has shown VMAT1 expression in adrenal chromaffin cells and other non-neural cells whereas VMAT2 occurs in the brain (PETER et al. 1995), suggesting that VMAT1 may localize to LDCVs and VMAT2 to SSVs.

To determine the subcellular location of the VMATs, we have again used the antibodies generated against peptides from the C-terminus of the proteins. In PC12 cells, endogenous VMAT1 colocalizes by immunofluorescence with the LDCV marker secretogranin rather than with the synaptic-like microvesicle (SLMV) marker synaptophysin (LIU et al. 1994). Sucrose density gradient fractionation also demonstrates preferential expression of VMAT1 in LDCVs. Smaller amounts of VMAT1 occur in lighter vesicle fractions, and velocity sedimentation through glycerol indicates that VMAT1 occurs at low levels on SLMVs. Consistent with these observations in PC12 cells, immunoelectron microscopy shows VMAT1 expression in the chromaffin granules of adrenal medullary cells, whereas synaptic-like microvesicles in these cells contain little VMAT1 (LIU et al. 1994).

To determine if VMAT2 differs in distribution from VMAT1, we have determined its subcellular location in PC12 cells and the brain. In contrast to our expectation that heterologous expression of VMAT2 in PC12 cells would confer localization to SLMVs, we found VMAT2 almost exclusively on LDCVs. Velocity sedimentation through glycerol showed no expression on SLMVs (Y. LIU, C. WAITES, P. TAN, R. EDWARDS, unpublished observations). Since PC12 cells derive from endocrine cells rather than neurons, they may not possess the machinery required to sort VMAT to SLMVs. However, the closely related vesicular ACh transporter does localize to SLMVs in PC12 cells as well as in vivo (Y. LIU, R. EDWARDS, in preparation; GILMOR et al. 1996), suggesting that these cells do have the machinery required to sort vesicular transport proteins to SLMVs. Further, immunoelectron microscopy performed to determine the localization of VMAT2 in vivo shows preferential expression in LDCVs of the rat solitary tract nuclei (NIRENBERG et al. 1995). However, VMAT2 also occurs in SSVs and in tubulovesicular membranes of midbrain dopamine cell bodies and dendrites (NIRENBERG et al. 1996). Classical studies have demonstrated storage of monoamines in these tubulovesicular structures and midbrain dopamine release does occur from a reserpine-sensitive compartment. Thus, VMAT2 shows preferential localization to vesicles other than SSVs in both PC12 cells and the brain.

VMATs may enter SSVs in brain or SLMVs in neurons through two distinct mechanisms. Both VMATs presumably sort to LDCVs on exit from the trans-Golgi network and stringent sorting of VMAT2 at this site presumably accounts for the exclusive localization of VMAT2 on LDCVs in PC12 cells. However, VMAT1 may not sort as stringently, enabling the transport protein to reach the plasma membrane through the constitutive secretory pathway and then to enter SLMVs after internalization, a route followed by other synaptic vesicle proteins (REGNIER-VIGOUROUX et al. 1991). Alternatively, both VMATs may enter SLMVs only after fusion of LDCVs with the plasma membrane. The absence of VMAT2 from SLMVs mitigates against this route for VMAT1 but supports this possibility in the case of VMAT2. Thus, fusion of LDCVs with the cell surface may trigger the appearance of VMATs in SSVs and so provides a mechanism for activity to alter the regulation, mode and site of monoamine release.

F. Structure/Function Studies of Vesicular Amine Transport

The alteration of amino acid residues by chemical modification and site-directed mutagenesis has been used to extract information about the role of specific residues in the transport mechanism. In terms of substrate recognition, studies have shown that the modifier N,N'-dicyclohexylcarbodiimide (DCCD), which reacts with carboxyl residues, inhibits monoamine transport as well as reserpine and tetrabenazine binding to the transporter expressed in bovine chromaffin granules, suggesting a role for acidic residues in the overall structure of the transporter (GASNIER et al. 1985; SUCHI et al. 1991). Consistent with this hypothesis, site-specific mutation of two aspartate residues in transmembrane domain 10 (TMD10) or TMD11 of VMAT1 to cysteine (D404C; D431C) abolishes transport activity (SCHULDINER et al. 1995).

In addition to the use of chemical modification by DCCD to guide site-directed mutagenesis, remote sequence homology of the VMATs to a class of bacterial drug resistance transporters (the tetracycline resistance genes of pBR322 and Tn10 and a bacterial multidrug resistance transporter) further suggests the involvement of an aspartate residue in the first transmembrane domain in substrate recognition. Interestingly, aspartate residues in the third transmembrane domain of receptors for monoamines are thought to bind to the cationic amino group of the monoamine ligand (STRADER et al. 1987, 1988). Using these observations to guide mutagenesis of the VMATs, we have shown that mutation of an aspartate in TMD1 of VMAT2 to asparagine (D34N) eliminates serotonin transport and the ability of serotonin to inhibit reserpine binding, indicating a specific deficit in substrate recognition (MERICKEL et al. 1995). The D34N mutant, however, exhibits equilibrium reserpine binding comparable to the wild-type protein, suggesting that the overall structure of D34N has not been perturbed.

Previous studies of substrate recognition by both monoamine receptors and plasma membrane transporters suggest that serine residues are central to substrate recognition. Mutation of two serines in TMD5 of the β-adrenergic receptor (STRADER et al. 1989) and two serines in TMD7 of the plasma membrane dopamine transporter (KITAYAMA et al. 1992) affect substrate recognition. More specifically, a compound lacking hydroxyl groups on the catechol ring is recognized equally well by mutant and wild-type β-adrenergic receptors (STRADER et al. 1989), suggesting that the serine residues interact with the hydroxyl groups of the normal substrate. Consistent with this hypothesis, mutation of three serines to alanine (S180, 181, 182) near TMD3 (Stmd3A) in VMAT2 completely abolishes transport activity (MERICKEL et al. 1995). Further, reserpine binding data with this mutant demonstrate that the protein folds normally and remains coupled to the proton-electrochemical gradient. Taken together, the results from the D34N and Stmd3A mutants suggest a model for substrate recognition in which, analogous to G-protein-coupled receptors, the cationic amino group of the substrate interacts with an aspartate in TMD1 and the hydroxyl group of the substrate interacts with serines in TMD3.

In addition to the substrate recognition event, chemical modification and site-directed mutations implicate specific residues in bioenergetic coupling to the driving force for transport. Diethylpyrocarbonate (DEPC) modifies histidine residues, and inhibits both serotonin transport and the acceleration of reserpine binding by a pH gradient (ISAMBERT and HENRY 1981; SUCHI et al. 1992), suggesting that modification of these sites either disrupts proton transport or prevents the resulting conformational change in the protein that occurs as part of the transport cycle. Further, site-directed mutagenesis shows that mutation of the only histidine residue conserved between VMAT1 and 2 to cysteine or arginine (H419C or H419R) abolishes transport activity. In this mutant, although equilibrium binding of reserpine is similar to the wild-type protein, a proton gradient no longer accelerates reserpine binding, suggesting a role for H419 in energy coupling (SHIRVAN et al. 1994). Finally, recent studies show that replacement of an aspartate in TMD10 at position 404 in VMAT1 with glutamate results in a downward shift in the pH optimum of transport, and replacement of an aspartate in transmembrane domain 11 at position 431 of VMAT1 with either glutamate or serine results in a transporter able to recognize substrate but unable to complete the entire transport cycle

(STEINER-MORDACH et al. 1996). Taken together, the site-directed mutagenesis studies indicate that multiple regions of the transporter are involved in the transport cycle.

A third method that has been used to deduce the structural basis for VMAT function involves the analysis of chimeric transporters (PETER et al. 1996). Chimeric analysis has several advantages over chemical mutagenesis and site-directed mutagenesis. It is less likely to perturb protein structure and does not require additional information to guide mutagenesis. Rather, the method relies on the analysis of functional chimerae that are derived from related parental proteins and has been used successfully to study the plasma membrane transporters for serotonin, dopamine and norepinephrine (BARKER et al. 1994; BUCK and AMARA 1994; GIROS et al. 1994).

The two vesicular amine transporters VMAT1 and VMAT2 are very similar in sequence yet exhibit multiple differences in substrate affinity and sensitivity to inhibitors (PETER et al. 1994). The sequence similarity allowed the construction and biochemical comparison of functional chimeric transporters of VMAT1 and VMAT2. Confirming results from the site-directed mutagenesis studies, the chimeric analysis suggests that multiple regions of the protein are involved in substrate recognition (for serotonin and histamine) and high-affinity interaction with the transport inhibitor tetrabenazine. By substituting domains of VMAT1 sequences with VMAT2, we found that high-affinity interactions with serotonin, histamine and tetrabenazine characteristic of VMAT2 require TMD5-8 and TMD9-12 of VMAT2, but neither domain alone suffices to confer the VMAT2 phenotype. Each domain requires the presence of an additional domain: a downstream domain in the case of TMD5-8 and an upstream domain in the case of TMD9-12. Presumably, each domain requires the other for the high-affinity interaction. In addition, a domain from the extreme N-terminus of VMAT2 when substituted alone into VMAT1 increases the affinity for substrates and the sensitivity to tetrabenazine; however, it is not required for the VMAT2 phenotype. Finally, a domain from TMD3-4 increases serotonin affinity but not histamine affinity or tetrabenazine sensitivity, and TMD5-7 of VMAT2 in the presence of N-terminal VMAT2 sequences reduces the affinity for serotonin. These results show that high-affinity serotonin and histamine recognition as well as high tetrabenazine sensitivity require similar regions of the protein, and that multiple regions of VMAT2 are required simultaneously to impart the VMAT2 phenotype. Interestingly, our chimera analysis suggests that the same domains of the transporter influence histamine recognition and tetrabenazine sensitivity. In vitro studies also show that histamine does not inhibit reserpine binding and reserpine does not inhibit histamine transport (MERIKEL and EDWARDS 1995), consistent with previous observations that tetrabenazine binds at a site distinct from the reserpine-binding site. Future site-directed mutagenesis of individual residues in these domains will further refine our knowledge of the critical residues involved in the transport mechanism.

References

Amara SG, Kuhar MJ (1993) Neurotransmitter transporters: recent progress. Ann Rev Neurosci 16:73–93

Angelides KJ (1980) Transport of catecholamines by native and reconstituted rat heart synaptic vesicles. J Neurochem 35:949–962

Barker EL, Kimmel HL, Blakely RD (1994) Chimeric human and rat serotonin transporters reveal domains involved in recognition of transporter ligands. Mol Pharmacol 46:799–807

Brodie BB, Olin JS, Kuntzman R, Shore PA (1957) Possible interrelationship between release of brain norepinephrine and serotonin by reserpine. Science 125:1293–1294

Buck KJ, Amara SG (1994) Chimeric dopamine-norepinephrine transporters delineate structural domains influencing selectivity for catecholamines and 1-methyl-4-phenylpyridinium. Proc Natl Acad Sci USA 91:12584–12588

Carboni E, Imperato A, Perezzani L, Di Chiara G (1989) Amphetamine cocaine phencyclidine and nomifensine increase extracellular DA concentrations preferentially in the NAcc of freely moving rats. Neuroscience 28:653–661

Carlson MD, Kish PE, Ueda T (1989) Characterization of the solubilized and reconstituted ATP-dependent vesicular glutamate uptake system. J Biol Chem 264:7369–7376

Carlsson A (1965) Drugs which block the storage of 5-hydroxytryptamine and related amines. Springer-Verlag, Berlin Heidelberg New York, pp 529–592 (Handbook of experimental pharmacology, vol 19)

Chaplin L, Cohen AH, Huettl P, Kennedy M, Njus D, Temperley SJ (1985) Reserpic acid as an inhibitor of norepinephrine transport into chromaffin vesicle ghosts. J Biol Chem 260:10981–10985

Daniels AJ, Reinhard JF (1988) Energy-driven uptake of the neurotoxin 1-methyl-4-phenylpyridinium into chromaffin granules via the catecholamine transporter. J Biol Chem 263:5034–5036

Darchen F, Scherman E, Henry JP (1989) Reserpine binding to chromaffin granules suggests the existence of two conformations of the monoamine transporter. Biochemistry 28:1692–1697

Di Chiara G, Imperato A (1988) Drugs abused by humans preferentially increase synaptic dopamine concentrations in the mesolimbic system of freely moving rats. Proc Natl Acad Sci USA 85:5274–5278

Erickson JD, Eiden LE (1993) Functional identification and molecular cloning of a human brain vesicle monoamine transporter. J Neurochem 61:2314–2317

Erickson JD, Eiden LE, Hoffman BJ (1992) Expression cloning of a reserpine-sensitive vesicular monoamine transporter. Proc Natl Acad Sci USA 89:10993–10997

Fischer JF, Cho AK (1979) Chemical release of DA from striatal homogenates: evidence for an exchange diffusion model. J Pharm Exp Ther 208:203–209

Floor E, Leventhal PS, Wang Y, Meng L, Chen W (1995) Dynamic storage of dopamine in rat brain synaptic vesicles in vitro. J Neurochem 64:689–699

Frize ED (1954) Mental depression in hypertensive patients treated for long periods with high doses of reserpine. N Engl J Med 251:1006–1008

Gasnier B, Scherman D, Henry J (1985) Dicyclohexylcarbodiimide inhibits the monoamine carrier of bovine chromaffin granule membranes. Biochemistry 24:1239–1244

Gilmor ML, Nash NR, Roghani A, Edwards RH, Yi H, Hersch SM, Levey AI (1996) Expression of the putative vesicular acetylcholine transporter in rat brain and localization in cholinergic synaptic vesicles. J Neurosci 16:2179–2190

Giros B, Wang YM, Suter S, McLeskey SB, Pifl C, Caron MG (1994) Delineation of discrete domains for substrate cocaine and tricyclic antidepressant interactions using chimeric dopamine-norepinephrine transporters. J Biol Chem 269:15985–15988

Giros B, Jaber M, Jones SR, Wightman RM, Caron MG (1996) Hyperlocomotion and indifference to cocaine and amphetamine in mice lacking the dopamine transporter. Nature 379:606–612

Heikkila RE, Manzino L, Cabbat FS, Duvoisin RC (1984) Protection against the dopaminergic neurotoxicity of 1-methyl-4-phenyl-1,2,5,6-tetrahydropyridine by monoamine oxidase inhibitors. Nature 311:467–469

Hell JW, Maycox PR, Jahn R (1990) Energy dependence and functional reconstitution of the gamma-aminobutyric acid carrier from synaptic vesicles. J Biol Chem 265:2111–2117

Hell JW, Edelmann L, Hartinger J, Jahn R (1991) Functional reconstitution of the gamma-aminobutyric acid transporter from synaptic vesicles using artificial ion gradients. Biochemistry 30 11795–11800

Howell M, Shirvan A, Stern-Bach Y, Steiner-Mordoch S, Strasser JE, Dean GE, Schuldiner S (1994) Cloning and functional expression of a tetrabenazine sensitive vesicular monoamine transporter from bovine chromaffin granules. FEBS Lett 338:16–22

Isambert MF, Henry JP (1981) Effect of diethylpyrocarbonate on pH-driven monoamine uptake by chromaffin granule ghosts. FEBS Lett 136:13–18

Isambert MF, Gasnier B, Botton D, Henry JP (1992) Characterization and purification of the monoamine transporter of bovine chromaffin granules Biochemistry 31:1980–1986

Jankovic J, Orman J (1988) Tetrabenazine therapy of dystonia chorea tics and other dyskinesias Neurology 38:391–394

Javitch J, D'Amato R, Nye J, Javitch J (1985) Parkinsonism-inducing neurotoxin N-methyl-4-phenyl-1,2,3,6-tetrahydropyridine: uptake of the metabolite N-methyl-4-phenylpyridine by dopamine neurons explains selective toxicity. Proc Natl Acad Sci USA 82:2173–2177

Jenner P, Rupniak NM, Rose S, Kelly E, Kilpatrick G, Lees A, Marsden CD (1984) 1-Methyl-4-phenyl-1.2.3.6-tetrahydropyridine-induced parkinsonism in the common marmoset. Neurosci Lett 50:85–90

Johnson RG (1988) Accumulation of biological amines into chromaffin granules: a model for hormone and neurotransmitter transport. Physiol Rev 68:232–307

Kaneko M, Yamaguchi A, Sawai T (1985) Energetics of tetracycline efflux system encoded by Tn10 in Escherichia coli. FEBS 193:194–198

Kanner BI, Schuldiner S (1987) Mechanisms of storage and transport of neurotransmitters. CRC Crit Rev Biochem 22:1–38

Kelly RB (1991) Secretory granule and synaptic vesicle formation. Curr Opin Cell Biol 3:654–660

Kirshner N (1962) Uptake of catecholamines by a particulate fraction of the adrenal medulla. Science 135:107–108

Kitayama S, Shimada S, Xu S, Markham L, Donovan DM, Uhl GR (1992) Dopamine transporter site-directed mutations differentially alter substrate transport and cocaine binding. ProcNatl Acad Sci USA 89:7782–7785

Knoth J, Zallakian M, Njus D (1981) Stoichiometry of H^+-linked dopamine transport in chromaffin granule ghosts. Biochemistry 20:6625–6629

Krueger MJ, Singer TP, Casida JE, Ramsay RR (1990) Evidence that the blockade of mitochondrial respiration by the neurotoxin 1-methyl-4-phenylpyridinium (MPP^+) involves binding at the same site as the respiratory inhibitor rotenone. Biochem Biophys Res Comm 169:123–128

Langston JW, Ballard P, Tetrud JW, Irwin I (1983) Chronic parkinsonism in humans due to a product of meperidine analog synthesis. Science 219:979–980

Langston JW, Forno LS, Revert CS, Irwin I (1984) Selective nigral toxicity after systemic administration of 1-methyl-4-phenyl-1,2,3,5-tetrahydropyridine (MPTP) in the monkey. Brain Res 292:390–394

Lesch KP, Gross J, Wolozin BL, Franzek E, Bengel D, Riederer P, Murphy DL (1994) Direct sequencing of the reserpine-sensitive vesicular monoamine transporter complementary DNA in unipolar depression and manic depressive illness. Pyschiatr Genet 4:153–160

Liange NY, Rutledge CO (1982) Comparison of the release of [³H]-dopamine from isolated corpus striatum by amphetamine fenfluramine and unlabeled dopamine. Biochem Pharmacol 31:983–992

Liu Y, Roghani A, Edwards RH (1992a) Gene transfer of a reserpine-sensitive mechanism of resistance to MPP⁺. Proc Natl Acad Sci USA 89:9074–9078

Liu Y, Peter D, Roghani A, Schuldiner S, Prive GG, Eisenberg D, Brecha N, Edwards RH (1992b) A cDNA that suppresses MPP⁺ toxicity encodes a vesicular amine transporter. Cell 70:539–551

Liu Y, Schweitzer ES, Nirenberg MJ, Pickel VM, Evans CJ, Edwards RH (1994) Preferential localization of a vesicular monoamine transporter to dense core vesicles in PC12 cells. J Cell Biol 127:1419–1433

Markey S, Johannessen J, Chiueh C, Burns R, Herkehnam M (1984) Intraneuronal generation of a pyridinium metabolite may cause drug-induced parkinsonism. Nature 311:464–467

Maron R, Stern Y, Kanner BI, Schuldiner S (1983) Functional asymmetry of the amine transporter from chromaffin granules. J Biol Chem 258:11476–11481

Martin JL, Magistretti PJ (1989) Pharmacological studies of the voltage-sensitive Ca²⁺ channels involved in the release of vasoactive intestinal peptide evoked by K⁺ in mouse cerebral cortical slices. Neuroscience 30:423–431

Matteoli M, Haimann C, Torri-Tarelli F, Polak JM, Ceccarelli B, De Camilli P (1988) Differential effect of alpha-latrotoxin on exocytosis of acetylcholine-containing small synaptic vesicles and CGRP-containing large dense-core vesicles at the frog neuromuscular junction. Proc Natl Acad Sci USA 85:7366–7370

Maycox PR, Deckwerth T, Hell JW, Jahn R (1988) Glutamate uptake by brain synaptic vesicles Energy dependence of transport and functional reconstitution in proteoliposomes. J Biol Chem 263:15423–15428

Merickel A, Edwards RH (1995) Transport of histamine by vesicular monoamine transporter-2. Neuropharmacology 34:1543–1547

Merickel A, Rosandich P, Peter D, Edwards RH (1995) Identification of residues involved in substrate recognition by a vesicular monoamine transporter. J Biol Chem 270:25798–25804

Michel PP, Hefti F (1990) Toxicity of 6-hydroxydopamine and dopamine for dopaminergic neurons in culture. J Neurosci Res 26:428–435

Neyfakh AA, Bidnenko VE, Chen LB (1991) Efflux-mediated multidrug resistance in Bacillus subtilis: similarities and dissimilarities with the mammalian system. Proc Natl Acad Sci USA 88:4781–4785

Nirenberg MJ, Liu Y, Peter D, Edwards RH, Edwards RH (1995) The vesicular monoamine transporter-2 is preferentailly localized to large dense core vesicles in the rat solitary tract nuclei. Proc Natl Acad Sci USA 92:8773–8777

Nirenberg MJ, Chan J, Liu Y, Edwards RH, Pickel VM (1996) Ultrastructural localization of the vesicular monoamine transporter-2 in midbrain dopaminergic neurons: potential sites for somatodendritic storage and release of dopamine. J Neurosci 16:4135–4145

Parkinson Study Group (1989) Effect of deprenyl on the progression of disability in early Parkinson's disease. N Engl J Med 321:1364–1371

Peter D, Finn JP, Klisak I, Liu Y, Kojis T, Heinzmann C, Roghani A, Sparkes RS, Edwards RH (1993) Chromosomal localization of the human vesicular amine transporter genes. Genomics 18:720–723

Peter D, Jimenez J, Liu Y, Kim J, Edwards RH (1994) The chromaffin granule and synaptic vesicle amine transporters differ in substrate recognition and sensitivity to inhibitors. J Biol Chem 269:7231–7237

Peter D, Liu Y, Sternini C, de Giorgio R, Brecha N, Edwards RH (1995) Differential expression of two vesicular monoamine transporters. J Neurosci 15:6179–6188

Peter D, Vu T, Edwards RH (1996) Chimeric vesicular monoamine transporters identify structural domains that influence substrate affinity and sensitivity to tetrabenazine. J Biol Chem 271:2979–2966

Rane S, Holz GG, Dunlap K (1987) Dihydropyridine inhibition of neuronal calcium current and substance P release. Pflugers Arch 409:361–366

Regnier-Vigouroux A, Tooze SA, Huttner WB (1991) Newly synthesized synaptophysin is transported to synaptic-like microvesicles via constitutive secretory vesicles and the plasma membrane. EMBO J 10:3589–3601

Rosenberg PA (1988) Catecholamine toxicity in cerebral cortex in dissociated cell culture. J Neurosci 8:2887–2894

Rudnick G, Wall SC (1992a) The molecular mechanism of "ecstasy" [3, 4-methylenedioxymethamphetamine (MDMA)]: serotonin transporters are targets for MDMA-induced serotonin release. Proc Natl Acad Sci USA 89:1817–1821

Rudnick G, Wall SC (1992b) p-Chloroamphetamine induces serotonin release through serotonin transporters. Biochemistry 31:6710–6718

Rudnick G, Wall SC (1992c) Non-neurotoxic amphetamine derivatives release serotonin through serotonin transporters. Mol Pharmacol 43:271–276

Rudnick G, Steiner-Mordoch SS, Fishkes H, Stern-Bach Y, Schuldiner S (1990) Energetics of reserpine binding and occlusion by the chromaffin granule biogenic amine transporter. Biochemistry 29:603–608

Schapira A, Cooper J, Dexter D, Clark J, et al (1990) Mitochondrial complex I deficiency in Parkinson's disease. J Neurochem 54:823–827

Scherman D, Henry JP (1984) Reserpine binding to bovine chromaffin granule membranes: characterization and comparison with dihydrotetrabenazine binding. Mol Pharmacol 25:113–122

Scherman D, Darchen F, Desnos C, Henry JP (1988) 1-Methyl-4-phenylpyridinium is a substrate of the vesicular monoamine transport system of chromaffin granules. Eur J Pharmacol 146:359–360

Schuldiner S, Liu Y, Edwards RH (1993a) Reserpine binding to a vesicular amine transporter expressed in Chinese hamster ovary fibroblasts. J Biol Chem 268:29–34

Schuldiner S, Steiner-Mordoch S, Yelin R, Wall SC, Rudnick G (1993b) Amphetamine derivatives interact with both plasma membrane and secretory vesicle biogenic amine transporters. Mol Pharmacol 44:1227–1231

Schuldiner S, Shirvan A, Linial M (1995) Vesicular neurotransmitter transporters: from bacteria to humans. Physiol Rev 75:369–392

Shirvan A, Laskar O, Steiner-Mordoch S, Schuldiner S (1994) Histidine-419 plays a role in energy coupling in the vesicular monoamine transporter from rat. FEBS Lett 356:145–150

Shoffner JM, Watts R, Juncos JL, Torroni A, Wallace DC (1991) Mitochondrial oxidative phosphorylation in Parkinson's disease. Ann Neurol 30:332–339

Smith AD (1992) Mechanisms involved in the release of noradrenaline from sympathetic nerve. Br Med Bull 29:123–129

Steiner-Mordach S, Shirvan A, Schuldiner S (1996) Modification of the pH profile and tetrabenazine sensitivity of rat VMAT1 by replacement of apspartate 404 with glutamate. J Biol Chem 271:13048–13054

Stern-Bach Y, Greenberg-Ofrath N, Flechner I, Schuldiner S (1990) Identification and purification of a functional amine transporter from bovine chromaffin granules. J Biol Chem 265:3961–3966

Stern-Bach Y, Keen JN, Bejerano M, Steiner-Mordoch S, Wallach M, Findlay JB, Schuldiner S (1992) Homology of a vesicular amine transporter to a gene conferring resistance to 1-methyl-4-phenylpyridinium. Proc Natl Acad Sci USA 89:9730–9733

Strader CD, Sigal IS, Register RB, Candelore MR, Rands E, Dixon RAF (1987) Identification of residues required for ligand binding to the beta-adrenergic receptor. Proc Natl Acad Sci USA 84:4384–4388

Strader CD, Sigal IS, Candelore MR, Rands E, Hill WS, Dixon RA (1988) Conserved aspartic acid residues 79 and 113 of the beta-adrenergic receptor have different roles in receptor function. J Biol Chem 263:10267–10271

Strader CD, Candelore MR, Hill WS, Sigal IS, Dixon RAF (1989) Identification of two serine residues involved in agonist activation of the beta-adrenergic receptor. J Biol Chem 264:13572–13578

Suchi R, Stern-Bach Y, Gabay T, Schuldiner S (1991) Covalent modification of the amine transporter with N,N'-dicyclohexylcarbodiimide. Biochemistry 30:6490–6494

Suchi R, Stern-Bach Y, Schuldiner S (1992) Modification of arginyl or histidinyl groups affects the energy coupling of the amine transporter. Biochemistry 31:12500–12503

Sudhof TC, Jahn R (1991) Proteins of synaptic vesicles involved in exocytosis and membrane recycling. Neuron 6:665–677

Sulzer D, Rayport S (1990) Amphetamine and other psychostimulants reduce pH gradients in midbrain dopaminergic neurons and chromaffin granules: a mechanism of action. Neuron 5:797–808

Sulzer D, Maidment NT, Rayport S (1993) Amphetamine and other weak bases act to promote reverse transport of dopamine in ventral midbrain neurons. J Neurochem 60:527–535

Sulzer D, St Remy C, Rayport S (1996) Reserpine inhibits amphetamine action in ventral midbrain culture. Mol Pharmacol 49:338–342

Surratt CK, Persico AM, Yang X-D, Edgar SR, Bird GS, Hawkins AL, Griffin CA, Li X, Jabs EW, Uhl GR (1993) A human synaptic vesicle monoamine transporter cDNA predicts posttranslational modifications reveals chromosome 10 gene localization and identifies TaqI RFLPs. FEBS Lett 318:325–330

Tanner CM, Langston JW (1990) Do environmental toxins cause Parkinson's disease? A critical review. Neurology 40:17–30

Thureson-Klein A (1983) Exocytosis from large and small dense cored vesicles in noradrenergic nerve terminals. Neuroscience 10:245–259

Vander Borght TMA, Sima AF, Kilbourn MR, Desmond TJ, Kuhl DE, Frey KA (1995) [^{3}H]Methoxytetrabenazine: a high specific activity ligand for estimating monoaminergic neuronal integrity. Neuroscience 68:955–962

Vincent MS, Near JA (1991) Identification of a [^{3}H]dihydrotetrabenazine-binding protein from bovine adrenal medulla. Mol Pharmacol 40:889–894

Weaver JH, Deupree JD (1982) Conditions required for reserpine binding to the catecholamine transporter on chromaffin granule ghosts. Eur J Pharmacol 80:437–438

Weihe E, Shafter MK, Erickson JD, Eiden LE (1994) Localization of vesicular monoamine transporter isoforms (VMAT1 and VMAT2) to endocrine cells and neurons in rat. J Mol Neurosci 5:149–164

CHAPTER 6

Regulation of the Environment of the Interior of Serotonin-Storing Vesicles

M.D. GERSHON and H. TAMIR

A. Introduction

Properties of the synaptic vesicles of serotoninergic neurons are of great interest, but very difficult to study. The importance of these neurons is highlighted by the successful use of 5-HT-selective reuptake inhibitors in the treatment of depression (JOFFE et al. 1996; KUZEL 1996), obsessive-compulsive behavior (BAER 1996), and panic disorders (KLEIN 1996), as well as the effectiveness of 5-HT$_{1A}$ agonists in treating anxiety (LUCKI 1996) and the controversial use of the serotonin-releasing drug fenfluramine to reduce appetite (NOACH 1994; BLUNDELL and LAWTON 1995). Clearly, the role played by 5-HT as a neurotransmitter cannot be thoroughly understood in the absence of knowledge about the characteristics of the synaptic vesicles of serotoninergic neurons. For investigators seeking to understand compounds that affect the storage of 5-HT, which, in addition to fenfluramine, include such drugs as reserpine, cocaine, and amphetamine (SULZER et al. 1995), information about these vesicles is even more critical. Since it is impossible to obtain a pure preparation of synaptic vesicles from serotoninergic neurons, investigators have sought to employ surrogates, which can be investigated as models that are relevant to serotoninergic neurons. Platelets, mast cells, and enterochromaffin cells are all peripheral cells that store 5-HT (ERSPAMER 1966); however, none of these are developmentally related to neurons. Platelets and mast cells are mesodermal derivatives and enterochromaffin cells develop from endoderm. There is thus no reason to believe that any of these cells are appropriate surrogates for serotoninergic neurons. Far better would seem to be a model cell the developmental origin of which is common to that of serotoninergic neurons. Such a surrogate is likely to share fundamental serotoninergic mechanisms with the neurons. For this reason, the secretory vesicles of serotoninergic paraneurons are of particular interest.

Paraneurons (parafollicular cells) that contain 5-HT are found in the thyroid gland (NUNEZ and GERSHON 1978). Parafollicular cells are endocrine; however, they are closely related to neurons (BARASCH et al. 1987a; RUSSO et al. 1992; JACOBS-COHEN et al. 1994; CLARK et al. 1995); moreover, parafollicular cells are accessible and can readily be isolated in pure form for experimental analysis (BERND et al. 1981; BARASCH et al. 1987b; CIDON et al. 1991). Serotoninergic neurons cannot similarly be isolated from the nuclei of the median raphe. The neural properties of parafollicular cells include their expression of voltage-gated Na⁺ channels (Deriemer et al. 1991; Tamir et al. 1994b) and their ability to "neuralize" when they are separated from thyroid follicular cells and cocultured with gut or exposed to nerve growth factor (Barasch et al. 1987a; Jacobs-Cohen et al. 1994; Clark et al. 1995). The following properties define the term "neuralize": (a) extend neurites, (b) synthesize neurofilament proteins, (c) change alternative splicing of mRNA encoded by the calcitonin gene to produce calcitonin gene-related peptide (CGRP) instead of calcitonin (Amara et al. 1982), and (d) express the specific plasmalemmal 5-HT transporter that is responsible for uptake of 5-HT (Clark et al. 1995). Both the secretory vesicles of parafollicular cells (Barasch et al. 1987b) and the synaptic vesicles of serotoninergic neurons (Gershon et al. 1983; Kirchgessner et al. 1988) contain a soluble 5-HT-binding protein (SBP). This protein is exclusively found in serotoninergic cells, but only in those, such as parafollicular cells and neurons, which are derived from neurectoderm (Gershon and Tamir 1985). Curiously, parafollicular cells even begin to "neuralize" when they are grown for extended periods of time in the absence of thyroid follicular cells; moreover, "neuralization" of parafollicular cells is antagonized by high concentrations of triiodothyronine, suggesting that the close contact of parafollicular cells with follicular cells in the thyroid may be responsible for maintaining these cells in an endocrine phenotype (Jacobs-Cohen et al. 1994). The default pathway for the precursor cells that give rise to parafollicular cells may thus be similar to that of the precursors of enteric serotoninergic neurons. Both are derived from the vagal region of the neural crest (Le Douarin 1982) and both are derived from cells that express the gene *mash-1* (Clark et al. 1995; Blaugrund et al. 1996). When such cells develop in the thyroid, the thyroid microenvironment, which includes the exposure of the cells to high concentrations of thyroid hormones, may keep them in an endocrine phenotype. When similar cells develop in the bowel, they may give rise to enteric serotoninergic neurons. In support of the idea that the endocrine state is a "less mature form" that requires the active intervention of an environmental factor, such as triiodothyronine, to prevent it from "neuralizing," is the fact that the expression of *mash-1* is permanent in parafollicular cells (in situ) (Clark et al. 1995), but is expressed only transiently during development by serotoninergic neurons (Guillemot and Joyner 1993; Blaugrund et al. 1996). Whether or not this hypothesis is correct, it is clear that parafollicular cells and enteric serotoninergic neurons are very closely related. Enteric serotoninergic neurons, moreover, are not known to be significantly different from their central counterparts. In discussing the specialized properties of the secretory vesicles of serotoninergic neurons or their parafollicular cell surrogates, it is useful to consider first what types of secretory vesicle are known to exist and how the vesicles of neurectodermally derived serotoninergic cells might be unique.

B. Secretory Pathways and the Vesicles Found at Synapses

Two secretory pathways have been identified in cells, the constitutive and the regulated (ROTHMAN and ORCI 1992; ROTHMAN and WARREN 1994). The constitutive pathway, found in almost every cell, consists of a continuous stream of transport vesicles that travel from the trans-Golgi network (TGN) to the plasma membrane. These vesicles deliver membrane proteins and lipids to the plasmalemma, while at the same time they secrete their soluble cargo to the extracellular space by exocytosis. The regulated pathway is found in a much more restricted subset of cells, which are specialized for the rapid signal-evoked secretion of products, such as enzymes, hormones, or neurotransmitters/neuromodulators. These products are sorted in the TGN and diverted from the constitutive stream of vesicles for delivery to specialized secretory vesicles in which they are stored in preparation for eventual stimulus-coupled secretion. The regulated secretory pathway is represented in nerve terminals by large, dense-cored vesicles, which are also found in neuroendocrine cells and paraneurons (DE CAMILLI and JAHN 1990; JAHN and SUDHOF 1994). Large, dense-cored vesicles contain proteins and/or peptides that are packaged in the TGN and transported to terminals for regulated secretion (THURESON-KLEIN and KLEIN 1990). Nerve endings also contain much smaller (~50nm) synaptic vesicles, which store small-molecule neurotransmitters, such as acetylcholine, glutamate, or GABA (JAHN and SUDHOF 1994). Synaptic vesicles differ from large, dense-cored vesicles in that they are derived from endosomes, not the TGN, and are specialized to permit exocytosis to recur at high frequencies (CAMERON et al. 1991; LINSTEDT and KELLY 1991). Since synaptic vesicles need not each be generated in the TGN and transported to terminals, they can instead be replenished locally, by endocytic recycling from the plasma membrane (HEUSER 1989; VALTORTA et al. 1990; FESCE et al. 1994). The membranous components of synaptic vesicles, including an H^+-translocating ATPase, capable of acidifying the internal environment of the vesicles, and transporters that import neurotransmitters from the cytosol, are initially delivered to the plasma membrane via the constitutive secretory pathway. The transporters enable synaptic vesicles to become filled with neurotransmitter, which is synthesized in the cytosol (LIU et al. 1992). The neurotransmitter-loaded synaptic vesicles translocate to the active zone of the plasma membrane (either by diffusion or by a mechanism aided by cytoskeletal components), where they dock and are primed to become competent for fast exocytosis triggered by Ca^{2+} (BAJJALIEH and SCHELLER 1995; SCHELLER 1995; SUDHOF 1995). The primed vesicles await a Ca^{2+} spike associated with action potentials, which triggers fusion of the vesicles with the plasmalemma and exocytosis. Following exocytosis, the synaptic vesicle membrane is re-internalized by endocytosis, probably by way of clathrin-coated pits, which become coated vesicles. The retrieved coated vesicles uncoat, acidify, reload with transmitter, and the cycle is repeated. In contrast to synaptic vesicles, there is no evidence that the large, dense-cored vesicles derived from the TGN recycle following exocytosis. The peptide contents of these vesicles have to be loaded with fresh peptide as the vesicles form in the TGN.

I. Secretory Vesicles of Parafollicular Cells

The vesicles of the regulated secretory pathway of neural crest-derived paraneurons seem to represent a third class of secretory vesicle, in that they share characteristics of both the large, TGN-derived, peptide-containing dense-cored vesicles and the small, endosome-derived, low-molecular-weight neurotransmitter-containing synaptic vesicles. Paraneurons include adrenal chromaffin cells and thyroid parafollicular cells (FUJITA 1987). The secretory vesicles of these cells are large, form initially within the TGN (THURESON-KLEIN and KLEIN 1990), and contain peptides (WEILER et al. 1989); however, like endosome-derived synaptic vesicles, they also contain small molecules (NUNEZ and GERSHON 1972; JOHNSON 1988; CIDON et al. 1991). Chromaffin granules costore catecholamines with endorphins and other peptides, while the secretory vesicles of parafollicular cells costore calcitonin (as well as somatostatin and other peptides) with 5-HT, which they obtain by transmembrane transport from the cytosol (NUNEZ and GERSHON 1972; CIDON et al. 1991). In fact, all of the secretory vesicles of parafollicular cells costore 5-HT with calcitonin and also take up ^{3}H-5-HT that has been synthesized in the cytosol from ^{3}H-5-hydroxytryptophan (NUNEZ and GERSHON 1972). Isolated parafollicular vesicles transport ^{3}H-5-HT in an ATP-dependent fashion (BARASCH et al. 1987b; CIDON et al. 1991). Moreover, the secretory vesicles of parafollicular cells recycle (TAMIR et al. 1994a) and their membranes contain the highly conserved set of proteins, which include synaptotagmin, synaptophysin, and synaptobrevin (WEILER et al. 1989; TAMIR et al. 1994c), that are present in synaptic vesicles and essential for exocytosis (ELFERIK et al. 1993; JAHN and SUDHOF 1994; LI et al. 1995; SUDHOF 1995).

The synaptic vesicles of serotoninergic neurons have not been well characterized, although, in the absence of specific evidence, they have been classified together with other small endosome-derived synaptic vesicles (JAHN and SUDHOF 1994). Because the information is lacking, however, it is unclear as to whether serotoninergic synaptic vesicles, like those which store acetylcholine, fall within the class of small molecule-storing synaptic vesicles, or whether instead 5-HT-containing synaptic vesicles resemble the serotoninergic secretory vesicles of parafollicular cells. This distinction is important, because the secretory vesicles of thyroid parafollicular cells display properties that have not yet been detected in synaptic vesicles, large, dense-cored vesicles, or even chromaffin granules. Not only is the secretion of the contents of these vesicles regulated, but their internal environment is also regulated by the same stimuli that initiate secretion (BARASCH et al. 1988; CIDON et al. 1991; TAMIR et al. 1994c). In contrast to the interior of chromaffin vesicles, which is constitutively

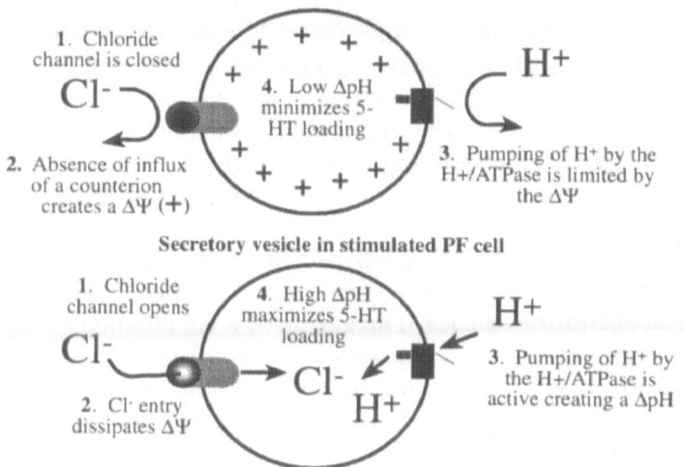

Fig. 1. The gating of a vesicular chloride channel regulates the acidity of the internal milieu of the secretory vesicles of parafollicular cells. *Top* In a nonstimulated cell, the chloride channel in the membrane of the vesicle is closed. As a result, a $\Delta\Psi$ develops in response to the constitutive action of the vesicular $H^+/ATPase$. The generation of the $\Delta\Psi$ limits transmembrane transport of H^+, and restricts the ΔpH that develops across the membrane. The low ΔpH provides a minimal drive for the vesicular uptake of 5-HT. *Bottom* In a stimulated cell, the chloride channel in the membrane of the vesicle is open. As a result, the $\Delta\Psi$ is dissipated. The vesicular $H^+/ATPase$ now effectively transports H^+ into the vesicles and a substantial ΔpH develops across the vesicular membrane. The high ΔpH provides a strong drive for the vesicular uptake of 5-HT

acidic (JOHNSON 1988), the interior of the secretory vesicles of parafollicular cells becomes acidic only when the cells are stimulated to secrete (BARASCH et al. 1988; CIDON et al. 1991). Acidification of the secretory vesicles of both chromaffin (JOHNSON 1988) and parafollicular (BARASCH et al. 1988; CIDON et al. 1991) cells occurs as a result of the action of an $H^+/ATPase$ in the vesicular membranes (MORIYAMA and NELSON 1989). In isolated chromaffin granules, H^+ transport gives rise to a transmembrane ΔpH, because Cl^- can permeate the membrane of the granules and act as a counterion, preventing the development of a transmembrane potential difference ($\Delta\Psi$; interior positive) that would otherwise limit continued transport of H^+ (JOHNSON 1988). In contrast (Fig. 1), Cl^- channels in the membranes of secretory vesicles of unstimulated parafollicular cells are probably closed (BARASCH et al. 1988) (see Sect. B.V). As a result, Cl^- cannot act as a counterion and the action of the $H^+/ATPase$ thus causes a transmembrane $\Delta\Psi$ to develop (CIDON et al. 1991). The $\Delta\Psi$, in turn, antagonizes transport of H^+ and prevents acidification of the vesicles. Stimulation of parafollicular cells with a secretogogue, however, leads to the opening of vesicular Cl^- channels and vesicle acidification in situ (BARASCH et al. 1988; TAMIR et al. 1994c). Acidification promotes the loading of vesicles with neurotransmitter because ΔpH provides a more potent driving force than $\Delta\Psi$ for the transmembrane transport of 5-HT into parafollicular vesicles (CIDON et al. 1991). It follows, therefore, that even the filling of parafollicular vesicles with 5-HT is probably regulated by secretogogue stimulation. Stimulus-induced acidification may thus control the 5-HT content/vesicle. If the uptake of 5-HT into vesicles is maximal during periods when parafollicular cells are stimulated with a secretogogue, then the 5-HT content of individual vesicles should increase in proportion to the duration or frequency of periods of stimulation.

Parafollicular cells are thus of intrinsic interest, if for no other reason than that they contain secretory vesicles that may have properties that are unlike those of other secretory vesicles. The gating of a Cl⁻ channel in the membrane of serotoninergic vesicles by a plasmalemmal receptor (see below for further details), moreover, is a novel mechanism that has not previously been described and may have applicability to serotoninergic neurons and perhaps to other cells of the nervous system.

II. Contents of Parafollicular Secretory Vesicles and Serotoninergic Neurons

The secretory vesicles of parafollicular cells (BERND et al. 1981; BARASCH et al. 1987b) and a tumor cell line derived from it (MTC cells) (TAMIR et al. 1989) resemble the synaptic vesicles of serotoninergic neurons (GERSHON et al. 1983) in that both contain a specific 5-HT-binding protein (SBP). SBP is only present in synaptic vesicles of central (GERSHON et al. 1983) and peripheral neurons that are serotoninergic (JONAKAIT et al. 1977; JONAKAIT et al. 1979; KIRCHGESSNER et al. 1988). Stimulated peripheral serotoninergic neurons and MTC cells secrete SBP together with 5-HT (JONAKAIT et al. 1979). Vesicles of the synaptic type (containing both 5-HT and SBP) are present in the growth cones of developing serotoninergic neurons prior to the formation of synapses (IVGY-MAY et al. 1994), indicating that this type of vesicle develops very early in serotoninergic cells and may enable serotoninergic neurons to influence the development of neurons that arise later than serotoninergic cells during ontogeny. In neurons, vesicles, marked by their content of SBP, move from the perikaryon to axon terminals by fast transport (GERSHON et al. 1983). This observation is consistent with the idea that the SBP-marked neuronal serotoninergic vesicles initially arise in the TGN, but this origin does not mean that they do not recycle.

III. Vesicle Recycling

Although SBP is a soluble constituent of the vesicular matrix and thus is secreted from stimulated cells, the secretion of 5-HT does not deplete cells of SBP. Although some SBP is released to the ambient medium, a portion of the vesicular SBP store remains membrane bound and is recaptured during vesicle recycling after exocytosis (TAMIR et al. 1994a). The recaptured membrane-bound SBP then dissociates from the membrane and is reutilized in recycled secretory vesicles. These observations indicate that although they are peptide containing and initially TGN derived, the secretory vesicles of parafollicular cells can also be recycled from endosomes. The reuse of SBP provides a convenient "tag" for studies of the recycling of parafollicular vesicles. Since, at the time of exocytosis, some SBP remains bound to what was the luminal face of the vesicular membrane, the bound SBP becomes exposed to the extracellular medium. When exocytosis is induced in the presence of antibodies to SBP, therefore, these antibodies bind to the exposed SBP and are recaptured with it when the vesicular membrane is retrieved (TAMIR et al. 1994a). The subsequent stimulation-dependent incorporation of antibodies to SBP into parafollicular secretory vesicles thus directly demonstrates vesicle recycling. When cells are incubated in the presence of antibodies to SBP, but are not stimulated, SBP is not internalized; moreover, antibodies in the extracellular medium that do not react with proteins of the luminal face of secretory vesicles fail to be incorporated into parafollicular cells, even when the cells are stimulated.

The presence of a vesicle-specific soluble protein (SBP) in the 5-HT-containing vesicles of both parafollicular cells and neurons suggests that the two vesicles are at least similar to one another and are more typical of a TGN than a synaptic vesicle that is purely endosome derived. When neuropeptides and small-molecule neurotransmitters are costored in the same *terminals*, the neuropeptides have usually been found in large, dense-cored vesicles and the small-molecule neurotransmitters in synaptic vesicles (DE POTTER et al. 1987; KONG et al. 1990; BAUERFEIND et al. 1993; MITCHELL and STAUBER 1993; ANNAERT et al. 1994). The TGN-derived large, dense-cored vesicles of most neurons, moreover, tend to lack the transporters that pump neurotransmitters into the vesicles from the cytosol and which appear to be necessary for the biosynthesis of endosome-derived synaptic vesicles (BAUERFEIND et al. 1993). The observations that 5-HT-containing synaptic vesicles contain a soluble protein (SBP), but are surrounded by a membrane that recycles and contains a transporter, thus suggest that the synaptic vesicles of serotoninergic neurons are "atypical" with respect to the category of small molecule-containing synaptic vesicles. They may, therefore, be more similar to the secretory vesicles of parafollicular cells than to either the large, dense-cored vesicles or synaptic vesicles of other types of neuron.

IV. Induction of Secretion

MTC cells are a serotoninergic parafollicular cell-derived tumor cell line with properties which resemble those of the parent parafollicular cells (TAMIR et al. 1989). The MTC line can be utilized to obtain large quantities of serotoninergic cells for biochemical analyses. They have thus been valuable for the characterization of the signal transduction mechanism leading to the secretion of 5-HT. Surprisingly, multiple signal transduction pathways were found to be involved. Although secretion is induced by increasing cytosolic cAMP, this second messenger is not affected by the natural secretogogues, which cause parafollicular cells to secrete, elevated $[Ca^{2+}]_e$ and thyrotropin, or by depolarizing parafollicular cells with high concentrations of $[K^+]_e$. In contrast, each of the natural secretogogues increases the levels of cytoplasmic free Ca^{2+} ($[Ca^{2+}]_i$), which can be detected by quantitative Fura-2 fluorescence with microscopic imaging (TAMIR et al. 1990). Activators of protein kinase C (PKC) also evoke secretion of 5-HT; moreover, downregulation of PKC blocks thyrotropin- or phorbol ester-induced secretion (TAMIR et al. 1992). Phorbol esters evoke 5-HT secretion by isolated parafollicular cells, as well as by MTC cells, and downregulation of PKC antagonizes the secretion of 5-HT evoked by a high concentration of $[Ca^{2+}]_e$; thus both elevated $[Ca^{2+}]_i$ and PKC contribute to the induction of secretion by parafollicular cells.

V. Stimulus-Coupled Acidification of Vesicles and Its Study

Sheep parafollicular cells have been studied most extensively because they can be isolated for experimental analysis in vitro (97% of the cells in the isolated preparations are parafollicular) (BERND et al. 1981; BARASCH et al. 1987b). Secretory vesicles, furthermore, can be isolated from sheep parafollicular cells (BARASCH et al. 1987b; CIDON et al. 1991; TAMIR et al. 1994c). The uptake of 5-HT by the secretory vesicles of sheep parafollicular cells is mediated by a reserpine-sensitive transporter in the vesicular membranes, is ATP-dependent, and is driven maximally by a transmembrane proton gradient and to a lesser extent by membrane potential ($\Delta\Psi$) (CIDON et al. 1991). As noted above, the interiors of the secretory vesicles of intact "resting" (unstimulated) parafollicular cells are rarely acidic, but almost all become acidic when parafollicular cells are exposed to a secretogogue. Secretory vesicles isolated from unstimulated parafollicular cells do not acidify in vitro when exposed to ATP; however, acidification proceeds if the isolated vesicles are exposed to ATP in the presence of the K^+ ionophore valinomycin in order to dissipate a $\Delta\Psi$ across the membranes of the vesicles (BARASCH et al. 1988). In the absence of the ionophore, the continued ATP-induced acidification of isolated vesicles is limited by the development of a transmembrane $\Delta\Psi$, which can be demonstrated with a voltage-sensitive dye, oxonolol VI; a ΔpH develops when this $\Delta\Psi$ is dissipated, either by exposing isolated vesicles to an ionophore, or by stimulating intact cells, which causes the opening of a conductance for a counterion in the vesicular membranes (CIDON et al. 1991). Secretory vesicles isolated from secretogogue-stimulated parafollicular cells have been stably changed and readily acidify when exposed to ATP; acidification is made possible by a Cl^- conductance in the membranes of the vesicles (BARASCH et al. 1988; CIDON et al. 1991; TAMIR et al. 1994c). The same vacuolar H^+-translocating ATPase that acidifies chromaffin granules and other Golgi-derived organelles is present in the membranes of the secretory vesicles of parafollicular cells (CIDON et al. 1991); these membranes contain p64, a 64-

kDa Cl⁻ channel, that is gated in response to secretogogue stimulation of intact parafollicular cells. Opening of this Cl⁻ channel is what naturally dissipates the ATP-dependent $\Delta\Psi$ and thus permits vesicles to acidify (BARASCH et al. 1987b; CIDON et al. 1991; TAMIR et al. 1994c). Demonstrable chloride actually enters the secretory vesicles of intact parafollicular cells in response to secretogogue stimulation and secretogogues also induce the phosphorylation of p64 (TAMIR et al. 1994c). These observations indicate that acidification of the secretory vesicles of parafollicular cells depends on the secretogogue-evoked gating of a vesicular Cl⁻ channel. The data also suggest that phosphorylation of the channel may be involved in its gating. As a result of the critical role in 5-HT uptake played by the transmembrane ΔpH in parafollicular secretory vesicles, the uptake and accumulation of 5-HT, as much as intravesicular acidity, is regulated by the secretogogue-induced gating of the vesicular Cl⁻ channel (CIDON et al. 1991; TAMIR et al. 1994c).

VI. Plasmalemmal Calcium Receptor of the Parafollicular Cell

Parafollicular cells secrete 5-HT and acidify their secretory vesicles in response to increases in extracellular Ca^{2+} ($[Ca^{2+}]_e$) (BARASCH et al. 1988). This remarkable property may be related to the presence of a plasma membrane calcium receptor (BROWN et al. 1993; MICHALCHIK and RUSSO 1994; GARRETT et al. 1995; TAMIR et al. 1996), a $[Ca^{2+}]_e$-binding integral membrane protein that is not a Ca^{2+} channel. Alternatively, the voltage-gated or other Ca^{2+} channels of the parafollicular cell plasma membrane might respond to elevated levels of $[Ca^{2+}]_e$ independently of the calcium receptor. Parathyroid chief cells, which, like parafollicular cells, respond to extracellular Ca^{2+}, express a plasmalemmal calcium receptor that is identical to that of parafollicular cells (BROWN et al. 1993). The distribution of the calcium receptor in the nervous system has recently been determined, and transcripts encoding the calcium receptor have been found in the brain and gut (ROGERS et al. 1994) (personal communication). Some of the cells that express the calcium receptor are serotoninergic raphe neurons; the receptor is thus expressed by serotoninergic neurons, although these cells are not the only neurons that express it. The calcium receptor is concentrated in axon terminals (RUAT et al. 1995) and is a member of the heptahelical (7-transmembrane domain) family of receptors. This type of receptor is invariably coupled to a heterotrimeric GTP-binding protein (G protein). In the case of the parafollicular cell, the calcium receptor is coupled primarily to Gαi and, to a lesser extent, to Gαq (TAMIR et al. 1996).

A substantial part of the transduction pathway to which the calcium receptor is coupled has now been ascertained. The parafollicular cell calcium receptor functions as a $[Ca^{2+}]_e$ sensor, which leads both to secretion and to vesicle acidification. The pathways responsible for stimulus-acidification and stimulus-secretion coupling diverge at a step distal to the G protein and, as a result, these two processes can be dissociated. Acidification of vesicles, moreover, is inhibited by compounds that antagonize a variety of protein kinases or phosphatases. The idea, noted above, that a covalent post-translational modification, such as phosphorylation, is responsible for gating vesicular Cl⁻ channels is supported by the observation that the phenomenon is sufficiently stable to permit vesicles to be isolated from stimulated cells with their Cl⁻ channels in the open configuration (BARASCH et al. 1988). All observations are consistent with the following working hypotheses: (a) p64 is the Cl⁻ channel in vesicular membranes that permits the stimulation-induced entry of Cl⁻ into vesicles; (b) the state and site of phosphorylation of p64 determine whether the channel is open or closed; and (c) signal transduction pathways couple the surface calcium receptor to protein kinases and phosphatases that determine the level of phosphorylation of the channel protein. It has been suggested that signal transduction for acidification from the calcium receptor follows the following pathway (TAMIR et al. 1996): calcium receptor $\rightarrow$ Gi $\rightarrow$ phospholipase C (PLC) $\rightarrow$ mobilization of $[Ca^{2+}]_i$ $\rightarrow$ Ca binds to calmodulin $\rightarrow$ activation of NO synthase (NOS) $\rightarrow$ stimulation of guanyl cyclase $\rightarrow$ increases cGMP $\rightarrow$ activation of protein kinase G $\rightarrow$ (participation of both protein kinases and phosphatases) $\rightarrow$ opening of the vesicular Cl⁻ channel (Fig. 2) (TAMIR et al.

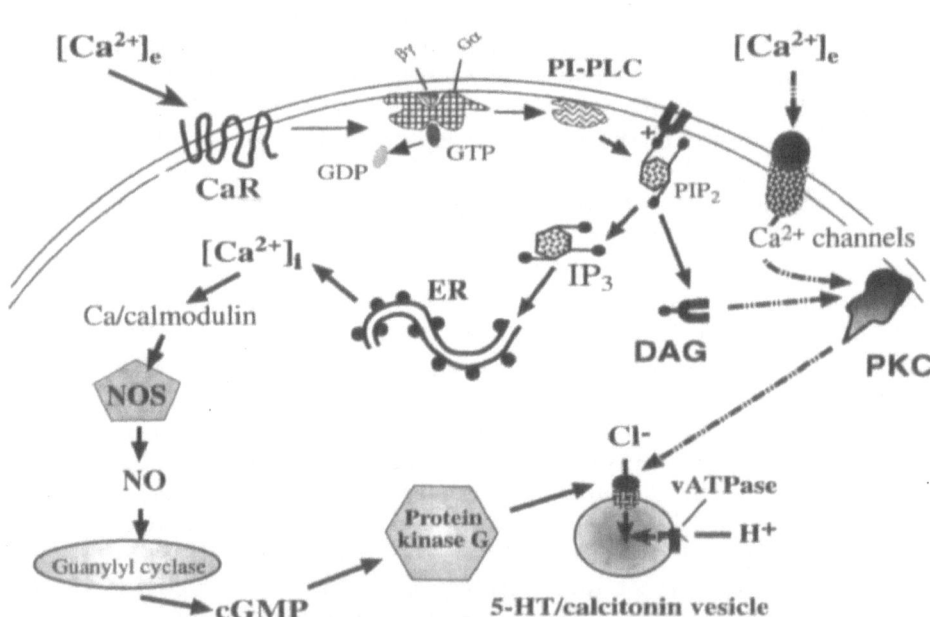

Fig. 2. A heuristic model showing the signal transduction pathway in parafollicular cells believed to be responsible for the gating of vesicular chloride channels in response to agonists that stimulate the plasmalemmal calcium receptor. The gating of this channel regulates the acidification of vesicles. It is possible that in serotoninergic neurons a similar or analogous pathway regulates vesicle acidification in response to any stimulus that activates the neurons

1996). A separate pathway that can also induce vesicle acidification is activated by Ca^{2+} entering parafollicular cells through plasmalemmal calcium channels. This pathway activates PKC. It now remains to summarize the evidence that has accumulated in support of this hypothetical pathway (TAMIR et al. 1996).

The most important natural secretogogue for parafollicular cells is increased $[Ca^{2+}]_e$. Stimulation by increased $[Ca^{2+}]_e$, as noted above, induces both the acidification of secretory vesicles and secretion. These phenomena suggested three questions that have now, at least partially, been answered. (a) Are vesicle acidification in, and secretion by, parafollicular cells the results of a single process or are they dissociable; could exocytosis occur if vesicle acidification were to be prevented? (b) What is the surface element of parafollicular cells that responds to elevations in $[Ca^{2+}]_e$; is it the calcium receptor, Ca^{2+} channel(s), or both? (c) How is the cellular perception of an increase in $[Ca^{2+}]_e$ transduced to gate vesicular Cl^- channels to the open configuration (how does either stimulation of the calcium receptor or entry of Ca^{2+} through a Ca^{2+}channel cause secretory vesicles to acidify)? The answers to these questions, summarized below, have served as the basis for constructing the model shown in Fig. 2 (TAMIR et al. 1996).

VII. Dissociation of Acidification of Parafollicular Secretory Vesicles from Secretion

Acidification of parafollicular secretory vesicles can be dissociated from secretion; exocytosis can occur even if acidification of secretory vesicles does not. Exposure of parafollicular cells to Ba^{2+} $(2-5\,mM)$, like exposure of these cells to increased $[Ca^{2+}]_e$, induces secretion of 5-HT. In fact, Ba^{2+} is a more potent secretogogue than is elevated $[Ca^{2+}]_e$, which does not induce significant secretion at concentrations below $5.0\,mM$. In contrast to increased $[Ca^{2+}]_e$, however, exposure of parafollicular cells to Ba^{2+} does not cause secretory vesicles to acidify. Ba^{2+} is known to enter cells through Ca^{2+} channels, but then to block these channels from the inside (PRZYWARA et al. 1993). Ba^{2+} also induces the release of Ca^{2+} from internal stores (PRZYWARA et al. 1993). As a result, Ba^{2+} powerfully evokes an increase in $[Ca^{2+}]_i$ (confirmed by measuring Fura-2 fluorescence), because the Ca^{2+} released internally by Ba^{2+} cannot diffuse out of cells through the Ba^{2+}-blocked plasmalemmal Ca^{2+} channels. Since the failure of vesicles to acidify does not interfere with Ba^{2+}-evoked secretion, vesicle acidification is clearly not a prerequisite for exocytosis.

C. Evoking of Vesicle Acidification by the Parafollicular Calcium Receptor

A role for the calcium receptor in secretogogue-induced acidification of parafollicular secretory vesicles has now been established (TAMIR et al. 1996). For this purpose, it was necessary to be able to distinguish the effects of stimulation of the calcium receptor from effects mediated solely by plasmalemmal Ca^{2+} channels. L-type Ca^{2+} channels appear to be able to contribute to the mediation of the acidification of vesicles, because dihydropyridines, such as nimodipine $(10\,\mu M)$, antagonize vesicle acidification when cells are exposed to very high concentrations of $[Ca^{2+}]_e$. Since the acidification induced by elevated $[Ca^{2+}]_e$ at concentrations $<2.0\,mM$, however, is unchanged by nimodipine, it seems likely that L-type Ca^{2+} channels are not involved in the vesicle acidification that occurs in response to exposure of parafollicular cells to low concentrations of $[Ca^{2+}]_e$. The response to $[Ca^{2+}]_e$ is thus complex; entry of Ca^{2+} through L-type Ca^{2+} channels evidently can provide a signal that causes vesicles to acidify, but this mechanism appears to be operative only under extreme conditions when the concentration of $[Ca^{2+}]_e$ is extraordinarily high. The resistance of the low end of the concentration-effect curve to inhibition by nimodipine is compatible with the idea that the calcium receptor, rather than a Ca^{2+} channel, mediates responses to low concentrations of $[Ca^{2+}]_e$. Although Ca^{2+} influx through L-type Ca^{2+} channels contributes to vesicle acidification only at higher concentrations of $[Ca^{2+}]_e$, the action of nimodipine alone does not rule out the participation of either Ca^{2+} channels that are insensitive to dihydropyridines, or the calcium receptor. To confirm the idea that the calcium channel is responsible for acidification of vesicles, it is necessary to expose parafollicular cells to a ligand that stimulates the calcium receptor, but does not increase the influx of Ca^{2+} through Ca^{2+} channels. In addition to Ca^{2+}, the calcium receptor binds to, and is activated by, lanthanides (Brown et al. 1993). Lanthanides block virtually all Ca^{2+} channels (Hess 1990). The lanthanide Gd^{3+} was therefore employed to stimulate the calcium receptor, while simultaneously blocking Ca^{2+} entry through

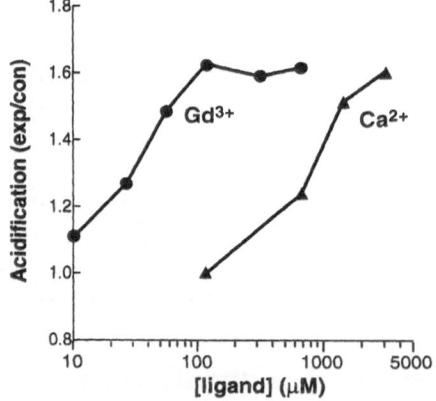

Fig. 3. Both Gd^{3+} and Ca^{2+} (in the presence of nimodipine to block L-type Ca^{2+} channels) stimulate vesicle acidification in parafollicular cells. Gd^{3+} is also a Ca^{2+} channel blocker. It is presumed that vesicle acidification in response to either agonist is due to stimulation of the plasmalemmal calcium receptor. Note that Gd^{3+} is more potent than Ca^{2+}; however, the concentration-effect curves for the two agonists are parallel

plasmalemmal Ca^{2+} channels. Gd^{3+} was found to be a potent inducer of vesicle acidification; the concentration-effect curve for Gd^{3+} is parallel to that of Ca^{2+}, suggesting an action on the same receptor; however, the potency of Gd^{3+} (ED_{50} = $52\mu M$) is ~tenfold greater than that of Ca^{2+} (ED_{50} = $636\mu M$) (Fig. 3). As expected, Gd^{3+} does not evoke 5-HT secretion, because entry of Ca^{2+} through calcium channels is absolutely required for exocytosis. Gd^{3+} causes vesicles to acidify, even in the presence of ethyleneglycoltetraacetic acid (EGTA) and nimodipine ($10\mu M$), confirming that its effects are independent of $[Ca^{2+}]_e$. These data suggest that the parafollicular cell calcium receptor can induce vesicle acidification, even if Ca^{2+} from the external solution is prevented from entering parafollicular cells; but since Gd^{3+} blocks Ca^{2+} channels, which are necessary for exocytosis, the action of Gd^{3+} does not, by itself, make clear whether activation of the calcium receptor is capable of inducing exocytosis as well as acidification. Conceivably, if calcium channels had not been blocked by Gd^{3+}, stimulation of the receptor might have induced secretion. The action of Gd^{3+} also establishes that if increased cytosolic $[Ca^{2+}]_i$ is a signal that leads to vesicle acidification the signaling pathway must include the relapse of $[Ca^{2+}]_i$ from an internal source.

Since stimulation of the calcium receptor evokes vesicle acidification, as the results with Gd^{3+} suggest, it is difficult to understand why Ba^{2+}, which binds to the calcium receptor and induces secretion (see above), does not cause vesicles to acidify. The answer to this seeming paradox is that Ba^{2+} evidently antagonizes signal transduction from the calcium receptor by an internal action. As a result, Ba^{2+} does induce vesicle acidification when its entry into parafollicular cells is blocked. When entry of Ba^{2+} into parafollicular cells is prevented by simultaneously exposing cells to nimodipine, Ba^{2+} causes vesicles to acidify. This observation suggests that Ba^{2+} activates the calcium receptor, but this action can be negated by an opposing intracellular action of Ba^{2+}. Indeed, the effects of submaximal concentrations of Ba^{2+} and Gd^{3+} on vesicle acidification are additive. The additive action can be explained by the ability of Gd^{3+} to block Ca^{2+} channels and prevent Ba^{2+} from entering parafollicular cells. If Ba^{2+} does not enter the cells, then its action is confined to stimulation of the calcium receptor on the cell surface, and vesicle acidification proceeds.

I. Induction of Binding of [³⁵S]GTPγS to Gαi and Gαq by Stimulation of Parafollicular Cells by the Calcium Receptor Agonists Ca^{2+} or Gd^{3+}

The G proteins that are coupled to the calcium receptor were identified by incubating isolated parafollicular cell membranes with [³⁵S]GTPγS in the presence of a ligand (Gd^{3+} or Ca^{2+}). The radioactivity immunoprecipitated by antibodies that react specifically with Gαi, Gαq, or all known Gα-subunits was then determined. Both Ca^{2+} and Gd^{3+} were found to stimulate the binding of [³⁵S]GTPγS to Gαi and, to a much lesser extent, to Gαq. Treatment of the membranes with pertussis toxin (PTx) abolished the binding of [³⁵S]GTPγS to

Gαi; however, the ligand-stimulated binding of [^{35}S]GTPγS to Gαq was unaffected. Since PTx blocks the agonist-induced binding of [^{35}S]GTPγS to Gαi, but not Gαq, the effects of PTx on vesicle acidification were determined. If the critical G protein involved in the transduction of calcium receptor stimulation is Gi, then PTx would be expected to inhibit the acidification of vesicles. In contrast, if a significant component of the pathway is dependent on the coupling of the receptor to Gq, then vesicle acidification should be at least relatively resistant to inhibition by PTx. In fact, PTx was able to virtually abolish agonist-induced vesicle acidification. It thus seems that Gi, rather than Gq, is essential for coupling the calcium receptor to the acidification of secretory vesicles.

Although not essential for acidification of secretory vesicles, entry of Ca^{2+} from the outside appears to contribute to signal transduction. Under physiological conditions, when parafollicular cells are stimulated by elevated $[Ca^{2+}]_e$ and not by Gd^{3+}, plasmalemmal voltage-sensitive Ca^{2+} channels do mediate Ca^{2+} entry and lead both to acidification of secretory vesicles and to secretion. At low concentrations, $[Ca^{2+}]_e$ can, as noted above, acidify vesicles even in the presence of nimodipine; however, the increment in acidification that occurs at concentrations of $[Ca^{2+}]_e$ above $1–2\,mM$ is nimodipine sensitive. In contrast to vesicle acidification, secretion of 5-HT evoked by increased $[Ca^{2+}]_e$ is abolished by nimodipine, suggesting that Ca^{2+} entry from the external medium is absolutely essential for exocytosis, even though it is not essential for vesicle acidification. This conclusion was confirmed by using thapsigargin ($10\,\mu M$; $60\,min$) to deplete Ca^{2+} from the endoplasmic reticulum. Thapsigargin irreversibly inhibits the transport of Ca^{2+} into the ER and other inositol trisphosphate (IP_3)-releasable Ca^{2+} pools, leading eventually to depletion of internal stores of Ca^{2+} (Ghosh et al. 1991; Lytton et al. 1991). Thapsigargin does not affect plasmalemmal Ca^{2+} pumps or channels. Exposure of isolated parafollicular cells to $\uparrow[Ca^{2+}]_e$ evoked 5-HT secretion, even when internal stores of Ca^{2+} were depleted by thapsigargin. Internal, IP_3-releasable Ca^{2+} stores are thus not essential for exocytosis, although, as we have seen, they probably are essential for vesicle acidification.

II. Release of Ca^{2+} from Internal Stores and Increase in Cytosolic $[Ca^{2+}]_i$ by Activation of the Calcium Receptor

The observations that ligands for the calcium receptor, such as Gd^{3+}, and the influx of Ca^{2+} through Ca^{2+} channels, can each induce vesicle acidification, suggest that $[Ca^{2+}]_i$ is an important signal in mediating the response. The calcium receptor could induce the release of Ca^{2+} from the endoplasmic reticulum, thereby enabling the calcium receptor to be effective even in the absence of $[Ca^{2+}]_e$. Fura-2 was therefore used to evaluate the effects of exposure to Gd^{3+} on $[Ca^{2+}]_i$. Gd^{3+} was found to cause an increase in $[Ca^{2+}]_i$. The source of the Ca^{2+} responsible for the Gd^{3+}-evoked increase in $[Ca^{2+}]_i$ was investigated by exposing isolated parafollicular cells for $60\,min$ to thapsigargin ($10\,\mu M$).

Thapsigargin, by itself, evoked an initial transient increase $[Ca^{2+}]_i$, which rapidly dissipated. Secretory vesicles acidified briefly during the transient increase of $[Ca^{2+}]_i$ induced by thapsigargin. This observation itself suggests that elevation of $[Ca^{2+}]_i$ is sufficient as signal to induce the acidification of secretory vesicles. The acidification in response to thapsigargin was rapidly lost as the $[Ca^{2+}]_i$ declined. Parafollicular cells were challenged with the calcium receptor agonist Gd^{3+} 60 min after application of thapsigargin, when the $[Ca^{2+}]_i$ was no longer elevated, and secretory vesicles were no longer acidic. Pretreatment of parafollicular cells with thapsigargin blocked the ability of Gd^{3+} to increase $[Ca^{2+}]_i$ and also to acidify secretory vesicles (TAMIR et al. 1996). These data suggest that stimulation of the calcium receptor by Gd^{3+} leads to the release of Ca^{2+} from the endoplasmic reticulum and that the consequent elevation of $[Ca^{2+}]_i$ is both necessary and sufficient to trigger acidification of parafollicular secretory vesicles. Similarly, low concentrations of $[Ca^{2+}]_e$ also evoked a rise in the concentration of $[Ca^{2+}]_i$, even in the presence of nimodipine. Again, the rise in $[Ca^{2+}]_i$ induced by low concentrations of $[Ca^{2+}]_e$ was prevented by pretreatment of cells with thapsigargin. These data are consistent with the idea that low concentrations of $[Ca^{2+}]_e$, like Gd^{3+}, activate the calcium receptor, which leads to the release of Ca^{2+} from the endoplasmic reticulum.

Since the calcium receptor is a G-protein-coupled receptor and its action evidently depends on the release of $[Ca^{2+}]_i$ from the endoplasmic reticulum, it is likely that the Gi activated by the calcium receptor stimulates a phosphoinositide-specific phospholipase C. The hydrolysis of phosphoinositides by a phosphoinositide-specific phospholipase C to produce IP_3 and diacylglycerol (DAG) is the most common mechanism of introducing to the cytosol an activator of $[Ca^{2+}]_i$-releasing receptors on the endoplasmic reticulum. The IP_3 resulting from the action of a PI-PLC could thus be the signal that evokes the release of Ca^{2+} from the endoplasmic reticulum. U73122, a selective phosphoinositide-specific phospholipase C inhibitor, but not its inactive congener, U733433, blocked both Ca^{2+}- and Gd^{3+}-induced vesicle acidification (TAMIR et al. 1996). These observations suggest that the Gi, which is activated by the calcium receptor, stimulates a phosphoinositide-specific phospholipase C to generate IP_3 and increase $[Ca^{2+}]_i$. The Gi could activate the phosphoinositide-specific phospholipase C either through its α- or $\beta\gamma$-subunits, each of which has previously been found to stimulate this enzyme (BLANK et al. 1991; MURTHY and MAKHLOUF 1995).

Further experiments were carried out in order to clarify the role played in the acidification signal transduction pathway by elevated $[Ca^{2+}]_i$. An unsurprising observation was that the calmodulin inhibitors W-7 ($50\,\mu M$) and trifluoperazine ($15\,\mu M$) each almost completely blocked agonist-induced acidification of secretory vesicles (TAMIR et al. 1994c). Parafollicular cells were also found to contain NADPH diaphorase, which was demonstrated histochemically, and nitric oxide synthase (NOS), which was demonstrated immunocytochemically (TAMIR et al. 1996). NOS has been shown to have NADPH diaphorase activity (DAWSON et al. 1991; HOPE et al. 1991). The constitutive form of NOS is known to be activated by increased $[Ca^{2+}]_i$; therefore, the involvement of NO production in the signal transduction pathway leading to acidification of secretory vesicles was tested. The idea that NO is interposed in the signaling pathway has been supported by four observations: (a) exposure of parafollicular cells to sodium nitroprusside, which generates NO (BERT and SNYDER 1989), causes parafollicular secretory vesicles to acidify (TAMIR et al. 1996). (b) L-N-5-(1-iminoethyl) ornithine (imino-ornithine), which is an irreversible inhibitor of NOS (McCALL et al. 1991), blocks Gd^{3+}-induced vesicle acidification (TAMIR et al. 1996). (c) If NO is involved in signaling, it is likely to activate guanylate cyclase (one of the NO receptors), thereby producing cGMP (YOUNG et al. 1993). LY83583, a guanylate cyclase inhibitor (PANDOL and SCHOEFFIELD-PAYNE 1990), does indeed antagonize Gd^{3+}-induced vesicle acidification (TAMIR et al. 1996); moreover, exposure of parafollicular cells to dibutyryl cGMP induces the acidification of secretory vesicles and this response to dibutyryl cGMP is not inhibited either by LY83583 or by imino-ornithine, as would be expected of agents that antagonize steps in the production of cGMP, but not its action. (d) RpcGMPS, an inhibitor of cGMP-dependent kinase, blocks the acidification of vesicles induced by calcium receptor agonists, implying that vesicle acidification depends on the phosphorylation of a protein by an enzyme activated by cGMP. These observations thus provide strong support for the idea that NO participates in the signal transduction pathway that couples the calcium receptor to the acidification of secretory granules in parafollicular cells.

Since the acidification of vesicles, induced by neither nitroprusside nor dibutyryl cGMP, is associated with an increase in $[Ca^{2+}]_i$, it is likely that NO acts at a step in the signal transduction pathway that is distal to that affected by $[Ca^{2+}]_i$. This conclusion is consistent with the hypothesis that Ca^{2+}/calmodulin activates NOS and thus acts prior to the generation of NO. NO activates guanylyl cyclase, which in turn activates cGMP-dependent kinase (G kinase). The G-kinase inhibitor Rp-cGMPS antagonizes vesicle acidification, suggesting that this enzyme gates the vesicular Cl⁻ channel, directly or via a regulatory intermediate. Phosphorylation of p64 is then associated with channel opening (TAMIR et al. 1994c).

Even if one accepts the evidence that NO, cGMP, and a G-kinase are responsible for gating the Cl⁻ channel in the membranes of parafollicular secretory vesicles that regulates the acidification of the vesicles, it does not necessarily follow that this transduction pathway is the only route from the calcium receptor to vesicle membrane. DAG is generated along with IP_3 as a result of activation of the calcium receptor. DAG is a well-known activator of PKC. It is therefore necessary to determine whether PKC also contributes to calcium receptor-mediated vesicle acidification. Strong evidence, in fact, suggests that it does (TAMIR et al. 1996). The PKC activator phorbol myristyl acetate is a potent inducer of vesicle acidification. In contrast to the calcium receptor agonists Gd^{3+} or increased $[Ca^{2+}]_e$, however, phorbol myristyl acetate does not increase $[Ca^{2+}]_i$. The acidification induced by phorbol myristyl acetate is also unaffected by inhibitors of NOS or guanylyl cyclase. The response of vesicles to phorbol myristyl acetate is blocked by staurosporin and by other, more selective PKC inhibitors, including calphostin C and chelerythrine. Neither staurosporin nor the other PKC inhibitors, however, inhibit the acidification of vesicles induced by the calcium receptor agonists Gd^{3+} or increased $[Ca^{2+}]_e$. Similarly, chronic downregulation of PKC (by prolonged exposure to a phorbol ester) blocks the acidification of secretory vesicles in response to phorbol myristyl acetate, but it fails to affect the acidification of vesicles that occurs in response to Gd^{3+} or $\uparrow[Ca^{2+}]_e$. These observations indicate that PKC can, when activated, cause parafollicular vesicles to acidify, but the enzyme is not likely to be a physiologically significant factor in transducing calcium receptor-related vesicle acidification.

Since there is a nimodipine-sensitive increment in vesicle acidification, which becomes active at concentrations of $[Ca^{2+}]_e$ higher than those needed to stimulate the calcium receptor, there is a second pathway that also opens Cl^- channels in the membranes of secretory vesicles. The fact that PKC is not physiologically involved in signal transduction of responses initiated by the calcium receptor does not rule out the possibility that PKC contributes physiologically to the Ca^{2+}-channel-related acidification of secretory vesicles. Neither the chronic downregulation of PKC nor the PKC inhibitor staurosporin affects the nimodipine-insensitive component of vesicle acidification (the calcium receptor-mediated response); nevertheless, both the chronic downregulation of PKC and staurosporin block the nimodipine-dependent increment in vesicle acidification that occurs at high concentrations of $[Ca^{2+}]_e$. These observations are consistent with the ideas that PKC does not contribute to the calcium receptor-induced acidification of vesicles, but PKC is responsible for an increment in vesicle acidification evoked by the entry of Ca^{2+} through nimodipine-sensitive calcium channels (TAMIR et al. 1996).

III. Electrophysiological Characterization of Plasmalemmal Ca^{2+} Channels of Parafollicular Cells

Electrophysiological studies have been carried out to characterize ion channels in the plasma membrane of isolated parafollicular cells (HSIUNG et al. 1993; MITHAL et al. 1994). Measurements of membrane potential, using whole-cell-perforated patch recordings, indicate that a high concentration of $[Ca^{2+}]_e$, $10\,mM$, induces a slow depolarization of the cells, with an associated increase in membrane conductance. This effect causes a rhythmic series of action potentials that continue until the $[Ca^{2+}]_e$ is returned to its resting level, $2.5\,mM$. Imaging of cells loaded with Fura-2 reveals that exposure of parafollicular cells to this high concentration of $[Ca^{2+}]_e$ causes a rhythmic series of oscillations in

$[Ca^{2+}]_i$ that coincides with the period during which $[Ca^{2+}]_e$ is elevated. Under voltage clamp conditions (V_{max} = –60mV), application of 10mM $[Ca^{2+}]_e$ is found to stimulate an inward current in parafollicular cells that does not return to baseline until the high concentration of $[Ca^{2+}]_e$ is washed out. Pretreatment of cells with non-subtype-selective Ca^{2+} channel blockers (La^{3+} or Gd^{3+}) prevents the depolarization induced by 10mM $[Ca^{2+}]_e$. Nimodipine, which interferes with the increase in $[Ca^{2+}]_i$, also inhibits the depolarization induced by 10mM $[Ca^{2+}]_e$, but not as strongly as does La^{3+}. These data are compatible with the idea that the changes in membrane conductance induced by exposing parafollicular cells to an elevation of $[Ca^{2+}]_e$ contribute to the ability of high concentrations of $[Ca^{2+}]_e$ to act as a secretogogue. The relationship between the calcium receptor, depolarization of parafollicular cells, and the opening of voltage-gated Ca^{2+} channels in their plasma membrane remains to be determined. Parafollicular cells also express voltage-dependent Na^+ channels in their plasma membrane (DERIEMER et al. 1991). These channels may play a role in secretory mechanisms since secretion induced by exposing parafollicular cells to elevated levels of $[Ca^{2+}]_e$ is partially inhibited by tetrodotoxin.

IV. Real-Time Measurement of 5-HT Secretion from Single Parafollicular Vesicles

The secretion of 5-HT by cultured parafollicular cells has recently been studied with an amperometric detection method that permits the release of 5-HT from single exocytotic events to be detected (personal observations, in collaboration with Dr. David Sulzer of Columbia University). This method employs a carbon fiber electrode to measure 5-HT. Single exocytotic events appear as spike-like oxidation currents, with a good time resolution. The mean quantal content of released 5-HT was found to be 100242 and the maximum was 265000 molecules. Preliminary observations suggest that repeated stimulation of parafollicular cells gives rise to a gradually increasing degree of acidity in the secretory vesicles. In parallel, the quantal content of secreted 5-HT also appears to increase. Blockade of the increment in vesicle acidification, for example by inhibition of the vesicular H^+ ATPase, prevents the increase in quantal content of secreted 5-HT that accompanies repeated stimulation. These observations suggest that the acidification of vesicles may serve to make them more efficient carriers of transmitter. As acidification increases, the vesicles would be expected, as noted above, to load increasing quantities of 5-HT, because the transmembrane transport of 5-HT into vesicles is driven by the ΔpH across the vesicular membrane (CIDON et al. 1991). In addition, the acidification of vesicles would be expected to enhance the dissociation of 5-HT from intravesicular (or membrane-bound) SBP, and thus promote its diffusion to the extracellular medium at the time of exocytosis. The binding of 5-HT to SBP is strongly inhibited when the ambient pH falls below 6.6 (BARASCH et al. 1987b).

References

Amara SG, Jonas V, Rosenfeld MG (1982) Alternative RNA processing in calcitonin gene expression generates mRNA encoding different polypeptide products. Nature 298:240–244

Annaert WG, Quatacker J, Liona I, De Potter WP (1994) Differences in the distribution of cytochrome b561 and synaptophysin in dog splenic nerve: a biochemical and immunocytochemical study. J Neurochem 62:265–274

Baer L (1996) Behavior therapy: endogenous serotonin therapy? J Clin Psychiatry 57 (Suppl 6):33–35

Bajjalieh S, Scheller R (1995) The biochemistry of neurotransmitter secretion. J Biol Chem 270:1971–1974

Barasch JM, Mackey H, Tamir H, Nunez EA, Gershon MD (1987a) Induction of a neural phenotype in a serotonergic endocrine cell derived from the neural crest. J Neurosci 7:2874–2883

Barasch JM, Tamir H, Nunez EA, Gershon MD (1987b) Serotonin-storing secretory granules from thyroid parafollicular cells. J Neurosci 7:4017–4033

Barasch JM, Gershon MD, Nunez EA, Tamir H, Al-Awqati Q (1988) Thyrotropin induces the acidification of the secretory granules of parafollicular cells by increasing the chloride conductance of the granular membrane. J Cell Biol 107:2137–2148

Bauerfeind R, Regnier-Vigouroux A, Flatmark T, Huttner WB (1993) Selective storage of acetylcholine but not catecholamine in neuroendocrine synaptic-like microvesicles of early endosomal origin. Neuron 11:105–121

Bernd P, Gershon MD, Nunez EA, Tamir H (1981) Separation of dissociated thyroid follicular and parafollicular cells: association of serotonin binding protein with parafollicular cells. J Cell Biol 88:499–508

Bert DS, Snyder SH (1989) Nitric oxide mediates glutamate-linked enhancement of cGMP levels in the cerebellum. Proc Natl Acad Sci USA 86:9030–9033

Blank JL, Brattain KA, Exton JH (1991) Activation of cytosolic phosphoinositide phospholipase C by G-protein beta gamma subunits. J Biol Chem 267:23069–23075

Blaugrund E, Pham TD, Tennyson VM, Lo L, Sommer L, Anderson DJ, Gershon MD (1996) Distinct subpopulations of enteric neuronal progenitors defined by time of development, sympathoadrenal lineage markers, and *Mash-1*-dependence. Development 122:309–320

Blundell JE, Lawton CL (1995) Serotonin and dietary fat intake: effects of dexfenfluramine. Metabolism 44:33–37

Brown EM, Gamba G, Riccardi D, Lombardi M, Butters R, Kifor O, Sun A, Hediger MA, Lytton J, Hebert SC (1993) Cloning and characterization of an extracellular Ca^{2+}-sensing receptor from bovine parathyroid. Nature 366:575–580

Cameron PL, Sudhof TC, Jahn R, De Camilli P (1991) Colocalization of synaptophysin with transferrin receptors: implication for synaptic vesicle biogenesis. J Cell Biol 115:151–164

Cidon S, Tamir H, Nunez EA, Gershon MD (1991) ATP-dependent uptake of 5-hydroxytryptamine by secretory granules isolated from thyroid parafollicular cells. J Biol Chem 266:4392–4400

Clark M, Lanigan T, Page N, Russo A (1995) Induction of a serotonergic and neuronal phenotype in thyroid C-cells. J Neurosci 15:6167–6178

Dawson TM, Bredt DS, Fotuhi M, Hwang PM, Snyder SH (1991) Nitric oxide synthase and neuronal NADPH diaphorase are identical in brain and peripheral tissue. Proc Natl Acad Sci USA 88:7797–7801

De Camilli P, Jahn R (1990) Pathways to the regulated exocytosis in neurons. Ann Rev Physiol 52:625–645

De Potter WP, Coen EP, De Potter RW (1987) Evidence for the coexistence and co-release of [Met]enkephalin and noradrenaline from sympathetic nerves of the bovine vas deferens. Neuroscience 20:855–866

Deriemer SA, Nelkin B, Tamir H (1991) Ras induction of sodium channels in medullary thyroid carcinoma. Physiologist 34:104

Elfcrink L, Pctcrson M, Schcller R (1993) A role for synaptotagmin (p65) in regulated exocytosis. Cell 72:153–159

Erspamer V (1966) Occurrence of indolealkylamines in nature . In: Erspamer V (eds): 5-Hydroxytryptamine and related indolealkylamines. Springer-Verlag, Berlin Heidelberg New York, pp 132–181 (Handbook of experimental pharmacology, vol 19)

Fesce R, Grohasz F, Valtorta F, Meldolesi J (1994) Neurotransmitter release: fusion or "kiss-and-run"? Trends Cell Biol 4:1–4

Fujita T (1987) Paraneurons. In: Adelman G (eds) Encyclopedia of neuroscience, vol II II. Birkhaüser, Boston, pp 921–923.

Garrett JE, Tamir H, Kifor O, Simin RT, Rogers KV, Mithal A, Gagel RF, Brown EM (1995) Calcitonin-secreting cells of the thyroid express and extracellular calcium sensing receptor gene. Endocrinology 136:5202–5211

Gershon MD, Tamir HT (1985) Peripheral sources of serotonin and serotonin-binding proteins. In: Vanhoutte PM (eds) Serotonin and the cardiovascular system. Raven, New York, pp 15–26

Gershon MD, Liu KP, Karpiak SE, Tamir H (1983) Storage of serotonin in vivo as a complex with serotonin binding protein in central and peripheral serotonergic neurons. J Neurosci 3:1901–1911

Ghosh TK, Bian J, Short AD, Rybak SL, Gill DL (1991) Persistent intracellular calcium pool depletion by thapsigargin and its influence on cell growth. J Biol Chem 266:24690–24697

Guillemot F, Joyner AL (1993) Dynamic expression of the murine achaete-scute homolog (MASH-1) in the developing nervous system. Mech Dev 42:171–185

Hess P (1990) Calcium channels in vertebrate cells. Annu Rev Neurosci 13:337–356

Heuser J (1989) The role of coated vesicle in recycling of synaptic vesicle membrane. Cell Biol Int Rep 13:1063–1076

Hope BT, Michael GJ, Knigge KM, Vincent SR (1991) Neuronal NADPH diaphorase is a nitric oxide synthase. Proc Natl Acad Sci USA 88:2811–2814

Hsiung SC, Tamir H, McGhee DS (1993) Extracellular calcium-induced changes in the membrane electrical properties of thyroid parafollicular cells. Neurosci Abst 19:1333

Ivgy-May N, Tamir H, Gershon MD (1994) Synaptic properties of serotonergic growth cones in developing rat brain. J Neurosci 14:1011–1029

Jacobs-Cohen R, Tamir H, Gershon M (1994) Expression of neuronal phenotype by neural crest-derived paraneurons (parafollicular cells) is antagonized by thyroid hormone (triiodothyronine; T3). Neurosci Abst 20:654

Jahn R, Sudhof TC (1994) Synaptic vesicles and exocytosis. Ann Rev Neurosci 17:219–246

Joffe RT, Levitt AJ, Sokolov ST (1996) Augmentation strategies: focus on anxiolytics. J Clin Psychiatry 57 (Suppl 7):25–31

Johnson RG (1988) Accumulation of biological amines into chromaffin granules: a model for hormone and neurotransmitter transport. Physiol Rev 68:232–307

Jonakait GM, Tamir H, Rapport MM, Gershon MD (1977) Detection of a soluble serotonin-binding protein in the mammalian myenteric plexus and other peripheral sites of serotonin storage. J Neurochem 28:277–284

Jonakait GM, Tamir H, Gintzler AR, Gershon MD (1979) Release of [³H] serotonin and its binding protein from enteric neurons. Brain Res 174:55–69

Kirchgessner AL, Gershon MD, Liu KP, Tamir H (1988) Co-storage of serotonin binding protein with serotonin in the rat CNS. J Neurosci 276: 607–621

Klein DF (1996) Panic disorder and agarophobia: hypothesis hothouse. J Clin Psychiatry 57 (Suppl 6):21–27

Kong JY, Thureson-Klein AK, Klein RL (1990) Are NPY and enkephalins co-stored in the same noradrenergic neurons and vesicles. Peptides 11:565–575

Kuzel R (1996) Management of depression. Current trends in primary care. Postgrad Med 99:179–180, 185–186, 189–192

Le Douarin NM (1982) The neural crest. Cambridge University Press, Cambridge

Li C, Ullrich B, Zhang J, Anderson R, Brose N, Sudhof T (1995) Ca^{2+}-dependent and -independent activities of neuronal and non-neuronal synaptotagmins. Nature 375:594–599

Linstedt AD, Kelly RB (1991) Synaptophysin is sorted from endocytic markers in neuroendocrine PC12 cells but not transfected fibroblasts. Neuron 7:309–317

Liu Y, Peter D, Roghani A, Schuldiner S, Privé GG, Eisenberg D, Brecha N, Edwards RH (1992) A cDNA that suppresses MPP^+ toxicity encodes a vesicular amine transporter. Cell 70:539–551

Lucki I (1996) Serotonin receptor specificity in anxiety disorders. J Clin Psychiatry 57 (Suppl 6):5–10

Lytton J, Westlin M, Hanley M (1991) Thapsigargin inhibits sarcoplasmal or endoplasmal reticulum calcium-ATPase family of calcium pumps. J Biol Chem 266:17067–17071

McCall TB, Feelish M, Palmer RM, Moncada S (1991) Identification of N-5-(1-iminoethyl) ornithine is an irreversible inhibitor of NO synthase in phagocytic cells. Br J Pharmacol 102:234–239

Michalchik MS, Russo AF (1994) Characterization of the extracellular calcium receptor in thyroid C-cells. Neurosci Abst 20:67

Mitchell BS, Stauber BV (1993) Localization of substance P and leucine enkephalin in nerve terminals of the guinea pig paracervical ganglion. Histochem J 25:144–149

Mithal A, Kifor O, Thun R, Krapcho K, Fuller F, Hebert S, Brown E, Tamir H (1994) Highly purified sheep C-cells express an extracellular Ca^{2+} receptor similar to that present in parathyroid. Am Soc Bone Mineral Res 9 (Suppl 1):S282

Moriyama Y, Nelson N (1989) H^+-translocating ATPase in Golgi apparatus. J Biol Chem 264:33–35

Murthy KS, Makhlouf GM (1995) Adenosine A1 receptor-mediated activation of phospholipase C-b3 in intestinal muscle: dual requirement for alpha and beta gamma subunits of Gi3. Mol Pharmacol 47:1172–1179

Noach EL (1994) Appetite regulation by serotoninergic mechanisms and effect of d-fenfluramine. Neth J Med 45:123–133

Nunez EA, Gershon MD (1972) Studies of the synthesis and storage of serotonin by parafollicular (C) cells of the thyroid gland of active, pre-hibernating and hibernating bats. Endocrinology 90:1008–1024

Nunez EA, Gershon MD (1978) The cytophysiology of thyroid parafollicular cells. Int Rev Cytol 52:1–80

Pandol S, Schoeffield-Payne M (1990) Cyclic GMP mediates the agonist stimulated increase in plasma membrane calcium entry in the pancreatic acinar cell. J Biol Chem 265:12846–12853

Przywara D, Pschowdhura P, Bhaze S, Wakade T, Wakade A (1993) Barium-induced exocytosis is due to internal calcium release and blocking of calcium efflux. Proc Natl Acad Sci USA 19:557–561

Rogers KV, Brown EM, Herbert SC (1994) Localization of calcium receptor mRNA in rat brain using in situ hybridization histochemistry. Neurosci Abst 20:1061

Rothman JE, Orci L (1992) Molecular dissection of the secretory pathway. Nature 30:409–415

Rothman JE, Warren G (1994) Implication of the SNARE hypothesis for intracellular membrane topology and dynamics. Curr Biol 1:220–233

Ruat M, Molliver M, Snowman A, Snyder S (1995) Calcium sensing receptor: molecular cloning in rat and localization to nerve terminals. Proc Natl Acad Sci USA 92:3161–3165

Russo AF, Lanigan TM, Sullivan BE (1992) Neuronal properties of a thyroid C-cell line: partial repression by dexamethasone and retinoic acid. Mol Endocrinol 6:207–218

Scheller R (1995) Membrane trafficking in the presynaptic nerve terminal. Neuron 14:893–897

Sudhof T (1995) The synaptic vesicle cycle: a cascade of protein-protein interactions. Nature 375:645–653

Sulzer D, Chen T, Lau Y, Kristensen H, Rayport S, Ewing A (1995) Amphetamine redistributes dopamine from synaptic vesicles to the cytosol and promotes reverse transport. J Neurosci 15:4102–4108

Tamir H, Liu K, Payette RF, Hsiung S, Adlersberg M, Nunez EA, Gershon MD (1989) Human medullary thyroid carcinoma: characterization of the serotonergic and neuronal properties of neurectodermally-derived cell line. J Neurosci 9:1199–1212

Tamir H, Liu KP, Hsiung SH, Adlersberg M, Nunez EA, Gershon MD (1990) Multiple signal transduction mechanisms leading to the secretion of 5-hydroxytryptamine by MTC cells, a neurectodermally derived cell line. J Neurosci 10:3743–3753

Tamir H, Hsiung S-C, Yu PY, Liu K-P, Adlersberg M, Nunez EA, Gershon MD (1992) Serotonergic signalling between thyroid cells: protein kinase C and 5-HT$_2$ receptors in the secretion and action of serotonin. Synapse 12:155–168

Tamir H, Liu K-P, Hsiung S-C, Adlersberg M, Gershon MD (1994a) Serotonin binding protein: synthesis, secretion, and recycling. J Neurochem 63:97–107

Tamir H, Liu KP, Heath M, Adlersberg M, Gershon MD (1994b) Stimulus-induced secretion and vesicle acidification in serotonergic paraneurons (parafollicular cells): the role of voltage-gated Ca^{2+} channels. Neurosci Abst 20:1719

Tamir H, Piscopo I, Liu K-P, Hsiung S, Adlersberg M, Nicolaides M, Al-Awqati Q, Gershon MD (1994c) Secretogogue-induced gating of chloride channels in the secretory vesicles of a paraneuron. Endocrinology 135:2045–2057

Tamir H, Liu K-P, Adlersberg M, Hsiung S-C, Gershon MD (1996) Acidification of serotonin-containing secretory vesicles induced by a plasma membrane calcium receptor. J Biol Chem 271:6441–6450

Thureson-Klein AK, Klein RL (1990) Exocytosis from neuronal large dense cored vesicles. Int Rev Cytol 121:67–126

Valtorta F, Fesce R, Grohovaz F, Haimann C, Hurlbut WP, Iezzi N, Tarelli FT, Villa A, Ceccarelli B (1990) Neurotransmitter release and synaptic vesicle recycling. Neuroscience 35:477–489

Weiler R, Cidon S, Gershon MD, Tamir H, Hogue-Angeletti R, Winkler H (1989) Adrenal chromaffin granules and secretory granules from thyroid parafollicular cells have several common antigens. FEBS Lett 257:457–459

Young HM, McConalogue K, Furness JB, De Vente J (1993) Nitric oxide targets in the guinea-pig intestine identified by induction of cyclic GMP immunoreactivity. Neuroscience 55:583–96

CHAPTER 7

Molecular Biology and Transductional Characteristics of 5-HT Receptors

P.R. Hartig

A. Serotonin, the Largest 7TM Neuroreceptor Family?

Serotonin is an ancient neurotransmitter. Phylogenetic comparisons suggest that the earliest serotonin receptor may have first appeared over 700 million years ago (Peroutka and Howell 1994). It is released from fiber tracts originating in primitive brainstem regions (the raphe nuclei of the brainstem reticular formation), and is found in species as ancient as *Aplysia*, *Caenorhabditis elegans*, and *Drosophila*. In addition, its synapses and receptors are widely distributed throughout the brain, serving many broad modulatory roles for diverse brain functions. It is not surprising, then, that serotonin receptors are primarily of the G-protein-coupled, 7-transmembrane receptor (7TM) class since these receptors couple to enzymatic and ion channel-modulating second messenger pathways (primarily adenylate cyclase, phosphoinositide turnover, and potassium channel modulation) which exhibit time constants in the second to millisecond range, consistent with broad signal modulation and gain-setting functions in the brain.

The broad distribution and diverse roles of serotonin are reflected in its large complement of 7TM receptor subtypes. These number 12 subtypes in man (13 in rodents) plus several pharmacologically unique species homologues (see below) plus a single ligand-gated ion channel subtype (5-HT$_3$). At present, this diversity of 7TM receptor subtypes appears to be the largest known for any neurotransmitter. This diversity may be a direct result of the ancient origin, broad regional distribution and wide-ranging roles of serotonin. These factors may have driven the evolutionary expansion and divergence of serotonin receptor subtypes by an overall process of gene duplication, evolutionary divergence, and fixation in the genome of new receptor subtypes (as occurred in the hemoglobin and homeobox gene families (Wills 1993)).

This review surveys the diversity of mammalian cDNA and genomic clones reported for serotonin receptors, information on second messenger coupling of these subtypes in transfected cell systems, and insights provided by molecular biological approaches such as receptor mutagenesis, chimeric receptor studies, gene knockouts, and antisense oligonucleotides. Only minimal comments will be provided on the distribution and roles of serotonin receptor subtypes since these are covered in other chapters. Additional reviews on the molecular biology of serotonin receptors include those by Julius (1991), Beer et al. (1993), BOESS and MARTIN (1994), and LUCAS and HEN (1995). Information on the chromosomal localization of both mouse and rat serotonin receptor genes can be found in the review by LUCAS and HEN (1995) and is not reviewed here.

B. Receptor Families and Subtypes

To date, 14 unique serotonin receptor subtypes and their related genes have been described in mammals, along with many additional species homologues (orthologous genes) that represent the "equivalent" receptor subtype in a different species [see HARTIG et al. 1992 for a discussion of species homologues (orthologues) of serotonin receptor genes]. With the exception of the 5-ht$_{5B}$ receptor (see below), all mammalian species studied to date appear to have the same set of functional serotonin receptor genes. This has led to a consistent molecular nomenclature for these mammalian serotonin receptor subtypes (HOYER et al. 1994; HARTIG et al. 1996) as shown in Table 1. The current nomenclature has evolved over the past 20 years from the surprisingly

Dendogram of the Serotonin Receptor Family

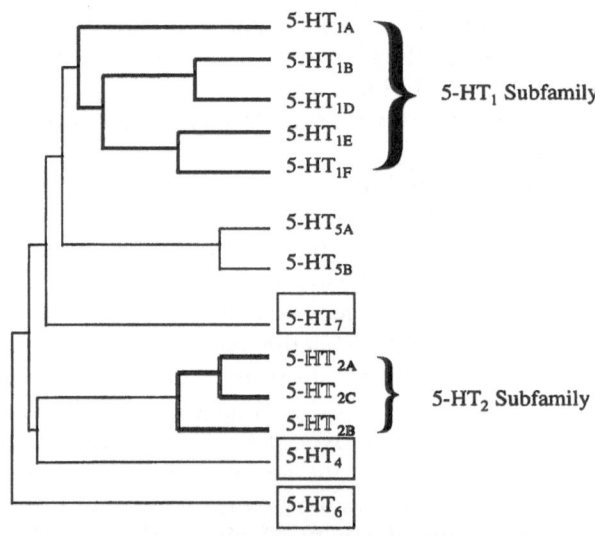

Fig. 1. Dendogram of the transmembrane amino acid identities between the 13 serotonin 7TM receptors. The lengths of line segments along the x-axis are proportional to the number of transmembrane amino acid differences between the receptor sequences shown in Fig. 2. All 5-HT$_1$ receptors are strongly coupled to G$_i$ or G$_o$ (adenylate cyclase inhibition) and all 5-HT$_2$ receptors are strongly coupled to G$_q$ or G$_{11}$ (activation of phosphoinositide hydrolysis). The subtypes *within boxes* (5-HT$_4$, 5-HT$_6$, 5-HT$_7$) are strongly coupled to G$_s$ (adenylate cyclase stimulation) but are not closely related and thus do not form a subfamily grouping. The second messenger coupling of the 5-ht$_5$ receptors is unknown

Table 1. The current recommended serotonin receptor nomenclature (Hoyer et al. 1994; Hartig et al. 1996) is shown below, utilizing a capital letter 5-HT designation for those subtypes that have fulfilled, and a small letter 5-ht designation for those subtypes that have not yet fulfilled, all of the operational, structural, and transductional criteria needed for full receptor subtype classification. Some of the previous names used for each serotonin subtype and their genes, which are found in the published literature, are also listed. The pharmacological properties of each receptor subtype may differ significantly between different species so the use of lowercase letters to designate the species from which the clone was obtained is encouraged in those cases (e.g., h5-HT$_{1D}$; see Hartig et al. 1996). All of these subtypes appear to be represented by single-copy genes in all studied mammalian genomes with the single possible exception of the 5-ht$_{5B}$ receptor, which may be an inactived gene in the human genome (see Section C. XII)

Preferred subtype name	Previous name(s)
5-HT$_{1A}$	
5-HT$_{1B}$	Human 5-HT$_{1D\beta}$, S12, rat 5-HT$_{1B\beta}$
5-HT$_{1D}$	Human 5-HT$_{1D\alpha}$, dog RDC4
5-ht$_{1E}$	S31
5-ht$_{1F}$	5-HT$_{1E\text{-like}}$, 5-HT$_{1E\beta}$
5-HT$_{2A}$	5-HT$_2$
5-HT$_{2B}$	5-HT$_{2F}$, SRL gene
5-HT$_{2C}$	5-HT$_{1C}$
5-HT$_3$	5-HT$_3$R-A
5-HT$_4$	
5-ht$_{5A}$	5-HT$_5$, 5-HT$_{5\alpha}$
5-ht$_{5B}$	5-HT$_{5\beta}$
5-HT$_6$	
5-HT$_7$	

successful initial efforts of pharmacologists to use biochemical and physiological data to deduce probable molecular subtypes without the benefit of cloning or receptor purification. This success may partly reflect the fact that most serotonin receptor subtypes are from the same 7TM superfamily and share many common features that have facilitated their study. Current nomenclature unites molecular and genomic information on serotonin receptors (one gene-one subtype) with an older literature on pharmacological (operational) and second messenger coupling (transductional) properties of these receptors (Hartig et al. 1996). We have generally found that any one of these criteria (structure, pharmacology, or coupling) will group 7TM receptor subtypes into the same subfamilies. This should not be considered surprising since the process of gene duplication, divergence through mutation, and preservation of favorable adaptations in the genome is believed to underlie the evolu-

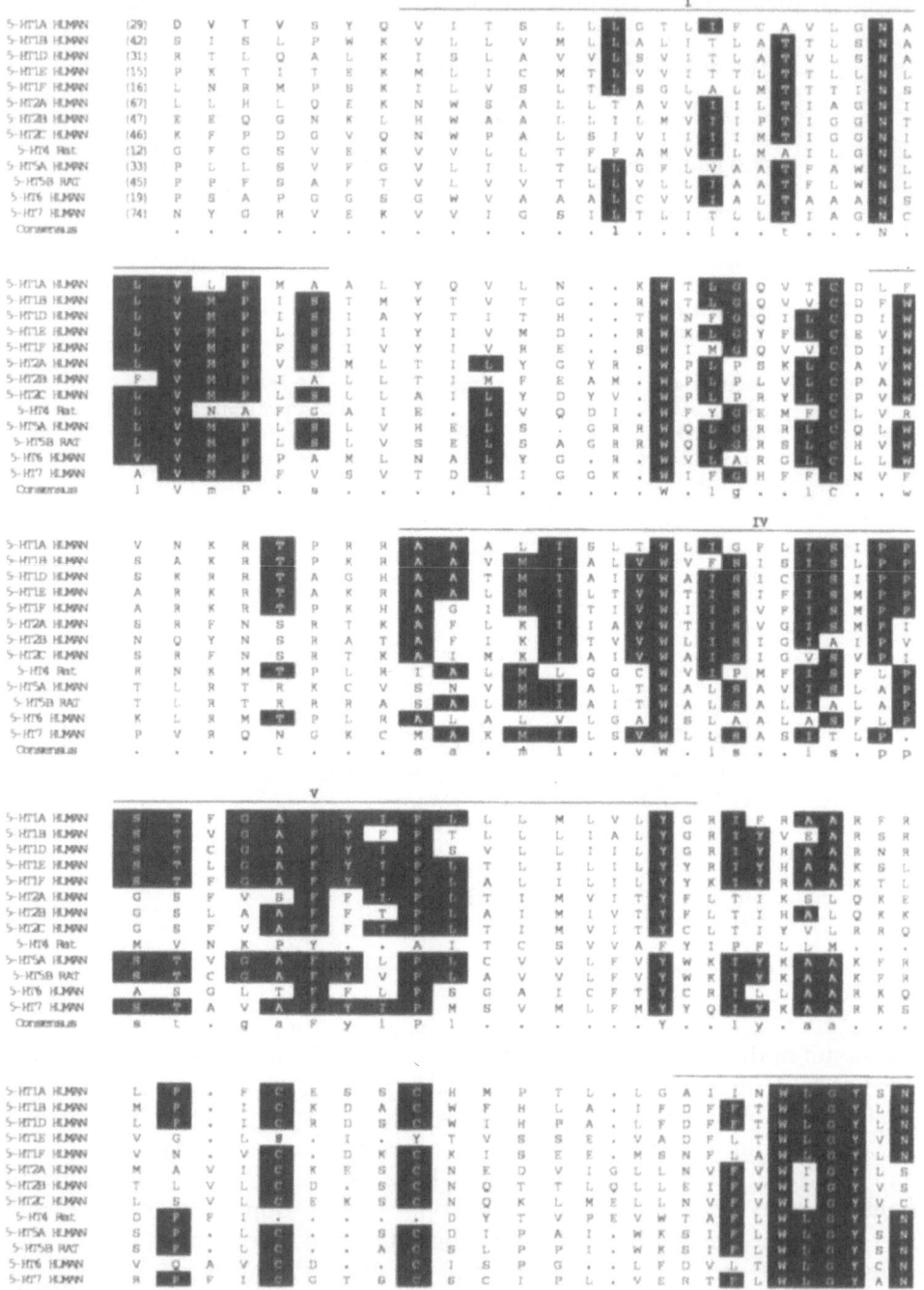

Fig. 2. Amino acid sequence alignment of 7TM serotonin receptors. Human receptor sequences obtained from the European Molecular Biology Laboratory (http://expasy.hcuge.ch/www/expasy-top.html) were used whenever available (see text for original references). Rat sequences were used for the 5-HT$_4$ and 5-ht$_{5B}$ receptors since no human sequences have been described. Sequences were aligned by the Multalin program (http://www.ibcp.fr/multalin.html). Conserved positions (amino acid identity)

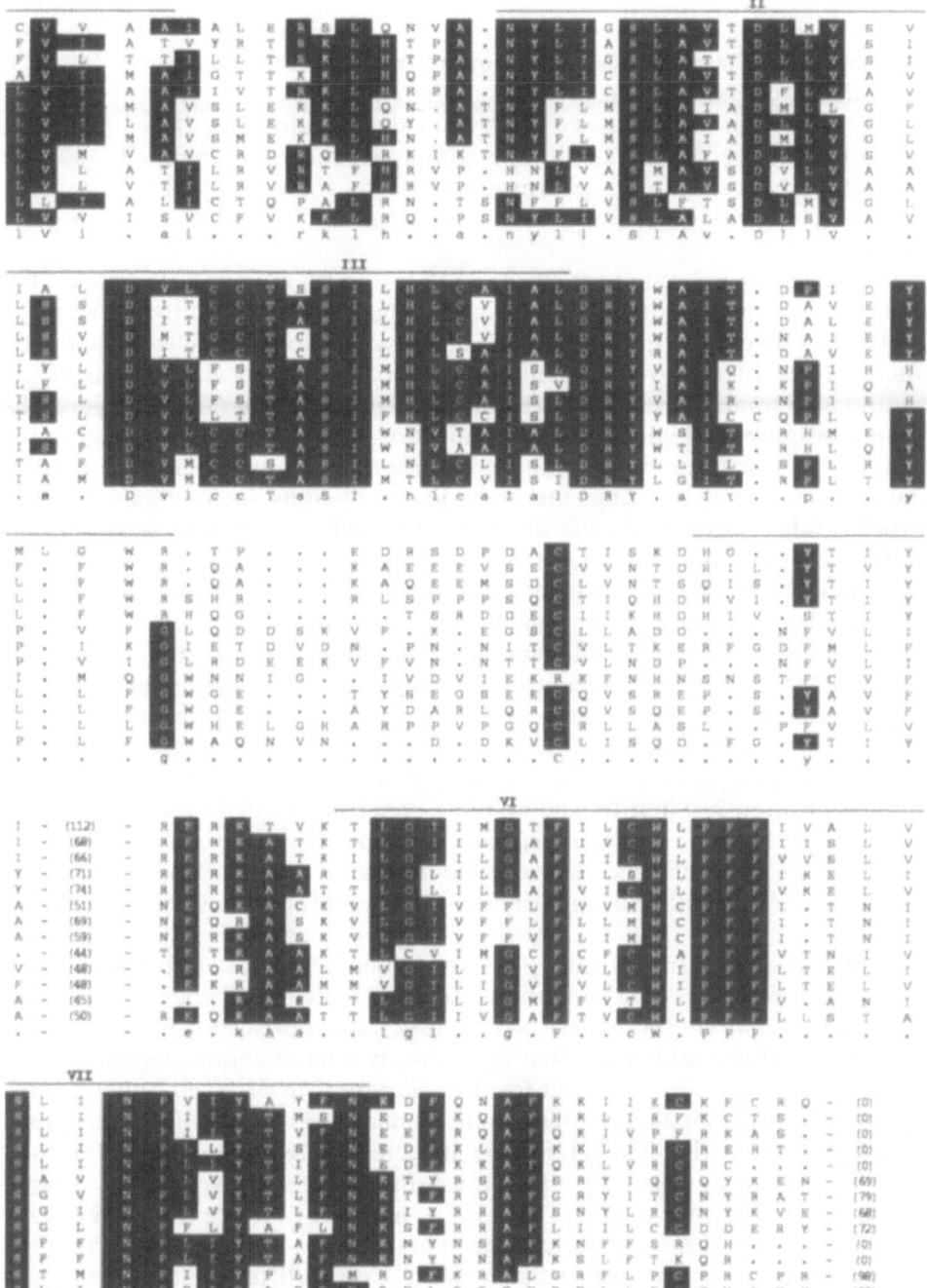

are shown by *inverted type (white on black)*. The start and stop positions of the transmembrane segments are not known with certainty and are reported in different positions by various authors. Note that the amino ends of the transmembrane segments alternate between extracellular and cytoplasmic faces of the plasma membrane as the receptor winds in and out of the lipid bilayer in a serpentine fashion

tion of protein families over evolutionary timescales (WILLS 1993). This process would be expected to produce subfamilies of closely related genes and receptor proteins, much as we observe in the current diversity of 7TM subtypes from the serotonin family, and many other receptor families. In the next section, these structural, pharmacological and coupling properties that have led to classification into the current serotonin receptor subfamilies are reviewed.

The transmembrane portions of 7TM serotonin receptors, the region believed to encompass the serotonin- and drug-binding site(s) (STRADER et al. 1994), can be used to group the known subtypes into subfamilies based upon their degree of amino acid similarity. Figure 1 shows a dendogram of the amino acid similarities among serotonin receptor subtypes. In this figure, the length of horizontal line segments connecting subtypes is proportional to the number of transmembrane amino acid differences found between receptor subtypes. Certain clusterings are immediately obvious, including the 5-HT_{2A}, 5-HT_{2B}, 5-HT_{2C} subfamily, and the group of 5-HT_1 subtypes, as shown in Fig. 1. Different authors have shown slightly different dendograms depending on how much of the receptor structure is included in the comparison, the assigned start and stop positions of transmembrane segments, and whether identity or similarity is examined, but all maps agree on the three basic groupings into 5-HT_1, 5-HT_2, and "other" receptor subfamilies. Detailed examination of the aligned transmembrane amino acids of serotonin 7TM receptor subtypes (Fig. 2) reveals several microdomains where this clustering into subtypes is reflected in nonconservative amino acid changes. For example, cysteine vs. phenylalanine at position 9 of TM III and cysteine or serine vs. methionine at position 12 of TM VI. These subfamily-specific microdomains may be interesting to examine for molecular mechanisms underlying certain subfamily differences (e.g., the low percentage of agonist high-affinity states formed by 5-HT_2 receptors).

Since the clustering of subtypes into subfamilies is based upon the amino acid structure of the transmembrane domains, which are believed to contain the drug-binding site(s) (STRADER et al. 1994), it is not surprising that related receptors within a subfamily often show closely related pharmacological properties that have made it difficult for drug designers to produce truly selective ligands. Prime examples are found in the human 5-HT_{1B} and 5-HT_{1D} receptors, which display nearly identical binding affinities for 18 compounds (HARTIG et al. 1992), although spiperone and ketanserin display some selectivity. A similar situation exists for the 5-HT_{2A}, 5-HT_{2B}, and 5-HT_{2C} receptors, which have been very difficult to separate pharmacologically. It is worth noting that the degree of amino acid homology between these difficult-to-separate subtypes is not unusual for subfamily members (typically around 75% transmembrane identity). What appears to distinguish these pharmacologically difficult-to-separate receptors is the lack of amino acid substitutions in those critical domains involved in drug binding. Site-directed mutagenesis studies have shown that single amino acid substitutions can dramatically alter the receptor-binding properties of serotonin receptors if they occur within or near the proposed drug-binding site (e.g., see KAO et al. 1992). For these "difficult" subtypes, it would appear that those random or selected amino acid substitutions that drove the separation and fixation of duplicated genes simply did not occur in the vicinity of the drug-binding site. Since the tiny natural ligand serotonin was probably the only significant ligand encountered by serotonin receptors in the course of their evolution, this should perhaps not be surprising. As far as classification schemes are concerned, it is difficult to define an appropriate ligand basis set for determining the degree of "relatedness" of receptor subtypes. Nevertheless the cases of difficult-to-separate subtypes and the frequent observation that similar chemical structures generally exhibit highest binding to related receptors within a subfamily serves to further validate the current groupings.

The classification of subtypes by amino acid similarity parallels the clustering of subtypes observed when quite different structural classification criteria are examined, such as gene structure (number and positions of introns) and the lengths of two receptor segments known to be critical for G protein coupling, the third intracellular loop, and the carboxy tail. All genes of the 5-HT_1 subfamily, for example, are composed of a single exon across the protein-coding region, whereas the coding region of the 5-HT_2 subfamily is interrupted by at least two introns in the coding region, one within the first extracellular loop (o1) joining transmembrane segments 2 and 3 (TM II to TM III) and one in the middle of TM IV (FOGUET et al. 1992a; STAM et al. 1992, 1994). All three 5-HT_2 genes contain these same two introns in the same two positions, suggesting a common evolutionary origin. The 5-HT_{2C} receptor contains at least one additional coding region intron which is not found in the other two subtypes. This additional intron is found in the amino-terminal segment of the 5-HT_{2C} receptor, 40 amino acids upstream of TM I (FOGUET et al. 1992a; STAM et al. 1994).The positions of all three coding region introns of the 5-HT_{2C} receptor are conserved in both the human (STAM et al. 1994) and mouse (FOGUET et al. 1992a) homologues, as are the positions of the two introns in the 5-HT_{2A} receptor gene in the rat (FOGUET et al. 1992a) and human (STAM et al. 1992) genomes. The remaining serotonin receptor subtypes, 5-HT_4, 5-HT_5, 5-HT_6, and 5-HT_7, all contain at least one coding region intron, but the positions are different in each receptor subtype, with the possible exception of the 5-HT_{5A}, 5-HT_{5B}, and 5-HT_6 receptors, which all appear to contain a splice site in the third intracellular loop (MATTHES et al. 1993; MONSMA et al. 1993). It is also interesting to note a strong overlap of certain serotonin receptor intron positions with a variety of dopaminergic and peptidergic receptor sites (STAM et al. 1994), perhaps pointing to a distant common ancestor gene.

The lengths of two protein segments critical to interaction with the α-subunit of the G protein, the i3 loop and the carboxy tail (STRADER et al. 1994), are also conserved in many 7TM receptor subfamilies. Among the serotonin receptor subtypes, this relationship also holds, although the lengths of these segments are not as well conserved between subfamily members as are other classification parameters. The 5-HT$_1$ subfamily members all exhibit long i3 loops but short carboxy tails, features commonly associated with Gi-coupled receptors (JIN et al. 1992; BEER et al. 1993). The i3 loop of 5-HT$_1$ receptors varies from 84 to 90 amino acids (with the single exception of 127 amino acids for 5-HT$_{1A}$), and the carboxy tail is 15–18 amino acids. The 5-HT$_2$ subfamily has long segments at both positions: 70–85 amino acids for i3 and 86–98 amino acids for the carboxy tail. The remaining subtypes do not resemble each other, especially in the carboxy tail region, where the length varies from 15 to 117 amino acids.

Signal transduction by these receptor subtypes follows the same subfamily patterns expressed by the protein and gene structural analyses. All members of the 5-HT$_1$ subfamily exhibit strong inhibition of adenylate cyclase activity in membrane assays, while all 5-HT$_2$ receptors are tightly coupled to enhancement of phosphoinositide turnover. The coupling of each 5-HT$_1$ and 5-HT$_2$ receptor to its primary second messenger pathway is generally or always tighter (more efficient) than coupling to other second messenger pathways, although alternative couplings can be observed, especially in the artificial environment of a transfected cell host where very high receptor expression densities can be obtained and receptors may be associated with G proteins not encountered in their native cell environment. For example, the 5-HT$_{1A}$ receptor can couple to inhibition of β-adrenergic agonist-stimulated and forskolin-stimulated adenylate cyclase activity when a 5-HT$_{1A}$ genomic clone is transfected into HeLa or COS-7 cells; yet it can also stimulate phospholipase C (PLC) activity in the HeLa cell host (FARGIN et al. 1989) and further stimulates phosphate uptake via a protein kinase C pathway (RAYMOND et al. 1989). The PLC-coupling pathway appears to be much weaker, however, than the coupling to Gi and adenylate cyclase inhibition as demonstrated by the much higher agonist concentration required to elicit PLC coupling (FARGIN et al. 1989). Furthermore, transfection of the 5-HT$_{1A}$ receptor into *Escherichia coli*, which is devoid of endogenous G proteins, demonstrated that this receptor couples efficiently to Gi α-subunits, but only very weakly to Go and not at all to Gs (BERTIN et al. 1992), indicating a high degree of selectivity for Gi pathways. These data that the 5-HT$_{1A}$ receptor shows a strong preference for Gi-coupling pathways in transfected systems are in agreement with studies showing a preferred link to adenylate cyclase inhibition in native tissues (HOYER et al 1994). Similar cross-reactivities of transfected receptors with different G proteins (promiscuity) have been observed for many 7TM receptors, although a preferred coupling is usually found to occur at a higher efficiency than any other, and to agree with the coupling normally observed in native tissues.

Although the predominant data in the 7TM field describe a strong preferential coupling of each receptor to a single G protein, clear exceptions to the rule have been described in the case of the calcitonin, thyrotropin, parathyroid hormone, and pituitary-adrenal cortex activating peptide (PACAP) receptors. For example, the PACAP receptor efficiently couples to both adenylate cyclase stimulation and phosphoinositide hydrolysis in native cell preparations (SPENGLER et al. 1993). Furthermore, different naturally occurring PACAP peptide agonists show quite different pharmacological properties and preferential activation of one second messenger pathway over the other (SPENGLER et al. 1993). This dual coupling might reflect a mixture of microdomains in one or more cell types, each one containing a different G-protein and coupling pathway. Although this degree of complexity in the coupling pathway does not yet appear to have been clearly documented for any 5-HT receptor, we should remain alert for its presence, especially in restricted anatomical domains or cell types.

A different type of second messenger promiscuity, downstream of Gi-protein activation, has been described for the 5-HT$_{1A}$ receptor and several other Gi-coupled receptors (KARSCHIN et al. 1991). This diversity of pathways arises from the ability of activated Giα-subunits to directly activate inwardly rectifying potassium channels independently of their ability to reduce adenylate cyclase activity (CODINA et al. 1987). Such an activity occurs naturally as a component of muscarinic cholinergic activity in cardiac atrial cells. KARSCHIN et al. (1991) used a recombinant vaccinia virus expression vector to express a human 5-HT$_{1A}$ receptor in primary cultures of rat atrial myocytes. They demonstrated agonist-induced serotoninergic activation of the same inwardly rectifying potassium current that is normally activated by the endogenous muscarinic cholinergic receptors. These data are in agreement with the demonstrated ability of 5-HT$_{1A}$ receptors to activate potassium channels in hippocampal neurons by a pertussis toxin-sensitive mechanism that is independent of any soluble cytoplasmic second messenger (ANDRADE et al. 1986). This direct coupling of 5-HT$_{1A}$ receptor activation to potassium channels must be considered as the possible primary signaling pathway of 5-HT$_{1A}$ receptors *in vivo* and perhaps also for other members of the 5-HT$_1$ receptor subfamily.

Another indicator of receptor-coupling activity for 7TM receptors is their ability to shift agonist high-affinity states to low affinity as they uncouple from G proteins in the presence of excess guanyl nucleotides. The presence of the G-protein-coupled, high-affinity agonist binding form of transfected cloned receptors is often difficult to demonstrate due to the high receptor densities typically achieved in transfection systems. This is especially a problem for transiently transfected cells where the receptor is probably expressed only by a subfraction of all cells, leading to very high local densities of receptor that may overwhelm the cell's ability to provide adequate levels of G protein. The choice of stably transfected cells containing an array of G proteins and expressing moderate to low levels of receptor often leads to better expression of agonist high-affinity states and an ability to demonstrate G coupling via guanyl nucleotide-induced shifts. This can be especially useful for demonstrating G protein coupling of receptors with no known second messenger response, such as the 5-ht$_5$ subtypes (see Sect. C. XI).

In summary, all these different receptor classification criteria: amino acid sequence similarity, lengths of protein segments associated with G protein coupling, gene structure, and preferred second messenger coupling, agree on the division of serotonin receptor subtypes into three main subfamilies: the 5-HT$_1$ group, the 5-HT$_2$ group, and an "others" class which includes all adenylate cyclase stimulatory (Gs-coupled) receptors. Mapping of these classification screens onto the dendogram of serotonin receptor relationships is shown in Fig. 1. Although there has been no serious challenge to these groupings, it remains unclear why the adenylate cyclase stimulatory serotonin receptors are unique in their diversity and failure to cluster into subfamilies, and why serotonin appears to be the most extreme case of 7TM receptor diversity among all 7TM receptor families. Finally, it is worth noting that the previous rapid pace of new receptor cloning has slowed considerably, which might indicate that we are near the end of serotonin receptor discovery, at least in the well-explored 7TM superfamily. A number of pharmacological subtypes remain to be fully understood, but it is possible that they may consist of mixtures of known subtypes (Hoyer et al. 1994). Whatever the truth of the matter, we should soon have all serotonin and other receptor subtypes available to us due to completion of the human genome project, which has currently produced portions of the sequence of an estimated 85% of all human cDNAs and is expected to provide nearly complete human genome information sometime near the year 2002 (Carey et al. 1995).

C. Properties of Each Subtype: Clones, Genes, and Transduction Pathways

In the sections that follow, each serotonin receptor subtype is discussed with regard to the available clones, orthologous genes in mammalian species, known splice variants and pseudogenes, and transductional (second messenger) properties. Brief comments on distribution and roles of the subtypes are provided, with a more thorough coverage of these topics available in companion chapters of this handbook.

I. 5-HT$_{1A}$ Receptor

The first 5-HT$_{1A}$ receptor clone (and the first clone of the serotonin receptor family to be isolated) was obtained from a human genomic library screened with a β_2-adrenergic receptor clone (Kobilka et al. 1987), although it was not identified as a serotonin 5-HT$_{1A}$ receptor until the following year (Fargin et al. 1988). A rat genomic clone has also been described, and shown to mediate agonist-induced inhibition of both basal and vasoactive intestinal polypeptide (VIP)-stimulated adenylate cyclase activity (Albert et al. 1990). In the rat, but not in humans, three RNA species were observed by Northern blot analysis, which led to speculation that different polyadenylation or transcriptional start sites might exist in the rat gene (Albert et al. 1990). Chanda et al. (1993) described an apparent sequencing error in the original G21 clone, in which they correct one and add one amino acid to the sequence near the start of TM IV of the human protein, which brings the sequence and number of amino acids (422) into agreement with the rat clone data.

The transmembrane amino acid sequence of the 5-HT_{1A} subtype is noteworthy for its high degree of homology to adrenergic receptor subtypes (approximately 45%) as well as to other 5-HT_1 receptors (approximately 55%, which might reflect a distant common ancestry with the adrenergic receptors. It is interesting to note that the α_2- and β-adrenergic receptors are colocalized on human chromosome 5 along with the 5-HT_{1A} receptor, although at distinct loci (OAKEY et al. 1991). The 5-HT_{1A} receptor gene has been localized to human chromosome 5, bands q11.2 to 13, in the vicinity of the glucocorticoid receptor (KOBILKA et al. 1987), whereas the adrenergic receptors are found near bands 5q31-34. In the mouse genome, these three receptors are localized to different chromosomes. Since similar conserved linkage groups appear around each gene in both genomes, the data suggest that the human organization may be closer to that of a primordial ancestor and that this set of receptor genes might have arisen from gene duplication followed by rearrangements and translocations of these linkage groups (OAKEY et al 1991). This relationship to the adrenergic receptors is also reflected in the pharmacology of the 5-HT_{1A} receptor, which is noteworthy for its tight binding of certain β-adrenergic receptor antagonists, such as pindolol. Site-directed mutagenesis has localized this adrenergic antagonist selectivity to a single residue in transmembrane domain 7, as discussed below.

The 5-HT_{1A} receptor, and its mRNA, are enriched in the hippocampus, septum, amygdala, and raphe nuclei (HOYER et al. 1994; ALBERT et al. 1990; MIQUEL et al. 1991), where they may modulate emotional responses via limbic pathways. This receptor has been shown to occur both presynaptically as a somatodendritic autoreceptor on raphe cells, and postsynaptically on target neurons in the hippocampus, although the degree of receptor reserve for the somatodendritic autoreceptor appears to be very high (MELLER et al. 1990), while there is no reserve for the postsynaptic receptor (YOCCA et al. 1992). This differential receptor reserve may underlie the observed therapeutic efficacy of the 5-HT_{1A}-selective partial agonist drug buspirone (Buspar) as an antianxiety medication (MELLER et al. 1990; YOCCA et al. 1992).

The 5-HT_{1A} receptor exhibits a preferred coupling to both inhibition of adenylate cyclase activity, and to direct activation of an inwardly rectifying potassium conductance as discussed in Sect. B. above. PAUWELS et al. (1993) extensively characterized the inhibition of forskolin-stimulated adenylate cyclase activity in HeLa cells stably transfected with the human 5-HT_{1A} receptor and found good agreement with previous tissue data. Coupling of this receptor to phosphoinositide hydrolysis, phosphate transport, and adenylate cyclase stimulatory pathways is much weaker and may or may not be physiologically relevant (see Sect. B.). Rapid desensitization of the cyclase response has been observed. VAN HUIZEN et al. (1993) observed a time-dependent downregulation and desensitization of adenyl cyclase inhibitory activity to

32% of preincubation values for the human 5-HT$_{1A}$ receptor transfected in Swiss 3T3 cells. HARRINGTON et al. (1994) examined the desensitization of the transfected human receptor in HeLa cells and found that the observed loss of high-affinity binding sites and loss of functional coupling was due to activation of both protein kinase C and phospholipase A$_2$ pathways. Finally, VARRAULT et al. (1994) used synthetic peptides derived from the second intracellular loop (i2) and the carboxy tail of the 5-HT$_{1A}$ receptor to demonstrate that these receptor fragments alone can strongly inhibit forskolin-stimulated adenylate cyclase activity and trigger guanyl nucleotide exchange of Gi/Go class proteins as does the serotonin-stimulated intact receptor protein.

II. 5-HT$_{1B}$ Receptor

The 5-HT$_{1B}$ receptor subtype, long known as a pharmacological binding site, was first cloned from rat brain (VOIGT et al. 1991) by polymerase chain reaction (PCR) with degenerate oligonucleotide primers. It was pharmacologically identified as a 5-HT$_{1B}$ subtype in binding assays. In parallel, independent studies (WEINSHANK et al. 1992; LEVY et al. 1992a; VELDMAN and BIENKOWSKY 1992), an orthologous gene was isolated from the human genome and shown to exhibit a quite different pharmacology, that of a 5-HT$_{1D}$-binding site as first described in bovine caudate. Due to this history, two closely related human genes with very similar pharmacological properties came to be named 5-HT$_{1D\alpha}$ and 5-HT$_{1D\beta}$, while the rodent homologue of the 5-HT$_{1D\beta}$ gene was most often called a 5-HT$_{1B}$ receptor. For a variety of reasons (including simplicity and a desire to use the same name for all species homologues), the Serotonin Nomenclature Committee has recently recommended that the 5-HT$_{1B}$ nomenclature be used for all species homologues of the rat 5-HT$_{1B}$ gene and that the human 5-HT$_{1D\alpha}$ gene and its homologues be associated with the 5-HT$_{1D}$ receptor (HARTIG et al. 1996).

The rat 5-HT$_{1B}$ receptor was first described as a rat cDNA clone obtained by PCR amplification of rat forebrain cDNA with degenerate primers designed against monoamine receptors (VOIGT et al. 1991). Although the binding properties of this transfected receptor clearly identified it as a 5-HT$_{1B}$ receptor, it was not possible to demonstrate coupling to any second messenger system in the transfection systems of the original study. Parallel independent reports on rat (ADHAM et al. 1992; HAMBLIN et al. 1992) and mouse (MAROTEAUX et al. 1992) genes also appeared and described adenylate cyclase inhibitory actions of this receptor in transfected systems. All of these reports described receptors with pharmacological properties in close agreement with the previously described rodent 5-HT$_{1B}$-binding site. In what may be the most exotic species homologue clone of the serotonin receptor family, CERUTIS et al. (1994) reported the cloning of a 5-HT$_{1B}$ cDNA from an opossum kidney (OK) cell line. The pharmacological and Gi-coupling properties were in agreement with those of the rat receptor, and it dispayed overall 82% and 54% amino acid identity to the rat and human clones, respectively.

Serotonin receptors with the pharmacological properties of this 5-HT$_{1B}$ site appear to be restricted to rat, mouse, hamster, and opossum among the animals studied so far, and are notably absent from human and most other mammals including dog, cow, and guinea pig (HOYER et al. 1994). As suggested earlier by HOYER and MIDDLEMISS (1989), cloning of both the human and rat homologues of this gene proved that the rat 5-HT$_{1B}$ pharmacology is expressed as a 5-HT$_{1D}$ pharmacology in the homologous human receptor (ADHAM et al. 1992).

The 5-HT$_{1B}$ receptor mRNA was found at its highest levels in Purkinje cells of the cerebellum, in the CA1 region of the hippocampus, and in the caudate-putamen (VOIGT et al. 1991; MAROTEAUX et al. 1992). Comparison of this distribution to receptor autoradiography of the 5-HT$_{1B}$ site suggests that the receptor may be primarily a presynaptic receptor (heteroreceptor controlling release of another transmitter) on striatal terminals in the substantia nigra and globus pallidus as well as on Purkinje cell terminals on deep nuclei in the cerebellum. Involvement of these pathways in motor control suggests a possible role for these receptors in the regulation of movement (MAROTEAUX et al. 1992). In addition, HAMEL et al. (1993) demonstrated the presence of 5-HT$_{1B}$ receptor transcripts in both bovine and human cerebral arteries, along with a pharmacological response profile suggesting that 5-HT$_{1B}$ receptors, but not 5-HT$_{1D}$ receptors. mediate contractile responses in these cerebral blood vessels.

The first 5-HT$_{1B}$ human clones were isolated from human genomic libraries using low-stringency homology screening with various probes (WEINSHANK et al. 1992; LEVY et al. 1992a; VELDMAN and BIENKOWSKY 1992) or PCR amplification with degenerate primers (DEMCHYSHYN et al. 1992; JIN et al. 1992). These identical clones, originally named 5-HT$_{1D\beta}$, human 5-HT$_{1B}$, or S12, encoded a receptor which displayed a 5-HT$_{1D}$ pharmacological profile and a large GTP-sensitive component of high-affinity agonist binding (LEVY et al. 1992a; DEMCHYSHYN et al. 1992). A restriction fragment length polymorphism has been reported for this human gene (DEMCHYSHYN et al. 1992), which could aid in exploring its role in human disease and normal physiology. In addition, the human 5-HT$_{1B}$ receptor contains a carboxy-terminal cysteine that has been implicated as a site for palmitoylation while the closely related human 5-HT$_{1D}$ receptor lacks this residue (DEMCHYSHYN et al. 1992). Northern blot analysis has shown two RNA transcripts in human brain, which may indicate the use of alternative transcription start sites (JIN et al. 1992).

The transfected human 5-HT$_{1B}$ receptor was found to potently inhibit adenylate cyclase activity in various assays and seemed inactive at stimulating intracellular calcium signaling in mouse L cells coexpressing the muscarinic M5 receptor (LEVY et al. 1992a). In contrast, ZGOMBICK et al. (1993) demonstrated strong coupling of the transfected human 5-HT$_{1B}$ receptor to elevation of intracellular calcium in their mouse fibroblast LM(tk-) cell line and a weak coupling to phosphoinositide hydrolysis. This group demonstrated that the majority of this intracellular calcium is released from intracellular stores and that similar dual coupling to adenylate cyclase inhibition and intracellular calcium elevation has been noted for several Gi-linked receptors (Zgombick et al. 1993).

Adham et al. (1993a) investigated the receptor reserve, EC_{50} values, and apparent intrinsic activities of a variety of partial agonists in a transfected rat 5-HT_{1B} receptor expression system. They demonstrated that a high degree of receptor reserve exists in these cells, which causes a strong leftward shift in dose-response relationships for agonists (compared to binding data) and causes some partial agonists to appear to be full agonists in this system. Since these experiments were done at receptor expression levels that are commonly found in transfection systems, and since information was presented that strong receptor reserve may exist even in some cell lines that express modest levels of receptors, this study indicates that extreme caution must accompany the interpretation of response data and intrinsic activity determinations in all transfected receptor studies. Furthermore, high degrees of receptor reserve, which may or may not reflect receptor responses in native tissues, can be expected to dominate response data from cloned receptors in expression systems.

In nonrodent mammalian species, the 5-HT_{1B} receptor appears to be significantly more abundant than the 5-HT_{1D} receptor (Hoyer et al. 1994), which would mean that human 5-HT_{1D} pharmacological responses may well be dominated by the 5-HT_{1B} gene product. The lack of highly selective ligands which can discriminate between the human 5-HT_{1D} and the human 5-HT_{1B} receptors has hampered investigation of these receptors. Since the 5-HT_{1B} receptor mRNA appears to be much more abundant, it is important to keep in mind the large species-specific pharmacological differences which exist between 5-HT_{1B} and 5-HT_{1D} receptors. This is especially important since both receptors may serve as autoreceptors and heteroreceptors in many overlapping brain regions (Hoyer et al. 1994).

The striking pharmacological differences between the rat and human 5-HT_{1B} receptors have been shown to reside primarily in a single transmembrane 7 asparagine residue that was shown to confer adrenergic ligand binding to the 5-HT_{1A} receptor (Metcalf et al. 1992; Oksenberg et al. 1992). This provides one of several examples in the serotonin receptor family in which a single transmembrane amino acid change in the course of species divergence during evolution has conferred markedly different ligand binding (pharmacological) properties to receptors which are otherwise quite equivalent in structure and function (see Sect. D).

III. 5-HT$_{1D}$ Receptor

The first 5-HT_{1D} clone to be isolated was RDC4, a dog cDNA clone obtained by degenerate PCR amplification of human thyroid cDNA followed by isolation of the full sequence from a dog thyroid cDNA library (Libert et al. 1989). This group noted that this 7TM receptor most closely resembled the 5-HT_{1A} sequence and might therefore be a serotonin receptor. Zgombick et al. (1991)

and MAENHAUT et al. (1991) expressed the canine RDC4 clone in mammalian cells and identified its pharmacological binding properties as those of a 5-HT$_{1D}$ subtype. ZGOMBICK et al. (1991) observed that serotonin decreased the rate of forskolin-stimulated cAMP accumulation in murine LM (tk-) fibroblasts transfected with RDC4. In contrast, MAENHAUT et al. (1991) observed a slight elevation of cAMP in COS-7 cells transfected with this clone, which remains a puzzling observation since the native 5-HT$_{1D}$ receptor is coupled to inhibition of cyclase and many other studies have demonstrated a strong coupling of various 5-HT$_{1D}$ clones to cyclase inhibition.

HAMBLIN and METCALF (1991) isolated the human homologue of RDC4 by homology screening a human genomic library with a PCR-derived RDC4 probe, and demonstrated that it exhibited a 5-HT$_{1D}$ pharmacological profile, coupled to a PTX-sensitive G protein, and inhibited forskolin-stimulated adenylate cyclase activity. WEINSHANK et al. (1992) obtained this human 5-HT$_{1D}$ clone by homology screening with RDC4, and demonstrated its 5-HT$_{1D}$-binding properties, coupling to adenylate cyclase inhibition and lack of coupling to phosphoinositide hydrolysis. NGUYEN et al. (1993) isolated a human pseudogene that most closely resembles the 5-HT$_{1D}$ receptor but is interrupted by several stop codons, frame shifts, and deletions along with an Alu repeat inserted in the coding region. BARD et al. (1995) showed that this pseudogene is transcribed, and exhibits a tissue distribution similar to the functional 5-HT$_{1D}$ gene. This 5-HT$_{1D}$ receptor pseudogene may represent a duplicated 5-HT$_{1D}$ gene (transposition including the promoter region) that diverged from its parent through subsequent mutations. Its function was disrupted, probably by acquiring a stop codon or other mutations in its coding region, thus ending its functional utility. Since this pseudogene would not be under evolutionary pressure to maintain its gene sequence, it would be expected to acquire mutations at an increased rate, leading to gradual change away from a recognizable 5-HT$_{1D}$ sequence. SHUCK et al. (1993) estimated that this critical mutation occurred 35–50 million years ago, and BARD et al. (1995) demonstrated the presence of the 5-HT$_{1D}$ pseudogene in the African green monkey, where it contains the same human Alu repeat sequence that may have been responsible for inactivating this gene.

A rat 5-HT$_{1D}$ receptor cDNA clone was reported by HAMBLIN et al. (1992) and by BACH et al. (1993). Highest levels of the mRNA for this subtype were reported in the pyramidal layer of the olfactory tubercle, caudate, and nucleus accumbens, and transcripts were located in the raphe nucleus, suggesting an autoreceptor role for this subtype in agreement with pharmacological studies (HOYER et al. (1994)). The density of mRNA and receptor protein for the 5-HT$_{1D}$ receptor appears to be much lower than that of the closely related 5-HT$_{1B}$ receptor, which has made it difficult to separately evaluate the role of the 5-HT$_{1D}$ receptor. The 5-HT$_{1D}$ receptor codistributes with the 5-HT$_{1B}$ receptor in many brain regions and also seems to provide both autoreceptor and heteroreceptor functions (HOYER et al. 1994). The majority of published studies on 5-HT$_{1D}$ receptor actions in nonrodent species appear to have primarily examined the role of the 5-HT$_{1B}$ component, since ketanserin and ritanserin (which exhibit high affinity for the 5-HT$_{1D}$ receptor) were generally inactive (HOYER et al. 1994).

Much confusion has existed in the literature regarding the pharmacology and nomenclature of these 5-HT$_{1D}$ and 5-HT$_{1B}$ subtypes (reviewed in HARTIG et al. 1992). This situation arose because these two closely related receptor subtypes surprisingly show two separate deviations from the usual conservation of form and function (amino acid sequence and pharmacological-binding properties) seen among other 7TM receptors. The human 5-HT$_{1D}$ and human 5-HT$_{1B}$ subtypes differ at 23% of their transmembrane amino acid positions (a fairly typical number for closely related subtypes, such as dopamine D_2 and D_4 receptors or the β-adrenergic receptors), and yet they display essentially identical binding properties for 19 different chemical structures (WEINSHANK et al. 1992). The second deviation from normal properties was for the human 5-HT$_{1B}$ versus the rat 5-HT$_{1B}$ receptor orthologues, where 93% overall amino acid identity exists between these clones and only seven amino acid differences exist in the ligand-binding transmembrane domains. In contrast to most rat vs. human species homologues for which ligand-binding properties are nearly identical, these two clones show markedly different binding profiles, especially for adrenergic ligands. It is not clear why these two unusual deviations occur in the same group of 5-HT$_{1D}$/5-HT$_{1B}$ receptors.

IV. 5-HT$_{1E}$ Receptor

LEVY et al. (1992b) were first to report the isolation of this serotonin 5-ht$_1$-like receptor clone, which they originally named S31. In a later study, this human genomic clone was shown to couple to adenylate cyclase and was pharmacologically characterized by this group (GUDERMANN et al. 1993). In parallel, independent studies, ZGOMBICK et al. (1992) and McALLISTER et al. (1992) isolated this same human clone and demonstrated that it displayed the pharmacological properties of a 5-ht$_{1E}$-binding site previously characterized by LEONHARDT et al. (1989), which is distinguished from other 5-HT$_1$ receptors by its unusually low affinity for 5-carboxamidotryptamine (5-CT). This receptor is found primarily in the striatum and cortex, but little is known about its possible physiological roles. Aside from the human genomic 5-ht$_{1E}$ clone, no other mammalian clones of the 5-ht$_{1E}$ receptor have been reported.

The human 5-ht$_{1E}$ gene was transfected into murine fibroblasts and shown to encode for a receptor with high-affinity guanine nucleotide-sensitive [^{3}H]5-HT binding (ZGOMBICK et al. 1992). When transfected into Y-1 adrenal cells, the receptor modestly inhibited (30%) forskolin-sensitive adenylate cyclase activity but was inactive at stimulating adenylate cyclase or increasing phosphoinositide hydrolysis (ZGOMBICK et al. 1992). When the human 5-ht$_{1E}$ receptor was stably transfected into human embryonic kidney cells (HEK

293), it exhibited modest inhibition (25%) of forskolin-stimulated adenylate cyclase, and guanine nucleotide sensitivity of [^{3}H]5-HT binding could not be demonstrated (MCALLISTER et al. 1992). Since the degree of adenylate cyclase inhibition in both of these studies was smaller than observed with most 5-HT$_1$ receptors, the 5-ht$_{1E}$ receptor may either display a high selectivity for a Gi isoform not abundant in either Y-1 or HEK 293 cells, or it may preferentially activate a different second messenger path *in vivo* (perhaps direct potassium channel modulation?). GUDERMANN et al. (1993) discovered that the degree of adenylate cyclase inhibition by the human 5-ht$_{1E}$ receptor is strongly dependent on the magnesium ion concentration used in the assay, and speculated that other ion channel actions, such as inhibition of calcium currents or stimulation of potassium channels, may be the primary signaling pathway of this channel. ADHAM et al. (1994a) transfected the human 5-ht$_{1E}$ receptor into African green monkey kidney cells (BS-C-1) and observed potent inhibition of forskolin-stimulated adenylate cyclase activity at low serotonin concentrations, but potentiation at higher concentrations. The pharmacological profiles of the two responses were similar but agonists were more potent at the inhibitory action, suggesting that inhibitory pathways are the more efficient and perhaps most relevant pathway.

V. 5-HT$_{1F}$ Receptor

AMLAIKY et al. (1992) were the first to report the cloning of a mouse brain cDNA clone that later was named the 5-ht$_{1F}$ receptor. This group reported a moderate degree of inhibition of forskolin-stimulated adenylate cyclase activity (35%) using NIH-3T3 cells as their transfection host and localized the low-abundance mRNA for this receptor primarily to forebrain including select regions of the hippocampus.

ADHAM et al. (1993b) and LOVENBERG et al. (1993a) reported human cDNA and genomic clones of the 5-ht$_{1F}$ receptor, and LOVENBERG et al. (1993a) reported a rat genomic clone. ADHAM et al. (1993b) demonstrated 74% inhibition of forskolin-stimulated adenylate cyclase activity in transfected NIH-3T3 cells and demonstrated inhibition of high-affinity agonist binding by guanyl nucleotides. They also demonstrated a high affinity of the antimigraine medication sumatriptan (K_i of 23 nM) for this site, implicating the 5-ht$_{1F}$ receptor, along with the 5-HT$_{1D}$ and 5-HT$_{1B}$ receptors, as possible contributors to the antimigraine action of this drug. ADHAM et al. (1993b) observed strong RNA hybridization signals for the 5-ht$_{1F}$ sequence in uterus, mesentery, and select regions of cortex and hippocampus. LOVENBERG et al. (1993a) demonstrated a moderate (32%) inhibition of β-adrenergic-stimulated adenylate cyclase activity in cells cotransfected with the β_2 receptor and the rat 5-ht$_{1F}$ receptor. They also described low-level expression of the receptor mRNA in rat cortex, striatum, hippocampus, thalamus, and pons and estimated its abundance at 0.0001% of total cortical mRNA. LOVENBERG et al. (1993a) described the presence of an apparent alternative splice site, indicating that an intron is present in the rat gene in the 5′-untranslated region, 42 nucleotides upstream of the translation initiation site.

Adham et al. (1993c) investigated the coupling of the human 5-ht$_{1F}$ receptor to multiple response pathways in two transfection hosts, NIH-3T3 cells and LM(tk-) fibroblasts. In NIH-3T3 cells, half-maximal inhibition of forskolin-stimulated adenylate cyclase activity required only 10% occupancy of receptor sites by 5-HT, indicating a high degree of receptor reserve and a potent coupling activity. Furthermore, in LM(tk-) cells, but not NIH-3T3 cells, moderate increases in phosphoinositide hydrolysis and intracellular calcium concentration were observed when slightly higher doses of 5-HT were applied. All three responses (cAMP, IP, and calcium) were sensitive to pertussis toxin, indicating a probable involvement of Gi or Go proteins.

VI. 5-HT$_{2A}$ Receptor

Pritchett et al. (1988) isolated the first 5-HT$_{2A}$ receptor clone from a rat brain cDNA library by homology screening with the 5-HT$_{2C}$ sequence. They observed a large increase (tenfold) in phosphoinositide (PI) hydrolysis and an increase in intracellular free calcium in transfected HEK 293 cells, in agreement with data on the native receptor (Hoyer et al. 1994). Injection of *in vitro* transcribed rat 5-HT$_{2A}$ RNA into *Xenopus* oocytes was shown by this group to induce a fast-desensitizing inward current response, apparently via serotoninergic activation of a calcium-sensitive chloride current as previously demonstrated for the 5-HT$_{2C}$ receptor (Lübbert et al. 1987). Julius et al. (1990) reported on the same clone with one correction to the nucleotide sequence that alters the reading frame and lengthens the amino terminus. Two distinct rat brain transcripts were described along with the ability of serotonin to activate cellular transformation and focus formation in transfected NIH 3T3 fibroblasts (Julius et al. 1990), as had been previously described for the 5-HT$_{2C}$ receptor (Julius et al. 1989). Van Obberghen-Schilling et al. (1991) reported on a hamster fibroblast cDNA clone and demonstrated that the mitogenic properties of this receptor require more than PI activation alone. Synergistic interactions between PI, Gi, and tyrosine kinase pathways were shown to be required for the growth-promoting actions of the 5-HT$_{2A}$ receptor. Segments of mouse genomic 5-HT$_{2A}$ clones were reported by Foguet et al. (1992a), indicating the presence of two coding regions introns also found in the 5-HT$_{2B}$ and 5-HT$_{2C}$ receptors, as discussed above (Sect. B).

Saltzman et al. (1991) first reported the cloning of a human 5-HT$_{2A}$ receptor from human brainstem cDNA. They noted that the amino acid sequences of the human and rat i3 loop are unusually well conserved (98% homology), suggesting a strong conservation of G protein coupling. Conservation of this region is not observed in other closely related systems (e.g., 5-HT$_{2C}$). Stam et al. (1992) constructed a full-length human 5-HT$_{2A}$ receptor gene from a combination of genomic and cDNA clones. They mapped the position of the two coding region introns to the same positions as the rat gene, and demonstrated that the first intron is 2.9 kb in length while the second is at least 3.7 kb. The receptor was transfected into Swiss 3T3 cells and the pharmacological binding properties determined along with the properties of the PI response. JOHNSON et al. (1995) isolated both pig and rhesus monkey cDNA clones of the 5-HT$_{2A}$ receptor, and demonstrated that the deduced amino acid sequence of both clones is identical to that of the human receptor within transmembrane regions, and exhibited 97% and over 99% overall identity, respectively. VAN OBBERGHEN-SCHILLING (1991) isolated a hamster fibroblast cDNA clone.

The availability of 5-HT$_{2A}$ receptor clones provided the opportunity to test a "two receptor" hypothesis, which proposed that the low density of agonist high-affinity sites labeled by the hallucinogenic agonist radioligand [^{3}H]DOB (4-bromo-2,5-dimethoxyphenylisopropylamine) was due to the presence of a new subtype of the 5-HT$_{2A}$ receptor with very low abundance. TEITLER et al. (1990) and BRANCHEK et al. (1990) demonstrated that the cloned, transfected 5-HT$_{2A}$ receptor expressed the binding sites for both the agonist hallucinogen and for classical 5-HT$_{2A}$ receptor antagonists, and that guanyl nucleotides could partially interconvert these sites. Thus, the "two state" hypothesis, which proposed that DOB binds to the agonist high-affinity conformation of the 5-HT$_{2A}$ receptor, was supported over the "two receptor" theory.

Site-directed mutagenesis studies have provided information on critical amino acids in the ligand- and drug-binding sites of the 5-HT$_{2A}$ receptor, as discussed in Sect. D below. Recently, work has begun on the promoter region of the 5-HT$_{2A}$ receptor gene. ZHU et al. (1995) have described Sp1, PEA3 and cyclic AMP response elements, and two novel transcription factors for the human gene. Multiple transcription initiation sites were detected, and both promoter and silencer regions were characterized. DU et al. (1995) also identified a number of positive and negative control regions, including a novel 17-nucleotide inverted repeat promoter element found only in the mouse and rat 5-HT$_{2A}$ genes. Dissection of the transcriptional control elements of this and other serotonin receptor genes will clearly be a major focus for future research, one that should eventually allow us to selectively control receptor levels in animal experiments, and eventually in human disease states.

VII. 5-HT$_{2B}$ Receptor

A preliminary report on a mouse genomic clone which encoded a stomach-fundus-like serotonin receptor (FOGUET et al. 1992a) was soon followed by full reports from several laboratories. FOGUET et al. (1992b) described a rat cDNA clone from a stomach fundus library that encoded a receptor which matched the binding properties of the stomach fundus serotonin receptor and triggered a modest chloride current increase when expressed in *Xenopus* oocytes. This chloride response was less efficient than for the 5-HT$_{2A}$ receptor and exhibited a delayed onset of action in response to serotonin. The gene for this receptor contains the same two introns found in the 5-HT$_{2A}$ receptor, but lacks the additional N-terminal intron of the 5-HT$_{2C}$ gene (Foguet et al. 1992a).

Kursar et al. (1992) described a rat stomach fundus cDNA clone which exhibited coupling to PI hydrolysis when transfected into AV-12 cells. Both groups reported that the receptor was expressed at high levels only in stomach fundus (Kursar et al. 1992; Foguet et al. 1992b). Wainscott et al. (1993) extensively characterized the pharmacological binding and PI response properties of the transfected receptor, and showed that agonists exhibit a strong temperature-dependent shift in binding constants (20- to 100-fold) while antagonist binding is little affected.

Loric et al. (1992) isolated a mouse 5-HT$_{2B}$ clone from a mouse brain library, demonstrating its presence in brain even though it appears to express at very low levels. Schmuck et al. (1994) isolated a human cDNA clone and characterized its regional distribution and the PI coupling of the transfected receptor. Two introns were described, at the same position as other 5-HT$_2$ family members. Relatively high levels of expression were found in peripheral tissues, including kidney, heart, and intestine. Modest expression was found in human stomach, which does not contain any tissue that is homologous to the rat stomach fundus.

VIII. 5-HT$_{2C}$ Receptor

The 5-HT$_{2C}$ receptor was the second serotonin receptor to be cloned, but the first clone to be identified with a previously characterized serotonin receptor subtype. A fragment of the 5-HT$_{2C}$ receptor was first obtained by a hybrid depletion and hybrid enrichment expression strategy in *Xenopus* oocytes (Lübbert et al. 1987), followed by a report on a full-length rat choroid plexus cDNA obtained by an RNA expression library-oocyte injection approach (Julius et al. 1988). The receptor was transfected in NIH 3T3 fibroblasts and shown to exhibit a 5-HT$_{2C}$ binding profile along with a serotonin-stimulated rise in intracellular calcium. The abundance of the mRNA for the 5-HT$_{2C}$ receptor was estimated at 0.02% of the choroid plexus messenger RNA population. Lower levels were noted throughout the brain (Julius et al. 1988), and were extensively mapped by *in situ* hybridization by Hoffman and Mezey (1989). Saltzman et al. (1991) first reported a human brainstem cDNA clone for this receptor. Stam et al. (1994) isolated both human genomic and human hippocampal cDNA clones of the 5-HT$_{2C}$ receptor. They mapped its three coding region introns and demonstrated that it is present in the human genome as a single-copy gene which spans over 20 kb. In transfected NIH 3T3 cells it displayed appropriate 5-HT$_{2C}$ ligand-binding properties and was efficiently coupled to PI turnover (Stam et al. 1994). Yu et al. (1991) obtained a mouse choroid plexus cDNA for the 5-HT$_{2C}$ receptor, demonstrated its functional activity in *Xenopus* oocytes, and mapped its location to the mouse X chromosome. Foguet et al. (1992a) reported a mouse genomic 5-HT$_{2C}$ clone and mapped its coding region introns.

JULIUS et al. (1989) demonstrated the ability of serotonin to generate foci in NIH 3T3 cells transfected with the 5-HT$_{2C}$ receptor. These foci were tumorigenic when injected into nude mice, but activation of PI pathways in these cells by other means is also known to induce foci, and the authors noted that the 5-HT$_{2C}$ receptor is unlikely to be tumorigenic in nondividing cells such as neurons. This study points to a possible involvement of the 5-HT$_{2C}$ receptor as a mitogen or growth factor receptor, especially in early brain development. TECOTT et al. (1995) generated 5-HT$_{2C}$ "knockout mice" by homologous re-combination in embryonic stem cells, which selectively inactivates the 5-HT$_{2C}$ gene. They observed a low frequency of spontaneous epileptic seizures and spontaneous death in these animals starting at the 5th week postnatally, ap-parently due to a lowered seizure threshold and a rapid progression of seizure activity. In addition, hemizygous mutant males were 13% overweight, with a 48% increase in perirenal white adipose tissue, with no change in brown fat or body length. The authors noted that a loss of 5-HT$_{2C}$ receptors in the paraventricular nucleus of the hypothalamus may be responsible for these actions.

IX. 5-HT$_3$ Receptor

Among the serotonin receptors, only one ligand-gated ion channel receptor (ionotropic receptor) has been described, the 5-HT$_3$ receptor. This places the serotonin receptor family in an intermediate position between the other monoamine families, which contain no ionotropic receptors (dopamine and norepinephrine), and certain amino acid receptors, which contain many ionotrophic receptor variants (e.g., glutamate and GABA). The 5-HT$_3$ recep-tor does, however, show many variations in properties and pharmacology among different tissues and different species. A major contributor to these variations is now thought to arise from species-specific differences in receptor structure and properties (PETERS et al. 1992).

MARICQ et al. (1991) first described the cloning of a 5-HT$_3$ receptor subunit cDNA from a mouse hybridoma NCB-20 cell line, which they named 5-HT$_3$R-A. The deduced receptor sequence contains four putative transmembrane segments (M1-M4) and is believed to form an active pentameric structure, by analogy to other ionotropic receptors (MARICQ et al. 1991). Formation of putative homo-oligomers of this receptor in oocytes injected with RNA tran-scribed from this clone produced a serotonin-gated cationic conductance which rapidly desensitized, exhibited a Hill slope of 1.6–1.8, and exhibited pharmacological properties in reasonable agreement with native tissue studies (MARICQ et al. 1991). Analysis of 5-HT$_3$ receptor pharmacology, current-voltage relationships, and single-channel conductances in transfected mamma-lian cell lines, N1E-115 cells, and superior cervical ganglion (SCG) neurons suggest that the formation of 5-HT$_3$R-A subunit homo-oligomers adequately accounts for the properties of 5-HT$_3$ receptors in N1E115 cells but not in the SCG preparation (HUSSY et al. 1994). These authors suggested that an undiscovered receptor subunit(s) may exist which modifies the properties of the 5-HT$_3$ receptor complex in the native SCG preparation.

Hope et al. (1993) isolated an apparent splice variant of the 5-HT$_3$R-A receptor which contains a deletion of six amino acids (GSDLLP) from the large cytoplasmic loop between M3 and M4 (and noted several apparent corrections to the sequence of Maricq et al. (1991)). This short-form variant encoded a receptor with identical pharmacological properties and Hill coefficients to the long form, with the single exception that the maximal response to 2-methyl-5-HT was much greater with the long-form homo-oligomer than with the short form (Downie et al. 1994). Similar short and long isoforms have been described in PCR-derived partial cDNA clones from rat nodose and superior cervical ganglia (Miquel et al. 1995). Interestingly, the rat clone apparently lacks the central aspartic acid of the putative alternatively spliced mouse exon since the rat isoforms differ by only five amino acids (GSLLP) (Miquel et al. 1995). In both mouse and rat native tissues, the short splice variant appears to be the dominant isoform, representing nearly 90% of the 5-HT$_3$R-A transcripts in the rat (Miquel et al. 1995).

The effects of cocaine (Fan et al. 1995), morphine (Fan 1995), alcohol, and volatile anesthetics (Machu and Harris 1994) on the cloned, transfected homo-oligomeric 5-HT$_3$ receptor have all been reported. In general, few differences were observed from previous work on the native receptors. Similarly, the allosteric interactions between agonists and antagonists (Bonhaus et al. 1995), as well as calcium current responses (Hargreaves et al. 1994), have been studied and were found to be in general agreement with native receptor studies. Finally, the short isoform of the human receptor has been isolated as a human hippocampal cDNA clone, and its tissue distribution, response, and pharmacological properties have been described (Miyake et al. 1995).

X. 5-HT$_4$ Receptor

The latest serotonin receptor subtype to be cloned, and the last of the pharmacologically well defined subtypes to be verified by cloning, was the 5-HT$_4$ receptor. Gerald et al. (1995) used a degenerate PCR strategy based on serotonin receptor-specific primers to isolate this very low abundance rat brain cDNA. They demonstrated that it encoded a receptor with the binding profile of the pharmacologically defined 5-HT$_4$ receptor. This receptor potently stimulated adenylate cyclase activity when transiently transfected into COS-7 cells and displayed antagonism by serotoninergic compounds that agreed with binding data from the same transfected receptor. The rank orders and affinities of agonists were, however, different in binding than in response assays (Gerald et al. 1995), as has been observed for many 7TM receptors.

Gerald et al. (1995) isolated two different isoforms of the 5-HT$_4$ receptor which differ in their carboxy-terminal domain. The deduced amino acid sequence of the two cDNA clones diverges at position 360 and contains a terminal 28 amino acid tail in the short form (5-HT$_{4S}$) and a 47 amino acid tail in the long form (5-HT$_{4L}$). These different isoforms appear to represent splice variants, with the long isoform widely expressed throughout the brain including striatum, thalamus, olfactory bulb, hippocampus, and brainstem but the short isoform restricted to the striatum. Peripheral tissues generally expressed both isoforms. Interestingly, the long isoform contains an additional protein kinase C phosphorylation site which might contribute to differences in desensitization phenomena or in G protein coupling, as was noted for similar carboxy-terminal splice isoforms of the PACAP and EP3 receptors (see GERALD et al. 1995 and LUCAS and HEN 1995). The two splice isoforms exhibited very similar binding profiles but differed in their adenylate cyclase responses to a series of agonists and partial agonists. Further work to map the cyclase response and tissue expression differences of these two isoforms should help to clarify their roles.

XI. 5-HT$_{5A}$ Receptor

A mouse brain cDNA encoding a receptor with high affinity for 5-CT and I-LSD (2-iodo-(+)lysergic acid diethylamide) but low affinity for sumatriptan was first described by PLASSAT et al. (1992). An orthologous rat hypothalamic cDNA clone has also been described (ERLANDER et al. 1993). Neither clone was able to activate cAMP or PI responses in any system tested, but a GTP-sensitive high-affinity binding site was described when the mouse clone was expressed at low density in NIH 3T3 cells (PLASSAT et al. 1992), suggesting that the receptor is capable of coupling to G proteins. This receptor is expressed widely and apparently exclusively in the brain, including hippocampus, cortex, and hypothalamus, and its pharmacological properties in many ways resemble those of the 5-HT$_{1B}$/5-HT$_{1D}$ receptors. In fact, it was suggested that certain 5-HT$_{1D}$-like responses such as inhibition of glutamate release from granule cells, or 5-HT$_{1D}$ responses in rat brain, may be mediated by 5-ht$_{5A}$ receptors (PLASSAT et al. 1992).

A human 5-ht$_{5A}$ receptor clone has been reported, which encodes a receptor with very similar pharmacological binding properties to the mouse receptor (REES et al. 1994). This clone is widely expressed throughout the adult CNS but no expression could be detected in peripheral tissues (REES et al. 1994) in agreement with studies on the mouse receptor (PLASSAT et al. 1992).

The mouse, rat, and human 5-ht$_{5A}$ genes all contain a single large coding region intron, approximately midway between TM V and TM VI on the i3 loop (MATTHES et al. 1993; REES et al. 1994). The human and mouse chromosomal locations have been mapped in the vicinity of a mouse reeler and human holoprosencephaly mutation, both of which are developmental brain mutations and might involve this receptor (MATTHES et al. 1993).

XII. 5-HT$_{5B}$ Receptor

The 5-ht$_{5B}$ receptor exhibits 77% amino acid homology to the 5-ht$_{5A}$ receptor in the transmembrane regions plus short connecting loops (excluding N and C terminus and i3) (Matthes et al. 1993), which is typical of other closely related monoamine receptor subtypes such as 5-HT$_{1B}$ and 5-HT$_{1D}$ or dopamine D$_1$ and D$_3$. Mouse brain and rat hypothalamic and forebrain cDNA clones have been described as well as a mouse genomic clone (Matthes et al. 1993; Erlander et al. 1993; Wisden et al. 1993). Like the 5-ht$_{5A}$ receptor, no second messenger coupling has yet been described, but again, coupling to G proteins was demonstrated from the ability of Gpp(NH)p to shift agonist binding in transfected COS1 cell membranes (Wisden et al. 1993). 8-Hydroxy-(2-N-dipropylamine)-tetralin (8-OH-DPAT) displays high affinity ($46\,nM$) for agonist (Wisden et al. 1993) but not antagonist binding (Matthes et al. 1993) to this receptor. The intron position of the 5-ht$_{5B}$ receptor matches that of the 5-ht$_{5A}$ receptor and their pharmacological properties are also closely matched (Matthes et al. 1993).

Recent preliminary observations (Grailhe et al. 1995; Rees et al. 1994) suggest that the 5-ht$_{5B}$ receptor is not expressed as a functional receptor in humans. This provides a very rare example of a receptor that was "lost" sometime following the evolutionary divergence between rodents and humans. Grailhe et al. (1995) have published a preliminary report that the coding sequence of the human 5-ht$_{5B}$ gene is interrupted by several stop codons and that no functional receptor protein is formed. These data suggest that the 5-ht$_{5B}$ gene was inactivated by acquiring a lethal nonsense mutation in its coding region sometime after the evolutionary divergence of rodents and humans. This is the only case of a serotonin receptor that is expressed in rodents but not humans and a very rare example of a receptor gene which is inactivated in the human genome but functional in other mammalian species. It will be very interesting to determine which other mammalian species contain active or inactive 5-ht$_{5B}$ genes and why we humans are able to do without it or to activate compensatory mechanisms for its replacement. Of course, it is possible that our inability to demonstrate functional consequences of 5-ht$_5$ receptor activation may indicate that one or both 5-ht$_5$ receptors is currently inactive or nonfunctional in multiple species, but the demonstrated ability of these receptors to couple to G proteins (agonist binding shifts) suggests otherwise.

XIII. 5-HT$_6$ Receptor

PCR amplification of 7TM sequences derived from rat striatal RNA led Monsma et al. (1993) to clone a novel serotonin receptor that couples to activation of adenylate cyclase. This receptor was found exclusively in the CNS, with highest expression levels in the striatum and in limbic and cortical regions. Monsma et al. (1993) detected one coding region intron at the carboxy terminus of TM VI, at a position not seen in any other serotonin receptor gene. RUAT et al. (1993a) isolated a rat striatal 5-HT$_6$ receptor cDNA in an independent homology screening effort based on the histamine H$_2$ receptor. This group detected a second intron in the middle of the i3 loop between TM V and TM VI, at a position shared by several serotonin and dopamine receptors. These same two coding sequence introns were found in the human 5-HT$_6$ receptor when it was recently isolated (KOHEN et al. 1996). A probable frame shift error in the original publication, caused by a single nucleotide sequencing error in a C-G-rich domain was detected during this study and led to correction of the original rat sequence data. Chromosomal mapping of the human gene identified a possible (silent) restriction length polymorphism (RFLP) and located the 5-HT$_6$ gene in close proximity to the human 5-HT$_{1D}$ receptor gene (KOHEN et al. 1996).

This receptor subtype has earned a high degree of interest from pharmaceutical researchers due to its high affinity for clozapine and the possibility that it may be a key site for "atypical" antipsychotic action with reduced extrapyramidal side effect liability. Since atypical antipsychotics with the lowest extrapyramidal side effects all exhibit high affinity for the human 5-HT$_6$ receptor (KOHEN et al. 1996), research on this receptor subtype and its ligands will remain an active field.

Recent antisense oligonucleotide studies on this subtype (see Sect. D, below) suggest an involvement of this subtype in the control of cholinergic transmission with no evidence of dopaminergic effects (BOURSON et al. 1995).

XIV. 5-HT$_7$ Receptor

In short order, six separate groups reported independent successes in isolating a serotonin 5-HT$_7$ receptor, the third member of a group of adenylate-cyclase-stimulatory serotonin receptors.

The first full report was submitted by MEYERHOF et al. (1993), who used a degenerate PCR strategy based on conserved 7TM sequences in the third and sixth transmembrane domains. The receptor they isolated is distantly related to other serotoninergic and monoamine receptors, and exhibits closest homology to a *Drosophila* serotonin receptor that also stimulates adenylate cyclase (WITZ et al. 1990). 5-HT$_7$ clones from rat (LOVENBERG et al. 1993b; RUAT et al. 1993b; SHEN et al. 1993), mouse (PLASSAT et al. 1993), human (BARD et al. 1993), and guinea pig (TSOU et al. 1994) have been reported. Two introns have been located in the coding region, one at the carboxy terminus of TM III (SHEN et al. 1993; RUAT et al. 1993b) and another close to the end of the carboxy tail (RUAT et al. 1993b). Several sequence differences appear in some 5-HT$_7$ publications, with the coding sequence found in the reports by MEYERHOF et al. (1993) and RUAT et al. (1993b) representing examples of the apparent consensus view. A pseudogene related to this receptor has also been identified (BARD et al. 1993).

Possible physiological functions for this receptor have been mentioned, including relaxation in several vascular preparations (Shen et al. 1993; Bard et al. 1993) and circadian rhythm control via the suprachiasmatic nucleus of the hypothalamus (Lovenberg et al. 1993b). Since 8-OH-DPAT was shown to exhibit relatively high affinity for this receptor, multiple functions formerly attributed to a 5-HT$_{1A}$-like receptor (including adenylate cyclase stimulation in gastrointestinal, cardiovascular, and hypothalamic preparations) may reside in the 5-HT$_7$ subtype. Finally, the high affinity of the atypical antipsychotic clozapine for this receptor made this subtype another candidate site for atypical antipsychotic actions, along with the 5-HT$_{2A}$, 5-HT$_6$, and dopamine D$_4$ subtypes.

D. Ligand- and Drug-Binding Sites: Mutagenesis and Chimeric Receptor Studies

In this section I have reviewed a subset of the serotonin mutagenesis studies which have given us a tentative picture of key amino acid residues forming the serotonin- and drug-binding site. These studies are in good agreement with similar studies on catecholamine receptors (reviewed in Strader et al. 1994) which preceded, and often guided, similar studies in the serotonin receptor field.

It is often very difficult to determine whether ligand-binding changes which result from mutation of a single receptor residue indicate direct bonding of that receptor residue to the ligand or a result from indirect effect (such as delocalized conformational changes, or alterations in the processing and insertion of the receptor). Nevertheless, certain mutations cause such consistent results across many 7TM receptors that tentative conclusions about ligand-binding sites have been drawn. These mutagenesis studies indicate that a single serotonin- and drug-binding site exists per receptor monomer, and that these sites overlap extensively. The key binding site residues are located mostly in the top (extracellular) one-third of the transmembrane regions of helices 3–7 (see Fig. 2) over 10 Å in from the extracellular membrane surface amid a ring of transmembrane domains (Strader et al. 1994). Apparently, most of the key amino acid positions involved in direct ligand interactions are the same in serotonin receptors as in the other monoamine receptors.

Key binding site amino acids have been tentatively identified in transmembrane domains 3, 5, 6, and 7 of serotonin receptors. With the possible exception of TM V (see below), they serve similar roles in all monoamine receptors and were previously identified in mutagenesis studies on adrenergic receptors. In the outer third of TM III, an aspartate residue (position 6 of TM III in Fig. 2) appears to serve as a counterion to the aliphatic primary amine of serotonin, critical for both agonist binding and activation. For the 5-HT$_{1A}$ receptor, the mutation D116N (aspartic acid at position 116 changed to asparagine) in TM III decreased agonist affinity and increased the EC_{50} for GTPase activation without affecting the affinity for the antagonist pindolol (Ho et al. 1992). In TM V, STRADER et al. (1989) proposed that a serine residue in the outer third of the transmembrane α-helix (position 9 or 10 of TM V in Fig. 2) may bind to the 5-hydroxyl group of serotonin. In the rat 5-HT_{2A} receptor, the mutation A242S in TM V increased the affinity of N(1)-unsubstituted ergolines and tryptamines but decreased the affinity of N(1)-alkylated compounds (JOHNSON et al. 1994). From this and complementary site-specific mutations, JOHNSON et al. (1994) suggested that a hydrogen bond forms between this TM V serine and the indole nitrogen of N(1)-unsubstituted ergolines and tryptamines. If serotonin binds in a similar manner, then it may be that the indole nitrogen of serotonin rather than the 5-hydroxyl group interacts with serine or threonine residues in the outer third of TM V. This deserves further study since such a binding mode for serotonin would deviate from the expectation (STRADER et al. 1989) set by known TM V–catechol ring hydroxyl interactions in the adrenoceptor family (STRADER et al. 1994). Similar results were observed with the 5-HT_{1A} receptor, where the mutations S199A and T200A (corrected amino acid numbering, see Sect. C.I) each reduced the receptor's affinity for agonist binding and activation but had no effect on antagonist binding (Ho et al. 1992).

TM VI is thought to form a stabilizing interaction with the aromatic ring of monoamine ligands (STRADER et al. 1994) via a key phenylalanine residue (position 17 of TM VI in Fig. 2). For the 5-HT_{2A} receptor, the mutation F340L in TM VI caused a decrease in affinity and functional activation for a wide range of ligands whereas the same change at the preceding residue (F339L) had no effect (CHOUDHARY et al. 1993). Finally, TM VII contains an asparagine residue in the 5-HT_{1A} receptor and the β-adrenergic receptors (position 4 of TM VII in Fig. 2) that is critical for antagonist binding. Conversion of this asparagine in the 5-HT_{1A} receptor to a valine (N386V) caused a dramatic reduction of affinity for pindolol and related β-adrenoceptor aryloxyalkylamine antagonists, with little effect on other ligands (GUAN et al. 1992). Conversely, mutation of other native residues in this position of the human 5-HT_{1B}, human 5-HT_{1D}, 5-HT_{1E}, and 5-HT_{1F} receptors to an asparagine induced a large increase (100- to 10000-fold) in the affinity of the host receptor for propranolol and related ligands (ADHAM et al. 1994b). The affinities for serotonin were essentially unchanged by these mutations, suggesting that the aryloxyalkylamine antagonists extend a contact interaction into this TM VII site but the natural ligand does not (partially overlapping binding sites).

Another important finding of these mutagenesis studies concerns the evolutionary differences between species homologues of a receptor subtype. In two cases, mutation of a single amino acid that differs between the human and rat homologues of the same serotonin receptor subtype caused dramatic changes in the pharmacology and essentially converted the pharmacology of one species homologue to the other. In the human 5-HT_{2A} receptor, 3 transmembrane and 38 additional amino acid differences separate the human from the rat sequence. Only one of these, S242 in transmembrane region V, is found near a key binding site residue as described above. KAO et al. (1992) and JOHNSON et al. (1994) demonstrated that conversion of the TM V human to rat (S242A) or rat to human (A242S) residue converted the pharmacology of the receptor to that of the other species, even though 40 other natural amino acid differences between the species homologues remained unchanged. Similarly, four laboratories have shown that conversion of a human 5-HT_{1B} threonine residue in TM VII to the asparagine residue found in rat (T355N) dramatically increases the receptor's affinity for pindolol and related ligands, and essentially converts the receptor from a human to a rat pharmacology (OKSENBERG et al. 1992; METCALF et al. 1992; PARKER et al. 1993; ADHAM et al. 1994b). These studies clearly demonstrate that large pharmacological differences can arise in species homologues of the same gene when a single amino acid residue sustains a mutation in the vicinity of a key ligand-binding domain. In both of these examples, it is interesting to note that the binding affinity for serotonin (presumably under strong evolutionary pressure) is little changed by these evolutionary changes while the affinities for certain artificial ligands (presumably no evolutionary constraint) are strongly altered.

A different type of natural mutation experiment occurs as a result of allelic polymorphism. Some genes exhibit natural variants in the population. In the case of the ApoE gene, different natural allelic variants have been shown to play a key role in determining our age of onset and susceptibility to Alzheimer's disease (STRITTMATTER and ROSES 1995). Among the serotonin receptors, two natural variants have been recently identified. An I128V natural variant of the human 5-HT_{1A} receptor was cloned from a schizophrenic patient and shown not to affect the pharmacological properties of the receptor, consistent with the location of this substitution on the extracellular aminoterminal segment of the receptor (BRÜSS et al. 1995). In contrast, a natural variant of the human 5-HT_{1B} receptor contains a cysteine substitution near the top of the transmembrane III α-helix (BÜHLEN et al. 1996). Both the position of this substitution near the ligand-binding pocket and its potential ability to form disulfide bonds suggest that it may affect ligand binding and agonist activity. This human receptor variant exhibits an increased affinity for several serotoninergic agonists, which might be physiologically important (BÜHLEN et al. 1996). Such studies of natural allelic variations may in the future provide important information on disease susceptibilities and disease subtyping. Such information could provide new directions for preventive medicine, disease prognosis, and individualized, genotype-specific therapies.

E. Gene Knockout and Antisense Approaches to Function

Two powerful new strategies, gene knockout by homologous recombination and *in vivo* application of antisense oligonucleotides, are showing great promise for revealing receptor function in living animals. Both strategies inactivate or reduce receptor expression, which should be functionally equivalent to blocking the receptor's activity with an antagonist medication. The antisense strategy involves injection of oligonucleotides which are complementary to the target sequence into the adult brain, causing a reduction of the corresponding target mRNA by RNAse H digestion and/or arrest of translation. Technical problems with delivery, distribution, and uptake of oligonucleotide into the appropriate tissues and cells, the general observation that only limited reductions of the target protein can be obtained (generally 50%–70% in vivo), and nonspecific toxic effects of oligonucleotides at doses only slightly above active concentrations are the current limitations of the approach (WAHLESTEDT 1994). In addition, receptors which exhibit high levels of receptor reserve or slow turnover rates may be little affected by current antisense treatments (BOURSON et al. 1995).

The gene knockout strategy introduces a mutated, disabled gene (nonsense mutation) in place of a normal gene by homologous recombination in cultured embryonic stem cells. The mutated cells are injected into blastocysts, leading to formation of chimeric mice which can be bred to homozygosity for the introduced mutation. The perfect specificity of the knockout and the complete absence of target protein product are clear advantages of the approach, which are balanced by the disadvantages that the gene is absent throughout development (which can lead to compensatory changes or lethal effects) and the fact that the technique is slow, laborious, and expensive. The absence of the gene throughout embryo development is particularly problematic since many receptors and neurotransmitters play varied and essential roles throughout development, which may be quite different from their roles in adult animals. It appears that lethal mutations are often produced by gene knockouts, and adaptive compensatory changes during development can produce changes in the adult animal that would lead to little information about the normal role of the protein in normal adult animals. Nevertheless, the techniques for both approaches continue to evolve, and useful data are available from current experiments as long as the limitations and cautions of each technique are understood.

The gene knockout approach has been applied to two serotonin receptors: 5-HT_{1B} and 5-HT_{2C}. The 5-HT_{1B} knockout mice were significantly more aggressive than wild-type mice of the same strain when assayed in the isolation-induced aggression test (SAUDOU et al. 1994). Serotonin is well known to be involved in modulating aggressive behavior in rodents and perhaps humans (BEVAN et al. 1989); thus these experiments suggest that the 5-HT_{1B} receptor may contribute to the antiaggressive properties of serotoninergic agonists. Indeed, several transgenic mouse lines which manifest aggressive behavior [overexpressors of transforming growth factor-α (HILAKIVI-CLARKE and GOLDBERG 1993), calcium-calmodulin-dependent kinase II knockouts (CHEN et al. 1994), monoamine oxidase A deletions (CASES et al. 1995)] also show alterations in serotonin levels or serotoninergic functions. The link of 5-HT_{1B}

function with aggression needs to be viewed as preliminary, however, both for the general concerns about gene knockouts discussed above, and due to the fact that alterations in pain, fear, anxiety, or cognition pathways could all induce secondary effects on aggression in animals.

Knockout mice lacking 5-HT$_{2C}$ receptors exhibit two noteworthy phenotypes: they are overweight (13% increase in body mass, 48% increase in white adipose tissue) due to increased food intake, and they exhibit spontaneous epileptic seizures (TECOTT et al. 1995). The knockout mice do not show reduced food intake upon administration of the nonselective serotonin agonist meta-chlorophenylpiperazine (mCPP), a drug that reduced food intake in wild-type mice by 78%. Thus the 5-HT$_{2C}$ receptor, which is present (as mRNA) in the paraventricular nucleus of the hippocampus, may mediate serotoninergic control of appetite (TECOTT et al. 1995). Spontaneous epileptic seizures that were observed in the knockout mice occasionally led to death and were accompanied by a lowering of seizure threshold and a more rapid progression of seizures, in comparison to wild type (TECOTT et al. 1995). Once again, the conclusion that the 5-HT$_{2C}$ receptor controls these activities must be viewed as preliminary due to the developmental cautions stated above. However, the observations that nonselective serotoninergic agents alter food intake and the induction of seizures in normal adult rats does support this conclusion.

One study on antisense oligonucleotides targeted against a serotonin receptor has been published. BOURSON et al. (1995) have shown that intracerebroventricular injection of antisense phosphorothioate 18-mer oligonucleotides complementary to the 5-HT$_6$ receptor induce a behavioral syndrome of yawning, stretching, and chewing, whereas scrambled control oligonucleotides had no effect. The behavior was antagonized by atropine but not haloperidol, suggesting that an increase in cholinergic transmission may be involved. The authors noted the tentative nature of this conclusion, due to the issues raised above.

Over the next 5–10 years we can expect to rapidly move from the era of receptor cloning and discovery to the genome era where all human gene sequences will be available to us, but the roles of only a small percentage of these genes will be known. At that time, the focus of many academic and pharmaceutical industry researchers will turn towards the characterization of these many "orphan" genes. Despite their technical limitations, both the gene knockout and antisense injection strategies promise to provide important leads on receptor and protein function well before specific agonist or antagonist drugs are available to probe these sites. This type of information cannot be reliably obtained by any other approach. These knockout strategies, along with related gene targeting and differential mRNA display methods, are expected to become major contributors to our neuroscience and pharmaceutical science advances as we increasingly gain understanding of the 100,000 or so proteins of the human genome, most of which are prominently expressed in the mammalian brain.

Acknowledgements. The author wishes to thank Dave Rominger for producing Fig. 2 and Shelley Van Horn for secretarial and literature support.

References

Adham N, Romanienko P, Hartig P, Weinshank RL, Branchek T (1992) The rat 5-hydroxytryptamine$_{1B}$ receptor is the species homologue of the human 5-hydroxytryptamine$_{1D\beta}$. Mol Pharmacol 41:1–7

Adham N, Ellerbrock B, Hartig P, Weinshank RL, Branchek T (1993a) Receptor reserve masks partial agonist activity of drugs in a cloned rat 5-hydroxytryptamine$_{1B}$ receptor expression system. Mol Pharmacol 43:427–433

Adham N, Kao H-T, Schechter LE, Bard J, Olsen M, Urquhart D, Durkin M, Hartig PR, Weinshank RL, Branchek TA (1993b) Cloning of another human serotonin receptor (5-HT$_{1F}$): a fifth 5-HT$_1$ receptor subtype coupled to the inhibition of adenylate cyclase. Proc Natl Acad Sci USA 90:408–412

Adham N, Borden LA, Schechter LE, Gustafon EL, Cochran TL, Pierre J-J, Weinshank RL, Branchek TA (1993c) Cell-specific coupling of the cloned human 5-HT$_{1F}$ receptor to multiple signal transduction pathways. Naunyn Schmiedebergs Arch Pharmacol 348:566–575

Adham N, Vaysse PJ-J, Weinshank RL, Branchek TA (1994a) The cloned human 5-HT$_{1E}$ receptor couples to inhibition and activation of adenylyl cyclase via two distinct pathways in transfected BS-C-1 cells. Neuropharmacology 33:403–410

Adham N, Tamm JA, Salon JA, Vaysse PJJ, Weinshank RL, Branchek TA (1994b) A single point mutation increases the affinity of serotonin 5-HT$_{1D\alpha}$, 5-HT$_{1D\beta}$, 5-HT$_{1E}$, and 5-HT$_{1F}$ receptors for β-adrenergic antagonists. Neuropharmacology 33:387–391

Albert PR, Zhou Q-Y, Van Tol HHM, Bunzow JR, Civelli O (1990) Cloning, functional expression, and mRNA tissue distribution of the rat 5-hydroxytryptamine$_{1A}$ receptor gene. J Biol Chem 265:5825–5832

Amlaiky N, Ramboz S, Boschert U, Plassat J-L, Hen R (1992) Isolation of a mouse "5-HT$_{1E}$-like serotonin receptor expressed predominantly in the hippocampus. J Biol Chem 267:19761–19765

Andrade R, Malenka RC, Nicoll RA (1986) A G protein couples serotonin and GABA$_B$ receptors to the same channels in hippocampus. Science 234:1261–1265

Bach AWJ, Unger L, Sprengel R, Mengod G, Palacios J, Seeburg PH, Voigt MM (1993) Structure, functional expression and spatial distribution of a cloned cDNA encoding a rat 5-HT$_{1D}$-like receptor. J Receptor Res 13:479–502

Bard JA, Zgombick J, Adham N, Vaysse P, Branchek TA Weinshank RL, (1993) Cloning of a novel human serotonin receptor (5-HT$_7$) positively linked to adenylate cyclase. J Biol Chem 268:23422–23426

Bard JA, Nawoschik SP, O'Dowd BF, George SR, Branchek TA, Weinshank RL (1995) The human serotonin 5-hydroxytryptamine$_{1D}$ receptor pseudogene is transcribed. Gene 153:295–296

Beer MS, Middlemiss DN, McAllister G (1993) 5-HT$_1$-like receptors: six down and still counting. Trends Pharmacol Sci 14:228–231

Bertin B, Freissmuth M, Breyer RM, Schutz W, Strosberg AD, Marullo S (1992) Functional expression of the human serotonin 5-HT$_{1A}$ receptor in Escherichia coli. J Biol Chem 267:8200–8206

Bevan P, Cools AR, Archer T (1989) Behavioural pharmacology of 5-HT. Erlbaum, Hillsdale, New Jersey

Boess FG, Martin IL (1994) Molecular biology of 5-HT receptors. Neuropharmacology 33:275–317

Bonhaus DW, Stefanich E, Loury DN, Hsu SA, Eglen RM, Wong EH (1995) Allosteric interactions among agonists and antagonists at 5-hydroxytryptamine$_3$ receptors. J Neurochem 65:104–110

Bourson A, Borroni E, Austin RH, Monsma FJ, Sleight AJ (1995) Determination of the role of the 5-HT$_6$ receptor in the rat brain: a study using antisense oligonucleotides. J Pharmacol Exp Ther 274:173–180

Branchek TA (1995) 5HT$_4$, 5HT$_6$, 5HT$_7$; molecular pharmacology of adenylate cyclase stimulating receptors. Neuroscience 7:375–382

Branchek TA, Adham N, Macchi M, Kao HT, Hartig PR (1990) [^{3}H]-DOB(4-bromo-2,5-dimethoxyphenylisopropylamine) and [^{3}H] ketanserin label two affinity states of the cloned human 5-hydroxytrypamine$_2$ receptor. Mol Pharmacol 38:604–609

Brüss M, Bühlen M, Erdmann J, Göthert M, Bönisch H (1995) Binding properties of the naturally occurring human 5-HT$_{1A}$ receptor variant with the Ile28Val substitution in the extracellular domain. Naunyn Schmiedebergs Arch Pharmacol 352:455–458

Bühlen M, Brüss M, Bönisch H, Göthert M (1966) Modified ligand binding properties of the naturally occurring Phe-124-CγS variant of the human 5-HT$_{1D\beta}$ receptor. Naunyn Schmiedebergs Arch Pharmacol 353(Suppl):R91

Carey J, Hamilton J O'C, Flynn J, Smith G (1995) The gene kings. Business Week, May 8, pp 72–78

Cases O, Seif I, Grimsby J, Gaspar P, Chen K, Pournin S (1995) Aggressive behavior and altered amounts of brain serotonin and norepinephrine in mice lacking MAOA. Science 268:1763–1766

Cerutis DR, Hass NA, Iversen LJ, Bylund DB (1994) The cloning and expression of an OK cell cDNA encoding a 5-hydroxytryptamine$_{1B}$ receptor. Mol Pharmacol 45:20–28

Chanda PK, Minchin MC, Davis AR, Greenberg L, Reilly Y, McGregor WH, Bhat R, Lubeck MD, Mizutani S, Hung PP (1993) Identification of residues important for ligand binding to the human 5-hydroxytryptamine$_{1A}$ serotonin receptor. Mol Pharmacol 43:516–520

Chen C, Rainnie DG, Greene RW, Tonegawa S (1994) Abnormal fear response and aggressive behavior in mutant mice deficient for alpha-calcium-calmodulin kinase II. Science 266:291–294

Choudhary MS, Craigo S, Roth BL (1993) A single point mutation (Phe340→Leu340) of a conserved phenylalanine abolishes 4-[^{125}I]iodo(2,5dimethoxy)-phenylisopropylamine and [^{3}H]mesulergine but not [^{3}H]ketanserin binding to 5-hydroxytryptamine$_2$ receptors. Mol Pharmacol 43:755–761

Codina J, Yatani A, Grenet D, Brown AM, Birnbaumer L (1987) The α-subunit of the GTP binding protein Gk opens atrial potassium channels. Science 236:442–445

Demchyshyn L, Sunahara RK, Miller K, Teitler M, Hoffman BJ, Kennedy JL, Seeman P, Van Tol HHM, Niznik HB (1992) A human serotonin $_{1D}$ receptor variant (5HT$_{1D\beta}$) encoded by an intronless gene on chromosome 6. Biochemistry 89:5522–5526

Downie DL, Hope AG, Lambert JJ, Peters JA, Blackburn TP, Jones BJ (1994) Pharmacological characterization of the apparent splice variants of the murine 5-HT$_3$ R-A subunit expressed in Xenopus laevis oocytes. Neuropharmacology 33:473–482

Du Y-L, Wilcox BD, Jeffrey JJ (1995) Regulation of rat 5-hydroxytryptamine type 2 receptor gene activity: identification of cis elements that mediate basal and 5-hydroxytryptamine-dependent gene activation. Mol Pharmacol 47:915–922

Erlander MG, Lovenberg TW, Baron BM, de Lecea L, Danielson PE, Racke M, Slone AL, Siegel BW, Foye PE, Cannon K, Burns JE, Sutcliffe JG (1993) Two members of a distinct subfamily of 5-hydroxytryptamine receptors differentially expressed in rat brain. Proc Natl Acad Sci USA 90:3452–3456

Fan P, Oz M, Zhang L, Weight FF (1995) Effect of cocaine on the 5-HT$_3$ receptor-mediated ion current in Xenopus oocytes. Brain Res 673:181–184

Fargin A, Raymond JR, Lohse MJ, Kobilka BK, Caron MC, Lefkowitz (1988) The genomic clone G-21 which resembles a β-adrenergic receptor sequence encodes the 5-HT$_{1A}$ receptor. Nature 335:358–360

Fargin A, Raymond JR, Regan JW, Cotecchia S, Lefkowitz RJ, Caron MG (1989) Effector coupling mechanisms of the cloned 5-HT$_{1A}$ receptor. J Biol Chem 264:14848–14852

Foguet M, Nguyen H, Le Huong, Lübbert H (1992a) Structure of the mouse 5-HT$_{1C}$, 5-HT$_2$ and stomach fundus serotonin receptor genes. Neuroreport 3:345–348

Foguet M, Hoyer D, Pardo LA, Parekh A, Kluxen FW, Kalkman HO, Stühmer W, Lübbert H (1992b) Cloning and functional characterization of the rat stomach fundus serotonin receptor. EMBO J 11:3481–3487

Gerald C, Adham N, Kao H-T, Olsen MA, Laz TM, Schechter LE, Bard JA, Vaysse PJJ, Hartig PR, Branchek TA, Weinshank RL (1995) The 5-HT$_4$ receptor: molecular cloning and pharmacological characterization of two splice variants. EMBO J 14:2806–2815

Grailhe R, Ramboz S, Boschert U, Hen R (1995) The 5-HT$_5$ receptors: characterization of the human 5-HT$_{5A}$ receptor; absence of the human 5-HT$_{5B}$ receptor; knockout of the mouse 5-HT$_{5A}$ receptor. Soc Neurosci Abst 21:1856

Guan XM, Peroutka SJ, Kobilka BK (1992) Identification of a single amino acid residue responsible for the binding of a class of β-adrenergic receptor antagonists to 5-hydroxytryptamine$_{1A}$ receptors. Mol Pharmacol 41:695–698

Gudermann T, Levy FO, Birnbaumer M, Birnbaumer L, Kaumann AJ (1993) Human S31 serotonin receptor clone encodes a 5-hydroxytryptamine$_{1E}$-like serotonin receptor. Mol Pharmacol 43:412–418

Hamblin MW, Metcalf MA (1991) Primary structure and functional characterization of a human 5HT$_{1D}$-type serotonin receptor. Mol Pharmacol 40:143–148

Hamblin MW, McGuffin RW, Metcalf MA, Dorsa DM, Merchang KM (1992) Distinct 5-HT$_{1B}$ and 5HT$_{1D}$ serotonin receptors in rat: structural and pharmacological comparison of the two cloned receptors. Mol Cell Neurosci 3:578–587

Hamel E, Gregoire L, Lau B (1993) 5-HT$_1$ receptors mediating contraction in bovine cerebral arteries: a model for human cerebrovascular "5-HT$_{1D\beta}$" receptors. Eur J Pharmacol 242:75–82

Hargreaves AC, Lummis SCR, Taylor CW (1994) Ca^{2+} permeability of cloned and native 5-hydroxytryptamine type 3 receptors. Mol Pharmacol 46:1120–1128

Harrington MA, Shaw K, Zhong P, Ciaranello RD (1994) Agonist-induced desensitization and loss of high-affinity binding sites of stably expressed human 5-HT$_{1A}$ receptors. J Pharmacol Exp Ther 268:1098–1106

Hartig PR, Branchek TA, Weinshank RL (1992) A subfamily of 5-HT$_{1D}$ receptor genes. Trends Pharm Sci 13:152–159

Hartig PR, Hoyer D, Humphrey PPA, Martin GR (1996) Alignment of receptor nomenclature with the human genome: classification of 5-HT$_{1B}$ and 5-HT$_{1D}$ receptor subtypes. Trends Pharm Sci 17:103–105

Hilakivi-Clarke LA, Goldberg R (1993) Effects of tryptophan and serotonin uptake inhibitors on behavior in male transgenic transforming growth factor alpha mice. Eur J Pharmacol 237:101–108

Ho BY, Karschin A, Branchek T, Davidson N, Lester HA (1992) The role of conserved aspartate and serine residues in ligand binding and in function of the 5-HT$_{1A}$ receptor: a site-directed mutation study. FEBS Lett 312:259–262

Hoffman BJ, Mezey E (1989) Distribution of serotonin 5-HT$_{1C}$ receptor mRNA in adult rat brain. FEBS Lett 247:453–462

Hope AG, Downie DL, Sutherland L, Lambert JJ, Peters JA, Burchell B (1993) Cloning and functional expression of an apparent splice variant of the murine 5-HT$_3$ receptor A subunit. Eur J Pharmacol 245:187–192

Hoyer D, Middlemiss DN (1989) Species differences in the pharmacology of terminal 5-HT autoreceptors in mammalian brain. Trends Pharmacol Sci 10:130–132

Hoyer D, Clarke DE, Fozard JR, Hartig PR, Martin GR, Mylecharane EJ, Saxena PR, Humphrey PPA (1994) VII. International union of pharmacology classification of receptors for 5-hydroxytryptamine (serotonin). Pharmacol Rev 46:157–203

Hussy N, Lukas W, Jones KA (1994) Functional properties of a cloned 5-hydroxytryptamine ionotropic receptor subunit: comparison with native mouse receptors. J Physiol 481:311–322

Jin H, Oksenberg D, Ashkenazi A, Peroutka SJ, Duncan AMV, Rozmahel R, Yang Y, Mengod G, Palacios JM, O'Dowd BF (1992) Characterization of the human 5-hydroxytryptamine$_{1B}$ receptor. J Biol Chem 267:5735–5738

Johnson MP, Loncharich RJ, Baez M, Nelson DL (1994) Species variations in transmembrane region V of the 5-hydroxytryptamine type 2A receptor alter the structure-activity relationship of certain ergolines and tryptamines, Mol Pharmacol 45:277–286

Johnson MP, Baez M, Kursar JD, Nelson DL (1995) Species differences in 5-HT$_{2a}$ receptors: cloned pig and rhesus monkey 5-HT$_{2a}$ receptors reveal conserved transmembrane homology to the human rather than rat sequence. Biochim Biophys Acta 1236:201–206

Julius D (1991) Molecular biology of serotonin receptors. Annu Rev Neurosci 14:335–360

Julius D, MacDermott AB, Axel R, Jessell TM (1988) Molecular characterization of a functional cDNA encoding the serotonin 1c receptor. Science 241:558–564

Julius D, Livelli TJ, Jessell TM, Axel R (1989) Ectopic expression of the serotonin 1c receptor and the triggering of malignant transformation. Science 244:1057–1062

Julius D, Huang KN, Livelli TJ, Axel R, Jessell TM (1990) The 5HT$_2$ receptor defines a family of structurally distinct but functionally conserved serotonin receptors. Proc Natl Acad Sci USA 87:928–932

Kao HT, Adham N, Olsen M, Weinshank RL, Branchek TA, Hartig P (1992) Site-directed mutagenesis of a single residue changes the binding properties of the serotonin 5-HT$_2$ receptor from a human to a rat pharmacology. FEBS Lett 307:324–328

Karschin A, Ho BY, LaBarca C, Elroy-Stein O, Moss B, Davidson N, Lester, HA (1991) Heterologously expressed serotonin $_{1A}$ receptors couple to muscarinic K$^+$ channels in heart. Proc Natl Acad Sci USA 88:5694–5698

Kobilka BK, Frielle T, Collins S, Yang-Feng T, Kobilka S, Francke U, Lefkowitz J, Caron MC (1987) An intronless gene encoding a potential member of the family of receptors coupled to guanine nucleotide regulatory proteins. Nature 329:75–79

Kohen R, Metcalf MA, Khan N, Druck T, Huebner K, Lachowicz JE, Meltzer HY, Sibley DR, Roth BL Hamblin MW (1996) Cloning, characterization, and chromosomal localization of a human 5-HT$_6$ serotonin receptor. J Neurochem 66:47–56

Kursar JD, Nelson DL, Wainscott DB, Cohen ML, Baez M (1992) Molecular cloning, functional expression, and pharmacological characterization of a novel serotonin receptor (5-hydroxytryptamine 2F) from rat stomach fundus. Mol Pharmacol 42:549–557

Leonhardt S, Herrick-Davis K, Titeler M (1989) Detection of a novel serotonin receptor subtype (5-HT$_{1E}$) in human brain: interaction with a GTP-binding protein. J Neurochem 53:465–471

Levy FO, Gudermann T, Perez-Reyes E, Birnbaumer M, Kaumann AJ, Birnbaumer L (1992a) Molecular cloning of a human serotonin receptor (S12) with a pharmacological profile resembling that of the 5-HT$_{1D}$ subtype. J Biol Chem 267:7553–7562

Levy FO, Gudermann T, Birnbaumer M, Kaumann AJ, Birnbaumer L (1992b) Molecular cloning of a human gene (S31) encoding a novel serotonin receptor mediating inhibition of adenylyl cyclase. FEBS Lett 296:201–206

Libert F, Parmentier M, Lefort A, Dinsart C, Van Sande J, Maenhaut C, Simons MJ, Dumont JE, Vassart G (1989) Selective amplification and cloning of four new members of the G-protein-coupled receptor family. Science 244:569–572

Loric S, Launay JM, Colas JF, Maroteaux L (1992) New mouse 5-HT$_2$ like receptor expression in brain, heart and intestine. FEBS Lett 312:203–207

Lovenberg TW, Erlander MG, Baron BM, Racke M, Slone AL, Siegel BW, Craft CM, Burns JE, Danielson PE, Sutcliffe (1993a) Molecular cloning and functional expression of 5-HT$_{1E}$-like rat and human 5-hydroxytryptamine receptor genes. Proc Natl Acad Sci USA 90:2184–2188

Lovenberg TW, Baron BM, deLecea L, Miller JD, Prosser RA, Rea MA, Foye PE, Racke M, Slone AL, Siegel BW, Danielson PE, Sutcliffe JG, Erlander MG (1993b) A novel adenylyl cyclase-activating serotonin receptor (5-HT$_7$) implicated in the regulation of mammalian circadian rhythms. Neuron 11:449–458

Lübbert H, Hoffman BJ, Snutch TP, van Dyke T, Levine AJ, Hartig PR, Lester HA, Davidson N (1987) cDNA cloning of a serotonin 5-HT$_{1C}$ receptor by electrophysiological assays of mRNA-injected Xenopus oocytes. Proc Natl Acad Sci USA 84:4332–4336

Lucas JJ, Hen R (1995) New players in the 5-HT receptor field: genes and knockouts. Trends Pharmacol Sci 16:246–252

Machu TK, Harris RA (1994) Alcohols and anesthetics enhance the function of 5-hydroxytryptamine$_3$ receptors expressed in Xenopus laevis oocytes. J Pharmacol Exp Ther 271:898–905

Maenhaut C, Van Sande J, Massart C, Dinsart C, Libert F, Monferini E, Giraldo E, Ladinsky H, Vassart G, Dumont JE (1991) The orphan receptor cDNA RDC4 encodes a 5-HT$_{1D}$ serotonin receptor. 180:1460–1468

Maricq AV, Peterson AS, Brake AJ, Myers RM, Julius D (1991) Primary structure and functional expression of the 5HT$_3$ receptor, a serotonin-gated ion channel. Science 254:432–437

Maroteaux L, Saudou F, Amlaiky N, Boschert U, Plassat JL, Hen R (1992) Mouse 5HT$_{1B}$ serotonin receptor: cloning, functional expression, and localization in motor control centers. Neurobiology 89:3020–3024

Matthes H, Boschert U, Amlaiky N, Grailhe R, Plassat JL, Muscatelli F, Mattei MG, Hen R (1993) Mouse 5-hydroxytryptamine$_{5A}$ and 5-hydroxytryptamine$_{5B}$ receptors define a new family of serotonin receptors: cloning, functional expression, and chromosomal localization. Mol Pharmacol 43:313–319

McAllister G, Charlesworth A, Snodin C, Beer MS, Noble AJ, Middlemiss DN, Iversen LL, Whiting P (1992) Molecular cloning of a serotonin receptor from human brain (5HT$_{1E}$): a fifth 5HT$_1$-like subtype. Neurobiology 89:5517–5521

Meller E, Goldstein M, Bohmaker K (1990) Receptor reserve for 5-hydroxytryptamine$_{1A}$-mediated inhibition of serotonin synthesis: possible relationship to anxiolytic properties of 5-hydroxytryptamine$_{1A}$ agonists. Mol Pharmacol 37:231–237

Metcalf MA, McGuffin RW, Hamblin MW (1992) Conversion of the human 5-HT$_{1D\beta}$ serotonin receptor to the rat 5-HT$_{1B}$ ligand-binding phenotype by Thr355Asn site directed mutagenesis. Biochem Pharmacol 44:1917–1920

Meyerhof W, Obermuller F, Fehr S, Richter D (1993) A novel serotonin receptor: primary structure, pharmacology, and expression pattern in distinct brain regions. DNA Cell Biol 12:402–409

Miquel MC, Doucet E, Bone C, Mestikawy S EL, Matthiessen L, Daval G, Verge D, Hamon M (1991) Central serotonin$_{1A}$ receptors: respective distribution of encoding mRNA, receptor protein and binding sites by in situ hybridization histochemistry, radioimmunohistochemistry and autoradiographic mapping in the rat brain. Neurochem Int 19:453–465

Miquel MC, Emerit MB, Gingrich JA, Nosjean A, Hamon M, Mestikawy S (1995) Developmental changes in the differential expression of two serotonin 5-HT$_3$ receptor splice variants in the rat. J Neurochem 65:475–483

Miyake A, Mochizuki S, Takemoto Y, Akuzawa S (1995) Molecular cloning of human 5-hydroxytryptamine$_3$ receptor: heterogeneity in distribution and function among species. Mol Pharmacol 48:407–416

Monsma FJ Jr, Shen Y, Ward RP, Hamblin MW, Sibley DR (1993) Cloning and expression of a novel serotonin receptor with high affinity for tricyclic psychotropic drugs. Mol Pharmacol 43:320–327

Nguyen T, Marchese A, Kennedy JL, Petronis A, Peroutka SJ, Wu PH, O'Dowd BF
   (1993) An Alu repeat interrupts a human 5-hydroxytryptamine$_{1D}$ receptor
   pseudogene. Gene 124:295–301
Oakey RJ, Caron MG, Lefkowitz RJ, Seldin MF (1991) Genomic organization of
   adrenergic and serotonin receptors in the mouse: linkage mapping of sequence-
   related genes provides a method for examining mammalian chromosome evolution, Genomics 10:338–344
Oksenberg D, Marsters SA, O'Dowd BF, Jin H, Havlik S, Peroutka SJ, Ashkenazi A
   (1992) A single amino-acid difference confers major pharmacological variation
   between human and rodent 5-HT$_{1B}$ receptors. Nature 360:161–163
Parker EM, Grisel DA, Iben LG, Shapiro RA (1993) A single amino acid difference
   accounts for the pharmacological distinctions between the rat and human 5-
   hydroxytryptamine$_{1B}$ receptors. J Neurochem 60:380–383
Pauwels PJ, Van Gompel P, Leysen JE (1993) Activity of serotonin (5-HT) receptor
   agonists, partial agonists and antagonists at cloned human 5-HT$_{1A}$ receptors that
   are negatively coupled to adenylate cyclase in permanently transfected HeLa cells.
   Biochem Pharmacol 45:375–383
Peroutka SJ, Howell TA (1994) The molecular evolution of G-protein-coupled
   receptors: focus on 5-hydroxytryptamine receptors. Neuropharmacology 33:319–
   324
Peters JA, Malone HM, Lambert JJ (1992) Recent advances in the electrophysiological
   characterization of 5-HT$_3$ receptors. Trends Pharm Sci 13:391–397
Plassat JL, Boschert U, Amlaiky N, Hen R (1992) The mouse 5HT$_5$ receptor reveals a
   remarkable heterogeneity within the 5HT$_{1D}$ receptor family. EMBO J 11:4779–
   4786
Plassat JL, Amlaiky N, Hen R, (1993) Molecular cloning of a mammalian serotonin
   receptor that activates adenylate cyclase. Mol Pharmacol 44:229–236
Pritchett DB, Bach AWJ, Wozny M, Taleb O, Dal Toso R, Shih JC, Seeburg PH (1988)
   Structure and functional expression of a cloned rat serotonin 5HT-2 receptor.
   EMBO J 7:4135–4140
Raymond JR, Fargin A, Middleton JP, Craff JM, Haupt DM, Caron MG, Lefkowitz
   RJ, Dennis VW (1989) The human 5-HT$_{1A}$ receptor expressed in HeLa cells
   stimulates sodium-dependent phosphate uptake via protein kinase C. J Biol Chem
   264:21943–21950
Rees S, den Daas I, Foord S, Goodson S, Bull D, Kilpatrick G, Lee M (1994) Cloning
   and characterisation of the human 5-HT$_{5A}$ serotonin receptor. FEBS Lett 355:242–
   246
Ruat M, Traiffort E, Arrang JM, Tardivel-Lacombe J, Diaz J, Leurs R, Schwartz JC
   (1993a) A novel rat serotonin (5-HT$_6$) receptor: molecular cloning, localization
   and stimulation of cAMP accumulation. Biochem Biophys Res Comm 193:268–
   276
Ruat M, Traiffort E, Leurs R, Tardivel-Lacombe J, Diaz J, Arrang JM, Schwartz JC
   (1993b) Molecular cloning, characterization, and localization of a high-affinity
   serotonin receptor (5-HT$_7$) activating cAMP formation. Proc Natl Acad Sci USA
   90:8547–8551
Saltzman AG, Morse B, Whitman MM, Ivanshchenko Y, Jaye M, Felder S (1991)
   Cloning of the human serotonin 5-HT$_2$ and 5-HT$_{1C}$ receptor subtypes. Biochem
   Biophys Res Comm 181:1469–1478
Saudou F, Amara DA, Dierich A, LeMeur M, Ramboz S, Segu L, Buhot MC, Hen R
   (1994) Enhanced aggressive behavior in mice lacking 5-HT$_{1B}$ receptor. Science
   265:1875–1878
Schmuck K, Ullmer C, Engles P, Lübbert H (1994) Cloning and functional character-
   ization of the human 5-HT$_{2B}$ serotonin receptor. FEBS Lett 342:85–90
Shen Y, Monsma FJ, Metcalf MA, Jose PA, Hamblin MW, Sibley DR (1993) Molecu-
   lar cloning and expression of a 5-hydroxytryptamine$_7$ serotonin receptor subtype.
   J Biol Chem 268:18200–18204

Shuck ME, Veldman SA, Bienkowski MJ (1993) Cloning, sequencing and phylogenetic analysis of a human 5-hydroxytryptamine$_{1D}$ receptor pseudogene. Gene 137:339–344

Spengler D, Waeber C, Pantaloni C, Holsboer F, Bockaert J, Seeburg PH, Journot L (1993) Differential signal transduction by five splice variants of the PACAP receptor. Nature 365:170–175

Spurlock G, Buckland P, O'Donovan M, McGuffin P (1994) Lack of effect of antidepressant drugs on the levels of mRNAs encoding serotonergic receptors, synthetic enzymes and 5HT transporter. Neuropharmacology 33:433–440

Stam NJ, Van Huizen F, Van Alebeek C, Brands J, Dijkema R, Tonnaer JADM, Olijve W (1992) Genomic organization coding sequence and functional expression of human 5-HT$_2$ and 5-HT$_{1A}$ receptor genes. Eur J Pharmacol Mol Pharmacol Section 227:153–162

Stam NJ, Vanderheyden P, van Alebeek C, Klomp J, deBoer T, van Delft AML, Olijve W (1994) Genomic organisation and functional expression of the gene encoding the human serotonin 5-HT$_{2c}$ receptor. Eur J Pharmacol 269:339–348

Strader CD, Candelore MR, Hill WS, Sigal RS, Dixon RAF (1989) Identification of two serine residues involved in agonist activation of the β-adrenergic receptor. J Biol Chem 264:13572–13578

Strader CD, Ming Fong T, Tota MR, Underwood D (1994) Structure and function of G protein-coupled receptors. Annu Rev Biochem 63:101–132

Strittmatter WJ, Roses AD (1995) Apolipoprotein E and Alzheimer's disease. Proc Natl Acad Sci USA 92:4725–4727

Tecott L, Sun LM, Akana SF, Strack AM, Lowenstein DH, Dallman MF, Julius D (1995) Eating disorder and epilepsy in mice lacking 5-HT$_{2c}$ serotonin receptors. Nature 374:542–546

Teitler M, Leonhardt S, Weisberg EL, Hoffman BJ (1990) 4-[125]Iodo-(2,5-dimethoxy)phenylisopropylamine and [³H]ketanserin labeling of 5-hydroxytryptamine$_2$ (5-HT$_2$) receptors in mammalian cells transfected with a rat 5-HT$_2$ cDNA: evidence for multiple states and not multiple 5-HT$_2$ receptor subtypes. Mol Pharmacol 38:594–598

Tsou AT, Kosaka A, Bach C, Zuppan P, Yee C, Tom L, Alvarez R, Ramsey S, Bonhaus DW, Stefanich E, et. al. (1994) Cloning and expression of a 5-hydroxytryptamine$_7$ receptor positively coupled to adenylyl cyclase. J Neurochem 63:456–464

van Huizen F, Bansse MT, Stam NJ (1993) Agonist-induced down-regulation of human 5-HT$_{1A}$ and 5-HT$_2$ receptors in Swiss 3T3 cells. Neuroreport 4:1327–1330

Van Obberghen-Schilling E, Vouret-Craviari V, Haslam RJ, Chambard J-C, Pouyssgur (1991) Cloning, functional expression and role in cell growth regulation of a hamster 5-HT$_2$ receptor subtype. Mol Endocrinol 5:881–889

Varrault A, Le Nguyen D, McClue S, Harris B, Jouin P, Bockaert J (1994) 5-Hydroxytryptamine$_{1a}$ receptor synthetic peptides. Mechanisms of adenylyl cyclase inhibition. J Biol Chem 269:16720–16725

Veldman SA, Bienkowski MJ (1992) Cloning and pharmacological characterization of a novel human 5-hydroxytryptamine$_{1D}$ receptor subtype. Mol Pharmacol 42:439–444

Voigt MM, Laurie DJ, Seeburg PH, Bach A (1991) Molecular cloning and characterization of a rat brain cDNA encoding a 5-hydroxytryptamine$_{1B}$ receptor. EMBO J 10:4017–4023

Wahlestedt C (1994) Antisense oligonucleotide strategies in neuropharmacology. Trends Pharmacol Sci 15:42–46

Wainscott DB, Cohen ML, Schenck KW, Audia JE, Nissen JS, Baez M, Kursar JD, Lucaites VL, Nelson DL (1993) Pharmacological characteristics of the newly cloned rat 5-hydroxytryptamine$_{2F}$ receptor. Mol Pharmacol 43:419–426

Weinshank RL, Zgombick JM, Macchi MJ, Branchek TA, Hartig PR (1992) Human serotonin$_{1D}$ receptor is encoded by a subfamily of two distinct genes: 5HT$_{1D\alpha}$ and 5-HT$_{1D\beta}$. Proc Natl Acad Sci USA 89:3630–3634

Wills C (1993) The runaway brain. Basic Books, New York

Wisden W, Parker EM, Mahle CD, Grisel DA, Nowak HP, Yocca FD, Felder CC, Seeburg PII, Voigt MM (1993) Cloning and characterization of the rat 5-HT$_{3B}$ receptor: evidence that the 5-HT$_{5B}$ receptor couples to a G protein in mammalian cell membranes. FEBS Lett 33:25–31

Witz P, Amlaiky N, Plassat JL, Maroteaux L, Borelli E, Hen R (1990) Cloning and characterization of a Drosophila serotonin receptor that activates adenylate cyclase. Proc Natl Acad Sci USA 87:8940–8944

Yocca FD, Iben L, Meller E (1992) Lack of apparent receptor reserve at postsynaptic 5-hydroxytryptamine$_{1A}$ receptors negatively coupled to adenylyl cyclase activity in rat hippocampal membranes. Mol Pharmacol 41:1066–1072

Yu L, Nguyen H, Le H, Bloem LJ, Kozak CA, Hoffman BJ, Snutch TP, Lester HA, Davidson N, Lübbert H (1991) The mouse 5-HT$_{1C}$ receptor contains eight hydrophobic domains and is X-linked. Mol Brain Res 11:143–149

Zgombick JM, Weinshank RL, Macchi M, Schecher LE, Branchek TA, Hartig PR (1991) Expression and pharmacological characterization of a canine 5-hydroxytryptamine$_{1D}$ receptor subtype. Mol Pharmacol 40:1036–1042

Zgombick JM, Schechter LE, Macchi M, Hartig PR, Branchek TA, Weinshank RL (1992) Human gene S31 encodes the pharmacologically defined serotonin 5-hydroxyptamine$_{1E}$ receptor. Mol Pharmacol 42:180–185

Zgombick JM, Borden LA, Cochran TL, Kucharewicz SA, Weinshank RL, Branchek TA (1993) Dual coupling of cloned human 5-hydroxytryptamine$_{1D\alpha}$ and 5-hydroxytryptamine$_{1D\beta}$ receptors stably expressed in murine fibroblasts: inhibition of adenylate cyclase and elevation of intracellular calcium concentrations via pertussis toxin-sensitive G proteins. Mol Pharmacol 44:575–582

Zhu QS, Chen K, Shih JC (1995) Characterization of the human 5-HT$_{2A}$ receptor gene promoter. J Neurosci 15(7Pt1):4885–4895

5-Hydroxytryptamine Receptor Histochemistry: Comparison of Receptor mRNA Distribution and Radioligand Autoradiography in the Brain

G. Mengod, J.M. Palacios, K.H. Wiederhold, and D. Hoyer

A. Introduction

Understanding neurotransmission begins with the anatomical and cellular localisation of neurotransmitters and their receptors. Ideally the question to be answered is: which neurone contains which receptor, and where are the receptors localised at the cellular level? It goes without saying that in many cases knowledge of this is only partial; the complexity of the 5-hydroxytryptamine (5-HT; for abbreviations, see "Appendix") receptor family with its seven subfamilies does not make the task simple (Hoyer et al. 1994). Fortunately, however, a number of radioligands which became available over the past decade have been used to characterise 5-HT binding sites; all of these sites were eventually recognised as true receptors since the corresponding cDNAs have been cloned. Receptor autoradiography emerged early as the method of choice to map brain receptors; it is indeed very gratifying to actually "see" the receptors labelled with a radioligand in many brain areas and nuclei, something that would simply be impossible to achieve using dissection techniques combined with membrane binding. However, there are a number of limitations to this methodology, which relate (a) to the potential lack of selectivity of the radioligands used (if at all available) and (b) to the technique, which in principle does not allow resolution at or below the cellular level.

In addition, even when electron microscopy resolution is reached with irreversible ligands it is difficult to associate the autoradiographic grains directly with specific cells and subcellular compartments, particularly in the brain. There are nevertheless exceptions, since evidence can be generated for the localisation of receptors in axons or even cell bodies; however, this is not the rule, because the receptor is often expressed in terminals which are at a significant distance from the cell body, which may not even be visible in the brain section studied. On the other hand, the recent molecular cloning of most members of the 5-HT receptor family has allowed the application of in situ hybridisation histochemistry to the study of 5-HT receptor subtype mRNA distribution, with high resolution and selectivity. In situ hybridisation can be extremely selective since the exactly matching "ligand" is used, for example, cRNA, cDNA or oligonucleotide of which the sequence is directly derived from the gene (Vilaro et al. 1995). In

Table 1. Summary of ligands used for 5-HT receptor autoradiography (modified from Waeber and Palacios 1993)

Receptor/ligand	Other receptors also labelled	Reference
5-HT$_{1A}$		
[³H]LSD	5-HT/DA receptors	Palacios et al. (1983)
[³H]5-HT	5-HT$_1$ receptors	Marcinkiewicz et al. (1984)
[³H]8-OH-DPAT	5-HT$_1$ receptors	Marcinkiewicz et al. (1984)
[³H]TVX Q 7821	5-HT$_1$ receptors	Glaser et al. (1985)
[³H]PAPP	5-HT$_1$ receptors	Ransom et al. (1986)
[¹²⁵I]BH-8-MeO-N-PAT	5-HT$_1$ receptors	Gozlan et al. (1988)
[³H]Buspirone	DA receptors	Brüning et al. (1989)
[³H]Eltoprazine	5-HT$_{1B}$ receptors	Sijbesma et al. (1990)
[³H]5-Methyl-urapidil	α-adrenoceptors	Laporte et al. (1991)
[³H]Tandospirone	α-adrenoceptors	Tanaka et al. (1991)
[³H]WAY 100635	α-adrenoceptors	Khawaja et al. (1995)
[³H]5-CT	5-HT$_{1B/7}$ receptors	Palacios et al. (1996), Waeber and Moskowitz (1995b)
5-HT$_{1B}$		
[³H]5-HT	5-HT$_1$ receptors	Pazos and Palacios (1985)
[¹²⁵I]I-CYP	β-adrenoceptors	Pazos et al. (1985b)
[¹²⁵I]Serotonin-O-CM-GTNH$_2$ (GTI)	5-HT$_{1D}$ receptors	Boulenguez et al. (1991), Palacios et al. (1996)
[³H]5-CT	5-HT$_{1A/7}$ receptors	Waeber and Moskowitz (1995b)
[³H]GR 125743	5-HT$_{1D}$ receptors	Mengod et al. (in preparation)
5-HT$_{1D}$		
[³H]5-HT	5-HT$_1$ receptors	Waeber et al. (1988a)
[¹²⁵I]Serotonin-O-CM-GTNH$_2$ (GTI)	5-HT$_{1B}$ receptors	Palacios et al. (1992)
[³H]GR 125743	5-HT$_{1B}$ receptors	Mengod et al. (in preparation)
5-HT$_{1E}$		
[³H]5-HT	Other non-5-HT $_{1A/1D/1C}$	Leonhard et al. (1989)
5-HT$_{1F}$		
[³H]Sumatriptan	5-HT$_{1B/1D}$ receptors	Waeber and Moskowitz (1995b)

5-HT$_{2A}$

Radioligand	Site	Reference
[3H]LSD	5-HT/DA receptors	Young and Kuhar (1980)
[3H]Spiperone	DA receptors	Palacios et al. (1981)
[3H]Ketanserin	"Tetrabenazine" sites	Schotte et al. (1983)
[3H]Mesulergine	5-HT$_{2C}$ receptors	Pazos et al. (1985a)
[125I]LSD	5-HT/DA receptors	Altar et al. (1986)
[125I]N-methyl-LSD	5-HT$_{2C}$ receptors	Hoffman et al. (1987)
8-[125I]7-NH$_2$-Ketanserin	"Tetrabenazine" sites	Schotte and Leysen (1988)
[125I]DOI	α-adrenoceptors	Appel et al. (1990)
[3H]MDL 100907	5-HT$_{2C}$ receptors	López-Giménez et al. (1997)

5-HT$_{2C}$

Radioligand	Site	Reference
[3H]LSD	5-HT/DA receptors	Palacios et al. (1983)
[3H]5-HT	5-HT$_1$ receptors	Pazos and Palacios (1985)
[3H]Mesulergine	5-HT$_{2A}$ receptors	Pazos et al. (1985a)
[125I]LSD	5-HT$_{2A}$ receptors	Yagaloff and Hartig (1985)
[3H]SCH 23390	DA receptors	Nicklaus et al. (1988)
[125I]DOI	5-HT$_{2A}$ receptors	Appel et al. (1990)
[3H]RP62203	5-HT$_{2A}$ receptors	Malgouris et al. (1993)

5-HT$_3$

Radioligand	Site	Reference
[3H]ICS 205930	—	Waeber et al. (1988b)
[3H]GR 65630	—	Kilpatrick et al. (1988)
[3H]BRL 43694	—	Reynolds et al. (1989)
[3H]Zacopride	—	Waeber et al. (1990b)
[3H]Quipazine	—	Perry (1990)
[3H]Iodo-zacopride	—	Koscielniak et al. (1990)
[3H]LY 278584	—	Gehlert et al. (1991)

5-HT$_4$

Radioligand	Site	Reference
[125I]SB 207710	—	Vilaró et al. (1996)
[3H]GR 113808	Sigma binding sites	Jakeman et al. (1994), Waeber et al. (1994)
[3H]BIMU-1	No data available	Jakeman et al. (1994)

5-HT$_5$ — No data available

5-HT$_6$

5-HT$_7$

Radioligand	Site	Reference
[3H]5-CT	5-HT$_{1A/1B}$ receptors	Palacios et al. (1996), Waeber and Moskowitz (1995b)

addition, labelling techniques allow the production of probes with very high specific radioactivity and thus enable detection of low levels of mRNA. Other advantages derive directly from these characteristics: it is even possible to study the distribution of the various subunits constitutive of a ligand-gated channel receptor, something which cannot be achieved using radioligand binding since subunit selective ligands are not commonplace. Further, given the resolution of the technique, different mRNAs can be co-expressed by and visualized in the same cell.

There are nevertheless limitations to in situ hybridisation. For example, the technique may not be as quantitative as initially anticipated; the relationship between receptor mRNA and protein may therefore not be quantitatively and spatially straightforward. Furthermore, there is no established recipe that warrants success: thus there is substantial trial and error with this method. However, radioligand receptor autoradiography and mRNA in situ hybridisation have produced such a wealth of complementary information in the 5-HT receptor field (Palacios et al. 1993) that it cannot be covered exhaustively in this short review.

B. Methodology

Radioligand receptor autoradiography and mRNA in situ hybridisation are normally performed with tissue sections of 10 to 20μm cut with the microtome cryostat and mounted onto microscope slides and kept at -20C until used. Radioligands can be tritiated or iodinated, and the incubation techniques, although adapted to autoradiography, are rather similar to those used for membrane binding. Table 1 summarises the ligands used to visualise 5-HT receptors in brain tissue. Also listed are the sites which may be labelled in addition to those targetted. In situ hybridisation can be performed using a range of tools: (a) Double-stranded DNA probes labelled by nick translation; however, reannealing can occur and the truly available antisense probe may be limited. (b) Single-stranded cDNA probes can be used in theory, but the labelling procedure is complex unless they are obtained by PCR. (c) Riboprobes or single-stranded cRNA probes, which combine the advantage of very high specific radioactivity and selectivity (RNA-RNA hybrids are very stable). (d) Oligonucleotide probes which are short (30–50 mers), single-stranded synthetic antisense DNA fragments which are easily synthesised or available commercially. These can be labelled at their 3' end with ^{32}P, ^{33}P or ^{35}S. Non-radioactive in situ hybridisation is feasible in most types of tissues but is successful only when the transcript density is high, which may not be the case for many receptor mRNAs. A number of controls need to be performed either when selecting the sequence of these probes and/or during the actual experiments (see Vilaro et al. 1995).

C. Distribution of 5-HT Receptor mRNA and Binding Sites in Brain (Mainly Rat)

I. 5-HT₁ Receptors

1. 5-HT₁ₐ Receptors

The 5-HT_{1A} receptor is the most commonly studied receptor of the 5-HT family; it was the first to be cloned, and good radioligands have been available for about 13 years starting, with [³H]8-OH-DPAT (GOZLAN et al. 1983; for chemical names see Table 1). There is a very good correspondence between the distribution of 5-HT_{1A} binding sites and receptor mRNA transcripts. High densities of 5-HT_{1A} sites have been found in the hippocampus, septum, raphe nuclei and interpeduncular nucleus. Corresponding mRNA signals have been located in the cell bodies of the hippocampus (pyramidal and granule cells), septal neurones, and 5-HT neurones of the raphe. Therefore it can be concluded that in a number of instances, 5-HT_{1A} receptors have a somatodendritic localisation (see POMPEIANO et al. 1992). This is also suggested from lesion studies using 5,7 dihydroxytryptamine, which causes a marked decrease in both transcripts and binding signals in the dorsal raphe. The septum shows some discrepancies between binding (low in the medial nucleus) and mRNA (very high).

One may suggest receptor transport or differences in turnover to account for these differences, although this is largely speculative (CHALMERS and WATSON 1991). In the hippocampus, for instance, binding in the dentate gyrus is very concentrated in the molecular layer, whereas mRNA transcripts are observed in the granule cell layer, suggesting a dendritic localisation for the receptor (see Fig. 1). Similar apparent mismatches occur in the olfactory bulb and primary olfactory cortex, where receptors are found on dendrites, whereas mRNA is on cells bodies. In the case of 5-HT_{1A} receptors one is in a rather fortunate situation, since studies carried out with antibodies directed against a synthetic fragment of the 5-HT_{1A} receptor have fully confirmed the distribution of 5-HT_{1A} sites reported previously by a number of investigators using ligands such as [³H]8-OH-DPAT or [³H]WAY 100635 (GOZLAN et al. 1983; KHAWAJA 1995). The distribution of sites seen with antagonist ligands is very similar to that seen with agonists, although the ratios of agonist/antagonist are not the same in every region studied.

2. 5-HT₁B/₁D Receptors

We limit this discussion principally to the rat brain, in which 5-HT_{1B} (1Dβ) and 5-HT_{1D} (1Dα) binding can be distinguished easily (see HARTIG et al. 1996 for the updated nomenclature on 5-HT_{1B} and 5-HT_{1D} receptors). Indeed, when [¹²⁵I]CYP is used as a radioligand in rat brain under appropriate conditions (HOYER et al. 1985b, 1986a), the binding is exclusively to 5-HT_{1B} receptors (as

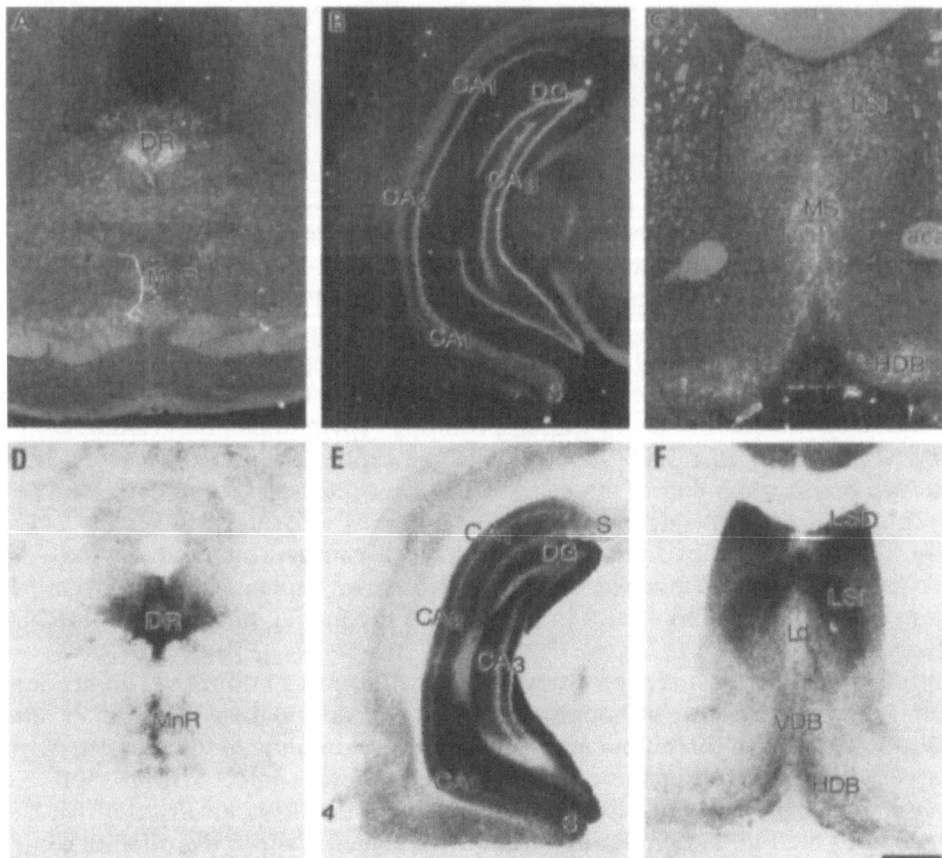

Fig. 1A–F. Regional distribution of 5-HT$_{1A}$ receptors and their mRNA in rat brain. Dark field photomicrographs from coronal sections of rat brain at the level of the raphe nucleus (**A**), hippocampus (**B**) and septum (**C**) show in situ hybridised sections with a ^{32}P-labelled oligonucleotide complementary to the rat 5-HT$_{1A}$ receptor mRNA. *White spots*, autoradiographic grains. Note the high density in the raphe nucleus, pyramidal and granular cell layer of the hippocampus, and the medial septum. **D–F** Adjacent sections displaying receptor autoradiography with the 5-HT$_{1A}$ receptor selective [^{3}H]8-OH-DPAT. The figure illustrates both the correspondence between mRNA transcript and binding sites colocalised in cell bodies (e.g. in the raphe nucleus), and presence of receptors in dendritic fields (see the hippocampus). *aca*, Anterior commissure anterior; *DG*, dentate gyrus; *DR*, dorsal raphe nucleus; *HDB*, nucleus horizontal limb diagonal band; *Ld*, lambdoid septal zone; *LSD*, lateral septal nucleus dorsal; *LSI*, lateral septal nucleus intermediate; *MnR*, median raphe nucleus; *MS*, medial septal nucleus; *S*, subiculum; *VDB*, nucleus vertical limb diagonal band. *Bar*, 1 mm

opposed to 5-HT$_{1D}$ receptors) since it labels only β-adrenoceptor sites in higher species (indole β blockers have low affinity for both 5-HT$_{1B}$ and 5-HT$_{1D}$ receptors, except for rat and mouse 5-HT$_{1B}$ receptors). When [^{125}I]GTI is used, the vast majority of the binding is also to 5-HT$_{1B}$ (1Dβ) sites, although a minor

component can be attributed to 5-HT$_{1D}$ binding (1Dα; BRUINVELS et al. 1993a;
Fig. 2). By contrast, in higher species, because of the very strong pharmacological similarity between 5-HT$_{1B}$ and 5-HT$_{1D}$ receptors, their respective contributions to [^{125}I]GTI binding is very difficult to establish (see previous
section), although the majority of the sites in most structures are 5-HT$_{1B}$. In the
rat brain 5-HT$_{1B}$ receptor mRNA is found in the hippocampus (pyramidal cells
of CA1), caudate putamen, Purkinje cell layer of the cerebellum and retinal
ganglion cells (see BRUINVELS et al. 1994a; Fig. 3). By contrast, 5-HT$_{1B}$ binding
is found in the globus pallidus, entopeduncular nucleus, substantia nigra,
subiculum and superficial grey layer of the superior colliculus. Thus binding
and mRNA at least in these regions appear to be mutually exclusive. One must
therefore consider that the receptors are located predominantly on axons.
Thus the 5-HT$_{1B}$ receptor of the substantia nigra originates in striatal cell
bodies.

   This hypothesis can and has in some cases been tested. The situation with
5-HT$_{1D}$ receptors is more complex: there is no detailed distribution of 5-HT$_{1D}$
binding in rat brain, although BRUINVELS and colleagues (1993a) using a rather
complex method estimated the non-5-HT$_{1B}$ component of [^{125}I]GTI binding in
rat brain which is quantitatively very limited and attributed essentially to 5-
HT$_{1D}$ sites (see Fig. 2). Interestingly, 5-HT$_{1D}$ mRNA levels are very low and

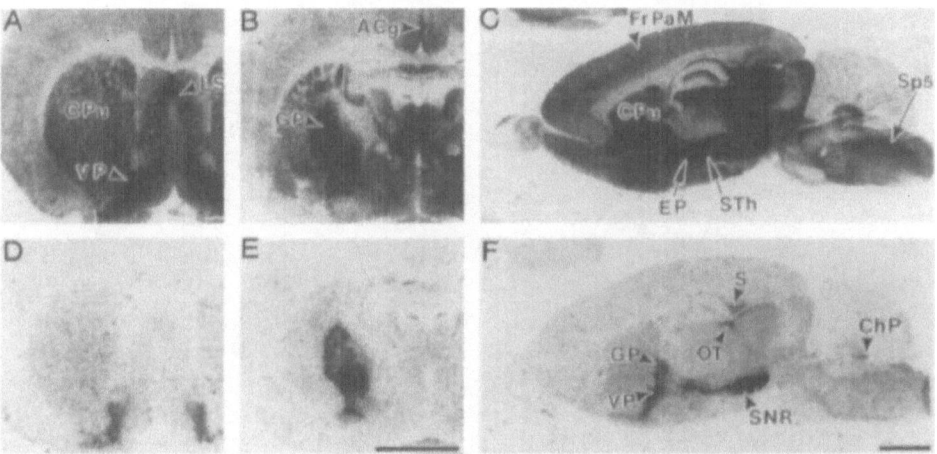

Fig. 2A–F. Regional distribution of 5-HT$_{1B}$ and 5-HT$_{1D}$ receptors in rat brain. 5-HT$_{1B}$
and 5-HT$_{1D}$ receptors were visualised with ^{125}I-GTI in coronal (**A,B**) and horizontal
section (**C**) of rat brain. **D–F** Consecutive sections to **A–C** where the incubation is
performed in the presence of an excess of CP 93129 to displace 5-HT$_{1B}$ binding. Thus
A–C represent a combination of 5-HT$_{1B}$ and 5-HT$_{1D}$ whereas **D–F** represent essentially
5-HT$_{1D}$ binding. *Acg*, Anterior cingulate cortex; *CPu*, caudate-putamen; *EP*,
entopeduncular nucleus; *FrPaM*, motor area frontoparietal cortex; *GP*, globus pallidus;
LS, lateral septal nucleus; *OT*, nucleus of the optic tract; *Pu*, putamen; *SNR*, substantia
nigra reticulata; *S*, subiculum; *Sp5*, spinal nucleus of the trigeminal nerve; *Sth*,
subthalamic nucleus; *VP*, ventral pallidum. *Bars*, 3mm

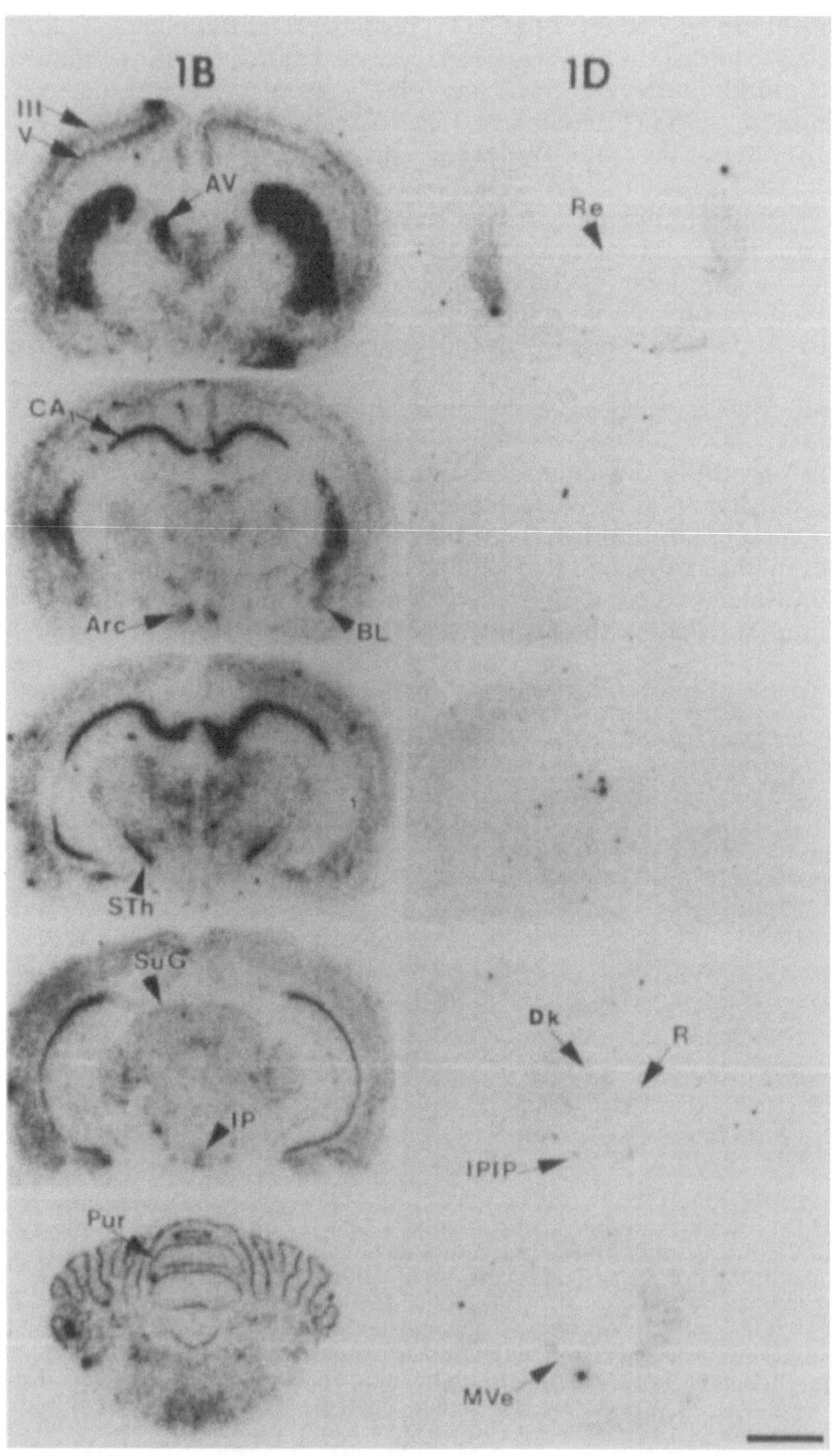

detected in only a few nuclei/structures in the rat brain (BRUINVELS et al. 1994b): olfactory bulb and primary olfactory cortex, basomedial amygdaloid nucleus, accumbens, caudate putamen, subthalamic nucleus, lateral mammilary nucleus, reuniens thalamic nucleus and red nucleus, deep layers of frontal and parietal cortex, and a number of nuclei in the medulla oblongata and pons (nucleus of the spinal tract of the trigeminal nerve, locus coeruleus, dorsal raphe and raphe magnus) and no transcripts in the cerebellum. The non-5-HT$_{1B}$ binding component of [^{125}I]GTI is found in globus pallidus, ventral pallidum, caudate putamen, subthalamic nucleus, entopeduncular nucleus, substantia nigra pars reticulata, frontoparietal cortex, optic tract, and some nuclei of the pons and medulla oblongata. The distributions of 5-HT$_{1B}$ and 5-HT$_{1D}$ receptors in rat brain are roughly comparable although in many cases 5-HT$_{1D}$ binding is only slightly above detection limit (see Fig. 2). The same comment can be made about their respective mRNAs, although the quantitative differences between 5-HT$_{1B}$ and 5-HT$_{1D}$ mRNA are even more pronounced (see Fig. 3). One might therefore speculate that 5-HT$_{1D}$ receptors are almost redundant for 5-HT$_{1B}$ receptors but at much lower concentrations and thus may function primarily as presynaptic auto- and heteroreceptors since they are located in projection regions.

3. 5-HT$_{1E}$ Receptors

The existence of the 5-HT$_{1E}$ receptor was first hypothesised by LEONHARD et al. (1989) based on observations that non-5-HT$_{1A,1D,1C}$ sites are labelled in human brain with [^{3}H]5-HT under conditions where the above-mentioned sites should in principle be entirely blocked (i.e. in the presence of an excess of 5-CT and mesulergine to block 5-HT$_{1A}$, 5-HT$_{1D}$, and 5-HT$_{2C}$ ligands). The 5-HT$_{1E}$ receptor has subsequently been cloned, but information on both distribution of binding sites and mRNA is still somewhat limited. There has been no extensive in situ study, nor is there a ligand and/or binding conditions which can be confidently described as 5-HT$_{1E}$ selective. Nevertheless, BARONE and colleagues (1993, 1994) reported the presence of 5-HT$_{1E}$ sites in rat, mouse and guinea-pig fronto-parietal and striate cortex, caudate-putamen, septum, claustrum and hippocampus (dentate gyrus, CA1, CA2, CA4). One should, however, keep in mind that such studies are performed with [^{3}H]5-HT in the

---

Fig. 3. Regional distribution of 5-HT$_{1B}$ and 5-HT$_{1D}$ receptor mRNA in rat brain. Photographs of autoradiograms from coronal sections at different levels of the rat brain showing in situ hybridised sections with ^{32}P-labelled oligonucleotides complementary to the rat 5-HT$_{1B}$ (*left*) and 5-HT$_{1D}$ receptor mRNA (*right*). III, V, Layers III and V of parietal motor cortex; *Arc*, arcuate hypothalamic nucleus; *AV*, anteroventral thalamic nucleus; *BL*, basolateral amygdaloid nucleus; *Dk*, nucleus of Darkschewitsch; *IP*, interpeduncular nucleus; *IPIP*, inner posterior subnucleus of the interpeduncular nucleus; *Mve*, medial vestibular nucleus; *Pur*, Purkinje cell layer of the cerebellum; *R*, red nucleus; *Re*, reuniens nucleus; *Sth*, subthalamic nucleus; *SuG*, superficial grey layer of the superior colliculus. *Bar*, 3mm

presence of 5-CT and other blocking drugs, conditions under which a number of other 5-HT receptors could be labelled. Miller and Teitler (1992) reported the presence of significant levels of 5-HT$_{1E}$ binding in human putamen, globus pallidus and frontal cortex. Bruinvels et al. (1993b, 1994a) investigated the distribution of 5-HT$_1$ sites which do not belong to the 5-HT$_{1A}$ or 5-HT$_{1B,1D}$ type in a number of species. Further, Bruinvels et al. (1994b) reported the distribution of 5-HT$_{1E}$ receptor mRNA in monkey and human brain; parietal cortex, caudate and putamen showed signals. Also labelled were the visual and entorhinal cortex and some hypothalamic nuclei, although a complete mapping was not performed.

4. 5-HT$_{1F}$ Receptors

There is at present no radioligand which allows the certain labelling 5-HT$_{1F}$ receptors, nor are there functional data providing definitive evidence for these receptors. However, Beer et al. (1993) using [^{3}H]5-HT described the distribution of 5-HT$_1$ sites in brain which are insensitive to 5-CT in rat and guinea-pig brain. These sites should therefore represent a mixture of 5-HT$_{1E/1F}$ sites present in olfactory tubercle, caudate putamen, accumbens and substantia nigra. Of note, the claustrum was more densely labelled in the guinea-pig than in the rat. There have been some reports on [^{3}H]sumatriptan binding which is in part associated with 5-HT$_{1F}$ binding in addition to 5-HT$_{1B/1D}$ binding in rat and guinea-pig brain (Waeber and Moskowitz 1995a). Interestingly, these authors also report that in the guinea-pig (but not rat) the claustrum shows dense 5-HT$_{1F}$ labelling; other regions include neocortical layers, mammilary and thalamic nuclei. It is suggested that in rat the caudate, lateral geniculate and spinal trigeminal nuclei contain both 5-HT$_{1B/1D}$ and 5-HT$_{1F}$ receptors.

Raurich et al. (manuscript in preparation) have extended these observations and combined [^{3}H]sumatriptan autoradiography with in situ hybridisation for 5-HT$_{1F}$ receptor mRNA. The data clearly suggest that both in rat and guinea pig an important component of [^{3}H]sumatriptan binding is to 5-HT$_{1F}$ sites, which co-localise with the mRNA for 5-HT$_{1F}$ receptors (see Mengod et al. 1996). On the other hand, Mills and Martin (1995) suggest that in cat brainstem, [^{3}H]sumatriptan labels 5-HT$_{1B/1D}$ sites in the nucleus tractus solitarius and trigeminal nucleus (since binding was sensitive to 5-CT and high concentrations of ketanserin but not to 8-OH-DPAT) but not 5-HT$_{1F}$ sites. We have determined the localisation of 5-HT$_{1F}$ receptor mRNA in guinea-pig brain (Bruinvels et al. 1994b). Transcripts were very prominently expressed in primary olfactory cortex, anterior olfactory nucleus, caudate-putamen, dentate gyrus, layers III–V of frontoparietal and cingulate cortex and claustrum. Significant in situ hybridisation was also observed in the central, medial and basomedial amygdaloid nuclei, pyramidal cell layers of CA1, CA2 and CA3, periventricular and supraoptic hypothalamic areas, some thalamic nuclei, ventrolateral geniculate nucleus, entorhinal cortex, principal sensory nucleus of the spinal tract, ventral tegmental nucleus, lateral vestibular nucleus and cochlear nucleus. The relatively good agreement between 5-CT insensitive [³H]sumatriptan binding and 5-HT$_{1F}$ transcripts could suggest that the 5-HT$_{1F}$ receptor is located predominantly in the somatodendritic compartment of the neurones expressing these receptors in contrast to 5-HT$_{1B/1D}$ receptors.

II. 5-HT$_2$ Receptors

1. 5-HT$_{2A}$ Receptors

5-HT$_{2A}$ receptor binding and mRNA distribution has been described in rat and human brain (see PAZOS et al. 1985a, 1987b; MENGOD et al. 1990b; POMPEIANO et al. 1994): both signals are densely represented across the cerebral cortex, especially in laminae I and IV–Va, the piriform and entorhinal cortex. They are also concentrated in the claustrum, endopiriform nucleus and olfactory bulb/anterior olfactory nucleus. A number of brainstem nuclei show high signals: pontine, motor trigeminal, facial and hypoglossal nuclei. Intermediate levels were observed in the limbic system and in the basal ganglia, for example, caudate nucleus and accumbens. There were no transcripts in the cerebellum and thalamic nuclei, whereas hippocampal expression was low (POMPEIANO et al. 1994). Of note, in the human brain the distribution of mRNA was similar to that in the rat, although mRNA was apparently absent from the striatum (BURNETT et al. 1995). Our data in human and monkey brain also showed a similar distribution of 5-HT$_{2A}$ receptor mRNA to that in the rat, and as in humans, mRNA is thus absent from the caudate and putamen nuclei (see MENGOD et al. 1996; Fig. 4). Note that the binding of [³H]ketanserin in the caudate and putamen nuclei is not to 5-HT$_{2A}$ sites but to tetrabenazine binding sites. There is generally good agreement between in situ hybridisation and binding data. Further, this distribution is compatible with immunocytochemical data using specific 5-HT$_{2A}$ receptor antibodies (GARLOW et al. 1993). Together the findings suggest that 5-HT$_{2A}$ receptors are expressed in pyramidal cells and interneurones in the neo-cortex.

2. 5-HT$_{2B}$ Receptors

The 5-HT$_{2B}$ receptor was initially referred to as the "fundus" receptor (originally called 5-HT$_{2F}$). It has been cloned from the rat (KURSAR et al. 1992; FOGUET et al. 1992a) and mouse (FOGUET et al. 1992b) stomach fundus and from various human libraries (KURSAR et al. 1994). The question of its presence in the brain is still controversial; it has been detected by reverse-transciptase polymerase chain reaction (RT-PCR) in several human brain regions but not in rat brain (KURSAR et al. 1994), but its presence is reported in the spinal cord of both species (HELTON et al. 1994). There are no reports on the visualisation of 5-HT$_{2B}$ binding sites due to the fact that there are no selective ligands for this receptor; very recently, however, CHOI and

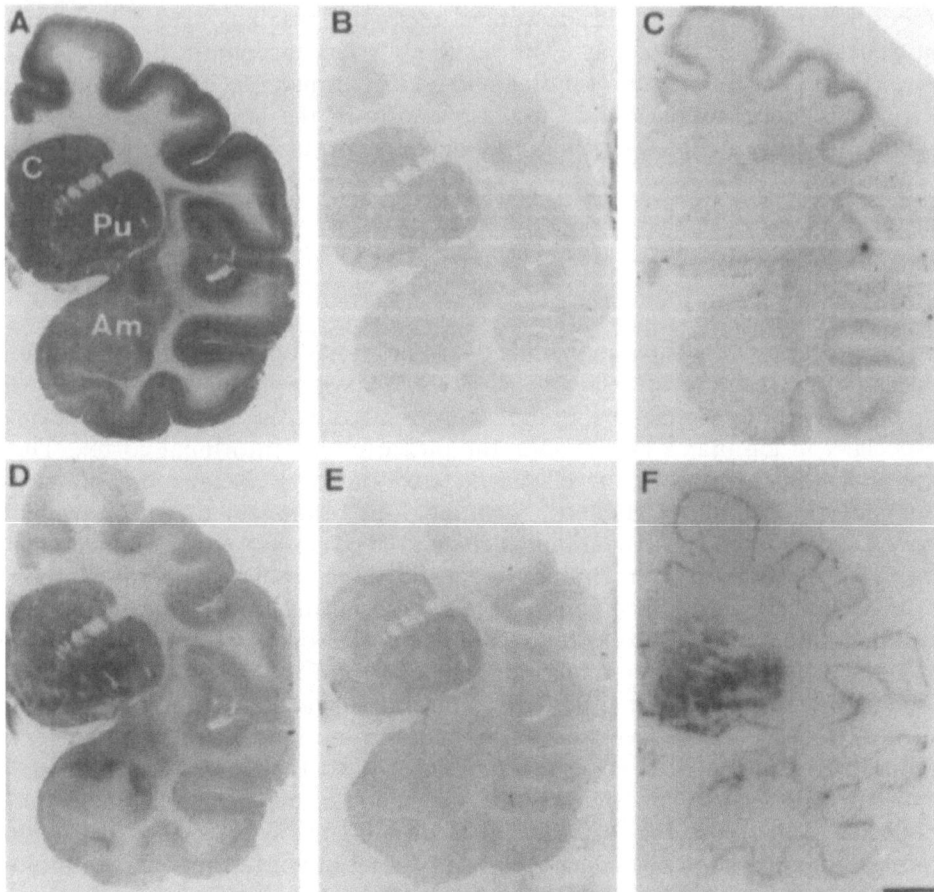

Fig. 4A–F. Visualisation of 5-HT$_{2A}$ and 5-HT$_{2C}$ receptors in primate brain. Photomicrographs from film autoradiograms of a coronal section of monkey (*Macaca fascicularis*) brain. **A** Tissue incubated with [^{3}H]ketanserin. **B** Incubation performed in the presence of 1 mM mianserin to establish non-specific binding. **C** In situ hybridised sections with a ^{32}P-labelled oligonucleotide complementary to the human 5-HT$_{2A}$ mRNA. **D** Tissue incubated with [^{3}H]mesulergine. **E** Non-specific binding in the presence of 1 mM mianserin. **F** Consecutive section hybridised with a ^{32}P-labelled oligonucleotide complementary to the human 5-HT$_{2C}$ receptor mRNA. *Am*, amygdala; *Cd*, caudate; *Hp*, hippocampus; *Pu*, putamen. *Bar*, 4mm

MAROTEAUX (1996), using antipeptide antibodies, reported limited binding in the mouse brain, especially the cerebellum.

3. 5-HT$_{2C}$ Receptors

The initial opinion that the 5-HT$_{2C}$ receptor should be essentially the "choroid plexus" receptor derived from radioligand binding studies that created this rather false impression (PAZOS et al. 1984) due to (a) the absence of truly 5-HT$_{2C}$-selective ligand and (b) the indeed very high concentration of binding sites in the plexus. Depending on the species, however, it is clear that 5-HT$_{2C}$ binding in brain is more widespread; indeed, in rat tissues [^{3}H]mesulergine is problematic since it also labels 5-HT$_{2C}$ and 5-HT$_{2A}$ receptors, whereas in higher species the ligand is more specific for 5-HT$_{2C}$ receptors (see Fig. 4). In situ hybridisation has corrected this unfortunate misrepresentation and confirmed that indeed 5-HT$_{2C}$ receptors are not limited to the choroid plexus (see HOFFMAN and MEZEY 1989; MENGOD et al. 1990a). MENGOD and colleagues compared 5-HT$_{2C}$ binding and mRNA expression in the rodent brain. Transcripts were, as expected, very highly expressed in the choroid plexus and concentrated in limbic structures: hippocampus (CA3), amygdala, anterior olfactory and endopiriform nuclei, cingulate and piriform cortex. In addition, 5-HT$_{2C}$ receptor mRNA is found in some thalamic nuclei, lateral habenula basal ganglia and particularly subthalamic nucleus and substantia nigra. Binding is generally comparable with mRNA, with a few exceptions, for example, subthalamic nucleus. In good agreement are also the low levels of binding and in situ signals in medulla and brainstem. Finally, these results are further supported by the use of anti-5-HT$_{2C}$ receptor peptide antibodies (ABRAMOWSKI et al. 1995) which, for instance, display strong labelling in rat and human substantia nigra. What is remarkable is the widespread distribution of 5-HT$_{2C}$ receptors in the brain, which is largely comparable to the 5-HT innervation.

III. 5-HT$_3$ Receptors

Initial attempts to localise 5-HT$_3$ receptors in the brain have been rather disappointing because of the low density of 5-HT$_3$ receptors in the CNS (KILPATRICK et al. 1988; WAEBER et al. 1988) and the low specific activity of the first radioligands used in 5-HT$_3$ receptor binding, although autoradiographic studies then revealed that 5-HT$_3$ binding sites can be found in a number of nuclei at very high concentrations in various species including mouse, rat and human (WAEBER et al. 1988, 1989). Although at present only one subunit of the 5-HT$_3$ multisubunit receptor has been cloned, there is no doubt that 5-HT$_3$ receptors are present in the brain, especially in the lower brainstem, i.e. the dorsal vagal complex, nucleus of the solitary tract, spinal trigeminal nucleus and around the area postrema. Limbic regions (hippocampus, amygdala, habenula, cerebral and entorhinal cortex) show lower but still significant densities. Binding is also present in the olfactory bulb. The dorsal horn of the spinal cord is densely labelled. The localisation of 5-HT$_3$ receptor mRNA has been carried out in detail in mouse brain (TECOTT et al. 1993) by in situ hybridisation. The cerebral cortex showed high densities of transcripts primarily in the piriform, cingulate and entorhinal cortex. Interneurones in the hippocampus (lacunosum moleculare of CA1, but also in CA3 and dentate gyrus) showed strong signals.

mRNA is also found in amygdaloid complex and olfactory bulb, dorsal tegmental area and in the hypothalamus (preoptic regions). Transcripts were present in the nuclei of the trochlear, facial and spinal tract of the trigeminal nerves and in the dorsal horn of the spinal cord. Combining the technique with immunocytochemistry, Morales et al. (1996) have recently reported that 90% of the 5-HT$_3$ mRNA is expressed in GABAergic cells in neocortex and hippocampus. In general mRNA and radioligand binding sites show similar distributions except in facial nerves and the hypothalamus, where binding is weak, and in the dorsal vagal complex, where no significant mRNA levels are detected, presumably because these receptors are presynaptic. Another possibility of course is that binding studies detect additional sites which do not correspond to the presently cloned 5-HT$_3$ receptor subunit.

IV. 5-HT$_4$ Receptors

Two splice variants of the 5-HT$_4$ receptor have been cloned (Gerald et al. 1995). There is at present no evidence to suggest that the currently used radioligands recognise preferentially either one or the other of these forms, or for that matter what the functional significance of the two splice variants is. The brain distribution of 5-HT$_4$ receptors has been rather extensively documented since 5-HT$_4$ receptor selective ligands (e.g. [^{3}H]GR113808, [^{3}H]BIMU-1 and [^{125}I]SB 207710; for abbreviations see "Appendix") made autoradiographic studies possible. In a number of species (rat, guinea-pig, human) 5-HT$_4$ receptors have a pronounced mesolimbic and nigrostriatal distribution (see Fig. 5). 5-HT$_4$ binding has been reported in olfactory tubercle, islands of Calleja, substantia nigra, ventral pallidum, striatum, septum, hippocampus and amygdala (Grossman et al. 1993; Jakeman et al. 1994; Waeber et al. 1994). RT-PCR studies suggest that the mRNA for the short isoform of the 5-HT$_4$ receptor of rat brain is expressed solely in the striatum, whereas the long form is found throughout the brain, especially in thalamus, striatum, hippocampus, brainstem and olfactory bulb (Gerald et al. 1995). However, in situ hybridisation shows that that both isoforms are expressed in the same regions. These findings underline that 5-HT$_4$ receptors, although widespread, would be part of two systems: (a) the septo-hippocampo-habenulo-peduncular pathway and (b) the striato-nigro-tectal pathway, suggesting two types of functions: limbic and visuo-motor. There is evidence that 5-HT$_4$ receptors and cholinergic terminals co-localise, and a modulation of acetylcholine release by 5-HT$_4$ receptors can be proposed. This would be in line with some electrophysiological findings reported in hippocampus.

V. 5-HT$_5$ Receptors

There are at present no autoradiographic data with which to compare the results of 5-HT$_5$ receptor in situ hybridisation since no radioligand has been identified to label sites equivalent to 5-HT$_5$ receptors, nor is there any func-

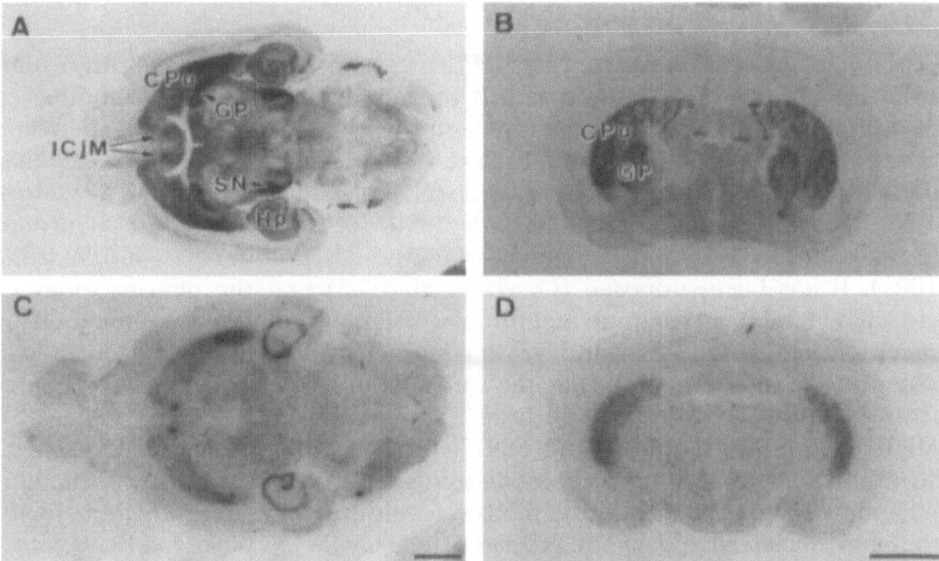

Fig. 5A–D. Regional distribution of 5-HT$_4$ receptors and their mRNA in rat brain. 5-HT$_4$ receptors were visualised with ^{125}I-SB 207710 in a horizontal (**A**) and coronal (**B**) section of rat brain. **C,D** Sections consecutive to **A** and **B** hybridised with an oligonucleotide probe that recognises both splice variants of the rat 5-HT$_4$ receptor mRNA. Note in **A,C** the high densities of receptor sites in the globus pallidus (*GP*) and substantia nigra (*SN*) together with the apparent absence of receptor mRNA in these regions. *Cd,* Caudate nucleus; *CPu,* caudate-putamen; *GP,* globus pallidus; *Hp,* hippocampus; *ICjM,* island of Calleja magna; *Pu,* putamen; *SN,* substantia nigra. *Bars,* 3 mm

tional evidence for the presence of 5-HT$_5$ receptors in the brain. Two subtypes of 5-HT$_5$ receptors have been cloned from rat, but in man only the 5-HT$_{5A}$ and not 5-HT$_{5B}$ cDNA can be identified. In rat brain 5-HT$_{5A}$ mRNA has been localised to cerebral cortex, hippocampus, habenula, olfactory bulb and cerebellum, whereas 5-HT$_{5B}$ mRNA is limited to CA1 (hippocampus) and the habenular complex (MATTHES et al. 1993). CARSON and colleagues (1996) showed that in the rat brain 5-HT$_{5A}$ receptors are expressed predominantly in astrocytes, as demonstrated by the use of receptor specific antisera. Labelled structures included hypothalamus, hippocampus, corpus callosum, fimbria, cerebral ventricles and glia. A similar distribution pattern was observed in mouse brain. The morphology and distribution of the cells labelled with antiserum are consistent with those of astrocytes (co-distribution with glial fibrillary acidic protein is prominent) except in the olfactory bulb and cortex where low levels of neurones are labelled. These observations are further supported by RT-PCR performed with cortical glial culture which showed marked expression of 5-HT$_{5A}$ receptor cDNA. It was also noted that receptor levels increased in reactive gliosis.

VI. 5-HT$_6$ Receptors

Although functional correlates for 5-HT$_6$ receptors have not been firmly established, there are some hints that this receptor is expressed in a number of neuronal tissues. The recombinant receptor mediates stimulation of adenylate cyclase activity. 5-HT stimulated cyclase activity with a 5-HT$_6$-like profile has been reported from neuroblastoma cells (N18-TG2, Unsworth and Molinoff 1994; NCB-20, Conner and Mansour 1990), cultured rat striatal neurones (Sebben et al. 1994) and pig caudate membranes (Schoeffter and Waeber 1994). It has been proposed (Glatt et al. 1995) that the pharmacology of clozapine binding in rat brain membranes is similar to that of the cloned 5-HT$_6$ receptor. However, no ligand yet has sufficient selectivity to label 5-HT$_6$ receptors in the brain, and thus the information on their distribution comes from in situ hybridisation in rat brain (Ruat et al. 1993; Ward et al. 1995). High levels of transcripts are reported in olfactory tubercle, islands of Calleja, striatum, nucleus accumbens, dentate gyrus, and CA1, CA2, CA3 of the hippocampus. Lower but significant mRNA levels are found in cerebellum, hypothalamic nuclei, amygdala and across cortical layers. As observed by Ward et al. (1995), most of the hybridisation signal for 5-HT$_6$ is located over the cell nuclei. It is not yet known what this location means. mRNA is not found in globus pallidus, septal nuclei and a number of thalamic areas, superior and inferior colliculus, and both dorsal and ventral raphe nuclei. The strong labelling in the striatum provides support for the 5-HT$_6$-like stimulation of adenylate cyclase observed in both striatal membranes and cell cultures (see above). That 5-HT$_6$ mRNA transcripts appear to be absent from serotonin-containing cell bodies but present in projection fields suggests that the 5-HT$_6$ receptor is expressed primarily in terminals.

VII. 5-HT$_7$ Receptors

It may not have been immediately clear when the 5-HT$_7$ receptor was cloned that this is a member of the 5-HT receptor family. On the one hand, the structure of the gene showed very low sequence similarity with the other 5-HT receptors cloned at that time, and this is still true now that the 5-HT$_4$ receptor has been cloned. On the other hand, the pharmacological profile of the 5-HT$_7$ receptor, although labelled by [^{3}H]5-HT and [^{3}H]5-CT, is still largely reminiscent of dopamine receptors. However, the 5-HT$_7$ receptor has been known for sometime as a relaxing receptor that inhibits, for instance, histamine-induced contraction in the guinea-pig ileum (Kalkman et al. 1986), although at that time the authors did not assign the receptor to one class or another. To et al. (1995) have cloned the guinea-pig 5-HT$_7$ receptor, established conditions for [^{3}H]5-CT binding to 5-HT$_7$ sites in brain, and performed side-by-side receptor autoradiography and in situ hybridisation in guinea-pig brain. Autoradiographic data reveal a high density of sites in the medial thalamic nuclei and related limbic and cortical regions with lower levels in sensory relay nuclei, substantia nigra, hypothalamus, central grey and dorsal raphe nuclei. In general there is a good correspondence between the areas and nuclei identified with the two methods, although there is an occasional mismatch; for instance, there are no mRNA transcripts in caudate-putamen, globus pallidus, ventral pallidum or substantia nigra, whereas 5-HT_7 binding is apparently present. GUSTAFSON et al. (1996) performed a similar comparative study in rat brain. $[^3\text{H}]$-5-CT binding under their conditions was present in layers 1–3 of cortex, septum, globus pallidus, thalamus, hypothalamus, centromedial amygdala, substantia nigra, periaqueductal grey, and superior colliculus. The authors proposed that the binding in the basal ganglia and substantia nigra is not to 5-HT_7 receptors since (a) methiothepin did not affect this binding and (b) no mRNA were found in these regions, in agreement with To et al. (1995). Otherwise, there was a rather similar distribution of binding sites and transcripts in rat brain, with a few exceptions – in the hippocampus mRNA was comparatively higher to binding whereas in the amygdala, the sub-regional distribution was inversely related. Overall, since the limbic system is particularly well represented, a potential role in sensory and affective processes may be suggested for the 5-HT_7 receptor.

D. Conclusion

This short overview demonstrates that in situ hybridisation and receptor autoradiography are powerful and complementary techniques which in combination provide substantial information about which 5-HT receptor subtype is located in which cells, assuming that sufficiently selective radioligands are available for any given receptor subtype. In principle, the final proof for many of these findings should come from "classical" immunochemistry applied to receptors, that is, from antibodies raised against carefully selected synthetic peptide fragments of the various 5-HT receptor subtypes. Initial work has been carried successfully out for 5-HT_{1A} receptors, and some reports deal with 5-HT_{1B}, 5-HT_{2A}, 5-HT_{2C}, 5-HT_{2B}, 5-HT_3 receptor localisation.

However, it should be realised that to obtain adequate antibodies (selective, specific, good titre, good signal/noise ratio) is not a trivial task. In most cases transmitter receptors represent only a very minor fraction of membrane proteins, and obtaining high-quality signals is not trivial. This is also a limitation to in situ hybridisation since mRNA levels can be extremely low. However, once at hand, the combination of these tools (selective radioligands, oligo- or riboprobes and antibodies) enable precise localisation of the receptor where expressed and where produced, at both the cellular and subcellular level. In the case of 5-HT receptors these techniques, when tools became available, made it possible to localise the receptors and provide support, for example, for the hypothesis that 5-HT_{2C} or 5-HT_3 receptors are indeed present in the brain at significant levels, although this subject was initially debated, that putative 5-HT_{1F} binding sites agree reasonably well with 5-HT_{1F} receptor mRNA transcripts, that 5-HT_{1B} receptors out-number 5-HT_{1D} receptors although their distribution is rather similar, and that 5-HT$_{1A}$ receptors are both dendritic and somato-dendritic. In the absence of selective radioligands, data from 5-HT$_6$ and 5-HT$_7$ in situ hybridisation studies have been very useful for establishing with reasonably confident conditions under which receptor auto-radiography could be performed.

In addition, there is good agreement between 5-HT$_3$ receptor binding and in situ hybridisation data. Since the latter was obtained with probes selective for a single subunit, does this mean that 5-HT$_3$ receptors in the brain are heterohomomers (since it is known that native 5-HT$_3$ receptors adopt a pentameric structure)? Or that all radioligands label only one subunit? It also remains to be seen whether studies with 5-HT$_{1E}$ and 5-HT$_{1F}$ receptors can be confirmed functionally, and this will be depend largely on selective tools. General questions such as the precise chemical phenotype of the neurones expressing 5-HT receptors or the cellular coexpression of more than one 5-HT receptor subtype per neurone remain to be answered for the large majority of these receptors. However, it is clear that studies such as those presented here pave the way for further developments in a field which due to its complexity was long considered exotic until it was realised that complexity is very common for most transmitter receptors.

Appendix: Glossary of Drug Names

1-NP:	1-(1-naphthyl)piperazine
5-CT:	5-carboxamidotryptamine
5-HT:	5-hydroxytryptamine (serotonin)
5-HTP-DP:	5-hydroxytryptophyl-5-hydroxytryptophan amide
5-MeOT:	5-methoxytryptamine HCl
5-OHIP:	5-hydroxyindalpine
6-OHIP:	6-hydroxyindalpine
8-OH-DPAT:	8-hydroxy-2-(di-*n*-propylamino)tetraline
AH 25086:	3-(2-aminoethyl)-*N*-methyl-1*H*-indole-5-acetamide
BIMU 1:	endo-*N*-(8-methyl-8-azabicyclo[3.2.1]oct-3-yl)-2,3-dihydro-3-ethyl-2-oxo-1*H*-benzimidazole-1-carboxamide
BIMU 8:	endo-*N*-(8-methyl-8-azabicyclo[3.2.1]oct-3-yl)-2,3-dihydro-(1-methyl)ethyl-2-oxo-1*H*-benzimidazole-1-carboxamide
BMY 7378:	8-[2-[4-(2-methoxyphenyl)-1-piperazinyl]ethyl]-8-azaspiro[4.5]-decane-7,9-dione dihydrochloride
BRL 20627:	(2a, 6b, 9aa)-(±)-4-amino-5-chloro-2-methoxy-*N*-(octahydro-6-methyl-2*H*-quinolizin-2-yl)benzamide
BW 501C67:	2-anilino-*N*-(2-(3-chlorophenoxy)propyl acetamidine HCl
CGS 12066B:	7-trifluoromethyl-4-(4-methyl-1-piperazinyl)-pyrrolo[1, 2-*a*]quinoxaline maleate
CP 93129:	3-(1,2,5,6-tetrahydropyrid-4-yl)pyrrolo[3, 2-*b*]pyrid-5-one
CP 96501:	3-(1,2,5,6-tetrahydropyrid-4-yl)5-*n*-propoxyindole

DAU 6285:   endo-6-methoxy-8-methyl-8-azabicyclo[3.2.1]oct-3-yl-2,3-
                 dihydro-2-oxo-1*H*-benzimidazole-1-carboxylate HCl
DHE: dihydroergotamine
DOB: 1-(2,5-dimethoxy-4-bromophenyl)-2-aminopropane
DOI: 1-(2,5-dimethoxy-4-iodophenyl)-2-aminopropane
DOM: 1-(2,5-dimethoxy-4-methylphenyl)-2-aminopropane
DP-5-CT: dipropyl-5-carboxamidotryptamine
GR 65630:   3-(5-methyl-1*H*-imidazol-4-yl)-1-(1-methyl-1*H*-indol-3-yl)-
                 1-propanone
GR 67330:   (±)-1,2,3,9-tetrahydro-9-methyl-3-[(5-methyl-1*H*-imidazol-
                 4-yl)methyl]-4*H*-carbazol-4-one
GR 113808:   [1-[2-(methylsulphonyl)amino]ethyl]-4-piperidinyl]methyl
                 1-methyl-1*H*-indole-3-carboxylate
GR 127935:   (*N*-[4-methoxy-3-(4-methyl-1-piperazinyl)phenyl]-2′-
                 methyl-4′-(5-methyl-1,2,4-oxadiazol-3-yl) [1, 1-biphenyl]-
                 4-carboxamide)
ICI 169369:   2-(2-dimethylaminoethylthio)-3-phenylquinoline HCl
[^{125}I]GTI: 5-*O*-carboxamidomethylglycyl[^{125}I]tyrosinamide-
                 tryptamine
[^{125}I]SCH 23982: *R*-(+)-7-hydroxy-8-[^{125}I]-3-methyl-1-phenyl-2,3,4,5-
                 tetrahydro(1*H*)-3-benzazepine HCl
LSD: (+)-lysergic acid diethylamide
LY 165163:   1-(2-(4-aminophenyl)ethyl)-4-(3-trifluoromethyl-
                 phenyl)piperazine (PAPP)
LY 278584:   1-methyl-*N*-(8-methyl-8-azabicyclo-[3.2.1]oct-3-yl)-1*H*-
                 indazole-3-carboxamide
LY 53857:   4-isopropyl-7-methyl-9-(2-hydroxy-1-
                 methylpropoxycarbonyl)-4,6,6A,7,8,9,10,10A-
                 octahydroindolo[4, 3-FG]quinolone maleate
mCPP: 1-(3-chlorophenyl)piperazine
MDL 72222:   1*α*H, 3*α*, 5*α*H-tropan-3-yl-3,5-dichlorobenzoate
MDL 72832:   8-(4-[1, 4-benzodioxan-2-ylmethylamino]butyl)-8-azaspiro
                 [4, 5]decane-7,9-dione
MDL 73005:   8-(2-[2, 3-dihydro-1, 4,benzodioxin-2-
                 ylmethylamino]ethyl)-8-azaspiro[4, 5]decane-7.9-dione
MDL 100907:   [*R*-(+)-*α*-(2, 3-dimethoxyphenyl)-1-[2-(4-fluorophenyl)
                 ethyl]-4-piperidinemethanol]
MK 212:   6-chloro-2-(1-piperazinyl)pyrazine HCl
NAN 190:   1-(2-methoxyphenyl)-4-[4-(2-phthalimmido)
                 butyl]piperzine HBr
PAPP: 1-(2-[4-aminophenyl]ethyl)-4-(3-trifluoromethylphenyl)
                 piperazine (LY 165163)
RP 62203:   (2-[3-(4-(4-fluorophenyl)-piperazinyl)propyl]naphto[1,8-
                 ca]isothiazole-1,1-dioxide)
RS 23597–190:   3-(piperidin-1-yl)propyl 2-methoxy-4-amino-5-
                 chlorobenzoate HCl

RU 24969:	5-methoxy-3(1,2,3,6-tetrahydro-4-pyridinyl)-1*H*-indole
SB 200646:	*N*-(1-methyl-5-indolyl)-*N*-(3-pyridyl urea hydrochloride)
SB 207710:	(1-butyl-4-piperidinylmethyl)-8-amino-7-iodo-1,4-benzodioxan-5-carboxylate
SC 53116:	exo-(1S,8S)-2-methoxy-4-amino-5-chloro-*N*-[(hexhydro-1*H*-pyrrolizin-1-yl)methyl]benzamide HCl
SCH 23390:	*R*-(+)-7-chloro-8-hydroxy-3-methyl-1-phenyl-2,3,4,5-tetrahydro(1*H*)-3-benzazepine HCl
SDZ 205557:	2-methoxy-4-amino-5-chlorobenzoic acid 2-(diethylamino)ethyl ester
SDZ 206830:	(3α-homotropanyl)-1-methyl-5-fluoro-indole-3-caroxylic acid ester
SDZ 21009:	4(3-terbutylamino-2-hydroxypropoxy)indol-2-carbonic-acid-isopropylester
SDZ 216525:	methyl-4(4-[4-(1, 1, 3-trioxo, 2*H*-1, 2-benziosothiazol-2-yl)butyl]-1-piperazinyl)1*H*-indole-2-carboxylate
TFMPP:	*N*-(3-trifluoromethylphenyl)piperazine
WAY 100135:	*N-tert*-butyl-3-(4-[2-methoxyphenyl]piperazin-1-yl)-2-phenylpropionamide dihydrochloride
WAY 100635:	(*N*-[2-[4(2-methoxyphenyl)-1-piperazinyl-1-piperazinyl]-*N*-2- pyridinyl) cyclohexanecarbonate
WB 4101:	2-(2,6-dimethoxyphenoxyethyl)aminomethyl-1,4-benzodioxane

Some other drugs were generally known by their code names until recently. The names of these drugs, as used in this review, and their previous code names are: ipsapirone (TVXQ 7821), granisetron (BRL 43694); ondansetron (GR 38032F); renzapride (BRL 24924); sumatriptan (GR 43175); tropisetron (ICS 205930).

References

Abramowski D, Rigo M, Duc D, Hoyer D, Staufenbiel M (1995) Generation of antisera against the 5-hydroxytryptamine$_{2c}$ receptor and its localization in human and rat brain. Neuropharmacology 34:1635–1645

Altar CA, Boyar WC, Marien MR (1986) ^{125}I-LSD autoradiography confirms the preferential localization of caudate-putamen S2 receptors to the caudal (peripallidal) region. Brain Res 372:130–136

Appel NM, Mitchell WM, Gralick RK, Glennon RA, Teitler M, De Souza EB (1990) Autoradiographic characterization of (+-)-1-(2,5-dimethoxy-4-[125I]iodophenyl)-2-aminopropane [125I]DOI) binding to 5-HT$_2$ and 5-HT$_{1C}$ receptors in rat brain. J Pharm Exp Ther 255:843–857

Barone P, Millet S, Moret C, Prudhomme N, Fillion G (1993) Quantitative autoradiography of 5-HT1E binding sites in rodent brains: effect of lesion of serotonergic neurones. Eur J Pharmacol 249:221–230

Barone P, Jordan D, Atger F, Kopp N, Fillion G (1994) Quantitative autoradiography of 5-HT1D and 5-HT1E binding sites labelled by [3H]5-HT, in frontal cortex and the hippocampal region of the human brain. Brain Res 638:85–94

Beer MS, Stanton JA, Hawkins LM, Middlemiss DN (1993) 5-Carboxamidotryptamine-insensitive 5-HT1-like receptors are concentrated in guinea pig but not rat, claustrum. Eur J Pharmacol 236:167–169

Boulenguez P, Chauveau J, Segu L, Morel A, Lanoir J, Delaage M (1991) A new 5-hydroxy-indole derivative with preferential affinity for 5-HT$_{1B}$ binding sites. Eur J Pharmacol 194:91–98

Bruinvels AT, Palacios JM, Hoyer D (1993a) Autoradiographic characterisation and localisation of 5-HT$_{1D}$ compared to 5-HT$_{1B}$ binding sites in rat brain. Naunyn Schmiedebergs Arch Pharmacol 347:569–582

Bruinvels AT, Palacios JM, Hoyer D (1993b) 5-Hydroxytryptamine$_1$ recognition sites in rat brain: heterogeneity of non 5-hydroxytryptamine$_{1A/1C}$ binding sites revealed by quantitative receptor autoradiography. Neuroscience 53:465–473

Bruinvels AT, Landwehrmeyer B, Probst A, Palacios JM, Hoyer D (1994a) A comparative autoradiographic study of 5-HT$_{1D}$ binding sites in human and guinea-pig brain using different radioligands. Mol Brain Res 21:19–29

Bruinvels AT, Landwehrmeyer B, Gustafson EL, Durkin MM, Mengod G, Branchek TA, Hoyer D, Palacios JM (1994b) Localization of 5-HT$_{1B}$, 5-HT$_{1Da}$, 5-HT$_{1E}$ and 5-HT$_{1F}$ receptor messenger RNA in rodent and primate brain. Neuropharmacol 33:367–386

Brüning G, Kaulen P, Schneider U, Baumgarten HG (1989) Quantitative autoradiographic distribution and pharmacological characterization of [^{3}H]buspirone binding to sections from rat, bovine, and marmoset brain. J Neural Transm 78:131–144

Burnet PW, Eastwood SL, Lacey K, Harrison PJ (1995) The distribution of 5-HT1A and 5-HT2A receptor mRNA in human brain. Brain Res 676:157–168

Carson MJ, Thomas EA, Danielson PE, Sutcliffe JG (1996) The 5-HT5A serotonin receptor is expressed predominantly in astrocytes in which it inhibits cAMP accumulation: a mechanism of reactive astrocytes. Glia 17:317–326

Chalmers DT, Watson SJ (1991) Comparative anatomical distribution of 5-HT1A receptor mRNA and 5-HT1A binding in rat brain-a combined in situ hybridisation/in vitro receptor autoradiographic study. Brain Res 561:51–60

Conner DA, Mansour TE (1990) Serotonin receptor mediated activation of adenylate cyclase in the neuroblastoma NCB.20: a novel 5 hydroxytryptamine receptor. Mol Pharmacol 37:742–751

Choi DS, Maroteaux L (1996) Immunohistochemical localisation of the serotonin 5-HT2B receptor in mouse gut, cardiovascular system and brain. FEBS Lett 391:45–61

Foguet M, Hoyer D, Pardo LA, Kluxen FW, Kalkman HO, Stuhmer W, Lübbert H (1992a) Cloning and functional characterization of the rat stomach fundus serotonin receptor. EMBO J 11:3481–3487

Foguet M, Nguyen H, Le H, Lübbert H (1992b) Structure of the mouse 5-HT$_{1C}$, 5-HT$_2$ and stomach fundus serotonin receptor genes. Neuroreport 3:345–348

Garlow SJ, Morilak DA, Dean RR, Roth BL, Ciaranello-RD (1993) Production and characterization of a specific 5-HT2 receptor antibody. Brain Res 615:113–120

Gehlert DR, Gackenheimer SL, Wong DT, Robertson DW (1991) Localization of 5-HT$_3$ receptors in the rat brain using [^{3}H]LY278584. Brain Res 553:149–154

Gerald C, Adham N, Hung-Teh K, Olsen MA, Laz TM, Schechter LE, Bard JA, Vaysse PJ-J, Hartig PR, Branchek TA, Weinshank, RL (1995) The 5-HT$_4$ receptor: molecular cloning and pharmacological characterization of two splice variants. EMBO J 14:2806–2815

Glaser T, Rath M, Traber J, Zilles K, Schleicher A (1985) Autoradiographic identification and topographical analysis of high affinity serotonin receptor subtypes as a target for the novel putative anxiolytic TVXQ7821. Brain Res 358:129–136

Glatt CE, Snowman AM, Sibley DR, Snyder S (1995) Clozapine: selective labeling of sites resembling 5HT$_6$ serotonin receptors may reflect psychoactive profile. Mol Med 1:398–406

Gozlan H, El Mestikawy S, Pichat L, Glowinski J, Hamon M (1983) Identification of presynaptic serotonin autoreceptors by a new ligand: 3H-PAT. Nature 305:140–142

Gozlan H, Ponchant M, Daval G, Vergé D, Ménard F, Vanhove A, Beaucourt JP, Hamon M (1988) [^{125}I] Bolton-Hunter-8-methoxy-2-[N-propyl-N-propylamino]tetralin as a new selective radioligand of 5-HT$_{1A}$ sites in the rat brain. In vitro binding and autoradiographic studies. J Pharmacol Exp Ther 244:751–759

Grossman CJ, Kilpatrick GJ, Bunce KT (1993) Development of a radioligand binding assay for 5-HT4 receptors in guinea-pig and rat brain. Br J Pharmacol 109:618–624

Gustafson EL; Durkin MM; Bard JA; Zgombick J; Branchek TA (1996) A receptor autoradiographic and in situ hybridization analysis of the distribution of the 5-ht7 receptor in rat brain. Br J Pharmacol 117:657–666

Hartig PR, Branchek TA, Weinshank RL (1992) A subfamily of 5-HT$_{1D}$ receptor genes. Trends Pharmacol Sci 13:152–159

Helton LA, Thor KB, Baez M (1994) 5-hydroxytryptamine$_{2A}$, 5-hydroxytryptamine$_{2B}$, and 5-hydroxytryptamine$_{2C}$ receptor mRNA expression in the spinal cord of rat, cat, monkey and human. Neuroreport 5:2617–2620

Hoffman BJ, Mezey E (1989) Distribution of serotonin 5 HT1C receptor mRNA in adult rat brain. Febs Lett 247:453–462

Hoffman BJ, Scheffel U, Lever JR, Karpa MD, Hartig PR (1987) N$_1$-methyl-2-^{125}I-lysergic acid diethylamide, a preferred ligand for in vitro and in vivo characterization of serotonin receptors. J Neurochem 48:115–124

Hoyer D, Engel G, Kalkman HO (1985) Characterization of the 5-HT$_{1B}$ recognition site in rat brain: binding studies with [^{125}I]iodocyanopindolol. Eur J Pharmacol 118:1–12

Hoyer D, Pazos A, Probst A, Palacios JM (1986a) Serotonin receptors in the human brain. I. Characterization and autoradiographic localization of 5-HT$_{1A}$ recognition sites. Apparent absence of 5-HT$_{1B}$ recognition sites. Brain Res 376:85–96

Hoyer D, Pazos A, Probst A, Palacios JM (1986b) Serotonin receptors in the human brain II. Characterization and autoradiographic localization of 5-HT$_{1C}$ and 5-HT$_2$ recognition sites. Brain Res 376:97–107

Hoyer D, Clarke DE, Fozard JR, Hartig PR, Martin GR, Mylecharane EJ, Saxena PR, Humphrey PPA (1994) International Union of Pharmacology classification of receptors for 5-hydroxytryptamine (serotonin) Pharmacol Rev 46:157–204

Jakeman LB, To ZP, Eglen RM, Wong EH, Bonhaus DW (1994) Quantitative autoradiography of 5-HT4 receptors in brains of three species using two structurally distinct radioligands, [3H]GR113808 and [3H]BIMU-1. Neuropharmacology 33:1027–1038

Kalkman HO, Engel G, Hoyer D (1986) Inhibition of 5-carboxamidotryptamine-induced relaxation of guinea-pig ileum correlates with [^{125}I]LSD binding. Eur J Pharmacol, 129:139-145

Khawaja X (1995) Quantitative autoradiographic characterization of the binding of [^{3}H]WAY-10065, a selective 5-HT$_{1A}$ receptor antagonist. Brain Res 67:217–225

Kilpatrick GJ, Jones BJ, Tyers MB (1988) The distribution of specific binding of the 5-HT$_3$ receptor ligand [^{3}H]GR65630 in rat brain using quantitative autoadioraphy. Neurosci Lett 94:156–160

Koscielniak T, Ponchant M, Laporte AM, Guminski Y, Vergé D, Hamon M, Gozlan H (1990) [^{125}I]iodo-zacopride: a new ligand for the autoradiographic study of the central 5-HT$_3$ receptors. CR Acad Sci Ser III 311:231–237

Kursar JD, Nelson DL, Wainscott DB, Cohen ML, Baez, M (1992) Molecular cloning, functional expression, and pharmacology of a novel serotonin receptor (5-hydroxytryptamine$_{2F}$) from rat stomach fundus. Mol Pharmacol 42:549–557

Kursar JD, Nelson DL, Wainscott DB, Baez M (1994) Molecular cloning, functional expression, and mRNA tissue distribution of the human 5-hydroxytryptamine$_{2B}$ receptor. Mol Pharmacol 46:227–234

Laporte AM, Schechter LE, Bolanos F, Vergé D, Hamon M, Gozlan H (1991) [³H] 5-methyl-urapidil labels 5-HT$_{1A}$ receptors and alpha-adrenoceptors in the rat CNS. In vitro binding and autoradiographic studies. Eur J Pharmacol 198:59–67

Leonhardt S, Herrick-Davis K, Titeler M (1989) Detection of a novel serotonin receptor subtype (5-HT1E) in human brain: interaction with a GTP-binding protein. J Neurochem 53:465–471

López-Giménez JF, Vilaró MT, Palacios JM, Mengod G. ³H-MDL100907 labels 5-HT$_{2A}$ receptors in the mammalian brain: pharmacological characterization by receptor autoradiography. (in preparation)

Malgouris C, Flamand F, Doble A (1993) Autoradiographic studies of RP 62203, a potent 5-HT$_2$ receptor antagonist. Pharmacological characterization of [³H]RP62203 binding in the rat brain. Eur J Pharmacol 233:37–45

Marcinkiewicz M, Vergé D, Gozlan H, Pichat L, Hamon M (1984) Autoradioraphic avidence for the heterogeneity of 5-HT$_1$ sites in the rat brain. Brain Res 291:159–163

Matthes H, Boschert U, Amlaiky N, Grailhe R, Plassat JL, Muscatelli F, Mattei MG, Hen R (1993) Mouse 5-hydroxytryptamine$_{5A}$ and 5-hydroxytryptamine$_{5B}$ define a new family of serotonin receptors: cloning, functional expression, and chromosal localization. Mol Pharmacol 43:313–319

Mengod G, Nguyen H, Le H, Waeber C, Lubbert H, Palacios JM (1990a) The distribution and cellular localization of the serotonin 1C receptor messenger mRNA in the rodent brain examined by in situ hybridization histochemistry comparison with receptor binding distribution. Neurosci 35:577–591

Mengod G, Pompeiano M, Martinez-Mir MI, Palacios JM (1990b) Localization of the mRNA for the 5-HT2 receptor by in situ hybridization histochemistry. Correlation with the distribution of receptor sites. Brain Res 524:139–143

Mengod G, Vilaró MT, Raurich A, López-Giménez JF, Cortés R, Palacios JM (1996) 5-HT receptors in mammalian brain: receptor autoradiography and in situ hybridization studies of the new ligands and newly identified receptors. Hist J (in press)

Miller KJ, Teitler M (1992) Quantitative autoradiography of 5-CT sensitive (5-HT$_{1D}$) and 5-CT insensitive (5-HT$_{1L}$) serotonin receptors in human brain. Neurosci Lett 136:223–226

Mills A, Martin GR (1995) Autoradiographic mapping of [3H]sumatriptan binding in cat brain stem and spinal cord. Eur J Pharmacol 280:175–178

Morales M, Battenberg E, DeLecea L, Bloom FE (1996) The type 3 serotonin receptor is expressed in a subpopulation of GABAergic neurons in the rat neocortex and hippocampus. Brain Res 731:199–202

Nicklaus KJ, McGonigle P, Molinoff PB (1988) [³H]SCH 23390 labels both dopamine-1 and 5-hydroxytryptamine(1C) receptors in the choroid plexus. J Pharmacol Exp Ther 247:343–348

Palacios JM, Niehoff DL, Kuhar MJ (1981) ³H-Spiperone binding sites in brain: Autoradiographic localization of receptors. Brain Res 213:277–289

Palacios JM, Probst A, Cortés R (1983) The distribution of serotonin in receptors in the human brain: high density of [³H]LSD binding sites in the raphe nuclei of the brainstem. Brain Res 274:150–155

Palacios JM, Waeber C, Bruinvels AT, Hoyer D (1992a) Direct visualization of serotonin $_{1D}$ receptors in the human brain using a new iodinated radioligand. Mol Brain Res 13:175–179

Palacios JM, Mengod G, Hoyer D (1993) Brain serotonin receptor subtypes: radioligand binding assays, second messenger, ligand autoradiography and in situ hybridization histochemistry. Methods Neurosci 12:238–261

Palacios JM, Raurich A, Mengod G, Hurt SD, Cortés R (1996) Autoradiographic analysis of 5-HT receptor subtypes labeled by ³H-5-CT (³H-5-carboxamidotryptamine) Behav Brain Res 295:271–274

Pazos A, Palacios JM (1985) Quantitative auto-radiographic mapping of serotonin receptors in the rat brain. I. Serotonin-1 receptors. Brain Res 346:205–230

Pazos A, Hoyer D, Palacios JM (1984) The binding of serotonergic ligand to the porcine choroid plexus: characterization of a new type of serotonin recognition site. Eur J Pharmacol 106:539–546

Pazos A, Cortés R, Palacios JM (1985a) Quantative autoradiographic mapping of serotonin receptors in the rat brain: II. Serotonin-2 receptors. Brain Res 346:231–245

Pazos A, Engel G, Palacios JM (1985b) Beta-adrenoceptor blocking agents recognize a subpopulation of serotonin receptors in brain. Brain Res 343:403–408

Pazos A, Probst A, Palacios JM (1987a) Serotonin receptors in the human brain: III. Autoradiographic mapping of serotonin-1 receptors. Neurosci 1:97–122

Pazos A, Probst A, Palacios JM (1987b) Serotonin receptors in the human brain. IV. Autoradiographic mapping of serotonin-2 receptors. Neurosci 21:123–139

Perry DC (1990) Autoradiography of [^{3}H] quipazine in rodent brain. Eur J Pharmacol 187:75–85

Pompeiano M, Palacios JM, Mengod G (1992) Distribution and cellular localization of mRNA coding for 5-HT1A receptor in the rat brain: correlation with receptor binding. J Neurosci 12:440–453

Pompeiano M, Palacios JM, Mengod G (1994) Distribution of the serotonin 5-HT2 receptor family mRNAs: comparison between 5-HT2A and 5-HT2C receptors. Brain Res Mol Brain Res 23:163–178

Ransom RW, Asarch KB, Shih JC (1986) [^{3}H]1-[2-(4-Amoniphenyl)ethyl]-4-(3-trufluoromethylphenyl)piperazine: a selective radioligand for 5-HT$_{1A}$ receptors in rat brain. J Neurochem 46:68–75

Reynolds DJM, Leslie RA, Grahame-Smith, Harvey JM (1989) Localization of 5-HT$_3$ receptor binding sites in human dorsal vagal complex. Eur J Pharmacol 174:127–130

Ruat M, Traiffort E, Arrang JM, Tardivel-Lacombe J, Diaz J, Leurs R, Schwartz JC (1993) A novel rat serotonin (5-HT6) receptor: molecular cloning, localization and stimulation of cAMP accumulation. Biochem Biophys Res Commun 193:268–276

Schoeffter P, Waeber C (1994) 5-Hydroxytryptamine receptors with a 5-HT$_6$ receptor-like profile stimulating adenylyl cyclase activity in pig caudate membranes. Naunyn Schmiedebergs Arch Pharmacol 350:356–60

Schotte A, Leysen JE (1988) Distinct autoradiographic labelling of serotonin 5-HT$_2$ receptors, a1-adrenoceptors and histamine-H1 receptors and of tetrabenazine displaceable ketanserin binding sites in rodent brain with [^{125}I] 7-amino-8-iodoketanserin. Eur J Pharmacol 145:213–216

Schotte A, Maloteaux JM, Laduron PM (1983) Characterization and regional distribution of serotonin S2 receptors in human brain. Brain Res 276:231–235

Sebben M, Ansanay H, Bockaert J, Dumuis A (1994) 5-HT6 receptors positively coupled to adenylyl cyclase in striatal neurones in culture. Neuroreport 5:2553–2557

Sijbesma H, Schipper J, DeKloet ER (1990) The anti-aggressive drug eltoprazine preferentially binds to 5-HT$_{1A}$ and 5-HT$_{1B}$ receptor subtypes in rat brain: sensitivity to guanine nucleotides. Eur J Pharmacol 187:209–223

Tanaka H, Shimiau H, Kumasaka Y, Hirose A, Tatsuno T, Nakamura M (1991) Autoradiographic localization and pharmacological characterization of [^{3}H] tandospirone binding sites in the rat brain. Brain Res 546:181–189

Tecott LH, Maricq AV, Julius D (1993) Nervous system distribution of the serotonin 5-HT3 receptor mRNA. Proc Natl Acad Sci USA 90:1430–1434

To ZP, Bonhaus DW, Eglen RM, Jakeman LB (1995) Characterization and distribution of putative 5-HT7 receptors in guinea-pig brain. Br J Pharmacol 115:107–116

Unsworth C D, Molinoff PB (1994) Characterization of a 5-hydroxytryptamine receptor in mouse neroblsatoma N18TG2 cells. J Pharmacol Exp Ther 269:246–255

Vilaro MT, Palacios JM, Mengod G (1995) Neurotransmitter receptor histochemistry: the contribution of in situ hybridization. Life Sci 57:1141–1154

Vilaró T, Cortés R, Palacios JM, Branchek T, Mengod G (1996) Localization of 5-HT4 receptor mRNA in the rat brain by in situ hybridization histochemistry. Mol Brain Res 43:356–360

Waeber C, Moskowitz MA (1995) [3H]sumatriptan labels both 5-HT1D and 5-HT1F receptor binding sites in the guinea pig brain: an autoradiographic study. Naunyn Schmiedebergs Arch Pharmacol 352:263–275

Waeber C, Moskowitz MA (1995) Autoradiographic visualization of [^{3}H] 5-carboxamidotryptamine binding sites in the guinea pig and rat brain. Eur J Pharmacol 283:31–46

Waeber C, Palacios JM (1993) Autoradiography of 5-HT receptors. In: Wharton J, Polak JM (eds) Receptor autoradiography: principles and practice. Oxford, p 195–210

Waeber C, Dietl MM, Hoyer D, Probst A, Palacios JM (1988a) Visualization of a novel serotonin recognition site (5-HT$_{1D}$) in the human brain by autoradiography. Neurosci Lett 88:11–16

Waeber C, Dixon K, Hoyer D, Palacios JM (1988b) Localization by autoradiography of neuronal 5-HT$_3$ receptors in mouse CNS. Eur J Pharmacol 151:351–352

Waeber C, Hoyer D, Palacios JM (1989) 5-HT$_3$ receptors in the human brain: autoradiographic visualization using [^{3}H]ICS 205–930. Neurosci 31:393–400

Waeber C, Sebben M, Grossman C, Javoy-Agid F, Bockaert J, Dumuis A (1993) [3H]-GR113808 labels 5-HT4 receptors in the human and guinea-pig brain. Neuroreport 4:1239–1242

Waeber C, Sebben M, Nieoullon A, Bockaert J, Dumuis A (1994) Regional distribution and ontogeny of 5-HT4 binding sites in rodent brain. Neuropharmacology 33:527–541

Ward RP, Hamblin MW, Lachowicz JE, Hoffman BJ, Sibley DR, Dorsa DM (1995) Localization of serotonin subtype 6 receptor messenger RNA in the rat brain by in situ hybridization histochemistry. Neuroscience 64:1105–1111

Yagaloff KA, Hartig PR (1985) [^{125}I] lysergic acid diethylamide binds to a novel serotoninergic site on rat choroid plexus epithelial cells. J Neurosci 5:3178–3183

Young WS, Kuhar MJ (1980) Serotonin receptor localization in rat brain by light microscopic autoradiography. Eur J Pharmacol 62:237–239

CHAPTER 9

The Main Features of Central 5-HT$_{1A}$ Receptors

M. Hamon

A. Introduction

Soon after the development of binding assays using [^{3}H]5-hydroxytryptamine ([^{3}H]5-HT) to label the specific recognition sites for serotonin on membrane-bound receptors in the central nervous system (CNS), differences from one brain area to another were noted which suggested the existence of several distinct classes of binding sites. In particular, comparison of the fate of [^{3}H]5-HT high-affinity binding sites in the rat brain after the selective degeneration of serotoninergic neurones due to the intra-raphe infusion of the neurotoxin 5,7-dihydroxytryptamine (5,7-DHT) showed that both the hippocampal and striatal sites are located on postsynaptic targets of serotoninergic projections (NELSON et al. 1978), where they are subjected, however, to differential adaptive changes after the lesion. Thus a significant increase in the density of [^{3}H]5-HT high-affinity binding sites was noted in the hippocampus but not in the striatum (NELSON et al. 1978). Studies of the pharmacological properties of [^{3}H]5-HT high-affinity binding also revealed that the radioactive indoleamine probably recognises several distinct high-affinity sites in brain membranes. Indeed, inhibition of [^{3}H]5-HT high-affinity binding by drugs such as methiothepin and quipazine yielded (apparent) Hill coefficients of less than 1.0, as expected of the heterogeneity of corresponding binding sites (NELSON et al. 1978). These observations were in fact the first to suggest the existence of several – at least two – different classes of high-affinity 5-HT$_1$ receptor binding sites, with distinct pharmacological properties and regional distributions in the rat brain.

Further investigations with other drugs, in particular some neuroleptics of the butyrophenone series, clearly confirmed the existence of two different classes of [^{3}H]5-HT high-affinity binding sites which were initially named 5-HT$_{1A}$ for those having a high (nM) affinity for spiperone and derivatives, and 5-HT$_{1B}$ for those having, in contrast, a low affinity for these drugs (PEDIGO et al. 1981). Although this so-called 5-HT$_{1B}$ class has subsequently been shown to correspond to a heterogeneous population of high-affinity (nM) sites for the indoleamine, the 5-HT$_{1A}$ class, initially identified by PEDIGO et al. (1981), turned out to be the 5-HT$_{1A}$ receptor, which was then cloned and sequenced a few years later (KOBILKA et al. 1987; FARGIN et al. 1988; ALBERT et al. 1990). In the early 1980s, a similar biphasic inhibition of [^{3}H]5-HT binding as obtained

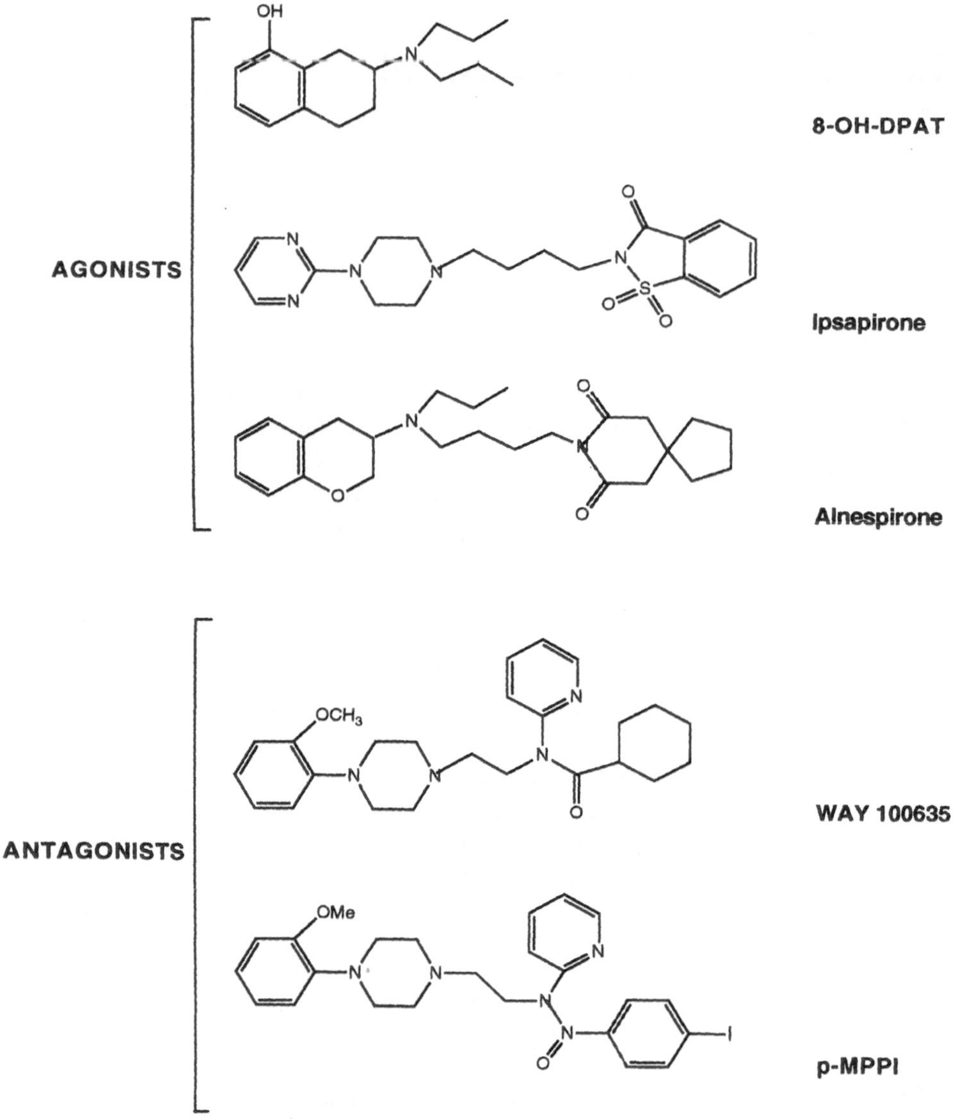

Fig. 1. Chemical structures of some agonists and antagonists at 5-HT$_{1A}$ receptors

with spiperone was reported with a novel tetralin derivative, 8-hydroxy-2-(di-*n*-propylamino)tetralin (8-OH-DPAT; Fig. 1; HJORTH et al. 1982), suggesting that the latter compound is also a selective high-affinity ligand of the 5-HT$_{1A}$ receptor (MIDDLEMISS and FOZARD 1983; HAMON et al. 1984). Interestingly, 8-OH-DPAT proved to be an agonist, whereas spiperone acted as an antagonist at 5-HT$_{1A}$ receptors. Furthermore, binding studies with [^{3}H]8-OH-DPAT (GOZLAN et al. 1983; HALL et al. 1985) subsequently showed that corresponding binding sites accounted only for a fraction of those specifically labelled by [^{3}H]5-HT, as expected of the recognition of the 5-HT$_{1A}$ subtype of the heterogeneous family of specific 5-HT$_1$ binding sites. Therefore an agonist, an antagonist and a selective radioligand were available for extensive studies of the 5-HT$_{1A}$ receptor as early as 1983. This explains why 13 years later this receptor is the best known among the 15 different 5-HT receptors which have been cloned, sequenced and pharmacologically characterised to date (HOYER et al. 1994).

B. Molecular Organisation of the 5-HT$_{1A}$ Receptor

In rodents as in man, the 5-HT$_{1A}$ receptor consists of a protein of 422 amino acids. The receptor has seven hydrophobic domains that probably correspond to membrane-spanning regions (KOBILKA et al. 1987; ALBERT et al. 1990; CHANDA et al. 1993; BOESS and MARTIN 1994). The intronless gene encoding the 5-HT$_{1A}$ receptor is located in the distal part of mouse chromosome 13 and on human chromosome 5 at the locus 5q11.2-q13. Three potentially glycosylated asparagine residues are located within the putative extracellular

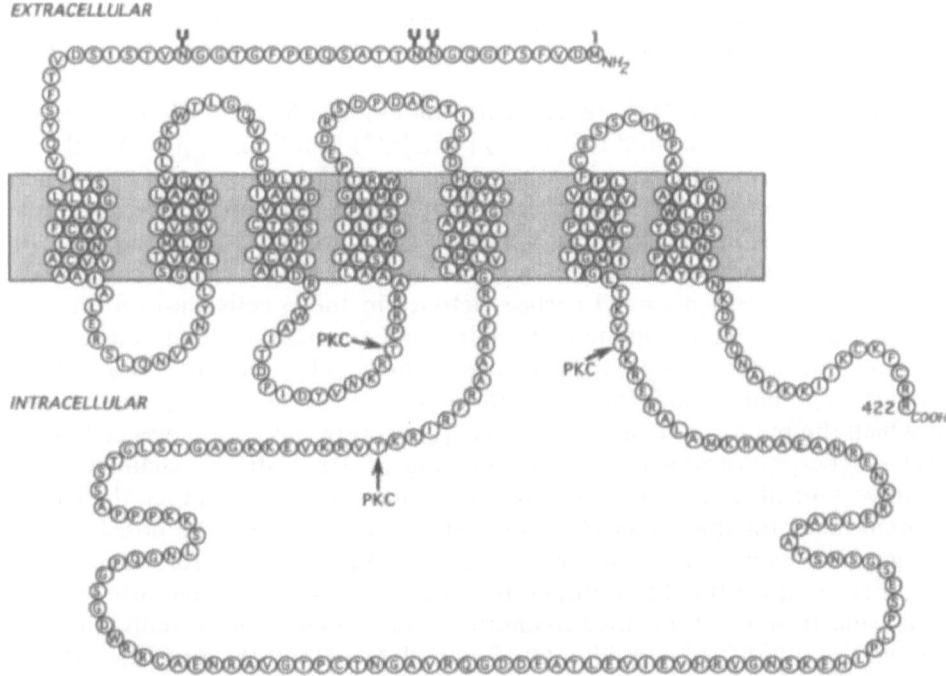

Fig. 2. Model of the transmembrane topology of the rat 5-HT$_{1A}$ receptor (ALBERT et al. 1990). Consensus sites for phosphorylation by protein kinase C (*PKC*) and N-glycosylation (*Y*) are marked

N terminal domain of the protein, whereas three threonine residues within consensus sequences for their phosphorylation by protein kinase C are present in the second and third intracellular loops (Fig. 2). This overall structure is typical of a G protein coupled receptor (EMERIT et al. 1990). Furthermore, the combined presence of a relatively long third intracellular loop (132 amino acids) and a short C terminal domain (18 amino acids) is a rather common feature of G protein coupled receptors negatively coupled with adenylyl cyclase, which is in fact the case of the $5\text{-}HT_{1A}$ receptor in brain tissues as well as in cell lines transfected with the corresponding cDNA coding sequence (BOESS and MARTIN 1994). Site-directed mutagenesis indicated that at least the aspartate residue at position 82, the serine residue at position 393 and the asparagine residue at position 396 are required for the binding of 8-OH-DPAT to the $5\text{-}HT_{1A}$ receptor (CHANDA et al. 1993).

Measurements of adenylyl cyclase activity of membranes from NIH-3T3 cells transfected with the $5\text{-}HT_{1A}$ receptor coding sequence showed that peptides corresponding to the second intracellular loop and the carboxyl end of the third intracellular loop mimicked the $5\text{-}HT_{1A}$ receptor-dependent activation of G protein(s) and the associated inhibition of the enzyme (VARRAULT et al. 1994). From these data, it can be inferred that these parts of the $5\text{-}HT_{1A}$ receptor are very likely those which interact with the functionally associated G protein(s). As expected from this inference, the binding of specific antibodies onto a portion of the receptor at a distance from these parts was shown to exert no effect on the $5\text{-}HT_{1A}$ receptor–G protein interaction (RIAD et al. 1991).

The blockade of $5\text{-}HT_{1A}$ receptor-dependent inhibition of adenylyl cyclase by pertussis toxin indicated that the G protein(s) functionally associated with the $5\text{-}HT_{1A}$ receptor belong(s) to the Gi/Go type (HAMON et al. 1990). Further experiments aimed at the identification of the latter protein(s) consisted of co-transfecting various cell lines with cDNAs encoding the $5\text{-}HT_{1A}$ receptor, on the one hand, and a given G protein, on the other. Measurements of $5\text{-}HT_{1A}$ receptor-dependent adenylyl cyclase activity in these cells showed that the larger responses were obtained with $Gi\alpha2$ and $Gi\alpha3$, suggesting that these G proteins are in fact those which couple to the $5\text{-}HT_{1A}$ receptor under physiological (optimal?) conditions (RAYMOND et al. 1993). However, other studies in which the measured response to $5\text{-}HT_{1A}$ receptor activation was cell membrane hyperpolarisaton due to the opening of a specific K^+ channel (HAJ-DAHMANE et al. 1991) favoured the idea that a Go (α_{o1}) rather than a Gi protein mediates this effect (OLESKEVICH 1995). Therefore the possible coupling of the same $5\text{-}HT_{1A}$ receptor binding subunit to different G proteins might account for the differential effects of $5\text{-}HT_{1A}$ receptor agonists on neuronal cells from one brain area to another. For instance, these compounds are known to inhibit forskolin-stimulated adenylyl cyclase activity in the rat hippocampus but not in the anterior raphe nuclei, although $5\text{-}HT_{1A}$ receptor binding sites are particularly concentrated in these two regions (VERGÉ et al. 1986). In contrast, the hyperpolarising response due to K^+ channel opening has been reported to occur upon 5-HT$_{1A}$ receptor activation in both the raphe area and the hippocampus (HAJ-DAHMANE et al. 1991; BECK et al. 1992). To date, however, the possible coupling of the 5-HT$_{1A}$ receptors to different G proteins from one brain area to another (i.e. from one cell type to another) has yet to be firmly demonstrated by direct biochemical studies.

C. Cellular Responses to 5-HT$_{1A}$ Receptor Activation

In addition to their negative coupling with adenylyl cyclase, which has been extensively used for the characterisaton of the pharmacological and functional properties of 5-HT$_{1A}$ receptors, the latter have also been shown to modulate the activity of phospholipase C in various cell lines transfected with the corresponding cDNA sequence. An increased hydrolysis of phosphatidyl inositol biphosphate, and a resulting intracellular accumulation of free Ca^{2+} have been reported to occur in transfected cells exposed to 5-HT$_{1A}$ receptor agonists (FARGIN et al. 1989; BODDEKE et al. 1992; RAYMOND et al. 1992). However, this mode of coupling apparently does not exist in brain tissues, even if 5-HT$_{1A}$ receptor agonists were found to exert a negative influence on the activation of phospholipase C due to muscarinic receptor stimulation in the hippocampus of young rats (MINISCLOU et al. 1995). Indeed, the effects of 5-HT$_{1A}$ receptor agonists are probably indirect, via some 5-HT$_{1A}$ receptor mediated alteration (possibly due to phosphorylation) of muscarinic receptors themselves or their effector mechanisms (MINISCLOU et al. 1995).

The functional coupling of 5-HT$_{1A}$ receptors with K$^+$ channels has been demonstrated both in the dorsal raphe nucleus where these receptors act as autoreceptors located on the somata and dendrites of serotoninergic neurones (VERGÉ et al. 1986; SOTELO et al. 1990), and in the hippocampus, septum and hypothalamus were they correspond to postsynaptic receptors with respect to serotoninergic afferent fibres (VERGÉ et al. 1986). In all these regions 5-HT$_{1A}$ receptor activation results in the opening of K$^+$ channels that produces both a hyperpolarisaton and a decrease in membrane resistance (AGHAJANIAN and LAKOSKI 1984; CORRADETTI et al. 1996). However, some differences exist from one area to another with regard to the efficiency of agonists and antagonists to affect these channels, which suggests that the molecular organisaton of the "5-HT$_{1A}$–G protein–K$^+$ channel" complex might exhibit some regional variations (CORRADETTI et al. 1996). In particular, possible differences in the nature of the coupled G protein(s) and the ratios of the respective concentrations of the 5-HT$_{1A}$ receptor binding subunit and the associated G protein(s) might explain why the same drugs can act as full agonists in some areas (i.e. the dorsal raphe nucleus) and only as partial agonists in other areas (i.e. the septum and the hippocampus; see HAMON et al. 1990).

At least in the dorsal raphe nucleus 5-HT$_{1A}$ autoreceptors have also been shown to be negatively coupled to Ca^{2+} channels (PENINGTON et al. 1991). The reduced Ca^{2+} conductance, in addition to the increased K$^+$ conductance, clearly contributes to the negative influence of 5-HT_{1A} receptor agonists on the firing rate of serotoninergic neurones.

Pharmacological characterisation of the K^+ channel functionally coupled with the 5-HT_{1A} autoreceptor on serotoninergic neurones in the dorsal raphe nucleus showed that it could be inhibited by 4-aminopyridine but not by apamin or charybdotoxin. These characteristics are in fact typical of K^+ channels directly controlled by G proteins (HAJ-DAHMANE et al. 1991).

D. Localisation of 5-HT_{1A} Receptors

I. Regional and Cellular Distributions of 5-HT_{1A} Receptors

Quantitative autoradiographic studies, performed as soon as [^{3}H]8-OH-DPAT became available, allowed the mapping of 5-HT_{1A} receptors in the rat CNS (MARCINKIEWICZ et al. 1984; PAZOS and PALACIOS 1985; VERGÉ et al. 1985, 1986). Subsequently, other selective 5-HT_{1A} radioligands were synthesised, such as: [^{125}I]Bolton-Hunter-8-methoxy-2-[*N*-propyl-*N*-propylamino]tetralin ([^{125}I]BH-8-MeO-*N*-PAT), [^{3}H]5-methoxy-3-(di-*n*-propylamino)chroman ([^{3}H]5-MeO-DPAC), [^{3}H]5-methyl-urapidil, [^{3}H]ipsapirone, [^{3}H]alnespirone, [^{3}H]1-[2-(4-fluorobenzoyl-amino)ethyl]-4-(7-methoxynaphtyl) piperazine ([^{3}H]*S* 14506), [^{3}H]*N*-[2-[4-(2-methoxyphenyl)-1-piperazinyl]ethyl]-*N*-(2-pyridinyl) cyclohexane carboxamide ([^{3}H]WAY 100635), [^{125}I]4-(2′-methoxyphenyl)-1-[2′-(*N*-2″-pyridinyl)-*p*-iodo-benzamido] ethylpiperazine ([^{125}I]p-

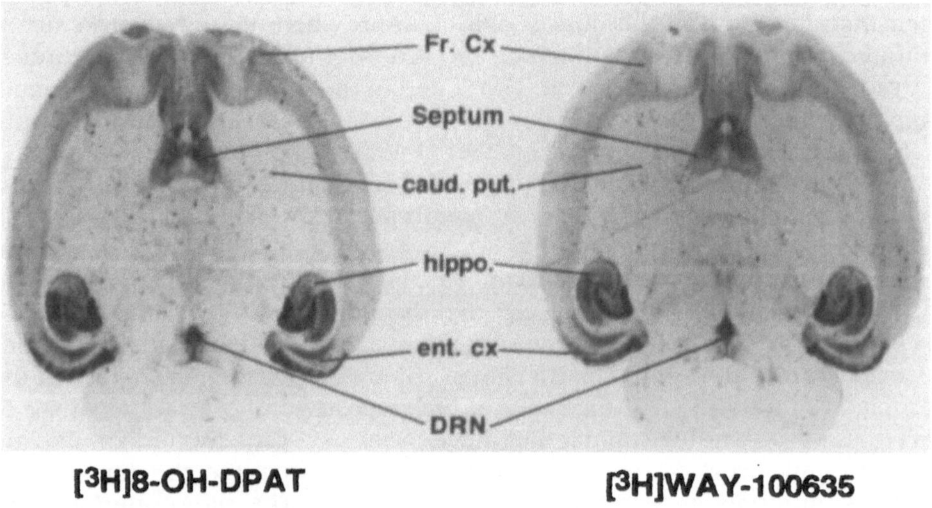

Fig. 3. Autoradiograms of the specific labelling of 5-HT_{1A} receptors by [^{3}H]8-OH-DPAT and [^{3}H]WAY 100,635 in the rat brain. Adjacent horizontal sections show that both radioligands bind precisely at the same areas. *Fr. Cx*, Frontal cortex; *caud. put.*, caudate putamen; *hippo.*, hippocampus; *ent. cx*, entorhinal cortex; *DRN*, dorsal raphe nucleus

MPPI), and the autoradiographic pictures obtained with these molecules fully confirmed the data obtained with [^{3}H]8-OH-DPAT (Cossery et al. 1987; Gozlan et al. 1988, 1995; Laporte et al. 1991; Radja et al. 1991; Lanfumey et al. 1993a, 1994; Kung et al. 1995; see Fig. 3). All these studies showed that 5-HT$_{1A}$ receptor binding sites are especially abundant in the hippocampus, notably in the dentate gyrus and the CA1 area of Ammon's horn, the lateral septum, the frontal and entorhinal cortex and the anterior raphe nuclei. In addition, the dorsal horn of the spinal cord, the amygdala, some thalamic and hypothalamic nuclei, and the lateral part of the caudeputamen also express 5-HT$_{1A}$ receptors but at a lower level. Finally, 5-HT$_{1A}$ receptors are hardly or not detectable in the globus pallidus, the substantia nigra and the cerebellum (Marcinkiewicz et al.1984; Vergé et al. 1986; Radja et al. 1991).

In general, the distribution of 5-HT$_{1A}$ receptor binding sites coincides with that of the 5-HT$_{1A}$ receptor mRNA visualised by in situ hybridisation of specific oligonucleotide or cRNA probes (Miquel et al. 1991; Pompeiano et al. 1992). Accordingly, it has been inferred that the 5-HT$_{1A}$ receptor protein is not transported over long distances from its site of synthesis, i.e. that distribution of the receptor may be restricted to the somato-dendritic compartment of the neurones. The first direct demonstration of such a phenomenon was obtained in the dorsal and median raphe nuclei where the selective destruction of the somata and dendrites of serotoninergic neurones consecutive to the local microinjection of the neurotoxin 5,7-DHT was shown to be associated with the disappearance of specific [^{3}H]8-OH-DPAT binding sites (Fig. 4; Vergé et al. 1985, 1986; Weissmann-Nanopoulos et al. 1985). Subsequently, double immunocytochemical staining procedures allowed the visualisaton of the 5-HT$_{1A}$ receptor protein on the somata and dendrites of neurones identified with anti-5-HT antibodies in both anterior raphe nuclei (Sotelo et al. 1990).

Another brain region in which the subcellular localisation of 5-HT$_{1A}$ receptors could be inferred from autoradiographic labelling with specific radioligands is the hippocampus. In this region [^{3}H]8-OH-DPAT specific binding is especially high in the stratum oriens, stratum radiatum and stratum lacunosum moleculare where the dendrites of pyramidal and granule cells are located (Vergé et al. 1986). However, at this level the addressing of 5-HT$_{1A}$ receptors seems to exclude the perikarya because [^{3}H]8-OH-DPAT specific binding is hardly detected in corresponding layers within the Ammon's horn and the dentate gyrus, where the 5-HT$_{1A}$ receptor mRNA is abundant (Miquel et al. 1991).

Observations in other regions of the rat CNS suggest that 5-HT$_{1A}$ receptors are not always confined to the somato-dendritic compartment of neurones but can also occur in axon terminals, for example, in the dorsal horn of the spinal cord. Similar to the opioid receptor binding sites, whose presynaptic location on the terminals of primary sensory fibres has been clearly demonstrated (Daval et al. 1987a; Besse et al. 1990), the selective degeneration of primary afferent terminals subsequent to dorsal rhizotomy results in a signifi-

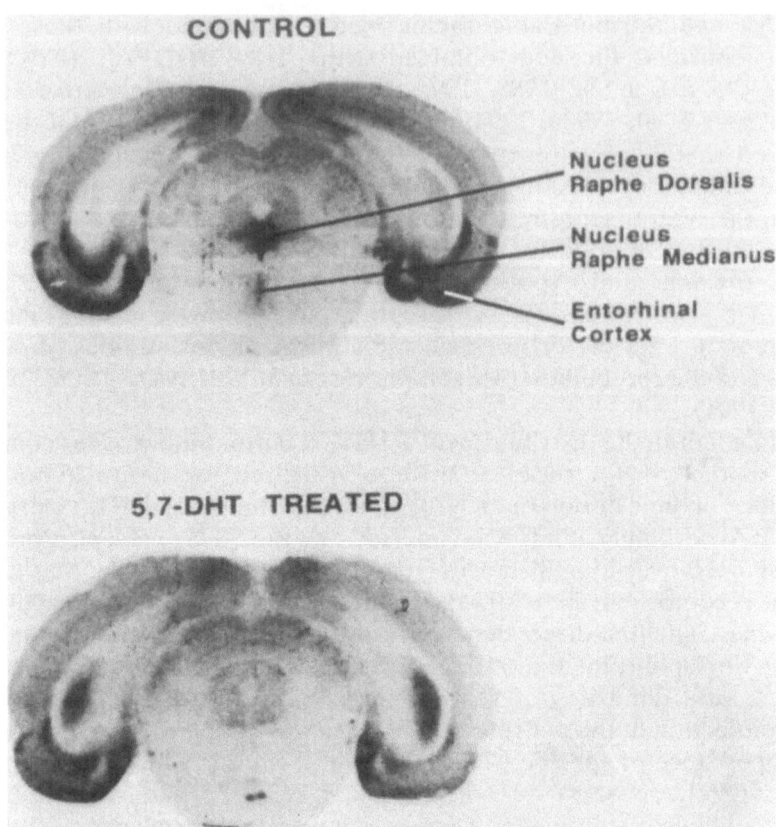

Fig. 4. Demonstration of the expression of somato-dendritic 5-HT$_{1A}$ autoreceptors by serotoninergic neurones in the rat brain. The selective lesion of serotoninergic neurones in the nucleus raphe dorsalis and the nucleus raphe medianus by the local microinjection of 5,7-DHT (VERGÉ et al. 1986) results in the disappearance of [^{3}H]8-OH-DPAT binding in these areas. In contrast, [^{3}H]8-OH-DPAT binding persists in the entorhinal cortex as expected of the location of 5-HT$_{1A}$ receptors on postsynaptic targets of serotoninergic neurones in this region

cant approx. 25% reduction in the density of [^{3}H]8-OH-DPAT specific binding sites within the superficial layers of the dorsal horn; such a finding could be expected provided a significant proportion of 5-HT$_{1A}$ receptors were located on terminals of primary afferent fibres originating from neurones in dorsal root ganglia (DRG; DAVAL et al. 1987; LAPORTE et al. 1995). Indeed, in situ hybridisaton studies have shown that the 5-HT$_{1A}$ receptor mRNA is expressed in some cells in DRG (POMPEIANO et al. 1992). Nevertheless, before a definitive conclusion can be reached, it must be excluded that the loss of 5-HT$_{1A}$ receptor binding sites following dorsal rhizotomy is due to plasticity phenomena which may in fact involve receptors located postsynaptically with respect to primary afferent fibres.

One important issue which derived from the cloning and sequencing of the 5-HT$_{1A}$ receptor has been the production of specific antibodies raised against either a synthetic peptide with the same sequence as that of a highly selective portion of the 5-HT$_{1A}$ receptor protein (EL MESTIKAWY et al. 1990; RIAD et al. 1991) or a recombinant fusion protein (GÉRARD et al. 1994). In the latter case the strategy consisted of forcing transformed bacteria to synthesise a "mixed" protein encoded by a plasmid where the sequence of a large selective portion of the 5-HT$_{1A}$ receptor cDNA (i.e. the coding region corresponding to the third intracellular loop of the receptor) had been fused to the gene of another protein whose purification could be rapidly achieved, such as the glutathione-S-transferase from *Schistosoma japonicum* (GÉRARD et al. 1994). Synthetic peptides coupled to protein carriers (such as bovine serum albumin or keyhole limpet hemocyanin) or the purified fusion protein have been injected to rabbits for the production of polyclonal antibodies. Western blotting experiments show that these antibodies recognise one major band of 63 kDa, corresponding to the MW of the native 5-HT$_{1A}$ receptor in brain tissues (EL MESTIKAWY et al. 1989). Furthermore, both antipeptide antibodies and anti-fusion protein antibodies are capable of immunoprecipitating 5-HT$_{1A}$ receptors in soluble extracts from rat hippocampal membranes (EL MESTIKAWY et al. 1990; RIAD et al. 1991; GÉRARD et al. 1994).

The incubation of rat brain sections with anti-peptide- or anti-fusion protein-antibodies, then [^{35}S]IgG-anti rabbit IgG, yielded immuno-autoradiograms which correspond exactly to autoradiograms obtained from sections labelled by [^{3}H]8-OH-DPAT or [^{3}H]WAY 100635 (EL MESTIKAWY et al. 1990; RIAD et al. 1991; GÉRARD et al. 1994; GOZLAN et al. 1995).

II. Addressing of 5-HT$_{1A}$ Receptors

In all brain areas where the ultrastructural localisation of 5-HT$_{1A}$ receptors has been examined by immunocytochemistry at the electron microscope level the protein appears to be located in the plasma membrane around the somata and dendrites of various neuronal types (KIA et al. 1996a,b). In contrast, no immunolabelling of glial cells is observed, in line with the lack of positive in situ hybridisation signal with a specific probe of 5-HT$_{1A}$ receptor mRNA in C6 glioma cells and astrocytes in primary culture (HILLION et al. 1994). Direct visualisation of such a somatodendritic addressing of the 5-HT$_{1A}$ receptor has been confirmed at the level of the dorsal and median raphe nuclei, the hippocampus, the septum and the cerebral cortex (KIA et al. 1996b). Detailed examination of the immunolabelling by anti-5-HT$_{1A}$ receptor antibodies using the immunogold technique shows that 5-HT$_{1A}$ receptors are distributed as patches in the plasma membrane, mainly outside synaptic differentiations (RIAD et al. 1995). Such an ultrastructural distribution is in line with the implication of 5-HT$_{1A}$ receptors in the mediation of neuromodulatory actions of 5-HT.

Direct investigations on the mechanisms responsible for the membrane addressing of 5-HT$_{1A}$ receptors to the somato-dendritic compartment of cen-

tral neurones have recently been performed using an epithelial cell line as a
model. Because several convergent sets of data had demonstrated that the
apical pole of epithelial cells is homologous to the terminal compartment of
neurones, whereas the baso-lateral pole of the former cells is homologous to
the somato-dendritic compartment, LANGLOIS et al. (1996) recently used the
pig kidney epithelial cell line LLC-PK1 to investigate the membrane addressing of 5-HT$_{1A}$ receptors. As expected from these homologies, the 5-HT$_{1A}$
receptor expressed in LLC-PK1 cells transfected with a plasmid encoding this
receptor was found to be located in the baso-lateral domain of the plasma
membrane (LANGLOIS et al. 1996). Interestingly, a chimera corresponding to
the 5-HT$_{1A}$ receptor in which the third intracytoplasmic loop is changed to that
of the 5-HT$_{1B}$ receptor (which is addressed to the terminal compartment of
neurones; RIAD et al. 1996) was no longer addressed to the baso-lateral pole of
cells transfected with the recombinant plasmid. Conversely, another chimera
made of the 5-HT$_{1B}$ receptor in which the third intracytoplasmic loop is
changed to that of the 5-HT$_{1A}$ receptor was addressed to the baso-lateral pole
(DARMON et al., unpublished observations), although the complete 5-HT$_{1B}$
receptor is normally not addressed to this domain of the cell membrane
(LANGLOIS et al. 1996). These data indicate that a signal very probably exists
within the third intracytoplasmic loop of the 5-HT$_{1A}$ receptor that is responsible for the receptor addressing to the baso-lateral pole of the plasma membrane in LLC-PK1 cells. Whether this signal sequence (which has still to be
clearly identified) is also implicated in the addressing of 5-HT$_{1A}$ receptors to
the somato-dendritic compartment of neurones has yet to be established.

E. Pharmacological Properties of 5-HT$_{1A}$ Receptors

To date 8-OH-DPAT is still considered as the prototypical 5-HT$_{1A}$ receptor
agonist although this compound has also some affinity for 5-HT$_7$ receptors
(HOYER et al. 1994). Indeed, the potency of 8-OH-DPAT is about 50 times
higher at 5-HT$_{1A}$ receptors than at 5-HT$_7$ receptors. Studies of the enantiomers
of 8-OH-DPAT showed that $(-)R$-8-OH-DPAT is about 10 times more potent
than $(+)S$-8-OH-DPAT. Numerous derivatives of 8-OH-DPAT have been
synthesised thanks to the substitution and addition of various chemical groups.
This approach led notably to the synthesis of (R)-5-fluoro-8-OH-DPAT [(R)-
UH-301] which is also a 5-HT$_{1A}$ receptor agonist (BJORK et al. 1991). In
contrast, the other enantiomer (S)-UH-301 is an antagonist at 5-HT$_{1A}$ receptors. However, its selectivity is much less than that of the parent compound
because (S)-UH-301 also exerts an agonist activity at dopamine D$_2$ receptors
(BJORK et al. 1991).

Numerous compounds of various chemical series (tetralins, chromans,
azapirones, etc.) have been synthesised among which potent and selective 5-
HT$_{1A}$ receptor agonists were found. In the chroman series 5-methoxy-3-(di-n-
propylamino)chroman (5-MeO-DPAC) exhibits a nanomolar affinity for

Table 1. Respective affinities (K_i values in nM) of selected
agonists and antagonists at 5-HT$_{1A}$ receptors (see text)

Compound	K_i (nM)
Agonists	
Alnespirone (S 20499)	0.20
N,N-di-propyl-5-carboxamido-tryptamine	0.20
5-Carboxamido-tryptamine	0.20
5-Methyl-urapidil	0.35
8-OH-DPAT	1.0
5-Methoxy-3-(di-n-propylamino)chroman	1.0
SR 57746 A	2.0
Flesinoxan	3.0
Ipsapirone	10
Buspirone	30
Lesopitron	50
Antagonists	
WAY 100,635	0.10
p-MPPI	0.32
SDZ 216,525	0.70
(–)Penbutolol	8
(–)Tertatolol	10
(–)Pindolol	15
WAY 100,135	20
(S)-UH-301	52
(–)Propranolol	55
Spiperone	60

S 20499, (+)-4-[N-(5-methoxy-chroman-3-yl)-N-propylam-
ino]butyl-8-azaspiro-(4,5)-decane-7,9-dione; SR 57746 A,
1-[2-(naphth-2-yl)ethyl]-4-(3-trifluoromethylphenyl)-1,2,
5,6-tetrahydropyridine; WAY 100,635, N-[2-[4-(2-
methoxyphenyl)-1-piperazinyl]ethyl]-N-(2-pyridinyl)
cyclohexane carboxamide; p-MPPI, 4-(2'-
methoxyphenyl)-1-[2'-(N-2''-pyridinyl)-p-iodobenzamido]
ethylpiperazine; SDZ 216,525, methyl 4-(4-(4-(1,1,3-
trioxo-2H-1,2-benziosothiazol-2-yl)butyl)-1-
piperazinyl)1H-indole-2-carboxylate; WAY 100,135, N-
$tert$-butyl-3,4-(2-methoxyphenyl)piperazin-1-yl-2-
phenylpropanamide; (S)-UH-301, (S)-5-fluoro-8-hydroxy-
2-(di-n-propylamino)tetralin.

5-HT$_{1A}$ receptors. Its high selectivity allows the use of a tritiated derivative of
this compound for the visualisation and quantification of 5-HT$_{1A}$ receptors in
brain (COSSERY et al. 1987). Another chroman derivative is alnespirone
(S 20499) which acts selectively, in the nanomolar range also, on 5-HT$_{1A}$
receptors (LANFUMEY et al. 1993; KIDD et al. 1993).

The first azapirone identified as a potent (partial) agonist at 5-HT$_{1A}$ receptors was buspirone (GOZLAN et al. 1983) (Table 1). However, this compound is
also an antagonist at presynaptic dopamine receptors, and its metabolism in
vivo leads to the synthesis of 1-pyrimidinyl-piperazine (1-PP) which acts as an
α_2-adrenergic antagonist. Therefore the effects produced by the in vivo admin-

istration of this drug are unavoidably complex and cannot be ascribed solely to
the (partial) stimulation of 5-HT$_{1A}$ receptors. Subsequently, other azapirones
were developed with less activity at dopamine receptors. For example,
ipsapirone, gepirone and tandospirone are more selective 5-HT$_{1A}$ receptor
agonists than buspirone (Hoyer et al. 1994). However, in vivo, these compounds also lead to the formation of 1-PP and therefore to the blockade of α_2-
adrenergic receptors.

Other agonists have been characterised such as 1-[2-(4-
fluorobenzoylamino)ethyl]-4-(7-methoxynaphtyl) piperazine (S 14506), 5-me-
thyl-urapidil, flesinoxan (Ahlenius et al. 1991), but their use in vivo for the
selective stimulation of 5-HT$_{1A}$ receptors is questionable because these drugs
also act generally, in rather low dose ranges, at other receptor types (5-HT$_{2C}$,
α_1-adrenergic receptors, etc.). In contrast, lesopitron (Haj-Dahmane et al.
1994) is a highly selective 5-HT$_{1A}$ receptor agonist but of rather low potency
both in vitro and in vivo.

Binding studies as well as electrophysiological and biochemical investigations with antagonists first characterised for their actions at other receptor
types led to the discovery that the dopamine antagonist spiperone and the β-
adrenergic antagonists (–)-propranolol, (–)-penbutolol, (–)-alprenolol, (–)-
pindolol and (–)-tertatotol also act as antagonists at 5-HT$_{1A}$ receptors (Lum
and Piercey 1988; Okskenberg and Peroutka 1988; Hamon et al. 1990;
Hjorth 1992; Middlemiss and Tricklebank 1992; Jolas et al. 1993; Langlois
et al. 1993, Prisco et al. 1993). This led to the development of derivatives with
enhanced efficacy and selectivity at 5-HT$_{1A}$ receptors such as (±)-1-(1H-indol-
4-yloxy)-3-(cyclohexylamino)-2-propanol (LY 206130; Engleman et al. 1996)
and pindobind [N^1-(bromoacetyl)-N^8-[3-(4-indolyloxy)-2-hydroxypropyl]-(Z)-
1,8-diamino-p-methane] (Liau et al. 1991). However, these drugs are generally
of limited interest for the blockade of the latter receptors, especially in vivo.
Other arylpiperazine drugs were then developed which were initially claimed
to act at 5-HT$_{1A}$ receptor antagonists, such as 1-(2-methoxyphenyl)-4-
[(phthalimido)butyl]piperazine (NAN-190; Glennon et al. 1989), 8-[2-[4-
(2-methoxyphenyl)-1-piperazinyl]ethyl]8-azaspiro(4,5)-decane-7,9-dione
(BMY 7378; Yocca et al. 1987), 1-(2-methoxyphenyl)-4-(4-succinimido)
butyl]piperazine (MM-77; Mokrosz et al. 1994), methyl 4-(4-(4-(1,1,3-trioxo-
2H-1,2-benziosothiazol-2-yl)butyl)-1-piperazinyl)1H-indole-2-carboxylate
(SDZ 216,525; Hoyer et al. 1992; Schoeffter et al. 1993) and N-tert-butyl-
3,4-(2-methoxyphenyl)piperazin-1-yl-2-phenylpropanamide   (WAY 100135;
Fletcher et al. 1993; Lanfumey et al. 1993b). In addition, benzodioxans such
as spiroxatrine and binospirone (MDL-73005EF: 8-[2-(2,3-dihydro-1,4-
benzodioxin-2-yl)methylamino]-8-azaspiro-(4,5)-decane-7,9-dione) have also
been proposed as 5-HT$_{1A}$ receptor antagonists (Van den Hooff and Galvan
1991; Cliffe et al. 1993). In fact, all these compounds, which are not really
selective for 5-HT$_{1A}$ receptors, act as partial agonists, leading to variable
effects from one cell type to another. For example, on serotoninergic neurones
in the dorsal raphe nucleus where a large reserve of 5-HT$_{1A}$ receptors exists

(MELLER et al. 1990) these drugs act as 5-HT$_{1A}$ agonists, whereas in the hippocampus where no 5-HT$_{1A}$ receptor reserve exists (YOCCA et al. 1992) they act essentially as antagonists. For instance, MDL 73005EF, BMY 73-78, SDZ 216,525 mimic, at least in part, the inhibitory effects of 8-OH-DPAT on the firing of serotoninergic neurones in the dorsal raphe nucleus and the associated release of 5-HT in their projection areas, whereas these drugs prevent the effects of the same agonist on adenylyl cyclase activity in the hippocampus and other postsynaptic targets (GARTSIDE et al. 1990; SPROUSE 1991; SHARP et al. 1990, 1993). In addition, at least in vivo, the effects of these drugs are often difficult to interpret because several of them (BMY 73-78, SDZ 216,525, NAN-190, etc.) are also potent α_1 adrenergic receptor blockers (CLAUSTRE et al. 1991; LANFUMEY et al. 1993b; ROUTLEDGE et al. 1994).

Actual antagonists now available include WAY 100635 [N-(2-(4-(2-methoxyphenyl)-1-piperazinyl)ethyl)-N-(2-pyridyl)cyclohexanecarboxamide] (FORSTER et al. 1995; FLETCHER et al. 1996; MUNDEY et al. 1996), p-MPPI [4-(2'-methoxyphenyl)-1-[2'-[N-(2''-pyridinyl-)-iodo-benzamido]ethyl]piperazine] and p-MPPF [4-(2'-methoxyphenyl)-1-[2'-[N-(2''-pyridinyl)-p-fluorobenzamido]ethyl]piperazine] (THIELEN et al. 1996). In all cases these drugs behave as silent antagonists on any cell type regardless of whether a receptor reserve is present. Thus they prevent the inhibitory effects of 5-HT$_{1A}$ receptor agonists on both the nerve impulse flow within serotoninergic neurones in the dorsal raphe nucleus and the forskolin-stimulated adenylyl cyclase activity in the hippocampus. In vivo they efficiently prevent the various behavioural effects of 5-HT$_{1A}$ receptor agonists, such as the flat-body posture and forepaw treading.

Because both WAY 100635 and MPPI are selective of 5-HT$_{1A}$ receptors within a relatively wide range of concentrations, radioactive derivatives of these molecules have been developed for the selective labelling of the receptors both in vitro and in vivo (LAPORTE et al. 1994; GOZLAN et al. 1995; KUNG et al. 1994, 1995). It appears that both [^{3}H]WAY 100635 and [^{125}I]p-MPPI bind to a greater number of specific sites than [^{3}H]8-OH-DPAT and other 5-HT$_{1A}$ agonist radioligands in all brain areas examined (GOZLAN et al. 1995; KHAWAJA 1995; KUNG et al. 1995). This discrepancy results simply from the fact that the antagonist radioligands bind equally, with the same high affinity, to 5-HT$_{1A}$ receptors that are physically associated with G proteins and to 5-HT$_{1A}$ receptor binding subunits which are free in membranes, whereas [^{3}H]8-OH-DPAT and other agonist radioligands have a high affinity solely for the 5-HT$_{1A}$ receptors coupled with G proteins (GOZLAN et al. 1995). Therefore measurements of the specific binding of both [^{3}H]8-OH-DPAT and [^{3}H]WAY 100635 in the same membrane preparation allow estimation of the total number of 5-HT$_{1A}$ receptors and of the proportion of those which are coupled to G proteins (i.e. the functional receptors; GOZLAN et al. 1995; NEVO et al. 1995).

Another important development from the availability of selective antagonist radioligands has been the labelling of 5-HT$_{1A}$ receptors in brain in vivo. Radioactive agonists cannot be used for this purpose because their binding to the receptors unavoidably produces the dissociation of the receptor–G protein complexes and therefore a drop in the receptor affinity for these molecules. Accordingly, at best only a very transient high-affinity labelling of 5-HT$_{1A}$ receptors can be expected with agonist radioligands. In contrast, high-affinity radioactive antagonists can theoretically bind for a long time to the receptors because the same high affinity is preserved regardless of whether the receptor is coupled to G protein. As expected from this reasoning, experiments in rats and mice showed that [^{3}H]WAY 100635 selectively binds to 5-HT$_{1A}$ receptors for at least 2 h after its intravenous administration. Measurements at 1 h after the injection indicated that the brain distribution of the accumulated radioactivity is perfectly correlated with the regional distribution of 5-HT$_{1A}$ receptors, with up to 10–16 times more radioactivity in the hippocampus where these receptors are especially abundant than in the cerebellum where they are expressed at hardly detectable levels (HUME et al. 1994; LAPORTE et al. 1994). Similarly, the other selective 5-HT$_{1A}$ receptor antagonist radioligand [^{123}I]p-MPPI has also been found to label 5-HT$_{1A}$ sites in the rat brain in vivo (KUNG et al. 1994). However, under optimal conditions the ratio of radioactivity accumulated in the hippocampus over the cerebellum reached a value of 3.3 (KUNG et al. 1994), indicating a higher level of non-specific binding than with [^{3}H]WAY 100635. Pretreatment with various 5-HT$_{1A}$ receptor ligands was shown to result in a marked decrease in the brain accumulation of [^{3}H]WAY 100635, giving access to the determination of 5-HT$_{1A}$ receptor occupancy in brain following the administration of a given dose of such ligands (HUME et al. 1994; LAPORTE et al. 1994).

Recently ^{11}C-radiolabelled derivatives of WAY 100635 have been synthesised for the in vivo labelling of 5-HT$_{1A}$ receptors in the human brain using positron emission tomography (PET). This approach led to the first visualisaton of central 5-HT$_{1A}$ receptors in vivo in man, including in discrete areas such as the dorsal raphe nucleus (PIKE et al. 1995; RITTER and JONES 1995). Obviously, important data can now be expected from PET investigations in man with this new radioligand, especially in the fields of anxiety, depression and schizophrenia where 5-HT$_{1A}$ receptors very probably play key roles.

F. Functional Roles of 5-HT$_{1A}$ Receptors

Because of the dual location of 5-HT$_{1A}$ receptors, both on serotoninergic neurones where they act as somato-dendritic autoreceptors, and on the targets of serotoninergic projections where they correspond to postsynaptic receptors (VERGÉ et al. 1986), very different effects can be expected from their stimulation in the raphe versus other regions, especially those of the limbic system (hippocampus, septum, frontal and entorhinal cortex, amygdala, etc.). Indeed, by acting at somato-dendritic autoreceptors, 5-HT$_{1A}$ agonists inhibit the electrical activity of serotoninergic neurones and thereby reduce serotoninergic

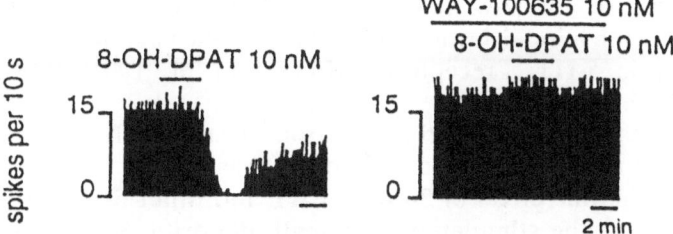

Fig. 5. Inhibitory effect of 5-HT$_{1A}$ autoreceptor stimulation on the firing of a serotoninergic neurone in the rat dorsal raphe nucleus. Prevention by WAY 100,635. Administration of 8-OH-DPAT (10nM) for 3 min onto a brainstem slice superfused with an artificial cerebro-spinal fluid (HAJ-DAHMANE et al. 1991) results in a complete transient inhibition of the firing (in spikes per 10s) of the recorded serotoninergic neurone. Treatment with the 5-HT$_{1A}$ receptor antagonist WAY 100,635 (10nM) completely suppresses this effect

neurotransmission (Fig. 5). In contrast, at the level of postsynaptic targets, 5-HT$_{1A}$ agonists reproduce the effect of 5-HT which is released from serotoninergic terminals and therefore enhance 5-HT$_{1A}$ receptor-dependent serotoninergic neurotransmission (HAMON et al. 1990). Conversely, 5-HT$_{1A}$ receptor antagonists block serotoninergic neurotransmission mediated through postsynaptic 5-HT$_{1A}$ receptors but can increase the electrical activity of serotoninergic neurones through the blockade of somato-dendritic 5-HT$_{1A}$ autoreceptors. However, not all 5-HT$_{1A}$ receptor antagonists produce the latter effect, probably because of their variable intrinsic efficacy to block the receptors. Indeed, only spiperone (FORNAL et al. 1994a), (−)-tertatolol (JOLAS et al. 1994) and WAY 100635 (FORNAL et al. 1994b) appeared to significantly increase the firing rate of serotoninergic neurones in the dorsal raphe nucleus, as expected of the prevention by these drugs of some inhibitory influence of endogenous 5-HT acting tonically at somato-dendritic 5-HT$_{1A}$ autoreceptors.

These data suggest that 5-HT agonists have dual effects on 5-HT-dependent mechanisms: inhibitory effects on serotonin neurone firing activity and serotonin neurotransmission via somato-dendritic autoreceptors and stimulatory effects on serotonin neurotransmission due to the activation of postsynaptic receptors on target neurones. The two effects may occur simultaneously. For instance, it is well established that acute treatment with 8-OH-DPAT and other 5-HT$_{1A}$ receptor agonists increase food consumption in rats. HUTSON et al. (1987) have shown that the stimulation of food consumption results from an inhibitory effect of the latter drugs on central serotoninergic neurotransmission through the stimulation of somato-dendritic 5-HT$_{1A}$ autoreceptors in anterior raphe nuclei. A similar effect on food consumption can be achieved by the blockade of postsynaptic 5-HT$_{2C}$ receptors located on satiety-mediating hypothalamic neurones, for example by administration of cyproheptadine or pizotifen. In contrast, parts of the behavioural "5-HT syndrome" which have been attributed to the stimulation of 5-HT$_{1A}$ receptors (flat-body posture, forepaw treading) as well as the "spontaneous" tail-flick triggered by acute administration of 5-HT$_{1A}$ receptor agonists are entirely due to postsynaptic 5-HT$_{1A}$ receptors (Millan et al. 1991; Wilkinson and Dourish 1991).

In some cases, however, the situation is still confused because interspecies variations apparently exist. Evidence has been reported, for example, that the hypothermic effect of 8-OH-DPAT and other 5-HT$_{1A}$ receptor agonists results from the stimulation of somato-dendritic 5-HT$_{1A}$ autoreceptors only in mice. In rats both the latter receptors and postsynaptic 5-HT$_{1A}$ receptors seem to be responsible for the hypothermic effect of these drugs (Wilkinson and Dourish 1991; Kleven et al. 1995).

In appropriate tests 5-HT$_{1A}$ receptor agonists exhibit both anxiolytic-like and antidepressant-like effects (Martin et al. 1990; Detke et al. 1995), which may seem to be paradoxical because anxiety-driven behaviour and depression-like behaviour are known to be associated with opposite (an increase and a decrease, respectively) changes in serotoninergic neurotransmission. Extensive studies with various approaches such as the selective lesion of serotoninergic neurones in anterior raphe nuclei in order to physically eliminate somato-dendritic 5-HT$_{1A}$ autoreceptors without altering postsynaptic 5-HT$_{1A}$ receptors (Vergé et al. 1985, 1986; Weissmann-Nanopoulos et al. 1985) as well as microinjections of 5-HT$_{1A}$ receptor ligands in various brain regions provided relatively convincing explanations to these puzzling observations. Thus it is now clearly established that the anxiolytic action of 8-OH-DPAT and other 5-HT$_{1A}$ receptor agonists results exclusively from the stimulation of somato-dendritic 5-HT$_{1A}$ autoreceptors by these drugs (Hogg et al. 1994; Jolas et al. 1995). Although some studies showed that the microinjection of 5-HT$_{1A}$ receptor agonists into the hippocampus exerts anxiolytic-like effects in rats, it has subsequently been demonstrated that these effects in fact result from the diffusion of the drugs from their injection sites to the dorsal raphe nucleus where they efficiently inhibit the firing of serotoninergic neurones (Jolas et al. 1995). Indeed, when diffusion is absent, an anxiogenic-like action can be triggered by the local administration of 5-HT$_{1A}$ receptor agonists into the hippocampus (File et al. 1996).

Interestingly, the "classical" anxiolytic drugs, i.e. benzodiazepines such as diazepam and its derivatives, also inhibit the firing of serotoninergic neurones in the dorsal raphe nucleus (Trulson et al. 1982). Accordingly, the reduction in serotoninergic neurotransmission (which can be increased by stressful and anxiogenic stimuli) probably plays a key role in the behavioural effects of both 5-HT$_{1A}$ receptor agonists and benzodiazepines as anxiolytics (Hamon 1994).

In contrast, the direct stimulation of somato-dendritic 5-HT$_{1A}$ autoreceptors in the dorsal raphe nucleus does not mimic the antidepressant-like effects of 5-HT$_{1A}$ receptor agonists in relevant behavioural paradigms (Martin et al. 1990). Furthermore, the selective lesion of serotoninergic neurones to eliminate these autoreceptors does not affect the antidepressant-like action of 5-HT$_{1A}$ receptor agonists. These data led to the proposal that postsynaptic 5-HT$_{1A}$ receptors, but not somato-dendritic 5-HT$_{1A}$ autoreceptors, are responsible for the antidepressant-like action of systemically administered 5-HT$_{1A}$ receptor agonists (MARTIN et al. 1990; WIELAND and LUCKI 1990; MATSUDA et al. 1995). Some of these postsynaptic 5-HT$_{1A}$ receptors appear to be located in the septum because the local administration of 8-OH-DPAT directly into this area exerts a clearcut antidepressant-like effect in the learned helplessness paradigm (MARTIN et al. 1990). However, 5-HT$_{1A}$ receptors in other limbic areas (frontal cortex, etc.) are probably also involved in this effect.

G. Regulation of 5-HT$_{1A}$ Receptors

In transfected cells the 5-HT$_{1A}$ receptor can be phosphorylated by protein kinase C, which results in a decreased potency of agonists to inhibit adenylyl cyclase activity functionally coupled to the receptor (RAYMOND 1991). Whether such a mechanism occurs in brain must still be established.

In vivo the chronic stimulation of 5-HT$_{1A}$ receptors by direct agonists or selective 5-HT reuptake inhibitors (SSRIs) has been shown to affect differently somato-dendritic 5-HT$_{1A}$ autoreceptors in the dorsal raphe nucleus and postsynaptic 5-HT$_{1A}$ receptors, notably in the hippocampus. Both in vivo and in vitro electrophysiological investigations (BLIER and DE MONTIGNY 1994; SCHECHTER et al. 1990; JOLAS et al. 1994; LE POUL et al. 1995) as well as biochemical studies have shown that a 2-to 3- week treatment with these drugs results in a functional desensitisation of somato-dendritic 5-HT$_{1A}$ autoreceptors without affecting postsynaptic 5-HT$_{1A}$ receptors. Thus the potency of 8-OH-DPAT and other 5-HT$_{1A}$ receptor agonists to inhibit the nerve impulse flow within serotoninergic neurones is markedly reduced after such treatments. Interestingly, binding studies generally revealed no change in the specific binding of [^{3}H]8-OH-DPAT and [^{3}H]WAY 100635 in any brain region examined, including the dorsal raphe nucleus, after a 2- to 3-week treatment with either a 5-HT$_{1A}$ receptor agonist or an SSRI (SCHECHTER et al. 1990; JOLAS et al. 1994; LE POUL et al. 1995). Accordingly, it can be concluded that the functional desensitisation of somato-dendritic 5-HT$_{1A}$ autoreceptors is associated neither with a down-regulation phenomenon nor with a possible dissociation of the 5-HT$_{1A}$ receptor–G protein complexes in the dorsal raphe nucleus. However, recent investigations at the mRNA level show that 3-week treatment with fluoxetine significantly reduces the concentration of 5-HT$_{1A}$ receptor mRNA measured by competitive RT-PCR in the anterior raphe area of adult rats (BONI et al., unpublished observations). Such a parallel between the functional desensitisation of somato-dendritic 5-HT$_{1A}$ autoreceptors and a decrease in the transcription of the corresponding gene suggests that binding data are not always relevant to assessing the actual physiological situation. Indeed, it can be speculated that only part of 5-HT$_{1A}$ receptor binding sites, i.e. a functional pool depending on the de novo synthesis from mRNA, are really functional, and that binding measurements concern not only this functional pool but also a large pool of non-functional receptor binding subunits. Possible changes in the number of receptors specifically in the functional pool may be masked by the bulk of receptor binding sites in the non-functional pool, therefore explaining the absence of measurable changes in [³H]8-OH-DPAT and [³H]WAY 100635 specific binding sites after chronic systemic administration of 5-HT$_{1A}$ receptor agonists or SSRIs. Although further investigations are needed to assess the relevance of this hypothesis, it must be stressed already at this stage that this scheme may apply especially to the dorsal raphe nucleus where a large reserve of 5-HT$_{1A}$ receptors exists (MELLER et al. 1990).

That 5-HT$_{1A}$ receptor mRNA is a more reliable marker of the dynamics of 5-HT receptor turnover is also supported by data obtained after the irreversible blockade of 5-HT$_{1A}$ receptors by the alkylating agent N-ethoxycarbonyl-2-ethoxy-1,2-dihydroquinoline (EEDQ). Administration of EEDQ produces an irreversible blockade of at least 80% of [³H]8-OH-DPAT and [³H]WAY 100635 specific binding sites within 12–24h after the treatment, which is then followed by a progressive recovery due to the de novo synthesis of 5-HT$_{1A}$ receptors (BOLAÑOS et al. 1991; GOZLAN et al. 1994). Interestingly, this treatment also induces a transient increase (i.e. at 12h after EEDQ administration) in the concentrations of 5-HT$_{1A}$ receptor mRNA in the anterior raphe area, the cerebral cortex and the hippocampus, which reaches up to 300% of the basal values in control rats (BONI et al. 1996; RAGHUPATHI et al. 1996). Therefore cells expressing 5-HT$_{1A}$ receptors appear to adapt to receptor blockade by increasing the rate of transcription of the corresponding gene so as to compensate rapidly for the loss of 5-HT signalling through the receptors.

Another example of changes at the transcription level concerns the relationships between central serotoninergic neurotransmission and the hypothalamo-pituitary-adrenal axis. In the hippocampus, but not in the raphe area, the synthesis of 5-HT$_{1A}$ receptors appears to be negatively controlled by corticosterone. Thus adrenalectomy has been repeatedly shown to increase both the density of [³H]8-OH-DPAT specific binding sites and the concentration of 5-HT$_{1A}$ receptor mRNA in selected subareas in the rat hippocampus (BURNET et al. 1992). Conversely, administration of corticosterone reduces these changes. Whether these observations are relevant to the physiological situation must still be established. In particular, whether chronic stress which elevates the plasma concentration of corticosterone is able to down-regulate hippocampal 5-HT$_{1A}$ receptors through a negative action of the steroid hormone at the transcription level is not yet known.

In addition to influencing the synthesis of 5-HT$_{1A}$ receptors at the transcription level, adrenal steroids also affect the functional properties of these receptors. Thus in slices of the hippocampus (JOËLS et al. 1991) and the dorsal raphe nucleus (LAARIS et al. 1995) that have been exposed to these hormones the electrophysiological responses to 5-HT$_{1A}$ receptor stimulation are of lower amplitude than under control conditions. However, different steroids are involved in these two regions since mineralocorticoids appear to be responsible for the functional desensitisation of 5-HT$_{1A}$ receptors in the hippocampus (i.e.

on pyramidal cells, JOËLS et al. 1991) whereas glucocorticoids induce this mechanism at somato-dendritic 5-HT$_{1A}$ autoreceptors in the dorsal raphe nucleus (LAARIS et al. 1995). At least in the latter area the in vitro effect of corticosterone is physiologically relevant since it can be reproduced by some stressful conditions known to markedly enhance the secretion of the endogenous hormone (LAARIS et al. 1996).

H. Clinical Perspectives

In addition to the well-known potential of 5-HT$_{1A}$ receptor agonists in the treatment of anxiety-related disorders (especially generalised anxiety; see CHARRIER et al. 1994; HAMON 1994) and depression (STAHL et al. 1992), some additional effects of these drugs may also be therapeutically relevant.

Developmental studies have shown that 5-HT$_{1A}$ receptors are expressed very early in brain during the embryonic life (HILLION et al. 1993). Within the cerebellum the highest density of 5-HT$_{1A}$ receptors is observed during the first week after birth, precisely at a period of very active development of this structure (DAVAL et al. 1987b; MATTHIESSEN et al. 1993; MIQUEL et al. 1994). These observations led to the proposal that 5-HT$_{1A}$ receptors play a role in the trophic action of 5-HT during brain maturation (LAUDER and KREBS 1978; HAMON et al. 1989; WHITAKER-AZMITIA 1991). Indeed, a stimulatory effect of 5-HT$_{1A}$ receptor stimulation on the differentiation of monoaminergic neurones in culture has been clearly established by DE VITRY et al. (1986). In addition, in culture of septal cells from rat embryos, ipsapirone and other 5-HT$_{1A}$ receptor agonists were found to enhance the growing and branching of neurites of cholinergic neurones (RIAD et al. 1994). Similar trophic effects also occur in vivo since subchronic treatment with the potent 5-HT$_{1A}$ receptor agonist 1-[2-(naphth-2-yl)ethyl]-4-(3-trifluoromethylphenyl)-1,2,5,6-tetra-hydro-pyridine (SR 57746A) reversed, at least partly, the loss of hippocampal choline acetyltransferase following the intra-septal injection of vincristine (to destroy septo-hippocampal cholinergic neurones) in both rats and marmosets (FOURNIER et al. 1993). In the same line, systemic administration of 5-HT$_{1A}$ receptor agonists has been shown to prevent neuronal degeneration in the cerebral cortex and hippocampus following global cerebral ischemia (BODE-GREUEL et al. 1990; PREHN et al. 1991). Whether such experimental data will have some clinical applications in various neurodegenerative diseases is a question which cannot yet be answered.

Another domain in which 5-HT$_{1A}$ receptors apparently play a modulatory role is nociception. In the spinal cord these receptors are especially concentrated within the superficial layers of the dorsal horn where primary afferent fibres that convey nociceptive signals terminate (DAVAL et al. 1987a; LAPORTE et al. 1996). Furthermore, part of these receptors are located on the terminals of primary afferent fibres (LAPORTE et al. 1995) and may therefore mediate some presynaptic modulatory action of 5-HT on the release of their neurotransmitters such as glutamate, substance P and/or calcitonin gene-related peptide.

A marked decrease in morphine-induced nociception has been reported in rats treated with 8-OH-DPAT and other 5-HT$_{1A}$ receptor agonists (Millan and Colpaert 1991). In light of the stimulatory effect of 5-HT$_{1A}$ receptor activation on the synthesis of dynorphin at the spinal level (Lucas et al. 1993), and the hyperalgesic effect of opioid κ receptor stimulation by endogenous dynorphin, one can speculate that, in contrast, 5-HT$_{1A}$ receptor antagonists are of interest for enhancing the analgesic effect of morphine especially when spinal dynorphinergic neurones are activated, such as, in chronic inflammatory pain (Pohl et al. 1996). Whether other pain syndromes that are largely reluctant to morphine and opioid-related drug actions, such as neuropathic pain, would benefit of the chronic blockade of 5-HT$_{1A}$ receptors because of a resulting enhancement in morphine efficacy is also another possibility to be explored in future investigations.

That 8-OH-DPAT facilitates sexual behaviour of male rats was first reported by Ahlenius et al. (1981) soon after the discovery of the selective 5-HT$_{1A}$ agonist properties of this drug. Other groups then showed that this compound and other 5-HT$_{1A}$ receptor agonists are able to reverse the sexual inhibition resulting from sexual exhaustion (Rodriguez-Manzo and Fernandez-Guasti 1994). Recently a novel 5-HT$_{1A}$ receptor agonist, 1-[2-(2-naphtyl)ethyl]-4-(3-trifluoromethylphenyl)-piperidine (SR 59026A), has been reported to facilitate the sexual behaviour of naive male rats characterised by a low level of sexual performance and to delay the time of sexual satiation (Arnone et al. 1995). Whether 5-HT$_{1A}$ receptor agonists are of interest for the treatment of certain sexual dysfunctions in man (notably in depressed patients, Baldwin 1996) is also another question to be addressed in future investigations.

In the field of schizophrenia 5-HT$_{1A}$ receptor ligands may also be of interest, especially for reducing the extrapyramidal motor dysfunction that can occur on long term treatment with anti-dopamine antipsychotic drugs. Using neuroleptic-induced catalepsy as a model, Wadenberg (1996) showed that this behaviour can be completely prevented by the administration of 5-HT$_{1A}$ receptor agonists acting at somato-dendritic 5-HT$_{1A}$ autoreceptors in the median raphe nucleus. In addition, some observations in rats also suggest that the combined treatment of a D$_2$ receptor antagonist (raclopride) with a 5-HT$_{1A}$ receptor agonist (8-OH-DPAT) would exert better antipsychotic-like effects than the former drug alone (Wadenberg and Ahlenius 1991). Indeed, buspirone was initially proposed as an antipsychotic drug, and its partial agonist effect at 5-HT$_{1A}$ receptors (Gozlan et al. 1983) might well contribute to this action. However, more potent and selective agonists must be tested to assess the potential antipsychotic effects of somato-dendritic 5-HT$_{1A}$ autoreceptor stimulation.

Chronic alcoholisation in rats is known to deeply affect central 5-HT neurotransmission, notably through functional alterations in pre- and postsynaptic 5-HT$_{1A}$ receptors (KLEVEN et al. 1995; NEVO et al. 1995). Convergent data have shown that 8-OH-DPAT, ipsapirone and other agonists reduce ethanol intake under conditions where neither food intake nor water drinking is affected (SVENSSON et al. 1993). Extensive studies by SCHREIBER et al. (1993) have demonstrated that the ethanol preference reducing effect of the systemic administration of 8-OH-DPAT can be reproduced by the local application of the same drug in the dorsal raphe nucleus, as expected of its mediation through the stimulation of somato-dendritic 5-HT$_{1A}$ autoreceptors. Accordingly, 5-HT$_{1A}$ receptor agonists acting preferentially at these sites might be of therapeutic value for reducing alcohol addiction in alcoholic patients.

To conclude, we emphasise the recent observation that the association of pindolol, to block somato-dendritic 5-HT$_{1A}$ receptors, with SSRIs, markedly enhances the antidepressant efficacy of the latter drugs (ARTIGAS et al. 1994; BLIER and BERGERON 1995). Extensive electrophysiological and microdialysis studies have clearly demonstrated that in the early phase of treatment with SSRIs these drugs do not enhance 5-HT neurotransmission. They provoke the activation of somato-dendritic 5-HT$_{1A}$ autoreceptors (through an increase in the extracellular concentration of 5-HT in the raphe area) which triggers inhibition of the firing of serotoninergic neurones, thereby reducing the release of 5-HT from their terminals in projection areas. During chronic treatment the resulting continuous stimulation of autoreceptors produces their desensitisation (see Sect. G), rendering these inhibitory influences ineffective. At this stage 5-HT neurotransmission is actually enhanced thanks to the persisting blockade of 5-HT reuptake by SSRIs. Because the desensitisation process requires 2–3 weeks to develop completely, which corresponds to the delay in the antidepressant action of SSRIs to occur (see BLIER and DE MONTIGNY 1994; GARDIER et al. 1996), it has been proposed that the blockade of somato-dendritic 5-HT$_{1A}$ autoreceptors should suppress this delay and improve the therapeutic efficacy of SSRIs, through the prevention of the inhibitory phenomena occurring in the early phase of the treatment. Using pindolol as a 5-HT$_{1A}$ autoreceptor antagonist, ARTIGAS et al. (1994) and BLIER and BERGERON (1995) successfully promoted the antidepressant efficacy of SSRIs and reduced their delay of action.

The latter example illustrates the interest of basic research on neurotransmission to design novel therapeutic strategies of psychiatric diseases. In the particular case of 5-HT$_{1A}$ receptors these data also emphasise the need of potent and specific ligands to interact with these targets. With regard to a better drug combination than that achieved with SSRI plus pindolol, the challenge will obviously be the development of an antagonist acting only at somato-dendritic 5-HT$_{1A}$ autoreceptors (because functional postsynaptic 5-HT$_{1A}$ receptors are needed for the antidepressant action of SSRIs). This will require further basic studies on the differential properties of 5-HT$_{1A}$ receptors expressed by serotoninergic neurones versus those expressed by their postsynaptic targets (see RADJA et al. 1992).

Acknowledgements. I am grateful to Mrs C. Sais for her excellent secretarial assistance, and to Drs J. Adrien, L. Lanfumey and A.M. Laporte for the illustrations.

References

Aghajanian GK, Lakoski JM (1984) Hyperpolarization of serotonergic neurons by serotonin and LSD: studies in brain slices showing increased K^+ conductance. Brain Res 305:181–185

Ahlenius S, Larsson K, Svensson L, Hjorth S, Carlsson A, Lindberg P, Wikström H, Sanchez D, Arvidsson L-E, Hacksell U, Nilsson JLG (1981) Effects of a new type of 5-HT receptor agonist on male rat sexual behavior. Pharmacol Biochem Behav 15:785–792

Ahlenius S, Larsson K, Wijkström A (1991) Behavioral and biochemical effects of the 5-HT$_{1A}$ receptor agonists flesinoxan and 8-OH-DPAT in the rat. Eur J Pharmacol 200:259–266

Albert PR, Zhou QY, Van Tol HH, Bunzow JR, Civelli O (1990) Cloning, functional expression and mRNA tissue distribution of the rat 5-HT$_{1A}$ receptor gene. J Biol Chem 265:5825–5832

Arnone M, Baroni M, Gai J, Guzzi U, Desclaux MF, Keane PE, Le Fur G, Soubrié P (1995) Effect of SR 59026A, a new 5-HT$_{1A}$ receptor agonist, on sexual activity in male rats. Behav Pharmacol 6:276–282

Artigas F, Perez V, Alvarez E (1994) Pindolol induces a rapid improvement of patients with depression treated with serotonin reuptake inhibitors. Arch Gen Psychiatry 51:248–251

Baldwin DS (1996) Depression and sexual function. J Psychopharmacol 10 [Suppl 1]:30–34

Beck SG, Choi KC, List TJ (1992) Comparison of 5-hydroxytryptamine$_{1A}$-mediated hyperpolarization in CA1 and CA3 hippocampal pyramidal cells. J Pharmacol Exp Ther 263:350–359

Besse D, Lombard MC, Zajac JM, Roques B, Besson JM (1990) Pre- and postsynaptic distribution of mu, delta and kappa opioid receptors in the superficial layers of the dorsal horn of the rat spinal cord. Brain Res 521:15–22

Bjork L, Cornfield LJ, Nelson DL, Hillver S-E, Anden N-E, Lewander T, Hacksell U (1991) Pharmacology of the novel 5-hydroxytryptamine$_{1A}$ receptor antagonist (S)-5-fluoro-8-hydroxy-2-(dipropylamino)tetralin: inhibition of (R)-8-hydroxy-2-(dipropylamino)tetralin-induced effects. J Pharmacol Exp Ther 258:58–65

Blier P, Bergeron R (1995) Effectiveness of pindolol with selected antidepressant drugs in the treatment of major depression. J Clin Psychopharmacol 15:217–222

Blier P, De Montigny C (1994) Current advances and trends in the treatment of depression. Trends Pharmacol Sci 15:220–226

Boddeke HWGM, Fargin A, Raymond JR, Schoeffter P, Hoyer D (1992) Agonist/antagonist interactions with cloned human 5-HT$_{1A}$ receptors: variations in intrinsic activity studied in transfected HeLa cells. Naunyn-Schmiedebergs Arch Pharmacol 345:257–263

Bode-Greuel KM, Klisch J, Glaser T, Traber J (1990) Serotonin(5-HT)$_{1A}$ receptor agonists as neuroprotective agents in cerebral ischemia, In: Krieglstein J, Oberpichler H (eds) Pharmacology of cerebral ischemia. Wissenschaftliche Verlagsgesellschaft, Stuttgart, pp 485–491

Boess FG, Martin JL (1994) Molecular biology of 5-HT receptors. Neuropharmacology 33:275–317

Bolaños FJ, Schechter LE, Laporte AM, Hamon M, Gozlan H (1991) Recovery of 5-HT$_{1A}$ receptors after irreversible blockade by N-ethoxycarbonyl-2-ethoxy-1,2-dihydroquinoline (EEDQ). Proc West Pharmacol Soc 34:387–393

Boni C, Fabre V, Lima L, Martres MP, Hamon M (1996) Transient increase in brain 5-HT$_{1A}$ receptor mRNA levels after EEDQ-induced irreversible blockade of 5-HT$_{1A}$ receptors. J Neurochem 66 [Suppl 2]:S36A

Burnet PWJ, Mefford IN, Smith CC, Gold PW, Sternberg EM (1992) Hippocampal [^{3}H]8-hydroxy-2-(di-n-propylamino)tetralin binding site densities, serotonin receptor (5-HT$_{1A}$) messenger ribonucleic acid abundance, and serotonin levels parallel the activity of the hypothalamo-pituitary-adrenal axis in rat. J Neurochem 59:1062–1070

Chanda PK, Minchin MCW, Davis AR, Greenberg L, Reilly Y, McGregor WH, Bhat R, Lubeck MD, Mizutani S, Hung PP (1993) Identification of residues important for ligand binding to the human 5-hydroxytryptamine$_{1A}$ serotonin receptor. Mol Pharmacol 43:516–520

Charrier D, Dangoumau L, Hamon M, Puech AJ, Thiébot MH (1994) Effects of 5-HT$_{1A}$ receptor ligands on a safety signal withdrawal procedure of conflict in the rat. Pharmacol Biochem Behav 48:281–289

Cliffe IA, Fletcher A, Dourish CT (1993) The evolution of selective, silent 5-HT$_{1A}$ receptor antagonists. Current Drugs Ltd ISSN 0961-4680, pp 1–25

Claustre Y, Rouquier L, Serrano A, Bénavidès J, Scatton B (1991) Effect of the putative 5-HT$_{1A}$ receptor antagonist NAN-190 on rat brain serotonergic transmission. Eur J Pharmacol 204:71–77

Corradetti R, Le Poul E, Laaris N, Hamon M, Lanfumey L (1996) Electrophysiological effects of N-(2-(4-(2-methoxyphenyl)-1-piperazinyl)ethyl)-N-(2-pyridinyl) cyclohexane carboxamide (WAY 100635) on dorsal raphe serotoninergic neurons and CA1 hippocampal pyramidal cells in vitro. J Pharmacol Exp Ther 278:679–688

Cossery JM, Gozlan H. Spampinato U, Perdicakis C, Guillaumet G, Pichat L, Hamon M (1987) The selective labelling of central 5-HT$_{1A}$ receptor binding sites by [^{3}H]5-methoxy-3-(di-n-propylamino)chroman. Eur J Pharmacol 140:143 155

Daval G, Vergé D, Basbaum AI, Bourgoin S, Hamon M (1987a) Autoradiographic evidence of serotonin$_1$ binding sites on primary afferent fibres in the dorsal horn of the rat spinal cord. Neurosci Lett 83:71–76

Daval G, Vergé D, Becerril A, Gozlan H, Spampinato U, Hamon M (1987b) Transient expression of 5-HT$_{1A}$ receptor binding sites in some areas of the rat CNS during postnatal development. Int J Devl Neurosci 5:171–180

Detlke MJ, Wieland S, Lucki I (1995) Blockade of the antidepressant-like effects of 8-OH-DPAT, buspirone and desipramine in the rat forced swim test by 5-HT$_{1A}$ receptor antagonists. Psychopharmacology 119:47–54

De Vitry F, Hamon M, Catelon J, Dubois M, Thibault J (1986) Serotonin initiates and autoamplifies its own synthesis during mouse central nervous system development. Proc Natl Acad Sci USA 83:8629–8633

El Mestikawy S, Taussig D, Gozlan H, Emerit MB, Ponchant M, Hamon M (1989) Chromatographic analyses of the serotonin 5-HT$_{1A}$ receptor solubilized from the rat hippocampus. J Neurochem 53:1555–1566

El Mestikawy S, Riad M, Laporte AM, Vergé D, Daval G, Gozlan H, Hamon M (1990) Production of specific anti-rat 5-HT$_{1A}$ receptor antibodies in rabbits injected with a synthetic peptide. Neurosci Lett 118:189–192

Emerit MB, El Mestikawy S, Gozlan H, Rouot B, Hamon M (1990) Physical evidence of the coupling of solubilized 5-HT$_{1A}$ binding sites with G regulatory proteins. Biochem Pharmacol 39:7–18

Engleman EA, Robertson DW, Thompson DC, Perry KW, Wong DT (1996) Antagonism of serotonin 5-HT$_{1A}$ receptors potentiates the increases in extracellular monoamines induced by duloxetine in rat hypothalamus. J Neurochem 66:599–603

Fargin A, Raymond JR, Lohse MJ, Kobilka BK, Caron MG, Lefkowitz RJ (1988) The genomic clone G-21 which resembles a β-adrenergic receptor sequence encodes the 5-HT$_{1A}$ receptor. Nature 335:358–360

Fargin A, Raymond JR, Regan JW, Cotecchia S, Lefkowitz RJ, Caron MG (1989) Effector coupling mechanisms of the cloned 5-HT$_{1A}$ receptor. J Biol Chem 264:14848–14852

File SE, Gonzalez LE, Andrews N (1996) Comparative study of pre- and postsynaptic 5-HT$_{1A}$ receptor modulation of anxiety in two ethological animal tests. J Neurosci 16:4810–4815

Fletcher A, Bill DJ, Bill SJ, Cliffe IA, Dover GM, Forster EA, Haskins TJ, Jones D, Mansell HL, Reilly Y (1993) WAY100135: a novel, selective antagonist at presynaptic and postsynaptic 5-HT$_{1A}$ receptors. Eur J Pharmacol 237:283–291

Fletcher A, Forster EA, Bill DJ, Brown G, Cliffe IA, Hartley JE, Jones DE, McLenachan A, Stanhope KJ, Critchley DJP, Childs KJ, Middlefell VC, Lanfumey L, Corradetti R, Laporte AM, Gozlan H, Hamon M, Dourish CT (1996) Electrophysiological, biochemical, neurohormonal and behavioural studies with WAY-100635, a potent selective, and silent 5-HT$_{1A}$ receptor antagonist. Behav Brain Res 73:337–353

Fornal CA, Marrosu F, Metzler CW, Tada K, Jacobs BL (1994a) Effects of the putative 5-hydroxytryptamine$_{1A}$ antagonists BMY 7378, NAN 190 and (−)-propranolol on serotonergic dorsal raphe unit activity in behaving cats. J Pharmacol Exp Ther 270:1359–1366

Fornal CA, Metzler CW, Gallegos R, Veasey SC, McCreary A, Jacobs BL (1994b) WAY-100635, a potent and selective 5-HT$_{1A}$ receptor antagonist, increases serotonergic neuronal activity and blocks the action of 8-OH-DPAT in behaving cats. Soc Neurosci 20:1544

Forster EA, Cliffe IA, Bill DJ, Dover GM, Jones D, Reilly Y, Fletcher A (1995) A pharmacological profile of the selective silent 5-HT$_{1A}$ receptor antagonist, WAY-100635. Eur J Pharmacol 281:81–88

Fournier J, Steinberg R, Gauthier T, Keane PE, Guzzi U, Coudé FX, Bougault I, Maffrand JP, Soubrié P, Le Fur G (1993) Protective effects of SR 57746A in central and peripheral models of neurodegenerative disorders in rodents and primates. Neuroscience 55:629–641

Gardier AM, Malagié I, Trillat AC, Jacquot C, Artigas F (1996) Role of 5-HT$_{1A}$ autoreceptors in the mechanism of action of serotoninergic antidepressant drugs: recent findings from in vivo microdialysis studies. Fundam Clin Pharmacol 10:16–27

Gartside SE, Cowen PJ, Hjorth S (1990) Effects of MDL 73005EF on central pre-and postsynaptic 5-HT$_{1A}$ receptor function in the rat in vivo. Eur J Pharmacol 191:391–400

Gérard C, Langlois X, Gingrich J, Doucet E, Vergé D, Kia HK, Raisman R, Gozlan H, El Mestikawy S, Hamon M (1994) Production and characterization of polyclonal antibodies recognizing the intracytoplasmic third loop of the 5-hydroxytryptamine$_{1A}$ receptor. Neuroscience 62:721–739

Glennon RA, Naiman NA, Peirson ME, Titeler M, Lyon RA, Herndon JL, Misenheimer B (1989) NAN-190, a potential 5-HT$_{1A}$ serotonin antagonist. Drug Dev Res 16:335–343

Gozlan H, El Mestikawy S, Pichat L, Glowinski J, Hamon M (1983) Identification of presynaptic serotonin autoreceptors using a new ligand: [^{3}H]PAT. Nature 305:140–142

Gozlan H, Ponchant M, Daval G, Vergé D, Ménard F, Vanhove A, Beaucourt JP, Hamon M (1988) [^{125}I]Bolton-Hunter-8-methoxy-2-[N-propyl-N-propylamino]tetralin as a new selective radioligand of 5-HT$_{1A}$ sites in the rat brain. In vitro binding and autoradiographic studies. J Pharmacol Exp Ther 244:751–759

Gozlan H, Laporte AM, Thibault S, Schechter LE, Bolaños F, Hamon M (1994) Differential effects of N-ethoxycarbonyl-2-ethoxy-1,2-dihydroquinoline (EEDQ) on various 5-HT receptor binding sites in the rat brain. Neuropharmacology 33:423–431

Gozlan H, Thibault S, Laporte AM, Lima L, Hamon M (1995) The selective 5-HT$_{1A}$ antagonist radioligand [^{3}H]WAY 100635 labels both G-protein-coupled and free 5-HT$_{1A}$ receptor binding sites in rat brain membranes. Eur J Pharmacol Mol Pharmacol 288:173–186

Haj-Dahmane S, Hamon M, Lanfumey L (1991) K$^+$ channel and 5-hydroxytryptamine$_{1A}$ autoreceptor interactions in the rat dorsal raphe nucleus: an in vitro electrophysiological study. Neuroscience 41:495–505

Haj-Dahmane S, Jolas T, Laporte AM, Gozlan H, Farré AJ, Hamon M, Lanfumey L (1994) Interactions of lesopitron (E-4424) with central 5-HT$_{1A}$ receptors: in vitro and in vivo studies in the rat. Eur J Pharmacol 255:185–196

Hall MD, El Mestikawy S, Emerit MB, Pichat L, Hamon M, Gozlan H (1985) [^{3}H]8-hydroxy-2-(di-n-propylamino)tetralin binding to pre- and postsynaptic 5-hydrox-ytryptamine sites in various regions of the rat brain. J Neurochem 44:1685–1696

Hamon M (1994) Neuropharmacology of anxiety: perspectives and prospects. Trends Pharmacol Sci 15:36–39

Hamon M, Bourgoin S, Gozlan H, Hall MD, Goetz C, Artaud F, Horn AS (1984) Biochemical evidence of the 5-HT agonist properties of PAT (8-hydroxy-2-(di-n-propylamino)tetralin) in the rat brain. Eur J Pharmacol 100:263–276

Hamon M, Bourgoin S, Chanez C, De Vitry F (1989) Do serotonin and other neu-rotransmitters exert a trophic influence on the immature brain? In: Evrard P, Minkowski A (eds) Developmental neurobiology. Raven, New York, pp 171–183

Hamon M, Gozlan H, El Mestikawy S, Emerit MB, Bolaños F, Schechter L (1990) The central 5-HT$_{1A}$ receptors: pharmacological, biochemical, functional and regulatory properties. Ann NY Acad Sci 600:114–131

Hillion J, Dumas Milne-Edwards JB, Catelon J, De Vitry F, Gros F, Hamon M (1993) Prenatal developmental expression of rat brain 5-HT$_{1A}$ receptor gene followed by PCR. Biochem Biophys Res Commun 191:991–997

Hillion J, Catelon J, Riad M, Hamon M, De Vitry F (1994) Neuronal localization of 5-HT$_{1A}$ receptor mRNA and protein in rat embryonic brain stem cultures. Devl Brain Res 79:195–202

Hjorth S (1992) (–)-Penbutolol as a blocker of central 5-HT$_{1A}$ receptor-mediated responses. Eur J Pharmacol 222:121–127

Hjorth S, Carlsson A, Lindberg P, Sanchez D, Wikström H, Arvidsson L-E, Hacksell U, Nilsson JLG (1982) 8-Hydroxy-2-(di-n-propylamino)tetralin, 8-OH-DPAT, a potent and selective simplified ergot congener with central 5-HT-receptor stimu-lating activity. J Neural Transm 55:169–188

Hogg S, Andrews N, File SE (1994) Contrasting behavioural effects of 8-OH-DPAT in the dorsal raphe nucleus and ventral hippocampus. Neuropharmacology 33:343–348

Hoyer D, Schoeffter P, Palacios JM, Kalkman HO, Bruinvels AT, Fozard JR, Seigl H, Seiler MP, Stoll A (1992) SDZ 216-525: a selective, potent and silent 5-HT$_{1A}$ receptor antagonist. Br J Pharmacol 105:29P

Hoyer D, Clarke D, Fozard JR, Hartig PR, Martin G, Mylecharane E, Saxena P, Humphrey PPA (1994) VII. International Union of Pharmacology Classification of receptors for 5-hydroxytryptamine (serotonin) Pharmacol Rev 46:157–203

Hume SP, Ashworth S, Opacka-Juffry J, Ahier RG, Lammertsma AA, Pike VW, Cliffe IA, Fletcher A, White AC (1994) Evaluation of [O-methyl-^{3}H]WAY-100635 as an in vivo radioligand for 5-HT$_{1A}$ receptors in rat brain. Eur J Pharmacol 271:515–523

Hutson PH, Dourish CT, Curzon G (1987) 8-Hydroxy-2-(di-n-propylamino)tetralin (8-OH-DPAT)-induced hyperphagia: neurochemical and pharmacological evidence for an involvement of 5-hydroxytryptamine somatodendritic autoreceptors. In: Dourish CT, Ahlenius S, Hutson PH (eds) Brain 5-HT$_{1A}$ receptors (eds) Ellis Horwood Series in Biomedicine, Chichester, pp 211–232

Joëls M, Hesen W, De Kloet ER (1991) Mineralocorticoid hormones suppress serotonin-induced hyperpolarization of rat hippocampal CA1 neurons. J Neurosci 11:2288–2294

Jolas T, Haj-Dahmane S, Lanfumey L, Fattaccini CM, Kidd EJ, Adrien J, Gozlan H, Guardiola-Lemaitre B, Hamon M (1993) (–)Tertatolol is a potent antagonist at pre- and postsynaptic serotonin 5-HT$_{1A}$ receptors in the rat brain. Naunyn-Schmiedebergs Arch Pharmacol 347:453–463

Jolas T, Haj-Dahmane S, Kidd EJ, Langlois X, Lanfumey L, Fattaccini CM, Vantalon V, Laporte AM, Adrien J, Gozlan H, Hamon M (1994) Central pre- and postsynaptic 5-HT$_{1A}$ receptors in rats treated chronically with a novel antidepressant, cericlamine. J Pharmacol Exp Ther 268:1432–1443

Jolas T, Schreiber R, Laporte AM, Chastanet M, De Vry J, Glaser T, Adrien J, Hamon M (1995) Are postsynaptic 5-HT$_{1A}$ receptors involved in the anxiolytic effects of 5-HT$_{1A}$ receptor agonists and in their inhibitory effects on the firing of serotonergic neurons in the rat? J Pharmacol Exp Ther 272:920–929

Khawaja X (1995) Quantitative autoradiographic characterisation of the binding of [^{3}H]WAY-100635, a selective 5-HT$_{1A}$ receptor antagonist. Brain Res 673:217–225

Kia, HK, Brisorgueil MJ, Hamon M, Calas A, Vergé D (1996a) Ultrastructural localization of serotonin1A receptors in the rat brain. J Neurosci Res 46:697–708

Kia HK, Miquel MC, Brisorgueil MJ, Daval G, Riad M, El Mestikawy, S, Hamon M, Vergé D (1996b) Immunocytochemical localization of 5-HT$_{1A}$ receptors in the rat central nervous system. J Comp Neurol 365:289–305

Kidd EJ, Haj-Dahmane S, Jolas T, Lanfumey L, Fattaccini CM, Guardiola-Lemaitre B, Gozlan H, Hamon M (1993) New methoxy-chroman derivatives, 4[N-(5-methoxy-chroman-3-yl)N-propylamino]butyl-8-azaspiro-(4,5)-decane-7,9-dione [(±)-S 20244] and its enantiomers, (+)-S 20499 and (–)-S 20500, with potent agonist properties at central 5-hydroxytryptamine$_{1A}$ receptors. J Pharmacol Exp Ther 264:863–872

Kleven M, Ybema C, Carilla E, Hamon M, Koek W (1995) Modification of behavioral effects of 8-hydroxy-2-(di-n-propylamino)tetralin following chronic ethanol consumption in the rat: evidence for the involvement of 5-HT$_{1A}$ receptors in ethanol dependence. Eur J Pharmacol 281:219–228

Kobilka BK, Frielle T, Collins S, Yang-Feng T, Kobilka TS, Francke U, Lefkowitz RJ, Caron MG (1987) An intron-less gene encoding a potential member of the family of receptors coupled to guanine nucleotide regulatory proteins. Nature 329:75–79

Kung M-P, Zhuang Z-P, Frederick D, Kung HK (1994) In vivo binding of [^{123}I]4-(2'-methoxy-phenyl)-1-[2'-(N-2''-pyridinyl)-p-iodobenzamido-]ethyl-piperazine, p-MPPI, to 5-HT$_{1A}$ receptors in rat brain. Synapse 18:359–366

Kung M-P, Frederick D, Mu M, Zhi-Ping Z, Kung HK (1995) 4-(2'-Methoxy-phenyl)-1-[2'-(n-2''-pyridinyl)-p-iodobenzamido]-ethyl-piperazine ([^{125}I]p-MPPI) as a new selective radioligand of serotonin-1A sites in rat brain: in vitro binding and autoradiographic studies. J Pharmacol Exp Ther 272:429–437

Laaris N, Haj-Dahmane S, Hamon M, Lanfumey L (1995) Glucocorticoid receptor-mediated inhibition by corticosterone of 5-HT$_{1A}$ autoreceptor functioning in the rat dorsal raphe nucleus. Neuropharmacology 24:1201–1210

Laaris N, Le Poul E, Laporte AM, Hamon M, Lanfumey L (1996) Effects of stress on the functional status of somatodendritic 5-HT$_{1A}$ autoreceptors in the rat dorsal raphe nucleus. Fundam Clin Pharmacol (in press)

Lanfumey L, Gaymard C, Fattaccini CM, Laporte AM, Lesourd M, Mocaer E, Hamon M (1993a) [^{3}H]S-20499: a new high affinity radioligand for the specific labelling of central 5-HT$_{1A}$ receptor. Eur J Neurosci Suppl 6:238

Lanfumey L, Haj-Dahmane S, Hamon M (1993b) Further assessment of the antagonist properties of the novel and selective 5-HT$_{1A}$ receptor ligands (+)-WAY 100135 and SDZ 216–525. Eur J Pharmacol 249:25–35

Lanfumey L, Gaymard C, Laporte AM, Lima L, Mocaer E, Hamon M (1994) [^{3}H]S-14506: a novel high affinity radioligand for the specific labelling of 5-HT$_{1A}$ receptors. Neuropsychopharmacology 10:93S

Langlois M, Brémont B, Rousselle D, Gaudy F (1993) Structural analysis by the comparative molecular field analysis method of the affinity of β-adrenoreceptor blocking agents for 5-HT$_{1A}$ and 5-HT$_{1B}$ receptors. Eur J Pharmacol 244:77–87

Langlois X, El Mestikawy S, Arpin M, Triller A, Hamon M, Darmon M (1996) Differential addressing of 5-HT$_{1A}$ and 5-HT$_{1B}$ receptors in transfected LLC-PK1 epithelial cells: a model of receptor targeting in neurons. Neuroscience 74:297–302

Laporte AM, Schechter LE, Bolaños FJ, Vergé D, Hamon M, Gozlan H (1991) [^{3}H]5-Methyl-urapidil labels 5-HT$_{1A}$ receptors and α_1 adrenoceptors in the rat CNS. In vitro binding and autoradiographic studies. Eur J Pharmacol 198:59–67

Laporte AM, Lima L, Gozlan H, Hamon M (1994) Selective in vivo labelling of brain 5-HT$_{1A}$ receptors by [^{3}H]WAY 100635 in the mouse. Eur J Pharmacol 271:505–514

Laporte AM, Fattaccini CM, Lombard MC, Chauveau J, Hamon M (1995) Effects of dorsal rhizotomy and selective lesion of serotonergic and noradrenergic systems on 5-HT$_{1A}$, 5-HT$_{1B}$ and 5-HT$_3$ receptors in the rat spinal cord. J Neural Transm 100:207–223

Laporte AM, Doyen C, Nevo IT, Chauveau J, Hauw JJ, Hamon M (1996) Autoradiographic mapping of serotonin 5-HT$_{1A}$, 5-HT$_{1D}$, 5-HT$_{2A}$, and 5-HT$_3$ receptors in the aged human spinal cord. J Chem Neuroanat 11:67–75

Lauder J, Krebs H (1978) Serotonin as a differentiation signal in early neurogenesis. Dev Neurosci 3:15–30

Le Poul E, Laaris N, Doucet E, Laporte AM, Hamon M, Lanfumey L (1995) Early desensitization of somato-dendritic 5-HT$_{1A}$ autoreceptors in rats treated with fluoxetine or paroxetine. Naunyn-Schmiedebergs Arch Pharmacol 352:141–148

Liau LM, Sleight AJ, Pitha J, Peroutka SJ (1991) Characterization of a novel and potent 5-hydroxytryptamine$_{1A}$ receptor antagonist. Pharmacol Biochem Behav 38:555–559

Lucas JJ, Mellström B, Colado MI, Naranjo JR (1993) Molecular mechanisms of pain: serotonin$_{1A}$ receptor agonists trigger transactivation by c-fos of the prodynorphin gene in spinal cord neurons. Neuron 10:599–611

Lum JT, Piercey MF (1988) Electrophysiological evidence that spiperone is an antagonist of 5-HT$_{1A}$ receptors in the dorsal raphe nucleus. Eur J Pharmacol 149:9–15

Marcinkiewicz M, Vergé D, Gozlan H, Pichat L, Hamon M (1984) Autoradiographic evidence for the heterogeneity of 5-HT$_1$ sites in the rat brain. Brain Res 291:159–163

Martin P, Beninger RJ, Hamon M, Puech AJ (1990) Antidepressant-like action of 8-OH-DPAT, a 5-HT$_{1A}$ agonist, in the learned helplessness paradigm: evidence for a postsynaptic mechanism. Behav Brain Res 38:135–144

Matsuda T, Somboonthum P, Suzuki M, Asano S, Baba A (1995) Antidepressant-like effect by postsynaptic 5-HT$_{1A}$ receptor activation in mice. Eur J Pharmacol 280:235–238

Matthiessen L, Kia HK, Daval G, Riad M, Hamon M, Vergé D (1993) Immunocytochemical localization of 5-HT$_{1A}$ receptors in the rat immature cerebellum. Neuroreport 4:763–766

Meller E, Goldstein M, Bohmaker K (1990) Receptor reserve for 5-hydroxytryptamine$_{1A}$-mediated inhibition of serotonin synthesis: possible relationship to anxiolytic properties of 5-hydroxytryptamine$_{1A}$ agonists. Mol Pharmacol 37:231–237

Middlemiss DN, Fozard JR (1983) 8-Hydroxy-2-(di-n-propylamino)tetralin discriminates between subtypes of the 5-HT$_1$ recognition site. Eur J Pharmacol 30:151–153

Middlemiss DN, Tricklebank MD (1992) Centrally active 5-HT receptor agonists and antagonists. Neurosci Biobehav Rev 16:75–82

Millan MJ, Colpaert FC (1991) 5-Hydroxytryptamine(HT)$_{1A}$ receptors and the tail-flick response. II. High efficacy 5-HT$_{1A}$ agonists attenuate morphine-induced antinociception in mice in a competitive-like manner. J Pharmacol Exp Ther 256:983–992

Millan MJ, Bervoets K, Colpaert FC (1991) 5-Hydroxytryptamine(5-HT)$_{1A}$ receptors and the tail-flick response. I. 8-Hydroxy-2-(di-n-propylamino)tetralin HBr-induced spontaneous tail-flicks in the rat as an in vivo model of 5-HT$_{1A}$ receptor-mediated activity. J Pharmacol Exp Ther 256:973–982

Minisclou C, Benavides J, Claustre Y (1995) Activation of 5-HT$_{1A}$ receptors inhibits carbachol-stimulated inositol 1,4,5-trisphosphate mass accumulation in the rodent hippocampus. Neurochem Res 20:977–983

Miquel MC, Doucet E, Boni C, El Mestikawy S, Matthiessen L, Daval G, Vergé D, Hamon M (1991) Central serotonin$_{1A}$ receptors: respective distributions of encoding mRNA, receptor protein and binding sites by in situ hybridization histochemistry, radioimmunohistochemistry and autoradiographic mapping in the rat brain. Neurochem Int 19:453–465

Miquel MC, Kia HK, Boni C, Doucet E, Daval G, Matthiessen L, Hamon M, Vergé D (1994) Postnatal development and localization of 5-HT$_{1A}$ receptor mRNA in rat forebrain and cerebellum. Devl Brain Res 80:149–157

Mokrosz MJ, Chojnacka-Wojcik E, Tatarczynska E, Klodzinska A, Filip M, Boksa J, Charakchieva-Minol S, Mokrosz JL (1994) 1-(2-methoxyphenyl)-4-(4-succinimido) butyl-piperazine (MM-77): a new, potent, postsynaptic antagonist of 5-HT$_{1A}$ receptors. Med Chem Res 4:161–169

Mundey MK, Fletcher A, Marsden CA (1996) Effects of 8-OH-DPAT and 5-HT$_{1A}$ antagonists WAY 100135 and WAY 100635, on guinea-pig behaviour and dorsal raphe 5-HT neurone firing. Br J Pharmacol 117:750–756

Nelson DL, Herbet A, Bourgoin S, Glowinski J, Hamon M (1978) Characteristics of central 5-HT receptors and their adaptive changes following intracerebral 5,7-dihydroxytryptamine administration in the rat. Mol Pharmacol 14:983–995

Nevo I, Langlois X, Laporte AM, Kleven M, Koek W, Lima L, Maudhuit C, Martres MP, Hamon M (1995) Chronic alcoholization alters the expression of 5-HT$_{1A}$ and 5-HT$_{1B}$ receptor subtypes in rat brain. Eur J Pharmacol 281:229–239

Oksenberg D, Peroutka SJ (1988) Antagonism of 5-hydroxytryptamine$_{1A}$ (5-HT$_{1A}$) receptor-mediated modulation of adenylate cyclase activity by pindolol and propranolol isomers. Biochem Pharmacol 37:3429–3433

Oleskevich S (1995) G$_{\alpha o1}$ decapeptide modulates the hippocampal 5-HT$_{1A}$ potassium current. J Neurophysiol 74:2189–2193

Pazos A, Palacios JM (1985) Quantitative autoradiographic mapping of serotonin receptors in the rat brain. I. Serotonin-1 receptors. Brain Res 346:205–230

Pedigo NW, Yamamura HI, Nelson DL (1981) Discrimination of multiple ^{3}H-5-hydroxytryptamine binding sites by the neuroleptic spiperone in rat brain. J Neurochem 36:220–226

Penington NJ, Kelly JS, Fow AP (1991) A study of the mechanism of Ca^{2+} current inhibition produced by serotonin in rat dorsal raphe neurons. J Neurosci 11:3594–3609

Pike VW, McCarron JA, Lammerstma AA, Hume SP, Poole K, Grasby PM, Malizia A, Cliffe IA, Fletcher A, Bench CJ (1995) First delineation of 5-HT$_{1A}$ receptors in human brain with PET and [^{11}C]WAY-100635. Eur J Pharmacol 283:R1-R3

Pohl M, Ballet S, Collin E, Mauborgne A, Bourgoin S, Benoliel JJ, Hamon M, Cesselin F (1997) Enkephalinergic and dynorphinergic neurons in the spinal cord and dorsal root ganglia of the polyarthritic rat – in vivo release and cDNA hybridization studies. Brain Res 749:18–28

Pompeiano M, Palacios JM, Mengod G (1992) Distribution and cellular localization of mRNA coding for 5-HT$_{1A}$ receptor in the rat brain: correlation with receptor binding. J Neurosci 12:440–453

Prehn JHM, Backhauß C, Karkoutly C, Nuglisch J, Peruche B, Roßberg C, Krieglstein J (1991) Neuroprotective properties of 5-HT$_{1A}$ receptor agonists in rodent models of focal and global cerebral ischemia. Eur J Pharmacol 203:213–222

Prisco S, Gagnotto A, Talone D, De Blasi A, Mennini T, Esposito E (1993) Tertatolol, a new β-blocker, is a serotonin (5-hydroxytryptamine$_{1A}$) receptor antagonist in rat brain. J Pharmacol Exp Ther 265:739–744

Radja F, Laporte AM, Daval G, Vergé D, Gozlan H, Hamon M (1991) Autoradiography of serotonin receptor subtypes in the central nervous system. Neurochem Int 18:1–15

Radja F, Daval G, Hamon M, Vergé D (1992) Pharmacological and physicochemical properties of pre- versus postsynaptic 5-hydroxytryptamine$_{1A}$ receptor binding sites in the rat brain: a quantitative autoradiographic study. J Neurochem 58:1338–1346

Raghupathi RK, Brousseau DA, McGonigle P (1996) Time-course of recovery of 5-HT$_{1A}$ receptors and changes in 5-HT$_{1A}$ receptor mRNA after irreversible inactivation with EEDQ. Mol Brain Res 38:233–242

Raymond JR (1991) Protein kinase C induces phosphorylation and desensitization of the human 5-HT$_{1A}$ receptor. J Biol Chem 266:14747–14753

Raymond JR, Albers FJ, Middleton JP (1992) Functional expression of human 5-HT$_{1A}$ receptors and differential coupling to second messengers in CHO cells. Naunyn-Schmiedebergs Arch Pharmacol 346:127–137

Raymond JR, Olsen CL, Gettys TW (1993) Cell-specific physical and functional coupling of human 5-HT$_{1A}$ receptors to inhibitory G protein α-subunits and lack of coupling to G$_{sα}$. Biochemistry 32:11064–11073

Riad M, El Mestikawy S, Vergé D, Gozlan H, Hamon M (1991) Visualization and quantification of central 5-HT$_{1A}$ receptors with specific antibodies. Neurochem Int 19:413–423

Riad M, Emerit MB, Hamon M (1994) Neurotrophic effects of ipsapirone and other 5-HT$_{1A}$ receptor agonists on septal cholinergic neurons in culture. Devl Brain Res 82:245–258

Riad M, Garcia S, Hamon M, Descarries L (1995) Somatodendritic localization of the serotonin 5-HT1A receptor in adult rat brain. Soc Neurosci 21:1368 (539.1)

Riad M, Jodoin N, Garcia S, Langlois X, Darmon M, Hamon M, Descarries L (1996) Axonal localization of the serotonin 5-HT1B receptor in brain. Soc Neurosci 22:1329 (528.2)

Ritter JM, Jones T (1995) PET: a symposium highlighting its clinical and pharmacological potential. Trends Pharmacol Sci 16:117–119

Rodriguez-Manzo G, Fernàndez-Guasti A (1994) Reversal of sexual exhaustion by serotonergic and noradrenergic agents. Behav Brain Res 62:127–134

Routledge C, Hartley J, Gurling J, Ashworth-Preece M, Brown G, Dourish CT (1994) In vivo characterization of the putative 5-HT$_{1A}$ receptor antagonist SDZ 216.525 using two models of somatodendritic 5-HT1A receptor function. Neuropharmacology 33:359–366

Schechter LE, Bolaños FJ, Gozlan H, Lanfumey L, Haj-Dahmane S, Laporte AM, Fattaccini CM, Hamon M (1990) Alterations of central serotoninergic and dopaminergic neurotransmission in rats chronically treated with ipsapirone – biochemical and electrophysiological studies. J Pharmacol Exp Ther 255:1335–1347

Schoeffter P, Fozard JR, Stoll A, Siegl H, Seiler MP, Hoyer D (1993) SDZ 216–525, a selective and potent 5-HT$_{1A}$ receptor antagonist. Eur J Pharmacol – Mol Pharmacol Sect 244:251–257

Schreiber R, Opitz K, Glaser T, De Vry J (1993) Ipsapirone and 8-OH-DPAT reduce ethanol preference in rats: involvement of presynaptic 5-HT$_{1A}$ receptors. Psychopharmacology 112:100–110

Sharp T, Backus LI, Hjorth S, Bramwell SR, Grahame-Smith DG (1990) Further investigation on the in vivo pharmacological properties of the putative 5-HT$_{1A}$ antagonist, BMY 7378. Eur J Pharmacol 176:331–340

Sharp T, McQuade R, Fozard JR, Hoyer D (1993) The novel 5-HT$_{1A}$ receptor antagonist, SDZ 216-525, decreases 5-HT release in rat hippocampus in vivo. Br J Pharmacol 109:699–702

Sotelo C, Cholley B, El Mestikawy S, Gozlan H, Hamon M (1990) Direct immunohistochemical evidence of the existence of 5-HT$_{1A}$ autoreceptors on serotoninergic neurons in the midbrain raphe nuclei. Eur J Neurosci 2:1144–1154

Sprouse JS (1991) Inhibition of dorsal raphe cell firing by MDL 73005EF, a novel 5-HT$_{1A}$ receptor ligand. Eur J Pharmacol 201:163–169

Stahl SM, Gastpar M, Keppel Hesselink JM, Traber J eds (1992) Serotonin 1A receptors in depression and anxiety, Raven, New York

Svensson L, Fahlke C, Hard E, Engel JA (1993) Involvement of the serotonergic system in ethanol intake in the rat. Alcohol 10:219–224

Thielen RJ, Fangon NB, Frazer A (1996) 4-(2'-Methoxyphenyl)-1-[2'-[N-(2''-pyridinyl)-p-iodobenzamido]ethyl]piperazine and 4-(2'-methoxyphenyl)-1-[2'-[N-(2''-pyridinyl)-p-fluorobenzamido]ethyl]piperazine, two new antagonists at pre- and postsynaptic serotonin-1A receptors. J Pharmacol Exp Ther 277:661–670

Trulson ME, Preussler DW, Gailyn A, Howell A, Frederickson CJ (1982) Raphe unit activity in freely moving cats: effects of benzodiazepines. Neuropharmacology 21:1045–1050

Van den Hoof P, Galvan M (1991) Electrophysiology of the 5-HT$_{1A}$ ligand MDL 73005EF in the rat hippocampal slice. Eur J Pharmacol 196:291–298

Varrault A, Le Nguyen D, McClue S, Harris B, Jouin P, Bockaert J (1994) 5-Hydroxytryptamine$_{1A}$ receptor synthetic peptides. Mechanisms of adenylyl cyclase inhibition. J Biol Chem 269:16720–16725

Vergé D, Daval G, Patey A, Gozlan H, El Mestikawy S, Hamon M (1985) Presynaptic 5-HT autoreceptors on serotonergic cell bodies and/or dendrites but not terminals are of the 5-HT$_{1A}$ subtype. Eur J Pharmacol 113:463–464

Vergé D, Daval G, Marcinkiewicz M, Patey A, El Mestikawy S, Gozlan H, Hamon M (1986) Quantitative autoradiography of multiple 5-HT$_1$ receptor subtypes in the brain of control or 5,7-dihydroxytryptamine-treated rats. J Neurosci 6:3474–3482

Wadenberg M-L (1996) Serotonergic mechanisms in neuroleptic-induced catalepsy in the rat. Neurosci Biobehav Rev 20:325–339

Wadenberg ML, Ahlenius S (1991) Antipsychotic-like profile of combined treatment with raclopride and 8-OH-DPAT in the rat: enhancement of antipsychotic-like effects without catalepsy. J Neural Transm 83:43–53

Weissmann-Nanopoulos D, Mach E, Magre J, Demassey Y, Pujol JF (1985) Evidence for the localization of 5-HT$_{1A}$ binding sites on serotonin containing neurons in the raphe dorsalis and raphe centralis nuclei of the rat brain. Neurochem Int 7:1061–1072

Whitaker-Azmitia PM (1991) Role of serotonin and other neurotransmitter receptors in brain development: basis for developmental pharmacology. Pharmacol Rev 43:553–561

Wieland S, Lucki I (1990) Antidepressant-like activity of 5-HT$_{1A}$ agonists measured with the forced swim test. Psychopharmacology 101:497–504

Wilkinson LO, Dourish CT (1991) Serotonin and animal behavior. In: Peroutka SJ (ed) Serotonin receptor subtypes: basic and clinical aspects. Wiley-Liss, New York, pp 147–210

Yocca FD, Hyslop DK, Smith DW, Maayani S (1987) BMY 7378, a buspirone analog with high affinity, selectivity and low intrinsic activity at the 5-HT$_{1A}$ receptor in rat and guinea pig hippocampal membranes. Eur J Pharmacol 137:293–294

Yocca FD, Iben L, Meller E (1992) Lack of apparent receptor reserve at postsynaptic 5-hydroxytryptamine$_{1A}$ receptors negatively coupled to adenylyl cyclase activity in rat hippocampus membranes. Mol Pharmacol 41:1066–1072

CHAPTER 10

Functional Neuropharmacology of Compounds Acting at 5-HT$_{1B/D}$ Receptors

M. Briley, P. Chopin, M. Marien, and C. Moret

A. Introduction

The neuropharmacology of serotonin (5-HT) has been fundamentally revised in recent years with the discovery of multiple serotonin receptor subtypes. Detailed descriptions of the pharmacology of the various receptor subtypes are to be found elsewhere in this book. We review here the available data on the in vitro and in vivo neuropharmacology of the 5-HT$_{1B/D}$ group of receptor subtypes and attempt to identify their possible functions and potential implications in psychiatric disorders.

The genes encoding the 5-HT$_{1B}$ receptor in rodents and the 5-HT$_{1D\beta}$ receptor in non-rodents, including humans, are species homologues (Hoyer and Middlemiss 1989; Hartig et al. 1992) with an overall homology of amino acid sequence of 92% (96% in the membrane-spanning domain). Although the pharmacological differences between these receptor subtypes are greater than usually seen between species homologues, the 5-HT$_{1B}$ and 5-HT$_{1D\beta}$ receptor subtypes have similar regional distributions (Waeber et al. 1989) and functions. Both subtypes are found in high densities in substantia nigra, globus pallidus, striatum and basal ganglia, where they function as 5-HT inhibitory autoreceptors on serotoninergic nerve terminals (Moret 1985; Engel et al. 1986; Middlemiss 1986, 1988; Schlicker et al. 1989; Starke et al. 1989; Hoyer et al. 1990) regulating the release of 5-HT. Other 5-HT$_{1B}$ or 5-HT$_{1D\beta}$ receptors are located, as heteroreceptors, on non-serotoninergic terminals while yet others are situated postsynaptically on the target neurons of 5-HT projections (Offord et al. 1988; Vergé et al. 1986).

Recently, Hartig et al. (1996) proposed a new classification of 5-HT$_{1B}$ and 5-HT$_{1D}$ receptor subtypes based on data from molecular biology. They suggest that the term 5-HT$_{1B}$ be used for the rodent 5-HT$_{1B}$ and human 5-HT$_{1D\beta}$, and that 5-HT$_{1D}$ should be used for the rodent 5-HT$_{1D\alpha}$ and human 5-HT$_{1D\beta}$. Species differences should be indicated by a prefix, e.g. "h" for human. Since our understanding of the function of the 5-HT$_{1D}$ and 5-HT$_{1D}$ receptors is still rudimentary, it is difficult to use the proposed nomenclature for functional studies. The terms 5-HT$_{1B}$ will thus be used to refer to the receptors in rodents and 5-HT$_{1D}$ for non-rodent receptors.

There are very few compounds that are selective for 5-HT$_{1B/D}$ receptors. Of the compounds having high affinity for the 5-HT$_{1B/D}$ receptors, 5-

carboxamidotryptamine (5-CT) has a tenfold greater affinity for the 5-HT_{1A} receptor, eltoprazine is equipotent on the 5-HT_{1B} and 5-HT_{1A} receptors, while trifluoromethylphenylpiperazine (TFMPP) is equipotent on the 5-HT_{1B} and 5-HT_{2C} receptors. RU 24969 (5-methoxy-3 (1,2,3,6-tetrahydropyridin-4-yl)-1*H*-indole) has some selectivity for 5-HT_{1B} receptors with a threefold higher affinity compared to 5-HT_{1A} receptors. *m*-Chlorophenylpiperazine (mCPP) is a non-selective 5-HT-receptor agonist with similar affinities for 5-HT_1, 5-HT_2 and 5-HT_3 receptors. Methiothepin (metitepine), which has come to be considered as a $5\text{-HT}_{1B/D}$ antagonist, has, in common with metergoline, similar affinities (in the 100 n*M* range) for most of the 5-HT_1- and 5-HT_2-receptor subtypes. In addition, methiothepin has low nanomolar affinity for the 5-HT_6 (MONSMA et al. 1993) and 5-HT_7 (TSOU et al. 1994; SLEIGHT et al. 1995) receptors. The β-adrenoceptor antagonists such as pindolol and cyanopindolol are equipotent at both 5-HT_{1B} and 5-HT_{1A} receptors.

Using such tools it is difficult to determine which receptor is responsible for their neurochemical or behavioural effects. It is only by combining data from various compounds that a tentative assignment can be made.

B. Biochemical Neuropharmacology

I. Regulation of Serotonin Release

1. In Vitro

Receptors of the 5-HT_{1B} subtype in rodents and the 5-HT_{1D} subtype in other species, including humans, have been shown to have an autoreceptor function regulating the release of 5-HT from brain slices in vitro. This subject is extensively reviewed elsewhere in this book.

Several studies have suggested that more than one type of 5-HT autoreceptor may exist. We (MORET and BRILEY 1986) have shown that whereas $0.1\,\mu M$ methiothepin was sufficient to shift the concentration-effect curve for the effect of lysergic acid diethylamide (LSD) on electrically evoked release of $[^3H]$5-HT from rat hypothalamic slices, $1\,\mu M$ methiothepin was required to antagonize the inhibitory effect of dihydroergocristine (DHEC). We suggested that LSD and DHEC might act on different subtypes of 5-HT autoreceptors, for which methiothepin had different affinities. In 1992, WILKINSON and MIDDLEMISS, using a similar approach in guinea pig hippocampal slices, found that the ability of methiothepin to antagonize the inhibitory effect of sumatriptan or 5-carboxamidotryptamine was less marked than for 5-HT. These authors also concluded a possible heterogeneity in the receptor mediating the inhibition of $[^3H]$5-HT release in guinea pig hippocampus. More recently, in 5-HT_{1B} "knock-out" mice, in which the gene for the 5-HT_{1B} receptor has been deleted, the relatively selective 5-HT_{1B} receptor agonist CP 93129 (3-(1,2,5,6-tetrahydropyrid-4-yl)pyrolol[3, 2, 6]pyrid-5-one) was no longer able to reduce the release of 5-HT from brain slices in vitro. 5-CT, however, still inhibited the electrically evoked release of $[^3H]$5-HT in frontal cortex and hippocampal slices. Since 5-CT binds with high affinity to multiple 5-HT receptors, the authors suggested that the inhibition by 5-CT may involve receptors other than the 5-HT$_{1B}$ (PINEYRO et al. 1995). 5-HT$_5$ and 5-HT$_7$ receptors are among the most likely candidates.

In vitro, fast cyclic voltammetry has been used to measure the modulation by 5-HT$_{1D}$ receptors of the release of 5-HT in slices of the guinea pig dorsal raphe nucleus. These studies demonstrated that agonists decreased, while antagonists by themselves had no effect, suggesting that although these receptors are functional they are not tonically activated (STARKEY and SKINGLE 1994; HUTSON et al. 1995) at least in these in vitro preparations.

2. In Vivo

The direct, in vivo, study of 5-HT autoreceptor function has become more accessible since the advent of intracerebral microdialysis, a technique which allows direct in vivo sampling and measurement of neurotransmitters and their metabolites in the extracellular fluid of the brain (UNGERSTEDT 1984; DI CHIARA 1990) of anaesthetized or freely moving animals (IMPERATO and DI CHIARA 1985; SHARP et al. 1986; IMPERATO et al. 1988).

a) Rat

RU 24969 has been reported to decrease extracellular 5-HT levels in the hippocampus of anaesthetized rat when administered systemically (SHARP et al. 1989a,b; MARTIN et al. 1992). A similar reduction of dialysate 5-HT has also been obtained with this compound in the frontal cortex of anaesthetized rats (BRAZELL et al. 1985; SLEIGHT et al. 1989) and in the diencephalon of awake rats (AUERBACH et al. 1991) following systemic administration of this compound. Since RU 24969 has significant affinity for 5-HT$_{1A}$ receptors, and 5-HT$_{1A}$ receptor agonists, such as 8-hydroxy-2-(di-n-propylamino)-tetralin (8-OH-DPAT), gepirone, ipsapirone and buspirone (SHARP et al. 1989a,b), produce similar effects, it is difficult to determine whether the effects of RU 24969 are due to stimulation of somatodendritic 5-HT$_{1A}$ receptors controlling the firing of raphe neurons (see elsewhere in this book for a detailed discussion of this point) or terminal release-controlling 5-HT$_{1B}$ receptors. However, unlike 5-HT$_{1A}$ agonists, local administration of RU 24969 onto 5-HT$_{1A}$ receptor-rich cell bodies in the raphe nucleus did not decrease the extracellular levels of the serotonin metabolite 5-hydroxyindole acetic acid (5-HIAA) (measured by voltammetry) (MARTIN and MARSDEN 1987; MARSDEN et al. 1987), suggesting that its effects on somatodendritic 5-HT$_{1A}$ receptors are weak. In addition, local administration of RU 24969 through the microdialysis probe via the perfusion medium directly into the terminal regions induced a decrease in extracellular 5-HT in the hippocampus of anaesthetized rat (HJORTH and TAO 1991; BOSKER et al. 1995) and in the diencephalon of awake rats, where locally applied TFMPP also decreased dialysate 5-HT levels (AUERBACH et al. 1991). It is worth noting that the effects of local administration of RU 24969 and TFMPP are only obtained in the presence of a 5-HT uptake inhibitor in the perfusion medium (Auerbach et al. 1991; Hjorth and Tao 1991). These compounds possess a non-negligible affinity for the 5-HT membrane carrier and are taken up into the nerve terminals where, by displacement, they induce the release of 5-HT, thus masking their autoreceptor-mediated inhibitory effects (Auerbach et al. 1991; Hjorth and Tao 1991). The effect of the local infusion of RU 24969 was attenuated by simultaneous infusion of the non-selective 5-HT autoreceptor antagonist methiothepin into the hippocampus (Martin et al. 1992).

Hjorth and Tao (1991) have shown that CP 93129, a 5-HT_{1B} receptor agonist with about a 100-fold higher affinity for 5-HT_{1B}- than for 5-HT_{1A}-binding sites (Koe et al. 1990; Macor et al. 1990), when administered via the dialysis perfusion medium, caused a reduction of 5-HT output in the hippocampus of anaesthetized rats. This effect was significantly antagonized by co-infusion of methiothepin. In contrast to RU 24969 and TFMPP, CP 93129 decreased dialysate 5-HT both in the presence and in the absence of a 5-HT uptake inhibitor, indicating that this compound is devoid of any action at the level of the 5-HT uptake site (Hjorth and Tao 1991). Thus CP 93129 is possibly the agonist of choice for studying 5-HT_{1B} receptors controlling 5-HT release from serotoninergic terminals in vivo. From the above results it would appear that 5-HT release in terminal areas in vivo is modulated by 5-HT_{1B} autoreceptors in the rat.

In the rat hypothalamus, methiothepin, when applied locally via the microdialysis probe, increased the extracellular levels of 5-HT both in the absence and in the presence of the 5-HT uptake inhibitor citalopram, suggesting that in awake animals 5-HT autoreceptors in the terminal projection areas are tonically activated and exert a potent inhibitory tone on the release of 5-HT (Moret and Briley 1996).

b) Guinea Pig

In guinea pigs, rabbits, pigs and humans, terminal serotoninergic autoreceptors are of the 5-HT_{1D} receptor subtype (Hoyer and Middlemiss 1989; Schlicker et al. 1989; Limberger et al. 1991; Galzin et al. 1992; Maura et al. 1993). Local (via microdialysis probe) administration of the $5\text{-HT}_{1A/B}$ agonist 5-CT into the frontal cortex of the freely moving guinea pig decreased extracellular 5-HT levels (Lawrence and Marsden 1992). Sumatriptan, when added to the perfusion medium, similarly reduced extracellular levels of 5-HT in the frontal cortex of anaesthetized guinea pigs (Sleight et al. 1990) presumably via an activation of 5-HT_{1D} autoreceptors. In guinea pig hypothalamus, local, through the probe, administration of naratriptan, which is somewhat more selective for 5-HT_{1D} compared to 5-HT_{1A} receptors, also decreased the extracellular levels of 5-HT (Fig. 1), an effect which was attenuated by the non-selective 5-HT antagonist methiothepin (Fig. 1) at $1\,\mu M$, a concentration which did not modify by itself the outflow of 5-HT (Moret, unpublished results).

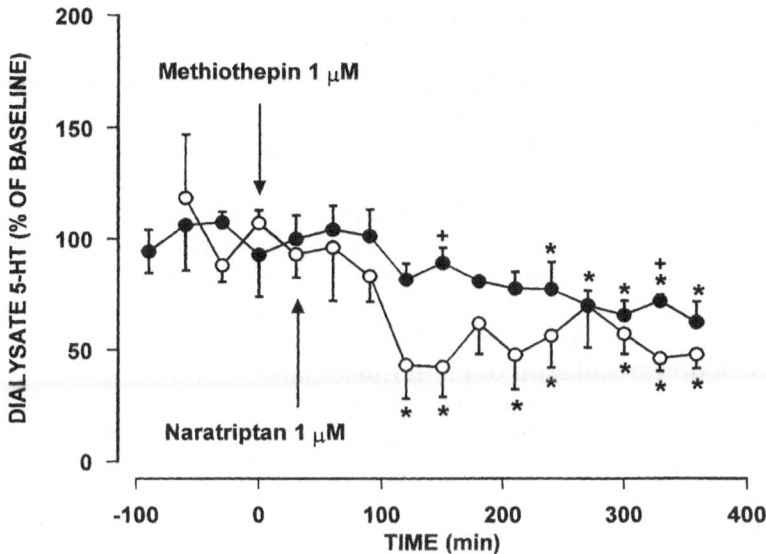

Fig. 1. Time course of microdialysis experiments studying the effect of naratriptan (1 μM) and its attenuation by methiothepin (1 μM) on extracellular levels of 5-HT, expressed as percentage of basal values, in the hypothalamus of freely moving guinea pigs. ○ Naratriptan, dissolved in Ringer's solution, was perfused from 30 min (*second arrow*) until the end of the experiment. ● Methiothepin, dissolved in Ringer's solution, was perfused (*first arrow*) from zero time, 30 min before the addition of naratriptan; both drugs were then present until the end of the experiment. Each point represents mean values ± SEM of four to five animals. *$P < 0.05$ in comparison to corresponding basal values (ANOVA followed by Dunnett's test). +$P < 0.05$ when compared to the respective value of naratriptan alone (two-way ANOVA for repeated measures followed by Tukey's test)

In vitro methiothepin has been shown to increase, by itself, the electrically evoked release of [^{3}H]5-HT from slices of rat hypothalamus (LANGER and MORET 1982) and guinea pig hypothalamus and substantia nigra (Moret, unpublished data). In freely moving guinea pig, methiothepin increases, in a concentration-dependent (10–100 μM) manner, the extracellular levels of 5-HT when applied through the probe into the hypothalamus (Moret, unpublished data). This suggests that, as in the rat (see above), 5-HT autoreceptors in the hypothalamus are tonically activated in the freely moving guinea pig and exert a potent inhibitory tone on the release of 5-HT. When this inhibition is removed by an antagonist, such as methiothepin, there is a major increase in 5-HT release. A similar effect has been found with methiothepin in guinea pig substantia nigra (BRILEY and MORET 1993).

3. Effects of Repeated Antidepressant Administration on Autoreceptor Sensitivity

Administration of citalopram (50 mg/kg p.o.) to rats for 21 days followed by a washout of 24 h resulted in an increased in vitro stimulation-induced release of

5-HT from hypothalamic slices preloaded with [³H]5-HT (Moret and Briley 1990). In addition, the concentration-effect curve of the agonist, LSD, was significantly shifted to the right compared with control animals, indicating a desensitization of the autoreceptor for the agonist. We suggested that repeated administration of the selective 5-HT reuptake inhibitor resulted in a decreased efficacy of the terminal autoreceptor, allowing an increased release of 5-HT (Moret and Briley 1990). A similar result has been obtained in rat hypothalamic slices following a long-term administration of amitriptyline (Schoups and De Potter 1988) once daily for 21 days at 10 mg/kg i.p. with 17 h withdrawal. Electrically evoked release of [³H]5-HT was increased and the modulatory effect of 5-HT itself and methiothepin was completely abolished, suggesting a downregulation of the terminal autoreceptor.

By using fast cyclic voltammetry, O'Connor and Kruk (1994) have shown that a repeated administration of the selective 5-HT reuptake inhibitor fluoxetine (5 mg/kg i.p.) for 21 days with 24 h washout resulted in an enhancement of electrically stimulated 5-HT overflow from brain slices containing suprachiasmatic nucleus and a significant shift to the right in the concentration-response curve for the 5-HT$_{1B}$ receptor agonist RU 24969 in comparison to brain slices from control (vehicle-treated) animals, again demonstrating a downregulation of the terminal 5-HT$_{1B}$ autoreceptors in this brain region.

In guinea pigs, Blier and Bouchard (1994) have found that the electrically induced release of [³H]5-HT was increased by a chronic treatment with paroxetine, a 5-HT reuptake inhibitor (14 days with 48 h withdrawal), in slices of hypothalamus, hippocampus and frontal cortex, and that the inhibitory effect of the non-selective 5-HT agonist 5-methoxytryptamine was attenuated in the hypothalamus and the hippocampus. Thus the terminal 5-HT$_{1D}$ autoreceptor also appears to be desensitized in the guinea pig after a long-term blockade of 5-HT uptake.

These findings are particularly interesting in the context of depressive illness. In spite of the rapid onset of uptake blockade in humans (within a few hours in platelets), the earliest signs of therapeutic improvement in depressive symptoms appear only after 2 weeks of treatment with selective 5-HT uptake inhibitors (for a general review of selective 5-HT uptake inhibitors, see Feighner and Boyer 1991). The latency of the therapeutic effects has been attributed to the need for adaptive changes to be brought about by long-term treatment (for review, see Briley and Moret 1993). One of these adaptive changes may be the desensitization of the terminal 5-HT$_{1B/D}$ autoreceptor as described above, with the subsequent rise of synaptic levels of 5-HT and the stimulation of one or more postsynaptic receptors, which is thought to be an essential long-term action of these antidepressants.

Recently, we (Moret and Briley 1996) attempted to demonstrate the increase in synaptic 5-HT using in vivo microdialysis on freely moving rats after a chronic administration of citalopram under exactly the same conditions as in the previous in vitro study (Moret and Briley 1990). Somewhat unexpectedly, no change was seen in the basal extracellular levels of endogenous 5-HT in chronic drug-treated animals. In addition, the enhancing effect of methiothepin, administered through the microdialysis probe, was similar in both control and chronically treated animals. These results suggest that under the conditions of this study, repeated administration of citalopram followed by a washout of 24 h does not lead to a desensitization of the terminal 5-HT autoreceptor of sufficient magnitude for it to be measured in vivo, in contrast to the effects shown in vitro. At present no clear explanation exists for the discrepancy between in vitro and in vivo findings. Methodological differences (slices preloaded with [^{3}H]5-HT in vitro compared to endogenous 5-HT in vivo; electrically stimulated release in vitro compared to basal release in vivo, etc.) may be important. It is possible, however, that in vivo other regulatory mechanisms come into play such as those controlling the firing rate of the raphe nucleus and the synthesis of 5-HT, both of which are under the control of the somatodendritic 5-HT$_{1A}$ autoreceptors. In the in vitro slice preparation where axon terminals have been physically separated from the cell bodies, these latter influences are absent. Thus it may not be possible to distinguish a change in the sensitivity of the terminal autoreceptor by studying the levels of extracellular 5-HT if there are concomitant changes in other parts of the system.

In contrast, when studied without washout, extracellular levels of 5-HT were increased by both acute and repeated citalopram administration (MORET and BRILEY 1996). In rats treated chronically without washout, methiothepin (administered locally via the probe) had a greater maximal effect on 5-HT outflow than in rats receiving acute citalopram treatment. This study shows that a 5-HT uptake inhibitor and an autoreceptor antagonist are both capable of increasing extracellular levels of 5-HT. Furthermore these two effects are additive or possibly synergistic, suggesting that a terminal 5-HT autoreceptor antagonist or a combination of such a drug with a 5-HT uptake inhibitor would produce a greater increase in extracellular levels of 5-HT in hypo-serotoninergic states and thus be potentially useful in the treatment of depressive disorders resistant to therapy by a single drug.

II. Regulation of 5-HT Synthesis

It is well established that activation of the somatodendritic 5-HT$_{1A}$ autoreceptors decrease 5-HT neuronal firing and, in turn, the synthesis, metabolism and release of the transmitter. Recently, however, HJORTH et al. (1995) showed, in the rat, that the 5-HT$_{1B/2C}$ receptor agonist TFMPP suppresses 5-HT synthesis in vivo as estimated by the accumulation of 5-hydroxytryptophan (5-HTP) after the inhibition of the amino acid decarboxylase, i.e. an index of tryptophan hydroxylation (CARLSSON et al. 1972). This suppression, which was evident in terminal projection areas such as the limbic forebrain and striatum, was also observed in axotomized animals, indicating that it was independent of neuronal firing. Furthermore, a similar inhibitory effect of TFMPP on 5-HT synthesis was found in vitro in slice preparations in the presence of depolarizing concentrations of potassium. In vitro the effect of TFMPP was attenuated by the non-selective 5-HT receptor antagonist methiothepin as well as the 5-HT$_{1B}$ receptor antagonists propranolol or cyanopindolol. By comparison, the decrease in 5-HT synthesis in forebrain regions induced in vivo by 8-OH-DPAT was prevented by transection of the brain. In addition, 8-OH-DPAT did not decrease 5-HT synthesis in vitro. These data thus suggest that the reduction of rat brain 5-HT synthesis by TFMPP is mediated by 5-HT autoreceptors located on the serotoninergic axon terminals, and that this is a direct effect and independent of 5-HT neuronal firing.

This finding confirms a number of earlier suggestions. Using a similar ex vivo protocol, MORET and BRILEY (1993) found that methiothepin administered systemically increased the synthesis of 5-HT in rat brain, whereas the 5-HT$_{1A}$/5-HT$_2$ receptor antagonist spiperone had no effect, suggesting that 5-HT$_{1B}$ autoreceptors may be involved in the modulation of 5-HT synthesis.

In guinea pig brain, systemic administration of GR 127935 (N-[4-methoxy-3-(4-methyl-1-piperazinyl)phenyl]-2′-methyl-4′-(5-methyl-1,2,4-oxadiazol-3-yl) [1, 1-biphenyl]-4-carboxamide), a reportedly selective 5-HT$_{1D}$ receptor antagonist (ROBERTS et al. 1994), increases 5-HT synthesis in the frontal cortex (maximum effect 233% of control). Smaller, non-significant increases (30%–50%) in 5-HT synthesis have also been found in other regions such as the hypothalamus, hippocampus and substantia nigra following systemic administration of similar doses of GR 127935 (Moret, unpublished data).

Taken together, these findings observed in both rat and guinea pig strongly suggest that the terminal 5-HT$_{1B/D}$ autoreceptors can play a role in the regulation of 5-HT synthesis. The future availability of specific agonists and antagonists for these receptors will help clarify and characterize their involvement in the control of this key step of 5-HT neurotransmission.

C. Behavioural Neuropharmacology

I. Locomotor Activity

The mixed 5-HT$_{1B/A}$ agonist RU 24969 dose dependently increases the spontaneous locomotor activity of rats and mice (GREEN et al. 1984). This effect is antagonized by the non-selective 5-HT$_{1B/A}$ antagonists (−)-pindolol and (−)-propranolol, but not by other 5-HT receptor antagonists, such as metergoline or methysergide (TRICKLEBANK et al. 1986; GREEN et al. 1984). CP 94253 (−(1,2,5,6-tetrahydro-4-pyridyl)-5-propoxypyrrolo[3, 2-b]pyridine), a 5-HT$_1$ receptor agonist exhibiting a greater affinity at 5-HT$_{1B}$ receptors than at 5-HT$_{1A}$ and 5-HT$_{2C}$ receptors, also produces hyperlocomotion in rats (KOE et al. 1992). In contrast, 5-HT$_{1A}$ receptor agonists (MITTMAN and GEYER 1989) and 5-HT$_{2A/2C}$ receptor agonists (WING et al. 1990) significantly reduce locomotor activity in rodents. The 5-HT$_{1B/2C}$ receptor agonists TFMPP and m-CPP also reduce the locomotor activity in rats (LUCKI and FRAZER 1982). This effect is blocked by antagonists with high affinity for 5-HT$_{2C}$ sites while antagonists at 5-HT$_{1A}$, 5-HT$_{1B}$, 5-HT$_{2A}$ and 5-HT$_3$ sites are without effect (KENNETT and CURZON 1988a), suggesting that the hypolocomotion induced by TFMPP and mCPP is mediated by 5-HT$_{2C}$ receptors and not by 5-HT$_{1B}$ receptors. However, TFMPP and m-CPP, like RU 24969, increase the spontaneous locomotor activity of mice previously isolated for one week (Chopin, unpublished results), a condition which is known to sensitize 5-HT$_{1B}$ receptors (FRANCÈS and MONIER 1991a) and would thus allow more complete expression of 5-HT$_{1B}$-receptor-mediated effects. Destruction of serotoninergic neurons with the specific neurotoxin 5,7-dihydroxytryptamine (5,7-DHT) potentiates the hyperactivity-inducing effects of RU 24969 (TRICKLEBANK et al. 1986), suggesting the involvement of receptors that are postsynaptic to serotoninergic terminals.

MDMA (3,4-methylenedioxymethamphetamine, "ecstasy"), a potent 5-HT-releasing agent, produces an increase in locomotor activity of rats that is antagonized by (−)-pindolol, (−)-propranolol (CALLAWAY et al. 1992) and the non-selective 5-HT receptor antagonist methiothepin, whereas the 5-HT$_{2A/C}$ receptor antagonists methysergide, cyproheptadine and ritanserin are ineffective (CALLAWAY et al. 1992). Chronic administration of RU 24969 attenuates the locomotor response to MDMA, suggesting 5-HT$_{1B}$ receptor subsensitivity (DE SOUZA et al. 1986; OBERLANDER et al. 1987). Chronic administration of MDMA also produces a tolerance to the hyperlocomotion induced by both acute MDMA and RU 24969 (CALLAWAY and GEYER 1992). The specificity of this cross-tolerance has been confirmed by the fact that chronic treatment with the 5-HT$_{1A}$ receptor agonist 8-OH-DPAT or with the 5-HT$_{2A/C}$ receptor agonist DOI does not reduce the hyperactivity induced by MDMA (CALLAWAY and GEYER 1992). These results are consistent with the MDMA-induced hyperactivity being mediated by stimulation of 5-HT$_{1B}$ receptors via the release of endogenous 5-HT (REMPEL et al. 1993).

II. Feeding Behaviour

Non-selective 5-HT receptor agonists and drugs that indirectly increase the synaptic availability of 5-HT, such as the 5-HT precursor 5-hydroxytryptophan (5-HTP), the 5-HT-releasing agent fenfluramine or the 5-HT uptake inhibitors fluoxetine and sertraline reduce feeding behaviour (for review, see CHOPIN et al. 1994). The 5-HT$_{1A/1B}$ antagonist cyanopindolol blocks the fenfluramine-induced decrease in eating rate, meal size and total food intake in freely feeding rats (GRIGNASCHI and SAMANIN 1992), suggesting that 5-HT$_1$ receptors are involved in the ability of fenfluramine to cause satiety. In contrast, 5-HT$_{1A}$ receptor agonists enhance food consumption in rats, whereas 5-HT$_{2A/C}$ receptor agonists inhibit it (for review, see CHOPIN et al. 1994).

The involvement of 5-HT$_{1B}$ receptors in the regulation of food intake is suggested by the ability of CP 94253 to inhibit food intake and decrease weight gain in rats (Koe et al. 1992). RU 24969, m-CPP and TFMPP also reduce food intake in both food-deprived and non-deprived rats (Garattini et al. 1989).

Pretreatment with the 5-HT$_{2A}$ receptor antagonists ritanserin and ketanserin, or with the 5-HT$_{1A/2A}$ receptor antagonist spiperone does not block anorexia induced by TFMPP, m-CPP or RU 24969, suggesting that 5-HT$_{2A}$ receptors are not involved (Kennett et al. 1987; Kennett and Curzon 1988a,b). The inhibitory effects on feeding behaviour of RU 24969 are blocked by pretreatment with the 5-HT$_{1A/1B}$ receptor antagonists, (−)pindolol, (−)propranolol and cyanopindolol, but not with the 5-HT$_{2A/C}$ receptor antagonists mesulergine or mianserin, suggesting an implication of 5-HT$_{1B}$ but not 5-HT$_{2C}$ receptors (Kennett and Curzon 1988b). RU 24969-induced hypophagia appears to be independent of endogenous levels of 5-HT since p-chlorophenylalanine (pCPA), a depletor of brain 5-HT, did not prevent the RU 24969-induced hypophagia (Kennett et al. 1987).

III. Sexual Behaviour

In general, 5-HT itself (Finberg and Vardi 1990) and drugs that increase the synaptic availability of 5-HT, such as its precursor 5-hydroxytryptophan (5-HTP) (Fernandez-Guasti and Rodriguez-Manzo 1992), have an inhibitory action on rat male sexual behaviour. The inhibitory effect of 5-HT on penile erectile function in the rat is antagonized by non-selective 5-HT receptor antagonists such as methysergide and methiothepin, but not by the 5-HT$_{2A}$ receptor antagonist ketanserin, or the 5-HT$_3$ receptor antagonist MDL 72222 (3-tropanyl-3,5-dichlorobenzoate) (Finberg and Vardi 1990), suggesting that 5-HT$_{2A}$ and 5-HT$_3$ receptors are not involved in penile erectile function. m-CPP produces dose-dependent penile erection (Bagdy et al. 1992), which is antagonized by the 5-HT$_{2A/2C}$ receptor antagonist mianserin but not by the selective 5-HT$_{2A}$ receptor antagonist ketanserin (Bagdy et al. 1992). The 5-HT$_{2A/C}$ receptor agonist DOI (1-(2,5-dimethoxy-4-indophenyl)-2-aminopropane) has no effect on penile erection, although, after ketanserin pretreatment, DOI is capable of inducing penile erection (Bagdy et al. 1992). Taken together these data suggest that penile erection may be induced by stimulation of 5-HT$_{2C}$ receptors.

RU 24969 and TFMPP inhibit sexual behaviour in male rats (Gorzalka et al. 1990) and potentiate the inhibitory effects of 5-HTP (Fernandez-Guasti and Rodriguez-Manzo 1992), suggesting an inhibitory role of 5-HT$_{1B}$ receptors on sexual behaviour in male rats. Moreover, the 5-HT synthesis inhibitor (already defined above) pCPA or the 5-HT neurotoxin 5,7-DHT facilitate sexual behaviour (copulation) of male rats without interfering with the inhibitory effects of TFMPP (Fernandez-Guasti and Escalante 1991). These results suggest that 5-HT postsynaptic receptors mediate the inhibitory action of TFMPP on copulation and further support the idea that endogenous 5-HT acts via the stimulation of 5-HT$_{1B}$ receptors to induce its inhibitory effects on male sexual behaviour.

IV. Aggressive Behaviour

Most, but not all, studies suggest that 5-HT inhibits aggression, more particularly predatory aggression, such as muricide in rats (OLIVIER and MOS 1992). Tryptophan-deficient diets facilitate muricidal behaviour in rats, whereas a supplement of tryptophan decreases this predatory aggression (GIBBONS et al. 1979). Similarly, fluvoxamine (OLIVIER et al. 1989) and zimelidine (ÖGREN et al. 1980), 5-HT uptake inhibitors that increase the synaptic availability of 5-HT, have been found to decrease aggressive behaviour. However, the destruction of 5-HT neurons with 5,7-DHT results in a reduction of offensive behaviour in a resident-intruder paradigm (SIJBESMA et al. 1991) and the inhibition of 5-HT synthesis by pCPA decreases isolation-induced aggression (MALICK and BARNETT 1976), suggesting that 5-HT might also have a stimulatory rather than an inhibitory influence on aggressive behaviour in certain situations.

5-HT$_1$ receptor agonists tend to inhibit aggressive behaviour (MCMILLEN et al. 1988; OLIVIER et al. 1989), whereas 5-HT$_{2A}$ and 5-HT$_3$ receptors do not seem to be involved in mediating aggressive behaviour (SÁNCHEZ et al. 1993). Isolation-induced offensive behaviour in male mice, resident-intruder offensive aggression and maternal aggression in rats are blocked by RU 24969, TFMPP and eltoprazine and also by 5-HT$_{1A}$ receptor agonists (OLIVIER and MOS 1992; SÁNCHEZ et al. 1993) although the effects of 5-HT$_{1A}$ receptor ligands on aggressive behaviour do not seem to be specific (for review, see CHOPIN et al. 1994). Furthermore, TFMPP and eltoprazine administered directly into the lateral ventricle of male rats block resident-intruder aggression without inducing sedation, whereas 8-OH-DPAT administered in the same way modifies neither aggression nor other behaviours (MOS et al. 1992). In addition, eltoprazine reduces resident-intruder aggression in both sham and 5,7-DHT-lesioned rats without inducing a decrease in exploration and social interest (SIJBESMA et al. 1991), suggesting that postsynaptic 5-HT$_{1B}$ receptors may specifically modulate aggressive behaviour.

It has also been shown that RU 24969, TFMPP and eltoprazine reduce aggression in muricidal rats as well as in isolation-induced aggression in mice (SÁNCHEZ et al. 1993). Both TFMPP and eltoprazine are 5-HT$_{1B}$ receptor agonists; however, the fact that TFMPP is also a 5-HT$_{2C}$ receptor agonist and eltoprazine a 5-HT$_{2C}$ receptor antagonist (OLIVIER and MOS 1992; RODGERS et al. 1992) suggests that 5-HT$_{2C}$ receptors are not involved in the regulation of aggression.

Interestingly the main behavioural characteristic of 5-HT$_{1B}$ "knock-out" mice, in which the gene for the 5-HT$_{1B}$ receptor has been specifically deleted, is aggression (see Chap. 15, this volume).

V. Isolation-Induced Social Behavioural Deficit in Mice

It has been shown that mice isolated for 1 week display a social behavioural deficit in relation to their motivation to escape from a Perspex cylinder when tested in the presence of either another isolated mouse or a grouped mouse (FRANCÈS 1988). The isolation-induced social behavioural deficit is reversed by RU 24969, TFMPP, m-CPP (FRANCÈS 1988), CGS 12066B (FRANCÈS and MONIER 1991a), CP 93129 and eltoprazine (Chopin, unpublished results). In contrast, 5-HT$_{1A}$ receptor agonists accentuate the behavioural deficit as well as decreasing exploration in the open field test (FRANCÈS 1988; FRANCÈS et al. 1990a). The putative 5-HT$_{1B/D}$ (but also 5-HT$_{1A}$) receptor agonist 5-CT also accentuates the deficit, whereas the 5-HT$_{2A/C}$ receptor agonist DOI does not (Chopin, unpublished results). Thus the pattern of agonist effects is consistent with the inhibition of this behavioural response by stimulation of 5-HT$_{1B}$ receptors.

Penbutolol, a 5-HT$_1$ receptor antagonist, dose-dependently antagonizes the effects of RU 24969, TFMPP, m-CPP (FRANCÈS et al. 1990b), CGS 12066B or eltoprazine (Chopin, unpublished results). The effect of anpirtoline, another 5-HT$_{1B}$ receptor agonist which reverses the social behavioural deficit induced by isolation, is also prevented by pretreatment with propranolol (SCHLICKER et al. 1992). In contrast, neither the 5-HT$_{2A}$ receptor antagonist ritanserin, the 5-HT$_3$ receptor antagonist ICS 205–930 nor the 5-HT$_{2A/2C}$ receptor antagonists mianserin and cyproheptadine modify the effects of TFMPP (FRANCÈS et al. 1990b). Thus the isolation-induced social behavioural deficit appears to be reversed by the stimulation of 5-HT$_1$ and more precisely 5-HT$_{1B}$ receptors, suggesting that this animal model may be useful for screening of 5-HT$_{1B}$ receptor ligands in vivo.

5-HT$_{1B}$ receptor agonists are inactive in grouped mice (FRANCÈS and MONIER 1991a), suggesting that isolated mice are more responsive than grouped mice and that isolation may lead to an increase in 5-HT$_{1B}$ receptor sensitivity. Repeated administration for 3 days with RU 24969 or CGS 12066B reduces the acute effects of RU 24969, TFMPP, m-CPP or CGS 12066B (FRANCÈS and MONIER 1991b) possibly through a decrease in the sensitivity of the 5-HT$_{1B}$ receptors. This could be therapeutically relevant, since chronic but not acute administration of antidepressant drugs also reverses the effects of TFMPP on isolation-induced social behavioural deficit (FRANCÈS and MONIER 1991b). The 5-HT uptake inhibitor fluoxetine or the monoamine oxidase inhibitor phenelzine, for example, are inactive on the isolation-induced deficit and do not modify the effects of TFMPP when given acutely, whereas they attenuate these effects after chronic administration (FRANCÈS and KHIDICHIAN 1990), suggesting a downregulation of 5-HT$_{1B}$ receptors resulting from the increased synaptic availability of 5-HT.

Finally, lesions of 5-HT nerve terminals by intracerebroventricular infusion of the specific serotoninergic neurotoxin 5,7-DHT does not modify the effects of TFMPP on isolation-induced deficit (FRANCÈS and MONIER 1991b), suggesting that the 5-HT$_{1B}$ receptors involved in social behavioural deficit are located postsynaptically.

VI. Animal Models of Anxiety

In general, compounds that decrease 5-HT neurotransmission tend to decrease the level of anxiety, whereas those that increase 5-HT stimulation tend to increase the level of anxiety (CHOPIN and BRILEY 1987; BRILEY and CHOPIN 1991, 1994). 5-HT$_{1A}$ receptor agonists, which reduce serotoninergic activity, have been found to be anxiolytic in a variety of animal models and in man, whereas 5-HT$_{2A}$ and 5-HT$_{2C}$ receptor agonists showed anxiogenic-like effects (CHOPIN and BRILEY 1987; BRILEY and CHOPIN 1991). Various agonists that exhibit a certain selectivity for 5-HT$_{1B}$ receptors, such as RU 24969, TFMPP and eltoprazine, show anxiogenic-like activity in animal models such as the shock probe conflict procedure (MEERT and COLPAERT 1986), the social interaction test and the elevated plus-maze test both in rats (PELLOW et al. 1987) and in mice (BENJAMIN et al. 1990; RODGERS et al. 1992), although results in conflict tests in rodents are more variable (DEACON and GARDNER 1986).

VII. Animal Models of Depression

5-HT is involved in the physiological processes of feeding, sleep, sexual behaviour, mood, vigilance and learning, all of which are modified to varying extents in human depression. However, the involvement of precise 5-HT receptor subtypes in depression, and in the action of antidepressant drugs, is still far from clear. The 5-HT$_{1A}$ receptor agonists 8-OH-DPAT, buspirone, ipsapirone and gepirone have been found active in certain behavioural tests in animals that are predictive of antidepressant effects such as the forced swimming test (WIELAND and LUCKI 1990) and the learned helplessness paradigm (GIRAL et al. 1988; MARTIN et al. 1990).

There is, however, little information available on the possible effects of direct 5-HT$_{1B}$ receptor stimulation. In the learned helplessness (LH) paradigm, rats that are exposed to uncontrollable footshocks fail to learn an escape response, such as a lever press in an operant cage or to move into another compartment in a shuttle box apparatus, whereas rats exposed to controllable shocks are able to acquire these escape responses. In this model, exposure to uncontrollable stress produces a number of signs seen in depressed patients such as weight loss, changes in sleep pattern, decreased locomotion as well as performance deficits in learning tasks (SHERMAN et al. 1979). These symptoms and specifically the performance deficits are reduced by a variety of clinically effective antidepressant drugs (MARTIN et al. 1990). In addition, the density of serotonin uptake sites as measured by ^{3}H-imipramine binding (in the cortex) (SHERMAN and PETTY 1984) or ^{3}H-paroxetine binding (in the hypothalamus) (EDWARDS et al. 1991) are reduced in LH rats as compared to controls. This closely resembles the clinical situation, where there have been numerous reports (for review see, for example, Briley 1985) of decreases in serotonin uptake sites labelled by ^{3}H-imipramine (and other more selective ligands) in platelets of depressed patients (Briley et al. 1980) and in the postmortem brain of suicide victims (Stanley et al. 1982).

In LH rats, 5-HT$_{1B}$ receptors are upregulated (increased receptor binding) in the cortex, hippocampus and septum and downregulated in the hypothalamus (Edwards et al. 1991). These results suggest that a change in 5-HT$_{1B}$ receptor responsiveness might be related to the escape deficit. 5-HT release measured in vivo by microdialysis in the cortex of LH rats is decreased (Petty et al. 1992), which is compatible with 5-HT$_{1B}$ autoreceptors being upregulated.

The 5-HT$_{1B}$ receptor agonist CGS 12066B does not modify the escape deficit in the learned helplessness paradigm, but it does reduce the ability of the 5-HT reuptake blockers citalopram and fluvoxamine to reverse the behavioural deficit resulting from uncontrollable shocks (Martin and Puech 1991), suggesting that 5-HT$_{1B}$ receptor agonists, by their inhibition of 5-HT release, might reduce the stimulation of 5-HT transmission induced by 5-HT reuptake blockers. Finally, a recent study has shown that methiothepin, a nonselective antagonist at the 5-HT$_{1B/1D}$ autoreceptor (Moret 1985), exhibited antidepressant-like activity in the olfactory bulbectomized rat model of depression (McNamara et al. 1995).

D. Clinical Neuropharmacology

The problem of drug selectivity is compounded in man by the limited number of investigational compounds available. Both neuroendocrine and behavioural challenge studies of 5-HT$_{1B/D}$ receptors have been essentially based on the use of sumatriptan. While sumatriptan is relatively selective for the 5-HT$_{1B/D}$ receptor, it is reputed to penetrate the blood-brain barrier only to a limited extent. The most specific neuroendocrine response induced in humans by the administration of sumatriptan is a major (greater than fivefold) increase in plasma levels of growth hormone (Franceschini et al. 1994; Herdman et al. 1994). This effect, which is prevented by prior administration of the non-selective 5-HT$_1$ receptor antagonist cyproheptadine (Franceschini et al. 1994), has been suggested to result from an inhibition of the release of somatostatin via 5-HT$_{1B}$ heteroreceptors (Mota et al. 1995).

The administration of the non-selective 5-HT agonist (already defined above) (mCPP) to untreated patients suffering from obsessive compulsive disorder (OCD) causes a marked and transient exacerbation of their symptoms (Zohar et al. 1987; Hollander et al. 1992), whereas the administration to healthy volunteers does not, in general, induce OCD symptomatology. This effect, which can be prevented by pretreatment with the non-selective 5-HT receptor antagonist metergoline (Pigott et al. 1991), has been suggested to result from stimulation of 5-HT receptors that are supersensitive in OCD patients. This idea has found support in the observation that the effects of mCPP are blunted in patients whose OCD has been successfully treated with clomipramine (ZOHAR et al. 1988), which presumably normalizes these super-sensitive receptors. mCPP has high affinity for 5-HT$_{1A}$, 5-HT$_{1D}$ and 5-HT$_{2C}$ receptors. Since more selective serotoninergic agonists, such as ipsapirone (5-HT$_{1A}$) and MK-212 (2-chloro-6-(1-piperazinyl) pyrazine) (5-HT$_{2C}$) (BASTANI et al. 1990), do not produce the exacerbation of OCD symptoms, it would appear that the receptors involved are probably of the 5-HT$_{1D}$ subtype. In addition, since OCD symptoms are unaltered by modification of synaptic 5-HT levels, through tryptophan loading (CHARNEY et al. 1988) or depletion (PIGOTT et al. 1993), the receptors involved are not 5-HT release-controlling autoreceptors but more probably postsynaptically located 5-HT$_{1D}$ receptors. The putative implication of 5-HT$_{1D}$ receptors has been recently tested by administering sumatriptan to OCD patients who reacted with a marked and transient aggravation of their OCD symptomatology but not their anxiety (DOLBERG et al. 1995). These studies suggest that supersensitive 5-HT$_{1D}$ receptors may indeed be involved in the pathophysiology of OCD and may represent a potential target for its treatment.

E. Conclusion

In spite of the lack of selective ligands, a careful analysis of the available data suggests the probable involvement of 5-HT$_{1B/D}$ receptors in the control of a certain number of behavioural activities. Feeding and male sexual activity are regulated by a large number of 5-HT receptor subtypes. Aggressive behaviour appears, however, to be more specifically mediated via 5-HT$_{1B}$ receptors in contrast to the antiaggressive effects produced by 5-HT$_{1A}$ receptor stimulation which result from general hypoactivity. Interestingly, all of the behavioural effects elicited by administration of 5-HT$_{1B}$ receptor agonists appear to be mediated through postsynaptic 5-HT$_{1B}$ receptors rather than release-controlling 5-HT$_{1B}$ autoreceptors.

5-HT$_{1B}$ receptor agonists have no effect on the escape behaviour of non-isolated mice, whereas they inhibit the social deficit in isolated animals. This suggests that isolation, which is often thought to produce a depressive-like state, increases 5-HT$_{1B}$ receptor sensitivity. The induction of the depressive-like state of learned helplessness also induces 5-HT$_{1B}$ receptor supersensitivity, which has been measured directly by binding studies (EDWARDS et al. 1991) in various regions. An increase in 5-HT$_{1B}$ autoreceptor sensitivity is consistent with the decreased release of 5-HT from learned helpless rats observed by microdialysis in the cortex (PETTY et al. 1992). Increased serotoninergic activity appears to be associated with an increased level of anxiety (CHOPIN and BRILEY 1987; BRILEY and CHOPIN 1994). The frequent co-existence of high levels of anxiety with depression, a supposedly hyposerotoninergic state, is nevertheless difficult to explain. In the case, however, of supersensitivity of both pre- and postsynaptic 5-HT$_{1B}$ receptors, a decreased release of 5-HT resulting from an increased autoinhibition and an increased level of anxiety resulting from activation of supersensitive postsynaptic 5-HT$_{1B}$ receptors would be expected. This corresponds to the situation found in the learned helplessness model of depression where pre- and postsynaptic 5-HT$_{1B}$ receptors appear to be supersensitive. Interestingly, rats exposed to repeated inescapable shocks, such as the "learned helpless rats", not only show a number of "depressive" signs (Sherman et al. 1979), but also exhibit behaviour associated with high levels of anxiety (Vandijken et al. 1992a,b). To date there is no indication as to whether 5-HT$_{1D}$ receptors are supersensitive in anxious or depressed patients, but the above animal data make this an attractive working hypothesis. Repeated administration of 5-HT uptake blocking antidepressants would be expected to desensitize both pre- and postsynaptic 5-HT$_{1D}$ receptors. Desensitization of presynaptic 5-HT$_{1B}$ autoreceptors has already been demonstrated in vitro in rats following chronic treatment with citalopram (Moret and Briley 1990). If pre- and postsynaptic 5-HT$_{1D}$ receptors are supersensitive in depression and anxiety, direct antagonism of 5-HT$_{1D}$ receptors may well produce both antidepressant and anxiolytic effects more rapidly than with selective 5-HT uptake blocking agents.

A totally independent line of reasoning has led Zohar and co-workers (Dolberg et al. 1995) to a similar conclusion in OCD, where 5-HT$_{1D}$ receptors (probably postsynaptic) appear to be supersensitive and their desensitization through long-term administration of 5-HT reuptake inhibitors is essential for a therapeutic effect. As in depression a significant gain in the delay of onset of action can be envisaged by the use of specific 5-HT$_{1D}$ receptor antagonists to treat OCD.

In conclusion, 5-HT$_{1B/D}$ receptors clearly play an important role in a number of basic behavioural activities in animals. Several lines of evidence suggest that changes in the sensitivity of pre- and/or postsynaptic 5-HT$_{1D}$ receptors may be fundamental to several psychiatric disorders. Further investigation into 5-HT$_{1B/D}$ receptor function in psychopathology would appear to be potentially rewarding. In addition the development of new selective agonists and antagonists could be important in both therapy and research.

References

Auerbach SB, Rutter JJ, Juliano PJ (1991) Substituted piperazine and indole compounds increase extracellular serotonin in rat diencephalon as determined by in vivo microdialysis. Neuropharmacology 30:307–311
Bagdy G, Kalogeras KT, Szemeredi K (1992) Effect of 5-HT$_{1C}$ and 5-HT$_2$ receptor stimulation on excessive grooming, penile erection and plasma oxytocin concentrations. Eur J Pharmacol 229:9–14
Bastani B, Nash F, Meltzer H (1990) Prolactin and cortisol responses to MK-212, a serotonin agonist, in obsessive-compulsive disorder. Arch Gen Psychiatry 47:946–951

Benjamin D, Lal H, Meyerson LR (1990) The effects of 5-HT$_{1B}$ characterizing agents in the mouse elevated plus-maze. Life Sci 47:195–203

Blier P, Bouchard C (1994) Modulation of 5-HT release in the guinea pig brain following long-term administration of antidepressant drugs. Br J Pharmacol 113:485–495

Bosker FJ, Van Esseveldt KE, Klompmakers AA, Westenberg HGM (1995) Chronic treatment with fluvoxamine by osmotic minipumps fails to induce persistent functional changes in central 5-HT$_{1A}$ and 5-HT$_{1B}$ receptors, as measured by in vivo microdialysis in dorsal hippocampus of conscious rats. Psychopharmacology 117:358–363

Brazell MP, Marsden CA, Nisbet AP, Routledge C (1985) The 5-HT$_1$ receptor agonist RU-24969 decreases 5-hydroxytryptamine (5-HT) release and metabolism in the rat frontal cortex in vitro and in vivo. Br J Pharmacol 86:209–216

Briley M (1985) Imipramine binding: its relationship with serotonin uptake and depression. In: Green R (ed) Neuropharmacology of serotonin. Oxford University Press, Oxford, pp 50–78

Briley M, Chopin P (1991) Serotonin in anxiety. Evidence from animal models. In: Sandler M, Coppen A, Harnett S (eds) 5-Hydroxytryptamine in psychiatry: a spectrum of ideas. Oxford University Press, pp 177–197

Briley M, Chopin P (1994) Is anxiety associated with a hyper- or hypo-serotonergic state? In: Palomo T, Archer T (eds) Strategies for studying brain disorders, vol 1: Depression, anxiety and drug abuse disorders. Editorial Complutence, Donoso Cortés, Madrid, pp 197–209

Briley M, Moret C (1993) Neurobiological mechanisms involved in antidepressant therapies. Clin Neuropharmacol 16:387–400

Briley M, Langer SZ, Raisman R, Sechter D, Zarifian E (1980) ^{3}H-imipramine binding sites are decreased in platelets of untreated depressed patients. Science 209:303–305

Callaway CW, Geyer MA (1992) Tolerance and cross-tolerance to the activating effects of 3,4-methylenedioxymethamphetamine and a 5-hydroxytryptamine$_{1B}$ agonist. J Pharmacol Exp Ther 263: 318–326

Callaway CW, Rempel N, Peng RY, Geyer MA (1992) Serotonin 5-HT(1)-like receptors mediate hyperactivity in rats induced by 3,4-methylenedioxymethamphetamine. Neuropsychopharmacology 7:113–127

Carlsson A, Davis JN, Kehr W, Lindqvist M, Atack CV (1972) Simultaneous measurement of tyrosine and tryptophan hydroxylase activities in brain in vivo using an inhibitor of the aromatic amino acid decarboxylase. Naunyn Schmiedebergs Arch Pharmacol 275:153–168

Charney DS, Goodman WK, Price LH, Woods SW, Rasmussen SA, Heninger GR (1988) Serotonin function in obsessive-compulsive disorder. Arch Gen Psychiatry 45:177–185

Chopin P, Briley M (1987) Animal models of anxiety: the effect of compounds that modify 5-HT neurotransmission. Trends Pharmacol Sci 8:383–388

Chopin P, Moret C, Briley M (1994) Neuropharmacology of 5-hydroxytryptamine$_{1B/D}$ receptor ligands. Pharmacol Ther 62:385–405

Deacon R, Gardner CR (1986) Benzodiazepine and 5-HT ligands in a rat conflict test. Br J Pharmacol 88:330P

De Souza RJ, Goodwin GM, Green AR, Heal DJ (1986) Effect of chronic treatment with 5-HT$_1$ agonist (8-OH-DPAT and RU 24969) and antagonist (ipsapirone) drugs on the behavioural responses of mice to 5-HT$_1$ and 5-HT$_2$ agonists. Br J Pharmacol 89:377–384

Di Chiara G (1990) In vivo brain dialysis of neurotransmitters. Trends Pharmacol Sci 11:116–121

Dolberg OT, Sasson Y, Cohen R, Zohar J (1995) The relevance of behavioral probes in obsessive-compulsive disorder. Eur Neuropsychopharmacol 5:161–162

Edwards E, Harkins K, Wright G, Henn FA (1991) 5-HT$_{1B}$ receptors in an animal model of depression. Neuropharmacology 30:101–105

Engel G, Göthert M, Hoyer D, Schlicker E, Hillenbrand K (1986) Identity of inhibitory presynaptic 5-hydroxytryptamine (5-HT) autoreceptors in the rat brain cortex with 5-HT$_{1B}$ binding sites. Naunyn Schmiedebergs Arch Pharmacol 332:1–7

Feighner JP, Boyer WF (1991) Selective serotonin reuptake inhibitors. Wiley, Chichester

Fernandez-Guasti A, Escalante A (1991) Role of presynaptic serotonergic receptors on the mechanism of action of 5-HT$_{1A}$ and 5-HT$_{1B}$ agonists on masculine sexual behaviour: physiological and pharmacological implications. J Neural Trans 85:95–107

Fernandez-Guasti A, Rodriguez-Manzo G (1992) Further evidence showing that the inhibitory action of serotonin on rat masculine sexual behavior is mediated after the stimulation of 5-HT$_{1B}$ receptor. Pharmacol Biochem Behav 42:529–533

Finberg JP, Vardi Y (1990) Inhibitory effect of 5-hydroxytryptamine on penile erectile function in the rat. Br J Pharmacol 101:698–702

Franceschini R, Cataldi A, Garibaldi A, Cianciosi P, Scordamaglia A, Barreca T, Rolandi E (1994) The effects of sumatriptan on pituitary secretion in man. Neuropharmacology 33:235–239

Francès H (1988) New animal model of social behavioural deficit: reversal by drugs. Pharmacol Biochem Behav 29:467–470

Francès H, Khidichian F (1990) Chronic but not acute antidepressants interfere with serotonin (5-HT$_{1B}$) receptors. Eur J Pharmacol 179:173–176

Francès H, Monier C (1991a) Isolation increases a behavioral response to the selective 5-HT$_{1B}$ agonist CGS 12066B. Pharmacol Biochem Behav 40:279–281

Francès H, Monier C (1991b) Tolerance to the effect of serotonergic (5-HT$_{1B}$) agonists in the isolation-induced social behavioral deficit test. Neuropharmacology 30:623–627

Francès H, Khidichian F, Monier C (1990a) Benzodiazepines impair a behavioral effect induced by stimulation of 5-HT$_{1B}$ receptors. Pharmacol Biochem Behav 35:841–845

Francès H, Khidichian F, Monier C (1990b) Increase in the isolation-induced social behavioral deficit by agonists at 5-HT$_{1A}$ receptors. Neuropharmacology 29:103–107

Galzin AM, Poirier MF, Lista A, Chodkiewicz JP, Blier P, Ramdine R, Loo H, Roux FX, Redondo A, Langer SZ (1992) Characterization of the 5-hydroxytryptamine receptor modulating the release of 5-[^{3}H]hydroxytryptamine in slices of the human neocortex. J Neurochem 59:1293–1301

Garattini S, Mennini T, Samanin R (1989) Reduction of food intake by manipulation of central serotonin. Current experimental results. Br J Psychiatry 8:41–51

Gibbons JL, Barr GA, Bridger WH, Leibowitz SF (1979) Manipulations of dietary tryptophan: effects on mouse killing and brain serotonin in the rat. Brain Res 169:139–153

Giral P, Martin P, Soubrié P, Simon P (1988) Reversal of helpless behavior in rats by putative 5-HT$_{1A}$ agonists. Biol Psychiatry 23:237–242

Gorzalka BB, Mendelson SD, Watson NV (1990) Serotonin receptor subtypes and sexual behavior. In: Whitaker-Azmitia PM, Peroutka SJ (eds) The neuropharmacology of serotonin. New York Academy of Sciences, New York, pp 435–446

Green AR, Guy AP, Gardner CR (1984) The behavioural effects of RU 24969, a suggested 5-HT$_1$ receptor agonist in rodents and the effect on the behaviour of treatments with antidepressants. Neuropharmacology 23:655–661

Grignaschi G, Samanin R (1992) Role of 5-HT receptors in the effect of d-fenfluramine on feeding patterns in the rat. Eur J Pharmacol 212:287–289

Hartig PR, Branchek TA, Weinshank RL (1992) A subfamily of 5-HT$_{1D}$ receptor genes. Trends Pharmacol Sci 13:152–159

Hartig PR, Hoyer D, Humphrey PPA, Martin GR (1996) Alignment of receptor nomenclature with the human genome: classification of 5-HT$_{1B}$ and 5-HT$_{1D}$ receptor subtypes. Trends Pharmacol Sci 17:103–105

Herdman JRE, Delva NJ, Hockney RE, Campling GM, Cowen PJ (1994) Neuroendocrine effects of sumatriptan. Psychopharmacology 113:561–564

Hjorth S, Tao R (1991) The putative 5-HT$_{1B}$ receptor agonist CP 93129 suppresses rat hippocampal 5-HT release in vivo – comparison with RU 24969. Eur J Pharmacol 209:249–252

Hjorth S, Suchowski CS, Galloway MP (1995) Evidence for 5-HT autoreceptor-mediated, nerve impulse-independent, control of 5-HT synthesis in the rat brain. Synapse 19:170–176

Hollander E, DeCaria CM, Nitescu A (1992) Serotonergic function in obsessive-compulsive disorder. Arch Gen Psychiatry 49:21–28

Hoyer D, Middlemiss DN (1989) Species differences in the pharmacology of terminal 5-HT autoreceptors in mammalian brain. Trends Pharmacol Sci 10:130–132

Hoyer D, Schoeffter P, Waeber C, Palacios JM (1990) Serotonin 5-HT$_{1D}$ receptors. Ann N Y Acad Sci 600:168–182

Hutson PH, Bristow LJ, Cunningham JR, Hogg JE, Longmore J, Murray F, Pearce D, Razzaque Z, Saywell K, Tricklebank MD, Young L (1995) The effects of GR127935, a putative 5-HT$_{1D}$ receptor antagonist, on brain 5-HT metabolism, extracellular 5-HT concentration and behaviour in the guinea pig. Neuropharmacology 34:383–392

Imperato A, Di Chiara G (1985) Dopamine release and metabolism in awake rats after systemic neuroleptics as studied by trans-striatal dialysis. J Neurosci 5:297–306

Imperato A, Tanda G, Frau R, Di Chiara G (1988) Pharmacological profile of dopamine receptor agonists as studied by brain dialysis in behaving rats. J Pharmacol Exp Ther 245:257–264

Kennett GA, Curzon G (1988a) Evidence that hypophagia induced by mCPP and TFMPP requires 5-HT$_{1C}$ and 5-HT$_{1B}$ receptors; hypophagia induced by RU 24969 only requires 5-HT$_{1B}$ receptors. Psychopharmacology 96:93–100

Kennett GA, Curzon G (1988b) Evidence that mCPP may have behavioural effects mediated by central 5-HT$_{1C}$ receptors. Br J Pharmacol 94:137–144

Kennett GA, Dourish CT, Curzon G (1987) 5-HT$_{1B}$ agonists induce anorexia at a post synaptic site. Eur J Pharmacol 141:429–435

Koe BK, Lebel LA, Burkhart CA, Macor JE (1990) CP-93,129, a new serotonergic ligand with marked affinity and high selectivity for 5-HT$_{1B}$ receptors. Soc Neurosci Abstr 16:1035

Koe BK, Nielsen JA, Macor JE, Heym J (1992) Biochemical and behavioral studies of the 5-HT$_{1B}$ receptor agonist, CP-94,253. Drug Dev Res 26:241–250

Langer SZ, Moret C (1982) Citalopram antagonizes the stimulation by lysergic acid diethylamide of presynaptic inhibitory serotonin autoreceptors in the rat hypothalamus. J Pharmacol Exp Ther 222:220–226

Lawrence AJ, Marsden CA (1992) Terminal autoreceptor control of 5-hydroxytryptamine release as measured by in vivo microdialysis in the conscious guinea pig. J Neurochem 58:142–146

Limberger N, Deicher R, Starke K (1991) Species differences in presynaptic serotonin autoreceptors: mainly 5-HT$_{1B}$ but possibly in addition 5-HT$_{1D}$ in the rat, 5-HT$_{1D}$ in the rabbit and guinea pig cortex. Naunyn Schmiedebergs Arch Pharmacol 343:353–364

Lucki I, Frazer A (1982) Behavioural effects of indole and piperazine-type serotonin receptor agonists. Soc Neurosci Abstr 8:101

Macor JE, Burkhart CA, Heym JH, Ives JL, Lebel LA, Newman ME, Nielsen JA, Rya B (1990) 3-(1,2,5,6-Tetrahydropyrid-4-yl)pyrrolo<3,2-b>pyrid-5-one: a potent and selective serotonin (5-HT$_{1B}$) agonist and rotationally restricted phenolic analogue of 5-methoxy-3-(1,2,5,6-tetrahydropyrid-4-yl)indole. J Med Chem 33: 2087–2093

Malick JB, Barnett A (1976) The role of serotonergic pathways in isolation-induced aggression in mice. Pharmacol Biochem Behav 5:55–61

Marsden CA, Martin KF, Brazell MP, Maidment NT (1987) In vivo voltammetry. Application to the identification of dopamine and 5-hydroxytryptamine receptors. In: Justice JB (ed) Voltammetry in the Neurosciences. Humana, Clifton, pp 209–237

Martin KF, Marsden CA (1987) In vivo voltammetry in the suprachiasmatic nucleus of the rat: effects of RU 24969, methiothepin and ketanserin. Eur J Pharmacol 121:135–139

Martin KF, Hannon S, Phillips I, Heal DJ (1992) Opposing roles for 5-HT$_{1B}$ and 5-HT$_3$ receptors in the control of 5-HT release in rat hippocampus in vivo. Br J Pharmacol 106:139–142

Martin P, Puech AJ (1991) Is there a relationship between 5-HT$_{1B}$ receptors and the mechanisms of action of antidepressant drugs in the learned helplessness paradigm in rats? Eur J Pharmacol 192:193–196

Martin P, Beninger RJ, Hamon M, Puech AJ (1990) Antidepressant-like action of 8-OH-DPAT, a 5-HT$_{1A}$ agonist, in the learned helplessness paradigm: evidence for a postsynaptic mechanism. Behav Brain Res 38:135–144

Maura G, Thellung S, Andrioli GC, Ruelle A, Raiteri M (1993) Release-regulating serotonin 5-HT$_{1D}$ autoreceptors in human cerebral cortex. J Neurochem 60:1179–1182

McMillen BA, Da Vanzo EA, Scott SM, Song AH (1988) N-Alkyl-substituted aryl-piperazine drugs: relationship between affinity for serotonin receptors and inhibition of aggression. Drug Dev Res 12:53–62

McNamara MG, Kelly JP, Leonard BE (1995) Some behavioural effects of methiothepin in the olfactory bulbectomised rat model of depression. Med Sci Res 23:583–585

Meert TF, Colpaert FC (1986) The shock probe conflict procedure. A new assay responsive to benzodiazepines, barbiturates and related compounds. Psychopharmacology 88:445–450

Middlemiss DN (1986) Blockade of the central 5-HT autoreceptor by β-adrenoceptor antagonists. Eur J Pharmacol 120:51–54

Middlemiss DN (1988) Autoreceptors regulating serotonin release. In: Sanders-Bush E (ed) The serotonin receptors. Humana, Clifton, pp 210–224

Mittman SM, Geyer MA (1989) Effects of 5-HT$_{1A}$ agonists on locomotor and investigatory behaviors in rats differ from those of hallucinogens. Psychopharmacology 98:321–329

Monsma FJ, Yong Shen JR, Ward RP, Hamblin MW, Sibley DR (1993) Cloning and expression of a novel serotonin receptor with high affinity for tricyclic psychotropic drugs. Mol Pharmacol 43:320–327

Moret C (1985) Pharmacology of the serotonin autoreceptor. In: Green AR (ed) Neuropharmacology of serotonin. Oxford University Press, Oxford, pp 21–49

Moret C, Briley M (1986) Dihydroergocristine-induced stimulation of the 5-HT autoreceptor in the hypothalamus of the rat. Neuropharmacology 25:169–174

Moret C, Briley M (1990) Serotonin autoreceptor subsensitivity and antidepressant activity. Eur J Pharmacol 180:351–356

Moret C, Briley M (1993) Which 5-HT receptors are involved in the modulation of 5-HT synthesis by the 5-HT uptake blocker, citalopram? Br J Pharmacol 108:95P

Moret C, Briley M (1996) Effects of acute and repeated administration of citalopram on extracellular levels of serotonin in rat brain. Eur J Pharmacol 295:189–197

Mos J, Olivier B, Poth M, Van Aken H (1992) The effects of intraventricular administration of eltoprazine, 1-(3-trifluoromethylphenyl)piperazine hydrochloride, and 8-hydroxy-2-(di-n-propylamino)tetralin on resident intruder aggression in the rat. Eur J Pharmacol 212:295–298

Mota A, Bento A, Peñalva A, Pombo M, Dieguez C (1995) Role of the serotonin receptor subtype 5-HT$_{1D}$ on basal and stimulated growth hormone secretion. J Clin Endocrinol Metab 80:1973–1977

Oberlander C, Demassey Y, Verdu A, van de Velde D, Bardelay C (1987) Tolerance to the serotonin 5-HT$_1$ agonist RU 24969 and effects on dopaminergic behaviour. Eur J Pharmacol 139:205–214

O'Connor JJ, Kruk ZL (1994) Effects of 21 days treatment with fluoxetine on stimulated endogenous 5-hydroxytryptamine overflow in the rat dorsal raphe and suprachiasmatic nucleus studied using fast cyclic voltammetry in vitro. Brain Res 640:328–335

Offord SJ, Ordway GA, Frazer A (1988) Application of [^{125}I]iodocyanopindolol to measure 5-hydroxytryptamine (1B) receptors in the brain of the rat. J Pharmacol Exp Ther 244:144–153

Ögren S-O, Holm A-C, Renyi AL, Ross SB (1980) Anti-aggressive effect of zimelidine in isolated mice. Acta Pharmacol Toxicol 47:71–74

Olivier B, Mos J (1992) Rodent models of aggressive behavior and serotonergic drugs. Prog Neuropsychopharmacol Biol Psychiatry 16:847–870

Olivier B, Mos J, van der Heyden JAM, Hartog J (1989) Serotonergic modulation of social interactions in isolated male mice. Psychopharmacology 97:154–156

Pellow S, Johnston AL, File SE (1987) Selective agonists and antagonists for 5-hydroxytryptamine receptor subtypes, and interactions with yohimbine and FG 7142 using the elevated plus-maze test in the rat. J Pharm Pharmacol 39:917–928

Petty F, Kramer G, Wilson L (1992) Prevention of learned helplessness – in vivo correlation with cortical serotonin. Pharmacol Biochem Behav 43:361–367

Pigott TA, Zohar J, Hill JL (1991) Metergoline blocks the behavioral and neuroendocrine effects of orally administered m-chlorophenylpiperazine in patients with obsessive-compulsive disorder. Biol Psychiatry 29:418–426

Pigott TA, Murphy DL, Brooks A (1993) Pharmacological probes in OCD: support for selective 5-HT dysregulation. Presented at the 1st international obsessive compulsive disorder congress, Capri, Italy, 12–13 March

Piñeyro G, Castanon N, Hen R, Blier P (1995) Regulation of 5-HT release in 5-HT$_{1B}$ knock-out mice: experiments in hippocampal, frontal cortex and midbrain raphe slices. Soc Neurosci Abstr 21:1368

Rempel N, Callaway CW, Geyer MA (1993) Serotonin-1B receptor activation mimics behavioural effects of presynaptic serotonin release. Neuropsychopharmacology 8:201–211

Roberts C, Thorn L, Price GW, Middlemiss DN, Jones BJ (1994) Effect of the selective 5-HT$_{1D}$ receptor antagonist, GR 127935, on in vivo 5-HT release, synthesis and turnover in the guinea pig frontal cortex. Br J Pharmacol 112:488P

Rodgers RJ, Cole JC, Cobain MR, Daly P, Doran PJ, Eells JR, Wallis P (1992) Anxiogenic-like effects of fluprazine and eltoprazine in the mouse elevated plus-maze: profile comparisons with 8-OH-DPAT, CGS 12066B, TFMPP and mCPP. Behav Pharmacol 3:621–634

Sánchez C, Arnt J, Hyttel J, Moltzen EK (1993) The role of serotonergic mechanisms in inhibition of isolation-induced aggression in male mice. Psychopharmacology 110:53–59

Schlicker E, Fink K, Göthert M, Hoyer D, Molderings G, Roschke I, Schoeffter P (1989) The pharmacological properties of the presynaptic serotonin autoreceptor in the pig brain cortex conform to the 5-HT$_{1D}$ receptor subtype. Naunyn Schmiedebergs Arch Pharmacol 340:45–51

Schlicker E, Werner U, Hamon M, Gozlan H, Nickel B, Szelenyi I, Göthert M (1992) Anpirtoline, a novel, highly potent 5-HT$_{1B}$ receptor agonist with antinociceptive/antidepressant-like actions in rodents. Br J Pharmacol 105:732–738

Schoups AA, De Potter WP (1988) Species dependence of adaptations at the pre- and postsynaptic serotonergic receptors following long-term antidepressant drug treatment. Biochem Pharmacol 37:4451–4460

Sharp T, Ljungberg T, Zetterström T, Ungerstedt U (1986) Intracerebral dialysis coupled to a novel activity box – a method to monitor dopamine release during behaviour. Pharmacol Biochem Behav 24:1755–1759

Sharp T, Bramwell SR, Hjorth S, Grahame-Smith DG (1989a) Pharmacological characterization of 8-OH-DPAT-induced inhibition of rat hippocampal 5-HT release in vivo as measured by microdialysis. Br J Pharmacol 98:989–997

Sharp T, Bramwell ST, Grahame-Smith DG (1989b) 5-HT$_1$ agonists reduce 5-hydroxytryptamine release in rat hippocampus in vivo as determined by brain microdialysis. Br J Pharmacol 96:283–290

Sherman AD, Petty F (1984) Learned helplessness decreases ^{3}H-imipramine binding in rat cortex. J Affect Disord 6:25–32

Sherman AD, Allers GL, Petty F, Henn FA (1979) A neuropharmacologically relevant animal model of depression. Neuropharmacology 18:891–894

Sijbesma H, Schipper J, De Kloet ER, Mos J, Van Aken H, Olivier B (1991) Postsynaptic 5-HT$_1$ receptors and offensive aggression in rats: a combined behavioural and autoradiographic study with eltoprazine. Pharmacol Biochem Behav 38:447–458

Sleight AJ, Smith RJ, Marsden CA, Palfreyman MG (1989) The effects of chronic treatment with amitriptyline and MDL 72394 on the control of 5-HT release in vivo. Neuropharmacology 28:477–480

Sleight AJ, Cervenka A, Peroutka SJ (1990) In vivo effects of sumatriptan (GR-43175) on extracellular levels of 5-HT in the guinea pig. Neuropharmacology 29:511–513

Sleight AJ, Carolo C, Petit N, Zwingelstein C, Bourson A (1995) Identification of 5-hydroxytryptamine$_7$ receptor binding sites in rat hypothalamus: sensitivity to chronic antidepressant treatment. Mol Pharmacol 47:99–103

Stanley M, Vigilio J, Gershon S (1982) Tritiated imipramine binding sites are decreased in the frontal cortex of suicides. Science 216:1337–1339

Starke K, Göthert M, Kilbinger H (1989) Modulation of neurotransmitter release by presynaptic autoreceptors. Physiol Rev 69:864–989

Starkey SJ, Skingle M (1994) 5-HT$_{1D}$ as well as 5-HT$_{1A}$ autoreceptors modulate 5-HT release in the guinea pig dorsal raphé nucleus. Neuropharmacology 33:393–402

Tricklebank MD, Middlemiss DN, Neill J (1986) Pharmacological analysis of the behavioural and thermoregulatory effects of the putative 5-HT$_1$ receptor agonist, RU 24969, in the rat. Neuropharmacology 25:877–886

Tsou A, Kosaka A, Bach C, Zuppan P, Yee C, Tom L, Alvarez R, Ramsey S, Bonhaus DW, Stefanich E, Jakeman L, Eglen RM, Chan HW (1994) Cloning and expression of a 5-hydroxytryptamine$_7$ receptor positively coupled to adenylyl cyclase. J Neurochem 63:456–464

Ungerstedt U (1984) Measurement of neurotransmitter release by intracranial dialysis. In: Marsden CA (ed) Measurement of neurotansmitter release in vivo. Wiley, New York, pp 81–105

Vandijken HH, Mos J, van der Heyden JAM, Tilders FJH (1992a) Characterization of stress-induced long-term behavioural changes in rats – evidence in favor of anxiety. Physiol Behav 52:945–951

Vandijken HH, van der Heyden JAM, Mos J, Tilders FJH (1992b) Inescapable footshocks induce progressive and long-lasting behavioural changes in male rats. Physiol Behav 51:787–794

Vergé D, Daval G, Marcinkiewicz M, Patey A, El Mestikawy S, Gozlan S, Hamon M (1986) Quantitative autoradiography of multiple 5-HT$_1$ receptor subtypes in the brain of control or 5,7-dihydroxytryptamine-treated rats. J Neurosci 6:3474–3482

Waeber C, Dietl MM, Hoyer D, Palacios JM (1989) 5-HT$_1$ receptors in the vertebrate brain: regional distribution examined by autoradiography. Naunyn Schmiedebergs Arch Pharmacol 340:486–494

Wieland S, Lucki I (1990) Antidepressant-like activity of 5-HT$_{1A}$ agonists measured with the forced swim test. Psychopharmacology 101:497–504

Wilkinson LO, Middlemiss DN (1992) Metitepine distinguishes two receptors mediating inhibition of [^{3}H]-5-hydroxytryptamine release in guinea pig hippocampus. Naunyn Schmiedebergs Arch Pharmacol 345:696–699

Wing LL, Tapson GS, Geyer MA (1990) 5-HT$_2$ mediation of acute behavioral effects of hallucinogens in rats. Psychopharmacology 100:417–425

Zohar J, Mueller EA, Insel TR (1987) Serotonin responsivity in obsessive compulsive disorder. Arch Gen Psychiatry 44:946–951

Zohar J, Insel TR, Zohar-Kadoush RC, Hill JL, Murphy DL (1988) Serotonergic responsivity in obsessive-compulsive disorder: effects of clomipramine treatment. Arch Gen Psychiatry 45:167–172

CHAPTER 11

5-HT-Moduline, an Endogenous Peptide Modulating Serotoninergic Activity via a Direct Interaction at 5-HT$_{1B/1D}$ Receptors

G. Fillion, O. Massot, J.C. Rousselle, M.P. Fillion, and N. Prudhomme

A. Introduction

The modulatory role of the serotoninergic system on central nervous system (CNS) activity is now well accepted (Jacobs and Fornal 1991; Jacobs and Azmitia 1992; Chap. 1, this volume) and the fact that it is implicated in numerous physiological events is in good agreement with this functional role. It is interesting to recall a few points which underline the importance of particular characteristics of the 5-HT system which are adapted to its modulatory function. First of all, this system is very centralized since all cellular bodies are located in the raphe area and, secondly, its projections are present in almost all areas of the brain (Baumgarten and Grozdanovic 1994). Interestingly also, it should be noted that the serotoninergic system uses two main types of neurons: one corresponds to a classical neuron having a cellular body, a single axon and a few terminals (varicosities) mainly originating from the median raphe and a second one centralized in the dorsal raphe corresponds to arborescent neurons presenting a very large number of varicosities (up to 500000 varicosities for a single neuron between raphe and cortex as described by Audet et al. 1989). These varicosities likely play an important role in the regulatory function of the serotoninergic system. Indeed, it can be considered that the serotoninergic system is an "oscillator" which delivers with a certain frequency (the frequency of discharge of the neuron) a signal of a certain amplitude (the amine released from the terminals). The principle of the oscillator corresponds to one of the best-adapted devices in the control of various functions and is largely applied in the domain of physics. The frequency of the serotoninergic oscillator varies from zero during paradoxical sleep to a few pulses per second during quiet waking and has a maximal rate of 5–7/s during active waking as reported by Jacobs and Fornal (1991) in the cat. The signal delivered by the oscillator is also characterized by its amplitude, which corresponds to the amount of the released 5-HT. Although various non-serotoninergic neurotransmissions may participate in the regulation of the functional parameters of the 5-HT oscillator, 5-HT autoreceptors have been shown to be crucially involved in the activity of the serotoninergic system: 5-HT$_{1A}$ autoreceptors located on soma and dendrites of the raphe serotoninergic neurons modulate the frequency of discharge of these neurons (Aghajanian 1978; Sprouse and Aghajanian 1987) whereas 5-HT$_{1B/1D}$ autoreceptors lo-

Table 1. Isolation procedure of 5-HT-moduline. Brain crude extract was prepared and purified by several chromatographic steps as described previously (ROUSSELLE et al. 1996). At each step of purification the collected fractions were tested for their ability to displace the binding of [³H]5-HT (30 nM) to its 5-HT$_{1nonA}$ receptors on rat brain cortical membranes. The elution or retention time of the active fraction was indicated at each step of the isolation procedure

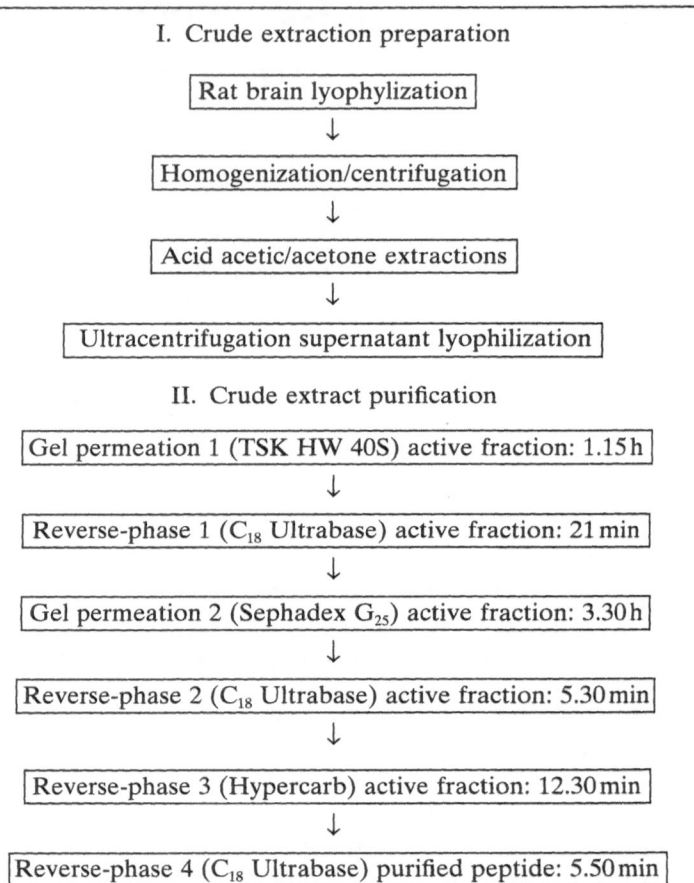

I. Crude extraction preparation

Rat brain lyophylization
↓
Homogenization/centrifugation
↓
Acid acetic/acetone extractions
↓
Ultracentrifugation supernatant lyophilization

II. Crude extract purification

Gel permeation 1 (TSK HW 40S) active fraction: 1.15 h
↓
Reverse-phase 1 (C$_{18}$ Ultrabase) active fraction: 21 min
↓
Gel permeation 2 (Sephadex G$_{25}$) active fraction: 3.30 h
↓
Reverse-phase 2 (C$_{18}$ Ultrabase) active fraction: 5.30 min
↓
Reverse-phase 3 (Hypercarb) active fraction: 12.30 min
↓
Reverse-phase 4 (C$_{18}$ Ultrabase) purified peptide: 5.50 min

cated on neuron terminals regulate the amount of 5-HT released from the neuron (MIDDLEMISS 1984; ENGEL et al. 1986; GÖTHERT et al. 1987; MIDDLEMISS et al. 1988; CHAPUT and DE MONTIGNY 1988).

Based on experimental observations, we previously proposed the hypothesis of the existence of an endogenous compound which could interact directly at 5-HT$_1$ receptors to affect their activity (FILLION and FILLION 1981). Therefore, if this hypothesis is confirmed, such compounds may exert an important role in the mechanisms regulating the activity of the "5-HT oscillator."

B. Isolation, Purification and Characterization of 5-HT-Moduline

The hypothesis of the existence of an endogenous ligand affecting the function of $5\text{-}HT_1$ receptors prompted us to examine whether or not various extracts of mammalian brain tissue may contain such a compound. Therefore, various extraction procedures (aqueous, acidic or organic) were carried out on bovine, horse and rat brain. These extracts were tested for their capacity to interact with the binding of $[^3H]5\text{-}HT$ to $5\text{-}HT_1$ receptors (ROUSSELLE et al. 1996). The extracts were then fractionated via several sequential procedures using gel permeation, reverse-phase chromatography with diverse mobile phases and matrices. Numerous trials were also carried out using ionic and normal-phase separation chromatography (Table 1).

Pharmacological specificity of the active fractions purified at each step of the chromatographic analysis was assessed, making it possible to show that the fractions were already reasonably specific for $5\text{-}HT_1$ receptor at early stages of purification (unpublished experiments). Numerous false-positive fractions had to be studied and then eliminated, reminding us of the numerous pitfalls described by LEE et al. (1987) in similar studies. Ultimately, a homogeneous fraction was isolated on C_{18} Hypercarb reverse-phase chromatography and shown to interact with the binding of $[^3H]5\text{-}HT$ to $5\text{-}HT_1$ sites.

The displacement was not total on rat brain membrane preparations containing various $5\text{-}HT_1$ receptor types; this result suggested that the fraction may specifically act at a single population of the $5\text{-}HT_1$ receptor subtype. This hypothesis was confirmed later (see below).

A major question which had to be answered was the nature of the active compound. The fact that this compound was metabolized in two main metabolites increased the complexity of the analysis. In any case, the NMR analysis of the active fraction was completed by amino acid analysis and protein sequencing. The identified compound was characterized as the tetrapeptide Leu-Ser-Ala-Leu and was later named 5-HT-moduline from its general properties (ROUSSELLE et al. 1996).

C. Molecular Interactions of 5-HT-Moduline with $5\text{-}HT_{1B/1D}$ Receptors

A first series of experiments was carried out to determine the properties of interaction of 5-HT-moduline with $5\text{-}HT_1$ receptors. It was soon clear that the peptide did not interact with all $5\text{-}HT_1$ receptor subtypes but solely with $5\text{-}HT_{1B/1D}$ receptors. The interaction consisted of a noncompetitive inhibitory effect affecting the binding of an agonist ($[^3H]5\text{-}HT$) or an antagonist ($[^{125}I]$cyanopindolol) to $5\text{-}HT_{1B}$ receptors in rat (Fig. 1). It is known that $5\text{-}HT_{1B}$ receptors in rat and mouse are the equivalent of $5\text{-}HT_{1D\beta}$ in other species (HARTIG et al. 1992; METCALF et al. 1992; OKSENBERG et al. 1992). Interestingly

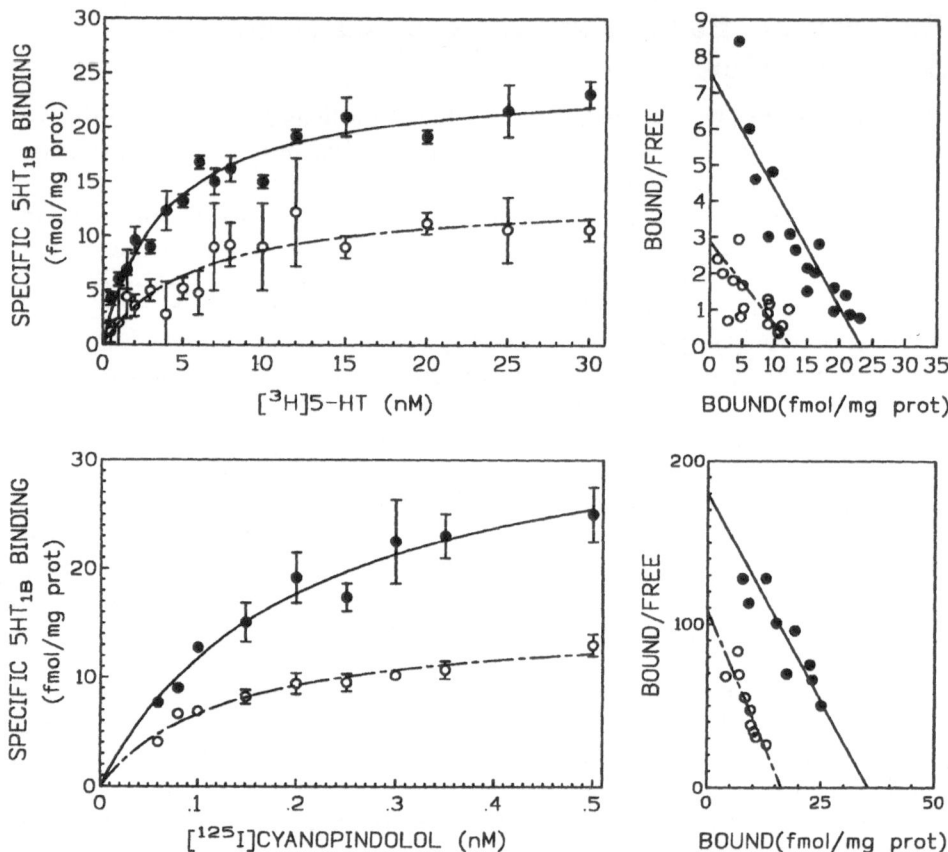

Fig. 1. Binding interactions of 5-HT-moduline on 5-HT$_{1B}$ receptors. Rat brain membranes were prepared as previously described (ROUSSELLE et al. 1996). *Upper*, [^{3}H]5-HT binding: aliquots of membrane preparation (500 μg protein per incubate) were incubated for 30 min at 25°C in a final volume of 500 μl with increasing concentrations (1–30 nM) of [^{3}H]5-HT (3.66 Tbq/mmol, Amersham), 0.1 μM 8-OH-DPAT (8-hydroxy-2 [di-*n*-propylamino]tetralin) and in the absence (*black circles*) or in the presence (*white circles*) of 5-HT-moduline (1 nM). The incubation medium consisted of a 50 mM Tris-HCl buffer, pH 7.4, containing 0.1% ascorbic acid, 4 mM CaCl$_2$ and 10 μM pargyline. Nonspecific binding was determined in the presence of 20 nM 5-CT (5-carboxytryptamine). At the end of the incubation, the tubes were cooled and filtered under vacuum on Whatman GF/B. Each value represents the mean ± SEM of triplicate determination. The presented data illustrate a typical experiment with the corresponding Scatchard plot. *Lower*, [^{125}I]Cyanopindolol binding: aliquots of rat brain membranes (25 μg protein per incubate) were incubated for 60 min (200 μl final volume) with increasing concentrations (0.06–0.5 nM) of [^{125}I]cyanopindolol (74 Tbq/mmol, Amersham) in the absence (*black circles*) or in the presence (*white circles*) of 5-HT-moduline (1 nM). The incubation medium consisted of a 10 mM Tris-HCl buffer, pH 7.7, containing 157 mM NaCl, 20 μM pargyline, 0.1 μM 8-OH-DPAT and 30 μM isoproterenol. Nonspecific binding was determined in the presence of 10 μM 5-HT. At the end of the incubation period, the tubes were cooled and filtered under vacuum on Whatman GF/B. Each point is the mean ± SEM of triplicate determinations. The presented data illustrate a typical experiment with the corresponding Scatchard plot

enough, 5-HT-moduline also had a noncompetitive antagonistic effect at 5-$HT_{1D\beta}$ receptors in guinea pig cortical membrane preparations. The apparent affinities in rat and guinea pig were very similar, the IC_{50} being close to 10^{-10} M. Bovine and human 5-HT_{1D} receptors were similarly sensitive to 5-HT-moduline as in guinea pig (Fig. 2). The interaction of the peptide with 5-$HT_{1B/1D}$ receptors was further demonstrated using cells transfected with the gene coding for 5-HT_{1B} or for 5-$HT_{1D\beta}$ receptor protein kindly given by René Hen, Columbia University, New York. The parameters of the interactions were very similar to those observed in rat brain membranes.

A second series of experiments was devoted to the specificity of action of 5-HT-moduline. It was shown that 5-HT-moduline at nanomolar concentrations exerted a maximal inhibitory effect on the binding of [^{3}H]5-HT to 5-$HT_{1B/1D}$, whereas it was inefficient, even at higher concentrations (up to $10^{-6}M$), in altering the binding of specific radioligands to other serotoninergic receptors: 5-HT_{1A}, 5-HT_{1E}, 5-HT_{1F}, 5-HT_3, 5-HT_6 and 5-HT_7. Moreover, it was also inactive at nonserotoninergic receptors: α-adrenergic, β-adrenergic, dopaminergic D_2, histaminergic H_1, muscarinic, benzodiazepine and opiate receptors (Table 2).

Therefore, these results indicate that 5-HT-moduline interacts with a high affinity solely with 5-$HT_{1B/1D}$ receptor and has no effect on other studied receptors (ROUSSELLE et al. 1996).

These properties of 5-HT-moduline were further confirmed and extended by direct binding experiments. 5-HT-moduline was tritiated and its binding

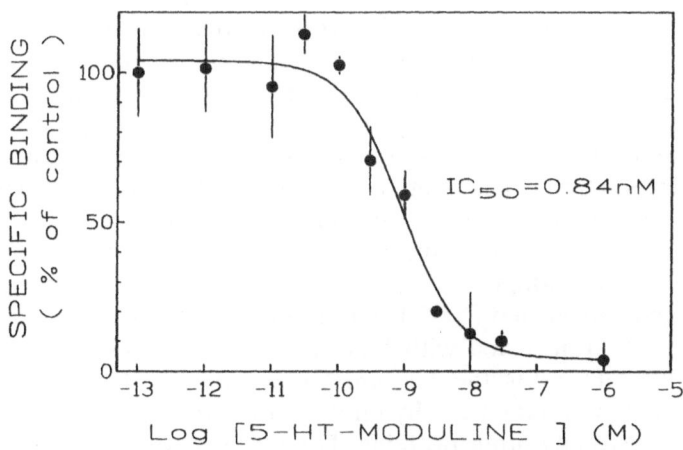

Fig. 2. Dose-response curve of 5-HT-moduline on [^{3}H]5-HT binding. Guinea pig cortical membranes (250 μg protein) were incubated for 30 min at 25°C with [^{3}H]5-HT (20 nM), 0.1 μM 8-OH-DPAT and increasing concentrations of 5-HT-moduline (0.1 pM to 1 μM). Nonspecific binding was determined in the presence of 10 μM 5-HT and represented about 50% of the total binding. Each point is the mean ± SEM of triplicate determinations

Table 2. Pharmacological properties of 5-HT-moduline. The capacity of 5-HT-moduline ($1\,nM$) to interact with the bindings of different radiolabeled ligands specific for serotoninergic receptors or other neurotransmitters receptors was examined on rat brain cortical membranes. Binding conditions were as previously described (Rousselle et al. 1996). Each value corresponds to the mean ± SEM of three independent experiments conducted in triplicate

Receptors	Ligands	Inhibitory activity (% of control)
5-HT$_{1A}$	[^{3}H]8-OH-DPAT ($3\,nM$)	0 ± 5
5-HT$_{1B}$	[^{125}I]Cyanopindolol ($0.3\,nM$) + $30\,\mu M$ isoproterenol + $0.1\,\mu M$ 8-OH-DPAT	70 ± 2
5-HT$_{1nonA}$	[^{3}H]5-HT ($30\,nM$) + $0.1\,\mu M$ 8-OH-DPAT	75 ± 3
5-HT$_{1E/1F}$	[^{3}H]5-HT ($30\,nM$) + $20\,nM$ 5-CT	0 ± 10
5-HT$_{2A}$	[^{3}H]Ketanserin ($5\,nM$)	10 ± 5
5-HT$_{2A}$	[^{3}H]DOB ($5\,nM$)	2 ± 8
5-HT$_3$	[^{3}H]BRL 43694 ($3\,nM$)	0 ± 10
α_1-Adrenergic	[^{3}H]Prazosin ($2\,nM$)	5 ± 3
β-Adrenergic	[^{3}H]Dihydroalprenolol ($3\,nM$) + $10\,\mu M$ 5-HT	10 ± 15
Dopaminergic D$_2$	[^{3}H]Spiperone ($2\,nM$) + $10\,\mu M$ 5-HT	2 ± 3
Histaminergic H$_1$	[^{3}H]Mepyramine ($5\,nM$)	10 ± 10
Muscarinic	[^{3}H]Quinuclidinyl benzylate ($3\,nM$)	0 ± 2
Opiate	[^{3}H]Naloxone ($2\,nM$)	10 ± 10
Benzodiazepine	[^{3}H]Flunitrazepam ($3\,nM$)	2 ± 2

studied in different membrane preparations (Massot et al. 1996). Rat and guinea pig membranes prepared from cortex or striatum were able to bind the labeled peptide in a saturable and reversible manner. The analysis of the saturation curves indicated that, apparently, a single population of sites was involved and recognized the peptide with affinity constants $K_D = 0.14$ in rat and $0.78\,nM$ in guinea pig, respectively (Fig. 3). These values are very similar to those corresponding to the EC$_{50}$ observed in displacement experiments, representing the interaction of the peptide with the 5-HT$_{1B/1D}$ binding. These results strongly suggest that the observed binding sites for 5-HT-moduline actually correspond to the sites involved in the interaction of the peptide with 5-HT$_{1B/1D}$ receptor binding.

It was also shown that [^{3}H]5-HT-moduline bound with the same affinity constant to cells transfected with 5-HT$_{1B}$ and with 5-HT$_{1D\beta}$ receptor genes, whereas it did not bind to the wild untransfected cells. These results demonstrate that 5-HT-moduline actually binds to the 5-HT$_{1B}$ and 5-HT$_{1D\beta}$ receptor protein and not to any other proteins. The fact that the observed interaction between the peptide and the binding of [^{3}H]5-HT obeys a noncompetitive mechanism strongly suggests that the site which recognizes 5-HT-moduline is located on the 5-HT$_{1B/1D}$ receptor protein and is distinct from that binding 5-HT itself.

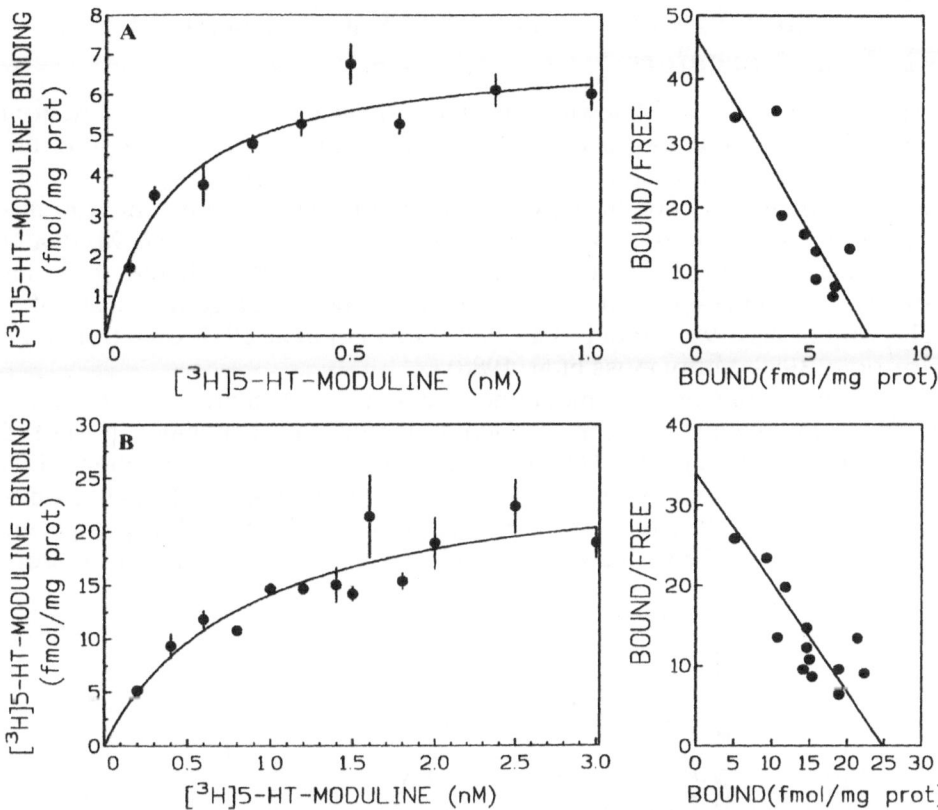

Fig. 3A,B. [³H]5-HT-moduline binding. Rat (*A*) or guinea pig (*B*) cortical membranes were pretreated for 10 min at 37°C with a mixture of antiproteases: 0.1 mM PMSF (phenylmethylsulfonyl fluoride), 2 mM ethylenediaminetetraacetate (EDTA), 5 IU/l aprotinin, 0.3 mg/ml bacitracine and 0.1% bovine serum albumin (BSA). Membranes (250 µg protein/tube) were then incubated overnight at 4°C in a 50 mM Tris-HCl buffer, pH 7.4, with increasing concentrations (0.1–3 nM) of [³H]5-HT-moduline (4.14 Tbq/mmol, CEA) in a volume of 1 ml. Nonspecific binding was determined in the presence of 1 µM nonlabeled 5-HT-moduline. At the end of the incubation period, the tubes were filtered under vacuum on Whatman GF/B. Each point is the mean ± SEM of triplicate determinations. Nonspecific binding represented 40% of the total binding. The presented data illustrate typical experiments with their corresponding Scatchard plot. Potential degradation of [³H]5-HT-moduline was examined under our experimental conditions (overnight incubation at 4°C and antiprotease pretreatment of the membranes); analysis by high-performance liquid chromatography (HPLC) on C$_{18}$ reverse-phase column of the labeled material present in the supernatant after centrifugation of the incubate indicated that 90% of the radiolabeled material was still the native radiolabeled compound

D. Functional Interactions of 5-HT-Moduline with 5-HT$_{1B/1D}$ Receptors

5-HT$_{1B/1D}$ receptors are known to be located on nonserotoninergic neuron terminals (heterologous presynaptic receptors) and on serotoninergic terminals (autologous presynaptic receptors) (Zifa and Fillion 1992). The former receptors mediate part of the activity of the 5-HT system on other neurotransmissions (Maura and Raiteri 1986; Molderings et al. 1987, 1990; Maura et al. 1989; Harel-Dupas et al. 1991; Van de Karr et al. 1989), whereas the latter directly control the serotoninergic activity itself by regulating the release of 5-HT (Middlemiss 1984; Engel et al. 1986; Göthert et al. 1987; Chaput and de Montigny 1988; Middlemiss et al. 1988).

The interaction of 5-HT-moduline with 5-HT$_{1B/1D}$ receptors was studied in in vitro experiments using preloaded synaptosomes containing [^{3}H]5-HT stimulated by a K$^+$ shock. 5-HT-moduline was shown to exert no effect of its own on the spontaneous release of the amine, whereas it had a clear antagonistic activity on the inhibitory effect induced by 5-HT (or by a 5-HT$_{1B/1D}$-specific agonist) on the evoked release of the amine. The antagonistic effect was dose

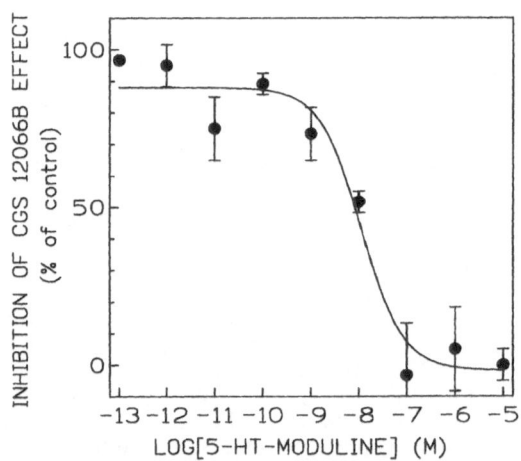

Fig. 4. Effect of 5-HT-moduline on the synaptosomal release of 5-HT. Hippocampal synaptosomes prepared according to Cotman and Matthews (1971) were loaded with 30 nM [^{3}H]5-HT and resuspended in a 37°C oxygenated Krebs-Ringer buffer, pH 7.4, and placed in 200-μl aliquots in a 96-well filtration plate (MAFB NOB glass fiber filter type B, Millipore). 5-HT-moduline or CGS 12066B or both were then added and incubated for 3 min with the loaded synaptosomes. At the end of this incubation period, a 5-min potassium stimulation (15 mM) was applied. The 96-well filtration plate was filtered under vacuum and the 96 filtrates recovered. The radioactivity contained in each filtrate was then measured by liquid scintillation counting. The presented data illustrate the dose response curve of 5-HT-moduline on the inhibition of [^{3}H]5-HT release induced by 50 μM CGS 12066B (CGS 12066B at 50 μM inhibits 40%–50% of the release of the [^{3}H]5-HT induced by the potassium stimulation). Each point is the mean ± SEM of quadruplicate determinations

dependent (Fig. 4). Under the experimental conditions used in these assays, the EC_{50} corresponded to higher values than those occurring in binding assays. This is likely due to the fact that at 37°C in the medium used to superfuse the synaptosomal preparation, the peptide is relatively rapidly metabolized whereas in binding assays carried out at 0°–2°C and in the presence of enzymatic inhibitors, the degradation of the peptide is markedly reduced; chromatographic controls performed in incubation medium support this hypothesis.

Thus, 5-HT-moduline appears capable of altering the reactivity of the receptor, which accordingly becomes less sensitive to a $5\text{-HT}_{1B/1D}$-specific agonist. The functional correlation of this result is that, under these conditions, 5-HT-moduline may induce an increase in the release of 5-HT from serotoninergic terminals.

E. In Vivo Activity of 5-HT-Moduline

The in vivo activity of 5-HT-moduline was studied in the social interaction test, an animal model known to involve 5-HT_{1B} receptors. In the test described by FRANCÈS (1988), the behavior of normal mice maintained in a group of ten animals is studied and compared with that of mice previously isolated for a week. The test discriminates between the behavior of the two types of animals placed for 6 min in the presence of a normal mouse (object mouse); the control

Table 3. Behavioral effect of 5-HT-moduline. 5-HT-moduline was tested in the social interaction test (FRANCÈS 1988). Mice (male Swiss NMRI, 3–4 weeks old, Iffa-Credo, France) were either housed in groups of five animals or isolated for 1 week. They were tested in pairs (one grouped and one isolated) under a transparent beaker inverted on a rough surface glass plate. The number of escape attempts was counted for 2 min and defined as one of the following: (1) the forepaws were placed against the beaker wall, (2) the mouse sniffed at the rim of the beaker or (3) the mouse scratched the glass floor. All mice were tested only once. Drugs (or sodium chloride for controls) were administered i.c.v., 45 min before the test, for 5-HT-moduline and i.p., 30 min before the test, for RU 24969. Results are expressed as means ± SEM of escape attempts per mouse. n represents the number of animals in each group

Swiss NMRI mice (20–25 g)	Number of escape attempts
Grouped mice	27 ± 1.5 ($n = 25$)
Isolated mice	14.8 ± 1.2 ($n = 23$)
Isolated mice treated with RU 24969 (4 mg/kg i.p.)	27 ± 2.7 ($n = 10$)
Isolated mice treated with 5-HT-moduline (50 µg i.c.v.)	14.3 ± 1.8 ($n = 10$)
Isolated mice treated with RU 24969 (4 mg/kg i.p.) and 5-HT-moduline (50 µg i.c.v.)	13 ± 1.2** ($n = 13$)

** indicates a significant difference ($P < 0.01$ using Student's t-test) between the group of isolated animals treated with RU 24969 and 5-HT-moduline and isolated mice treated with RU 24969.

animal has a "normal" behavior characterized by a certain locomotor activity whereas the isolated mouse exhibits a deficit of this activity. Administration of RU 24969, acting at 5-HT$_{1B}$ receptors, restores the normal locomotor behavior in the isolated animal. 5-HT-moduline administered dose-dependently i.c.v. antagonizes this effect whereas it has no effect of its own (Table 3). Here again, it is shown that 5-HT-moduline has an antagonistic property on the 5-HT$_{1B}$-induced activity.

F. Release of 5-HT-Moduline

Assays were performed to tentatively determine whether 5-HT-moduline could be released from cerebral tissue. Synaptosomes were prepared from rat brain cortex and submitted to a K$^+$ shock (30 mM) for 3 min in the presence and absence of Ca^{2+} (122 mM). The incubate was then centrifuged and the supernatant processed by chromatographic separation to isolate fractions containing 5-HT-moduline. These fractions then were tested by a radioreceptor assay to determine quantitatively their peptide content. The results show that 5-HT-moduline was released from cortical synaptosomes in a K$^+$-Ca^{2+}-dependent manner, suggesting that it originates from excitable tissue (Fig. 5). Although it cannot be ruled out that glial cells are involved, the hypothesis that the peptide is stored in neurons is further considered in our current experiments.

G. Conclusion

The results obtained in our recent research demonstrate the existence of 5-HT-moduline in the CNS and also in peripheral tissues (kidney, lung, stomach, blood) (Rousselle et al. 1996). It is also clear that this peptide possesses a very high affinity for the 5-HT$_{1B/1D}$ receptor protein and specifically recognizes it. The interaction of the peptide with the receptor induces a dose-dependent decrease in the binding of 5-HT using a noncompetitive mechanism of action. In parallel, the peptide induces a desensitization of the receptor which is evidenced in vitro on the cellular function of 5-HT$_{1B/1D}$ receptor; indeed, it corresponds to the decrease in efficacy of the 5-HT$_{1B/1D}$ receptor in inhibiting the evoked release of 5-HT from neuron terminals. Moreover, this antagonistic effect is evidenced in vivo in an animal model.

The noncompetitive interaction of 5-HT-moduline at the 5-HT$_{1B/1D}$ receptor leading to its desensitization suggests that the action of the peptide involves an allosteric mechanism. A number of experimental observations strongly suggest, if not demonstrate, that G-protein-coupled receptors are allosteric proteins able to exist under distinct structural conformations (Kent et al. 1980; Levitzki 1988; Stiles et al. 1984; Tucek and Proska 1995). It is proposed that 5-HT-moduline is able to alter the equilibrium existing between various conformational states of the 5-HT$_{1B/1D}$ receptor and ultimately pro-

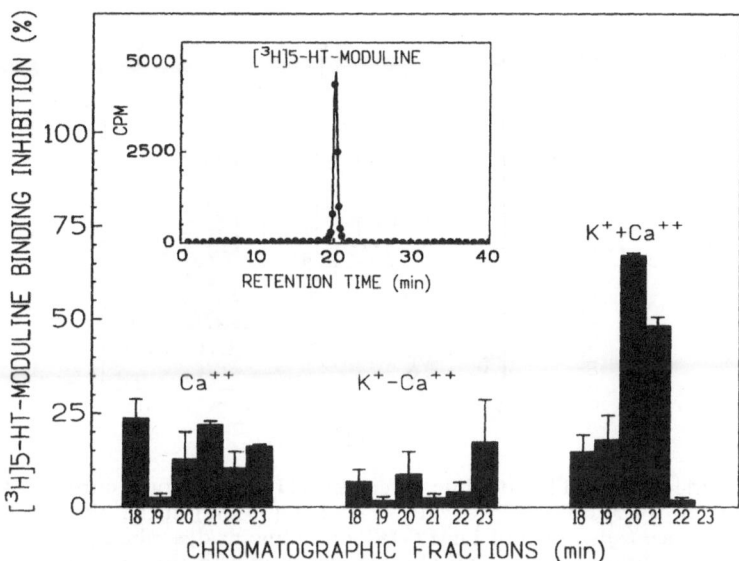

Fig. 5. Brain release of 5-HT-moduline. Synaptosomes from six rat brains were prepared according to the method of COTMAN and MATTHEWS (1971) and resuspended at 37°C in 30 ml of an oxygenated Krebs-Ringer buffer, pH 7.4, without $CaCl_2$. A 4-min potassium stimulation (30 mM) was then applied to the synaptosomes in the presence or absence of $CaCl_2$ (1.22 mM). A control without potassium stimulation but with $CaCl_2$ was also set up. The tubes were then immediatedly centrifuged (17 500 × g/25 min/4°C) and the supernatants lyophilized. The dried extracts were resuspended in 1 ml 50 mM ammonium acetate buffer, pH 5, and injected into a C_{18} Ultrabase reverse-phase column (250 × 10 mm). Elution was run at 4 ml/min with the following gradient (isocratic step of 3 min, 30 min linear gradient from 0% to 30% of acetonitrile and a 5-min step gradient at 50% acetonitrile) and 40 fractions of 1 min were collected. For each extract, fractions 18–23 were lyophilized (retention time of [³H]5-HT-moduline = 20.5 min), resuspended in 400 μl incubation buffer and then tested for their ability to displace the [³H]5-HT-moduline binding (1.5 nM) on guinea pig brain membranes. *Each bar* is the mean ± SEM of three independent experiments conducted in quadruplicate. *Inset*, elution profile of [³H]5-HT-moduline on C_{18} Ultrabase reverse-phase column

mote a "desensitized" conformation of the receptor protein unable to bind 5-HT (Fig. 1). Current experiments support this hypothesis; in particular, it has been shown that GTP restores the binding of 5-HT to the receptor after 5-HT-moduline had suppressed it (Fig. 6). Furthermore, it is also possible that 5-HT-moduline induces a transitory stabilization of the receptor conformation corresponding to the coupling with G-protein. Under these conditions, the peptide would facilitate the activation of the receptor by 5-HT. Additional work is needed to fully clarify the functional relationship of the interaction of 5-HT-moduline with the receptor protein; however, up to now, all results suggest an allosteric interaction of the peptide with the receptor.

The functional consequences of the interaction are of interest since it appears that 5-HT-moduline can control two important steps in the action of

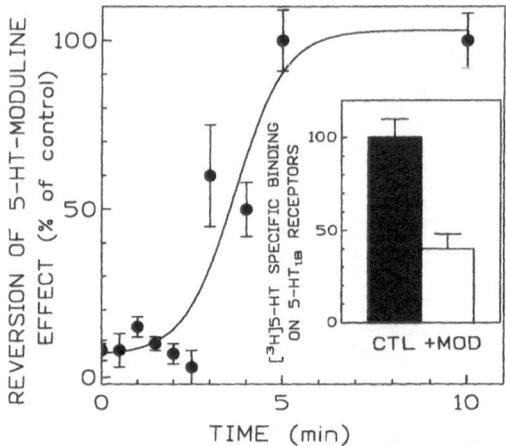

Fig. 6. Reversion of 5-HT-moduline effect by GTP. Rat brain cortical membranes were incubated for 30 min at 25°C with [³H]5-HT (30 nM) in the absence (*filled bar*) or in the presence (*open bar*) of 1 nM 5-HT-moduline as described previously. At the end of the incubation period, GTP (1 mM) was added to the membranes incubated in the presence of 5-HT-moduline. The incubates were then filtered at different times (0–10 min) after addition of GTP. *Each point* is the mean ± SEM of triplicate determinations

the serotoninergic system. A first one is directly related to the control of the release of 5-HT from terminals. According to the model of the serotoninergic "oscillator," 5-HT-moduline can regulate the amplitude of the signal delivered by the oscillator. The second step is related to the effect of 5-HT in controlling the release of other neurotransmitters (Ach, NA, GABA, prolactine, etc.) via heterologous presynaptic receptors, and here again 5-HT-moduline can affect the efficacy of the 5-HT$_{1B/1D}$ receptors involved in that control.

It is important to speculate that 5-HT-moduline, potentially released by a neuronal system, may affect almost individually the serotoninergic terminals. Accordingly, the functional consequences of the electrical activity of a serotoninergic neuron would be different on the release of the amine at each varicosity and would depend on the state of sensitivity of the 5-HT$_{1B/1D}$ receptors. Indeed, the serotoninergic negative feedback regulating the release of 5-HT would be controlled by 5-HT-moduline.

The 5-HT neurotransmission is one of the most extended and efficient systems regulating other neurotransmissions involved in the elaboration of the response of the CNS to a given stimulus. It is accepted that it plays a role in the homeostasis of the brain. The existence of 5-HT-moduline may add a new concept in this hypothesis. Indeed, 5-HT-moduline appears to be able to control the regulatory activity of the serotoninergic system on the CNS activity; therefore, it may define (with other factors) a level of 5-HT activity which corresponds to a given state of homeostasis of CNS and, accordingly, it might be important in defining personality traits, or possibly also might be involved in psychiatric pathologies.

The need for additional work is evident but promising. Finally, the fact that we have identified 5-HT-moduline specifically acting at the serotoninergic system is in favor of the existence of analogous peptides acting specifically at other regulatory systems. Current experiments in our laboratory are being carried out to explore this hypothesis.

Acknowledgements. This work was supported by "Direction des Recherches et Techniques" (DRET, contract No. 93/062).

References

Aghajanian GK (1978) Feedback of central monoaminergic neurons: evidence from single-cell recording studies. In: Youdim MB, Lornberg Y, Sharman DF, Lagnado JR (ed) Essays in neurochemistry and neuropharmacology. Wiley, New York, pp 1–32

Audet MA, Descarries L, Doucet G (1989) Quantified regional and laminar distribution of the serotonin innervation in the anterior half of adult rat cerebral cortex. J Chem Neuroanat 2:29–44

Baumgarten HG, Grozdanovic Z (1994) Neuroanatomy and neurophysiology of the central serotonergic system. J Serot Res 3:171–179

Chaput Y, De Montigny C (1988) Effects of the 5-hydroxytryptamine-1 receptor antagonist, BMY 7378, on 5-hydroxytryptamine neurotransmission: electrophysiological studies in the rat central nervous system. J Pharmacol Exp Ther 246:359–370

Cotman CW, Matthews DA (1971) Synaptic plasma membranes from rat brain synaptosomes: isolation and partial characterization. Biochim Biophys Acta 249:380–394

Engel G, Göthert M, Hoyer D, Schlicker E, Hillenbrand K (1986) Identity of inhibitory presynaptic 5-hydroxytryptamine (5-HT) autoreceptors in the rat brain cortex with 5-HT$_{1B}$ binding sites. Naunyn Schmiedebergs Arch Pharmacol 332:1–7

Fillion G, Fillion MP (1981) Modulation of affinity of postsynaptic serotonin receptors by antidepressant drugs. Nature 292:349–351

Francès H (1988) New animal model of social behavioural deficit: reversal by drugs. Pharmacol Biochem Behav 29:467–470

Göthert M, Schlicker E, Fink K, Classen K (1987) Effects of RU 24969 on serotonin release in rat brain cortex: further support for the identify of serotonin autoreceptors with 5-HT$_{1B}$ sites. Arch Int Pharmacodyn Ther 288:31–42

Harel-Dupas C, Cloëz, I, Fillion G (1991) The inhibitory effect of TFMPP of [^{3}H]acetylcholine release in guinea pig hippocampal synaptosomes is mediated by a 5-HT$_1$ receptor distinct from 1A-subtype, 1B-subtype, and 1C-subtype. J Neurochem 56:221–227

Hartig PR, Branchek TA, Weinshank RL (1992) A subfamily of 5-HT$_{1D}$ receptor genes. Trends Pharmacol 13:152–159

Jacobs BL, Azmitia EC (1992) Structure and function of the brain serotonin system. Physiol Rev 72:165–229

Jacobs BL, Fornal CA (1991) V. Activity of brain serotonergic neurons in the behaving animal. Pharmacol Rev 43:563–578

Kent RS, DeLean A, Lefkowitz RJ (1980) A quantitative analysis of beta-adrenergic receptor interactions: resolution of high and affinity states of the receptor by computer modeling of ligand binding data. Mol Pharmacol 17:14–23

Lee CR, Galzin AM, Taranger MA, Langer SZ (1987) Pitfalls in demonstrating an endogenous ligand of imipramine recognition sites. Biochem Pharmacol 36:945–949

Levitzki A (1988) From epinephrine to cyclic AMP. Science 241:800–806

Massoto, Rousselle JC, Fillion MP, Grinaldi B, Cloez-Tayarani I, Fugelli A, Prudhomme N, Seguin L, Rousseau B, Plantegol M, Hen R, Fillion G (1996) 5-hydroxytryptamine-moduline, a New Endogenous Cerebral Peptide, controls the Serotonergic activity via its specific Interaction with 5-Hydroxytryptamine$_{1B/1D}$ Receptors. Mol Pharmacol 50:752–762

Maura G, Raiteri M (1986) Cholinergic terminals in rat hippocampus possess 5-HT$_{1B}$ receptors mediating inhibition of acetylcholine release. Eur J Pharmacol 129:333–337

Maura G, Fedele E, Raiteri M (1989) Acetylcholine release from rat hippocampal slices is modulated by 5-HT. Eur J Pharmacol 165:173–179

Metcalf MA, McGuffin RW, Hamblin MW (1992) Conversion of the human 5-HT$_{1D}$ β serotonin receptor to the rat 5-HT$_{1B}$ ligand-binding phenotype by Thr^{355} Asn site-directed mutagenesis. Biochem Pharmacol 44:1917–1920

Middlemiss DN (1984) Stereoselective blockade at [^{3}H]5-HT binding sites and at the 5-HT autoreceptor by propanol. Eur J Pharmacol 101:289–293

Middlemiss DN, Bremer ME, Smith SM (1988) A pharmacological analysis of the 5-HT receptor mediating inhibition of 5-HT release in the guinea-pig frontal cortex. Eur J Pharmacol 157:101–107

Molderings GJ, Fink K, Schlicker E, Göthert M (1987) Inhibition of noradrenaline release via presynaptic 5-HT$_{1B}$ receptors of the rat vena cava. Naunyn Schmiedebergs Arch Pharmacol 336:245–250

Molderings GJ, Likungu J, Göthert M (1990) Inhibition of noradrenaline release from the sympathetic nerves of the human saphenous vein via presynaptic 5-HT receptors similar to the 5-HT$_{1D}$ subtype. Naunyn Schmiedebergs Arch Pharmacol 342:371–377

Oksenberg D, Marsters SA, O'Dowd BF, Jin H, Havlik S, Peroutka SJ, Ashkenazi A (1992) A single amino-acid difference confers a major pharmacological variation between human and rodent 5-HT$_{1B}$ receptors. Nature 360:161–163

Rousselle JC, Massot O, Delepierre M, Zifa E, Rousseau B, Fillion G (1996) Isolation and characterization of an endogenous peptide from rat brain interacting specifically with the serotonergic$_{1B}$ receptor subtypes. J Biol Chem 271:1–10

Sprouse JS, Aghajanian GK (1987) Electrophysiological responses of serotoninergic dorsal raphe neurons to 5-HT$_{1A}$ and 5-HT$_{1B}$ agonists. Synapse 1:3–9

Stiles GL, Caron MG, Lefkowitz RJ (1984) Beta-adrenergic receptors: biochemical mechanisms of physiological regulation. Physiol Rev 64:661–743

Tucek S, Proska J (1995) Allosteric modulation of muscarinic acetylcholine receptors. TIPS 16:205–212

Van de Kar LD, Carnes M, Maslowski RJ, Bonadonna AM, Rittenhouse PA, Kunimoto K, Piechowski RA, Bethea CL (1989) Neuroendocrine evidence for denervation supersensitivity of serotonin receptors: effects of the 5-HT agonist RU 24969 on corticotropin, corticosterone, prolactine and renin secretion. J Pharmacol Exp Ther 251:428–434

Zifa E, Fillion G (1992) 5-Hydroxytryptamine receptors. Pharmacol Rev 44:401–458

CHAPTER 12

Regulation of 5-HT Release in the CNS by Presynaptic 5-HT Autoreceptors and by 5-HT Heteroreceptors

M. Göthert and E. Schlicker

A. Introduction: Definitions and Scope

The amount of serotonin (5-hydroxytryptamine; 5-HT) released from the varicosities of the serotoninergic axon terminals in response to invading action potentials at a given frequency is by no means constant. As generally accepted now, exocytotic release of 5-HT can be significantly modified by pre-synaptic receptors, i.e. receptors located on the serotoninergic axon terminals (for reviews, see MORET 1985; MIDDLEMISS 1988; STARKE et al. 1989; GÖTHERT 1990). When such receptors are stimulated by 5-HT released from the serotoninergic axon terminals, they are termed presynaptic 5-HT au-toreceptors. When receptors on the serotoninergic axon terminals are acti-vated by other transmitters released from non-serotoninergic neurons, they are denoted as presynaptic heteroreceptors. In the latter case, the axon termi-nals of these neurons may form axo-axonic synapses with the serotoninergic terminals, or the interacting neighbouring neurons may release their transmit-ter non-synaptically into the extracellular space and, thus, may influence larger neuronal territories including the serotoninergic axon terminals (BEAUDET and DESCARRIES 1978; TÖRK 1990).

The present report will focus mainly on inhibitory presynaptic 5-HT autoreceptors. Such receptors, via an ultrashort inhibitory feedback loop, probably play an important physiological role in the fine regulation of 5-HT release. The presynaptic 5-HT autoreceptors have to be distinguished from 5-HT autoreceptors on the cell bodies and dendrites of the serotoninergic neurons, i.e. the somadendritic 5-HT autoreceptors (SPROUSE and AGHAJANIAN 1986, 1987; HJORTH and MAGNUSSON 1988; SCHECHTER et al. 1990; PINEYRO et al. 1995a). The latter can be activated by 5-HT released from the somadendritic area or recurrent branches of the serotoninergic axon of the same neuron or from axons of other serotoninergic neurons innervating the cell bodies and dendrites of the neuron under consideration. According to the location of these receptors, their stimulation induces a decrease in neuronal cell firing (5-HT$_{1A}$ receptors; SPROUSE and AGHAJANIAN 1986, 1987; HJORTH and MAGNUSSON 1988; SCHECHTER et al. 1990) or in somadendritic 5-HT release (5-HT$_{1D}$ receptors; PINEYRO et al. 1995a). In this report, these receptors, which certainly are also of high functional importance, will not be reviewed in detail, but, when relevant, will be considered in comparison with the presynaptic 5-HT autoreceptors.

Whereas the cell bodies and dendrites of the serotoninergic neurons are located in a limited brain area, namely the raphe nuclei of the brainstem, their axons project into virtually all parts of the CNS (TÖRK 1990; Chap. 2, this volume) and, accordingly, presynaptic 5-HT autoreceptors are widely distributed within the brain and spinal cord (see, e.g. STARKE et al. 1989). It may be deduced from the presynaptic location of heteroreceptors on the serotoninergic axon terminals (e.g. presynaptic α_2-heteroreceptors; GÖTHERT and HUTH 1980; FRANKHUYZEN and MULDER 1980; GÖTHERT et al. 1981) that signals can directly be conducted from other brain areas via non-serotoninergic nerves (e.g. from the locus coeruleus via noradrenergic neurons) to the serotoninergic varicosities. At these sites the transmitter released shares the property of 5-HT itself to modify the strength of the chemical signal (i.e. the amount of 5-HT available in the synaptic cleft) at the location where this signal is formed. In other words, such a modulatory-type action of non-serotoninergic transmitters on the release of serotonin exerted via presynaptic heteroreceptors resembles the restraining effect of extracellular 5-HT acting on its autoreceptors.

The presynaptic heteroreceptors on the serotoninergic nerves will not be considered here in detail, since their occurrence and function has recently been reviewed (GÖTHERT and SCHLICKER 1991). However, presynaptic heteroreptors may also come into play in the context of the main topic of the present report, namely the autoregulation of 5-HT release. This process may not only be operative via 5-HT autoreceptors, but may also involve neuronal circuits with at least one interneuron. In the case of *one* interneuron only, it should be activated by 5-HT via a 5-HT receptor, resulting in the release of a transmitter which acts via its heteroreceptor on the serotoninergic axon terminals. If *several* neurons are involved in such a feedback loop, the first one would be endowed with the 5-HT receptor; the last one would release the transmitter activating its presynaptic heteroreceptor on the serotoninergic terminal. Activation of such a feedback loop may result in a facilitation or inhibition of 5-HT release and, thus, may mimic the function of a facilitatory or inhibitory presynaptic 5-HT autoreceptor, respectively. Whereas the occurrence of an inhibitory neuronal circuit consisting of several neurons will only briefly be considered in Sect. D.I.4, the probable involvement of such a feedback loop will be more extensively discussed in the context of, e.g. 5-HT receptors mediating facilitation of 5-HT release (GALZIN et al. 1990; GALZIN and LANGER 1991; BLIER and BOUCHARD 1993).

5-HT receptors are also present as presynaptic 5-HT heteroreceptors on non-serotoninergic axon terminals (see, e.g. GÖTHERT 1990; CASSEL and JELTSCH 1995). These receptors, which are, as a rule, less well investigated (except 5-HT heteroreceptors on cholinergic axon terminals; CASSEL and JELTSCH 1995), will not be dealt with in this report.

B. Methodological Approaches

The operation of 5-HT receptors mediating inhibition or facilitation of 5-HT release has mainly been proved by experiments in which the effects of 5-HT receptor ligands on the release of 5-HT were investigated either in vitro or in vivo. According to the concept underlying such investigations, a receptor agonist should inhibit or increase 5-HT release in a manner susceptible to blockade by an appropriate 5-HT receptor antagonist. If the receptor is tonically activated by 5-HT, the antagonist, given alone, should produce an effect opposite to that of the agonist. Such functional studies cannot only prove the existence of presynaptic 5-HT receptors, but also provide valuable information about their pharmacological properties, function, potential physiological role and, by application of appropriate pharmacological tools, the site and mechanism by which they influence neurotransmitter release.

I. In Vitro Studies

Most investigations available have been carried out with slices or synaptosomes. As a rule, the preparations were preincubated with radioactively labelled neurotransmitter, and the stimulation-evoked overflow of radioactivity or (after column chromatography) labelled 5-HT from the superfused (in a few cases, incubated) preparations has been determined (for review, see, e.g. MIDDLEMISS 1988; STARKE et al. 1989). In some investigations, slices were incubated with a radioactive precursor leading to the synthesis of labelled 5-HT. whose overflow in response to stimulation was studied (HAMON et al. 1974). It is generally accepted that stimulation-evoked overflow of tritium or labelled 5-HT, in particular in the presence of an inhibitor of neuronal 5-HT uptake, reflects the release of endogenous 5-HT. Therefore, in these cases, the term "5-HT release" will be used throughout the subsequent sections. Very few experiments were based on the investigation of the stimulation-evoked overflow of endogenous 5-HT; as a rule, the latter has been determined by means of fast cyclic voltammetry. This technique has been applied in slices of the rat dorsal raphe nucleus (STARKEY and SKINGLE 1994; DAVIDSON and STAMFORD 1995). In most investigations, either electrical field stimulation or high K^+ was used to depolarize the cell membrane of the respective neurons and, thus, to stimulate transmitter release. Recently, N-methyl-D-aspartate (NMDA) has also been applied for this purpose (FINK et al. 1996).

The advantage of the use of superfused synaptosomes, consisting mainly of isolated varicosities, for such experiments is that the data obtained provide evidence for the presynaptic location of the receptors involved (CERRITO and RAITERI 1979). Furthermore, such synaptosomal experiments make it possible to study the function of presynaptic receptors undisturbed of the influence of its endogenous ligand. In conventional experiments on slices, the endogenous agonist must be assumed to be present in the biophase of the presynaptic receptor under consideration at sufficiently high concentration to interact with an additional exogenous agonist. However, the interaction of the endogenous neurotransmitter can also be eliminated in experiments in which the overflow of radioactivity or a labelled transmitter is elicited by a small number of electrical impulses (as a rule up to four) at high frequency (100 Hz). In such "pseudo-one-pulse" ("POP") experiments, the concentration of the endogenous transmitter in the biophase of the corresponding presynaptic receptor is too low to stimulate it, and, on the other hand, the radioactive signal is strong enough to be measurable (SINGER 1988).

II. In Vivo Studies

Intracerebral microdialysis has been the most frequently applied technique in anaesthetized or freely moving animals. 5-HT and its metabolites in the dialysate have been measured by means of high-pressure liquid chromatography (HPLC) (BRAZELL et al. 1985; SHARP et al. 1989). This technique makes it possible to directly apply 5-HT receptor agonists and/or antagonists to the brain area under investigation and to simultaneously determine their influence on the overflow of 5-HT and its metabolites. The effect of direct application can be compared to that of systemic administration of the drugs (ASSIE and KOEK 1996). This comparison provides hints at whether or not the drug passes the blood-brain barrier, and allows the determination of the overall importance of the presynaptic 5-HT autoreceptor in the control of 5-HT release. In studies in which the drugs are exclusively injected systemically, it is difficult to decide whether an effect observed is due to a direct action at the serotoninergic nerve endings or whether the drugs primarily act in the somadendritic area of this neuron. A further possibility is that the drugs may influence the serotoninergic neuron indirectly by acting at non-serotoninergic nerves endowed with 5-HT receptors of a subtype identical to, or different from, the presynaptic 5-HT autoreceptors (see Sect. C.II, D.II).

In some of the in vivo studies, fast cyclic voltammetry has been used to determine 5-HT release (MARSDEN et al. 1986; PINEYRO et al. 1995a). The disadvantage of this technique is the lower specificity compared to HPLC, the main advantage being the high topographic resolution, which makes it possible to measure release in very small brain areas.

Finally, electrophysiological methods have been successfully applied to study presynaptic 5-HT autoreceptor function in vivo (CHAPUT et al. 1986; CHAPUT and DE MONTIGNY 1988). This approach is based on a comparison of the depressant response of hippocampal pyramidal neurons to directly applied exogenous 5-HT and of endogenous 5-HT, released in response to electrical stimulation of the ascending serotoninergic axons. Application of a 5-HT receptor antagonist which leaves the postsynaptic pyramidal 5-HT receptor at least largely unaffected but blocks the presynaptic 5-HT autoreceptors electively enhances the depressant response to stimulation of the afferent serotoninergic neurons without influencing the response to directly applied 5-HT.

C. Identification, Location and Physiological Role of 5-HT Receptors Mediating Inhibition or Facilitation of 5-HT Release

Up to 1990 presynaptic 5-HT autoreceptors were reviewed a number of times, either specifically (MORET 1985; MIDDLEMISS 1988) or in a more general context of various presynaptic receptors (GÖTHERT 1982; TIMMERMANS and THOOLEN 1987; STARKE et al. 1989; GÖTHERT 1990). In order to avoid redundance, the present article will focus mainly on recent developments not yet covered in the review by STARKE et al. (1989) coauthored by one of us. However, key findings and important conclusions relevant to presynaptic 5-HT autoreceptors will be briefly summarized here, irrespective of whether or not they have already been extensively reviewed.

I. Inhibitory Presynaptic 5-HT Autoreceptors

1. Identification and Location

Evidence for the existence of inhibitory presynaptic 5-HT autoreceptors was presented first by FARNEBO and HAMBERGER (1971). These authors showed that the non-selective 5-HT receptor agonist, LSD, inhibited 5-HT release in rat brain cortex slices, a finding which was confirmed by HAMON et al. (1974). FARNEBO and HAMBERGER (1974) demonstrated that the effect of LSD was shared by other 5-HT receptor agonists and that the non-selective 5-HT receptor antagonist methiothepin increased 5-HT release. On the basis of these findings, the concept of an inhibitory feedback regulation of 5-HT release was developed which was supported by CERRITO and RAITERI (1979) using rat hypothalamic synaptosomes and by GÖTHERT and WEINHEIMER (1979) using rat brain cortex slices. These studies revealed that in the presence of an inhibitor of the neuronal 5-HT transporter, 5-HT itself when applied exogenously inhibited 5-HT release and that the effect of 5-HT was counteracted by methiothepin; furthermore, the results of the experiments on cortical slices confirmed those by FARNEBO and HAMBERGER (1974) in that methiothepin, given alone, disinhibited, i.e. increased, 5-HT release.

In the following years, inhibitory 5-HT autoreceptors were also identified by means of basically identical techniques in other regions of the rat brain and in the brain of other species, including mouse, guinea pig, rabbit, pig and man (Fig. 1; Table 1). Generally speaking, presynaptic 5-HT autoreceptors have been identified in every region of the mammalian CNS which has been investigated for this purpose (Table 1). During the last 8 years, the occurrence and operation of these receptors in the brain of the species mentioned so far have been repeatedly confirmed (WICHMAN et al. 1989; FEUERSTEIN et al. 1992;

Table 1. Occurrence of inhibitory 5-HT autoreceptors in various brain regions of the rat and other species. P, presynaptic location has been proved; E, indirect evidence for activation by endogenous 5-HT has been presented; C, the receptor has been classified according to its pharmacological properties in terms of 5-HT$_{1B}$ and "5-HT$_{1D}$". More recent findings not yet considered in the review by STARKE et al. (1989; including brain regions and species in which autoreceptors had not yet been identified) are underlined. References for autoreceptors characterized or classified during the last 8 years: cortex of guinea pig: HOYER and MIDDLEMISS (1989), LIMBERGER et al. (1991), BÜHLEN et al. (1996a); cortex of rabbit: LIMBERGER et al. (1991); cortex of mouse and rhesus monkey: SCHIPPER (1990); cortex and hypothalamus of mouse: SCHLICKER, FINK and GÖTHERT (unpublished); hippocampus of guinea pig: WILKINSON and MIDDLEMISS (1992); periaqueductal gray of the rat: VERSTEEG et al. (1991); superior colliculus of the rabbit: WICHMANN et al. (1989); hypothalamus and substantia nigra of the guinea pig: MORET and BRILEY (1995); corpus striatum (caudate nucleus) of rabbit: FEUERSTEIN et al. (1992); human hippocampus: SCHLICKER et al. (1996)

Cerebral cortex		Hypothalamus	
Rat	P, E, C	Rat	P, E, C
Mouse	E, C	Mouse	E
Guinea pig	P, E, C	Guinea pig	E
Rabbit	E, C	Rabbit	E
Pig	P, E, C		
Rhesus monkey	C	**Superior colliculus**	
Man	P, E, C	Rabbit	E
Hippocampus		**Periaqueductal grey**	
Rat	P, E, C	Rat	
Guinea pig	E		
Rabbit	E, C	**Cerebellum**	
Man	E	Rat	P, C
		Mouse	E
Nucleus accumbens			
Rat	P	**Medulla oblongata**	
		Rat	E
Corpus striatum			
Rat		**Spinal cord**	
Rabbit	E, C	Rat	P, E, C
Substantia nigra			
Guinea pig	C		

WILKINSON and MIDDLEMISS 1992; MORET and BRILEY 1995; BÜHLEN et al. 1996a; FINK et al. 1996) and have, in addition, been demonstrated in the cerebral cortex of another species, namely the rhesus monkey (SCHIPPER 1990).

In view of the potential role of brain serotoninergic neurotransmission in the pathogenesis and drug therapy of neuropsychiatric diseases (see Sect. G and, e.g. GÖTHERT 1991; PEROUTKA 1991; CHOPIN et al. 1994; Chap. 10, this volume), it was of particular interest to investigate whether presynaptic 5-HT autoreceptors also occur in the human hippocampus. This has recently been proven in experiments on human hippocampal slices (SCHLICKER et al. 1996).

The occurrence of inhibitory 5-HT autoreceptors has been shown not only in the in vitro experiments discussed so far, but also in vivo in the brain of anaesthetized and freely moving animals. As summarized in Table 2, 5-HT

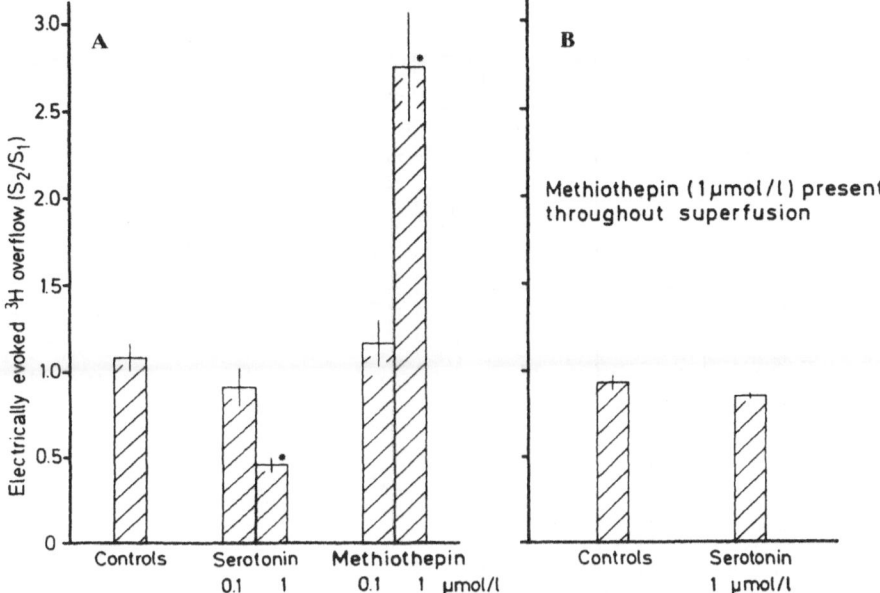

Fig. 1A,B. Inhibition by unlabelled serotonin (5-HT) and facilitation by methiothepin of the electrically (3 Hz) evoked overflow of tritium (consisting of more than 80% [³H]5-HT) from superfused human cerebral cortical slices **(A)** and abolition of the inhibitory effect of 5-HT by methiothepin **(B)**. The slices were preincubated with [³H]5-HT and superfused in the presence of the 5-HT uptake inhibitor, paroxetine (3.2 μM). Two periods of stimulation were applied after 40 and 90 min of superfusion (S_1, S_2) and the ratios of the overflow evoked by S_2 over that evoked by S_1 (S_2/S_1) are given (means ± SEM; $n = 4$–7; *$P < 0.001$). **A** 5-HT or methiothepin was present from the 65th min of superfusion onward: **B** methiothepin was present throughout superfusion and 5-HT from the 65th min of superfusion onward. (Modified from SCHLICKER et al. 1985)

receptor agonists inhibited 5-HT release in a manner sensitive to blockade by methiothepin, which given alone facilitated 5-HT release. In these studies the technique of in vivo microdialysis was used and 5-HT release was quantified by HPLC. Table 2 exclusively contains data from studies in which the drugs were administered directly via the dialysis probe.

Investigations of 5-HT release in the raphe nuclei were not included since in this region it is difficult to distinguish unequivocally between responses mediated by presynaptic autoreceptors on recurrent axon terminals, by somadendritic autoreceptors or by both (STARKEY and SKINGLE 1994; DAVIDSON and STAMFORD 1995). Exclusion of 5-HT receptors modulating 5-HT release in the raphe nuclei from a review on *presynaptic* 5-HT autoreceptors is the more justified as results obtained in this brain area of the rat with subtype-selective 5-HT receptor agonists and antagonists strongly support the view that *somadendritic* 5-HT autoreceptors modulate 5-HT release from the cell bodies and dendrites (PINEYRO et al. 1995a). These somadendritic 5-HT autoreceptors belong to the r5-HT$_{1D}$ subtype (PINEYRO et

Table 2. Effects of 5-HT receptor agonists and antagonists on 5-HT release in the rat and guinea pig brain in vivo. 5-HT release was determined by the microdialysis technique. 5-HT receptor ligands and 5-HT uptake blockers were administered locally through the dialysis probe (the concentration obtained in the brain tissue is about 20-fold lower). Note that the 5-HT receptor ligands indicated have not been examined in each of the studies

Species, CNS region	References (anaesthetized or freely moving animal? 5-HT uptake blocker present?)	5-HT receptor ligands (μmol/l) causing		Comment
		Inhibition	Facilitation	
Rat, cerebral cortex	Auerbach and Hjorth (1995; CH-anaesthetized; citalopram)	RU 24969 0.1–10 CP-93, 129 10		
Rat, hippocampus	Claustre et al. (1991; CH-anaesthetized) Hjorth and Tao (1991; CH-anaesthetized; citalopram) Martin et al. (1992; CH-anaesthetized; citalopram) Hjorth and Sharp (1993; CH-anaesthetized; citalopram) Auerbach and Hjorth (1995; CH-anaesthetized; citalopram) Bosker et al. (1995; freely moving; fluvoxamine)	RU 24969 0.1–10 TFMPP 10 CP-93, 129 3–10	Methiothepin 10 (–)-Penbutolol 1	No effect of 8-OH-DPAT (up to 100 μmol/l); effect of RU 24969 and CP-93, 129 antagonized by methiothepin and/or (–)-penbutolol; data are compatible with a 5-HT$_{1B}$ receptor[a]

Preparation	Reference	Agonist		Antagonist		Comments
Rat, diencephalon	Auerbach et al. (1991; freely moving; fluoxetine)	RU 24969	1			No effect of TFMPP 10 µmol/l
Guinea pig, cerebral cortex	Sleight et al. (1990; CH-anaesthetized); Claustre et al. (1991; CH-anaesthetized); Lawrence and Marsden (1992; freely moving); Hutson et al. (1995; isoflurane-anaesthetized); Skingle et al. (1995; CH-anaesthetized)	5-CT Oxymetazoline Sumatriptan	0.1–10 10 0.01–10	Methiothepin	33–100	No effect of RU 24969; facilitatory effect of GR 127935 0.1 µmol/l in the study by Skingle et al. (1995) but no effect of the same drug (10–100 µmol/l) in the study by Hutson et al. (1995)
Guinea pig, substantia nigra	Briley and Moret (1993; freely moving); Moret and Briley (1995; freely moving)			Methiothepin 1-(1-Naphthyl)-piperazine	100 1–100	

[a] Suggested both by the effectiveness of the selective 5-HT$_{1A/B}$ receptor agonist CP-93, 129 and the lack of effect of 8-OH-DPAT (a 5-HT$_{1A}$ receptor agonist with moderate affinity for 5-HT$_{1D}$ and virtually no affinity for r5-HT$_{1B}$ receptors). CH, chloral hydrate; TFMPP, m-trifluoromethylphenylpiperazine; 5-CT, 5-carboxamidotryptamine; 8-OH-DPAT, 8-hydroxy-2(di-n-propylamino)tetralin; RU 24969, 5 methoxy-3(1,2,3,6-tetrahydropyridin-4-yl)-1H-indole.

al. 1995a; new nomenclature of 5-HT$_{1B/1D}$ receptors; see Sect. D.I.3). Via these receptors, release can be controlled independently of firing activity and the firing-regulating 5-HT$_{1A}$ autoreceptors (Starkey and Skingle 1994; Pineyro et al. 1996). 5-HT$_{1D}$ autoreceptors also modulate 5-HT release in the mesencephalic raphe of guinea pigs (El Mansari and Blier 1996); it may be assumed that these receptors are also located in the somadendritic area of the serotoninergic neurons.

As already mentioned above (Sect. B), the presynaptic location of receptors can be proved by experiments in synaptosomes. Another approach consists of the functional isolation of the axon terminals in brain slices by interrupting the impulse traffic along the axons by tetrodotoxin. Under this condition K$^+$-evoked 5-HT release cannot be influenced by short neuronal circuits within the slices. However, a theoretical, but very improbable, possibility which is not excluded by such experiments is that 5-HT, via a presynaptic 5-HT heteroreceptor, primarily stimulates the release of a neurotransmitter from a non-serotoninergic axon terminal. This, in turn, modifies 5-HT release via a presynaptic heteroreceptor for its transmitter. Both slices superfused in the presence of tetrodotoxin and synaptosomes have been used to provide evidence for the presynaptic location of inhibitory 5-HT autoreceptors in several areas of the rat CNS (Starke et al. 1989). In Table 1, the areas in which the presynaptic location of the inhibitory autoreceptors has been proved are marked by "P".

Recently, the presynaptic location of inhibitory 5-HT autoreceptors has also been demonstrated in the human (Maura et al. 1993; Fink et al. 1995; Fig. 2) and guinea pig cerebral cortex (Bühlen et al. 1996a) by means of experiments on synaptosomes. In this context, it should be noted that the guinea pig brain represents an important model for the development of ligands of presynaptic 5-HT autoreceptors, which may become useful for the treatment of human neuropsychiatric diseases (Chopin et al. 1994; Chap. 10, this volume). Furthermore, in experiments on slices superfused with solution containing tetrodotoxin, the presynaptic location of the inhibitory 5-HT autoreceptors in the human hippocampus was made probable, since under this condition 5-carboxamidotryptamine 5-CT still inhibited the K$^+$-evoked 5-HT release in a manner sensitive to antagonism by methiothepin (Schlicker et al. 1996).

As in previous studies (for review see Starke et al. 1989), attempts have been made in rats to more directly prove the location of the presynaptic 5-HT autoreceptors which belong to the 5-HT$_{1B}$ subtype (see Sect. D.1) on the serotoninergic axon terminals themselves. In these experiments, the density of 5-HT$_{1B}$-binding sites has been determined after chemical axotomy with 5,7-dihydroxytryptamine. Either no change or even an increase in 5-HT$_{1D}$ receptor density occurred (Frankfurt et al. 1993, 1994), suggesting that the presynaptic 5-HT$_{1B}$ autoreceptors do not account for a major proportion of the total number of these receptors and that 5-HT$_{1B}$ receptors which are not located on 5-HT neurons appear to be upregulated. Comparison of the levels of 5-HT$_{1B}$ receptor messenger RNA with the densities of 5-HT$_{1B}$-binding sites in mouse

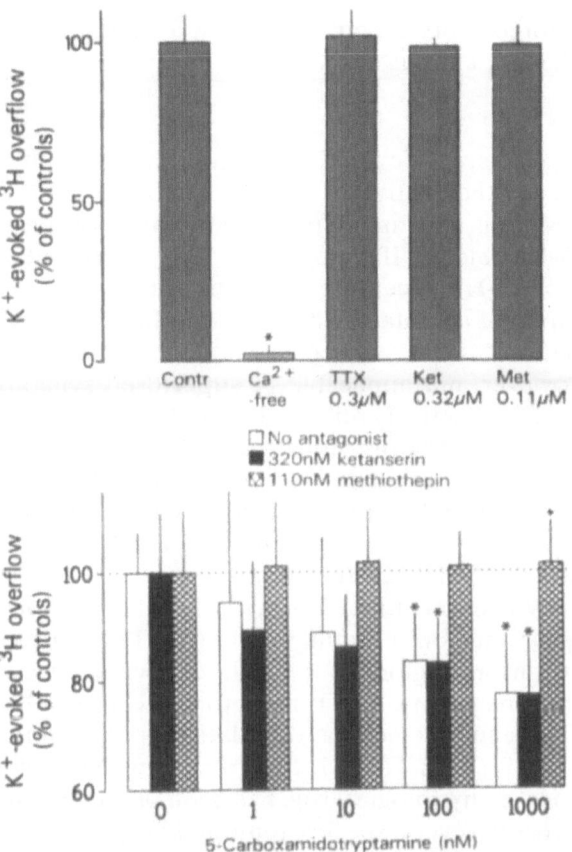

Fig. 2. Effects of serotonin (5-HT) receptor ligands on the K^+ (25 mM)-evoked tritium overflow from superfused human cerebral cortical synaptosomes preincubated by [^{3}H]5-HT. *Upper panel,* lack of influence of ketanserin (*Ket*), methiothepin (*Met*) or tetrodotoxin (*TTX*) on the evoked tritium overflow and abolition of overflow by omission of Ca^{2+} from the superfusion fluid; *Contr,* drug-free controls. *Lower panel,* inhibition by 5-carboxamidotryptamine (5-CT) of evoked tritium overflow and interaction with 5-HT receptor antagonists: effect of 5-CT not changed by Ket, but abolished by Met. Ket 320 nM corresponds to 4.4 times its K_i at cloned h5-HT$_{1D}$ receptors and Met 110 nM corresponds to 4.4 times its K_i at cloned h5-HT$_{1B}$ receptors (K_i values from WEINSHANK et al. 1991). The synaptosomes were superfused in the presence of the 5-HT uptake inhibitor fluvoxamine (10 μM). Ca^{2+} was omitted from, and the drugs were added to, the superfusion fluid from the 20th min before stimulation onward. Means ± SEM of 6–14 experiments. *$P < 0.05$, compared to the corresponding controls without 5-CT. $^+P < 0.05$, compared to the corresponding effect of 5-CT in the absence of Met. (Modified from FINK et al. 1995)

brain areas containing cell bodies and in areas to which they project revealed that the 5-HT$_{1B}$ receptor is localized predominantly on axon terminals (BOSCHERT et al. 1994); although this study referred to presynaptic 5-HT$_{1B}$ heteroreceptors, the data were compatible with the above suggestion, namely that 5-HT$_{1B}$ autoreceptors account for only a minor fraction of the whole population of 5-HT$_{1B}$-binding sites.

2. Physiological Role

Do the inhibitory presynaptic 5-HT autoreceptors play a physiological role or, in other words, are they activated by endogenous 5-HT? The usual way to answer this question is to examine the effect of an appropriate 5-HT receptor antagonist on 5-HT release. If endogenous 5-HT is present in the biophase of such an inhibitory 5-HT receptor at a sufficient amount to produce a tonic activation, addition of an antagonist should modify 5-HT release in a manner opposite to that of 5-HT itself, i.e. it should facilitate 5-HT release. This approach is, however, not applicable to superfused synaptosomes, in which endogenous 5-HT is effectively removed from the biophase of the receptors by the superfusion stream. The procedure can also not be used in slices in which 5-HT release is evoked by a single electrical pulse or by the pseudo-one-pulse stimulation (few pulses administered at a very high frequency; SINGER 1988; see Sect. B.1) since in this case the amount of 5-HT released is too low to tonically activate the 5-HT autoreceptor.

It has already been mentioned previously that, in experiments on slices, the 5-HT receptor antagonist methiothepin $(0.1-1\,\mu M)$ given alone increased 5-HT release in various regions of the CNS of several species. These findings suggest that the presynaptic 5-HT receptors are tonically activated at the physiological frequency of electrical stimulation applied in the numerous studies (STARKE et al. 1989). In Table 1, the areas in which this has been demonstrated and, hence, a physiological role has been proved are marked by "E". In some of these areas of several species, tonic activation of the inhibitory presynaptic 5-HT autoreceptors has been found only recently. This holds true for rat spinal cord (YANG et al. 1994), mouse cerebral cortex and hypothalamus (SCHLICKER, FINK and GÖTHERT, unpublished), guinea pig cerebral cortex (WILKINSON et al. 1993), hippocampus, hypothalamus and substantia nigra (WILKINSON and MIDDLEMISS 1992; CHOPIN et al. 1994), rabbit colliculus superior and caudate nucleus (WICHMANN et al. 1989; FEUERSTEIN et al. 1992) and human hippocampus (SCHLICKER et al. 1996). Taken together it may be stated that methiothepin facilitated 5-HT release in any region of the CNS in which it has been studied. Such an effect has also been observed in in vivo experiments in which methiothepin was administered to the rat hippocampus or guinea pig cerebral cortex or substantia nigra via the dialysis probe (Table 2). These data support the view that the inhibitory presynaptic 5-HT autoreceptors play a physiological role under in vivo conditions.

II. 5-HT Receptors Mediating Facilitation of 5-HT Release

1. Identification and Location

A facilitatory feedback loop in which 5-HT receptors are involved has been identified only recently. Using superfused slices from the rat hypothalamus and guinea pig cerebral cortex, GALZIN and LANGER (1991) showed that the 5-HT$_3$ receptor agonist 2-methyl-5-HT facilitated the electrically evoked 5-HT release. This effect was counteracted by the 5-HT$_3$ receptor antagonists ondansetron and tropisetron. Basically the same results were obtained in guinea pig hypothalamic slices (BLIER and BOUCHARD 1992). In a more detailed study by BLIER and BOUCHARD (1993) on slices of the cerebral cortex, hypothalamus and hippocampus of guinea pigs, the facilitation of the electrically (1, 3 or 5Hz) evoked 5-HT release by 2-methyl-5-HT was confirmed, as was the sensitivity of this effect to blockade by various 5-HT$_3$ receptor antagonists (among others, ondansetron, tropisetron and S-zacopride). The enhancing effect of 2-methyl-5-HT in guinea pig hypothalamic slices was the more pronounced, the higher the frequency of stimulation with a constant number of pulses (3Hz compared with 1Hz) and the shorter the duration of stimulation at a constant frequency (BLIER and BOUCHARD 1993). In guinea pig hypothalamic slices, 2-methyl-5-HT also increased the K$^+$ (15–35mM)-evoked 5-HT release (BLIER et al. 1993). The facilitatory effect of 2-methyl-5-HT on the 5-HT release evoked by 25mM K$^+$ was antagonized by ondansetron (BLIER et al. 1993). At high concentrations, 2-methyl-5-HT even produced direct stimulation of 5-HT release from guinea pig hypothalamic slices, which was sensitive to blockade by S-zacopride (BLIER and BOUCHARD 1993).

A facilitatory effect of 2-methyl-5-HT on the electrically evoked 5-HT release was observed in rat frontal cortical and hippocampal slices as well (HADDJERI and BLIER 1995); this effect was blocked by the 5-HT$_3$ receptor antagonists BRL 46470A [(endo-N-(8-methyl-8-azabicyclo[3.2.1.]oct-3-yl)2,3-dihydro-3,3-dimethylindole-1-carboxamide; ricasetion] and S-zacopride.

The receptors involved in these effects are also operative in vivo (MARTIN et al. 1992). In the ventral hippocampus of rats anaesthetized with chloral hydrate, administration of 2-methyl-5-HT via the microdialysis probe increased dialysate 5-HT levels in a concentration-related manner. This effect was counteracted by the 5-HT$_3$ receptor antagonist MDL 72222 (endo-8-methyl-8-azabicyclo[3.2.1.]oct-3-yl-3,5-dichlorobenzoate).

The receptors mediating facilitation of 5-HT release are not autoreceptors (BLIER et al. 1993) but are probably located on an interneuron. This conclusion can be drawn from data of BLIER et al. (1993), who found in guinea pig hypothalamic synaptosomes that 2-methyl-5-HT did not facilitate the K$^+$-evoked 5-HT release, nor did it stimulate 5-HT release by itself. In agreement with this, no enhancement of the K$^+$-evoked 5-HT release by 2-methyl-5-HT was observed in hypothalamic slices, when tetrodotoxin was present in the superfusion fluid (BLIER et al. 1993). Hence, a neuronal circuit comprising one or more interneurons and an unknown presynaptic heteroreceptor on the serotoninergic axon terminals should be involved (see Sect. A).

2. Physiological Role

Relatively little information is so far available with respect to the potential tonic serotoninergic input to the facilitatory 5-HT receptors. In this case, 5-

HT$_3$ receptor antagonists should cause an inhibition of 5-HT release, but Blier and Bouchard (1993) found in guinea pig hypothalamic slices that, as a rule (e.g. with ondansetron, tropisetron, MDL 72222), this did not occur. Nevertheless, their experiments in which the neuronal 5-HT transporter was blocked by paroxetine left the possibility open that the facilitatory 5-HT receptors can be activated by endogenous 5-HT when, under certain conditions, its concentration in the synaptic cleft is increased: paroxetine mimicked the effect of 2-methyl-5-HT in that it enhanced the electrically evoked 5-HT release in guinea pig hypothalamic slices, an inhibition which was sensitive to blockade by the 5-HT$_3$ receptor antagonist tropisetron.

In the in vivo study of Martin et al. (1992) on the rat ventral hippocampus, the 5-HT$_3$ receptor antagonist MDL 72222 did not affect 5-HT release, thus arguing against the possibility that the 5-HT receptors involved are activated by endogenous 5-HT under the conditions applied.

D. Classification

I. Inhibitory Presynaptic 5-HT Autoreceptors

1. Classification in Terms of 5-HT$_{1B}$ and "5-HT$_{1D}$" Receptors

Until 1990, classification and nomenclature of 5-HT receptors had mainly been based on their operational/transductional behaviour and their pharmacological properties (Bradley et al. 1986; Leff and Martin 1988; Peroutka 1988; Frazer et al. 1990). At that time, little information was available on 5-HT receptor structure, since only few genes coding for these receptors had been sequenced; however, the results available from molecular biological studies roughly supported the classification scheme derived from operational data.

Due to the lack of selective ligands at certain 5-HT receptors, these 5-HT receptors, among them the presynaptic 5-HT autoreceptor (see Sect. C.I), could only be classified by application of a large number of 5-HT receptor agonists and antagonists (e.g. Engel et al. 1986). In such attempts to classify the presynaptic 5-HT autoreceptors, the potencies of agonists in inhibiting 5-HT release from rat hypothalamic synaptosomes and in rat and pig cerebral cortical slices and of antagonists in antagonizing this inhibition were compared with their potencies in inhibiting binding of appropriate [^{3}H]ligands and [^{125}I]ligands to the various 5-HT-binding sites in membranes from the CNS of several species (Martin and Sanders-Bush 1982; Engel et al. 1983, 1986; Schlicker et al. 1989). In subsequent studies which were based on part of these results, relatively few ligands which exhibited a certain, yet limited degree of selectivity for 5-HT receptor subtypes have been applied for classification of the presynaptic 5-HT autoreceptors (e.g. Bonanno et al. 1986; Maura et al. 1986; for review, see Starke et al. 1989). Surprisingly, it was found that the presynaptic 5-HT autoreceptor belongs to another subtype than the firing-regulating somadendritic 5-HT autoreceptor: whereas the latter had been characterized as being of the 5-HT$_{1A}$ subtype (Sprouse and Aghajanian

1986, 1987), the presynaptic 5-HT autoreceptor was classified as 5-HT_{1B} in various regions of the rat CNS and as "5-HT_{1D}" (former nomenclature, labelled by quotation marks) in the pig cerebral cortex (for review and references, see STARKE et al. 1989; nomenclature of "5-HT_{1D}" receptors recently revised, see HARTIG et al. 1996 and below, Sect. D.I.3); furthermore, there were hints that in the guinea pig and rabbit brain the presynaptic 5-HT autoreceptors also possess "5-HT_{1D}" properties.

Although several pharmacological similarities were recognized between rat 5-HT_{1B}-binding sites and/or receptors on the one hand and guinea pig, pig, bovine and human "5-HT_{1D}" sites/receptors on the other, there were some fundamental differences between these subtypes. The most striking one was observed with certain β-adrenoceptor antagonists. Thus, propranolol and cyanopindolol also exhibited rather high affinity for rat 5-HT_{1B}-binding sites ($pK_i > 7$; see footnote to Table 3) and stereoselectively acted as partial agonists exhibiting antagonistic property at rat presynaptic 5-HT autoreceptors [the $(-)$-enantiomers being more potent]; however, β-adrenoceptor antagonists had a clearly lower affinity for "5-HT_{1D}"-binding sites in the brain of the guinea pig, pig, cow and man (see footnote to Table 3; HEURING and PEROUTKA 1987; WAEBER et al. 1988; 1989). Furthermore, propranolol did not antagonize the inhibitory effect of the 5-HT receptor agonist 5-methoxytryptamine on 5-HT release in the pig cerebral cortex (for review, see STARKE et al. 1989; for details of the investigation in pig cerebral cortex, see SCHLICKER et al. 1989). Another difference between the rat 5-HT_{1B}-binding sites and the "5-HT_{1D}" sites of the other species mentioned was that the 5-HT_{1A} receptor agonist 8-OH-DPAT [8-hydroxy-2(di-n-propylamino)tetralin] had very low affinity for the former, but moderate affinity ($pK_i \geq 6$) for "5-HT_{1D}" sites (see footnote to Table 3). Accordingly, 8-OH-DPAT did not inhibit 5-HT release in the rat CNS (see, however, LIMBERGER et al. 1991 and Sect. D.I.4), but behaved as a low-potency agonist at the inhibitory presynaptic 5-HT autoreceptors of the pig cerebral cortex (see STARKE et al. 1989). Other ligands exhibiting a preference in favour of "5-HT_{1D}" receptors are the two rauwolfia alkaloids rauwolscine and yohimbine as well as dipropyl-5-carboxamidotryptamine (see footnotes to Table 3). In recent years, the inhibitory presynaptic 5-HT autoreceptors in various regions of the brain of various species other than rat and pig, including man, have been classified in terms of 5-HT_{1B} and "5-HT_{1D}" (see Tables 1, 3). The similarities in distribution and function between the 5-HT_{1B}- and "5-HT_{1D}"-binding sites/receptors in the various species led HOYER and MIDDLEMISS (1989) to propose that these receptors are species homologues.

2. Subclassification of "5-HT_{1D}" Receptors

Molecular biological studies which revealed that the "5-HT_{1D}" receptors are not homogeneous made it necessary to extend the nomenclature of these receptors (for review, see HARTIG et al. 1992). Two different genes which code for receptors with binding properties virtually identical to those previously

Table 3. Pharmacological properties of the inhibitory presynaptic 5-HT autoreceptors in the brain of various species as a basis for the classification of these receptors in terms of 5-HT$_{1B}$ and "5-HT$_{1D}$" (last but one column). This "classical" nomenclature was primarily based on the pharmacological differences between 5-HT$_{1B}$-binding sites in rat brain membranes and "5-HT$_{1D}$"-binding sites in bovine, guinea pig, porcine and human cerebral membranes (for references, see Sect. D.I.1). These differences were most pronounced with the drugs considered in this table (for references, see footnotes). The classification scheme of the 5-HT$_{1B}$/"5-HT$_{1D}$" receptors has recently been revised on

Species	CNS region	References	Potency of 8-OH-DPAT[a]	Potency of yohimbine (Y) or rauwolscine (R)[b]	Potency of dipropyl-5-CT[c]
Rat	Cerebral cortex	Middlemiss and Hutson (1990)[h]			
	Spinal cord	Matsumoto et al. (1992)	pD$_2$ < 6		
		Yang et al. (1994)	(0.1 μM inactive)		
Mouse	Cerebral cortex	Schipper (1990)	("8-OH-DPAT inactive")[j]		
Guinea pig	Cerebral cortex	Ormandy (1993)			pD$_2$ 6.8
		Bühlen et al. (1996a) Synaptosomes Slices			
		Limberger et al. (1991)	pD$_2$ 7.06 (partial agonism)	Y: pK$_B$ 6.66 (partial agonism)	
Rabbit	Cerebral cortex	Limberger et al. (1991)	pD$_2$ 6.82 (partial agonism)	Y: pD$_2$ 7.01 (partial agonism)	
	Caudate nucleus	Feuerstein et al. (1992)	(agonism at 0.1 and 0.3 μM)	R: pD$_2$ about 6.5 (partial agonism)	
Rhesus monkey	Cerebral cortex	Schipper (1990)	pD$_2$ 7.1		
Man	Cerebral cortex	Galzin et al. (1992)	(tendency towards agonism at 0.1 μM)	(agonism by Y 1 μM)	
		Maura et al. (1993)	(agonism at 0.3 and 1 μM)		

[a] pK_i values of 8-OH-DPAT [8-hydroxy-2(di-n-propylamino)tetralin] in binding studies: 5-HT$_{1B}$ 4.22–5.75; "5-HT$_{1D}$" 5.90–7.55 (compiled by Zifa and Fillion 1992).
[b] pK_i values of yohimbine and rauwolscine in binding studies: 5-HT$_{1B}$ 5.46 and 5.25, respectively; "5-HT$_{1D}$" 7.12 and 7.65, respectively (Schoeffter and Hoyer 1990).
[c] pK_i values of dipropyl-5-CT in binding studies: 5-HT$_{1B}$ 4.88; "5-HT$_{1D}$" 6.92 (Schoeffter and Hoyer 1990).
[d] pK_i values of cyanopindolol (CYP) and (−)-propranolol (PRP) in binding studies: 5-HT$_{1B}$ 8.28 and 7.33, respectively; "5-HT$_{1D}$" 6.85 and 5.49, respectively (Schoeffter and Hoyer 1990).
[e] The potencies (pD$_2$) of agonists [in the study by Middlemiss and Hutson (1990), also of antagonists; pA$_2$] at the presynaptic autoreceptors were compared by linear regression analysis with their affinities (pK_i) at 5-HT$_{1B}$- and "5-HT$_{1D}$"-binding sites (n, number of drugs). Either the regression analyses were

the basis of the primacy of the human genome (last column; new nomenclature according to Hartig et al. 1996), taking into account that all presynaptic 5-HT autoreceptors listed in this table are probably encoded by orthologous genes (see Sects. D.I.3, 4). This table is restricted to more recent investigations not yet considered in the review by Starke et al. (1989). Slices or synaptosomes preincubated with [³H] 5-HT were superfused and [³H] 5-HT release was evoked by electrical field stimulation or high K⁺. In each study, the inhibitory effect of 5-HT or 5-carboxamidotryptamine 5-CT on the evoked [³H] 5-HT release was counteracted by methiothepin or metergoline

Potency of cyanopindolol (CYP) or propranolol (PRP)[d]	Correlation coefficients (r): comparison of potencies of ligands with affinities for binding sites[e]		Classification	
			Primacy of pharmacology[f]	Primacy of genome[g]
	5-HT$_{1B}$	"5-HT$_{1D}$"		
CYP: pA$_2$ 8.4	0.87 (n = 12)	0.32 (n = 11)	5-HT$_{1B}$[i]	r5-HT$_{1B}$
	0.79 (n = 8)	0.71 (n = 7)	5-HT$_{1B}$	r5-HT$_{1B}$
CYP: pA$_2$ 8.3			5-HT$_{1B}$	m5-HT$_{1B}$
	0.87 (n = 5)	0.88 (n = 5)	"5-HT$_{1D}$"	gp5-HT$_{1B}$
	0.78 (n = 6)	0.86 (n = 6)		
	0.55 (n = 6)	0.82 (n = 6)		
PRP: pA$_2$ < 6; pD$_2$ < 5[k]	0.76 (n = 4)	0.95 (n = 4)		
PRP: pA$_2$ < 6; pD$_2$ < 6[k]	0.84 (n = 5)	0.94 (n = 5)	"5-HT$_{1D}$"	rb5-HT$_{1B}$
(agonism by CYP 1 and 10 μM)			"5-HT$_{1D}$"	rb5-HT$_{1B}$
CYP: pA$_2$ 6.5			"5-HT$_{1D}$"	mk5-HT$_{1B}$
(antagonism by PRP 1 μM)			"5-HT$_{1D}$"	h5-HT$_{1B}$

carried out by the authors themselves or by us, using the potencies given by the authors quoted in this table and the pK$_i$ values reported by Schoeffter and Hoyer (1990; 1991), Hoyer and Schoeffter (1991) and Beer et al. (1993).
[f] "Classical" nomenclature in terms of 5-HT$_{1B}$ and "5-HT$_{1D}$" (see Sect. D.I.1).
[g] New nomenclature (see Sects. D.I.3, 4).
[h] pD$_2$, pA$_2$ and pK$_i$ values were assessed from Fig. 4 of that paper.
[i] The same conclusion was previously drawn by Engel et al. (1986). The study by Limberger et al. (1991), however, provides evidence that, in addition, an r5-HT$_{1D}$ receptor is involved (see Sect. D.I.4).
[j] No exact concentration range given in that paper.
[k] pD$_2$ for partial agonism.

determined for "5-HT$_{1D}$"-binding sites in human cerebral membranes, but different from rat 5-HT$_{1B}$ sites, were isolated from the human brain. These receptors were called "5-HT$_{1D\alpha}$" and "5-HT$_{1D\beta}$". Whereas they exhibited almost identical binding characteristics for 19 different 5-HT receptor agonists and antagonists (see HARTIG et al. 1992), they can be distinguished by ketanserin and, less well, by ritanserin, which have 75- and 24-fold higher affinity, respectively, for cloned human "5-HT$_{1D\alpha}$" than "5-HT$_{1D\beta}$" receptors (ZGOMBICK et al. 1995).

Accordingly, ketanserin has been used to determine the "5-HT$_{1D}$" receptor subtype to which the presynaptic 5-HT autoreceptors belong in the human (FINK et al. 1995 and Fig. 2) and guinea pig cerebral cortex (BÜHLEN et al. 1996a): they were subclassified as "5-HT$_{1D\beta}$" and "5-HT$_{1D\beta}$-like", respectively (for more details see below, Sect. D.I.4).

The gene which encodes the human "5-HT$_{1D\beta}$" receptor is highly homologous to a gene which was isolated from the rat brain and which was found to code for a receptor with the pharmacological characteristics of the 5-HT$_{1B}$-binding sites in rat cerebral membranes (HARTIG et al. 1992). These findings strongly supported the suggestion that, in spite of the pharmacological differences mentioned above, the presynaptic 5-HT$_{1B}$ autoreceptor in the rat and the presynaptic "5-HT$_{1D}$" autoreceptor in other species such as the guinea pig and man are species homologues (see above; HOYER and MIDDLEMISS 1989): only few differences in the amino acid sequence are responsible for the differences in pharmacological profile (METCALF et al. 1992; OKSENBERG et al. 1992; PARKER et al. 1993).

In addition to the gene encoding the rat 5-HT$_{1B}$ receptor, a rat homologue of the gene encoding the human 5-HT$_{1D\alpha}$ receptor was identified; this rat gene coded for a receptor with the binding properties of a "5-HT$_{1D}$" receptor (BACH et al. 1993).

3. Revised Nomenclature of 5-HT$_{1B/1D}$ Receptors on the Basis of the Primacy of the Human Genome

As a result of the history of the discovery of genes, binding sites and receptors of the "5-HT$_{1B/1D}$" subtype, a complex and confusing nomenclature of this receptor subfamily has developed (GÖTHERT 1992; ZIFA and FILLION 1992; BOESS and MARTIN 1994; HOYER et al. 1994; MARTIN and HUMPHREY 1994). Therefore, the nomenclature has recently been revised and simplified on the basis of the primacy of the human genome (HARTIG et al. 1996) taking the following general rules, accepted by the International Union of Pharmacology Committee for Receptor Nomenclature, into account: a name should first be established for the amino acid sequence of a receptor subtype deduced from a certain human gene. The same name should be used for all species homologues of this receptor subtype or, in other words, for all receptors which are the products of orthologous genes, irrespective of whether or not the pharmacological properties of these receptors are identical. The species should be specified by brief letter prefixes (e.g. "h" for human, "mk" for monkey, "r" for rat, "m" for mouse, "gp" for guinea pig, "rb" for rabbit, "p" for pig; VANHOUTTE et al. 1996). For instance, taking these principles into account, the human "5-HT$_{1D\beta}$" and "5-HT$_{1D\alpha}$" receptors have been renamed h5-HT$_{1B}$ and h5-HT$_{1D}$, respectively, and analogously the rat 5-HT$_{1B}$ and "5-HT$_{1D}$" receptors are termed r5-HT$_{1B}$ and r5-HT$_{1D}$, respectively. Thus, it becomes obvious at first sight that the h5-HT$_{1B}$ and r5-HT$_{1B}$ receptors belong to the same subtype in spite of their pharmacological differences.

It is also clear from the previous sections that the new nomenclature has more far-reaching consequences for the terminology of the presynaptic 5-HT autoreceptors than for any other functional 5-HT receptors described so far. Therefore, the development of the nomenclature had to be described here in some detail in order to enable access from this review to the underlying original work on these receptors.

4. Subclassification and Nomenclature in Terms of the Revised Nomenclature

As outlined above and in Table 3, the presynaptic 5-HT autoreceptors have originally been classified as 5-HT$_{1B}$ in rats and mice and as "5-HT$_{1D}$" (according to the nomenclature introduced before 1989) in the other species investigated up to now. It has also been mentioned that, in view of the heterogeneity of the latter receptors, attempts have recently been made in guinea pig and human brain slices and synaptosomes to subclassify the presynaptic 5-HT autoreceptors (Sect. D.I.2). Some details of these studies, which appeared before the new nomenclature was published, are worthwhile to be reported here. In order to avoid confusion, the primacy of the new nomenclature is emphasized by mentioning the new name either alone or first, the old nomenclature being added in some cases in quotation marks.

It has already been stated above (Sect. D.I.2) that subclassification of human presynaptic 5-HT autoreceptors was complicated by the lack of compounds which discriminate between h5-HT$_{1B}$ ("5-HT$_{1D\beta}$") and h5-HT$_{1D}$ ("5-HT$_{1D\alpha}$") receptors and that among the drugs available ketanserin was most suitable in this respect. In a study in human cortical synaptosomes (FINK et al. 1995; GÖTHERT et al. 1996), ketanserin at a concentration 4.4 times higher than its K_i at cloned h5-HT$_{1D}$ receptors but more than 16 times lower than its K_i at cloned h5-HT$_{1B}$ receptors failed to antagonize the inhibitory effect of 5-CT on the K$^+$-evoked [^{3}H]5-HT release (Fig. 2). In contrast, the non-selective antagonist at h5-HT$_{1B/1D}$ receptors, methiothepin, at a concentration 4.4 times higher than its K_i at cloned h5-HT$_{1B}$ receptors abolished the 5-HT-induced inhibition (Fig. 2). This pattern of effects of the two antagonists made it possible to conclude that the presynaptic inhibitory 5-HT autoreceptor of the human cerebral cortex belongs to the h5-HT$_{1B}$ ("5-HT$_{1D\beta}$"; Sect. D.I.2) subtype.

Very recently two new drugs SB 216641 (N-[3-(2-dimethylamino)ethoxy-4-methoxyphenyl]-2'-methyl-4'-(5-methyl-1,2,4-oxadiazol-3-yl)-(1,1'-

biphenyl)-4-carboxamide) and BRL 15572 (1-phenyl-3-[4-(3-chlorophenyl) piperazin-1-yl]phenylpropan-2-ol), which discriminate between h5-HT$_{1B}$ and h5-HT$_{1D}$ receptors, have become available: SB 216641 exhibited 30-fold higher affinity for cloned h5-HT$_{1B}$ receptors than for cloned h5-HT$_{1D}$ receptors, whereas BRL 15572 had over 100-fold greater affinity for cloned h5-HT$_{1D}$ than for cloned h5-HT$_{1B}$ receptors (PRICE et al. 1996). In human cerebral cortical synaptosomes, the inhibitory effect of 5-HT on the K$^+$-evoked 5-HT release was antagonized by SB 216641, but not affected by BRL 15572 (both drugs applied at concentrations 15 times higher than their K_i at cloned h5-HT$_{1B}$ and h5-HT$_{1D}$ receptors, respectively; SCHLICKER et al. 1997). These data strongly support the conclusion that the presynaptic 5-HT autoreceptor in the human cerebral cortex is of the 5-HT$_{1B}$ subtype. In agreement with this, SB 216641, but not BRL 15572, increased the electrically evoked 5-HT release from human cerebral cortical slices.

Basically the same approach as in human brain preparations was used to subclassify the presynaptic 5-HT autoreceptor in the guinea pig cerebral cortex, although the gp5-HT$_{1B}$ and gp5-HT$_{1D}$ receptors have not yet been cloned. However, the pharmacological properties of the presynaptic 5-HT autoreceptors in the guinea pig cerebral cortex (LIMBERGER et al. 1991; BÜHLEN et al. 1996a; ROBERTS et al. 1996) are similar to those in the human cerebral cortex (SCHLICKER et al. 1985; GALZIN et al. 1992; MAURA et al. 1993) and the same holds true for the "5-HT$_{1D}$"-binding sites in cerebral membranes from both species (WAEBER et al. 1988, 1989; BEER et al. 1992; BRUINVELS et al. 1992). Therefore, it is generally accepted that the guinea pig brain is suitable as an experimental model for the development of autoreceptor ligands which may become useful as therapeutic compounds in human diseases (see below). To provide evidence that in the guinea pig cerebral cortex the presynaptic 5-HT autoreceptor is similar to the h5-HT$_{1B}$ receptor ("5-HT$_{1D\beta}$"-like), a study was carried out in guinea pig cortical synaptosomes in which ketanserin and methiothepin were applied at appropriate concentrations (BÜHLEN et al. 1996a). In fact it was found that the inhibitory effect of 5-CT on 5-HT release was not affected by ketanserin at a concentration 4 times higher than its K_i at cloned h5-HT$_{1D}$ ("5-HT$_{1D\alpha}$") receptors, but that methiothepin, at a concentration 4 times higher than its K_i value at h5-HT$_{1B}$ ("5-HT$_{1D\beta}$") receptors, caused a rightward shift of the concentration-response curve of 5-CT.

Further support for the suggestion that the presynaptic 5-HT autoreceptors in the guinea pig brain predominantly are h5-HT$_{1B}$-like came from binding studies in cortical synaptosomes and membranes from this species. In the synaptosomal preparation, the fraction of presynaptic membranes related to the total quantity may be assumed to be higher than in conventionally prepared membranes. Whereas the pharmacological properties of the [^{3}H]5-HT binding sites in membranes correlated with those of both h5-HT$_{1B}$ and h5-HT$_{1D}$ receptors, the pharmacological characteristics of the [^{3}H]5-HT-binding sites in synaptosomes correlated with those of h5-HT$_{1B}$ ("5-HT$_{1D\beta}$") only, but not with those of h5-HT$_{1D}$ receptors ("5-HT$_{1D\alpha}$"; BÜHLEN et al.

1996a). In both preparations there was no correlation with the properties of the other 5-HT receptors within the seven 5-HT receptor families which have so far been identified in the guinea pig or other species.

On the basis of the observation that the presynaptic 5-HT autoreceptor belongs to the $5-HT_{1B}$ subclass of the new classification scheme not only in rat and mouse but also in guinea pig and man, it may be suggested that this probably also holds true for the other species in which they have previously been classified as "$5-HT_{1D}$", i.e. for pig (SCHLICKER et al. 1989), rhesus monkey and rabbit (Table 3; the $rb5-HT_{1B}$ receptor has recently been cloned; BARD et al. 1996).

However, this general rule does not implicate that all presynaptic 5-HT receptors are of the $5-HT_{1B}$ subtype. Thus, the presynaptic inhibitory 5-HT heteroreceptors on the sympathetic nerve fibres innervating the human atrial appendages have been classified by means of ketanserin as $h5-HT_{1D}$ ("$5-HT_{1D\alpha}$"; GÖTHERT et al. 1996; MOLDERINGS et al. 1996). Another example refers to the presynaptic 5-HT autoreceptors in the rat brain. According to experiments by LIMBERGER et al. (1991) in cerebral cortical slices, not all of these receptors belong to the $r5-HT_{1B}$ subtype, but a part of them could be subclassified as $r5-HT_{1D}$: whereas the effects of most drugs conformed to the $r5-HT_{1B}$ character of the presynaptic 5-HT autoreceptor (e.g. the antagonism by the preferential $5-HT_{1B}$ receptor antagonist isamoltane of the inhibitory effect of RU 24969 [5-methoxy-3(1,2,3,6 tetrahydropyridin-4-yl)-1H-indole] on [^{3}H]5-HT release), the inhibitory effect of 8-OH-DPAT in the high nanomolar range did not. The potency of this compound was virtually identical to that in guinea pig or rabbit cortical slices examined in the same study. The effect of 8-OH-DPAT was counteracted by methiothepin but not by isamoltane. These findings with 8-OH-DPAT are compatible with the involvement of $r5-HT_{1D}$ receptors.

In the guinea pig hippocampus, two different presynaptic 5-HT autoreceptors appear to occur as well (WILKINSON and MIDDLEMISS 1992): in slices of this brain region methiothepin antagonized the effect of 5-HT at higher potency (apparent pA_2 value 7.6) than the effects of 5-CT and sumatriptan (pA_2 6.7–7.0). It is by no means clear whether these findings can be accounted for by the occurrence of both the $gp5-HT_{1B}$ and the $gp5-HT_{1D}$ receptor or of another 5-HT receptor in addition to the $gp5-HT_{1B}$ receptor. The operation of multiple presynaptic 5-HT autoreceptors has recently been confirmed in the guinea pig cerebral cortex by analysis of the concentration response curves for 5-HT and 5-CT and of the antagonistic property of GR 127935 [2'-methyl-4'-(5-methyl-[1,2,4]oxadiazol-3-yl)-biphenyl-4-carboxylic acid [4-methoxy-3-(4-methyl-piperazin-1-yl)-phenyl]-amide] (ROBERTS et al. 1996).

Heterogeneity of 5-HT receptors modulating 5-HT release has recently been proved in cortical and hippocampal slices from the $m5-HT_{1B}$ receptor "knock-out" mice (PINEYRO et al. 1995b) generated by SAUDOU et al. (1994). It was found that the $r5-HT_{1B}$ receptor agonist CP-93129 [3-(1,2,5,6-tetrahydropyrid-4-yl)pyrrolo[3,2-b]pyrid-5-one] and the $5-HT_{1B/1D}$ receptor agonist sumatriptan did not inhibit the electrically evoked 5-HT release, whereas the less selective compound 5-CT was still capable of decreasing 5-HT release. In fact, 5-CT has relatively high affinity not only for 5-HT$_{1B}$ and 5-HT$_{1D}$ receptors but also for 5-HT$_5$ and 5-HT$_7$ receptors (Hoyer et al. 1994). Therefore, the results with 5-CT suggest that, at least in the absence of m5-HT$_{1B}$ receptors, autoinhibition of 5-HT release may be mediated by m5-HT$_5$ and/or m5-HT$_7$ receptors (Pineyro et al. 1995b). It is conceivable that the inhibitory effect mediated via other than m5-HT$_{1B}$ receptors may play a more important role in m5-HT$_{1B}$ receptor "knock-out" animals than under normal conditions: these receptors may be expressed at higher density as a compensatory mechanism. The location of such non-5-HT$_{1B}$ receptors is not clear. They may either represent true presynaptic 5-HT autoreceptors or they may be located on non-serotoninergic neurons within a short regulatory neuronal loop (see Sect. A).

The new nomenclature of 5-HT$_{1B/1D}$ receptors offers a further advantage, namely that allelic polymorphism as a basis of structural variants of receptor subtypes can easily be dealt with in this classification scheme (Hartig et al. 1996). Recently, a coding mutation in nucleotide position 371 (T→G substitution) of the h5-HT$_{1B}$ receptor gene was found in single-strand conformation analysis of the h5-HT$_{1B}$ receptor gene from healthy persons for DNA sequence variation. This mutation leads to an amino acid exchange (Phe→Cys) in position 124 of the receptor protein in the third transmembrane domain close to the junction with the extracellular loop (Nöthen et al. 1995). In membranes from cultured COS-7 cells transfected with the cDNA of the wild-type or mutant h5-HT$_{1B}$ receptor, binding of [^{3}H]5-CT was determined (Bühlen et al. 1996b,c). In saturation experiments the affinity of [^{3}H]5-CT for the mutant h5-HT$_{1B}$ receptor was higher, whereas B_{max} of this receptor was lower compared to the wild-type receptor. Competition experiments with eight full or partial 5-HT$_{1B}$ receptor agonists including 5-HT, dihydroergotamine, sumatriptan and methysergide revealed that their inhibitory potencies were 0.3–0.5 log units higher than for the wild-type receptor. Corresponding changes in the pharmacological properties of the h5-HT$_{1B}$ receptor may be assumed to occur in individuals in whom the mutant h5-HT$_{1B}$ gene is expressed as presynaptic 5-HT autoreceptor: the increased affinity of the endogenous agonist 5-HT and of other 5-HT$_{1B}$ receptor ligands may lead to a decrease in 5-HT release (by increased autoreceptor function) and to pharmacogenetic differences in the action of 5-HT$_{1B}$ receptor ligands.

II. 5-HT Receptors Mediating Facilitation of 5-HT Release

The facilitatory 5-HT receptors identified in the guinea pig and rat brain in vitro (Galzin and Langer 1991; Blier and Bouchard 1993) and in vivo (Martin et al. 1992) belong to the 5-HT$_3$ receptor class. This can be deduced mainly from the findings with the 5-HT$_3$ receptor agonists and antagonists mentioned above (Sect. C.II.1). When taking the species differences in the

pharmacological properties of 5-HT$_3$ receptors into account (for review see
KILPATRICK et al. 1990; HOYER et al. 1994; Chap. 18, this volume), the facilitatory 5-HT receptor in the guinea pig brain (cerebral cortex, hippocampus and
hypothalamus) also fulfilled the criteria characteristic for the 5-HT$_3$ receptor
of this species (BLIER and BOUCHARD 1993): the agonists phenylbiguanide and
m-chlorophenylbiguanide, which have been shown to be effective 5-HT$_3$ receptor agonists in rat but not guinea pig tissues (KILPATRICK et al. 1991;
ROBERTSON and BEVAN 1991; BUTLER et al. 1990), failed to mimic the facilitatory effect of 2-methyl-5-HT; in agreement with the general properties of 5-
HT$_3$ receptors in guinea pig tissues (for references, see above), the potencies of
the 5-HT$_3$ receptor antagonists were generally low, (+)-tubocurarine even
being ineffective.

In agreement with the property of 5-HT$_3$ receptors to desensitize in response to prolonged stimulation (PETERS and LAMBERT 1989; YAKEL 1992;
Chap. 18, this volume), it was observed in rat hypothalamic and guinea pig
cerebral cortical and hypothalamic slices that 2-methyl-5-HT facilitated 5-HT
release only if the time of exposure was 8 min but not if it was extended to
20 min (GALZIN and LANGER 1991; BLIER and BOUCHARD 1993). However, in
this context it should be noted that in electrophysiological experiments, 5-HT$_3$
receptors proved to desensitize much faster (YAKEL and JACKSON 1988; YAKEL
1992).

E. Site and Mechanism of Action

I. Inhibitory Presynaptic 5-HT Autoreceptors

From early work carried out before 1989 only few, rather general conclusions
could be drawn concerning the ionic or biochemical events induced in the cell
membrane by activation of inhibitory presynaptic 5-HT autoreceptors
(for review and references, see STARKE et al. 1989). Propagation of action
potentials seems not to be affected, but stimulus-release coupling, in particular
transmembrane Ca^{2+} influx into the serotoninergic axon terminals, appears
to be inhibited. The results of experiments designed to investigate the
involvement of G proteins in autoreceptor-mediated effects were contradictory. Whereas PASSARELLI et al. (1988) found the effect of two
5-HT receptor agonists on 5-HT release in rat hippocampal slices to be
pertussis toxin-sensitive, the data obtained in interaction experiments of
5-HT receptor ligands with pharmacological tools known to influence the
adenylyl cyclase (SCHLICKER et al. 1987, rat brain cortex slices) or protein
kinase C second messenger system (FEUERSTEIN et al. 1987, rabbit hippocampal slices; WANG and FRIEDMAN 1987, rat brain cortex slices) argued against
inhibitory coupling of the autoreceptors to these biochemical systems via G
proteins.

Although it is now clear that the 5-HT$_{1B}$ receptor and, accordingly, the
presynaptic 5-HT autoreceptor belongs to the superfamily of G-protein-

coupled receptors (Hoyer et al. 1994), the experimental evidence for the involvement of G proteins and adenylyl cyclase or protein kinase C remained equivocal. Blier (1991) studied the effect of two inhibitors of $G_{i/o}$ proteins, pertussis toxin and N-ethylmaleimide (NEM), and of a stimulator of G_s proteins, cholera toxin, in rat hippocampal slices. Pertussis and cholera toxin, with which the rats were pretreated, and which by themselves did not change 5-HT release, failed to modify the effect of 5-methoxytryptamine, 5-CT and/or methiothepin on 5-HT release. NEM, which by itself increased 5-HT release, also did not change the effect of 5-methoxytryptamine. In guinea pig brain cortex slices, forskolin (an activator of adenylyl cyclase), rolipram (an inhibitor of cAMP phosphodiesterase) and 8-Br-cAMP (a lipid-soluble analogue of cAMP) did not affect 5-HT release when given alone, but attenuated the modulatory effect of 5-CT and/or sumatriptan on 5-HT release (Ormandy 1993). 1,9-Dideoxyforskolin, which fails to activate adenylyl cyclase (but, like forskolin, blocks K^+ channels), did not attenuate the effect of 5-CT. In rat hypothalamic slices, forskolin, 8-Br-cAMP and an inhibitor of phospodiesterase (IBMX) both increased 5-HT release, when given alone, and did not attenuate the inhibitory effects of 5-methoxytryptamine and RU 24969 on 5-HT release (Ramdine et al. 1989). In contrast, inhibition of protein kinase C by phorbol-11,12-dibutyrate (which by itself also increased 5-HT release) attenuated the modulatory effects of 5-methoxytryptamine and methiothepine on 5-HT release.

Taken together, the inconsistencies observed in these experiments on cerebral slices are probably due to the fact that the methods applied so far have been too indirect. More direct approaches in future studies will probably provide more unequivocal results.

II. 5-HT Receptors Mediating Facilitation of 5-HT Release

The ionic and biochemical effects in the cell membrane induced by activation of the 5-HT receptor mediating facilitation of 5-HT release are even less well known. This is already evident when considering that the exact location of the receptor is still unclear. As mentioned above (Sects. A, C.II.1), the 5-HT receptor is probably located on a non-serotoninergic interneuron. Hence, in a cerebral slice preparation, an event induced by activation of such a receptor may occur in any of the nerves involved in the facilitatory neuronal circuit.

On the basis of the investigation by Blier and Bouchard (1993) on guinea pig hypothalamic slices, it appears that an increase in transmembrane influx of Ca^{2+} ions is a prerequisite for the 5-HT$_3$ receptor-mediated facilitation: the facilitatory effect of 2-methyl-5-HT on 5-HT release only occurred when 5-HT release was stimulated by Ca^{2+}-dependent methods of stimulation such as electrical impulses or high K^+, whereas the Ca^{2+}-independent fenfluramine-induced 5-HT release was not altered.

F. Plasticity of Inhibitory Presynaptic 5-HT Autoreceptors

Very recently it has been shown that the functional activity of the presynaptic 5-HT autoreceptor can be modulated by an endogenous tetrapeptide, Leu-Ser-Ala-Leu (LSAL), which was called 5-HT-moduline (MASSOT et al. 1996; Chap. 11, this volume). This peptide is released from synaptosomes and binds to specific recognition sites in mammalian brain and in transfected cells expressing 5-HT$_{1B}$ receptors. In rat hippocampal synaptosomes, interaction of 5-HT-moduline with presynaptic 5-HT$_{1B}$ autoreceptors was shown to decrease, and at appropriate concentration even abolish, the inhibitory effect of a 5-HT$_{1B}$ receptor agonist on evoked 5-HT release.

In numerous investigations (see below), the possibility has been examined that the sensitivity of inhibitory 5-HT autoreceptors may change with the light-dark cycle, aging and chronic treatment with drugs, in particular antidepressant drugs, or electroconvulsive shocks (ECS).

I. Lack of Circadian Variations

Many of the steps involved in the synthesis and release of 5-HT are subject to circadian variations (MARTIN 1991). For example, the release of 5-HT measured by in vivo microdialysis in the hippocampus of freely moving rats undergoes changes during the light-dark cycle (KALEN et al. 1989), a phenomenon which might be related to variations of the sensitivity of the inhibitory 5-HT autoreceptor. This possibility was examined in experiments of SINGH and REDFERN (1994a,b), who determined 5-HT release and its autoreceptor-mediated modulation in slices from the rat cerebral cortex and hippocampus and from the guinea pig cerebral cortex at four equally spaced time points in the 12-h light/12-h dark cycle (end-dark, mid-light, end-light, mid-dark). The potencies of exogenous 5-HT in inhibiting 5-HT release and of methiothepin in antagonizing the effect of endogenous or exogenous 5-HT did not differ between the four time points. In another study, BLIER et al. (1989) compared 5-HT release and its autoreceptor-mediated modulation in hypothalamic slices prepared from two groups of rats subjected to a 12-h light/12-h dark cycle. In the first group, in which the lights were turned on at 7h and off at 19h, 5-HT release was higher than in the second group, in which the lights were turned on at 19h and off at 7h. However, both groups did not differ with respect to the inhibitory effect of 5-methoxytryptamine and RU 24969 and the facilitatory effect of methiothepin on 5-HT release. Thus, circadian variations of the sensitivity of inhibitory 5-HT autoreceptors do not appear to occur.

II. Age-Dependent Variations

The density of various types of 5-HT-binding sites in the brain has been reported to undergo alterations during aging (for review, see AMENTA et al.

1991). Therefore, the possibility that the inhibitory 5-HT autoreceptor may also undergo age-dependent alterations was examined in two in vitro studies. In cerebral cortex slices from 3-month-old rats, 5-HT release was higher than in slices from 2.5-year-old animals, but the potency of exogenous 5-HT in inhibiting 5-HT release and the potency of methiothepin in antagonizing the inhibitory effect of endogenous and exogenously 5-HT was identical in both groups (Schlicker et al. 1989). In spinal cord slices from 3-month-old rats, 5-HT and mCPP [1-(3-chlorophenyl)piperazine] inhibited 5-HT release, whereas in slices from 18- to 20-month-old animals, both compounds facilitated 5-HT release (release in the absence of the 5-HT receptor agonists not mentioned; Murphy and Zemlan 1989).

III. Changes by Drugs and Electroconvulsive Shocks

Long-term treatment of patients suffering from affective disorders with tricyclic antidepressants, selective serotonin reuptake inhibitors (SSRIs), reversible or irreversible monoamine oxidase (MAO) inhibitors and electroconvulsive shocks may cause alterations of the sensitivity of the inhibitory 5-HT autoreceptor. Such changes might be due to the increased concentration of 5-HT in the synaptic cleft as a consequence of reduced inactivation or increased release of 5-HT. This possibility was studied in animals treated as mentioned, usually for 2–3 weeks (see Table 4 for such in vitro studies; as an exception, this table also contains data obtained before 1989). Furthermore, the possibility was investigated in animals that blockade of the presynaptic 5-HT autoreceptors with methiothepin or treatment with nimodipine, a Ca^{2+} channel blocker potentially leading to a decreased 5-HT release, might cause an increased sensitivity of the autoreceptors (Table 4).

The sensitivity of the autoreceptor was examined in functional experiments in vitro, i.e. by investigating the effects of agonists and/or antagonists on 5-HT release. Therefore, an important prerequisite for such investigations is that, at the time of determination of autoreceptor function, no residual compound used for long-term treatment is left in the tissue. Accordingly, in all studies entered into Table 4, administration of the drug was stopped 17–72h before sacrifice and in some of them the absence of residual compound was proven by demonstrating that 5-HT uptake or MAO was no longer inhibited. In order to make sure that a change in sensitivity after long-term treatment with a drug is different from its acute effect, the sensitivity of autoreceptors should also be examined after 1- or 2-day administration. This has been examined in some, but not all, studies listed in Table 4. Ideally, the recovery from long-term treatment-induced change in sensitivity after withdrawal of the drug should also be demonstrated [as was done by Maura and Raiteri (1984) only; Table 4].

In analogy to the time schedule applied in the experiments with long-term treatment with drugs, the sensitivity of the release-modulating receptors was

Table 4. Effect of long-term treatment with various drugs or electroconvulsive shocks on the sensitivity of inhibitory 5-HT autoreceptors. "0", "+" and "−" means that the long-term treatment did not affect, increase or decrease, respectively, the sensitivity of the 5-HT receptors. "Acute treatment" means that the drug was given for 1 or 2 day(s) only

Species, brain region Authors; preparation; stimulus	Treatment schedule	Change in sensitivity of the autoreceptor (in parentheses. drug(s) used to test sensitivity)	Comment
Rat, cerebral cortex			
SCHLICKER et al. (1991); slices; electrical pulses	Nimodipine 65 mg/kg in food pellets for at least 6 weeks	0 (5-HT)	5-HT release not affected by chronic nimodipine
MAURA and RAITERI (1984); synaptosomes; high K⁺	CGP 6085A 10 mg/kg i.p. plus clorgyline 1 mg/kg i.p. for 15 days	− (5-HT)	5-HT release not affected by the two chronic treatment schedules
	Methiothepin 10 mg/kg i.p. for 15 days	+ (5-HT)	Acute treatment with CGP 6085A plus clorgyline or methiothepin did not affect autoreceptor sensitivity
			Autoreceptor sensitivity recovered 10 days after withdrawal from chronic CGP 6085A + clorgyline
Rat, hypothalamus			
HAGAN and HUGHES (1983); slices; electrical pulses	Methiothepine 10 mg/kg i.p. for 21 days	0 (5-HT, methiothepin)	5-HT release and uptake not affected by chronic methiothepin
OXFORD and WARWICK (1987); slices; high K⁺	Nialamide 40 mg/kg i.p. or clomipramine 10 mg/kg i.p. for 13 days each	− (5-MeOT) 0 (5-MeOT)	5-HT release increased after chronic nialamide but not affected after chronic clomipramine
			Acute treatment with nialamide did not affect autoreceptor sensitivity
MORET and BRILEY (1990); slices; electrical pulses	Citalopram 10 mg/kg in the drinking water or citalopram 50 mg/kg in food pellets or milnacipran 50 mg/kg in food pellets for 21 days each	− (LSD) − (LSD) 0 (LSD)	5-HT release increased after chronic citalopram (50 > 10 mg/kg) but not affected after chronic milnacipran
			Effect of acute treatment with citalopram on autoreceptor sensitivity not examined

Table 4. *Continued*

Species, brain region Authors; preparation; stimulus	Treatment schedule	Change in sensitivity of the autoreceptor (in parentheses, drug(s) used to test sensitivity)	Comment
Schoups and de Potter (1988); slices; electrical pulses	Amitriptyline 10mg/kg i.p. for 21 days	– (5-HT, methiothepin)	5-HT release increased by chronic amitriptyline. Effect of acute treatment with amiptriptyline on autoreceptor sensitivity not examined
Guinea pig, frontal cortex, hippocampus and hypothalamus; Blier and Bouchard (1994); slices; electrical pulses	Paroxetine 10mg/kg[a] for 21 days; Befloxatone 0.75mg/kg[a] for 21 days	0 or –[b]; 0 (5-MeOT, methiothepin)	Residual 5-HT uptake or MAO blockade after chronic paroxetine and befloxatone excluded. 5-HT release increased after chronic paroxetine or befloxatone. Effect of acute treatment with paroxetine on the sensitivity of the inhibitory autoreceptor has not been studied
Guinea pig, frontal cortex; El Mansari et al. (1995); slices; electrical pulses	Paroxetine 10mg/kg[a] for 3 or 8 weeks	0 (5-MeOT)	5-HT release increased by chronic paroxetine (3 or 8 weeks). 5-HT transporter desensitized by chronic paroxetine (8 weeks). No residual inhibition of 5-HT uptake after chronic paroxetine (8 weeks)

Tissue; reference; method	Treatment	Autoreceptor sensitivity (test drug)	Effect
Guinea pig, orbitofrontal cortex EL MANSARI et al. (1995); slices; electrical pulses	Paroxetine 10mg/kg[a] for 3 weeks; for 8 weeks; Fluoxetine 5 mg/kg[a] for 3 or 8 weeks	0 (5-MeOT) – (5-MeOT) 0 (5-MeOT)	5-HT release increased by chronic paroxetine (8 weeks) but not by chronic paroxetine (3 weeks) and chronic fluoxetine (3 or 8 weeks) 5-HT transporter not desensitized by chronic paroxetine or fluoxetine (8 weeks each)
Guinea pig, caudate nucleus EL MANSARI et al. (1995); slices; electrical pulses	Paroxetine 10mg/kg[a] or fluoxetine 5mg/kg[a] for 3 or 8 weeks each	0 (5-MeOT)	5-HT release not increased by chronic paroxetine or fluoxetine (3 or 8 weeks each)
Guinea pig, hypothalamus BLIER and BOUCHARD (1992); slices; electrical pulses	Electroconvulsive shock three times per week for 14 days	0 (5-CT, methiothepin)	5-HT release not affected after repeated administration of electroconvulsive shocks Same treatment schedule attenuated the 8-OH-DPAT-induced hypothermia in rats in the same study
Rabbit, hypothalamus SCHOUPS and DE POTTER (1988); slices; electrical pulses	Amitriptyline 20mg/kg i.p. or clomipramine 10mg/kg i.p. or imipramine 20mg/kg i.p. or mianserin 20mg/kg i.p. for 21 days each	0 (5-HT) – (methiothepin)	5-HT release not affected by chronic amitriptyline, clomipramine or imipramine but decreased by chronic mianserin Effect of acute treatment with the four drugs on autoreceptor sensitivity not examined Amitriptyline 10mg/kg i.p. for 21 days reduced the autoreceptor sensitivity in the rat hypothalamus (see above)

[a] Administered via osmotic minipump.

[b] Sensitivity of the inhibitory autoreceptor was not changed in the cerebral cortex (test drug: 5-MeOT) but decreased in the hippocampus (test drugs: 5-MeOT, methiothepin) and hypothalamus (test drugs: 5-MeOT, 5-CT, methiothepin).

CGP 6085A, 4-(5,6-dimethyl-2-benzofuranyl)piperidine; 5-CT, 5-carboxamidotryptamine; 5-MeOT, 5-methoxytryptamine; 8-OH-DPAT, 8-hydroxy-2-(di-n-propylamino)tetralin.

examined 48 h after the last of a series of six electroconvulsive shocks over a 2-week period (Table 4; BLIER and BOUCHARD 1992).

Table 4 indicates that in some, but not all, studies of long-term treatment with tricyclics, SSRI, MAO inhibitors and electroconvulsive shocks, the expected decrease in sensitivity of inhibitory 5-HT autoreceptors was observed. An increase in sensitivity of the inhibitory 5-HT autoreceptors which was expected after their long-term blockade with methetiothepin was found only by MAURA and RAITERI (1984), but not by HAGAN and HUGHES (1983; see Table 4).

An increase in 5-HT release is another effect which could be predicted for experiments in slices (but not in synaptosomes; see Sect. C.I.2) if long-term treatment of animals with antidepressant drugs led to a decreased function of the inhibitory 5-HT autoreceptors. Since the autoreceptors are normally tonically activated, this should occur because of the impaired negative feedback loop. Indeed such an increase in 5-HT release was found in several of the studies summarized in Table 4 (OXFORD and WARWICK 1987; SCHOUPS and DE POTTER 1988; MORET and BRILEY 1990; BLIER and BOUCHARD 1994; EL MANSARI and BLIER 1996), but not in synaptosomes (MAURA and RAITERI 1984).

The effect of long-term treatment with SSRI and/or dexfenfluramine on the sensitivity of inhibitory 5-HT autoreceptors was also examined in vivo. GARDIER et al. (1992) injected rats with fluoxetine 30 mg/kg i.p. or dexfenfluramine 7.5 mg/kg i.p. for 3 days, and 24 h after the last injection they determined the effect of methiothepin administered via the dialysis probe on the basal and high K^+-induced 5-HT release in the frontal cortex of the animals anaesthetized with α-chloralose/urethane. After both fluoxetine and dexfenfluramine treatment, the basal and K^+-induced 5-HT release was decreased; the facilitatory effect of methiothepin on the basal 5-HT release was not affected whereas the facilitatory effect on the evoked 5-HT release was attenuated (effect of an acute administration of the drugs not examined). In the study of BOSKER et al. (1995), the rats received fluvoxamine 6.7 mg/kg daily via an osmotic minipump for 21 days, leading to a fluvoxamine brain level corresponding to its IC_{50} for inhibition of the synaptosomal 5-HT transporter. Three days after removal of the minipump, they measured the effect of RU 24969, administered via the dialysis probe, on basal 5-HT release in the dorsal hippocampus of the freely moving animals. Both 5-HT release and its inhibition by RU 24969 were unaltered. AUERBACH and HJORTH (1995) treated rats with citalopram, either with 2×5 mg/kg s.c. for 2, 7 or 14 days, with 2×20 mg/kg s.c. for 14 days, or with 10 mg/kg via an osmotic minipump for 14 days. Twenty-four hours after the last injection or removal of the minipump, citalopram and later, in addition, RU 24969 or CP-93,129 were administered via the dialysis probe to the frontal cortex or dorsal hippocampus of the chloral hydrate-anaesthetized animals. Citalopram (treatment for 14 days) given via the dialysis probe caused a more marked increase in 5-HT release (compared to the saline-treated control animals) when administered to the dorsal hippocampus but not when injected into the frontal cortex. Basal 5-HT release (prior to the injection of citalopram) and the inhibitory effect of RU 24969 or CP-93,129 on 5-HT release (after injection of citalopram) did not differ between citalopram- and saline-treated animals in both brain regions.

The possibility that *lithium* given for up to 21 days affects the sensitivity of the inhibitory 5-HT autoreceptor was also examined. After administration of lithium to rats in the food for 3 weeks, the sensitivity of the inhibitory 5-HT autoreceptor in slices of the cerebral cortex, hippocampus and hypothalamus was decreased (FRIEDMAN and WANG 1988, WANG and FRIEDMAN 1988). When rats were injected i.p. with lithium for 3 days only, the sensitivity of the autoreceptor was decreased in hippocampal, but not in cortical slices (HOTTA and YAMAWAKI 1988). However, a shortcoming was that in the former two studies, the rats received food pellets containing lithium until sacrifice and in the latter, the last injection of lithium was administered only 1 h before sacrifice. Thus, lithium was probably not yet washed out of the tissue. This is a drawback since lithium, directly applied to the tissue in superfusion experiments, was shown to abolish the inhibitory effect of exogenously added 5-HT on 5-HT release (HIDE and YAMAWAKI 1989).

Taken together, evidence in support of the view that the sensitivity of the inhibitory presynaptic 5-HT autoreceptors is modified by long-term treatment with drugs used in affective disorders is equivocal.

Whether or not long-term treatment with such drugs may affect the sensitivity of 5-HT receptors mediating *facilitation* of 5-HT release has almost not yet been investigated at all. In the only study available, pretreatment of guinea pigs with paroxetine 10 mg/kg for 3 weeks decreased the facilitatory effect of 2-methyl-5-HT on the electrically evoked 5-HT overflow in frontal cortical, hippocampal and hypothalamic slices (BLIER and BOUCHARD 1994).

G. Potential Pathophysiological and Therapeutic Roles of Inhibitory Presynaptic 5-HT Autoreceptors

Since 5-HT is involved in the central regulation of blood pressure (for review, see WOLF et al. 1985; COOTE 1990), it was conceivable that alterations of inhibitory 5-HT autoreceptors in the CNS might play a role in the pathogenesis of hypertension. Therefore, the sensitivity of these receptors in the hypothalamus, nucleus tractus solitarii and cerebral cortex of spontaneously hypertensive (SHR) was compared to that in age-matched Wistar-Kyoto (WKY) rats (SCHLICKER et al. 1988). However, irrespective of the age of the rats and the brain region, neither 5-HT release in the absence of drugs nor modulation by exogenous 5-HT and methiothepin differed between the two strains.

5-HT probably plays an even more important role in various other functions of the CNS, including modulation of pain, feeding behaviour, neuroendocrinology, sleep, locomotor activity, sexual activity, mood and cognitive function. Furthermore, psychiatric and neurological diseases such as depres-

Depression

No depression

Before
treatment

Acute
effect

Long-term
treatment

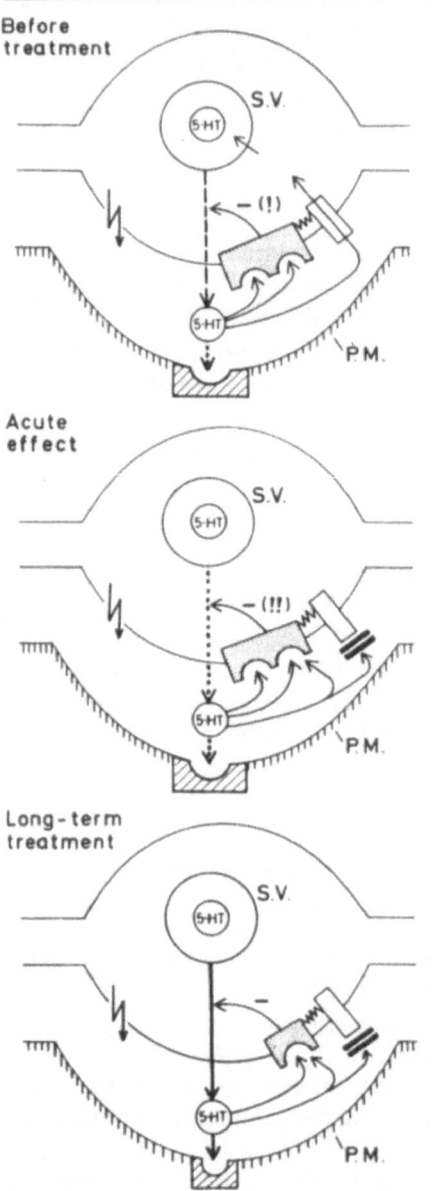

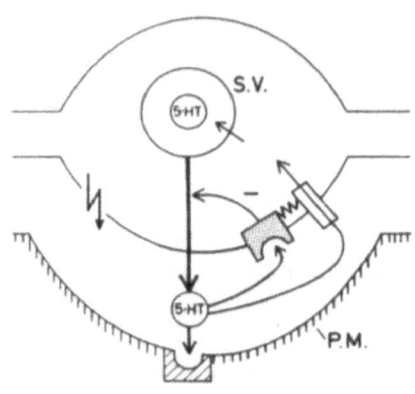

S.V. : Storage vesicle
P.M. : Postsynaptic membrane

sion, anxiety, obsessive-compulsive disorder and myoclonus may be caused or accompanied by alterations in the serotoninergic system (for review, see SIEVER et al. 1991). In particular, reduced serotoninergic neurotransmission has been hypothesized to be involved in the pathogenesis of major depression (COPPEN 1967). It is conceivable that a decrease in 5-HT release is due to an increased sensitivity of the inhibitory presynaptic h5-HT$_{1B}$ autoreceptors (GÖTHERT 1991; GÖTHERT and SCHLICKER 1993; CHOPIN et al. 1994; see Fig. 3).

This view is supported in an animal model of depression, namely the learned helplessness paradigm test in rats (for detailed discussion see CHOPIN et al. 1994; Chap. 10, this volume). The depression-like state of learned help-lessness was induced by prolonged exposure of rats to repetitive electrical footshocks from which the animals could not escape. As a symptom of the depression-like state, it was observed that the rats reacted apathic and did not try to escape from this stress, when, subsequent to the "training session" with unescapable shocks, they had the opportunity to avoid the footshocks. This depression-like state of learned helplessness was associated with an upregulation of 5-HT$_{1B}$ receptors, as suggested by an increase in density of [^{125}I]cyanopindolol-binding sites in the cortex, hippocampus and septum (EDWARDS et al. 1991). Such an upregulation may refer not only to postsynaptic 5-HT$_{1B}$ receptors on the somadendritic or terminal areas of non-serotoninergic nerves but also to the presynaptic 5-HT$_{1B}$ autoreceptors. In agreement with an increased responsiveness of presynaptic 5-HT$_{1B}$ autoreceptors resulting from such an upregulation, a decrease in 5-HT release has been measured in microdialysis studies in depressed-like learned helpless rats (PETTY et al. 1992).

Fig. 3. Function of the inhibitory presynaptic 5-HT$_{1B}$ autoreceptor on a serotoninergic varicosity in the brain under normal conditions (*right panel*) and hypothesized function in depressed patients before and after treatment with selective or non-selective 5-HT uptake inhibitors (*left panel*). *Stippled area* in the membrane of the serotoninergic varicosity, 5-HT$_{1B}$ autoreceptor; ⩗ action potential. *Solid vertical arrows*, normal 5-HT release resulting in normal 5-HT concentration at postsynaptic 5-HT receptors; *broken vertical arrows*, decreased release and decreased 5-HT concentration at postsynaptic 5-HT receptors; *thin solid arrows* within the synaptic cleft, binding of 5-HT to presynaptic autoreceptors and 5-HT uptake by the 5-HT transporter (*rectangle* within the cell membrane). Autoreceptor-mediated inhibition of 5-HT release: -, normal inhibition; -(!) and -(!!), reinforced inhibition. In untreated, depressed patients, the number of autoreceptors is assumed to be increased, symbolized by the increased *stippled area*, by the two recognition sites for 5-HT and by -(!). Blockade of the transporter by certain antidepressants (∥) initially results in an increased concentration of 5-HT at the autoreceptor, leading immediately to further reinforcement of autoreceptor function, symbolized by -(!!). As a result of both effects (uptake blockade and increased autoreceptor function), the 5-HT concentration at postsynaptic receptors remains de-creased. After long-term treatment with such antidepressants, downregulation of the autoreceptor may occur, resulting in normal serotoninergic synaptic transmission. *Hatched area* in the postsynaptic membrane, non-5-HT$_{1B}$ (e.g. 5-HT$_{2A}$) receptor. (From GÖTHERT and SCHLICKER 1993)

On the other hand, a supersensitivity of the postsynaptic receptors may be associated with an increased level of anxiety which has basically been related to an elevated serotoninergic "activity" (CHOPIN and BRILEY 1987). The neurons innervated by serotoninergic axon terminals and endowed with an increased density of postsynaptic 5-HT_{1B} receptors may be hyperstimulated even when, due to the simultaneous supersensitivity of presynaptic 5-HT_{1B} autoreceptors, the release of 5-HT is to a certain, yet relatively less, extent decreased. In agreement with these considerations and the results of the binding studies mentioned above, learned helpless rats exhibited behavioural signs of anxiety (VANDIJKEN et al. 1992a,b) combined with depression-like symptoms (SHERMAN et al. 1979). Analogously, supersensitivity of pre- and postsynaptic 5-HT_{1B} receptors in man would be expected to be associated with depression and anxiety, respectively, a combination which is frequently observed in patients.

It is well known that the therapeutic effect of antidepressant drugs (tricyclics, SSRIs) develops within 2–3 weeks of treatment whereas the blockade of the 5-HT (and noradrenaline) transporter(s) occurs immediately. The reason for this discrepancy probably is that not the inhibition of the transporter(s) per se but rather adaptive changes of pre- and postsynaptic receptors, including the presynaptic 5-HT_{1B} autoreceptors (see Fig. 3 and Sect. F.III in this context), account for the therapeutic effect. In depressed patients the presynaptic 5-HT_{1B} autoreceptors may be assumed to be supersensitive and, as a consequence, release of 5-HT from the serotoninergic axon terminals may be hypothesized to be diminished. Under these conditions, acute administration of a drug which inhibits 5-HT reuptake does probably not increase serotoninergic neurotransmission: the decreased inactivation of 5-HT may be assumed to lead to an initial increase in 5-HT concentration in the synaptic cleft which, in turn, causes an increased stimulation of the inhibitory autoreceptors and, as consequence, a diminished 5-HT release (GÖTHERT 1991; GÖTHERT and SCHLICKER 1993). Thus, these opposite effects may outweigh each other (Fig. 3). However, after long-term treatment, the sensitivity of the presynaptic 5-HT autoreceptors may decrease, leading to an increase in 5-HT release. This hypothesis is supported by some (although not all) studies in experimental animals which revealed a downregulation of the presynaptic 5-HT autoreceptors and an increase in 5-HT release after administration of antidepressant drugs for 2–3 weeks (see Table 4 and Sect. F.III). An additional mechanism may contribute to the increase in 5-HT release, namely the desensitization of the inhibitory somadendritic 5-HT_{1A} autoreceptors (for reviews, see BLIER and DE MONTIGNY 1994; GARDIER et al. 1996).

On the basis of the hypothesis that an increased serotoninergic neurotransmission relieves certain symptoms of depression, blockade of inhibitory somadendritic or presynaptic 5-HT autoreceptors, leading to an immediate increase in 5-HT release (for in vitro evidence for such an effect, see Fig. 1), has become an attractive concept for the development of novel antidepressant drugs (GÖTHERT 1982, 1991; GÖTHERT and SCHLICKER 1993; ARTIGAS 1993; BRILEY and MORET 1993a,b; BLIER and DE MONTIGNY 1994;

CHOPIN et al. 1994). Such h5-HT$_{1A}$ or h5-HT$_{1B}$ receptor antagonists, respectively, in particular when combined with an inhibitor of the 5-HT transporter, may be assumed to exhibit the advantage of inducing an immediate antidepressant effect without the latency period characteristic for the currently available antidepressant drugs. In fact, clinical evidence has been obtained for an increased onset of antidepressant effect when an SSRI was combined with the h5-HT$_{1A}$ receptor antagonist pindolol (ARTIGAS et al. 1994; BLIER and BERGERON 1995). In agreement with this effect of somadendritic autoreceptor blockade, it has recently been proven in an in vivo microdialysis study in guinea pig hypothalamus (ROLLEMA et al. 1996) that blockade of presynaptic 5-HT$_{1D}$ autoreceptors by GR 127935 [(N-[4-methoxy-3-(4-methyl-1-piperazinyl)phenyl]-2'-methyl-4'-(5-methyl-1,2,4-oxadiazol-3-yl)[1,1-biphenyl]-4-carboxamide)] potentiates the weak increasing effect of sertraline, an SSRI, on 5-HT release.

H. Concluding Remarks

Inhibitory presynaptic 5-HT autoreceptors have been shown to play a physiological role in the regulation of 5-HT release in the central nervous system. They have been identified in any region of the mammalian brain and spinal cord which has been investigated for this purpose. Furthermore, 5-HT release in the guinea pig and rat brain cortex, hippocampus and hypothalamus is modulated via facilitatory 5-HT receptors. These are not autoreceptors but may be assumed to be located on an interneuron which is involved in a neuronal circuit impinging on the serotoninergic axon terminals. The latter terminals should be endowed with a heteroreceptor for a neurotransmitter released from the innervating interneuron.

Whereas the facilitatory 5-HT receptors, whose physiological role in the regulation of 5-HT release is uncertain, belong to the 5-HT$_3$ receptor class, the inhibitory 5-HT autoreceptors have been classified as 5-HT$_{1B}$. The latter receptors differ considerably in their pharmacological properties among various species, e.g. between rat and mouse on the one hand and guinea pig, rabbit, pig and man on the other. Accordingly, guinea pig cerebral preparations have proven to represent excellent models for the development of selective ligands at inhibitory presynaptic 5-HT autoreceptors in the human brain.

As shown in some, but not all, animal experiments, long-term treatment with tricyclic antidepressants and SSRIs, both of which inhibit neuronal 5-HT uptake, induces a downregulation of the inhibitory presynaptic 5-HT autoreceptors. By attenuating the autoinhibitory feedback loop this may be assumed to result in an increase in serotoninergic neurotransmission, thus probably contributing to the mechanism of the antidepressant action of those drugs.

In view of the marked heterogeneity of 5-HT receptors and their specific distribution and function in the human CNS, the classification of the inhibitory 5-HT autoreceptors as h5-HT$_{1B}$ offers a promising chance for the development of drugs with beneficial effects in certain neuropsychiatric disorders. Thus, selective antagonists at human 5-HT$_{1B}$ receptors, which immediately and considerably increase the release of 5-HT, may be expected to possess more favourable properties as, e.g. antidepressant drugs than the available ones. This may refer to their main actions in relation to potential side effects and to a more rapid onset of action.

Acknowledgement. The studies on inhibitory presynaptic 5-HT autoreceptors in the authors' own laboratory were supported by the Deutsche Forschungsgemeinschaft (Go 343/1, Go 343/2–6 and SFB 400).

References

Amenta F, Zaccheo D, Collier WL (1991) Neurotransmitters, neuroreceptors and aging. Mech Age Dev 61:249–273

Artigas F (1993) 5-HT and antidepressants: new views from microdialysis studies. Trends Pharmacol Sci 14:262

Artigas F, Perez V, Alvarez E (1994) Pindolol induces a rapid improvement of depressed patients treated with serotonin reuptake inhibitors. Arch Gen Psychiatry 51:248–251

Assie MB, Koek W (1996) (-)-Pindolol and (+)-tertatolol affect rat hippocampal 5-HT levels through mechanism involving not only 5-HT$_{1A}$, but also 5-HT$_{1B}$ receptors. Neuropharmacology 35:213–222

Auerbach SB, Hjorth S (1995) Effect of chronic administration of the selective serotonin (5-HT) uptake inhibitor citalopram on extracellular 5-HT and apparent autoreceptor sensitivity in rat forebrain in vivo. Naunyn-Schmiedeberg's Arch Pharmacol 352:597–606

Auerbach SB, Rutter JJ, Juliano J (1991) Substituted piperazine and indole compounds increase extracellular serotonin in rat diencephalon as determined by in vivo microdialysis. Neuropharmacology 30:307–311

Bach AWJ, Unger L, Sprengel R, Mengod G, Palacios J, Seeburg PH, Voigt MM (1993) Structure, functional expression and spatial distribution of a cloned cDNA encoding a rat 5-HT$_{1D}$-like receptor. J Receptor Res 13:479–502

Bard JA, Kucharewicz SA, Zgombick JM, Weinshank RL, Branchek TA, Cohen ML (1996) Differences in ligand binding profiles between cloned rabbit and human 5-HT$_{1D\alpha}$ and 5-HT$_{1D\beta}$ receptors: ketanserin and methiothepin distinguish rabbit 5-HT$_{1D}$ receptor subtypes. Naunyn-Schmiedeberg's Arch Pharmacol 354:237–244

Beaudet A, Descarries L (1978) The monoamine innervation of rat cerebral cortex: synaptic and nonsynaptic axon terminals. Neuroscience 3:851–860

Beer MS, Stanton JA, Bevan Y, Chauhan NS, Middlemiss DN (1992) An investigation of the 5-HT$_{1D}$ receptor binding affinity of 5-hydroxytryptamine, 5-carboxamidotryptamine and sumatriptan in the central nervous system of seven species. Eur J Pharmacol 213:193–197

Beer MS, Stanton JA, Bevan Y, Heald A, Reeve AJ, Street LJ, Matassa VG, Hargreaves RJ, Middlemiss DN (1993) L-694,247: a potent 5-HT$_{1D}$ receptor agonist. Br J Pharmacol 110:1196–1200

Blier P (1991) Terminal serotonin autoreceptor function in the rat hippocampus is not modified by pertussis and cholera toxins. Naunyn-Schmiedeberg's Arch Pharmacol 344:160–166

Blier P, Bouchard C (1992) Effect of repeated electroconvulsive shocks on serotonergic neurons. Eur J Pharmacol 211:365–373

Blier P, Bouchard C (1993) Functional characterization of a 5-HT$_3$ receptor which modulates the release of 5-HT in the guinea-pig brain. Br J Pharmacol 108:13–22

Blier P, Bouchard C (1994) Modulation of 5-HT release in the guinea-pig brain following long-term administration of antidepressant drugs. Br J Pharmacol 113:485–495

Blier P, de Montigny C (1994) Current advances and trends in the treatment of depression. Trends Pharmacol Sci 15:220–226

Blier P, Bergeron R (1995) Effectiveness of pindolol with selected antidepressant drugs in the treatment of major depression. J Clin Psychopharmacol 15:217–222

Blier P, Galzin AM, Langer SZ (1989) Diurnal variation in the function of serotonin terminals in the rat hypothalamus. J Neurochem 52:453–459

Blier P, Monroe PJ, Bouchard C, Smith DL, Smith DJ (1993) 5-HT$_3$ receptors which modulate [^{3}H]5-HT release in the guinea pig hypothalamus are not autoreceptors. Synapse 15:143–148

Bonanno G, Maura G, Raiteri M (1986) Pharmacological characterization of release-regulating serotonin autoreceptors in rat cerebellum. Eur J Pharmacol 126:317–321

Boess FG, Martin IL (1994) Molecular biology of 5-HT receptors. Neuropharmacology 33:275–317

Boschert U, Amara DA, Segu L, Hen R (1994) The mouse 5-hydroxytryptamine$_{1B}$ receptor is localized predominantly on axon terminals. Neuroscience 58:167–182

Bosker FJ, van Esseveldt KE, Klompmakers AA, Westenberg HGM (1995) Chronic treatment with fluvoxamine by osmotic minipumps fails to induce persistent functional changes in central 5-HT$_{1A}$ and 5-HT$_{1B}$ receptors, as measured by in vivo microdialysis in dorsal hippocampus of conscious rats. Psychopharmacology 117:358–363

Bradley PB, Engel G, Feniuk W, Fozard JR, Humphrey PPA, Middlemiss DN, Mylecharane EJ, Richardson BP, Saxena PR (1986) Proposals for the classification and nomenclature of functional receptors for 5-hydroxytryptamine. Neuropharmacology 25:563–576

Brazell MP, Marsden CA, Nisbet AP, Routledge C (1985) The 5-HT$_1$ receptor agonist RU-24969 decreases 5-hydroxytryptamine (5-HT) release and metabolism in the rat frontal cortex in vitro and in vivo. Br J Pharmacol 86:209–216

Briley M, Moret C (1993a) Neurobiological mechanisms involved in antidepressant therapies. Clin Neuropharmacol 16:387–400

Briley M, Moret C (1993b) 5-HT and antidepressants: in vitro and in vivo release studies. Trends Pharmacol Sci 14:396–397

Bruinvels AT, Lery H, Nozulak J, Palacios JM, Hoyer D (1992) 5-HT$_{1D}$ binding sites in various species: similar pharmacological profile in dog, monkey, calf, guinea-pig and human brain membranes. Naunyn-Schmiedeberg's Arch Pharmacol 346:243–248

Bühlen M, Fink K, Böing C, Göthert M (1996a) Evidence for presynaptic location of inhibitory 5-HT$_{1D\beta}$-like autoreceptors in the guinea-pig brain cortex. Naunyn-Schmiedeberg's Arch Pharmacol 353:281–289

Bühlen M, Brüss M, Bönisch H, Göthert M (1996b) Modified ligand binding properties of the naturally occurring PhE-124-Cys variant of the human 5-HT$_{1D\beta}$ receptor. Naunyn-Schmiedeberg's Arch Pharmacol 353(Suppl):R91

Bühlen M, Brüss M, Bönisch H, Göthert M (1996c) Modified ligand binding to the naturally occurring Phe-124-Cys variant of the human 5-HT$_{1B}$ receptor. Soc Neurosci Abstr 22, part 2, p 1329

Butler A, Elswood CJ, Burridge J, Ireland SJ, Bunce KT, Kilpatrick GJ, Tyers MB (1990) The pharmacological characterization of 5-HT$_3$ receptors in three isolated preparations derived from guinea-pig tissues. Br J Pharmacol 101:591–598

Cassel J-C, Jeltsch H (1995) Serotonergic modulation of cholinergic function in the central nervous system: cognitive implications. Neuroscience 69:1–41

Cerrito F, Raiteri M (1979) Serotonin release is modulated by presynaptic autoreceptors. Eur J Pharmacol 57:427–430

Chaput Y, Blier P, de Montigny C (1986) In vivo electrophysiological evidence for the regulatory role of autoreceptors on serotonergic terminals. J Neurosci 6:2796–2801

Chaput Y, de Montigny (1988) Effects of the 5-hydroxytryptamine₁ receptor antagonist, BMY 7378, on 5-hydroxytryptamine neurotransmission: electrophysiological studies in the rat central nervous system. J Pharmacol Exp Ther 246:359–370

Chopin P, Briley M (1987) Animal models of anxiety: the effect of compounds that modify 5-HT neurotransmission. Trends Pharmacol Sci 8:383–388

Chopin P, Moret C, Briley M (1994) Neuropharmacology of 5-hydroxytryptamine$_{1B/D}$ receptor ligands. Pharmacol Ther 62:385–405

Claustre Y, Rouqier L, Bonvento G, Benavides J, Scatton B (1991) In vivo regulation of serotonin release by 5-HT autoreceptors and α_1-adrenoceptors in the rat and guinea-pig brain. In: Rollema H, Westerink B, Drijfhout WJ (eds) Monitoring molecules in neuroscience. Krips Repro, Meppel, The Netherlands, pp 247–249

Coote JH (1990) Bulbospinal serotonergic pathways in the control of blood pressure. J Cardiovasc Pharmacol 15(Suppl 7):S35–S41

Coppen A (1967) The biochemistry of affective diseases. Br J Psychiatry 113:1237–1264

Davidson C, Stamford J (1995) The effect of paroxetine and 5-HT-efflux in the rat dorsal raphe nucleus is potentiated by both 5-HT$_{1A}$ and 5-HT$_{1B}$ receptor antagonists. Neurosci Lett 188:41–44

Edwards E, Harkins K, Wright G, Henn FA (1991) 5-HT$_{1B}$ receptors in an animal model of depression. Neuropharmacology 30:101–105

El Mansari M, Bouchard C, Blier P (1995) Alteration of serotonin release in the guinea pig orbito-frontal cortex by selective serotonin reuptake inhibitors – relevance to obsessive-compulsive disorder. Neuropsychopharmacology 13:117–127

El Mansari M, Blier P (1996) Functional characterization of 5-HT$_{1D}$ autoreceptors on the modulation of 5-HT release in guinea-pig mesencephalic raphe, hippocampus and frontal cortex. Br J Pharmacol 118:681–689

Engel G, Göthert M, Müller-Schweinitzer E, Schlicker E, Sistonen L, Stadler PA (1983) Evidence for common pharmacological properties of [³H]5-hydroxytryptamine binding sites, presynaptic 5-hydroxytryptamine autoreceptors in CNS and inhibitory presynaptic 5-hydroxytryptamine receptors on sympathetic nerves. Naunyn-Schmiedeberg's Arch Pharmacol 324:116–124

Engel G, Göthert M, Hoyer D, Schlicker E, Hillenbrand K (1986) Identity of inhibitory presynaptic 5-hydroxytryptamine (5-HT) autoreceptors in the rat brain cortex with 5-HT$_{1B}$ binding sites. Naunyn-Schmiedeberg's Arch Pharmacol 332:1–7

Farnebo LO, Hamberger B (1971) Drug-induced changes in the release of ³H-monoamines from field stimulated rat brain slices. Acta Physiol Scand Suppl 371:35–44

Farnebo LO, Hamberger B (1974) Regulation of [³H]5-hydroxytryptamine release from rat brain slices. J Pharm Pharmacol 26:642–644

Feuerstein TJ, Allgaier C, Hertting G (1987) Possible involvement of protein kinase C (PKC) in the regulation of electrically evoked serotonin (5-HT) release from rabbit hippocampal slices. Eur J Pharmacol 139:267–272

Feuerstein TJ, Lupp A, Hertting G (1992) Quantitative evaluation of the autoinhibitory feedback of release of 5-HT in the caudate nucleus of the rabbit where an endogenous tone on α_2-adrenoceptors does not exist. Neuropharmacology 31:15–23

Fink K, Zentner J, Göthert M (1995) Subclassification of presynaptic 5-HT autoreceptors in the human cerebral cortex as 5-HT$_{1D\beta}$ receptors. Naunyn-Schmiedeberg's Arch Pharmacol 352:451–454

Fink K, Böing C, Göthert M (1996) Presynaptic 5-HT autoreceptors modulate N-methyl-D-aspartate-evoked 5-hydroxytryptamine release in the guinea-pig brain cortex. Eur J Pharmacol 300:79–82

Frankfurt M, Mendelson SD, McKittrick CR, McEwen BS (1993) Alterations of serotonin receptor binding in the hypothalamus following acute denervation. Brain Res 601:349–352

Frankfurt M, McKittrick CR, Mendelson SD, McEwen BS (1994) Effect of 5,7-dihydroxytryptamine, ovariectomy and gonadal steroids on serotonin receptor binding in rat brain. Neuroendocrinology 59:245–250

Frankhuyzen AL, Mulder AH (1980) Noradrenaline inhibits depolarisation-induced ^{3}H-serotonin release from slices of rat hippocampus. Eur J Pharmacol 63:179–182

Frazer A, Maayani S, Wolfe BB (1990) Subtypes of receptors for serotonin. Annu Rev Pharmacol Toxicol 30:307–348

Friedman E, Wang HY (1988) Effect of chronic lithium treatment on 5-hydroxytryptamine autoreceptors and release of 5-[^{3}H]hydroxytryptamine from rat brain cortical, hippocampal and hypothalamic slices. J Neurochem 50:195–201

Galzin AM, Poncet V, Langer SZ (1990) 5-HT$_3$ receptor agonists enhance the electrically-evoked release of [^{3}H]-5-HT in guinea-pig frontal cortex slices. Br J Pharmacol 101:307P

Galzin AM, Langer SZ (1991) Modulation of 5-HT release by presynaptic inhibitory and facilitatory 5-HT receptors in brain slices. In: Langer SZ, Galzin AM, Costentin J (eds) Presynaptic receptors and neuronal transporters. (Advances in the biosciences, vol 82 Pergamon, Oxford, pp 59–62)

Galzin AM, Poirier MF, Lista A, Chodkiewicz JP, Blier P, Ramdine R, Loo H, Roux FX, Redondo A, Langer SZ (1992) Characterization of the 5-hydroxytryptamine receptor modulating the release of 5-[^{3}H]hydroxytryptamine in slices of the human neocortex. J Neurochem 59:1293–1301

Gardier AM, Kaakola S, Erfurth A, Wurtman RJ (1992) Effects of methiothepin on changes in brain serotonin release induced by repeated administration of high doses of anoretic serotoninergic drugs. Brain Res 588:67–74

Gardier AM, Malagié I, Trillat AC, Jacquot C, Artigas F (1996) Role of 5-HT$_{1A}$ autoreceptors in the mechanism of action of serotoninergic antidepressant drugs: recent findings from in vivo microdialysis studies. Fundam Clin Pharmacol 10:16–27

Göthert M (1982) Modulation of serotonin release in the brain via presynaptic receptors. Trends Pharmacol Sci 3:437–440

Göthert M (1990) Presynaptic serotonin receptors in the CNS. Ann NY Acad Sci 604:102–112

Göthert M (1991) Presynaptic effects of 5-HT. In: Stone TW (ed) Aspects of synaptic transmission, LTP, galanin, autonomic, 5-HT, vol. I. Taylor and Francis, London, pp 314–329

Göthert M (1992) 5-Hydroxytryptamine receptors. An example for the complexity of chemical transmission of information. Arzneim Forsch/Drug Res 42:238–246

Göthert M, Weinheimer G (1979) Extracellular 5-hydroxytryptamine inhibits 5-hydroxtryptamine release from rat brain cortex slices. Naunyn-Schmiedeberg's Arch Pharmacol 310:93–96

Göthert M, Huth H (1980) alpha-Adrenoceptor-mediated modulation of 5-hydroxytryptamine release from rat brain cortex slices. Naunyn-Schmiedeberg's Arch Pharmacol 313:21–26

Göthert M, Schlicker E (1991) Regulation of serotonin release in the CNS by presynaptic heteroreceptors. In: Feigenbaum J, Hanani M (eds) A handbook. Presynaptic regulation of neurotransmitter release, vol II. Freund, Tel Aviv, pp 845–876

Göthert M, Schlicker E (1993) Relevance of 5-HT autoreceptors for psychotropic drug action. In: Gram LF, Balant LP, Meltzer HY, Dahl SG (eds) Clinical pharmacology in psychiatry. Springer, Berlin Heidelberg New York, pp 38–51

Göthert M, Huth H, Schlicker E (1981) Characterization of the receptor subtype involved in alpha-adrenoceptor-mediated modulation of serotonin release from rat brain cortex slices. Naunyn-Schmiedeberg's Arch Pharmacol 317:199–203

Göthert M, Fink K, Frölich D, Likungu J, Molderings G, Schlicker E, Zentner J (1996) Presynaptic 5-HT auto- and heteroreceptors in the human central and peripheral nervous system. Behav Brain Res 73:89–92

Haddjeri N, Blier P (1995) Pre- and post-synaptic effects of the 5-HT₃ agonist 2-methyl-5-HT on the 5-HT system in the rat brain. Synapse 20:54–67

Hagan RM, Hughes IE (1983) Lack of effect of chronic methiothepin treatment on 5-hydroxytryptamine autoreceptors. Br J Pharmacol 80:513P

Hamon M, Bourgoin S, Jagger J, Glowinski J (1974) Effects of LSD on synthesis and release of 5-HT in rat brain slices. Brain Res 69:265–280

Hartig PR, Branchek TA, Weinshank RL (1992) A subfamily of 5-HT$_{1D}$ receptor genes. Trends Pharmacol Sci 13:152–159

Hartig PR, Hoyer D, Humphrey PPA, Martin GR (1996) Alignment of receptor nomenclature with the human genome: classification of 5-HT$_{1B}$ and 5-HT$_{1D}$ receptor subtypes. Trends Pharmacol Sci 17:103–105

Heuring RE, Peroutka SJ (1987) Characterization of a novel ³H-5-hydroxytryptamine binding site subtype in bovine brain membranes. J Neurosci 7:894–903

Hide I, Yamawaki S (1989) Inactivation of presynaptic 5-HT autoreceptors by lithium in rat hippocampus. Neurosci Lett 107:323–326

Hjorth S, Magnusson T (1988) The 5-HT$_{1A}$ receptor agonist, 8-OH-DPAT, preferentially activates cell body 5-HT autoreceptors in rat brain in vivo. Naunyn-Schmiedeberg's Arch Pharmacol 338:463–471

Hjorth S, Tao R (1991) The putative 5-HT$_{1B}$ receptor agonist CP-93,129 suppresses rat hippocampal 5-HT release in vivo: comparison with RU 24969. Eur J Pharmacol 209:249–252

Hjorth S, Sharp T (1993) In vivo microdialysis evidence for central serotonin$_{1A}$ and serotonin$_{1B}$ autoreceptor blocking properties of the beta adrenoceptor antagonist (-)penbutolol. J Pharmacol Exp Ther 265:707–712

Hotta J, Yamawaki S (1988) Possible involvement of presynaptic 5-HT autoreceptors in effect of lithium on 5-HT release in hippocampus of rat. Neuropharmacology 27:987–992

Hoyer D, Middlemiss DN (1989) Species differences in the pharmacology of terminal 5-HT autoreceptors in mammalian brain. Trends Pharmacol Sci 10:130–132

Hoyer D, Schoeffter P (1991) 5-HT receptors: subtypes and second messengers. J Recep Res 11:197–214

Hoyer D, Clarke DE, Fozard JR, Hartig PR, Martin GR, Mylecharane EJ, Saxena PR, Humphrey PPA (1994) VII. International Union of Pharmacology classification of receptors for 5-hydroxytryptamine (serotonin). Pharmacol Rev 46:157–203

Hutson PH, Bristow LJ, Cunningham JR, Hogg JE, Longmore J, Murray F, Pearce D, Razzague Z, Saywell K, Tricklebank MD, Young L (1995) The effects of GR 127935, a putative 5-HT$_{1D}$ receptor antagonist, on brain 5-HT$_{1D}$ metabolism, extracellular 5-HT concentration and behaviour in the guinea pig. Neuropharmacology 34:383–392

Kalen P, Rosengren E, Lindvall O, Bjorklund A (1989) Hippocampal noradrenaline and serotonin release over 24 hours as measured by the dialysis technique in freely moving rats: correlation to behavioural activity state, effect of handling and tail-pinch. Eur J Neurosci 1:181–188

Kilpatrick GJ, Bunce KT, Tyers MB (1990) 5-HT₃ receptors. Med Res Rev 10:441–475

Kilpatrick GJ, Barnes NM, Cheng CHK, Costall B, Naylor RJ, Tyers MB (1991) The pharmacological characterization of 5-HT₃ receptor binding sites in rabbit ileum: comparison with those in rat ileum and rat brain. Neurochem Int 4:389–396

Lawrence AJ, Marsden CA (1992) Terminal autoreceptor control of 5-hydroxytryptamine release as measured by in vivo microdialysis in the conscious guinea-pig. J Neurochem 58:142–146

Leff P, Martin GR (1988) The classification of 5-hydroxytryptamine receptors. Med Res Rev 8:187–202

Limberger N, Deicher R, Starke K (1991) Species differences in presynaptic serotonin autoreceptors: mainly 5-HT$_{1B}$ but possibly in addition 5-HT$_{1D}$ in the rat, 5-HT$_{1D}$ in the rabbit and guinea-pig brain cortex. Naunyn-Schmiedeberg's Arch Pharmacol 343:353–364

Marsden CA, Martin KF, Routledge C, Brazell MP, Maidment NT (1986) Application of intracerebral dialysis and in vivo voltammetry to pharmacological and physiological studies of amine neurotransmitters. Ann NY Acad Sci 473:106–124

Martin KF (1991) Rhythms in neurotransmitter turnover: focus on the serotonergic system. Pharmacol Ther 51:421–429

Martin LL, Sanders-Bush E (1982) Comparison of the pharmacological characteristics of 5-HT₁ and 5-HT₂ binding sites with those of serotonin autoreceptors which modulate serotonin release. Naunyn-Schmiedeberg's Arch Pharmacol 321:165–170

Martin GR, Humphrey PPA (1994) Receptors for 5-hydroxytryptamine: current perspective on classification and nomenclature. Neuropharmacology 33:261–273

Martin KF, Hannon S, Phillips I, Heal DJ (1992) Opposing roles for 5-HT$_{1B}$ and 5-HT$_3$ receptors in the control of 5-HT release in rat hippocampus in vivo. Br J Pharmacol 106:139–142

Massot O, Rousselle JC, Fillion MP, Grimaldi B, Cloëz-Tayarani I, Fugelli A, Prudhomme N, Seguin L, Rousseau B, Plantefol M, Hen R, Fillion G (1996) 5-Hydroxytryptamine-moduline, a new endogenous cerebral peptide, controls the serotonergic activity via its specific interaction with 5-hydroxytryptamine$_{1B/1D}$ receptors. Mol Pharmacol 50:752–762

Matsumoto I, Combs MR, Jones DJ (1992) Characterization of 5-hydroxytryptamine$_{1B}$ receptors in rat spinal cord via [^{125}I]-iodocyanopindolol binding and inhibition of [^{3}H]-5-hydroxytryptamine release. J Pharmacol Exp Ther 260: 614–626

Maura G, Raiteri M (1984) Functional evidence that chronic drugs induce adaptive changes of central autoreceptors regulating serotonin release. Eur J Pharmacol 97:309–313

Maura G, Roccatagliata E, Raiteri M (1986) Serotonin autoreceptor in rat hippocampus: pharmacological characterization as a subtype of the 5-HT₁ receptor. Naunyn-Schmiedeberg's Arch Pharmacol 334:323–326

Maura G, Thellung S, Andriolo GC, Ruelle A, Raiteri M (1993) Release-regulating serotonin 5-HT$_{1D}$ autoreceptors in human cerebral cortex. J Neurochem 60:1179–1182

Metcalf MA, McGuffin RW, Hamblin MW (1992) Conversion of the human 5-HT$_{1D\beta}$ serotonin receptor to the rat 5-HT$_{1B}$ ligand-binding phenotype by Thr^{355}Asn site directed mutagenesis. Biochem Pharmacol 44:1917–1920

Middlemiss DN (1988) Autoreceptors regulating serotonin release. In: Sanders-Bush E (ed) The serotonin receptors. Humana, Clifton, NJ, pp 201–224

Middlemiss DN, Hutson P (1990) The 5-HT$_{1B}$ receptors. Ann NY Acad Sci 600:132–147

Molderings GJ, Frölich D, Likungu J, Göthert M (1996) Inhibition of noradrenaline release via presynaptic 5-HT$_{1D\alpha}$ receptors in human atrium. Naunyn-Schmiedeberg's Arch Pharmacol 353:272–280

Moret C (1985) Pharmacology of the serotonin autoreceptor. In: Green AR (ed) Neuropharmacology of serotonin. Oxford University Press, Oxford, pp 21–49

Moret C, Briley M (1990) Serotonin autoreceptor subsensitivity and antidepressant activity. Eur J Pharmacol 180:351–356

Moret C, Briley M (1993) The unique effect of methiothepin on the terminal serotonin autoreceptor in the rat hypothalamus could be an example of inverse agonism. J Psychopharmacol 7:331–337

Moret C, Briley M (1995) In vitro and in vivo activity of 1-(1-naphthyl)piperazine at terminal 5-HT autoreceptors in guinea-pig brain. Naunyn-Schmiedeberg's Arch Pharmacol 351:377–384

Murphy RM, Zemlan FP (1989) Functional change in the 5-HT presynaptic receptor in spinal cord of aged rats. Neurobiol Aging 10:95–97

Nöthen MM, Erdmann J, Shimron-Abarbanell D, Propping P (1995) Identification of genetic variation in the human serotonin 1Dβ receptor gene. Biochem Biophys Res Comm 202:1194–1200

Oksenberg D, Marsters SA, O'Dowd BF, Jun H, Havlik S, Peroutka SJ, Ashkenazi A (1992) A single amino-acid difference confers major pharmacological variation between human and rodent 5-HT$_{1B}$ receptors. Nature 360:161–163

Ormandy GC (1993) Increased cyclic AMP reduces 5-HT$_{1D}$ receptor-mediated inhibition of [^{3}H]5-hydroxytryptamine release from guinea-pig cortical slices. Eur J Pharmacol Mol Pharmacol Sect 244:189–192

Oxford SJ, Warwick RO (1987) Differential effects of nialamide and clomipramine on serotonin efflux and autoreceptors. Pharmacol Biochem Behav 26:593–600

Parker EM, Grisel DA, Ibhen LG, Shapiro RA (1993) A single amino acid difference accounts for the pharmacological distinctions between the rat and human 5-hydroxytryptamine$_{1B}$ receptors. J Neurochem 60:380–383

Passarelli F, Costa T, Almeida OFX (1988) Pertussis toxin inactivates the presynaptic serotonin autoreceptor in the hippocampus. Eur J Pharmacol 155:297–299

Peroutka SJ (1988) 5-Hydroxytryptamine receptor subtypes. Annu Rev Neurosci 11:45–60

Peroutka SJ (1991) VI. Serotonin receptor subtypes and neuropsychiatric diseases: Focus on 5-HT$_{1D}$ and 5-HT$_3$ receptor agents. Pharmacol Rev 43:579–586

Peters JA, Lambert JJ (1989) Electrophysiology of 5-HT$_3$ receptors in neuronal cell lines. Trends Pharmacol Sci 10:172–175

Petty F, Kramer G, Wilson L (1992) Prevention of learned helplessness – In vivo correlation with cortical serotonin. Pharmacol Biochem Behav 43:361–367

Pineyro G, de Montigny C, Blier P (1995a) 5-HT$_{1D}$ receptors regulate 5-HT release in the rat raphe nuclei – in vivo voltammetry and in vitro superfusion studies. Neuropsychopharmacology 13:249–260

Pineyro G, Castanon N, Hen R, Blier P (1995b) Regulation of [^{3}H]5-HT release in raphe, frontal cortex and hippocampus of 5-HT$_{1B}$ knock-out mice. NeuroReport 7:353–359

Pineyro G, de Montigny C, Weiss M, Blier P (1996) Autoregulatory properties of dorsal raphe 5-HT neurons. Possible role of electronic coupling and 5-HT$_{1D}$ receptors in the rat brain. Synapse 22:54–62

Price GW, Burton MJ, Roberts C, Watson J, Duckworth M, Gaster L, Middlemiss DN, Jones BJ (1996) SB 216641 and BRL 15572 pharmacologically discriminate between 5-HT$_{1B}$ and 5-HT$_{1D}$ receptors. Br J Pharmacol 119:301P

Ramdine R, Galzin AM, Langer SZ (1989) Phorbol-12,13-dibutyrate antagonizes while forskolin potentiates the presynaptic autoreceptor-mediated inhibition of [^{3}H]-5-hydroxytryptamine release in rat hypothalamic slices. Synapse 3:173–181

Roberts C, Watson J, Burton M, Price GW, Jones BJ (1996) Functional characterization of the 5-HT terminal autoreceptor in the guinea-pig brain cortex. Br J Pharmacol 117:384–388

Robertson B, Bevan S (1991) Properties of 5-hydroxytryptamine$_3$ receptor-gated currents in adult rat dorsal root ganglion neurones. Br J Pharmacol 102:272–276

Rollema H, Clarke T, Sprouse JS, Schulz DW (1996) Combined administration of a 5-hydroxytryptamine (5-HT)$_{1D}$ antagonist and a 5-HT reuptake inhibitor synergistically increases 5-HT release in guinea pig hypothalamus in vivo. J Neurochem 67:2204–2207

Saudou F, Amara DA, Dierich A, Le Meur M, Ramboz S, Segu L, Buhot MC, Hen R (1994) Enhanced aggressive behavior in mice lacking 5-HT$_{1B}$ receptor. Science 265:1875–1878

Schechter LE, Bolanos FJ, Gozlan H, Lanfumey L, Haj-Dahmane S, Laporte AM, Fattaccini CM, Hamon M (1990) Alterations of central serotoninergic and dopaminergic neurotransmission in rats chronically treated with ipsapirone: biochemical and electrophysiological studies. J Pharmacol Exp Ther 255:1335–1347

Schipper J (1990) Pharmacological characterization of serotonin autoreceptors. Pharmacol Toxicol 66 (Suppl 3):149

Schipper J, Tulp MTM (1988) Serotonin autoreceptors in guinea pig cortex slices resemble the 5-HT$_{1D}$ binding site. Soc Neurosci Abstr 14:552

Schlicker E, Brandt F, Classen K, Göthert M (1985) Serotonin release in human cerebral cortex and its modulation via serotonin receptors. Brain Res 331:337–341

Schlicker E, Fink K, Classen K, Göthert M (1987) Facilitation of serotonin (5-HT) release in the rat brain cortex by cAMP and probable inhibition of adenylate cyclase in 5-HT nerve terminals by presynaptic α_2-adrenoceptors. Naunyn-Schmiedeberg's Arch Pharmacol 336:251–256

Schlicker E, Classen K, Göthert M (1988) Presynaptic serotonin receptors and α-adrenoceptors on central serotoninergic and noradrenergic neurons of normotensive and spontaneously hypertensive rats. J Cardiovasc Pharmacol 11:518–528

Schlicker E, Betz R, Göthert M (1989) Investigation into the age-dependence of release of serotonin and noradrenaline in the rat brain cortex and of autoreceptor-mediated modulation of release. Neuropharmacology 28:811–815

Schlicker E, Glaser T, Lümmen G, Neise A, Göthert M (1991) Serotonin and histamine receptor-mediated inhibition of serotonin and noradrenaline release in rat brain cortex under nimodipine treatment. Neurochem Int 19:437–444

Schlicker E, Fink K, Zentner J, Göthert M (1996) Presynaptic inhibitory serotonin autoreceptors in the human hippocampus. Naunyn-Schmiedeberg's Arch Pharmacol 354:393–396

Schlicker E, Fink K, Molderings GJ, Price GW, Middlemiss DN, Zentner J, Likungu J, Göthert M (1997) Effects of SB 216641 and BRL 15572 (selective h5-HT$_{1B}$ and h5-HT$_{1D}$ receptor antagonists, respectively) on guinea-pig and human 5-HT auto- and heteroreceptors. Br J Pharmacol 120:143P

Schoeffter P, Hoyer D (1990) 5-Hydroxytryptamine (5-HT)-induced endothelium-dependent relaxation of pig coronary arteries is mediated by 5-HT receptors similar to the 5-HT$_{1D}$ receptor subtype. J Pharmacol Exp Ther 252:387–395

Schoeffter P, Hoyer D (1991) Interaction of the α-adrenoceptor agonist oxymetazoline with serotonin 5-HT$_{1A}$, 5-HT$_{1B}$, 5-HT$_{1C}$ and 5-HT$_{1D}$ receptors. Eur J Pharmacol 196:213–216

Schoups AA, De Potter WP (1988) Species dependence of adaptations at the pre- and postsynaptic serotonergic receptors following long-term antidepressant drug treatment. Biochem Pharmacol 37:4451–4460

Sharp T, Bramwell SR, Grahame-Smith DG (1989) 5-HT$_1$ agonists reduce 5-hydroxytryptamine release in rat hippocampus in vivo as determined by brain microdialysis. Br J Pharmacol 96:283–290

Sherman AD, Allers GL, Petty F, Henn FA (1979) A neuropharmacologically relevant animal model of depression. Neuropharmacology 18:891–894

Siever LJ, Kahn RS, Lawlor BA, Trestman RL, Lawrence TL, Coccaro EF (1991) Critical issues in defining the role of serotonin in psychiatric diseases. Pharmacol Rev 43:509–525

Singer EA (1988) Transmitter release from brain slices elicited by single pulses: a powerful method to study presynaptic mechanisms. Trends Pharmacol Sci 9:274–276

Singh A, Redfern P (1994a) Lack of circadian variation in the sensitivity of rat terminal 5-HT$_{1B}$ autoreceptors. J Pharm Pharmacol 46:366–370

Singh A, Redfern P (1994b) Guinea pig terminal 5-HT$_{1D}$ autoreceptors do not display a circadian variation in their responsiveness to serotonin. Chronobiol Int 11:165–172

Skingle M, Sleight AJ, Feniuk W (1995) Effects of the 5-HT$_{1D}$ receptor antagonist GR 127935 on extracellular levels of 5-HT in the guinea-pig frontal cortex as measured by microdialysis. Neuropharmacology 34:377–382

Sleight AJ, Cervenka A, Peroutka SJ (1990) In vivo effects of sumatriptan (GR 43175) on extracellular levels of 5-HT in the guinea pig. Neuropharmacology 29:511–513

Sprouse JS, Aghajanian GK (1986) (-)-Propranolol blocks the inhibition of serotonergic dorsal raphe cell firing by 5-HT$_{1A}$ selective agonists. Eur J Pharmacol 128:295–298

Sprouse JS, Aghajanian GK (1987) Electrophysiological responses of serotoninergic dorsal raphe neurons to 5-HT$_{1A}$ and 5-HT$_{1B}$ agonists. Synapse 1:3–9

Starke K (1981) Presynaptic receptors. Annu Rev Pharmacol Toxicol 21:7–30

Starke K, Göthert M, Kilbinger H (1989) Modulation of neurotransmitter release by presynaptic autoreceptors. Physiol Rev 69:864–989

Starkey SJ, Skingle M (1994) 5-HT$_{1D}$ as well as 5-HT$_{1A}$ autoreceptors modulate 5-HT release in the guinea-pig dorsal raphe nucleus. Neuropharmacology 33:393–402

Timmermans PBMWM, Thoolen MJC (1987) Autoreceptors in the central nervous system. Med Res Rev 7:307–332

Törk I (1990) Anatomy of the serotonergic system. Ann NY Acad Sci 600:9–34

Vandijken HH, Mos J, van der Heyden JAM, Tilders FJH (1992a) Characterization of stress-induced long-term behavioural changes in rats – Evidence in favour of anxiety. Physiol Behav 52:945–951

Vandijken HH, van der Heyden JAM, Mos J, Tilders FJH (1992b) Inescapable footshocks induce progressive and long-lasting behavioural changes in male rats. Physiol Behav 51:787–794

Vanhoutte PM, Humphrey PPA, Spedding M (1996) X. International Union of Pharmacology Recommendations for nomenclature of new receptor subtypes. Pharmacol Rev 48:1–2

Versteeg DHG, Csikós T, Spierenburg H (1991) Stimulus-evoked release of tritiated monoamines from rat periaqueductal gray slices in vitro and its receptor-mediated modulation. Naunyn-Schmiedeberg's Arch Pharmacol 343:595–602

Waeber C, Schoeffter P, Palacios JM, Hoyer D (1988) Molecular pharmacology of 5-HT$_{1D}$ recognition sites: radioligand binding studies in human, pig and calf brain membranes. Naunyn-Schmiedeberg's Arch Pharmacol 337:595–601

Waeber C, Schoeffter P, Palacios JM, Hoyer D (1989) 5-HT$_{1D}$ receptors in guinea-pig and pigeon brain. Naunyn-Schmiedeberg's Arch Pharmacol 340:479–485

Wang HY, Friedman E (1987) Protein kinase C: regulation of serotonin release from rat brain cortical slices. Eur J Pharmacol 141:15–21

Wang HY, Friedman E (1988) Chronic lithium: desensitization of autoreceptors mediating serotonin release. Psychopharmacology 94:312–314

Weinshank RL, Branchek T, Hartig PR (1991) International Application Patent, International Publication Number WO 91/17174, November 14, 1991

Wichmann T, Limberger N, Starke K (1989) Release and modulation of release of serotonin in rabbit superior colliculus. Neuroscience 32:141–151

Wilkinson LO, Middlemiss DN (1992) Metitepine distinguishes two receptors mediating inhibition of [^{3}H]-5-hydroxytryptamine release in guinea pig hippocampus. Naunyn-Schmiedeberg's Arch Pharmacol 345:696–699

Wilkinson LO, Hawkins LM, Beer MS, Hibert MF, Middlemiss DN (1993) Stereoselective actions of the isomers of metitepine at 5-HT$_{1D}$ receptors in the guinea pig brain. Neuropharmacology 32:205–208

Wolf WA, Kuhn DM, Lovenberg W (1985) Serotonin and central regulation of arterial blood pressure. In: Vanhoutte PM (ed) Serotonin and the cardiovascular system. Raven, New York, pp 63–73

Yakel JL, Jackson MB (1988) 5-HT$_3$ receptors mediate rapid responses in cultured hippocampus and a clonal cell line. Neuron 1:615–621

Yakel JL (1992) 5-HT$_3$ receptors as cation channels. In: Hamon M (ed) Central and peripheral 5-HT$_3$ receptors. Academic, London pp 103–128

Yang L, Jacocks HM, Helke CJ (1994) Release of [^{3}H]5-hydroxytryptamine from the intermediate area of rat thoracic spinal cord is modulated by presynaptic autoreceptors. Synapse 18:198–204

Zgombick JM, Schechter LE, Kucharewicz SA, Weinshank RL, Branchek TA (1995) Ketanserin and ritanserin discriminate between recombinant human 5-HT$_{1D\alpha}$ and 5-HT$_{1D\beta}$ receptor subtypes. Eur J Pharmacol 291:9–15

Zifa E, Fillion G (1992) 5-Hydroxytryptamine receptors. Pharmacol Rev 44:401–458

CHAPTER 13

Behavioral Consequences of 5-HT$_{1B}$ Receptor Gene Deletion

N. Castanon, S. Ramboz, F. Saudou, and R. Hen

A. Introduction

Serotonin (5-hydroxytryptamine, 5-HT) is a biogenic amine which is involved in a wide range of physiological functions including sleep, appetite, pain perception, sexual activity, memory and mood control (for a review see Wilkinson et al. 1991). A central serotonin deficit has been associated with behaviors such as suicidality, impulsive violence (Higley et al. 1992), depression and alcoholism (Eichelmann et al. 1992), and serotoninergic drugs are used in the treatment of a number of pathological states including migraine, depression and anxiety (Sleight et al. 1991). The multiple actions of serotonin are mediated by the interaction of this amine with at least 14 receptors (for a review see Saudou and Hen 1994), most of which belong to the G-protein-coupled receptor family.

The purpose of this study was to determine the contribution of one of these receptors, the 5-HT$_{1B}$ subtype, to the various behavioral responses elicited by serotonin and serotoninergic drugs. The 5-HT$_{1B}$ receptor, which is the rodent homologue of the human 5-HT$_{1D\beta}$ receptor, is expressed in a variety of brain regions including motor control centers such as the basal ganglia as well as structures involved in mood control such as the central gray, hippocampus and raphe nuclei (Boschert et al. 1992; Bruinvels et al. 1993; Maroteaux et al. 1992). Pharmacological studies using poorly specific agonists have suggested that activation of 5-HT$_{1B}$ receptors might lead to an increase in anxiety and locomotion (Green et al. 1984; Griebel et al. 1990; Oberlander et al. 1986, 1987; Pellow et al. 1987) and to a decrease in food intake, sexual activity and aggressive behavior (Fernandez-Guasti et al. 1992; Kennett et al. 1987; Koe et al. 1992; Olivier et al. 1986). In particular, a class of 5-HT$_1$ agonists have been termed "serenics," because of their antiaggressive properties in several rodent aggression models (Flannelly et al. 1985; Mos et al. 1992; Olivier 1980; Olivier et al. 1986, 1989). However, it is not clear to what extent the effects of the serenics are mediated by 5-HT$_{1B}$ receptors because these drugs also activate 5-HT$_{1A}$ receptors and possibly some of the recently discovered 5-HT receptors. In order to study the function of the 5-HT$_{1B}$ receptor, we have generated by homologous recombination homozygous mutant mice lacking both copies of the gene encoding this receptor. These mice are viable and fertile. They were analyzed for a variety of behaviors that are thought to be modulated by 5-HT$_{1B}$ receptors such as locomotion, anxiety and aggression.

B. 5-HT$_{1B}$ Receptor Gene Targeting

The 5-HT$_{1B}$ receptor gene was disrupted by homologous recombination (Capecchi et al. 1989 and Fig. 1). The JA construct consisted of 6.0 kilobasepairs (kb) of genomic sequence in which part of the 5-HT$_{1B}$ coding sequence was replaced by a neomycin phosphotransferase gene (neo) under the control of the GTI-II enhancer (Lufkin et al. 1991 and Fig. 1A). In the JB construct, the neo cassette was inserted in the coding sequence of the 5-HT$_{1B}$ gene (Fig. 1B). The two linearized targeting vectors were electroporated into D3 embryonic stem (ES) cells and G418-resistant colonies were screened by Southern blotting (Fig. 1). The EX400 probe identified positive clones with the expected 10- and 8.5-kb *Kpn*I fragments for the JA and JB constructs, respectively (Fig. 1). Four positive clones were obtained with both constructs, yielding a targeting frequency of 1/15 (JA) and 1/12 (JB). Southern analyses using *Xba*I digests and the E2A1 probe or the neo probe confirmed that accurate targeting occurred and that no additional integration took place (data not shown). Cells from the positive clones JA7 and JB13 were microinjected into 3.5-day C57BL/6 mouse blastocysts. The two clones gave rise to highly chimaeric mice which were bred with C57BL/6 females in order to test for germline transmission of the mutated 5-HT$_{1B}$ receptor gene. The positive

$\longrightarrow$

Fig. 1A,B. Homologous recombination at the 5-HT$_{1B}$ locus. Schematic representation of the targeting event using JA **(A)** and JB **(B)** constructs and Southern analysis of JA7 **(A)** and JB13 **(B)** mutant mice. *Upper panels* correspond to the targeting vectors, genomic structure of the 5-HT$_{1B}$ gene and predicted structures of the mutated alleles after homologous recombination. *Black box* corresponds to the coding sequence of the 5-HT$_{1B}$ receptor and *hatched box* to the neo cassette. *Arrows on the neo cassette and on the 5-HT$_{1B}$ gene* indicate the direction of transcription (*from left to right*). The locations of the probes E2A1 and EX400 used in Southern analysis are shown. E2A1 and EX400 probes were used to screen neomycin-resistant clones after *Xba*I and *Kpn*I digests, respectively. *Bottom,* Southern blot analysis. Tail DNA from wild-type, heterozygous and homozygous JA7 **(A)** and JB13 **(B)** mutant mice were cut by *Kpn*I and hybridized with the 3′-end EX400 probe. *E, Eco*RI; *B, Bal*I; *X, Xba*I; *K, Kpn*I; *V, Eco*RV; +/+, wild type; +/−, heterozygous; −/−, homozygous mutant.

Methods: JA and JB targeting vectors were constructed from phage 172 containing the 5-HT$_{1B}$ coding sequence (Maroteaux et al 1992). The JA vector contains a 6-kb *Eco*RI genomic fragment subcloned into pBluescript SK- (Stratagene) in which the two 860-bp *Bal*I fragments were replaced by a 1.7-kb *Sma*I fragment containing the GTI-II neo-cassette lacking a poly(A) tail [purified from p581 (Lufkin et al 1991)]. The JB vector consisted of the same genomic fragment in which the 1.7-kb *Sma*I fragment containing the GTI-II neo-cassette was inserted into the *Eco*RV site of the 5-HT$_{1B}$ gene. Before electroporation of ES cells, JA and JB targeting vectors were linearized by *Xho*I and *Spe*I, respectively. Electroporation of D3 ES cells (a gift from R Kemler), cell culture, G418 selection, DNA preparation, Southern analysis and generation of chimaeric mice were as described by Lufkin et al (1991)

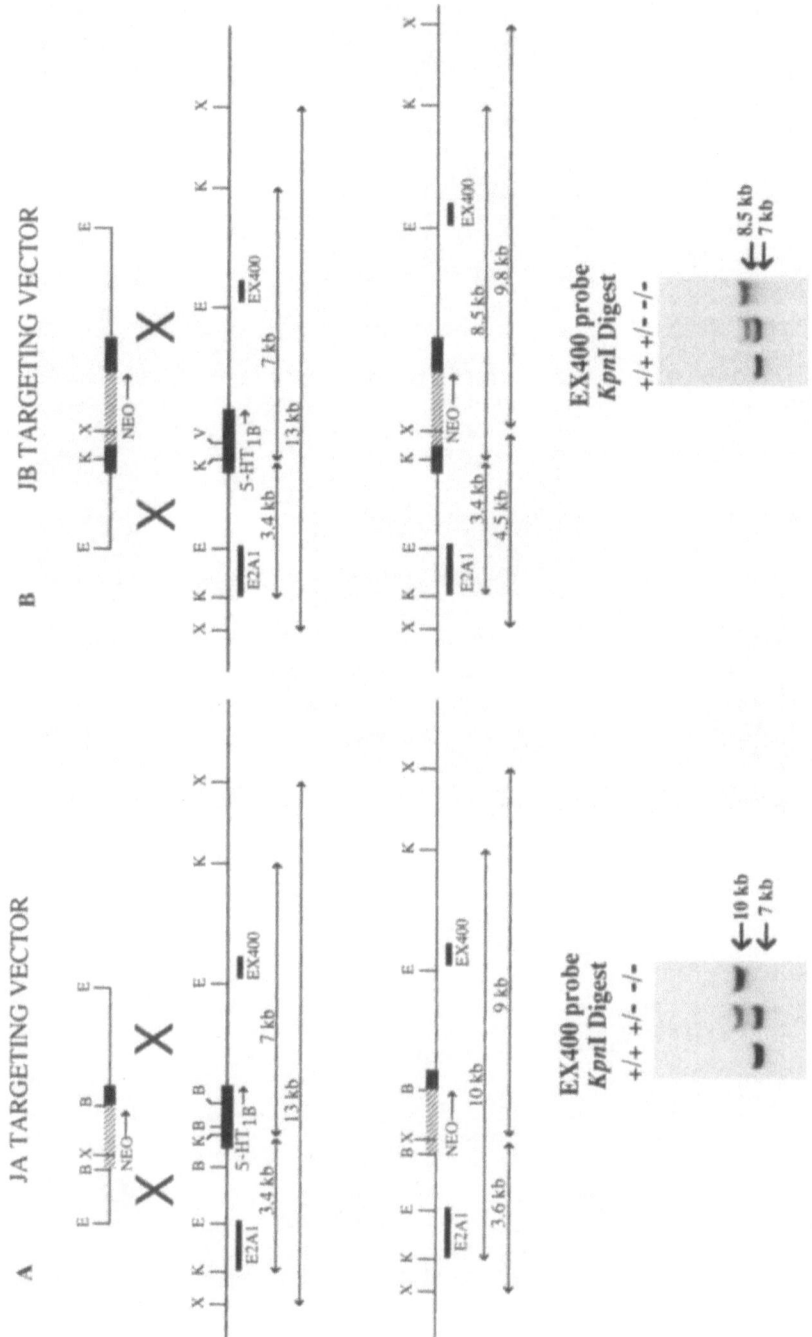

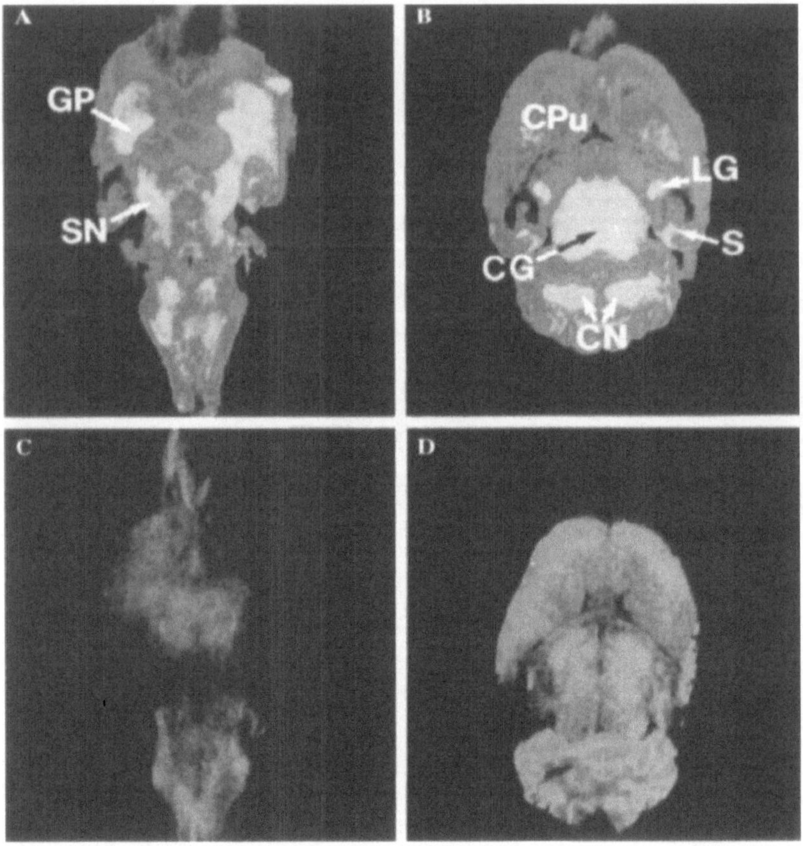

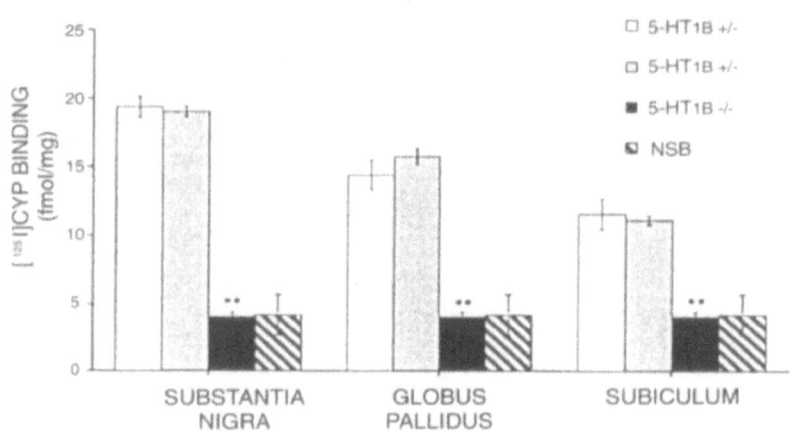

chimaeras were bred with females from the 129/Sv-ter inbred strain to obtain heterozygotes on the 129/Sv-ter genetic background. Heterozygous mice were phenotypically normal and fertile. Homozygous animals deriving from both cell lines were generated by heterozygote crossings. In 243 JA offsprings and 213 JB offsprings, the expected 1:2:1 ratio of wild-type, heterozygous and homozygous mutant progeny was observed. The homozygous mutants did not display any obvious developmental or behavioral abnormality and were fertile. Although the average lifespan has not yet been determined, no spontaneous deaths occurred during the first 12 months of life. All the analyses presented here were performed on animals having a pure 129/Sv genetic background. There were no differences between the mice derived from the JA7 and JB13 targeted cell lines.

C. Effectiveness of 5-HT$_{1B}$ Receptor Ablation

In order to ensure that disruption of the 5-HT$_{1B}$ receptor gene was effective, we performed autoradiographic studies (SEGU et al. 1990) on brains of wild-type, heterozygous and homozygous mutants using the radiolabeled ligand $3[^{125}I]$iodocyanopindolol ($[^{125}I]$CYP). When used in the presence of appropriate masking agents, this radioligand binds specifically to the 5-HT$_{1B}$ receptor (OFFORD et al. 1988; PAZOS and PALACIOS 1985). No specific binding was observed in homozygous mutants, demonstrating that effective disruption of the 5-HT$_{1B}$ gene occurred (SAUDOU et al. 1994). We also performed autoradiographic studies with S-CM-G$[^{125}I]$TNH$_2$ ($[^{125}I]$GTI), which is specific for 5-HT$_{1B}$ and 5-HT$_{1D}$ receptors (BOULENGUER et al. 1991; SEGU et al. 1991). In wild-type mice, $[^{125}I]$-GTI-binding sites were found in the globus pallidus, substantia nigra, cerebellar nuclei, subiculum, lateral geniculate nucleus, central gray and colliculi, while a low level of specific binding was observed in homozygous mutants in the globus pallidus and substantia nigra (Fig. 2C). Preliminary pharmacological studies indicate that these residual sites correspond to 5-HT$_{1D\alpha}$ receptors (not shown).

◄

Fig. 2A–D. Levels of 5-HT$_{1B}$ and 5-HT$_{1D\alpha}$ receptors in wild-type and homozygous 5-HT$_{1B}$-minus mice analyzed by autoradiography with S-CM-G$[^{125}I]$TNH$_2$ ($[^{125}I]$GTI). Autoradiograms of horizontal sections of adult wild-type mice (**A, B**) or 5-HT$_{1B}$-minus mice (**C, D**) incubated with 0.3 nM $[^{125}I]$-GTI. *CPu*, caudate putamen; *CG*, central gray; *LG*, lateral geniculate nucleus; *SN*, substantia nigra; *GP*, globus pallidus; *CN*, cerebellar nuclei, *TGN*, trigeminal nucleus; *NSB*, nonspecific binding.

Methods: Preparation of the brains, horizontal sections and preincubation of the sections were as described in BOSCHERT et al. (1992). Incubations were performed with 0.3 nM of S-CM-G$[^{125}I]$TNH$_2$. Nonspecific binding was determined with 10 mM 5-HT. Washing conditions and exposure to films were as described by BOSCHERT et al. (1992). Quantitative analyses of the autoradiograms was carried out for the different anatomical structures with a computer device for image analysis (SEGU et al. 1990)

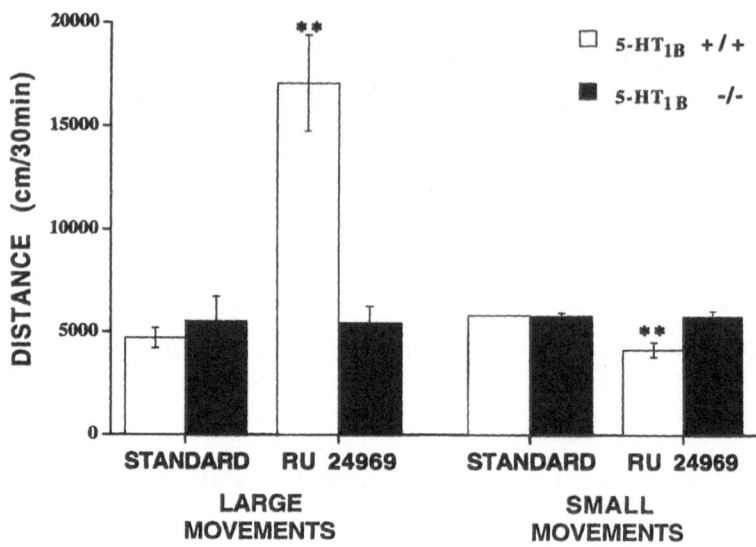

Fig. 3. Locomotor activity of the mutant mice and effect of RU 24969 (mean ± SEM). A video tracking device was used to measure the distance travelled by the animals during a period of 30 min in an open field. Locomotor activity in control conditions of wild type ($n = 12$) and mutant mice ($n = 10$) and effect of RU 24969 injected in the same mice 10 days after the first test. There was no significant difference between the wild-type mice and the mutant mice in control conditions as revealed by t-test analysis. However, after RU 24969 treatment there was a significant difference between the two groups (**$P < 0.01$, *$P < 0.05$).

Methods: Male mice were 12 weeks old at time of testing. They were housed alone in a standard cage with food and water, and kept on a 12/12-h light-dark cycle with light onset at 0700 hours. The mice were tested between 1000 and 1600 hours during the light phase. Mice were placed in a circular open field (70 cm in diameter) and the distance traveled by the animal was recorded by a videotracking apparatus (Videotrack, Viewpoint, Lyon, France). Ten days after the first test, mice were injected with RU 24969 prior to the second test. RU 24969 (5-methoxy-3-(1,2,5,6-tetrahydropyrid-4-yl)-1*H*-indole) was dissolved in saline and administered intraperitoneally, 40 min before testing, at a concentration of 5 mg/kg body weight in a volume of 10 ml/kg body weight

D. Locomotion

As shown in Fig. 2, the 5-HT_{1B} receptor is localized in motor control centers such as the globus pallidus, substantia nigra and deep cerebellar nuclei. Furthermore, pharmacological studies have suggested an involvement of 5-HT_{1B} receptors in the control of locomotor activity (GREEN et al. 1984; OBERLANDER et al. 1986, 1987). The activity of the mice in an open field was analyzed with a video-tracking device. No significant differences were detected between the mutant mice and their wild-type littermates (Fig. 3). Administration of the 5-HT_1 agonist RU 24969 stimulated locomotor activity in the wild-type mice while it had no effect in the mutants (Fig. 3). These results indicate that the hyperlocomotor effect of RU 24969 is mediated by 5-HT_{1B} receptors.

E. Anxiety

5-HT$_1$ agonists such as RU 24969, eltoprazine, fluprazine or 1-(m-trifluoromethylphenyl)piperazine (TFMPP) have been reported to induce anxiogenic responses in rats and mice (GRIEBEL et al. 1990; PELLOW et al. 1987). To evaluate the level of anxiety of the 5-HT$_{1B}$-minus mice, we used the light/dark choice test (CRAWLEY and GOODWIN 1980 and Fig. 4). The time spent in the lit compartment, as well as the number of transitions between the dark and the lit compartments, have been considered as indices of anxiety since they are increased by anxiolytic drugs (CRAWLEY and GOODWIN 1980; MISSLIN et al 1989). There were no significant differences between mutants

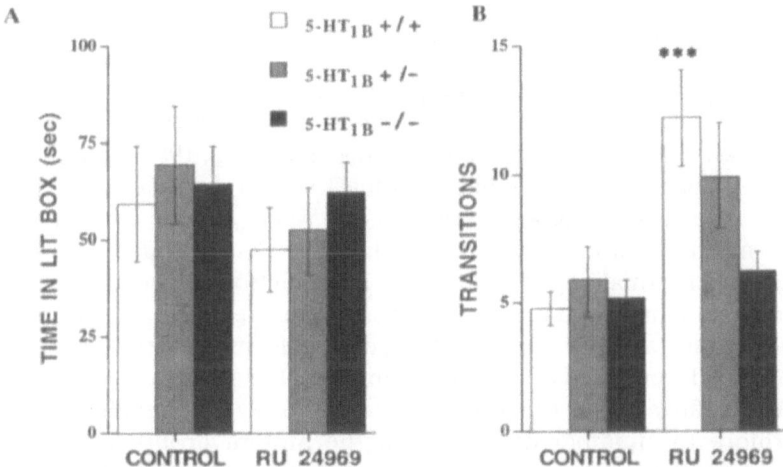

Fig. 4A,B. Behavior of wild-type and mutant mice in the light-dark choice test. **A** Time spent by mice in the lit box (time in lit box: mean ± SEM); **B** Number of tunnel crossings from the dark box to the lit box (transitions: mean ± SEM). *Left columns* (control) correspond to wild type ($n = 22$), heterozygous ($n = 12$) and homozygous mutant mice ($n = 26$) injected with saline vehicle and the right columns correspond to a different series of wild type ($n = 23$), heterozygous ($n = 12$) and mutant mice ($n = 26$) injected with RU 24969. ANOVA (factors = genotype and treatment) revealed no significant differences among groups for time spent by mice in the lit box **(A)**: genotype [$F(2, 115) = 0.9$, NS]; treatment [$F(1, 115) = 1.31$, NS]; genotype × treatment interaction [$F(2, 115) = 0.25$, NS]. In contrast, there were significant differences among groups for the number of transitions **(B)**: genotype [$F(2, 115) = 3.38, P < 0.05$]; treatment [$F(1, 115) = 15.62, P < 0.0001$]; genotype × treatment interaction [$F(2, 115) = 4.12, P < 0.05$]. There were no significant differences among control groups for the transitions [$F(2, 58) = 0.49$, NS] but the treatment with RU 24969 increased significantly the number of transitions in wild-type mice [$t(42) = 3.54, P < 0.001$], not quite significantly in heterozygous mice [$t(22) = 1.63, P = 0.11$] and not in homozygous mutant mice [$t(50) = 1.01$, NS].

Methods: Male mice were housed under the conditions described in Fig. 3 except that the mice were tested during the dark phase. Procedure and apparatus were as described by CRAWLEY and GOODWIN (1980) and MISSLIN et al. (1989). For the control conditions, mice were injected with saline. Injection of saline or RU 24969 was as in Fig. 3

and their wild-type littermates for either parameter, suggesting that the mutants have the same level of anxiety as the wild-type mice in this test (Fig. 4). In the wild-type mice, RU 24969 increased significantly the number of transitions between the dark box and the lit box, but had no significant effect on the time spent in the lit box (Fig. 4). In the mutants RU 24969 did not modify these parameters. The larger number of transitions displayed by the wild-type mice are probably a result of the increase in locomotion induced by RU 24969. The lack of response of the mutants demonstrates that the effect of RU 24969 is mediated by 5-HT$_{1B}$ receptors. In contrast with earlier reports, RU 24969 had no anxiogenic effect. Such an effect might have been masked in the present experiment by the hyperlocomotor effect of this drug. A dose response curve as well as other anxiety tests might allow us to reveal an effect of RU 24969.

F. Aggression

A class of 5-HT$_1$ agonists including eltoprazine and fluprazine have been termed "serenics" because of their antiaggressive properties (FLANNELLY et al. 1985; Mos et al. 1992; OLIVIER 1980; OLIVIER et al. 1986, 1989), and their effects have been suggested to be mediated, at least in part, by 5-HT$_{1B}$ receptors. We therefore investigated the aggressiveness of 5-HT$_{1B}$-minus mice in a classical aggression test. After an isolation period of 4 weeks, test mice (resident) were analyzed for intermale aggression after introduction in their cage of a wild-type mouse that had been reared in a group (intruder) (Fig. 5). In this test, the latency of attack and the number of attacks performed by the resident during a 3-min period were used as aggression indices. The mutant residents attacked the intruder faster than the wild-type or heterozygous residents (Fig. 5A). Furthermore the number of attacks in the mutant group was significantly higher than in the wild-type or heterozygote groups (Fig. 5B). In addition, the intensity of attacks of the mutant residents was higher, as well as the number of tail rattlings preceding the attacks (not shown). Similar results were obtained in two tests performed 1 week apart. The level of aggressiveness was higher in the second test with both the wild-type and the mutant animals, in good agreement with previous reports showing that aggression increases with fighting experience (LAGERSPETZ and LAGERSPETZ 1971). A qualitative analysis of the attacks during the 3-min test period revealed additional marked differences between wild-type and mutant mice (Fig. 6). In the first test, 29% of the mutant residents attacked the intruder within less than 10s after introduction of the intruder in the cage (impulsive attacks, Fig. 6A) while no wild-type or heterozygous mice attacked the intruder during that time interval. Conversely, 75% of the wild-type mice and only 21% of the mutants did not attack during the 3-min test. In the second test, the percentage of mutants displaying impulsive attacks was even higher (46%) while still no wild-type animals performed such short-latency attacks (Fig. 6B). These results indicate that the 5-HT$_{1B}$-minus mice are more aggressive than their wild-type or heterozygous litter-

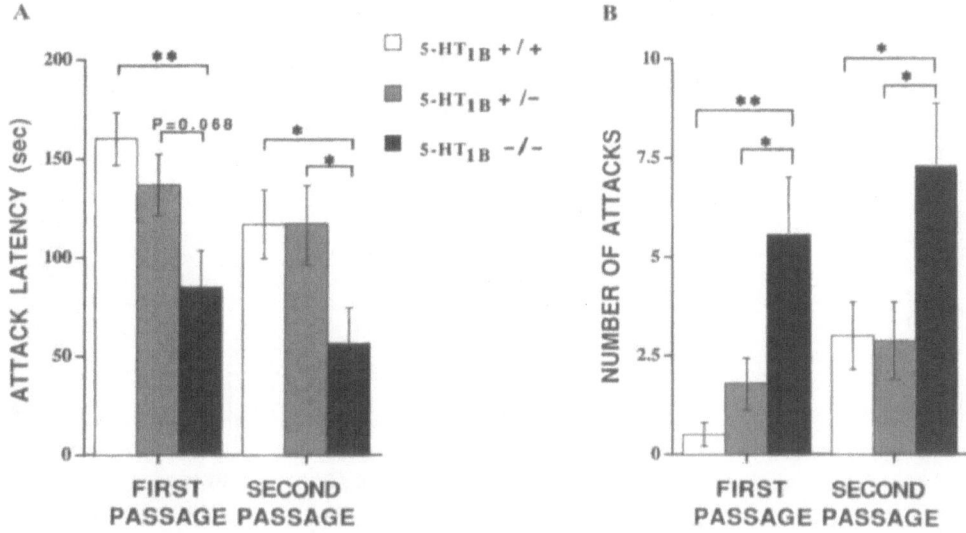

Fig. 5A,B. Resident-intruder aggression test. Resident mice were: wild-type ($n = 12$), heterozygote ($n = 16$) and mutant mice ($n = 14$). **A** Attack latency (mean ± SEM): time between the introduction of the intruder and the first attack by the resident. ANOVA revealed significant differences for the attack latency both in the first test [$F(2, 39) =$ 5.38, $P < 0.01$] and in the second test [$F(2, 37) = 3.49$, $P < 0.05$]. Further statistical analyses revealed significant differences between wild-type and mutant mice [first test, $t(24) = 3.19$, $P < 0.01$; second test, $t(23) = 2.38$, $P < 0.05$], heterozygotes and mutant mice [first test, $t(28) = 2.17$, $P < 0.05$; second test, $t(26) = 2.26$, $P < 0.05$], but not between wild-type and heterozygous mice [first test, $t(26) = 1.10$, NS; second test, $t(25) = 0.01$, NS]. **B** Number of attacks (means ± SEM) during the session. ANOVA: first test, $F(2, 39) =$ 7.39, $P < 0.01$; second test, $F(2, 37) = 4.48$, $P < 0.02$. t-tests: wild-type vs. mutants (first test [$t(24) = 3.19$, $P < 0.01$]; second test [$t(23) = 2.32$, $P < 0.05$]), heterozygotes vs. mutants (first test [$t(28) = 2.16$, $P < 0.05$]; second test [$t(26) = 2.44$, $P < 0.05$]), wild-type vs. heterozygotes (first test [$t(26) = 1.72$, NS]; second test [$t(25) = 0.99$, NS]). *$P < 0.05$; **$P < 0.01$.

Methods: Male mice, 12–14 weeks old at time of testing, were isolated during 4 weeks in transparent cages ($22 \times 16 \times 13$ cm). Litter was changed once a week, without moving the animal, but not during the week preceding the test. Food, water and light-dark cycle was as in Fig. 3. Wild-type male mice of the same strain (129/Sv-ter), 8 weeks old at time of testing and housed eight per cage were used as intruders. The intruder was then introduced into the resident cage and attack latency and the number of attacks were measured during the 3-min session. The latency of mice that did not attack was scored as 180 s. One week later, the same mice were tested again (second test)

mates and suggest that the 5-HT$_{1B}$ receptor is, at least in part, responsible for the antiaggressive properties of the serenics. In order to determine whether 5-HT$_{1B}$ receptors are the preferential target of these drugs, we studied the effect of an eltoprazine injection on the aggression test. As expected, eltoprazine abolished completely the aggressive behavior in the wild-type mice (Fig. 7). However, eltoprazine also decreased the aggressiveness of the mutant mice. These results confirm the antiaggressive properties of the eltoprazine

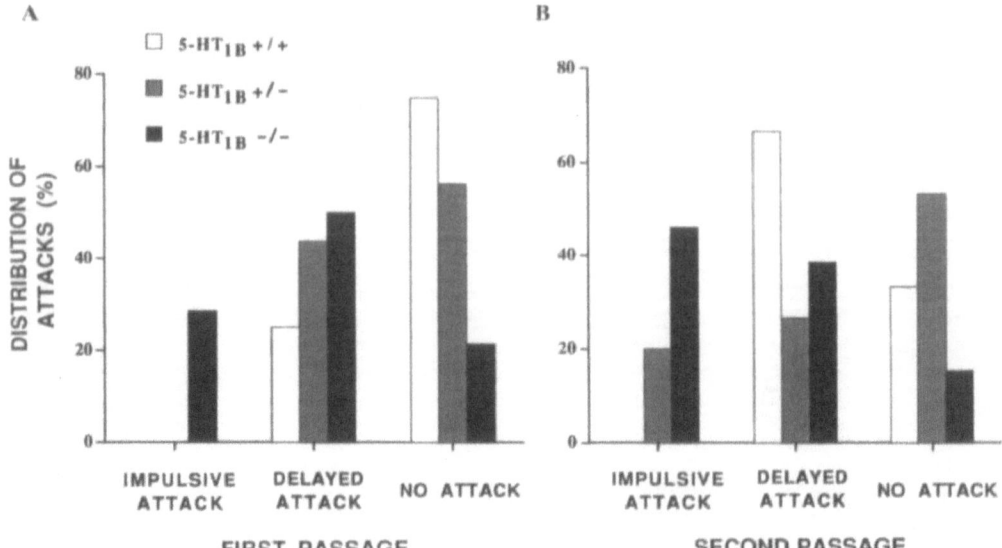

Fig. 6A,B. Resident-intruder aggression test: distribution of attacks. Attacks were categorized as impulsive attacks (attacks within less than 10s), delayed attacks (attacks displayed between 10 and 180s) and no attacks during the 3-min test session. **A** First test; **B** second test. The values presented here derive from the experiment described in Fig. 5

(Flannelly et al. 1985; Mos et al. 1992; Olivier 1980; Olivier et al. 1986, 1989) and suggest that the $5-HT_{1B}$ receptor is not the only target since an antiaggressive effect remains in the absence of this receptor. Eltoprazine is also an effective agonist of $5-HT_{1A}$ receptors and activation of $5-HT_{1A}$ receptors has been shown to decrease aggressive behavior. It is therefore likely that the remaining effect of eltoprazine in the mutant mice is due to activation of $5-HT_{1A}$ receptors.

G. Discussion and Conclusions

We have generated by homologous recombination mice lacking the $5-HT_{1B}$ receptor. Our autoradiographic data demonstrated the absence of $5-HT_{1B}$ receptors in the homozygous mutants. Such mice develop and live apparently normally. Preliminary histological analyses of their central nervous system did not reveal any obvious defect (not shown).

We analyzed a number of behaviors that were thought to be modulated by activation of $5-HT_{1B}$ receptors, such as locomotion, anxiety and aggression (Fernandez-Guasti et al. 1992; Green et al. 1984; Griebel et al. 1990; Kennett et al. 1987; Koe et al. 1992; Oberlander et al. 1986, 1987; Olivier et al. 1987; Pellow et al. 1987). Surprisingly, in our test conditions, we did not

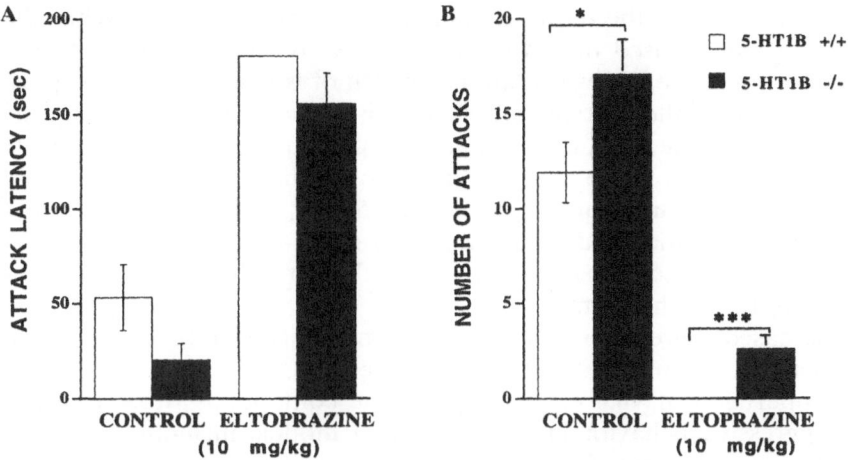

Fig. 7A,B. Resident-intruder aggression test: effect of eltoprazine on the attack latency **(A)** (means ± SEM) and the number of attacks **(B)** (means ± SEM). *Left columns* (control) correspond to wild-type (n = 11) and homozygous mutant mice (n = 11) injected with saline vehicle and *right columns* correspond to a different series of wild type (n = 12) and mutant mice (n = 10) injected with eltoprazine (10 mg/kg, s.c.). ANOVA (factors = genotype and treatment) revealed significant differences among groups for the attack latency **(A)**: genotype [$F(1, 40) = 5.49$, $P < 0.05$]; treatment [$F(1, 40) = 112.25$, $P < 0.001$]; genotype × treatment interaction [$F(2, 40) = 0.10$, NS]; and for the number of attacks **(B)**: genotype [$F(1, 40) = 9.50$, $P < 0.005$]; treatment [$F(1, 40) = 109.35$, $P < 0.001$]; genotype × treatment interaction [$F(2, 40) = 1.05$, NS]. There were no significant differences among control groups [$F(1, 21) = 2.87$, NS] and among eltoprazine groups [$F(1, 21) = 1272.5$, NS] for the attack latency. In contrast, the number of attacks is significantly higher in mutant mice than in wild-type mice both with [$F(1, 21) = 36.87$, $P < 0.001$] and without [$F(1, 21) = 147.68$, $P < 0.05$] eltoprazine. *Methods:* Male mice (5 months old) were housed in the conditions described in Fig. 5. Eltoprazine was dissolved in saline and administered subcutaneously, 30 min before testing, at a concentration of 10 mg/kg body weight

detect any differences in the levels of basal locomotor activity or anxiety in the mutant mice. However, the hyperlocomotor effect of the 5-HT$_1$ agonist RU 24969 was totally absent in the mutants, demonstrating that this effect is mediated by 5-HT$_{1B}$ receptors. We might therefore have expected a decrease in locomotor activity in the 5-HT$_{1B}$-minus mice. The absence of such a motor effect in our experimental conditions suggests either that compensatory mechanisms occurred during development or, alternatively, that in normal "baseline" conditions, the 5-HT$_{1B}$ receptor is not activated. Our preliminary results indicated that the levels of the 5-HT$_{1A}$ and 5-HT$_{1D\alpha}$ receptors were not altered in the mutants (not shown). We are currently analyzing the levels of 5-HT and catecholamines as well as the levels of aminergic receptors which are involved in motor control and which might have compensated for the absence of the 5-HT$_{1B}$ receptor. An alternative possibility, that 5-HT$_{1B}$ receptors are activated in response to environmental changes such as stressful situations, is appealing in the light of the results obtained in the aggression test. When the mutants are group housed they do not appear to be more aggressive than grouped wild-type mice. However, after a month of isolation and in the presence of an intruder, the mutants are significantly more aggressive than the wild-type mice. In male mice, isolation and the presence of a conspecific male intruder have been shown to increase aggressive behavior (LAGERSPETZ and LAGERSPETZ 1971). Our results indicate that 5-HT_{1B}-minus mice are more responsive to the isolation and the stress or fear generated by the intruder and suggest that 5-HT_{1B} receptors might be activated in stressful situations such as those encountered in this test.

The increased aggressiveness of 5-HT_{1B}-minus mice is in good agreement with the fact that a family of 5-HT_{1B} agonists termed "serenics" have antiaggressive properties (OLIVIER et al. 1987). These compounds were shown to decrease aggressive behavior in several animal models including isolation-induced aggression in mice (OLIVIER et al. 1989), resident-intruder aggression in rats (FLANNELLY et al. 1985; MOS et al. 1992) and maternal aggression in rats (MOS et al. 1992). Our results suggest that the 5-HT_{1B} receptor is in part responsible for the antiaggressive properties of the serenics, but it is not the only target of these drugs. Therefore, our results emphasize a participation of other receptors with a high affinity for these compounds such as the 5-HT_{1A} receptor. We are currently testing the effects of smaller doses of serenics in the 5-HT_{1B}-minus mice as well as a combination of serenics and 5-HT_{1A} antagonist.

Several studies have revealed an association between aggressive behavior and a reduction in the activity of the serotoninergic system. In rodents and primates, aggressiveness is increased after inhibition of 5-HT synthesis (VERGNES et al. 1986) or destruction of serotoninergic neurons (MOLINA et al. 1987). Mouse strains that display increased aggressiveness have low brain 5-HT levels (BOURGAULT et al 1963; MAAS 1962). In humans, impulsive aggressive behaviors have been associated with a deficit in central serotonin (COCCARO et al. 1989). Cerebrospinal fluid (CSF) concentrations of the 5-HT metabolite 5-HIAA (5-hydroxyindoleacetic acid) are reduced in the brain of violent offenders (BROWN et al. 1979), arsonists (VIRKKUNEN et al. 1987) and people who committed violent suicide (COCCARO et al. 1989; MANN et al. 1989). Interestingly, in one recent study, impulsive violent offenders had low CSF 5-HIAA levels while offenders who premeditated their acts had high CSF 5-HIAA levels (VIRKKUNEN et al. 1994). Similarly, in studies of suicide victims, only those who performed violent suicides exhibited low CSF 5-HIAA levels (COCCARO et al. 1989; MANN et al. 1989). These findings suggest a link between low serotonin levels and a lack of impulse control. The aggressive behavior displayed by the 5-HT_{1B}-minus mice might be considered impulsive, since the mutants attacked much faster than the wild-type mice. In the light of these results it is tempting to speculate that low serotonergic activity would result in a decreased activation of 5-HT_{1B} receptors which might trigger aggressive behavior.

The 5-HT$_{1B}$ receptor is localized both presynaptically on serotoninergic terminals, where it inhibits the release of 5-HT, and postsynaptically on other nerve endings, where it might inhibit the release of various neurotransmitters (for a review see SAUDOU and HEN 1994). The antiaggressive effect of serenics is most likely mediated by postsynaptic receptors since they are not affected by lesions of serotoninergic neurons (SIJBESMA et al. 1991). Such postsynaptic receptors might be localized in the central gray, a brain structure involved in defensive behavior and response to fear (FANSELOW 1991) and containing moderate densities of 5-HT$_{1B}$ receptors (BOSCHERT et al. 1992; BRUINVELS et al. 1993). Activation of the 5-HT$_{1B}$ receptor might be a component of the adaptation to fearful stimuli. Interestingly, several behavioral responses elicited by fear such as those observed in a flight situation: increased locomotion and decreased aggressiveness, sexual activity and food intake, are also induced by 5-HT$_{1B}$ agonists. We are currently analyzing these behaviors as well as the level of stress hormones in order to determine whether adaptation to stress or fear is altered in the absence of the 5-HT$_{1B}$ receptor.

Acknowledgements. We thank M. Digelmann and E. Blondelle for ES culturing; M. Duval and P. Mellul for blastocyst injection; P. Charles, F. Tixier and R. Matyas for animal care; E. Vogel for behavioral analyses; J-L Vonesch for image analysis; and B. Boulay and C. Lappi for preparing the photographs and the manuscript. For helpful discussions we thank J. Crabbe as well as our colleagues N. Amlaiky, U. Boschert, A. Ghavami, R. Grailhe and J. Lucas. This work was supported by grants from CNRS, INSERM, Ministère de la Recherche, Rhône Poulenc Rorer and Association pour la Recherche contre le Cancer.

References

Boschert U, Ait Amara D, Segu L, Hen R (1992) The mouse 5-HT$_{1B}$ receptor is localized predominantly on axon terminals. Neuroscience 58:167–182

Boulenguer P, Chauveau J, Segu L, Morel A, Lanoir J, Delaage M (1991) A new 5-hydroxy-indole derivative with preferential affinity for 5HT$_{1B}$ binding sites. Eur J Pharmacol 194:91–98

Bourgault PC, Karczmar AG, Scudder CL (1963) Contrasting behavioral, pharmacological, neurophysiological, and biochemical profiles of C57 B1/6 and SC-I strains of mice. Life Sci 8:533–537

Brown GL, Goodwin FK, Ballenger JC, Goyer PF, Major LF (1979) Aggression in humans correlates with cerebrospinal fluid amine metabolites. Psychiatry Res 1:131–139

Bruinvels AT, Palacios JM, Hoyer D (1993) Pharmacological characterization and distribution of serotonin 5-HT$_{1D}$-like and 5H$_{1D}$ binding sites in rat brain. Naunyn Schmiedebergs Arch Pharmacol 347:569–582

Capecchi MR (1989) Altering the genome by homologous recombination. Science 244:1288–1292

Coccaro EF (1989) Central serotonin and impulsive aggression. Br J Psychiatry 155(Suppl):52–62

Crawley JN, Goodwin FK (1980) Preliminary report of a simple animal model for the behavioral actions of benzodiazepines. Pharmacol Biochem Behav 13:167–170

Eichelmann B (1992) Aggressive behavior: from laboratory to clinic quo vadit? Arch Gen Psychiatry 49:488–492

Fanselow MS (1991) The midbrain periaqueductal gray as a coordinator of action in response to fear and anxiety. In: Depaulis A, Bandler R (eds) The midbrain periaqueductal gray matter: functional, anatomical, and neurochemical organization. Plenum, New York, pp 151–173

Fernandez-Guasti A, Escalante AL, Ahlenius S, Hillegaart V, Larsson K (1992) Stimulation of 5-HT$_{1A}$ and 5-HT$_{1B}$ receptors in brain regions and its effects on male rat sexual behaviour. Eur J Pharmacol 210:121–129

Flannelly KJ, Muraoka MY, Blanchard DC, Blanchard RJ (1985) Specific antiaggressive effects of fluprazine hydrocloride. Psychopharmacology 87:86–89

Green AR, Guy AP, Gardner CR (1984) The behavioural effects of RU 24969, a suggested 5-HT$_1$ receptor agonist in rodents and the effect on the behavior of treatment with antidepressants. Neuropharmacology 23:655–661

Griebel G, Saffroy-Spittler M, Misslin R, Vogel E, Martin JR (1990) Serenics fluprazine (DU 27716) and eltoprazine (DU 28853) enhance neophobic and emotional behaviour in mice. Psychopharmacology 102:498–502

Higley JD, Mehlman PT, Taub DM, Higley SB, Suomi SJ, Vickers JH, Linnoila M (1992) Cerebrospinal fluid monoamine and adrenal correlates of aggression in free-ranging rhesus monkeys. Arch Gen Psychiatry 49:436–441

Kennett GA, Dourish CT, Curzon G (1987) 5-HT$_{1B}$ agonists induce anorexia at a postsynaptic site. Eur J Pharmacol 141:429–435

Koe BK, Nielsen JA, Macor JE, Heym J (1992) Biochemical and behavioral analyses of the 5-HT$_{1B}$ receptor agonist, CP-94. Drug Dev Res 26:241–250

Lagerspetz KMJ, Lagerspetz KYH (1971) Changes in the aggressiveness of mice resulting from selective breeding, learning, and social isolation. Scand J Psychol 12:241–248

Lufkin T, Dierich A, LeMeur M, Mark M, Chambon P (1991) Disruption of the Hox-1.6 Homeobox gene results in defects in a region corresponding to its rostral domain of expression. Cell 66:1105–1119

Maas JW (1962) Neurochemical differences between two strains of mice. Science 137:621–625

Mann JJ, Arango V, Marzuk PM, Theccanat S, Reis DJ (1989) Evidence of the 5-HT hypothesis of suicide. A review of post-mortem studies. Br J Psychiatry 155(Suppl):7–14

Maroteaux L, Saudou F, Amlaiky N, Boschert U, Plassat JL, Hen R (1992) Mouse 5HT$_{1B}$ serotonin receptor: cloning, functional expression, and localization in motor control centers. Proc Natl Acad Sci USA 89:3020–3024

Misslin R, Belzung C, Vogel E (1989) Behavioural validation of a light/dark choice procedure for testing anti-anxiety agents. Behav Proc 18:119–132

Molina V, Ciesielski L, Gobaille S, Isel F, Mandel P (1987) Inhibition of mouse killing behavior by serotonin-mimetic drugs: effect of partial alterations of serotonin neurotransmisssion. Pharmacol Biochem Behav 27:123–131

Mos J, Olivier B, Tulp MThM (1992) Ethopharmacological studies differentiated the effects of various serotoninergic compounds on aggression in rats. Drug Develop Res 26:343–360

Oberlander C, Blaquière B, Pujol JF (1986) Distinct function for dopamine and serotonin in locomotor behavior: evidence using the 5-HT$_1$ agonist RU 24969 in globus pallidus-lesioned rats. Neurosci Lett 67:113–118

Oberlander C, Demassey Y, Verdu A, van de Velde D, Bardeley C (1987) Tolerance to the serotonin 5-HT$_1$ agonist RU24969 and effects on dopaminergic behavior. Eur J Pharmacol 139:205–214

Offord SJ, Odway GA, Frazer A (1988) Application of [^{125}I] iodocyanopindolol to measure 5-hydroxytryptamine 1$_B$ receptors in the brain of rat. J Pharmacol Exp Ther 244:144–153

Olivier B (1980) A new antiaggressive compound. Ethological studies. Aggress Behav 6:262–263

Olivier B, VanDalen D, Hartog J (1986) A new class of psychoactive drugs, serenics. Drugs of the Future 11:473–499

Olivier B, Mos J, van der Heyden J, Schipper J, Tulp M, Berkelmans B, Bevan P (1987) Serotonin modulation of agonistic behaviour. In: Olivier B, Mos J, Brain PF (eds) Ethopharmacology of agonistic behaviour in animals and humans. Nijhoff, Dordrecht, pp 162–186

Olivier B, Mos J, van der Heyden J, Hartog J (1989) Serotonergic modulation of social interactions in isolated male mice. Psychopharmacology 97:154–156

Pazos A, Palacios JM (1985) Quantitative autoradiographic mapping of serotonin receptors in the rat brain. l. Serotonin-1 receptors. Brain Res 346:205–230

Pellow S, Johnston AL, File SE (1987) Selective agonists and antagonists for 5-hydroxytryptamine receptor subtypes, and interactions with yohimbine and FG 7142 using the elevated plus-maze in the rat. J Pharm Pharmacol 39:917–922

Saudou F, Hen R (1994) 5-Hydroxytryptamine receptor subtypes in vertebrates and invertebrates. Neurochem Int 25:503–532

Saudou F, Ait Amara D, Dierich A, Lemeur M, Ramboz S, Segu L, Buhot MC, Hen R (1994) Enhanced aggressive behavior in mice lacking 5-HT$_{1B}$ receptor. Science 265:1875–1878

Segu L, Rage P, Boulenguer P (1990) A new system for computer assisted quantitative receptor autoradiography. J Neurosci Methods 31:197–205

Segu L, Chauveau J, Boulenguer P, Morel A, Lanoir J, Delaage M (1991) Synthesis and pharmacological study of radioiondinated serotonin derivative specific of 5-HT$_{1B}$ and 5-HT$_{1D}$ binding sites of central nervous system. C R Acad Sci III 312:655–661

Sijbesma H, Schipper J, De Kloet ER, Mos J, van Aken H, Olivier B (1991) Postsynaptic 5-HT$_1$ receptors and offensive aggression in rats: a combined behavioural and autoradiographic study with eltoprazine. Pharmacol Biochem Behav 38:447–458

Sleight AJ, Pierce PA, Schmidt AW, Hekmatpanah CR, Peroutka SJ (1991) The clinical utility of serotonin receptor active agents in neuropsychiatric Disease. In: Peroutka S (ed) Serotonin receptor subtypes: basic and clinical aspects. Wiley, New York, pp 211–227

Vergnes M, Depaulis A, Boehrer A (1986) Parachlorophenylalanine-induced serotonin depletion increases offensive but not defensive aggression in male rats. Physiol Behav 36:653–658

Virkkunen M, Nuutila A, Goodwin FK, Linnoila M (1987) Cerebrospinal fluid monoamine metabolite levels in male arsonists. Arch Gen Psychiatry 44:241–247

Virkkunen M, Rawlings R, Tokola R, Poland RE, Guidotti A, Nemeroff C, Bissette G, Kalogeras K, Karonen SL, Linnoila M (1994) CSF biochemistries, glucose metabolism, and diurnal activity rhythms in alcoholic, violent offenders, fire setters, and healthy volunteers. Arch Gen Psychiatry 51:20–27

Wilkinson LO, Dourish CT (1991) Serotonin and animal behavior. In: Peroutka S (ed) Serotonin receptor subtypes: basic and clinical aspects. Wiley, New York, pp 147–210

CHAPTER 14

Pharmacology of 5-HT$_2$ Receptors

B.L. ROTH and E.G. HYDE

A. Introduction

The molecular mechanisms by which drugs bind to and regulate G-protein-coupled receptors (GPCR) remains a major unsolved problem for modern molecular biologists, biochemists and structural biologists. Ideally, we would like to ultimately define drug binding at the atomic level. An ideal molecular model would also be able to demonstrate why it is that agonists activate receptors while antagonists, which may or may not bind in overlapping domains, do not activate receptors. A perfect molecular explanation of drug action would elucidate the molecular determinants responsible for cellular regulation processes (e.g. desensitization, internalization and down-regulation). At the present time, there are no verifiable models which elucidate these properties for any G-protein-coupled receptor.

In the absence of direct structural information, most studies directed at identifying drug actions have focused on structure-activity studies of ligands, site-directed mutagenesis studies of receptors and molecular modelling of drug-receptor interactions. This chapter will summarize information obtained via these methods for various 5-HT receptors, using the 5-HT$_{2A}$ receptor as a convenient model system.

The organization of the chapter will be as follows. First, we will discuss published structure-activity studies which focus on the 5-HT$_2$ receptor family. Second, we will review molecular modelling approaches to understanding drug actions at 5-HT receptors. Third, we will discuss site-directed mutagenesis studies which complement the molecular modelling studies. In this way, a unified approach to understanding how drugs might bind to and activate 5-HT$_2$ receptors will be presented.

B. Structure-Activity Studies of 5-HT$_2$ Receptor Ligands

I. Introduction

The first relatively selective 5-HT$_2$ antagonist was ketanserin (LEYSEN et al. 1982, Fig. 1). [^{3}H]-Ketanserin continues to be widely used as the radioligand of choice for 5-HT$_{2A}$ receptors, particularly with cloned receptor populations. [^{3}H]-Ketanserin binds to several other receptor sites including the α_1-

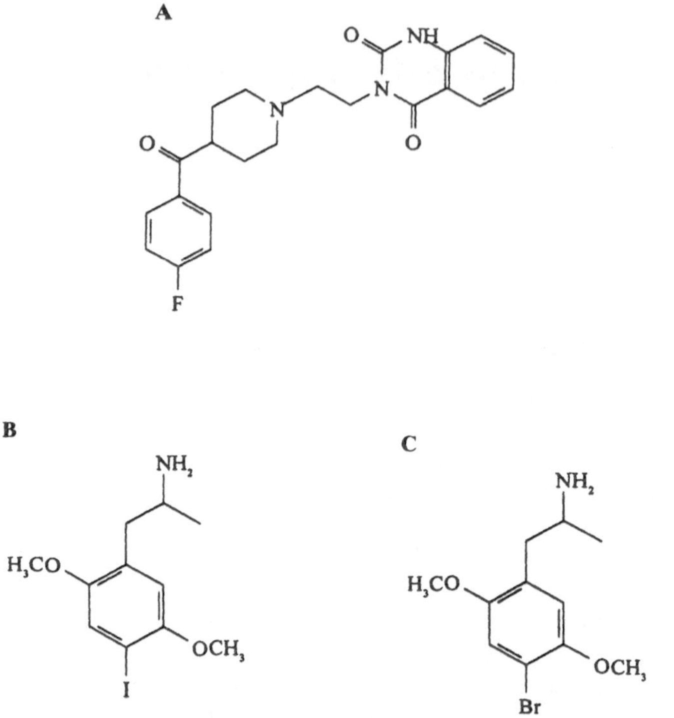

Fig. 1A–C. Selective antagonist and agonists active at 5-HT$_2$ receptors include ketanserin **(A)**, DOI **(B)** and DOB **(C)**

adrenergic receptor (Leysen et al. 1982), the 5-HT$_{2C}$ receptor (Roth et al. 1992) and a tetrabenezine-sensitive site identified as the vesicular amine pump (Leysen et al. 1987; Roth et al. 1987). The phenylalkylamines 4-iodo-2,5-dimethoxyphenylisopropylamine (DOI) and 4-bromo-2,5-dimethoxyphenylisopropylamine (DOB) were the first "selective" 5-HT$_2$ agonists discovered (for review see Nichols and Glennon 1984; Glennon et al. 1984, 1983; Fig. 1). DOI and DOB bind well to 5-HT$_{2A}$ and 5-HT$_{2C}$ receptors (Sanders-Bush and Breeding 1991). A recent review by Glennon and Dukat (1993) divided 5-HT$_2$-active ligands into six major classes: (1) indolealkylamines, (2) N-alkylpiperidines, (3) piperazines, (4) phenylalkylamines, (5) tricyclics and (6) miscellaneous agents. This classification scheme will be adhered to in this review and new information regarding representative members of each group will be briefly summarized.

II. Indolealkylamines

5-Hydroxytryptamine (5-HT, Fig. 2) is a non-selective indolealkylamine which, by definition, binds to all 5-HT receptors. 5-HT has variable affinities for 5-HT$_2$ receptor subtypes (see Table 1) with relatively low affinity for 5-

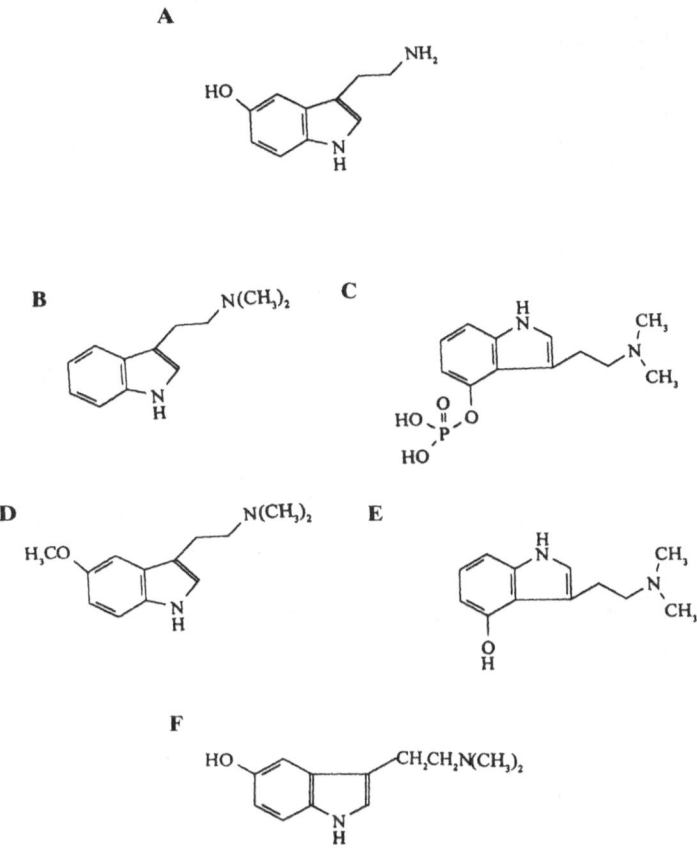

Fig. 2A–F. Representative indolealkylamine compounds that are 5-HT₂ ligands are 5-hydroxytryptamine **(A)**, *N,N′*-dimethyltryptamine **(B)**, psilocybin **(C)**, 5-*O*-methyl-DMT **(D)**, psilocin **(E)** and bufotenine **(F)**

HT$_{2A}$ receptors, moderate affinity for 5-HT$_{2B}$ receptors and high affinity for 5-HT$_{2C}$ receptors (NELSON 1993). 5-HT appears to bind to low- and high-affinity states of the 5-HT$_{2A}$ receptor (BRANCHEK et al. 1990; TEITLER et al. 1990).

III. Tryptamines

Indolealkylamines may be divided structurally into tryptamines, ergolines and β-carbolines. Tryptamine derivatives have been extensively characterized based on their hallucinogenic potential at 5-HT$_{2A}$ receptors. Thus, for instance, the following tryptamine derivates (Fig. 2) have been classed as hallucinogens: *N,N′*-dimethyltryptamine (DMT), 5-*O*-methyl-DMT, bufotenine, psilocibin and its active derivative psilocin (NICHOLS and GLENNON 1984; GLENNON, in press). We have recently discovered that these compounds all behave as agonists at cloned 5-HT$_{2A}$ receptors (Roth and Khan, unpublished observa-

Table 1. Comparison of mutagenesis data and theoretical contribution of various residues to 5-HT and ketanserin binding to the 5-HT$_{2A}$ receptor

Residue/ helix	Dahl, ketanserin	Holtje, ketanserin	Dahl, 5-HT	Weinstein, N-5-HT[a]	Höltje, Doi[b]	Weinstein N-DOM[c]	Mutagenesis data, 5-HT[d]	Mutagenesis data, ketanserin[e]
Thr88/TMI	-0.9		-5.3					
Asn92/TMI	-1.2		-7.8					
Asp120/TMII	-13.1		-54.3				+/-	
Leu123/TMII	-3.1							
Gly124/TMII	-0.8							
Val1271/TMII	-2.8							
Met128/TMII	-3.3							
Ser131/TMII	-2.3							
Trp151/TMIII	-5.0	+						
Asp155/TMIII	-42.2	+	-3.0	+	+	+	+	
Ser159/TMIII	-1.8	+		+		+	+	+
Trp200/TMIV	-1.0	+				+		
Ser203/TMIV	-2.1							
Ile206/TMIV	-0.7	+						
Ser207/TMIV	-1.2							
Ile210/TMIV	-0.4							
Ser239/TMV	-1.6	+				+		
Phe240/TMV	-9.0		-3.5	+		+		
Phe243/TMV	-4.8		-2.1	+				
Phe244/TMV	-1.9		-0.8	+				
Leu247/TMV	-0.9							
Met250/TMV	-0.7							

Residue	$\Delta\Delta G$					
Met335/TMVI	−0.6					
Trp336/TMVI	−3.6	−4.4	+		+	+
Phe339/TMVI	−0.7	+	+		+	
Phe340/TMVI	+	+			+	
Asn343/TMVI	−0.9	+	+		+	+
W367			+	+	+	+
I368			+	+	+	
G369			+	+	+	
Y370			+	+	+	+
L371			+	+	+	+
S372			+	+	+	
V375			+			

Shown are the calculated $\Delta\Delta G$ values from molecular modelling studies for the relative contribution of each residue for ligand binding. $\Delta\Delta G$ values are in kcal/mole. For molecular modelling studies in which $\Delta\Delta G$ values are not given "+" denotes a positive interaction between the ligand and the receptor at the residue indicated. For mutagenesis studies, "+" denotes a large decrease in ligand-binding affinity for a particular mutation of tenfold or greater, "−" denotes no significant change in ligand-binding affinity for a particular mutation (<twofold decrease in binding affinity), "+/−" denotes a modest change in ligand-binding affinity for a particular mutation (two- to tenfold decrease in binding affinity).

[a] Weinstein model also postulates the following additional residues for 5-HT docking: S156, F158 (both helix III), V204 (helix IV) and V241 (helix V).

[b] Höltje model also postulates the following residues for ketanserin binding: F213 (helix IV) and F158 (helix III).

[c] Weinstein model also postulates the following additional residues for DOM binding: F158, S162 and I163 (helix III), T201, V204 (helix IV), L236 (helix V).

[d] Mutagenesis data from tests of the following additional residues which have no apparent effect on 5-HT or ketanserin binding: W76, T81, T82 (helix I), F125, M132, T134 (helix III), F385 (helix VII).

[e] Mutagenesis data implicating the following additional residue as being important for 5-HT and ketanserin binding: F365 (helix VII).

tions). Glennon and Dukat (1993) reported that N1 alkylation and C7 alkylation enhance the affinity and selectivity of tryptamines for 5-HT$_2$-type receptors.

A recent study of 5-HT$_{2A}$ receptor mutations reported on structural features of the 5-HT$_{2A}$ receptor which are essential for tryptamine binding (Johnson et al. 1994). In this paper, the authors discovered that a point mutation (S242A) of a non-conserved serine in helix V selectively altered the affinity of N1-substituted tryptamines and ergolines for the 5-HT$_{2A}$ receptor. It appears that the higher affinity of rat vs. human 5-HT$_{2A}$ receptors for N1-substituted ergolines and tryptamines is primarily due to the presence of a Ser residue in the rat receptor which can form favourable hydrogen bonds with the N1-substituted group (Kao et al. 1992; Johnson et al. 1994).

IV. Ergolines

Tritiated lysergic acid diethylamide (LSD, Fig. 3) was the first successful radioligand for studying 5-HT receptors (Peroutka et al. 1981). Initial studies suggested that LSD bound selectively to 5-HT$_2$ receptors but later findings

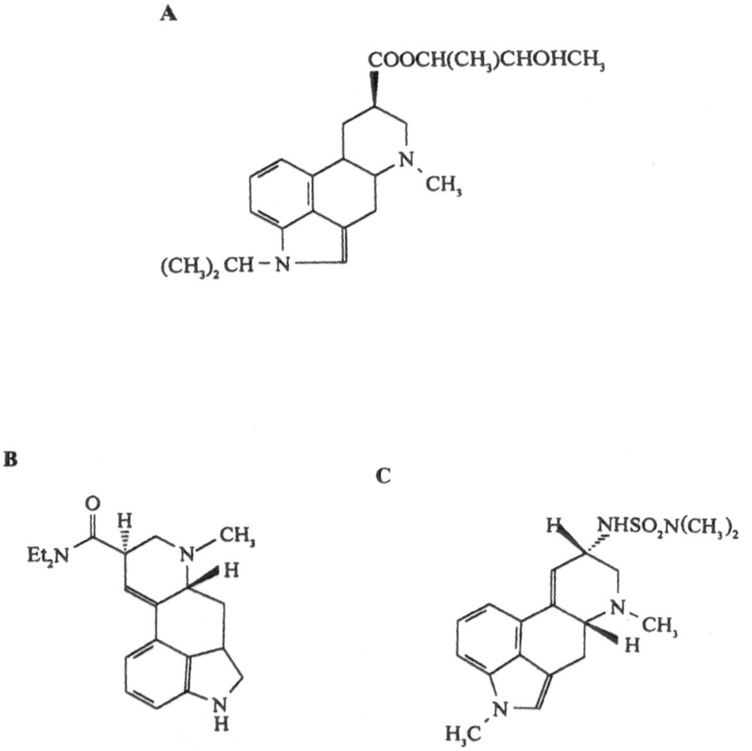

Fig. 3A–C. Representatives of the ergoline class of 5-HT$_2$ ligands are LY53857 (**A**), lysergic acid diethylamide (**B**) and mesulergine (**C**)

have demonstrated that LSD binds to most major 5-HT receptor families (BURRIS et al. 1991). A large number of ergoline and ergopeptine derivatives have been synthesized which have high affinity for the 5-HT$_{2A}$ and 5-HT$_{2C}$ receptors. Most ergoline and ergopeptines bind to other 5-HT and non-5-HT receptors (e.g. dopaminergic and α-adrenergic) so they are generally not selective.

A number of relatively selective ergoline derivatives have recently been synthesized by researchers at Eli Lilly & Company including LY53857 (Fig. 3), which is relatively 5-HT$_2$ selective but which binds with high affinity to 5-HT$_{2A}$, 5-HT$_{2B}$ and 5-HT$_{2C}$ receptors (COHEN et al. 1981). Mesulergine also has relative selectivity for 5-HT$_2$-family receptors, but also appears to bind to 5-HT$_6$ receptors with high affinity (MONSMA et al. 1993; ROTH et al. 1994). Studies which have used [^{3}H]-mesulergine to label 5-HT$_{2C}$ receptors (with spiperone to block the 5-HT$_{2A}$ receptor component of mesulergine binding) may have been inadvertently labelled 5-HT$_6$ receptors as well (MONSMA et al. 1993).

In general, ergolines function as partial agonists or antagonists at 5-HT$_2$ receptors, although few detailed studies of their agonist actions have been reported. In this regard, it is interesting to note that Sanders-Bush's group has recently suggested that mesulergine functions as a neutral agonist at 5-HT$_{2C}$ receptors while compounds such as mianserin behave as inverse agonists (BARKER et al. 1994; WESTPHAL and SANDERS-BUSH 1994). Identifying the structural features responsible for inverse agonist and neutral antagonist activity could represent a profitable direction for drug development research.

The structural features essential for ergoline binding include an aromatic residue at position 340 for the 5-HT$_{2A}$ receptor in helix VI, a negatively charged group at position 155 in helix III and, for N1-substituted ergolines, a Ser at position 242 in helix V (CHOUDHARY et al. 1994, 1995; JOHNSON et al. 1994; WANG et al. 1993). Results from these studies suggest that several amino acids located in neighboring helices combine to form the binding pocket. Additionally, it is clear that ergopeptines differ from ergolines in the structural requirements essential for optimal binding to 5-HT$_{2A}$ receptors (Choudhary et al. 1995).

V. N-Alkylpiperidines

Haloperidol (Fig. 4), the first N-alkylpiperidine used clinically, possesses modest affinity ($K_i = 15$–30 nM) for 5-HT$_{2A}$ receptors compared to spiperone ($K_i = 0.3$ nM; LEYSEN et al. 1978). In general, these compounds have high affinity for D$_2$ dopamine receptors as well. A number of other N-alkylpiperidines (Fig. 4) with high affinity for 5-HT$_{2A}$ receptors include melperone ($K_i = 79$ nM) and pimozide ($K_i = 9$ nM), which are also clinically effective antipsychotic drugs (ROTH et al. 1995b). Additional drugs used as antipsychotic agents include risperidone, sertindole, pirenperone, ziprasidone and MDL100907 (see FATEMI et al. 1996, for recent review). All of the later compounds are

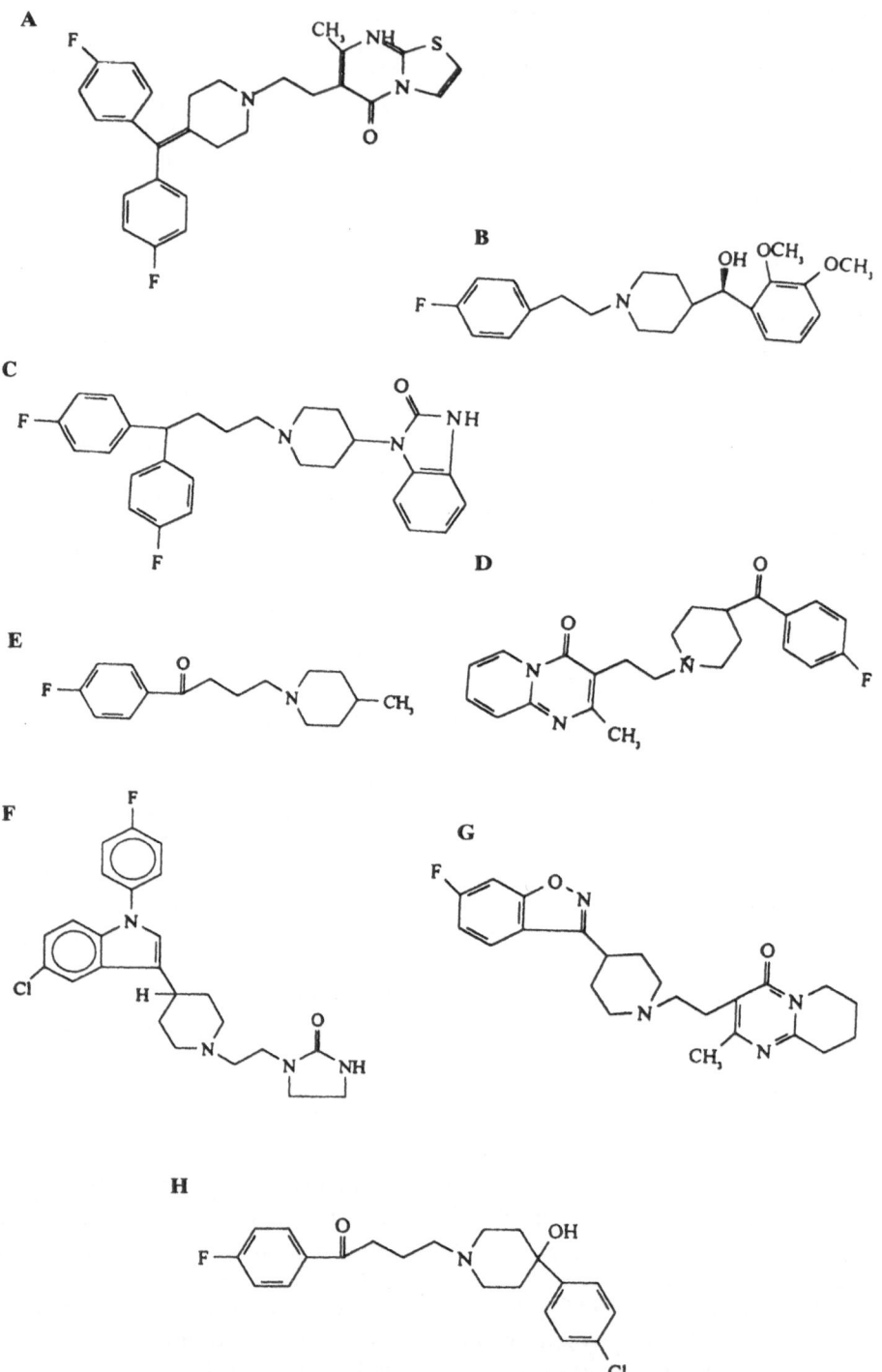

Fig. 4A–H. The *n*-alkylpiperidines are represented by ritanserin (**A**), MDL 100907 (**B**), pimozide (**C**), pirenperone (**D**), melperone (**E**), sertindole (**F**), riperidone (**G**) and the first clinically used *n*-alkylpiperidine, haloperidol (**H**)

characterized as having higher affinities for 5-HT$_{2A}$ receptors than for 5-HT$_{2C}$ and, with the exception of pirenperone, for D$_2$ dopamine receptors.

Ketanserin, ritanserin, pirenperone and MDL100907, unlike the others, possess little affinity for D$_2$ dopamine receptors (LEYSEN et al. 1993). Because at least two of these compounds (ritanserin and MDL100907) possess potential antipsychotic activity, some investigators have proposed that 5-HT$_{2A}$ receptor antagonist activity alone is sufficient for antipsychotic effects (SORENSON et al. 1993). It is beyond the scope of the present chapter to present evidence for and against this idea; the interested reader is directed to a recent review on the subject (FATEMI et al. 1996).

Attempts to synthesize 5-HT$_2$-subtype-selective agents have been marginally successful when it comes to N-alkylpiperidines. MDL100907 has been reported to be 100- to 300-fold selective for 5-HT$_{2A}$ receptors, while GLENNON and colleagues recently synthesized spiperone derivatives which were more than 1000-fold selective for 5-HT$_{2A}$ vs. 5-HT$_{2C}$ receptors (ISMAIEL et al. 1993). Ketanserin is between 10- and 30-fold selective for 5-HT$_{2A}$ receptors.

The structural features essential for N-alkylpiperidine binding to 5-HT$_{2A}$ receptors include aromatic residues at positions 339 and/or 340 and a negatively charged residue at position 155 of the 5-HT$_{2A}$ receptor (ROTH et al. 1993). Multiple structural domains appear essential for determining the subtype selectivity of drugs such as ketanserin and spiperone (CHOUDHARY et al. 1992). Preliminary findings suggest that one or more non-conserved threonine residues in helix III (Thr134) might be important for the relatively higher affinity haloperidol has for the 5-HT$_{2A}$ vs. 5-HT$_{2C}$ receptor (ROTH et al. 1993).

In general N-alkylpiperidines function as antagonists at 5-HT$_{2A}$ and 5-HT$_{2C}$ receptors. Ketanserin has been reported to be an inverse agonist at 5-HT$_{2C}$ receptors. Whether other N-alkylpiperidines function as neutral antagonists or inverse agonists is unknown.

VI. Piperazines

A number of piperazines (Fig. 5) have been discovered which function as agonists or antagonists at 5-HT$_{2A}$ receptors. Quipazine and MK-212 are full agonists at 5-HT$_{2A}$ and 5-HT$_{2C}$ receptors while m-chlorophenylpiperazine (mCPP) and trifluorophenylpiperazine (TFMPP) are partial agonists at 5-HT$_{2A}$ receptors and full agonists at 5-HT$_{2C}$ receptors (GROTEWEIL et al. 1994).

In addition to piperazines, a number of arylpiperazines have been evaluated as potential antipsychotic drugs because of their high 5-HT$_{2A}$/D$_2$ affinity ratios. These include tiospirone and amperozide (for review see FATEMI et al. 1996). Both drugs apparantly failed clinical testing because of side effects and/or lack of efficacy. Amperozide and tiospirone are both relatively selective for 5-HT$_{2A}$ vs. 5-HT$_{2C}$ receptors (ALBINSSON et al. 1990; AXELSSON et al. 1991; CHRISTENSSON and BJORK 1990), while MK-212 prefers the 5-HT$_{2C}$ site. mCPP, TFMPP and quipazine are nearly eqipotent at 5-HT$_{2A}$ and 5-HT$_{2C}$ sites (ROTH et al. 1992).

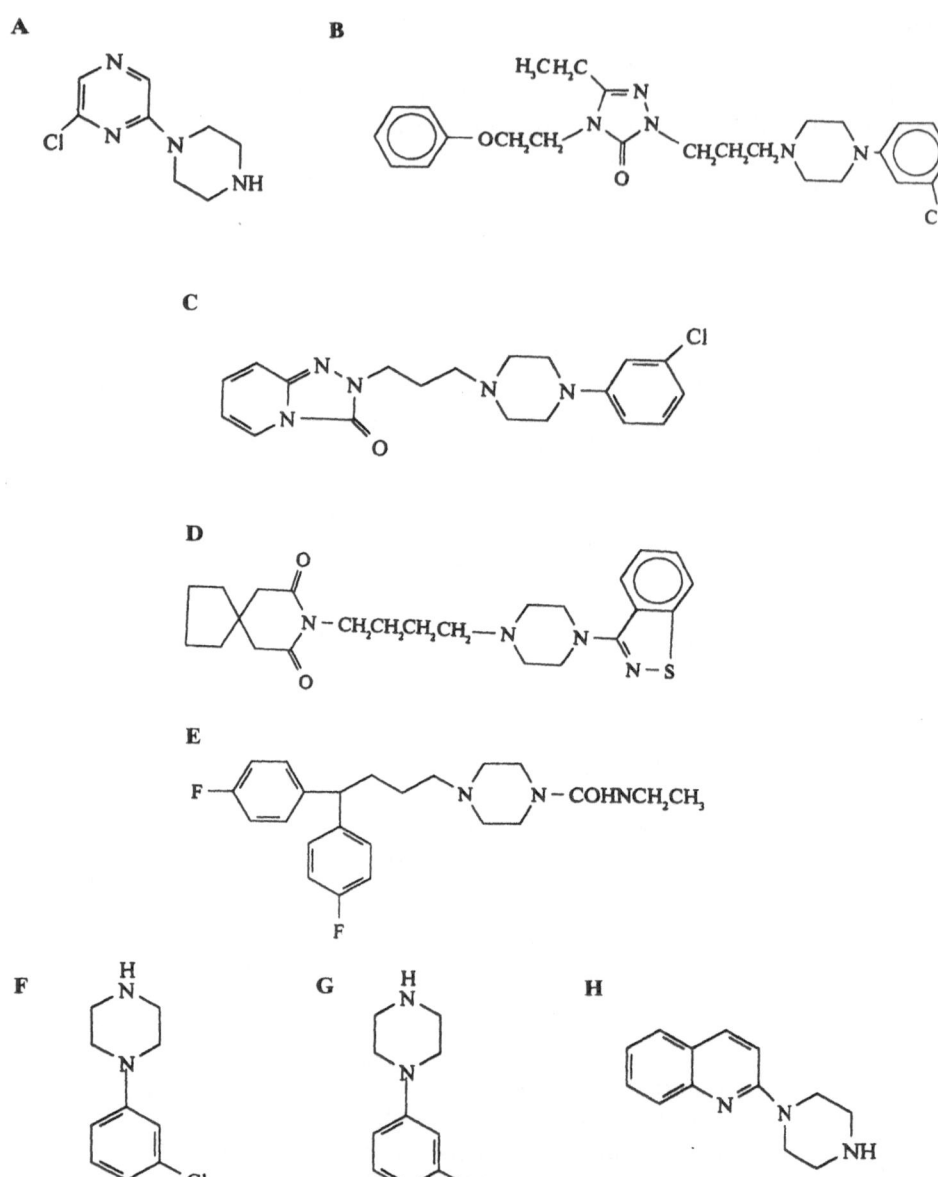

Fig. 5A–H. Piperazine compounds active at 5-HT$_2$ receptors include MK-212 **(A)**, nefazodone **(B)**, trazodone **(C)**, tiospirone **(D)**, amperiozide **(E)**, mCPP **(F)**, TFMPP **(G)** and quipazine **(H)**

The structural requirements essential for arylpiperazine binding have been incompletely studied. MK212, mCPP and TFMPP all appear to require an aromatic residue at position 340 for high-affinity binding and agonist efficacy while quipazine does not (CHOUDHARY et al. 1993). An aromatic residue at position 340 is essential for quipazine's efficacy but not binding affinity at 5-HT$_{2A}$ receptors (ROTH et al. 1995a).

Several arylpiperazine compounds possess appreciable antidepressant activity. These include trazadone and nefazadone, which are potent 5-HT$_{2A}$ receptor partial agonists (Roth and Khan, unpublished observations). It is likely that the antidepressant activity of these compounds is related to their 5-HT2A receptor activity (Peroutka and Snyder 1980; Maes and Meltzer 1995).

VII. Tricyclic Compounds

A large number of tricyclic antipsychotic and antidepressant drugs possess high affinities for both 5-HT$_{2A}$ and 5-HT$_{2C}$ receptors (ROTH et al. 1992; PALVIMAKKI et al. 1996; FATEMI et al. 1996, Fig. 6). These include typical antipsychotic drugs such as thioridazine, chlorpromazine and fluphenazine among many others (ROTH et al. 1992). Atypical antipsychotic drugs which possess a tricyclic core also have high affinities for 5-HT$_{2A}$ and/or 5-HT$_{2C}$ receptors including clozapine, zotepine, fluperlapine and olanzepine. Most tricyclic antidepressants have high affinities for both 5-HT$_{2A}$ and 5-HT$_{2C}$ receptors (PALVIMAKKI et al. 1996).

In general, atypical antipsychotic drugs with a tricyclic nucleus possess higher 5-HT$_{2A}$/D$_2$ affinity ratios than typical antipsychotic drugs (MELTZER et al. 1989). A vast number of these compounds have been synthesized and characterized as typical and atypical antipsychotic drugs. Typical antipsychotic drugs, as a rule, have higher affinities for the D$_2$ receptor than the 5-HT$_{2A}$ receptor, although some exceptions do exist (e.g. loxapine, amoxapine). Many of these compounds also possess appreciable affinity for the 5-HT$_{2C}$ receptor as well (CANTON et al. 1990; ROTH and MELTZER 1995; ROTH et al. 1992). although in general 5-HT$_{2C}$ receptor binding affinity does not predict whether a drug will possess antipsychotic activity. Additionally, 5-HT$_{2C}$ binding affinity does not predict the atypical nature of antipsychotic drugs (for discussion see FATEMI et al. 1996).

The other main therapeutic class of tricyclic compounds are antidepressants. For some time it has been known that many tricyclic antidepressants have high affinities for 5-HT$_{2A}$ receptors (PEROUTKA et al. 1980), although we have recently noted equally high affinities for 5-HT$_{2C}$ sites (PALVIMAKKI et al. 1996). Mianserin, a tricyclic compound with high affinity for 5-HT$_{2A}$, 5-HT$_{2C}$ and histamine receptors, is a potent antidepressant (BRUNELLO et al. 1982). The efficacy of mianserin as an antidepressant suggests that 5-HT$_{2A/2C}$ blocking activity may be an important mediator of the actions of many tricyclic antidepressants.

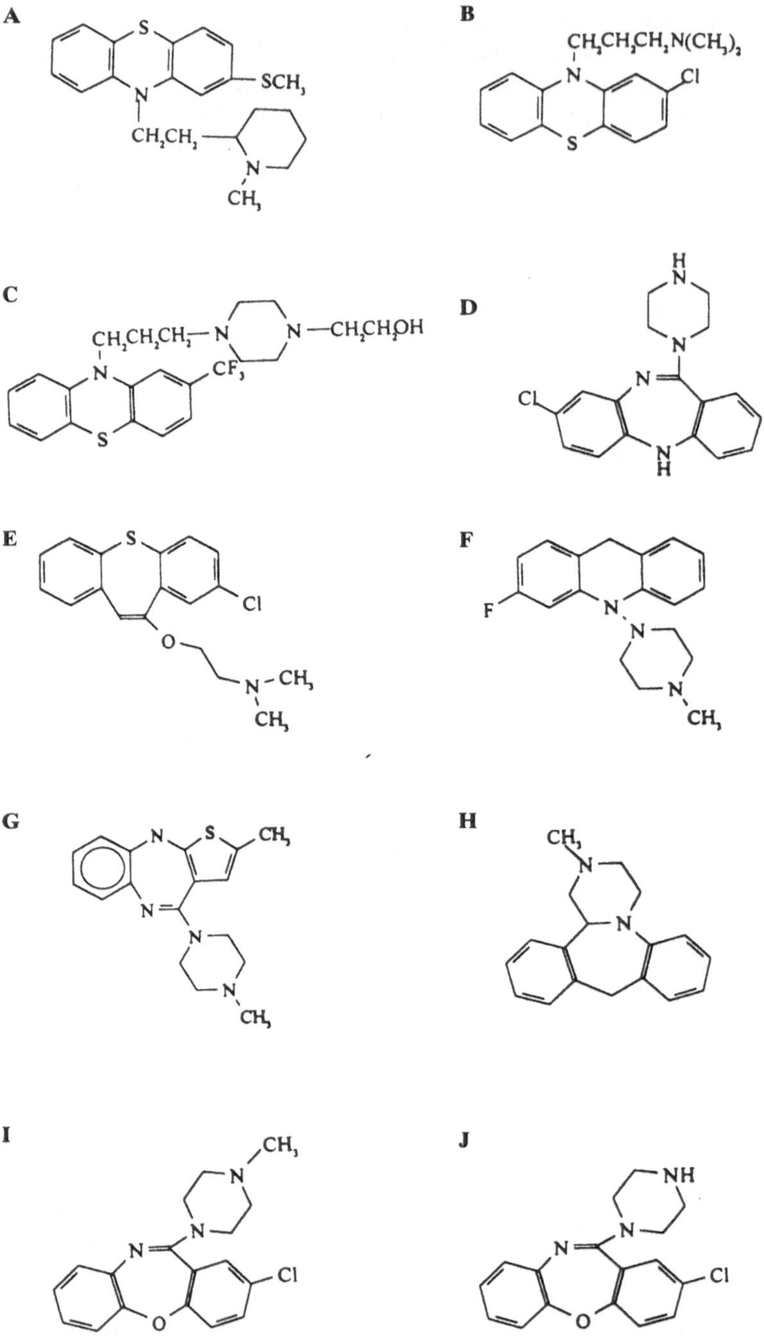

Fig. 6A–J. Representative tricyclic compounds active at 5-HT$_2$ receptors include thioridazine (**A**), chlorpromzine (**B**), fluphenzaine (**C**), clozapine (**D**), zotepine (**E**), fluperlapine (**F**), olanzapine (**G**), mianserin (**H**), loxapine (**I**) and amoxapine (**J**)

VIII. Phenylisopropylamines

A large number of phenylisopropylamines (Fig. 7) have been synthesized and characterized as 5-HT$_{2A}$ and 5-HT$_{2C}$ agonists (for review see GLENNON, in press). The prototypical agent in this class is 4-iodo-3,5-dimethoxy-phenylisopropylamine (DOI; see NICHOLS and GLENNON 1984 and GLENNON, in press). A large number of structurally related compounds have been synthesized primarily through the efforts of two groups of investigators (see GLENNON and DUKAT, 1993). In general, phenylisopropylamines behave as 5-HT$_{2A/2C}$ receptor agonists. DOI is a partial agonist while 4-bromo-3,5-dimethoxyphenylisopropylamine (DOB) appears to be a full agonist (SEGAL et al. 1987, 1990). Although most phenylisopropylamines are agonists, one compound, DOPP, behaves as an antagonist in some test systems (SEGAL et al. 1987). Typically, the phenylisopropylamines act as hallucinogenic agents with DOB being the prototypical hallucinogen (GLENNON 1990). A high correlation has been noted comparing the potency for induction of hallucinogenic activity of these compounds in man and their affinities at 5-HT$_{2A}$ receptors (GLENNON 1990).

Phenylisopropylamine agonists appear to selectively label a high-affinity agonist state of the 5-HT$_{2A}$ receptor (TEITLER et al. 1990; BRANCHEK et al. 1990). Thus [^{125}I]-DOI and [^{3}H]-DOB label a subpopulation of 5-HT$_{2A}$ receptors which have a high affinity for agonists. Agonist inhibition of [^{3}H]-ketanserin binding, which labels all 5-HT$_{2A}$ receptors, typically reveals high- and low-affinity binding sites. The high-affinity binding site/state appears to represent the same site labelled by [^{125}I]-DOI and [^{3}H]-DOB, while the low-affinity site/state probably represents the uncoupled form of the receptor. In general [^{125}I]-DOI and [^{3}H]-DOB binding is sensitive to guanine nucleotides while [^{3}H]-ketanserin binding is not.

[^{125}I]-DOI and [^{3}H]-DOB also have high affinity for 5-HT$_{2C}$ receptors (LEONHARDT et al. 1992). Phenylisopropylamines, in general, appear to be potent 5-HT$_{2C}$ receptor agonists (SANDERS-BUSH et al. 1988; SANDERS-BUSH

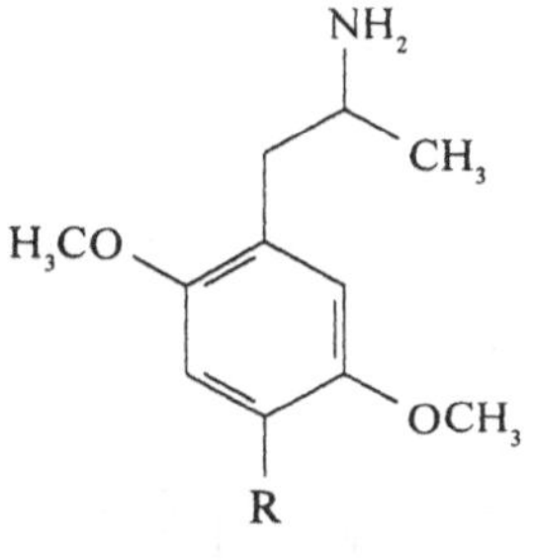

R = I, Br or Phenylpiperazine

Fig. 7. General structure of the phenylisopropylamines; DOI has an iodine at R, DOB, a bromine, and DOPP has a phenylpiperazine group at R

and BREEDING 1991). In general, their potency at 5-HT$_{2C}$ receptors is greater than that seen for 5-HT$_{2A}$ receptors.

IX. Miscellaneous Agents

Several interesting 5-HT$_{2A}$ antagonists have recently been developed including ICI 169369 and seroquel (FATEMI et al. 1996, Fig. 8). These drugs have been evaluated as atypical antipsychotics and, at least in the case of seroquel, show

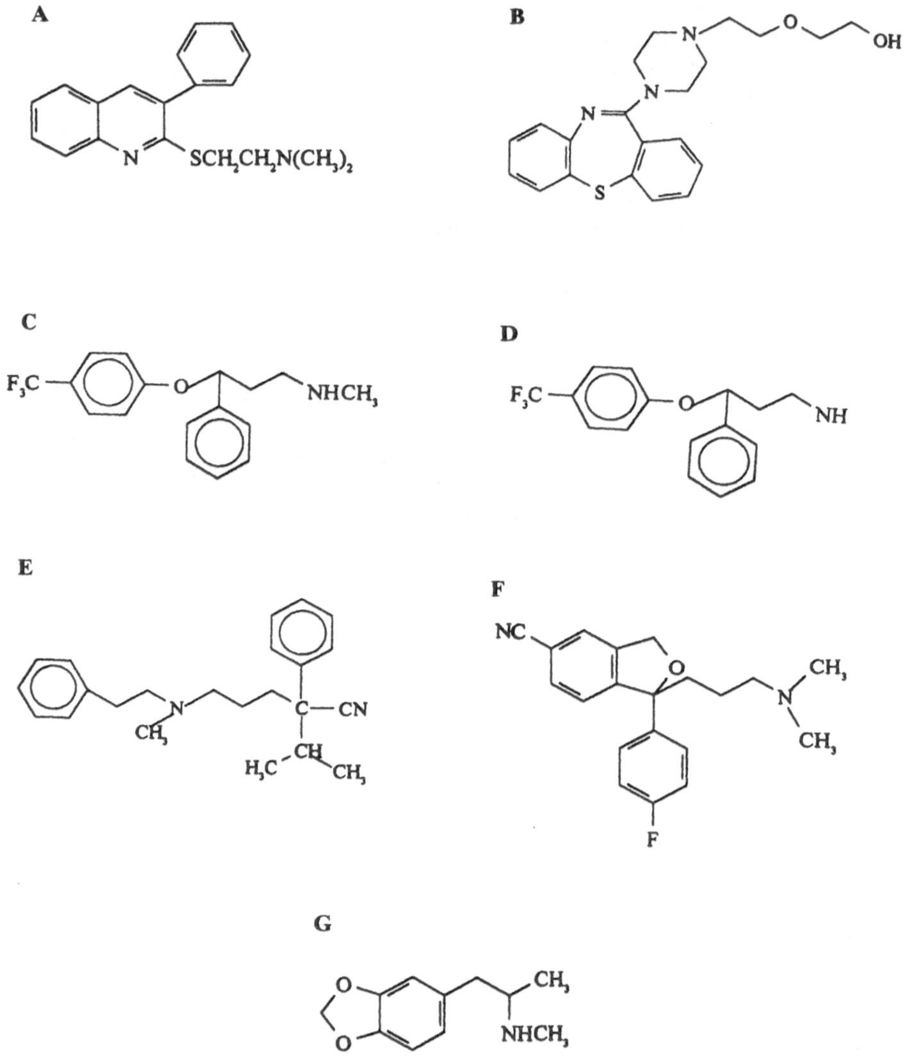

Fig. 8A–G. Drugs active at 5-HT$_2$ receptors that cannot be classified into the previous categories include ICI 169369 **(A)**, seroquel **(B)**, fluoxetine and norfluoxetine **(C, D)**, emopamil **(E)**, citalopram **(F)** and MDMA **(G)**

promise clinically. GLENNON and DUKAT (1993) included in this group of compounds emopamil and others. We recently discovered that citalopram and fluoxetine also had appreciable 5-HT$_{2A}$ and 5-HT$_{2C}$ affinities. Fluoxetine, norfluoxetine and citalopram all behaved as 5-HT$_{2C}$ receptor antagonists (PALVIMAKI et al. 1996). Because of the success of compounds such as ritanserin and risperidone as atypical antipsychotic drugs and the success of mianserin, trazadone and nefazadone as antidepressants, it is clear that 5-HT$_{2A}$ receptors might be involved in the therapeutic actions of atypical antipsychotic and antidepressant agents.

Compounds such as methylene-dioxyamphetamine (MDMA) also have appreciable affinity for 5-HT$_{2A}$ and 5-HT$_{2C}$ receptors. We recently discovered that MDMA and a related compound (NASH et al. 1994) behaved as partial agonists at both 5-HT$_{2A}$ and 5-HT$_{2C}$ receptors. It is conceivable that certain of the pharmacological properties of these compounds are related to 5-HT$_{2A/2C}$ receptor activation.

C. Structure-Activity Studies of 5-HT$_2$ Receptors

I. Molecular Modelling Overview

A large number of investigators have published models of 5-HT$_{2A}$ receptors and the general approaches for model building will be summarized here. All but the most recent models are based on bacteriorhodopsin templates (HENDERSON et al. 1990). The general approach to model building has been adequately summarized in some detail by others (BALLESTEROS and WEINSTEIN 1995) and only a minimum of information will be given here. Typically, alignments based on hydrophobicity plots are made with the 7α-helical regions of bacteriorhodopsin (see, for example, WESTKAEMPER and GLENNON 1993a; HIBERT et al. 1991; HOLTJE and JENDRETZKI 1995; TRUMPP-KALLMEYER et al. 1992). With the exception of one group of investigators (PARDO et al. 1992), the first putative α-helical region of the 5-HT$_{2A}$ receptor is assumed to be homologous with helix I of bacteriorhodopsin. In the same way, the second α-helical region is generally assumed to be homologous with helix II of bacteriorhodopsin and so on up to the seventh α-helix.

One of the major problems with this method of model construction is that there is little, if any, amino acid sequence homology between bacteriorhodopsin and 5-HT$_2$ receptors. Thus, the homology, if any really exists, must be structural rather than sequence based. Although this approach appears somewhat speculative, abundant examples exist in which greatly different protein sequences displayed remarkable structural similarities (CHOUDHARY et al. 1995), suggesting that this general approach may not be totally problematic. Additionally, recent high-resolution cryomicroscopy studies of mammalian rhodopsin (SCHERTLER et al. 1993; SCHERTLER and HARGRAVE 1996; UNGER and SCHERTLER 1995) suggest that the general schema of 7α-helical regions for G-protein-coupled receptors is probably correct. Important

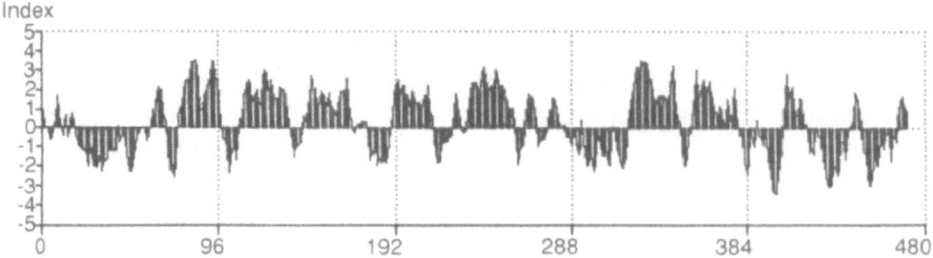

Fig. 9. Hydrophobicity plot of 5-HT$_{2A}$ receptor. Shown is a typical hydrophobicity plot calculated using the Kyte-Doolittle algorithmn (see Westkaemper and Glennon 1993a for example). The plot displays seven potential hydrophobic regions

differences in the helical packing and helical arrangement have been found comparing rhodopsin (both bovine and frog) and bacteriorhodopsin. These differences will be expanded upon below.

Figure 9 shows a hydrophobicity plot and a-helical predictions for the 5-HT$_{2A}$ receptor. As can be seen, there are seven hydrophobic regions, each of which has a strong α-helical character. The α-helical, hydrophobic regions are then typically used together with the bacteriorhodopsin atomic coordinates to construct an initial model. The initial model is then inspected for "unnatural" conformations and subjected to extensive energy minimization. Energy minimization is usually done in vacuo, although some investigators have attempted to add a dielectric constant similar to that of the lipid bilayer to approximate the postulated native environment of the receptor. Because of the tremendous theoretical degrees of freedom, energy minimization is ideally done on supercomputers or high-end graphics systems (see Edvardsen et al. 1992).

II. Bacteriorhodopsin- and Rhodopsin-Based Models

1. Dahl Model

Dahl and colleagues were the first to publish an energy-minimized model based on bacteriorhodopsin coordinates for the 5-HT$_{2A}$ receptor. They have used their model to simulate the interactions of ketanserin, ritanserin and 5-HT at the 5-HT$_{2A}$ receptor (Edvardsen et al. 1992; Kristianssen et al. 1993). In their computer simulations, they have based the structures of the test compounds (e.g. ketanserin and 5-HT) on crystallographic data, where possible. For testing both agonist and antagonist binding, they have postulated ionic interactions with either D120 (in helix II) or D155 (in helix III). Mutagenesis data (see below) suggest that D155 forms the counter ion for the positively charged amine moiety of serotoninergic compounds. Table 1 summarizes the postulated residues essential for ketanserin and 5-HT binding and compares this data with experimentally determined findings. Figures 10 and 11 display the residues found experimentally to alter agonist binding and those

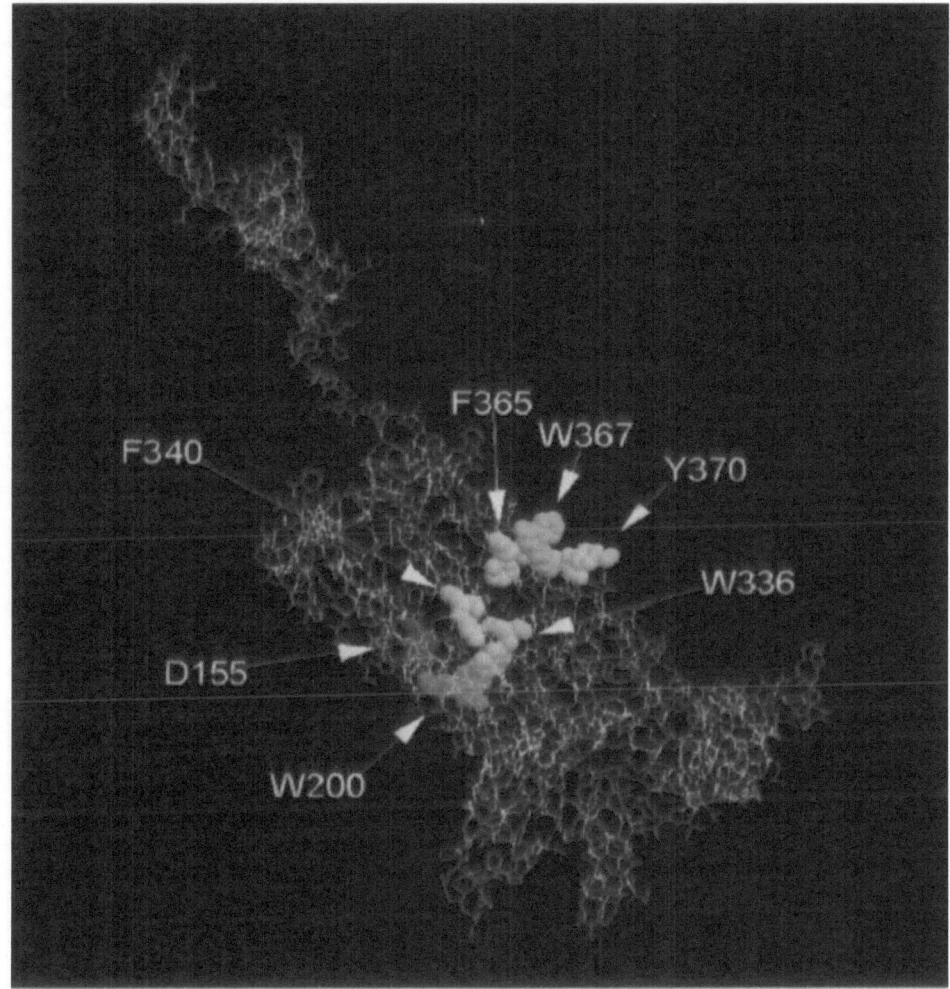

Fig. 10. Identification of residues experimentally determined to affect agonist binding

found not to alter agonist binding (CHOUDHARY et al. 1992, 1993, 1995; ROTH et al. 1993; SEALFON et al. 1995; WANG et al. 1993).

As shown in Table 1, the Dahl model explicitly implicates several residues for ketanserin binding. Several of the residues identified by Dahl's group have been tested by site-directed mutagenesis experiments including: D120, D155, W200, W336, F339, F340 and W367. Results from Roth et al. (unpublished observations) and Fig. 10 implicated the following residues for ketanserin binding: D155, W200, W336, F339 and W367. Additionally (Roth et al., unpublished observations) Y370 was also identified as being important for ketanserin binding. Interestingly, F340 was not implicated by Dahl's group,

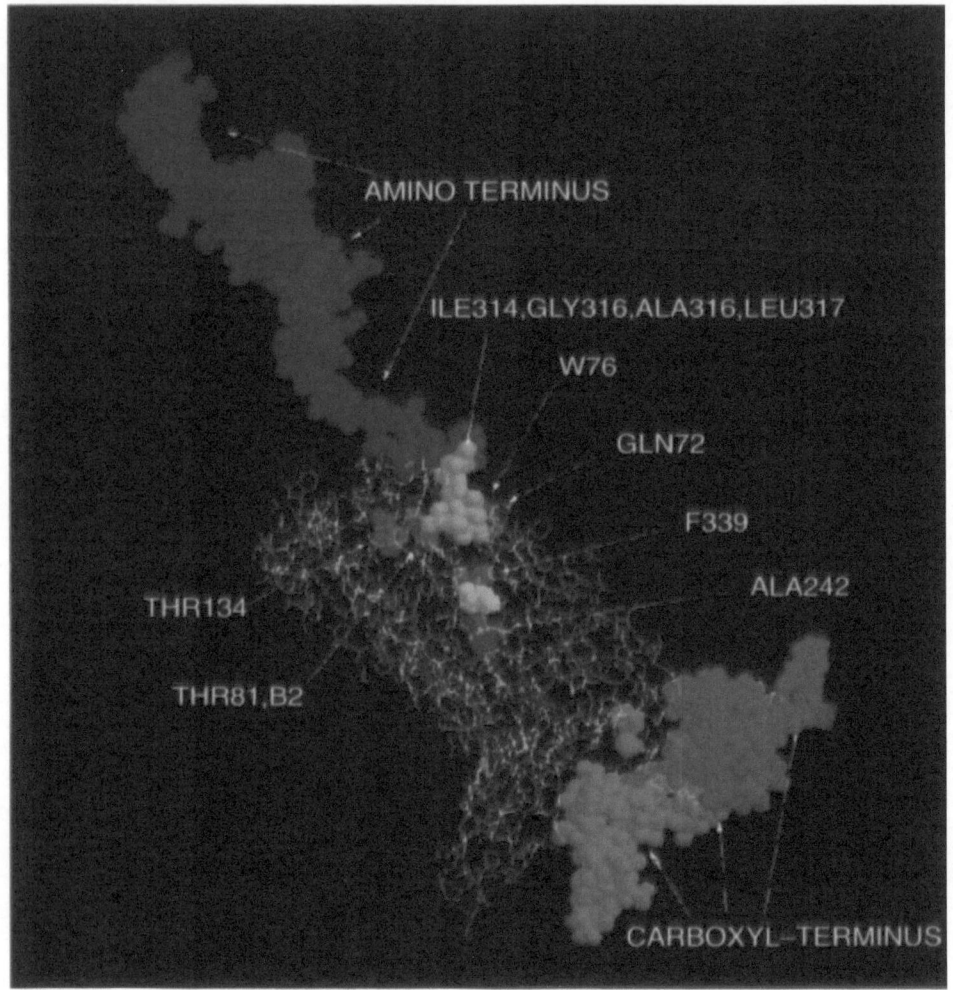

Fig. 11. Identification of residues not determined to affect agonist binding

and mutagenesis results (Choudhary et al. 1993, 1995) showed it not to be important for ketanserin binding.

Dahl's group has also modelled the interaction of the agonist 5-HT with their 5-HT$_{2A}$ receptor model (Kristianssen et al. 1993). Their initial studies (Edvardsen et al. 1992) suggested that 5-HT primarily interacted with D120 (TM II) and not D155 (TM III). These results are inconsistent with mutagenesis results, which have consistently implicated the conserved aspartic acid in helix III as being essential for agonist docking. Other residues implicated by this group for 5-HT binding included: Thr88, Asn92 (TM II), Ile206 (TM IV), Ser239 (TMV), W336, F340 and Asn343 (TM VI). Mutagensis results have implicated W200, W336, F340, F365, W367 and Y370 in 5-HT binding and efficacy (Roth et al., unpublished observations). Thus, there is only a small degree of correspondence comparing the mutagenesis results with the predicted model of EDVARDSEN et al. (1992).

2. Höltje and Jendretski Model

Another recent bacteriorhodopsin-based model also attempting to model ketanserin and DOI binding to the 5-HT$_{2A}$ receptor was reported by HÖLTJE and JENDRETSKI (1995), who used a conventional bacteriorhodopsin-based template with minimal energy minimization and no dynamics simulations. They predicted the following residues to be important for ketanserin binding: W151, D155, F158, Ser159 (helix III), Phe213 (helix IV), W336, F340, Asn343 (helix VI) and F365 (helix VII). The residues implicated previously by mutagenesis experiments only partially overlap with those shown by HÖLTJE and JENDRETSKI (1995) and include D155 and W336. Interestingly, F340 and F365 have been found to be non-essential for ketanserin binding (CHOUDHARY et al. 1993, 1995; Roth et al., unpublished observations).

The same group also identified potential residues for DOI binding including: D155 (helix III), W336, F340 and Asn343 (helix VI). In general, these residues identified agree quite well with the mutagenesis data and strengthen the predictive power of the HÖLTJE and JENDRETZKI (1995) model.

3. Weinstein Model

Weinstein's group has published a large number of studies using both bacteriorhodopsin- and rhodopsin-based models. In their initial studies, PARDO et al. (1992) suggested, based on their sequence analysis of bacteriorhodopsin vs. various GPCRs, that a different helical arrangement was possible. These authors proposed that exon shuffling has occurred during evolution so that, in fact, helix II of BR was homologous to helix IV for GPCR, for example. The following helical homology was suggested: BRII = GPCRIV; BRIII = GPCRV; BRIV = GPCRVI; BRV = GPCRI; BRVI = GPCRII; BRVII = GPCRIII. No apparent homology was seen for GPCRVII or BRI. The PRADO et al. (1992) model proposes that a binding pocket composed of helices II, III, IV and V could exist with helices I, VI and VII apparently too far from helix III for significant interactions to occur. Additionally, GPCR helices II and VII appear to be too far from each other for functional interactions to occur (SCHLERTLER and HARGRAVES 1996; BALDWIN 1993).

Another model of helical connectivity was proposed by WEINSTEIN and ZHANG (1994). This model proposes quite a different topology from that of PRADO et al. (1992) or later models. In particular, TMs I, II and III appear to have been shuffled. TMIII is some distance from TMVI. Specific interactions with many residues were also proposed. Since this model has been abandoned, it will not be further discussed here.

More recently, ZHOU et al. (1995) proposed a somewhat different helical arrangement based on rhodopsin coordinates. The rationale behind this particular model has been adequately summarized elsewhere (BALLESTEROS and WEINSTEIN 1995). This model proposes a helical arrangement in which BRI = GPCRI; BRII = GPCRII; and so on. The reasons for abandoning the prior helical arrangements are not given (BALLESTEROS and WEINSTEIN 1995; LUO et al. 1994). In particular, this model proposes a close approximation of the following helical pairs: II-VII; III-VI and IV-V. In general, this more recent model agrees with the most recent 6-Å model of frog rhodopsin by SCHERTLER and HARGRAVE (1996) as well as Baldwin's model based on a 9-Å map of bovine rhodopsin (BALDWIN 1993).

Evidence in favour of this helical connectivity model has come from recent experiments in which reciprocal mutations were made of conserved residues found in helices II and VII (SEALFON et al. 1995; ZHOU et al. 1995). In these experiments, Asp120 (helix II) and Asn376 (helix VII) were mutated as follows: D120N; N376D; and the double mutant D120N/N376D. These authors found that both the D120N mutation abolished signal transduction without altering ligand binding and the N376D mutation functioned at 95% relative E_{max} when compared with the native receptor. The double mutant D120N/N376D functioned at a level of approximately 50% of the native receptor. Despite the fact that the D120N mutant was apparently uncoupled from signal transduction, no change in ligand affinity for either agonist or antagonist was measured. Based on these findings, these authors suggested that a helical-helical interaction between helices II and VII occurs, which is essential for signal transduction.

These results differ from those reported by WANG et al. (1993), in which a significant decrease in agonist and antagonist affinity was noted following a D120N mutation of the rat 5-HT_{2A} receptor. WANG et al. (1993) also discovered that the D120N mutation was not coupled to PI hydrolysis. The reason for these discrepant results is not immediately obvious, although SEALFON et al. (1995) proposed that the human 5-HT_{2A} receptor differs from the rat receptor in the effects of the D120N mutation on agonist- and antagonist-binding affinities (WANG et al. 1993).

4. Westkaemper, Glennon and Roth Models

CHOUDHARY et al. (1995) recently proposed a model to account for the effects of various point mutations on ergoline and ergopeptine binding to 5-HT_{2A} receptors. These authors discovered that a point mutation of the highly conserved phenylalanine F340 (F340L; F340A) dramatically altered the affinities of selected ergolines for the 5-HT_{2A} receptor while mutations of an adjacent phenylalanine (F339L; F339A) had no effect on ergoline binding. Interestingly, F340Y, which maintains the aromatic nature at the 340 position, did not alter ergoline binding. These authors proposed that the aromatic ring of ergolines and not ergopeptines was oriented close to the aromatic ring of F340 but not F339. A model based on bacteriorhodopsin coordinates supported these findings (CHOUDHARY et al. 1995).

WESTKAEMPER and GLENNON (1991, 1993) have also proposed additional residues as being essential for agonist binding to 5-HT$_{2A}$ receptors. These findings are in general agreement with some, but not all, of the mutagenesis results. For instance, F340 and D155 are suggested to be essential for agonist binding as also suggested by mutagenesis findings (WANG et al. 1993; CHOUDHARY et al. 1993). Further studies are currently in progress to refine these models using a rhodopsin-based template (Roth, Glennon and Westkaemper, in progress).

5. Other 5-HT$_{2A}$ Receptor Models

Recently, a number of 5-HT$_{2A}$ receptor models were summarized. MOEREELS and JANSSEN (1993) used a ligand-based strategy for building a 5-HT$_{2A}$ receptor model. In this strategy, the receptor is modelled around the agonist in a step-by-step process and weaker (e.g. hydrogen-type bonds) were considered. The authors proposed that Ser239, Asp155 and phenylalanines in TMV were essential for 5-HT binding.

SHIH et al. (1992) also proposed a model for 5-HT binding to 5-HT$_{2A}$ receptors. They suggested as well that Ser residues in TMV (Ser239, Ser242) as well as Asp155 were essential for 5-HT binding to 5-HT$_{2A}$ receptors. This hypothesis was supported by mutagenesis data suggesting that Ser242 affects N1-substituted ergoline and tryptamine binding to rat and human 5-HT$_{2A}$ receptors (JOHNSON et al. 1994).

III. Mutations Involving Aromatic Residues

CHOUDHARY et al. (1993, 1995) found that mutation of a highly conserved phenylalanine residue (F340L) significantly affected the binding of a large number of agonists to the 5-HT$_{2A}$ receptor. In their initial study, CHOUDHARY et al. (1993) examined three mutations: F125L, F339L and F340L. They found that only the F340L mutation significantly altered agonist binding. By contrast, the F339L mutation significantly diminished ketanserin and spiperone binding but did not alter agonist or ergoline binding. The F125L mutation had no effect on agonist or antagonist binding. Interestingly, the F340L mutation also diminished the affinities of three model ergolines for the 5-HT$_{2A}$ receptor. These results suggested that certain agonists and ergolines may share overlapping binding sites on the 5-HT$_{2A}$ receptor (CHOUDHARY et al. 1993, 1995).

CHOUDHARY et al. (1993) also examined agonist efficacies at the various phenylalanine mutations. They found that F340L, but not F339L or F125L, mutations significantly diminished agonist efficacies at the 5-HT$_{2A}$ receptor. Thus agonist affinities (K_{act} values) and efficacies (V_{max} values) were attenuated by the F340L mutations. The F339L and F125L mutations had no significant effect on K_{act} or V_{max} values for selected agonists (DOI, 5-HT and quipazine). These results suggest that F340, but not F339 or F125, are essential for agonist efficacy.

Roth et al. (1995a) subsequently investigated the regulatory properties of the F340L receptor. They discovered that the F340L mutant was able to be desensitized by DOI but not quipazine. By contrast, the native receptor was desensitized by both quipazine and DOI. These results suggested that mere binding to the receptor was insufficient for agonist-mediated desensitization.

In follow-up studies, Choudhary et al. (1995) examined the effects of various types of substitutions at the F339 and F340 positions on ergoline and ergopeptine binding. These findings have been summarized previously. Choudhary et al. (1995) also examined the effects of these various mutations on ketanserin binding. They discovered, as predicted by Kristiansen et al. (1993), that ketanserin-binding affinity was diminished by F339 but not F340 mutations. These results suggest that ketanserin binds slightly differently from the ergolines and ergopeptines.

IV. Mutations of Non-Conserved Residues

Choudhary et al. (1992) investigated the roles of non-conserved domains of the 5-HT$_{2A}$ and 5-HT$_{2C}$ receptors for subtype-selective binding properties. Choudhary et al. (1992) discovered that structurally distinct antagonists apparently preferred different domains for subtype-selective binding. Thus, haloperidol's high-affinity binding to the 5-HT$_{2A}$ receptor was due to residues

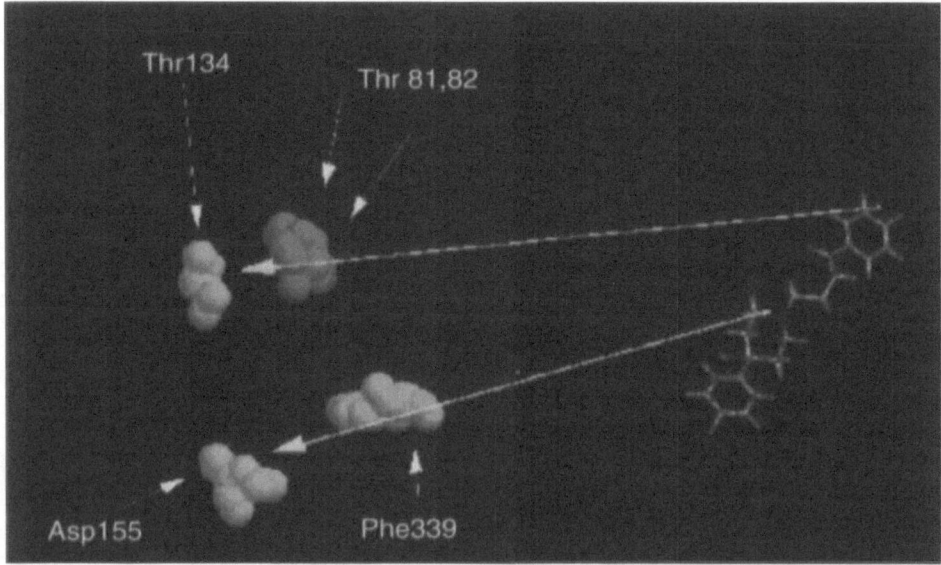

Fig. 12. Model of haloperidol docking. Site-directed mutagenesis studies (Roth et al. 1993) demonstrated that Thr134, Thr81, Thr82, Asp155 and Phe339 all affected haloperidol binding to the 5-HT$_{2A}$ receptor. The figure shows a potential model which can account for these findings

located in helices I-II. By contrast, 5-HT showed higher affinity for the 5-HT2C receptor due to residues located in helices VI-VII.

Follow-up studies of individual amino acids were performed for haloperidol (ROTH et al. 1993). In these studies individual non-conserved amino acids were mutated in helices I-II and the affinity of haloperidol determined. The Thr134Ala mutation appeared to represent the amino acid in helix II which contributed most to the special pharmacological properties of haloperidol for the 5-HT$_{2A}$ vs. 5-HT$_{2C}$ receptor. A preliminary model (Fig. 12) proposes potential interactions essential for the high-affinity binding of haloperidol for the 5-HT$_{2A}$ receptor.

V. Mutagenesis Overview

In addition to molecular modelling, a number of investigators have attempted to investigate ligand-receptor interactions utilizing molecular biological approaches. These primarily include mutagenesis studies. The strengths of this approach are: first, empirical results rather than theoretical constructs are analysed; second, specific binding models can be empirically tested provided appropriate mutagenic constructs are made; and third, appropriate controls can be done to verify that the mutations do not perturb the overall structure of the receptor.

Mutagenesis approaches are not without their pitfalls. The major problem is that one is never entirely sure that the mutation did not introduce subtle changes in the three-dimensional folding of the protein. Examples of this problem abound in the crystallographic literature, where it has been shown that even minor modifications can lead to striking changes in folding and, as a consequence, alterations in activity. Additionally, particular mutations can destroy the ability of the protein to be synthesized, so that independent means of determining receptor protein production also need to be formulated.

Certain control experiments can be easily performed which serve to verify that receptor protein is being produced and properly inserted into the plasma membrane. The most appropriate control is the visualization of receptor protein on the cell surface using an antireceptor antibody as was recently demonstrated by VON ZASTROW and KOBILKA (1992). Some investigators have relied on the identification of receptor mRNA, but this is an inadequate control since receptor mRNA production may in many cases have very little to do with production of a mutant protein. FERSHT and colleagues have devised a number of criteria they have used to determine whether a particular mutation has altered the three-dimensional structure of a protein. FERSHT's group proposed, first, that the replacement residue be smaller than the native residue and that additional functional groups not be added (FERSHT et al. 1987). They also proposed that point mutations which affected the binding of certain substrates (but not others) were unlikely to grossly affect the conformation of the protein. FERSHT et al. (1988) and LOWE et al. (1987) also proposed that the $\Delta\Delta G$ values for mutations should be less than 4 kcal/mol for mutations which do not affect the overall conformation of the protein. Finally, where mutant receptors are still coupled to second messenger production this indicates that the functional properties of the receptor are not altered.

VI. Mutations Involving Charged Residues

Following the initial work of STRADER et al. (1989a,b) on β-adrenergic receptors in which a highly conserved aspartic acid residue in helix III was found to be essential for ligand binding, WANG et al. (1993) mutated D155 (helix III) and D120 (helix II). They discovered that the D155N mutation had lower affinity for agonists (30-fold) and antagonists (14- to 75-fold), but only slightly diminished affinity for ^{125}I-LSD (2.8nM vs. 0.6nM for the native receptor). Additionally, the D155N mutation had dramatically decreased K_{act} values for 5-HT (7800 vs. 128nM) and DOI (6800 vs. 4.3nM) for PI hydrolysis as well as somewhat lower percentage maximum responses of 78% compared to the native receptor.

The D120N mutation, by contrast, abolished agonist-mediated signal transduction (WANG et al. 1993; SEALFON et al. 1995). Additionally, WANG et al. (1993) found that the GTP-induced shift in agonist affinity was also abolished by the D120N mutation, but not by the D155N nor by a D172N mutation. WANG et al. (1993) also found that the D120N mutation diminished agonist and antagonist affinities for the 5-HT$_{2A}$ receptor. SEALFON et al. (1995) had somewhat different results for the D120N mutation, which they discovered did not alter agonist or antagonist affinities, suggesting that these differences could be due to the fact that WANG et al. (1993) used a rat clone while SEALFON et al. (1995) used the human clone.

References

Albinsson A, Eriksson E, Andersson G (1990) Amperozide-effect on prolactin release in the rat. Pharmacol Toxicol 66:49–51

Axelsson R, Nilsson A, Christensson E, Bjork A (1991) Effects of amperozide in schizophrenia: an open study of a potent 5-HT$_2$ receptor antagonist. Psychopharmacology (Berl) 104:287–292

Baldwin JM (1993) The probable arrangement of the helices in G protein-coupled receptors. EMBO J 12:1693–1703

Ballesteros JA, Weinstein H (1995) Integrated methods for the construction of three-dimensional models and computational probing of structure-function relations in G protein-coupled receptors. Methods Neurosci 25:366–428

Barker EL, Westphal RS, Schmidt D, Sanders-Bush E (1994) Constitutively active 5-hydroxytryptamine2C receptors reveal novel inverse agonist activity of receptor ligands. J Biol Chem 269:11687–11690

Branchek T, Adham N, Macchi M, Kao H-T, Hartig PR (1990) [^{3}H]-DOB (4-bromo-25-dimethoxyphenylisopropylamine) and [^{3}H]-ketanserin label two affinity states of the cloned human 5-hydroxytryptamine2 receptor. Mol Pharmacol 38:604–609

Brunello N, Chuang D-M, Costa E (1982) Different synaptic location of mianserin and imipramine binding sites. Science 215:1112–1115

Burris KD, Breeding M, Sanders-Bush E (1991) (+)Lysergic acid diethylamide, but not its nonhallucinogenic congeners, is a potent serotonin 5HT1c receptor agonist. J Pharmacol Exp Ther 258:891–896

Canton H, Verriele L, Colpaert FC (1990) Binding of typical and atypical antipsychotics to 5-HT$_{1c}$ and 5-HT$_2$ sites: clozapine potently interacts with 5-HT$_{1c}$ sites. Eur J Pharmacol 191:93–96

Choudhary S, Craigo S, Roth BL (1992) Identification of domains which modify serotonin receptor pharmacology. Soc Neurosci Abstr 18:100.9

Choudhary MS, Craigo S, Roth BL (1993) A single point mutation (Phe340->Leu340) of a conserved phenylalanine abolishes 4-[^{125}I]-iodo-(25-dimethoxy)phenylisopropylamine and [^{3}H]-mesulergine but not [^{3}H]-ketanserin binding to 5-hydroxytryptamine2 receptors. Mol Pharmacol 43:755–763

Choudhary MS, Sachs N, Uluer A, Glennon RB, Westkaemper RA, Roth BL (1995) Differential ergoline and ergopeptine binding to 5-hydroxytryptamine2A (5-HT$_{2A}$) receptors: ergolines require an aromatic residue at position 340 for high affinity binding. Mol Pharmacol 47:450–457

Christensson E, Bjork A (1990) Amperozide: a new pharmacological approach in the treatment of schizophrenia. Pharmacol Toxicol 66:5–7

Cohen ML, Fuller RW, Wiley KS (1981) Evidence for 5-HT$_2$ receptors mediating contraction in vascular smooth muscle. J Pharmacol Exp Ther 218:421–425

Evardsen O, Sylte I, Dahl SG (1992) Molecular dynamics of serotonin and ritanserin interacting with 5-HT$_2$ receptor. Mol Brain Res 14:166-178

Fatemi SH, Meltzer HY, Roth BL (1996) Interaction of atypical antipsychotic drugs with non-dopaminergic systems. Csernansky JH (ed) Antipsychotic drugs. Springer, Berlin Heidelberg New York (Handbook of experimental pharmacology) pp 77–115

Fersht AR (1988) Relationships between apparent binding energies measured in site-directed mutagenesis experiments and energetics of binding and catalysis. Biochemistry 27:1577–1580

Fersht AR, Leatherbarrow RJ, Wells TNC (1987) Structure-activity relationships in engineered proteins: analysis of use of binding energy by linear free energy relationships. Biochemistry 26:6030–6038

Glennon RA (1990) Do classical hallucinogens act as 5-HT$_2$ agonists or antagonists? Neuropsychopharmacology 3:509–517

Glennon RA, Dukat M (1993) 5-HT receptor ligands – update 1992. Curr Drugs, pp 1–45

Glennon RA, Young R, Rosencrans JA (1983) Antagonism of the effects of the hallucinogen DOM, and the purported 5-HT agonist quipazine by 5-HT$_2$ antagonists. Eur J Pharmacol 91:189–193

Glennon RA, Titler M, McKenney JD (1984) Evidence for 5-HT$_2$ involvement in the mechanism of action of hallucinogenic agents. Life Sci 35:2505–2511

Grotewiel MS, Chu H, Sanders-Bush E (1994) m-Chlorophenylpiperazine and m-trifluoromethylphenylpiperazine are partial agonists at cloned 5-HT$_{2A}$ receptors expressed in fibroblasts. J Pharmacol Exp Ther 271:1122–1126

Henderson R, Baldwin J, Ceska TH, Zemlin F, Beckmann E, Downing K (1990) Model for the structure of bacteriorhodopsin based on high resolution electron cryomicroscopy. J Mol Biol 213:899–929

Hibert MF, Trumpp-Kallmeyer S, Bruinvels A, Hoflack J (1991) Three-dimensional models of neurotransmitter G-binding protein-coupled receptors. Mol Pharmacol 40:8–15

Höltje HD, Briem H (1991) Theoretical determination of the putative receptor-bound conformations of 5-HT$_2$ receptor agonists. Quant Struct Activ Relat 10:193–197

Höltje HD, Jendretzki UW (1995) Construction of a detailed serotoninergic 5-HTsA receptor mode. Arch Pharm 328:577–584

Ismaiel AM, De Los Angeles J, Teitler M, Ingher S, Glennon RA (1993) Antagonism of 1-(3,5-dimethoxy-4-methylphenyl)-2-aminopropane stimulus with a newly identified 5-HT$_2$- versus 5-HT$_{1C}$-selective antagonist. J Med Chem 36:2519–2525

Johnson MP, Loncharich RJ, Baez M, Nelson DL (1994) Species variations in transmembrane region V of the 5-hydroxytryptamine type 2A receptor alter the structure-activity relationship of certain ergolines and tryptamines. Mol Pharmacol 45:277–286

Kao H-T, Adhan N, Olsen MA, Weinshank RL, Branchek TA, Hartig PR (1992) A single amino acid distinguishes human from rat 5-HT$_2$ receptors. FEBS Lett 307:324–326

Kristianssen K, Edvardsen O, Dahl SG (1993) Molecular modelling of ketanserin and its interactions with the 5-HT$_2$ receptor. Med Chem Res 3:370–385

Leonhardt S, Gorospe E, Hoffman BJ, Teitler M (1992) Molecular pharmacological differences in the interaction of serotonin with 5-hydroxytryptamine$_{1c}$ and 5-hydroxytryptamine$_2$ receptors. Mol Pharmacol 42:328–335

Leysen JE, Niemegeers CJE, Tollenaere JP, Laduron PM (1978) Serotonergic component of neuroleptic receptors. Nature 272:168–171

Leysen JE, Niemegeers CJE, van Nueten JM, Laduron PM (1982) [^{3}H]-ketanserin (R 41 468) a selective ^{3}H-ligand for serotonin2 receptor binding sites. Mol Pharmacol 21:301–314

Leysen JE, Gommeren W, van Gompel P, Wynants J, Janssen PAJ (1987) Non-serotonergic [^{3}H]-ketanserin binding sites in striatal membranes are associated with a dopac release system on dopaminergic nerve endings. Eur J Pharmacol 134:373–375

Leysen JE, Janssen PMF, Schotte A, Luyten WHML, Megens AAHP (1993) Interaction of antipsychotic drugs with neurotransmitter receptor sites in vitro and in vivo in relation to pharmacological and clinical effects: role of 5HT$_2$ receptors. Psychopharmacology 112:S40–S54

Lowe DM, Winter G, Fersht AR (1987) Structure-activity relationships in engineered proteins: characterization of disrupted deletions in the alpha-ammonium group binding site of tyrosyl-tRNA synthetase. Biochemistry 26:6038–6043

Luo XZ, Hang D, Weinstein H (1994) Ligand-induced domain motion in the activation mechanism of a G-protein-coupled receptor. Protein Eng 7:1441–1448

Maes M, Meltzer HY (1995) The serotonin hypothesis of major depression. In: Bloom FE, Kupfer DJ (eds) Psychopharmacology: the 4th generation of progress. Raven Press, New York, pp 933–944

Meltzer HY, Matsubara S, Lee J-C (1989) Classification of typical and atypical antipsychotic drugs on the basis of dopamine D-1, D-2 and serotonin$_2$ pKi values. J Pharmacol Exp Ther 251:238–246

Moereels H, Janssen PAJ (1993) Molecular modelling of G-protein coupled receptors: going step by step. Med Chem Res 3:335–343

Monsma FJ, Shen Y, Ward RP, Hamblin MW, Sibley DR (1993) Cloning and expression of a novel serotonin receptor with high affinity for tricyclic psychotropic drugs. Mol Pharmacol 43:320–327

Nash JF, Roth BL, Brodkin JD, Nichols DE, Gudelsky GA (1994) Effect of the R(−) and S(+) isomers of MDA and MDMA on phosphatidylinositol turnover in cultured cells expressing 5-HT$_{2A}$ or 5-HT$_{2C}$ receptors. Neurosci Lett 177:111–115

Nelson DL (1993) The serotonin2 (5-HT$_2$) subfamily of receptors: pharmacological considerations. Med Chem Res 3:306–316

Nichols DE, Glennon RA (1984) Medicinal chemistry and structure-activity relationships of hallucinogens. In: Jacobs BL (ed) Hallucinogens: neurochemical behavioral and clinical perspectives. Raven, New York, pp 95–142

Palvimaki EP, Roth BL, Majasuo H, Laakso A, Kuoppamaki M, Syvlanti E, Hietala J (1996) Interactions of selective serotonin reuptake inhibitors with the serotonin 5-HT$_{2C}$ receptor. Psychopharmacology 126:234–240

Pardo L, Ballesteros JA, Osman R, Weinstein H (1992) On the use of the transmembrane domain of bacteriorhodopsin as a template for modelling the three-dimensional structure of guanine nucleotide-binding regulatory protein-coupled receptors. Proc Natl Acad Sci USA 89:4009–4012

Pehak EA, Meltzer HY, Yamamoto BK (1993) The atypical antipsychotic drug amperozide enhances rat cortical and striatal dopamine efflux. Eur J Pharmacol 240:1993

Peroutka SJ, Snyder SH (1980) Long-term antidepressant treatment decreases spiroperidol-labelled serotonin receptor binding. Science 210:86–90 1980

Peroutka SJ, Lebovitz RM, Snyder SH (1981) Two distinct serotonin receptors with distinct physiological functions. Science 212:827–829

Roth BL, Meltzer HY (1995) Psychopharmacology: the 4th generation of progress. Raven, New York

Roth BL, McLean S, Zhu X-Z, Chuang D-M (1987) Characterization of two [³H]-ketanserin recognition sites in rat striatum. J Neurochem 49:1833–1838

Roth BL, Ciaranello RD, Meltzer HY (1992) Binding of typical and atypical antipsychotic agents with transiently expressed 5-HT$_{1c}$ receptors. J Pharmacol Exp Ther 260:1361–1365

Roth BL, Choudhary MS, Craigo S (1993) Mutagenesis of serotonin receptors: what does an analysis of many mutant serotonin receptors tell us? Med Chem Res 3:407–418

Roth BL, Craigo SC, Choudhary MS, Monsma FJ, Shen Y, Meltzer HY, Sibley DR (1994) Binding of typical and atypical antipsychotic agents to 5-hydroxytryptamine6 (5-HT6) and 5-hydroxytryptamine7 (5-HT7) receptors. J Pharmacol Exp Ther 256:1403–1410

Roth BL, Pekka-Palvimaki E, Berry SA, Khan N, Sachs N, Uluer A, Choudhary MS (1995a) 5-Hydroxytryptamine2A (5-HT$_{2A}$) receptor desensitization can occur without down-regulation. J Pharmacol Exp Ther 275:1638–1645

Roth BL, Tandra S, Burgess LH, Sibley DR, Meltzer HY (1995b) D$_4$ dopamine receptor binding affinity does not distinguish between typical and atypical antipsychotic drugs. Psychopharmacology 120:365–368

Sanders-Bush E, Breeding M (1991) Choroid plexus epithelial cells in primary culture – a model of 5HT$_{1c}$ receptor activation by hallucinogenic drugs. Psychopharmacology 105:340–346

Sanders-Bush E, Burris KD, Knoth K (1988) Lysergic acid diethylamide and 2,5-dimethoxy-4-methylamphetamine are partial agonists at serotonin receptors linked to phosphoinositide hydrolysis. J Pharmacol Exp Ther 246:924–928

Schertler GFX, Billa C, Henderson R (1993) Projection structure of rhodopsin. Nature 262:770–772

Schertler GFX, Hargrave PA (1995) Projection structure of frog rhodopsin in two crystal forms. Proc Natl Acad Sci USA 92:11578–11582

Sealfon SC, Chi L, Ebersole BJ, Rodic V, Zhang D, Ballesteros JA, Weinstein H (1995) Related contribution of specific helix 2 and helix 7 residues to conformational activation of the serotonin 5-HT$_{2A}$ receptor. J Biol Chem 270:16683–16688

Segal MR, Youssif MY, Lyons RA, Titler M, Roth BL, Suba EA, Glennon RA (1990) A structure-affinity study of the binding of 4-substituted analogues of 1-(25-dimethoxyphenyl)-2-aminopropane at the 5-HT$_2$ serotonin receptors. J Med Chem 33:1032–1036

Serrano L, Bycroft M, Fersht AR (1991) Aromatic-aromatic interactions and protein stability: investigation by double mutant cycles. J Mol Biol 218:465–475

Shih JC, Gallaher T, Wang C-D, Chen K (1992) Site-directed mutagenesis of serotonin 5-HT$_2$ receptors. J Chem Neuroanat 5:218–282

Sorensen SM, Kehne JH, Fadayl EM, Humphreys TM, Ketteler HJ, Sullivan C, Taylor VL, Schmidt CJ (1993) Characterization of the 5-HT$_2$ antagonist MDL 100907 as a putative atypical antipsychotic: behavioral, electrophysiological and neurochemical studies. J Pharmacol Exp Ther 266:684–691

Strader CD, Candelore MR, Hill WS, Sigal IS, Dixon RAF (1989a) Identification of two serine residues involved in agonist activation of the beta-adrenergic receptor. J Biol Chem 264:13572–13578

Strader CD, Sigal IS, Dixon RAF (1989b) Structural basis of beta-adrenergic receptor function. FASEB J 3:1825–1831

Teitler M, Leonhardt S, Weisberg EL, Hoffman BJ (1990) 4-[^{125}I]-(25-Dimethoxy)phenylisopropylamine and [^{3}H]-ketanserin labeling of 5-hydroxytryptamine2 (5HT$_2$) receptors in mammalian cells transfected with a rat 5HT$_2$ cDNA: evidence for multiple states and not multiple 5HT$_2$ receptor subtypes. Mol Pharmacol 38:594–598

Trumpp-Kallmeyer S, Hoflack J, Bruinvels A, Hibert M (1992) Modeling of G-protein-coupled receptors: application to dopamine, adrenaline, serotonin, acetylcholine and mammalian opsin receptors. J Med Chem 35:3448–3462

Unger VM, Schertler GFX (1995) Low resolution structure of bovine rhodopsin determined by electron cryo-microscopy. Biophys J 68:1776–1786

Von Zastrow M, Kobilka B (1992) Ligand-regulated internalization and recycling of humand beta 2-adrenergic receptors between the plasma membrane and endosomes containing transferrin receptors. J Biol Chem 267:3530–3538

Wang C-D, Gallaher TK, Shih JC (1993) Site-directed mutagenesis of the serotonin 5-hydroxytryptamine2 receptor: identification of amino acids necessary for ligand binding and receptor activation. Mol Pharmacol 43:931–940

Weinstein H, Zhang D (in press) Receptor models and ligand-induced responses. QSAR and molecular modeling: New insights for structure activity relations Prous Science, Barcelona pp 497–507

Westkaemper RB, Glennon RA (1991) Approaches to molecular modeling studies and specific application to serotonin ligands and receptors. Pharmacol Biochem Behav 40:1019–1031

Westkaemper RB, Glennon RA (1993a) Molecular graphics models of members of the 5-HT$_2$ subfamily: 5-HT$_{2A}$, 5-HT$_{2B}$, and 5-HT$_{2C}$ receptors. Med Chem Res 3:317–334

Westkaemper RB, Glennon RA (1993b) Molecular modelling of the interaction of LSD and other classical hallucinogens with 5-HT$_2$ receptors. In: Lin G, Glennon RA (eds) NIDA research monograph, Classical Hallucinogens: An Update, 146:263–283

Westphal RS, Sanders-Bush E (1994) Reciprocal binding properties of 5-hydroxytryptamine type 2C receptor agonists and inverse agonsits. Mol Pharmacol 46:937–942

Zhou W, Rodic V, Kitanovic S, Flanagan CA, Chi L, Weinstein H, Maayani S, Millar RP, Sealfon SC (1995) A locus of the gonadotropin-releasing hormone receptor that differentiates agonist and antagonist binding sites. J Biol Chem 270:18853–18857

Regulation of 5-HT$_{2A}$ Receptor at the Molecular Level

M. Toth and D. Benjamin

A. Introduction

The serotonin$_{2A}$ (5-HT$_{2A}$) receptor belongs to the family of G-protein-coupled receptors. The number of available 5-HT$_{2A}$ receptors in the brain is influenced by several factors. 5-HT$_{2A}$ receptor shows a marked increase in number during perinatal and early postnatal development (Roth et al. 1991). Changes in the 5-HT$_{2A}$-receptor-binding sites have also been described in diseased brains compared to controls; it has been repeatedly shown that the level of 5-HT$_{2A}$ receptor is increased in postmortem brains of depressed patients (Cooper et al. 1986; McKeith et al. 1987; Yates et al. 1990). Drugs also modify the number of 5-HT$_{2A}$ receptors (Peroutka and Snyder 1980; Conn and Sander-Bush 1986). In particular, antidepressants downregulate 5-HT$_{2A}$ receptor level, raising the possibility of a causal relationship between receptor level and attenuation of symptoms in depression. These examples demonstrate the importance of studying regulatory processes involving the 5-HT$_{2A}$ receptor. The elucidation of these processes has been facilitated by recent advances in molecular biology techniques and here we have summarized the knowledge accumulated in the last few years regarding the regulation of the 5-HT$_{2A}$ receptor.

B. Cell-Type-Specific Regulation of the 5-HT$_{2A}$ Receptor

I. Cells Expressing 5-HT$_{2A}$ Receptor in the Nervous System

Since the 5-HT$_{2A}$ receptor has been implicated in a variety of behavioral and physiological processes, as well as in a number of neuropsychiatric disorders (Jacobs et al. 1990; Fuller 1990; Leysen 1990; Murphy 1990), it is important to specify the brain regions and cell types expressing the receptor.

In the neocortex the receptor is heterogeneously distributed across the different layers forming three tangential bands at the level of layers 1, 5a and 5b (Mengod et al. 1990). In the olfactory bulb, 5-HT$_{2A}$ sites are very abundant in the external plexiform layer, where the dendrites of the mitral cells are located. Immunolabeling with antibody raised against the 5-HT$_{2A}$ receptor showed a pattern in forebrain similar to the pattern obtained by receptor autoradiography (Garlow et al. 1993; Morilak et al. 1993). Based upon the distribution, localization and morphology of immunoreactive neurons, it is likely that subpopulations of receptor-containing cells may be GABAergic interneurons. Cholinergic neurons in the laterodorsal and pedunculopontine tegmental nuclei in the basal forebrain represent a second class of neurotransmitter-specific neurons that possess 5-HT$_{2A}$ receptors (Morilak and Ciaranello 1993a).

The distribution of receptor-specific mRNA visualized by in situ hybridization is similar but not identical to that of the expression of the receptor. 5-HT$_{2A}$ receptor mRNA appears concentrated in few areas in the CNS, including the cerebral cortex and some nuclei of the brainstem (Pompeiano et al. 1994). In the cortex, 5-HT$_{2A}$ receptor mRNA is present on interneurons located at the border of layers II and III. In Ammon's horn, the receptor mRNA is expressed at high levels in the pyramidal cell layer of the ventral and caudal parts of CA3. Receptor mRNA expression is restricted to a few nuclei in the thalamus and hypothalamus. The receptor mRNA is present at intermediate levels in the caudate putamen, nucleus accumbens, fundus striati, claustrum and substantia nigra pars compacta. The olfactory bulb presents a low level of the transcript with an enrichment of 5-HT$_{2A}$ receptor mRNA in the mitral cell layer.

In addition to neurons, the 5-HT$_{2A}$ receptor is also expressed in astrocytes (Deecher et al. 1993). Radioligand binding, Northern blot analysis and changes in [Ca^{2+}]$_i$ demonstrated the presence of 5-HT$_{2A}$ receptor in cultured cortical astrocytes. The 5-HT-induced, spiperone- and ketanserin-sensitive increases in free [Ca^{2+}]$_i$, as measured by FURA-2, demonstrated that the 5-HT$_{2A}$ receptors were functional in these cells.

II. Reconstruction of Cell-Type-Specific Expression in Cultured Cells

The widespread expression of the 5-HT$_{2A}$ receptor in brain raises the question of how the receptor gene is regulated in different neuron types and in astroglial cells. The mechanism most commonly utilized to determine cell-type-specific expression is transcriptional regulation. Cell-specific transcriptional regulation can be achieved through positive regulatory mechanisms in which *trans*-acting factors activate the basal transcription machinery (Mitchell et al. 1987; Struhl 1991). Alternatively, transcription can be repressed in nonexpressing tissues through the action of a repressor that acts in some cell types and not in others (Muglia et al. 1986; Levine and Manley 1989; Maue et al. 1990; Kraner et al. 1992; Wuenschell et al. 1990; Mori et al. 1990, 1992) or by a repressor that functions in all cell types with subsequent reactivation in the appropriate cell by cell-specific activators (Baniahmad et al. 1987; Evans et al. 1988; Emerson et al. 1989; Herbst et al. 1989; Seal et al. 1991). Cloning the 5'-flanking sequences of the 5-HT$_{2A}$ receptor from several species facilitated the identification of potential positive and negative elements involved in the tissue-specific regulation of the receptor (Ding et al. 1993; Du et al. 1994; Garlow et al. 1994; Zhu et al. 1995). It has been demonstrated that the promoters of the mouse, rat and human 5-HT$_{2A}$ receptors utilize multiple start sites. In the approximately 5-kb rat sequence, DU et al. (1994) found two GC boxes (SP-1-binding sites) and several AP-2-binding sites upstream of the transcription initiation sites; however, their role in the tissue-specific receptor expression was not investigated. Transient transfection of a plasmid construct containing a 1.4-kb 5-HT$_{2A}$ receptor sequence upstream of the dominant initiation site into rat myometrial smooth muscle cells revealed a basal promoter activity. In another study, transfection of chimeric growth hormone plasmids containing various DNA fragments of a 1.6-kb human 5-HT$_{2A}$ promoter sequence into receptor expressing human cell lines showed that a 0.74-kb fragment 5' of the downstream initiation site exhibited significant promoter activity (ZHU et al. 1995). In the 1.6-kb promoter sequence, ZHU et al. (1995) identified multiple elements binding transcription factors, including Sp1, PEA3, cyclic AMP response element (CRE)-like sequence, and E-boxes, but their contribution in gene activity was not studied. DING et al. (1993) isolated a 6.9-kb segment of the 5'-flanking region of the mouse 5-HT$_{2A}$ receptor. Constructs containing different segments of this 5'-flanking region were tested in a variety of cell lines, including neuroblastoma, glial, fibroblast and epithelial cell lines (DING et al. 1993). A basal promoter was identified between −0.6 kb and −2.3 kb which was functional in both neuronal and non-neuronal cells. Upstream of the basal promoter, between −2.3 and −4.2 kb, two repressor domains were found within the 5'-flanking sequence of the receptor gene. These sequences repressed gene activity in all cells except cells of neuronal origin; thus the repressor domains were the primary determinants to generate neuronal cell-specific transcription of the 5-HT$_{2A}$ receptor gene. Presumably, non-neuronal cells contain factors that bind within these regions to inhibit transcription. Cells of neuronal origin either lack these repressors or express additional factors that modify their function. Since these neuronal cells did not express the 5-HT$_{2A}$ receptor, the expression of the exogenous gene was neuronal specific rather than 5-HT$_{2A}$ receptor specific. The transcriptional regulatory functions of the 5'-flanking region of the mouse 5-HT$_{2A}$ receptor were also tested in glial cells (DING et al. 1993). The repressor domains between −2.3 and −4.2 kb inhibited the transcription of the transfected 5-HT$_{2A}$ receptor gene in C6 glioma cells, which expresses the 5-HT$_{2A}$ receptor. However, a far upstream domain within the 5'-flanking region relieved repression and reactivated transcription of the gene specifically within C6 glioma cells but not within glial cells where the 5-HT$_{2A}$ receptor was not expressed. In the rat 5-HT$_{2A}$ receptor promoter, GARLOW et al. (1994) found a minimal promoter sequence within 200 bp of the major site of transcription initiation. They identified negative attenuating elements between 2.5 and 2.3 kb and positive elements between 1.1 and 0.2 kb upstream of the transcription initiation, suggesting again that the expression of the 5-HT$_{2A}$ receptor gene is regulated by a combination of positive and negative elements operating through the minimal promoter. However, there is a discrepancy between the studies by DING et al. (1993) and GARLOW et al. (1994) on the actual positions of the positive and negative elements within the 5-HT$_{2A}$ receptor promoter. Since these studies used different cell lines, it is possible that the regulation of the receptor is unique for each cell type, which is likely determined by the available transcription factors. For example, a silencer which interacts with a promoter element in one cell type may not exist in another; instead, a different silencer and its binding site become dominant in regulating gene expression.

In summary, both activation by positive regulatory elements and inhibition by repressors have been described as a mechanism functioning in the cell-specific regulation of the 5-HT$_{2A}$ receptor. The combinatorial action of positively and negatively acting transcriptional regulatory proteins and mechanisms is presumably essential to provide complex regulation and flexibility in the expression of the 5-HT$_{2A}$ receptor.

III. Transgenic Strategy to Study the Regulation of the 5-HT$_{2A}$ Promoter

Although in vitro experiments provide an insight into the cell-type-specific regulation of the 5-HT$_{2A}$ receptor gene, the developmental and complex, tissue-specific regulation of the receptor gene can be studied only in transgenic animals. TOTH et al. (1994) produced mice carrying chloramphenicol-acetyltransferase (CAT) reporter genes whose expression was directed by 5-HT$_{2A}$ receptor 5'-flanking sequences by microinjection of fertilized oocytes. A sequence extending from the translational initiation codon to −5.6kb in the 5'-flanking region of the receptor gene was found to be sufficient to target gene expression to the brain in transgenic animals. However, the region-specific pattern of the transgene expression was different in individual mouse lines and none of them matched precisely the endogenous receptor expression pattern, suggesting that the position of the transgene had a major influence on expression. It is also possible that additional elements, upstream or downstream, are required for authentic cell-type-specific expression in the brain.

C. Regulation of 5-HT$_{2A}$ Receptor by Agonists and Antagonists

I. Regulation by Agonists

For most G-protein-linked receptors, agonist stimulation can initiate desensitization, a process that causes them to become progressively less able to trigger signaling mechanisms (SIBLEY et al. 1987). The molecular mechanisms underlying desensitization, which is essentially complete within minutes, include phosphorylation and sequestration of the receptor (HAUSDORFF et al. 1989). More prolonged exposure to an agonist can result in a progressive reduction in receptor density, and this may be associated with a decrease in the level of receptor mRNA (HADCOCK et al. 1989; COLLINS et al. 1992). Indeed, chronic treatment with 5-HT$_{2A}$ receptor agonists such as 1-(2,5-dimethoxy-4-iodophenyl)-2-aminopropane HCl (DOI), 1-(2,5-dimethoxy-4-bromophenyl)-2-aminopropane (DOB), 1-(2,5-dimethoxy-4-methylphenyl)-2-aminopropane (DOM), quipazine and lysergic acid diethylamide (LSD) reduces 5-HT$_{2A}$ receptor densities (BUCKHOLTZ et al. 1985, 1988; EISON et al. 1989; McKENNA et al. 1989; LEYSEN et al. 1989).

P11 cells, which constitutively express a high density of 5-HT$_{2A}$ receptors, were used to study the regulation of receptors in vitro by agonists and partial agonists (FERRY et al. 1993). 5-HT, DOI and LSD caused marked reductions in the density of 5-HT$_{2A}$ receptors as has been observed in vivo. Downregulation was prevented by coincubation with ketanserin, a 5-HT$_{2A}$ receptor antagonist. Interestingly, 5-HT treatment resulted in a transient increase in levels of 5-HT$_{2A}$ receptor mRNA prior to the downregulation (FERRY et al. 1994). This transient upregulation of the mRNA was mediated by a post-translation mechanism. When agonist-promoted increases in receptor mRNA were prevented by the protein kinase C inhibitor bisindolylmaleimide, the rate of agonist-induced downregulation was accelerated (FERRY et al. 1994). RYDELEK-FITZGERALD et al. (1993) found an approximately fourfold increase in the levels of 5-HT$_{2A}$ receptor mRNA in cultured smooth muscle cells within 2 h after administration of 5-HT with maximal levels occurring after 12 h. Nuclear run-on analysis revealed that 5-HT increased the initiation of 5-HT$_{2A}$ receptor mRNA synthesis and the 5-HT-dependent increase in 5-HT$_{2A}$ receptor transcription required de novo protein synthesis. A similar, but more prolonged upregulation of both 5-HT$_{2A}$ receptor and its mRNA was found in cerebellar granule cells after 5-HT or DOI treatment (AKIYOSHI et al. 1993). The increase in 5-HT$_{2A}$ receptor mRNA was detected within 2 h after 5-HT or DOI prestimulation, reached a maximum around 4 h and remained at a plateau for at least 24 h. TOTH and SHENK (1994) studied the effect of long-term (2–6 days) 5-HT treatment on the regulation of the 5-HT$_{2A}$ receptor in C6 cells. The density of 5-HT$_{2A}$ receptor-binding sites was reduced by approximately 40% in the presence of 1 or 10 μM 5-HT. Long (2–6 days) but not short (1 h) treatment altered receptor density, consistent with earlier work performed in animals (CONN and SANDERS-BUSH 1986; SANDERS-BUSH 1990). Treatment of C6 cells with 5-HT caused a decrease in the level of 5-HT$_{2A}$-receptor-specific RNA as well, and the magnitude of this reduction corresponded well with the decrease observed for receptor-binding activity. It is unlikely that transcriptional mechanisms played a role in the downregulation of the 5-HT$_{2A}$ receptor mRNA in C6 cells, since 5-HT treatment caused little if any reduction in enzyme activity when plasmids containing portions of the 5-HT$_{2A}$ receptor gene 5′-flanking sequence and the CAT reporter gene were transfected into C6 cells (TOTH and SHENK 1994).

In summary, prolonged exposure with agonists downregulates the 5-HT$_{2A}$ receptor in vivo that is generally reproducible in different in vitro systems, such as P11 and C6 cells. However, immediately after receptor stimulation,

receptor expression can be upregulated transiently (P11) or in a more prolonged fashion (smooth muscle cells, cerebellar granule cells).

II. Receptor Downregulation by Antagonists

It is somewhat paradoxical that antagonists elicit the downregulation of the 5-
HT_{2A} receptor (Blackshear and Sanders-Bush 1982). Downregulation of
most G-protein-linked receptors is generally induced by agonist stimulation,
while chronic treatment with antagonists induces disuse supersensitivity, a
state characterized by an increase in receptor density (Samanin et al. 1980).
However, chronic administration of 5-HT_{2A} receptor antagonists such as
methysergide, mianserin, ritanserin, setoperone and ketanserin decreases receptor density in rats (Leysen et al. 1986; May et al. 1986; Eison et al. 1989;
Roth and Ciaranello 1991). The downregulation of the 5-HT_{2A} receptor was
not accompanied by a reduction of the 5-HT_{2A}-receptor-specific mRNA level
in rat brain (Roth and Ciaranello 1991).

A number of studies investigated the mechanism of receptor downregulation in cell cultures. In P11 cells, ketanserin and mianserin did not alter
the density of 5-HT_{2A} receptors, even after prolonged incubation with drug
(Ferry et al. 1993). On the other hand, in C6 cells, long-term treatment with
mianserin caused a reduction in the level of 5-HT_{2A} receptor and its mRNA
(Toth and Shenk 1994). The discrepancy between the two studies may be
explained by the different origin of the P11 (neuronal) and C6 cells (glial),
suggesting that the 5-HT_{2A} receptor is not regulated in the same fashion in all
cells. In C6 cells, mianserin downregulated the 5-HT_{2A} receptor gene expression at the level of transcription, a conclusion based on transient transfection
experiments with CAT reporter plasmids containing 5'-flanking regions of the
5-HT_{2A} receptor (Toth and Shenk 1994).

The downregulation of the 5-HT_{2A} receptor by mianserin both in vivo and
in vitro could be based on the ability of antagonists to interrupt a basal,
constitutive level signaling process that maintains a normal level of receptor
gene transcription. It has been demonstrated that the 5-HT_{2C} receptor, which
is closely related to the 5-HT_{2A} receptor, exhibits constitutive receptor activation, defined as agonist-independent receptor activation of the signal transduction pathway (Westphal et al. 1995). Alternatively, antagonists may be able to
act as "inverse agonists" as has been demonstrated with the 5-HT_{2C} receptor
(Barker et al. 1994). Mianserin can be classified as an inverse agonist on the
5-HT_{2C} receptor, since it exhibits a negative intrinsic activity, defined as
a decrease in agonist-independent, receptor-mediated, phosphoinositide
hydrolysis.

III. Antidepressant-Induced Downregulation of the 5-HT_{2A} Receptor

Studies have shown that the density of 5-HT_{2A} receptor is altered during
antidepressant drug treatment (Peroutka and Snyder 1980; Blackshear and

SANDERS-BUSH 1982; GOODWIN et al. 1984). A variety of pharmacologically distinct antidepressants downregulate the 5-HT$_{2A}$ receptor such as monoamine oxidase (MAO) inhibitors and 5-HT uptake blockers. These drugs can be considered indirect agonists, since they increase the availability of 5-HT in synapses. As discussed above, mianserin, a 5-HT$_{2A}$ receptor antagonist and atypical antidepressant, also elicits a decrease in receptor density (PEROUTKA and SNYDER 1980; BENJAMIN et al. 1992). Since both agonists and antagonists can produce antidepressant effect and receptor downregulation, it is intriguing to hypothesize that the downregulation of the 5-HT$_{2A}$ receptor is involved in the development of the therapeutic effect of antidepressants.

D. Glucocorticoid Regulation of 5-HT$_{2A}$ Receptor Expression

I. Glucocorticoids Upregulate 5-HT$_{2A}$ Receptor In Vivo

Animal studies have shown that corticosteroids can alter several elements of the serotoninergic neurotransmission including 5-HT metabolism, turnover and receptors themselves (CURZON 1971; BIEGON et al. 1985; DE KLOET et al. 1986). KURODA et al. (1992, 1993) found that cortical 5-HT$_{2A}$ receptor is upregulated by administration of adrenocorticotropic hormone (ACTH) and dexamethasone in rats. Stress-induced increases in corticosteroid levels can also lead to the upregulation of 5-HT$_{2A}$ receptor (TORDA et al. 1990; BENJAMIN et al. 1993; MCKITTRICK et al. 1995).

II. Glucocorticoid Regulation of the 5-HT$_{2A}$ Receptor in Cultured Cells

The glucocorticoid-induced regulation of the 5-HT$_{2A}$ receptor has been studied in cell lines of both neuronal and glial origins (SAIFF et al. 1995). In agreement with the in vivo data, the 5-HT$_{2A}$ receptor was upregulated by dexamethasone in P11 neuronal cells. However in C6 glial cells, the receptor was actually downregulated, indicating that the same gene can be differentially regulated depending on the cell context. As discussed above, the mechanism of cell-type-specific expression of the 5-HT$_{2A}$ receptor is distinct in neuronal and glial cells, and that may consequently influence the glucocorticoid-induced regulation. Based on these results, it is possible that both upregulation (in neurons) and downregulation (in astocytes) of the 5-HT$_{2A}$ receptor occurs in brain, but because the majority of the 5-HT$_{2A}$ receptors are expressed by neurons, the glial-specific regulation is masked.

The ability of glucocorticoids to influence transcription of the rat 5-HT$_{2A}$ receptor gene was tested by using a promoter-reporter plasmid by GARLOW and CIARANELLO (1995). Sequence analysis revealed a putative glucocorticoid response element (GRE), which consists of a palindromic sequence of 5'

AGAACA 3' separated by a random 3-bp spacer. In the neuronal RS1 and CCL-39 cells, the dexamethasone treatment caused an inhibition of transcription of the 5-HT$_{2A}$ receptor promoter, whereas in the Neuro-2a cells the dexamethasone treatment stimulated transcription from the 5-HT$_{2A}$ promoter. The glucocorticoid response seemed to be indirect, as sequence analysis of the 4.2 kb preceding the site of transcription initiation revealed only an 11/15-nt match to a putative GRE, and deletion of this sequence did not alter the response to dexamethasone. However, the promoter region for the rat 5-HT$_{2A}$ receptor gene contains a number of partial glucocorticoid response elements of 5' AGAACA 3'. Because of several presently uncharacterized interactions of transcription factors and the current uncertainty as to the exact requirements for activation when a response element is not complete, it will be necessary to empirically determine the efficiency of these partial response elements to activate or inhibit 5-HT$_{2A}$ receptor gene expression in neuronal and glial cells. Nevertheless, it is likely that glucocorticoids through glucocorticoid receptors can either inhibit or enhance transcription of the 5-HT$_{2A}$ receptor, depending on their interaction with other transcription factors expressed in the specific cell type.

E. Clinical Relevance

I. Upregulation of the 5-HT$_{2A}$ Receptor in a Group of Depressed Patients

It has been repeatedly shown that the level of 5-HT$_{2A}$ receptor is increased in postmortem brains of depressed patients (Cooper et al. 1986; McKeith et al. 1987; Yates et al. 1990) and in suicide victims (Stanley and Mann 1983; Mann et al. 1986; Arora and Meltzer 1989; Arango et al. 1990); the latter finding, however, was not confirmed by other investigators (Owen et al. 1983; Cheetham et al. 1988). Recently, an in vivo study confirmed these postmortem findings. A higher uptake of ^{123}I-ketanserin was observed in the parietal cortex of depressed patients by single photon emission computed tomography (D'Haenen et al. 1992). However, the in vivo study did not find any correlation between receptor density and suicidal preoccupation. As mentioned above, most chronic antidepressant treatments downregulate the receptor (Peroutka and Snyder 1980; Blackshear and Sanders-Bush 1982; Goodwin et al. 1984), again suggesting a positive correlation between receptor density and disease symptoms.

II. Upregulation of the 5-HT$_{2A}$ Receptor in Depression May Be Related to Stress

Since a hyperfunction of the 5-HT$_{2A}$ receptor may exist in depression, the molecular mechanisms underlying the upregulation of 5-HT$_{2A}$ receptor are of basic interest. It has been shown that depressed patients exhibit adrenocortical hypertrophy (NEMEROFF 1992; RUBIN et al. 1995) and hypercortisolemia (NEMEROFF et al. 1989), and are generally unresponsive to feedback inhibition of the hypothalamic-pituitary-adrenal (HPA) axis (RUPPRECHT et al. 1991). Consistent with a causal role of the HPA axis, depression is frequently observed in patients with Cushing's syndrome, a primary endocrine disorder of high glucocorticoid levels (KATHOL et al. 1985). It is known that stress may underlie and often precipitates depression and it is possible that hypercortisolemia in depression, that is due to a dysregulation of the HPA axis, is an exaggerated response to stress. As described above, glucocorticoids can upregulate the 5-HT$_{2A}$ receptor (KURODA et al. 1993); therefore, elevated levels of corticosteroids induced by stress could lead to the upregulation of the 5-HT$_{2A}$ receptor. Indeed, MCKITTRICK et al. (1995) demonstrated that in the colony model of chronic social stress, subordinate male rats were severely stressed, and had elevated basal corticosterone levels and increased levels of 5-HT$_{2A}$ receptors. Similarly, the dominance status in triad-housed rats correlated inversely with cortical 5-HT$_{2A}$ receptor binding (BENJAMIN et al. 1993). However, glucocorticoids may not be the sole route to upregulate the 5-HT$_{2A}$ receptor in stress. TORDA et al. (1990) described that acute immobilization stress increased the number of 5-HT$_{2A}$ receptors in the rat frontal cortex which were abolished by peripheral administration of propranolol or central administration of 6-hydroxydopamine, suggesting that the noradrenergic system, presumably through stimulation of β-adrenoreceptors, may control the regulation of 5-HT$_{2A}$ receptors during acute stress.

References

Akiyoshi J, Hough C, Chuang DM (1993) Paradoxical increase of 5-hydroxytryptamine$_2$ receptors and 5-hydroxytryptamine$_2$ receptor mRNA in cerebellar granule cells after persistent 5-hydroxytryptamine$_2$ receptor stimulation. Mol Pharmacol 43:349–55

Arango V, Ernsberger P, Marzuk PM, Chen JS, Tierney H, Stanley M, Reis DJ, Mann JJ (1990) Autoradiographic demonstration of increased serotonin 5-HT$_2$ and beta-adrenergic receptor binding sites in the brain of suicide victims. Arch Gen Psychiatry 47:1038–1047

Arora RC, Meltzer HY (1989) Increased serotonin$_2$ (5-HT$_2$) receptor binding as measured by ^{3}H-lysergic acid diethylamide (^{3}H-LSD) in the blood platelets of depressed patients. Life Sci 44:725–734

Baniahmad A, Muller M, Steiner C, Renkawitz R (1987) Activity of two different silencer elements of the chicken lysozyme gene can be compensated by enhancer elements. EMBO J 6:2297–2302

Barker EL, Westphal RS, Schmidt D, Sanders-Bush E (1994) Constitutively active 5-hydroxytryptamine$_{2C}$ receptors reveal novel inverse agonist activity of receptor ligands. J Biol Chem 269:11687–11690

Benjamin D, Saiff EI, Nevins T, Lal H (1992) Mianserin-induced 5-HT$_2$ receptor downregulation results in anxiolytic effects in the elevated plus-maze test. Drug Dev Res 26:287–297

Benjamin D, Saiff EI, Goldstein KR, Larson SA, De Adrianza A, Pohorecky LA (1993) Cortical 5-HT$_2$ receptor binding is inversely related to dominance status in triad-housed rats. Soc Neurosci Abstr 19:663–668

Biegon A, Rainbow TC, McEwen BS (1985) Corticosterone modulation of neurotransmitter receptors in rat hippocampus: a quantitative autoradiographic study. Brain
  Res 332:309–314
Blackshear MA, Sanders-Bush E (1982) Serotonin receptor sensitivity after acute and
  chronic treatment with mianserin. J Pharmacol Exp Ther 221:303–308
Buckholtz NS, Freedman DX, Middaugh LD (1985) Daily LSD administration selectively decreases serotonin$_2$ receptor binding in rat brain. Eur J Pharmacol 109:421–
  425
Buckholtz NS, Zhou D, Freedman DX (1988) Serotonin$_2$ agonist administration down-
  regulates rat brain serotonin$_2$ receptors. Life Sci 42:2439–2445
Cheetham SC, Crompton MR, Katona CL, Horton RW (1988) Brain 5-HT$_2$ receptor
  binding sites in depressed suicide victims. Brain Res 443:272–280
Collins S, Caron MG, Lefkowitz RJ (1992) From ligand binding to gene expression:
  new insights into the regulation of G-protein-coupled receptors. Trends Biochem
  Sci 17:37–39
Conn PJ, Sanders-Bush E (1986) Regulation of serotonin-stimulated phosphoinositide
  hydrolysis: relation to the serotonin 5-HT-2 binding site. J Neurosci 6:3669–
  3675
Cooper SJ, Owen F, Chambers DR, Crow TJ, Johnson J, Poulter M (1986) Post-
  mortem neurochemical findings in suicide and depression: a study of the
  serotoninergic system and imipramine binding in suicide victims. In: Deakin JFW
  (ed), The biology of depression. Gaskell, London pp 53-70
Curzon G (1971) Effects of adrenal hormones and stress on brain serotonin. Am J Clin
  Nutr 24:830–834
D'Haenen H, Bossuyt A, Mertens J, Bossuyt-Piron C, Gijsemans M, Kaufman L (1992)
  SPECT imaging of serotonin$_2$ receptors in depression. Psychiatry Res 45:227–237
De Kloet ER, Sybesma H, Reul HM (1986) Selective control by corticosterone of
  serotonin$_1$ receptor capacity in raphe-hippocampal system. Neuroendocrinology
  42:513–521
Deecher DC, Wilcox BD, Dave V, Rossman PA, Kimelberg HK (1993) Detection of 5-
  hydroxytryptamine$_2$ receptors by radioligand binding, northern blot analysis, and
  Ca^{2+} responses in rat primary astrocyte cultures. J Neurosci Res 35:246–256
Ding D, Toth M, Zhou Z, Parks C, Hoffman B, Shenk T Glial (1993) Cell-specific
  expression of the serotonin 2 receptor gene: selective reactivation of a repressed
  promoter. Mol Brain Res 20:181–191
Du YL, Wilcox BD, Teitler M, Jeffrey JJ (1994) Isolation and characterization of the
  rat 5-hydroxytryptamine type 2 receptor promoter: constitutive and inducible
  activity in myometrial smooth muscle cells. Mol Pharmacol 45:1125–1131
Eison AS, Eison MS, Yocca FD, Gianutsos G (1989) Effects of imipramine and
  serotonin-2 agonists and antagonists on serotonin-2 and beta-adrenergic receptors
  following noradrenergic or serotonergic denervation. Life Sci 44:1419–1427
Emerson BM, Nickol JM, Fong TC (1989) Erythroid-specific activation and derepression of the chick β-globin promoter in vitro. Cell 57:1189–1200
Evans T, Reitman M, Felsenfeld G (1988) An erythroid-specific DNA-binding factor
  recognizes a regulatory sequence common to all chicken globin gene. Proc Natl
  Acad Sci USA 85:5976–5980
Ferry RC, Unsworth CD, Molinoff PB (1993) Effects of agonists, partial agonists, and
  antagonists on the regulation of 5-hydroxytryptamine receptors in P11 cells. Mol
  Pharmacol 43:726–733
Ferry RC, Unsworth CD, Artymyshyn RP, Molinoff PB (1994) Regulation of mRNA
  encoding 5-HT$_{2a}$ receptors in P11 cells through a post-transcriptional mechanism
  requiring activation of protein kinase. J Biol Chem 269:31850–31857
Fuller RW (1990) Serotonin receptors and neuroendocrine responses. Neuropsychopharmacology 3:495–502
Garlow SJ, Ciaranello RD (1995) Transcriptional control of the rat serotonin-2 receptor gene. Brain Res Mol Brain Res 31:201–209

Garlow SJ, Morilak DA, Dean RR, Roth BL, Ciaranello RD (1993) Production and characterization of a specific 5-HT$_2$ receptor antibody. Brain Res 615:113–120

Garlow SJ, Chin AC, Marinovich AM, Heller MR, Ciaranello RD (1994) Cloning and functional promoter mapping of the rat serotonin-2 receptor gene. Mol Cell Neurosci 5:291–300

Goodwin GM, Green AR, Johnson P (1984) 5-HT$_2$ receptor characteristics in frontal cortex and 5-HT$_2$ receptor-mediated head-twitch behaviour following antidepressant treatment to mice. Br J Pharmacol 83:235–242

Hadcock JR, Wang H, Malbon CC (1989) Agonist-induced destabilization of β-adrenergic receptor mRNA. Attenuation of glucocorticoid-induced up-regulation of β-adrenergic receptors. J Biol Chem 264:19928–19933

Hausdorff WP, Bouvier M, O'Dowd BF, Irons GP, Caron MG, Lefkowitz RJ (1989) Phosphorylation sites on two domains of the β_2-adrenergic receptor are involved in distinct pathways of receptor desensitization. J Biol Chem 264:12657–12666

Herbst RS, Friedman N, Darnell JE Jr, Babiss LE (1989) Positive and negative regulatory elements in the mouse albumin enhancer. Proc Natl Acad Sci USA 86:1553–1557

Jacobs BL, Wilkinson LO, Fornal CA (1990) The role of brain serotonin. A neurophysiologic perspective. Neuropsychopharmacology 3:473–479

Kathol RG, Delahunt JW, Hannah L (1985) Transition from bipolar affective disorder to intermittent Cushing's syndrome: case report. J Clin Psychiatry 46:194–196

Kraner SD, Chong JA, Tsay H-J, Mandel G (1992) Silencing the type II sodium channel gene: a model for neural-specific gene regulation. Neuron 9:37–44

Kuroda Y, Mikuni M, Ogawa T, Takahashi K (1992) Effect of ACTH, adrenalectomy and the combination treatment on the density of 5-HT$_2$ receptor binding sites in neocortex of rat forebrain and 5-HT$_2$ receptor-mediated wet-dog shake behaviors. Psychopharmacology 108:27–32

Kuroda Y, Mikuni M, Nomura N, Takahashi K (1993) Differential effect of subchronic dexamethasone treatment on serotonin-2 and beta-adrenergic receptors in the rat cerebral cortex and hippocampus. Neurosci Lett 155:195–198

Levine M, Manley JL (1989) Transcriptional repression of eukaryotic promoters. Cell 59:405–408

Leysen JE (1990) Gaps and peculiarities in 5-HT$_2$ receptor studies. Neuropsychopharmacology 3:361–369

Leysen JE, VanGompel P, Gommeren W, Woestenborghs R, Janssen PFM (1986) Down regulation of serotonin-S2 receptor sites in rat brain by chronic treatment with the serotonin-S2 antagonists: ritanserin and setoperone. Psychopharmacology 88:434–444

Leysen JE, Janssen PFM, Niemegeers CJE (1989) Rapid desensitization and down-regulation of 5-HT$_2$ receptors by DOM treatment. Eur J Pharmacol 163:145–149

Mann JJ, Stanley M, McBride PA, McEwen BS (1986) Increased serotonin$_2$ and β-adrenergic receptor binding in the frontal cortices of suicide victims. Arch Gen Psychiatry 43:954–959

Maue RA, Kraner SD, Goodman RH, Mandel G (1990) Neuron-specific expression of the rat brain type II Na$^+$ channels by nerve growth factor. Proc Natl Acad Sci USA 85:924–928

May PC, Morgan DG, Finch CE (1986) Regional serotonin receptor studies: chronic methysergide treatment induces a selective and dose-dependent decrease in serotonin-2 receptors in mouse cerebral cortex. Life Sci 38:1741–1747

McKeith IG, Marshall EF, Ferrier IN, Armstrong MM, Kennedy WN, Perry RH, Perry EK, Eccleston D (1987) 5-HT receptor binding in post-mortem brain from patients with affective disorder. J Affect Disord 13:67–74

McKenna DJ, Nazarali AJ, Himeno A, Saavedra JM (1989) Chronic treatment with ($\pm$) DOI, a psychotomimetic 5-HT$_2$ agonist, downregulates 5-HT$_2$ receptors in rat brain. Neuropsychopharmacology 2:81–87

McKittrick CR, Blanchard DC, Blanchard RJ, McEwen BS, Sakai RR (1995) Serotonin receptor binding in a colony model of chronic social stress. Biol Psychiatry 37:383–393

Mengod G, Pompeiano M, Martinez-Mir MI, Palacios JM (1990) Localization of the mRNA for the 5-HT$_2$ receptor by in situ hybridization histochemistry. Correlation with the distribution of receptor sites. Brain Res 524:139–143

Mitchell PJ, Wang C, Tjian R (1987) Positive and negative regulation of transcription in vitro: enhancer binding protein AP2 is inhibited by SV40 T antigen. Cell 50:847–861

Mori N, Schoenherr C, Vandenbergh DJ, Anderson DJ (1992) A common silencer element in the SCG10 and type II Na$^+$ channel genes binds a factor present in non-neuronal but not in neuronal cells. Neuron 9:45–54

Mori N, Stein R, Sigmund O, Anderson DJ (1990) A cell type-preferred silencer element that controls the neural-specific expression of the SCG10 gene. Neuron 4:583–594

Morilak DA, Ciaranello RD (1993) 5-HT$_2$ receptor immunoreactivity on cholinergic neurons of the pontomesencephalic tegmentum shown by double immunofluorescence. Brain Res 627:49–54

Morilak DA, Garlow SJ, Ciaranello RD (1993) Immunocytochemical localization and description of neurons expressing serotonin$_2$ receptors in the rat brain. Neuroscience 54:701–717

Muglia L, Rothman-Denes LB (1986) Cell type-specific negative regulatory element in the control region of the rat alpha-fetoprotein gene. Proc Natl Acad Sci USA 83:7653–7657

Murphy DL (1990) Neuropsychiatric disorders and the multiple human brain serotonin receptor subtypes and subsystems. Neuropsychopharmacology 3:457–471

Nemeroff CB (1989) Clinical significance of psychoneuroendocrinology in psychiatry: focus on the thyroid and adrenal. J Clin Psychiatry 50:13–19

Nemeroff CB, Krishnan RR, Reed D, Leder R, Beam C, Dunnick NR (1992) Adrenal gland enlargement in major depression. Arch Gen Psychiatry 49:384–390

Owen F, Cross AJ, Crow TJ, Deakin JF, Ferrier IN, Lofthouse R, Poulter M (1983) Brain 5-HT-2 receptors and suicide. Lancet ii:1256

Peroutka SJ, Snyder SH (1980) Long-term antidepressant treatment decreases spiroperidol-labeled serotonin receptor binding. Science 210:88–90

Pompeiano M, Palacios JM, Mengod G (1994) Distribution of the serotonin 5-HT$_2$ receptor family mRNAs: comparison between 5-HT$_{2A}$ and 5-HT$_{2C}$ receptors. Mol Brain Res 23:163–178

Roth BL, Ciaranello RD (1991) Chronic mianserin treatment decreases 5-HT$_2$ receptor binding without altering 5-HT$_2$ receptor mRNA levels. Eur J Pharmacol 207:169–172

Roth BL, Hamblin M, Ciaranello RD (1991) Developmental regulation of 5-HT$_2$ and 5-HT$_{1C}$ mRNA and receptor levels. Dev Brain Res 58:51–58

Rubin RT, Phillips JJ, Sadow TF, McCracken JT (1995) Adrenal gland volume in major depression: increase during the depressive episode and decrease with successful treatment. Arch Gen Psychiatry 52:213–218

Rupprecht R, Kornhuber J, Wodarz N, Lugauer J, Gobel C, Haack D, Beck G, Muller OA, Rieder P, Beckman H (1991) Disturbed glucocorticoid receptor auto-regulation and corticotropin response to dexamethasone in depressives pretreated with metyrapone. Biol Psychiatry 29:1099–1109

Rydelek-Fitzgerald L, Wilcox BD, Teitler M, Jeffrey JJ (1993) Serotonin-mediated 5-HT$_2$ receptor gene regulation in rat myometrial smooth muscle cells. Mol Cell Endocrinol 92:253–259

Saiff E, Goska J, Sibille E, Toth M, Pohorecky LA, Benjamin D (1995) Differential regulation by corticosteroids of 5-HT$_{2A}$ receptors in P-11 and C-6 cells. Soc Neurosci Abstr 21:443.18

Samanin R, Mennini J, Ferraris A, Bendotti C, Borsini R (1980) Hyper- and hyposensitivity of central serotonin receptors: [^{3}H]-serotonin binding and functional studies in the rat. Brain Res 189:449–457

Sanders-Bush E (1990) Adaptive regulation of central serotonin receptors linked to phosphoinositide hydrolysis. Neuropsychopharmacology 3:411–416

Seal SN, Davis DL, Burch JBE (1991) Mutational studies reveal a complex set of positive and negative control elements within the chicken vitellogenin II promoter. Mol Cell Biol 11:2704–2717

Sibley DR, Benovic JL, Caron MG, Lefkowitz RJ (1987) Regulation of transmembrane signaling by receptor phosphorylation. Cell 48:913–922

Stanley M, Mann JJ (1983) Increased serotonin-2 binding sites in frontal cortex of suicide victims. Lancet i:214–216

Struhl K (1991) Mechanism for diversity in gene expression patterns. Neuron 7:177–181

Torda T, Murgas K, Cechova E, Kiss A, Saavedra JM (1990) Adrenergic regulation of [^{3}H]ketanserin binding sites during immobilization stress in the rat frontal cortex. Brain Res 527:198–203

Toth M, Shenk T (1994) Antagonist-mediated downregulation of serotonin$_2$ receptor gene expression: modulation of transcription. Mol Pharmacol 145:1095–1100

Toth M, Ding D, Shenk T (1994) The 5′ flanking region of the serotonin$_2$ receptor gene directs brain specific expression in transgenic animals. Mol Brain Res 27:315–319

Westphal RS, Backstrom JR, Sanders-Bush E (1995) Increased basal phosphorylation of the constitutively active serotonin 2C receptor accompanies agonist-mediated desensitization. Mol Pharmacol 48:200–205

Wuenschell CW, Mori N, Anderson DJ (1990) Analysis of SCG10 gene expression in transgenic mice reveals that neural specificity is achieved through selective derepression. Neuron 4:4595–4602

Yates M, Leake A, Candy JM, Fairbairn AF, McKeith IG, Ferrier IN (1990) 5HT$_2$ receptor changes in major depression. Biol Psychiatry 27:489–496

Zhu QS, Chen K, Shih JC (1995) Characterization of the human 5-HT$_{2A}$ receptor gene promoter. J Neurosci 15:4885–48895

CHAPTER 16

Neuropharmacology of 5-HT₃ Receptor Ligands

B. Costall and R.J. Naylor

A. Introduction

Gaddum and Picarelli (1957), using the guinea pig ileum, were the first researchers to show that 5-hydroxytryptamine (5-HT) can mediate a neuronal depolarising response and release of acetylcholine. A clarification of the nature and function of the receptor mediating the response to 5-HT has subsequently taken some 40 years. The present chapter reviews the beginnings of a pharmacological understanding of this receptor and the therapeutic implications of drugs which have affinity for it.

B. Electrophysiological Studies of the 5-HT₃ Receptors

I. The 5-HT₃ Receptor as a Member of the Ligand-Gated Ion Channel Superfamily

In common with other agonist-gated channels such as the nicotinic and glycine receptors, the 5-HT₃ receptor has probably evolved to mediate rapid synaptic events in the nervous system (Sugita et al. 1992). Intracellular microelectrode recordings performed on central and peripheral autonomic, sensory and enteric neurons have shown that 5-HT₃ receptors mediate a rapid depolarising response, associated with an increase in membrane conductance consequent on the opening of cation selective channels (see reviews by Fozard 1984a; Wallis 1989; Peters et al. 1994). Such responses also occur in a number of neuronal clonal cell lines, primary cultures and slice preparations; a summary of the electrophysiological properties of 5-HT₃ receptors obtained from studies employing voltage clamp and single channel recording techniques is given in Peters et al. (1994). Neither G proteins nor soluble second messengers appear necessary for channel activation and the response to 5-HT is usually described in terms of a cooperative receptor model in which the occupation by 5-HT of one receptor subunit enhances the binding of other agonist molecules to other subunits (Yakel et al. 1991). At least two or three molecules of 5-HT must bind to activate the 5-HT₃ receptor, the response in all test systems being characterised by rapid and pronounced desensitisation (Wallis and North 1978; see Peters et al. 1994). Desensitisation is recorded as a progressive decline in the amplitude of depolarisation in response to repeated challenge with 5-HT; the precise nature of the desensitisation process may differ between and within cell types (Yakel 1992).

Single cell studies employing intracellular recording of 5-HT$_3$ receptor mediated depolarisation in several preparations have indicated that an influx of sodium and potassium ions contributes to the response (Wallis and North 1978; Higashi and Nishi 1982). The channels gated by 5-HT discriminate weakly between small monovalent cations ($Cs^+ > K^+ > Li^+ > Na^+ > Rb^+$, see Peters et al. 1994), but they remain permeable to divalent cations such as Ca^{2+} and Mg^{2+}, and a small influx of Ca^{2+} may occur under physiological conditions (Yang et al. 1992). One consequence of increasing the extracellular concentration of Ca^{2+} or Mg^{2+} is a reduction in the amplitude and duration of 5-HT$_3$ mediated responses (Peters et al. 1989; Yakel et al. 1990; Robertson and Bevan 1991). Zn^{2+}, Cd^{2+} and Cu^{2+} ions also inhibit 5-HT$_3$ receptor mediated responses, and the mechanisms involved, e.g. a reduction in the affinity of the 5-HT$_3$ receptors for agonists, are being investigated (see Peters et al. 1994). Further studies are also required to establish the significance of notable differences in single-channel conductance between different preparations (Peters et al. 1994).

II. Agonist Pharmacology

2-Methyl-5-HT and the amidine derivatives 1-phenyl biguanide and meta-chlorophenylbiguanide have relatively selective affinity for the 5-HT$_3$ receptor (Richardson et al. 1985; Wallis and Nash 1981; Kilpatrick et al. 1990a). Their agonist potency and efficacy has been established in many different preparations from many species (e.g. vagus nerve of guinea pig, rabbit and rat, superior cervical ganglia of guinea pig, mouse and rat, N1E-115 and NG108-15 cells) and varies between preparations, displaying agonist or partial agonist effects (see Peters et al. 1994). There are also species differences: 1-phenyl biguanide has no action on peripheral neurons of the guinea pig (Butler et al. 1990) but is a full (Fortune et al. 1983) or partial (Kilpatrick et al. 1990a) agonist in the rat and mouse.

III. Antagonist Pharmacology

The first selective 5-HT$_3$ receptor antagonist to be identified was MDL 72222 (Fozard 1984b); this compound was shown in low concentrations to inhibit the effects of 5-HT on the rabbit nodose ganglia in vitro by competitive antagonism: higher concentrations of MDL 72222 caused an insurmountable antagonism (Azami et al. 1985). In the superior cervical ganglion or the rat isolated vagus nerve the antagonism afforded by MDL 72222 at any concentration was insurmountable (Azami et al. 1985; Ireland and Tyers 1987). But in the guinea pig isolated vagus nerve and in the rabbit preganglionic cervical sympathetic nerves, blockade by MDL 72222 demonstrates a simple competitive antagonism (Elliott and Wallis 1988; Butler et al. 1990). A similar situation has been shown to occur for tropisetron (ICS 205930). Thus the apparent competitive blockade by tropisetron of the depressant effect of 5-HT upon the compound action potential amplitude of the rabbit vagus nerve (RICHARDSON et al. 1985) contrasts with the insurmountable blockade on rat isolated vagus nerve (IRELAND and TYERS 1987; TATTERSALL et al. 1992). In the rabbit isolated nodose and superior cervical ganglia, the antagonism by tropisetron of the 5-HT induced depolarisation response was essentially insurmountable: only in the nodose ganglia and at low concentrations did the antagonism achieve a competitive nature (ROUND and WALLIS 1986). Also in the mouse, rat and guinea pig superior cervical ganglia, the antagonism afforded by tropisetron of the effect of 2-Me-5-HT was non-competitive (NEWBERRY et al. 1991). Ondansetron exerts a competitive blockade of the 5-HT response on the rat vagus nerve and superior cervical ganglion (BUTLER et al. 1990) whereas zacopride, granisetron and BRL 46470A exert a non-competitive blockade (see PETERS et al. 1994). However, it should be noted that even the antagonism afforded by ondansetron in the rabbit vagus nerve was not entirely overcome by 5-HT (BUTLER et al. 1988).

From the use of in vitro models it becomes clear that, with the possible exception of ondansetron, 5-HT$_3$ receptor antagonists may attenuate the effects of 5-HT through competitive and non-competitive mechanisms. This may relate to variations in response between tissues, species, differing drug pretreatment times and differences in receptor reserve, or to the development of receptor desensitisation. The latter, in particular, will seriously interfere with equilibrium conditions necessary for the accurate measurement of agonist-induced responses.

C. Molecular Characterisation of 5-HT$_3$ Receptors

The archetypal member of the ligand-gated ion channel receptor is the nicotinic acetylcholine receptor (nAChR), which consists of five individual subunits forming a pentameric ion channel complex (see HERTLING-JAWEED et al. 1990). Electron microscopic images of purified 5-HT$_3$ receptors show a similar structure, indicating that the 5-HT$_3$ receptor is also composed of five subunits arranged around a central cavity, the vestibule of an ion channel (see BOESS et al. 1992; BOESS and MARTIN 1994).

I. Biochemical Characterisation of 5-HT$_3$ Receptors

Attempts to purify and biochemically characterise the 5-HT$_3$ receptor were the essential first step in the partial amino acid sequencing of the subunits to facilitate cloning. Sequence information allows an examination of the localisation of receptor mRNA (with specific RNA or DNA probes) and receptor protein with subtype specific antibodies. Substantial amounts of receptor protein are required for such studies and whilst 5-HT$_3$ receptor levels in brain are relatively low, a number of immortal cell lines of neuronal origin express 5-HT$_3$ receptors in high density. NG108-15 (mouse neuroblastoma × rat glioma hybrid), NIE-115 (mouse neuroblastoma) and NCB20 (mouse neuroblastoma × Chinese hamster brain cell hybrid) cell lines allowed the purification of 5-HT$_3$ receptors which appeared similar to receptors found in the mammalian brain (BOLANOS et al. 1990; HOYER and NEIJT 1988a,b; MCKERNAN et al. 1990; LUMMIS and MARTIN 1992; LUMMIS et al. 1994; BOESS et al. 1992).

II. Cloning of the 5-HT$_3$ Receptor

The first cDNA containing the coding sequence of a 5-HT gated ion channel (5-HT$_3$ R-A) was obtained by screening an NCB20 expression library for 5-HT activated currents in *Xenopus* oocytes (MARICQ et al. 1991). The cDNA sequence predicted a protein of 487 amino acids long including the signal peptide. JOHNSON and HEINEMANN (1992) reported the isolation of a 5-HT$_3$ receptor cDNA from a rat superior cervical ganglion library encoding a protein of 461 amino acids. The rat 5-HT$_3$ receptor shows significant homology of amino acid sequence with the mouse 5-HT$_3$ receptor and other members of the ligand gated ion channel superfamily. Features include four hydrophobic regions (M1–M4), and residues with negative charges bracket the M2 region in both the nAChR and 5-HT$_3$ R-A. It would be predicted that modification of these residues by site directed mutagenesis would affect conductance and ion selectivity (IMOTO et al. 1988; KONNO et al. 1991).

III. Splice Variant of 5-HT$_3$ R-A

A splice variant of 5-HT$_3$ R-A termed 5-HT$_3$ R-As has been cloned from NIE-115 neuroblastoma cells and NG108-15 hybridoma cells (HOPE et al. 1993; WERNER et al. 1993). 5-HT$_3$ R-As demonstrates a deletion of six consecutive amino acids in the putative intracellular loop between M3 and M4. The absent stretch contains a putative casein kinase II phosphorylation site but no clear functional differences have been reported between 5-HT$_3$ R-A and 5-HT$_3$-R-As. When expressed in *Xenopus* oocytes or NEK293 cells, homo-oligomeric complexes formed from either 5-HT$_3$ R-A or 5-HT$_3$ R-As exhibit inward current responses to 5-HT and many of the pharmacological properties recorded for 5-HT$_3$ in cell lines. Indeed, only 2-Me-5-HT was found to produce a lesser maximal response at the 5-HT$_3$ R-As (see BOESS and MARTIN 1994; DOWNIE et al. 1994).

The cloning of the murine 5-HT$_3$ receptor subunit 5-HT$_3$ R-A demonstrated the pharmacological profile of the receptor endogenous to mouse tissue. The cloning and characterisation of species homologues of the 5-HT$_3$ R-A provides a powerful tool in identifying interspecies receptor heterogeneity. HOPE et al. (1995) and BELELLI et al. (1995a) have recently reported that a cDNA encoding a 5-HT$_3$ receptor subunit isolated from a human amygdala cDNA library was cloned into a eukaryotic expression vector and introduced into HEK293 cells by lipid mediated transfection; RNA transcripts of the human cDNA were also injected into *Xenopus* oocytes (BELELLI et al. 1995b). Radioligand binding assays using [³H]granisetron indicated high nanomolar affinity for ondansetron and micromolar affinity for (+)tubocurarine. 5-HT elicited an inward current response that desensitised; 2-methyl-5-HT, phenylbiguanide and *m*-chlorophenylbiguanide mimicked the effects of 5-HT, achieving 87%, 68% and 81% of the maximal response attained by 5-HT. The response to 5-HT was antagonised by picomolar concentrations of ondansetron and micromolar concentrations of (+)tubocurarine.

These responses obtained to the human 5-HT₃ R-A receptor show clear differences to those obtained in the murine 5-HT₃ R-A receptor (GILL et al. 1995); in the human system there was a 1500– to 1800-fold reduced potency of (+)tubocurarine and greater efficacy of 2-methyl-5-HT. Also, low concentrations of Zn^{2+} enhanced the effects of 5-HT in the murine but not in the human preparation. BELELLI et al. (1995a,b) suggest that in view of the small number of sequence substitutions that occur between the human and mouse homologues of the 5-HT₃ R-As in the extracellularly located N-terminal domain, compounds such as (+)tubocurarine in conjunction with site directed mutagenesis may prove valuable in locating amino acid residues that contribute to the ligand binding site(s) of the 5-HT₃ receptor.

D. Distribution of 5-HT₃ Receptors

The distribution of 5-HT₃ receptors was initially established using traditional radioligand binding techniques in tissue homogenates and autoradiography. It is ironic to note that the key compounds which established the functional role of 5-HT₃ receptors (tropisetron, ondansetron and MDL 72222) were ineffectual or of limited use as receptor ligands in binding assays. It would have been concluded from a reliance on these three agents as potential ligands that 5-HT₃ receptors did not exist! KILPATRICK et al. (1987) were the first to demonstrate the presence of 5-HT₃ binding sites using the radioligand [³H]GR 65630 and a rat brain homogenate preparation. Subsequently, many authors using numerous ligands have firmly established the location of 5-HT₃ receptors in body tissues in many species; the literature has been extensively reviewed (see LESLIE et al. 1994).

Briefly, 5-HT₃ receptor density in the brain of many species including man is highest in the subnucleus gelatinosus of the solitary nucleus, nucleus of the solitary tract and dorsal motor nucleus of the vagus nerve. Whilst forebrain 5-HT₃ receptor density varies a little between species, 5-HT₃ receptors are located throughout cortical and limbic brain regions, substantia gelatinosa of the spinal trigeminal nucleus and dorsal horn of the spinal cord (see KILPATRICK et al. 1987, 1988; WAEBER et al. 1989; BARNES et al. 1988; GEHLERT et al. 1991). 5-HT₃ receptors have also been located in the enteric nervous system (PINKUS et al. 1989; KILPATRICK et al. 1991; CHAMPANERIA et al. 1992), in the sympathetic nervous system from cat and rabbit superior cervical ganglia (HOYER et al. 1989), and on cat, rat and rabbit vagus nerve (HOYER et al. 1989; KILPATRICK et al. 1990b). Future developments in positron emission tomography may prove helpful in the localisation of 5-HT$_3$ receptors and in their activity in living brain.

Future studies will inevitably build on advances in molecular biology to develop antibodies directed against the 5-HT$_3$ receptor itself. Immunohistochemistry, in conjunction with electron microscopy, should allow a precise definition of the 5-HT$_3$ receptors at the subcellular level. Already in situ hybridisation techniques have revealed the presence of 5-HT$_3$ receptor mRNA throughout cortical and limbic structures, the hypothalamus, facial nerve nucleus, nucleus of the spinal tract of the trigeminal nerve and spinal cord dorsal horn. In the peripheral nervous system, 5-HT$_3$ receptor mRNA is present in the parasympathetic and sympathetic ganglion neurons (TECOTT et al. 1993; JOHNSON and HEINMANN 1993). It remains an interesting observation that no mRNA was detected in the area postrema/nucleus tractus solitarius, the area of highest 5-HT$_3$ receptor density. This suggests a presynaptic location of 5-HT$_3$ receptors on peripheral afferents, which is in agreement with lesion studies (PRATT and BOWERY 1989).

In situ hybridisation techniques have also been used in developmental studies where the distribution pattern of 5-HT$_3$ receptor mRNA was examined during mouse embryogeneus (SCHTROM et al. 1993). It was suggested that 5-HT$_3$ receptors may influence cell migration or differentiation.

E. 5-HT$_3$ Receptor Heterogeneity

The first indication of a heterogeneity between neuronally located 5-HT$_3$ receptors derived from observations that antagonist affinities of nor-(−)-cocaine and benzoic acid esters of tropine, most notably MDL 72222 and tropisetron, were much higher in the rabbit heart than in the guinea pig ileum (FOZARD 1983, 1984; FOZARD and GITTOS 1983; DONATSCH et al. 1984; RICHARDSON et al. 1985). Subsequently, RICHARDSON and ENGEL (1986) hypothesised the existence of 5-HT$_{3A}$, 5-HT$_{3B}$ and 5-HT$_{3C}$ receptor subtypes in the rabbit heart, rabbit vagus nerve and guinea pig ileum respectively to explain the differences in affinity. However, this pioneering delineation has been viewed with caution, with respect to the apparent non-competitive antagonism exerted by tropisetron in some preparations (KILPATRICK and TYERS 1992).

Following on from the hypothesis of RICHARDSON and ENGEL, it is now well accepted that there are major species variations in the properties of 5-HT$_3$ receptors. A comparison of 5-HT$_3$ receptor antagonist affinities in preparations from guinea pig, rabbit, rat and mouse is shown in Table 1. It is quite clear that antagonist affinities in the guinea pig are much lower than in the rat or rabbit. However, antagonist affinities in different tissues within a given species are very similar, giving no evidence of tissue dependence.

Table 1. Antagonist potencies of 5-HT$_3$ receptor antagonists in various preparations and species. The table was taken from PETERS et al. (1994). The table shows pA_2 (or apparent pA_2) values for antagonists acting against the extracellularly recorded depolarisation elicited by 5-HT (or 2-methyl-5-HT) in vagus nerve (VN), superior cervical ganglia (SCG) and nodose ganglia (NG). For the heart, ileum and colon, positive chronotropic and contractile effects were the quantified agonist-induced responses

Species preparation	Guinea pig				Rat		Rabbit				Mouse
Compound	Ileum	Colon	VN	SCG	VCN	SCG	VN	SCG	NG	Heart	SCG
Metoclopramide	5.5[a,2]	5.7[a,2]	5.4[a,2]	—	6.5[g,2]	6.3[k]	7.4[l]	7.2[o]	7.1[o]	7.2[q]	—
MDL 72222	6.7[a,2]	6.7[a,2]	6.4[a,2]	—	7.9[g]	—	8.0[m]	7.8[o]	7.7[o]	9.3[r]	—
(+)-Tubocurarine	—	—	—	4.8[f]	7.2[h,2]	7.1[f]	—	—	—	—	8.1[f]
Quipazine	—	—	—	—	8.5[g]	—	—	7.6[o]	7.5[o]	—	—
Ondansetron	7.3[b,2]	7.1[a,2]	7.0[a,2]	—	8.6[g,2]	8.1[b]	9.4[b]	—	—	10.5[b]	—
GR 65630	7.5[c,2]	7.5[c,2]	7.2[c,2]	—	9.9[c,2]	—	—	—	—	—	—
Renzapride	7.6[d]	—	—	—	—	—	8.5[n]	—	—	8.9[d]	—
SDZ 206792	7.7[e]	—	—	—	—	—	8.9[e,1]	—	—	9.8[e]	—
GR 80284	7.8[c,2]	7.6[c,2]	7.2[c,2]	—	7.9[c,2]	—	—	—	—	—	—
ICS 205930	8.0[c,2]	8.0[a,2]	7.8[a,2]	7.2[f]	11.0[g]	9.4[f]	10.2[c,l]	10.4[p]	10.2[p]	10.6[e]	8.7[f]
L 683877	—	—	—	—	9.3[i,2]	8.3[i,2]	10.1[l]	—	—	10.1[l,2]	—
Granisetron	8.1[a,2]	8.1[a,2]	7.9[a,2]	—	9.8[a,2]	—	—	—	—	10.7[s]	—
Zacopride	8.1[a,2]	8.3[a,2]	8.0[a,2]	—	9.9[a,2]	—	10.1[m]	—	—	—	—
GR 67330	9.0[a,2]	8.7[a,2]	8.0[a,2]	—	10.2[j]	—	—	—	—	—	—
SDZ 206830	9.1[e]	—	—	—	—	—	13.1[e,l]	—	—	10.1[e]	—

[1] Agonist-induced responses quantified from depression of "C" fibre spike amplitude rather than membrane depolarisation.
[2] pK_B value.

The data were collated from the following sources: [a] BULTER et al. 1990; [b] BUTLER et al. 1988; [c] KILPATRICK and TYERS 1992; [d] SANGER 1987; [e] RICHARDSON et al. 1985; [f] NEWBERRY et al. 1991; [g] IRELAND and TYERS 1987; [h] NEWBERRY et al. 1992; [i] TATTERSALL et al. 1992; [j] KILPATRICK et al. 1990b; [k] IRELAND et al. 1990; [l] ELLIOT et al. 1987; [m] SMITH et al. 1988; [n] BUCHHEIT, cited by FOZARD 1989; [o] ROUND and WALLIS 1987; [p] ROUND and WALLIS 1986; [q] FOZARD and MOBAROK ALI 1978; [r] FOZARD 1984b; [s] SANGER and NELSON 1989.

Table 2. Antagonist potencies of 5-HT$_3$ receptor antagonists reported from single cell studies. The table was taken from PETERS et al. (1994). The table shows the $_pIC_{50}$ values [i.e. $-\log_{10} IC_{50}(M)$] determined for the antagonists acting against 5-HT-induced depolarising responses, or 5-HT-evoked inward membrane currents (*) recorded under voltage-clamp conditions

Species preparation	Mouse				Rabbit	Guinea pig			Rat
	N1E-115 cells	NG108-15 cells	Hippocampal neurons	Nodose ganglion cells	Nodose ganglion cells	Nodose ganglion cells	Coeliac ganglion cells	Submucous plexus neurons	Lateral amygdala neurons
Compound									
Cocaine	5.1[a]*	—	—	5.3[bh]*	7.1[a]*	5.7[a]*	—	5.5[j]	—
Metoclopramide	7.6[a]*	6.7[f]*	6.4[f]*	—	7.9[a]*	5.9[a]*	—	—	—
MDL 72222	8.7[b]*	—	—	—	9.4[a]*	6.5[a]*	5.8[j]	5.5[j]	—
(+)-Tubocurarine	9.1[c]*	9.1[f]*	8.8[f]*	8.8[h]*	6.8[a]*	5.0[a]*	4.4[i]	4.7[j]	4.2[k]
Ondansetron	9.6[d]*	—	—	9.4[h]*	10.2[a]*	7.7[a]*	—	7.0[j]	6.3[k]
ICS 205930	9.7[b]	—	—	—	10.3[a]*	—	—	7.9[j]	7.3[k]
Granisetron	9.9[c]*	—	—	—	—	—	—	—	—

The data were obtained from the following sources: [a] PETERS et al. 1991; [b] NEIJT et al. 1988; [c] PETERS et al. 1990; [d] LAMBERT et al. 1989; [e] SEPULVEDA et al. 1991; [f] YAKEL and JACKSON 1988; [g] MALONE, PETERS and LAMBERT, unpublished observations; [h] MALONE et al. 1991; [i] WALLIS and DUN 1988; [j] VANNER and SURPRENANT 1990; [k] SUGITA et al. 1992.

Single cell studies determining the pIC_{50} values for antagonists acting against 5-HT induced depolarising responses, or 5-HT evoked inward membrane currents recorded under voltage clamp conditions, are shown in Table 2. The data support the results obtained from other functional studies, i.e. the potency of the 5-HT₃ receptor antagonists is considerably less in the guinea pig than in the rabbit or mouse tissues. The differences detected with (+)tubocurarine are particularly striking, reflecting a 100- to 10000-fold potency difference. Radioligand binding studies also reveal an apparent low affinity of antagonist ligands at 5-HT₃ receptors on guinea pig enteric neurons (PINKUS and GORDON 1991; WONG et al. 1992).

There remain two further points. The 5-HT₃ receptors in mouse tissue show a consistent and particularly high affinity to (+)tubocurarine: the receptor expressed in this species may reflect an interspecies variant of the 5-HT₃ receptor (see Sect. C.III). Secondly, and perhaps of even greater interest, are the unexpectedly low potency values of ondansetron and tropisetron as inhibitors of the 5-HT induced depolarising response in rat lateral amygdala neurons (SUGITA et al. 1992). If the data can be confirmed and extended to other central neuronal systems, and if the low values are not a peculiarity of the brain slice technique (obtained using a transient application of 5-HT), then the results may have important implications for an understanding of the central actions of 5-HT₃ receptor antagonists in vivo. It would be particularly instructive to extend such studies to primate tissue.

The possibility of 5-HT receptor heterogeneity is also derived from the electrophysiological studies of WANG and colleagues. They have reported inhibitory actions of iontophoretically applied 5-HT and 5-HT₃ receptor agonists, 2-methyl-5-HT, phenylbiguanide and SR 57227A, upon spontaneously firing rat medial prefrontal cortical neurons and glutamate activated firing of CA1 hippocampal pyramidal cells (ASHBY et al. 1991, 1992; WANG et al. 1991a,b, 1992; ZHANG et al. 1994). The inhibitory effect is slow in onset, does not desensitise and is selectively antagonised by 5-HT₃ receptor antagonists. The inhibitory response profile is of a diametrical nature to the classical 5-HT₃ mediated fast excitatory response. The response does not appear to be GABA mediated (WANG et al. 1991a) and, in concomitant studies, EDWARDS et al. (1991) have shown that 5-HT₃ receptor activation in the entorhinal and frontocingulate cortex increases phosphoinositide turnover. If such changes can be dissociated unequivocally from the release of intermediate transmitter substance(s), the possibility of 5-HT₃ receptor subtypes would be strengthened.

F. Involvement of 5-HT₃ Receptors in Neurotransmitter Release

The neuronal depolarising effect of 5-HT₃ receptor stimulation would be predicted to enhance neurotransmitter release and this has been noted for at least five neurotransmitter/modulatory substances.

It is probable that the results obtained from in vitro preparations are critically dependent on the selectivity of agonist action to a receptor(s) sub-type, that antagonist action in its own right, administered peripherally or centrally, is without effect to moderate neurotransmitter release, that endogenous 5-HT tone may greatly influence agonist or antagonist action, and that rapid 5-HT$_3$ receptor desensitisation may curtail drug action. Indeed, it is perhaps puzzling that any effects have been observed to 5-HT$_3$ receptor manipulation; given that the effect of agonist drug treatment observed over minutes or hours would be expected to desensitise the receptor within seconds.

With these constraints in mind, HAGAN et al. (1987) originally showed, using in vivo studies, that the increase in endogenous dopamine turnover in the rat mesolimbic nucleus accumbens caused by stimulation of the ventral tegmental area was attenuated by ondansetron. Stress or the administration of ethanol, nicotine or morphine can also enhance limbic dopamine release; 5-HT$_3$ receptor antagonists block such effects (IMPERATO and ANGELUCCI 1989; CARBONI et al. 1988; IMPERATO et al. 1990; WOZNIAK et al. 1990). 5-HT$_3$ agonists administered centrally or to slices of rat striatum increase dopamine release which is attenuated by 5-HT$_3$ receptor antagonists (BLANDINA et al. 1988; JIANG et al. 1990; CHEN et al. 1991). The ability of 5-HT$_3$ receptor antagonists to moderate dopamine release may relate to their attenuation of the behavioural consequences of a raised mesolimbic dopamine function (see Sect. G.II).

5-HT$_3$ receptor agonists are also reported to enhance 5-HT release in vitro using guinea pig frontal cortex slices (GALZIN et al. 1990) and cholecystokinin (CCK) in the rat brain (PAUDICE and RAITERI 1991) and to inhibit noradrenaline release in the rat hypothalamus (BLANDINA et al. 1991). The ability of 5-HT$_3$ receptor ligands to moderate Ach release is discussed in Sect. G.III.

It is clear that further in vitro and in vivo studies are required in different species and brain regions to definitively characterise the role of 5-HT$_3$ receptors in neurotransmitter release.

G. Behavioural Effects of 5-HT$_3$ Agonists and Antagonists

A decade of experimentation has shown unequivocally that 5-HT$_3$ receptor antagonists fail to cause any overt change in the behaviour of normal animals. The administration of millions of doses of ondansetron and other 5-HT$_3$ receptor antagonists has also been without behavioural effect in the cancer patient or postoperatively (Sect. H). Yet in animals, 5-HT$_3$ receptor antagonists have been shown to profoundly affect behaviourally induced changes that may be relevant to the treatment of psychiatric disorders. It is the unusual profile of a failure to modify normal behaviour yet to correct a disturbed behaviour that has created intense interest. This has focused on the reversal of behavioural suppression, a reduction of disturbed limbic dopamine activity and an enhancement of cognitive performance. It has been hypothesised that these effects are relevant to the treatment of anxiety, psychoses and drug dependence, and cognitive disorders, respectively.

I. Reversal of Behavioural Suppression

The importance of 5-HT has been a major theme in the understanding of the physiological mechanisms involved in behavioural suppression in animals and anxiety in man. It has also provided the basis for pharmacological manipulations designed to modify anxiety and anxiety-like responding. It is an area of increasing complexity and an extensive literature (see reviews by IVERSEN 1984; JOHNSTON and FILE 1986; CHOPIN and BRILEY 1987; MURPHY et al. 1993). Briefly, 5-HT₁, 5-HT₂, 5-HT₃ and 5-HT₄ receptors have all been implicated in anxiety related behaviours. Thus, 5-HT₁ₐ and 5-HT₄ receptor agonist action mediates behavioural disinhibition whilst agonist action at the 5-HT₂ and 5-HT₃ receptor predisposes to behavioural inhibition (BARRAT and GLEESON 1991; KENNETT et al. 1989; CHENG et al. 1994; COSTALL and NAYLOR 1991).

It is important to recognise that drug action at the 5-HT receptor subtypes to modify 5-HT neurotransmission will be critically dependent on basal, and possibly changing, endogenous 5-HT tone. This, in turn, may vary markedly between species and strains of animals and also the animal holding conditions. Simple handling of animals can markedly elevate 5-HT release and modify the responsiveness to subsequent challenge with anxiolytic-like agents (FILE et al. 1992; ANDREWS and FILE 1993); this is likely to vary between experimenters. A further major challenge to the use of ondansetron and the earlier 5-HT₃ receptor antagonists was the development of a bell-shaped dose response curve (JONES et al. 1988). The loss of efficacy with increasing dose has been confirmed in many studies and remains to be explained: it is not observed with second generation agents such as alosetron (GR 68755) and BRL 47460 (HAGAN et al. 1991; KENNETT and BLACKBURN 1990).

Increasingly, there is a tendency to use animal models that have a face validity rather than those that involve punishment. In contrast to the benzodiazepines, the 5-HT₃ receptor antagonists are clearly inactive in the "conflict" tests and in tests of defensive burying and rat pup ultrasonic vocalisation (see OLIVIER et al. 1992; JONES and PIPER 1994).

In the mouse, rat, gerbil, marmoset and cynomolgus monkey, using a battery of tests, the two compartment light/dark test, social interaction, elevated X-maze, ethological test procedures, passive avoidance paradigms, defeat analgesia and a startle paradigm, there is a consensus of data obtained from many laboratories that 5-HT₃ receptor antagonists have an anxiolytic profile of action (see reviews by COSTALL and NAYLOR 1991; JONES and PIPER 1994). In these many and varied species and tests the 5-HT₃ receptor antagonists were as efficacious as diazepam but lacked its sedative potential and the anxiogenic-like behaviours associated from withdrawal of a repeated treatment (OAKLEY et al. 1988). Furthermore, treatment with ondansetron can prevent the anxiogenic profile of withdrawal from drugs of abuse – diazepam, alcohol, nicotine and cocaine – in the rat and marmoset (COSTALL et al. 1990; ONAIVI et al. 1989; OAKLEY et al. 1988).

The location of the sites of action of the 5-HT$_3$ receptor antagonists has been investigated in the mouse and rat using the intracerebral injection technique. Ondansetron and other 5-HT$_3$ antagonists were effective on injection into the rat and mouse amygdala and into the mouse dorsal raphe nucleus to reduce behavioural responding to aversive situations in their own right as well as that induced following withdrawal from drugs of abuse (COSTALL et al. 1989, 1990; TOMKINS et al. 1990; HIGGINS et al. 1991).

This encouraging preclinical data has prompted trials of the 5-HT$_3$ receptor antagonists ondansetron, tropisetron and zacopride in patients suffering from anxiety. The largest study, a multicentre, general practice placebo controlled trial in 400 patients, reported that ondansetron had a statistically significant effect at the lowest dose of 1 mg three times daily, notwithstanding a high (40%–45%) placebo response (LADER 1991). LECRUBIER et al. (1991) conducted a double blind trial in 92 outpatients receiving placebo or 3 doses of tropisetron and measured the response after 1 and 3 weeks. In contrast to the ondansetron study, there was no bell-shaped dose response curve, the highest dose (25 mg/day) producing the greatest effect. ABUZZAHAB (1991) reported an ongoing small placebo controlled trial using high doses of ondansetron (1–16 mg three times daily) and analyses are still awaited. However, in a small placebo controlled double blind study zacopride could be shown by combination of data from group analyses, including a "high" dose of 0.1 mg four times a day, to have a significant effect (PECKNOLD 1990). The studies provide encouragement to the design of larger trials, although an anecdotal comment from SCHWEIZER and RICKELS (1991) was doubtful that data from other zacopride and ondansetron studies would prove positive. It is clear that further studies are required to establish the clinical usefulness of the 5-HT$_3$ receptor antagonist in anxiety, to better understand the predictive value of the animal tests, the role of 5-HT and 5-HT$_3$ receptors in anxiety, and the nature of anxiety in man that might prove sensitive to treatment with such agents.

II. 5-HT$_3$ Receptor Involvement in the Interaction Between Limbic Serotoninergic and Dopaminergic Systems

The earliest studies indicating a 5-HT$_3$ receptor involvement with limbic dopamine came from behavioural experiments. Ondansetron injected i.p., directly into the nucleus accumbens or the amygdala of the rat or marmoset, antagonised the hyperactivity induced by amphetamine, a combination of amphetamine/2-methyl-5-HT or the infusion of dopamine (COSTALL et al. 1987). Other 5-HT$_3$ receptor antagonists were also shown to block the "limbic" hyperactivity response (see review by COSTALL and NAYLOR 1994a). This profile mimicked the inhibitory effects of neuroleptic drugs. Yet this profile was different in a crucial way: the 5-HT$_3$ receptor antagonists were quite devoid of classical neuroleptic effects such as catalepsy, antagonism of stereotyped behaviour or inhibition of a conditioned avoidance response (PALFREYMAN et al. 1992). The data indicated that 5-HT$_3$ receptors may have a permissive role in the regulation of limbic dopamine function and this hypothesis has support from studies showing that zatosetron can decrease the number of spontaneously active cells in the ventral tegmental area, with some evidence of a selective effect (MINABE et al. 1991; RASMUSSEN et al. 1991; WANG et al. 1991). In addition, rewarding effects of nicotine and morphine in conditioned place preference conditioning are abolished by 5-HT$_3$ receptor antagonists, which also attenuate the increase in dopamine release caused by these agents or stress (CARBONI et al. 1988; HIGGINS et al. 1992; IMPERATO et al. 1990; IMPERATO and ANGELUCCI 1989). There is also neurochemical evidence that continuous cocaine administration in rats involves an alteration of 5-HT$_3$ receptor-mediated effects (KING et al. 1995). However, electrical brain self-stimulation or rewarding drug (cocaine, amphetamine, nicotine) effects thought to be mediated via dopamine release in the nucleus accumbens are reported to be resistant to blockade by 5-HT$_3$ receptor antagonists in the rodent (GREENSHAW 1992; DUNN et al. 1991; HERBERG et al. 1992; MONTGOMERY et al. 1993). Also, cocaine self-administration in the rat and the discriminative stimulus effects of cocaine are resistant to 5-HT$_3$ receptor antagonism (LANE et al. 1992).

The preclinical evidence that 5-HT$_3$ receptor antagonists selectively inhibit mesolimbic dopamine function prompted their use in schizophrenia, their important advantage being the absence of endocrine and severe extrapyramidal side effects routinely associated with traditional neuroleptic therapy. Early open trials indicated a beneficial effect (WHITE et al. 1991). Subsequently, two open, uncontrolled multicentre studies were undertaken in hospitalised patients with a DSM-III diagnosis of schizophrenia, using doses of ondansetron of either 4 or 8 mg twice daily in one study, whilst the other employed a low dose regimen with a maximum possible dose of 8 mg twice daily and a high dose regimen with a maximum possible dose of 16 mg twice daily. In these studies, ondansetron appeared to possess antipsychotic activity, but the drug's efficacy appeared to be inversely related to dose. For this reason, a larger study was undertaken at a dose of 4 mg given twice daily.

This was a double blind placebo controlled study in acute schizophrenia. Schizophrenic patients satisfying DSM-III criteria were randomly assigned to ondansetron at the 4-mg dose level, or to placebo, for a 4-week double blind trial. Chlorpromazine (CPZ) in multiples of doses of 50–100 mg or diazepam was permitted for additional clinical management, if needed. Analysis of the total Brief Psychiatric Rating Scale (BPRS), BPRS positive and negative symptoms, and the clinical global impression (CGI) did not show a significant drug treatment effect or drug versus CPZ interaction. There was also no difference in the proportions of the entire group of patients who responded to ondansetron or placebo. However, several measures suggested that the group of patients treated with ondansetron alone may have had a more favourable outcome than the other three groups, e.g. in final CGI scores and final BPRS positive and negative symptoms. Ondansetron failed to induce extrapyramidal or other side effects and was well tolerated. It was concluded that further study of the antipsychotic efficacy of 5-HT$_3$ receptor antagonists using more conventional research designs was indeed warranted (Meltzer 1991).

One single blind trial of zacopride indicated no clinical benefit in the treatment of schizophrenia (Newcomer et al. 1992). Further trials have been in progress since these initial observations but they remain unpublished. This is disappointing: it effectively precludes comment on the validity of the animal models and disadvantages drug research. It is undoubtedly an exceptionally difficult area of clinical research with major variables. The heterogeneity of the syndrome of schizophrenia and the questions this raises contribute to the lack of outcome: different symptoms ranging from the "positive", e.g. hallucination/delusions, to the "negative", e.g. affective flattening, loss of drive; change over time; the dosage of neuroleptic required, particularly in view of the preclinical data; assessment criteria and whether they are optimal to reveal drug effects on discrete disease components; whether drug action limited to a proportion of patients would have a commercial future.

With respect to the treatment of drug abuse, the situation is even more problematic. The closely defined experiments that can be achieved in animals are almost impossible to design in man. Thus multisubstance drug abuse is the rule rather than the exception and this may have occurred for years prior to a subject being entered for a clinical trial. This pattern of abuse of many substances may cause unknown short- and long-term neurochemical or pathological changes which have not been investigated in animal models. Also, although the craving for drugs can be mimicked in animals, the plethora of social and other cues related to drug abuse have received little attention in animal models.

Notwithstanding these many problems, in a placebo controlled study Sellers et al. (1994) examined the effects of ondansetron in the treatment of alcohol dependent men: a significant reduction in alcohol intake was evident. Johnson and colleagues (see Johnson and Cowen 1993) also report that, following ondansetron administration, there is a reduction in the pleasurable effect of a small dose of ethanol and a reduced desire to drink, although this could not be extended to the use of higher doses (Doty et al. 1994). Ondansetron also failed to reduce the craving and withdrawal symptoms in opiate addicts when exposed to a video containing drug related cues (Sell et al. 1995), or to reduce cigarette smoking (Zacny et al. 1993). The effectiveness of other 5-HT$_3$ receptors in the treatment of drug abuse is unknown.

III. 5-HT$_3$ Receptors and Cognition

The potential role of 5-HT$_3$ receptors in cognitive behaviour has provided one of the most exciting aspects of 5-HT$_3$ receptor research. It is also highly complex. The term "cognition" includes the processes of attention, the gating of sensory input, recognition and interpretation, consolidation and encoding of data, storage and retrieval. Emphasis on the cognitive defects of Alzheimer's disease has focused on memory impairments and its chronic and irretrievable neurodegenerative nature, with numerous neurochemical deficits. The treatment of cognitive disorders has been characterised by attempts to replace the declining function in cholinergic transmission, with limited success and many side effects (see HOLTUM and GERSHON 1992). Manipulation of the 5-HT system may provide an alternative approach.

Activation of the 5-HT cerebral systems impairs performance in various learning and memory tasks in animals (see review by COSTALL and NAYLOR 1994b). Therefore, it could be predicted that a reduced 5-HT function may enhance performance and it is of note that 5-HT₃ receptors are conserved in Alzheimer's disease (BARNES et al. 1990).

A 5-HT₃ receptor involvement in cognition was first revealed by behavioural studies: ondansetron prevented the impairment in performance of mice in a habituation test or rats in a T maze reinforced alternation task caused by scopolamine, old age or lesions of the nucleus basalis magnocellularis, and significantly decreased the number of trials to criteria in a marmoset object discrimination and reversal task (BARNES et al. 1990; DOMENEY et al. 1991; CAREY et al. 1992). In similar and other paradigms, a passive avoidance task, a continuous operant delayed non-matching to position task and the Morris water maze task, granisetron, tropisetron, zacopride and DAU 6215 in microgram per kilogram doses have been reported to improve rodent cognitive performance in aged or scopolamine treated animals, with lower doses generally being more effective than higher doses (CHUGH et al. 1991a,b; JAKAL et al. 1993; PITSIKAS et al. 1993, 1994). Accepting the importance of the hippocampus to learning and memory, and that LTP may provide a substrate for certain forms of memory, ondansetron in the rat was found to induce a reliable and dose-dependent increase in hippocampal theta rhythm, a significant increase in the magnitude and duration of LTP and an improved retention in an odor matching problem and in a spatial task (STAUBLI and XU 1995). The actions of the 5-HT₃ receptor antagonists may be task dependent since ondansetron failed to attenuate a scopolamine induced impairment in the stone maze (BRATT et al. 1994). Also, repeated treatments may be required, for example, DAU 6215 was only effective on chronic treatment (PITSIKAS et al. 1993).

A potential neurochemical basis for the actions of ondansetron and other 5-HT₃ receptor antagonists was revealed by their ability to prevent the effect of 2-methyl-5-HT to decrease Ach release in rat entorhinal cortex (BARNES et al. 1989), a finding extended to the rat hippocampus (Barnes, personal communication) and the guinea pig and human cortex (BIANCHI et al. 1990; MAURA et al. 1992). However, the situation is one of considerable complexity since 2-methyl-5-HT could also increase Ach release as a dominant effect mediated via a 5-HT₂ receptor in the entorhinal cortex (BARNES et al. 1989) and also the dorsal hippocampus (CONSOLO et al. 1994). The multiplicity of 5-HT receptors regulating Ach release indicates a potential plurality of effects and Johnson et al. (1993) have failed to record a 5-HT$_3$ receptor involvement in Ach release from slices of rat entorhinal cortex. Following the positive preclinical findings, the effect of ondansetron and alosetron have been investigated in man using healthy volunteers, in subjects with age-related memory impairment and in patients with Alzheimer's disease.

In treating healthy male subjects, in a randomised double blind, double dummy, four-way cross-over study, each subject received placebo, scopolamine (0.4mg i.m.), scopolamine + alosetron (10mg i.v.) or scopolamine (250mg i.v.) (Preston et al. 1991; Preston 1994). Assessments of verbal and spatial memory, sedation and sustained attention were performed prior to and post treatment. The main results from the study were that scopolamine induced robust deficits on all primary variables measured, the reduction in verbal and spatial memories being attenuated by 10-μg and 250-μg doses of alosetron respectively, with no effect on the sedation or on changes in attention.

In a study of age-associated memory impairment, in a double blind placebo controlled trial using three doses of ondansetron (10, 250 or 1000μg p.o. b.d.), patients were treated for 12 weeks, followed by a 2-week washout period, with assessments being made at the initiation, during and at the termination of treatment. Behavioural rating scales and a computerised battery of tests related to learning and memory tasks of daily life were used to assess changes in cognitive performance. Ondansetron caused dose related effects to enhance name-face association, acquisition, name-face association – delayed recall and facial recognition – number recognised before first error, with the intermediate dose being more efficacious than the highest dose. It was concluded that ondansetron merited further study (Crook and Lakin 1991).

In a study designed to measure the modulation of anticholinergic effects on cognition and behaviour in elderly humans, ten elderly normal subjects received, by infusion, placebo, ondansetron (0.15mg/kg i.v.), scopolamine (0.4mg i.v.), scopolamine plus ondansetron, and scopolamine plus m-chlorophenylpiperazine (0.08mg/kg i.v.), the five study days being separated by at least 72h (Little et al. 1995). Cognitive measures examined episodic and semantic memory, lexical search and retrieval, processing speed and behavioural ratings were undertaken. Scopolamine caused impairment in a host of tests and was exacerbated by m-chlorophenylpiperazine in areas of letter fluency and serial visual search. Ondansetron at the high dose failed to modify the scopolamine induced impairments on cognitive, physiological or behavioural measures. The authors, using a single high dose of ondansetron and a single pretreatment time, emphasised that careful dose-response studies may be required before ruling out possible cholinergic potentiating effects of ondansetron in humans.

Overall, these preliminary studies, whilst emphasising the difficulties of dosage selection, offered encouragement to the design of more extensive trials in dementia of the Alzheimer's type. However, a large programme of trials has apparently failed to demonstrate a convincing treatment effect on the core cognitive symptoms of dementia or the patient's clinical status (Corn, personnel communication). It is hoped that such trials will be published to allow a better understanding of the effect of 5-HT$_3$ receptor antagonists in dementia and to better understand the validity of animal models of cognition. Possibly in cognition, as with anxiety, there are subsets of responders in the patient populations examined.

H. Role of 5-HT$_3$ Receptors in Nociception and Migraine

There is an extensive literature to support the concept that endogenous 5-HT, acting through a number of 5-HT receptor subtypes at a multiplicity of sites (primary nociceptive afferent nerve terminals in peripheral tissues, and centrally in the dorsal horn, spinothalamic tract, brain stem nuclei and thalamus) can moderate the perception and processing of pain (see reviews by ROBERTS 1984; RICHARDSON 1992; FOZARD 1994).

In the periphery, tissue injury causing the release of 5-HT produces pain per se and potentiates the algetic effects of other substances such as bradykinin (see RICHARDSON 1992). In animal models there is convincing evidence that 5-HT$_3$ receptors located on peripheral sensory neurons are involved in inflammatory (but not mechanical or thermal) pain, several 5-HT$_3$ receptor antagonists producing an inhibition of nociceptive effects (GIORDANO and DYCHE 1989; GIORDANO and ROGERS 1989; ESCHALIER et al. 1989; FOZARD 1994). However, 5-HT$_3$ receptor antagonism in the central nervous system may complicate the peripherally mediated effects. Thus, 5-HT has antinociceptive effects via influence on the CNS, the analgesic effects of 5-HT being mimicked by 2-methyl 5-HT, and the effects of both agonists being blocked by tropisetron and MDL 72222 (GLAUM et al. 1988, 1990; ALHAIDER et al. 1991; GIORDANO 1991). Drug action at the spinal level may be an important locus of such action, where 5-HT$_3$ receptor interactions may be mediating their effects via a GABAergic interneuron (ALHAIDER et al. 1991; GIORDANO 1991). Such actions may compromise the peripherally mediated analgesic effects of the 5-HT$_3$ receptor antagonists. However, FOZARD (1994) has emphasised that the centrally mediated effects are exclusively the result of exogenously applied pharmacological manipulations; the physiological relevance of 5-HT remains to be established.

In man, use of the blister base technique has established that 5-HT induces pain and sensitises the nociceptive neuron to other algesic agents such as bradykinin, that 2-methyl-5-HT mimics the effects of 5-HT and that tropisetron blocks the algesic effects (KEELE and ARMSTRONG 1964; RICHARDSON et al. 1985). A flare response was also recorded by RICHARDSON et al. (1985), which could be reduced by treatment with tropisetron or MDL 72222 (ORWIN and FOZARD 1986) and subsequently many other 5-HT$_3$ receptor antagonists (see FOZARD 1994).

MDL 72222 was originally developed as an antimigraine drug, on the hypothesis that 5-HT released within the cranial vasculature would activate 5-HT_3 receptors on the perivascular sensory neurons causing pain, and sensitise to other nociceptive stimuli (FOZARD 1989; FOZARD et al. 1985). Initial clinical trials confirmed the hypothesis in a double blind placebo and an open study, MDL 72222 and granisetron proving effective to treat the acute attack (LOISY et al. 1985; COUTURIER et al. 1991). In a second double blind placebo controlled trial granisetron appeared to offer some relief from the symptoms, but the differences were not statistically significant (PEWAT et al. 1991). Given prophylactically, tropisetron failed to significantly reduce attack frequency and the best results were obtained with the use of the lowest dose (FERRARI et al. 1991). Clearly, further well designed studies are required to establish the value of 5-HT_3 receptor antagonists in a therapy for migraine, either as a treatment in its own right or as an adjunct to existing treatment.

With the exception of migraine, the only other use of the 5-HT_3 receptor antagonists in the treatment of inflammatory pain has been in irritable bowel syndrome, a topic beyond the scope of this chapter, but which is reported elsewhere (see SANGER et al. 1994).

I. Role of 5-HT_3 Receptors in Emesis

The nausea and vomiting associated with chemotherapy and radiation in the cancer patient seriously reduces the quality of life and may be so severe as to cause the patient to discontinue therapy (LAZLO and LUCAS 1981; HOAGLAND et al. 1983). MINER and SANGER (1986) and COSTALL et al. (1986) were the first to show that the 5-HT_3 receptor antagonists can inhibit emesis, reporting that MDL 72222 and tropisetron could antagonise cisplatin-induced emesis in the ferret. The findings were rapidly confirmed by other groups and in other species, shown to occur in man (LEIBUNDGUT and LANCRANJAN 1987) and extended to many other 5-HT_3 receptor antagonists and different chemotherapeutic regimens as well as to radiation. These findings caused a renaissance of emesis research and revolutionised the treatment of emesis in the cancer patient (the topic has been reviewed in detail – ANDREWS 1994; NAYLOR and RUDD 1994; BUTCHER 1993).

There is a consensus that the 5-HT_3 receptor antagonists block emesis by 5-HT_3 receptor antagonism at central sites, i.e. in the nucleus tractus solitarius, area postrema and dorsal motor vagal nucleus, and in the periphery at the 5-HT_3 receptors on the afferent vagus nerve terminals. It is hypothesised that chemotherapy and radiation disrupt serotoninergic function in the gut, causing a release of 5-HT from enterochromaffin cells to stimulate 5-HT_3 receptors on the vagus nerve (CUBEDDU et al. 1992; ANDREWS 1994). Certainly, 5-HT_3 agonists phenylbiguanide, m-chlorophenylbiguanide and 2-methyl-5-HT have been shown to induce emesis in animals, which is blocked by 5-HT_3 receptor antagonists in many species (ANDREWS 1994). Yet it remains curious that it has proven difficult to induce emesis using 5-HT itself, that cyclophosphamide is a potent emetogen yet fails to induce changes in 5-HT release, and that the carcinoid syndrome rarely encompasses emesis (see ANDREWS 1994).

Whilst the value of the 5-HT$_3$ receptor antagonists as antiemetics is undisputed, an extensive clinical usage has revealed that "delayed" emesis, i.e. emesis occurring after the 1st day of chemotherapy, is less well controlled, and that dexamethasone will enhance the antiemetic efficacy of 5-HT$_3$ receptor antagonists (CUNNINGHAM et al. 1989). The nature of the synergy between 5-HT receptor antagonists and dexamethasone is now being further investigated in animal models of acute and delayed emesis (RUDD et al. 1994; Rudd, unpublished data). Also worthy of further study is the action of granisetron, and perhaps other 5-HT$_3$ receptor antagonists, to alleviate the emesis of migraine (ROWAT et al. 1991), and the use of ondansetron to attenuate emesis associated with symptoms of abdominal pain and gastrointestinal disorder (EVANS 1993). Finally, it has been demonstrated that ondansetron is clinically the preferred treatment to reduce postoperative nausea and vomiting, with a notable absence of significant side effects (LARIJANI et al. 1991; DERSHWITZ et al. 1992; MCKENZIE et al. 1993). This may constitute a future major use, the ability of ondansetron to reduce nausea and vomiting induced by morphine (KOCH and BINGAMAN 1993) probably contributing to its success. The successful use of ondansetron in postoperative nausea and vomiting indicates an ability to reduce emesis in the presence of a multifactorial aetiology and there is preliminary evidence that ondansetron can relieve nausea and vomiting induced by baclofen (BROGGI et al. 1995), antibiotics (GOMPELS et al. 1993), the serotonin reuptake inhibitors (BAILEY et al. 1994), hyperemesis gravidarum (SULLIVAN et al. 1996; GUIKONTES et al. 1993), uraemia (ANDREWS et al. 1995) and neurological trauma (KLEINERMAN et al. 1993).

References

Abuzzahab FS (1991) Ondansetron: a novel anti-anxiety agent. In: Briley M, File SE (eds) New concepts in anxiety. MacMillan, Basingstoke, pp 185–189

Alhaider AA, Lei SZ, Wilcox GL (1991) Spinal 5-HT$_3$ receptor-mediated antinociception: possible release of GABA. J Neurosci 11:1881–1887

Andrews PLR (1994) 5-HT$_3$ receptor antagonists and antiemesis. In: King FD, Jones BJ, Sanger GJ (eds) 5-Hydroxytryptamine-3 receptor antagonists. CRC, Boca Raton, pp 255–317

Andrews N, File SE (1993) Handling history of rats modifies behavioural effects of drugs in the elevated plus-maze test of anxiety. Eur J Pharmacol 235:109–112

Andrews PA, Quan V, Ogg CS (1995) Ondansetron for symptomatic relief in terminal uraemia. Nephrol Dial Transplant 10: 140

Ashby CR, Minabe Y, Edwards E, Wang RY (1991) 5-HT$_3$-like receptors in the rat medial prefrontal cortex: an electrophysiological study. Brain Res 550:181–191

Ashby CR, Edwards E, Wang RY (1992) Action of serotonin in the medial prefrontal cortex: mediation by serotonin$_3$-like receptors. Synapse 10:7–15

Azami J, Fozard JR, Round AA, Wallis DI (1985) The depolarising action of 5-hydroxytryptamine on rabbit vagal primary afferent and sympathetic neurones and its selective blockade by MDL 72222. Naunyn Schmiedebergs Arch Pharmacol 328:762–763

Bailey J, Potokar J, Nutt D (1994) Can the GI disturbance produced by the SSRIs be attenuated by a 5-HT$_3$ antagonist? Neuropsychopharmacology 10:220

Barnes JM, Barnes NM, Costall B, Naylor RJ, Tyers MB (1989) 5-HT$_3$ receptors mediate inhibition of acetylcholine release in cortical tissue. Nature 338:762–763

Barnes JM, Costall B, Coughlan J, Domeney AM, Gerrard PA, Kelly ME, Naylor RJ, Onaivi ES, Tomkins DM, Tyers MB (1990a) The effects of ondansetron, a 5-HT$_3$ receptor antagonist, on cognition in rodents and primates. Pharmacol Biochem Behav 35:955–962

Barnes NM, Costall B, Naylor RJ (1988) [^{3}H]Zacopride: ligand for the identification of 5-HT$_3$ recognition sites. J Pharm Pharmacol 40:548–551

Barnes NM, Costall B, Naylor RJ, Williams TJ, Wischik CM (1990b) Normal densities of 5-HT$_3$ receptor recognition sites in Alzheimer's disease. Neuroreport 1:253–254

Barrat JE, Gleeson S (1991) Anxiolytic effects of 5-HT$_{1A}$ agonists, 5-HT$_3$ antagonists and benzodiazepines: conflict and drug discrimination studies. In: Rogers RJ, Cooper SJ (eds) 5-HT$_{1A}$ agonists, 5-HT$_3$ antagonists and benzodiazepines, their comparative behavioural pharmacology. Wiley, Chichester, pp 59–105

Belelli D, Balcarek JM, Hope AG, Peters JA, Lambert JT, Blackburn TP (1995a) Cloning and functional expression of a human 5-hydroxytryptamine type 3A$_S$ receptor (5-HT$_3$R-A$_S$) subunit. Mol Pharmacol 48:1054–1062

Belelli D, Hope AG, Peters JA, Lambert JL, Blackburn TP, Balcarek JM (1995b) Functional properties of a human recombinant 5-HT$_3$ receptor expressed in Xenopus laevis oocytes. Br J Pharmacol 116:229P

Bianchi C, Siniscalchi A, Beani L (1990) 5-HT$_{1A}$ agonists increase and 5-HT$_3$ agonists decrease acetylcholine efflux from the cerebral cortex of freely-moving guinea pigs. Br J Pharmacol 101:448–452

Blandina P, Goldfarb J, Green JP (1988) Activation of 5-HT$_3$ receptor releases dopamine from rat striatal slice. Eur J Pharmacol 155:349–350

Blandina P, Goldfarb J, Walcott J, Green JP (1991) Serotonergic modulation of the release of endogenous norepinephrine from rat hypothalamic slices. J Pharmacol Exp Ther 256:341–347

Boess FG, Martin IL (1994) Molecular biology of 5-HT receptors. Neuropharmacology 33:275–317

Boess FG, Lummus SCR, Martin IL (1992) Molecular properties of 5-hydroxytryptamine$_3$ receptor type binding sites purified from NG108-15 cells. J Neurochem 59:1692–1701

Bolanos FJ, Schechter LE, Miguel MC, Emerit MB, Rumigny JF, Hamon M, Gozlan H (1990) Common pharmacological and physiological properties of 5-HT$_3$ binding sites in the rat cerebral cortex and NG108-15 clonal cells. Biochem Pharmacol 40:1541–1550

Bradley PB, Engel G, Feniuk W, Fozard JR, Humphrey PPA, Middlemiss DN, Mylecharane EJ, Richardson BP, Saxena PR (1986) Proposals for the classification and nomenclature of functional receptors for 5-hydroxytryptamine. Neuro-pharmacology 25:563–576

Bratt AM, Kelly ME, Domeney AM, Naylor RJ, Costall B (1994) Failure of ondansetron to attenuate a scopolamine-induced deficit in a stone maze task. Neuroreport 5:1921–1924

Brett RR, Prat JA (1990) Chronic handling modifies the anxiolytic effect of diazepam in the elevated plus-maze. Eur J Pharmacol 178:135–138

Broggi U, Dones I, Servello D, Ferrazza C (1996) A possible pharmacological treatment of baclofen overdosage. Italian J Neurol Sci 17:179–180

Butcher ME (1993) Global experience with ondansetron and future potential. Oncology 50:191–197

Butler A, Hill JM, Ireland SJ, Jordan CC, Tyers MB (1988) Pharmacological properties of GR38032F, a novel antagonist at 5-HT$_3$ receptors. Br J Pharmacol 94:397–412

Butler A, Elswood CJ, Burridge J, Ireland SJ, Bunce KT, Kilpatrick GJ, Tyers MB (1990) The pharmacological characterisation of 5-HT$_3$ receptors in three isolated preparations derived from guinea-pig tissues. Br J Pharmacol 101:591–598

Carboni E, Acquas E, Leone P, Perrezzani L, Di Chiara G (1988) 5-HT$_3$ receptor antagonists block morphine and nicotine induced place preference conditioning. Eur J Pharmacol 151:1590–160

Carey GJ, Costall B, Domeney AM, Gerrard PA, Jones DNC, Naylor RJ, Tyers MB (1992) Ondansetron and arecoline prevent scopolamine-induced cognitive deficits in the marmoset. Pharmacol Biochem Behav 42:75–83

Champaneria S, Costall B, Naylor RJ, Robertson DW (1992) Identification and distribution of 5-HT$_3$ recognition sites in the rat gastrointestinal tract. Br J Pharmacol 106:693–696

Chen J, Van Praag HM, Gardner EL (1991) Activation of 5-HT$_3$ receptor by 1-phenylbiguanide increases dopamine release in the rat nucleus accumbens. Brain Res 543:354–357

Cheng CHK, Costall B, Kelly ME, Naylor RJ (1994) Actions of 5-hydroxytryptophan to inhibit and disinhibit mouse behaviour in the light dark test. Eur J Pharmacol 255:39–49

Chopin P, Briley M (1987) Animal models of anxiety: the effect of compounds that modify 5-HT neurotransmission. Trends Pharmacol Sci 8:383–388

Chugh Y, Saha N, Sankaranarayanan A, Sharma PL (1991a) Memory enhancing effects of granisetron (BRL 43694) in a passive avoidance task. Eur J Pharmacol 203:121–123

Chugh Y, Saha N, Sankaranarayanan A, Datta H (1991b) Enhancement of memory retrieval and attenuation of scopolamine induced amnesia following administration of 5-HT$_3$ antagonist ICS205-930. Pharmacol Toxicol 69:105–106

Consolo S, Bertorelli R, Russi G, Zambelli M, Ladinsky H (1994) Serotonergic facilitation of acetylcholine release in vivo from rat dorsal hippocampus via serotonin 5-HT$_3$ receptors. J Neurochem 62:2254–2261

Costall B, Naylor RJ (1991) Anxiolytic effects of 5-HT$_3$ antagonists in animals. In: Rodgers RJ, Cooper SJ (eds) 5-HT$_{1A}$ agonists, 5-HT$_3$ antagonists and benzodiazepines, their comparative behavioural pharmacology. Wiley, Chichester, pp 133–157

Costall B, Naylor RJ (1994a) 5-HT$_3$ receptor antagonism and schizophrenia. In: King FD, Jones BJ, Sanger GJ (eds) 5-Hydroxytryptamine-3 receptor antagonists. CRC, Boca Raton, pp 183–219

Costall B, Naylor RJ (1994b) 5-HT$_3$ receptor antagonists in the treatment of cognitive disorders. In: King FD, Jones BJ, Sanger GJ (eds) 5-Hydroxytryptamine-3 receptor antagonists. CRC, Boca Raton, pp 203–219

Costall B, Domeney AM, Naylor RJ (1986) 5-Hydroxytryptamine M-receptor antagonism to prevent cisplatin-induced emesis. Neuropharmacology 25:959–961

Costall B, Domeney AM, Naylor RJ, Tyers MB (1987) Effects of the 5-HT$_3$ receptor antagonist GR38032F, on raised dopaminergic activity in the mesolimbic system of the rat and marmoset brain. Br J Pharmacol 92:881–894

Costall B, Kelly ME, Naylor RJ, Onaivi ES, Tyers MB (1989) Neuroanatomical sites of action of 5-HT$_3$ receptor agonist and antagonists for alteration of aversive behaviour in the mouse. Br J Pharmacol 96:325–332

Costall B, Naylor RJ, Tyers MB (1990) The psychopharmacology of 5-HT$_3$ receptors. Pharmacol Ther 47:181–202

Couturier EGM, Hering R, Foster CA, Steiner TJ, Rose FC (1991) First clinical study of the selective 5-HT$_3$ antagonist, granisetron (BRL43694), in the acute treatment of migraine headache. Headache 31:296–297

Crook TH, Lakin M (1991) Effects of ondansetron image-associated memory impairment. Proceedings of the 5th world congress of biological psychiatry, Florence

Cubeddu LX, Hoffman IS, Fuenmayor NT, Malave JJ (1992) Changes in serotonin metabolism in cancer patients: its relationship to nausea and vomiting induced by chemotherapeutic agents. Br J Cancer 66:198–203

Cunningham D, Turner A, Hawthorn J, Rosin RD (1989) Ondansetron with and without dexamethasone to treate chemotherapy. Lancet 1:1323

Dershwitz M, Rosow S-E, Dibiase PM, Joslyn AF, Sanderson PE (1992) Ondansetron is effective in decreasing postoperative nausea and vomiting. Clin Pharmacol Ther 52:96–101

Domeney AM, Costall B, Gerrard PA, Jones DN, Naylor RJ, Tyers MD (1991) The effect of ondansetron on cognitive performance in the marmoset. Pharmacol Biochem Behav 38:169–175

Donatsch P, Engel G, Richardson BP, Stadler P (1984) ICS205-930:a highly selective and potent antagonist at peripheral neuronal 5-hydroxytryptamine (5-HT) receptors. Br J Pharmacol 81:34P

Doty P, Zacny JP, de Wit H (1994) Effects of ondansetron pretreatment on acute responses to ethanol in social drinkers. Behav Pharmacol 5:461–469

Downie DL, Hope AG, Lambert JJ, Peters JA, Blackburn TP, Jones BJ (1994) Pharmacological characterisation of the apparent splice variants of the murine 5-HT$_3$ R-A subunit expressed in Xenopus laevis oocytes. Neuropharmacology 33:473–482

Dunn RW, Carlezon WA, Corbett R (1991) Preclinical anxiolytic versus antipsychotic profiles of the 5-HT$_3$ antagonists ondansetron, zacopride, 3-tropanyl-1H-indole-3-carboxylic acid ester, and 1-H, 3, 5H-tropan-3-yl-3,5-dichlorobenzoate. Drug Dev Res 23:289–300

Edwards E, Harkins KL, Ashby CR, Wang RY (1991) The effects of 5-HT$_3$ receptor agonists on phosphoinositide hydrolysis in the rat frontocingulate and entorhinal cortices. J Pharmacol Exp Ther 256:1025–1032

Elliott P, Wallis DI (1988) The depolarising action of 5-hydroxtryptamine on rabbit isolated preganglionic cervical sympathetic nerves. Naunyn Schmiedebergs Arch Pharmacol 338:608–615

Elliott P, Seemungal BM, Wallis DI (1990) Antagonism of the effects of 5-hydroxytryptamine on the rabbit isolated vagus nerve by BRL43694 and metoclopramide. Naunyn Schmiedbergs Arch Pharmacol 341:503–509

Eschalier A, Kayser V, Guilband G (1989) Influence of a specific 5-HT$_3$ antagonist on carrageenan-induced hyperalgesia in rats. Pain 36:249–255

Evans JE (1993) Nausea, abdominal pain and diarrhoea of uncertain cause responding to ondansetron. Med J Aust 159:125–127

Ferrari MD, Wilkinson M, Hirt D, Lataste X, Notter M (1991) Efficacy of ICS205-930 a novel 5-hydroxytryptamine-3 5-HT$_3$ receptor antagonist in the prevention of migraine attacks: a complex answer to a simple question. Pain 45:283–291

File SE, Andrews N, Wu PY, Zharkovsky A, Zangrossi H (1992) Modification of chlordiazepoxide's behavioural and neurochemical effects by handling and plus-maze experience. Eur J Pharmacol 218:9–14

Fortune DH, Ireland SJ, Tyers MB (1983) Phenylbiguanide mimics the effects of 5-hydroxytryptamine on the rat isolated vagus nerve and superior cervical ganglion. Br J Pharmacol 79:298P

Fozard JR (1983) Differences between receptors for 5-hydroxytryptamine on autonomic neurones revealed by nor-(−)cocaine. J Auton Pharmacol 3:21–216

Fozard JR (1984a) MDL72222: a potent and highly selective antagonist at neuronal 5-hydroxytryptamine receptors. Naunyn Schmiedebergs Arch Pharmacol 326:36–44

Fozard JR (1984b) Neuronal 5-HT receptors in the periphery. Neuropharmacology 23:1473–1486

Fozard JR (1984c) 5-Hydroxytryptamine in the pathophysiology of migraine. In: Bevan JA, Godfraind T, Maxwell RA, Stoclet JC, Worcel M (eds) Vascular neuroeffector mechanisms. Elsevier, Amsterdam, pp 321–329

Fozard JR (1989) The development and early clinical evaluation of selective 5-HT$_3$ receptor antagonists. In: Fozard JR (ed) The peripheral actions of 5-hydroxytryptamine. Oxford University Press, New York, pp 354–376

Fozard JR (1994) Role of 5-HT$_3$ receptors in nociception. In: King FD, Jones BJ, Sanger GJ (eds) 5-Hydroxytryptamine$_3$ receptor antagonists. CRC, Boca Raton, pp 241–253

Fozard JR, Gittos MW (1983) Selective blockade of 5-hydroxytryptamine neuronal receptors by benzoic acid esters of tropine. Br J Pharmacol 80:511P

Fozard JR, Loisy C, Tell GP (1985) Blockade of neuronal 5-hydroxytryptamine receptors with MDL72222: a novel approach to the symptomatic treatment of migraine. In: Rose FC (ed) Migraine clinical and research advances. Skarger, Basle, pp 264–272

Fozard JR, Mobarok Ali ATM (1978) Blockade of neuronal tryptamine receptors by metoclopramide. Eur J Pharmacol 49:109–112

Gaddum JHR, Picarelli ZP (1957) Two kinds of tryptamine receptor. Br J Pharmacol 12:323–328

Galzin AM, Poncet V, Langer SZ (1990) 5-HT$_3$ receptor agonists enhance the electrically evoked release of [^{3}H]5-HT in guinea-pig frontal cortex slices. Br J Pharmacol 100:307P

Gehlert DR, Gackenheimer SL, Wong DT, Robertson DW (1991) Localisation of 5-HT$_3$ receptors in the rat brain using [^{3}H]LY 278584. Brain Res 553:149–154

Gill CH, Peters JA, Lambert JJ (1995) An electrophysiological investigation of the properties of a murine recombinant 5-HT$_3$ receptor stably expressed in HEK293 cells. Br J Pharmacol 114:1211–1221

Giordano J (1991) Analgesic profile of centrally administered 2-methylserotonin against acute pain in rats. Eur J Pharmacol 199:233–236

Giordani J, Dyche J (1989) Differential analgesic actions of serotonin 5-HT$_3$ receptor antagonists in the mouse. Neuropharmacology 28:423–427

Giordani J, Rogers L (1989) Peripherally administered serotonin 5-HT$_3$ receptor antagonists reduce inflammatory pain. Eur J Pharmacol 170:83–86

Glaum SR, Proudfit HK, Anderson EG (1988) Reversal of the antinociceptive effects of intrathecally administered serotonin in the rat by a selective 5-HT$_3$ receptor antagonist. Neurosci Lett 95:313–317

Glaum SR, Proudfit HK, Anderson EG (1990) 5-HT$_3$ receptors modulate spinal nociceptive reflexes. Brain Res 510:12–18

Gompels M, McWilliams S, O'Hare M, Harris JR, Pinching FJ, Main J (1993) Ondansetron usage in HIV positive patients – a pilot study on the control of nausea and vomiting in paitents on high dose co-trimoxazole for Pneumocytis carinii pneumonia. Int J Std Aids 4: 293–296

Greenshaw AJ (1992) Differential effects of antipsychotics and the 5-HT$_3$ antagonist ondansetron on electrical self-stimulation of the VTA in rats. Clin Neuropsychopharmacol 15:88–92

Guikontes E, Spantideas A, Diakakis J (1992) Ondansetron and hyperemesis gravidarum. Lancet 340:1223

Hagan RM, Butler A, Hill JM, Jordan CC, Ireland SJ, Tyers MB (1987) Effect of the 5-HT$_3$ receptor antagonist GR38032F on responses to injection of a neurokinin agonist into the ventral tegmental area of the rat brain. Eur J Pharmacol 138:303–305

Hagan RM, Oakley NR, Burridge J, Kilpatrick GJ, Tyers MB (1991) Effects of the 5-HT$_3$ receptor antagonist, GR68755, in models of anxiety and raised mesolimbic dopaminergic activity in the rat. Proceedings of the international meeting on serotonin, Birmingham, p 115

Herberg LJ, De Belleroche JS, Rose IC, Montgomer AMJ (1992) Effect of the 5-HT$_3$ receptor antagonist ondansetron on hypothalamic self-stimulation in rats and its interaction with the CCK analogue caerulin. Neurosci Lett 140:16–18

Hertling-Jaweed S, Bandini G, Hucho F (1990) Purification of nicotinic receptors. In: Hulme EC (ed) Receptor biochemistry, a practical approach. IRL Press, London, pp 163–176

Higashi H, Nishi S (1982) 5-Hydroxytryptamine receptors on visceral primary afferent neurones of rabbit nodose ganglia. J Physiol (Lond) 323:543–567

Higgins GA, Jones BJ, Oakley NR, Tyers MB (1991) Evidence that the amygdala is involved in the disinhibitory effects of 5-HT$_3$ receptor antagonists. Psychopharmacology 104:545–551

Higgins GA, Joharchi N, Nguyen P, Sellers EM (1992) Effect of the 5-HT₃ receptor antagonists, MDL72222 and ondansetron on morphine place conditioning. Psychopharmacology 106:315–320

Hoagland AC, Bennet JM, Morrow GR, Carnrike CLM (1983) Oncologist's view of cancer patients' non-compliance. Am J Oncol 6: 239–244

Holtum JR, Gershon S (1992) The cholinergic model of dementia, Alzheimer type: progression from the unitary transmitter concept. Dementia 3:174–185

Hope AG, Downie DL, Sutherland L, Lambert JL, Peters JA, Burchell B (1993) Cloning and functional expression of an apparent splice variant of the murine 5-HT₃ receptor A subunit. Eur J Pharmacol 245:187–192

Hope AG, Brown AM, Peters JA, Lambert JJ, Balcarek JM, Blackburn TP (1995) Characterisation of a cloned human 5-HT₃ receptor subunit stably expressed in HEK293 cells. Br J Pharmacol 116:82P

Hoyer D, Neijt HC (1988a) Identification of serotonin 5-HT₃ recognition sites by radioligand binding in NG108-15 neuroblastoma-glioma cells. Eur J Pharmacol 143:291–292

Hoyer D, Neijt HC (1988b) Identification of serotonin 5-HT₃ recognition sites in membranes of N1E-115 neuroblastoma cells by radioligand binding. Mol Pharmacol 33:303–309

Hoyer D, Waeber C, Karpf A, Neijt H, Palacios JM (1989) [³H]CS205-930 labels 5-HT₃ recognition sites in membranes of cat and rabbit vagus nerve and superior cervical ganglion. Naunyn Schmiedebergs Arch Pharmacol 340:396–402

Imoto K, Busch C, Sakmann B, Mishina M, Konno T, Nakai J, Bujo H, Mori Y, Fukuda K, Numa S (1988) Rings of negatively charged amino acids determine the acetylcholine receptor channel conductance. Nature 335:645–648

Imperato A, Angelucci L (1989) 5-HT₃ receptor controls dopamine release in nucleus accumbens of freely moving rats. Neurosci Lett 101:214–217

Imperato A, Puglisi-Allergra S, Zocchi A, Scrocco MG, Casolini P, Angelucci L (1990) Stress activation of limbic and cortical dopamine release is prevented by ICS205-930 but not diazepam. Eur J Pharmacol 175:211–214

Ireland SJ, Tyers MB (1987) Pharmacological characterisation of 5-hydroxytryptamine-induced depolarisations of the rat isolated vagus nerve. Br J Pharmacol 90:229–238

Ireland SJ, Straughan DW, Tyers MB (1987) Influence of 5-hydroxytryptamine uptake on the apparent 5-hydroxytryptamine antagonist potency of metoclopramide in the rat isolated superior cervical ganglion. Br J Pharmacol 90:151–160

Iversen SD (1984) 5-HT and anxiety. Neuropharmacology 23:1553–1560

Jiang LH, Ashby CR, Kasser RJ, Wang RY (1990) The effect of intraventricular administration of the 5-HT₃ receptor agonist 2-methylserotonin on the release of dopamine in the nucleus accumbens: an in vivo chronocoulometric study. Brain Res 513:156–160

Jakala P, Sirvio J, Riekkinen PJ (1993) The effects of tacrine and zacopride on the performance of adult rats in the working memory task. Gen Pharmacol 24:675–679

Johnson BA, Cowen PJ (1993) Alcohol-induced reinforcement: dopamine and 5-HT₃ receptor interactions in animals and humans. Drug Dev Res 30:153–169

Johnson DS, Heinemann SF (1992) Cloning and expression of the rat 5-HT₃ receptor reveals species-specific sensitivity to curare antagonism. Soc Neurosci Abstr 18:249

Johnson DS, Heinemann SF (1993) Rat 5-HT₃ receptor expression in the central and peripheral nervous system – an in situ hybridization study. Soc Neurosci Abstr 19:632

Johnson RM, Inouye GT, Eglen RM, Wong EH (1993) 5-HT₃ receptor ligands lack modulatory influence on acetylcholine release in rat entorhinal cortex. Naunyn Schmiedebergs Arch Pharmacol 347:241–247

Johnston AL, File SE (1984) 5-HT and anxiety: promises and pitfalls. Pharmacol Biochem Behav 24:1467–1470

Jones BJ, Piper DC (1994) 5-HT₃ receptor antagonists in anxiety. In: King FD, Jones BJ, Sanger GJ (eds) 5-Hydroxytryptamine-3 receptor antagonists. CRC, Boca Raton, pp 155–181

Jones BJ, Costall B, Domeney AM, Kelly ME, Naylor RJ, Oakley NR, Tyers MB (1988) The potential anxiolytic activity of GR38032F, a 5-HT₃ receptor antagonist. Br J Pharmacol 93:985–993

Keele CA, Armstrong D (1964) Substances producing pain and itch. Williams and Wilkins, Baltimore, pp 152–167

Kennett GA, Blackburn TP (1990) Anxiolytic-like actions of BRL 46470, a novel 5-HT₃ antagonist. Proceedings of the British Association for Psychopharmacology, annual meeting, Cambridge, England, 17th July, abstract 117

Kennett GA, Whitton P, Shah K, Curzon G (1989) Anxiogenic-like effects of mCPP and TFMPP in animal models are opposed by 5-HT₁C receptor antagonists. Eur J Pharmacol 164:445–454

Kilpatrick GJ, Tyers MB (1992) Interspecies variants of the 5-HT₃ receptor. Biochem Soc Trans 20:118–121

Kilpatrick GJ, Jones BJ, Tyers MB (1987) Identification and distribution of 5-HT₃ receptors in rat brain using radioligand binding. Nature 330:746–748

Kilpatrick GJ, Jones BJ, Tyers MB (1988) The distribution of specific binding of the 5-HT₃ receptor ligand [³H]GR65650 in rat brain using quantitative autoradiography. Neurosci Lett 94:156–160

Kilpatrick GJ, Butler A, Burridge J, Oxford AW (1990a) 1-(m-Chlorophenyl)-biguanide, a potent high affinity 5-HT₃ receptor agonist. Eur J Pharmacol 182:193–197

Kilpatrick GJ, Butler A, Hagan RM, Jones BJ, Tyers MB (1990b) [³H]GR67330, a very high affinity ligand for 5-HT₃ receptors. Naunyn Schmiedebergs Arch Pharmacol 342:22–30

Kilpatrick GJ, Barnes NM, Cheng CHK, Costall B, Naylor RJ, Tyers MB (1991) The pharmacological characterisation of 5-HT₃ receptor binding sites in rabbit ileum: comparison with those in rat ileum and rat brain. Neurochem Int 19:389–396

King GR, Xue Z, Calvi C, Ellinwood EH (1995) 5-HT₃ agonist-induced dopamine overflow during withdrawal from continuous or intermittent cocaine administration. Psychopharmacology (Berl) 117: 458–465

Kleinerman KB, Deppe SA, Sargent AI (1993) Use of ondansetron for control of projectile vomiting in patients with neurosurgical trauma: two case reports. Ann Pharmacother 27: 566–568

Koch KL, Bingaman S (1993) Effects of ondansetron on morphine-induced nausea, vasopressin and gastric myoelectrical activity in healthy humans. Gastroenterology 104:A535

Konno T, Busch C, von Kitzing E, Imoto K, Wang F, Nakai J, Mishina M, Numa S, Sakmann B (1991) Rings of anionic amino acids as structural determinants of ion selectivity in the acetylcholine receptor channel. Proc R Soc Lond B 244:67–79

Lader MH (1991) Ondansetron in the treatment of anxiety. In: Racagni G, Brunello N, Fukuda T (eds) Biological psychiatry. Excerpta Medica, Amsterdam, pp 885–887

Lambert JJ, Peters JA, Hales TG, Dempster J (1989) The properties of 5-HT₃ receptors in clonal cell lines studied by patch-clamp techniques. Br J Pharmacol 97:27–40

Lane JD, Pickering CL, Hooper ML, Fagan K, Tyers MB, Emmett-Oglesby MW (1992) Failure of ondansetron to block the discriminative or reinforcing stimulus properties of cocaine in the rat. Drug Alcohol Depend 30:151–162

Larijani GE, Gratz I, Afshar M, Minassian S (1991) Treatment of postoperative nausea and vomiting with ondansetron – a randomised, double-blind comparison with placebo. Anaesth Analg 73:246–249

Lazlo J, Lucas VSJ (1981) Emesis as a critical problem in chemotherapy. N Engl J Med 305: 948–949

Lecrubier Y, Puech AJ, Azcona A (1991) 5-HT₃ receptors in anxiety disorders. Proceedings of the British Association for Psychopharmacology, annual meeting, Cambridge, England, 17th July

Leibundgut U, Lancranjan I (1987) First results with ICS205-930 (5-HT$_3$ receptor antagonist) in prevention of chemotherapy-induced emesis. Lancet i:1198

Leslie RA, Reynolds DJM, Newberry NR (1994) Localisation of 5-HT$_3$ receptors. In: King FD, Jones BJ, Sanger GJ (eds) 5-Hydroxytryptamine-3 receptor antagonists. CRC, Boca Raton, pp 79–96

Little JT, Broocks A, Martin A, Hill JL, Tune LE, Mack C, Cantillon M, Molchan S, Murphy DL, Sunderland T (1995) Serotonergic modulation of anticholinergic effects on cognition and behaviour in elderly humans. Psychopharmacology 120:280–288

Loisy C, Beorchia S, Centonze V, Fozard JR, Schechter PJ, Tell GP (1985) Effects on migraine headache of MDL72222, an antagonist at neuronal 5-HT receptors, double-blind, placebo-controlled study. Cephalalgia 5:79–82

Lummis SCR, Martin IL (1992) Solubilisation, purification and functional reconstitution of 5-hydroxytryptamine$_3$ receptors from N1E-115 neuroblastoma cells. Mol Pharmacol 41:18–23

Lummis SCR, Kilpatrick GJ, Martin IL (1994) Molecular characterisation of 5-HT$_3$ receptors. In: King FD, Jones BJ, Sanger GJ (eds) 5-Hydroxytryptamine-3 receptor antagonists. CRC, Boca Raton, pp 97–113

Malone HM, Peters JA, Lambert JJ (1991) (+)-Tubocurarine and cocaine reveal species differences in the 5-HT$_3$ receptors of rabbit, mouse and guinea pig nodose ganglion neurones. Br J Pharmacol 104:68P

Maricq AV, Petersen AS, Brake AJ, Myers RM, Julius D (1991) Primary structure and functional expression of the 5-HT$_3$ receptor, a serotonin-gated ion channel. Science 254:432–437

Maura G, Andrioli GC, Cavazzani P, Raiteri M (1992) 5-Hydroxytryptamine$_3$ receptor sites on cholinergic axon terminals of human cerebral cortex mediate inhibition of acetylcholine release. J Neurochem 58:2334–2337

McKenzie R, Kovac A, O'Connor T, Duncalf D, Angel J, Gratz I, Tolpin E, McLeskey C, Joslyn A (1993) Comparison of ondansetron versus placebo to prevent post operative nausea and vomiting in women undergoing ambulatory gynecological surgery. Anesthiology 78:21–28

McKernan RM, Gillard NP, Quirk K, Kneen CO, Stevenson GI, Swain CJ, Ragan CI (1990) Purification of the 5-hydroxytryptamine 5-HT$_3$ receptor from NCB20 cells. J Biol Chem 265:13572–13577

Meltzer HY (1991) Studies of ondansetron in schizophrenia. Satellite symposium on ondansetron, 5th world congress of biological psychiatry, Florence

Minabe Y, Asby CR, Schwartz JE, Wang RY (1991) The 5-HT$_3$ receptor antagonists LY277359 and granisetron potentiate the suppressant action of apomorphine on the basal firing rate of ventral tegmental dopamine cells. Eur J Pharmacol 209:143–150

Miner WD, Sanger GJ (1986) Inhibition of cisplatin-induced vomiting by selective 5-hydroxytryptamine M-receptor antagonism. Br J Pharmacol 88:497–499

Montgomery AMJ, Rose IC, Herberg LJ (1993) The effect of a 5-HT$_3$ receptor antagonist, ondansetron, on brain stimulation reward, and its interaction with direct and indirect stimulants of central dopaminergic transmission. J Neural Trans 91:1–11

Murphy DL, Broocks A, Aulakh C, Pigott TA (1993) Anxiolytic effects of drugs acting on 5-HT receptor subtypes. In: Vanhoutte PM (ed) Serotonin. Kluwer Academic, Fondazione Giovanni Lrenzini, Amsterdam, pp 223-230

Naylor RJ, Rudd JA (1994) Emesis and antiemesis. In: Hanks GW, Sidebottom E (eds) Palliative medicine problem areas in pain and symptom management. Cold Spring Harbor Laboratory Press, New York, pp 117–135

Neijt HC, Te Duits IJ, Vijverberg HPM (1988) Pharmacological characterisation of serotonin 5-HT$_3$ receptor-mediated electrical response in cultured mouse neuroblastoma cells. Neuropharmacology 27:301–307

Newberry NR, Cheshire SH, Gilbert MJ (1991) Evidence that the 5-HT$_3$ receptors of the rat, mouse and guinea pig cervical ganglion may be different. Br J Pharmacol 102:615–620

Newberry NR, Watkins CJ, Sprosen TS (1992) BRL 46470A antagonises 5-HT₃ receptor-mediated responses on the rat vagus and on NG 108–15 cells. Br J Pharmacol 105:276P

Newcomer JW, Faustman WO, Zipursky RB, Csernansky JG (1992) Zacopride in schizphrenia: a single-blind serotonin type 3 antagonist trial [letter]. Arch Gen Psychiatry 49:751–752

Oakley NR, Jones BJ, Tyers MB (1988) Tolerance and withdrawal studies with diazepam and GR38032F in the rat. Br J Pharmacol 95:764P

Olivier B, Mos J, Van der Heyden J, Van der Poel G, Tulp M, Slangen J, De Jong R (1992) Preclinical evidence for the anxiolytic activity of 5-HT₃ receptor antagonists – a review. Stress Med 8:117–136

Onaivi ES, Todd S, Martin BR (1989) Behavioural effects in the mouse during and following withdrawal from ethanol ingestion and or nicotine administration. Drug Alcohol Depend 24:205–211

Orwin JM, Fozard JR (1986) Blockade of the flare response to intradermal 5-hydroxytryptamine in man by MDL72222, a selective antagonist at neuronal 5-hydroxytryptamine receptors. Eur J Clin Pharmacol 30:209–212

Palfreyman MG, Sorenson SM, Baron BM (1992) Antipsychotic potential of 5-HT₃ antagonists. In: Meltzer HY (ed) Novel antipsychotic drugs. Raven, New York, pp 211–223

Paudice P, Raiteri M (1991) Cholecystokinin release mediated by 5-HT₃ receptors in rat cerebral cortex and nucleus accumbens. Br J Pharmacol 103:1790–1794

Pecknold JC (1990) Platelet [³H]parocetine and [³H]imipramine binding in zacopride treated patients with generalised anxiety disorder: preliminary results, presented at the International Symposium on New Concepts in Anxiety, Castres, France, April 1990

Peters JA, Hales TG, Lambert JJ (1989) Electrophysiology of 5-HT₃ receptors in neuronal cell lines. Trends Pharmacol Sci 10:172–175

Peters JA, Malone HM, Lambert JJ (1990) Antagonism of 5-HT₃ receptor mediated currents in murine N1E-115 neuroblastoma cells by (+)-tubocurarine. Neurosci Lett 110:107–112

Peters JA, Malone HM, Lambert JJ (1991) Characterisation of 5-HT₃ receptor mediated electrical responses in nodose ganglion neurones and clonal neuroblastoma cells maintained in culture. In: Fozard JR, Saxena PR (eds) Serotonin: molecular biology, receptors and functional effects. Birkhauser, Basel, pp 84–94

Peters JA, Lambert JJ, Malone HM (1994) Electrophysiological studies of 5-HT₃ receptors. In: King FD, Jones BJ, Sanger GJ (eds) 5-Hydroxytryptamine-3 receptor antagonists. CRC, Boca Raton, pp 115–153

Pinkus LM, Gordon JC (1991) Utilisation of zacopride and its R- and S-enantiomers in studies of 5-HT₃ receptor subtypes. In: Fozard JR, Saxena PR (eds) Serotonin: receptors, molecular biology and functional effects. Birkhauser, Basel, pp 439–448

Pinkus LM, Sarbin NS, Barefoot DS, Gordon JC (1989) Association of [³H]zacopride with 5-HT₃ binding sites. Eur J Pharmacol 163:355–362

Pitsikas N, Brambilla A, Borsini F (1993) DAU 6215, a novel 5-HT₃ receptor antagonist, improves performance in the aged rat in the Morris water maze task. Neurobiol Aging 14:561–564

Pitsikas N, Brambilla A, Borsini F (1994) Effect of DAU 6215, a novel 5-HT₃ receptor antagonist, on scopolamine induced amnesia in the rat in a spatial learning task. Pharmacol Biochem Behav 47:95–99

Pratt GD, Bowery NG (1989) The 5-HT₃ receptor ligand, [³H]BRL43694, binds to presynaptic sites in the nucleus tractus solitarius of the rat. Neuropharmacology 28:1367–1376

Preston GC (1994) 5-HT₃ antagonists and disorders of cognition. In: Rascagni G, Brunello N, Langer SZ (eds) Recent advances in the treatment of neuro-degenerative disorders and cognitive dysfunction. Karger, Basel. Int Acad Biomed Drug Res 7:89–93

Preston GC, Millson DS, Ceuppens PR, Warburton DM (1991) Effects of the 5-HT$_3$ receptor antagonist GR68755 on a scopolamine-induced cognitive deficit in healthy subjects. Br J Clin Pharmacol 32:546P

Rasmussen K, Stockton ME, Czachura JR (1991) The 5-HT$_3$ receptor antagonist zatosetron decreases the number of spontaneously active A10 dopamine neurons. Eur J Pharmacol 205:103–116

Richardson BP (1992) Organisation of 5-HT neurones regulating central pain. In: Bradley PB, Handley SL, Cooper SJ, Key BJ, Barnes NM, Coote JH (eds) Serotonin, CNS receptors and brain function. Pergamon, Oxford, pp 335–347

Richardson BP, Engel G (1986) The pharmacology and function of 5-HT$_3$ receptors. Trends Neurosci 9:424–428

Richardson BP, Engel G, Donatsch P, Stadler PA (1985) Identification of serotonin M-receptor subtypes and their specific blockade by a new class of drugs. Nature 316:126–131

Roberts MHT (1984) 5-Hydroxytryptamine and antinociception. Neuropharmacology 23:1529–1536

Robertson B, Bevan S (1991) Properties of 5-hydroxytryptamine$_3$ receptor-gated currents in adult rat dorsal route ganglion neurones. Br J Pharmacol 102:272–276

Rodgers RJ, Shepherd JK (1992) Attenuation of defensive analgesia in male mice by 5-HT$_3$ receptor antagonists ICS205-930, MDL72222, MDL73147F and MDL72699. Neuropharmacology 31:553–560

Round AA, Wallis DI (1986) The depolarising action of 5-hydroxytryptamine on rabbit vagal afferent and sympathetic neurones and its selective blockade by ICS205-930. Br J Pharmacol 88:485–494

Round A, Wallis DI (1987) Further studies on the blockade of 5-HT depolarisations of rabbit vagal afferent and sympathetic ganglion cells by MDL72222 and other antagonists. Neuropharmacology 26:39–48

Rowat BMT, Merrill CF, Davies A, South V (1991) A double-blind comparison of granisetron and placebo for the treatment of acute migraine in the emergency department. Cephalgia 11:207–213

Rudd JA, Jordan CC, Naylor RJ (1994) Profiles of emetic action of cisplatin in the ferret: a potential model of acute and delayed emesis. Eur J Pharmacol 262:R1–R2

Sanger GJ (1987) Increased gut cholinergic activity and antagonism of 5-hydroxytryptamine M-receptors by BRL24924: potential clinical importance of BRL24924. Br J Pharmacol 91:77–87

Sanger GJ, Nelson DR (1989) Selective and functional 5-hydroxytryptamine$_3$ receptor antagonism by BRL43694 (granisetron). Eur J Pharmacol 159:113–124

Sanger GJ, Banner SE, Wardle KA (1994) The role of the 5-HT$_3$ receptor in normal and disordered gastrointestinal function. In: King FD, Jones BJ, Sanger GJ (eds) 5-Hydroxytryptamine-3 receptor antagonists. CRC, Boca Raton, pp 319–358

Schtrom SS, Tecot LH, Julius DJ (1993) Developmental expression of 5-HT$_3$ receptor in RNA. Soc Neurosci Abstr 19:632

Schweizer E, Rickels K (1991) Serotonergic anxiolytics: a review of their clinical efficacy. In: Rodgers RJ, Cooper SJ (eds) 5-H$_{1A}$ agonists, 5-HT$_3$ antagonists and benzodiazepines: their comparative behavioural pharmacology. John Wiley and Sons, Chichester, pp 365–376

Sell LA, Cowan PJ, Robson PJ (1995) Ondansetron and opiate craving: a novel pharmacological approach to addition. Br J Pharmacol 166:511–514

Sellers EM, Toneatto T, Romach MK (1994) Clinical efficacy of the 5-HT$_3$ antagonist ondansetron in alcohol abuse and dependence. Alcohol Clin Exp Res 18:879–885

Sepulveda MI, Martin IL, Lummis SCR (1991) Complex actions of BRL 43694 in N1E-115 neuroblastoma cells. Symposium 5-hydroxytryptamine – CNS receptors and brain function, Birmingham, UK, abstract, p 76

Smith WW, Sancilio LF, Owera-Atepo JB, Naylor RJ, Lambert LL (1988) Zacopride, a potent 5-HT$_3$ antagonist. J Pharm Pharmacol 40:301–302

Staubli U, Xu FB (1995) Effects of 5-HT₃ receptor antagonism on hippocampal theta rhythm, memory, and LTP induction in the freely moving rat. J Neurosci 15:2445–2452

Sugita S, Shen KZ, North RA (1992) 5-Hydroxytryptamine is a fast excitatory transmitter at 5-HT₃ receptors in rat amygdala. Neuron 8:199–203

Sullivan CA, Johnson CA, Roach H, Martin RW, Stewart DK, Morrison JC (1996) A pilot study of intravenous ondansetron for hyperemesis gravidarum. Am J Obstet Gynecol 174:1565–1568

Tattersall D, Newberry N, Beer MS, Rigby M, Gilbert M, Maguire JJ, Mudunktotuwa N, Duchnowski M, McKight AT, Swain CJ, Keen C, Dourish C (1992) L683,677: pharmacological profile of a novel 5-HT₃ receptor antagonist. Drug Dev Res 25:17–28

Tecott LH, Maricq AV, Julius D (1993) Nervous system distribution of the serotonin 5-HT₃ receptor in RNA. Proc Natl Acad Sci USA 90:1430–1434

Tomkins DM, Costall B, Kelly ME (1990) Release of suppressed behaviour of rat on the elevated X-maze by 5-HT₃ receptor antagonists injected into the basolateral amygdala. British Association for Psychopharmacology Conference, Cambridge, abstract 68

Vanner S, Suprenant A (1990) Effect of 5-HT₃ receptor antagonists on 5-HT and nicotinic depolarisations in guinea pig submucosal neurones. Br J Pharmacol 99:840–844

Waeber C, Hoyer D, Palacios JM (1989) 5-Hydroxytryptamine₃ receptors in the human brain: autoradiographic visualisation using [³H]ICS205-930. Neuroscience 31:393–400

Wallis DI (1989) Interaction of 5-hydroxytryptamine with autonomic and sensory neurones. In: Fozard JR (ed) The peripheral actions of 5-hydroxytryptamine. Oxford University Press, Oxford, pp 220–246

Wallis DI, Dun DJ (1988) A comparison of fast and slow depolarisations evoked by 5-HT in guinea pig coeliac ganglion cells in vitro. Br J Pharmacol 93:110–120

Wallis DI, Nash H (1981) Relative activities of substances related to 5-hydroxytryptamine as depolarising agents of superior cervical ganglion cells. Eur J Pharmacol 70:381–392

Wallis DI, North RA (1978) The action of 5-hydroxytryptamine on single neurones of the rabbit superior cervical ganglion. Neuropharmacology 17:1023–1028

Wang RY, Ashby CR, Edwards E (1991a) Characterisation of 5-HT₃-like receptors in the rat cortex: electrophysiological and biochemical studies. In: Fozard JR, Saxena PR (eds) Serotonin: molecular biology, receptors and functional effects. Birkhauser, Basel, pp 174–185

Wang RY, Ashby CR, Minabe Y (1991b) Antipsychotic potential of granisetron: electrophysiological studies. Presented at Serotonin 1991, Birmingham, UK, 14th July, abstract, p 74

Wang RY, Ashby CR, Zhang JY (1992) Functional roles of 5-HT₃-like receptors in the medial prefrontal cortex. In: Bradley PB, Handley SL, Cooper SJ, Key BJ, Barnes NM, Cote JH (eds) Serotonin, CNS receptors and brain function. Pergammon, New York, pp 81–96 (Advances in biosciences, vol 85)

Werner P, Humbert Y, Boess F, Reid J, Jones K, Kawashima E (1993) Organisation of the 5-HT₃ₐ receptor gene and investigation of its splice variants. Soc Neurosci Abstr 19:1164

White A, Corn TH, Feetham C, Faulconbridge C (1991) Ondansetron in the treatment of schizophrenia. Lancet 337:1173

Wong EHF, Wu I, Eglen RM, Whiting RL (1992) Labelling of species variants of 5-HT₃ receptors by a novel 5-HT₃ receptor ligand. Br J Pharmacol 105:33P

Wozniak KM, Pert A, Linnoila M (1990) Antagonism of 5-HT₃ receptors attenuates the effects of ethanol on extracellular dopamine. Eur J Pharmacol 187:287–289

Yakel JL (1992) 5-HT₃ receptors as cation channels. In: Hamon M (ed) Central and peripheral 5-HT₃ receptors. Academic, New York, pp 163–184

Yakel JL, Jackson MB (1988) 5-HT₃ receptors mediate rapid responses in cultured hippocampus and a clonal cell line. Neuron 1:615–621

Yakel JL, Shao XM, Jackson SM (1990) The selectivity of the channel coupled to the 5-HT₃ receptor. Brain Res 533:46–52

Yakel JL, Shao XM, Jackson MB (1991) Activation and desensitisation of the 5-HT₃ receptor in a rat glioma × mouse neuroblastoma hybrid cell. J Physiol (Lond) 436:293–308

Yang J, Mathie A, Hille B (1992) 5-HT₃ receptor channels in dissociated rat superior cervical ganglion neurones. J Physiol (Lond) 448:237–256

Zacny JP, Apfelbaum JL, Lichtor JL, Zaragoza JG (1993) Effects of 5-hydroxytryptamine-3 antagonist ondansetron, on cigarette smoking, smoke exposure and mood in humans. Pharmacol Biochem Behav 44:387–391

Zhang JY, Zeise ML, Wang RY (1994) Serotonin₃ receptor agonists attenuate glutamate-induced firing in rat hippocampal CA1 pyramidal cells. Neuro-pharmacology 33:483–491

CHAPTER 17

5-HT$_4$ Receptors: An Update

J. BOCKAERT, L. FAGNI, and A. DUMUIS

A. Introduction: A Review

SHENKER et al. (1987) showed, in guinea pig hippocampal membranes, that stimulation of adenylyl cyclase (AC) by 5-hydroxytryptamine (5-HT) involves two receptors: the 5-HT (RH) receptor, which displays a high affinity for 5-HT, is characterized as a 5-HT$_{1A}$-like receptor and is now recognized as a 5-HT$_7$ receptor (HOYER et al. 1994), and the 5-HT (RL) receptor, which has a low affinity for 5-HT and which has not been identified. A clear biphasic dose-activation curve was obtained only with 5-carboxamidotryptamine (5-CT). Indeed, this agonist had high (13 nM) and low (3000 nM) affinities for 5-HT (RH) and 5-HT (RL) receptors, respectively.

One year later we found, in mouse colliculi neurons, a 5-HT receptor that stimulates AC and which has a pharmacology different from the well-known 5-HT$_1$, 5-HT$_2$ and 5-HT$_3$ receptors (DUMUIS et al. 1988a). We proposed the name of 5-HT$_4$ receptor, immediately recognizing that this receptor shared with the 5-HT (RL) receptor defined by SHENKER et al. (1987) similar potencies for a series of agonists: 5-HT = 5-MeOT > bufotenine > 5-CT > tryptamine. Our conviction that the 5-HT$_4$ receptor was different from the 5-HT$_{1,2,3}$ receptors (BRADLEY et al. 1986) came from the observation that highly potent and specific 5-HT$_{1,2,3}$ antagonists were unable to inhibit the 5-HT$_4$ receptors in colliculi neurons (DUMUIS et al. 1988b). Later we found a weak (millimolar potency) but competitive inhibitor of 5-HT$_4$ receptors: tropisetron. Shortly after, we demonstrated that a series of gastrointestinal prokinetic benzamide derivatives, including metoclopramide, renzapride, cisapride and zacopride, act as agonists at 5-HT$_4$ receptors (DUMUIS et al. 1989a, b). These observations immediately linked the 5-HT$_4$ receptors described in colliculi neurons and an unclassified 5-HT receptor in the enteric nervous system which was postulated to mediate the gastrokinetic actions of these compounds (CRAIG and CLARKE 1990; SANGER 1987; SCHUURKES et al. 1985). It was shown that, in guinea pig ileum, this non-5-HT$_{1,2,3}$ receptor was located in neurons, had high affinity for 5-methoxytryptamine (5-MeOT) and low affinity for 5-CT, was stimulated by benzamides and displayed low affinity for tropisetron (millimolar) (for review see BOCKAERT et al. 1992; CLARKE et al. 1989; FORD and CLARKE 1993).

It became likely that this 5-HT receptor in guinea pig ileum was similar, if not identical, to the 5-HT$_4$ receptor present in guinea pig hippocampus and

Table 1. 5-HT$_4$ receptor-mediated responses

Species	Target (organ-tissue cell)	Mechanisms	Effects	References
Neuronal tissues				
Mouse	Colliculi neurons	↗ cAMP	↘ K$^+$ current (long-term effect)	Dumuis et al. (1988a) Fagni et al. (1992) Ansanay et al. (1995)
Guinea pig	Hippocampus	↗ cAMP		Bockaert et al. (1990)
Human	Frontal cortex	↗ cAMP		Monferini et al. (1993)
Rat	Hippocampus (CA$_1$)	↗ cAMP	↘ K$^+$ current slow depolarization	Torres et al. (1995)
Rat	Vagus nerve		Depolarization	Rhodes et al. (1992) Coleman and Rhodes (1995) Bley et al. (1994)
Guinea pig	Myenteric neurons		↗ Fast EPSP (indirect)	Nemeth and Gullickson (1989) Tonini et al. (1991)
Rat	CNS		↗ EEG	Boddeke and Kalkman (1990)
Rat	CNS		↘ Spatial memory deficit induced by atropine	Eglen et al. (1995a)
Rat	CNS		↘ In hypoxia-induced amnesia	Ghelardini et al. (1994)
Rat	CNS		↗ Social olfactory memory	Letty et al. (1996)
Mice	CNS		Analgesic effects	Ghelardini et al. (1996)
Gastrointestinal tract				
Guinea pig	Ileum	↗ Acetylcholine release ↗ NANC release ↗ Ascending excitatory reflex	↗ Contraction ↗ Peristaltic reflex ↗	Craig and Clarke (1990) Rizzi et al. (1992) Ford and Clarke (1993) Yuan et al. (1994)

Species	Tissue	Response	Second messenger	References
	Myenteric neurons	Fast EPSP (indirect)		NEMETH and GULLICKSON (1989), TONINI et al. (1991)
Rat	Ileum	Relaxation		TULADHAR et al. (1991)
Rat	Ileum, colon	↗ Secretory response (neuronal)		FRANKS et al. (1995)
Guinea pig	Ileal muscosa	↗ Cl⁻ secretion (neuronal and non-neuronal)		LEUNG et al. (1995), SCOTT et al. (1992)
Mice	Colon	5-HTP-induced diarrhoea		HEDGE et al. (1994)
Human	Ileum, Jejunum	↗ Cl⁻ secretion (non-neuronal)		BORMAN and BURLEIGH (1993), BURLEIGH and BORMAN (1993)
Guinea pig	Colon	Contraction (neuronal)		ELSWOOD et al. (1991), BRIEJER et al. (1993), WARDLE and SANGER (1993)
Human	Colon	Circular muscle relaxation		TAM et al. (1995)
Guinea pig	Stomach	Contraction		BUCHHEIT and BUHL (1994)
Rat	Stomach	Gastroprokinetic (gastric emptying)		HEDGE et al. (1995b), RIZZI et al. (1994)
Human	Stomach	Gastroprokinetic (gastric emptying)		WISEMAN and FAULDS (1994)
Dog	Stomach	5-HT-induced contraction		BINGHAM et al. (1995), GULLIKSON et al. (1993)
Rat	Oesophagal smooth muscle	Relaxation	↗ cAMP	BIEGER and TRIGGLE (1985), BAXTER et al. (1991), REEVES et al. (1991), FORD et al. (1992), RONDÉ et al. (1995), FORD et al. (1992)

Table 1. *Continued*

Species	Target (organ-tissue cell)	Mechanisms	Effects	References
Urinary bladder				
Monkey	Smooth muscle		Relaxation	WAIKAR et al. (1994)
Human	Neurons		Contraction (neuronal)	CORSI et al. (1991)
				TONINI et al. (1994)
Adrenal gland				
Frog	Adrenocortical cells (homologous to glomerulosa cells)	↗ cAMP	↗ Glucorticoid secretion	IDRES et al. (1991)
				CONTESSE et al. (1994)
				CONTESSE et al. (1996)
Human			↗ Cortisol	LEFÉBVRE et al. (1992)
			↗ Aldosterone (in vivo, in vitro)	LEFÉBVRE et al. (1993)
				LEFÉBVRE et al. (1994)
Heart				
Pig, piglet	Atria		↗ Heart rate	BOM et al. (1988)
				VILLALÒN et al. (1990)
				SCHOEMAKER et al. (1992)
				SAXENA and VILLALÒN (1990)
				KAUMANN (1990)
Human	Atria	↗ cAMP	↗ Atria contractile force	KAUMANN et al. (1990)
			↗ Ca²⁺ channels (L-type)	KAUMANN et al. (1991)
				SANDERS and KAUMANN (1992)
				OUADID et al. (1992)
			Induction of rate-dependent arrhythmia	JAHNEL et al. (1992)
				KAUMANN and SANDERS (1994)

mouse colliculi neurons (CLARKE et al. 1989; TONINI et al. 1991). In addition to brain and gastrointestinal tract, 5-HT$_4$ receptors are expressed in heart, adrenal gland and urinary bladder of several species including humans (Table 1). The pharmacology of 5-HT$_4$ receptors now includes highly specific and potent antagonists but few specific agonists (Figs. 1,2). Since 1993, it has been possible to label the 5-HT$_4$ receptors with selective ligands [^{3}H]GR 113808 and [^{125}I]SB 207710 (Fig. 3; BROWN et al. 1993; GASTER and SANGER 1994; GROSSMAN et al. 1993; WAEBER et al. 1993). Finally, the 5-HT$_4$ receptor gene has now been cloned from rat brain (GERALD et al. 1995).

B. Structure of 5-HT$_4$ Receptors

Recently, molecular studies have provided primary amino acid sequence and signal transduction data for a great number of serotonin receptor subtypes. These include five 5-HT$_1$-like receptor subtypes (5-HT$_{1A}$, 5-HT$_{1B}$/$_{1Db}$, 5-HT$_{1Da}$, 5-HT$_{1E}$, 5-HT$_{1F}$), all negatively coupled to AC, three 5-HT$_2$ receptor subtypes (5-HT$_{2A}$, 5-HT$_{2B}$, 5-HT$_{2C}$) all coupled to phospholipase C, two 5-HT$_5$ receptor subtypes (5-HT$_{5A}$, 5-HT$_{5B}$) for which no signal transduction has been described, and a 5-HT$_6$ and a 5-HT$_7$ receptor positively coupled to AC (for reviews see BOESS and MARTIN 1994; HOYER et al. 1994; LUCAS and HEN 1995; MARTIN and HUMPHREY 1994). For unclear reasons, but perhaps because of the low abundancy of the corresponding mRNA as well as low sequence similarities with the other 5-HT receptor types, cloning of the 5-HT$_4$ gene was delayed until 1995. Using a strategy based on nucleotide homologies between transmembrane domain III and V of 5-HT receptors, two cDNAs were isolated which likely corresponded to two splice variants of a G-protein-coupled receptor having seven putative membrane-spanning domains, the 5-HT$_{4L}$ and the 5-HT$_{4S}$ receptors. They differ in the second half of their C-termini at position 360 (GERALD et al. 1995) (Fig. 4). The 5-HT$_{4S}$ and 5-HT$_{4L}$ cDNAs encode 387 and 406 amino acids, respectively. The generation of alternative splicing in G-protein-coupled receptors as a source of functional diversity of coupling is particularly evident in pituitary adenylyl cyclase activating polypeptide (PACAP) receptors and prostaglandin EP$_3$ receptors (NAMBA et al. 1993; SPENGLER et al. 1993). The possibility that each isoform of 5-HT$_4$ receptor is coupled to distinct G protein coupling remains to be demonstrated. Note that their sensitivity to GTPγS is slightly different (GERALD et al. 1995). The 5-HT$_4$ receptor is typical of the first group of heptahelice receptors, which includes most of the receptors for small-sized neurotransmitters (with the exception of metabotropic glutamate receptors; BOCKAERT 1995; COUGHLIN 1994). 5-HT$_4$ receptors have typical signatures of receptors belonging to group 1 such as an aspartic acid (D) in the transmembrane domain II important in the coupling to G proteins and its regulation by Na$^+$ (D66), the sequence DRY at the N-terminal end of the second intracellular loop and a putative S-S bridge

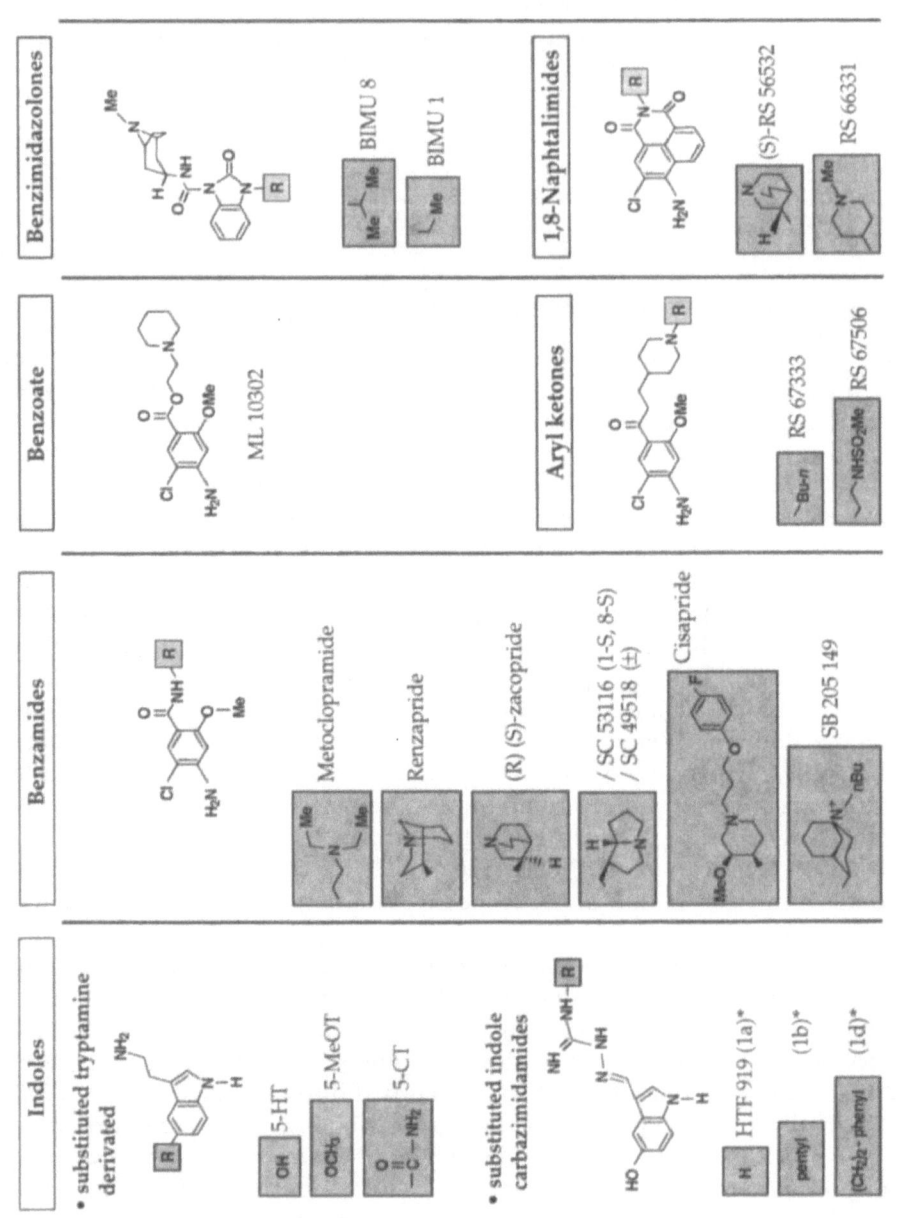

Fig. 1. Chemical structure of key 5-HT$_4$ receptor agonists. These molecules are representative of six active chemical classes including indoles, benzamides, benzoate, aryl ketones, benzimidazolones and 1,8-naphtalimides. *5-HT$_4$-receptor agonists of substituted indole carbazimidamides are taken from BUCHHEIT et al. (1995a, b)

5-HT4 antagonists

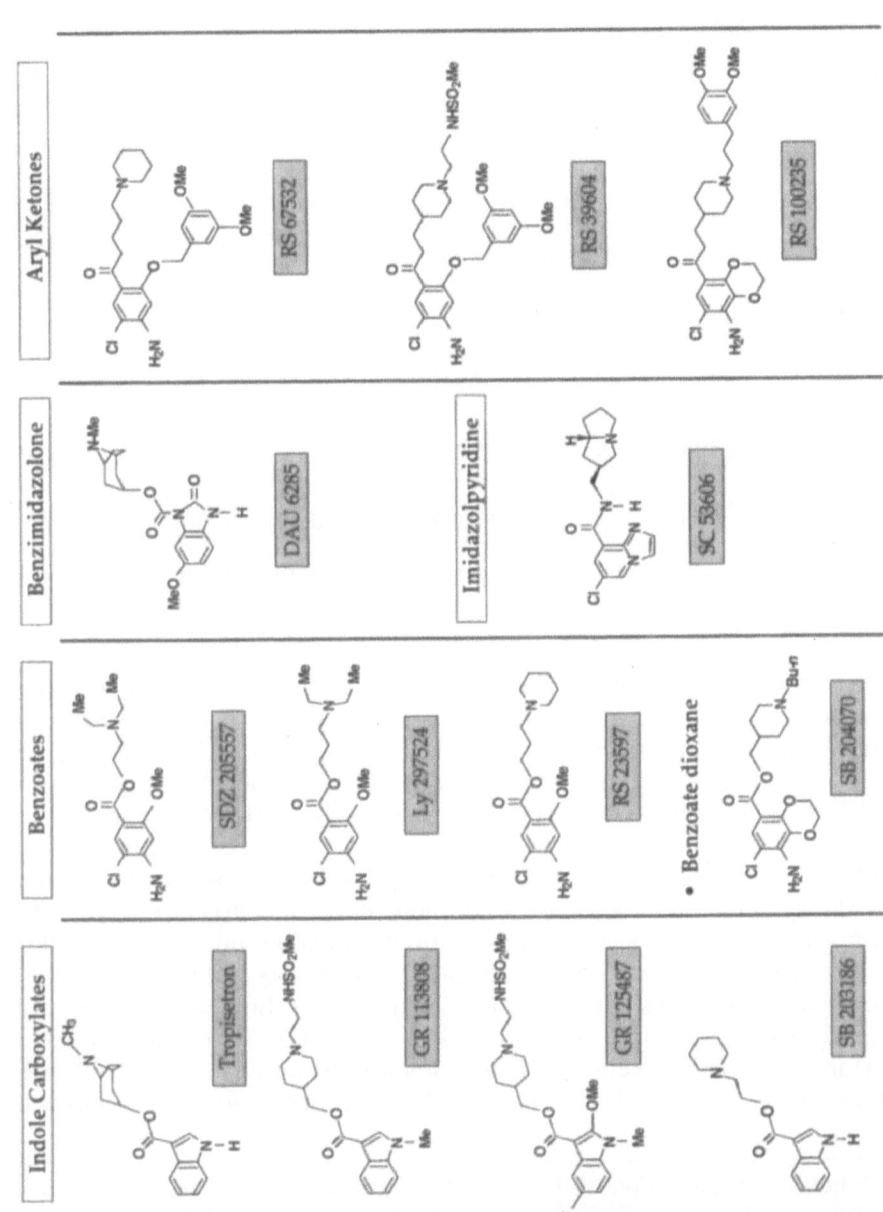

Fig. 2. Chemical structure of key 5-HT₄ receptor antagonists. These molecules are representative of five active chemical classes including indoles, benzoates and benzoate dioxane, benzimidazolone, imidazolpyridine and aryl ketones

Most selective and potent
5HT4 receptor antagonists used as radioligands

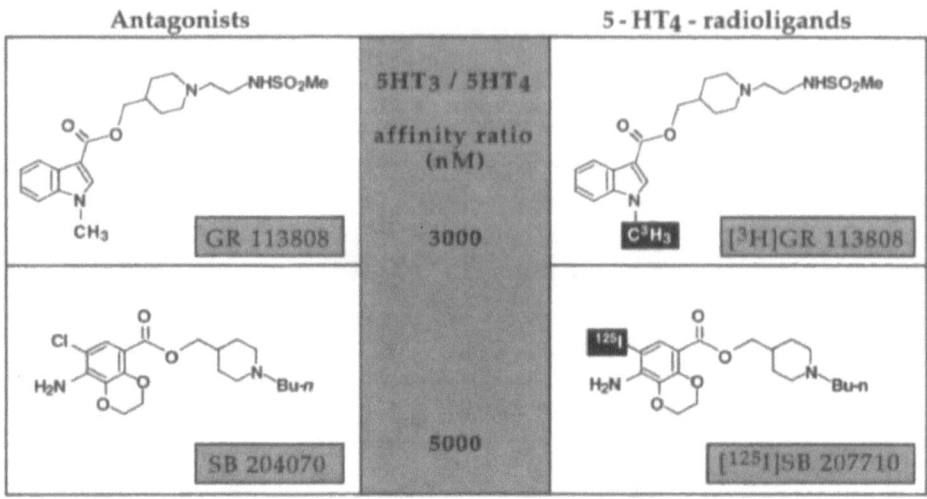

Fig. 3. Chemical structure of the most selective and highly potent 5-HT$_4$ receptor radioligands. For each molecule the nature and the site at which the compound has been radiolabelled are indicated. The increase in affinity ratio indicates the selectivity for 5-HT$_4$ receptor

between the first and second extracellular loops (BOCKAERT 1991) (Fig. 4). An aspartic acid in the transmembrane domain III (D100) is conserved in all neurotransmitter receptors having a cationic amino group, including the 5-HT$_4$ receptors, and which may serve as the primary counterion for this receptor group (BOCKAERT 1991). The 5-HT$_{4L}$ and 5-HT$_{4S}$ have three and four putative protein kinase C phosphorylation sites (Fig. 4). Threonine 218 is also a putative casein-kinase II phosphorylation site. Cysteine 328 is a potential palmitoylation site. Many G-protein-coupled receptors carry a cysteine at a similar position, which may play a role in functional coupling (NG et al. 1993; O'DOWD et al. 1989). The 5-HT$_{4L}$ and 5-HT$_{4S}$ receptors possess, in their C-termini, 16 and 11 serines and threonines, respectively. Some of them may be putative phosphorylation sites for G-protein-coupled-receptor kinases (GRKs) involved in the well-described homologous desensitization of 5-HT$_4$ receptors (ANSANAY et al. 1992). The dendrogram obtained from amino acid

Fig. 4. Schematic organization of the two splice variants of rat 5-HT$_4$ receptor. The amino acid sequences were taken from GERALD et al. (1995). The positions for the putative phosphorylation sites by protein kinase C, G-protein-coupled receptor kinase (GRK) or casein-kinase II are indicated. Potential palmitoylation and glycosylation sites are also shown. Key amino acids present in all G-protein-coupled receptors of the first group (BOCKAERT 1995) are indicated

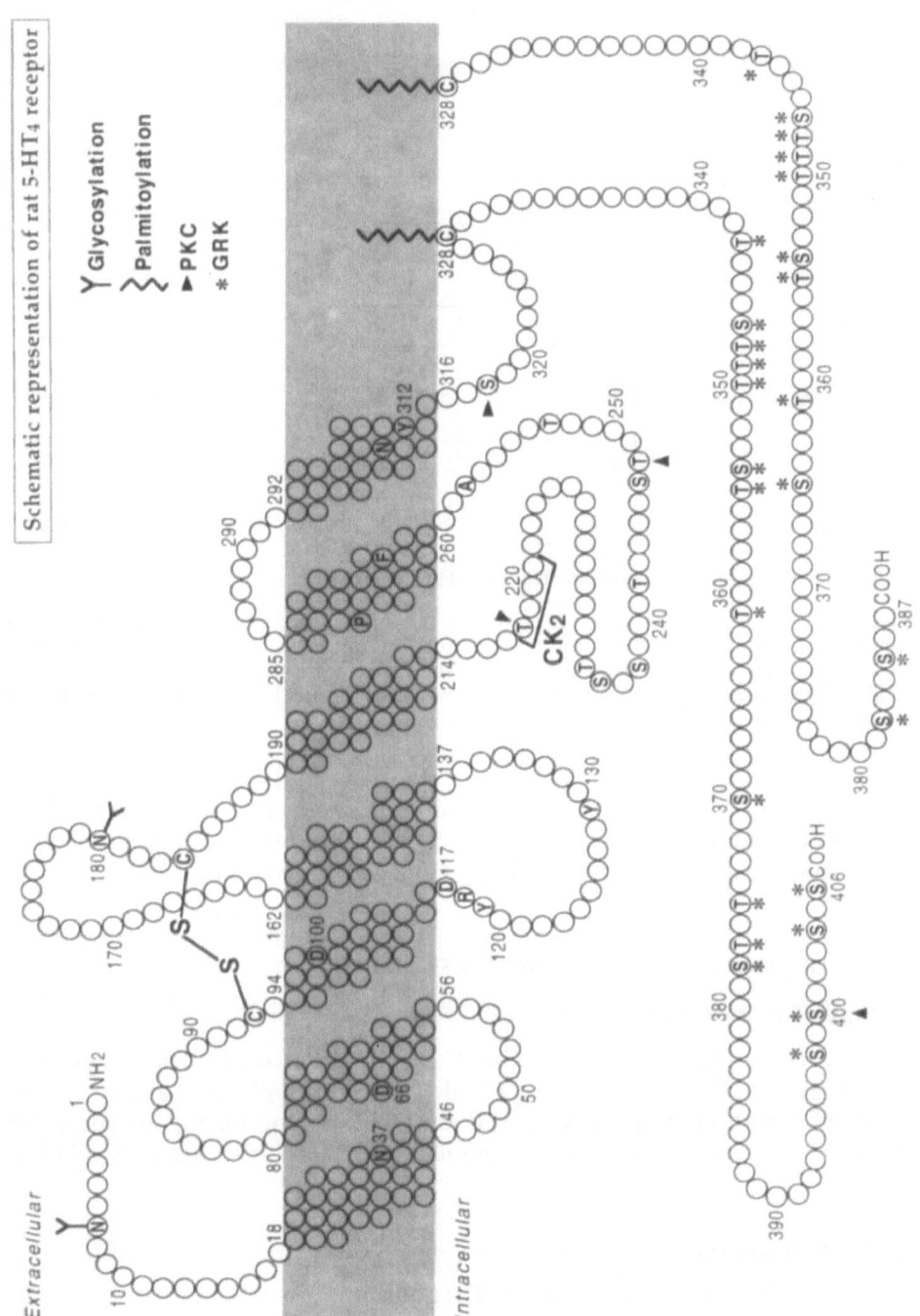

Schematic representation of rat 5-HT$_4$ receptor

Y Glycosylation
〜 Palmitoylation
► PKC
* GRK

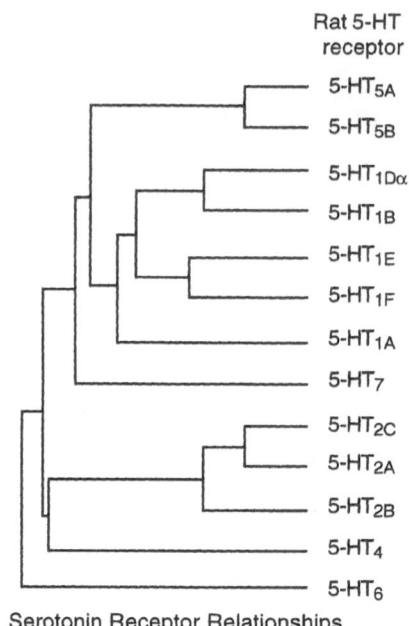

Fig. 5. Dendrogram of amino acid sequences of 13 rat G-protein-coupled 5-HT receptors cloned. The amino acid sequences corresponding to the transmembrane domains were compared and clustered according to sequence homology as reported by GERALD et al. (1995). *Length of horizontal lines* is inversely proportional to percentage homology between receptors or groups of receptors

homologies within the transmembrane domains indicates a weak (60%–70%) overall homology between the 5-HT$_4$ receptors and other 5-HT receptors, with some clustering with the 5-HT$_2$ family (Fig. 5) (GERALD et al. 1995).

C. Pharmacological Characterization

I. Agonists and Antagonists

Most 5-HT$_4$ receptor ligands have been discovered using functional models (colliculi neurons, guinea pig ileum, rat oesophagus) due to the late availability of a binding assay as well as cDNA coding for this receptor. The chemical structures of several key agonists and antagonists are shown in Figs. 1, 2 and 3.

1. 5-HT$_4$ Agonists

There are six distinct classes of 5-HT$_4$ agonists:

1. Indole-based 5-HT analogues including the substituted tryptamines and the substituted indole carbazimidamides. Among the tryptamines, 5-MeOT is a

full agonist in all assays. It is also interesting because it lacks affinity for 5-HT$_3$ receptors. It has been used to discriminate between 5-HT$_3$ and 5-HT$_4$ receptor-mediated responses (CRAIG et al. 1990), although it should be noted that it is also a potent agonist at 5-HT$_1$ and 5-HT$_2$ receptors. No selective tryptamine exists for the 5-HT$_4$ receptors (BOCKAERT et al. 1992; FORD and CLARKE 1993). Recently, a new class of potent and selective 5-HT$_4$ receptor agonists, containing an indole linked to a carbazimidamide, has been described (Fig. 1; BUCHHEIT et al. 1995a, b). These compounds have been designed based on the putative pharmacophore of 5-HT$_4$ receptors. They incorporate the aromatic ring of 5-HT, a basic nitrogen on the side chain, giving the possibility of the formation of a hydrogen bond between the indole NH of serotonin and the receptor. These compounds, including HTF 919 (compound 1a) (Fig. 1), incorporate near the basic nitrogen of serotonin and zacopride (which are 2 Å apart in the pharmacophore) a guanidine function which can interact in a twin nitrogen-twin oxygen fashion with the putative carboxylic acid counterpart. HTF 919 is more potent than 5-HT (pD_2 = 8.8 and 8.4, respectively) on 5-HT$_4$ receptors and is the most selective agonist known so far. It has a moderate affinity (K_i = 63 nM) for 5-HT$_{1A}$ receptors and a very low affinity (K_i > 250 nM) for the other 5-HT receptors. Addition of hydrophobic groups near the basic nitrogen (where the pharmacophore predicts a hydrophobic pocket) such as a pentyl (compound 1b) or a phenyl (compound 1d) (Fig. 1) increases the potency of the drug (pD_2 = 9.3 and 9.1, respectively; BUCHHEIT et al. 1995a,b).

2. The benzamides bearing the 2-methoxy-4-amino-5-chloro substitution are agonists (Fig. 1; DUMUIS et al. 1989b). The benzamides of the first generation have antagonistic activity at 5-HT$_3$ receptors (zacopride, renzapride, metoclopramide), 5-HT$_2$ receptors (cisapride) and dopamine D$_2$ receptors (metoclopramide). New compounds such as SB 205149 (quaternized renzapride; BAXTER et al. 1993) and SC 53116, the active isomer of the racemic mixture SC 49518, appear to be among the most selective 5-HT$_4$ receptor agonists (FLYNN et al. 1992; GULLIKSON et al. 1993).

3. The benzoate (ML 10302) bearing the same substitution as the benzamide (2-methoxy-4-amino-5-chloro; LANGLOIS et al. 1994) (Fig. 1) is a potent, although partial agonist (nanomolar) at some 5-HT$_4$ receptor-mediated responses. For example, ML 10302 is a potent agonist in inducing contractions of guinea pig ileum and relaxation of rat oesophagus (LANGLOIS et al. 1994) (pEC_{50} = 8.4), whereas it is a potent antagonist (K_i = 10 nM) at 5-HT$_4$ receptors expressed in colliculi neurons (unpublished data). In ML 10302, addition of a carbon between the benzoate and the piperidyl ring gives the potent 5-HT$_4$ receptor antagonist RS 23597 (CLARK et al. 1994) (Fig. 2).

4. Aryl ketones including RS 67333 and RS 67506 are new, selective 5-HT$_4$ agonists related to 2-methoxy-4-amino-5-chloro-substituted benzoate (EGLEN et al. 1995a,c). They exhibit similar high affinities to ML 10302 but

the replacement of the ester group by a ketone moiety gave to the aryl ketones (RS 67333 and RS 67506) an action duration significantly longer than that of the related benzoate ester.

5. Azabicycloalkyl benzimidazolones were found to be 5-HT$_4$ receptor agonists. Indeed, we thought that drugs which have gastrointestinal prokinetic activity, and for which a clear mechanism of action was not yet known, could be good candidates. The azabicycloalkyl benzimidazolone derivatives were of interest because the electrically stimulated guinea pig ileum was blocked by millimolar concentrations of tropisetron (a 5-HT$_4$ characteristic) but not by ondansetron, a selective 5-HT$_3$ antagonist (Rizzi et al. 1992; Tonini et al. 1991). Indeed, we found, in colliculi neurons, that BIMU 1 and BIMU 8, which are alkyl-substituted benzimidazolones at position 3 (Fig. 1), were potent and efficacious agonists at 5-HT$_4$ receptors in colliculi neurons (Dumuis et al. 1991). Similar effects of this class of agonists were observed on other 5-HT$_4$ receptor-mediated responses (Bockaert et al. 1992). As with many benzamides, the benzimidazolones are also potent 5-HT$_3$ antagonists. Note that BIMU 1 is also a sigma-2 ligand (Bonhaus et al. 1993).

6. 1, 8-Naphthalimides constitute a new class of potent in vivo and in vitro active 5-HT$_4$ agonist: S-RS 56532 and RS 66331 (pEC$_{50}$ = 7.9 and 7.6, respectively; Eglen et al. 1994, 1995c). Interestingly, RS 56532 exhibits opposing enantiomery selectivity at 5-HT$_3$ and 5-HT$_4$ receptors, the (R)-enantiomer is a potent 5-HT$_3$ ligand whereas the (S)-enantiomer is a potent 5-HT$_4$ receptor agonist devoid of 5-HT$_3$ activity.

Table 2. 5-HT$_4$ receptor antagonists: affinities and efficacies of 13 5-HT$_4$ receptor antagonists. Their selectivities are indicated by their 5-HT$_3$/ 5-HT$_4$ affinity (nM) ratios

Compounds	Antagonistic activity (pK$_b$)	Binding (pK$_i$)	5-HT$_3$/5-HT$_4$ affinity ratio (nM)	References
Tropisetron	6	6.8	0.003	Dumuis et al. (1988a)
DAU 6285	6.8	7.5	3	Dumuis et al. (1992)
SDZ 205557	7.2	8.2	25	Buchheit et al. (1992)
LY 297524	7.7			Krushinski et al. (1992)
SC 53606	7.91	7.82	21	Yang et al. (1993)
RS 23597	8.4	7.8		Bonhaus et al. (1993)
RS 67532	8.8	8.7		Clark et al. (1994)
SB 203186	9		50	Parker et al. (1993)
RS 39604	9.3	9.1	1000	Hedge et al. (1995a)
GR 113808	10.2	9.3	3000	Grossman et al. (1993)
GR 125487	10.2	10.42	>3000	Gale et al. (1994a)
SB 204070	10.9	10.8	5000	Gaster and Sanger (1994)
RS 100235	11.2	10	>5000	Clark et al. (1995)

For references see also Eglen et al. (1995c), Schiavi et al. (1994) and Ansanay et al. (1996).

2. 5-HT$_4$ Antagonists

The chemical structures of the 5-HT$_4$ antagonists are illustrated in Fig. 2. Their pK$_i$, pK$_b$ and 5HT$_3$/5-HT$_4$ receptor selectivity are shown in Table 2. They belong to five different classes:

1. The indole carboxylate class includes the first-described 5-HT$_4$ receptor antagonist, tropisetron. This is a weak antagonist (pK$_i$ = 6) and a potent 5-HT$_3$ antagonist (pK$_i$ = 9.4). This class also contains the most potent and selective 5-HT$_4$ antagonists GR 125487 (GALE et al. 1994a) and GR 113808 (GALE et al. 1994b; GROSSMAN et al. 1993; Figs. 2, 3). Both compounds are esters with short half-lives; however, GR 125487 possesses an action of longer duration than GR 113808.
2. The benzoates are also esters and potent but not selective 5-HT$_3$ over 5-HT$_4$ receptor antagonists (for a review see EGLEN et al. 1995c) (Fig. 2, Table 2). In addition, the derivative RS 23597 is a potent sigma-1 ligand(pK$_i$ = 8.4; CLARK et al. 1994; DOUGLAS et al. 1994). The benzoate dioxane (SB 204070), which belongs to the same class, also presents an ester moiety but it is one of the most potent and selective antagonists (Table 2). SB 204070 is, when iodinated, an excellent radiolabelled ligand (BROWN et al. 1993; GASTER and SANGER 1994) (under the name of SB 207710) (Figs. 2, 3, Table 2).
3. The benzimidazolone class contains a single compound, DAU 6285. It exibits similar affinity at both 5-HT$_3$ and 5-HT$_4$ receptors (Fig. 2, Table 2) but has the advantage of being stable in vivo (DUMUIS et al. 1992).
4. The imidazolpyridine class also contains a single compound: SC 53606, which does not discriminate well between 5-HT$_3$ and 5-HT$_4$ receptors (Table 2) but has the advantage of being stable in vivo (YANG et al. 1993).
5. The aryl ketone class contains phenyl ketone compounds substituted at position 2 by an O-phenyl group: RS 67532 and RS 39604 (Fig. 2; EGLEN et al. 1995a; HEDGE et al. 1995a), which are very potent and specific antagonists orally active with relatively long half-lives. This class also contains a benzoate dioxanyl ketone, RS 100235, related to the benzoate dioxane (SB 204070). It is to date the most potent and the most stable 5-HT$_4$ receptor antagonist (half-life after an i.v. dose of 3 mg/kg > 360 min; CLARK et al. 1995). This drug gives unsurmountable antagonism.

II. Are There 5-HT$_4$ Receptor Subtypes?

Correlation plots of agonist potencies in different assays often analysed in different species have revealed no evidence for 5-HT$_4$ receptor heterogeneity (BOCKAERT et al. 1992). Similarly, with the advent of specific radioligand binding, no discrepancies have been observed between the pharmacology of 5-HT$_4$-binding sites labelled with [^{3}H]GR 113808 and [^{125}I]SB 207710 and the pharmacology of 5-HT$_4$ receptors determined on functional pharmacological analyses (ANSANAY et al. 1996; BROWN et al. 1993; GROSSMAN et al. 1993; WAEBER et al. 1993). However, several unexplained observations deserve some comment. Benzamide and benzimidazolone derivatives, unlike many indoles, behave as partial agonists in guinea pig hippocampus (BOCKAERT et al. 1990), rat oesophagus, guinea pig ileum, guinea pig distal colon and human right atria and as antagonists in rat hippocampal CA_1 neurons (ANDRADE and CHAPUT 1991; BAXTER et al. 1991; CHAPUT et al. 1990; KAUMANN et al. 1990; OUADID et al. 1992; WARDLE and SANGER 1993). By contrast, in colliculi neurons, benzamides are full agonists and some of them, such as cisapride and renzapride, are super agonists (efficacy greater than that of 5-HT). A difference in coupling efficiency may explain these differences. It is surprising that benzamides are partial agonists in the most efficiently coupled systems (having high potency for 5-HT). For example, the EC_{50} for 5-HT is 100 times lower in distal guinea pig colon than in rat oesophagus, while benzamides are still partial agonists (WARDLE and SANGER 1993). Some difference in potency of a given drug was noted depending on the system. For example, 5-MeOT is as potent as 5-HT in most preparations but a weak agonist in rat oesophagus and guinea pig ileum (ELSWOOD et al. 1991; REEVES et al. 1991). Cisapride, which is the most potent agonist in colliculi neurons, is a relatively weak agonist in pig heart and rat oesophagus (for review see FORD and CLARKE 1993). Cloning has also revealed interesting differences between the EC_{50} for several drugs depending on the nature of the splice variant studied (GERALD et al. 1995). The splice variants varied in their intracellular domains. Such an influence of the intracellular domains of G-protein-coupled receptors has already been reported (SPENGLER et al. 1993). Finally, several "5-HT$_4$-like" receptors having non-orthodox 5-HT$_4$ receptor pharmacology have been described in bovine pulmonary artery and in isolated strips of the human urinary bladder detrusor muscle (BECKER et al. 1992; CORSI et al. 1991).

D. Transduction Mechanisms

I. cAMP as the Second Messenger Mediating 5-HT$_4$ Receptor Action

5-HT$_4$ receptors coupled to AC were first described in cultured colliculi neurons (DUMUIS et al. 1988a). The coupling of 5-HT$_4$ receptors to AC has been reported in many tissues such as guinea pig hippocampus (BOCKAERT et al. 1990), human cortex (MONFERINI et al. 1993), tunica muscularis mucosae (TMM) of oesophagus (FORD et al. 1992; RONDÉ et al. 1995) and human and piglet atria (KAUMANN 1990; KAUMANN et al. 1990, 1991). Such a preferential coupling of 5-HT$_4$ receptors to cAMP transduction systems has been confirmed when cloned 5-HT$_{4S}$ and 5-HT$_{4L}$ receptors could be expressed in COS-7 cells (GERALD et al. 1995). Many G-protein-coupled receptors can affect more than one signal transduction system. Therefore, it will not be surprising that some cellular functions of 5-HT$_4$ receptors may involve other second messengers. Such a hypothesis has been proposed for the release of acetylcholine in guinea pig myenteric plexus (KILBINGER et al. 1995) and for 5-HT$_4$-receptor-induced vagus nerve depolarization (discussed in the review by FORD and CLARKE 1993). However, in these two latter cases only indirect and preliminary data have been published. Following cAMP production, activation of protein kinase A (PKA) has been directly demonstrated in atrial tissues (SANDERS and KAUMANN 1992) and in neurons (ANSANAY et al. 1995; FAGNI et al. 1992).

II. Control of K$^+$ Channels

Voltage-clamp experiments show that in colliculi neurons, 5-HT$_4$ receptors inhibit a delayed rectifier voltage-dependent K$^+$ current (ANSANAY et al. 1995; FAGNI et al. 1992). This effect was mimicked by cAMP and other agents able to increase cAMP (forskolin, isoproterenol, PACAP) and blocked by PKA inhibitors. In current clamp experiments, 5-HT$_4$ receptors and cAMP have been shown to slow the falling phase of the action potential (Fig. 6A, C), to inhibit the afterhyperpolarization (AHP) which follows the action potential (Fig. 6A) and spike accommodation during a train of action potentials (Fig. 6B; ANSANAY et al. 1995). All these effects can be explained by inhibition of K$^+$ channels including Ca^{2+}-activated K$^+$ channels. In colliculi neurons a remarkable observation is that after a short application of 5-HT or 8-Br-cAMP (for a few seconds) all the effects described above slowly develop and remain stable for more than 30 min (Fig. 6). Indeed, a complete recovery of the blockade occurs only after 2 h (Fig. 6F). The long-term effect of 5-HT$_4$ receptor and 8-Br-cAMP stimulations can be prolonged up to 4 h in the presence of okadaic acid, an inhibitor of phosphatases 1 and 2A (ANSANAY et al. 1995). It is not due to a prolonged activation of PKA but to a prolonged inhibition of phosphatases. The reduced phosphatase activity could result from a sustained activity of cAMP-regulated phosphatase inhibitors such as phosphatase inhibitor 1 or DARPP-32. How the sustained activity of these inhibitors could be maintained after the transient activation of PKA remains to be demonstrated. In addition to the reduction of phosphatase activity, a long-lived phosphorylation of the K$^+$ channels cannot be excluded. Figure 7 is a schematic representation of the data in colliculi neurons. Also illustrated is the hypothesis that the long-lasting broadening of the action potential and reduction of the AHP may result in a prolonged increase in neurotransmitter release. 5-HT$_4$ receptors are indeed implicated in stimulating: (1) acetylcholine release in gastrointestinal tract as well as in rat frontal cortex (CONSOLO et al. 1994; KILBINGER et al. 1995; KILBINGER and WOLF 1992); (2) dopamine release in striatal slices and in vivo (BENLOUCIF et al. 1993; BONHOMME et al. 1995; GE and BARNES 1995b; STEWARD et al. 1995; STOWE and BARNES 1995); and (3) 5-HT release in rat hippocampus in vivo (GE and BARNES 1995a).

In CA$_1$ neurons of hippocampus, activation of 5-HT$_4$ receptors increases membrane excitability by reducing the Ca^{2+}-activated potassium current responsible for the slow AHP obtained in these cells. Although benzamides seem to be partial agonists or even antagonists in this preparation (see above), the 5-HT$_4$ antagonists are very potent and the transduction pathway includes

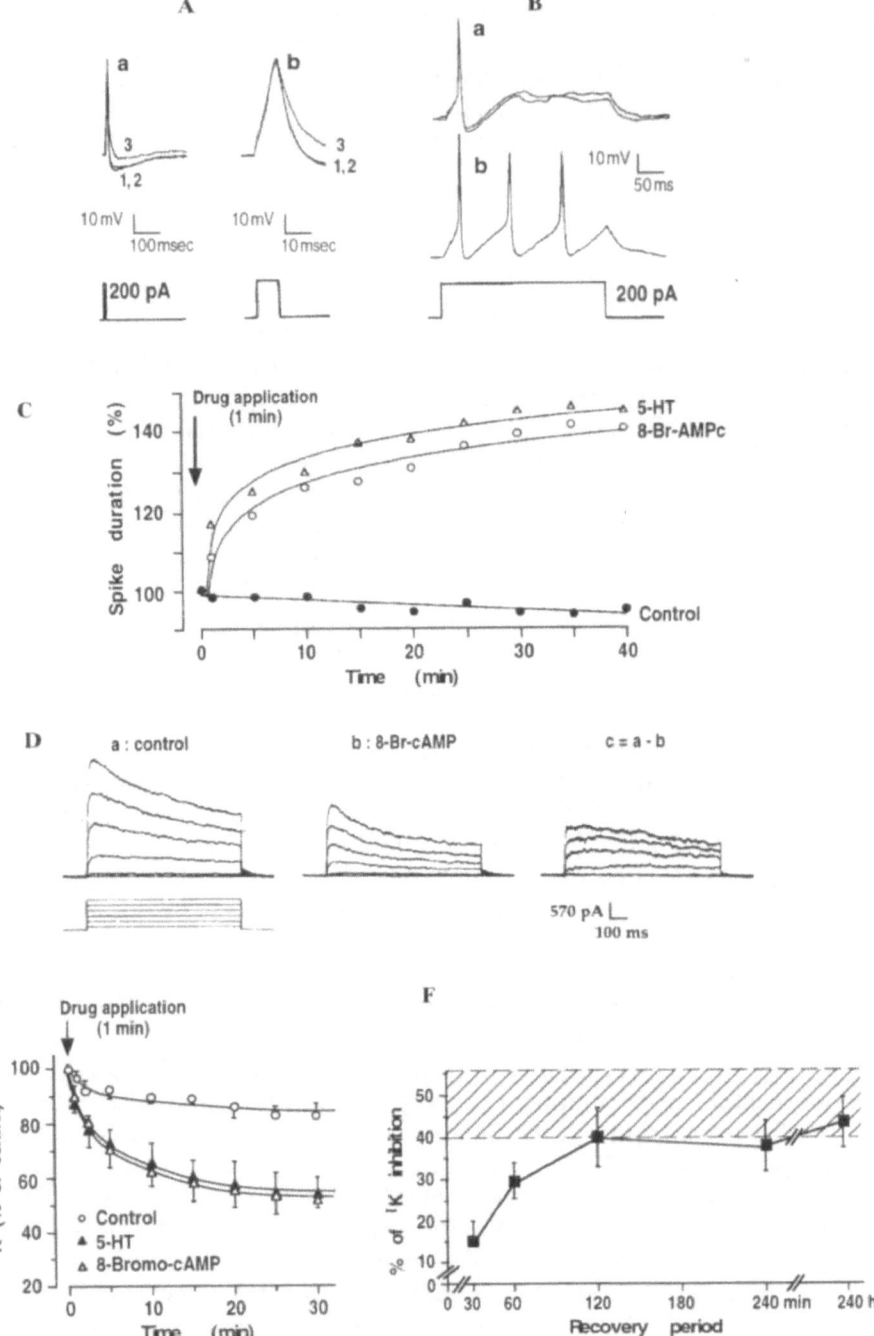

cAMP-mediated events (ANDRADE and CHAPUT 1991; CHAPUT et al. 1990; TORRES et al. 1994, 1995). Interestingly, the effect of 5-HT$_4$ receptor stimulation is modulated by short-term and long-term treatments with corticosterone and mineralocorticoids (BIRNSTIEL and BECK 1995). Whether or not the modulation of adrenal medulla steroid release by 5-HT$_4$ agonists (see below) influences central 5-HT$_4$ receptor effects remains to be demonstrated.

III. Control of Ca^{2+} Channels

In atrial myocytes, a 5-HT$_4$-receptor-induced increase in cAMP is followed by facilitation of the L-type Ca^{2+} current to the same extent as in catecholamines,

◄ ──

Fig. 6A–F. cAMP- and 5-HT-induced changes in action potential, AHP, accommodation and K$^+$ current in cultured mouse colliculi neurons. In **A** and **B** recordings were obtained in the whole-cell current-clamp mode. **A** Traces b are the expanded version of traces a (*horizontal calibration bar* is 100 ms for traces a and 10 ms for traces b). The superimposed traces 1 and 2 were recorded at 30-min intervals before a 1-min application of 1 μM 5-HT. Traces 3 were recorded after a 30-min washout of the drug. Similar results were obtained with 100 μM 8-Br-cAMP. **B** *Panel a* represents two superimposed recordings obtained at a 30-min interval before a 1-min application of 1 μM 5-HT. *Trace b* was recorded after 30-min washout of the drug. Similar results were obtained with 100 μM 8-Br-cAMP. **C** Spike duration measured from the beginning of the EPSP to the time at which the resting potential was restored (just before the beginning of AHP) and expressed as percentage of the control value, under different conditions: without drug application (control, *filled circles*) and after 1 min application of 100 μM 8-Br-cAMP or 100 μM 5-HT. *Arrow* indicates the time at which drug was applied and therefore does not apply for the control condition. **D** In this and the following figures, recordings were obtained in the voltage-clamp mode. K$^+$ currents evoked by successive depolarizing pulses before (*a*) and 30 min after a brief application of 100 μM 8-Br-cAMP (*b*), in a similar cell. *Traces in c* represent the 8-Br-cAMP-inhibited currents obtained by subtracting those of *b* from those of *a*. Similar results were obtained with 100 μM 5-HT. The voltage traces in *a* apply also for panels *b* and *c* and represent voltage steps from − 60 mV (holding potential) to −50, −30, −10, +10, +30 and +50 mV. **E** Potassium current amplitude measured after a 1-min application of either 5-HT (1 μM; *n* = 7) or 8-Br-cAMP (100 μM, *n* = 9; mean ± SEM) and during washout of the drugs, and expressed as percentage of the K$^+$ current measured before each drug application. Stable control recordings were taken for 15 min before drug application (not shown). *Arrow* indicating the time of the drug applications does not apply to the control curve (*n* = 7). **F** In these experiments, after a first application of 8-Br-cAMP (100 μM), which inhibited K$^+$ current, subsequent applications of this drug or 5-HT (1 μM) in the same cell did not induce further inhibition of the current, at least within a 30-min period, probably because of a long-lasting and full inhibition of the current by the agonists. We took advantage of this particular characteristic to measure the period necessary for the K$^+$ current to recover from the 5-HT$_4$/cAMP-induced inhibition (recovery period; *x*-axes). Thus, colliculi cultures underwent a 1-min conditioning treatment with 8-Br-cAMP (100 μM) and the effect of an identical application of the drug (test application) was measured 30 min to 2 h later in several neurons (*n* = 9). Indeed, current amplitude was measured before (*a*) and 20 min after (*b*) the test application. The *y*-axis is the percentage inhibition of the K$^+$ current calculated as follows: [(*a*-*b*)/*a*] × 100]. *Shaded area* represents K$^+$ current inhibition achieved by cAMP stimulation in naive cells. Note that the inhibitory effect of cAMP lasted for up to 2 h, in the absence of drug

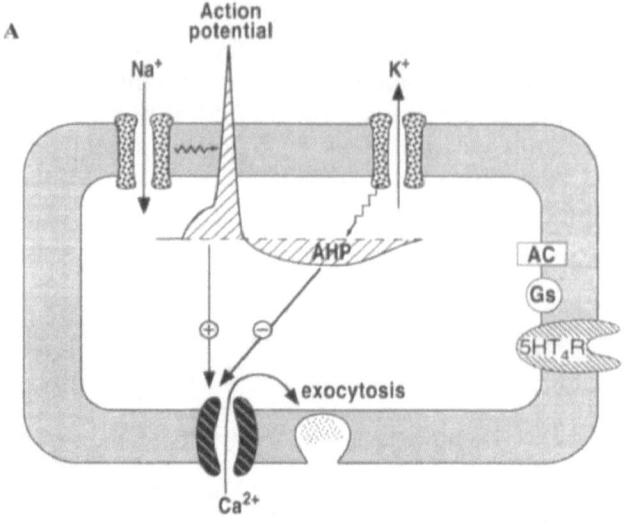

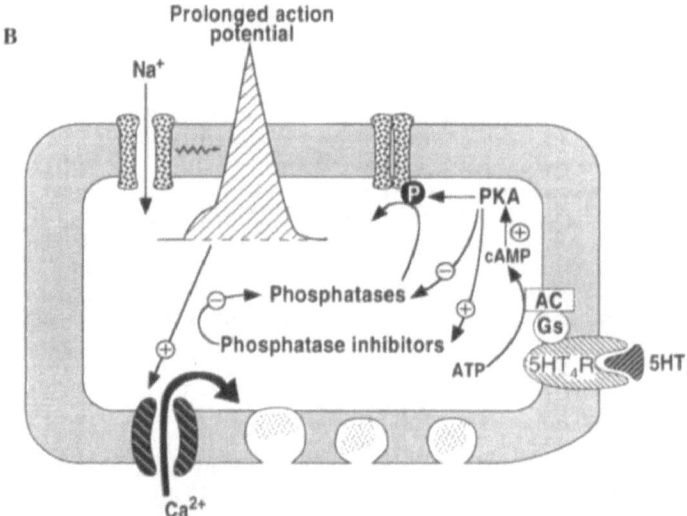

Fig. 7A,B. Tentative hypothesis relative to the effect of 5-HT$_4$ receptor stimulation on neurotransmitter release. **A** Action potential opens voltage-dependent Ca^{2+} channels. The resulting Ca^{2+} influx will trigger neurotransmitter exocytosis. The AHP that follows action potential results from activation of voltage-dependent K$^+$ channels and shuts off the Ca^{2+}channels. This will stop neurotransmitter exocytosis. **B** Activation of 5-HT$_4$ receptors and consequent phosphorylation of voltage-activated K$^+$ channels will block AHP. As a consequence, the action potential is significantly prolonged. This should increase the Ca^{2+} influx and neurotransmitter release. This synaptic modulation should last as long as K$^+$ channels remain phosphorylated: 2h in our model. Such a prolonged phosphorylation state of K$^+$channels in colliculi neurons might result from PKA-mediated sustained activation of protein phosphatase inhibitors or sustained inhibition of phosphatases an effect involving PKA activation (JAHNEL et al. 1992; OUADID et al. 1992). A similar cAMP, PKA-mediated pathway is involved in activation of T-type Ca^{2+} channels in frog adrenocortical cells (CONTESSE et al. 1996).

IV. Desensitization

5-HT₄ receptors desensitize very rapidly in various preparations (CRAIG et al. 1990; DUMUIS et al. 1989b, 1991; RONDÉ et al. 1995). In colliculi neurons and rat oesophagus, we have analysed the molecular mechanisms of this desensitization (ANSANAY et al. 1992; RONDÉ et al. 1995). More than 50% of desensitization occurred following 5-min exposure to 5-HT. This rapid desensitization was not mediated or produced by cAMP or other agonists that increase cAMP in these cells. This suggested that the desensitization was homologous. In the β-adrenergic system, homologous desensitization is dependent upon phosphorylation of the agonist-occupied receptor by a specific kinase: the βARK (BENOVIC et al. 1989). A good correlation is observed between the affinities (pEC_{50}) of a series of agonists and the percentage of desensitization they are inducing. No correlation was obtained between the efficacies (E_{max}) of 5-HT₄ agonists and their abilities to desensitize 5-HT₄ receptors (ANSANAY et al. 1992). This indicates that homologous desensitization is a function of mean time of receptor occupancy by the agonist. This is consistent with the fact that receptor kinases (such as βARK) phosphorylate only occupied receptors. One consequence of this observation is that following activation with agonists having a low potency (such as metoclopramide) the receptor desensitization has a much slower time course than in the presence of potent agonists (such as cisapride). TURCONI et al. (1991) reported that low desensitizing properties of the partial agonist BIMU 1 may explain its good clinical efficacy in the alimentary tract compared to BIMU 8, which induces rapid desensitization. This may have important therapeutic consequences.

During rapid desensitization (5-HT stimulatory period shorter than 30 min), the homologous desensitization is not associated with any loss in 5-HT₄-binding sites labelled with [³H]GR 113808 (ANSANAY et al. 1996). However, if the 5-HT₄ receptor stimulation is prolonged over 30 min, a loss of binding sites is apparent (ANSANAY et al. 1996).

In vivo, lesions of 5-HT neurons induced an upregulation of 5-HT₄ receptor expression in caudate putamen, nucleus accumbens and hippocampus (COMPAN et al. 1996). This suggests that 5-HT₄ receptor density is regulated by a tonic activation by 5-HT in these brain areas. A tonic activation of 5-HT₄ receptors is also suggested by the inhibition of basal 5-HT and dopamine release by the 5-HT₄ receptor antagonist GR 113808 (GE and BARNES 1995a, b).

Contraction of guinea pig ileum by 5-HT is due to activation of both 5-HT₃ and 5-HT₄ receptors. In order to study separately the effect of these two receptors, it is possible to desensitize either 5-HT₃ or 5-HT₄ receptors. Indeed, 5-MeOT is able to completely desensitize 5-HT₄ responses without affecting

5-HT$_3$ responses, whereas 2-Me-5-HT desensitizes 5-HT$_3$ responses without affecting the 5-HT$_4$ responses (Craig et al. 1990). A potent desensitization of 5-HT$_4$ receptors following prolonged exposure to 5-HT and zacopride has also been observed in frog and human adrenocortical tissues (Idres et al. 1991; Lefèbvre et al. 1992), whereas 5-HT$_4$ receptors in heart seem to be less susceptible to desensitization (Kaumann 1994).

E. 5-HT$_4$ Receptor Distribution and Functions

I. Central Nervous System

The precise regional distribution of 5-HT$_4$ receptors has only been obtained recently, using radioligand and autoradiographic analysis of [^{3}H]GR 113808 and [^{125}I]SB 207710 binding (Fig. 8). They have been performed in rat (Grossman et al. 1993; Jakeman et al. 1994; Waeber et al. 1994), mouse (Waeber et al. 1994), guinea pig (Domenech et al. 1994; Grossman et al. 1993; Jakeman et al. 1994; Waeber et al. 1993, 1994), pig (Schiavi et al. 1994), calf (Domenech et al. 1994), monkey (Jakeman et al. 1994) and humans (Domenech et al. 1994; Reynolds et al. 1995; Waeber et al. 1993). The main finding is a heterogeneous and comparable distribution in adult brain of different species (Fig. 8) with few interspecies differences in globus pallidus, substantia nigra and interpeduncular nucleus (Waeber et al. 1994). The highest densities (100–400 fmol/mg protein) are found in several regions belonging to the limbic system (islands of Calleja, olfactory tubercle, fundus striati, ventral pallidum, septal region, hippocampus, amygdala) or known to be components of different pathways, such as hippocampo-habenulo-interpeduncular and striato-nigro tectal pathways (Fig. 8). All brain regions express the 5-HT$_{4L}$ form of the receptor whereas the 5-HT$_{4S}$ form is mainly expressed in striatum (Gerald et al. 1995). Lesion studies in rat as well as postmortem studies of patients suffering from Parkinson's disease indicate that in the striato-nigral pathway the 5-HT$_4$ receptors are not located on dopaminergic neurons (Compan et al. 1996; Mengod et al. 1995; Patel et al. 1995; Reynolds et al. 1995). In contrast, lesions of neurons having their cell bodies in striatum (with kainic acid or ibotenic acid) result in a loss of 5-HT$_4$ receptors in striatum. In addition, following kainic acid lesions, a loss of binding sites in the dorsolateral part of the substantia nigra has been found (Compan et al. 1995; Patel et al. 1995). In addition, Compan et al. (1996) have found a loss of [^{3}H]GR 113808-binding sites in globus pallidus which was not found by Patel et al. (1995). This indicates a localization of 5-HT$_4$ receptors in cell bodies of striatal neurons that project to the substantia nigra and/or globus pallidus.

Such a localization is in agreement with a subtantial loss of 5-HT$_4$ receptors in putamen of patients suffering from Huntington's disease (Reynolds et al. 1995). In vitro studies using rat striatal slices or in vivo microdialysis experiments also suggest that 5-HT$_4$ receptors facilitate dopamine release

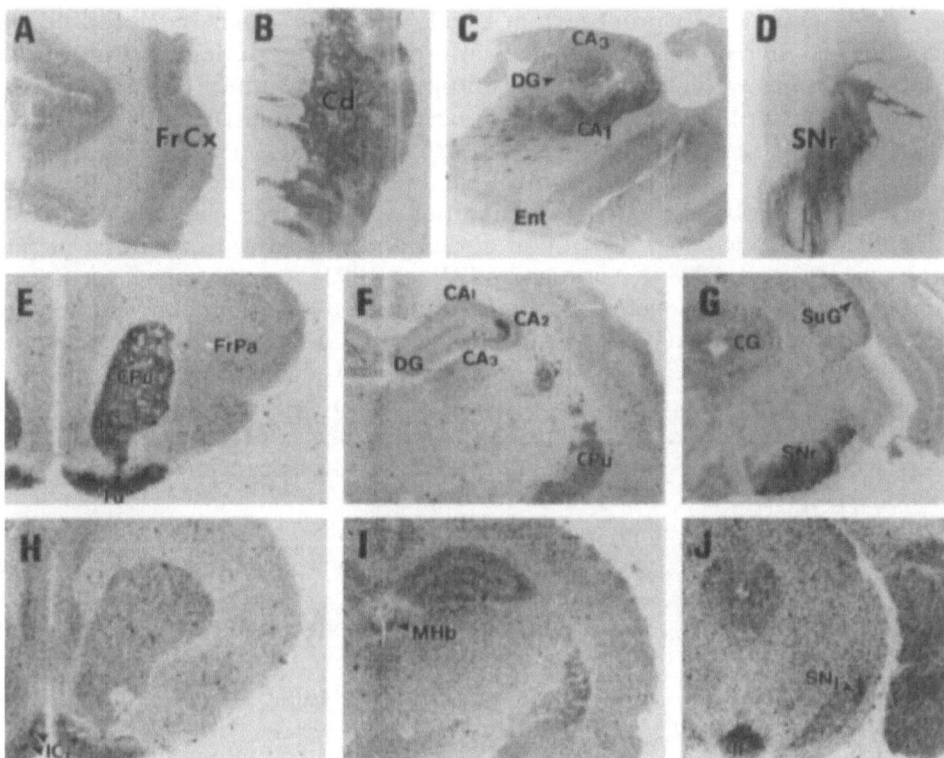

Fig. 8A–J. Autoradiographic localization of [³H]GR 113808-binding sites at different levels of human, guinea pig and rat brain. Pictures from autoradiograms generated by incubating human (**A–D**) guinea pig (**E–G**) or rat (**H–J**) brain sections in the presence of the 5-HT₄ receptor antagonist: [³H]GR 113808. The distribution of 5-HT₄-binding sites is comparable in all species: high densities of sites being found in the caudate (*Cd*), caudate-putamen (*CPu*), hippocampal formation (*CA₁*, *CA₂*, *CA₃* subfields of Ammon's horn), dentate gyrus (*DG*) and substantia nigra (*SNr* or *SN₁*) while lower densities are observed in cortical areas [frontal cortex (*FrCx*), frontoparietal cortex (*FrPa*)]. Some species differences are nevertheless noteworthy. The labelling in the substantia nigra is concentrated over the reticular part in human and guinea pig, in the lateral part in rat. The interpeduncular nucleus (*IP*) contains high densities of sites in rats, but not in guinea pig (this region has not been studied in humans). *Tu*, olfactory tubercle; *ICj*, islands of Calleja; *Ent*, entorhinal cortex; *MHb*, medial habenula; *SuG*, superficial grey layer of the superior colliculus; *CG*, periaqueductal central grey. (Data taken from WAEBER et al. 1993, 1994)

(BENLOUCIF et al. 1993; BONHOMME et al. 1995; GE and BARNES 1995b; STEWARD et al. 1995; STOWE and BARNES 1995). In vivo, intrastriatal injections of GR 113808 reduced basal dopamine release in one study (GE and BARNES 1995b) but not in another (POZZI et al. 1995). One report shows that blockade of 5-HT₄ receptors in substantia nigra attenuate morphine-stimulated but not basal dopamine release in rat striatum. Since 5-HT₄ receptors are not localized on dopamine neurons, the mechanisms by which 5-HT₄ ligands regulate dopamine release in striatum remain unexplained.

It is tempting to propose that 5-HT$_4$ receptors might be localized on cholinergic nerve endings in the interpeduncular nucleus arising from habenula since we have found that ontogenic expression of [^{3}H]GR 113808-binding sites in the interpeduncular nucleus is concomitant with activity of choline acetyltransferase in this nucleus (CONTESTABILE et al. 1987). The possibility that 5-HT$_4$ receptors control cholinergic neuron activity is also suggested by the observation that stimulation of the 5-HT$_4$ receptor increases acetylcholine release in rat frontal cortex and guinea pig myenteric plexus (KILBINGER et al. 1995; KILBINGER and WOLF 1992). In addition, there is evidence for clear neuroanatomical substrates of a serotoninergic/cholinergic interaction in brain (CASSEL and JELTSCH 1995). Finally, 5-HT$_4$ receptor agonists such as zacopride and renzapride reverse decreases in electro-encephalographic energy induced by scopolamine (BODDEKE and KALKMAN 1990). A problem of unambiguously attributing these effects to 5-HT$_4$ receptors was the lack of stereospecificity of (S)- and (R)-zacopride (BODDEKE and KALKMAN 1992).

We have discussed above the long-term blockade of K$^+$ channels in colliculi neurons triggered by a brief stimulation of 5-HT$_4$ receptors (ANSANAY et al. 1995). This process mirrors the one seen in *Aplysia californica* in which 5-HT-induced cAMP generation inhibits K$^+$ currents, resulting in a persistent facilitation of evoked transmitter release. These events provide a mechanistic basis for the gill withdrawal reflex in *Aplysia*, an elementary model of learning (HOCHNER and KANDEL 1992; KANDEL and SCHWARTZ 1982). For this reason we proposed several years ago that 5-HT$_4$ receptors might play a role in memory of vertebrates (BOCKAERT et al. 1992, 1994). Recently the important role of cAMP in invertebrate and vertebrate learning has been recognized (BOURTCHULADZE et al. 1994; KANDEL and ADEL 1995; YIN et al. 1994; ZHONG 1995). The direct role of 5-HT$_4$ receptors in cognition has recently been investigated. Preliminary data indicate that RS 67333, a potent and selective 5-HT$_4$ agonist (Fig. 1), reverses the rat performance deficit induced by atropine in the Morris Water Maze learning test, an effect reversed by the selective 5-HT$_4$ receptor antagonist RS 67532 (EGLEN et al. 1995a, c). None of these drugs have any effect on the performances of rats not treated with atropine (EGLEN et al. 1995a). Similarly, BIMU 1 and BIMU 8 improve performance after hypoxia-induced amnesia in mice (GHELARDINI et al. 1994). We have also demonstrated that BIMU 1, a mixed 5-HT$_3$ receptor antagonist/5-HT$_4$ receptor agonist, improves short-term olfactory memory. This effect was mediated by 5-HT$_4$ receptors since it was blocked by doses of GR 125487 necessary for blocking central 5-HT$_4$ receptors. Ondansetron, a specific 5-HT$_3$ antagonist, was inactive in inhibiting the BIMU 1 effect. A role of acetylcholine in this 5-HT$_4$-mediated effect was suggested because arecoline, a muscarinic agonist, mimicked the BIMU 1 effect and scopolamine, a muscarinic antagonist, blocked both BIMU 1 and arecoline effects (LETTY et al. 1996). Three other functions of 5-HT$_4$ receptors possibly mediated by central effects have been reported but need further confirmation: (1) an anti-anxiolytic effect of the 5-

HT₄ receptor antagonist SDZ 205557 observed under high serotoninergic tone (Costall and Naylor 1993), (2) an analgesic response (Ghelardini et al. 1990, 1996) mediated by 5-HT₄ agonists and (3) a reduction in ethanol intake observed in alcohol-preferring rats (Panocka et al. 1995). The BIMU 1- and BIMU 8-induced analgesic effects are blocked by anticholinergic drugs such as atropine and hemicholinium but not by naloxone. Morever, analgesic effects are obtained by intracerebroventricular injection of BIMU 1 at doses which are ineffective following parental routes (Ghelardini et al. 1996). This suggests a central-mediated effect. Developmental studies in rodent indicate that expression of 5-HT₄ receptors is essentially prenatal in the pons area and postnatal in most other brain areas. Although the density of 5-HT₄ receptors is stable during postnatal life, a decline is observed in globus pallidus and pons (Waeber et al. 1994).

II. Peripheral Nervous System

The 5-HT₄-receptor-induced contraction of guinea pig ileum is blocked by atropine as well as tetrodotoxin (Buchheit et al. 1985; Craig and Clarke 1990). In this preparation, 5-HT₄ receptors stimulate acetylcholine release (Kilbinger et al. 1995; Kilbinger and Wolf 1992). These results indicate involvement of presynaptic cholinergic neurons on which 5-HT₄ receptors are presumably localized (Ford and Clarke 1993; Tonini et al. 1991). A role for nonadrenergic, noncholinergic (NANC) neurotransmission as well as neurokinin B (Ramirez et al. 1994) in 5-HT₄ receptor-mediated prokinesis has also been proposed (Ford and Clarke 1993). In alimentary tract, some other 5-HT₄ receptor-mediated responses are also likely to be mediated by an increase in neurotransmitter release. There is contraction of guinea pig ascending colon (Briejer et al. 1993; Elswood et al. 1991; Wardle and Sanger 1993) and stimulation of Cl⁻ secretion in guinea pig ileal mucosa (Leung et al. 1995). In guinea pig myenteric neurons, 5-HT₄ receptor agonists potentiate the nicotinic fast excitatory postsynaptic potential (EPSP) via an increase in acetylcholine release (Nemeth and Gullickson 1989; Tonini et al. 1991), which facilitates cholinergic transmission between myenteric neurons (Rizzi et al. 1992). Rat vagus nerve depolarization can be triggered by 5-HT₄ receptor activation (Bley et al. 1994; Coleman and Rhodes 1995; Rhodes et al. 1992). Enterochromaffin cells express a 5-HT₄ receptor which inhibits 5-HT release (Gebauer et al. 1993). In human urinary bladder, 5-HT₄ receptors facilitate a cholinergic-mediated contractile response to electrical stimulation (Corsi et al. 1991).

III. Alimentary Tract

More than 80% of the 5-HT cell bodies is present in enterochromaffin cells or enteric neurons and therefore one would expect an important role of this amine in alimentary tract (Costa et al. 1982; Erspamer 1966). One characteristic of the cellular distribution of 5-HT$_4$ receptors in different parts of the alimentary tract is the high species variation. Table 1 gives the different effects of 5-HT$_4$ receptors according to the species and organ considered.

1. Excitatory Responses

We have already discussed the discovery by CRAIG and CLARKE that 5-HT$_4$ receptors mediate, together with 5-HT$_3$ receptors, the 5-HT-induced stimulation of guinea pig ileum contraction and how they have studied each receptor separately by inducing receptor-specific desensitization (CRAIG et al. 1990). Soon after, it was demonstrated that 5-HT$_4$ receptor stimulation facilitates the peristaltic reflex in isolated segments of guinea pig ileum (BUCHHEIT and BUHL 1991; CRAIG and CLARKE 1991). It seems, however, that 5-HT$_4$ receptors are not tonically active on the endogenous ascending and descending reflex pathways (YUAN et al. 1994). 5-HT$_4$ receptor-mediated contractions have also been demonstrated in guinea pig colon, in which the sensitivity to 5-HT is 100 times greater than in rat oesophagus (see the discussion on 5-HT$_4$ receptor subtypes; BRIEJER et al. 1993; ELSWOOD et al. 1991; WARDLE and SANGER 1993). In guinea pig, rat, human and dog stomach, 5-HT$_4$ receptors induce gastroprokinetic effects and gastric emptying (BINGHAM et al. 1995; BUCHHEIT and BUHL 1994; GULLIKSON et al. 1993; HEDGE et al. 1995b; RIZZI et al. 1994).

2. Inhibitory Responses

One of the most-studied 5-HT$_4$ receptor effects is certainly the relaxation of rat oesophagus (for a review see FORD and CLARKE 1993). After the pioneering work of BIEGER and TRIGGLE (1985), several groups unequivocally demonstrated that 5-HT-induced relaxation of the rat oesophagus was mediated via 5-HT$_4$ receptors displaying the same pharmacology as those described in colliculi, hippocampal neurons and guinea pig ileum (see above; BAXTER et al. 1991; REEVES et al. 1991). 5-HT$_4$ receptors are expressed by the smooth muscle itself and their relaxing effect is mediated via cAMP accumulation (see above). It seems that guinea pig, rabbit and dog oesophagal smooth muscles do not express 5-HT$_4$ receptors (COHEN et al. 1994). It is possible that many parts of the gastrointestinal tract relax in the presence of 5-HT via 5-HT$_4$ receptors present in smooth muscle. Of these, we can cite the rat ileum (MC LEAN et al. 1995; TULADHAR et al. 1991) and human colon (McLEAN et al. 1995; TAM et al. 1995).

3. Secretory Responses in the Alimentary Tract

5-HT is a potent secretagogue. In some species, 5-HT increases, in several regions of the alimentary tract, short-circuit current (I_{sc}) due to an increase in Cl$^-$ secretion. In guinea pig ileum, the receptors involved appear to be 5-HT$_3$ and 5-HT$_4$ receptors (LEUNG et al. 1995; SCOTT et al. 1992). They are localized on neuronal and non-neuronal elements. In rat, 5-HT$_4$ receptors do not make any contribution to the electrically neuronal-mediated 5-HT intestinal secretory response in jejunum in vivo, but may play a role in the ileum and colon (FRANKS et al. 1995). In human ileum and jejunum, a 5-HT$_4$ receptor mediates a non-neuronal increase in secretory response (BORMAN and BURLEIGH 1993; BURLEIGH and BORMAN 1993). In mice, 5-hydroxytryptophane (5-HTP), the endogenous precursor of 5-HT, increases faecal pellet output and sustained diarrhoea. In both cases, inhibition of 5-HT$_4$ receptors with specific 5-HT$_4$ antagonists greatly reduces the 5-HPT effect (BANNER et al. 1993; HEDGE et al. 1994).

IV. Urinary Bladder

In isolated strips of rhesus and cynomolgous monkey urinary bladder, and under conditions designed to favour pharmacological isolation of 5-HT$_4$ receptors, 5-HT$_4$ agonists inhibit electrical and cholinergic-stimulated contractions. The inhibitory 5-HT$_4$ receptor-mediated effect appears to be located postjunctionally on smooth muscles (FORD and CLARKE 1993; WAIKAR et al. 1994). In contrast, "atypical" 5-HT$_4$ receptors (see above), likely present in neurons, stimulate contractions of the detrusor muscle of human urinary bladder. Further work is needed to fully characterize the receptor involved.

V. Adrenal Gland

In frog adrenal gland, 5-HT$_4$ receptors stimulate both corticosterone and aldosterone release (CONTESSE et al. 1994; IDRES et al. 1991). This effect appears to be mediated by cAMP production, PKA activation and stimulation of Ca^{2+} channels of the T type (see above; CONTESSE et al. 1996). In human adrenocortical slices, 5-HT most likely released from intra-adrenal mast-like cells stimulates cortisol secretion (LEFÈBVRE et al. 1992). In vitro, zacopride, a 5-HT$_4$ receptor agonist, mimicked this effect, but this drug was much more potent at stimulating aldosterone release (LEFÈBVRE et al. 1993). In healthy volunteers, metoclopramide, zacopride and cisapride stimulate aldosterone release but are much less potent stimulators of cortisol release (LEFÈBVRE et al. 1993, 1994). This effect appears to be direct on glomerulosa cells rather than centrally mediated. Such an effect on aldosterone release is expected to induce hypertension in patients chronically treated with 5-HT$_4$ agonists. However, such a hypertension has never been reported. A more rapid desensitization of this 5-HT$_4$ response compared to the gastrointestinal response may explain this result (LEFÈBVRE et al. 1993).

VI. Heart

Tachycardia is induced by 5-HT in many species including man. However, the 5-HT receptors involved are almost as diverse as the species investigated (SAXENA and VILLALÒN 1991). SAXENA and coworkers reported that tachycardia induced by 5-HT in pigs was mediated by a receptor different from $5-HT_1$, $5-HT_2$ and $5-HT_3$ receptors and subsequently demonstrated its identity with the $5-HT_4$ receptors (Bom et al. 1988; Villalòn et al. 1990). Recently, they demonstrated that $5-HT_4$ receptors are located solely in the atria, where they have a stimulatory effect on heart rate (Saxena et al. 1992; Schoemaker et al. 1992). In parallel with those studies, Kaumann and coworkers demonstrated, in right atrial appendages obtained from patients undergoing open heart surgery, that $5-HT_4$ receptors increase contractile force and arrhythmias (Kaumann 1993; Kaumann and Sanders 1994; Kaumann et al. 1990, 1991). Interestingly, the arrhythmias induced in vitro were more severe in strips obtained from patients treated with β-adrenergic-blocking agents (Kaumann and Sanders 1994; Kaumann 1990; Kaumann et al. 1990; Ouadid et al. 1992). As in pig, $5-HT_4$ receptor expression in human heart is restricted to the atria and they are not present in ventricular muscles (Jahnel et al. 1992). In piglet atria, as well as in human atria, the $5-HT_4$ effects are mediated by cAMP accumulation, PKA activation and increase in L-type Ca^{2+} current (see above; Kaumann 1990; Kaumann et al. 1990; Ouadid et al. 1992).

F. Present and Future Therapeutic Drugs Acting at 5-HT₄ Receptors

I. Alimentary Tract

Metoclopramide and cisapride have been widely used in the treatment of dyspepsia, gastro-oesophageal reflux disease and, to some extent, gastroparesis (Bockaert et al. 1994; Ford and Clarke 1993). The mechanism of the therapeutic effects of these drugs is not fully understood. It certainly includes oesophageal relaxation combined with gastrokinesis and enhanced gastric emptying (see above). The putative use of $5-HT_4$ agonists in the treatment of constipation is not well established since metoclopramide and cisapride also have a $5-HT_3$ antagonistic activity known to cause constipation. However, cisapride has been shown to overcome the colonic stasis that occurs during constipation and to cause diarrhoea in healthy volunteers (Müller-Lissner 1987; Sanger and Gaster 1994). The observations that 5-HT induces diarrhoea in humans, and secretory responses in human jujenum and ileum, suggest that $5-HT_4$ receptor antagonists could be useful in treatments of specific diseases created by an abnormally high turnover of endogenous 5-HT (e.g. in diarrhoea predominant irritable syndrome, IBS). The fact that $5-HT_4$ antagonists inhibit 5-HTP-induced diarrhoea in mice is also encouraging for such a therapeutic action (Hedge et al. 1994; Sanger and Gaster 1994). The presence of $5-HT_4$ receptors in the vagus nerve (see above) could be at the origin of the emetic action of copper sulphate and of zacopride in ferret and dog (Andrews et al. 1988; Bhandari and Andrews 1991; Fukui et al. 1994).

II. Central Nervous System

For reviews on possible therapeutic usefulness of 5-HT$_4$ ligands see EGLEN et al. (1995c) and SANGER and GASTER (1994). We have discussed evidence for a role of 5-HT$_4$ receptors in memory and in reversing some deficits in memory performance in rodents. Thus, 5-HT$_4$ agonists may provide a novel therapeutic approach for cognitive disorders. 5-HT$_4$ antagonists may be useful as anxiolytics. Finally, a central reduction in pain perception may be obtained with 5-HT$_4$ agonists (GHELARDINI et al. 1990, 1996).

III. Cardiovascular System

Cisapride, a weak partial agonist at human atrial 5-HT$_4$ receptors, has been shown to generate supraventricular arrhythmias and tachycardia in humans (OLSSEN and EDWARDS 1992; for a review see KAUMANN 1994). It has been proposed by KAUMANN that 5-HT, released from platelet aggregation following stagnant circulation (observed for example in elderly patients), can induce atrial fibrillation. Chronic atrial fibrillation may then facilitate intra-atrial blood clotting, which in turn would generate more 5-HT, inducing a self-perpetuation of the arrhythmia. 5-HT$_4$ antagonists may therefore be able to brake this positive feedback cycle and attenuate thromboembolic events associated with atrial arrhythmia KAUMANN (1994).

IV. Urinary System

Cisapride and metoclopramide have been shown to increase urinary incontinence in humans (PILLANS and WOOD 1994). If tonic 5-HT$_4$-receptor-mediated contractions of the urinary bladder exist, 5-HT$_4$ receptor antagonists may be of interest for the treatment of incontinence.

G. Conclusion

Within less than 10 years, 5-HT$_4$ receptors have gained a birth certificate, pharmacological legitimacy and a gene structure. Numerous specific ligands have been synthesized, although more specific agonists crossing the blood-brain barrier are still needed. 5-HT$_4$ receptor agonists were introduced to gastrointestinal therapeutics before the identification of their target. New 5-HT$_4$ agonists and antagonists should soon be introduced for the treatment of other peripheral and central diseases.

References

Andrade R, Chaput Y (1991) 5-Hydroxytryptamine$_4$ receptors mediate the slow excitatory response to serotonin in the rat hippocampus. J Pharmacol Exp Ther 257:930–937

Andrews PL, Rapeport NG, Sanger GJ (1988) Neuropharmacology of emesis induced by anticancer therapy. Trends Pharmacol Sci 9:334–341

Ansanay H, Sebben M, Bockaert J, Dumuis A (1992) Characterization of homologous 5-HT$_4$ receptor desensitization in colliculi neurons. Mol Pharmacol 42:808–816

Ansanay H, Dumuis A, Sebben M, Bockaert J, Fagni L (1995) A cyclic AMP-dependent, long-lasting inhibition of a K$^+$ current in mammalian neurons. Proc Natl Acad Sci USA 92:6635–6639

Ansanay H, Sebben M, Bockaert J, Dumuis A (1996) Pharmacological comparison between [^{3}H]GR 113808 binding sites and functional 5-HT$_4$ receptors in neurons. Eur J Pharmacol 298:165–174

Banner SE, Smith MI, Bywater D, Sanger GJ (1993) 5-HT$_4$ receptor antagonism by SB 204070 inhibits 5-hydroxytryptophan-evoked defaecation in mice. Br J Pharmacol 110:16P

Baxter GS, Craig DA, Clarke DE (1991) 5-HT$_4$ receptors mediate relaxation of the rat oesophageal tunica muscularis mucosae. Naunyn Schmiedeberg's Arch Pharmacol 343:439–446

Baxter GS, Boyland P, Gaster LM, King FD (1993) Quaternised renzapride as a potent and selective 5-HT$_4$ receptor agonist. Biorg Med Chem Lett 3:633–634

Becker BN, Gettys TW, Middleton JP, Olsen CL, Albers FJ, Lee SL, Fabburg BL, Raymond JR (1992) 8-Hydroxy-2-(di-n-propylamino)tetralin-responsive 5-hydroxytryptamine$_4$-like receptor expressed in bovine pulmonary artery smooth muscle cells. Mol Pharmacol 42:817–825

Benloucif S, Keegan MJ, Galloway MP (1993) Serotonin-facilitated dopamine release in vivo: pharmacological characterization. J Pharmacol Exp Ther 265:373–377

Benovic JL, DeBlasi A, Stone WC, Caron MG, Lefkowitz RJ (1989) β-Adrenergic receptor kinase: primary structure delineates a multiple family. Science 246:235–240

Bhandari P, Andrews PLR (1991) Preliminary evidence for the involvement of the putative 5-HT$_4$ receptor in zacopride and copper sulphate induced vomiting in the ferret. Eur J Pharmacol 204:273–280

Bieger D, Triggle C (1985) Pharmacological properties of mechanical responses of the rat oesophagal muscularis mucosae to vagal and field stimulation. Br J Pharmacol 84:93–106

Bingham S, King BF, Rushant B, Smith MI, Gaster L, Sanger GJ (1995) Antagonism by SB 204070 of 5-HT-evoked contractions in the dog stomach: an in vivo model of 5-HT$_4$ receptor function. J Pharm Pharmacol 47:219–222

Birnstiel S, Beck SG (1995) Modulation of the hydroxytryptamine 4 receptor-mediated response by short-term and long-term administration of corticosterone in rat CA1 hippocampal neurons. J Pharmacol Exp Ther 273:1132–1138

Bley KR, Eglen RM, Wong EHF (1994) Characterization of 5-hydroxytryptamine-induced depolarization in rat isolated vagus nerve. Eur J Pharmacol 260:139–147

Bockaert J (1991) G proteins and G-protein-coupled receptors: structure, function and interactions. Curr Opin Neurobiol 1:32–42

Bockaert J (1995) Les récepteurs à sept domaines transmembranaires: physiologie et pathologie de la transduction. Medecine/Science 11:382–394

Bockaert J, Sebben M, Dumuis A (1990) Pharmacological characterization of 5-HT$_4$ receptors positively coupled to adenylate cyclase in adult guinea pig hippocampal membranes: effect of substituted benzamide derivatives. Mol Pharmacol 37:408–411

Bockaert J, Fozard J, Dumuis A, Clarke D (1992) The 5-HT$_4$ receptor: a place in the sun. Trends Pharmacol Sci 13:141–145

Bockaert J, Ansanay H, Waeber C, Sebben M, Fagni L, Dumuis A (1994) 5-HT$_4$ receptors. Potential therapeutic implications in neurology and psychiatry. CNS Drugs 1:6–15

Boddeke HWGM, Kalkman HO (1990) Zacopride and BRL 24924 induce an increase in EEG-energy in rats. Br J Pharmacol 101:281–284

Boddeke HWGM, Kalkman HO (1992) Agonist effects at putative central 5-HT₄ receptors in rat hippocampus by R(+)- and S(-)-zacopride: no evidence for stereoselectivity. Neurosci Lett 134:261–263

Boess FG, Martin IL (1994) Molecular biology of 5-HT receptors. Neuropharmacology 33:275–317

Bom AH, Dunker DJ, Saxena PR, Verdouw PD (1988) 5-HT-induced tachycardia in the pig: possible involvement of a new type of 5-HT receptor. Br J Pharmacol 93:663–671

Bonhaus DW, Loury DN, Jakeman LB, To Z, DeSouza A, Eglen RM, Wong EHF (1993) [³H]BIMU-1, a 5-hydroxytryptamine₃ receptor ligand in NG-108 cells, selectively labels sigma-2 binding sites in guinea-pig hippocampus. J Pharmacol Exp Ther 267:961–970

Bonhomme N, De Deurwaërdere P, Le Moal M, Spampinato U (1995) Evidence for 5-HT₄ receptor subtype involvement in the enhancement of striatal dopamine release induced by serotonin: a microdialysis study in the halothane-anesthetized rat. Neuropharmacology 34:269–279

Borman RA, Burleigh DE (1993) Evidence for the involvement of a 5-HT₄ receptor in the secretory response of the human small intestine to 5-HT. Br J Pharmacol 110:927–928

Bourtchuladze R, Frenguelli B, Blendy J, Cioffi D, Schütz G, Silva A (1994) Deficient long-term memory in mice with a targeted mutation of the cAMP-responsive element-binding protein. Cell 79:59–68

Bradley PB, Engel G, Feniuk W, Fozard JR, Humphrey PPA, Middlemiss DN, Myelcharane EJ, Richardson BP, Saxena PR (1986) Proposals for the classification and nomenclature of functional receptors for 5-hydroxytryptamine. Neuropharmacology 25:563–576

Briejer MR, Akkermans LMA, Meulemans AL, Lefrevre RA, Schuurkes JAJ (1993) Cisapride and a structural analogue, R 76186, are 5-hydroxytryptamine₄ (5-HT₄) receptor agonists on the guinea-pig colon ascendens. Naunyn Schmiedeberg's Arch Pharmacol 347:464–470

Brown AM, Young TJ, Patch TL, Cheung CW, Kauman A, Gaster L, King FD (1993) [¹²⁵I]SB 207710, a potent, selective radioligand for 5-HT₄ receptors. Br J Pharmacol 110:10P

Buchheit KH, Games R, Pfannküche H (1992) SDZ 205-557, a selective surmountable antagonist for 5-HT₄ receptors in the isolated guinea-pig ileum. Naunyn Schmiedeberg's Arch Pharmacol 345:387–393

Buchheit K, Gamse R, Giger R, Hoyer D, Klein F, Klöppner E, Pfannküche H, Mattes H (1995a) The serotonin 5-HT₄ receptor. 1. Design of a new class of agonists and receptor map of the agonist recognition site. J Med Chem 38:2326–2330

Buchheit K, Gamse R, Giger R, Hoyer D, Klein F, Klöppner E, Pfannküche H, Mattes H (1995b) The serotonin 5-HT₄ receptor. 2. Structure-activity studies of the indole carbazimidamide class of agonists. J Med Chem 38:2331–2338

Buchheit KH, Buhl T (1991) Prokinetic benzamides stimulate peristaltic activity in the isolated guinea pig ileum by activation of 5-HT₄ receptors. Eur J Pharmacol 205:203–208

Buchheit KH, Buhl T (1994) Stimulant effects of 5-hydroxytryptamine on guinea pig stomach preparation in vitro. Eur J Pharmacol 262:91–97

Buchheit KH, Engel G, Mutschler E, Richardson BP (1985) Study of the contractile effect of 5-hydroxytryptamine (5-HT) in isolated longitudinal muscle strip from guinea pig ileum. Evidence for two distinct release mechanisms. Naunyn Schmiedeberg's Arch Pharmacol 329:36–41

Burleigh DE, Borman RA (1993) Short-circuit current responses to 5-hydroxytryptamine in human ileal mucosa are mediated by a 5-HT₄ receptor. Eur J Pharmacol 241:125–128

Cassel JC, Jeltsch H (1995) Serotonergic modulation of cholinergic function in the central nervous system: cognitive implications. Neuroscience 69:1–41

Chaput Y, Araneda RC, Andrade R (1990) Pharmacological and functional analysis of a novel serotonin receptor in the rat hippocampus. Eur J Pharmacol 182:441–456

Clark RD, Jahangir A, Langston JA, Weinhardt KK, Miller AB, Leung E, Eglen RM (1994) Ketones related to the benzoate 5-HT$_4$ receptor antagonist RS-23597 are high affinity partial agonists. BioOrg Med Chem Lett 4:2481–2484

Clark RD, Jahangir A, Flippin LA, Langston JA, Leung E (1995) RS-100235: a high affinity 5-HT$_4$ receptor antagonist. Bio Org Med Chem Lett 5:2119–2122

Clarke DE, Craig DA, Fozard JR (1989) The 5-HT$_4$ receptor: naughty but nice. Trends Pharmacol Sci 10:385–386

Cohen ML, Susemichel AD, Bloomquist W, Robertson DW (1994) 5-HT$_4$ receptors in rat but not guinea pig, rabbit or dog esophageal smooth muscle. Gen Pharmacol 25:1143–1148

Coleman J, Rhodes KF (1995) Further characterization of the putative 5-HT$_4$ receptor mediating depolarization of the rat isolated vagus nerve. Naunyn Schmiedeberg's Arch Pharmacol 352:74–78

Compan V, Dusticier N, Nieoullon A, Daszuta A (1995) Serotonergic influence on striatal neurons containing neuropeptide Y, substance P and Met-enkephalin. Society for Neuroscience, 25th annual meeting, San Diego, abstract 559.17

Compan V, Daszuta A, Salin P, Sebben M, Bockaert J, Dumuis A (1996) Lesion study of the distribution of serotonin 5-HT$_4$ receptors in rat basal ganglia and hippocampus. Eur J Neurosci 8:2591–2598

Consolo S, Arnaboldi S, Giorgi S, Russi G, Ladinsky H (1994) 5-HT$_4$ receptor stimulation facilitates acetylcholine release in rat frontal cortex. Neuro Report 5:1230–1232

Contesse V, Hamel C, Delarue C, Lefèbvre H, Vaudry H (1994) Effect of a series of 5-HT$_4$ receptor agonists and antagonists on steroid secretion by adrenal gland in vitro. Eur J Pharmacol 265:27–33

Contesse V, Hamel C, Lefèbvre H, Dumuis A, Vaudry H, Delarue C (1996) Activation of 5-hydroxytryptamine$_4$ receptors causes calcium influx in adrenocortical cells: involvement of calcium in 5-HT-induced steroid secretion. Mol Pharmacol 49:481–493

Contestabile A, Virgili M, Bernabei O (1987) Development profiles of cholinergic activity in the habenulae and interpeduncular nucleus of the rat. Int J Dev Neurosci 8:561–564

Corsi M, Pietra C, Toson G, Trist D, Tuccitto G, Artibani W (1991) Pharmacological analysis of 5-hydroxytryptamine effects on electrically stimulated human isolated urinary bladder. Br J Pharmacol 104:719–725

Costa M, Furness JB, Cuello AC, Verhofstad AAJ, Steinbush HWJ, Elde RP (1982) Neurones with 5-hydroxytryptamine-like immunoreactivity in the enteric nervous system: their visualization and reactions to drug treatment. Neuroscience 7:351–363

Costall B, Naylor RJ (1993) The pharmacology of the 5-HT$_4$ receptor. Int Clin Psychopharmacol 8 [Suppl 2]:11–18

Coughlin SR (1994) Expanding horizons for receptors coupled to G proteins: diversity and disease. Curr Opin Biol 6:191–197

Craig DA, Clarke DE (1990) Pharmacological characterization of a neuronal receptor for 5-HT in guinea pig ileum with properties similar to the 5-HT$_4$ receptor. J Pharmacol Exp Ther 252:1378–1386

Craig DA, Clarke DE (1991) Peristalsis evoked by 5-HT and renzapride: evidence for putative 5-HT$_4$ receptor activation. Br J Pharmacol 102:563–564

Craig DA, Eglen RM, Walsh LKM, Perkins LA, Whiting RL, Clarke DE (1990) 5-Methoxytryptamine and 2-methyl-5-hydroxytryptamine-induced desensitization as a discriminative tool for the 5-HT$_3$ and putative 5-HT$_4$ receptors in guinea pig ileum. Naunyn Schmiedeberg's Arch Pharmacol 342:9–16

Domenech T, Beleta J, Fernandez AG, Gristwood RW, Cruz Sanchez F, Tolosa E, Palacios JM (1994) Identification and characterization of serotonin 5-HT$_4$ receptor binding sites in human brain: comparison with other mammalian species. Mol Brain Res 21:176–180

Douglas W, Bonhaus DN, Loury DN, Jakeman LB, Hsu SAO, To ZP, Leung E, Zeitung KD, Eglen RM, Wong EHF (1994) [^{3}H]RS-23597, a potent 5-hydroxytryptamine$_4$ antagonist, labels Sigma-1 but Sigma-2 binding sites in guinea pig brain. J Pharmacol Exp Ther 271:484–493

Dumuis A, Bouhelal R, Sebben M, Cory R, Bockaert J (1988a) A non classical 5-hydroxytryptamine receptor positively coupled with adenylate cyclase in the central nervous system. Mol Pharmacol 34:880–887

Dumuis A, Sebben M, Bockaert J (1988b) Pharmacology of 5-hydroxytryptamine$_{1A}$ receptors which inhibit cAMP production in hippocampal and cortical neurons in primary culture. Mol Pharmacol 33:178–186

Dumuis A, Sebben M, Bockaert J (1989a) BRL 24924: a potent agonist at a non-classical 5-HT receptor positively coupled with adenylate cyclase in colliculi neurons. Eur J Pharmacol 162:381–384

Dumuis A, Sebben M, Bockaert J (1989b) The gastrointestinal prokinetic benzamide derivatives are agonists at the non-classical 5-HT receptor (5-HT$_4$) positively coupled to adenylate cyclase in neurons. Naunyn Schmiedeberg's Arch Pharmacol 340:403–410

Dumuis A, Sebben M, Monferini E, Nicola M, Ladinsky H, Bockaert J (1991) Azabicycloalkylbenzimidazolone derivatives as a novel class of potent agonists at the 5-HT$_4$ receptor positively coupled to adenylate cyclase in brain. Naunyn Schmiedeberg's Arch Pharmacol 343:245–251

Dumuis A, Gozlan H, Sebben M, Ansanay H, Rizzi C, Turconi M, Monferini E, Giraldo E, Schiantarelli P, Ladinsky H, Bockaert J (1992) Characterization of a novel 5-HT$_4$ receptor antagonist of the azabicycloalkylbenzimidazolone class: DAU 6285. Naunyn Schmiedeberg's Arch Pharmacol 345:264–269

Eglen RM, Bonhaus DW, Clark RD, Johson LG, Lee CH, Leung E, Smith WL, Wong EHF, Whiting RL (1994) (R) and (S) RS 56532: mixed 5-HT$_3$ and 5-HT$_4$ receptor ligands with opposing enantiomeric selectivity. Neuropharmacology 33:515–526

Eglen RM, Bonhaus DW, Clark RD, Daniels S, Leung E, Wong EHF, Fontana DJ (1995a) Effects of a selective and potent 5-HT$_4$ receptor agonist, RS-67333, and antagonist, RS-67532, in a rodent model of spatial learning and memory. Br J Pharmacol 116:235p

Eglen RM, Bonhaus DW, Johnson LG, Leung E, Clark RD (1995b) Pharmacological characterization of two novel and potent 5-HT$_4$ receptor agonists, RS 67333 and RS 67506, in vitro and in vivo. Br J Pharmacol 115:1387–1392

Eglen RM, Wong EHF, Dumuis A, Bockaert J (1995c) Central 5-HT$_4$ receptors. Trends Pharmacol Sci 16:391–398

Elswood CJ, Bunce KT, Humphrey PPA (1991) Identification of putative 5-HT$_4$ receptors in guinea pig ascending colon. Eur J Pharmacol 196:149–155

Erspamer V (1966) 5-Hydroxytryptamine and related indole-alkylamines. Springer, Berlin Heidelberg New York, pp 32–181 (Handbook of Experimental Pharmacology, vol 19)

Fagni L, Dumuis A, Sebben M, Bockaert J (1992) The 5-HT$_4$ receptor subtype inhibits K$^+$ current in colliculi neurons via activation of a cyclic AMP-dependent protein kinase. Br J Pharmacol 105:973–979

Flynn DL, Zabrowski DL, Becker DP, Nosal R, Villamil CI, Gullikson GW, Moummi C, Yang DC (1992) SC 53116: The first selective agonist at the newly identified serotonin 5-HT$_4$ receptor subtype. J Med Chem 35:1486–1489

Ford APDW, Clarke DE (1993) The 5-HT$_4$ receptor. Med Res Rev 13:633–662

Ford APDW, Baxter GS, Eglen RM, Clarke DE (1992) 5-Hydroxytryptamine stimulates cyclic AMP formation in the tunica muscularis mucosae of the rat oesophagus via 5-HT$_4$ receptors. Eur J Pharmacol 211:117–120

Franks CM, Hardcastle J, Hardcastle PT, Sanger GJ (1995) Do 5-HT$_4$ receptors medi-ate the intestinal secretory response to 5-HT in rat in vivo? J Pharm Pharmacol 47:213–218

Fukui H, Yamanoto M, Sasaki S, Sato S (1994) Possible involvement of peripheral 5-HT$_4$ receptors in copper sulfate-induced vomiting in dogs. Eur J Pharmacol 257:47–52

Gale JD, Green A, Darton J, Sargant RS, Clayton NM, Bunce KT (1994a) GR125487: a 5-HT$_4$ receptor antagonist with long duration of action in vivo. Br J Pharmacol 113:119P

Gale JD, Grossman CJ, Whitehead JWF, Oxford AW, Bunce KT, Humphrey PPA (1994b) GR113808: a novel, selective antagonist with high affinity at the 5-HT$_4$ receptor. Br J Pharmacol 111:332–338

Gaster LM, Sanger GJ (1994) SB 204070: 5-HT$_4$ receptor antagonists and their poten-tial therapeutic utility. Drugs Fut 19:1109–1121

Ge J, Barnes NM (1995a) 5-HT$_4$ receptor-mediated modulation of extracellular levels of 5-HT in the rat hippocampus in vivo. Br J Pharmacol 116:232P

Ge J, Barnes NM (1995b) Further characterization of the 5-HT$_4$ receptor modulating extracellular levels of dopamine in the rat striatum in vivo. Br J Pharmacol 116:233P

Gebauer A, Merger M, Kilbinger H (1993) Modulation by 5-HT$_3$ and 5-HT$_4$ receptors of the release of 5-hydroxytryptamine from the guinea-pig small intestine. Naunyn Schmiedeberg's Arch Pharmacol 347:137–140

Gerald C, Adham N, Kao H-T, Olsen M, Laz TM, Schechter LE, Bard JA, Vaysse PJJ, Hartig PR, Branchek TA, Weinshank RL (1995) The 5-HT$_4$ receptor: molecular cloning and pharmacological characterization of two splice variants. EMBO J 14:2806–2815

Ghelardini C, Mamberg-Aiello P, Malcangio M, Bartolini A (1990) Investigation into atropine-induced antinociception. Br J Pharmacol 101:49–54

Ghelardini C, Meoni P, Galleotti N, Malmberg-Aiello P, Rizzi CA, Bartolini A (1994) Effect of the two benzimidazolone derivates: BIMU 1 and BIMU 8 on a model of hypoxia-induced amnesia in the mouse. Proceedings, third IUPHAR satellite meeting on serotonin, Chicago, 30 July – 3 Aug, p 55

Ghelardini C, Galeotti N, Casamenti F, Malmberg-Aiello P, Pepeu G, Gualtieri F, Bartolini A (1996) Central cholinergic antinociception induced by 5-HT$_4$ agonists: BIMU 1 and BIMU 8. Life Sci 58:2297–2309

Grossman CJ, Kilpatrick GJ, Bunce KT (1993) Development of a radioligand binding assay for 5-HT$_4$ receptors in guinea-pig and rat brain. Br J Pharmacol 109:618–624

Gullikson GW, Virina MA, Loeffler RF, Yang DC, Goldstin B, Wang SX, Moummi C, Flynn DL, Zabrowski DL (1993) SC 49518 enhances gastric emptying of solid and liquid meals and stimulates gastrointestinal motility in dogs by a 5-hydroxytrytamine$_4$ receptor mechanism. J Pharmacol Exp Ther 264:240–248

Hedge SS, Moy TM, Perry MR, Loeb M, Eglen RM (1994) Evidence for the involve-ment of 5-hydroxytryptamine$_4$ receptors in the 5-hydroxytryptophan-induced diar-rhea in mice. J Pharmacol Exp Ther 271:741–747

Hedge SS, Bonhaus DW, Johson LG, Leung E, Clark RD, Eglen RM (1995a) RS 39604: a potent, selective and orally active 5-HT$_4$ receptor antagonist. Br J Pharmacol 115:1087–1095

Hedge SS, Wong AG, Perry MR, Ku P, Moy TM, Loeb M, Eglen RM (1995b) 5-HT$_4$ receptor mediated stimulation of gastric emptying in rats. Naunyn Schmiedeberg's Arch Pharmacol 351:589–595

Hochner B, Kandel ER (1992) Modulation of a transient K$^+$ current in the pleural sensory neurons of *Aplysia* by serotonin and cAMP: implications for spike broad-ening. Proc Natl Acad Sci USA 89:11476–11480

Hoyer D, Clarke DE, Fozard JR, Hartig PR, Martin GR, Mylecharane EJ, Saxena PR, Humphrey PPA (1994) International union of pharmacology classification of re-ceptors for 5-hydroxytryptamine (serotonin). Pharmacol Rev 46:157–203

Idres S, Delarue C, Lefèbrve H, Vaudry H (1991) Benzamide derivatives provide evidence for the involvement of a 5-HT₄ receptor type in the mechanism of action of serotonin in frog adrenocortical cells. Mol Brain Res 10:251–258

Jahnel U, Rupp J, Ertl R, Nawrath H (1992) Positive inotropic response to 5-HT in human atrial but not in ventricular heart muscle. Naunyn Schmiedeberg's Arch Pharmacol 346:346–485

Jakeman LB, To ZP, Eglen RM, Wong EHP, Bonhaus DW (1994) Quantitative autoradiography of 5-HT₄ receptors in brains of three species using two structurally [³H] GR113808 and [³H] BIMU 1. Neuropharmacology 33:1027–1038

Kandel E, Adel T (1995) Neuropeptides, adenylyl cyclase and memory storage. Science 268:825–826

Kandel ER, Schwartz JH (1982) Molecular biology of learning. Science 218:433–438

Kaumann A (1993) Blockade of human atrial 5-HT receptors by GR 113808. Br J Pharmacol 110:1172–1174

Kaumann A, Sanders L (1994) 5-Hydroxytryptamine causes rate-dependent arrhythmias through 5-HT₄ receptors in human atrium: facilitation of chronic β-adrenergic blockade. Naunyn Schmiedeberg's Arch Pharmacol 349:331–337

Kaumann AJ (1990) Piglet sinoatrial 5-HT receptors resemble human atrial 5-HT₄-like receptors. Naunyn Schmiedeberg's Arch Pharmacol 342:619–622

Kaumann AJ (1994) Do human atrial 5-HT₄ receptors mediate arrhythmias? Trends Pharmacol Sci 15:451–455

Kaumann AJ, Sanders L, Brown AM, Murray KJ, Brown MJ (1990) A 5-hydroxytryptamine receptor in human atrium. Br J Pharmacol 100:879–885

Kaumann AJ, Sanders L, Brown AM, Murray KJ, Brown MJ (1991) A 5-HT₄-like receptor in human right atrium. Naunyn Schmiedeberg's Arch Pharmacol 344:150–159

Kilbinger H, Wolf D (1992) Effect of 5-HT₄ receptor stimulation on basal and electrically evoked release of acetylcholine from guinea pig myenteric plexus. Naunyn Schmiedeberg's Arch Pharmacol 345:270–275

Kilbinger H, Gebauer A, Hass J, Ladinsky H, Rizzi CA (1995) Benzimidazolones and renzapride facilitate acetylcholine release from guinea-pig myenteric plexus via 5-HT₄ receptors. Naunyn Schmiedeberg's Arch Pharmacol 351:229–236

Krushinski JH, Susemichel A, Robertson DW, Cohen ML (1992) Interaction of metoclopramide analogues with 5-HT₄ receptors. American Chemical Society, Division Medical Chemistry, 203rd national meeting

Langlois M, Zhang L, Yang D, Brémont B, Shen S, Manara L, Croci T (1994) Design of a potent 5-HT₄ receptor agonist with nanomolar affinity. Bioorg Med Chem Lett 4:1433–1436

Lefèbvre H, Contesse V, Delarve C, Fevilloley M, Héry F, Grise P, Raynaud G, Verhofstad AAJ, Wolf LM, Vaudry H (1992) Serotonin-induced stimulation of cortisol secretion from human adrenocortical tissue is mediated through activation of a serotonin-₄ receptor subtype. Neuroscience 47:999–1007

Lefèbvre H, Contesse V, Delarue C, Soubrane C, Legrand A, Kuhn JM, Wolf LM, Vaudry H (1993) Effect of the Serotonin-₄ receptor agonist zacopride on aldosterone secretion from the human adrenal cortex: in vivo and in vitro studies. J Clin Endocrinol Metabol 77:1662–1666

Lefèbvre H, Contesse V, Delarue C, Legrand A, Kuhn JM, Vaudry H, Wolf L (1994) The serotonin-4 receptor agonist cisapride and angiotensin II exert additive effects on aldosterone secretion in normal man. J Clin Endocrinol Metabol 80:504–507

Letty S, Child R, Gale JD, Dumuis A, Bockaert J, Rondouin G (1996) Central 5-HT₄ receptors improve short-term social memory in rat. Neuropharmacology (in press)

Leung E, Blissard D, Jett MF, Eglen RM (1995) Investigation of the 5-hydroxytryptamine receptor mediating the "transient" short-circuit current response in guinea-pig ileal mucosa. Naunyn Schmiedeberg's Arch Pharmacol 351:596–602

Lucas JJ, Hen R (1995) New players in the 5-HT receptor field: genes and knockouts. Trends Pharmacol Sci 16:246–252

Martin GR, Humphrey PPA (1994) Receptors for 5-hydroxytryptamine: current perspectives on classification and nomenclature. Neuropharmacology 33:261–273

McLean PG, Coupar I, Molenaar P (1995) A comparative study of functional 5-HT$_4$ receptors in human colon, rat oesophagus and rat ileum. Br J Pharmacol 115:47–56

Mengod G, Cortes R, Salcedo C, Palacios JM (1995) 5-Hyroxytryptamine$_4$ receptors (5-HT$_4$R) are located presynaptically in the striatonigral pathway and co-distributed with 5-HT$_{1D}$ receptors. Abstr Soc Neurosci 21:P1856

Monferini E, Gaetani P, Rodriguez Y, Baena R, Giraldo E, Parentini M, Zocchetti A, Rizzi CA (1993) Pharmacological characterization of the 5-hydroxytrytamine receptor coupled to adenylyl cyclase stimulation in human brain. Life Sci 52:61–65

Müller-Lissner S (1987) Bavarian constipation study group: treatment of chronic constipation with cisapride and placebo. Gut 28:1033

Namba T, Sugimoto Y, Negishi F, Irie A, Ushikubi A, Kakizuka A, Ito S, Ichikawa A, Narumiya S (1993) Alternative splicing of the C-terminal tail of prostaglandin E receptor subtype EP3 determines G-protein specificity. Nature 365:166–170

Nemeth PR, Gullickson GW (1989) Gastrointestinal mobility stimulating drugs and 5-HT receptors on myenteric neurons. Eur J Pharmacol 166:387–391

Ng GY, George SR, Zastawny RI, Caron M, Bouvier M, Dennis M, O'Dowd BF (1993) Human serotonin 1B receptor expression in Sf9 cells – phosphorylation, palmitoylation and adenylyl cyclase inhibition. Biochemistry 32:11727–11733

O'Dowd BF, Hnatowich M, Caron MG, Lefkowitz RJ, Bouvier M (1989) Palmitoylation of the human β_2-adrenergic receptors. Mutation of Cys 341 in carboxy tail leads to an uncoupled non-palmitoylated form of the receptor. J Biol Chem 264:7564–7569

Olssen S, Edwards IR (1992) Tachycardia during cisapride treatment. Br Med Chem 305:748–749

Ouadid H, Seguin J, Dumuis A, Bockaert J, Nargeot J (1992) Serotonin increases Ca^{2+} current in human atrial myocytes via the newly described 5-hydroxytryptamine$_4$ receptors. Mol Pharmacol 41:346–351

Panocka I, Ciccociopo R, Polidori C, Pompei P, Massi M (1995) The 5-HT$_4$ receptor antagonist, GR 113808, reduces ethanol intake in alcohol-preferring rats. Pharmacol Biochem Behavior 52:255–259

Patel S, Roberts J, Moorman J, Reavill C (1995) Localization of serotonin-4 receptors in the striato-nigral pathway in rat brain. Neuroscience 69:1159–1167

Parker SG, Hamburger S, Taylor EM, Kaumann AJ (1993) SB 203186, a potent 5-HT$_4$ receptor antagonist in porcine sinoatrial node and human and porcine atrium. Br J Pharmacol 108:68

Pillans PI, Wood SM (1994) Cisapride increases micturition frequency. J Clin Gastroenterol 19:336–338

Pozzi L, Trabace L, Invernizzi R, Samanin R (1995) Intracellular GR 113808, a selective 5-HT$_4$ antagonist, attenuates morphine-stimulated release in rat striatum. Brain Res 692:265–268

Ramirez MJ, Cenarruzabeitia E, Del Rio J, Lasheras B (1994) Involvement of neurokinins in the non-cholinergic response to activation of 5-HT$_3$ and 5-HT$_4$ receptors in guinea-pig ileum. Br J Pharmacol 111:419–424

Reeves JJ, Bunce KT, Humphrey PPA (1991) Investigation into the 5-hydroxytryptamine receptor mediating smooth muscle relaxation in rat oesophagus. Br J Pharmacol 103:1067–1072

Reynolds GP, Mason SL, Meldrum A, De Keczer S, Parnes H, Eglen RM, Wong EHF (1995) 5-Hydroxytryptamine (5-HT$_4$) receptors in post mortem human brain tissue: distribution, pharmacology and effects of neurodegenerative diseases. Br J Pharmacol 114:993–998

Rhodes KF, Coleman J, Lattimer N (1992) A component of 5-HT-evoked depolarization of the rat isolated vagus nerve is mediated by a putative 5-HT₄ receptor. Naunyn Schmiedeberg's Arch Pharmacol 346:496–503

Rizzi CA, Coccini T, Onori L, Manzo L, Tonini M (1992) Benzimidazolone derivatives: a new class of 5-hydroxytryptamine₄ receptor agonists with prokinetic and acetylcholine releasing properties in guinea-pig ileum. J Pharmacol Exp Ther 261:412–419

Rizzi CA, Sagrada A, Schiavone A, Schiantarelli P, Cesana R, Schiavi GB, Ladinsky H, Donetti A (1994) Gastroprokinetic properties of benzimidazolone derivative BIMU 1, an agonist at 5-hydroxytryptamine₄ and antagonist at 5-hydroxytryptamine₃ receptors. Naunyn Schmiedeberg's Arch Pharmacol 349:338–345

Rondé P, Ansanay H, Dumuis A, Miller R, Bockaert J (1995) Homologous desensitization of 5-hydroxytryptamine₄ receptors in rat oesophagus: functional and second messenger studies. J Pharmacol Exp Ther 272:977–983

Sanders L, Kaumann AJ (1992) A 5-HT₄-like receptor in human left atrium. Naunyn Schmiedeberg's Arch Pharmacol 345:382–386

Sanger GJ (1987) Activation of a myenteric 5-hydroxytryptamine-like receptor by metoclopramide. J Pharm Pharmacol 39: 449–453

Sanger GJ, Gaster L (1994) Central and peripheral nervous system. 5-HT₄ receptor antagonists. Exp Opin Ther Patents 4:323–334

Saxena PR, Villalon CM (1991) 5-Hydroxytryptamine: a chameleon in the heart. Trends Pharmacol 12:223–227

Saxena PR, Villalon CM, Dhasmana KM, Verdouw PD (1992) 5-Hydroxytryptamine-induced increase in left ventricular dP/dt$_{max}$ does not suggest the presence of ventricular 5-HT₄ receptors in the pig. Naunyn Schmiedeberg's Arch Pharmacol 346:629–636

Schiavi GB, Brunet S, Rizzi CA, Ladinsky H (1994) Identification of serotonin 5-HT₄ receptor sites in the porcine caudate nucleus by radioligand binding. Neuropharmacology 33:543–549

Schoemaker RG, Du XY, Bax WA, Saxena PR (1992) 5-Hydroxytryptamine increases contractile force in porcine right atrium but not in left ventricle. Naunyn Schmiedeberg's Arch Pharmacol 346:486–489

Schuurkes JAJ, van Nueten JM, van Daele PGH, Reyntjens AJ, Janssen PAJ (1985) Motor stimulating properties of cisapride, and isolated gastrointestinal preparations of the guinea pig. J Pharmacol Exp Ther 234:775-783

Scott CM, Bunce KT, Spraggs CF (1992) Investigation of the 5-hydroxytryptamine receptor mediating the "maintained" short circuit current response in guinea pig ileal mucosa. Br J Pharmacol 106:877–882

Shenker A, Maayani S, Weinstein H, Green JP (1987) Pharmacological characterization of two 5-hydroxytryptamine receptors coupled to adenylate cyclase in guinea pig hippocampal membranes. Mol Pharmacol 31:357–367

Spengler D, Waeber C, Pantaloni C, Holsboer F, Bockaert J, Seeburg PH, Journot L (1993) Differential signal transduction by five splice variants of the PACAP receptor. Nature 365:170–175

Steward LJ, Ge J, Barnes NM (1995) Ability of 5-HT₄ receptor ligands to modify rat striatal dopamine release in vitro and in vivo. Br J Pharmacol 114:381P

Stowe RL, Barnes NM (1995) Further characterization of the 5-HT₄ receptor modulating dopamine release from rat striatal slices. Br J Pharmacol 116:234P

Tam FSH, Hillier K, Bunce KT, Grossman C (1995) Differences in response to 5-HT₄ receptor agonists and antagonists of the 5-HT₄-like receptor in human colon circular smooth muscle. Br J Pharmacol 115:172–176

Tonini M, Rizzi CA, Manzo L, Oniri L (1991) Novel enteric 5-HT₄ receptors and gastrointestinal prokinetic action. Pharmacol Res 24:5–14

Tonini M, Messori E, Franceschetti GP, Rizzi CA, Castoldi AF, Coccini T, Candura SM (1994) Characterization of the 5-HT receptor potentiating neuromuscular cholinergic transmission in strips of human isolated detrusor muscle. Br J Pharmacol 113:1-2

Torres G, Chaput Y, Andrade R (1995) Cyclic AMP and protein kinase A mediate 5-hydroxytryptamine type 4 receptor regulation of calcium-activated potassium current in adult hippocampal neurons. Mol Pharmacol 47:191–197

Torres GE, Holt IL, Andrade R (1994) Antagonists of 5-HT$_4$ receptor-mediated responses in adult hippocampal neurons. J Pharmacol Exp Ther 271:255–261

Tuladhar BR, Costall B, Naylor RJ (1991) Putative 5-HT$_4$ receptor involvement in the relaxation induced by 5-HT in rat ileum. Br J Pharmacol 104:151P

Turconi M, Schiantarelli P, Borsini F, Rizzi CA, Ladinsky H, Donetti A (1991) Azabicycloalkyl benzimidazolones: interaction with serotonergic 5-HT$_3$ and 5-HT$_4$ receptors and potential therapeutic implications. Drugs Fut 16:1011–1026

Villalòn CM, Den Boer MO, Heiligers JPC, Saxena PR (1990) Mediation of 5-hydroxytryptamine-induced tachycardia in the pig by the putative 5-HT$_4$ receptor. Br J Pharmacol 100:665–667

Waeber C, Sebben M, Grossman C, Javoy-Agid F, Bockaert J, Dumuis A (1993) [^{3}H] GR 113808 labels 5-HT$_4$ receptors in the human and guinea pig brain. NeuRoreport 4:1239–1242

Waeber C, Sebben M, Nieoullon A, Bockaert J, Dumuis A (1994) Regional distribution and ontogeny of 5-HT$_4$ binding sites in rodent brain. Neuropharmacology 33:527–541.

Waikar MV, Ford APDW, Clarke DE (1994) Evidence for an inhibitory 5-HT$_4$ receptor in urinary bladder of rhesus and cynomolgous monkeys. Br J Pharmacol 111:213–218

Wardle KA, Sanger GJ (1993) The guinea-pig distal colon, a sensitive preparation for the investigation of 5-HT$_4$ receptor-mediated contraction. Br J Pharmacol 110:1593–1599

Wiseman LR, Faulds D (1994) Cisapride: an updated review of its pharmacology and therapeutic efficacy as a prokinetic agent in gastrointestinal mobility disorders. Drugs 47:116-152

Yuan SY, Bornstein J, Furness JB (1994) Investigation of the role of 5-HT$_3$ and 5-HT$_4$ receptors in ascending and descending reflexes to circular muscle of guinea-pig small intestine. Br J Pharmacol 112:1095–1100

Yang D, Goldstin B, Moormann AE, Flynn DL, Gullikson GW (1993) SC-53606, a potent and selective antagonist of 5-hydroxytryptamine$_4$ receptors in isolated rat esophageal tunica muscularis mucosae. J Pharmacol Exp Ther 266:1339–1346

Yin JCP, Wallach JS, Del Vecchio M, Wilder EL, Zhou H, Quinn WG, Tully T (1994) Induction of a dominant negative CREB transgene specifically blocks long-term memory in Drosophila. Cell 79:49–58

Zhong Y (1995) Mediation of PACAP-like neuropeptide transmission by coactivation of Ras/Raf and cAMP signal transduction pathways in Drosophila. Nature 375:588–591

Molecular Biology and Potential Functional Role of 5-HT$_5$, 5-HT$_6$, and 5-HT$_7$ Receptors

T.A. BRANCHEK and J.M. ZGOMBICK

A. Introduction

The supposition that multiple receptor subtypes mediate the numerous physiological actions of 5-HT has been elegantly substantiated by molecular cloning, including the discovery of "novel" receptor subtypes such as 5-HT$_{5A}$, 5-HT$_{5B}$, 5-HT$_6$, and 5-HT$_7$. All four subtypes are members of the G-protein-coupled receptor (GPCR) superfamily. Detailed characterization of the pharmacological profiles of these cloned receptor subtypes facilitates the interpretation of complex binding data as well as *in vitro* and *in vivo* responses to 5-HT in native systems. This review focuses on the structural features, binding properties, and mRNA distribution of these novel 5-HT receptor subtypes, as well as an evaluation of the tool sets available to link these cloned receptors to function and therapeutic potential.

B. Molecular Biology of the 5-HT$_5$, 5-HT$_6$, and 5-HT$_7$ Receptors

I. 5-HT$_5$ Receptors

The 5-HT$_5$ receptor subfamily includes the 5-HT$_{5A}$ and 5-HT$_{5B}$ subtypes in mouse (PLASSAT et al. 1992; MATTHES et al. 1993) and rat (ERLANDER et al. 1993; WISDEN et al. 1993). The deduced amino acid sequences of the rat receptors indicate that they encode proteins containing 357 and 371 amino acids, respectively (Fig. 1). They display >75% amino acid identity to each other in the transmembrane (TM) regions and lower TM identities (<50%) to 5-HT$_6$ and 5-HT$_7$ receptors (Fig. 2). Their sequences are not closely related to any other known member of the 5-HT receptor family; the closest relationship is to the *Drosophila* cyclase inhibitory receptors 5-HT$_{Dro2}$ (SAUDOU et al. 1992). Analysis of the genomic structure of the mouse sequences indicates that they are interrupted by an intron between TM V and TM VI, in the middle of the third intracellular loop, at the same position in the two genes (MATTHES et al. 1993). Chromosomal localization of the mouse genes reveals a single locus for each found on chromosomes 5 (position 5B) and 1 (position 1F), respectively (MATTHES et al. 1993). The mouse 5-HT$_{5A}$ receptor sequence contains two sites for potential N-linked glycosylation in the amino terminus and it has several

```
Rat 5-HT5A  . . . . . . . . . . . . . . . . . . . . . . . . . . . . . . . . . . . . . . M D . L P I N L T S     9
Rat 5-HT5B  . . . . . . . . . . . . . . . . . . . . . . . . . . . M E V S N L S G A T P G I A . F P P G P E S    21
Rat 5-HT6   . . . . . . . . . . . . . . . . . . . . . . . . . . . . . . . . . . . . . . . . . . . . . . . .     0
Rat 5-HT7   M M D V N S S G R P D L Y G H L R S L I L P E V G R G L Q D L S P D G G A H P V V S S W M P H L L S    50

                                                                           I
Rat 5-HT5A  F S L S T P S T L E P N R S L D T E A L R T S Q S F L S . . . A F R V L V L T L L G F L A A A T F T    56
Rat 5-HT5B  C S D S P S S G R S M G S T P G G L I L S G R E P P F S . . . A F T V L V V T L L V L L I A A T F L    68
Rat 5-HT6   . . . . . . . . M V P E P G P V N S S T P A W G P G P P P A P G G S G W V A A A L C V V I V L T A A    42
Rat 5-HT7   G F L E V T A S P A P T W D A P P D N V S G C G E Q I N Y G R V E K V V I G S I L T L I T L L T I A   100

                                                            II
Rat 5-HT5A  W N L L V L A T I L R V R T F H R V P H N L V A S M A I S D V L V A V L V M P L S L V H E L S . G R   105
Rat 5-HT5B  W N L L V L V T I L R V R A F H R V P H N L V A S T A V S D V L V A A L V M P L S L V S E L S A G R   118
Rat 5-HT6   A N S L L I V L I C T Q P A L R N T S N F F L V S L F T S D L M V G L V V M P P A M L N A L Y G . .    90
Rat 5-HT7   G N C L V V I S V C F V K K L R Q P S N Y L I V S L A L A D L S V A V A V M P F V S V T D L I G G .   149

                                           III
Rat 5-HT5A  R W Q L G R R L C Q L W I A C D V L C C T A S I W N V T A I A L D R Y W S I T R H L E Y T L R A R K   155
Rat 5-HT5B  R W Q L G R S L C H V W I S F D V L C C T A S I W N V A A I A L D R Y W T I T R H L Q Y T L R T R R   168
Rat 5-HT6   R W V L A R G L C L L W T A F D V M C C S A S I L N L C L I S L D R Y L L I L S P L R Y K L R M T A   140
Rat 5-HT7   K W I F G H F F C N V F I A M D V M C C T A S I M T L C V I S I D R Y L G I T R P L T Y P V R Q N G   199

                               IV
Rat 5-HT5A  R V S N V M I L L T W A L S A V I S L A P L L F G W . . . G E T Y S E L S E E C Q V S R E P S Y T V   202
Rat 5-HT5B  R A S A L M I A I T W A L S A L I A L A P L L F G W . . . G E A Y D A R L Q R C Q V S Q E P S Y A V   215
Rat 5-HT6   P R A L A L I L G A W S L A A L A S F L P L L L G W H E L G K A R T P A P G Q C R L L A S L P F V L   190
Rat 5-HT7   K C M A K M I L S V W L L S A S I T L P P . L F G W . . . A Q N V N D . D K V C L I S Q D F G Y T I   244

                    V
Rat 5-HT5A  F S T V G A F Y L P L C V V L F V Y W K I Y K A A K F R M G S R K . . . . . . . . . T N S V S P I P E   244
Rat 5-HT5B  F S T C G A F Y V P L A V V L F V Y W K I Y K A A K F R F G R R R . . . . . . . . . . R A V V P L P A   256
Rat 5-HT6   V A S G V T F F L P S G A I C F T Y C R I L L A A R K Q A V Q . . . . . . . . . . . . . V A S L T T   227
Rat 5-HT7   Y S T A V A F Y I P M S V M L F M Y Y Q I Y K A A R K S A A K H K F P G F P R V Q P E S V I S L N G   294

Rat 5-HT5A  A V E V K D A S Q H P Q M V F T V R . H A T V T F Q T E G D T W R E Q K E Q R A A L M V G I L I G V   293
Rat 5-HT5B  T T Q A K E A P Q E S E T V F T A R C R A T V A F Q T S G D S W R E Q K E K R A A M M V G I L I G V   306
Rat 5-HT6   G T . A G Q A L E T L Q V P R T P R P G M E S A D S R R L A T K H S R K A L K A S L T L G I L L G M   276
Rat 5-HT7   V V K L Q K E V E E C A N L . . . . . . S R L L K H E R K N I S I F K R E Q K A A T T L G I I V G A   338
```

Fig. 1. Alignment of the deduced amino acid sequences of recombinant rat 5-HT_{5A}, 5-HT_{5B}, 5-HT_6, and 5-HT_7 receptor subtypes. The seven transmembrane domains are *defined by brackets*. Sequence alignment was performed using the software program PileUp (Genetics Computer Inc., Madison, WI)

VI
VII

Rat 5-HT₅ₐ	FVLCWFPFFVTE . . . . LISPLCSWDIPALWKSIFLWLGYSNSFFNPLIYT	357
Rat 5-HT₅ᵦ	FVLCWIPFFLTE . . . . LVSPLCACSLPPIWKSIFLWLGYSNSFFNPLIYT	370
Rat 5-HT₆	FFVTWLPFFVAN . . ARPFI . IAQAVCDCISPGLFEDVLTWLGGYCNSTMNPIIYP	371
Rat 5-HT₇	FTVCWLPFFLLSTARPFICGTSCSCIPLWVERTCLWLGYANSLINPFIYA	435

Rat 5-HT₅ₐ	AFNRSYSSAFKVFFSKQQ . . .	357
Rat 5-HT₅ᵦ	AFNKNYNNAFKSLFTKQR . . . . GARP . . . . QQVLP	370
Rat 5-HT₆	LFMRDEKRALGRFLPCVHC . . PEHRASPASPSMWTSHSGARPGLSLQQVLP	371
Rat 5-HT₇	FFNRDLRTTYRSLLQCQYKNINRKLSAAGMHEALKLAERPERSEFVL . .	435

Rat 5-HT₅ₐ	. . . . . SGGTSGLQLTAQLLPGEATRDPPPPTRATTVVNFFVT .	357
Rat 5-HT₅ᵦ	. . . SDSDSASGGTSG . . .	370
Rat 5-HT₆	VPVPPNSDSDSASGGTSG .	421
Rat 5-HT₇	. . .	435

Rat 5-HT₅ₐ	. . . . . EIRPHPLSSPVN . . .	351
Rat 5-HT₅ᵦ	. . . EIRPHPLSSPVN . . .	370
Rat 5-HT₆	DSVEPEIRPHPLSSPVN . . .	438
Rat 5-HT₇	. . .	435

Fig. 1. *Continued*

sites for protein kinase A and C phosphorylation in the intracellular domains. The mouse 5-HT$_{5B}$ receptor sequence contains a single site for N-linked glycosylation and multiple phosphorylation sites. Both receptors have modest size third intracellular loops (~60 amino acids) and short carboxylterminal tails (~15 amino acids). Functional coupling to a signal transduction system has not yet been reported.

A cDNA encoding a 5-HT$_{5A}$ receptor has been cloned from the human (REES et al. 1994) and it has 84% overall amino acid identify to its murine homologue. The human 5-HT$_{5A}$ receptor is also 357 amino acids and the sequence contains an intron at the same position as in the rodent. The human 5-HT$_{5A}$ gene is located on chromosome 7 at position 7q36 (MATTHES et al. 1993). In contrast to the mouse and rat, only an untranscribed fragment (one of the two exons) of the human 5-HT$_{5B}$ receptor has been detected (REES et al. 1994) and its locus was mapped to human chromosome 2 (position 2q11–13) (MATTHES et al. 1993). Quite recently, the second exon of the human sequence has been sequenced and was found to be interrupted by several stop codons (GRAILHE et al. 1995). Both Southern blot and polymerase chain reaction (PCR) experiments indicate that this is the only homologue of the rodent 5-HT$_{5B}$ gene and, thus, there will be no functional 5-HT$_{5B}$ receptor in man. This appears to be the first GPCR found in lower mammalian species but which is absent in man.

II. 5-HT$_6$ Receptors

The rat 5-HT$_6$ receptor has been reported to encode a protein of 436 amino acids (MONSMA et al. 1993; RUAT et al. 1993a). However, this has recently been reevaluated and it is now deduced to form a protein of 438 amino acids (Fig. 1) while the human homologue contains 440 amino acids (KOHEN et al. 1996). The human sequence is 89% identical to the rat amino acid sequence. No additional subtypes have been cloned. Key structural elements of this receptor sequence include a short third intracellular loop (50 amino acids, rat; 57 amino acids, human) and a long carboxyl tail (~120 amino acids), common to some other GPCRs which couple to adenylate cyclase stimulation (LEFKOWITZ and CARON 1988). The 5-HT$_6$ receptor is characterized by a single glycosylation site in the amino terminus, and by multiple potential phosphorylation sites for protein kinase C (PKC) in the cytoplasmic domains. The sequence also contains a consensus "leucine zipper" motif in TM III (RUAT et al. 1993a). The sequence has the highest identity (37%) to the *Drosophila* cyclase stimulatory serotonin receptor (5-HT$_{DRO1}$, WITZ et al. 1990) and the histamine H$_2$ receptor (RUAT et al. 1991); lower identities are observed with 5-HT$_5$ and 5-HT$_7$ receptors (Figs. 1, 2). Both the rat and human sequences contain two introns, one in the third intracellular loop and the other in the third extracellular loop (KOHEN et al. 1996). The intron in the third cytoplasmic loop appears at the same location as that described for the 5-HT$_{5A}$ and 5-HT$_{5B}$ receptors, as well as the dopamine D$_2$ and D$_3$ receptors (see KOHEN et al. 1996). The gene for the

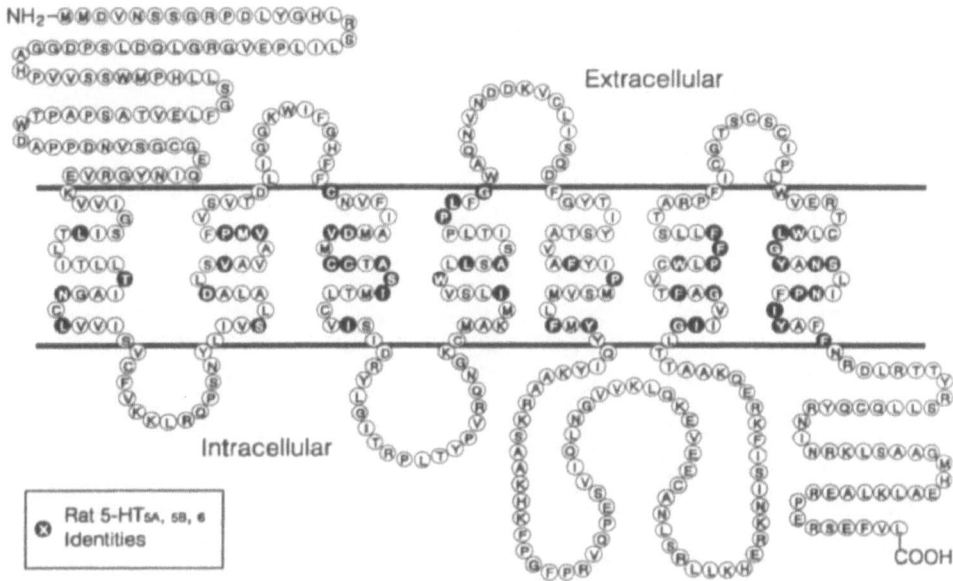

Fig. 2. Structural relationship of the transmembrane domains of recombinant rat 5-HT$_{5A}$, 5-HT$_{5B}$, 5-HT$_6$, and 5-HT$_7$ receptor subtypes. Identical amino acids are *denoted by solid symbols*. The deduced amino acid sequence of the recombinant rat 5-HT$_7$ receptor served as the structural template for comparisons

human 5-HT$_6$ receptor maps to chromosome region 1p35-p36, thus overlapping with the gene locus for the 5-HT$_{1D}$ receptor (Kohen et al. 1996). The 5-HT$_6$ receptor sequence contains an *Rsa*I restriction length polymorphism in the first extracellular loop. The 5-HT$_6$ receptor couples to the stimulation of adenylate cyclase activity, although no detailed functional analysis has been reported.

III. 5-HT$_7$ Receptors

The human 5-HT$_7$ receptor sequence encodes a protein of 445 amino acids (Bard et al. 1993). 5-HT$_7$ cDNAs have also been cloned from mouse (Plassat et al. 1993), rat (Lovenberg et al. 1993; Shen et al. 1993; Ruat et al. 1993b; Meyerhof et al. 1993), guinea pig (Tsou et al. 1994), and *Xenopus* (Nelson et al. 1995). Pseudogenes for the 5-HT$_7$ have been observed (Shen et al. 1993; Bard et al. 1993; R. Hen, personal communication). Key structural features which distinguish this subtype include a relatively short third intracellular loop (55 amino acids) and a moderately long carboxyl terminus (58 amino acids). There are two potential sites for N-linked glycosylation in the amino terminus and multiple sites for protein kinase A and C phosphorylation in the intracellular domains. The amino acid sequence of the rat 5-HT$_7$ receptor (Fig. 1) displays the highest identity to the 5-HT$_{Drol}$ (57% TM identity) and lower TM

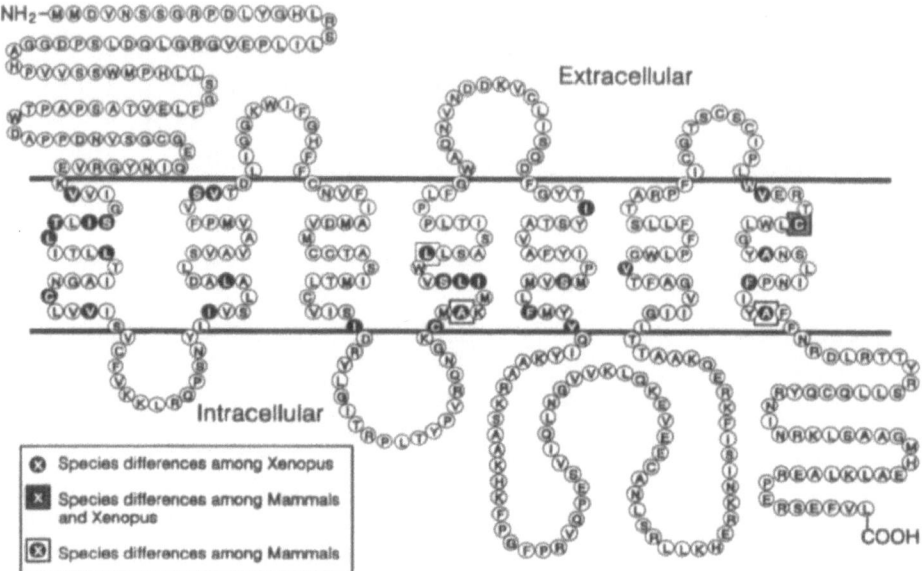

Fig. 3. Structural relationship of the transmembrane domains of recombinant mammalian and *Xenopus* 5-HT₇ receptors. Nonconserved amino acids are *denoted by solid symbols*. The deduced amino acid sequence of the recombinant rat 5-HT₇ receptor served as the structural template for comparisons

identities (<50%) to 5-HT$_5$ and 5-HT$_6$ receptors (Fig. 2). The 5-HT$_7$ receptor couples to the stimulation of adenylate cyclase activity (Bard et al. 1993), although it lacks any homology to the other cloned 5-HT stimulatory receptors (only 39% TM identity to the 5-HT$_6$ receptor and only 46% TM identity to the 5-HT$_4$ receptor, Gerald et al. 1995). No inhibition of adenylate cyclase activity or coupling to phospholipase C has been demonstrated for the 5-HT$_7$ receptor.

Sequences of mammalian 5HT$_7$ receptors range in size from 435 to 448 amino acids due to differences in the size of their carboxyl termini. *Xenopus* receptors are the shortest at 425 amino acids due to a shorter smaller amino terminus. In addition, varying lengths have been reported for the rat sequence by different groups. Such differences appear to be due to alternative Kozak consensus sequences for the initiating methionine as well as to possible splice sites in the carboxyl terminus. In general, there is a high conservation of amino acid identities in the TM regions (99%) of mammalian 5-HT$_7$ receptors (Fig. 3). A nonconservative substitution of leucine for proline in TM IV of the mouse sequence may account for the >tenfold lower binding affinities of methiothepin, 5-methoxytryptamine (5-MeOT), and 8-hydroxy-2-(di-*n*-propylamino)tetralin (8-OH-DPAT) for the mouse relative to the rat homologue. Site-directed mutagenesis experiments to evaluate these possible interactions have not been reported. In contrast, the *Xenopus* 5-HT$_7$ receptor displays lower TM amino acid identities (84%) to mammalian sequences. However, a direct comparison of mammalian pharmacology to the *Xenopus* receptor is not possible since binding affinities were not reported.

C. Molecular Pharmacology of the 5-HT$_5$, 5-HT$_6$, and 5-HT$_7$ Receptors

I. 5-HT$_5$ Receptors

The discriminating properties of the pharmacology of the 5-HT$_{5A}$ and 5-HT$_{5B}$ receptor subtypes include a rank order of ligand affinities from binding assays: 5-carboxamidotryptamine (5-CT) > 5-HT > 8-OH-DPAT > sumatriptan (Table 1). No compounds have been reported which can discriminate effectively (>tenfold) between the closely related 5-HT$_5$ subtypes (ERLANDER et al. 1993; MATTHES et al. 1993). Both [^{125}I]LSD (ERLANDER et al. 1993; MATTHES et al. 1993) and [^{3}H]5-CT (WISDEN et al. 1993; GRAILHE et al. 1994) have been used to determine the binding properties of the 5-HT$_5$ receptors. Competition experiments using [^{125}I]LSD have yielded biphasic curves for compounds which traditionally behave as agonists in systems expressing native or recombinant 5-HT receptors including 5-HT, 5-CT, 8-OH-DPAT, and sumatriptan (PLASSAT et al. 1992). These results would imply that [^{125}I]LSD is an antagonist at the 5-HT$_5$ receptors. The high-affinity site was guanine nucleotide-sensitive, indicating coupling of 5-HT$_5$ receptors to G protein(s). Interestingly, the K_L value of these compounds determined from [^{125}I]LSD competition experiments displayed better correspondence with the K_i value determined from [^{3}H]5-CT displacement studies, consistent with the small proportion (25%) of receptors existing in the high-affinity state. Few significant differences in binding properties have been observed among the rat, mouse, and human 5-HT$_5$ receptors with the exception of methiothepin, which exhibits significantly higher affinity (71-fold) for the human 5-HT$_{5A}$ receptor relative to the mouse homologue; the rat subtype has intermediate affinity for this compound.

II. 5-HT$_6$ Receptors

The discriminating properties of the pharmacology of the cloned rat 5-HT$_6$ receptor are the high affinity for a series of antipsychotic compounds, including clozapine and loxapine, as well as affinity for a number of tricyclic antidepressants such as amoxapine, clomipramine, and amitriptyline (MONSMA et al. 1993; ROTH et al. 1994). Analysis of competition binding studies using [^{125}I]LSD as a radioligand gives the rank order of binding affinities: methiothepin > 5-MeOT > 5-HT > tryptamine > 5-CT > sumatriptan ≫ 8-OH-DPAT (Table 1). A detailed functional analysis has not been reported for the cloned 5-HT$_6$ receptor. However, a similar but native receptor in N8TG2 cells, a neuroblastoma line, displays a rank order of agonist potency in both

Table 1. Affinity (pK$_i$) of serotonergic ligands for recombinant rat 5-HT$_{5A}$, 5-HT$_{5B}$, 5-HT$_6$, and 5-HT$_7$ receptors

Structural class	Drug	5-HT$_{5A}$[a]	5-HT$_{5B}$[a]	5-HT$_{5B}$[b]	5-HT$_6$[c]	5-HT$_6$[d]	5-HT$_7$[e]	5-HT$_7$[f]
Tryptamines	5-HT	5.9	6.6	8.1	6.8	7.3	8.7	8.8
	5-CT	6.6	7.9	8.9	6.1	6.6	9.5	9.8
	5-MeOT				7.4	7.7	8.7	9.2
	5-MeO-DMT				7.1		7.7	8.1
	Sumatriptan			6.1			6.2	6.6
	Tryptamine						7.5	7.8
Ergolines	DHE			7.5	6.4	8.3		
	Ergotamine	7.8	8.2	8.5	7.9			
	LSD				8.9			8.0[g]
	Mesulergine	<6.0	<6.0	<6.0	5.8		8.1	7.7
	Methysergide	<6.0	6.7	7.4	6.4		7.9	7.9
	Metergoline	<6.0	<6.0		7.5	7.2	8.7	8.2
Piperazine	TFMPP				6.3		6.3	6.6
Aminotetralin	8-OH-DPAT	<6.0	<6.0	7.3	<5.0		7.5	7.5
Benzoyl piperidine	Ketanserin			<6.0	<5.0		6.7	6.6
Bisphenylmethylenepiperidine	Ritanserin				7.4	7.8	7.7	7.8
Aryloxypropanolamine	Pindolol			<6.0	<5.0		<6.0	<5.6[g]
Dibenzoheptene	Cyproheptadine			6.9			7.3	7.5
Triazospirodecanone	Spiperone			<6.0	<5.0			7.7[g]
Dibenzodiazepine	Clozapine			<6.0	7.9	7.7	7.9	7.4
Rauwolfia alkaloid	Yohimbine	<6.0		<6.0				
Dibenzothiepine	Methiothepin	6.8	7.5	7.4	8.7	9.4	9.0	9.4
Benzamide	Zacopride	<6.0	<6.0	<6.0	<5.0		<6.0	<6.0

TFMPP, N-(3-Trifluoromethylphenyl) piperazine.
[a] Erlander et al. 1993. [125I]LSD used as the radioligand.
[b] Wisden et al. 1993. [3H]5-CT used as the radioligand.
[c] Monsma et al. 1993. [125I]LSD used as the radioligand.
[d] Monsma et al. 1993. [3H]5-HT used as the radioligand.
[e] Shen et al. 1993. [125I]LSD used as the radioligand.
[f] Shen et al. 1993. [3H]5-HT used as the radioligand.
[g] Ruat et al. 1993b. [3H]5-HT used as the radioligand.

radioligand binding and cAMP assays: 5-MeOT > 5-HT > tryptamine > 2-Me tryptamine > 5-CT > α-methyl-5-HT (UNSWORTH and MOLINOFF 1994). Responses to 5-HT in this cell line are also antagonized by clozapine. Affinities of compounds for the human cloned 5-HT_6 receptor are similar to those determined for the rat, with the exception of four compounds. Methiothepin exhibits a fourfold higher affinity for the human receptor, while metergoline and the atypical antipsychotics tiopyrone and amperozide display > tenfold higher affinity for the rat receptor. Clozapine acts as a high-affinity antagonist at both human and rat 5-HT_6 receptors. The rank order of antagonist binding affinities is: methiothepin > clozapine = olanzapine > ritanserin >> risperidone. A nonconserved amino acid substitution of threonine (rat) for leucine (human) in TM III may contribute to the differences in binding affinities observed between the species homologues.

III. 5-HT_7 Receptors

The discriminating properties of the molecular pharmacology of the 5-HT_7 receptor include a rank order of agonist affinities in binding assays: 5-CT > 5-HT ≥ 5-MeOT > 8-OH-DPAT > sumatriptan (Table 1). Agonist potencies determined from studies of the activation of adenylate cyclase activity generally displayed a tenfold lower activity than those derived from competition binding studies (PLASSAT et al. 1993; ZGOMBICK et al. 1995), without a change in their rank order. Antagonists for this subtype are methiothepin, metergoline, mesulergine, methysergide, ritanserin, clozapine, and LSD. In addition, a series of antipsychotic and tricyclic antidepressant compounds have high affinity for 5-HT_7 (ROTH et al. 1994). However, specific compounds possess unique 5-HT_6 and 5-HT_7 receptor profiles and discriminate between these two subtypes. For example, olanzapine displays significant 5-HT_6 selectivity (40-fold) while pimozide, risperidone, and spiperone exhibit marked 5-HT_7 selectivity (>150-fold; ROTH et al. 1994).

D. Distribution of 5-HT_5, 5-HT_6, and 5-HT_7 Receptors

I. 5-HT_5 Receptors

In the mouse, the distribution of mRNA for the 5-HT_{5A} receptor sequence has been evaluated by multiple techniques (PLASSAT et al. 1992). Using Northern analysis, three transcripts were detected in the brain and cerebellum. These multiple transcripts may indicate different polyadenylation lengths, possible pseudogenes, or even homologous genes not yet cloned. No signal was observed in the liver or kidney. By RT-PCR (reverse transcription polymerase chain reaction), a more sensitive detection method, a similar pattern was observed with labeling of the spinal cord and brain. No signal was amplified from mRNA derived from spleen, liver, heart, kidney, or lung. Therefore, the CNS distribution was more carefully mapped using *in situ* hybridization his-

Table 2. Localization of 5-HT$_5$, 5-HT$_6$, and 5-HT$_7$ receptor mRNA by *in situ* hybridization histochemistry

Region	5-HT$_{5A}$[a,b,c]	5-HT$_{5B}$[a,b,d]	5-HT$_6$[e,f]	5-HT$_7$[g,h,i]	Comments
Olfactory bulb	+	+	+	−	5-HT$_{5A}$ data sketchy for many regions
Cortex	+	+	+	+	
Septal area	+	−	−	+	
Hippocampus:					
CA1	+	+	+	−	
CA2	+	−	+	+	
CA3	+	−	+	+	
Dentate gyrus, Granule cell	+	+	+	−	Some see 5-HT$_7$ signals here but AS = S
Amygdala	+	−	+	+	
Basal ganglia:					
Caudate putamen	−	−	+	−	
Nucleus accumbens	−	−	+	−	
Olfactory tubercle	−	−	+	−	
Thalamus	+	−	+	+	5-HT$_7$ >> 5-HT$_6$
Habenula	+	+	+	−	
Hypothalamus	−	+	+	+	5-HT$_7$ most widely distributed
Superior colliculus	−	−	−	+	
Dorsal raphé	−	+	−	−	
Periaquaductal gray	−	−	−	+	
Pontine nucleus	−	−	−	+	
Dorsal tegmental nucleus	−	−	−	+	
Nucleus of solitary tract	−	−	−	+	
Cerebellum:					
Granule cell layer	+	−	+	−	5-HT$_6$ sense high
Spinal cord	−	−	−	−	

[a] Erlander et al. 1993; [b] Matthes et al. 1993; [c] Plassat et al. 1992; [d] Wisden et al. 1993; [e] Ruat et al. 1993a; [f] Ward et al. 1995; [g] Lovenberg et al. 1993; [h] Ruat et al. 1993b; [i] Gustafson et al. 1996. Detectable levels of receptor subtype mRNA are denoted by plus (+); undetectable levels of receptor subtype mRNA are denoted by minus (−).

tochemistry (Table 2). The predominant localization of 5-HT$_{5A}$ transcripts was detected in the hippocampus (CA1, CA2, CA3), cerebral cortex, habenula, olfactory bulb, and granule layer of cerebellum. These data suggest overlap with radiolabeled binding sites previously attributed to 5-HT$_{1D}$ sites in the mouse cortex, cerebellum, and hippocampus (Plassat et al. 1992). The distribution of message for the 5-HT$_{5B}$ was studied using Northern analysis and *in situ* hybridization histochemistry (Matthes et al. 1993). By Northern analysis, no signal was detected in brain, heart, kidney, lung, liver, or intestine. In contrast, signals were detected in very discreet and limited brain regions which included the CA1 region of the hippocampus, the habenula, and dorsal raphé nucleus by *in situ* hybridization histochemistry. Although the distribution of the proteins encoding 5-HT$_{5A}$ and 5-HT$_{5B}$ receptors has not been studied directly, it is possible that some 5-HT$_5$ binding sites have been embedded in the study of other 5-HT receptor binding sites. For example, [^{3}H]5-CT binding sites have been identified in guinea pig frontal cortex (Mahle et al.

1991). Some of these [^{3}H]5-CT-labeled sites may be related to 5-HT$_5$ receptor subtypes since sumatriptan displacement curves were clearly biphasic. The high-affinity component is due to interaction with 5-HT$_{1B/1D}$ subtypes while the low-affinity component may comprise 5-HT$_5$ and/or 5-HT$_7$ receptors.

In the rat (ERLANDER et al. 1993), 5-HT$_{5A}$ transcripts were detected by Northern analysis, with the hippocampus giving the strongest signal followed by the cortex, thalamus, pons, striatum, and medulla. No signal was obtained from the heart, kidney, or liver. Using *in situ* hybridization histochemistry, 5-HT$_{5A}$ mRNA was detected in the piriform cortex, hippocampus, amygdala, septum, and several thalamic nuclei. Transcripts for the 5-HT$_{5B}$ (ERLANDER et al. 1993) were detected by Northern analysis only in the hippocampus. Although the 5-HT$_{5B}$ cDNA was derived from a hypothalamic library, no transcript was detected in the hypothalamic mRNA source. In addition, signals were absent in heart, kidney, and liver. This absence of peripheral distribution was confirmed by WISDEN and colleagues (1993) for liver, heart, and kidney and extended to include muscle, lung, and spleen. Using *in situ* hybridization, 5-HT$_{5B}$ mRNA was found in pyramidal cells of the CA1 region of the hippocampus and subiculum, in the medial and lateral habenular nuclei, and in the median raphé nucleus. A weak signal was obtained in the piriform cortex and supraoptic nucleus (ERLANDER et al. 1993; WISDEN et al. 1993). Localization in the hippocampus suggests a possible role for the 5-HT$_5$ receptors in learning and memory consolidation while the habenular locus may implicate the 5-HT$_5$ receptor in the acquisition of adaptive behavior under stressful situations (THORNTON and DAVIES 1991).

Studies of localization of the human 5-HT$_{5A}$ receptor have been accomplished by RT-PCR (REES et al. 1994). Sites of representation included the amygdala, caudate, hypothalamus, cerebellum, substantia nigra, and thalamus. No signal was detected from aorta, kidney, heart, liver, small intestine, spleen, or uterus. There was only a low level of expression in the fetal brain. These data are consistent with the mouse and rat receptor distribution. In general, the 5-HT$_5$ subtypes appear to have an exclusive, or at least predominate, CNS distribution.

II. 5-HT$_6$ Receptors

The distribution of mRNA encoding the 5-HT$_6$ receptor has been determined by Northern analysis in the rat brain and peripheral tissues (MONSMA et al. 1993). The region with highest expression was the striatum, with lower density signals detected in the amygdala, cerebral cortex, and olfactory tubercle. Tissues in which mRNA was undetectable included cerebellum, hippocampus, hypothalamus, medulla, olfactory bulb, pituitary, retina, thalamus, heart, lung, kidney, liver, spleen, pancreas, skeletal or smooth muscle, stomach, ovary, prostate, and testes. A second group observed 5-HT$_6$ mRNA in the hippocampus, hypothalamus, stomach, and adrenal (RUAT et al. 1993a). Initial *in situ* hybridization studies in the rat brain detected a high level of mRNA in the striatum and olfactory tubercles (RUAT et al. 1993a). Other positive structures included the nucleus accumbens, olfactory bulb, and hippocampus. A comprehensive examination of the mRNA distribution for the 5-HT$_6$ receptor in the rat brain using *in situ* hybridization (WARD et al. 1995) has confirmed the highest abundance of message in the olfactory tubercle, striatum, nucleus accumbens, dentate gyrus, and CA1, CA2, and CA3 fields of the hippocampus. Lower intensity labeling was obtained in the cerebellum, some diencephalic nuclei, amygdala, and several cortical layers (2, 3, 4, and 6; see Table 2). In the rat brain, GLATT and colleagues (1995) have employed [^{3}H]clozapine as a label for 5-HT$_6$ receptors using rat brain membranes. They have reported that 40% of the sites which they detected exhibit a 5-HT$_6$ profile. No apparent differences in the density of these sites were detected when comparisons were made between cerebral cortex, striatum, and hippocampus. Thus the [^{3}H]clozapine binding is consistent with data from *in situ* hybridization studies. Perhaps future studies using [^{3}H]clozapine for receptor autoradiography will provide a detailed map of 5-HT$_6$ receptors. Methiothepin is another ligand which has even higher affinity for the 5-HT$_6$ receptor. It has previously been radiolabeled with tritium, but it has not been found to be a suitable radioligand in the brain (NELSON et al. 1979) due to its physico-chemical properties (e.g., lipophilicity) and low receptor subtype selectivity.

Recently, the distribution of the human 5-HT$_6$ mRNA has been evaluated by Northern blot analysis (KOHEN et al. 1996). It parallels the distribution in the rat, with the highest expression detected in the caudate nucleus. In the human brain, lower expression of 5-HT$_6$ mRNA was found in the hippocampus, amygdala, and thalamus.

III. 5-HT$_7$ Receptors

The distribution of mRNA for the 5-HT$_7$ receptor has been evaluated by several methods (LOVENBERG et al. 1993; MEYERHOF et al. 1993; PLASSAT et al. 1993; BARD et al. 1993; TSOU et al. 1994; GUSTAFSON et al. 1996). Using Northern analysis, LOVENBERG and colleagues (1993) detected transcripts with a predominant distribution in the rat thalamus and hypothalamus, with low levels detected in the hippocampus, cortex, and medulla. 5-HT$_7$ mRNA was not detected in the cerebellum, striatum, heart, liver, kidney, adrenal, spleen, testes, or ovaries. Similar results were obtained by SHEN and colleagues (1993), with the notable exception that they detected a robust transcript in the spleen. In the guinea pig (TSOU et al. 1994), a strong signal was observed in hippocampus and cortex but not in the cerebellum. The ileum and spleen were weakly positive but other peripheral tissues such as lung, liver, and adrenal were negative.

In the mouse, transcripts for the 5-HT$_7$ gene were undetectable by Northern analysis and the distribution was determined by RT-PCR (PLASSAT et al. 1993). In this species, positive signals were obtained in the brainstem, forebrain, cerebellum, embryonic collicular neurons, heart, and intestine. In the human, 5-HT$_7$ mRNA was detected, also by RT-PCR, at relatively high levels in total brain, coronary artery, and gastrointestinal tract and at low levels in kidney, liver, pancreas, and spleen. The peripheral distribution is consistent with the pharmacological relationship of the 5-HT$_7$ subtype to 5-HT-induced relaxation responses of several smooth muscle preparations (BARD et al. 1993; see Sect. E).

In situ hybridization studies in the adult rat brain (LOVENBERG et al. 1993) displayed intense labeling of the anterioventral and paraventricular thalamic nuclei, and less intense labeling in the hippocampal pyramidal cells in layers CA2 and CA3. Relatively weak levels of message were detected in the piriform and retrosplenial cortices and in the neocortex in layers 2 and 3. In the hypothalamus, sites of labeling included the arcuate and anterior nuclei (Table 2). Expression in the suprachiasmatic nucleus (SCN) was not definitive by either *in situ* hybridization or by PCR analysis, in spite of the proposed functional link of this subtype to the control of circadian rhythms (LOVENBERG et al. 1993). In the guinea pig CNS, the expression pattern was similar to that described in the rat (TSOU et al. 1994). The predominant loci of 5-HT$_7$ mRNA were the hippocampus, periventricular nucleus of the thalamus, and superficial layers of the cortex. These brain regions receive a diffuse projection from serotoninergic fibers from the midbrain and dorsal raphé nuclei (which do not appear to contain this transcript). The distribution of these signals indicates a potential role in information processing in the hippocampus and modulation of the corticothalamolimbic system (TSOU et al. 1994).

The distribution of the 5-HT$_7$ receptor protein in the rat and guinea pig brain has been studied using [^{3}H]5-CT as a radioligand (in the presence of various sets of compounds to mask other subtypes labeled by this ligand), including 5-HT$_{1A}$, 5-HT$_{1B}$ (rat), 5-HT$_{1D}$, and 5-HT$_5$ (GUSTAFSON et al. 1996; To et al. 1995; RAURICH et al. 1995; WAEBER and MOSKOWITZ 1995). The distribution of the receptor binding sites using this method was largely consistent with that reported for *in situ* hybridization studies and included labeling of septum, anterior thalamic nuclei, hypothalamus, and medial amygdala. This predominantly limbic distribution is consistent with suggestions of a role for this receptor subtype in emotion and in sensory processes (MEYERHOF et al. 1993). Regions in which receptor binding but no *in situ* hybridization signal was obtained were the globus pallidus, substantia innominata, and substantia nigra. The binding in the globus pallidus and substantia nigra was resistant to methiothepin and thus is probably not due to 5-HT$_7$ sites (GUSTAFSON et al. 1996). In the guinea pig brain, To and colleagues (1995) have also observed residual binding in these brain regions. It is possible that these are sites of 5-HT$_5$ binding since the 5-HT$_5$ subtypes have lower affinity for methiothepin than does the 5-HT$_7$ receptor. In general, the guinea pig and rat had very similar distributions of both mRNA and binding for 5-HT$_7$ receptors.

E. Potential Functional Roles of the 5-HT$_5$, 5-HT$_6$, and 5-HT$_7$ Receptors

I. 5-HT$_5$ Receptors

At the present time there are no selective ligands to evaluate 5-HT$_5$ function. Possible functions of the 5-HT$_5$ receptors can be projected based on localization studies. For example, the similarity in the molecular pharmacology of 5-HT$_5$ sites in the brain to 5-HT$_{1B/1D}$ sites indicates a possible involvement of the 5-HT$_5$ receptor in motor control, feeding, anxiety, and depression (PLASSAT et al. 1992; WILKINSON and DOURISH 1991). In addition, the limbic distribution in the mouse brain indicates a possible role in learning and in mood (PLASSAT et al. 1992) while the distribution of 5-HT$_{5A}$ mRNA in the cerebellum may suggest a role in the coordination of fine motor skills.

The distribution of 5-HT$_5$ message in the habenula is interesting. This region is a relay between the limbic system and midbrain and projects to the interpeduncular nucleus, raphé nuclei, substantia nigra, and ventral tegmental areas. The stimulation of GABAergic neurons in the lateral habenula inhibits both 5-HT neurons in the raphé and dopamine neurons in the substantia nigra and ventral tegmentum. Therefore, it is possible that modulation of 5-HT and dopamine release through 5-HT$_5$ receptors may have behavioral consequences. Habenular lesions have been shown to lead to enhanced exploratory behavior. Together, these data indicate a possible role for the 5-HT$_5$ receptors in the inhibition of behaviors related to emotional states. Additionally, since the habenular-raphé pathway is implicated in sleep cycle and in perception of pain, 5-HT$_5$ receptors may be involved in these activities. These are possible areas for assessment using transgenic mice with targeted disruption of the 5-HT$_5$ genes. Such knock-out mice have been made for the 5-HT$_5$ receptor, but no behavioral data have been reported (GRAILHE et al. 1995).

A possible functional correlate of 5-HT$_5$ activity may be embedded in earlier reports of adenylate cyclase stimulatory responses from the rat cortex (FAYOLLE et al. 1988). The high-affinity portion of the response to antagonists somewhat matches the rank order for 5HT$_5$ sites: dihydroergotamine (DHE) > methiothepin > methysergide >> ketanserin (inactive). However, the agonist rank order did not match: 5-CT was not active and bufotenin was more active than 5-HT. In the guinea pig airway, activation of an atypical 5-HT receptor mediates inhibition of a non-cholinergic, non-adrenergic bronchoconstriction (WARD et al. 1994). In general the rank order of agonist potency of this response is consistent with a 5-HT$_5$ receptor profile: 5-CT > 5-HT >> 8-OH-DPAT > α-Me-5-HT; sumatriptan and 2-Me-5-HT are inactive. The only inconsistency is that 5-MeOT was inactive in the guinea pig but similar in affinity to 5-HT in the cloned receptor systems, although the guinea pig 5-HT$_5$ receptors may display species variations in pharmacology due to potential differences in their deduced amino acid sequences.

II. 5-HT$_6$ Receptors

Possible *in vivo* roles of the 5-HT$_6$ receptor may involve neuropsychiatric functions, based on the high affinity of atypical antipsychotic drugs and several tricyclic antidepressants for this receptor (MONSMA et al. 1993; ROTH et al. 1994). Analysis of the clinical profiles of atypical antipsychotics underscores the potential impact of 5-HT$_6$ activity. Risperidone, a mixed 5-HT$_{2A}$/D$_2$ antagonist with low 5-HT$_6$ activity, has been shown to have a substantial trend toward causing extrapyramidal symptoms whereas olanzapine and zotepine, which have similar affinities at 5-HT$_{2A}$ and 5-HT$_6$, do not (KOHEN et al. 1996). Thus 5-HT$_6$ receptor antagonism may have a beneficial effect on the clinical profile of atypical antipsychotics. The distribution of 5-HT$_6$ mRNA in limbic pathways (WARD et al. 1995) also indicates that the 5-HT$_6$ receptor may modulate the affective state. The localization of mRNA for the 5-HT$_6$ receptor in the nucleus accumbens may also suggest a role in reinforcement/reward. The ability of 5-HT$_6$ receptors to modulate the mesolimbic dopaminergic projections to the nucleus accumbens remains to be evaluated experimentally. Preliminary data indicate that mRNA for the 5-HT$_6$ receptor is preferentially downregulated in rats in certain brain regions after a 2-week treatment with clozapine or haloperidol (FREDERICK et al. 1995).

The first behavioral studies of possible 5-HT$_6$-mediated function have been attempted using antisense oligonucleotides targeted to the 5-HT$_6$ receptor subtype (BOURSON et al. 1995). In these studies, the rats exhibited a behavioral phenotype consisting of an increased number of yawns and stretches. This behavior was blocked by atropine, suggesting a role of the 5-HT$_6$ receptor in the control of cholinergic neurotransmission. If so, then a 5-HT$_6$ antagonist might be useful in the treatment of depression, anxiety, and/or memory disorders (BOURSON et al. 1995). Further developments in this area await additional antisense studies, knock-out mice, or the development of selective agonists and antagonists for this receptor subtype.

Functional correlates of possible effects of 5-HT$_6$ receptors have also been observed *in vitro* (QUACH et al. 1982). A study of glycogenolysis in rat cortical slices may reflect a 5-HT$_6$-like profile. In this slice preparation, 5-HT, 5-MeOT, and tryptamine stimulated glycogen hydrolysis; tricyclic antidepressants were among the best competitive antagonists of the response. However, methiothepin was weak in antagonizing the response although physico-chemical properties of the compound may have limited its efficacy. *N,N*-Dimethyltryptamine (*N,N*-DMT) was an antagonist of this response with greater efficacy than methiothepin. At the human 5-HT$_6$ receptor, *N,N*-DMT was equal in effect to 5-HT ($pK_i = 7.2$).

There have been several early reports of cyclase-stimulatory serotonin receptors in cells lines, particular NCB.20 cells. This cell line is a fusion between a mouse neuroblastoma line, N18TG2, and an embryonic hamster brain explant (BERRY-KRAVITZ and DAWSON 1983). In this hybrid cell line, cAMP accumulation was stimulated by 5-HT, 5-MeOT, and methysergide.

The response was antagonized by clozapine as well as by spiperone, although at much higher concentrations. This response was reinvestigated using more recently available tools for characterization (Connor and Mansour 1990; Cossery et al. 1990). The response was inhibited by metergoline ($K_b = 50 \text{n}M$), but not by ICS 205,930 (Cossery et al. 1990), consistent with a 5-HT$_6$, but not a 5-HT$_4$ or 5-HT$_7$, response profile. The parental mouse cell line, N18TG2, has also been evaluated for 5-HT-stimulated cAMP responses. It appears to display a pharmacology similar to the cloned 5-HT$_6$ receptor based on evaluation of the rank order of agonist potency in both radioligand binding and second messnger assays: 5-MeOT > 5-HT > tryptamine > 2-Me-5-HT >> 5-CT > α-Me-5-HT (Unsworth and Molinoff 1994). In binding assays methiothepin displayed higher affinity than clozapine while in second messenger assays methiothepin, clozapine, and mianserin exhibited similar antagonistic potencies (pA$_2$ ~ 6.5). A molecular analysis of the N18TG2 cells to evaluate the presence of mRNA for serotonin receptor subtypes has not been reported.

Stimulatory cyclase responses mediated via a 5-HT$_6$-like receptor in striatal neurons have been described (Schoeffter and Waeber 1994; Sebben et al. 1994). In cultured mouse striatal neurons, that rank order of agonist potencies to stimulate cAMP production was: 5-HT > LSD > 5-MeOT > 5-CT; 8-OH-DPAT, sumatriptan, and cisapride were inactive. Antagonists of this response included methiothepin, nortriptyline, clozapine, and amitriptyline. In pig caudate membranes, a similar rank order of agonist potencies was observed: 5-HT ≥ 5-MeOT > 5-CT; 8-OH-DPAT, sumatriptan, and renzapride were also not active. The antagonist rank order in this preparation was: methiothepin > clozapine >> ketanserin. No 5-HT$_4$ receptor responses were detected, although receptor autoradiographic studies (Grossman et al. 1993), as well as mRNA distribution (Gerard et al. 1995), indicate a strong representation of 5-HT$_4$ in the striatum in rat, guinea pig, and human. Neither of these receptor profiles derived from striatal preparations matches the rank order of potencies in either the N18TG2 cell line or the rank order of binding affinities from the cloned rat receptor. However, it is possible that cross-species comparisons or methodological differences may obscure the true relationships. These reports argue for a strong similarity of the striatal responses to the cloned 5-HT$_6$ receptor, but corroborative evidence of the presence of mRNA encoding this subtype in these preparations, cloning and characterization of mouse and pig 5-HT$_6$ receptors, and/or receptor-selective antibodies or compounds is required before these responses can be characterized definitively.

III. 5-HT$_7$ Receptors

There are no selective ligands available to study the 5-HT$_7$ receptor. However, based on the activity profiles of a set of clinically useful compounds (which have activity at several other receptor subtypes), the 5-HT$_7$ receptor may play a role in antipsychotic or antidepressive therapy (Sleight et al. 1995; Roth et al. 1994). Like the 5-HT$_6$ receptor, the 5-HT$_7$ subtype has high affinity for several antipsychotic compounds and antidepressants such as pimozide and clozapine, suggesting that the therapeutic actions of these drugs may be exerted partially via this receptor (ROTH et al. 1994). Chronic antidepressant treatment with the 5-HT uptake inhibitor fluoxetine substantially reduces the binding of [^{3}H]5-HT to 5-HT$_7$ receptors in the hypothalamus (SLEIGHT et al. 1995), suggesting a potential role for the 5-HT$_7$ receptor in depression. Predictions of possible receptor function based on studies of the localization of both mRNA and protein for the 5-HT$_7$ receptor in the septum, hypothalamus, and amygdala further include a role in so-called limbic processes. Additional brain regions with 5-HT$_7$ binding sites including layers 1–3 of cortex, thalamus, and periaquaductal gray indicate the potential for roles of the 5-HT$_7$ receptor in pain processing. It has been suggested that the 5-HT$_7$ receptor may play a role in circadian function (see below). However, neither the 5-HT$_7$ receptor nor its mRNA has been localized to the SCN (LOVENBERG et al. 1993; GUSTAFSON et al. 1996). The distribution of 5-HT$_7$ mRNA in the anterior hypothalamic area, immediately lateral and dorsal to the SCN, suggests the possibility that the receptor may be localized on neurons whose dendritic fields extend into the SCN, and/or that the anterior hypothalamic neurons may have axonal projections into the SCN. Similar observations have been made in the guinea pig (To et al. 1995).

The regulation of mammalian circadian rhythms by the 5-HT$_7$ receptor was initially suggested by LOVENBERG and colleagues (1993). This suggestion was based on the observation that 5-CT and 8-OH-DPAT, compounds with affinity for both 5-HT$_{1A}$ and 5-HT$_7$ receptor subtypes, are able to phase advance neuronal firing activity in rat hypothalamic slices *in vitro* (MEDANIC and GILLETTE 1992) and to inhibit spontaneous activity and photic responses in the hamster SCN *in vivo* (YING and RUSAK 1994). Ritanserin, an antagonist with moderate affinity for the 5-HT$_7$ subtype, but not pindolol, a 5-HT$_{1A}$ antagonist, reverses the phase advance induced by 8-OH-DPAT in the hypothalamic slice preparation (LOVENBERG et al. 1993). Additional evidence implicating the 5-HT$_7$ receptor in circadian function has been provided using whole cell voltage clamp recordings from rat SCN neurons *in vitro* (KAWAHARA et al. 1994). 5-HT, 5-CT, and 8-OH-DPAT all inhibited the GABA$_A$ currents and this effect was blocked by ritanserin, but not by pindolol. The inhibition of the GABA current could be mimicked by agents that directly elevate cAMP, such as forskolin and 8-Br-cAMP. In the isolated rat SCN, 8-OH-DPAT-mediated phase shifts are blocked by cAMP-dependent protein kinase inhibitors, supporting a role for a G$_s$-coupled receptor (PROSSER et al. 1994). These authors demonstrated that potassium channel blockers antagonize, while potassium channel activators mimic, the 8-OH-DPAT response, suggesting the 5-HT$_7$ receptor may also couple to potassium channels in addition to adenylate cyclase stimulation. Taken together, these results are consistent with the involvement of the 5-HT$_7$ receptor in circadian processes. However, definitive proof of the role of 5-HT$_7$ receptors awaits the development of subtype-selective compounds.

    Additional functional correlates of 5-HT$_7$ receptor activity can be found in
the modulation of vasculature tone. Serotonin-induced smooth muscle relaxation has been reported for a variety of isolated tissue preparations including
the porcine vena cava (Trevethick et al. 1984, 1986; Sumner et al. 1989), cat
saphenous vein (Feniuk et al. 1983), guinea pig ileum (Fenuik et al. 1983;
Carter et al. 1995), canine coronary artery (Cushing and Cohen 1992), porcine pial veins (Lee et al. 1994; Ueno et al. 1995), guinea pig stomach fundus
(Kojima et al. 1992), and guinea pig trachea (Pype et al. 1994). In most of these
tissues, elevations of intracellular cAMP levels have been detected. The relaxant response is resistant to tetrodotoxin, suggesting a postjunctional receptor
localization except in the guinea pig trachea, in which a prejunctional 5-HT$_7$
receptor mediates relaxation (Pype et al. 1994). Pharmacological properties
characteristic of these preparations include a high sensitivity to 5-CT with a
rank order of agonist potencies: 5-CT > 5-HT $\geq$ 5-MeOT > 8-OH-DPAT
>> sumatriptan (inactive). The relaxant responses are antagonized by
methiothepin, methysergide, mesulergine, spiperone, and clozapine. Antagonists selective for other 5-HT receptor subtypes are inactive (e.g., ketanserin,
GR 113808, ondansetron). Further evidence for the mediation of this response
via the 5-HT$_7$ receptor is provided by the localization by RT-PCR of mRNA
encoding the 5-HT$_7$ receptor in many of these blood vessels (Ullmer et al.
1995). Taken together, these data indicate that the 5-HT$_7$ receptor may play a
role in smooth muscle relaxation and thus may be involved in diseases such as
irritable bowel syndrome or angina.
    In other *in vitro* and *in vivo* preparations, 5-HT$_7$ responses are intermixed
with one or more additional serotonin responses. For example, Shenker and
colleagues (1987) described two 5-HT receptors coupled to adenylate cyclase
stimulation in guinea pig hippocampal membranes designated R$_H$ and R$_L$. The
R$_H$ response component was originally classified as 5-HT$_{1A}$-like due to the high
potency of DPAT and competitive antagonism by spiperone (K_b = 24nM).
Subsequent studies indicate that the R$_H$ receptor displays a pharmacological
profile consistent with the 5-HT$_7$ subtype (Tsou et al. 1994) while the R$_L$
corresponds to the 5-HT$_4$ receptor (Dumuis et al. 1990). In the rabbit jugular
vein, vascular relaxation is mediated indirectly via an endothelial receptor
with properties consistent with a 5-HT$_{2B}$ receptor (Ellis et al. 1995), and
directly by a smooth muscle receptor with characteristics of a 5-HT$_7$ receptor
(Martin et al. 1987). In the isolated sheep pulmonary vein, both 5-HT$_4$- and 5-
HT$_7$-like receptors appear to mediate endothelium-independent relaxations
(Cocks and Arnold 1992). In intact dogs, a receptor resembling the 5-HT$_7$
receptor mediates vasorelaxation of the coronary vascular bed, while 5-HT$_{1D}$-
like receptors produce vasoconstriction of the renal and carotid vascular beds
(Cambridge et al. 1995).

F. Summary and Conclusions

At the time of the last issue of this series considering 5-HT, it was already clear
that there were a multitude of responses to indolealkylamines (Erspamer

1966; GYERMEK 1966) and that 5-HT had significant clinical effects in man (STACEY 1966). Since that time, remarkable progress has been made in the elucidation of serotonin receptor subtypes, in the design of more refined pharmacological tools with which to probe receptor function, and in the therapeutic application of such tools to human disease. However, even now there are many serotonin responses that cannot be readily assigned to a particular receptor subtype. The search for more selective pharmacological and genetic tools continues. The ability to clone receptor subtypes and to perform structure activity analysis and receptor profiling in heterologous expression systems containing a single receptor subtype has greatly enhanced the ability to dissect subtypes in various preparations. It has become possible to compare the pharmacology of cloned receptor subtypes and isolated tissue preparations using exact species matches (BRANCHEK et al. 1995). Subtypes such as 5-HT$_5$ have been uncovered which were formerly embedded in the overlapping pharmacological profiles of other receptor subtypes such as the 5-HT$_{1D}$ receptors. In addition, apparently "novel" profiles such as 5-HT$_6$ have been uncovered. However, retrospective analysis of the literature points to examples of this subtype in native preparations (SCHOEFFTER and WAEBER 1994; SEBBEN et al. 1994). The 5-HT$_7$ receptor represents a "novel" subtype which had been observed in many functional assays where it mediated smooth muscle relaxation. It has similar operational properties to 5-HT$_1$ receptors (BRADLEY et al. 1986) but was never detailed in binding studies. However, it could have been 5-HT$_7$ sites embedded in assays of high-affinity [^{3}H]5-HT binding to 5-CT-sensitive but sumatriptan-insensitive sites along with 5-HT$_5$ receptors.

At present, the least information is available concerning the 5-HT$_5$ and 5-HT$_6$ receptor subtypes. For the 5-HT$_6$ receptor, the activity of several major compounds which are in clinical use will certainly generate many studies of its potential role in psychosis and depression. For the 5-HT$_{5A}$ receptor, careful studies of transgenic animals with targeted disruption of the 5-HT$_{5A}$ gene (GRAILHE et al. 1995) may soon yield valuable insight into the functions of this subtype. In parallel, further refinement of chemical tools and the generation of antibodies to these subtypes (e.g., 5-HT$_5$; CARSON et al. 1995) may further refine our understanding of the functional roles of these novel 5-HT receptor subtypes. Although the pace of receptor cloning appears to have slowed, the task of identifying pathophysiological states in which these receptor subtypes are involved, as well as the development of subtype-selective compounds for the 5-HT$_5$, 5-HT$_6$, and 5-HT$_7$ receptors for clinical applications, represent major challenges for the future.

References

Bard JA, Zgombick J, Adham N, Vaysse P, Branchek TA, Weinshank RL (1993) Cloning of a novel human serotonin receptor (5-HT$_7$) positively linked to adenylate cyclase. J Biol Chem 268:23422–23426

Berry-Kravis E, Dawson G (1983) Characterization of an adenylate cyclase-linked serotonin (5-HT$_1$) receptor in a neuroblastoma brain explant hybrid cell line (NCB-20). J Neurochem 40:977–985

Bourson A, Boroni E, Austin RH, Monsma FJ Jr, Sleight AJ (1995) Determination of the role of the 5-HT$_6$ receptor in rat brain: a study using antisense oligonucleotides. J Pharmacol Exp Ther 274:173–180

Bradley PB, Engel G, Feniuk W, Fozard JR, Humphrey PPA, Middlemiss DN, Mylecharane EJ, Richardson BP, Saxena PR (1986) Proposals for the classification and nomenclature of functional receptors for 5-hydroxytryptamine. Neuropharmacol 25:563–576

Branchek TA, Bard JA, Kucharewicz SA, Zgombick JM, Weinshank RL, Cohen ML (1995) Migraine: relationship to cloned canine and human 5-HT$_{1D}$ receptors. In: Olesen J, Moskowitz MA (eds) Experimental headache models. Lippincott-Raven, Philadelphia, pp 125–134

Cambridge D, Whiting MV, Butterfield LJ, Marston C (1995) Vascular 5-HT$_1$-like receptors mediating vasoconstriction and vasodilation: their characterization and distribution in intact canine cardiovascular system. Br J Pharmacol 114:961–968

Carson MJ, Danielson PE, Thomas EA, Sutcliffe JG (1995) The 5-HT$_{5A}$ receptor is expressed predominantly on astrocytes within the developing and adult rat CNS. Soc Neurosci Abstr 21:1856

Carter D, Champney M, Hwang B, Englen RM (1995) Characterization of a postjunctional 5-HT receptor mediating relaxation of guinea pig isolated ileum. Eur J Pharmacol 280:243–250

Cocks TM, Arnold PJ (1992) 5-Hydroxytryptamine (5-HT) mediates potent relaxation in the sheep isolated pulmonary vein via activation of 5-HT$_4$ receptors. Br J Pharmacol 107:591–596

Connor DA, Mansour TE (1990) Serotonin receptor-mediated activation of adenylate cyclase in the neuroblastoma NCB.20: a novel 5-hydroxytryptamine receptor. Mol Pharmacol 37:742–751

Cossery JM, Mienville J-M, Sheehy PA, Mellow AM, Chuang D-M (1990) Characterization of two distinct 5-HT receptors coupled to adenylate cyclase activation and ion current generation in NCB-20 cells. Neurosci Lett 108:149–154

Cushing DJ, Cohen ML (1992) Serotonin-induced relaxation in canine coronary artery smooth muscle. J Pharmacol Exp Ther 263:123–129

Dumuis A, Bouhelal R, Sebben M, Cory R, Bockaet J (1990) A nonclassical 5-hydroxytryptamine receptor positively coupled with adenylate cyclase in the central nervous system. Mol Pharmacol 34:880–887

Ellis ES, Byrne C, Murphy OE, Tilford NS, Baxter GS (1995) Mediation by 5-hydroxytryptamine$_{2B}$ receptors of endothelium-dependent relaxation in the rat jugular vein. Br J Pharmacol 114:400–404

Erlander MG, Lovenberg TW, Baron BM, De Lecea L, Danielson PE, Racke M, Slone AL, Siegel BW, Foye PE, Cannon K, Burns JE, Sutcliffe JG (1993) Two members of a distinct subfamily of 5-hydroxytryptamine receptors differentially expressed in rat brain. Proc Natl Acad Sci USA 90:3452–3456

Erspamer V (1966) Peripheral physiological and pharmacological actions of indolealkylamines. In: Erspamer V (ed) 5-Hydroxytryptamine and related indolealkylamines. Springer, Berlin Heidelberg New York, pp 245–359 (Handbook of experimental pharmacology, vol 19)

Fayolle C, Fillion M-P, Barone P, Oudar P, Rousselle J-C, Fillion G (1988) 5-Hydroxytryptamine stimulates two distinct adenylate cyclase activities in rat brain: high affinity activation is related to the 5-HT$_1$ subtype different from 5-HT$_{1A}$, 5-HT$_{1B}$, and 5-HT$_{1C}$. Fund Clin Pharmacol 2:195–214

Feniuk W, Humphrey PPA, Watts AD (1983) 5-Hydroxytryptamine-induced relaxation of isolated mammalian smooth muscle. Eur J Pharmacol 96:71–78

Frederick JA, Lopez JF, Meador-Woodruff JH (1995) Effects of clozapine and haloperidol on expression of 5-HT$_6$ and 5-HT$_7$ receptors. Soc Neurosci Abstr 21:1857

Gerard C, Adham N, Kao H-T, Olsen MA, Laz TM, Schechter LE, Bard JA, Vaysse PJ-J, Hartig PR, Branchek TA, Weinshank RL (1995) The 5-HT$_4$ receptor: molecular cloning and pharmacological characterization of two splice variants. EMBO J 14:2806–2815

Glatt CE, Snowman A, Sibley DR, Snyder SH (1995) Clozapine: selective labeling of sites resembling 5-HT$_6$ serotonin receptors. Mol Med 1:398–406

Grailhe R, Amlaiky A, Ghavami A, Ramboz S, Yocca F, Mahle C, Margouris C, Perrot F, Hen R (1994) Human and mouse 5-HT$_{5A}$ and 5-HT$_{5B}$ receptors: cloning and functional expression. Soc Neurosci Abstr 20:1160

Grailhe R, Ramboz S, Boschert U, Hen R (1995) The 5-HT$_5$ receptors: characterization of the human 5-HT$_{5A}$ receptor: absence of the human 5-HT$_{5B}$ receptor; knockout of the mouse 5-HT$_{5A}$ receptor. Soc Neurosci Abstr 21:1856

Grossman CJ, Kilpatrick GJ, Bunce KT (1993) Development of a radioligand binding assay for 5-HT$_4$ receptors in guinea pig and rat brain. Br J Pharmacol 109:618–624

Gustafson EL, Durkin MM, Bard JA, Zgombick JM, Branchek TA (1996) A receptor autoradiographic and *in situ* hybridization analysis of the distribution of the 5-HT$_7$ receptor in rat brain. Br J Pharmacol 117:657–666

Gyermek L (1966) Drugs which antagonize 5-hydroxytryptamine and related indolealylamines. In: Erspamer V (ed) 5-Hydroxytryptamine and related indolealkylamines. Springer, Berlin Heidelberg New York, pp 471–528 (Handbook of experimental pharmacology, vol 19)

Kawahara F, Saito H, Katsuki H (1994) Inhibition by 5-HT$_7$ receptor stimulation of GABA$_A$ receptor-activated current in cultured rat suprachiasmatic neurones. J Physiol (Lond) 478:67–73

Kohen R, Metcalf MA, Khan N, Druck T, Huebner K, Lachowicz JE, Meltzer HY, Sibley DR, Roth BL, Hamblin MW (1996) Cloning, characterization, and chromosomal localization of a human 5-HT$_6$ serotonin receptor. J Neurochem 66:47–56

Kojima S-I, Ishizaki R, Shimo Y (1992) Investigation into the 5-hydroxytryptamine-induced relaxation of the circular smooth muscle of guinea pig stomach fundus. Eur J Pharmacol 224:45–49

Lee T J-F, Ueno M, Sunagane N, Sun M-H (1994) Serotonin relaxes porcine pial vessels. Am J Physiol 266:H1000–H1006

Lefkowitz RJ, Caron MG (1988) Adrenergic receptors. Models for the study of receptors coupled to guanine nucleotide regulatory proteins. J Biol Chem 263:4993–4996

Lovenberg TW, Baron BM, de Lecea L, Miller JD, Prosser RA, Rea MA, Foye PE, Racke M, Slone AL, Siegel BW, Danielson PE, Sutcliffe JG, Erlander MG (1993) A novel adenylyl cyclase-activating serotonin receptor (5-HT$_7$) implicated in the regulation of mammalian circadian rhythms. Neuron 11:449–458

Mahle CD, Nowak HP, Mattson RJ, Hurt SD, Yocca FD (1991) [^{3}H]5-Carboxamidotryptamine labels multiple high affinity 5-HT$_{1D}$-like sites in guinea pig brain. Eur J Pharmacol 205:323–324

Martin GR, Leff P, Cambridge D, Barrett VJ (1987) Comparative analysis of two types of 5-hydroxytryptamine receptor mediating vasorelaxation: differential classification using tryptamines. Naunyn-Schmiedeberg's Arch Pharmacol 336:365–373

Matthes H, Boschert U, Amaiky N, Grailhe R, Plassat J-L, Muscatelli F, Mattei M-G, Hen R (1993) Mouse 5-hydroxytryptamine$_{5A}$ and 5-hydroxytryptamine$_{5B}$ receptors define a new family of serotonin receptors: cloning, functional expression, and chromosomal localization. Mol Pharmacol 43:313–319

Medanic M, Gillette MU (1992) Serotonin regulates the phase of the rat suprachiasmatic circadian pacemaker *in vitro* only during the subjective day. J Physiol (Lond) 450:629–642

Meyerhof W, Obermuller F, Fehr S, Richter D (1993) A novel rat serotonin receptor: primary structure, pharmacology, and expression pattern in distinct brain regions. DNA Cell Biol 12:401–409

Monsma FJ Jr, Shen Y, Ward RP, Hamblin MW, Sibley DR (1993) Cloning and expression of a novel serotonin receptor with high affinity for tricyclic psychotropic drugs. Mol Pharmacol 43:320–327

Nelson CS, Cone RD, Robbins LS, Allen CN, Adelman JP (1995) Cloning and expression of a 5-HT$_7$ receptor from *Xenopus laevis*. Recept Chann 3:61–70

Nelson DL, Herbet A, Pichat L, Glowinski J, Hamon M (1979). *In vitro* and *in vivo* disposition of [^{3}H]methiothepin in brain tissues: relationship to the effects of acute treatment with methiothepin on central serotonergic receptors. Naunyn-Schmiedeberg's Arch Pharmacol 310:25–33

Plassat J-L, Boschert U, Amlaiky N, Hen R (1992) The mouse 5-HT$_5$ receptor reveals a remarkable heterogeneity within the 5-HT$_{1D}$ receptor family. EMBO J 11:4779–4786

Plassat J-L, Amlaiky N, Hen R (1993) Molecular cloning of a mammalian serotonin receptor that activates adenylate cyclase. Mol Pharmacol 44:229–236

Prosser RA, Heller HC, Miller JD (1994) Serotonergic phase advances of the mammalian circadian clock involve protein kinase A and K$^+$ channel opening. Brain Res 644:67–73

Pype JL, Verleden GM, Demedts MG (1994) 5-HT modulates noncholinergic contraction in guinea pig airways *in vitro* by a prejunctional 5-HT$_1$-like receptor. J Appl Physiol 77:1135–1141

Quach TT, Rose C, Duchemin AM, Schartz JC (1982) Glycogenolysis induced by serotonin in brain: identification of a new class of receptor. Nature 298:373–375

Raurich A, Mengod G, Hurt S, Palacios JM, Cortes R (1995) Correlation between 5-HT$_7$ receptor binding and its mRNA in the rat and guinea pig brain visualized autoradiographically. Soc Neurosci Abstr 21:1857

Rees S, den Daas I, Foord S, Goodson S, Bull D, Kilpatrick G, Lee M (1994) Cloning and characterization of the human 5-HT$_{5A}$ serotonin receptor. FEBS Lett 355:242–246

Roth BL, Craigo SC, Choudray MS, Uluer A, Monsma FJ Jr, Shen Y, Meltzer HY, Sibley DR (1994) Binding of typical and atypical antipsychotic agents to 5-hydroxytryptamine-6 and 5-hydroxytryptamine-7 receptors. J Pharmacol Exp Ther 268:1403–1410

Ruat M, Traiffort E, Arrang J-M, Leurs R, Schartz J-C (1991) Cloning and tissue expression of a rat histamine H$_2$-receptor gene. Biochem Biophys Res Commun 179:1470–1478

Ruat M, Traiffort E, Arrang J-M, Tardivel-Lacombe J, Diaz J, Leurs R, Schartz J-C (1993a) A novel rat serotonin (5-HT$_6$) receptor: molecular cloning, localization, and stimulation of cAMP accumulation. Biochem Biophys Res Commun 193:268–276

Ruat M, Traiffort E, Leurs R, Tardivel-Lacombe J, Diaz J, Arrang J-M, Schartz J-C (1993b) Molecular cloning, characterization, and localization of a high-affinity serotonin receptor (5-HT$_7$) activating cAMP formation. Proc Natl Acad Sci USA 90:8547–8551

Saudou F, Boschert U, Amlaiky N, Plassat J-L, Hen R (1992) A family of *Drosophila* serotonin receptors with distinct intracellular signaling properties and expression patterns. EMBO J 11:7–17

Schoeffter P, Waeber C (1994) 5-Hydroxytryptamine receptors with a 5-HT$_6$ receptor-like profile stimulating adenylyl cyclase activity in pig caudate membranes. Naunyn-Schmiedeberg's Arch Pharmacol 350:356–360

Sebben M, Ansanay H, Bockaert J, Dumuis A (1994) 5-HT$_6$ receptors positively coupled to adenylyl cyclase in striatal neurones in culture. Neuroreport 5:2553–2557

Shen Y, Monsma FJ Jr, Metcalf MA, Jose PA, Hamblin MW, Sibley DR (1993) Molecular cloning and expression of a 5-hydroxytryptamine$_7$ serotonin receptor subtype. J Biol Chem 68:18200–18204

Shenker A, Maayani S, Weinstein H, Green JP (1987) Pharmacological characterization of two 5-hydroxytryptamine receptors coupled to adenylyl cyclase in guinea pig hippocampal membranes. Mol Pharmacol 31:357–367

Sleight AJ, Carolo C, Petit N, Zwingelstein C, Bourson A (1995) Identification of 5-hydroxytryptamine$_7$ receptor binding sites in rat hypothalamus: sensitivity to chronic antidepressant treatment. Mol Pharmacol 47:99–103

Stacey RS (1966) Clinical aspects of cerebral and extracerebral 5-hydroxytryptamine. In: Erspamer V (ed) 5-Hydroxytryptamine and related indolealkylamines. Springer, Berlin Heidelberg New York, pp 744–774 (Handbook of experimental pharmacology, vol 19)

Sumner MJ, Feniuk W, Humphrey PPA (1989) Further characterization of the 5-HT receptor mediating relaxation and elevation of cyclic AMP in porcine vena cava. Br J Pharmacol 97:292–300

Thornton EW, Davies C (1991) A water-maze discrimination learning deficit in the rat following lesion of the habenula. Physiol Behav 49:819–822

To ZP, Bonhaus DW, Eglen RM, Jakeman LB (1995) Characterization and distribution of putative 5-HT$_7$ receptors in guinea pig brain. Br J Pharmacol 115:107–116

Trevethick MA, Feniuk W, Humphrey PPA (1984) 5-Hydroxytryptamine-induced relaxation of porcine vena cava in vitro. Life Sci 35:477–486

Trevethick MA, Feniuk W, Humphrey PPA (1986) 5-Carboxamidotryptamine: a potent agonist mediating relaxation and elevation of cyclic AMP in the isolated neonatal porcine vena cava. Life Sci 38:1521–1528

Tsou A-P, Kosaka A, Bach C, Zuppan P, Yee C, Tom L, Alvarez R, Ramsey S, Bonhaus DW, Stefanich E, Jakeman L, Eglen RM, Chan HW (1994) Cloning and expression of a 5-hydroxytryptamine$_7$ receptor positively coupled to adenylyl cyclase. J Neurochem 63:456–464

Ueno M, Ishine T, Lee T J-F (1995) A novel 5-HT$_1$-like receptor subtype mediates cAMP synthesis in porcine pial vein. Am J Physiol 268:H1383–H1389

Ullmer C, Schmuck K, Kalkman HO, Lübbert H (1995) Expression of serotonin receptor mRNAs in blood vessels. FEBS Lett 370:215–221

Unsworth CD, Molinoff PB (1994) Characterization of a 5-hydroxytryptamine receptor in mouse neuroblastoma N18TG2 cells. J Pharmacol Exp Ther 269:246–255

Waeber C, Moskowitz MA (1995) Autoradiographic visualization of [^{3}H]5-carboxamidotryptamine binding sites in the guinea pig and rat brain. Eur J Pharmacol 283:31–46

Ward JK, Fox AJ, Barnes PJ, Belvisi MG (1994) Inhibition of excitatory non-adrenergic, non-cholinergic bronchoconstriction in guinea pig airways in vitro by activation of an atypical 5-HT receptor. Br J Pharmacol 111:1095–1102

Ward RP, Hamblin MW, Lachowicz JE, Hoffman BJ, Sibley DR, Dorsa DM (1995) Localization of serotonin subtype 6 receptor messenger RNA in the rat brain by in situ hybridization histochemistry. Neuroscience 64:1105–1111

Wilkinson LO, Dourish CT (1991) Serotonin and animal behavior. In: Peroutka SJ (ed) Serotonin receptor subtypes: basic and clinical aspects. Wiley-Liss, New York, pp 147–210

Wisden W, Parker EM, Mahle CD, Grisel DA, Nowak HP, Yocca FD, Felder C, Seeburg PH, Voight MM (1993) Cloning and characterization of the rat 5-HT$_{5B}$ receptor: evidence that the 5-HT$_{5B}$ receptor couples to a G protein in mammalian cell membranes. FEBS 333:25–31

Witz P, Amlaiky N, Plassat JL, Maroteaux L, Borrelli E, Hen R (1990) Cloning and characterization of a Drosophila serotonin receptor that activates adenylate cyclase. Proc Natl Acad Sci USA 87:8940–8944

Ying S-W, Rusack B (1994) Effects of serotonergic agonists on firing rates of photically responsive cells in the hamster suprachiasmatic nucleus. Brain Res 651:37–46

Zgombick JM, Adham N, Bard JA, Vaysse PJ-J, Weinshank RL, Branchek TA (1995) Pharmacological characterization of the recombinant human 5-HT$_7$ receptor subtype coupled to adenylate cyclase stimulation in a clonal cell line. Soc Neurosci Abstr 21:1364

CHAPTER 19
Electrophysiology of 5-HT Receptors

G.K. Aghajanian and R. Andrade

A. Introduction

Within the past decade, molecular cloning techniques have confirmed that 5-hydroxytryptamine (5-HT) receptor subtypes, originally predicted from radioligand binding and functional studies (e.g., $5\text{-}HT_1$, $5\text{-}HT_2$, $5\text{-}HT_3$, $5\text{-}HT_4$), represent separate and distinct gene products. This knowledge has had a crucial impact on electrophysiological approaches to the 5-HT system in two important ways: (1) studies on previously *recognized* 5-HT receptors are now being directed, through the use of in situ mRNA hybridization and immunocytochemical maps, more precisely toward neurons that express these specific 5-HT receptor subtypes and (2) the functional role of previously *unrecognized* receptors (e.g., $5\text{-}HT_5$, $5\text{-}HT_6$, $5\text{-}HT_7$) can now be explored. Depending on the expression pattern for each type of neuron, the various 5-HT receptor subtypes can interact with their own set of G proteins, second messengers, and ion channels, to give rise to the wide range of electrophysiological actions produced by 5-HT throughout the brain and spinal cord. In addition, it is becoming evident that more than one 5-HT receptor sub-type may be expressed by the same neuron or by different neurons within the same region. Thus, while the following review is organized primarily according to individual 5-HT receptor subtypes, interactions between different receptor subtypes within a single neuron or region are also discussed where appropriate.

B. $5\text{-}HT_1$ Receptors

Many electrophysiological studies on the $5\text{-}HT_1$ receptor subtype have been conducted in areas with a dense concentration of $5\text{-}HT_{1A}$-binding sites and a high level of $5\text{-}HT_{1A}$ mRNA expression such as the dorsal raphe nucleus, the hippocampal pyramidal cell layer, and the entorhinal cortex (CHALMERS and WATSON 1991; MIQUEL et al. 1991; PAZOS and PALACIOS 1985; POMPEIANO et al. 1992). Studies in these and other regions will be reviewed in the following sections.

I. Raphe Nuclei

1. Physiology and Pharmacology

Because serotonergic neurons of the raphe nuclei are inhibited by the local (microiontophoretic) application of 5-HT to their cell body region, the receptor mediating this effect has been termed a somatodendritic autoreceptor (as opposed to the prejunctional autoreceptor). Early studies in the dorsal raphe nucleus showed that lysergic acid diethylamide (LSD) and other indoleamine hallucinogens are powerful agonists at the somatodendritic 5-HT autoreceptor (AGHAJANIAN et al. 1968; AGHAJANIAN et al. 1972). Functionally, the somatodendritic 5-HT autoreceptor has been shown to mediate collateral inhibition (WANG and AGHAJANIAN 1977). Studies in the brain slice preparation have revealed that the ionic basis for the autoreceptor-mediated inhibition, by either 5-HT or LSD, is an opening of K^+ channels to produce a hyperpolarization (AGHAJANIAN and LAKOSKI 1984); these channels are characterized by their inwardly rectifying properties (WILLIAMS et al. 1988). As in the dorsal raphe nucleus, it has been shown recently that serotonergic neurons of the nucleus raphe magnus are also hyperpolarized by 5-HT via the opening of inwardly rectifying K^+ channels (PAN et al. 1993). Furthermore, similar findings have been obtained in acutely isolated (PENINGTON et al. 1993a) and individually microcultured (JOHNSON 1994) dorsal raphe neurons, underscoring the fact that autoreceptor inhibition is independent of any inputs to the raphe nucleus. Patch clamp recordings in the cell-attached and outside-out configuration from such acutely isolated dorsal raphe neurons show that the increase in K^+ current results from a greater probability of opening of unitary resting K^+ channel activity (PENINGTON et al. 1993b).

The somatodendritic autoreceptors of serotonergic neurons in both the dorsal raphe nucleus and the nucleus raphe magnus appear to be predominantly of the 5-HT$_{1A}$ subtype as a variety of drugs with 5-HT$_{1A}$ selectivity [e.g., 8-hydroxy-2-(di-n-propylamino)-tetralin (8-OH-DPAT) and the anxiolytic drugs buspirone and ipsapirone] share the ability to potently inhibit raphe cell firing in a dose-dependent manner (PAN et al. 1993; SPROUSE and AGHAJANIAN 1987 1988). Furthermore, intracellular recordings from dorsal raphe neurons in brain slices show that 5-HT$_{1A}$ agonists such as 8-OH-DPAT and ipsipirone fully mimic 5-HT in producing hyperpolarization and decreasing input resistance (Fig. 1; SPROUSE and AGHAJANIAN 1987). For a series of 5-HT$_{1A}$ agonists, treatment with the irreversible receptor inactivator EEDQ (N-ethoxycarbonyl-2-ethoxy-1,2-dihydroquinoline) shifts agonist dose-response curves to the right but does not reduce the maximal inhibitory response, suggesting that a considerable autoreceptor reserve exists in the raphe (Cox et al. 1993).

While classical 5-HT antagonists have proven ineffective in blocking the electrophysiological effects at the 5-HT autoreceptor, the acute intravenous administration of the D_2 receptor antagonist spiperone, which has moderate affinity for the 5-HT$_1$-receptor-binding site, rapidly blocks 5-HT$_{1A}$-mediated

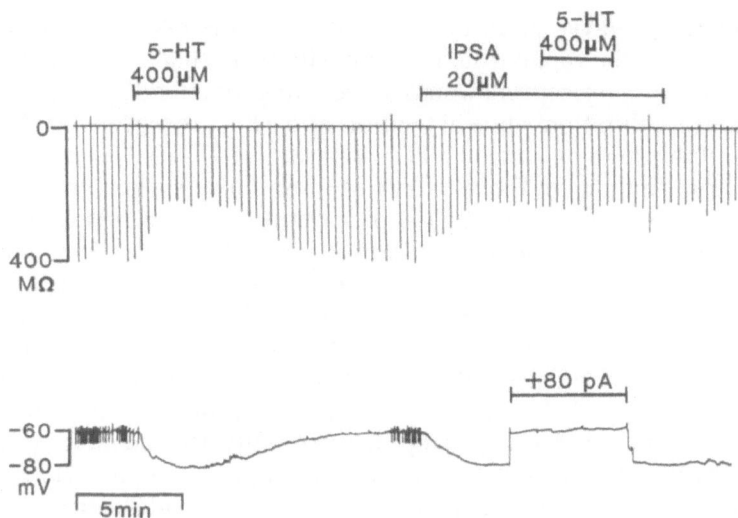

Fig. 1. Hyperpolarizing effects of 5-HT and the 5-HT$_{1A}$ agonist ipsapirone (*IPSA*) on a serotonergic neuron in the dorsal raphe nucleus. Note *in the lower trace* that both ipsapirone and 5-HT, given at supramaximal concentrations, induce an identical degree of hyperpolarization in this cell (~20 mV); also note that both agonists reduce input resistance to a similar degree (*top trace*). A shared receptor/transduction pathway is indicated by the fact that, in the presence of ipsapirone, 5-HT produces no further effect on membrane potential or input resistance even when original resting membrane potential is restored by the injection of +80 pA current. (Provided by G.K. AGHAJANIAN)

inhibitions in the dorsal raphe (BLIER et al. 1989; LUM and PIERCEY 1987) but not postsynaptically in CA3 neurons of the hippocampus, suggesting the possibility that pre- and postsynaptic 5-HT$_{1A}$ receptors are not identical (Blier et al. 1993a). The basis for this difference is unclear since to date only one 5-HT$_{1A}$ clone has been reported. Very recently, a highly selective 5-HT$_{1A}$ antagonist (WAY 100635) has been found which potently blocks the direct inhibition of dorsal raphe serotonergic neurons by both 5-HT and selective 5-HT$_{1A}$ agonists (CRAVEN et al. 1994; FLETCHER et al. 1996; FORSTER et al. 1995). WAY 100635 also blocks the indirect inhibition of dorsal raphe neurons induced by selective 5-HT reuptake inhibitors (GARTSIDE et al. 1995). After chronic treatment with a selective 5-HT reuptake inhibitor, there is a recovery of the basal firing rates of raphe neurons; under these conditions a 5-HT$_{1A}$ antagonist induces an even further increase in firing rate, suggesting that compensatory mechanisms have developed to maintain basal firing rates (ARBORELIUS et al. 1995).

In contrast to WAY 100635, a selective 5-HT$_{1D}$ antagonist (GR 127935) was found to be ineffective in blocking suppression of dorsal raphe firing by 5-HT (CRAVEN et al. 1994), suggesting that 5-HT$_{1D}$ receptors do not contribute to the inhibitory effect of exogenous 5-HT on 5-HT neurons. On the other hand, it has been proposed that an activation of 5-HT$_{1D}$ receptors reduces collateral inhibition within the dorsal raphe nucleus, possibly by inhibiting 5-HT release from dendrites (PIÑEYRO et al. 1996). If this were the case, the function of 5-HT$_{1D}$ receptors in the raphe would be opposite from that of 5-HT$_{1A}$ receptors: activation of 5-HT$_{1A}$ receptors would serve a negative feedback role by mediating collateral inhibition while activation of 5-HT$_{1D}$ receptors would oppose collateral inhibition by reducing 5-HT release.

Receptor binding and behavioral studies have suggested that the β-adrenoceptor antagonists such as (–)-propranolol possess 5-HT$_{1A}$ antagonist properties. Electrophysiological studies have shown that low microiontophoretic currents of β-blockers such as (–)-propranolol or (–)-tertatolol effectively block the suppressant effects of the 5-HT$_{1A}$ agonists (e.g., ipsapirone and 8-OH-DPAT) on raphe cell firing (JOLAS et al. 1993; SPROUSE and AGHAJANIAN 1986; PRISCO et al. 1993). These results fit with cloning data, which reveal a remarkable degree of sequence homology between the β_2-adrenoceptor and the 5-HT$_{1A}$ receptor, especially the membrane-spanning domains that are characteristic of G-protein-coupled receptors (FARGIN et al. 1988). These findings provide a molecular basis for the interaction between β-adrenoceptor antagonists and 5-HT$_{1A}$ agonists.

In addition to opening of K$^+$ channels, whole cell recordings from acutely dissociated raphe neurons have shown that 5-HT decreases high-threshold calcium currents, probably via 5-HT$_{1A}$ receptors since the effect of 5-HT is mimicked by 8-OH-DPAT (PENINGTON and KELLY 1990). This calcium current is virtually insensitive to L-type calcium channel blockers but is partially sensitive to ϖ-conotoxin, an N-type channel channel blocker (PENINGTON et al. 1991).

2. Signal Transduction Pathways

It has been shown that the opening of K$^+$ channels via 5-HT$_{1A}$ receptors in dorsal raphe neurons is mediated by pertussis-toxin-sensitive G proteins. Pertussis toxin catalyzes the ADP ribosylation of the α-subunit of certain G proteins (e.g., G$_i$ and G$_o$), causing an irreversible uncoupling of the G protein from its receptor. Extracellular and intracellular experiments in the dorsal raphe nucleus have shown that a 48-h preinjection with pertussis toxin (local or intracerebroventricular) causes an almost total blockade of the inhibitory and hyperpolarizing effect of 5-HT (INNIS et al. 1988). Consistent with their 5HT$_{1A}$-binding properties, the inhibitory effects of ipsapirone, 8-OH-DPAT, and LSD in the dorsal raphe are also blocked by pertussis toxin (INNIS et al. 1988; BLIER et al. 1993b). The coupling to 5-HT$_1$ receptors to G proteins in the dorsal raphe is also shown by the fact that intracellular injection of GTPγS, a nonhydrolyzable analog of GTP which induces an irreversible activation of G proteins, mimics and is nonadditive with the hyperpolarizing action of 5-HT (INNIS et al. 1988). Intracellular GTPγS also renders irreversible the suppression of high-threshold calcium currents in dorsal raphe neurons (PENINGTON et al. 1991). Other areas where pertussis toxin blockade of inhibitory responses to 5-HT has been demonstrated include the dorsal root ganglia (CRAIN et al.

1987), the ventromedial hypothalamus (NEWBERRY and PRIESTLEY 1988), and the hippocampus (see below).

II. Other Subcortical Regions

Inhibitory or hyperpolarizing responses to 5-HT have been reported in a wide variety of neurons in the spinal cord, brain stem, and diencephalon. In general, such responses have been attributed to mediation by 5-HT_1 receptors. In sensory neurons of dorsal root ganglia, a 5-HT_1-like receptor has been reported to reduce the calcium component of action potentials and to produce hyperpolarizations which can be mimicked by 5-HT_{1A} agonists such as 8-OH-DPAT (SCROGGS and ANDERSON 1990; TODOROVIC and ANDERSON 1990). In cerebellar Purkinje cells, 5-HT-induced inhibition but not excitation is mediated through 5-HT_{1A} receptors (DARROW et al. 1990). In brain slices of the nucleus prepositus hypoglossi, focal electrical stimulation evokes inhibitory postsynaptic potentials (IPSPs) that are mediated by 5-HT_{1A} receptors to activate an inwardly rectifying K^+ conductance (BOBKER and WILLIAMS 1989a) and a novel outwardly rectifying K^+ conductance (BOBKER and WILLIAMS 1995); focal electrical stimulation in the nucleus prepositus hypoglossi also induces a late excitatory postsynaptic potential (EPSP) which is mediated by 5-HT_2 receptors (BOBKER 1994). In the midbrain periaqueductal gray, a region known to be involved in pain modulation and fear responses, approximately half the cells are inhibited/hyperpolarized by 8-OH-DPAT, suggesting mediation by 5-HT_{1A} receptors (BEHBEHANI et al. 1993). In the ventromedial hypothalamus (NEWBERRY 1992) and lateral septum (JOËLS et al. 1987; VAN DEN HOOFF and GALVAN 1992), 5-HT and 5-HT_{1A} agonists produce inhibitory effects also by activating a potassium conductance. In the locus coeruleus, 5-HT suppresses depolarizing synaptic potentials, apparently through both 5-HT_{1A} and 5-HT_{1B} receptors located presynaptically (BOBKER and WILLIAMS 1989b). Also in the locus coeruleus, 5-HT appears to selectively suppress the postsynaptic excitatory response to locally applied glutamate through a 5-HT_{1A} receptor (CHARLETY et al. 1991). In the rat laterodorsal tegmental nucleus (LDT), bursting cholinergic neurons are hyperpolarized by 5-HT via 5-HT_1 receptors (LUEBKE et al. 1992). It has been suggested that during REM sleep the removal of a tonic inhibitory 5-HT influence from these cholinergic neurons may be responsible for the emergence of ponto-geniculo-occipital (PGO) spikes during this behavioral state.

III. Hippocampus

1. Physiology and Pharmacology

Pyramidal cells of the CA1 region express high levels of 5-HT_{1A} receptor mRNA and 5-HT_{1A} receptor binding (POMPEIANO et al. 1992). Early electrophysiological studies using in vivo extracellular recording and iontophoresis reported that the predominant effect of 5-HT on these cells was an inhibition of firing activity (BISCOE and STRAUGHAN 1966). Intracellular recordings in brain slices showed that the 5-HT-induced inhibition was due to a hyperpolarization resulting from an opening of K^+ channels (SEGAL 1980). Subsequent work, using a variety of pharmacological approaches in vitro in brain slices, showed that the 5-HT-induced inhibition in both CA1 and CA3 pyramidal cells was mediated by the activation of receptors of the 5-HT$_{1A}$ subtype (ANDRADE and NICOLL 1987; OKUHARA and BECK 1994; SEGAL et al.1989; ZGOMBICK et al. 1989). As in the dorsal raphe, these receptors activate a perussis-toxin-sensitive G protein which couples to the opening of inwardly rectifying potassium channels through membrane-delimited pathways (ANDRADE et al. 1986, ANDRADE and NICOLL 1987). The 5-HT$_{1A}$-mediated increase in potassium conductance in hippocampus is but one example of a widespread cellular effector mechanism used by numerous neurotransmitter receptors in diverse preparations (NICOLL 1988; ANDRADE et al. 1986), the best known of which is the muscarinic M_2 receptor inhibition of the heart (PFAFFINGER et al. 1985).

In addition to the above direct effects on pyramidal cells, 5-HT has been shown to depress both excitatory and inhibitory synaptic potentials in the hippocampus. Relatively high concentrations of 5-HT cause a reduction in electrically evoked EPSPs in CA1 pyramidal cells (SCHMITZ et al. 1995), an effect that is mimicked by 8-OH-DPAT, suggesting mediation by 5-HT$_{1A}$ receptors. Indirect measures indicate that 5-HT acts presynaptically to reduce Ca^{2+} entry and thereby gluamatergic synaptic transmission. In addition, there is a 5-HT$_{1A}$-mediated inhibitory effect on putative inhibitory interneurons of the hippocampus (SEGAL 1990; SCHMITZ et al. 1995a,b). Consistent with an opening of potassium channels, the inhibitory effects of 5-HT on interneurons result from a hyperpolarization associated with a reduction in input resistance. Functionally, the 5-HT$_{1A}$-mediated inhibition of GABAergic interneurons in the hippocampus leads to a disinhibition of pyramidal cells in CA1. Clearly, the effects of 5-HT in the hippocampus are highly complex, involving both pre- and postsynaptic actions which may, to varying degrees, be inhibitory or disinhibitory, facilitatory or disfacilitatory.

2. Signal Transduction Pathways

The molecular mechanisms underlying the opening of potassium channels are most likely common to all neurotransmitter receptors, including 5-HT$_{1A}$ receptors, that couple through the G_i/G_o family of G proteins. Muscarinic receptors in the heart activate potassium channels through a G protein of the G_i/G_o family (BREITWIESER and SZABO 1985; PFAFFINGER et al. 1985). The mechanism involved in this muscarinic response is highly homologous to that used by 5-HT$_{1A}$ receptors in hippocampus (ANDRADE 1992; ANDRADE et al. 1986), and 5-HT$_{1A}$ receptors transfected onto heart cells can substitute for the muscarinic receptors (KARSCHIN et al. 1991). The mechanisms by which G proteins couple these receptors to the channels have been the subject of considerable work and controversy. The initial work of PFAFFINGER et al. (1985) and of BREITWIESER and SZABO (1985) demonstrated that muscarinic receptors coupled to potassium channels in the heart through a membrane-delimited pathway involving a G protein of the G_i/G_o family. Subsequent work showed that a similar mechanism mediated the effect of the 5-HT_{1A} receptors in the CA1 region of the hippocampus (ANDRADE et al. 1986, ZGOMBICK et al. 1989). Initial work in isolated heart cells suggested that it was the α_i-subunit of the G protein that gated the opening of the potassium channels (YATANI et al. 1988). However, more recently this conclusion has been questioned, leading to the current view that the $\beta\gamma$-subunits regulate the channels (WICKMAN et al. 1994; REUVENY et al. 1994; KOFUJI et al. 1995). There seems to be little selectivity of different $\beta\gamma$-subunit combinations (WICKMAN et al. 1994), raising questions about how cells maintain selective receptor regulation of the potassium channels. One possible explanation derives from recent studies in which it was found that $\beta\gamma$-subunits directly bind to both the N- and C-terminus of G-protein-activated inwardly rectifying K^+ channels and that trimeric $G_{\alpha\beta\gamma}$ binds to the N-terminal domain of these channels (HUANG et al. 1995). Based on these observations, it was suggested that the binding of $G_{\alpha\beta\gamma}$ to the K^+ channel confers specificity due to a compartmentalization of the receptor/G protein/channel complex.

The effector mechanism that ultimately mediates the inhibitory effect signaled by M_2 or 5-HT_{1A} receptors is the potassium channel. Molecular biological approaches have made it possible to identify several of the subunits comprising these channels (KRAPIVINSKY et al. 1995; LESAGE et al. 1994; DASCAL et al. 1995; KUBO et al. 1993). All of them are part of the growing family of inwardly rectifying potassium channel subunits. This name refers to the unusual property of the channels to allow potassium to flow into the cell much more readily than out of the cell, a property that confers upon these channels an unmistakable electrophysiological signature. Interestingly, at least one of the potassium channel subunits identified in heart, GIRK1, is expressed at high levels in hippocampus (KARSCHIN et al. 1994), suggesting that it might be involved in mediating the 5-HT_{1A}-receptor-induced hyperpolarization in this region. Consistent with this possibility, the potassium current activated by 5-HT_{1A} receptors in the CA1 region does show the characteristic signature of this potassium channel family; namely it shows inward rectification (ANDRADE and NICOLL 1987). However, the degree of inward rectification in situ is considerably less than that displayed by channels formed by GIRK1 (DASCAL et al. 1995; KUBO et al. 1993). Since the channels in situ are most likely heteromultimers (KOFUJI at al. 1995; KRAPIVINSKY et al. 1995) and inward rectification varies between subunits (WIBLE et al. 1994), a possible explanation for these findings is that additional subunits coassemble with GIRK1 to form the potassium channels that are activated by 5-HT_{1A} receptor in the CA1 region of hippocampus. In one instance, it has been shown that the coexpression of GIRK4 with GIRK1 in *Xenopus* oocytes enhances the in-

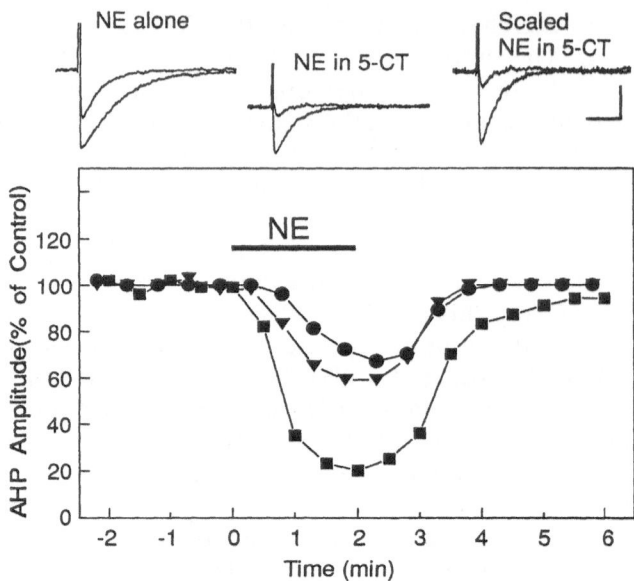

Fig. 2. Enhancement of adenylate cyclase/cAMP signaling by 5-HT$_{1A}$ receptors in the hippocampus. Administration of norepinephrine (*NE*) inhibits the afterhyperpolarization (*AHP*) in pyramidal cells of the CA1 region. This effect is mediated by β-adrenergic receptors and is signaled by the adenylate cyclase/cAMP/protein kinase A cascade. Activation of 5-HT$_{1A}$ receptors by 5-carboxyamidotryptamine (*5-CT*) enhances the ability of NE to reduce the AHP. *Calibration bar* corresponds to 5 mV and 1 s. Control, 300 nM 5-CT, recovery. *Lower panel: square*, control; *triangle*, NE alone; *circle*, NE in 5-CT. (Provided by R. Andrade)

wardly rectifying K$^+$ current induced by coexpressed 5-HT$_{1A}$ receptors (Spauschus et al. 1996).

5-HT$_{1A}$ receptors inhibit forskolin-stimulated adenylate cyclase in biochemical assays (DeVivo and Maayani 1986). Thus a potential pathway for these receptors to affect electrophysiological functions might be to inhibit signaling through the adenylate cyclase/cAMP-signaling pathway. In the CA1 region, stimulation of adenylate cyclase results in the reduction of a specific calcium-activated potassium current responsible for the slow afterhyperpolarization (AHP) present in these cells (Madison and Nicoll 1982). The presence of this robust cAMP-mediated response has made it possible to examine the effect of 5-HI$_{1A}$ receptor activation on responses signaled through the adenylate cyclase/cAMP-signaling cascade. Surprisingly, activation of 5-HT$_{1A}$ receptors not only fails to inhibit but actually enhances the β-adrenergic-induced reduction of the AHP (Fig. 2; Andrade 1993). In contrast, no effect of 5-HT$_{1A}$ receptor activation is seen on the 8-bromo-cAMP-induced reduction of the AHP. This indicates that the 5-HT$_{1A}$ receptors enhanced signaling along this pathway by acting at the level of the adenylate cyclase. A possible molecular explanation for these paradoxical results may be found in the observation that some adenylate cyclase isoforms, including adenylate cyclase type II, are synergistically activated by $\beta\gamma$-subunits (FEDERMAN et al. 1992; TANG and GILMAN 1991). Since pyramidal cells of the CA1 region are highly enriched in type II cyclase (Mons et al. 1993), 5-HT$_{1A}$ receptor activation could enhance cAMP signaling by making $\beta\gamma$-subunits available to synergize with G$_s$ in activating adenylate cyclase (ANDRADE 1993). These results suggest that the effects of 5-HT$_{1A}$ receptor activation on cAMP signaling should be critically dependent on the specific isoform(s) of adenylate cyclase expressed by each cell type.

IV. Cerebral Cortex

5-HT$_{1A}$-induced hyperpolarizing responses in pyramidal cells of the cerebral cortex have been described in a number of studies (ARANEDA and ANDRADE 1991; DAVIES et al. 1987; SHELDON and AGHAJANIAN 1991; TANAKA and NORTH 1993). However, all of these studies also show that cortical neurons typically display mixed inhibitory and excitatory responses to 5-HT, involving actions at multiple 5-HT receptor subtypes, including 5-HT$_{1A}$ and 5-HT$_{2A/2C}$, that are expressed by the same pyramidal cells. As in the dorsal raphe and other regions, pretreatment with intracerebral injections of the GiGo inactivator pertussis toxin attenuates responses to the 5-HT$_{1A}$ agonist 8-OH-DPAT but not to 5-HT$_{2A/2C}$ or 5-HT3 agonists (ZHANG et al. 1994). Inhibitory responses mediated by 5-HT$_{1A}$ receptors are often unmasked or enhanced in the presence of 5-HT$_2$ antagonists, consistent with the idea that there is an interaction between 5-HT$_{1A}$ and 5-HT$_{2A}$ receptors at an individual neuronal level (ARANEDA and ANDRADE 1991; ASHBY et al. 1994; LAKOSKI and AGHAJANIAN 1985). Consistent with the idea that 5-HT$_2$ receptor blockade might unmask or potentiate 5-HT$_{1A}$ inhibitory responses in the cerebral cortex, the systemic administration of a drug which possesses both 5-HT$_{1A}$ agonist and 5-HT$_2$ antagonist activity (BIMT 17) has been shown to be especially effective in suppressing the firing of neurons in prefrontal cortex in vivo (BORSINI et al 1995). A similar suggestion of a shift in the balance between 5-HT-mediated excitation and inhibition comes from another in vivo study in which both the systemic and local application of 5-HT$_2$ antagonists was shown to prevent an enhancement of unit activity (and cortical desynchronization) that normally occurs in response to noxious stimuli (tail compression) in anesthetized rats (NEUMAN and ZEBROWSKA 1992).

In the medial prefrontal cortex in vivo, both spontaneous unit activity and evoked unit activity (i.e., noxious stimuli or electrical stimulation of the mediodorsal nucleus of the thalamus) is suppressed by electrical stimulation of the midbrain raphe nuclei (MANTZ et al. 1990); surprisingly, the ability of midbrain raphe stimulation to suppress spontaneous activity is blocked rather than enhanced by the systemic administration of 5-HT$_2$ antagonists. Thus, these results do not fit with the general concept, as described above, that an unmasking of 5-HT$_{1A}$ inhibitory effects occurs in the presence of 5-HT$_2$ antagonists. While the reason for this apparent discrepancy is not clear at this time, one possible explanation would be that there is a preferential excitation of inhibitory interneurons via 5-HT$_2$ receptors in response to electrical stimulation of the raphe nuclei, resulting in a transynaptic GABAergic inhibition of pyramidal cells. Alternatively, since the 5-HT$_2$ antagonists were administered systemically, they could be acting at a subcortical site such as the thalamus.

As in the dorsal raphe nucleus, the activation of 5-HT$_{1A}$ receptors has been found to reduce high-voltage-activated calcium currents in acutely dissociated neocortical pyramidal cells (Foehring 1996). The 5-HT$_{1A}$ modulation is reduced by the N-type blocker ω-conotoxin and the P-type blocker ω-agatoxin but not the L-type blocker nifedipine, suggesting that N- and P- but not L-type calcium channels are involved. The reduction in calcium currents was shown to be G protein mediated since the inclusion of GTPγS in the recording pipettes rendered the responses irreversible. Moreover, this effect was shown to be membrane delimited rather than through a diffusible cytosolic pathway since the response to 5-HT$_{1A}$ receptor activation did not spread to on-cell patches. However, in contrast to the dorsal raphe nucleus, the 5-HT$_{1A}$-mediated reduction in calcium currents in cortical pyramidal cells does not appear to be blocked by pertussis toxin. The reason for this apparent regional difference remains to be determined (Foehring 1996).

In addition to the above postsynaptic effects, there are presynaptic effects mediated by 5-HT$_1$ receptors in the cerebral cortex. In cingulate cortex, 5-HT reduces the amplitude of electrically evoked excitatory synaptic potentials, including both N-methyl-D-aspartate (NMDA) and non-NMDA components (Tanaka and North 1993); selective agonists and antagonists were used to show that this inhibition resulted from the activation of presynaptic 5-HT$_{1B}$ receptors, probably through the reduction in release of glutamate from cortico-cortical excitatory fibers. Similar findings have been reported for medial prefrontal (Read et al. 1994) and somatosensory cortex (Liu et al. 1993).

C. 5-HT$_2$ Receptors

Quantitative autoradiographic studies show high concentrations of 5-HT$_2$-binding sites and mRNA expression in certain regions of the forebrain such as the neocortex (layers IV/V), piriform cortex, claustrum, and olfactory tubercle (Mengod et al. 1990a). With few exceptions (e.g., motor nuclei and the nucleus tractus solitarius), relatively low concentrations of 5-HT$_2$ receptors or mRNA expression are found in the brain stem and spinal cord. Studies aimed at examining the physiological role of 5-HT$_2$ receptors in several of these regions are discussed in the following sections.

I. Motoneurons

1. Physiology and Pharmacology

Motoneurons of the facial and other cranial motor nuclei have a high density of 5-HT$_2$-receptor-binding sites. Consistent with the binding data, in situ hybridization shows a high level of 5-HT$_2$ receptor mRNA in these neurons (MENGOD et al. 1990b). Early studies in vivo showed that 5-HT applied microiontophoretically does not by itself induce firing in the normally quiescent facial motoneurons but does facilitate the subthreshold and threshold excitatory effects of glutamate (McCALL and AGHAJANIAN 1979). Intracellular recordings from facial motoneurons in vivo or in brain slices in vitro (AGHAJANIAN and RASMUSSEN 1989; LARKMAN et al. 1989) show that 5-HT induces a slow, subthreshold depolarization associated with an increase in

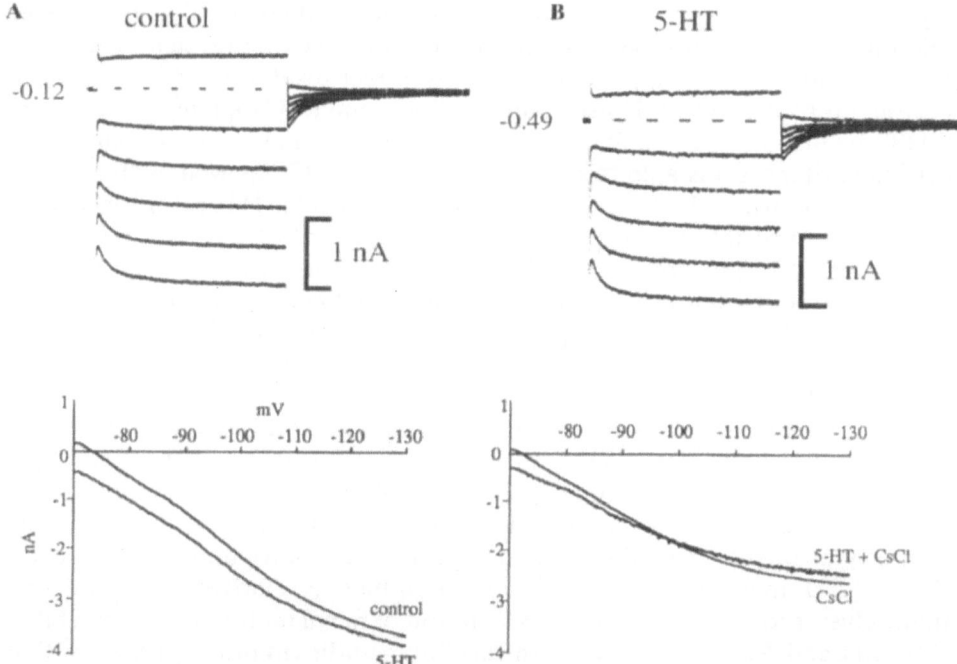

Fig. 3A,B. Voltage clamp recordings showing the dual effects of 5-HT on a rat facial motoneuron: (1) depolarization of the resting membrane potential (as indicated by the induction of an inward current) and (2) enhancement of the hyperpolarization-activated cationic current I_h. *In the top traces*, 5-HT can be seen to induce an inward current of approximately 370 pA (holding potential, − 70 mV). In addition, time-dependent inward currents induced by depolarizing steps (−80 to −120) are increased by 5-HT. *Lower traces* represent slow IV ramps (−70 to −130 mV) to determine the reversal potential of the 5-HT-induced inward current: (1) *left panel* shows that in normal perfusate the IV curves do not cross at the potassium reversal potential (∼− 90 mV); *right panel* shows that in the presence of Cs$^+$, which blocks I_h, there is crossing at expected potassium reversal potential. (Modified from GARRATT et al. 1993)

input resistance, indicating a decrease in resting K^+ conductance. Similar effects of 5-HT also have been described in medullary respiratory neurons (Lalley et al. 1995) and spinal motoneurons (White and Fung 1989; Yamazaki et al. 1992). Recently, it has been shown that 5-HT also increases the excitability of facial motoneurons by enhancing a hyperpolarization-activated cationic current I_h (Fig. 3; Garratt et al. 1993; Larkman and Kelly 1992). In contrast to the dual action of 5-HT (i.e., both a reduction in I_K^+ and an enhancement of I_h), the depolarizing effect of norepinephrine on facial motoneurons appears to involve only a closure of K^+ channels (Larkman and Kelly 1992).

The 5-HT$_2$ antagonist ritanserin is able to selectively block the excitatory effects of 5-HT in facial motoneurons (Rasmussen and Aghajanian 1990). On the other hand, the selective 5-HT$_{1A}$ agonist 8-OH-DPAT, although it increases facial motoneuron excitability when given in vivo by systemic injection, fails to produce excitation when applied locally (by either microiontophoresis or bath application) in brain slices. Thus, this selective 5-HT$_{1A}$ ligand does not appear to have any *direct* excitatory effect on facial motoneurons. However, 5-carboxamidotryptamine (5-CT), a broad-spectrum 5-HT$_1$ agonist, is able to enhance directly facial motoneuron excitability. Surprisingly, ritanserin is able to block the effects of 5-CT as well as those of 5-HT. Since ritanserin has extremely low affinity for all 5-HT$_1$ receptors except 5-HT$_{2C}$, it is possible that it is the 5-HT$_{2C}$ component of the 5-CT's receptor profile that is responsible for its effect on facial motoneurons. Alternatively, since 5-CT and ritanserin both have relatively high affinity for certain of the newly cloned 5-HT receptors such as 5-HT$_7$ (Bard et al. 1993; Lovenberg et al. 1993), it is possible that a receptor outside the 5-HT$_2$ family also is involved.

A large number of studies, using behavioral, ligand-binding, and electrophysiological techniques, have shown that indoleamine (e.g., LSD and psilocin) and phenethylamine [e.g., mescaline and 1-(2,5-dimethoxy-4-iodophenyl)-2-aminopropane (DOI)] hallucinogens have in common the property of interacting with 5-HT$_2$ receptors. The iontophoretic administration of LSD, mescaline, or psilocin, although having relatively little effect by themselves, produces a prolonged facilitation of facial motoneuron excitability (McCall and Aghajanian 1980). Intracellular studies in brain slices show that the enhancement is due in part to a small but persistent depolarizing effect of the hallucinogens (Garratt et al. 1993; Rasmussen and Aghajanian 1990). In addition, LSD and the phenethylamine hallucinogen DOI enhance the cationic current I_h to an even greater degree than does 5-HT itself, suggesting that this action of the drugs may be more important quantitatively than their ability to close K^+ channels (Garratt et al. 1993). All of these effects of the hallucinogens are reversed by spiperone and ritanserin, consistent with mediation by 5-HT$_2$ receptors. However, it is now evident that the drugs employed in these studies (e.g., LSD, spiperone, and ritanserin) also have relatively high affinity for the 5-HT$_7$ receptor (Roth et al 1994; Ruat et al. 1993), again raising the possibility that a receptor outside the 5-HT$_2$ family may be involved. This possible involvement of a 5-HT receptor that is positively coupled to adenylate cyclase such as 5-HT$_7$ is especially intriguing since in other neurons in the CNS 5-HT has been suggested to increase I_h through a cAMP-coupled mechanism (e.g., BOBKER and WILLIAMS 1989b; PAPE and McCORMICK 1989).

2. Signal Transduction Pathways

The role of G proteins in mediating the 5-HT$_2$-induced slow inward current that results from K$^+$ channel closure has been evaluated in facial motoneurons by using the hydrolysis-resistant guanine nucleotide analogs GTPγS and GTPβS (AGHAJANIAN 1990). The 5-HT-induced inward current becomes largely irreversible in the presence of intracellular GTPγS. Mediation by G proteins is also suggested by the fact that the inward current is reduced by intracellular GTPβS, which prevents G protein activation. Although the identity of the G protein(s) mediating the electrophysiological responses remains to be determined, the 5-HT$_2$ family of receptors are known to be coupled to phospholipase C. Thus, a member of the G$_q$ family may be involved as the latter can directly activate phospholipase C (SMRCKA et al. 1991).

The activation of phospholipase C leads to an increase in phosphoinoside (PI) hydrolysis, yielding inositol trisphosphate and diacylglycerol (DAG) (CONN and SANDERS-BUSH 1984; SANDERS-BUSH 1988). DAG, by activating protein kinase C (PKC), would be expected to affect many long-term cellular responses through protein phosphorylation. Hence, the effect of protein kinase inhibitors on the response of facial motoneurons to 5-HT has been tested (AGHAJANIAN 1990). Two protein kinase inhibitors with different mechanisms of action, 1-(5-isoquinolylsulfonyl)-2-methylpiperazine (H7), a nonselective protein kinase inhibitor, and sphingosine, a selective PKC inhibitor, in concentrations that have no effect of their own (100 and 10 mM, respectively), both markedly enhance and prolong the excitation of facial motoneurons induced by 5-HT. Conversely, phorbol esters that are known to activate PKC reduce the excitatory effect of 5-HT. These results suggest that activation of PI turnover, perhaps through receptor phosphorylation, has a negative feedback effect on 5-HT-induced excitations in the facial nucleus. Similar findings have also been obtained in the cerebral cortex using the grease-gap method for assessing the 5-HT enhancement of NMDA-induced depolarizations (RAHMAN and NEUMAN 1993). In the latter studies, phorbol esters and diacylglycerol, rather than mimicking the ability of 5-HT to enhance NMDA depolarizations, promoted 5-HT desensitization. On the other hand, in this same preparation, Ca^{2+}-mobilizing agents (e.g., cyclopiazonic acid or thapsigarin) mimic the 5-HT$_{2A}$ facilitation of NMDA depolarization (NEUMAN AND RAHMAN 1996). A possibly related finding is that in cultured interneurons from mouse olfactory bulb, 5-HT has been shown directly to mobilize intracellular free calcium through a 5-HT$_2$-like receptor (TANI et al. 1992).

II. Locus Coeruleus

Systemically administered mescaline or LSD induces a simultaneous *decrease* in spontaneous activity and an *increase* in sensory responsivity of noradrenergic neurons in the locus coeruleus (LC) (AGHAJANIAN 1980). The effects of LSD and mescaline (and other phenethylamine hallucinogens) on LC neurons can be reversed by low doses of 5-HT$_2$ antagonists such as ritanserin and LY 53857 (RASMUSSEN and AGHAJANIAN 1986). In addition to reversing the effects of hallucinogens, the 5-HT$_2$ antagonists induce a small but significant increase in basal LC firing rates, suggesting the existence of a tonic 5-HT$_2$ inhibitory influence. The latter findings are paralleled by voltammetric studies which show an increase in the catechol metabolite 3,4-dihyroxyphenylacetic acid (DOPAC) in the LC following systemic injections of the 5-HT$_2$ antagonist ritanserin (CLEMENT et al. 1992). In correlation with their affinity for 5-HT$_2$-binding sites, antipsychotic drugs are also able to reverse the actions of hallucinogens in the locus coeruleus independently of their affinity for dopamine and other types of receptors (RASMUSSEN and AGHAJANIAN 1988). Similarly, the relative potencies of hallucinogens in affecting LC neurons correlates with their affinity for 5-HT$_2$ receptors. However, the effects of hallucinogens in the LC are not direct as they are not mimicked by the local, iontophoretic application onto LC cell bodies. Thus, the hallucinogens are likely to be acting indirectly on LC neurons via afferents to this nucleus. Consistent with the idea of an indirect action is the finding that the decrease in LC neuronal firing induced by a 5-HT$_2$ agonist (DOI) is blocked by the local infusion into the locus coeruleus of the GABA antagonists bicuculline or picrotoxin (CHIANG and ASTON-JONES 1993).

III. Other Subcortical Regions

In brain slices of the medial pontine reticular formation, 5-HT induces a hyperpolarization in some cells (34%) and a depolarization in other cells (56%) (STEVENS et al. 1992). The hyperpolarizing responses are associated with an increase in membrane conductance and have a 5-HT$_1$ pharmacological profile. The depolarizing responses have a 5-HT$_2$ pharmacology and are associated with a decrease in membrane conductance resulting from a decrease in an outward K$^+$ current. These two actions of 5-HT do not appear to coexist in the same neurons since none of the cells display dual responses to selective agonists. In brain slices of the substantia nigra pars reticulata, a majority of neurons are excited by 5-HT; pharmacological analysis using a variety of antagonists indicates that the excitation is mediated by 5-HT$_2$ receptors (PESSIA et al. 1994), possibly of the 5-HT$_{2C}$ rather than the 5-HT$_{2A}$ subtype (RICK et al.1995). Neurons in the inferior olivary nucleus are excited by 5-HT via 5-HT$_{2A}$ receptors, thereby altering the oscillatory frequency of input to cerebellar Purkinje cells (SUGIHARA et al. 1995). In the nucleus accumbens the great majority of neurons are depolarized by 5-HT, inducing them to fire

(NORTH and UCHIMURA 1989). This depolarization is associated with an increase in input resistance due to a reduction in an inward rectifier K^+ conductance. Pharmacological analysis shows that the depolarization is mediated by a $5\text{-}HT_2$ rather than a $5\text{-}HT_1$ or $5\text{-}HT_3$ type receptor.

In GABAergic neurons of the nucleus reticularis thalami, marked depolarizing responses to 5-HT, associated with a decrease in a resting or "leak" K^+ conductance, have been reported; these excitatory responses are blocked by the $5\text{-}HT_2$ antagonists ketanserin and ritanserin (MCCORMICK and WANG 1991). The 5-HT-induced slow depolarization potently inhibits burst firing in these cells and promotes single-spike activity. It has been suggested that this 5-HT-induced switch in firing mode from rhythmic oscillation to single-spike activity, which occurs during states of arousal and attentiveness, contributes to the enhancement of information transfer through the thalamus during these states.

There are also GABAergic neurons within the medial septal nucleus that are excited by 5-HT via $5\text{-}HT_2$ receptors (ALREJA 1996). GABAergic neurons of the medial septum form one of the two major components of the septohippocampal pathway and innervate a large number of GABAergic neurons within the hippocampus. It has been suggested that 5-HT-induced excitation of septohippocampal GABAergic neurons would inhibit a large number of GABAergic neurons in the hippocampus, reducing both feedback and feedforward inhibition within this stucture, and thereby promote the induction of long-term potentiation (ALREJA 1996). Interestingly, in the dentate gyrus of the hippocampus there appears to be a subpopulation of GABAergic neurons that are activated via $5\text{-}HT_{2A}$ receptors, as evidenced by an increase in IPSP frequency in granule cells in the dentate gyrus (PIGUET an GALVAN 1994). In contrast, 5-HT has a direct hyperpolarizing effect on granule cells mediated by postsynaptic $5\text{-}HT_{1A}$ receptors. These observations in the dentate gyrus closely resemble similar findings in the piriform cortex, where a subpopulation of GABAergic interneurons is excited by 5-HT via $5\text{-}HT_{2A}$ receptors (see below). There is also indirect evidence that 5-HT-induced inhibition of dentate/interpositus neurons of the deep cerebellar nuclei is mediated indirectly by the activation of GABAergic interneurons through $5\text{-}HT_2$ receptors (CUMMING-HOOD et al. 1993). Taken together, these findings suggest that in multiple locations within the CNS there are subpopulations of interneurons that are excited by 5-HT via $5\text{-}HT_2$ receptors, giving rise to paradoxical, indirect inhibitory effects.

IV. Cerebral Cortex

The electrophysiological effects of 5-HT have been studied in several cortical regions. In vivo, $5\text{-}HT_2$ agonists applied by microiontophoresis have been reported to have primarily inhibitory effects on the firing of unidentified neurons in prefrontal cortex (ASHBY et al. 1989a); these inhibitions are blocked by $5\text{-}HT_2$ antagonists. (Note that the inhibitions produced by 5-HT itself are not blocked by 5-HT$_2$ antagonists but are blocked by 5-HT$_3$ antagonists; see below.) In contrast, in vitro studies in the brain slice preparation have shown that pyramidal cells in various regions of the cerebral cortex respond to 5-HT by a small hyperpolarization, depolarization, or no change in potential (Araneda and Andrade 1991; Davies et al. 1987; McCormick and Williamson 1989; Sheldon and Aghajanian 1990). Depending on the region of cortex under study, as described below, the depolarizations appear to be mediated by 5-HT$_{2A}$ or 5-HT$_{2C}$ receptors.

1. Piriform Cortex

In brain slices of rat piriform cortex, excitatory responses to 5-HT in pyramidal cells are mediated by a reduction in K$^+$ conductances (Sheldon and

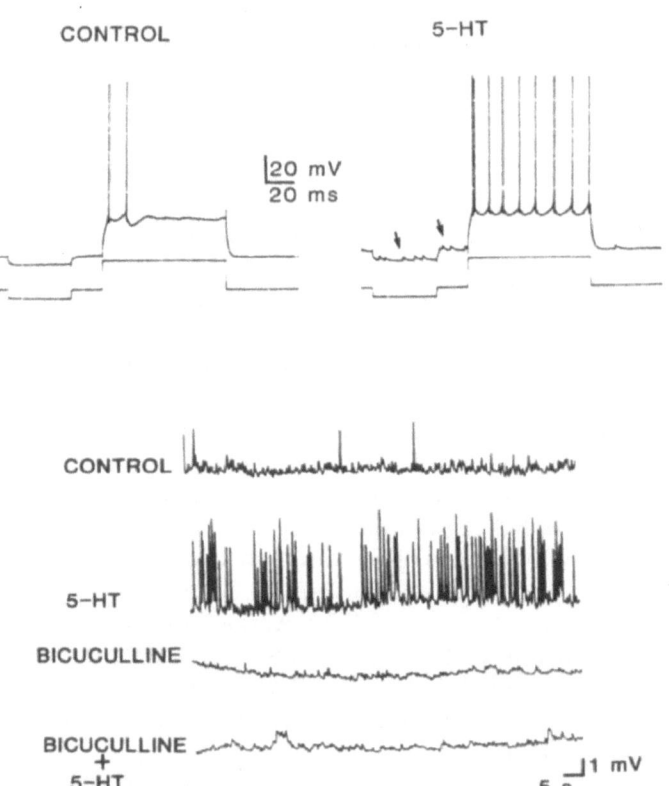

Fig. 4. A 5-HT-induced increase in bicuculline-sensitive spontaneous inhibitory postsynaptic potentials (IPSPs) in layer II pyramidal cells of the piriform cortex. *Top traces* are from an intracellular recording showing that 5-HT, in addition to increasing the electrical excitability of a pyramidal cell (increased spikes in response to a constant depolarizing pulse of 0.5 nA), induces an increase in synaptic potentials (*arrows*). *Bottom traces* at higher gain from another piriform pyramidal cell show that 5-HT-induced IPSPs are blocked by the GABA$_A$ antagonist bicuculline. (Modified from Sheldon and Aghajanian 1993)

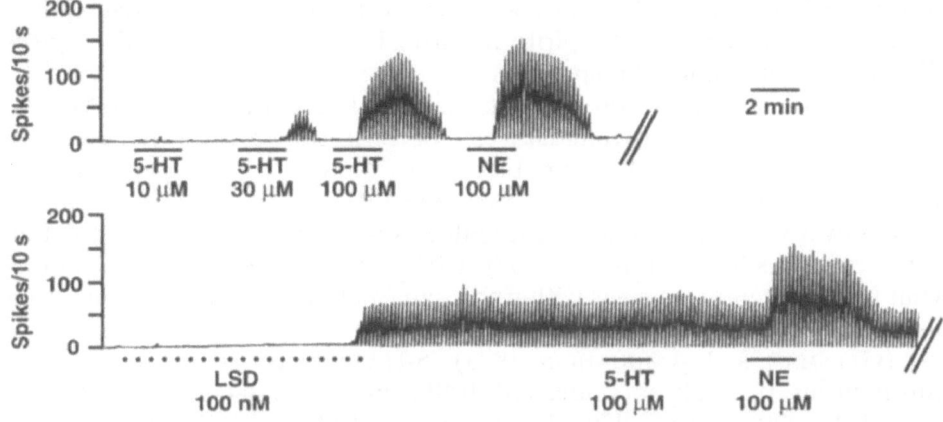

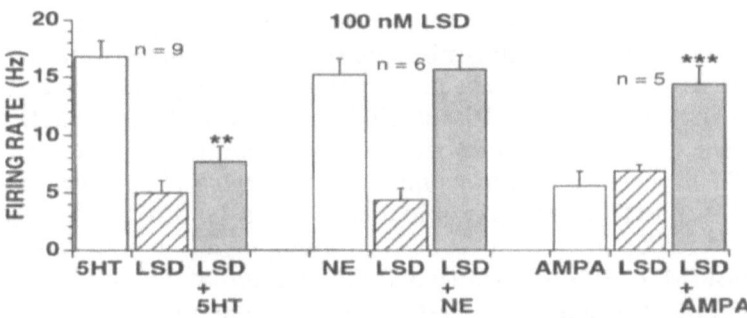

Fig. 5. Firing-rate histograms (*top traces*) showing the partial agonist effect of LSD with respect to 5-HT on the firing of an interneuron in piriform cortex. Note that the activation of firing produced by a maximal concentration of LSD (100nM) is only ~50% of the maximal effect of 5-HT itself; when the LSD effect is ongoing (*lower trace*) the maximal effect of 5-HT is blocked. The graph (below) summarizes data showing the blockade by LSD of the maximal effect of 5-HT but not norepinephrine (*NE*) or the excitatory amino acid agonist AMPA. (Modified from MAREK and AGHAJANIAN 1996)

AGHAJANIAN 1990). Three different types of conductances appear to be involved: a resting K^+ conductance, a depolarization-activated K^+ conductance (M current), and a Ca^{2+}-activated K^+ conductance. There is pharmacological evidence that receptors mediating the excitatory effects of 5-HT on pyramidal cells in piriform cortex are predominantly of the 5-HT$_{2C}$ rather than the 5-HT$_{2A}$ type (SHELDON and AGHAJANIAN 1991). These electrophysiological results are consistent with in situ hybridization studies showing a high level of 5-HT$_{2C}$ receptor expression in the pyramidal cell layer of piriform cortex (MENGOD et al. 1990a). Recent studies show that nearly all pyramidal cells in the piriform cortex express *both* 5-H$_{2A}$ and 5-HT$_{2C}$ receptors (WRIGHT et al. 1995). Thus, it remains to be determined, with the help of more selective pharmacological agents than were available at the time of the initial studies, whether pyramidal cells of the piriform cortex have a mixture of 5-HT$_{2A}$ and 5-HT$_{2C}$ electrophysiological responses.

In addition to these direct postsynaptic effects, there is an induction by 5-HT of IPSPs in pyramidal cells of piriform cortex (SHELDON and AGHAJANIAN 1990). The IPSPs are blocked by the GABA antagonist bicuculline, suggesting that the IPSPs arise from GABAergic interneurons that are excited by 5-HT (Fig. 4). Consistent with this interpretation, a subpopulation of interneurons has been found in layer III that are excited by 5-HT. Somewhat less frequently, neurons within this same subpopulation of interneurons also tend to be excited by norepinephrine and much less frequently by dopamine (GELLMAN and AGHAJANIAN 1993). 5-HT-induced activation of these interneurons (as well as associated IPSPs in pyramidal cells) is blocked by 5-HT$_2$ antagonists. The hallucinogens LSD and 2,5-dimethoxy-4-methylamphetamine (DOM) behave as potent partial agonists in this system, producing a modest activation by themselves but, at higher concentrations, blocking the full effect of 5-HT (Fig. 5; MAREK and AGHAJANIAN 1996).

In piriform cortex, the 5-HT$_2$ antagonist ritanserin blocks the activation of interneurons more readily than it blocks the depolarization of pyramidal cells (SHELDON and AGHAJANIAN 1990). Ritanserin has a nearly tenfold higher affinity for the 5-HT$_{2A}$ receptor than for the 5-HT$_{2C}$ receptor, suggesting that the action of 5-HT on these interneurons might be through 5-HT$_{2A}$ receptors whereas the action of 5-HT on the pyramidal cells might be through 5-HT$_{2C}$ receptors. Recent studies in the piriform cortex employing a new, highly selective antagonist (MDL 100,907) with a 300-fold greater affinity for 5-HT$_{2A}$ than 5-HT$_{2C}$ receptors also suggest that 5-HT$_{2A}$ rather than 5-HT$_{2C}$ receptors are mainly responsible for the 5-HT excitation of interneurons (MAREK and AGHAJANIAN 1994). These results are consistent with recent findings that 5-HT$_{2A}$-receptor immunoreactivity (MORILAK et al. 1993) and mRNA (BURNET et al. 1995) is located in a subpopulation of interneurons in the piriform cortex.

In contrast to piriform cortex, we have found that 5-HT produces primarily an increase in spontaneous EPSPs rather than IPSPs in pyramidal cells within other regions of the cortex, including frontal, parietal, temporal, and occipital (AGHAJANIAN and MAREK, in preparation). The ability of 5-HT to induce EPSPs was blocked by the selective 5-HT$_2$A antagonist MDL 100907 but not 5-HT$_{1A}$, 5-HT$_3$, or 5-HT$_4$ antagonists; this pharmacology is consistent with the abundance of 5-HT$_{2A}$ receptor mRNA in many pyramidal cells throughout the neocortex (BURNET et al. 1995; RAHMAN et al. 1995; WRIGHT et al. 1995). In addition, the EPSPs are totally blocked by the AMPA (α-amino-3-hydroxy-5-methylisoxazole-4-propionic acid) antagonists, indicating that the EPSPs are primarily mediated by AMPA-type excitatory amino acid receptors. The 5-HT-induced EPSPs are eliminated by tetrodotoxin (TTX) treatment, indicating that fast Na$^+$ channels are involved in their transmission.

Stimulation of 5-HT$_{2A}$ receptors in tissue from cerebral cortex activates phospholipase C, resulting in increased phosphoinositide (PI) turnover (CONN and SANDERS-BUSH 1984; KENDALL and NAHORSKI 1985), leading to an activation and translocation of protein kinase C (PKC) from the cytosol to the membrane (WANG and FRIEDMAN 1990). To investigate the possibility that an activation of PKC is responsible for the excitation of interneurons in piriform cortex, the effect of inhibitors of this enzyme on these excitatory responses was examined (MAREK and AGHAJANIAN 1995). Interestingly, both selective (bisindolylmalemide and chelerythrine) and nonselective (H7) inhibitors of PKC enhance rather than block the excitation of piriform interneurons by 5-HT. Conversely, the PKC activator phorbol 12,13-diacetate decreases the excitatory effect of 5-HT: this decrease was rapidly reversed by the PKC inhibitor H7. Thus, it appears that PKC has a negative feedback rather than a mediating role in modulating the excitation by 5-HT of interneurons in piriform cortex. The identity of the transduction mechanism used by 5-HT$_{2A}$ receptors to produce excitatory effects on piriform interneurons remains to be determined. Presumably, the transduction mechanism involves either a membrane-delimited G-protein gating of an ion channel (CLAPHAM 1994) or a diffusable second messenger such as myoinositol 1,4,5-trisphosphate and/or Ca^{2+} (BERRIDGE 1984).

2. Prefrontal Cortex

In situ hybridization studies have reported high levels of expression of 5-HT$_{2A}$, but not 5-HT$_{2B}$ or 5-HT$_{2C}$, receptor mRNA in the rat medial prefrontal cortex (HOFFMAN and MEZEY 1989; MENGOD et al. 1990a,b). Consistent with these observations, administration of serotonin to pyramidal cells of layer V in rat prefrontal cortex slices in many cases results in a delayed membrane depolarization and the appearance of a slow depolarizing afterpotential (ARANEDA and ANDRADE 1991), effects that appear to be common to G$_q$-coupled receptors in this region (ANDRADE 1991; ARANEDA and ANDRADE 1991). These effects are blocked by ketanserin and also by high nanomolar concentrations of spiperone. Since spiperone's affinity for the 5-HT$_{2A}$ receptor is almost 1000-fold higher than that for the 5-HT$_{2C}$ receptor (HOYER 1988), and high nanomolar concentrations of spiperone can be expected to distinguish between 5-HT$_{2A}$ and 5-HT$_{2C}$ sites (ARANEDA and ANDRADE 1991), these results identified the receptor responsible for these excitatory responses as belonging to the 5-HT$_{2A}$ subtype.

5-HT$_{2A}$ responses in prefrontal cortex show desensitization upon repeated agonist administration (ARANEDA and ANDRADE 1991). This has made it difficult to determine the cellular and molecular mechanisms underlying the cellular effects signaled by these receptors. However, other G$_q$-coupled receptors capable of signaling essentially identical physiological responses, such as muscarinic or α_1-adrenergic receptors, show little if any desensitization upon repeated agonist exposure (ANDRADE 1991; ARANEDA and ANDRADE 1991). This has made it possible to address the ionic mechanisms underlying these physiological responses in the context of other G_q-coupled receptors. In this case, both the depolarization and the slow depolarizing afterpotential result from the activation of nonselective cationic channels (HAJ-DAHMANE and ANDRADE 1995a,b). It seems likely that a similar or identical mechanism mediates 5-HT_{2A} responses in rat prefrontal cortex.

D. 5-HT₃ Receptors

There are rapidly desensitizing depolarizing responses to 5-HT in the periphery that are mediated by 5-HT_3 receptors (formerly known as M receptors). In brain, excitatory responses to 5-HT have been found in cultured mouse hippocampal and striatal neurons, which have many of the characteristics of peripheral 5-HT_3 responses: rapid onset and rapid desensitization, features that are typical of ligand-gated ion channels rather than G-protein-coupled receptor responses (YAKEL and JACKSON 1988; YAKEL et al. 1988, 1990). In cultured NG 108-15 cells the permeation properties of the 5-HT_3 channel are indicative of a cation channel with relatively high permeability to Na^+ and K^+ and low permeability to Ca^{2+} (YAKEL et al. 1990). Recently, a 5-HT-gated ion channel has been cloned which has physiological and pharmacological properties appropriate for a 5-HT_3 receptor (MARICQ et al. 1991). In the oocyte expression system, this receptor shows rapid desensitization and is blocked by 5-HT_3 antagonists (e.g., ICS 205–930 and MDL 72222). Because of its sequence homology with the nicotinic acetylcholine receptor (27%), the β_1-subunit of the $GABA_A$ receptor (22%), and the 48K subunit of the glycine receptor (22%), it is likely that this 5-HT_3 receptor clone is a member of the ligand-gated ion channel superfamily.

Rapidly desensitizing 5-HT_3 responses have also been reported in brain slices. In slices containing the lateral nucleus of the amygdala, 5-HT_3-mediated fast excitatory synaptic responses to focal electrical stimulation can be demonstrated when glutamate receptors are blocked; these synaptic responses show rapid cross-desensitization with bath-applied 5-HT (SUGITA et al. 1992). In hippocampal slices, 5-HT has been reported to increase spontaneous GABAergic IPSPs, most likely through a 5-HT_3-receptor-mediated excitation of inhibitory interneurons; these responses also show fading with time (ROPERT 1988; ROPERT and GUY 1991). On the other hand, the 5-HT_3 agonist 2-methyl-5-HT has been reported to reduce both excitatory and inhibitory electrically evoked synaptic potentials via orthodromic stimulation of Schaffer collarerals onto CA1 pyramidal cells in the hippocampus (ZEISE et al. 1994). It is not known whether these effects are through a presynaptic or postsynaptic action.

While fast, rapidly inactivating excitation has generally become accepted as being characteristic of 5-HT_3 receptors, nondesensitizing responses have also been reported. In dorsal root ganglion cells a relatively rapid but *noninactivating* depolarizing response has been described that has a 5-HT$_3$ pharmacological profile (Todorovic and Anderson 1990). In neurons of nucleus tractus solitarius brain slices, there is a postsynaptic depolarizing response to 5-HT$_3$ agonists which does not appear to be rapidly desensitizing (Glaum et al. 1992). In the latter preparation there are also 5-HT3-mediated enhancements of presynaptic responses (both IPSPs and EPSPs). In medial prefrontal cortex a slow *inhibitory* response to microiontophoretically applied 5-HT has been described which also has a 5-HT$_3$-like pharmacology (Ashby et al. 1989a). This effect is mimicked by the 5-HT$_3$ agonist 2-methylserotonin and is blocked by the 5-HT$_3$ antagonists BRL 43693 and ICS 205930. At this juncture, it is not clear whether there is more than just a superficial pharmacological relationship among these various so-called 5-HT$_3$ responses.

E. G$_s$-Coupled Serotonin Receptors: the 5-HT$_4$, 5-HT$_6$, and 5-HT$_7$ Subtypes

The existence of the 5-HT$_4$ receptor in the central nervous system was first suspected on the basis of biochemical data showing positive coupling of 5-HT responses to adenylyl cyclase (see Bockaert et al. 1992). Thus, this receptor differs from the 5-HT$_1$ subfamily, all of whose members are negatively coupled to adenylyl cyclase. Very recently, two novel 5-HT receptors positively coupled to adenylyl cyclase have been cloned; since their pharmacology differs from that of the previously described 5-HT$_4$ site, they have been tentatively designated as 5-HT$_6$ and 5-HT$_7$ receptors (Lovenberg et al. 1993; Monsma et al. 1993; SHEN et al. 1993). Physiological studies are available at this time only for the 5-HT$_4$ and 5-HT$_7$ receptors, as described below.

I. 5-HT$_4$ Receptors

Binding studies using GR 113808 indicate that 5-HT$_4$ receptors are expressed in several discrete regions of the mammalian brain including the striatum, substantia nigra, olfactory tubercle, and hippocampus (Grossman et al. 1993). As these regions also express 5-HT$_4$ receptor mRNA, it appears likely that these receptors function postsynaptically to mediate certain actions of 5-HT. The best studied of these regions is the hippocampus, where both biochemical and electrophysiological studies have provided us with a uniquely detailed picture of the actions of 5-HT at 5-HT$_4$ receptors.

Early physiological studies on hippocampal pyramidal cells of the CA1 region (Andrade and Nicoll 1987; Beck 1989) provided evidence for the existence of an orphan 5-HT receptor capable of mediating "excitatory" responses to 5-HT. These studies identified two separate mechanisms that combined could account for the observed increase in membrane excitability. First,

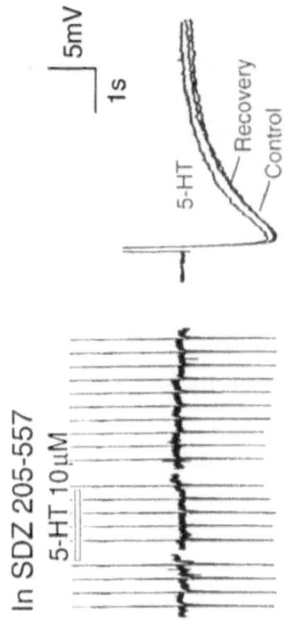

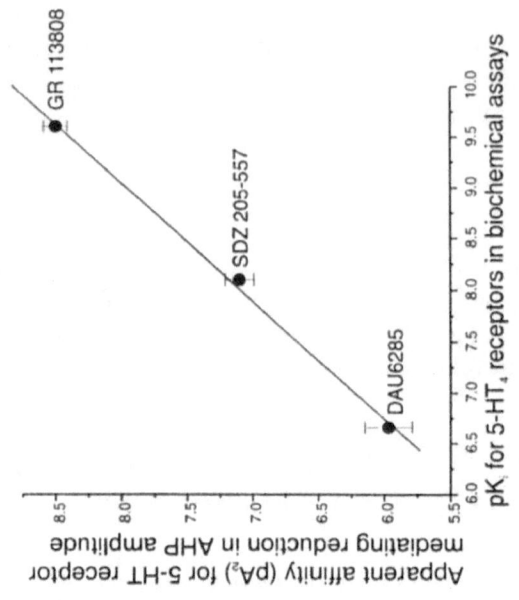

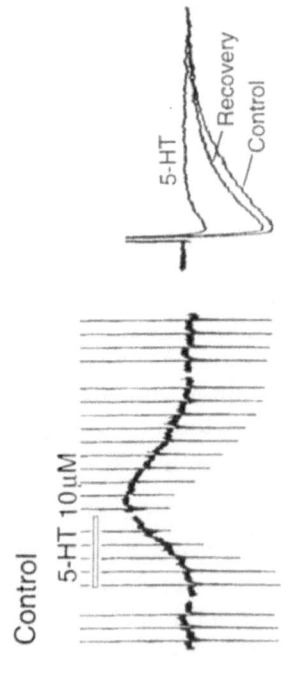

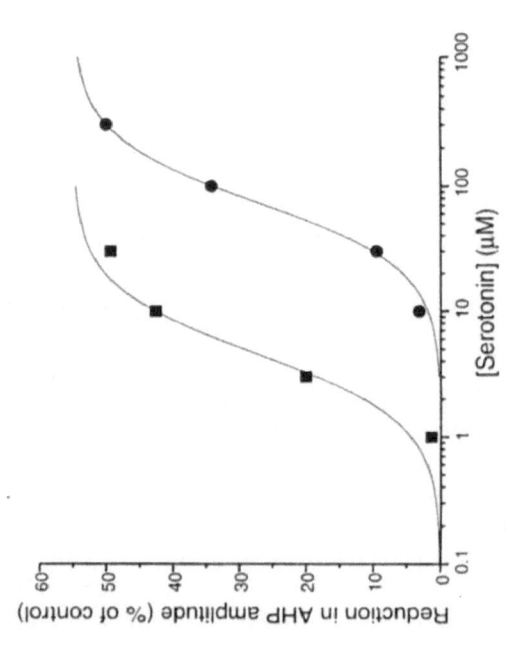

5-HT activation of this novel receptor elicited a slow membrane depolarization, most likely through a reduction of the resting potassium permeability responsible for the maintenance of membrane potential. Second, serotonin inhibited the calcium-activated potassium current which in these cells is responsible for the generation of a slow calcium-activated afterhyperpolarization (AHP). The combined action of both of these effects could markedly increase membrane excitability. The first of these actions brought the membrane potential close to threshold while the second facilitated the ability of these cells to respond to excitatory inputs with robust spiking activity.

The initial pharmacological characterization of this novel receptor indicated that it was insensitive to most of the then available serotonergic antagonists, but exhibited a reasonably high affinity for the benzamides renzapride, cisapride, and zacopride (ANDRADE and CHAPUT 1991; CHAPUT et al. 1990a,b; ANDRADE and NICOLL 1987). This pharmacological profile closely paralleled that of the then newly described 5-HT$_4$ receptor subtype (DUMUIS et al. 1988; BOCKAERT et al. 1990), and led to the tentative identification of this receptor as belonging to this receptor subclass. Consistent with this provisional identification, the serotonin-induced depolarization and reduction of the AHP proved to be highly sensitive to the selective 5-HT$_4$ receptor antagonist developed subsequently (TORRES et al. 1994) (Fig. 6).

Consistent with the biochemical data, the ability of 5-HT$_4$ receptor activation to inhibit the AHP is mimicked by manipulations that increase intracellular cAMP and is enhanced by phosphodiesterase inhibitors (TORRES et al. 1995). Moreover, the effects of 5-HT on the AHP are inhibited by the protein kinase inhibitor staurosporine and the selective protein kinase A inhibitor (Rp)cAMPs (Fig. 7). Together these results indicate that 5-HT$_4$ receptors in the CA1 region reduce the AHP by stimulating adenylate cyclase, increasing intracellular cAMP, and activating protein kinase A. Similarly, activation of a cAMP-dependent protein kinase has been implicated in the suppression of a voltage-activated K$^+$ current in cultured neurons from the superior colliculus (FAGNI et al. 1992).

◀
──

Fig. 6A–C. Pharmacology of the serotonin-induced depolarization and reduction of the afterhyperpolarization (AHP) in the CA1 region of the hippocampus. **A** Administration of serotonin in the presence of a 5-HT$_{1A}$ antagonist produces a slow membrane depolarization. This depolarization is accompanied by a decrease in the AHP (downward deflections) that follows calcium influx into the cell. AHPs taken at a constant membrane potential in the presence and absence of serotonin are illustrated *to the right of the membrane potential trace using an expanded time scale*. The depolarization and reduction of the AHP are both antagonized by administration of the selective 5-HT$_4$ receptor antagonist SDZ 205–557 (10 mM). **B** Bath administration of SDZ 205–557 (1 mM) causes a parallel shift to the right of the concentration response curve for the serotonin reduction of the afterhyperpolarization (*AHP*). **C** Correlation between the apparent affinities (pA$_2$) of the 5-HT$_4$ antagonists DAU 6285, SDZ 205–557 and GR 113808 for the serotonin receptor acting on the AHP and their affinity for 5-HT$_4$ receptors in model systems. (Provided by R. ANDRADE)

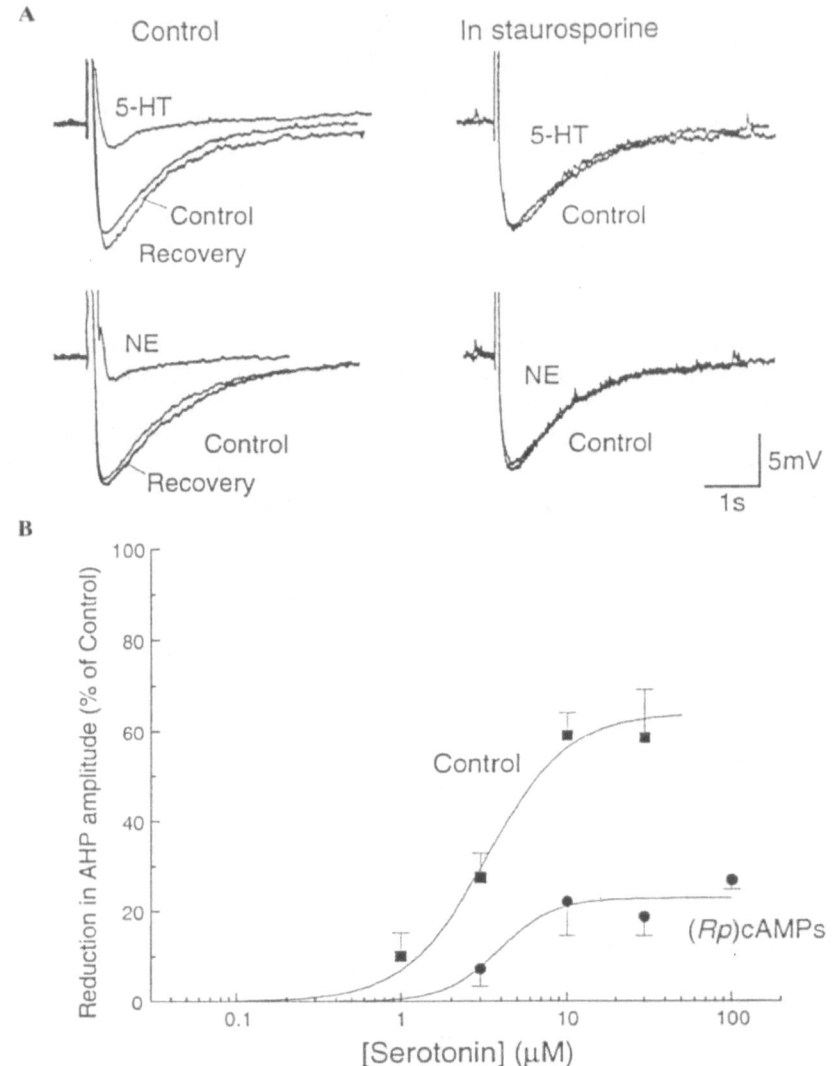

Fig. 7A,B. Effect of PKA inhibitors on the 5-HT₄-mediated reduction of the afterhyperpolarization (*AHP*). **A** Pretreatment with the protein kinase inhibitor staurosporin (1–10 mM) blocks the ability of serotonin and norepinephrine to reduce the AHP. **B** Intracellular injection of the selective PKA inhibitor (Rp)cAMPs noncompetitively inhibits the ability of serotonin to reduce the afterhyperpolarization (*AHP*). (Provided by R. Andrade)

II. 5-HT₇ Receptors

The circadian rhythm in mammals is set by a pacemaker located primarily in the susprachiasmatic nucleus of the hypothalamus. This pacemaker activity can be maintained in in vitro hypothalamic slices, where suprachiasmatic neurons display diurnal changes in neuronal firing rate. Administration of 5-

HT produces a phase shift in this activity (MEDANIC and GILLETTE 1995) by acting on a receptor of the 5-HT$_7$ subtype (LOVENBERG et al. 1993). This shift appears to be mediated by stimulation of adenylate cyclase as it is mimicked by increasing intracellular cAMP (PROSSER and GILLETTE 1989) and blocked by inhibitors of protein kinase A (PROSSER et al. 1994). However, the precise mechanism by which 5-HT$_7$ receptors act is not presently known since it is unclear whether suprachiasmatic neurons themselves express the 5-HT$_7$ receptors (LOVENBERG et al. 1993). Furthermore, the effect of 5-HT on the membrane properties of these cells has not been examined. 5-HT$_7$ receptor activation has been reported to inhibit GABA$_A$ currents on suprachiasmatic neurons in culture (KAWAHARA et al. 1995), but the relationship, if any, between these observations and 5-HT changes in ciracadian activity remains to be determined.

A more general kind of electrophysiological effect that may be mediated through 5-HT$_7$ (or other 5-HT receptors that are positively coupled to adenylate cyclase) is the enhancement of the hyperpolarizing-activated nonselective cationic current I_h. This possibility, which has been raised for the 5-HT-induced increase in I_h in motoneurons (see above), remains to be evaluated in other neurons of the CNS, such as in the thalamus (PAPE and McCORMICK 1989), prepositus hypoglossi (BOBKER and WILLIAMS 1989b), or substantia nigra zona compacta (NEDERGAARD et al. 1991), where 5-HT has been shown to enhance I_h.

F. Summary and Conclusions

I. 5-HT Receptor Subtypes, Ion Channels, and Second Messenger Systems

The extraordinarily diverse electrophysiological actions of 5-HT in the central nervous system can now be understood according to various receptor subtypes and their respective effector mechanisms. As summarized in Table 1, the following generalizations are emerging: (1) inhibitory effects of 5-HT are mediated by 5-HT$_1$ receptors linked to the opening of K$^+$ channels or the closing of Ca^{2+} channels, both via pertussis-toxin-sensitive G proteins; (2) certain facilitatory effects of 5-HT involve a closing of K$^+$ channels mediated by 5-HT$_2$ receptors, with the PI second messenger system and PKC acting as a negative feedback loop; (3) other facilitatory effects of 5-HT are mediated by 5-HT$_4$ (and possibly by other Gs-linked) 5-HT receptors through a reduction in $I_{K(Ca^{2+})}$ or voltage-dependent K$^+$ current through the cAMP pathway; and (4) fast excitations are mediated by 5-HT$_3$ receptors through a ligand-gated cationic ion channel which does not require coupling with a G protein or a second messenger. Thus, the range of electrophysiological actions of 5-HT is remarkably broad, encompassing the two major neurotransmitter superfamilies: the G-protein-coupled receptors (i.e., 5-HT$_1$, 5-HT$_2$, 5-HT$_4$) and the ligand-gated channels (5-HT$_3$).

Table 1. 5-HT receptor subtypes: physiological transduction pathways

Receptor subtype	G protein	Ion channel	Physiological response
$5\text{-}HT_1$	G_i/G_o	$\uparrow K^+, \downarrow Ca^{2+}$	Slow inhibition
$5\text{-}HT_2$	$G_{q/11}$	$\downarrow K^+$	Slow excitation
$5\text{-}HT_3$	Ligand-gated	$\uparrow Na^+/K^+$	Fast excitation
$5\text{-}HT_4$	G_s	$\downarrow I_K, I_{K^+(Ca^{2+})}$	Slow excitation
$5\text{-}HT_5$	?	?	?
$5\text{-}HT_{6\text{-}7}$	G_s	?	?

II. Role of 5-HT in Neuronal Networks

The direct actions of 5-HT at a cellular level, as detailed in this review, do not have a simple relationship to the overall input-output relations. For example, while $5\text{-}HT_{1A}$ agonists are directly inhibitory upon serotonergic neurons in the raphe nuclei (possibly at doses insufficient to affect postsynaptic $5\text{-}HT_{1A}$ receptors), the net effect of this inhibition may be disinhibitory for postsynaptic neurons that express $5\text{-}HT_{1A}$-receptors and simultaneously dysfacilitatory for postsynaptic neurons that express $5\text{-}HT_2$, $5\text{-}HT_3$, and $5\text{-}HT_4$ receptors. Similarly, while $5\text{-}HT_2$ agonists may be directly excitatory at a subpopulation of GABAergic interneurons in piriform cortex and dentate gyrus that express $5\text{-}HT_2$ receptors, the net effect of this excitation may be inhibitory for the principal neurons that receive inputs from these interneurons. In general, to understand the functional consequences of the discrete cellular actions of 5-HT, these must be viewed within the context of the neuronal networks where they occur.

III. Integrative Action of 5-HT

This review has described mainly the isolated cellular actions of 5-HT at individual receptor subtypes that are found in many different regions of the brain and spinal cord. Do these discrete and disparate effects of 5-HT have an integrated function greater than the sum of the parts? Suggestive of this possibility is the fact that: (1) serotonergic neurons are clustered in a relatively small number of nuclei within the brain stem (i.e., the raphe nuclei), projecting diffusely to almost every other region of the neuraxis and (2), in unanesthetized animals, the tonic firing of serotonergic neurons *as a group* varies according to behavioral state, such that activity is greatest during behavioral arousal, diminished during slow wave sleep, and virually absent during REM sleep (see Chap. 3, this volume).

The integrative function of 5-HT can be viewed at a cellular as well as at a systems level. A number of examples have been given where more than one 5-HT receptor subtype is expressed functionally within the same neuron (e.g., $5\text{-}HT_1$ and $5\text{-}HT_2$ receptors in cortical pyramidal cells and $5\text{-}HT_1$ and $5\text{-}HT_4$

receptors in hippocampal pyramidal cells). Given the contrasting electrophysiological effects mediated by these different receptor subtypes, the general question arises as to how these disparate individual subtype actions become integrated into a meaningful overall effect of 5-HT upon a target neuron. In approaching this question, both temporal and spatial issues must be addressed. For example, the 5-HT$_{1A}$ inhibitory effect of 5-HT upon hippocampal pyramidal cells tends to manifest itself at an earlier phase in time than does the 5-HT$_4$ excitatory effect (ANDRADE and CHAPUT 1991). Spatial issues must also be considered in terms of a segregated cytoarchitectural localization of the various receptor subtypes. For example, 5-HT$_1$ receptors appear to be preferentially localized to the initial axon segment (axon hillock) of hippocampal and cortical pyramidal cells (AZMITIA et al. 1996). The localization of 5-HT$_1$ receptors to the axon hillock would place them in a strategic position to regulate the output of cortical pyramidal cells. In contrast, 5-HT$_{2A}$ receptors appear to be localized to the apical rather than the basilar aspect of layer V cortical pyramidal cells (BLUE et al. 1988), and therefore would be in an ideal position to regulate inputs to pyramidal cells. Such temporal and spatial distinctions undoubtedly contribute, at a cellular level, to the integrative action of 5-HT in many regions of the CNS.

Do the individual cellular actions of 5-HT in different regions of the central nervous system serve to coordinate overall behavioral states? The possibility of such an integrative action of 5-HT can be illustrated for four disparate sets of neurons as follows. During the waking state, when serotonergic neurons are in a tonic firing mode, the following conditions would prevail: (1) motoneurons would be in a relatively depolarized, excitable state (via 5-HT$_2$ receptors), and would thus be receptive to the initiation of movement; (2) neurons of the nucleus reticularis thalami would be in a depolarized, single-spike mode (via 5-HT$_2$ receptors), and would thus be conducive to thalamocortical sensory information transfer (McCORMICK and WANG 1991; PAPE and McCORMICK 1989); (3) GABAergic neurons of the septohippocampal pathway would be activated (in part via 5-HT$_{2A}$ receptors), potentially enhancing long-term potentiation by inhibiting GABAergic neurons of the hippocampus (ALREJA 1996); and (4) neurons of the laterodorsal tegmental nucleus would be hyperpolarized (via 5-HT$_1$ receptors) and therefore not able to generate the bursting activity of REM sleep (LUEBKE et al. 1992). Conversely, with a reduction in serotonergic activity during various stages of sleep, the above conditions would switch such that motoneurons would become less excitable, thalamocortical sensory information transfer would be diminished, hippocampal function would be reduced, and sleep spindles and PGO waves would emerge.

Thus, serotonergic neurons can be seen as simultaneously modulating, in a complex but coordinated fashion, motor, sensory, and other systems to promote a given behavioral state or function. It remains to be seen whether *all* the diverse cellular actions of 5-HT can be incorporated into this kind of unified scheme or whether there are subsets of serotonergic neurons that influence different groups of postsynaptic neurons, both temporally and spatially, in an independent fashion.

Acknowledgements. G.K.A. was supported by USPHS grants MH 17871, MH 15642, and the State of Connecticut; R.A. was supported by USPHS grant MH 43985.

References

Aghajanian GK (1980) Mescaline and LSD facilitate the activation of locus coeruleus neurons by peripheral stimuli. Brain Res 186:492–498

Aghajanian GK (1990) Serotonin-induced current in rat facial motoneurons: evidence for mediation by G proteins but not protein kinase C. Brain Res 524:171–174

Aghajanian GK, Lakoski JM (1984) Hyperpolarization of serotonergic neurons by serotonin and LSD: studies in brain slices showing increased K⁺ conductance. Brain Res 305:181–185

Aghajanian GK, Rasmussen K (1989) Intracellular studies in the facial nucleus illustrating a simple new method for obtaining viable motoneurons in adult rat brain slices. Synapse 3:331–338

Aghajanian GK, Foote WE, Sheard MH (1968) Lysergic acid diethylamide: sensitive neuronal units in the midbrain raphe. Science 161:706–708

Aghajanian GK, Haigler HJ, Bloom FE (1972) Lysergic acid diethylamide and serotonin: direct actions on serotonin-containing neurons in rat brain. Life Sci 11:615–622

Alreja MA (1996) Excitatory actions of serotonin on GABAergic neurons of the medial septum and diagonal band of Broca. Synapse 22:15–27

Andrade R (1991) Cell excitation enhances muscarinic cholinergic responses in rat association cortex. Brain Res 548:81–93

Andrade R (1992) The electrophysiology of 5-HT$_{1A}$ receptors in the hippocampus and cortex. Drug Dev Res 26:275–286

Andrade R (1993) Enhancement of β-adrenergic responses by G$_i$-linked receptors in rat hippocampus. Neuron 10:83–88

Andrade R, Chaput Y (1991a) 5-HT$_4$-like receptors mediate the slow excitatory response to serotonin in the rat hippocampus. J Pharmacol Exp Ther 40:399–412

Andrade R, Chaput Y (1991b) 5-HT$_4$-like receptors mediate the slow excitatory response to serotonin in the rat hippocampus. J Pharmacol Exp Ther 257:930–937

Andrade R, Nicoll RA (1987) Pharmacologically distinct actions of serotonin on single pyramidal neurones of the rat hippocampus recorded in vitro. J Physiol (Lond) 394:99–124

Andrade R, Malenka R, Nicoll A (1986) A G protein couples serotonin and GABA$_B$ receptors to the same channels in hippocampus. Science 234:1261–1265

Araneda R, Andrade R (1991) 5-Hydroxytryptamine$_2$ and 5-hydroxytryptamine$_{1A}$ receptors mediate opposing responses on membrane excitability in rat association cortex. Neuroscience 40:399–412

Arborelius L, Nomikos GG, Brillner P, Hertel P, Höök BB, Hacksell U, Svensson TH (1995) 5-HT$_{1A}$ receptor antagonists increase the activity of serotonergic cells in the dorsal raphe nucleus in rats treated acutely or chronically with citalopram. Naunyn Schmiedebergs Arch Pharmacol 352:157–165

Ashby CR Jr, Edwards E, Harkins K, Wang RY (1989a) Characterization of 5-hydroxytryptamine-3 receptors in the medial prefrontal cortex: a microiontophoretic study. Eur J Pharmacol 173:193–196

Ashby CR Jr, Jiang LH, Kasser RJ, Wang RY (1989b) Electrophysiological characterization of 5-hydroxytryptamine-2 receptors in rat medial prefrontal cortex. J Pharmacol Exp Ther 252:171–178

Ashby CR Jr, Edwards E, Wang RY (1994) Electrophysiological evidence for a functional interaction between 5-HT$_{1A}$ and 5-HT$_{2A}$ receptors in the rat medial prefrontal cortex: an iontophoretic study. Synapse 17:173–181

Azmitia EC, Gannon PJ, Kheck NM, Whitaker-Azmitia PM (1996) Cellular localization of the 5-HT$_{1A}$ receptor in primate brain neurons and glial cells. Neuropsychopharmacology 14:35–46

Bard JA, Zgombick J, Adham N, Vaysse P, Branchek TA, Weinshank RL (1993) Cloning of a novel human serotonin receptor (5-HT$_7$) positively linked to adenylate cyclase J Biol Chem 268:23422–23426

Beck SG (1989) 5-Carboxyamidotryptamine mimics only the 5-HT elicited hyperpolarization of hippocampal pyramidal cells via 5-HT$_{1A}$ receptor. Neurosci Letts 99:101–106

Behbehani MM, Liu H, Jiang M, Pun RYK, Shipley MT (1993) Activation of serotonin$_{1A}$ receptors inhibits midbrain periaqueductal gray neurons of the rat. Brain Res 612:56–60

Berridge MJ (1984) Inositol trisphosphate and diacylglycerol as second messengers. Biochem J 220:345–360

Biscoe TJ, Straughan DW (1966) Micro-electrophoretic studies of neurones in the cat hippocampus. J Physiol (Lond) 183:341–359

Blier P, Steinberg S, Chaput Y, deMontigny C (1989) Electrophysiological assessment of putative antagonists of 5-hydroxytryptamine receptors: a single-cell study in the dorsal raphe nucleus. Can J Physiol Pharmacol 67:98–105

Blier P, Lista A, deMontigny C (1993a) Differential properties of pre- and postsynaptic 5-hydroxytryptamine$_{1A}$ receptors in the dorsal raphe and hippocampus: I. Effect of spiperone. J Pharmacol Exp Ther 265:7–15

Blier P, Lista A, deMontigny C (1993b) Differential properties of pre- and postsynaptic 5-hydroxytryptamine$_{1A}$ receptors in the dorsal raphe and hippocampus: II Effect of pertussis and cholera toxins. J Pharmacol Exp Ther 265:16–23

Blue M, Yagaloff KA, Mamounas LA, Hartig PR, Molliver ME (1988) Correspondence between 5-HT$_2$ receptors and serotonergic axons in rat neocortex. Brain Res. 453:314–328

Bobker DH (1994) A slow excitatory postsynaptic potential mediated by 5-HT$_2$ receptors in nucleus prepositus hypoglossi. J Neurosci 14:2428–2434

Bobker DH, Williams JT (1989a) Serotonin agonists inhibit synaptic potentials in the rat locus ceruleus in vitro via 5-hydroxytryptamine$_{1A}$ and 5-hydroxytryptamine$_{1B}$ receptors. J Pharmacol Exp Ther 250:37–43

Bobker DH, Williams JT (1989b) Serotonin augments the cationic current I$_h$ in central neurons. Neuron 2:1535–1540

Bobker DH, Williams JT (1995) The serotonergic inhibitory postsynaptic potential in prepositus hypoglossi is mediated by two potassium currents. J Neurosci 15:223–229

Bockaert J, Sebben M, Dumuis A (1990) Pharmacological characterization of 5-hydroxytryptamine$_4$ (5-HT$_4$) receptors positively coupled to adenylate cyclase in adult guinea pig hippocampal membranes: effect of substituted benzamide derivatives. Mol Pharmacol 347:408–411

Bockaert J, Fozard JR, Dumuis A, Clarke DE (1992) The 5-HT$_4$ receptor: a place in the sun. TIPS 13:141–145

Borsini F, Ceci A, Bietti G, Donetti A (1995) BIMT 17, a 5-HT$_{1A}$ receptor agonist/5-HT$_{2A}$ receptor antagonist, directly activates postsynaptic 5-HT inhibitory responses in the rat cerebral cortex. Naunyn Schmiedebergs Arch Pharmacol 353:283–290

Breitwieser GE, Szabo G (1985) Uncoupling of cardiac muscarinic and β-adrenergic receptors from ion-channels by a guanine nucleotide analogue. Nature 317:538–540

Burnet PWJ, Eastwood SL, Lacey K, Harrison PJ (1995) The distribution of 5-HT$_{1A}$ and 5-HT$_{2A}$ receptor mRNA in human brain. Brain Res 676:157–168

Chalmers DT, Watson SJ (1991) Comparative anatomical distribution of 5-HT$_{1A}$ receptor mRNA and 5-HT$_{1A}$ binding in rat brain – a combined in situ hybridisation/in vitro receptor autoradiographic study. Brain Res 561:51–60

Chaput Y, Araneda RC, Andrade R (1990a) Pharmacological and functional analysis of a novel serotonin receptor in the rat hippocampus. Eur J Pharmacol 182:441–456

Chaput Y, Araneda RC, Andrade R (1990b) Pharmacological and functional analysis of a novel serotonin receptor in the rat hippocampus. Eur J Pharmacol 182:441–456

Charlety PJ, Aston-Jones G, Akaoka H, Buda M, Chouvet G (1991) 5-HT decreases glutamate-evoked activation of locus coeruleus neurons through 5-HT$_{1A}$ receptors. Neurophysiology 421–426

Chiang C, Aston-Jones G (1993) A 5-hydroxytryptamine$_2$ agonist augments γ-aminobutyric acid and excitatory amino acid inputs to noradrenergic locus coeruleus neurons. Neuroscience 54:409–420

Clapham DE (1994) Direct G protein activation of ion channels. Ann Rev Neurosci 17:441–464

Clement HW, Gemsa D, Wesemann W (1992) Serotonin-norepinephrine interactions: a voltammetric study on the effect of serotonin receptor stimulation followed in the N. raphe dorsalis and the locus coeruleus of the rat. J Neural Trans 88:11–23

Conn PJ, Sanders-Bush E (1984) Selective 5-HT-2 antagonists inhibit serotonin stimulated phosphatidylinositol metabolism in cerebral cortex. Neuropharmacology 23:993–996

Cox RF, Meller E, Waszczak BL (1993) Electrophysiological evidence for a large receptor reserve for inhibition of dorsal raphe neuronal firing by 5-HT$_{1A}$ agonists. Synapse 14:298–304

Crain SM, Crain B, Makman MH (1987) Pertussis toxin blocks depressant effects of opioid, monoaminergic and muscarinic agonists on dorsal-horn network responses in spinal cord-ganglion cultures. Brain Res 400:185–190

Craven, R, Graham-Smith, Newberry, N (1994) WAY-100635 and GR127935: effects on 5-hydroxytryptamine-containing neurons. Eur J Pharmacol 271:R1-R3

Cumming-Hood PA, Strahlendorf HK, Strahlendorf JC (1993) Effects of serotonin and 5-HT$_{2/1C}$ receptor agonist DOI on neurons of the cerebral dentate/interpositus nuclei: possible involvement of a GABAergic interneuron. Eur J Pharmacol 236:457–465

Darrow EJ, Strahlendorf HK, Strahlendorf JC (1990) Response of cerebellar Purkinje cells to serotonin and the 5-HT$_{1A}$ agonists 8-OH-DPAT and ipsapirone in vitro. Eur J Pharmacol 175:145–153

Dascal N, Schreibmayer W, Lim NF, Wnag W, Chavkin C, DiMagno L, Labarca C, Kieffer B, Gaveriaux RC, Trollinger D, Lester HA, Davidson N (1995) Atrial G protein-activated K$^+$ channel: expression cloning and molecular properties. Proc Natl Acad Sci USA 90:10235–10239

Davies MF, Deisz RA, Prince DA, Peroutka SJ (1987) Two distinct effects of 5-hydroxytryptamine on single cortical neurons. Brain Res 423:347–352

DeVivo M, Maayani S (1986) Characterization of the 5-hydroxytryptamine$_{1A}$ receptor-mediated inhibition of forskolin-stimulated adenylate cyclase activity in guinea pig and rat hippocampal membranes. J Pharmacol Exp Ther 238:248–253

Dumuis A, Bouhelal R, Sebben M, Cory R, Bockaert J (1988) A nonclassical 5-hydroxytryptamine receptor positively coupled with adenylate cyclase in the central nervous system. Mol. Pharmacol 34:880–887

Fagni L, Dumuis A, Sebben M, Bockaert J (1992) The 5-HT$_4$ receptor subtype inhibits K$^+$ current in colliculi neurones via activation of a cyclic AMP-dependent protein kinase. Br J Pharmacol 105:973–979

Fargin A, Raymond JR, Lohse MJ, Kobilka BK, Caron MG, Lefkowitz RJ (1988) The genomic clone G-21 which resembles a β-adrenergic receptor sequence encodes the 5-HT$_{1A}$ receptor. 335:358–360

Federman AD, Conklin BR, Schrader KA, Reed RR, Bourne HR (1992) Hormonal stimulation of adenylyl cyclase via G_i protein $\beta\gamma$ subunits. Nature 356:159–161

Fletcher A, Forster EA, Bill DJ, Brown G, Cliffe IA, Gartley JE, Jones DI, McLenachan A, Stanhope KJ, Critchley DJP, Childs, KJ, Middlefell VC, Lanfumey L, Corradetti R, Laporte A-M, Gozlan H, Hamon M, Dourish CT (1996) Behav Brain Res 73:337–353

Foehring RC (1996) Serotonin modulates N-and P-type calcium currents in neocortical pyramidal neurons via a membrane-delimited pathway. J. Neurophysiol (in press)

Forster EA, Cliffe IA, Bill DJ, Dover GM, Jones D, Reilly Y, Fletcher A (1995) A pharmacological profile of the selective silent 5-HT$_{1A}$ receptor antagonist, WAY-10635. Eur J Pharmacol 281:81–88

Garratt JC, Alreja M, Aghajanian GK (1993) LSD has high efficacy relative to serotonin in enhancing the cationic current I_h: intracellular studies in rat facial motoneurons. Synapse 13:123–134

Gartside SE, Umbers V, Hajós M, Sharp T (1995) Interaction between a selective 5-HT$_{1A}$ receptor antagonist and an SSRI in vivo: effects on 5-HT cell firing and extracellular 5-HT. Br J Pharmacol 115:1064–1070

Gellman RL, Aghajanian GK (1993) Pyramidal cells in piriform cortex receive a convergence of inputs from monoamine activated GABAergic interneurons. Brain Res 600:63–73

Glaum SR, Brooks PA, Spyer KM, Miller RJ (1992) 5-Hydroxytryptamine-3 receptors modulate synaptic activity in the rat nucleus tractus solitarius in vitro. Brain Res 589:62–68

Grossman CJ, Killpatrick GJ, Bunce KT (1993) Development of a radioligand binding assay for 5-HT$_4$ receptors in guinea-pig and rat brain. Br J Pharmacol 109:618–624

Haj-Dahmane S, Andrade R (1995a) Muscarinic activation of a voltage-dependent cation non-selective current in rat association cortex (in press)

Haj-Dahmane S, Andrade R (1995b) A voltage and calcium-dependent cation current underlies the depolarization and depolarizing afterpotential elicited by muscarinic activation in rat cortex. Abst Soc Neurosci 21:2038

Hoffman BJ, Mezey E (1989) Distribution of serotonin 5-HT$_{1C}$ receptor mRNA in adult rat brain. FEBS Lett 247:453–462

Hoyer D (1988) Functional correlates of serotonin 5-HT$_1$ recognition sites. J Recept Res 8:59

Huang CL, Slesinger PA, Casey PJ, Jan YN, Jan LY (1995) Evidence that direct binding of $G\beta\gamma$ to the GIRK1 G protein-gated inwardly rectifying K^+ channel is important for channel activation. Neuron 15:1133–1143

Innis RB, Nestler EJ, Aghajanian GK (1988) Evidence for G protein mediation of serotonin- and GABA$_B$-induced hyperpolarization of rat dorsal raphe neurons. Brain Res 459:27–36

Joëls M, Shinnick-Gallagher P, Gallagher JP (1987) Effect of serotonin and serotonin analogues on passive membrane properties of lateral septal neurons in vitro. Brain Res 417:99–107

Johnson MD (1994) Electrophysiological and histochemical properties of postnatal rat serotonergic neurons in dissociated cell culture. Neuroscience 63:775–787

Jolas T, Haj-Dahmane S, Lanfumey L, Fattaccini CM, Kidd EJ, Adrien J, Gozlan H, Guardiola-Lemaitre B, Hamon M (1993) (−)Tertatolol is a potent antagonist at pre- and postsynaptic serotonin 5-HT$_{1A}$ receptors in the rat brain. Naunyn Schmiedebergs Arch Pharmacol 347:453–463

Karschin A, Ho BY, Labarca C, Elroy-Stein O, Moss B, Davidson N, and Lester H (1991) Heterologously expressed serotonin$_{1A}$ receptors couple to muscarinic K^+ channels in heart. Proc Natl Acad Sci USA 88:5694–5698

Karschin C, Schreibmayer W, Dascal N, Lester H, Davidson N, Karschin A (1994) Distribution and localization of a G protein-coupled inwardly rectifying K^+ channel in the rat. FEBS Lett 348:139–144

Kawahara F. Saito H, Katsuki H (1995) Inhibition by 5-HT$_7$ receptor stimulation of GABA$_A$ receptor-activated current in cultured rat suprachiasmatic neurones. J Physiol (Lond) 478:67–73

Kendall DA, Nahorski SR (1985) 5-Hydroxytryptamine-stimulated inositol phospholipid hydrolysis in rat cerebral cortex slices: pharmacological characterization and effects of antidepressants. J Pharmacol Exp Ther 233:473–479

Kofuji P, Davidson N, Lester HA (1995). Evidence that neuronal G-protein-gated inwardly rectifying K$^+$ channels are activated by G$\beta\gamma$ subunits and function as heteromultimers. Proc Natl Acad Sci USA 92:6542–6546

Krapivinsky G, Gordon EA, Wickman K, Velimirovic B, Krapivinsky L, Claplham DE (1995) The G-protein-gated atrial K$^+$ channel 1$_{KACH}$ is a heteromultimer of two inwardly rectifying K$^+$-channel proteins. Nature 374:135–141

Kubo Y, Reuveny E, Slesinger PA, Jan YN, Jan LY (1993) Primary structure and functional expression of a rat G-protein-coupled muscarinic potassium channel. Nature 364:802–806

Lakoski, JM, Aghajanian, GK (1985) Effects of ketanserin on neuronal responses to serotonin in the prefrontal cortex, lateral geniculate and dorsal raphe nucleus. Neuropharmacology 24:265–273

Lalley PM, Bischoff AM, Schwarzacher SW, Richter DW (1995) 5-HT$_2$ receptor-controlled modulation of medullary respiratory neurones in the cat. J Physiol (Lond) 487:653–661

Larkman PM, Kelly JS (1992) Ionic mechanisms mediating 5-hydroxytryptamine- and noradrenaline-evoked depolarization of adult rat facial motoneurones. J Physiol (Lond) 456:473–490

Larkman PM, Penington NJ, Kelly JS (1989) Electrophysiology of adult rat facial motoneurones: the effect of serotonin (5-HT) in a novel in vitro brainstem slice. J Neurosci Meth 28:133–146

Lesage F, Duprat F, Fink M, Guillemare E, Coppola T, Lazdunski M, Hugnot J-P (1994) Cloning provides evidence for a family of inward rectifier and G-protein coupled K$^+$ channels in brain. FEBS Lett 353:37–42

Lovenberg TW, Baron BM, deLecea L, Miller JD, Prosser RA, Rea MA, Foye PE, Racke M, Slone AL, Siegel BW, Danielson PE, Sutcliffe JG, Erlander MG (1993) A novel adenylyl cyclase-activating serotonin receptor (5-HT$_7$) implicated the regulation of mammalian circadian rhythms. Neuron 11:449–458

Luebke JI, Greene RW, Semba K, Kamond A, McCarley RW, Reiner PB (1992) Serotonin hyperpolarizes cholinergic low-threshold burst neurons in the rat laterodorsal tegmental nucleus in vitro. Proc Natl Acad Sci USA 89:743–747

Liu W, Sessler JM, Lin RCS, Waterhouse BD (1993) Serotonergic effects on the response properties of morphologically identified layer II/III neurons of rat barrel field cortex. Soc Neurosci Abst 19:742

Lum JT, Piercey MF (1987) Electrophysiological evidence that spiperone is an antagonist of 5-HT$_{1A}$ receptors in the dorsal raphe nucleus. Eur J Pharmacol 149:9–15

Madison DV, Nicoll R (1982) Noradrenaline blocks accommodation of pyramidal cell discharge in the hippocampus. Nature 299:636–638

Mantz J, Godbout J-P, Tassin J, Glowinski J, Thierry A-M (1990) Inhibition of spontaneous and evoked unit activity in the rat medial prefrontal cortex by mesencephalic raphe nuclei. Brain Res 524:22–30

Marek GJ, Aghajanian GK (1994) Excitation of interneurons in piriform cortex by serotonin (5-HT): blockade by MDL 100,907, a highly selective 5-HT$_{2A}$ antagonist. Eur J Pharmacol 259:137–141

Marek GJ, Aghajanian GK (1995) Protein kinase C inhibitors enhance the 5-HT$_{2A}$ receptor-mediated excitatory effects of serotonin on interneurons in rat piriform cortex. Synapse 21:123–130

Marek GJ, Aghajanian GK (1996) LSD and the phenethylamine hallucinogen DOI are potent partial agonists at 5-HT$_{2A}$ receptors on interneurons in the rat piriform cortex. J Pharmacol Exp Ther in press

Maricq AV, Peterson AS, Brake AJ, Myers R, Julius D (1991) Primary source and functional expression of the 5-HT$_3$ receptor, a serotonin-gated ion channel. Science 254:432–437

McCall RB, Aghajanian GK (1979) Serotonergic facilitation of facial motoneuron excitation. Brain Res 169:11–27

McCall RB, Aghajanian GK (1980) Hallucinogens potentiate responses to serotonin and norepinephrine in the facial motor nucleus. Life Sci 26:1149–1156

McCormick DA, Wang Z (1991) Serotonin and noradrenaline excite GABAergic neurones of the guinea-pig and cat nucleus reticularis thalami. J Physiol (Lond) 442:235–255

McCormick DA, Williamson A (1989) Convergence and divergence of neurotransmitter action in human cerebral cortex. Proc Natl Acad Sci USA 86:8098–8102

Medanic, M, Gillette MU (1995) Serotonin regulates the phase of the rat suprachiasmatic circadian pacemaker in vitro only during the subjective day. J Physiol (Lond) 450:629–642

Mengod G, Nguyen H, Lee H, Waeber C, Lubbert H, Palacios JM (1990a) The distribution and cellular localization of 5-HT$_{1C}$ receptor mRNA in the rodent brain examined by in situ hybridization histochemistry. Comparison with receptor binding distribution. Neuroscience 35:577–592

Mengod G, Pompeiano M, Martinez-Mir MI, Palacios JM (1990b) Localization of the mRNA for the 5-HT$_2$ receptor by in situ hybridization histochemistry. Correlation with the distribution of receptor sites. Brain Res 524:139–143

Miquel MC, Doucet E, Boni C, El Mestikawy S, Matthiessen L, Daval G, Verge D, Hamon M (1991) Central serotonin$_{1A}$ receptors: respective distributions of encoding mRNA, receptor protein and binding sites by in situ hybridization histochemistry, radioimmunohistochemistry and autoradiographic mapping in the rat brain. Neurochem Int 19:453–465

Mons N, Yoshimsura M, Cooper DMF (1993). Discrete expression of Ca^{2+}/calmodulin-sensitive and Ca^{2+}-insensitive adenylyl cyclases in the rat brain. Synapse 14:51–59

Monsma FJ Jr, Shen Y, Ward RP, Hamblin MW, Sibley DR (1993) Cloning and expression of a novel serotonin receptor with high affinity for tricyclic psychotropic drugs. Mol Pharmacol 43:320–327

Morilak DA, Garlow SJ, Ciaranello RK (1993) Immunocytochemical localization and description of neurons expressing serotonin receptors in the rat brain. Neuroscience 54:701–717

Nedergaard S, Flatman JA, Engberg I (1991) Excitation of substantia nigra pars compacta neurones by 5-hydroxytryptamine in vitro. NeuroReport 2:329–332

Neuman RS, Rahman S (1996) Ca^{2+} mobilizing agents mimic serotonin 5-HT$_{2A}$ facilitation of N-methyl-D-aspartate depolarization. Behav Brain Res 73:273–275

Neuman RS, Zebrowska G (1992) Serotonin (5-HT$_2$) receptor mediated enhancement of cortical unit activity. Can J Physiol Pharmacol 70:1604–1609

Newberry NR (1992) 5-HT$_{1A}$ receptors activate a potassium conductance in rat ventromedial hypothalamic neurones. Eur J Pharmacol 210:209–212

Newberry NR, Priestley T (1988) A 5-HT$_1$-like receptor mediates a pertussis toxin-sensitive inhibition of rat ventromedial hypothalamic neurons in vitro. Br J Pharmacol 95:6–8

Nicoll RA (1988) The coupling of neurotransmitter receptors to ion channels in the brain. Science 241:545–551

North RA, Uchimura N (1989) 5-Hydroxytryptamine acts at 5-HT$_2$ receptors to decrease potassium conductance in rat nucleus accumbens neurones. J Physiol (Lond) 417:1–12

Okuhara DY, Beck SG (1994) 5-HT$_{1A}$ receptor linked to inward-rectifying potassium current in hippocampal CA3 pyramidal cells. J Neurophysiol 71:2161–2167

Pan ZZ, Wessendorf MW, Williams JT (1993) Modulation by serotonin of the neurons in rat nucleus raphe magnus in vitro. Neuroscience 54:421–429

Pape HC, McCormick DA (1989) Noradrenaline and serotonin selectively modulate thalamic burst firing by enhancing a hyperpolarization-activated cation current. Nature 340:715–718

Pazos A, Palacios JM (1985) Quantitative autoradiographic mapping of serotonin receptors in the rat brain. I. Serotonin-I receptors. Brain Res 346:205–230

Penington NJ, Kelly JS (1990) Serotonin receptor activation reduces calcium current in an acutely dissociated adult central neuron. Neuron 4:751–758

Penington NJ, Kelly JS, Fox AP (1991) A study of the mechanism of Ca^{2+} current inhibition produced by serotonin in rat dorsal raphe neurons. J Neurosci 11:3594–3609

Penington NJ, Kelly JS, Fox AP (1993a) Whole-cell recordings of inwardly rectifying K^+ currents activated by 5-HT_{1A} receptors on dorsal raphe neurones in the adult rat. J Physiol (Lond) 469:387–405

Penington NJ, Kelly JS, Fox AP (1993b) Unitary properties of potassium channels activated by 5-HT in acutely isolated rat dorsal raphe neurones. J Physiol (Lond) 469:407–426

Pessia M, Jiang Z-G, North RA, Johnson SW (1994) Actions of 5-hydroxytryptamine on ventral tegmental area neurons of the rat in vitro. Brain Res 654:324–330

Pfaffinger PJ, Martin JM, Hunter DD, Nathanson NM, Hille B (1985). GTP-binding proteins couple cardiac muscarinic receptors to a K channel. Nature 317:536–538

Piguet P, Galvan M (1994) Transient and long-lasting action of 5-HT on rat dentate gyrus neurones in vitro. J Physiol (Lond) 481:629–639

Piñeyro G, deMontigny C, Weiss M, Blier P (1996) Autoregulatory properties of dorsal raphe 5-HT neurones: possible role of electronic coupling and 5-HT_{1D} receptors in the rat brain. Synapse 22:54–62

Pompeiano M, Palacios JM, Mengod G (1992) Distribution and cellular localization of mRNA coding for 5-HT_{1A} receptor in the rat brain: correlation with receptor binding. J Neursci 12:440–453

Prisco S, Cagnotto A, Talone D, DeBlasi A, Mennini T, Esposito E (1993) Tertatolol, a new β-blocker, is a serotonin (5-hydroxytryptamine$_{1A}$) receptor antagonist in rat brain. J Pharmacol Exp Ther 265:739–744

Prosser RA, Gillette MU (1989) The mammalian circadian clock in the suprachiasmatic nuclei is reset in vitro by cAMP. J Neurosci 9:1073–1081

Prosser RA, Heller HC, Miller JD (1994) Serotonergic phase advances of the mammalian circadian clock involve protein kinase A and K^+ channel opening. Brain Res 644:67–73

Rahman S, Neuman RS (1993) Activation of 5-HT_2 receptors facilitates depolarization of neocortical neurons by N-methyl-D-aspartate. Eur J Pharmacol 231:347–354

Rahman S, McLean JH, Darby-King A, Paterno G, Reynolds JN, Neuman RS (1995) Loss of cortical serotonin$_{2A}$ signal transduction in senescent rats: reversal following inhibition of protein kinase C. Neuroscience 55:891–901

Rasmussen K, Aghajanian GK (1986) Effects of hallucinogens on spontaneous and sensory-evoked locus coeruleus unit activity in the rat: reversal by selective 5-HT_2 antagonists. Brain Res 385:395–400

Rasmussen K, Aghajanian GK (1988) Potency of antipsychotics in reversing the effects of a hallucinogenic drug on locus coeruleus neurons correlates with 5-HT_2 binding affinity. Neuropsychopharmacology 1:101–107

Rasmussen K, Aghajanian GK (1990) Serotonin excitation of facial motoneurons: receptor subtype characterization. Synapse 5:324–332

Read HL, Beck SG, Dun NJ (1994) Serotonergic suppression of interhemispheric cortical synaptic potentials. Brain Res 643:17–28

Reuveny E, Slesinger PA, Inglese J, Morales JM, Iniguez-Lluhi JA, Lefkowitz RJ, Bourne HR, Jan YN, Jan LY (1994) Activation of the cloned muscarinic potassium channel by G protein $\beta\gamma$ subunits. Nature 370:143–146

Rick CE, Stanford IM, Lacey MG (1995) Excitation of rat substantia nigra pars reticulata mediated by 5-hydroxytryptamine$_{2C}$ receptors. Neuroscience 69:903–913

Ropert N (1988) Inhibitory action of serotonin in CA1 hippocampal neurons in vitro. Neuroscience 26:69–81

Ropert N, Guy N (1991) Serotonin facilitates GABAergic transmission in the CA1 region of rat hippocampus in vitro. J Physiol (Lond) 441:121–136

Roth BL, Craigo SC, Choudhary MS, Uluer A, Monsma FJ Jr, Shen Y, Meltzer HY, Sibley DR (1994) Binding of typical and atypical antipsychotic agents to 5-hydroxytryptamine-6 and 5-hydroxytryptamine-7 receptors. J Pharmacol Exp Ther 268:1403–1410

Ruat M, Traiffort E, Leurs R, Tardivel-Lacombe J, Diaz J, Arrang JM Schwartz JC (1993) Molecular cloning, characterization, and localization of a high-affinity serotonin receptor (5-HT$_7$) activating cAMP formation. Proc Natl Acad Sci USA 90:8547–8551

Sanders-Bush E, Burris KD, Knoth K (1988) Lysergic acid diethylamide and 2,5-dimethoxy-4-methylamphetamine are partial agonists at serotonin receptors linked to phosphoinositide hydrolysis. J Pharmacol Exp Ther 246:924–928

Schmitz D, Empson RM, Heinemann U (1995a) Serotonin and 8-OH-DPAT reduce excitatory transmission in rat hippocampal area CA1 via reduction in presumed presynaptic Ca^{2+} entry. Brain Res 701:249–254

Schmitz D, Empson RM, Heinemann U (1995b) Serotonin reduces inhibition via 5-HT$_{1A}$ receptors in area CA1 of rat hippocampal slices in vitro. J Neurosci 15:7217–7225

Scroggs RS, Anderson EG (1990) 5-HT$_1$ receptor agonists reduce the Ca^{++} component of sensory neuron action potentials. Eur J Pharmacol 178:229–232

Segal M (1980) The action of serotonin in the rat hippocampal slice preparation. J Physiol (Lond) 303:423–439

Segal M (1990) Serotonin attenuates a slow inhibitory postsynaptic potential in rat hippocampal neurons. Neuroscience 36:631–641

Segal M, Azmitia EC, Whitaker-Azmitia PM (1989) Physiological effects of selective 5-HT$_{1A}$ and 5-HT$_{1B}$ ligands in rat hippocampus: comparison to 5-HT. Brain Res 502:67–74

Sheldon PW, Aghajanian GK (1990) Serotonin (5-HT) induces IPSPs in pyramidal layer cells of rat piriform cortex: evidence for the involvement of a 5-HT$_2$-activated interneuron. Brain Res 506:62–69

Sheldon PW, Aghajanian GK (1991) Excitatory responses to serotonin (5-HT) in neurons of the rat piriform cortex: evidence for mediation by 5-HT$_{1C}$ receptors in pyramidal cells and 5-HT$_2$ receptors in interneurons. Synapse 9:208–218

Shen Y, Monsma FJ Jr, Metcalf MA, Jose PA, Hamblin MW, Sibley DR (1993) Molecular cloning and expression of a 5-hydroxtryptamine$_7$ serotonin receptor subtype. J Biol Chem 268:18200–18204

Smrcka AV, Hepler JR, Brown KO, Sternweis PC (1991) Regulation of polyphosphoinositide-specific phospholipase C activity by purified G$_q$. Science 251:804–808

Spauschus A, Lentes K-U, Wischmeyer E, Dibmann E, Karschin C, Karschin A (1996) A G-protein-activated inwardly rectifying K$^+$ channel (GIRK4) from human hippocampus associates with other GIRK channels. J Neurosci 16:930–938

Sprouse JS, Aghajanian GK (1986) (-)-Propranolol blocks the inhibition of serotonergic dorsal raphe cell firing by 5-HT$_{1A}$ and 5-HT$_{1B}$ selective agonists. Eur J Pharmacol 128:295–298

Sprouse JS, Aghajanian GK (1987) Electrophysiological responses of serotonergic dorsal raphe neurons to 5-HT$_{1A}$ and 5-HT$_{1B}$ agonists. Synapse 1:3–9

Sprouse JS, Aghajanian GK (1988) Responses of hippocampal pyramidal cells to putative serotonin 5-HT$_{1A}$ and 5-HT$_{1B}$ agonists: a comparative study with dorsal raphe neurons. Neuropharmacology 27:707–715

Stevens DR, McCarley RW, Greene RW (1992) Serotonin$_1$ and serotonin$_2$ receptors hyperpolarize and depolarize separate populations of medial pontine reticular formation neurons in vitro. Neuroscience 47:545–553

Sugihara I, Lang EJ, Llinás R (1995) Serotonin modulation of inferior olivary oscillations and synchronicity: a multiple-electrode study in the rat cerebellum. 7:521–534

Sugita S, Shen KZ, North RA (1992) 5-Hydroxytryptamine is a fast excitatory transmitter at 5-HT$_3$ receptors in rat amygdala. Neuron 8:199–203

Tanaka E, North RA (1993) Actions of 5-hydroxytryptamine on neurons of the rat cingulate cortex. J Neurophysiol 69:1749–1757

Tang W, Gilman AG (1991) Type specific regulation of adenylyl cyclase by G protein $\beta\gamma$ subunits. Science 254:1500–1503

Tani A, Yoshihara Y, Mori K (1992) Increase in cytoplasmic free Ca^{2+} elicited by noradrenaline and serotonin in cultured local interneurons of mouse olfactory bulb. Neuroscience 49:193–199

Todorovic S, Anderson EG (1990) 5-HT$_2$ and 5-HT$_3$ receptors mediate two distinct depolarizing responses in rat dorsal root ganglion neurons. Brain Res 511:71–79

Torres GE, Holt IL, Andrade R (1994) Antagonists of 5-HT$_4$ receptor-mediated responses in adult hippocampal neurons. J Pharmacol Exp Ther 271:255–261

Torres GE, Chaput Y, Andrade R (1995) cAMP and PKA mediate 5-HT$_4$ receptor regulation of calcium-activated potassium current in adult hipocampal neurons. Mol Pharmacol 47:191–197

Van den Hooff P, Galvan M (1992) Actions of 5-hydroxytryptamine and 5-HT$_{1A}$ receptor ligands on rat dorsolateral septal neurones in vitro. Br J Pharmacol 106:893–899

Wang RY, Aghajanian GK (1977) Antidromically identified neurons in the rat midbrain raphe: evidence for collateral inhibition. Brain Res 132:186–193

Wang HY, Friedman E (1990) Central 5-hydroxytryptamine receptor-linked protein kinase C translocation: a functional postsynaptic transduction system. Mol Pharmacol 37:75–79

White SR, Fung SJ (1989) Serotonin depolarizes cat spinal motoneurons in situ and decreases motoneuron afterhyperpolarizing potentials. Brain Res 502:205–213

Wible TM, Ficker E, Brown AM (1994) Gating in inwardly rectifying K$^+$ channels localized to a single negatively charged residue. Nature 371:246–249

Wickman KD, Iniguez-Lluhi JA, Davenport PA, Taussig R, Krapivinsky G, Linder ME, Gilman AG, Clapham DE (1994) Recombinant G-protein protein $\beta\gamma$ subunits activate the muscarinic-gated atrial potassium channel. Nature 368:255–25

Williams JT, Colmers WF, Pan ZZ (1988) Voltage- and ligand-activated inwardly rectifying currents in dorsal raphe neurons in vitro. J Neurosci 8:3499–3506

Wright DE, Seroogy KH, Lundgren KH, Davis BM, Jennes L (1995) Comparative localization of serotonin$_{1A, 1C}$, and $_2$ receptor subtype mRNAs in rat brain. J Comp Neurol 351:357–373

Yakel JL, Jackson MB (1988) 5-HT$_3$ receptors mediate rapid responses in cultured hippocampus and a clonal cell line. Neuron 1:615–621

Yakel JL, Trussell LO, Jackson MB (1988) Three serotonin responses in cultured mouse hipocampal and striatal neuron. J Neurosci 8:1273–1285

Yakel JL, Shao XM, Jackson MB (1990) The selectivity of the channel coupled to the 5-HT$_3$ receptor. Brain Res 533:46–52

Yamazaki J, Fukuda H, Nagao T, Ono H (1992) 5-HT$_2$/5-HT$_{1C}$ receptor-mediated facilitatory action on unit activity of ventral horn cells in rat spinal cord slices. Eur J Pharmacol 220:237–242

Yatani A, Mattera R, Codina J, Graf R, Okabe K, Padrell E, Iyengar R, Brown AM, Birnbaumer L (1988) The G protein-gated atrial channel is stimulated by three distinct G$_i\alpha$-subunits. Nature 336:680–682

Zeise ML, Batsche K, Wang RY (1994) The 5-HT$_3$ receptor agonist 2-methyl-5-HT reduced postsynaptic potentials in rat CA1 pyramidal neurons of the hippocampus in vitro. Brain Res 651:337–341

Zgombick JM, Beck SG, Mahle CD, Craddock-Royal B, Maayani S (1989) Pertussis toxin-sensitive guanine nucleotide-binding protein(s) couple adenosine A$_1$ and 5-hydroxtryptamine$_{1A}$ receptors to the same effector system in rat hippocampus: biochemical and electrophysiological studies. Mol Pharmacol 35:484–494

Zhang JY, Ashby CR Jr, Wang RY (1994) Effect of pertussis toxin on the response of rat medial prefrontal cortex cells to the iontophoresis of serotonin receptor agonists. J Neural Transm 95:165–172

CHAPTER 20
5-HT Receptors Involved in the Regulation of Hormone Secretion

L.D. VAN DE KAR

A. Organization of the Neuroendocrine System

The hypothalamus, particularly the paraventricular nucleus (PVN), plays a central role in coordinating and regulating neuroendocrine function. Axons from cells originating in the PVN release corticotropin-releasing hormone (CRH) into the hypophysial portal vessels. CRH is then transported to the anterior pituitary gland, where it controls the release of corticotropin (adrenocorticotropic hormone, ACTH), leading to increased secretion of adrenal glucocorticoids (cortisol in humans, corticosterone in rats) (SWANSON and SAWCHENKO 1983; SAWCHENKO and SWANSON 1985). Activation of other cells in the PVN increases prolactin release from the anterior lobe of the pituitary gland (BAGDY and MAKARA 1994, 1995; BLUET PAJOT et al. 1995; RITTENHOUSE et al. 1993; AREY and FREEMAN 1992; MINAMITANI et al. 1987; KISS et al. 1986). Cells in the PVN and supraoptic nucleus (SON) that synthesize oxytocin and vasopressin release these peptides into the circulation from their nerve terminals in the posterior (neural) lobe of the pituitary gland (SWANSON and SAWCHENKO 1983). Finally, cells in the PVN are also important in the regulation of renin release from the kidneys (RITTENHOUSE et al. 1992b; RICHARDSON MORTON et al. 1989). Serotoninergic neurons originating in the midbrain dorsal and median raphe nuclei and the ventrolateral B9 cell group send axons into the hypothalamus, where they innervate neuroendocrine control sites (AZMITIA and SEGAL 1978; VAN DE KAR and LORENS 1979; SAWCHENKO et al. 1983) (Fig. 1). Serotoninergic synapses have been identified on CRH cells in the PVN (LIPOSITS et al. 1987). No such histological evidence is currently available for serotoninergic synapses on oxytocin, vasopressin or other, as yet unidentified, cells that regulate the secretion of prolactin and renin. However, lesion studies have indicated that the dorsal raphe nucleus contains the serotoninergic perikarya involved in stimulating the secretion of prolactin (QUATTRONE et al. 1979; VAN DE KAR and BETHEA 1982; BAROFSKY et al. 1983) and renin (VAN DE KAR et al. 1982, 1984). Other studies also indicate that stimulation of serotoninergic neurons in the dorsal raphe nucleus increases the activity of oxytocin-containing cells in the PVN (SAPHIER 1991; KAWANO et al. 1992). There is inconsistent evidence for a serotoninergic stimulation of growth hormone (GH) secretion (SELETTI et al. 1995; COWEN et al. 1994; MELLER and BOHMAKER 1994). Since this topic has been previously

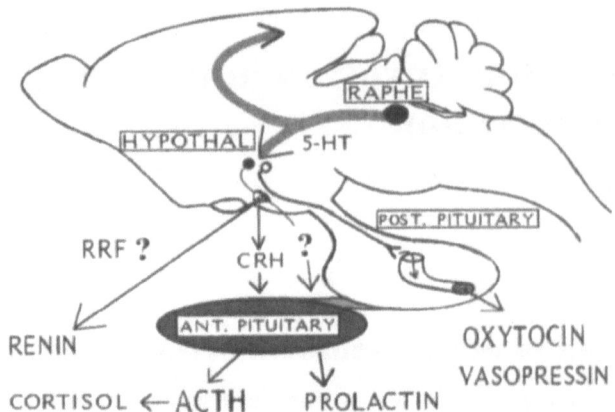

Fig. 1. Schematic illustration of serotoninergic input into the hypothalamic centers that control the secretion of hormones. *Ant*, anterior; *CRH*, corticotropin-releasing hormone; *Hypothal*, hypothalamus; *post*, posterior; *RRF*, renin-releasing factor

reviewed (Levy et al. 1994; Van de Kar 1991; Yatham and Steiner 1993), it will not be addressed in this chapter. Serotoninergic neurons in the dorsal raphe send collaterals to both the hypothalamic PVN and other limbic regions such as the central amygdaloid nucleus, prefrontal cortex and nucleus accumbens (Van Bockstaele et al. 1993; Petrov et al. 1994). Hence, changes in the function of dorsal raphe neurons could lead to alterations in both neuroendocrine and limbic function.

B. 5-HT Receptor Subtypes

As reviewed in other chapters of this book, 5-HT receptors are currently classified into seven main groups, some of which are subdivided into subgroups: 5-HT_{1A-F}, 5-HT_{2A-C}, 5-HT_3, 5-HT_4, 5-HT_{5a-b}, 5-HT_6 and 5-HT_7. Table 1 summarizes the known neuroendocrine effects of activation of 5-HT receptor subtypes. To be complete, catecholamine receptors are also included in Table 1.

C. Neuroendocrine Effects of 5-HT_{1A} Agonists

I. Hypothalamic Pituitary Adrenal Axis

Administration to rats of 5-HT_{1A} agonists, such as 8-hydroxy-2-(dipropylamino)tetralin (8-OH-DPAT), flesinoxan, ipsapirone, buspirone, gepirone, alnespirone and tandospirone, increases the secretion of ACTH and corticosterone (Lesch 1991; Groenink et al. 1995; Levy et al. 1995; Koenig et al. 1988; Matheson et al. 1989; Seletti et al. 1995). Several 5-HT_{1A} antagonists inhibit the stimulatory effects of 5-HT_{1A} agonists on the hypothalamic pitu-

Table 1. Summary of monoamine receptor influence on hormone secretion

Hormone	Serotonin									Dopamine		Norepinephrine		
	1A	1B/D	2A	2C	3	4	5	6	7	D_1	D_2	α_1	α_2	β
ACTH	↑	0	↑	↑	↑	?	?	?	0	↑	↑	↑	↓?	?
Corticosterone/cortisol	↑	0	↑	↑	↑	↑	?	?	↑?	↑	↑	↑	↑?	↑?
Prolactin	↑?	?	↑	↑	↑	?	?	?	0	↑	↓	↑	↑	↑
Oxytocin	↑	0	↑	↑	0	?	?	?	0	↑	↓	↑	?	↓
Vasopressin	0	0	↑	↑	0	?	?	?	?	↑	↓	↑	?	↓
Renin	0	0	↑	↑	0	?	?	?	↑?	0	0	↓	↓	↑

↑, Stimulatory influence; 0, no influence; ↓, inhibitory influence; ?, unknown influence or inconclusive data.

itary adrenal (HPA) axis. These include: (−)pindolol and (−)tertatolol (β-adrenoceptor and 5-HT$_{1A}$ antagonists), spiperone (also a dopamine D$_2$, 5-HT$_{2A}$ and 5-HT$_{1A}$ antagonist) and the more selective 5-HT$_{1A}$ (partial) antagonists NAN 190, WAY 100,135 and the full antagonist WAY 100,635 (Critchley et al. 1994; Lejeune et al. 1993; Przegalinski et al. 1990; Pan and Gilbert 1992; Lesch 1991; Levy et al. 1995). Administration of buspirone and ipsapirone to humans also increases plasma ACTH and/or cortisol (Anderson and Cowen 1992; Lesch et al. 1991). A receptor reserve exists for the 5-HT$_{1A}$-receptor-mediated increase in plasma ACTH and corticosterone (Meller and Bohmaker 1994; Pinto et al. 1994). Consequently, changes in brain 5-HT$_{1A}$ receptors will not be accurately reflected in the magnitude of the ACTH response to a challenge with 5-HT$_{1A}$ agonists.

Direct injection of 8-OH-DPAT into the hypothalamic PVN increased plasma ACTH (Pan and Gilbert 1992). However, a recent study compared lower doses of 8-OH-DPAT, injected directly into the PVN or into the dorsal raphe nucleus, and concluded that the location of the 5-HT$_{1A}$ receptors involved in ACTH secretion is in the dorsal raphe nucleus (Bluet Pajot et al. 1995). Additional evidence against a postsynaptic location was obtained when depletion of 5-HT stores using p-chlorophenylalanine (PCPA) reduced the ACTH response to i.v.-injected 8-OH-DPAT (Bluet Pajot et al. 1995). However, other studies found that PCPA potentiated, or at least did not reduce, the effect of 8-OH-DPAT on plasma corticosterone (Cowen et al. 1990; Kelder and Ross 1992). Also, addition of 8-OH-DPAT to hypothalamic explants dose dependently increased CRH release in vitro, with a maximal increase observed at 0.1 nM (Calogero et al. 1989). Hence, it is unclear whether the ACTH response to 5-HT$_{1A}$ agonists represents activation of postsynaptic 5-HT$_{1A}$ receptors in the hypothalamus or autoreceptors in the midbrain, or both.

The 5-HT$_{1A}$ agonist 8-OH-DPAT elevates plasma concentrations of β-endorphin, an effect that is inhibited by spiperone and pindolol (Koenig et al. 1987). The effect of 8-OH-DPAT on plasma β-endorphin concentration is blocked by dexamethasone, a CRH antiserum, or by transection of the infundibulum (Bagdy et al. 1990). Addition of 8-OH-DPAT to pituitary cells in vitro does not alter β-endorphin secretion, suggesting that 5-HT$_{1A}$ receptors stimulate the release of β-endorphin by a CRH-dependent mechanism. Finally, an intracerebroventricular (i.c.v.) injection of 8-OH-DPAT (5 μg) increases the plasma concentration of β-endorphin, suggesting that the receptors mediating the secretion of β-endorphin are located in the brain (Di Sciullo et al. 1990).

II. Oxytocin

The 5-HT$_{1A}$ agonists 8-OH-DPAT, buspirone, ipsapirone and alnespirone also increase the secretion of oxytocin (Bagdy and Makara 1994; Bagdy and Kalogeras 1993; Li et al. 1993b, 1994; Van de Kar et al., unpublished observations). The 5-HT$_{1A}$ antagonists NAN 190, (–)pindolol and spiperone also reduce the oxytocin response to 8-OH-DPAT or alnespirone (BAGDY and KALOGERAS 1993; Van de Kar et al., unpublished observations). In a recent experiment in my laboratory, we found that the full 5-HT$_{1A}$ antagonist WAY 100,635 inhibits the oxytocin response to 8-OH-DPAT (0.5 mg/kg s.c.). The minimum WAY 100,635 dose of 0.1 mg/kg s.c. inhibited, while a complete blockade of the oxytocin response was observed at a WAY 100,635 dose of 1 mg/kg s.c. (saline–saline, 8.5 ± 1.5; saline–8-OH-DPAT, 78.0 ± 9.6; WAY 100,635–saline, 8.8 ± 1.2; WAY 100,635–8-OH-DPAT, 6.7 ± 1.1 pg/ml). There is no receptor reserve for the 5-HT$_{1A}$-receptor-mediated increase in plasma oxytocin (PINTO et al. 1994). No information is available on the oxytocin response to 5-HT$_{1A}$ agonists in humans. Vasopressin and renin secretion cannot be increased by 5-HT$_{1A}$ agonists (BROWNFIELD et al. 1992; VAN DE KAR and BROWNFIELD 1993).

The location of the 5-HT$_{1A}$ receptors involved in the secretion of oxytocin has not been established. Mechanical destruction of the PVN prevents the oxytocin response to ipsapirone (BAGDY and MAKARA 1994). Cell-selective lesions with ibotenic acid injections into the PVN also inhibit the oxytocin response to the 5-HT releaser p-chloroamphetamine. Similar injections of ibotenic acid into the SON do not inhibit the oxytocin response to p-chloroamphetamine (VAN DE KAR et al. 1995b). Nevertheless, these data do not provide direct evidence that the 5-HT$_{1A}$ receptors are located on oxytocin-containing neurons in the PVN. It cannot be excluded that, after peripheral injection of a 5-HT$_{1A}$ agonist, receptors in other sites could be activated, and in turn activate the oxytocin-containing cells in the PVN. Furthermore, destruction of the PVN also prevents the effect of the 5-HT$_2$ agonist 1-(2,5-dimethoxy-4-iodophenyl)-2-aminopropane (DOI) on plasma oxytocin (BAGDY and MAKARA 1994), suggesting that the PVN is not selectively activated by only one 5-HT receptor subtype. Destruction of serotoninergic neurons, using i.c.v. injections of 5,7-DHT, reduces the maximal oxytocin response to 8-OH-DPAT, which could suggest that the 5-HT$_{1A}$ receptors might be located presynaptically (Li and Van de Kar, unpublished observations). However, the reduction in the maximal oxytocin response is probably due to the ability of this dose of i.c.v.-injected 5,7-DHT to reduce the concentration of oxytocin in the hypothalamus, by about 38% (SAYDOFF et al. 1993). There is insufficient evidence to conclude that the 5-HT$_{1A}$ receptors are post- or presynaptic and their location remains to be determined.

III. Prolactin

Administration of 8-OH-DPAT and ipsapirone to rats produces a very short-lived increase in plasma prolactin, subsiding within 30 min (SIMONOVIC et al. 1984; NASH and MELTZER 1989; VAN DE KAR et al. 1989b; KELLAR et al. 1992; DI SCIULLO et al. 1990). This short duration is unusual, compared with the prolactin response to other 5-HT agonists and when comparing prolactin with other hormones after injection of 5-HT$_{1A}$ agonists. Although buspirone also increases prolactin secretion, this effect is more likely due to antagonism of dopamine D$_2$ receptors in the pituitary gland, rather than due to activation of 5-HT$_{1A}$ receptors in the hypothalamus (ANDERSON and COWEN 1992; COWEN et al. 1990). Ipsapirone does not increase prolactin secretion in humans (COWEN et al. 1990). Flesinoxan (1 mg i.v.) increases plasma prolactin as well as ACTH in humans. However, its effect on plasma prolactin is not inhibited by pindolol, a dose (30 mg p.o.) that inhibits the ACTH response to flesinoxan (SELETTI et al. 1995). Furthermore, methysergide inhibits both the prolactin and ACTH responses to flesinoxan, suggesting that other 5-HT receptors might mediate the prolactin response to this 5-HT agonist. Recent findings in my laboratory placed additional doubt on the nature of the receptor mediating the effect of 8-OH-DPAT on plasma prolactin. We found that the full 5-HT$_{1A}$ antagonist WAY 100,635 cannot inhibit the effect of 8-OH-DPAT on plasma prolactin. A dose (1 mg/kg s.c.) that completely blocks the ACTH and oxytocin responses to 8-OH-DPAT does not inhibit the prolactin response (saline–saline, 2.2 ± 0.2; saline–8-OH-DPAT, 6.7 ± 1.5; WAY 100,635–saline, 1.1 ± 0.1; WAY 100,635–8-OH-DPAT, 5.8 ± 0.9 ng/ml). To confuse the issue further, pertussis toxin, which inactivates Gi and Go proteins, prevents the effect of 8-OH-DPAT (50 µg/kg s.c.) on plasma prolactin, suggesting that Gi or Go proteins mediate this effect (VAN DE KAR et al. 1995a). In addition, i.c.v. injection of forskolin also prevents the effect of i.c.v.-injected 8-OH-DPAT (0.1, 1 and 10 µg/kg i.c.v.) on plasma prolactin (VAN DE KAR et al. 1995a). Since forskolin directly activates adenylyl cyclase, elevating cellular concentrations of cAMP, these data suggest that the effect of 8-OH-DPAT on plasma prolactin involves Gi- or Go-protein-induced inhibition of adenylyl cyclase. This fits with our knowledge of the coupling of 5-HT$_{1A}$ receptors to their second messenger system (BOESS and MARTIN 1994), but could also be related to other receptors that are coupled via Gi or Go proteins to adenylyl cyclase. The data are insufficient to conclude that 5-HT$_{1A}$ receptors play an important role in the secretion of prolactin.

   In contrast with the HPA axis, the prolactin response to 8-OH-DPAT is potentiated in rats pretreated with PCPA, consistent with the hypothesis of denervation supersensitivity of postsynaptic 5-HT receptors (BLUET PAJOT et al. 1995; SIMONOVIC et al. 1984). I.c.v. injection of much lower doses of 8-OH-DPAT than peripherally effective doses (0.1–10 µg/kg) increases plasma prolactin to much higher levels than peripheral injections of much higher doses (VAN DE KAR et al. 1995a). These data suggest that peripheral mechanisms activated by 8-OH-DPAT might mask some of its central activating effects. We also observed that destruction of serotoninergic pathways using i.c.v. injection of 5,7-DHT potentiated the prolactin response to 8-OH-DPAT (10–200 µg/kg s.c.) (VAN DE KAR et al. 1995a). Injection of 8-OH-DPAT into the dorsal raphe nucleus increased plasma ACTH but did not alter plasma prolactin, while injections of 8-OH-DPAT into the PVN increased prolactin secretion (BLUET PAJOT et al. 1995). Taken together, the data suggest that activation of 5-HT receptors in the hypothalamic PVN increases the secretion of prolactin. However, PVN lesions do not inhibit the prolactin response to ipsapirone (BAGDY and MAKARA 1994). Clearly, the data are not conclusive with respect to the identity or location of 5-HT$_{1A}$ receptors that stimulate the secretion of prolactin.

IV. Inhibition by 5-HT$_{1A}$ Agonists of the Neuroendocrine Responses to Stress

Inhibition of serotoninergic neurotransmission can produce anxiolytic effects. Anxiolytic drugs of the benzodiazepine class inhibit serotoninergic transmission and act in the dorsal raphe nucleus to reduce anxiety-related behaviors (NISHIKAWA and SCATTON 1986; THIEBOT et al. 1982). The 5-HT$_{1A}$ partial agonists buspirone, gepirone, tandospirone and ipsapirone are clinically effective anxiolytic drugs (EISON 1989; BARRETT and VANOVER 1993). These 5-HT$_{1A}$ agonists appear to produce their anxiolytic effect by activating somatodendritic 5-HT$_{1A}$ autoreceptors on serotoninergic cells in the dorsal raphe nucleus, inhibiting the firing rate of serotoninergic neurons and reducing serotoninergic neurotransmission (DOURISH et al. 1986; VANDERMAELEN et al. 1986; MATHESON et al. 1994).

While the hormones ACTH, corticosterone, prolactin, oxytocin and renin are released by activation of 5-HT receptors, these hormones (except for vasopressin) are also secreted by exposure to most stressors and can be considered stress-markers (FELDMAN et al. 1995; CHAOULOFF 1993; BLAIR et al. 1976; PARIS et al. 1987; VAN DE KAR et al. 1991; LANG et al. 1983; CALLAHAN et al. 1992; VAN DE KAR 1996). Anxiolytic drugs of the benzodiazepine class suppress stress-induced corticosterone (GRAM and CHRISTENSEN 1986; LE FUR et al. 1979) and prolactin (VAN DE KAR et al. 1985a; FEKETE et al. 1981) secretion. In contrast, these drugs do not inhibit the renin response to conditioned stress (VAN DE KAR et al. 1985a). Because of the importance of the renin-angiotensin system in the regulation of blood pressure, the lack of effect of benzodiazepines could be clinically important and suggests that other drugs should be explored for the treatment of anxiety disorders in hypertensive patients.

The 5-HT$_{1A}$ agonist 8-OH-DPAT and partial 5-HT$_{1A}$ agonists buspirone and ipsapirone, at doses lower than the doses that increase the secretion of hormones, inhibit the effect of stress on ACTH, corticosterone, prolactin and renin secretion (VAN DE KAR et al. 1985b, 1991; URBAN et al. 1986; RITTENHOUSE et al. 1992a; SAPHIER et al. 1995). In my laboratory, Dr. Peter A. Rittenhouse observed that the oxytocin response to immobilization stress can be reduced by ipsapirone at doses below those that increase the secretion of oxytocin (Table 2). Since the somatodendritic 5-HT$_{1A}$ autoreceptors are more sensitive than postsynaptic receptors to the effects of 5-HT$_{1A}$ agonists, these

Table 2. Inhibition by ipsapirone of the effect of immobilization stress on plasma oxytocin

Dose of ipsapirone (mg/kg i.p.)	Plasma oxytocin (pg/ml)	
	Control	Immobilization stress
0 (saline)	5.1 ± 1.3	14.0 ± 2.9**
0.05	7.6 ± 1.4	10.2 ± 0.9
0.1	8.1 ± 1.5	10.3 ± 1.7
0.5	6.6 ± 0.7	11.7 ± 1.0
1.0	4.9 ± 0.9	9.5 ± 0.9

Data represent mean ± SEM of eight rats per group.
** Significant difference from nonstressed rats, $P < 0.01$ (one-way ANOVA and Newman Keuls' test).

observations suggest that the anxiolytic effects of buspirone and ipsapirone on stress-induced increase in secretion of hormones are due to inhibition of serotoninergic transmission. Consistent with this hypothesis, electrolytic lesions in the dorsal raphe nucleus prevent the effect of conditioned fear stress on renin and corticosterone secretion (Van de Kar et al. 1984; Richardson Morton et al. 1986). These findings are similar to observations by other investigators, who found that injections of 5,7-DHT into the raphe or into the hypothalamic PVN inhibit the effect of several stressors on corticosterone secretion (Feldman et al. 1984, 1987, 1991; Feldman and Weidenfeld 1995). Together, the observations suggest that plasma concentrations of ACTH, prolactin, oxytocin and renin can be useful as neuroendocrine markers of anxiety in experimental animals, and in humans. Drugs with anxiolytic effects can be characterized using their ability to suppress the neuroendocrine manifestations of anxiety.

D. Neuroendocrine Effects of 5-HT$_{1B/1D}$ Agonists

There is insufficient convincing evidence to support a role for 5-HT$_{1B}$ receptors in stimulating hormone secretion. A stimulatory role for 5-HT$_{1B}$ receptors in the secretion of prolactin has been postulated (Van de Kar et al. 1989b), but subsequent studies do not support this hypothesis. It is unlikely that 5-HT$_{1B}$ receptors influence the secretion of ACTH, corticosterone, oxytocin, vasopressin or renin because no antagonists for this receptor can inhibit the effects of the 5-HT$_{1B}$ agonist CP 93,129 on these hormones (Van de Kar et al. 1994). The increase in prolactin also was only inhibited by metergoline but not by propranolol, suggesting that other 5-HT receptors mediate the prolactin response to CP 93,129 (Van de Kar et al. 1994). In humans, the 5-HT$_{1D}$ agonist sumatriptan does not increase plasma ACTH, cortisol, β-endorphin or prolactin (Franceschini et al. 1994; Entwisle et al. 1995).

E. Neuroendocrine Effects of 5-HT₂ Agonists

I. Hypothalamic Pituitary Adrenal Axis

Administration to rats of several 5-HT$_2$ agonists, such as DOI, and 1-(-4-bromo-2,5-dimethoxyphenyl)-2-aminopropane (DOB), which activate both 5-HT$_{2A}$ and 5-HT$_{2C}$ receptors, and the less selective agonists m-chlorophenylpiperazine (mCPP), MK-212, 1-(2,5-dimethoxy-4-methylphenyl)-2-aminopropane (DOM), RU 24969 and quipazine (which mainly activate 5-HT$_{2C}$ receptors) increases the secretion of ACTH and corticosterone (BAGDY et al. 1993; KING et al. 1989; FULLER and SNODDY 1990; OWENS et al. 1991; GARCIA-BORREGUERO et al. 1995; HALBREICH et al. 1995; RITTENHOUSE et al. 1994a; KRYSTAL et al. 1993; HOLLANDER et al. 1992). The effects of DOI are inhibited in rats pretreated with 5-HT$_{2A}$ antagonists such as spiperone (dopamine D$_2$, 5-HT$_{2A}$ and 5-HT$_{1A}$ antagonist) (RITTENHOUSE et al. 1994a). The effects of DOI, MK-212, DOM, mCPP and RU 24969 are inhibited by the 5-HT$_{2A/2C}$ antagonists ritanserin, ketanserin, altanserin, metergoline, mianserin, mesulergine, metergoline and LY 53857 (AULAKH et al. 1992, 1994; KING et al. 1989; RITTENHOUSE et al. 1994a; VAN DE KAR et al. 1989a; GARTSIDE et al. 1992). In humans, the cortisol response to mCPP is inhibited by metergoline, suggesting involvement of 5-HT$_{2C}$ receptors (MUELLER et al. 1985). MK-212 also increases plasma cortisol in humans, an effect that is not inhibited by pindolol (LOWY and MELTZER 1988; BASTANI et al. 1990; MELTZER and MAES 1995). Injection of DOI into the lateral cerebral ventricles of conscious rats increases plasma ACTH, an effect that can be inhibited by i.c.v. injection of ritanserin or LY 53857 (RITTENHOUSE et al. 1994a). Superfusion of hypothalamic explants with DOI increases the release of CRH into the perfusing buffer (CALOGERO et al. 1989). In addition, 5-HT increases CRH release from hypothalamic explants and this effect is inhibited by ketanserin, metergoline and ritanserin (CALOGERO et al. 1989). Thus, it is clear that the ACTH response to 5-HT$_2$ agonists represents activation of 5-HT$_2$ receptors in the hypothalamus. Activation of 5-HT$_2$ receptors also stimulates β-endorphin secretion. DOI, mCPP and MK-212 increase plasma β-endorphin in rats (KOENIG et al. 1987; BAGDY et al. 1990). The responses to MK-212 are inhibited by ketanserin, altanserin, ritanserin and metergoline (KOENIG et al. 1987). The effect of DOI on β-endorphin is attenuated by transection of the infundibulum, suggesting that it is at least partially mediated by a CRH-dependent mechanism (BAGDY et al. 1990).

II. Oxytocin

Most 5-HT$_2$ agonists also increase the secretion of oxytocin. Among them are DOI, mCPP, MK-212 and, with lower efficacy, RU 24969 (SAYDOFF et al. 1991; BAGDY and MAKARA 1994; BAGDY et al. 1992a; BAGDY and KALOGERAS 1993). RU 24969 is a more potent 5-HT$_{1A/1B}$ agonist than a 5-HT$_{2C}$ agonist (TITELER et al. 1987; HOYER 1988) and its low efficacy in increasing oxytocin release is surprising, considering the effectiveness of 5-HT$_{1A}$ agonists. However, RU 24969 also has a moderate affinity for β-adrenoceptors (VAN WIJNGAARDEN et al. 1990) which exert an inhibitory effect on the secretion of oxytocin (CROWLEY and ARMSTRONG 1992). The 5-HT$_2$ antagonists mentioned for the HPA axis (ritanserin, spiperone and ketanserin) also inhibit the oxytocin response to DOI (SAYDOFF et al. 1991; BAGDY et al. 1992a; BROWNFIELD et al. 1992; BAGDY and KALOGERAS 1993). Ritanserin at a high dose (2.5 mg/kg s.c.) also inhibits (but does not completely block) the oxytocin response to MK-212 (SAYDOFF et al. 1991), suggesting involvement of additional receptors. It is important to note that the effect of mCPP on plasma oxytocin is not inhibited by the 5-HT$_{2A}$ antagonist ketanserin, while antagonists with similar affinity for both 5-HT$_{2A}$ and 5-HT$_{2C}$ receptors, such as mianserin and LY 53857, can inhibit the effects of mCPP. In contrast, ketanserin and spiperone (which have a higher affinity for 5-HT$_{2A}$ than for 5-HT$_{2C}$ receptors) inhibit the oxytocin responses to DOI (SAYDOFF et al. 1991; BAGDY et al. 1992a), suggesting also a role for 5-HT$_{2A}$ receptors.

III. Prolactin

DOI and MK-212 dose dependently increase plasma prolactin concentration in rats (LI et al. 1993a; LEVY et al. 1992; RITTENHOUSE et al. 1993). The effects of DOI and MK-212 are inhibited by the 5-HT$_{2A/2C}$ antagonists ritanserin and ketanserin (ALBINSSON et al. 1994; PAN and TAI 1992; RITTENHOUSE et al. 1993). The relative contribution of 5-HT$_{2A}$ vs. 5-HT$_{2C}$ receptors in this response has not been established. At a high dose, ketanserin (5 mg/kg) could not differentiate between these receptors because ketanserin only has an approximately tenfold higher affinity for 5-HT$_{2A}$ than for 5-HT$_{2C}$ receptors (BOESS and MARTIN 1994). The prolactin response to i.c.v. injection of DOI was inhibited by i.c.v. infusion of a low dose of ritanserin (2 μg/kg i.c.v.) and the prolactin response to i.c.v. injection of RU 24969 was inhibited by i.c.v. injection of LY 53857 (50 μg/kg i.c.v.) (RITTENHOUSE et al. 1993). Since RU 24969 has a lower affinity for 5-HT$_{2A}$ than for 5-HT$_{2C}$ receptors (ZIFA and FILLION 1992), these data suggest that activation of brain 5-HT$_{2C}$ receptors increases the secretion of prolactin. The prolactin response to MK-212 (which has a very low affinity for 5-HT$_{2A}$ receptors) is inhibited by the 5-HT$_{2A/2C}$ antagonist LY 53857, further supporting a stimulatory role of 5-HT$_{2C}$ receptors in prolactin secretion (VAN DE KAR et al. 1989b). However, the prolactin response to MK-212 (10 mg/kg, i.p.) was only partially attenuated by LY 53857 (0.1–1 mg/kg), whereas some other hormone responses were completely blocked (LORENS and VAN DE KAR 1987; VAN DE KAR et al. 1989b). These data suggest that MK-212 increases prolactin secretion partially via activation of 5-HT$_{2C}$ receptors, and partially through other mechanisms. Consistent with this suggestion, recent data in humans indicate that pindolol inhibits the prolactin response to MK-212, suggesting that MK-212 also activates 5-HT$_{1A}$ receptors to increase the secretion of prolactin (MELTZER and MAES 1995). mCPP also increases prolactin secretion in humans and in rats (QUATTRONE et al. 1981; MUELLER et al. 1985; BAGDY et al. 1989; SEIBYL et al. 1991; AULAKH et al. 1992; KAHN et al. 1994). In rats, the prolactin response to mCPP is inhibited by several 5-HT$_2$ antagonists such as metergoline, clozapine, mianserin and mesulergine, but not by ritanserin or xylamidine (which does not cross the blood-brain barrier; BAGDY et al. 1989; AULAKH et al. 1992; KAHN et al. 1994). However, in humans, ritanserin does inhibit the prolactin response to mCPP (SEIBYL et al. 1991). To further confuse the issue, propranolol also inhibits the prolactin response to mCPP, suggesting that either this is due to the ability of propranolol to antagonize 5-HT$_{2C}$ receptors or is due to effects of mCPP on other receptors (AULAKH et al. 1992). In conclusion, most of the data agree that activation of 5-HT$_{2C}$ receptors increases the secretion of prolactin.

IV. Renin and Vasopressin

ACTH, oxytocin and prolactin also can be released after activation of other 5-HT receptor subtypes, notably 5-HT$_{1A}$ and 5-HT$_3$ (LEVY et al. 1994); renin secretion can only be increased by administration of 5-HT$_2$ agonists, such as MK-212, DOI, mCPP, quipazine and RU 24969 (BAGDY et al. 1992b, 1993; ALPER and SNIDER 1987; LORENS and VAN DE KAR 1987; VAN DE KAR et al. 1981, 1989a; ALPER 1990; RITTENHOUSE et al. 1991, 1994b; LI et al. 1992). The ability of DOI to increase renin release is antagonized by ritanserin and spiperone (RITTENHOUSE et al. 1991). The effects of mCPP, RU 24969 or quipazine on renin secretion are antagonized by ritanserin, LY 53857, spiperone and xylamidine (LORENS and VAN DE KAR 1987; VAN DE KAR et al. 1989a; BAGDY et al. 1992b; ZINK et al. 1990). Because spiperone has a higher affinity for 5-HT$_{2A}$ than for 5-HT$_{2C}$ receptors, these data suggest that 5-HT$_{2A}$ receptors primarily mediate the renin response to DOI, although a contribution of 5-HT$_{2C}$ receptors is also possible, particularly to the agonists with lower affinity for 5-HT$_{2A}$ receptors (i.e., MK-212, mCPP and RU 24969). Because xylamidine is a 5-HT$_2$ antagonist that does not readily cross the blood-brain barrier, peripheral 5-HT$_2$ receptors have been postulated to contribute to the renin response to DOI and quipazine (ZINK et al. 1990; RITTENHOUSE et al. 1991). However, a central site of 5-HT$_{2C}$ receptors is evident from the renin response to i.c.v.-injected RU 24969 that is blocked by i.c.v. administration of LY 53857 (RITTENHOUSE et al. 1994b).

Central serotoninergic neurons also stimulate the secretion of vasopressin by activating 5-HT$_2$ receptors. Injection of 5-HT (i.c.v.) to conscious rats increases plasma vasopressin, an effect that can be inhibited by i.c.v. or peripheral administration of the 5-HT$_{2A/2C}$ antagonist LY 53857 (PERGOLA et al. 1993; SAYDOFF et al. 1992). mCPP (2.5 mg/kg i.v.) and MK-212, but not DOI (1 mg/kg i.v.), increase vasopressin secretion (STEARDO and IOVINO 1986; BROWNFIELD et al. 1988, 1992; BAGDY et al. 1990, 1992b). In rats withdrawn from cocaine (15 mg/kg b.i.d. for 7 days), the effect of MK-212 on plasma vasopressin is inhibited (VAN DE KAR et al. 1992). There is a report that mCPP does not elevate plasma vasopressin in humans (KAHN et al. 1992). However, the dose used (0.25–0.5 mg/kg, p.o.) might have been too low to be able to observe an effect, although this dose and lower i.v. doses of mCPP (0.1 mg/kg) were capable of increasing plasma cortisol and prolactin (GARCIA-BORREGUERO et al. 1995; HOLLANDER et al. 1994). A problem exists in identifying the effects of peripherally administered 5-HT agonists on the secretion of both renin and vasopressin. The secretion of both hormones can be suppressed when blood pressure is increased (SCHOLZ et al. 1993; ANDERSEN et al. 1994). Because of the hypertensive effects of peripherally injected 5-HT$_2$ agonists, neither renin nor vasopressin can be used reliably as indices of CNS 5-HT$_2$ receptor function if the 5-HT$_2$ agonists are injected peripherally (LEVY et al. 1994; RITTENHOUSE et al. 1991). In addition, the hypertensive effects that are mediated by sympathetic activation and α_1-adrenoceptor-induced vasoconstriction can mask the effect of 5-HT-releasing drugs or 5-HT agonists. Prazosin unmasks a vasopressin response to 5-HT agonists (BROWNFIELD et al. 1992).

Evidence suggests that the serotoninergic stimulation of vasopressin secretion is mediated by the renin angiotensin system. Several 5-HT$_2$ antagonists inhibit the effects of 5-HT$_2$ agonists on both plasma vasopressin (BROWNFIELD et al. 1988) and renin (LORENS and VAN DE KAR 1987; RITTENHOUSE et al. 1991, 1994b). Furthermore, the effect of fenfluramine on plasma renin activity can be inhibited by LY 53857 (LORENS and VAN DE KAR 1987), suggesting that centrally released 5-HT activates 5-HT$_2$ receptors to stimulate the release of renin from the kidneys. Treatment with the angiotensin-converting enzyme inhibitor enalapril inhibits the vasopressin response to fenfluramine and the angiotensin II (AT1) antagonist losartan (i.c.v.) inhibits the vasopressin response to i.c.v.-injected 5-HT (SAYDOFF et al. 1996). These observations suggest that brain angiotensin II (AT1) receptor activation mediates the stimulation of vasopressin secretion after activation of brain 5-HT$_2$ receptors.

V. Location of the 5-HT$_2$ Receptors Involved

Little doubt exists that the 5-HT$_2$ receptors involved in hormone secretion are postsynaptic. Depletion of 5-HT with PCPA and destruction of serotoninergic pathways using i.c.v. injections of 5,7-dihydroxytryptamine (5,7-DHT) potentiate ACTH, corticosterone, prolactin and renin responses to 5-HT$_2$ agonists (MELTZER et al. 1976; QUATTRONE et al. 1979; RITTENHOUSE et al. 1994a; VAN DE KAR et al. 1989a). Mechanical lesions in the hypothalamic PVN, using a triangular knife, prevent the effect of DOI on plasma oxytocin, prolactin and corticosterone (BAGDY and MAKARA 1994), and inhibit the corticosterone and prolactin but not oxytocin responses to mCPP (BAGDY and MAKARA 1995). Cell-specific lesions in the PVN, using ibotenic acid, inhibit the ACTH, corticosterone, prolactin and renin responses to RU 24969 and to p-chloroamphetamine (RITTENHOUSE et al. 1992b, 1993, 1994a). Direct injections of RU 24969 into the PVN of conscious rats increases plasma renin concentration (VAN DE KAR et al. 1990). Together, these studies suggest that activation of 5-HT$_2$ receptors in the hypothalamus increases the secretion of these hormones by activating sites in the PVN.

F. Neuroendocrine Effects of 5-HT$_3$ Agonists

A role for brain 5-HT$_3$ receptors in stimulating the secretion of prolactin and ACTH/corticosterone has been suggested by several studies (JORGENSEN et al. 1992; SAPHIER et al. 1995; SAPHIER and WELCH 1994). While it is clear that activation of 5-HT$_3$ receptors increases the release of these hormones, it is not clear that these receptors reside inside the brain. Studies suggesting a central role in the secretion of corticosterone, although very well designed, did not present ACTH data, only plasma corticosterone data. Therefore, it is not clear whether the increase in plasma corticosterone was only due to increased plasma ACTH or due to additional mechanisms. With respect to ACTH, the evidence suggests that activation of 5-HT$_3$ receptors in the anterior pituitary directly stimulates the release of ACTH in vitro (CALOGERO et al. 1995). Systemic administration of 5-HT, DOM or p-chloroamphetamine increases prolactin secretion, which is attenuated by pretreatment with the 5-HT$_3$ antagonist ondansetron, ICS 205–930, or MDL 7222 (AULAKH et al. 1994; JORGENSEN et al. 1992; LEVY et al. 1993). The studies suggesting a role for 5-HT$_3$ receptors in the secretion of prolactin utilized peripheral administration of agonists and antagonists, including i.v. injection of 5-HT (JORGENSEN et al. 1992). Since 5-HT does not readily cross the blood-brain barrier, it is not clear that these effects are due to mechanisms inside the brain. Indeed, when 5-HT (30 μg/kg) or the putative 5-HT$_3$ agonist 2-methylserotonin (1, 20 or 200 μg/kg) were injected i.c.v. into conscious rats, they increased the release of ACTH, prolactin and renin (LEVY et al. 1993) and of oxytocin and vasopressin (Van de Kar et al., unpublished observations). However, peripheral administration of the 5-HT$_3$ antagonist ondansetron (1 mg/kg) did not inhibit any of the hormone responses to i.c.v.-administered 5-HT or 2-methylserotonin (LEVY et al. 1993). Consequently, it appears that the 5-HT$_3$ receptors involved in stimulating the secretion of ACTH and prolactin are located outside the brain and that the i.c.v.-injected 5-HT and 2-methylserotonin activate other 5-HT-receptor subtypes in the brain to stimulate the secretion of hormones. It is worth noting that while 2-methylserotonin is often mentioned as a 5-HT$_3$ agonist; it has a higher affinity for 5-HT$_{2C}$ receptors (VAN WIJNGAARDEN et al. 1990).

G. Neuroendocrine Effects of 5-HT$_4$ Agonists

No central effect of 5-HT$_4$ receptors has been demonstrated so far. However, in vitro studies with human adrenocortical cells have demonstrated that drugs with high affinity for 5-HT$_4$ receptors (zacopride, cisapride and BRL 24924) increase the release of both aldosterone and cortisol (Idres et al. 1991; Lefebvre et al. 1993). Administration of zacopride to humans (400 μg orally) increases plasma aldosterone without altering plasma ACTH, cortisol or renin (Lefebvre et al. 1993). Furthermore, metoclopramide might also increase aldosterone secretion by activating 5-HT$_4$ receptors on adrenal cortical cells (Rizzi 1994). Together, the data so far suggest only a role for peripheral 5-HT$_4$ receptors in altering the secretion of hormones.

I. Hormones as Markers of Serotoninergic Function ("Neuroendocrine Challenge Tests")

Hormone responses to challenges with specific 5-HT agonists can be used as peripheral indicators of central serotoninergic receptor function. It is important to measure several hormones concomitantly, instead of only one hormone, because each hormone is regulated by several neurotransmitters (see Table 1). If only one hormone were measured after a challenge with a 5-HT agonist, the side effects of such an agonist could make the interpretation of the results difficult and possibly lead to wrong conclusions. mCPP and MK-212 are used in challenge tests to examine the functional state of hypothalamic 5-HT$_2$ receptors in humans. Both drugs increase ACTH/cortisol and prolactin secretion and their effects are believed to be mediated by activation of 5-HT$_{2C}$ receptors (Murphy et al. 1991; Roth et al. 1992; Zifa and Fillion 1992; Mueller et al. 1986; Lowy and Meltzer 1988; Seibyl et al. 1991). Patients with seasonal affective disorder (SAD) show a higher prolactin and cortisol response to mCPP (0.1 mg/kg i.v.) when compared with control subjects, suggesting sensitization of 5-HT$_{2C}$ receptors (Garcia-Borreguero et al. 1995). The effects of MK-212 on plasma prolactin and cortisol are blunted in patients with obsessive compulsive disorder (OCD), suggesting desensitization of 5-HT$_{2C}$ receptors (Bastani et al. 1990). These 5-HT agonists, though not selective, can provide useful information on the status of hypothalamic 5-HT$_{2C}$ receptors. The most selective 5-HT$_{2A/2C}$ agonists, DOI and DOB, are not approved for human use.

Most human neuroendocrine challenge tests with 5-HT$_{1A}$ agonists involve buspirone (which also is a dopamine D$_2$ antagonist), ipsapirone or gepirone. All increase plasma ACTH and cortisol (Gelfin et al. 1995; Cowen et al. 1990, 1994; Lesch et al. 1990a,b, 1991; Lesch 1991). As mentioned above, there are problems with the measurement of plasma prolactin in such tests. The effects of buspirone are likely mediated by antagonism of D$_2$ receptors on lactotrophs in the pituitary, and ipsapirone did not elevate plasma prolactin in humans (Lesch et al. 1989; Cowen et al. 1990; Maskall et al. 1995; Meltzer and Maes 1994; Anderson and Cowen 1992; Levy and Van de Kar 1992; Nash and Meltzer 1989). The ACTH response to ipsapirone is blunted in depressed patients, suggesting that 5-HT$_{1A}$ receptors are desensitized (Lesch 1991; Lesch et al. 1990a).

Plasma oxytocin has not been measured yet in humans after a challenge with 5-HT agonists, but it is likely to be a more sensitive marker of 5-HT receptor function than plasma ACTH/cortisol, prolactin and renin. The cells that synthesize oxytocin are located in the hypothalamus while their axons terminate on blood vessels in the hypophysial lobe outside the brain. Unlike ACTH, there is no receptor reserve for the oxytocin response to 5-HT$_{1A}$ agonists (PINTO et al. 1994; MELLER and BOHMAKER 1994). Oxytocin is less sensitive than ACTH to stressors and would not be as easily triggered by the phenomenon of "white coat" stress of meeting physicians. No hormonal feedback is known to exist as there is for cortisol affecting the secretion of ACTH. Also, no evidence exists, in depressed patients, for elevations in basal plasma oxytocin as there is for basal cortisol (THAKORE and DINAN 1995). Serotoninergic neurons make direct synaptic contacts with oxytocin cells in the PVN (KAWANO et al. 1992; SAPHIER 1991), and activation of both 5-HT$_{1A}$ and 5-HT$_2$ receptors increases plasma oxytocin. Thus, oxytocin has a strong potential as a marker of the functional state of specific 5-HT receptors.

H. Effects of 5-HT Uptake Blockers on 5-HT$_{1A}$ and 5-HT$_2$-Receptor-Mediated Hormone Secretion

A delay of about 2 weeks exists between the onset of antidepressant therapy and the first appearance of clinical improvement, suggesting that adaptive processes need to run their course before improvement can be observed (BRILEY and MORET 1993; RICHELSON 1991). One study observed that OCD patients who previously had a normal ACTH response to ipsapirone (0.3 mg/kg p.o.) had a reduced ACTH response to ipsapirone after chronic administration of fluoxetine (LESCH 1991). Since previous studies confirmed that ipsapirone increases the secretion of ACTH in humans by activating 5-HT$_{1A}$ receptors (COWEN et al. 1990; LESCH et al. 1990b), these observations suggest that chronic exposure of humans to fluoxetine produces subsensitive 5-HT$_{1A}$ receptors.

Our studies in rats confirm the observations by LESCH et al. (1990b) in humans and indicate that daily injections of fluoxetine for 21 days reduce the ACTH and corticosterone and, in addition, the oxytocin responses to 8-OH-DPAT and ipsapirone (LI et al. 1993b, 1994). A single injection of fluoxetine did not have an effect on the hormone responses to 8-OH-DPAT (LI et al. 1993b, 1995). It is not likely that the effects of fluoxetine were directly exerted at the level of the cells that produce and release ACTH or oxytocin because the same exposure to fluoxetine potentiated the response of both ACTH and oxytocin to a challenge with the 5-HT$_2$ agonist DOI (LI et al. 1993a). This potentiated neuroendocrine response to DOI was associated with increased density of agonist-state (^{125}I-DOI binding) of 5-HT$_2$ receptors in the hypothalamus (LI et al. 1993a). To determine whether the reductions in the sensitivity of 5-HT$_{1A}$ receptors were an adaptive change, in a recent study by Qian

Li, daily administration of fluoxetine (3, 7, 14 or 22 days) and paroxetine (1, 3, 7 or 14 days) was followed by a challenge with the 5-HT$_{1A}$ agonist 8-OH-DPAT, 18 h after the last injection. Both fluoxetine and paroxetine reduced the oxytocin, ACTH and corticosterone responses to 8-OH-DPAT in a gradual manner. The first significant reduction in hormone responses was observed during the 3rd day, and this became maximal after 14 daily injections of both fluoxetine and paroxetine (Li et al. 1995). These reductions in neuroendocrine responses were accompanied by a parallel reduction in the hypothalamic levels of Giα_1 and Giα_3 proteins and in the concentration of Goα and Giα_2 proteins in the midbrain (all of which are hypothesized to couple 5-HT$_{1A}$ receptors to their second messenger systems) (Li et al. 1995). The correlation, in rats, of biochemical changes in the hypothalamus with neuroendocrine changes may lead to more accurate interpretation of neuroendocrine challenge tests in humans.

I. Concluding Remarks

This chapter has reviewed evidence in support of specific and independent roles of 5-HT$_{1A}$ and 5-HT$_2$ receptors in stimulating the secretion of specific

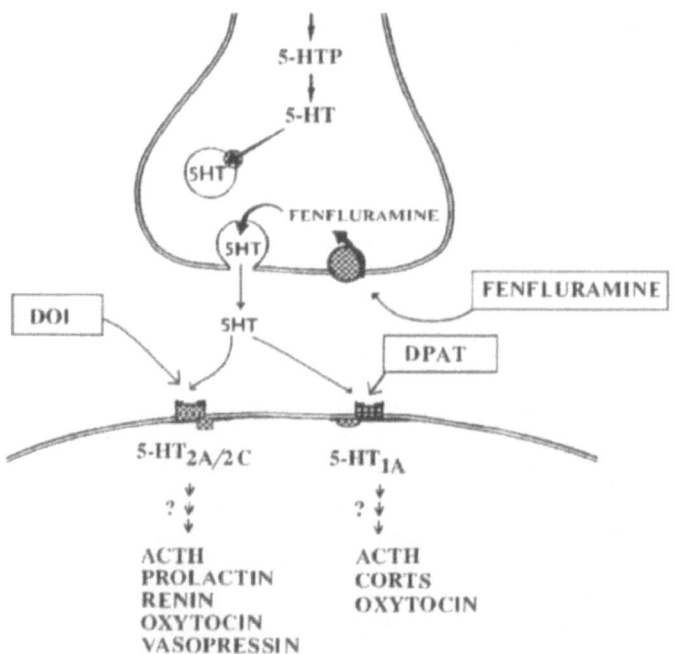

Fig. 2. Schematic illustration of the serotoninergic synapse and hormones released in response to activation of 5-HT$_{1A}$ and 5-HT$_2$ receptors. Note that prolactin has been omitted from the list of hormones released by activation of 5-HT$_{1A}$ receptors until more definitive evidence supports this conclusion. *DPAT*, 8-OH-DPAT

hormones as summarized in Table 1 and Fig. 2. Activation of 5-HT_{1A} receptors increases the secretion of ACTH, corticosterone/cortisol and oxytocin while activation of 5-HT_2 receptors also increases the secretion of these hormones and, in addition, increases the secretion of prolactin, vasopressin and renin. Although there is little doubt that these 5-HT receptors are located in the brain, their exact location (i.e., hypothalamus or elsewhere) remains to be determined. These latter determinations may lead to neuroendocrine challenge tests in which the hormone responses to specific 5-HT agonists will provide specific information about the functional status of specific 5-HT receptors and their signal transduction systems in the brain.

Acknowledgements. The studies reported in this chapter were supported in part by United States Public Health Service Grants MH45812, NS34153 and DA04865. The authors thank Dr. Lanny C. Keil (Sunnyvale, CA) for the generous supply of the oxytocin antiserum.

References

Albinsson A, Palazidou E, Stephenson J, Andersson G (1994) Involvement of the 5-HT_2 receptor in the 5-HT receptor-mediated stimulation of prolactin release. Eur J Pharmacol 251:157–161

Alper RH (1990) Hemodynamic and renin response to (+)-DOI, a selective 5-HT_2 receptor agonist, in conscious rats. Eur J Pharmacol 175:323–332

Alper RH, Snider JM (1987) Activation of serotonin$_2$ (5-HT_2) receptors by quipazine increases arterial pressure and renin secretion in conscious rats. J Pharmacol Exp Ther 243:829–833

Andersen JL, Andersen LJ, Thrasher TN, Keil LC, Ramsay DJ (1994) Left heart and arterial baroreceptors interact in control of plasma vasopressin, renin, and cortisol in awake dogs. Am J Physiol Regul Integr Comp Physiol 266:R879–R888

Anderson IM, Cowen PJ (1992) Effect of pindolol on endocrine and temperature responses to buspirone in healthy volunteers. Psychopharmacology 106:428–432

Arey BJ, Freeman ME (1992) Activity of oxytocinergic neurons in the paraventricular nucleus mirrors the periodicity of the endogenous stimulatory rhythm regulating prolactin secretion. Endocrinology 130:126–132

Aulakh CS, Hill JL, Murphy DL (1992) Effects of various serotonin receptor subtype-selective antagonists alone and on m-chlorophenylpiperazine-induced neuroendocrine changes in rats. J Pharmacol Exp Ther 263:588–595

Aulakh CS, Mazzola-Pomietto P, Hill JL, Murphy DL (1994) Role of various 5-HT receptor subtypes in mediating neuroendocrine effects of 1-(2,5-dimethoxy-4-methylphenyl)-2-aminopropane (DOM) in rats. J Pharmacol Exp Ther 271:143–148

Azmitia EC, Segal M (1978) An autoradiographic analysis of the differential ascending projections of the dorsal and median raphe nuclei in the rat. J Comp Neurol 179:641–668

Bagdy G, Szemeredi K, Kanyicska B, Murphy DL (1989) Different serotonin receptors mediate blood pressure, heart rate, plasma catecholamine and prolactin responses to m-chlorophenylpiperazine in conscious rats. J Pharmacol Exp Ther 250:72–78

Bagdy G, Calogero AE, Szemeredi K, Gomez MT, Murphy DL, Chrousos GP, Gold PW (1990) β-Endorphin responses to different serotonin agonists: involvement of corticotropin-releasing hormone, vasopressin and direct pituitary action. Brain Res 537:227–232

Bagdy G, Kalogeras KT, Szemeredi K (1992a) Effect of 5-HT$_{1C}$ and 5-HT$_2$ receptor stimulation on excessive grooming, penile erection and plasma oxytocin concentrations. Eur J Pharmacol 229:9–14

Bagdy G, Sved AF, Murphy DL, Szemeredi K (1992b) Pharmacological characterization of serotonin receptor subtypes involved in vasopressin and plasma renin activity responses to serotonin agonists. Eur J Pharmacol 210:285–289

Bagdy G, Szemeredi K, Listwak SJ, Keiser HR, Goldstein DS (1993) Plasma catecholamine, renin activity, and ACTH responses to the serotonin receptor agonist DOI in juvenile spontaneously hypertensive rats. Life Sci 53:1573–1582

Bagdy G, Kalogeras KT (1993) Stimulation of 5-HT$_{1A}$ and 5-HT$_2$/5-HT$_{1C}$ receptors induce oxytocin release in the male rat. Brain Res 611:330–332

Bagdy G, Makara GB (1994) Hypothalamic paraventricular nucleus lesions differentially affect serotonin-1A (5-HT$_{1A}$) and 5-HT$_2$ receptor agonist-induced oxytocin, prolactin, and corticosterone responses. Endocrinology 134:1127–1131

Bagdy G, Makara GB (1995) Paraventricular nucleus controls 5-HT$_{2C}$ receptor-mediated corticosterone and prolactin but not oxytocin and penile erection responses. Eur J Pharmacol 275:301–305

Barofsky AL, Taylor J, Massari VJ (1983) Dorsal raphe-hypothalamic projections provide the stimulatory serotoninergic input to suckling-induced prolactin release. Endocrinology 113:1894–1903

Barrett JE, Vanover KE (1993) 5-HT receptors as targets for the development of novel anxiolytic drugs: models, mechanisms and future directions. Psychopharmacology (Berl) 112:1–12

Bastani B, Nash JF, Meltzer HY (1990) Prolactin and cortisol responses to MK-212, a serotonin agonist, in obsessive-compulsive disorder. Arch Gen Psychiatry 47:833–839

Blair ML, Feigl EO, Smith OA (1976) Elevation of plasma renin activity during avoidance performance in baboons. Am J Physiol 231:772–776

Bluet Pajot MT, Mounier F, Di Sciullo A, Schmidt B, Kordon C (1995) Differential sites of action of 8-OH-DPAT, a 5-HT$_{1A}$ agonist, on ACTH and PRL secretion in the rat. Neuroendocrinology 61:159–166

Boess FG, Martin IL (1994) Molecular biology of 5-HT receptors. Neuropharmacology 33:275–317

Briley M, Moret C (1993) Neurobiological mechanisms involved in antidepressant therapies. Clin Neuropharmacol 16:387–400

Brownfield MS, Greathouse J, Lorens SA, Armstrong J, Urban JH, Van de Kar LD (1988) Neuropharmacological characterization of serotoninergic stimulation of vasopressin secretion in conscious rats. Neuroendocrinology 47:277–283

Brownfield MS, Armstrong J, Rittenhouse PA, Li Q, Levy AD, Van de Kar LD (1992) Pharmacological differentiation of serotonergic (5-HT) stimulation of oxytocin and vasopressin secretion in the conscious rat. Neurosci Abstr 18:823 (#346.7)

Callahan MF, Thore CR, Sundberg DK, Gruber KA, O'Steen K, Morris M (1992) Excitotoxin paraventricular nucleus lesions: stress and endocrine reactivity and oxytocin mRNA levels. Brain Res 597:8–15

Calogero AE, Bernardini R, Margioris AN, Bagdy G, Gallucci WT, Tamarkin L, Tomai TP (1989) Effect of serotonergic agonists and antagonists on corticotropin-releasing hormone secretion by explanted rat hypothalami. Peptides 10:189–200

Calogero AE, Bagdy G, Burrello N, Polosa P, D'Agata R (1995) Role for serotonin$_3$ receptors in the control of adrenocorticotropic hormone release from rat pituitary cell cultures. Eur J Endocrinol 133:251–254

Chaouloff F (1993) Physiopharmacological interactions between stress hormones and central serotonergic systems. Brain Res Rev 18:1–32

Cowen PJ, Anderson IM, Grahame-Smith DG (1990) Neuroendocrine effects of azapirones. J Clin Psychopharmacol 10 [Suppl]:21S–25S

Cowen PJ, Power AC, Ware CJ, Anderson IM (1994) 5-HT$_{1A}$ receptor sensitivity in major depression. A neuroendocrine study with buspirone. Br J Psychiatry 164:372–379

Critchley DJP, Childs KJ, Middlefell VC, Dourish CT (1994) Inhibition of 8-OH-DPAT-induced elevation of plasma corticotrophin by the 5-HT$_{1A}$ receptor antagonist WAY100635. Eur J Pharmacol 264:95–97

Crowley WR, Armstrong WE (1992) Neurochemical regulation of oxytocin secretion in lactation. Endocr Rev 13:33–65

Di Sciullo A, Bluet Pajot MT, Mounier F, Oliver C, Schmidt B, Kordon C (1990) Changes in anterior pituitary hormone levels after serotonin 1A receptor stimulation. Endocrinology 127:567–572

Dourish CT, Hutson PH, Curzon G (1986) Putative anxiolytics 8-OH-DPAT, buspirone and TVX Q 7821 are agonists at 5-HT$_{1A}$ autoreceptors in the raphe nuclei. Trends Pharmacol Sci 7:212–214

Eison MS (1989) The new generation of serotonergic anxiolytics: possible clinical roles. Psychopathology 22 [Suppl 1]:13–20

Entwisle SJ, Fowler PA, Thomas M, Eckland DJA, Lettis S, York M, Freedman PS (1995) The effects of oral sumatriptan, a 5-HT$_1$ receptor agonist, on circulating ACTH and cortisol concentrations in man. Br J Clin Pharmacol 39:389–395

Fekete MIK, Szentendrei T, Kanyicska B, Palkovits M (1981) Effects of anxiolytic drugs on the catecholamine and dopac (3, 4-dihydroxyphenylacetic acid) levels in brain cortical areas and on corticosterone and prolactin secretion in rats subjected to stress. Psychoneuroendocrinology 6:113–120

Feldman S, Weidenfeld J (1995) Posterior hypothalamic deafferentiation or 5,7-dihydroxytryptamine inhibit corticotropin-releasing hormone, ACTH and corticosterone responses following photic stimulation. Neurosci Lett 198:143–145

Feldman S, Melamed E, Conforti N, Weidenfeld J (1984) Effect of central serotonin depletion on adrenocortical responses to neural stimuli. Exp Neurol 85:661–666

Feldman S, Conforti N, Melamed E (1987) Paraventricular nucleus serotonin mediates neurally stimulated adrenocortical secretion. Brain Res Bull 18:165–168

Feldman S, Weidenfeld J, Conforti N, Saphier D (1991) Differential recovery of adrenocortical responses to neural stimuli following administration of 5,7-dihydroxytryptamine into the hypothalamus. Exp Brain Res 85:144–148

Feldman S, Conforti N, Weidenfeld J (1995) Limbic pathways and hypothalamic neurotransmitters mediating adrenocortical responses to neural stimuli. Neurosci Biobehav Rev 19:235–240

Franceschini R, Cataldi A, Garibaldi A, Cianciosi P, Scordamaglia A, Barreca T, Rolandi E (1994) The effects of sumatriptan on pituitary secretion in man. Neuropharmacology 33:235–239

Fuller RW, Snoddy HD (1990) Serotonin receptor subtypes involved in the elevation of serum corticosterone concentration in rats by direct- and indirect-acting serotonin agonists. Neuroendocrinology 52:206–211

Garcia-Borreguero D, Jacobsen FM, Murphy DL, Joseph-Vanderpool JR, Chiara A, Rosenthal NE (1995) Hormonal responses to the administration of m-chlorophenylpiperazine in patients with seasonal affective disorder and controls. Biol Psychiatry 37:740–749

Gartside SE, Ellis PM, Sharp T, Cowen PJ (1992) Selective 5-HT$_{1A}$ and 5-HT$_2$ receptor-mediated adrenocorticotropin release in the rat: effect of repeated antidepressant treatments. Eur J Pharmacol 221:27–33

Gelfin Y, Lerer B, Lesch KP, Gorfine M, Allolio B (1995) Complex effects of age and gender on hypothermic, adrenocorticotrophic hormone and cortisol responses to ipsapirone challenge in normal subjects. Psychopharmacology (Berl) 120:356–364

Gram LF, Christensen P (1986) Benzodiazepine suppression of cortisol secretion: a measure of anxiolytic activity. Pharmacopsychiatry 19:19–22

Groenink L, Van der Gugten J, Verdouw PM, Maes RAA, Olivier B (1995) The anxiolytic effects of flesinoxan, a 5-HT$_{1A}$ receptor agonist, are not related to its neuroendocrine effects. Eur J Pharmacol 280:185–193

Halbreich U, Rojansky N, Palter S, Tworek H, Hissin P, Wang K (1995) Estrogen augments serotonergic activity in postmenopausal women. Biol Psychiatry 37:434–441

Hollander E, DeCaria CM, Nitescu A, Gully R, Suckow RF, Cooper TB, Gorman JM, Klein DF, Liebowitz MR (1992) Serotonergic function in obsessive-compulsive disorder: behavioral and neuroendocrine responses to oral m-chlorophenylpiperazine and fenfluramine in patients and healthy volunteers. Arch Gen Psychiatry 49:21–28

Hollander E, Stein DJ, DeCaria CM, Cohen L, Saoud JB, Skodol AE, Kellman D, Rosnick L, Oldham JM (1994) Serotonergic sensitivity in borderline personality disorder: preliminary findings. Am J Psychiatry 151:277–280

Hoyer D (1988) Functional correlates of serotonin 5-HT$_1$ recognition sites. J Recept Res 8:59–81

Idres S, Delarue C, Lefebvre H, Vaudry H (1991) Benzamide derivatives provide evidence for the involvement of a 5-HT$_4$ receptor type in the mechanism of action of serotonin in frog adrenocortical cells. Mol Brain Res 10:251–258

Jorgensen H, Knigge U, Warberg J (1992) Involvement of 5-HT$_1$, 5-HT$_2$, and 5-HT$_3$ receptors in the mediation of the prolactin responses to serotonin and 5-hydroxytryptophan. Neuroendocrinology 55:336–343

Kahn RS, Kling MA, Wetzler S, Asnis GM, Van Praag H (1992) Effect of m-chlorophenylpiperazine on plasma arginine-vasopressin concentrations in healthy subjects. Psychopharmacology (Berl) 108:225–228

Kahn RS, Davidson M, Siever LJ, Sevy S, Davis KL (1994) Clozapine treatment and its effect on neuroendocrine responses induced by the serotonin agonist, m-chlorophenylpiperazine. Biol Psychiatry 35:909–912

Kawano S, Osaka T, Kannan H, Yamashita H (1992) Excitation of hypothalamic paraventricular neurons by stimulation of the raphe nuclei. Brain Res Bull 28:573–579

Kelder D, Ross SB (1992) Long lasting attenuation of 8-OH-DPAT-induced corticosterone secretion after a single injection of a 5-HT$_{1A}$ receptor agonist. Naunyn Schmiedebergs Arch Pharmacol 346:121–126

Kellar KJ, Hulihan-Giblin BA, Mulroney SE, Lumpkin MD, Flores CM (1992) Stimulation of serotonin$_{1A}$ receptors increases release of prolactin in the rat. Neuropharmacology 31:643–647

King BH, Brazell C, Dourish CT, Middlemiss DN (1989) MK-212 increases rat plasma ACTH concentration by activation of the 5-HT$_{1C}$ receptor subtype. Neurosci Lett 105:174–176

Kiss JZ, Kanycska B, Nagy GY (1986) Hypothalamic paraventricular nucleus has a pivotal role in regulation of prolactin release in lactating rats. Endocrinology 119:870–873

Koenig JI, Gudelsky GA, Meltzer HY (1987) Stimulation of corticosterone and beta-endorphin secretion in the rat by selective 5-HT receptor subtype activation. Eur J Pharmacol 137:1–8

Koenig JI, Meltzer HY, Gudelsky GA (1988) 5-Hydroxytryptamine$_{1A}$ receptor-mediated effects of buspirone, gepirone and ipsapirone. Pharmacol Biochem Behav 29:711–715

Krystal JH, Seibyl JP, Price LH, Woods SW, Heninger GR, Aghajanian GK, Charney DS (1993) m-Chlorophenylpiperazine effects in neuroleptic-free schizophrenic patients: evidence implicating serotonergic systems in the positive symptoms of schizophrenia. Arch Gen Psychiatry 50:624–635

Lang RE, Heil JWE, Ganten D, Hermann K, Unger T, Rascher W (1983) Oxytocin unlike vasopressin is a stress hormone in the rat. Neuroendocrinology 37:314–316

Lefebvre H, Contesse V, Delarue C, Soubrane C, Legrand A, Kuhn J-M, Wolf L-M, Vaudry H (1993) Effect of the serotonin-4 receptor agonist zacopride on aldosterone secretion from the human adrenal cortex: in vivo and in vitro studies. J Clin Endocrinol Metab 77:1662–1666

Le Fur G, Guilloux F, Mitrani N, Miazoule J, Uzan A (1979) Relationship between plasma corticosteroids and benzodiazepines in stress. J Pharmacol Exp Ther 211:305–308

Lejeune F, Rivet J-M, Gobert A, Canton H, Millan MJ (1993) WAY 100,135 and (−)-tertatolol act as antagonists at both 5-HT$_{1A}$ autoreceptors and postsynaptic 5-HT$_{1A}$ receptors in vivo. Eur J Pharmacol 240:307–310

Lesch KP (1991) 5-HT$_{1A}$ receptor responsivity in anxiety disorders and depression. Prog Neuropsychopharmacol Biol Psychiatry 15:723–733

Lesch KP, Rupprecht R, Poten B, Muller U, Sohnle K, Fritze J (1989) Endocrine responses to 5-hydroxytryptamine-1A receptor activation by ipsapirone in humans. Biol Psychiatry 26:203–205

Lesch KP, Mayer M, Disselkamp-Tietze J, Hoh A, Wiesmann M, Osterheider M, Schulte HM (1990a) 5-HT$_{1A}$ receptor responsivity in unipolar depression. Evaluation of ipsapirone-induced ACTH and cortisol secretion in patients and controls. Biol Psychiatry 28:620–628

Lesch KP, Sohnle K, Poten B, Schoellnhammer G, Rupprecht R, Schulte HM (1990b) Corticotropin and cortisol secretion after central 5-hydroxytryptamine-1A (5-HT$_{1A}$) receptor activation: effects of 5-HT receptor and β-adrenoceptor antagonists. J Clin Endocrinol Metab 70:670–674

Lesch KP, Schulte HM, Osterheider M, Muller T (1991) Long-term fluoxetine treatment decreases 5-HT$_{1A}$ receptor responsivity in obsessive-compulsive disorder. Psychopharmacology 105:415–420

Levy AD, Van de Kar LD (1992) Endocrine and receptor pharmacology of serotonergic anxiolytics, antipsychotics and antidepressants. Life Sci 51:83–94

Levy AD, Li Q, Alvarez Sanz MC, Rittenhouse PA, Brownfield MS, Van de Kar LD (1992) Repeated cocaine modifies the neuroendocrine responses to the 5-HT$_{1C}$/5-HT$_2$ receptor agonist DOI. Eur J Pharmacol 221:121–127

Levy AD, Li Q, Rittenhouse PA, Van de Kar LD (1993) Investigation of the role of 5-HT$_3$ receptors in the secretion of prolactin, ACTH and renin. Neuroendocrinology 58:65–70

Levy AD, Baumann MH, Van de Kar LD (1994) Monoaminergic regulation of neuroendocrine function and its modification by cocaine. Front Neuroendocrinol 15:1–72

Levy AD, Li Q, Gustafson M, Van de Kar LD (1995) Neuroendocrine profile of the potential anxiolytic drug S-20499. Eur J Pharmacol 274:141–149

Li Q, Rittenhouse PA, Levy AD, Alvarez Sanz MC, Van de Kar LD (1992) Neuroendocrine responses to the serotonin$_2$ agonist DOI are differentially modified by three 5-HT$_{1A}$ agonists. Neuropharmacology 31:983–989

Li Q, Brownfield MS, Battaglia G, Cabrera TM, Levy AD, Rittenhouse PA, Van de Kar LD (1993a) Long-term treatment with the antidepressants fluoxetine and desipramine potentiates endocrine responses to the serotonin agonists 6-chloro-2-(1-piperazinyl)-pyrazine (MK-212) and (±)-1-(2,5-dimethoxy-4-iodophenyl)-2-aminopropane HCl (DOI). J Pharmacol Exp Ther 266:836–844

Li Q, Levy AD, Cabrera TM, Brownfield MS, Battaglia G, Van de Kar LD (1993b) Long-term fluoxetine, but not desipramine, inhibits the ACTH and oxytocin responses to the 5-HT$_{1A}$ agonist 8-OH-DPAT in male rats. Brain Res 630:148–156

Li Q, Brownfield MS, Levy AD, Battaglia G, Cabrera TM, Van de Kar LD (1994) Attenuation of hormone responses to the 5-HT$_{1A}$ agonist ipsapirone by long-term treatment with fluoxetine, but not desipramine, in male rats. Biol Psychiatry 36:300–308

Li Q, Battaglia G, Pinto W, Vicentic A, Chambers CB, Van de Kar LD (1995) Repeated injections of fluoxetine produce a delayed and gradual reduction in the

ACTH and oxytocin responses to the 5-HT$_{1A}$ agonist 8-OH-DPAT. Soc Neurosci Abstr 21:343.13

Liposits Z, Phelix C, Paull WK (1987) Synaptic interaction of serotonergic axons and corticotropin releasing factor (CRF) synthesizing neurons in the hypothalamic paraventricular nucleus of the rat. A light and electron microscopic immunocytochemical study. Histochemistry 86:541–549

Lorens SA, Van de Kar LD (1987) Differential effects of serotonin (5-HT$_{1A}$ and 5-HT$_2$) agonists and antagonists on renin and corticosterone secretion. Neuroendocrinology 45:305–310

Lowy MT, Meltzer HY (1988) Stimulation of serum cortisol and prolactin secretion in humans by MK-212, a centrally active serotonin agonist. Biol Psychiatry 23:818–828

Maskall DD, Zis AP, Lam RW, Clark CM, Kuan AJ (1995) Prolactin response to buspirone challenge in the presence of dopaminergic blockade. Biol Psychiatry 38:235–239

Matheson GK, Gage-White D, White G, Guthrie D, Rhoades J, Dixon V (1989) The effects of gepirone and 1-(2-pyrimidinyl)-piperazine on levels of corticosterone in rat plasma. Neuropharmacology 28:329–334

Matheson GK, Pfeifer DM, Weiberg MB, Michel C (1994) The effects of azapirones on serotonin$_{1A}$ neurons of the dorsal raphe. Gen Pharmacol 25:675–683

Meller E, Bohmaker K (1994) Differential receptor reserve for 5-HT$_{1A}$ receptor-mediated regulation of plasma neuroendocrine hormones. J Pharmacol Exp Ther 271:1246–1252

Meltzer HY, Maes M (1994) Effects of buspirone on plasma prolactin and cortisol levels in major depressed and normal subjects. Biol Psychiatry 35:316–323

Meltzer HY, Maes M (1995) Effect of pindolol pretreatment on MK-212-induced plasma cortisol and prolactin responses in normal men. Biol Psychiatry 38:310–318

Meltzer HY, Fang VS, Paul SM, Kaluskar R (1976) Effect of quipazine on rat plasma prolactin levels. Life Sci 19:1073–1078

Minamitani N, Minamitani T, Lechan RM, Bollinger-Gruber J (1987) Paraventricular nucleus mediates prolactin secretory responses to restraint stress, ether stress, and 5-hydroxy-l-tryptophan injection in the rat. Endocrinology 120:860–867

Mueller EA, Murphy DL, Sunderland T (1985) Neuroendocrine effects of m-chlorophenylpiperazine, a serotonin agonist, in humans. J Clin Endocrinol Metab 61:1179–1184

Mueller EA, Murphy DL, Sunderland T (1986) Further studies of the putative serotonin agonist, m-chlorophenylpiperazine: evidence for a serotonin receptor mediated mechanism of action in humans. Psychopharmacology 89:388–391

Murphy DL, Lesch KP, Aulakh CS, Pigott TA (1991) Serotonin-selective arylpiperazines with neuroendocrine, behavioral, temperature, and cardiovascular effects in humans. Pharmacol Rev 43:527–552

Nash JF, Meltzer HY (1989) Effect of gepirone and ipsapirone on the stimulated and unstimulated secretion of prolactin in the rat. J Pharmacol Exp Ther 249:236–241

Nishikawa T, Scatton B (1986) Neuroanatomical site of the inhibitory influence of anxiolytic drugs on central serotonergic transmission. Brain Res 271:123–132

Owens MJ, Knight DL, Ritchie JC, Nemeroff CB (1991) The 5-hydroxytryptamine$_2$ agonist, (±)-1-(2,5-dimethoxy-4-bromophenyl)-2-aminopropane stimulates the hypothalamic-pituitary-adrenal (HPA) axis. II. Biochemical and physiological evidence for the development of tolerance after chronic administration. J Pharmacol Exp Ther 256:795–800

Pan J-T, Tai M-H (1992) Effects of ketanserin on DOI-, MCPP- and TRH-induced prolactin secretion in estrogen-treated rats. Life Sci 51:839–845

Pan L, Gilbert F (1992) Activation of 5-HT$_{1A}$ receptor subtype in the paraventricular nuclei of the hypothalamus induces CRH ACTH release in the rat. Neuroendocrinology 56:797–802

Paris JM, Lorens SA, Van de Kar LD, Urban JH, Richardson-Morton KD, Bethea CL (1987) A comparison of acute stress paradigms: hormonal responses and hypothalamic serotonin. Physiol Behav 39:33–43

Pergola PE, Sved AF, Voogt JL, Alper RH (1993) Effect of serotonin on vasopressin release: a comparison to corticosterone, prolactin and renin. Neuroendocrinology 57:550–558

Petrov T, Krukoff TL, Jhamandas JH (1994) Chemically defined collateral projections from the pons to the central nucleus of the amygdala and hypothalamic paraventricular nucleus in the rat. Cell Tissue Res 277:289–295

Pinto W, Cabrera TM, Li Q, Van de Kar LD, Battaglia G (1994) Receptor reserve for 5-HT$_{1A}$ receptors in rat brain as assessed by neuroendocrine responses to 8-OH-DPAT stimulation. Soc Neurosci Abstr 19:1539 (no 632.4)

Przegalinski E, Ismaiel AM, Chojnacka-Wojcik E, Budziszewska B, Tatarczynska E, Blaszczynska E (1990) The behavioural, but not the hypothermic or corticosterone, response to 8-hydroxy-2-(di-n-propylamino)-tetralin, is antagonized by NAN-190 in the rat. Neuropharmacology 29:521–526

Quattrone A, Schettini G, Di Renzo G, Tedeschi G, Preziosi P (1979) Effect of midbrain raphe lesion or 5,7-dihydroxytryptamine treatment on the prolactin-releasing action of quipazine and D-fenfluramine in rats. Brain Res 174:71–79

Quattrone A, Schettini G, Annunziato L, Di Renzo G (1981) Pharmacological evidence of supersensitivity of central serotonergic receptors involved in the control of prolactin secretion. Eur J Pharmacol 76:9–13

Richardson Morton KD, Van de Kar LD, Lorens SA, Paris JM, Urban JH, Kunimoto K (1986) The effect of stress on renin and corticosterone secretion is blocked by electrolytic lesions in the mesencephalic dorsal raphe and hypothalamic paraventricular nuclei. Soc Neurosci Abstr 12:290.15

Richardson Morton KD, Van de Kar LD, Brownfield MS, Bethea CL (1989) Neuronal cell bodies in the hypothalamic paraventricular nucleus mediate stress-induced renin and corticosterone secretion. Neuroendocrinology 50:73–80

Richelson E (1991) Biological basis of depression and therapeutic relevance. J Clin Psychiatry 52 [6, Suppl]:4–10

Rittenhouse PA, Bakkum EA, Van de Kar LD (1991) Evidence that the serotonin agonist, DOI, increases renin secretion and blood pressure through both central and peripheral 5-HT$_2$ receptors. J Pharmacol Exp Ther 259:58–65

Rittenhouse PA, Bakkum EA, O'Connor PA, Carnes M, Bethea CL, Van de Kar LD (1992a) Comparison of neuroendocrine and behavioral effects of ipsapirone, a 5-HT$_{1A}$ agonist, in three stress paradigms: immobilization, forced swim and conditioned fear. Brain Res 580:205–214

Rittenhouse PA, Li Q, Levy AD, Van de Kar LD (1992b) Neurons in the hypothalamic paraventricular nucleus mediate the serotonergic stimulation of renin secretion. Brain Res 593:105–113

Rittenhouse PA, Levy AD, Li Q, Bethea CL, Van de Kar LD (1993) Neurons in the hypothalamic paraventricular nucleus mediate the serotonergic stimulation of prolactin secretion via 5-HT$_{1C/2}$ receptors. Endocrinology 133:661–667

Rittenhouse PA, Bakkum EA, Levy AD, Li Q, Carnes M, Van de Kar LD (1994a) Evidence that ACTH secretion is regulated by serotonin$_{2A/2C}$ (5-HT$_{2A/2C}$) receptors. J Pharmacol Exp Ther 271:1647–1655

Rittenhouse PA, Bakkum EA, Levy AD, Li Q, Yracheta JM, Kunimoto K, Van de Kar LD (1994b) Central stimulation of renin secretion through serotonergic, noncardiovascular mechanisms. Neuroendocrinology 60:205–214

Rizzi CA (1994) A serotonergic mechanism for the metoclopramide-induced increase in aldosterone level. Eur J Clin Pharmacol 47:377–378

Roth BL, Ciaranello RD, Meltzer HY (1992) Binding of typical and atypical antipsychotic agents to transiently expressed 5-HT$_{1C}$ receptors. J Pharmacol Exp Ther 260:1361–1365

Saphier D (1991) Paraventricular nucleus magnocellular neuronal responses following electrical stimulation of the midbrain dorsal raphe. Exp Brain Res 85:359–363

Saphier D, Welch JE (1994) 5-HT$_3$ receptor activation in the rat increases adrenocortical secretion at the level of the central nervous system. Neurosci Res Commun 14:167–173

Saphier D, Farrar GE, Welch JE (1995) Differential inhibition of stress-induced adrenocortical responses by 5-HT$_{1A}$ agonists and by 5-HT$_2$ and 5-HT$_3$ antagonists. Psychoneuroendocrinology 20:239–257

Sawchenko PE, Swanson LW (1985) Localization, colocalization and plasticity of corticotropin-releasing factor immunoreactivity in rat brain. Fed Proc 44:221–227

Sawchenko PE, Swanson LW, Steinbusch HWM, Verhofstad AAJ (1983) The distribution and cells of origin of serotonergic inputs to the paraventricular and supraoptic nuclei of the rat. Brain Res 277:355–360

Saydoff JA, Rittenhouse PA, Van de Kar LD, Brownfield MS (1991) Enhanced serotonergic transmission stimulates oxytocin secretion in conscious male rats. J Pharmacol Exp Ther 257:95–99

Saydoff JA, Leanza F, Carnes M, Brownfield MS (1992) Central serotonin induces vasopressin and oxytocin secretion and a specific pattern of c-Fos expression in rat brain. Soc Neurosci Abstr 18:823 (no 346.5)

Saydoff JA, Carnes M, Brownfield MS (1993) The role of serotonergic neurons in intravenous hypertonic saline-induced secretion of vasopressin, oxytocin, and ACTH. Brain Res Bull 32:567–572

Saydoff JA, Rittenhouse PA, Carnes M, Armstrong J, Van de Kar LD, Brownfield MS (1996) Neuroendocrine and cardiovascular activation by serotonin: selective role of brain angiotensin in vasopressin secretion. Am J Physiol Regul Integr Comp Physiol 33:E513–E521

Scholz H, Vogel U, Kurtz A (1993) Interrelation between baroreceptor and macula densa mechanisms in the control of renin secretion. J Physiol (Lond) 469:511–524

Seibyl JP, Krystal JH, Price LH, Woods SW, D'Amico C, Heninger GR, Charney DS (1991) Effects of ritanserin on the behavioral, neuroendocrine, and cardiovascular responses to meta-chlorophenylpiperazine in healthy human subjects. Psychiatry Res 38:227–236

Seletti B, Benkelfat C, Blier P, Annable L, Gilbert F, de Montigny C (1995) Serotonin$_{1A}$ receptor activation by flesinoxan in humans. Body temperature and neuroendocrine responses. Neuropsychopharmacology 13:93–104

Simonovic M, Gudelsky GA, Meltzer HY (1984) Effect of 8-hydroxy-2-(di-n-propylamino)tetralin on rat prolactin secretion. J Neural Transm 59:143–149

Steardo L, Iovino M (1986) Vasopressin release after enhanced serotonergic transmission is not due to activation of the peripheral renin-angiotensin system. Brain Res 382:145–148

Swanson LW, Sawchenko PE (1983) Hypothalamic integration: organization of the paraventricular and supraoptic nuclei. Annu Rev Neurosci 6:269–324

Thakore JH, Dinan TG (1995) Cortisol synthesis inhibition: a new treatment strategy for the clinical and endocrine manifestations of depression. Biol Psychiatry 37:364–368

Thiebot MH, Hamon M, Soubrie P (1982) Attenuation of induced-anxiety in rats by chlordiazepoxide: role of raphe dorsalis benzodiazepine binding sites and serotoninergic neurons. Neuroscience 7:2287–2294

Titeler M, Lyon RA, Davis KH, Glennon RA (1987) Selectivity of serotonergic drugs for multiple brain serotonin receptors. Biochem Pharmacol 36:3265–3271

Urban JH, Van de Kar LD, Lorens SA, Bethea CL (1986) Effect of the anxiolytic drug buspirone on prolactin and corticosterone secretion in stressed and unstressed rats. Pharmacol Biochem Behav 25:457–462

Van Bockstaele EJ, Biswas A, Pickel VM (1993) Topography of serotonin neurons in the dorsal raphe nucleus that send axon collaterals to the rat prefrontal cortex and nucleus accumbens. Brain Res 624:188–198

Van de Kar LD (1991) Neuroendocrine pharmacology of serotonergic (5-HT) neurons. Annu Rev Pharmacol Toxicol 31:289–320

Van de Kar LD (1996) Forebrain pathways mediating stress-induced renin secretion. Clin Exp Pharmacol Physiol 23: 166–170

Van de Kar LD, Bethea CL (1982) Pharmacological evidence that serotonergic stimulation of prolactin secretion is mediated via the dorsal raphe nucleus. Neuroendocrinology 35:225–230

Van de Kar LD, Brownfield MS (1993) Serotonergic neurons and neuroendocrine function. NIPS 8:202–207

Van de Kar LD, Lorens SA (1979) Differential serotonergic innervation of individual hypothalamic nuclei and other forebrain regions by the dorsal and median raphe nuclei. Brain Res 162:45–54

Van de Kar LD, Wilkinson CW, Ganong WF (1981) Pharmacological evidence for a role of brain serotonin in the maintenance of plasma renin activity in unanesthetized rats. J Pharmacol Exp Ther 219:85–90

Van de Kar LD, Wilkinson CW, Skrobik Y, Brownfield MS, Ganong WF (1982) Evidence that serotonergic neurons in the dorsal raphe nucleus exert a stimulatory effect on the secretion of renin but not of corticosterone. Brain Res 235:233–243

Van de Kar LD, Lorens SA, McWilliams CR, Kunimoto K, Urban JH, Bethea CL (1984) Role of midbrain raphe in stress-induced renin and prolactin secretion. Brain Res 311:333–341

Van de Kar LD, Lorens SA, Urban JH, Richardson KD, Paris J (1985a) Pharmacological studies on stress-induced renin and prolactin secretion: effects of benzodiazepines, naloxone, propranolol and diisopropylfluorophosphate (DFP). Brain Res 345:257–263

Van de Kar LD, Urban JH, Lorens SA, Richardson KD (1985b) The non-benzodiazepine anxiolytic buspirone inhibits stress-induced renin secretion and lowers heart rate. Life Sci 36:1149–1155

Van de Kar LD, Carnes M, Maslowski RJ, Bonadonna AM, Rittenhouse PA, Kunimoto K, Piechowski RA, Bethea CL (1989a) Neuroendocrine evidence for denervation supersensitivity of serotonin receptors: effects of the 5-HT agonist RU 24969 on corticotropin, corticosterone, prolactin and renin secretion. J Pharmacol Exp Ther 251:428–434

Van de Kar LD, Lorens SA, Urban JH, Bethea CL (1989b) Effect of selective serotonin (5-HT) agonists and 5-HT$_2$ antagonist on prolactin secretion. Neuropharmacology 28:299–305

Van de Kar LD, Urban JH, Brownfield MS (1990) Serotonergic regulation of renin and vasopressin secretion. In: Paoletti R, Vanhoutte PM, Brunello N, Maggi FM (eds) Serotonin. From cell biology to pharmacology and therapeutics. Kluwer Academic, Dordrecht, Netherlands, pp 123–129

Van de Kar LD, Richardson Morton KD, Rittenhouse PA (1991) Stress: neuroendocrine and pharmacological mechanisms. In: Jasmin G, Cantin M (eds) Stress revisited. 1. Neuroendocrinology of stress. Methods and achievements in experimental pathology. Karger, Basel, pp 133–173

Van de Kar LD, Rittenhouse PA, O'Connor P, Palionis T, Brownfield MS, Lent SJ, Carnes M, Bethea CL (1992) Repeated cocaine injections reduce the magnitude of the effect of the serotonin agonist MK-212 on prolactin, vasopressin, oxytocin and renin but not ACTH or corticosterone secretion. Biol Psychiatry 32:258–269

Van de Kar LD, Alvarez Sanz MC, Yracheta JM, Kunimoto K, Li Q, Levy AD, Rittenhouse PA (1994) ICV injection of the serotonin 5-HT$_{1B}$ agonist CP-93,129 increases the secretion of ACTH, prolactin, and renin and increases blood pressure by nonserotonergic mechanisms. Pharmacol Biochem Behav 48:429–436

Van de Kar LD, Li Q, Vicentini A, Battaglia G (1995a) Evidence that the prolactin response to the 5-HT$_{1A}$ agonist 8-OH-DPAT is mediated by Gi proteins. Soc Neurosci Abstr 21:343.14

Van de Kar LD, Rittenhouse PA, Li Q, Levy AD, Brownfield MS (1995b) Hypothalamic paraventricular, but not supraoptic neurons, mediate the serotonergic stimulation of oxytocin secretion. Brain Res Bull 36:45–50

Van Wijngaarden I, Tulp MTM, Soudijn W (1990) The concept of selectivity in 5-HT receptor research. Eur J Pharmacol 188:301–312

VanderMaelen CP, Matheson GK, Wilderman RC, Patterson LA (1986) Inhibition of serotonergic dorsal raphe neurons by systemic and iontophoretic administration of buspirone, a non-benzodiazepine anxiolytic drug. Eur J Pharmacol 129:123–130

Yatham LN, Steiner M (1993) Neuroendocrine probes of serotonergic function: a critical review. Life Sci 53:447–463

Zifa E, Fillion G (1992) 5-Hydroxytryptamine receptors. Pharmacol Rev 44:401–458

Zink MH,III, Pergola PE, Doane JF, Sved AF, Alper RH (1990) Quipazine increases renin release by a peripheral hemodynamic mechanism. J Cardiovasc Pharmacol 15:1–9

Role of Serotoninergic Neurons and 5-HT Receptors in the Action of Hallucinogens

D.E. NICHOLS

A. Introduction

Brain serotonin receptors and serotoninergic pathways have received increasing attention as targets for a wide variety of therapeutic agents. Perhaps peculiar to this realm, however, are the so-called hallucinogenic drugs, which presently lack demonstrated therapeutic utility, and still remain, as they have for at least the past 50 years, pharmacological curiosities. Research into their mechanism of action is generally poorly funded, and we know relatively little about how they affect the brain, despite their continued popularity as recreational drugs among a significant proportion of the population.

In contrast to virtually every other type of substance that affects the central nervous system, modern man is at a loss to provide either a succinct description of the effects of hallucinogenic drugs or even a widely acceptable name for the drug class itself. The amount of factual information about hallucinogenic drugs circulating among the public is dismally inadequate. Even in the scientific world there is now a whole generation of neuroscientists who generally know very little about hallucinogens other than the fact that they are subject to very strict legal controls. Enforcement authorities consider these substances extremely dangerous, and the applications and procedures to gain approval to study them, especially in humans, are burdensome nearly everywhere in the world. What is it, exactly, that makes these pharmacological curiosities so fearsome? It is necessary for this chapter to devote just a little time to set the background for a discussion of these substances.

First of all, the name "hallucinogen" is a misnomer, because these drugs do not reliably produce hallucinations. It is only one of many terms that have been proposed as a name for this drug class. The German toxicologist Louis Lewin coined the name "phantastica" earlier in this century, but perhaps the term most widely used has been "psychotomimetic," which incorrectly implies that these drugs produce a state that mimics psychosis. In the lay press "psychedelic" has found favor for nearly 3 decades. This word, which implies that these substances manifest beneficial qualities of mind, has so far not gained acceptance in mainstream camps. Even more recently, the name "entheogen" has been adopted in some counterculture circles, a term denoting that these drugs produce or manifest a god within, but it seems very unlikely that this name will ever be accepted in formal usage. The fact that there is no consensus even on the best name for this class perhaps sets the tone for any discussion of the present topic.

What we do know is that naturally occurring hallucinogenic drugs have played a significant role in the development of philosophy and religious thought in many cultures. There are those who argue, quite persuasively, that hallucinogenic drugs were in fact responsible for the stirrings of humankind's earliest theologies. Important examples of the use of hallucinogenic substances in other cultures include the *soma* of ancient India, to which numerous Vedic hymns were written, *teonanacatl*, "god's flesh" used by the Aztec shaman, and *peyote*, taken as a sacrament during services of the Native American Church. Further, in the village of Eleusis in ancient Greece, for more than 2000 years it was a greatly treasured opportunity for any Greek citizen to participate in the secret ceremony held each September that involved the drinking of a special hallucinogenic potion, κψκεον. The ritual was partially described in the second century AD: ". . . of all the divine things that exist among men, it is both the most awesome and the most luminous" (WASSON et al. 1977). And in Brazil today, there is a religious order that uses as a sacrament *ayahuasca*, an hallucinogenic plant extract employed for millenia in rituals by Indians of the Amazon valley. It might be noted in this context that one doctoral dissertation has even provided evidence that psilocybin-induced mystical-religious experiences could not be distinguished, by objective criteria, from spontaneously occurring ones (PAHNKE 1963).

What exactly are these substances, feared by modern man, yet revered by many ancient cultures? JAFFE (1990) recently provided a definition consistent with the ritual use of these drugs in other societies. Arguing that the term "psychedelic" is better either than "hallucinogen" or "psychotomimetic," he states that, ". . . the feature that distinguishes the psychedelic agents from other classes of drugs is their capacity reliably to induce states of altered perception, thought, and feeling that are not experienced otherwise except in dreams or at times of religious exaltation." FREEDMAN (1968) probably agreed with this assessment, because he has stated that, ". . . one basic dimension of behavior . . . compellingly revealed in LSD states is 'portentousness' – the capacity of the mind to see more than it can tell, to experience more than it can explicate, to believe in and be impressed with more than it can rationally justify, to experience boundlessness and 'boundaryless' events, from the banal to the profound." These definitions do focus on the more spectacular effects that these substances are capable of producing, whereas low doses of hallucinogens generally elicit less dramatic results.

It is also a unique feature of the hallucinogens that they do not obey regular dose-response relationships. That is, high doses do not simply produce effects similar to low doses, but at greater intensity. Furthermore, identical doses given to the same individual on different occasions may provoke dramatically different responses. There are several important variables that determine the nature of the drug effect. The most important of these are probably what have been called the "set" and "setting," i.e., the frame of mind and expectations of the subject: the "set," and the environment within which the experience takes place: the "setting." BARR et al. (1972) have further shown that certain features of the LSD effect vary with the personality type of the subject, which in some sense one might consider to be included as a determinant of set.

Dr. Stanislov Grof, who supervised more clinical LSD sessions than any other individual, has opined, "I consider LSD to be a powerful unspecific amplifier or catalyst of biochemical and physiological processes in the brain" (GROF 1975). These thoughts are echoed by BARR et al. (1972), who state, "... the phenomena induced by LSD (and probably by any similar drug) cannot be predicted or understood in purely pharmacological terms; the personality of the drug taker plays an enormous and critical role in determining how much effect there will be and of what particular type.

It is precisely this unpredictability of effect that makes clinical research with LSD and other hallucinogens so difficult. What criteria does one use to quantitate drug effect? How can one establish baseline values? This unpredictability is no doubt a primary factor in adverse reactions ("bad trips") that occur with recreational use of these drugs. At the extremes, a user might on one occasion experience ecstasy and mystical union with the cosmos, while on another he or she might endure a hellish nightmare, extreme paranoia, and the like.

Although there is less written today about psychedelics, LSD use has continued at a relatively constant level among high school youth over the past several decades, with some modest increase reported in recent years. The major difference between present recreational usage and that of prior years seems to be that dosages currently available on the clandestine market are typically in the 40–60 µg range, rather than the 100–300 µg more characteristic of the 1960s and 1970s. At lower dosages, the psychological effects of the psychedelics are generally not overwhelming, with less likelihood of an adverse reaction that might require medical intervention and be brought to the attention of reporting agencies or the press. Although there is also high variability even in low dosage responses, intoxication would typically include many of the following (HOLLISTER 1984):

1. Somatic symptoms: dizziness, weakness, tremors, nausea, drowsiness, paresthesias, and blurred vision.
2. Perceptual symptoms: altered shapes and colors, difficulty in focusing on objects, a sharpened sense of hearing, and, rarely, synesthesias.
3. Psychic symptoms: alternations in mood, tension, distorted time sense, difficulty in expressing thoughts, depersonalization, dreamlike feeling, and visual hallucinations.

One gets the clear impression that the substrates in the brain that are affected by hallucinogenic drugs must be very important to us as conscious beings in defining exactly who we are in relation to the rest of the world. One is not actually concerned so much with the structures of the receptors themselves with which hallucinogens interact. Rather, interest lies in understanding the role of those receptors in transducing effects in relevant neuronal systems that provide for the perception of ordinary states of reality ("consensus reality") under one set of biochemical circumstances, but which under another set allow an experience of what can only be described as an ineffable state of mystical consciousness.

Finally, although present interest in hallucinogenic drugs is, for the most part, focused on purely basic science goals such as identification of the receptor types involved in their action or the anatomical substrates for their effects, ultimately one must not lose sight of the fact that these drugs only came to our attention in the first place because of their powerful and *unique* effects upon the psyche. The driving force for the investigation of hallucinogens must be the desire for a greater understanding of the human mind. It is this relevance of hallucinogenic drugs to the ageless quest of our species for meaning in life that makes research into their mechanism of action important, and that makes current legal strictures on their availability for research so paradoxical. One must not, of course, ignore the possibility that hallucinogens might also have great therapeutic importance to present day society, were they to be employed in an appropriate and constructive manner.

The failure of modern science and society to envision hallucinogens as key ingredients in our efforts to understand the nature of the human mind is what allows these agents to be viewed simply as drugs of abuse. One sees hopeful signs that this perspective may be gradually changing.

B. Chemical Classes of Hallucinogens

The actual chemical structures of hallucinogens can be classified into two broad categories: (1) the tryptamines, and (2) the phenethylamines. Within the tryptamines, however, one should probably include two subsets, the simple tryptamines such as *N,N*-dimethyltryptamine (DMT) and psilocin, which possess considerable conformational flexibility, and the ergolines, relatively rigid analogs including LSD and a few closely related compounds.

The phenethylamines have as their prototype the naturally occuring compound mescaline, the principal active component in the peyote cactus, *Lophophora williamsii*. Extensive structure-activity relationship studies carried out over many years principally in the laboratories of A.T. Shulgin (see SHULGIN and SHULGIN 1991), D.E. Nichols, and R.A. Glennon (for recent reviews see NICHOLS et al. 1991; NICHOLS 1994) have led to extremely potent substituted phenethylamine derivatives. The structures of some of the most important hallucinogens are presented in Fig. 1.

Tryptamines

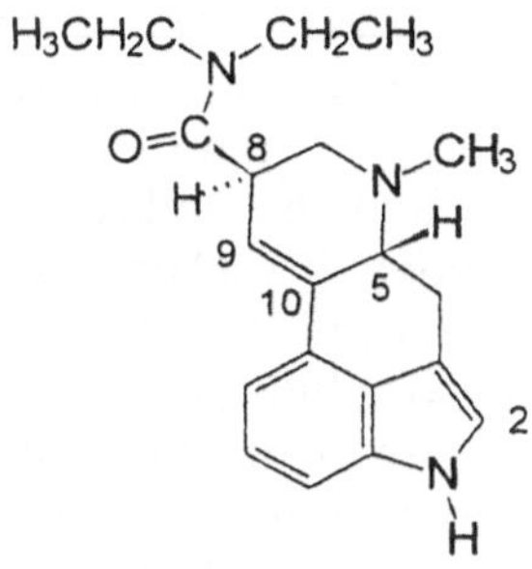

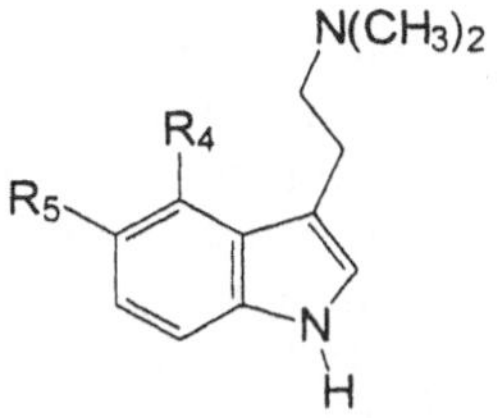

H3CH2C N CH2CH3

O=C 8 N–CH3
H·· H
9
10 5
N 2
H

(+)-LSD (LSD-25: Delysid)

N(CH3)2

R4
R5
N
H

R4 = R5 = H: N,N-Dimethyltryptamine (DMT)
R4 = OH, R5 = H; Psilocin
R4 = OPO3-, R5 = H; Psilocybin
R4 = H, R5 = OCH3; 5-Methoxy-DMT

Phenethylamines

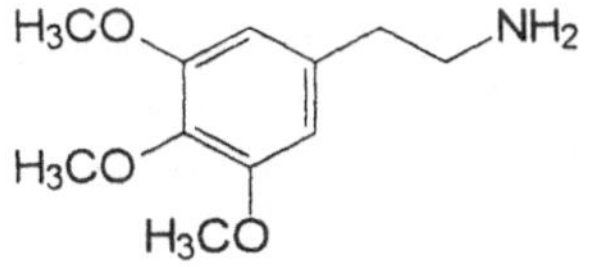

H3CO NH2

H3CO

H3CO

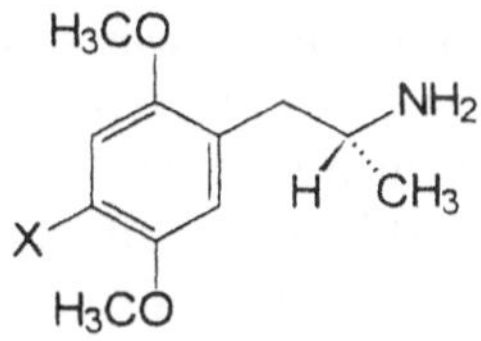

Mescaline

H3CO

X

H3CO

NH2

H CH3

(-)-2.5-Dimethoxy-4X-amphetamines

X = CH3; DOM
X = Br; DOB
X = I: DOI
X = CF3; DOTFM

Fig. 1. Chemical structures of the most important representatives of hallucinogen molecules. (+)-LSD is commonly available as its (+)-tartaric acid salt. While the phenethylamines include substituted "amphetamines," e.g., DOM, these are more properly named as 1-(2,5-dimethoxy-4X-phenyl)-2-aminopropanes. All the phenethylamines are shown here in the form of their free bases, but are typically available either as a sulfate or hydrochloride salt

C. Historical Relationship Between Serotonin and Hallucinogen Action

Following the isolation and identification of serotonin, interest in its role as a neurotransmitter and its possible relevance to behavior was greatly stimulated by, and was intertwined with, the virtually contemporaneous discovery of LSD and recognition that this potent psychoactive substance had the ability to interact with serotonin systems. From a chemist's perspective, it was easy to recognize the tryptamine template within the ergoline framework of LSD and to appreciate that the much simpler molecule of serotonin was built upon the same molecular scaffold. It was only a decade after the discovery of LSD that TWAROG and PAGE (1953) used a sensitive bioassay to demonstrate the presence of serotonin in brain extracts, and GADDUM (1953) reported that LSD was an antagonist of the actions of serotonin in the rat's uterus. A year later GADDUM and HAMEED (1954) reported details of experiments describing LSD's specific antagonism of the effects of serotonin in peripheral tissues, while WOOLLEY and SHAW (1954) hypothesized "... that the mental disturbances caused by lysergic acid diethylamide were to be attributed to an interference with the action of serotonin in the brain." In the same year UDENFRIEND et al. (1954) reported new sensitive assays for serotonin and its metabolites, a development that made possible the analysis of various neuronal tissues for these neurochemicals and ushered in an era of intense research on the role of serotoninergic systems in the brain.

The idea that the effects of LSD could be attributed to the blockade of central serotonin receptors was short lived, however. A brominated derivative of LSD, 2-bromo-LSD (BOL), a potent serotonin antagonist in peripheral tissues (CERLETTI and DOEPFNER 1958), was found to be essentially devoid of LSD-like effects (CERLETTI and ROTHLIN 1955). BOL pretreatment actually prevented the effects of a subsequently administered dose of LSD (GINZEL and MAYER-GROSS 1956), although at high doses BOL itself does have psychoactive properties (ISBELL et al. 1959). In addition, while LSD-like activity appeared to be generally correlated with serotonin antagonist activity (CERLETTI and DOEPFNER 1958), the morpholide analogue of LSD had only about 8% of the antiserotonin activity of LSD, yet was about 75% of the potency of LSD as an hallucinogen (GOGERTY and DILLE 1957). In addition, a series of cycloalkyl monosubstituted amides of lysergic acid had serotonin antagonist effects in the rat intestine 30% greater than LSD itself, yet lacked LSD-like behavioral effects (VOTAVA et al. 1958).

Nevertheless, it was evident that even if LSD was not a central antagonist of serotonin, it did have effects on serotoninergic function in the CNS. FREEDMAN (1961) showed that systemic LSD administration elevated brain 5-HT levels, an effect not shared by the nonhallucinogenic BOL or by the behaviorally inactive *levo* isomer of LSD. Later, ROSECRANS et al. (1967) reported that LSD also reduced brain levels of the serotonin metabolite 5-hydroxyindoleacetic acid (5-HIAA).

ANDÉN et al. (1968) were among the first to suggest that LSD might have a direct *agonist* effect at serotonin receptors in the CNS. Within a few years studies had been reported showing that drugs from all the various classes of hallucinogens, including psilocybin, *N*,*N*-dimethyltryptamine (DMT), 5-methoxy-DMT, and 2,5-dimethoxy-4-methylamphetamine (DOM), all increased brain serotonin levels and/or decreased the turnover of serotonin (ANDÉN et al. 1971, 1974; FUXE et al. 1972; FREEDMAN et al. 1970; LEONARD 1973; RANDIC and PADJEN 1971), phenomena consistent with an agonist effect of these drugs.

D. Early Hypothesis for a Presynaptic Agonist Mechanism of Action

It was found early on that LSD had the potent ability to suppress the firing of cells in the dorsal raphe nucleus, whether given systemically (AGHAJANIAN et al. 1968, 1970) or applied directly to the cell bodies by microiontophoresis (AGHAJANIAN et al. 1972). The tryptamine hallucinogens DMT (AGHAJANIAN et al. 1970), as well as psilocin and 5-methoxy-DMT (AGHAJANIAN and HAIGLER 1975; DEMONTIGNY and AGHAJANIAN 1977), also inhibit the firing of dorsal raphe cells. These observations led AGHAJANIAN and HAIGLER (1975) to hypothesize that this presynaptic action in the raphe might in fact be the underlying basis for the action of hallucinogens.

Serious problems soon developed with regard to this hypothesis, however, in view of the fact that phenethylamine hallucinogens lacked this effect. For example, systemic administration of mescaline or DOM failed to inhibit raphe cell firing (HAIGLER and AGHAJANIAN 1973; TRULSON et al. 1981). Furthermore, a nonhallucinogenic ergoline, lisuride, also potently suppressed raphe cell firing (ROGAWSKI and AGHAJANIAN 1979). In addition, in a cat behavioral model the suppression of raphe firing by hallucinogens outlasted the behavioral effects (TRULSON et al. 1981). Once the rate suppressant effect on raphe cell firing had been established to be mediated by stimulation of 5-HT$_{1A}$ somatodendritic receptors, 5-HT$_{1A}$ agonists were identified that suppressed raphe firing, but which were not hallucinogenic (SPROUSE and AGHAJANIAN 1987, 1988). This hypothesis for the mechanism of action was, therefore, not tenable, although as discussed below it is certainly possible that the importance of this effect to the overall pharmacology of tryptamine hallucinogens may at the present time be underestimated.

E. Evidence for Agonist Activity at the 5-HT$_{2A}$ Serotonin Receptor Subtype

Unfortunately, the vast majority of mechanistic studies have been carried out in rodents. What little we know about the human pharmacology of hallucinogens essentially dates from a period 30–40 years ago. Nevertheless, it will serve no useful purpose to summarize the various animal behavioral paradigms that have been employed over the years. The primary choice of an animal model today appears to be the two-lever drug discrimination procedure, wherein rats are trained to discriminate between the effects of an injection of saline and a training drug, typically LSD or an hallucinogenic amphetamine derivative such as DOM or 1-(2,5-dimethoxy-4-iodophenyl)-2-aminopropane (DOI). Using a variety of antagonists of the training drug effect, as well as substitution tests with various compounds of known pharmacology, one is able to determine the underlying pharmacological basis for the training drug stimulus in the rat.

Although one cannot be certain that studies carried out with this paradigm actually reflect what would happen in humans if similar experiments were performed, the model has certainly shown the best correlation with extant human data. A minor deficiency of the model is that from time to time it produces false positives, i.e., gives data predicting a compound to be hallucinogenic in man, when in fact it is known that the compound is not clinically active. Nevertheless, these are probably the best data available, and most of what we know about the in vivo mechanism of action of hallucinogens is based on drug discrimination studies in rats. The use of drug discrimination to study hallucinogens has recently been reviewed by WINTER (1994).

The earliest hypothesis that hallucinogenic drugs acted specifically at 5-HT_2 receptor subtypes was offered by GLENNON et al. (1983), following drug discrimination studies showing that the 5-HT_2 antagonists ketanserin and pirenperone blocked the discriminative stimulus effects of phenethylamine and tryptamine hallucinogens, including LSD (COLPAERT et al. 1982; LEYSEN et al. 1982; COLPAERT and JANSSEN 1983; GLENNON et al. 1983). At the present time, there seems to be a fairly clear consensus that the key site for hallucinogen action is the 5-HT_{2A} receptor subtype (BRANCHEK et al. 1990; MCKENNA and SAAVEDRA 1987; PIERCE and PEROUTKA 1989; SADZOT et al. 1989; TITELER et al. 1988; TEITLER et al. 1990). This conclusion was largely developed by correlation of the rat behavioral activity of hallucinogenic amphetamines with their affinities and efficacies at the 5-HT_2 receptor (GLENNON et al. 1983, 1984a,b, 1986; SANDERS-BUSH et al. 1988). The most potent substituted phenethylamine hallucinogen analogue reported to date, DOTFM, had 5-$HT_{2A/2C}$ affinity ($K_i =$ 1.5 nM at the $[^{125}I]$DOI-labeled site) slightly greater than that of DOI, and both DOI and DOTFM were full agonists at the cloned 5-HT_{2A} and 5-HT_{2C} receptors (NICHOLS et al. 1994).

Although it is not yet proven that agonist activity at the 5-HT_{2A} receptor is the essential feature of hallucinogen action *in humans*, there is one report that the mixed 5-$HT_{2A/2C}$ antagonist cyproheptadine was able to antagonize the subjective effects of DMT in some subjects (MELTZER et al. 1982).

Because all of the hallucinogens have nearly equal potency in binding to the 5-HT_{2A} and 5-HT_{2C} receptor subtypes, there has been some uncertainty as to which of these receptors was more important to the mechanism of action.

Although ketanserin, used in earlier studies, is clearly more selective for 5-HT$_{2A}$ sites, ISMAIEL et al. (1993) developed a spiperone analogue with about 2000-fold selectivity for the rat 5-HT$_{2A}$ over the 5-HT$_{2C}$ receptor that was able to block the discriminative cue of the hallucinogenic amphetamine DOM in the two-lever drug discrimination paradigm in rats.

SCHREIBER et al. (1994) were able to abolish the discriminative cue of the hallucinogenic amphetamine derivative DOI in rats with the highly selective 5-HT$_{2A}$ receptor antagonist MDL 100,907 (SCHMIDT et al. 1992). This antagonist has greater than 200-fold selectivity for 5-HT$_{2A}$ versus 5-HT$_{2C}$ receptors. A selective 5-HT$_{2C}$ receptor antagonist, SB 200,646 (KENNETT et al. 1994), did not block the stimulus effect of DOI at relatively high doses that were effective in antagonizing behavioral effects caused by 5-HT$_{2C}$ receptor activation.

FIORELLA et al. (1995b) have recently provided additional compelling evidence for 5-HT$_{2A}$ receptor mediation of the hallucinogen stimulus in rats using antagonist correlation analysis. The ability of a series of nonselective 5-HT$_{2A/2C}$ receptor antagonists to block the discriminative stimulus properties of (–)-DOM was correlated with their affinity for 5-HT$_{2A}$ sites, and not with 5-HT$_{2C}$ affinity. The data reported by those authors have been plotted in Fig. 2 to illustrate this conclusion.

The head-twitch in rats also appears to be mediated by an agonist action at 5-HT$_{2A}$ receptors. Recently SCHREIBER et al. (1995) showed that head twitches induced by DOI were abolished by low doses of the 5-HT$_{2A}$ selective antagonist MDL 100,907. The selective 5-HT$_{2C}$ antagonist SB 200,646A failed to block DOI-induced head twitch.

Although the preponderance of evidence suggests that hallucinogens are agonists, the issue is clouded by studies that show LSD to be a partial agonist (MCCLUE et al. 1989; SANDERS-BUSH et al. 1988), or even an antagonist (NORMAN et al. 1989; PIERCE and PEROUTKA 1990) at the 5-HT$_{2A}$ receptor. GLENNON (1990) has reviewed this controversy and has concluded that hallu-

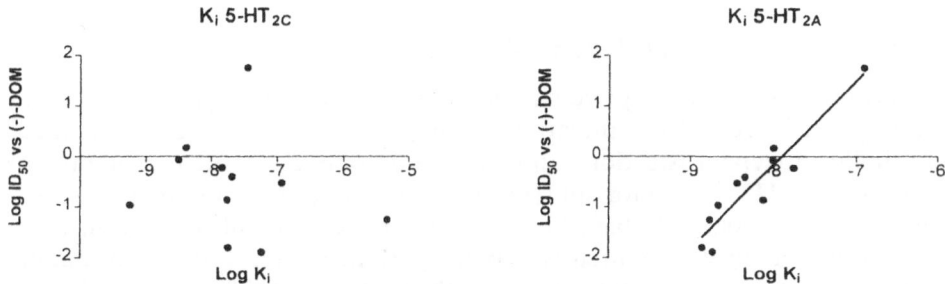

Fig. 2. Data for antagonist correlation analysis, taken from the report of FIORELLA et al. (1995b). They illustrate that the blockade of the discriminative cue of (–)-DOM in LSD-trained rats by a series of mixed 5-HT$_{2A/2C}$ antagonists is highly correlated only with affinity of the antagonist for the 5-HT$_{2A}$ receptor, and not for the 5-HT$_{2C}$ receptor subtype

cinogens are either agonists or in some cases may be partial agonists, but are not 5-HT$_{2A}$ antagonists.

Some caution must still remain regarding the extrapolation of rat drug discrimination data to humans. It is particularly troubling that the 5-HT$_2$ antagonists ketanserin or pirenperone failed to block the LSD cue in drug discrimination studies with LSD-trained monkeys (Nielsen 1985). Mescaline, another hallucinogenic 5-HT$_2$ agonist, did not substitute for LSD in monkeys, but 5-methoxy-DMT, a mixed 5-HT$_{1A}$/5-HT$_2$ agonist (Martin and Sanders-Bush 1982), did.

I. Mechanistic Basis for Rapid Tolerance to the Hallucinogens

The administration of hallucinogens to humans results in extremely rapid development of tolerance. For example, LSD treatment on sequential days leads to tolerance and, by the 4th day, there is an almost complete loss of sensitivity to the effects of LSD (Cholden et al. 1955; Isbell et al. 1956). This development of tolerance to LSD and other hallucinogens (Freedman et al. 1958) has been of great interest because the elucidation of a basis for this tolerance would have relevance to understanding the mechanism of action of hallucinogens.

There is evidence to support the idea that the phenomenon of tolerance is also mediated by 5-HT$_2$ receptors. Leysen et al. (1989) and Leysen and Pauwels (1990) observed rapid desensitization and downregulation of central 5-HT$_2$ receptors following repeated treatment of rats with DOM. A significant reduction in the total number of cortical 5-HT$_2$ sites was observed after only two drug treatments (2.5 mg/kg s.c. every 8 h). After four injections in 24 h, the receptor number recovered only very slowly ($T_{1/2} = 5$ days). Chronic treatment of rats for 7 or 8 days with either DOB (Buckholtz et al. 1988) or DOI (McKenna et al. 1989), respectively, also produced downregulation of brain 5-HT$_2$ receptors in rat. There is no present evidence that repeated administration of hallucinogens leads to downregulation of other receptor types.

II. Possible Species Differences?

Even though the interoceptive cue produced by hallucinogens may be mediated by 5-HT$_{2A}$ receptor stimulation, the hypothesis that this effect also mediates hallucinogen intoxication in humans is confounded by the fact that the rat and human 5-HT$_{2A}$ receptor subtypes have slightly different structure-activity relationships, apparently based on a single amino acid substitution in transmembrane region V (see discussion below). Hence, effects of various hallucinogens on rat behavior may not be strictly analogous to those in humans.

Kao et al. (1992) first showed that residue 242 in transmembrane helix 5 (TM V), which is serine in the human and alanine in the rat 5-HT$_{2A}$ receptor, could explain different binding affinities of these two receptors for mesulergine, an N-methylated ergoline 5-HT$_{2C}$ antagonist ligand. Mutation of

Ser^{242} in the human receptor to Ala gave a mutant receptor with binding characteristics similar to the rat wild type.

Following a similar line of inquiry, JOHNSON et al. (1994) and NELSON et al. (1993) have shown that differential affinities of ergolines and their $N(1)$-alkyl derivatives for the rat and human 5-HT$_{2A}$ receptors can be attributed to the difference in this same amino acid. Ergoline and tryptamine ligands had higher affinity for the human receptor (Ser^{242}), while $N(1)$-alkylated ergolines and tryptamines had higher affinity for the rat receptor (Ala^{242}). JOHNSON et al. (1994) also prepared the mutant rat receptor Ala$^{242} \rightarrow$ Ser^{242}, complementing the work of KAO et al. (1992), and showed that this converted the rat receptor into one having binding characteristics virtually identical to the human receptor. These workers speculated that the serine in the human receptor serves as a hydrogen bond acceptor for the indole NH, while in the rat receptor, the alanine residue interacts with the N-alkyl group of the alkylated ergolines.

GALLAHER et al. (1993) found that psilocin had 15-fold higher affinity for the human than the rat 5-HT$_{2A}$ receptor and speculated that Ser^{242} in the human receptor might hydrogen bond to the 4-hydroxy of psilocin, leading to a binding orientation where the indole ring was rotated 180° from its orientation in that of N,N-dimethyltryptamine, serotonin, or bufotenin (5-hydroxy-N,N-dimethyltryptamine). These studies emphasize the probability that this residue is part of the ligand recognition site in the 5-HT$_{2A}$ receptor, and also point out that the binding orientations of ligands with only subtle structural differences (i.e., 4- versus 5-hydroxy) may differ dramatically and may not be intuitively obvious. Clearly, these observations create some uncertainty regarding the degree to which rat data will parallel the human situation.

F. Evidence for Involvement of the 5-HT$_{2C}$ Receptor Subtype

Even though there may be compelling evidence that the most salient feature of the interoceptive cue of hallucinogens in rats is produced by stimulation of the 5-HT$_{2A}$ receptor, and that by inference this event may be a key component of hallucinogen action in man, it is still possible that activation of the 5-HT$_{2C}$ receptor may play a role in the overall intoxication process. Lisuride, for example, is generally considered to be nonhallucinogenic. FIORELLA et al. (1995c) have recently reexamined the stimulus properties of lisuride, and provided evidence that lisuride substitution in hallucinogen-trained rats is mediated by stimulation of 5-HT$_{2A}$ receptors. In addition, lisuride is known to suppress the firing of dorsal raphe cells, as does LSD (ROGAWSKI and AGHAJANIAN 1979). Further, both lisuride and LSD are potent dopamine D$_2$ agonists. Hence, lisuride appears to possess the pharmacological components believed necessary for hallucinogenic effects in man. In contrast to LSD and other hallucinogens, however, lisuride is not an agonist at 5-HT$_{2C}$ receptors. Rather, it can actually block the agonist actions of 5-HT at this receptor

(SANDERS-BUSH 1994). Indeed, other known hallucinogens are potent serotonin $5\text{-}HT_{2C}$ agonists (SANDERS-BUSH and BREEDING 1991; BURRIS et al. 1991), while nonhallucinogenic analogues such as BOL and lisuride lack agonist properties.

Additional supporting animal data have been reported by FIORELLA et al. (1995a), who observed that p-chlorophenylalanine (PCPA) depletion of central serotonin resulted in supersensitivity to the stimulus effects of LSD in rats. While PCPA had no effect on the maximal level of PI hydrolysis mediated by $5\text{-}HT_{2A}$ receptors, the maximal phosphoinositide (PI) hydrolysis in response to $5\text{-}HT_{2C}$ receptor stimulation was increased by 46%. It has also been reported that humans are more sensitive to the effects of LSD following treatment with reserpine (RESNICK et al. 1965), a regimen that would deplete CNS serotonin (among other effects). It seems possible, therefore, that stimulation of the $5\text{-}HT_{2A}$ receptor may be a necessary, but not sufficient, condition for hallucinogenesis. This issue cannot be resolved until an hallucinogen analogue is tested in man that possesses selective $5\text{-}HT_{2A}$ agonist effects but lacks any action at the $5\text{-}HT_{2C}$ receptor.

The possible significance of the $5\text{-}HT_{2C}$ receptor has not been widely appreciated, probably due to the lack of ligands, both agonists and antagonists, specific for this subtype, as well as a perception by many that it is primarily localized only in the choroid plexus. However, POMPEIANO et al. (1994), using in situ hybridization histochemistry, have shown that the $5\text{-}HT_{2C}$ receptor subtype is a principal 5-HT receptor in the rat brain. mRNA for the $5\text{-}HT_{2C}$ receptor was expressed at high levels in many brain regions other than choroid plexus. The $5\text{-}HT_{2A}$ and $5\text{-}HT_{2C}$ receptor mRNAs appeared to be expressed in distinct (but overlapping) sets of brain regions.

G. Evidence for $5\text{-}HT_{1A}$ Receptor Involvement

The earliest hypothesis for the mechanism of action of hallucinogens at the cellular level was based on a number of studies carried out by George Aghajanian and his colleagues. As discussed earlier, those experiments showed that LSD or psilocybin caused a reduction in the firing rate of cell bodies in the dorsal raphe nucleus (AGHAJANIAN et al. 1968, 1970; HAIGLER and AGHAJANIAN 1975). DMT also inhibits the firing rate of cells in the dorsal raphe (AGHAJANIAN et al. 1970). These observations led to the hypothesis that this pharmacological effect might be the underlying mechanism for hallucinogenesis (AGHAJANIAN and HAIGLER 1975). DeMONTIGNY and AGHAJANIAN (1977) also showed that administration of 5-methoxy-DMT produced a rapid and dose-dependent inhibition of unit activity in the dorsal raphe.

These observations could not, however, be extended to the phenethylamine hallucinogens. For example, direct iontophoresis of mescaline did not produce inhibition of raphe unit activity (HAIGLER and AGHAJANIAN 1973). Thus, the ability to inhibit raphe cell firing was not tenable as a comprehensive explanation for the mechanism of action of hallucinogens. Furthermore, as noted earlier, a variety of 5-HT$_{1A}$ agonists that suppressed raphe firing were not hallucinogenic (SPROUSE and AGHAJANIAN 1987, 1988). Immunohistochemical studies have shown that the 5-HT$_{1A}$ receptors located in the midbrain raphe nuclei are almost exclusively localized on the cell membranes of serotonin neurons (SOTELO et al. 1990) and agonist drugs at the 5-HT$_{1A}$ autoreceptor strongly inhibit central serotonin neuron activity (WILLIAMS et al. 1988). That hallucinogens of the phenethylamine class do not suppress raphe firing can then be rationalized by the fact that this class of molecule lacks significant affinity for the 5-HT$_{1A}$ receptor subtype (TITELER et al. 1988).

LSD has high affinity for 5-HT$_{1A}$ receptors, as do potent tryptamine hallucinogens such as 5-methoxy-DMT and psilocin (MCKENNA et al. 1990). LSD has been shown to be a full agonist at central 5-HT$_{1A}$ receptors linked to inhibition of adenylate cyclase (DEVIVO and MAAYANI 1986). DELIGANIS et al. (1991) also found DMT to bind to 5-HT$_{1A}$ receptors ($K_i = 130\,nM$), with affinity reduced ($K_i = 464\,nM$) by addition of guanyl nucleotides, suggesting that DMT was an agonist. Using forskolin-stimulated cAMP formation in rat hippocampal homogenate, these workers reported that DMT was equally efficacious to the 5-HT$_{1A}$ agonist 8-hydroxy-2-(di-n-propylamino)tetralin (8-OH-DPAT) in inhibiting forskolin-stimulated cAMP formation. DMT also enhances the acoustic startle response in rats (DAVIS and SHEARD 1974), an effect now attributed to 5-HT$_{1A}$ receptor activation (NANRY and TILSON 1989). 5-Methoxy-DMT is also a full agonist at 5-HT$_{1A}$ receptors (DUMUIS et al. 1988).

In addition to their role as somatodendritic autoreceptors, there are also postsynaptic 5-HT$_{1A}$ receptors. Their highest density is found in limbic regions of the brain (HAMON et al. 1990). While no studies have differentiated the effects of hallucinogens at postsynaptic versus somatodendritic receptors, there are data suggesting that the 5-HT$_{1A}$ effects of LSD may be detectable, at least in some animal behavioral assays. Drug discrimination studies in pigeons suggest that LSD may produce 5-HT$_{1A}$ receptor-mediated effects (WALKER et al. 1991; YAMAMOTO et al. 1991). The selective 5-HT$_{1A}$ receptor agonist 8-OH-DPAT is able to mimic LSD in rats trained to discriminate 0.16 mg/kg LSD from saline (WINTER and RABIN 1988; MEERT et al. 1990), but LSD can only partially mimic 8-OH-DPAT in rats trained to discriminate the latter agent (CUNNINGHAM and APPEL 1987).

As noted earlier, NIELSEN (1985) reported that ketanserin or pirenperone failed to block the LSD cue in LSD-trained monkeys. Further, while mescaline failed to substitute for LSD in these monkeys, substitution did occur with 5-methoxy-DMT, a mixed 5-HT$_{1A/2}$ agonist. These results, and others, prompted NIELSEN to conclude that there might be a "preferential role for 5-HT$_1$ sites in mediating the LSD stimulus effect." While previous studies, cited earlier, have suggested that the nature of the LSD cue may be essentially expressed by 5-HT$_{2A}$ receptor activation, it is clear that the cue can be modulated by effects of

LSD at 5-HT$_{1A}$ and at other monoamine neurotransmitter receptors (e.g., MARONA-LEWICKA and NICHOLS 1995).

Recently, several studies have reported evidence of functional interactions between 5-HT$_{1A}$ and 5-HT$_2$ receptors. For example, in vivo, 5-HT$_{1A}$ agonists may appear to be 5-HT$_2$ antagonists (e.g., ARNT and HYTTEL 1989; GLENNON 1991). Further, the behavioral syndrome induced by 5-methoxy-DMT can be stereoselectively antagonized by (–)-pindolol or propranolol (LUCKI et al. 1984; TRICKLEBANK et al. 1985). In addition, an early study by DIXON (1968) showed that propranolol could block the disruptive behavior induced by LSD in rats. At that time, this finding was related to a possible involvement of β-adrenergic receptors. In view of present knowledge, however, this result may have been a consequence of 5-HT$_{1A}$ receptor antagonism, a pharmacological property of propranolol that was unknown at the time.

The 5-HT$_{1A}$ agonist 8-OH-DPAT has been shown to inhibit the DOI-induced head twitch in rats (ARNT and HYTTEL 1989; BERENDSEN 1991). In their studies of the rat head twitch induced by DOI, SCHREIBER et al. (1995) replicated this finding and also reported that the ability of the 5-HT$_{1A}$ agonist 8-OH-DPAT to block the DOI-induced head twitch was antagonized by the 5-HT$_{1A}$ antagonist (–)-tertatolol. SCHREIBER et al. (1995) concluded that activation of 5-HT$_{1A}$ receptors inhibited functional effects mediated by 5-HT$_{2A}$ receptors. This conclusion is consistent with other studies suggesting functional interactions between these two receptor types in the modulation of various behaviors (BERENDSEN and BROEKKAMP 1990; DARMANI et al. 1990; YOCCA et al. 1990). GLENNON (1991) has reviewed many of the studies presenting results suggesting that effects mediated through one subtype of serotonin receptor may be modulated by other serotonin receptor subtypes.

Similarly, BACKUS et al. (1990) reported that 5-HT$_2$ antagonists enhanced the behavioral syndrome induced by 8-OH-DPAT, 5-methoxy-DMT, or gepirone. These workers suggested that 5-HT$_2$ receptors may modulate 5-HT$_{1A}$ receptors. Functional interactions clearly exist between 5-HT$_{1A}$ and 5-HT$_{2A}$ receptors, evident not only in various behaviors, but also observable at the neuronal level, at least in rodent pharmacology (ASHBY et al. 1994; BRANDAO et al. 1991).

While the observation that tryptamine hallucinogens can suppress raphe cell activity now seems largely ignored, simply because it did not cut across all the structural types of hallucinogens, it is still possible that this pharmacological effect may be relevant, at least for the actions of hallucinogens of the tryptamine type.

What might be the consequences of 5-HT$_{1A}$ agonist activity in tryptamine hallucinogens? Clinical data are lacking that might clarify this issue. In general, however, LSD and tryptamines such as DMT or psilocybin are described as being particularly able to provoke highly colored and dynamic displays of visual imagery. By contrast, anecdotal data for many of the phenethylamine type hallucinogens, as provided for example by SHULGIN and SHULGIN (1991), seem to indicate more of an effect on mood and emotion, and less production of visual imagery. The author offers the hypothesis that hallucinogens that have a significant agonist action at 5-HT_{1A} receptors may elicit a greater effect on central visual processes than compounds that are purely $5\text{-HT}_{2A/2C}$ agonists. This feature may be responsible, in part, for qualitative differences between the tryptamine and phenethylamine type hallucinogens. Although mescaline is a phenethylamine, its ability to produce impressive and colorful visual imagery is also well known. Like other phenethylamines, however, its affinity for 5-HT_{1A} receptors is in the low micromolar range. Nevertheless, the quantitative potency of mescaline is very low, and at effective doses in man (200–400 mg) there would be significant interaction with the 5-HT_{1A} receptor subtype.

H. Potentiating Effects of Interactions at Other Receptor Subtypes

In the study by FIORELLA et al. (1995b), only 56% of the variability in the potency of a given antagonist to block the interoceptive cue produced by LSD could be accounted for by 5-HT_{2A} affinity alone. Further, except for pirenperone, all of the antagonists tested in this study were more potent in blocking responding elicited by (–)-DOM in LSD-trained rats than in blocking the responding to LSD itself. These results are consistent with the hypothesis that the primary cue of LSD can be modulated by interactions at other receptor types.

Although the essential nature of the LSD cue in rats appears to be expressed through 5-HT_{2A} receptor activation, this cue may be modulated by effects of LSD at other monoamine receptors. These ancillary monoamine interactions could be important, particularly in conferring the extraordinarily high potency possessed by LSD. The affinity of LSD for the 5-HT_{2A} and 5-HT_{2C} receptors is not significantly different from, and may actually be slightly less than, that of phenethylamine compounds such as DOB or DOI. Similarly, the hydrophobicity of LSD is also of the same magnitude as these amphetamine derivatives. Nevertheless, the in vivo potency of LSD is about one order of magnitude greater than for these phenethylamines. These facts would seem to suggest that some other pharmacological property of LSD may be potentiating its effects. It is difficult to know what this action might be, however, since LSD binds with high affinity to a variety of receptors, including 5-HT_{1A}, 5-HT_{1B}, 5-HT_{1D}, 5-HT_{2A}, 5-HT_{2C}, 5-HT_6, 5-HT_7, dopamine D_1 and D_2 receptors, and α_1- and α_2-adrenergic receptors (BURT et al. 1976; CREESE et al. 1976; HOYER 1988; LEYSEN 1985; MARONA-LEWICKA and NICHOLS 1995; MEIBACH et al. 1980; U'PRICHARD et al. 1977; WATTS et al. 1995). By contrast, DOM lacks affinity for most of these receptors (LEYSEN et al. 1989), including 5-HT_{1A} receptors, where it has micromolar affinity, and seems to have significant effects only at the 5-HT_2 receptors (GLENNON et al. 1984a; TITELER et al. 1988).

Which of these other sites, if any, might be responsible for synergism with 5-HT$_{2A}$ agonism? It seems unlikely that the 5-HT$_{1A}$ agonist effects of LSD could be responsible for its remarkable potency, since the tryptamines in general (e.g., DMT, psilocin) possess 5-HT$_2$ and 5-HT$_{1A}$ affinities comparable to LSD, but are a good deal less behaviorally potent. Indeed, most present evidence suggests that 5-HT$_{1A}$ agonists may functionally antagonize 5-HT$_{2A}$ agonists (ARNT and HYTTEL 1989; BERENDSEN 1991; DARMANI et al. 1990; SCHREIBER et al. 1995). Hence, if anything, 5-HT$_{1A}$ agonist effects might be expected to *attenuate* rather than potentiate 5-HT$_{2A}$ agonist actions.

Although there is no evidence to suggest that interactions with other 5-HT$_1$ subtypes are responsible for potentiation of 5-HT$_{2A}$ effects, these cannot presently be ruled out as possibilities. Almost nothing is known about the functional significance of 5-HT$_6$ or 5-HT$_7$ receptors, where LSD also has high affinity.

LSD has modest affinity for α-adrenergic receptors. It was recently shown that stimulation of α_2-adrenoceptors by clonidine was able to potentiate the stimulus properties of LSD in the two-lever drug discrimination paradigm (MARONA-LEWICKA and NICHOLS 1995). This is a possible potentiating interaction, although the affinity of LSD itself for α_2-receptors, reported in the same study ($K_i = 37\,nM$ at [^{3}H]clonidine sites), may not be great enough to make a significant contribution at relevant doses of LSD.

LSD also has high affinity for dopamine D$_2$ receptors (WATTS et al. 1995). The affinity of LSD for D$_2$ receptors ($K_{0.5} = 6.4\,nM$) is certainly in a range where receptor activation might occur at behaviorally relevant doses. The dopamine D$_1$ affinity of LSD ($K_{0.5} = 27\,nM$) is somewhat lower, similar to its affinity for α_2-receptors, and would appear to be less important as a possibility, but cannot be excluded. Although a potentiating effect of D$_2$-receptor stimulation has not been extensively investigated for hallucinogens, risperidone, a mixed 5-HT$_2$ and D$_2$ antagonist, was more potent in antagonizing the discriminative cue of LSD in the rat than was ritanserin, a pure 5-HT$_{2A/2C}$ antagonist (MEERT et al. 1989). Those authors argued for a catecholamine component of the discriminative cue of LSD. Further, SCHREIBER et al. (1995) reported that the DOI head twitch response in rats, in addition to being blocked by 5-HT$_{2A}$ antagonists, was also potently blocked by both D$_1$ and D$_2$ dopamine receptor antagonists. In addition, it has been recently shown that stimulation of the 5-HT$_{2A}$ receptor can enhance dopaminergic function (HUANG and NICHOLS 1993; ICHIKAWA and MELTZER 1995; SCHMIDT et al. 1992).

There is a large body of evidence now developing on the interaction between dopamine D$_2$ and serotonin 5-HT$_{2A}$ receptors, with much of the focus being on the discovery of atypical antipsychotic agents (MELTZER 1991). Thus, the possibility that dopamine D$_2$ receptor stimulation may potentiate 5-HT$_{2A}$ agonism seems to be a promising area for further research, and would lead the investigation of halllucinogens full circle, back to the premise that they may be useful tools for investigating schizophrenia and psychosis.

I. Conclusions

At the present time there is compelling evidence that the most salient feature of the pharmacology of hallucinogens, at least in rats, is an agonist interaction with serotonin 5-HT$_{2A}$ receptors. There is also some evidence that this is a necessary, but perhaps not sufficient, mechanism, and that the concurrent stimulation of serotonin 5-HT$_{2C}$ receptors may also be required. There is no other distinguishing feature, at least of the phenethylamine hallucinogens, to suggest that additional pharmacological components are required. Tryptamine hallucinogens, including LSD, also potently activate serotonin 5-HT$_{1A}$ receptors, and the author hypothesizes that this may distinguish qualitative features of the actions of tryptamines from those of the phenethylamines.

The exceptional potency of LSD, compared with other types of hallucinogens, cannot be explained simply by its affinity for, or interactions with, serotonin 5-HT$_{2A}$, 5-HT$_{2C}$, or 5-HT$_{1A}$ receptors. It is proposed here that the high affinity of LSD for dopamine D$_2$ receptors may be a potentiating interaction to explain this high potency; this possibility would have relevance to current interest in the role of both dopamine and serotonin pathways in mediating various psychiatric disorders. It is also possible that the high potency of LSD may be a consequence of synergistic interactions resulting from the interplay of LSD with many of the receptors for which it has high affinity. Because of this complex interaction at so many different monoamine receptors, it seems likely that LSD will remain a unique psychopharmacological agent, with effects on the psyche that are never completely mimicked by any other substance.

Finally, it must be kept in mind that all these comments should be considered in the context of animal pharmacology studies. It is very unfortunate that legal restrictions have kept these extremely interesting substances from receiving more clinical study in the past 4 decades. It is to be hoped that renewed interest in these molecules in the last few years will continue, eventually leading to greater knowledge of their *human* pharmacology, and providing us with a better understanding of why these unique molecules have such a profound effect upon the human psyche.

Acknowledgement. Great appreciation is expressed for funding (grant DA 02189 from the National Institute on Drug Abuse) that has supported the study of hallucinogenic drugs in the author's laboratory for many years.

References

Aghajanian GK, Foote WE, Sheard MH (1968) Lysergic acid diethylamide: sensitive neuronal units in the midbrain raphe. Science 161:706–708

Aghajanian GK, Foote WE, Sheard MH (1970) Action of psychotogenic drugs on single midbrain raphe neurons. J Pharmacol Exp Ther 171:178–187

Aghajanian GK, Haigler HJ (1975) Hallucinogenic indoleamines: preferential action upon presynaptic serotonin receptors. Psychopharmacol Commun 1:619–629

Aghajanian GK, Haigler HJ, Bloom FE (1972) Lysergic acid diethylamide and serotonin: direct actions on serotonin-containing neurons. Life Sci 11:615–622
Andén NE, Corrodi H, Fuxe K, Hokfelt T (1968) Evidence for a central 5-
   hydroxytryptamine receptor stimulation by lysergic acid diethylamide. Br J
   Pharmacol Chemother 34:1–7
Andén NE, Corrodi H, Fuxe K (1971) Hallucinogenic drugs of the indolealkylamine
   type and central monoamine neurons. J Pharmacol Exp Ther 179:236–249
Andén NE, Corrodi H, Fuxe K, Meek JL (1974) Hallucinogenic phenethylamines:
   interactions with serotonin turnover and receptors. Eur J Pharmacol 25:176–184
Arnt J, Hyttel J (1989) Facilitation of 8-OH-DPAT-induced forepaw treading of rats by
   the 5-HT$_2$ agonist DOI. Eur J Pharmacol 161:45–51
Ashby CR Jr, Edwards E, Wang RY (1994) Electrophysiological evidence for a functional interaction between 5-HT$_{1A}$ and 5-HT$_{2A}$ receptors in the rat medial prefrontal cortex: an iontophoretic study. Synapse 17:173–181
Backus LI, Sharp T, Grahame-Smith DG (1990) Behavioural evidence for a functional
   interaction between central 5-HT$_2$ and 5-HT$_{1A}$ receptors. Br J Pharmacol 100:793–
   799
Barr HL, Langs RJ, Holt RR, Goldberger L, Klein GS (1972) LSD: personality and
   experience. Wiley-Interscience, New York
Berendsen HHG (1991) Behavioural consequences of selective activation of 5-HT
   receptor subtypes; possible implications for the mode of action of antidepressants.
   PhD thesis, University of Groningen, The Netherlands
Berendsen HHG, Broekkamp CLE (1990) Behavioural evidence for functional interactions between 5-HT receptor subtypes in rats and mice. Br J Pharmacol 101:667–
   673
Branchek T, Adham N, Macchi M, Kao H-T, Hartig PR (1990) [^{3}H]-DOB (4-bromo-
   2,5-dimethoxyphenylisopropylamine) and [^{3}H]ketanserin label two affinity states
   of the cloned human 5-hydroxytryptamine$_2$ receptor. Mol Pharmacol 38:604–609
Brandao ML, Lopez-Garcia JA, Graeff FG, Roberts MHT (1991) Electrophysiological
   evidence for excitatory 5-HT$_2$ and depressant 5-HT$_{1A}$ receptors on neurones of the
   rat midbrain tectum. Brain Res 556:259–266
Buckholtz NS, Zhou D, Freedman DX (1988) Serotonin$_2$ agonist administration downregulates rat brain serotonin$_2$ receptors in rat brain. Life Sci 42:2439–2445
Burris KD, Breeding M, Sanders-Bush E (1991) (+)Lysergic acid diethylamide, but not
   its nonhallucinogenic congeners, is a potent serotonin 5HT$_{1C}$ receptor agonist. J
   Pharmacol Exp Ther 258:891–896
Burt DR, Creese I, Snyder SH (1976) Binding interactions of lysergic acid diethylamide
   and related agents with dopaminergic receptors in brain. Mol Pharmacol 12:631–
   638
Cerletti A, Doepfner W (1958) Comparative study on the serotonin antagonism of
   amide derivatives of lysergic acid and of ergot alkaloids. J Pharmacol Exp Ther
   122:124–136
Cerletti A, Rothlin E (1955) Role of 5-hydroxytryptamine in mental disease and its
   antagonism to lysergic acid derivatives. Nature 176:785–786
Cholden LS, Kurland A, Savage C (1955) Clinical reactions and tolerance to LSD in
   chronic schizophrenia. J Nerv Ment Dis 122:211–221
Colpaert FC, Janssen PAJ (1983) A characterization of LSD-antagonist effects of
   pirenperone in the rat. Neuropharmacology 22:1001–1005
Colpaert FC, Niemegeers CJE, Janssen PAJ (1982) A drug discrimination analysis of
   lysergic acid diethylamide (LSD): in vivo agonist and antagonist effects of pur-
   ported 5-hydroxytryptamine antagonists and of pirenperone, an LSD-antagonist. J
   Pharmacol Exp Ther 221:206–214
Creese I, Burt DR, Snyder SH (1976) The dopamine receptor: differential binding of d-
   LSD and related agents to agonist and antagonist states. Life Sci 17:1715–1720
Cunningham KA, Appel JB (1987) Neuropharmacological reassessment of the discriminative stimulus properties of d-lysergic acid diethylamide (LSD). Psychop-
   harmacology 91:67–73

Darmani NR, Martin BR, Pandey U, Glennon RA (1990) Do functional relationships exist between 5-HT$_{1A}$ and 5-HT$_2$ receptors? Pharmacol Biochem Behav 36:901–906

Davis M, Sheard MH (1974) Biphasic dose-response effects of N-N-dimethyltryptamine on the rat startle reflex. Pharmacol Biochem Behav 2:827–829

Deliganis AV, Pierce PA, Peroutka SJ (1991) Differential interactions of dimethyltryptamine (DMT) with 5-HT$_{1A}$ and 5-HT$_2$ receptors. Biochem Pharmacol 41:1739–1744

DeMontigny C, Aghajanian GK (1977) Preferential action of 5-methoxytryptamine and 5-methoxydimethyltryptamine on presynaptic serotonin receptors: a comparative iontophoretic study with LSD and serotonin. Neuropharmacology 16:811–818

DeVivo M, Maayani S (1986) Characterization of the 5-hydroxytryptamine$_{1A}$ receptor-mediated inhibition of forskolin-stimulated adenylate cyclase activity in guinea pig and rat hippocampus membranes. J Pharmacol Exp Ther 238:248–253

Dixon AK (1968) Evidence of catecholaminergic mediation in the "aberrant" behavior induced by lysergic acid diethylamide (LSD) in the rat. Experientia 15:743–747

Dumuis A, Sebben M, Bockaert J (1988) Pharmacology of 5-hydroxytryptamine-1A receptors which inhibit cAMP production in hippocampal and cortical neurons in primary culture. Mol Pharmacol 33:178–186

Fiorella D, Helsley S, Lorrain DS, Rabin RA, Winter JC (1995a) The role of the 5-HT$_{2A}$ and 5-HT$_{2C}$ receptors in the stimulus effects of hallucinogenic drugs III: the mechanistic basis for the supersensitivity to the LSD stimulus following serotonin depletion. Psychopharmacology 121:364–372

Fiorella D, Rabin RA, Winter JC (1995b) The role of the 5-HT$_{2A}$ and 5-HT$_{2C}$ receptors in the stimulus effects of hallucinogenic drugs I: antagonist correlation analysis. Psychopharmacology 121:347–356

Fiorella D, Rabin RA, Winter JC (1995c) Role of 5-HT$_{2A}$ and 5-HT$_{2C}$ receptors in the stimulus effects of hallucinogenic drugs II: reassessment of LSD false positives. Psychopharmacology 121:357–363

Freedman DX (1961) Effects of LSD-25 on brain serotonin. J Pharmacol Exp Ther 134:160–166

Freedman DX (1968) On the use and abuse of LSD. Arch Gen Psychiatry 18:330–347

Freedman DX, Aghajanian GK, Ornitz EM, Rosner BS (1958) Patterns of tolerance to lysergic acid diethylamide and mescaline in rats. Science 127:1173–1174

Freedman DX, Gottleib R, Lovell RA (1970) Psychotomimetic drugs and brain 5-hydroxytryptamine metabolism. Biochem Pharmacol 19:1181–1188

Fuxe K, Holmstedt B, Jonsson G (1972) Effects of 5-methoxy-N,N-dimethyltryptamine on central monoamine neurons. Eur J Pharmacol 19:25–34

Gaddum JH (1953) Antagonism between lysergic acid diethylamide and 5-hydroxytryptamine. J Physiol (Lond) 121:15P

Gaddum JH, Hameed KA (1954) Drugs which antagonize 5-hydroxytryptamine. Br J Pharmacol 9:240–248

Gallaher TK, Chen K, Shih J (1993) Higher affinity of psilocin for human than rat 5-HT$_2$ receptor indicates binding site structure. Med Chem Res 3:52–66

Ginzel KH, Mayer-Gross W (1956) Prevention of psychological effects of d-lysergic acid diethylamide (LSD 25) by its 2-brom derivative (BOL 148). Nature 178:210

Glennon RA (1990) Do classical hallucinogens act as 5-HT$_2$ agonists or antagonists? Neuropsychopharmacology 3:509–517

Glennon RA (1991) Multiple populations of serotonin receptors may modulate the behavioral effects of serotoninergic agents. Life Sci 48:2493–2498

Glennon RA, Young R, Rosecrans JA (1983) Antagonism of the stimulus effects of the hallucinogen DOM and the purported serotonin agonist quipazine by 5-HT$_2$ antagonists. Eur J Pharmacol 91:189–192

Glennon RA, Young R, Hauck AE, McKenney JD (1984a) Structure activity studies on amphetamine analogues using drug discrimination methodology. Pharmacol Biochem Behav 21:895–901

Glennon RA, Titeler M, McKenney JD (1984b) Evidence for 5-HT$_2$ involvement in the mechanism of action of hallucinogenic agents. Eur J Pharmacol 35:2505–2511

Glennon RA, Titeler M, Young R (1986) Structure-activity relationships and mechanisms of action of hallucinogenic agents based on drug discrimination and radioligand binding studies. Psychopharmacol Bull 22:953–958

Gogerty JH, Dille JM (1957) Pharmacology of d-lysergic acid morpholide (LSM). J Pharmacol Exp Ther 120:340–348

Grof S (1975) Realms of the human unconscious. Observations from LSD research. Viking, New York

Haigler HJ, Aghajanian GK (1975) Mescaline and LSD: direct and indirect effects on serotonin-containing neurons in brain. Eur J Pharmacol 21:53–60

Hamon M, Gozlan H, Mestikawy SEL, Emerit MB, Bolanol F, Schechter L (1990) The central 5-HT$_{1A}$ receptors: pharmacological, biochemical, functional, and regulatory properties. Ann N Y Acad Sci 600:114–131

Hollister LE (1984) Effects of hallucinogens in humans. In: Jacobs BL (ed) Hallucinogens: neurochemical, behavioral, and clinical perspectives. Raven, New York, pp 19–33

Hoyer D (1988) Functional correlates of serotonin 5-HT$_1$ recognition sites. J Recept Res 8:59–81

Huang X, Nichols DE (1993) 5-HT$_2$ receptor-mediated potentiation of dopamine synthesis and central serotonergic deficits. Eur J Pharmacol 238:291–296

Ichikawa J, Meltzer HY (1995) DOI, a 5-HT$_{2A/2C}$ receptor agonist, potentiates amphetamine-induced dopamine release in rat striatum. Brain Res 698:204–208

Isbell H, Belleville RE, Fraser HF, Wikler A, Logan CR (1956) Studies on lysergic acid diethylamide (LSD-25). I. Effects in former morphine addicts and development of tolerance during chronic intoxication. Arch Neurol Psychiatty 76:468–478

Isbell H, Miner EJ, Logan CR (1959) Relationships of psychotomimetic to anti-serotonin potencies of congeners of lysergic acid diethylamide (LSD-25). Psychopharmacology 1:20–28

Ismaiel AM, De Los Angeles J, Teitler M, Ingher S, Glennon RA (1993) Antagonism of 1-(2,5-dimethoxy-4-methyl)-2-aminopropane stimulus with a newly identified 5-HT$_2$ versus 5-HT$_{1C}$-selective antagonist. J Med Chem 36:2519–2525

Jaffe JH (1990) Drug addiction and drug abuse. In: Gilman AG, Rall TW, Nies AS, Taylor P (eds) Goodman and Gilman's The pharmacological basis of therapeutics, 8th edn. McGraw-Hill, New York, pp 522–573

Johnson MP, Loncharich RJ, Baez M, Nelson DL (1994) Species variations in transmembrane region V of the 5-hydroxytryptamine Type 2A receptor alter the structure-activity relationship of certain ergolines and tryptamines. Mol Pharmacol 45:277–286

Kao H-T, Adham N, Olsen MA, Weinshank RL, Branchek TA, Hartig PR (1992) Site-directed mutagenesis of a single residue changes the binding properties of the serotonin 5-HT$_2$ receptor from a human to a rat pharmacology. FEBS Lett 307:324–328

Kennett GA, Wood MD, Glen A, Grewal S, Forbes I, Gadre A, Blackburn TP (1994) In vivo properties of SB 200,646A, a novel 5-HT$_{2C/2B}$ receptor antagonist. Br J Pharmacol 111:797–802

Leonard BE (1973) Some effects of the hallucinogenic drug 2,5-dimethoxy-4-methylamphetamine on the metabolism of biogenic amines in the rat brain. Psychopharmacologia 32:33–49

Leysen JE (1985) Serotonin receptor binding sites. In: Green AR (ed) Neuropharmacology of serotonin. Oxford Press, Oxford, pp 86–87

Leysen JE, Pauwels PJ (1990) 5-HT$_2$ receptors, roles and regulation. In: Whitaker-Azmitia PM, Peroutka SJ (eds) The neuropharmacology of serotonin. Ann N Y Acad Sci 600:183–193

Leysen JE, Niemegeers CJE, Van Nueten JM, Laduron PM (1982) [^{3}H]-ketanserin (R 41,468), a selective ^{3}H-ligand for serotonin$_2$ receptor binding sites. Mol Pharmacol 21:301–304

Leysen JE, Janssen PFM, Niemegeers CJE (1989) Rapid desensitization and down-regulation of 5-HT$_2$ receptors by DOM treatment. Eur J Pharmacol 163:145–149

Lucki I, Nobler MS, Frazer A (1984) Differential actions of serotonin antagonists on two behavioral models of serotonin receptor activation in the rat. J Pharmacol Exp Ther 228:1133–1139

Marona-Lewicka D, Nichols DE (1995) Complex stimulus properties of LSD: a drug discrimination study with α_2-adrenoceptor agonists and antagonists. Psychopharmacology 120:384–391

Martin LL, Sanders-Bush E (1982) Comparison of the pharmacological characteristics of 5-HT$_1$ and 5-HT$_2$ binding sites with those of serotonin autoreceptors which modulate serotonin release. Naunyn Schmiedebergs Arch Pharmacol 321:165–170

McClue SJ, Brazell C, Stahl SM (1989) Hallucinogenic drugs are partial agonists of the human platelet shape change response: a physiological model of the 5-HT$_2$ receptor. Biol Psychiatry 26:297–302

McKenna DJ, Saavedra JM (1987) Autoradiography of LSD and 2,5-dimethoxy-phenylisopropylamine psychotomimetics demonstrates regional, specific cross-displacement in the rat brain. Eur J Pharmacol 142:313–315

McKenna DJ, Nazarali AJ, Himeno A, Saavedra JM (1989) Chronic treatment with ($\exists$)-DOI, a psychotomimetic 5-HT$_2$ agonist, downregulates 5-HT$_2$ receptors in rat brain. Neuropsychopharmacology 2:81–87

McKenna DJ, Repke DB, Peroutka SJ (1990) Differential interactions of indolealkylamines with 5-hydroxytryptamine receptor subtypes. Neuropharmacology 29:193–198

Meert TF, de Haes P, Janssen PAJ (1989) Risperidone (R 64,766), a potent and complete LSD antagonist in drug discrimination by rats. Psychopharmacology 97:206–212

Meert TF, De Haes PLAJ, Vermote PCM (1990) The discriminative stimulus properties of LSD: serotonergic and catecholaminergic interactions. Psychopharmacology 101:S71(271)

Meibach RC, Maayani S, Green JP (1980) Characterization and radioautography of [^{3}H]LSD binding by rat brain slices in vitro: the effect of 5-hydroxytryptamine. Eur J Pharmacol 67:371–382

Meltzer HY (1991) The mechanism of action of novel antipsychotic drugs. In: Kane JM (ed) New developments in the pharmacologic treatment of schizophrenia. NIMH Public Health Service, Washington DC, pp 71–95

Meltzer HY, Wiita B, Tricou BJ, Simonovic M, Fang VS, Manov G (1982) Effects of serotonin precursors and serotonin agonists on plasma hormone levels. In: Ho BT, Schoolar JC, Usdin E (eds) Serotonin in biological psychiatry. Raven Press, New York

Nanry KP, Tilson HA (1989) The role of 5-HT$_{1A}$ receptors in the modulation of the acoustic startle reflex in rats. Psychopharmacology (Berlin) 97:507–513

Nelson DL, Lucaites VL, Audia JE, Nissen JS, Wainscott DB (1993) Species differences in the pharmacology of the 5-hydroxytrypamine$_2$ receptor: structurally specific differentiation by ergolines and tryptamines. J Pharmacol Exp Ther 265:1271–1279

Nichols DE (1994) Medicinal chemistry and structure-activity relationships. In: Cho AK, Segal DS (eds) Amphetamine and its analogs. Academic, New York, pp 3–41

Nichols DE, Oberlender R, McKenna DJ (1991) Stereochemical aspects of hallucinogenesis. In: Watson RR (ed) Biochemistry and physiology of substance abuse, vol III. CRC, Boca Raton, FL, pp 1–39

Nichols DE, Frescas S, Marona-Lewicka D, Huang X, Roth BL, Gudelsky GA, Nash JF (1994) 1-(2,5-dimethoxy-4-(trifluoromethyl)phenyl)-2-aminopropane: a potent serotonin 5-HT$_{2A/2C}$ agonist. J Med Chem 37:4346–4351

Nielsen EB (1985) Discriminative stimulus properties of lysergic acid diethylamide in the monkey. J Pharmacol Exp Ther 234:244–249

Norman AB, Nash DR, Sanberg PR (1989) [^{3}H]Lysergic acid diethylamide (LSD): differential agonist and antagonist binding properties at 5-HT receptor subtypes in rat brain. Neurochem Int 14:497–504

Pahnke (1963) Drugs and mysticism. An analysis of the relationship between psychedelic drugs and the mystical consciousness. PhD thesis, Harvard University

Pierce PA, Peroutka SJ (1989) Hallucinogenic drug interactions with neurotransmitter receptor binding sites in human cortex. Psychopharmacology 97:118–122

Pierce PA, Peroutka SJ (1990) Antagonist properties of d-LSD at 5-hydroxytryptamine$_2$ receptors. Neuropsychopharmacology 3:503–508

Pompeiano M, Palacios JM, Mengod G (1994) Distribution of the serotonin 5-HT$_2$ receptor family mRNAs: comparison between 5-HT$_{2A}$ and 5-HT$_{2C}$ receptors. Mol Brain Res 23:163–178

Randic M, Padjen A (1971) Effect of N,N-dimethyltryptamine and d-lysergic acid diethylamide on the release of 5-hydroxyindoles in rat forebrain. Nature 230:532–533

Resnick O, Krus DM, Raskin M (1965) Accentuation of the psychological effects of LSD-25 in normal subjects treated with reserpine. Life Sci 4:1433–1437

Rogawski MA, Aghajanian GK (1979) Response of central monoaminergic neurons to lisuride: comparison with LSD. Life Sci 24:1289–1298

Rosecrans JA, Lovell RA, Freedman DX (1967) Effects of lysergic acid diethylamide on the metabolism of brain 5-hydroxytryptamine. Biochem Pharmacol 16:2011–2021

Sadzot B, Baraban JM, Glennon RA, Lyon RA, Leonhardt S, Jan C-R, Titeler M (1989) Hallucinogenic drug interactions at human brain 5-HT$_2$ receptors: implications for treating LSD-induced hallucinogenesis. Psychopharmacology 98:495–499

Sanders-Bush E (1994) Neurochemical evidence that hallucinogenic drugs are 5-HT$_{1C}$ receptor agonists: what next? In: Lin GC, Glennon RA (eds) Hallucinogens: an update. US-DHHS, Rockville, MD, pp 203–213 (NIDA research monograph series, 146)

Sanders-Bush E, Breeding M (1991) Choroid plexus epithelial cells in primary culture: a model of 5-HT$_{1C}$ receptor activation by hallucinogenic drugs. Psychopharmacology 105:340–346

Sanders-Bush E, Burris KD, Knoth K (1988) Lysergic acid diethylamide and 2,5-dimethoxy-4-methylamphetamine are partial agonists at serotonin receptors linked to phosphoinositide hydrolysis. J Pharmacol Exp Ther 246:924–928

Schmidt CJ, Fadayel GM, Sullivan CK, Taylor VL (1992) 5-HT$_2$ receptors exert a state-dependent regulation of dopaminergic function: studies with MDL 199,907 and the amphetamine analogue, 3,4-methylenedioxymethamphetamine. Eur J Pharmacol 223:65–74

Schreiber R, Brocco M, Millan MJ (1994) Blockade of the discriminative stimulus effects of DOI by MDL 100,907 and the 'atypical' antipsychotics, clozapine and risperidone. Eur J Pharmacol 264:99–102

Schreiber R, Brocco M, Audinot V, Gobert A, Veiga S, Millan MJ (1995) (1-(2,5-Dimethoxy-4-iodophenyl)-2-aminopropane)-induced head-twitches in the rat are mediated by 5-hydroxytryptamine (5-HT)$_{2A}$ receptors: modulation by novel 5-HT$_{2A/2C}$ antagonists, D$_1$ antagonists and 5-HT$_{1A}$ agonists. J Pharmacol Exp Ther 273:101–112

Shulgin AT, Shulgin A (1991) PIHKAL. A chemical love story. Transform, Berkeley, CA

Sotelo C, Cholley B, El Mestikawy S, Gozlan H, Hamon M (1990) Direct immunohistochemical evidence of the existence of 5-HT$_{1A}$ autoreceptors on serotoninergic neurons in the midbrain raphe nuclei. Eur J Neurosci 2:1144–1154

Sprouse JS, Aghajanian GK (1987) Electrophysiological responses of serotonergic dorsal raphe neurons to 5-HT$_{1A}$ and 5-HT$_{1B}$ agonists. Synapse 1:3–9

Sprouse JS, Aghajanian GK (1988) Responses of hippocampal pyramidal cells to putative serotonin 5-HT$_{1A}$ and 5-HT$_{1B}$ agonists: a comparative study with dorsal raphe neurons. Neuropharmacology 27:707–715

Teitler M, Leonhardt S, Weisberg EL, Hoffman BJ (1990) 4-[^{125}I]Iodo-(2,5-dimethoxy)phenylisopropylamine and [^{3}H]ketanserin labeling of 5-hydroxytryptamine$_2$ (5HT$_2$) receptors in mammalian cells transfected with a rat 5HT$_2$ cDNA: evidence for multiple states and not multiple 5HT$_2$ receptor subtypes. Mol Pharmacol 38:594–598

Titeler M, Lyon RA, Glennon RA (1988) Radioligand binding evidence implicates the brain 5-HT$_2$ receptor as site of action for LSD and phenylisopropylamine hallucinogens. Psychopharmacology 94:213–216

Tricklebank MD, Forler C, Middlemiss DN, Fozard JR (1985) Subtypes of the 5-HT receptor mediating the behavioral responses to 5-methoxy-N,N-dimethyltryptamine in the rat. Eur J Pharmacol 117:15–24

Trulson ME, Heym J, Jacobs BL (1981) Dissociations between the effects of hallucinogenic drugs on behavior and raphe unit activity in freely moving cats. Brain Res 215:275–293

Twarog BM, Page I (1953) Serotonin content of some mammalian tissues and urine and a method for its determination. Am J Physiol 175:157–161

Udenfriend S, Weissbach H, Clark CT (1954) The estimation of 5-hydroxytryptamine (serotonin) in biological tissues. J Biol Chem 215:337–344

U'Prichard DC, Greenberg DA, Snyder SH (1977) Binding characteristics of radiolabeled agonist and antagonists at central nervous system alpha noradrenergic receptors. Mol Pharmacol 13:454–473

Votava Z, Podvolava I, Semonsky M (1958) Studies on the pharmacology of d-lysergic acid cycloalkylamides. Arch Int Pharmacodyn Ther 115:114–130

Walker EA, Yamamoto T, Hollingsworth PJ, Smith DB, Woods JH (1991) Discriminative stimulus effects of quipazine and l-5-hydroxytryptophan in relation to serotonin binding sites in pigeons. J Pharmacol Exp Ther 259:772–782

Wasson RG, Hofmann A, Ruck AP (1977) The road to eleusis. Harcourt Brace Jovanovich, New York, pp 17–18

Watts VJ, Lawler CP, Fox DR, Neve KA, Nichols DE, Mailman RB (1995) LSD and structural analogs: pharmacological evaluation at D$_1$ dopamine receptors. Psychopharmacology 118:401–409

Williams JT, Colmers WF, Pan ZZ (1988) Voltage and ligand activated inwardly rectifying currents in rat dorsal raphe neurons in vitro. J Neurosci 8:3499–3506

Winter JC (1994) The stimulus effects of serotonergic hallucinogens in animals. In: Lin GC, Glennon RA (eds) Hallucinogens: an update. US-DHHS, Rockville, MD, pp 157–182 (NIDA research monograph series 146)

Winter JC, Rabin RA (1988) Interactions between serotonergic agonist and antagonists in rats trained with LSD as a discriminative stimulus. Pharmacol Biochem Behav 30:617–624

Woolley DW, Shaw E (1954) A biochemical and pharmacological suggestion about certain mental disorders. Proc Natl Acad Sci USA 40:228–231

Yamamoto T, Walker EA, Woods JH (1991) Agonist and antagonist properties of serotonergic compounds in pigeons trained to discriminate either quipazine or l-5-hydroxytryptophan. J Pharmacol Exp Ther 258:999–1007

Yocca FD, Wright RN, Margraf RR, Eison AS (1990) 8-OH-DPAT and buspirone analogs inhibit the ketanserin-sensitive quipazine-induced head shakes response in rats. Pharmacol Biochem Behav 35:251–254

The Interaction of the Serotoninergic and Other Neuroregulatory Systems in Alcohol Dependence

E.M. SELLERS, M.K. ROMACH, and A.D. Lê

A. Introduction

Substantial progress in the treatment of mental disorders over the past 40 years can be ascribed to the development of psychotropic medications for anxiety, sedation, depression, and psychosis. However, from a public health perspective, alcohol abuse and dependence have an even larger social, health, and economic impact than these other disorders.

Since many of the patients (>50%) who enter an alcohol rehabilitation treatment program relapse within the first 3 months of treatment (NATHAN 1986), improved pharmacotherapeutic or psychosocial approaches are needed to improve a patient's likelihood of maintaining abstinence (VOLPICELLI et al. 1995). Available medications for alcohol dependence show quite limited efficacy; therefore, strategies to improve their efficacy are needed. While more selective and potent agents may be found that have greater efficacy than currently marketed agents, evidence suggests that agents selective for several neuropharmacologic systems might also be effective and possibly safer.

In this chapter, we first examine studies which evaluate the general role of serotonin (5-HT) and specific 5-HT receptor types and/or subtypes on alcohol consumption in experimental animals and humans. Where applicable, we compare the effects of specific 5-HT receptor types on self-administration with other drugs. In addition, the possible interaction between 5-HT system and/or specific receptors with other neurotransmitters in the regulation of alcohol intake is discussed. The chapter concludes with a rationale for combined pharmacologic approaches that could target several neural systems concurrently based on the clinically important role that these systems play in the basic and clinical phenomena of craving, impulse regulation, and regulation of internal dysphoric mood states in initiating, maintaining, and causing relapse to alcohol dependence.

B. Role of the 5-HT System in Regulating Alcohol Consumption

I. Anatomical Pathways

The anatomy of 5-HT pathways and their various receptor types and subtypes are described in detail in other chapters in this book and will not be reviewed here. Briefly, the ascending 5-HT system is comprised mainly of 5-HT neurons originating from median and dorsal raphe nuclei which are aligned along the midline, dorsal to the fourth ventricle. Serotoninergic innervation of the striatum, substantia nigra, and amygdala derives mainly from neurons originating from the dorsal raphe nucleus, whereas septohippocampal 5-HT derives mainly from 5-HT neurons originating from the median raphe. A number of structures such as hypothalamus, cortex, and nucleus accumbens receive 5-HT innervation from both nuclei.

II. Role of Serotonin in Regulating Ethanol Consumption

1. Preclinical Studies

Myers and Veale (1968) were the first to report that general depletion of 5-HT by administration of p-chlorophenylalanine, a 5-HT synthesis inhibitor, reduces alcohol consumption in rats. However, research during the 1970s resulted in conflicting results concerning the effects of various 5-HT manipulations on ethanol consumption in experimental animals. The development of better pharmacological agents for modifying central 5-HT activity and better models for assessing ethanol intake in experimental animals resulted in a consistent pattern of 5-HT effects on ethanol intake becoming apparent in the 1980s. In general, manipulations which elevate 5-HT neurotransmission reduce ethanol consumption. Examples of such drugs include the 5-HT releaser and re-uptake inhibitor dexfenfluramine (Higgins et al. 1992; Rowland and Morian 1992) and the selective 5-HT uptake inhibitors such as sertraline (Gill et al. 1988; Higgins et al. 1992), fluoxetine (Haraguchi et al. 1990), and zimelidine (Amit et al. 1984). A recent review by Sellers et al. (1992) summarizes the effects of various 5-HT agents on alcohol-related behaviors.

In contrast, treatments which reduce 5-HT function, however, do not necessarily lead to an increase in ethanol consumption. In fact, in some circumstances, reducing 5-HT function might produce a similar effect as that of stimulating 5-HT activity. Such apparent paradoxical effects of the 5-HT system on alcohol intake are related to the system's complexity. Over the last 10 years, much progress has been made on identifying various 5-HT receptor types and subtypes as well as their functional significances. As will be discussed, ethanol consumption not only depends on the particular 5-HT receptor types or subtypes, but also their location.

2. Human Experimental and Clinical Studies

In humans, there is evidence that 5-HT and its metabolites are decreased in the cerebrospinal fluid of many alcoholics (Borg et al. 1985; Linnoila 1990), suggesting a central 5-HT deficiency in alcohol dependence (Romach and Tomkins 1995). Variations in 5-HT function have been implicated in a large number of psychiatric disorders (Coccaro and Murphy 1990). These include affective disorders, eating disorders, impulsivity, and violent behaviors. An analogy is often drawn between the role of the 5-HT system in the appetite for food and for alcohol, and it has been suggested that excesses of eating and alcohol consumption represent a form of self-treatment for a disorder in 5-HT functioning (BREWERTON et al. 1990).

a) Serotonin Uptake Inhibitors (SSRIs)

The most extensively studied group of serotoninergic drugs have been the 5-HT uptake inhibitors. A number of short-term clinical trials, 2–4 weeks in duration and placebo-controlled, have been conducted in mild to moderately alcohol-dependent individuals (DSM-III-R criteria) (NARANJO and SELLERS 1989). Zimelidine (NARANJO et al. 1984), citalopram (NARANJO et al. 1987, 1992), viqualine (NARANJO et al. 1989), and fluoxetine (NARANJO et al. 1990) reduced alcohol intake by an average of 9%–17%. There appeared to be some minor differences in response patterns amongst the drugs. Zimelidine and citalopram increased the number of abstinent days, whereas viqualine, a 5-HT releaser and uptake inhibitor, and fluoxetine decreased the number of drinks on drinking days. The effects seen were dose-related and in contrast to an antidepressant effect; the onset of effect was rapid. A similar trial with sertraline, another uptake inhibitor, has now been completed (NARANJO and SELLERS, unpublished data). The clinical patterns of response are consistent with observations in rats studied with a drinkometer in a continuous access paradigm in which SSRIs increased the latency to drinking and decreased the probability of initiating drinking (HIGGINS et al. 1992).

Recent human experimental ethanol self-administration studies with citalopram (NARANJO et al. 1992) and fluoxetine (GORELICK and PAREDES 1992) support the earlier findings of rapid onset within the first week of treatment and modest decreases in alcohol intake, 18% and 14%, respectively. Furthermore, these studies reported diminution in craving and interest to use ethanol.

SSRIs are clinically efficacious antidepressants and depression is common in alcoholics. However, the effectiveness of 5-HT uptake inhibitors in the treatment of depressed alcoholics has not been systematically investigated. One open trial of fluoxetine in 12 detoxified alcoholics showed significant improvements in measures of depressive symptomatology and alcohol intake (CORNELIUS et al. 1992). A recently completed 6-month, double-blind, placebo-controlled trial of fluoxetine in alcohol-dependent individuals showed decreases in scores on scales of depression (KRANZLER et al. 1995).

b) Dexfenfluramine

One clinical trial has been conducted (ROMACH et al. 1996) in which 7.5–30 mg dexfenfluramine (a 5-HT releaser and uptake inhibitor) twice daily was administered together with structured psychosocial treatment in chronic alcoholics in an attempt to follow up on the prediction that a drug with several mechanisms of action could be more effective than the SSRIs. Due to a large placebo-control response (i.e., 50% reduction in drinking), no main drug effect was found. The issue of whether dexfenfluramine has efficacy in the presence of minimal psychosocial treatment is left open by this trial.

c) Role of General Serotoninergic Agents

Despite consistent preclinical data, agents which generally increase 5-HT have not proven to have sufficiently large clinical effects to be of therapeutic importance. Unfortunately, the dose ranges studied and duration of the trials have been very limited.

C. Involvement of Selective 5-HT Receptor Types and Subtypes in Alcohol Consumption

I. 5-HT Receptor Types and Subtypes

At present, 5-HT receptor cloning work has identified at least seven different types of 5-HT receptors, which have been named 5-HT_1 to 5-HT_7. A number of receptor subtypes also exists for these various 5-HT receptors. Readers are referred to an excellent review by HOYER et al. (1994) for a detailed description of these various receptor types, subtypes, their distribution, and their functional significances. At present, selective ligands, however, are available for certain numbers of 5-HT receptor types and subtypes. Experimental studies with alcohol or other behavioral aspects of these receptor functions are therefore limited to such availability.

II. 5-HT$_1$ Receptor Type and Subtypes

At present, various 5-HT_1 subtypes, namely, 5-HT_{1A}, 5-HT_{1B}, 5-HT_{1D}, 5-HT_{1E}, and 5-HT_{1F}, have been cloned. The 5-HT_{1C} receptor has been renamed as the 5-HT_{2C} receptor on the basis of operational, transductional, and structural criteria (HUMPHREY et al. 1993). Among these various 5-HT_1 receptor subtypes, the involvement of 5-HT_{1A} and 5-HT_{1B} on alcohol consumption has been examined.

1. 5-HT$_{1A}$ Receptors

a) Preclinical Studies

Depending on their location, 5-HT_{1A} receptors can be classified into somatodendritic and postsynaptic receptors. The receptors located postsynaptically to 5-HT axons in the forebrains are referred to as postsynaptic 5-HT_{1A} receptors and are found in the hippocampus, septum, and amygdala, and in some cortical layers (HOYER et al. 1994). The 5-HT_{1A} receptors located in the median and dorsal raphe are referred to as somatodendritic autoreceptors. Activation of the somatodendritic receptors causes a reduction in 5-HT synthesis, and in the release and firing activity of 5-HT neurons (HJORTH and

SHARP 1991; INVERNIZZI et al. 1991). Activation of 5-HT$_{1A}$ receptors by systemic administration of an 5-HT$_{1A}$ agonist or antagonist would produce a complex pattern of responses that depends on the relative activation of both somatodendritic 5-HT$_{1A}$ autoreceptors and postsynaptic 5-HT$_{1A}$ receptors induced by the doses of the agonists or antagonists employed.

8-Hydroxy-2-(di-n-propylamino) tetralin (8-OH-DPAT) is one of the 5-HT$_{1A}$ agonists that has been most commonly employed to investigate the involvement of 5-HT$_{1A}$ receptors in alcohol consumption. In general, it can be said that the reported effects of 5-HT$_{1A}$ agonists from various laboratories on alcohol consumption are rather inconsistent, with decrease, increase, or no effects on alcohol consumption all having been reported. This is not surprising considering the complex action of 5-HT$_{1A}$ receptors mentioned above. A close examination of the data, however, reveals a number of critical factors which might account for the discrepancies observed from these various studies.

With respect to studies which employed systemic administration of 8-OH-DPAT, the doses of 8-OH-DPAT and the duration of alcohol drinking appear to play a critical role in determining whether an increase or decrease in ethanol consumption is observed.

When access to alcohol consumption was restricted to short duration or when alcohol consumption was monitored at repeated intervals, doses of 8-OH-DPAT ranging from 0.5 to 2.0 mg/kg suppressed alcohol intake (MURPHY et al. 1987a; McBRIDE et al. 1990; HIGGINS et al. 1992). However, an increase in alcohol consumption was observed when low doses of 8-OH-DPAT (0.03–0.06 mg/kg) were administered or when alcohol intake was measured several hours after administration of higher doses of 8-OH-DPAT (TOMKINS et al. 1994a; MURPHY et al. 1987b; HIGGINS et al. 1992). A biphasic effect was observed with gepirone: low doses stimulate, whereas higher doses decrease or have no effect on alcohol drinking (KNAPP et al. 1992; TOMKINS et al. 1993).

The biphasic effect of 8-OH-DPAT with respect to doses and duration might be related to its differential action on somatodendritic and postsynaptic 5-HT$_{1A}$ receptors. Low doses of 8-OH-DPAT might suppress 5-HT activity by preferentially acting at somatodendritic 5-HT$_{1A}$ receptors, thereby stimulating alcohol intake. Higher doses of 8-OH-DPAT have been shown to produce 5-HT syndromes which cause motor impairment, at least during the early period following its administration. Such impairment would interfere with the animal's ability to consume alcohol and thereby produce a decrease in alcohol intake. The subsequent enhancement of alcohol consumption induced by higher doses at later times may be due to the elimination of 8-OH-DPAT, which results in a lower level of 8-OH-DPAT.

This notion of the preferential activation of somatodendritic 5-HT$_{1A}$ receptors induced by low doses of 8-OH-DPAT which produces an increase in ethanol intake is consistent with other studies employing microinjection of 8-OH-DPAT into the raphe nuclei. Focal infusion of 5 or 2.5 fg of 8-OH-DPAT into the median or dorsal raphe nucleus, which specifically activate somatodendritic 5-HT$_{1A}$ receptors in these areas, enhances alcohol intake.

Such effects of intraraphe injection of 8-OH-DPAT can be reversed by peripheral administration of WAY100135, a 5-HT$_{1A}$ receptor antagonist (Tomkins and Fletcher 1996). Intradorsal raphe injection of a higher dose of 8-OH-DPAT (10 μg), however, has been reported to decrease alcohol preference but not the amount of alcohol consumption in Alko-alcohol (AA) rats which have been selectively bred for high ethanol consumption (Schreiber et al. 1993). It should be pointed out that the dose of 8-OH-DPAT employed in this study was four times higher than that employed in the Tomkins et al. study (1994a), and much higher than that required to induce changes in other behaviors (Higgins et al. 1988; Fletcher 1991).

The effects of various partial 5-HT$_{1A}$ receptor agonists such as buspirone (Privette et al. 1988; Kostowski and Dyr 1992; Rezvani et al. 1991), ipsapirone (Engel et al. 1990; Schreiber et al. 1993), and gepirone (Knapp et al. 1992; Tomkins et al. 1993) on alcohol consumption have also been examined. Ipsapirone in doses ranging from 2.5 mg to 10 mg/kg reduced alcohol intake in rats. A similar finding has also been observed for buspirone and gepirone, with the exception of the study by Tomkins et al. (1993) which showed that gepirone in doses from 1–3 mg/kg s.c. increased alcohol in rats that have a low preference for alcohol. On the other hand, doses of gepirone greater than 3 mg/kg reduced alcohol intake in high-alcohol-preferring rats.

These studies with 8-OH-DPAT and various partial 5-HT$_{1A}$ agonists, indicate that 5-HT$_{1A}$ receptors play a role in the regulation of alcohol consumption. The stimulation of ethanol intake induced by suppression of 5-HT activity through peripheral administration of low doses or intraraphe injection of 8-OH-DPAT is consistent with observations that certain alcohol-preferring strains of rats such as the preferring (P), high-alcohol drinking (HAD), and the Fawn-hooded rats have lower 5-HT levels than their lower alcohol-preferring counterparts (McBride et al. 1990; Wong et al. 1990). The involvement of postsynaptic 5-HT$_{1A}$ receptors in the regulation of alcohol consumption is not clearly understood at present. Due to the complex nature of the 5-HT$_{1A}$ receptor with its somatodendritic and postsynaptic locations, studies that utilize direct administration of 5-HT$_{1A}$ receptor agonists into the relevant brain sites as well as an examination of the effects of 5HT$_{1A}$ agonists in animals with neurotoxin lesions of raphe nuclei are required to clarify the involvement of the postsynaptic 5-HT$_{1A}$ receptors in alcohol intake.

b) Human Experimental and Clinical Studies

Preclinical studies do not suggest 5-HT$_{1A}$ (partial) agonists to have a useful therapeutic role in directly decreasing alcohol consumption to clinically safe doses. However, the lack of any clinical evidence for a potential role, particularly in anxious alcoholics, has been reported. For example, an initial study with buspirone, an anxiolytic 5-HT$_{1A}$ partial agonist, in alcoholics suggested some decrease in drinking (Bruno 1989). In a larger 12-week, placebo-controlled trial in 61 anxious alcoholics, buspirone therapy was associated with greater retention during the trial period, reduced anxiety, a slower return to heavy alcohol consumption, and fewer drinking days during the follow-up period (KRANZLER et al. 1994). In 51 dually diagnosed patients (generalized anxiety plus alcohol abuse/dependence), buspirone was superior to placebo as an anxiolytic and was associated with a reduction in the number of days of desiring alcohol and with an overall clinical improvement (TOLLEFSON et al. 1992). In contrast, in a well-conducted, double-blind trial of buspirone versus placebo in highly anxious alcoholics who had recently completed inpatient detoxification for alcoholism, no benefit over placebo was observed in a number of anxiety and alcohol use tests (MALCOLM et al. 1992). Clearly, further larger studies with careful characterization of anxiety in alcohol-dependent patients will be needed to establish a role for $5-HT_{1A}$ agents.

2. $5-HT_{1B}$ Receptors

a) Preclinical Studies

Few selective ligands are available for $5-HT_{1B}$ receptors. A number of compounds such as 5-methoxy-3-(1,2,3,6-tetrahydropyridin-4-yl)-1H-indole (RU 24969), m-trifluoromethylphenylpiperazine (TFMPP) and m-chloro-phenylpiperazine (mCPP) have been claimed to act as agonists at $5-HT_{1B}$ receptors (HOYER et al. 1994). Among these compounds, the effects of mCPP on alcohol consumption have been examined. HIGGINS et al. (1992) demonstrated that, in rats which have been maintained on continuous access drinking, mCPP suppressed alcohol intake during the first hour post-injection; however, alcohol intake increased 6–12h post injection. In limited access drinking, mCPP produces dose-dependent (0.1–1 mg/kg) suppression of alcohol intake in the rat. The effects of the highest dose of mCPP appeared to be nonspecific, as a suppression of water intake was also observed (BUCZEK et al. 1994). Although mCPP has a good affinity as an agonist at $5-HT_{1B}$ receptors, it is difficult to pinpoint the precise primary site(s) of mCPP action on ethanol drinking, and although mCPP has an agonist action at $5-HT_{1B}$ receptors, it also has comparable potency as agonist at $5-HT_{2B}$ and $5-HT_{2C}$ receptors (HOYER et al. 1994). In fact, in the same study, BUCZEK et al. (1994) also found that the effects of mCPP on ethanol is reversed by pretreatment with metergoline but not by ritanserin. Since metergoline has a better affinity than mCPP as an agonist at $5-HT_{1B}$ receptors, it is also a highly selective $5-HT_{2C}$ antagonist, whereas ritanserin is a good antagonist at both $5-HT_{2A}$ and $5HT_{2C}$ receptors. This raises the issues of whether the effects of mCPP on ethanol intake can be attributed to its action on $5-HT_{1B}$ or $5-HT_{2C}$ receptors.

Probably the strongest evidence for the involvement of $5-HT_{1B}$ receptors in alcohol intake is derived from a recent study by CRABBE et al. (1996) with $5-HT_{1B}$ knockout mice. $5-HT_{1B}$ receptor-deficient mice consume much more alcohol than the wild-type control. In fact, some of the mutants consume as much as 26g/kg of ethanol per day, which is almost twice that of the C57BL/

6 mice which have a well-documented high preference for alcohol (CRABBE et al. 1996).

b) Human Experimental and Clinical Studies

KRYSTAL and colleagues (1994) presented data from a human study with the pharmacological probe mCPP, a drug which interacts with several 5-HT receptor subtypes, which implicated the serotoninergic system in the discriminative properties of ethanol and suggested a serotoninergic contribution to craving. mCPP infusions generated dysphoric effects, and this was felt to have contributed to its capacity to stimulate craving for alcohol.

III. 5-HT$_{2A}$ Receptors

1. Preclinical Studies

Ritanserin and amperozide are the two prototype compounds that have been used quite extensively over the last several years to examine the involvement of 5-HT$_{2A}$ receptor antagonists on alcohol intake. The effects of ritanserin on alcohol consumption have been rather inconsistent with reduction (MEERT et al. 1991; MEERT 1993; PANOCKA and MASSI 1992; PANOCKA et al. 1993b,c) and no effects (HIGGINS et al. 1992; MYERS et al. 1993; PANOCKA et al. 1993a) both having been reported.

MEERT et al. (1991) were the first to report that treatment with ritanserin markedly decreased preference for 3% w/v ethanol solution in a dose-dependent manner(0.01–0.16 mg/kg), as measured in a 24 h/two-bottle choice drinking paradigm. In a subsequent study, they found that doses of ritanserin ranging from 0.4 to 2.5 mg/kg also reduced preference for alcohol. Doses of 10 mg/kg of ritanserin, however, produced no effect on alcohol preference and slightly increased fluid intake (MEERT 1993). PANOCKA and co-workers (1993a,b,c; PANOCKA and MASSI 1992) have carried out extensive investigation into the role and mechanisms of ritanserin effects on alcohol preference. Employing a similar paradigm as that of MEERT (1993), they showed that daily treatment with 1 or 10 mg/kg of ritanserin for a period of 9 days produced a marked suppression of alcohol preference. The suppression of alcohol preference by ritanserin was found to be long lasting, as it was still observed at 20 days after the cessation of the treatment. These investigators have also shown that the effects of ritanserin on alcohol intake were different from that of haloperidol, a dopamine D$_2$ receptor antagonist, and risperidone, a mixed serotoninergic and dopaminergic receptor antagonist. The suppression of alcohol preference by haloperidol or by a high dose of risperidone (10 mg/kg) was rapid in its onset but of short duration, whereas the effects of ritanserin or a low dose of risperidone (0.1 mg/kg), which preferentially binds to 5-HT$_2$ receptor on alcohol preference, were of slow onset and long-lasting effect (PANOCKA and MASSI 1992). These results suggest that the effects of ritanserin on alcohol preference are likely to be due to its antagonism of 5-HT$_{2A}$ receptor rather than its possible action on dopamine receptors. The effects of ritanserin on alcohol preference have also been suggested by the same group of investigators to be mediated centrally and possibly through its antagonism of 5-HT$_{2A}$ receptors located in the nucleus accumbens, as intracerebroventricular or intra-accumbens injection of ritanserin also suppresses alcohol preference (Panocka et al. 1993b).

Failure to observe any effects of ritanserin on alcohol drinking and/or alcohol preference has also been reported by Panocka et al. (1993a) as well as by other investigators (Myers and Lankford 1993; Higgins et al. 1992; Svensson et al. 1993). In all these studies, ritanserin in doses ranging from 0.3 to 10 mg/kg was found to have no effects on alcohol preference (Panocka et al. 1993a; Svensson et al. 1993; Myers and Lankford 1993) or alcohol consumption (Higgins et al. 1992; Myers and Lankford 1993). Although Higgins et al. (1992) found no effect of ritanserin on alcohol consumption by itself, pretreatment with 0.3 or 3 mg/kg of ritanserin attenuated the suppression of alcohol intake induced by dexfenfluramine.

The effects of amperozide on alcohol consumption and/or preference has been derived mainly from Myers' laboratory. Amperozide has been found to decrease alcohol preference and/alcohol intake in Sprague Dawley rats (Myers et al. 1992; McMillen et al. 1993) or the P rats (McMillen et al. 1994), which have been selectively bred for high alcohol consumption. McMillen et al. (1994) have also shown that while amperozide was effective in reducing alcohol intake in the P rats, similar treatment with trazodone, another 5-HT$_{2A}$ receptor antagonist, was ineffective in reducing alcohol consumption.

It is difficult to reconcile various findings concerning the effects of ritanserin on alcohol consumption. In all these studies, the same paradigm of alcohol consumption, 24 h/two-bottle choice, was employed to assess the effect of ritanserin. It is possible, however, that the effects of ritanserin might be related to the amount of alcohol consumed by the animals. In studies which reported a positive effect of ritanserin on alcohol consumption, the amount of alcohol consumed by the animals was relatively low (2–3 g/kg per day) as estimated from the preference, total fluid consumption, and the percentage of alcohol solution employed (Meert et al. 1991; Meert 1993; Panocka and Massi 1992; Panocka et al. 1993a,c). Negative effects of ritanserin were observed in studies that employed high-alcohol-preferring rats (Panocka et al. 1993b) or normal rats which consumed high amounts such as 5–6 g/kg per day (Myers and Lankford 1993). It is clear that further work is required before any definite conclusion regarding the effects of ritanserin on alcohol consumption can be evaluated. Data on the effects of ritanserin treatment in different paradigms of alcohol drinking, such as an operant paradigm or a limited access drinking paradigm where the amount and/or pattern of alcohol consumption during a particular period of time are known, would be of importance to evaluate the effect of ritanserin on alcohol consumption.

Although ritanserin and amperozide are both 5-HT$_{2A}$ receptor antagonists, amperozide has also been shown to inhibit 5-HT re-uptake in vitro and
also to increase extracellular dopamine in the nucleus accumbens (McMILLEN
et al. 1993). It is possible that the effects of amperozide on alcohol consumption might be operated through more mechanisms than that induced by 5-HT
receptor blockade alone.

2. Human Experimental and Clinical Studies

Ritanserin is a selective 5-HT$_{2A/2C}$ antagonist which in several animal paradigms of alcohol dependence has been suggested, but not strongly, to play a
role in alcoholism treatment. Three clinical trials are reported to be underway
(JANSSEN 1994). To date only one human experimental study has been reported. The short-term effects of ritanserin in 39 (35 male, four female) heavy
social drinkers (consuming at least 28 drinks/week) aged 19–63 years and who
were not seeking treatment have been studied (NARANJO et al. 1995). After
an intake assessment, they received placebo for 7 days in a single-blind
baseline. They were then randomly assigned to one of three double-blind
treatments for 14 days: ritanserin 5 mg/day ($n = 12$), ritanserin 10 mg/day ($n =$
13) or placebo ($n = 14$). Experimental drinking sessions were conducted after
baseline (EDS1) and treatment (EDS2); in each session, subjects were offered
18 mini-drinks (total = six standard) and rated their desire to drink, intoxication and mood (POMS). The outpatient results were as follows: ritanserin
5 mg/day decreased desire and craving for alcohol (versus baseline, $p < 0.05$),
but not alcohol intake. Liking of alcohol decreased from baseline with
ritanserin 10 mg/day ($p = 0.01$) and placebo ($p = 0.05$). Changes in alcohol
intake from baseline with ritanserin 10 mg/day (increase, $p > 0.05$) and placebo
(decrease, $p > 0.05$) were different ($p < 0.05$). Both ritanserin 5 mg/day and
10 mg/day enhanced alcohol-induced decreases in fatigue compared with placebo ($p < 0.05$). These results indicate that ritanserin may have differential
effects on alcohol intake, desire, craving, and liking, as well as intoxication and
some of alcohol's effects on mood. However, they suggest that ritanserin has
limited efficacy in reducing alcohol intake in heavy drinkers (NARANJO et al.
1995).

A double-blind crossover design trial was carried out to investigate the
effect of simultaneous administration of alcohol (0.5 g/kg) and ritanserin
(10 mg) on biological and behavioral functioning. Twenty healthy volunteers
were selected to participate in the study. Psychomotor performance and vital
signs during the ritanserin session did not differ significantly from the placebo
session. Similar results were obtained with respect to alcohol intoxication,
euphoria, and mood, except for tiredness and alertness, which were significantly different compared to placebo. The findings tend to indicate that
ritanserin neither enhances the central nervous system depressant effect of
alcohol nor produces a pharmacokinetic interaction during acute alcohol ingestion (ESTEVEZ et al. 1995).

A trial has also been carried out to determine the effect of ritanserin or a placebo on sleep and mood in two groups of abstinent alcoholic patients (MONTI et al. 1993). Ritanserin was given at a daily dose of 10 mg for 28 days and was preceded (10 days) and followed (2 days) by a placebo. In the ritanserin group, there was a reduction of total waking time. Total sleep time increase was associated with significantly larger amounts of nonrapid eye movement sleep. Slow wave sleep and rapid eye movement sleep were not significantly modified. Compared to the placebo group, a statistically significant decrease in the Hamilton Rating Scale for Depression and Hamilton Rating Scale for Anxiety was found in the ritanserin group after 1 week of treatment. The absence of an effect in the placebo-treated group suggests that the clinical response and sleep improvement were mainly related to ritanserin administration (MONTI et al. 1993).

IV. 5-HT$_3$ Receptors

1. Preclinical Studies

In contrast to the reported effects of 5-HT$_2$ receptor antagonists, pretreatment with 5-HT$_3$ receptor antagonists has been shown to produce a consistent reduction in ethanol intake in various paradigms of alcohol self-administration. Various 5-HT$_3$ receptor antagonists such as ondansetron (TOMKINS et al. 1995), zacropride (KNAPP and POHORECKY 1992), MDL72222 (FADDA et al. 1991), tropisetron, or ICS 205–930 (HODGE et al. 1993; KOSTOWSKI et al. 1993) have been reported to reduce alcohol consumption or decrease alcohol preference in rats. SVENSSON et al. (1993), however, failed to observed any effect of ondansetron on ethanol preference. This disparity may be partly due to the short length of the ethanol acquisition procedure employed in this study as compared to other studies. In addition, the absence of the effects of ondansetron on ethanol intake in the SVENSSON et al. (1993) study might be related to the duration of ondansetron treatment. Depending on the experimental conditions, the effects of 5-HT$_3$ receptor antagonists on ethanol intake might require several days of treatment to become manifest.

5-HT$_3$ receptor antagonists might reduce ethanol intake by modifying its reinforcing properties. It has been proposed that ethanol-induced release of dopamine in the nucleus accumbens may play a major role in ethanol reinforcement (DI CHIARA and IMPERATO 1988; KOOB 1992). Pretreatment with 5-HT$_3$ receptor antagonists has been shown to attenuate such effects of ethanol (CARBONI et al. 1989; WOZNIAK et al. 1990; YOSHIMOTO et al. 1991). Recent studies by JANKOWSKA and KOWTOWSKI (1995) and JANKOWSKA et al. (1995) also support this view. These workers demonstrated that bilateral infusion of tropisetron into the nucleus accumbens can reduce ethanol intake (JANKOWSKA and KOWTOWSKI 1995). Moreover, they also showed that the reduction of ethanol intake by peripheral administration of tropisetron is attenuated in rats subjected to dopamine lesions by intracerebral administration of 6-hydroxydopamine (JANKOWSKA et al. 1995).

Since 5-HT$_3$ receptors also exist in the terminals of sympathetic and para-sympathetic of the peripheral nervous system and have also been implicated in the control of gastric motility (Cohen 1992), one could argue that 5-HT$_3$ receptors might alter alcohol intake by a pharmacokinetic mechanism. Indeed, one of the 5-HT$_3$ receptor antagonists, MDL 72222, has been shown to reduce ethanol absorption in the rats when ethanol is administered by the oral route (Grant 1995) but not when ethanol is administered by the intraperitoneal route. However, other 5-HT$_3$ antagonists such as ondansetron or tropisetron did not affect alcohol absorption (Grant and Colombo 1993; Tomkins et al. 1995) in rats or in humans (Johnson et al. 1993a), indicating that the observed effects of 5-HT$_3$ receptor antagonists on ethanol consumption are unlikely to be attributed to an alteration in ethanol kinetics.

5-HT$_3$ receptor antagonists appear to selectively reduce ethanol self-administration, but not other reinforcers such as saccharin and food. In addition, pretreatment with 5-HT$_3$ receptor antagonists did not affect cocaine (Lane et al. 1992; Peltier and Schenk 1991), heroin (Higgins et al. 1994), or nicotine (Corrigall and Coen 1994) self-administration.

2. Clinical Studies

Ondansetron, a 5-HT$_3$ antagonist, has also been shown to have significant but modest efficacy in decreasing alcohol consumption in a treatment trial of alcohol dependence (Sellers et al. 1991, 1994). In a randomized, placebo-controlled study, the efficacy of 6 weeks of ondansetron (0.25 mg b.i.d. or 2.0 mg b.i.d.) was tested in the treatment of 71 non-severely alcohol-dependent males. The results showed a reduction in drinking, with the differences steadily increasing toward the end of the treatment period and approaching significance at week 7 in the 0.25 mg group ($p = 0.06$). Twice as many patients in this group showed a 2 standard deviations decrease in drinking compared with the other groups. When patients drinking more than ten drinks per drinking day at baseline ($n = 11$) were excluded from the analysis, significant group differences were found at both treatment and follow-up, with the lower ondansetron dose producing the greatest reduction from baseline (i.e., 2.8 standard drinks; 35% compared with baseline and 21% compared with placebo; $p < 0.02$–0.001). Within this group, there was an almost fourfold greater number of patients showing a clinically meaningful decrease in drinking.

In an experimental study, the effect of the 5-HT$_3$ receptor antagonist ondansetron (4 mg administered orally) on some of the psychological effects of a small dose of alcohol (580 ml of 3.6% alcohol content by volume of lager) was studied in 16 healthy male volunteers using a double-blind placebo controlled, Latin Square crossover design. Pretreatment with ondansetron significantly attenuated several of the subjective pleasurable effects of alcohol, and also decreased the subjective desire to drink. These findings are consistent with preclinical studies, suggesting that the reinforcing properties of alcohol may be attenuated by 5-HT$_3$ receptor blockade (Johnson et al. 1993b).

V. 5-HT₄ Receptors

1. Preclinical Studies

So far there is only one study which has examined the effects of 5-HT$_4$ receptor antagonists on ethanol intake. GR113808, which has been shown to have 3000-fold selectivity for 5-HT$_4$ compared to 5-HT$_3$ receptors, has been shown to reduce ethanol intake in water-sated or water-deprived Sardinian alcohol preferring (sP) rats. The inhibition of ethanol intake induced by pretreatment with GR-113808 (1–10 mg/kg subcutaneously) was observed during the 2 h daily access to ethanol solution (PANOCKA et al. 1995). Indirect evidence for the involvement of 5-HT$_4$ receptors on ethanol intake can also be derived from a study which employed high doses of tropisetron (HODGE et al. 1995). Doses of tropisetron (10 and 17 mg/kg) have been shown to attenuate the stimulation of ethanol intake induced by morphine, while having no effect on ethanol intake on their own. High doses of tropisetron (>1.0 mg/kg), however, have been shown to have antagonist activity at 5-HT$_4$ receptors (DUMUIS et al. 1988). No studies in humans have been conducted.

D. Interaction Between 5-HT and Other Neurotransmitters in the Regulation of Alcohol Intake

Pharmacologic approaches to understanding neurobiological mechanisms underlying alcohol abuse have typically focused on manipulating the activity of a single neurochemical system. Neurochemical systems, however, do not act singly but interact with one another to exert various physiological or behavioral effects. It is not surprising, therefore, that a number of neurotransmitters have been shown to affect alcohol consumption. For example, beside the 5-HT system, other neurotransmitters such as dopamine, GABA, and opioids have been shown to play a role in alcohol consumption. In this section we will examine the interaction between 5-HT and such other neurochemical systems in the regulation of alcohol consumption.

I. 5-HT and Opioid Systems

1. Preclinical Studies

Acute administration of 5-HT agonists such as lysergic acid diethylamide (LSD) and mCPP has been shown to decrease the level of immunoreactive L-enkephalin in various brain regions (KMIECIAK-KOLADA and KOWALSKI 1986). Similarly, a number of studies have shown a significant increase in the levels of met- and leu-enkephalins or β-endorphin 24 h after acute administration of fenfluarmine or dexfenfluramine (GROPETTI et al. 1984; HARSING et al. 1982a,b; MAJEED et al. 1986). Administration of morphine, on the other hand, has been shown to increase 5-HT utilization (BRASE 1977). Similarly, administration of enkephalinase inhibitors produces an increase in 5-HT turnover (LLORENS-

Cortes and Schwartz 1984). Behaviorally, it was shown that combined administration of 5-hydroxytryptophan and naloxone reduced food intake at doses that had no effect on food intake when administered separately (Fernandez-Tome et al. 1988). Similarly, tropisetron, which does not have any effect on food intake on its own, can potentiate naloxone-induced hypophagia in rats (Beczowska et al. 1991).

Despite such evidence for the interaction between the opiate and 5-HT systems, little attention has been focused on the role of such interactions in ethanol intake. We have examined the interaction between the 5-HT and opiate systems on alcohol over the last few years. Low doses of fluoxetine and naltrexone, which had no effects on alcohol intake on their own, produced a strong suppression of alcohol intake when given together (Lê and Sellers, unpublished results). Similarly, we have also found that the 5-HT$_3$ antagonist ondansetron can potentiate the reduction of ethanol intake induced by naltrexone (Lê and Sellers 1994). Given that both 5-HT and opiate compounds have been used clinically for possible treatment of alcohol dependence on their own, the combined use of these agents might be of potential importance in the development of medication for the treatment of alcohol abuse.

2. Clinical Studies

Two clinical trials with the opiate antagonist naltrexone have shown the medication to be useful in the treatment of alcoholism. In a double-blind study, Volpicelli et al. (1992) treated detoxified alcoholics with 50 mg of naltrexone or placebo for 12 weeks. Naltrexone-treated subjects had significantly lower relapse rates, fewer drinking days, and significantly less craving for alcohol than placebo-treated subjects. Naltrexone-treated subjects felt less "high" than usual after drinking alcohol. Only 17% of the naltrexone-treated subjects relapsed to clinically significant drinking after treatment compared with nearly 66% of placebo-treated patients. O'Malley et al. (1992) found that patients who received naltrexone and supportive psychotherapy for 12 weeks abstained from drinking alcohol longer than subjects who received placebo and supportive therapy. Naltrexone-treated subjects who also received coping skills therapy to prevent alcohol relapse were less likely to progress from a slip to excessive drinking than placebo-treated subjects who received similar coping skills therapy. In both these studies, naltrexone-treated patients who did drink reported that they consumed fewer drinks per drinking day. Sixteen alcoholic patients treated with naltrexone and 27 treated with placebo who participated in a 12-week clinical trial reported retrospectively on their subjective responses to their first episode of a lapse into alcohol consumption and on their reasons for terminating the drinking episode. Compared to the subjects who received placebo, the subjects who received naltrexone reported lower levels of craving for alcohol and were more likely to give reasons for terminating drinking that were consistent with decreased incentive to drink. These findings support the hypothesis that one central effect of naltrexone is the modification of alcohol-induced craving. A pilot study with an experimental opioid antagonist, nalmefene, has shown similar effects in alcohol-dependent individuals compared to placebo, a significantly lower rate of relapse, and a significant decrease in the number of drinks per drinking day (MASON et al. 1994).

BOHN et al. (1994) studied the effects of adding naltrexone to brief counseling in 14 nondependent heavy drinkers. Naltrexone significantly decreased desire for alcohol, drinking frequency, frequency of heavy alcohol consumption, and total alcohol consumption. In an experimental double-blind crossover study of the effects of naltrexone versus placebo in social drinkers, SWIFT et al. (1994) examined a number of subjective alcohol intoxication measures. When subjects were treated with naltrexone before they consumed alcohol, they reported reduced positive reinforcing stimulant effects and augmented sedative and "central" effects (e.g., feeling down, slow thoughts, head spinning) compared to subjects pretreated with placebo.

Recently, DAVIDSON et al. (1996) studied drinking by 16 non-alcohol-dependent young men and women in an experimental cocktail bar in a double-blind trial of placebo versus naltrexone versus no drug. Naltrexone significantly increased the time to sip the first drink and consume the second drink and had an effect on the self-reported urge to drink. Interestingly, naltrexone also increased predrinking tension and fatigue on the POMS and increased nausea during drinking.

II. 5-HT and GABA Systems

1. Preclinical Studies

Drugs acting at the GABAergic system have been shown to modulate alcohol intake. Peripheral administration of $GABA_A$ agonists increased both acquisition and maintenance of alcohol intake (SMITH et al. 1992; BOYLE et al. 1992, 1993), whereas GABA antagonists such as picrotoxin and the partial benzodiazepine inverse agonist Ro 15–4513 have been shown to reduce alcohol intake (SAMSON et al. 1987; RASSNICK et al. 1993).

Part of the neurochemical and behavioral effects of GABAergic agents have been shown to be mediated through their action on 5-HT neurons located in the raphe nuclei. Injection of muscimol into the dorsal raphe nucleus decreased 5-HT release in several brain areas (NISHIKAWA and SCATTON 1985). Recently, TOMKINS et al. (1994b; TOMKINS and FLETCHER 1996) have shown that injection of muscimol into the median and dorsal raphe increases alcohol consumption in rats. The stimulation of alcohol intake induced by injection of muscimol into the dorsal raphe is specific for alcohol as such injection did not produce any changes in water consumption. On the other hand, the effect of muscimol in the median raphe appears to be due to an enhancement in behavioral activation, as an increase in water consumption was also observed (TOMKINS et al. 1994a; TOMKINS and FLETCHER 1996).

III. Alcohol, 5-HT and Mood Disorders

Alterations in 5-HT have also been reported consistently in depression (5-HT decreased). Therefore, the linkage of mood disorders and alcohol dependence is of special importance. Data from the NIMH Epidemiologic Catchment Area (ECA) Study (REGIER et al. 1990) indicate that 37% of alcohol-abusing individuals meet DSM-III-R criteria for at least one other mental disorder. This is important because comorbidity may alter the time of onset and clinical course of alcoholism, affect the therapeutic intervention selected, and alter treatment outcome, and because there may be gender differences (ROUNSAVILLE et al. 1987; KRANZLER et al. 1992). The prevalence rates of depression vary considerably due to heterogeneous patient populations, the diagnostic criteria used, the chronology of disorders, and the temporal relation of drinking or withdrawal. The reported frequency ranges from 8% to 53% in clinical samples (MERIKANGAS and GELERNTER 1990; KRANZLER and LIEBOWITZ 1988), but is also common in the general population (GRANT and HARFORD 1995). Depression is a serious disorder accompanied by extensive distress and disability (BROADHEAD et al. 1990; VON KORFF et al. 1992) and an increased risk (10%–15%) of suicide completion in alcoholism (SCHUCKIT 1986). In a recently completed 12-week controlled clinical trial of dexfenfluramine versus placebo, 34% of alcoholics presenting for treatment had a lifetime history of major depression as determined using a validated structured diagnostic interview (Structured Clinical Interview for DSM-III-R Diagnoses, SCID) (SPITZER et al. 1989) and extensive depressive symptomatology irrespective of diagnostic category (ROMACH et al. 1995).

Even though SSRIs do decrease alcohol consumption, they are foremost highly effective antidepressants. However, surprisingly, the utility of SSRIs and other serotoninergic antidepressants in treating alcohol dependence in depressed patients, for either immediate efficacy or relapse prevention has been understudied (LITTEN and ALLEN 1995). JAFFE and CIRAULO (1986) reviewed the usefulness of the tricyclic antidepressants for the treatment of depression in alcoholics and concluded that, although a few (imipramine, doxepin, amitriptyline) showed some positive effect, the poor methodology precluded any firm conclusions. A recent study of desipramine (DMI) ($n = 71$) incorporated multidimensional treatment outcome measures (e.g., DMI plasma level monitoring, depressive symptoms, and alcohol consumption) and provided support for DMI treatment efficacy of depression secondary to alcoholism. Depressed alcoholics who received DMI showed significant improvements in mood and a trend to maintain sobriety for a longer period of time, compared to the placebo group (MASON and KOCSIS 1991; MASON et al. 1996). A similar trial with imipramine and placebo in 69 alcoholic patients with primary depressive disorders showed that in patients whose mood improved, their alcohol consumption had also decreased. This reduction was more marked in the imipramine group (McGRATH et al. 1996). MASON et al. (1996) showed that antidepressant therapy produced greater treatment satisfaction in alcoholics with major depression. In another trial in alcoholics, those with a diagnosis of depression had HAM-D scores which declined significantly with treatment with fluoxetine compared to placebo, but this trial had small numbers of patients with depression and retention was not specifically examined (KRANZLER et al. 1995). In a trial of dexfenfluramine in alcohol dependence, the HAM-D in those patients completing week 4 was significantly lower ($p < 0.01$) than in those who were early drop outs (5.8 ± 4.3 versus 9.4 ± 6.4) irrespective of a diagnosis of major depression. Similar data were found with the Beck Depression Inventory (BDI) and Symptom Checklist-90 (SCL-90) depression subscale, suggesting that depressive symptoms are associated with early dropout (ROMACH et al. 1995).

While it has been traditional to consider the role of depression and alcohol dependence using diagnostic approaches and to separate depression in alcoholism into primary and secondary (SCHUCKIT 1986), this temporal convenience of classification fails to account for the complex and intertwined nature of mood disorders with the drinking behavior and largely ignores the observation that "diagnosis" of major depression seems to have little to do with symptoms and function (GOTLIEB et al. 1995). Negative affective states such as anxiety and depression appear to initiate and maintain maladaptive drinking patterns (MARLATT and GORDON 1985) and, therefore, modification of these dysphoric states pharmacologically and/or psychologically could lead to a decrease in or cessation of excessive drinking. Modification of these symptoms and drinking triggers could have a beneficial effect in decreasing drinking if drinking and depressive symptoms are functionally linked (McGRATH et al. 1996).

IV. Alcohol and Craving

Urges, craving, and compulsion to use alcohol, tobacco and other drugs are believed to contribute importantly to the maintenance of dependence and to relapse after abstinence. As a result, treatment approaches as diverse as cognitive–behavioral and pharmacotherapy either target the decreasing of such urges as their primary goal or look for evidence post hoc that the urge or compulsion to use has decreased. A larger literature, reviewed by BLUNDELL (1991), supports an important role of 5-HT in the modulation of feeding. Detailed analysis of feeding sequences suggest that satiety is earlier with 5-HT increases, and that the microstructure of feeding is markedly altered by drugs such as dexfenfluramine and 5-HT uptake inhibitors. In a 12-h access, two-bottle ethanol–water choice paradigm, HIGGINS et al. (1992) examined the impact of sertraline, dexfenfluramine, and mCPP on ethanol-drinking behavior. These authors observed that sertraline and dexfenfluramine increased latency to first ethanol drink, delayed the time to drink 50% of the ethanol, and decreased the number of drinking bouts. These data suggest that initiation of drinking (initial drink) or between bout interval are prolonged by a general increase in 5-HT levels. Whether this altered behavioral control is a model for a decrease in craving is unknown. The data do, however, suggest that 5-HT agents may have a modulatory role in the behavior consequent to "craving". Earlier in this chapter, we included clinical and experimental evidence that naltrexone and selected serotoninergic agents can decrease "craving" in the context of alcohol use.

V. Alcohol, 5-HT and Impulsivity

1. Preclinical Studies

Epidemological studies have found a high correlation between high impulsivity and alcohol abuse (LACEY and EVANS 1986; DULIT et al. 1990). In fact a number of clinical studies have identified impulsivity as a defining feature of alcohol abuse in certain types of alcoholics (CLONINGER 1987; MILLER and BROWN 1991). An animal study also showed a strong correlation between level of impulsivity and alcohol consumption, with high level of impulsivity corresponding to high level of alcohol consumption (POULOS et al. 1995).

There is substantial evidence that the serotoninergic system plays an important role in regulating behavioral inhibitions that underlie impulse control (DEAKIN 1983; SOUBRIÉ 1986; FLETCHER 1993). It is likely that the reduction in alcohol consumption induced by serotoninergic agents might be mediated through their effects on impulsivity. Pharmacological manipulations that enhance 5-HT activity reduce impulsivity as well as alcohol consumption (THIÉBOT et al. 1985; SELLERS et al. 1992). This correlation has also been recently shown by POULOS et al. (1996) with 8-OH-DPAT. Low doses of 8-OH-DPAT which augmented alcohol consumption also increased impulsivity, whereas high doses of 8-OH-DPAT which suppressed alcohol consumption also decreased impulsivity as measured in a delay-reward paradigm (POULOS et al. 1996), in a similar manner to that observed for its effect on alcohol intake (TOMKINS et al. 1994a).

E. Conclusions

The preclinical behavioral pharmacology of the serotoninergic agents suggest that they alter the reinforcing properties of alcohol, or the regulation of the behavior controlling self-administration (cf. craving or impulsivity). Clinically, mechanisms affecting craving, impulsivity, or mood could be important. In the case of opiate antagonists, similar pharmacologic properties exist, except that there is no evidence for an important effect of opiate antagonists on mood. Since none of the existing agents has major clinical effects, the combination of 5-HT agents and opiate antagonists could produce major clinical effects by variously affecting ethanol reinforcing properties, craving, and dysphoric mood.

Other synergistic interactions can be postulated on the basis of the known neuroanatomical localization and role of 5-HT receptor and opioid receptor subtypes, because the two drugs decrease alcohol consumption by different mechanisms. Side effects should not be more severe than for either drug alone, because the mechanisms of producing side effects differ from the mechanism of decreasing alcohol consumption. The greatest potential for combination therapy would seem to be in those clinical conditions for which serotoninergic agents are already used and effective (e.g., major depression and anxiety disorders such as panic or agoraphobia). The basis for the interaction, therefore, would not only be the primary treatment of the alcohol dependence disorder, but also the amelioration of the trigger symptoms for drinking, such as dysphoric mood.

Alcohol dependence is a heterogeneous disorder in which a variety of factors contribute to excessive alcohol consumption. It is unlikely that any one pharmacologic agent or any one psychosocial intervention will be effective in treating all types of alcohol-dependent patients. Rather than focusing on discovering medications with high selectivity and specificity, the identification of agents with selective but several mechanisms of action may meet with considerable success.

References

Amit Z, Sutherland EA, Gill K, Ogren S (1984) Zimelidine. A review of its effects on ethanol consumption. Neurosci Biobehav Rev 8:35–54

Beczowska IW, Koch JE, Bodnar RJ (1992) Naltrexone, serotonin receptor subtype antagonist, and glucoprivic intake: 1.2-deoxyglucose. Pharmacol Biochem Behav 42:661–670

Blundell J (1991) Pharmacological approaches to appetite suppression. Trends Pharmacol Sci 12:147–157

Bohn MJ, Kranzler HR, Beazoglou D, Staehler BA (1994) Naltrexone and brief counselling to reduce heavy drinking. Am J Addict 3:91–99

Borg S, Kvande H, Liljeberg P, Mossberg D, Valverius P (1985) 5-Hydroxyindoleacetic acid in cerebrospinal fluid in alcoholic patients under different clinical conditions. Alcohol 2:415–418

Boyle AE, Smith BR, Amit Z (1992) Microstructural analysis of the effects of THIP, a $GABA_A$ agonist, on voluntary ethanol intake in laboratory rats. Pharmacol Biochem Behav 43:1121–1127

Boyle AE, Segal R, Smith BR, Amit Z (1993) Bidirectional effects of GABAergic agonists and antagonists on maintenance of voluntary ethanol intake in rats. Pharmacol Behav 46:179–182

Brase DA (1977) Roles of serotonin and aminobutyric acid in opiod effects. Adv Biochem Psychopharmacol 20:409–428

Brewerton TD, Brandt HA, Lessem MD, Murphy DL, Jimerson DC (1990) Serotonin in eating disorders. In: Coccaro EF, Murphy DL (eds) Serotonin in major psychiatric disorders. American Psychiatric Press, Washington, DC, pp 153–184

Broadhead WE, Blazer DG, George LK, Tse CK (1990) Depression, disability days and days lost from work in a prospective epidemiologic survey. JAMA 264:2524–2528

Bruno F (1989) Buspirone in the treatment of alcoholic patients. Psychopathology 22 [Suppl 1]:49–59

Buczek Y, Tomkins DM, Higgins GA, Sellers EM (1994) Dissociation of serotonergic regulation of anxiety and ethanol self-administration: a study with mCPP. Behav Pharmacol 5(4/5):470–484

Carboni E, Acquas E, Frau R, Di Chiara G (1989) Differential effects of a 5-HT₃ antagonist on drug-induced stimulation of dopamine release. Eur J Pharmacol 164:515–519

Cloninger CR (1987) Neurogenetic adaptive mechanisms in alcoholism. Sciense 236:410–416

Coccaro EF, Murphy DL (eds) (1990) Serotonin in major psychiatric disorders. American Psychiatric Press, Washington, DC

Cohen ML (1992) 5-HT₃ receptors in the periphery. In: Hamon M (ed), Central and peripheral 5-HT₃ receptors. Academic, London, pp 19–32

Cornelius JR, Fisher BW, Salloum IM, Cornelius MD, Ehler JG (1992) Fluoxetine trial in depressed alcoholics. Alcohol Clin Exp Res 16(2):362

Corrigall WA, Coen KM (1994) Nicotine self-administration and locomotor activity are not modified by the 5-HT₃ antagonists ICS 205–930 and MDL 72222. Pharmacol Biochem Behav 49(1):67–71

Crabbe JC, Phillips TJ, Feller DJ, Wenger CM, Lessor CM, Schafer GL (1996) Increased alcohol preference and reduced ataxia in null mutant mice missing the serotonin 5-HT₁ᴮ. Alcohol Clin Exp Res 20(2):23A

Davidson D, Swift R, Fitz E (1996) Naltrexone increases the latency to drink alcohol in social drinkers. Alcohol Clin Exp Res 20(4):732–739

Deakin JFW (1983) Roles of serotonergic systems in escape, avoidance and other behaviours. In: Cooper SJ (ed) Theory in psychopharmacology, vol. 2. Academic, New York

Di Chiara G, Imperato A (1988) Drugs of abuse preferentially stimulate dopamine release in the mesolimbic system of freely moving rats. Proc Natl Acad Sci USA 85:5274–5278

Dulit RA, Fryer M, Haas G, Sullivan T, Frances A (1990) Substance use in borderline personality disorder. Am J Psychiatry 147:1002–1007

Dumuis A, Bouhelal R, Sebben M, Bockaert JA (1988) 5-HT receptor in the central nervous system, positively coupled with adenylate cyclase, is antagonized by ICS 205–930. Eur J Pharmacol 146:187–188

Engel JA, Falke C, Hard E, Svensson L (1990) Effects of 5-HT₁ₐ receptor agonists on ethanol preference in the rat. Pharmacol Toxicol 67 [Suppl 19]:21

Estevez F, Parrillo S, Giusti M, Monti JM (1995) Single-dose ritanserin and alcohol in healthy volunteers: a placebo-controlled trial. Alcohol 12(6):541–545

Fadda F, Garau B, Marchei F, Colmbo G, Gessa GL (1991) MDL 72222, a selective 5-HT₃ receptor antagonist, suppresses voluntary ethanol consumption in alcohol-preferring rats. Alcohol Alcohol 26:107–110

Fernandez-Tome MP, Gonzalez Y, DelRio J (1988) Interaction between opioid agonists or naloxone and 5-HTP on feeding behavior in food-deprived rats. Pharmacol Biochem Behav 29:387–392

Fletcher PJ (1991) Dopamine receptor blockade in nucleus accumbens or caudate nucleus differentially affects feeding induced by 8-OH-DPAT injected into dorsal or median raphe. Brain Res 552:181–189

Fletcher PJ (1993) A comparison of the effects of dorsal or median raphe injections in three operant tasks measuring response inhibition. Behav Brain Res 54:187–197

Gill K, Amit Z, Koe BK (1988) Treatment with sertraline, a new serotonin uptake inhibitor, reduces voluntary ethanol consumption in rats. Alcohol 5:349–354

Gorelick DA, Paredes A (1992) Effect of fluoxetine on alcohol consumption in male alcoholics. Alcohol Clin Exp Res 16(2):261–265

Gotlieb IH, Lewinsohn PM, Seeley JR (1995) Symptoms as a diagnosis of depression: differences in psychosocial functioning. J Consult Clin Psychol 63:90–100

Grant KA, Colombo G (1993) The ability of 5-HT₃ antagonists to block an ethanol discrimination: effect of route of administration. Alcohol Clin Exp Res 17:497

Grant BF, Harford TC (1995) Comorbidity between DSM-IV alcohol use disorders and major depression: results of a national survey. Drug Alcohol Depend 39:197–206

Grant KA (1995) The role of 5-HT$_3$ receptors in drug dependence. Drug Alcohol Depend 38:155–171

Gropetti A, Parenti M, Dellavedova L, Tirone F (1984) Possible involvement of endogenous opiates in tolerance to the anorectic effect of fenfluramine. J Pharmacol Exp Ther 228:446–453

Haraguchi M, Samson HH, Tolliver GA (1990) Reduction in oral ethanol self-administration in the rat by the 5-HT uptake blocker fluoxetine. Pharmacol Biochem Behav 35:259–262

Harsing LL, Yang HYT, Costa E (1982a) Accumulation of hypothalamic endorphins after repeated injections of anorectics which release serotonin. J Pharmacol Exp Ther 223:689–694

Harsing LL, Yang HYT, Govoni S, Costa E (1982b) Elevation of met-enkephalin and β-endorphin hypothalamic content in rats receiving anorectic drugs: differences between d-fenfluramine and d-amphetamine. Neuropharmacology 21:141–145

Higgins GA, Bradbury AJ, Jones BJ, Oakley NR (1988) Behavioural and biochemical consequences following activation of 5-HT$_1$-like and GABA receptors in the dorsal raphe nucleus of the rat. Neuropharmacology 27:993–1001

Higgins GA, Tomkins DM, Fletcher PJ, Sellers EM (1992) Effect of drugs influencing 5-HT function on ethanol drinking and feeding behaviour in rats: studies using a drinkometer system. Neurosci Biobehav Rev 16(4):535–552

Higgins GA, Wang Y, Corrigall WA, Sellers EM (1994) Influence of 5-HT$_3$ receptor antagonists and the indirect 5-HT agonist, dexfenfluramine, on heroin self-administration in rats. Psychopharmacology 114:611–619

Hjorth S, Sharp T (1991) Effect of the 5-HT$_{1A}$ receptor agonist 8-OH-DPAT on the release of 5-HT in dorsal and median raphe-innervated rat brain regions as measured by in vivo microdialysis. Life Sci 48:1779–1786

Hodge CW, Samson HH, Lewis RS, Erickson HL (1993) Specific decreases in ethanol-but not water-reinforced responding produced by the 5-HT$_3$ antagonist ICS 205–930. Alcohol 10:191–196

Hodge CW, Niehus JS, Samson HH (1995) Morphine induced changes in ethanol-and water-intake are attenuated by the 5-HT$_{3/4}$ antagonist tropisetron (ICS 205–930). Psychopharmacology 119:186–192

Hoyer D, Clarke DE, Fozard JR, Hartig PR, Martin GR, Mylecharane EJ, Saxena PR, Humphrey PPA (1994) VII International Union of Pharmacology classification of receptors for 5-hydroxytryptamine (serotonin). Pharmacol Rev 46:157–203

Humphrey PPA, Harting P, Hoyer D (1993) A proposed new nomenclature for 5-HT receptors. Trends Pharmacol Sci 14:233–236

Invernizzi R, Carli M, Di Clemente A, Samanin R (1991) Administration of 8-hydroxy-2-(di-n-propylamino)tetralin in raphe nuclei dorsalis and medianus reduces serotonin synthesis in the rat brain: differences in potency and regional sensitivity. J Neurochem 56:143–147

Jaffe J, Ciraulo D (1986) Alcoholism and depression. In: Meyer RE (ed) Psychopathology and addictive disorders. Guilford, New York, pp 293–320

Jankowska E, Kowtowski W (1995) The effect of tropisetron injected into the nucleus accumbens septi on ethanol consumption in rats. Alcohol 12(3):195–198

Jankowska E, Bidzinski A, Kostowski W (1995) Alcohol drinking in rats injected ICV with 6-OHDA: effect of 8-OHDPAT and tropisetron (ICS 205930). Alcohol 12(2):121–126

Janssen PA (1994) Addiction and the potential for therapeutic drug development [review]. EXS 71:361–70

Johnson BA, Rue J, Cowen PJ (1993a) Ondansetron and alcohol pharmacokinetics. Psychopharmacology 112:145–147

Johnson BA, Campling GM, Griffiths P, Cowen PJ (1993b) Attenuation of some alcohol-induced mood changes and the desire to drink by 5-HT3 receptor blockade. Psychopharmacology 112:111–115

Kmieciak-Kolada K, Kowalski J (1986) Involvement of the central serotonergic system in the changes of leu-enkephalin level in discrete rat brain areas. Neuropeptides 7(4):351–360

Knapp DJ, Pohorecky LA (1992) Zacopride, a 5-HT$_3$ receptor antagonist, reduces voluntary ethanol consumption in rats. Pharmacol Biochem Behav 41:847–850

Knapp DJ, Benjamin D, Pohorecky LA (1992) Effects of gepirone on ethanol consumption, exploratory behavior, and motor performance in rats. Drug Dev Res 26:319–341

Koob GF (1992) Neural mechanisms of drug reinforcement. In: Kalivas PW, Samson HH (eds) The neurobiology of drug and alcohol addiction. Ann NY Acad Sci 654:171–191

Kostowski W, Dyr W (1992) Effect of 5-HT$_{1A}$ receptor agonists on ethanol preference in the rat. Alcohol 9:283–286

Kostowski W, Dyr W, Krzaścik P (1993) The abilities of 5-HT$_3$ receptor antagonist ICS 205–930 to inhibit alcohol preference and withdrawal seizures in rats. Alcohol 10:369–373

Kranzler HR, Liebowitz NR (1988) Anxiety and depression in substance abuse: clinical implications. Med Clin North Am 72:867–885

Kranzler H, Del Boca F, Rounsaville B (1992) Psychopathology as a predictor of outcome three years after alcoholism treatment. Alcohol Clin Exp Res 16(2):363

Kranzler HR, Burleson JA, Del Boca FK, Babor TF, Korner P, Brown J, Bohn MJ (1994) Buspirone treatment of anxious alcoholics. A placebo-controlled trial. Arch Gen Psychiatry 51(9):720–731

Kranzler HR, Burleson JA, Korner P et al (1995) Placebo-controlled trial of fluoxetine as an adjunct to relapse prevention in alcoholics. Am J Psychiatry 152:391–397

Krystal JH, Webb E, Cooney N, Kranzler HR, Charney DS (1994) Specificity of ethanol-like effects elicited by serotonergic and noradrenergic mechanisms. Arch Gen Psychiatry 51(11):898–911

Lacey J, Evans C (1986) The impulsivist: a multi-impulsive personality disorder. Br J Addict 81:641

Lane JD, Pickering CL, Hooper ML, Fagan K, Tyers MB, Emmett-Oglesby MW (1992) Failure of ondansetorn to block the discriminative or reinforcing stimulus effects of cocaine in the rat. Drug Alcohol Depend 30(2):151–162

Lê AD, Sellers EM (1994) Interactions between opiate and 5-HT$_3$ receptor antagonists in the regulation of alcohol intake. Alcohol Alcoholism Suppl. 2:545–549

Linnoila M (1990) Monoamines and impulse control. In: Swinkels JA, Blijleven W (eds) Depression, anxiety and aggression: factors that influence the course. Medidact b.v.i.o., Houten, Netherlands, pp 167–169

Litten RZ, Allen JP (1995) Pharmacotherapy for alcoholics with collateral depression or anxiety: an update of research findings. Exp Clin Psychopharmacol 3(2):87–93

Llorens-Cortes C, Schwartz JC (1984) Changes in turnover of monoamines following inhibition of enkephalin metabolism by thiorphan and betastin. Eur J Pharmacol 104:369–374

Majeed NH, Lason W, Przewlocka B, Przewlocki R (1986) Involvement of endogenous opioid peptides in fenfluramine anorexia. Pharmacol Biochem Behav 25:967–972

Malcolm R, Anton RF, Randall CL, Johnston A, Brady K, Thevos A (1992) A placebo-controlled trial of buspirone in anxious inpatient alcoholics. Alcohol Clin Exp Res 16(6):1007–1013

Marlatt GA, Gordon JR (1985) Determinants of relapse: implications for the maintenance of behavior change. In: Davidson PO, Davidson SM (eds) Behavioral medicine: changing health lifestyles. Brunner-Mazel, New York, pp 410–452

Mason BJ, Kocsis JH (1991) Desipramine treatment of alcoholism. Psychopharmacol Bull 27:155–161

Mason BJ, Ritvo EC, Morgan RO, Salvato FR, Goldberg G, Welch B, Mantero-Atienza E (1994) A double-blind, placebo-controlled pilot study to evaluate the efficacy and safety of oral nalmefene HCl for alcohol dependence. Alcohol Clin Exp Res 18:1162–1167

Mason BJ, Kocsis JH, Ritvo EC, Cutler RB (1996) A double-blind placebo-controlled trial of desipramine for primary alcohol dependence stratified on the presence or absence of major depression. JAMA 275:761–767

McBride WJ, Murphy JM, Lumeng L, Li T-K (1990) Serotonin, dopamine and GABA involvement in alcohol drinking of selectively bred rats. Alcohol 7:199–205

McGrath PJ, Nunes EV, Stewart JW, Goldman D, Agosti V, Ocepek-Welikson K, Quitkin F (1996) Imipramine treatment of alcoholics with primary depression. Arch Gen Psychiatry 53:232–240

McMillen BA, Jones EA, Hill LJ, Williams HL, Björk A, Myers RD (1993) Amperozide, a 5-HT$_2$ antagonist, attenuates craving for cocaine by rats. Pharmacol Biochem Behav 46(1):125–129

McMillen BA, Walter S, Williams HL, Myers RD (1994) Comparison of the action of the 5-HT$_2$ antagonists amperozide and trazodone on preference for alcohol in rats. Alcohol 11(3):203–206

Meert TF (1993) Effects of various serotonergic agents on alcohol intake and alcohol preference in Wistar rats selected at two different levels of alcohol prefernece. Alcohol Alcoholism 28:157–170

Meert TF, Awouters F, Niemegeers CJE, Schellekens KHL, Janssen PAJ (1991) Ritanserin reduces abuse of alcohol, cocaine, and fentanyl in rats. Pharmacopsychiat 24:159–163

Merikangas KR, Gelernter CS (1990) Comorbidity for alcoholism and depression. Psychiatr Clin N Am 13(4):613–632

Miller WR, Brown JM (1991) Self-regulation as a conceptual basis for the prevention and treatment of addictive behaviors. In: Heather N, Miller WR, Greeley J (eds) Self-control and the addictive behaviors. Maxwell Macmillan, Australia, pp 3–79

Monti JM, Alterwain P, Estevez F, Alvarino F, Giusti M, Olivera S, Labraga P (1993) The effects of ritanserin on mood and sleep in abstinent alcoholic patients. Sleep 16(7):647–654

Murphy JM, McBride WJ, Lumeng L, Li T-K (1987a) Contents of monoamines in forebrain regions of alcohol-preferring (P) and -nonpreferring (NP) lines of rats. Pharmacol Biochem Behav 26:389–392

Murphy JM, McBride WJ, Lumeng L, Li T-K (1987b) Effects of serotonergic agents on ethanol intake on the high alcohol drinking (HAD) line of rats. Alcohol Clin Exp Res 11:208

Myers RD, Lankford MF (1993) Failure of the 5-HT$_2$ receptor antagonist, ritanserin, to alter preference for alcohol in drinking rats. Pharmacol Biochem Behav 45(1):233–237

Myers RD, Veale W (1968) Alcohol preference in the rat: reduction following depletion of brain serotonin. Science 160:1469–1471

Myers RD, Lankford MF, Björk A (1993) 5-HT$_2$ receptor blockade by amperozide suppresses ethanol drinking in genetically preferring rats. Pharmacol Biochem Behav 45:741–747

Myers RD, Lankford M, Björk A (1992) Selective reduction by the 5-HT antagonist amperozide of alcohol preference induced in rats by systemic cyanamide. Pharmacol Biochem Behav 43:661–667

Naranjo CA, Sellers EM (1989) Serotonin uptake inhibitors attenuate ethanol intake in problem drinkers. In: Galanter M (ed) Recent developments in alcoholism, vol. 7. Plenum, New York, pp 255–266

Naranjo CA, Sellers EM, Roach CA, Woodley DV, Sanchez-Craig M, Sykora K (1984) Zimelidine-induced variations in alcohol intake by non-depressed heavy drinkers. Clin Pharmacol Ther 35:374–381

Naranjo CA, Sellers EM, Sullivan JT, Woodley DV, Kadlec K, Sykora K (1987) The serotonin uptake inhibitor citalopram attenuates ethanol intake. Clin Pharmacol Ther 41(3):266–274

Naranjo CA, Sullivan JT, Kadlec KE, Woodley-Remus DV, Kennedy G, Sellers EM (1989) Differential effects of viqualine on alcohol intake and other consummatory behaviors. Clin Pharmacol Ther 46(3):301–309

Naranjo CA, Kadlec KE, Sanhueza P, Woodley-Remus D, Sellers EM (1990) Fluoxetine differentially alters alcohol intake and other consummatory behaviors in problem drinkers. Clin Pharmacol Ther 47(4):490–498

Naranjo CA, Poulos CV, Bremner KE, Lanctôt KL (1992) Citalopram decreases desirability, liking and consumption of alcohol in alcohol dependent drinkers. Clin Pharmacol Ther 51:729–739

Naranjo CA, Poulos CX, Lanctôt KL, Bremner KE, Kwok M, Umana M (1995) Ritanserin, a central 5-HT$_2$ antagonist, in heavy social drinkers: desire to drink, alcohol intake and related effects. Addiction 90(7):893–905

Nathan PE (1986) Outcomes of treatment for alcoholism: current data. Ann Behav Med 8:40–46

Nishikawa T, Scatton B (1985) Inhibitory influence of GABA on central serotonergic transmission. Raphe nuclei as the neuroanatomical site of the GABAergic inhibition of cerebral serotonergic neurons. Brain Res 331:91–103

O'Malley SS, Jaffe AJ, Chang G et al (1992) Naltrexone and coping skills therapy for alcohol dependence: a controlled study. Arch Gen Psychiatry 49:881–887

Panocka I, Massi M (1992) Long-lasting suppression of alcohol preference in rats following serotonin receptor blockade by ritanserin. Brain Res Bull 28(3):493–496

Panocka I, Ciccocioppo R, Pompei P, Massi M (1993a) 5-HT$_2$ receptor antagonists do not reduce ethanol preference in Sardinian alcohol-preferring (sP) rats. Pharmacol Biochem Behav 46:853–856

Panocka I, Ciccocioppo R, Polidori C, Massi M (1993b) The nucleus accumbens is a site of action for the inhibitory effect of ritanserin on ethanol intake in rats. Pharmacol Biochem Behav 46:857–862

Panocka I, Pompei P, Massi M (1993c) Suppression of alcohol preference in rats induced by risperidone, a serotonin 5-HT$_2$ and dopamine D2 receptor antagonist. Brain Res Bull 31(5):595–599

Panocka I, Ciccocioppo R, Polidori C, Pompei P, Massi M (1995) The 5-HT$_4$ receptor antagonist, GR113808, reduces ethanol intake in alcohol-preferring rats. Pharmacol Biochem Behav 52(2):255–259

Peltier R, Schenk S (1991) Effects of serotonergic manipulations on cocaine self-administration in rats. Psychopharmacology 110:390–394

Poulos CX, Lê AD, Parker JL (1995) Impulsivity predicts individual susceptibility to high levels of alcohol self-administration. Behav Pharmacol 6:810–814

Poulos CX, Parker JL, Lê AD (1996) Dexfenfluramine and 8-OH-DPAT modulate impulsivity in a delay-of-reward paradigm: implications for a correspondence with alcohol consumption. Behav Pharmacol 7:395–399

Privette TH, Hornsby RL, Myers RD (1988) Buspirone alters alcohol drinking induced in rats by tetrahydropapaveroline injected into brain monoaminergic pathways. Alcohol 5:147–152

Rassnick S, D'Amico E, Riley E, Koob GF (1993) GABA antagonist and benzodiazepine partial inverse agonist reduce motivated responding for ethanol. Alcohol Clin Exp Res 17:124–130

Regier DA, Farmer ME, Rae DS, Locke BZ, Keith SJ, Judd LL, Goodwin FK (1990) Comorbidity of mental disorders with alcohol and other drug abuse. JAMA 264:2511–2518

Rezvani AH, Overstreet DH, Janowsky DS (1991) Drug-induced reductions in etahnol intake in alcohol-preferring and Fawn-Hooded rats. Alcohol [Suppl 1]:433–437

Romach MK, Tomkins DM (1995) Clinical application of findings from animal research on alcohol self-administration and dependence. In: Kranzler HR (ed) The pharmacology of alcohol abuse. Springer, Berlin Heidelberg New York, (Handbook of experimental pharmacology, vol. 114) pp 261–295

Romach MK, Somer G, Kaplan HL, Magnatta A, Baylon GJ, Sellers EM, Sobell L, Toneatto T (1995) The three faces of depression and alcohol dependence. Abstracts of Science Day, Addiction Research Foundation, Toronto, Ontario

Romach MK, Sellers EM, Kaplan HL, Somer G, Sobell MB, Sobell LC (1996) Efficacy of dexfenfluramine (DEX) in the treatment of alcohol dependence. Alcohol Clin Exp Res 20(2):90A

Rounsaville BJ, Dolinsky ZS, Babor TF et al (1987) Psychopathology as a predictor of treatment outcome in alcoholics. Arch Gen Psychiatry 44:505–513

Rowland NE, Morian KR (1992) Effect of dexfenfluramine on alcohol intake in alcohol-preferring "P" rats. Alcohol 9:559–561

Samson HH, Tolliver GA, Pfeffer AO, Sadeghi KG, Mills FG (1987) Oral ethanol reinforcement in the rat: effect of the partial inverse benzodiazepine agonist Ro 15–4513. Pharmacol Biochem Behav 27:517–519

Schreiber R, Opitz K, Glaser T, De Vry J (1993) Isapirone and 8-OH-DPAT reduce ethanol preference in rats: involvement of presynaptic 5-HT$_{1A}$ receptors. Psychopharmacology 112:100–110

Schuckit M (1986) Genetic and clinical implications of alcoholism and affective disorder. Am J Psychiatry 143:140–147

Sellers EM, Higgins GA, Tomkins DM, Romach MK, Toneatto T (1991) Opportunities for treatment of psychoactive substance use disorders with serotonergic medications. J Clin Psychiatry 52 [Suppl]:49–54

Sellers EM, Higgins GA, Sobell MB (1992) 5-HT and alcohol abuse. Trends Pharmacol Sci 13:69–75

Sellers EM, Toneatto T, Romach MK, Somer GR, Sobell LC, Sobell MB (1994) Clinical efficacy of the 5-HT$_3$ antagonist ondansetron in alcohol abuse and dependence. Alcohol Clin Exp Res 8(4):879–885

Smith BR, Robidoux J, Amit Z (1992) GABAergic involvement in the acquisition of voluntary ethanol intake in laboratory rats. Alcohol Alcoholism 27:227–231

Soubrié P (1986) Reconciling the role of central serotonin neurons in human and animal behavior. Behav Brain Sci 9:319–364

Spitzer RL, Willams JB, Gibbon M (eds) (1989) Structured Clinical Interview for DSM-IIIR. Biometrics Department, New York State Psychiatric Institute, New York

Svensson L, Fahlke C, Hård E, Engel JA (1993) Involvement of the serotonergic system in ethanol intake in the rat. Alcohol 10:219–224

Swift RM, Whelihan W, Kuznetsov O et al (1994) Naltrexone induced alterations in human ethanol intoxication. Am J Psychiatry 151:1463–1467

Thiébot MH, Le Bihan C, Soubrié P, Simon P (1985) Benzodiazepines reduce the tolerant to reward delay in rats. Psychopharmacology 86:147–162

Tollefson GD, MOntague-Clouse J, Tollefson SL (1992) Treatment of comorbid generalized anxiety in a recently detoxified alcoholic population with a slective serotonergic drug (buspirone). J Clin Psychopharmacol 12(1):192–26

Tomkins DM, Fletcher PJ (1996) Evidence that GABA$_A$ but not GABA$_B$ receptor activation in the dorsal raphe nucleus modulates ethanol intake in Wistar rats. Behav Pharamcol 7:85–93

Tomkins DM, Higgins GA, Sellers EM (1993) Enhancement of ethanol intake in Wistar rats following pre-treatment with low doses of 8-OH DPAT and gepirone using a limited access paradigm. Br J Pharmacol 108 (Proceedings Supplement):93P

Tomkins DM, Sellers EM, Fletcher PJ (1994a) Median and dorsal raphe injections of the 5-HT$_{1A}$ agonist, 8-OH-DPAT, and the GABA$_A$ agonist, muscimol, increase voluntary ethanol intake in Wistar rats. Neuropharmacology 33(3/4):349–358

Tomkins DM, Higgins GA, Sellers EM (1994b) Low doses of the 5-HT$_{1A}$ agonist, 8-hydroxy-2-(Di-N-propylamino)tetralin (8-OH-DPAT), increase ethanol intake in the rat. Psychopharmacology 115:173–179

Tomkins DM, Lê AD, Sellers EM (1995) Effect of the 5-HT$_3$ antagonist ondansetron on voluntary ethanol intake in rats and mice maintained on a limited access procedure. Psychopharmacology 117:479–485

Volpicelli JR, Alterman AI, Hayashida M et al (1992) Naltrexone in the treatment of alcohol dependence. Arch Gen Psychiatry 49:876–880

Volpicelli JR, Clay KL, Watson NT, O'Brien CP (1995) Naltrexone in the treatment of alcoholism: predicting response to naltrexone. J Clin Psychiatry 56 [Suppl 7]:39–44

Von Korff M, Ormel J, Katon W, Lin EHB (1992) Disability and depression among high utilizers of of health care. Arch Gen Psychiatry 49:91–100

Wong DT, Threlkeld PG, Lumeng L, Li T-K (1990) Higher density of serotonin$_{1A}$ receptors in the hippocampus and cerebral cortex of alcohol preferring P rats. Life Sci 46:231–235

Wozniak KM, Pert A, Linnoila M (1990) Antagonism of 5-HT$_3$ receptors attenuates the effects of ethanol on extracellular dopamine. Eur J Pharmacol 187:287–289

Yoshimoto K, McBride WJ, Lumeng L, Li TK (1991) Alcohol stimulates the release of dopamine and serotonin in the nucleus accumbens. Alcohol 9:17–22

CHAPTER 23

Neuronal Pathophysiology of Migraine as a Basis for Acute Treatment with 5-HT Receptor Ligands*

M.A. Moskowitz and C. Waeber

A. Introduction

Migraine, which afflicts 12%–15% of the general population, is acknowledged as the first neurological condition which can be treated successfully by administering relatively selective drugs targeted to specific 5-hydroxytryptamine receptor subtypes. The receptor subtypes involved in the effects of anti-migraine drugs might include 5-HT_{1B}, 5-HT_{1D} and possibly 5-HT_{1F}, as sumatriptan binds to each with high affinity. However, there is still no universal agreement as to sumatriptan's specific mechanism of action in migraine. Some propose the importance of vascular smooth muscle constriction whereas others suggest that sumatriptan binding to neuronal receptors on trigeminovascular fibers is important for its headache-relieving action. We believe that the significance of each mechanism will become clearer as more selective 5-HT ligands are developed and applied to the human condition. Major discoveries at the preclinical level have already advanced and await testing in humans.

This chapter reviews the preclinical data supporting an important role for neuronal receptors in the actions of sumatriptan and related compounds. Most of the evidence has been developed in animal models, which include neurogenic inflammation within the meninges and c-Fos expression within trigeminal nucleus caudalis following noxious meningeal stimulation. Molecular evidence will also be reviewed which points to the expression of selective 5-HT receptor subtypes in human and animal tissues relevant to migraine. Results from animal models complemented by human studies provide powerful information about migraine which will likely advance our understanding of human pathophysiology and treatment into the next millennium.

---

* Note concerning the nomenclature of 5-HT_{1B} and 5-HT_{1D} receptors: Receptors previously known as $5\text{-HT}_{1D\alpha}$ and $5\text{-HT}_{1D\beta}$ have been named 5-HT_{1D} and 5-HT_{1B} throughout this chapter (as recommended by Hartig et al. 1996). The prefixes h, gp, r and m are used to specify the species (human, guinea pig, rat and mouse, respectively). "5-HT_{1B}" (with no prefix) denotes a species-specific pharmacology (high affinity for β-blockers and CP-93,129).

B. In Vivo Animal Models

I. Dural Plasma Extravasation

Neurogenic inflammation has been proposed as a mechanism relevant to vascular headache pathogenesis and treatment (Moskowitz 1992). This process involves vasodilation and plasma protein extravasation and is mediated by the release of vasoactive neuropeptides [substance P, neurokinin A, calcitonin gene-related protein (CGRP)] from trigeminovascular sensory fibers. The tachykinins induce both an endothelium-dependent vasodilation and enhanced permeability via receptors located on the vascular endothelium. CGRP induces vasodilation by activating receptors on vascular smooth muscle cells. While the factors leading to activation of the sensory trigeminovascular fibers during migraine are unknown, it is possible to mimic this activation in anesthetized rodents by stimulating the trigeminal ganglion electrically (Markowitz et al. 1987). Alternatively, sensory fibers can be activated by an intravenous administration of capsaicin (or stereotactic microinjection of capsaicin into the ganglion), which releases vasoactive neuropeptides (Buzzi and Moskowitz 1990).

Unilateral electrical stimulation of the trigeminal ganglion (5 Hz, 5 ms, 0.6–1 mA, 5 min) is usually preceded by a femoral vein injection of [^{125}I]albumin. Five minutes after stimulation, the animals are perfused with saline to remove residual tracer from the vessel lumen and the dural tissues from both sides are harvested. The radioactivity on the stimulated side is then compared with that on the unstimulated side and expressed as a ratio (1.6–1.7 in vehicle-treated rats or guinea pigs). Test drugs are usually administered 10–30 min prior to electrical stimulation.

To determine whether drugs act on receptors located on the trigeminal neurons or on postjunctional sites, it is possible to elicit extravasation by intravenous injection of substance P (1 nmol/kg) and compare the drug response to that observed after electrical trigeminal stimulation (Lee and Moskowitz 1993; Lee et al. 1995). Drugs acting on prejunctional sites are expected to inhibit extravasation induced by electrical (or capsaicin) stimulation, but not that induced by substance P injection.

Light and electron microscopy studies have shown that following electrical stimulation of the trigeminal ganglion, striking changes occur in the appearance of endothelial and mast cells as well as in the platelets surrounding or within postcapillary venules (Dimitriadou et al. 1991, 1992). Platelet aggregates appeared within the lumen of the stimulated side shortly after stimulation; some were found adhering to the endothelium. Endothelial cells later developed numerous clear cytoplasmic vesicles and vacuoles; the endothelium hypertrophied and microvilli formed. Mast cells also showed signs of secretion and degranulation. All of these changes depended on the presence of small, unmyelinated fibers, as they were not observed in rats neonatally treated with capsaicin.

Table 1. Potency of a series of 5-HT$_1$ receptor agonists at inhibiting of plasma protein extravasation in the rodent dura mater

Drug	Species	Threshold dose	Reference
Ergotamine	Rat	75 nmol/kg	Saito et al. (1988)
Dihydroergotamine	Rat	75 nmol/kg	Saito et al. (1988)
Methysergide	Rat	100 nmol/kg[a]	Saito et al. (1988)
5-CT	Rat	3 pmol/kg	Buzzi et al. (1991)
5-benzyloxytryptamine	Rat	35 nmol/kg	Buzzi et al. (1991)
8-OH-DPAT	Rat	900 nmol/kg	Buzzi et al. (1991)
Metergoline	Rat	2.5 μmol/kg	Buzzi et al. (1991)
Methiothepin	Rat	2.2 μmol/kg	Buzzi et al. (1991)
Sumatriptan	Rat	250 nmol/kg	Buzzi et al. (1991)
Sumatriptan	Rat	14 nmol/kg[b]	Beattie and Connor (1995)
CP-93,129	Rat	140 nmol/kg	Matsubara et al. (1991)
CP-93,129	Mouse	200 nmol/kg	Yu et al. (1996)
LY334370	Rat	20 pg/kg[b]	Phebus et al. (1996)
LY334370	Guinea pig	30 pg/kg[b]	Phebus et al. (1996)
GTI	Guinea pig	150 nmol/kg	Yu et al. (in preparation)
Sumatriptan	Guinea pig	1 nmol/kg	Yu et al. (1997)
CP-93,129	Guinea pig	inactive at 1400 nmol/kg	Matsubara et al. (1991)
CP-122,288	Guinea pig	1 pmol/kg	Lee and Moskowitz (1993)
CP-122,288	Rat	0.3 pmol/kg[b]	Beattie and Connor (1995)
CP-122,288	Rat	6 pmol/kg	Gupta et al. (1995)
CP-122,638	Guinea pig	0.1 pmol/kg	Lee and Moskowitz (1993)
MK-462	Rat	75 nmol/kg[b]	Shepheard et al. (1995a)

[a] Chronic treatment (four times a day for 3 days).
[b] ID$_{50}$ value.

Several factors indicate the relevance of neurogenic inflammation to the acute treatment of migraine. Ergotamine and sumatriptan have been shown to block [^{125}I]albumin extravasation in the dura mater as well as endothelial and mast cell activation (Saito et al. 1988; Buzzi and Moskowitz 1990). The dosages required to block neurogenic inflammation in rodents are comparable to those efficacious in relieving migraine headache. These drugs block the inflammatory response even when administered 45 min after electrical stimulation (Huang et al. 1993). Finally, other medications effective in the treatment of migraine attacks but inactive at 5-HT$_1$ receptors block plasma protein extravasation: opiates (Saito et al. 1988), valproate (Lee et al. 1995) and ketorolac (Lee and Moskowitz, unpublished).

1. Studies Using Agonists

The potency of a series of 5-HT receptor agonists at inhibiting plasma protein extravasation after electrical stimulation is summarized in Table 1. With the exception of 8-hydroxy-dipropylaminotetralin (8-OH-DPAT), CP-93,129 and LY334370, all the agonists listed in Table 1 have a high affinity for both 5-HT$_{1B}$ and 5-HT$_{1D}$ receptors in all species. 8-OH-DPAT is a selective 5-HT$_{1A}$ agonist;

its weak potency in the model rules out a contribution of this subtype and is compatible with the very low potency of this compound at inhibiting adenylate cyclase in rat substantia nigra (an effect mediated by r5-HT_{1B} receptors). It should be mentioned that at this dose, 8-OH-DPAT only partially blocked plasma extravasation in rats (BUZZI et al. 1991). CP-93,129 has been shown to recognize specifically r5-HT_{1B}, and not h5-HT_{1B} or h5-HT_{1D} receptors (MACOR et al. 1990). Its potency in rats (species where 5-HT_{1B} receptors show a distinct pharmacological profile) and lack of effect in guinea pig indicate the involvement of r5-HT_{1B} in rats and gp5-HT_{1B} (or gp5-HT_{1D}) receptors in guinea pigs (MATSUBARA et al. 1991).

While these data strongly imply that 5-$HT_{1B/1D}$ receptors inhibit plasma extravasation, they do not rule out a role of other receptors. In particular, sumatriptan's and CP-122,288's effects could be mediated by 5-HT_{1F} receptors (WAEBER and MOSKOWITZ 1995a). Indeed, JOHNSON et al. (1996) have recently shown that the potency of drugs at inhibiting dural plasma extravasation in guinea pig correlated most significantly with their binding affinity at 5-HT_{1F} receptors. Interestingly, LY334370 is a selective, high-affinity 5-HT_{1F} receptor agonist ($K_i = 1.6\,nM$), suggesting that these receptors, in addition to 5-$HT_{1B/1D}$ sites, might also be involved in the inhibition of neurogenic inflammation. Furthermore, 5-CT might act at 5-HT_5 or 5-HT_7 receptors (HOYER et al. 1994). The most striking feature of the profile in Table 1 is the fact that 5-carboxamidotryptamine (5-CT), CP-122,288 and CP-122,638 (the latter two compounds are conformationally restricted sumatriptan analogues) are 100–1000 times more potent than expected from their affinity at 5-$HT_{1B/1D}$ receptors. This suggests that these agonists might act at an additional site to inhibit plasma extravasation (see below).

The possible functional significance of 5-HT_{1E}, 5-HT_3, 5-HT_4 and 5-HT_6 receptors in this system has not been explored, mainly because of the lack of selective agonist at the time the studies were performed. Similarly, the role of subtypes of the class (5-$HT_{2A/2B/2C}$) has not been investigated; the 5-HT_2 receptor agonist α-methyl-5-HT stimulates endothelial cell leakage directly, and this effect is blocked by pretreatment with the 5-HT_2 antagonist pimozide (MOSKOWITZ 1993). The opposing effects of 5-HT_1 and 5-HT_2 receptors on dural plasma extravasation might explain why 5-HT itself is inactive in the model.

The partial inhibition of plasma extravasation in rat dura by methiothepine is unlikely to be accounted for by an action at r5-HT_{1B} receptors, as this effect is also observed in mice lacking m5-HT_{1B} receptors (see below). Also methiothepine is an antagonist at 5-HT_{1A}, 5-$HT_{1B/1D}$, 5-HT_{1F}, 5-HT_5, 5-HT_6 and 5-HT_7 receptors, as well as at α_2-adrenoceptors (the stimulation of the latter sites also inhibits plasma extravasation; MATSUBARA et al. 1992). Methiothepin's effect on extravasation might be explained by inhibition of a receptor stimulating vascular leakage. Alternatively, methiothepin might inhibit extravasation via a receptor similar to the site where it mediates hypoalgesia in 5-HT-depleted mice (EIDE et al. 1988).

2. Studies Using Antagonists

The interpretation of pharmacological data based on agonist potencies can be misleading, particularly in the presence of receptor reserves or when multiple receptors are involved in the observed response (KENAKIN 1993). Antagonists provide more reliable data, which can directly be compared to binding affinities obtained in vitro. Unfortunately, a selective 5-HT$_{1D}$ antagonist has been developed only recently (SKINGLE et al. 1996) and the antagonists used in the original plasma extravasation studies (metergoline and methiothepine) lacked specificity (BUZZI et al. 1991). In rats, pretreatment with metergoline only partially reversed the response to sumatriptan, while methiothepine did not reverse the response at all; the effects of 5-CT were not blocked by either antagonist. In guinea pigs, pretreatment with metergoline partially antagonized the effect of CP-122,288 (LEE and MOSKOWITZ 1993). However, methiothepine blocks the inhibitory effect of sumatriptan on nicotine-induced basilar artery relaxation in vitro, a response mediated predominantly via substance P release from trigeminal nerve terminals (O'SHAUGHNESSY and CONNOR 1994).

While methiothepin (also called metitepine) is consistently reported as an antagonist at both 5-HT$_{1B}$ and 5-HT$_{1D}$ receptors, metergoline seems to behave as an agonist in most systems containing nonrodent 5-HT$_{1B/1D}$ receptors (SCHOEFFTER et al. 1988; SCHOEFFTER and HOYER 1989; MILLER et al. 1992), with some exceptions, however (MIDDLEMISS et al. 1988; SCHOEFFTER and HOYER 1990; SCHLICKER et al. 1989). Its behavior at 5-HT$_{1B}$ receptors is also complex, as it has been described as an agonist by some authors (SCHOEFFTER and HOYER 1989; PAUWELS and PALMIER 1994) and an antagonist by others (BOUHELAL et al. 1988), even in very similar experimental conditions (inhibition of adenylate cyclase). An additional level of complexity can be expected with methiothepine, as this compound has been suggested to be an inverse agonist at recombinant h5-HT$_{1B}$ and h5-HT$_{1D}$ receptors (WATSON et al. 1995; THOMAS et al. 1995) as well as at the rat terminal 5-HT autoreceptor (MORET and BRILEY 1993).

GR-127,935 has been shown to be a potent antagonist against hypothermia induced by 5-HT$_{1B/1D}$ receptor agonists in guinea pig (SKINGLE et al. 1996). In in vitro experiments, GR-127,935 has a 100-fold selectivity for 5-HT$_{1D}$ (both h5-HT$_{1B}$ and h5-HT$_{1D}$) receptors over 5-HT$_{1A}$, 5-HT$_{1E}$ and 5-HT$_{2C}$ receptors, and negligible affinity for 5-HT$_3$, 5-HT$_4$, 5-HT$_5$ and 5-HT uptake, and α_1- and α_2-adrenoceptor, dopamine D$_{1-4}$ and muscarinic M$_{1-3}$ binding sites (SKINGLE et al. 1996). It has a somewhat higher affinity for 5-HT$_{2A}$ receptors. PAUWELS and COLPAERT (1995) have also shown that GR-127,935 behaves as a partial agonist in cell lines transfected with h5-HT$_{1D}$ receptors, but does not show any intrinsic activity in cells transfected with h5-HT$_{1B}$ receptors.

We have used GR-127,935 in order to characterize further the profile of the receptors mediating the effects of sumatriptan, 5-CT and CP-122,288 in the plasma extravasation model (YU et al. 1997). As described in other paradigms,

1 mg/kg GR-127,935 was a partial agonist in guinea pigs; however, it seemed to be a silent antagonist in rats. In rats, a dose of 0.1 mg/kg GR-127,935 was sufficient to shift sumatriptan's dose/response curve to the right; in guinea pigs, a significant shift of sumatriptan's dose/response curve was only observed with 1 mg/kg GR-127,935. Interestingly, this dose did not reverse the inhibitory effects of CP-122,288 and 5-CT.

These results indicate that sumatriptan is likely to act via $5\text{-HT}_{1B/1D}$ receptors to inhibit extravasation (although sumatriptan also displays a high affinity for 5-HT_{1F} receptors, GR-127,395 is not expected to block this subtype, for which it only shows a low affinity; Yu et al. 1997). The higher potency of GR-127,935 in rats than in guinea pigs, taken together with its partial agonist activity in the latter, but not the former species, suggests that sumatriptan acts via the species homologue of $h5\text{-HT}_{1D}$ receptors in guinea pigs, via $r5\text{-HT}_{1B}$ receptors in rats. The lack of effect of the antagonist against 5-CT and CP-122,288 indicates that these agonists act at another receptor (see next section).

3. Studies in Mice Lacking 5-HT_{1B} Receptors

As mentioned earlier, a selective $5\text{-HT}_{1B/1D}$ antagonist was lacking until recently, and even GR-127,935 seems to be of limited use in view of its partial agonist activity in guinea pigs. René Hen and his group have developed a mouse "knockout" lacking $m5\text{-HT}_{1B}$ (pharmacologically similar to $r5\text{-HT}_{1B}$) receptors (see Chap. 15, this volume). These animals display enhanced aggressiveness, but do not seem to synthesize more $m5\text{-HT}_{1D}$ receptors as a compensatory mechanism.

In order to use these animals in the extravasation model, we have characterized the profile of the 5-HT receptors inhibiting neurogenic inflammation in SV-129 wild-type mice (Yu et al. 1996). As expected, this profile is identical to that found in rats, with 5-CT and CP-122,288 showing enhanced potency, and the 5-HT_{1B}-selective agonist CP-93,129 displaying a higher potency than sumatriptan. Both CP-93,129 and sumatriptan were ineffective in mice lacking $m5\text{-HT}_{1B}$ receptors (as was the $5\text{-HT}_{1B/1D}$-selective serotonin-5-O-carboxymethyl-glycyl-tyrosinamide, also called GTI). The lack of effect of sumatriptan indicates that, in mice, $m5\text{-HT}_{1D}$ and $m5\text{-HT}_{1F}$ receptors are not involved in the response to a significant extent.

Interestingly, 5-CT and CP-122,288 were fully effective in knockout mice in agreement with their effects in guinea pigs in the presence of GR-127,935. Perhaps more surprisingly, dihydroergotamine retained its potency as well. The drug-binding profile of CP-122,288 at 5-HT receptor subtypes is very similar to that of sumatriptan (Waeber and Moskowitz 1995a). Exclusion criteria thus suggest that CP-122,288 acts at a presently uncharacterized 5-HT receptor (it cannot act at 5-HT_{1B}, 5-HT_{1D} and 5-HT_{1F}, and has a much lower affinity for the other subtypes). The situation with 5-CT is more complex, as it possesses a high affinity for 5-HT_{1A}, 5-HT_5 and 5-HT_7 receptors (Hoyer et al. 1994). However, methiothepin is a potent antagonist at these three subtypes, and does not block the effects of 5-CT in rats. Also, 8-OH-DPAT is very weak in the rat model, ruling out 5-HT_{1A} receptors. 5-HT_{5A} sites are unlikely to be involved as 5-CT keeps its potency in 5-HT_{5A}-knockout mice (unpublished observation, in collaboration with René Hen, Columbia University, New York).

Data obtained with the antagonist GR-127,935 in guinea pigs agree with the observations in 5-HT_{1B} knockout mice. They indicate that a receptor other than $5\text{-HT}_{1B/1D}$ mediates the inhibition of plasma extravasation in the dura mater. This receptor also appears to be located prejunctionally (i.e., on trigeminovascular fibers), as CP-122,288 does not inhibit the extravasation induced by substance P injection (LEE and MOSKOWITZ 1993). Importantly, CP-122,288 is devoid of vasoconstrictor activity in either the dog carotid or coronary artery at concentrations which fully inhibit neurogenic inflammation in the dura (GUPTA et al. 1995). Molecules with a similar profile might constitute the next generation of antimigraine drugs, with an improved safety profile over sumatriptan.

II. Neuropeptide Release

When trigeminovascular fibers depolarize, impulses are transmitted centrally toward the superficial laminae of the trigeminal nucleus caudalis (STRASSMAN et al. 1986; DAVIS and DOSTROVSKY 1986). In addition, depolarization and antidromic conduction cause the release of sensory neuropeptides from widely branching perivascular trigeminal axons. These neuropeptides (substance P, neurokinin A, CGRP) mediate neurogenic inflammation. It has been demonstrated that substance P, neurokinin A and CGRP immunoreactivities are colocalized in neurons of the trigeminal ganglion of several species, including humans (UDDMAN and EDVINSSON 1989).

After electrical stimulation of the trigeminal ganglion in rats (BUZZI et al. 1991), CGRP levels increased in the draining venous effluents (sagittal sinus). The effect was maximal after 1 min. Pretreatment with dihydroergotamine (75 nmol/kg, i.e., a dose similar to that inhibiting plasma extravasation) did not change baseline CGRP concentrations but decreased the levels of CGRP during stimulation by 55% at 1 min and by 50% at 3 min. Sumatriptan (700 nmol/kg) attenuated the increase by 57% at 3 min. Increased CGRP and VIP levels were also observed in a slightly different paradigm (electrical stimulation of the superior sagittal sinus in cats; ZAGAMI et al. 1990). GOADSBY and EDVINSSON (1994) recently reported that stimulation of the trigeminal ganglion in the cat led to increases in both CGRP and vasoactive intestinal polypeptide (VIP) levels in the external jugular vein. These were both attenuated markedly by administration of the antimigraine drug 311C90 in a dose of 100 mg/kg (about 200 nmol/kg). KNYIHAR-CSILLIK et al. (1995) confirmed the trigeminal origin of increased blood CGRP content after trigeminal stimulation by showing increased CGRP immunoreactivity in fibers innervating the dura mater after brief electrical stimulation; longer-lasting stimulation produced irregular terminals with corroded outlines, indicative of release.

In humans, neurogenic inflammation develops in the facial skin during thermocoagulation of the trigeminal ganglion (Sweet and Wepsic 1974) and is associated with increased CGRP and substance P levels in the jugular vein (Goadsby et al. 1988). Interestingly, high venous levels of CGRP have been found in patients with migraine (Goadsby et al. 1990), and this increase was reversed after sumatriptan administration (Goadsby and Edvinsson 1991). CGRP levels are also elevated in patients with subarachnoid hemorrhage (Juul et al. 1990), as is tachykinin gene expression within trigeminal ganglia (Linnik et al. 1989), suggesting that this condition is also associated with activation of the trigeminovascular system. In fact, blood placed in the subarachnoid space in rats induces c-Fos expression (a marker of cell activation, see Sect. B.III) in the trigeminal nucleus caudalis, which can be blocked by pretreating the animals with sumatriptan and dihydroergotamine (Nozaki et al. 1992a). These data indicate that $5\text{-}HT_1$ agonists may be useful for the alleviation of pain in headache associated with meningeal activation (e.g., bacterial and viral meningitis) and that a response to treatment is not reliable diagnostically.

Although CGRP is a potent vasodilator (McCulloch et al. 1986), substance P, but not CGRP, causes plasma extravasation in dura mater or extracranial tissues innervated by the trigeminal ganglion (Markowitz et al. 1987). Increased blood flow in the medial meningeal artery after electrical stimulation of the rat dura mater is blocked by the CGRP receptor antagonist $CGRP_{8\text{-}37}$, but not by the NK1 receptor antagonist RP-67,580 (Messlinger et al. 1995). Conversely, NK1 receptor antagonists inhibit neurogenic plasma extravasation completely, suggesting the importance of substance P in this process (Lee et al. 1994). Although substance P levels in venous meningeal effluents have not been measured directly during migraine attacks, increased concentrations of this peptide have been found in the external jugular vein of both humans and cats after trigeminal ganglion stimulation (Goadsby et al. 1988). An in vitro study also reported that substance P is released from afferent nerve endings in bovine pia arachnoid following excitation by chemical or depolarizing stimuli (Moskowitz et al. 1983). It is not known whether substance P release would be blocked using $5\text{-}HT_1$ agonists, but this seems likely, in view of the reported costorage of substance P with CGRP in the trigeminal ganglion (Lee et al. 1985) and the simultaneous release of CGRP with several tachykinins in rat spinal cord slices after capsaicin stimulation (Saria et al. 1986).

III. c-Fos Expression in the Trigeminal Nucleus Caudalis

The nuclear phosphoprotein c-Fos is the product of an immediate early gene (i.e., it is expressed rapidly and transiently in stimulated cells without the requirement for de novo protein synthesis) which may play a role in long-term alterations of cellular function (Sheng and Greenberg 1990). Trauma, ischemia, seizures and more subtle forms of neuronal stimulation induce c-Fos expression in neuronal tissues (DRAGUNOW and FAULL 1989; ONODERA et al. 1989). Detection of the c-Fos protein by immunohistochemistry can thus be used to identify areas of neuronal activity and map functionally related neuronal pathways. The induction of c-Fos protein-like immunoreactivity occurs in the dorsal horn of the spinal cord and the trigeminal nucleus caudalis within 1–2 h following peripheral noxious and non-noxious stimulation (HUNT et al. 1987; ANTON et al. 1991).

NOZAKI et al. (1992a,b) noted the appearance of c-Fos immunoreactive cells within the trigeminal nucleus caudalis (laminae I and IIo) in response to autologous blood injected in the subarachnoid space of the rat. The number of positive cells corresponded to the amount of administered blood. Morphine decreased the number of labeled cells by 63%, while sumatriptan (720 nmol/kg, twice), dihydroergotamine (86 nmol/kg, twice) and the selective r5-HT$_{1B}$ agonist CP-93,129 (460 nmol/kg, twice) decreased this number by 31%, 33% and 39%, respectively. No decrease was observed in the nucleus of the solitary tract and area postrema. Drug-induced blockade of c-Fos expression was likely to be mediated by an action on primary afferent fibers, since CP-93,129 did not reduce the number of expressing cells in rats treated as neonates with capsaicin.

Capsaicin has been extensively employed as an activating stimulus for c-Fos expression (PELTO-HUIKKO et al. 1991). STRASSMAN and VOS (1993) showed that injection of capsaicin into the supraorbital region of rats evoked c-Fos expression in laminae I and II of the trigeminal nucleus caudalis. Capsaicin evoked a more robust c-Fos response than subarachnoid blood injection (CUTRER et al. 1995a,b), but the laminar distribution of labeled cells was similar with both stimulants. Pretreatment with the conformationally restricted sumatriptan analogue CP-122,288 (100 pmol/kg, i.v.) decreased the number of positive cells by 50%–60% in laminae I and II (CUTRER et al. 1995a). This dose is 1000 times lower than the effective dose of dihydroergotamine and almost 10000 times lower than that of sumatriptan (NOZAKI et al. 1992a). Although the models are slightly different (capsaicin in guinea pigs versus autologous blood in rats), the enhanced potency of CP-122,288 at inhibiting c-Fos expression seems to mirror its high potency in the plasma extravasation paradigm (LEE and MOSKOWITZ 1993).

Apparently discrepant results were recently published using sumatriptan to inhibit c-*fos* mRNA expression in the trigeminal nucleus of rats (SHEPHEARD et al. 1995b): sumatriptan (1 mg/kg, i.e., about 2.5 mmol/kg) pretreatment only resulted in a nonsignificant 26% ± 8% reduction of c-Fos mRNA expression. Interestingly, this dose of sumatriptan reduced the level of c-Fos hybridization signal by 65% when the blood-brain barrier was disrupted by prior hyperosmolar mannitol injection. These results suggest that the antimigraine action of sumatriptan is predominantly peripherally mediated (inhibition of dural neurogenic inflammation or vasoconstriction). The variance between these findings and the efficacy of sumatriptan at inhibiting subarachnoid blood-induced c-Fos expression might be caused by experimental factors (electrical stimulation might induce a stronger, i.e., less prone to inhibition, c-Fos expression than subarachnoid blood; also, c-*fos* mRNA was measured 30 min after stimulation, whereas immunohistochemistry was performed 2 h after blood administration). Furthermore, electrical trigeminal stimulation may cause widespread c-Fos expression by discharging many more afferents than those innervating the meninges. Alternatively, the subarachnoid hemorrhage might have altered the blood-brain barrier to some extent, allowing sumatriptan to access central sites. In line with these data, HOSKIN et al. (1995) recently reported that dihydroergotamine produced a stronger inhibition than sumatriptan of c-Fos expression after superior sagittal sinus stimulation, correlating this finding with the higher brain penetration of dihydroergotamine (GOADSBY and GUNDLACH 1991). The brain penetration properties of CP-122,288 have not been reported; however, the incorporation of the C3-aminoethyl side chain of sumatriptan into a C3-(*R*)-pyrrolidine-2-*yl*-methyl group in CP-122,288 might be expected to increase the lipophilicity of the molecule and hence its brain barrier permeation.

Activation of sensory trigeminovascular fibers causes the release of neuropeptides (including substance P, neurokinin A and CGRP) both locally, as part of the meningeal neurogenic inflammatory response, and centrally, in the dorsal horn of the lower medulla. Substance P has been shown to coexist with glutamate in some nociceptive neurons (BATTAGLIA and RUSTIONI 1988), and both molecules act synergistically on dorsal horn neurons in response to noxious as well as innocuous stimuli (DOUGHERTY and WILLIS 1991). The NK1 receptor antagonist RPR-100,893 has been shown to reduce the number of c-Fos-positive cells in the trigeminal nucleus caudalis after intracisternal capsaicin injection by up to 50% (CUTRER et al. 1995b). Similar results were found with another NK1 antagonist, CP-99,994 (SHEPHEARD et al. 1995b). Hence, it might be hypothesized that 5-HT$_1$ agonists inhibit c-Fos expression by blocking substance P release from primary afferent neurons.

In a search for a more physiological inducer of c-Fos expression, we have found that recurrent neocortical spreading depression (an excitatory wave of depolarization followed by inhibition, moving over neocortical gray matter at a rate of 2–6 mm/min), increases c-Fos labeling within the ipsilateral superficial lamina of the rat trigeminal nucleus caudalis (MOSKOWITZ et al. 1993); this expression is dependent upon trigeminovascular activation and is inhibited by sumatriptan (700 nmol/kg). Spreading depression itself, however, is not blocked by sumatriptan. The fact that spreading depression induces trigeminal nucleus activation suggests that endogenous neurophysiological events in the neocortex are capable of stimulating trigeminovascular fibers innervating the surrounding meninges and penetrating blood vessels.

This finding also suggests that sumatriptan does not block migraine etiology per se, but inhibits selectively the response of a common pathway. The site of sumatriptan's effect in spreading depression is most likely to be the primary afferent fiber, since spreading depression is not known to alter the properties of the blood-brain barrier within brain stem. It is not known whether spreading depression occurs in humans, although some investigators have postulated that it underlies the focal neurological events during the aura or initial stages of migraine headache.

C. Data from In Vitro Experiments

I. Autoradiographic Data

The functional data mentioned in the previous sections point to a role for 5-$HT_{1B/1D}$ receptors in the inhibition of neurogenic inflammation by sumatriptan and ergot alkaloids. However, the significance of other receptors, mainly 5-HT_{1F} and 5-HT_7, but also 5-HT_{1A} and 5-HT_5, cannot be completely ruled out. Moreover, there is evidence that 5-CT and CP-122,288 might inhibit plasma extravasation and c-Fos expression via a novel receptor subtype (see plasma extravasation studies with GR-127,935 and 5-HT_{1B} knockout mice). Most of the functional models listed above are elaborate and do not lend themselves to easy pharmacological characterization (one animal is needed for each data point in the extravasation and c-Fos models). Radioligand binding studies can test a substantial number of drugs in a much shorter period of time, but are mostly performed on homogenates from transfected cells or easily accessible brain tissues (such as cortex or striatum), but irrelevant to migraine. The advantage of in vitro autoradiography is that binding studies can be performed on tissues that are actually involved in inhibition of neurogenic inflammation.

Cerebral blood vessels, the dura mater, trigeminal ganglion and trigeminal nuclear complex in the lower medulla are all likely to contain target receptors for antimigraine drugs. Although binding sites for some receptors (histamine H_1, α- and β-adrenergic and muscarinic) have been detected on cerebral vessels and meninges (OTTOSON et al. 1990; NAKAI et al. 1986; LASBENNES et al. 1992; DAUPHIN and HAMEL 1992), studies of 5-HT receptors are scarce and contradictory (PEROUTKA and KUHAR 1984; MIYAMOTO et al. 1994; DE KEYSER et al. 1993). [^{3}H]Sumatriptan-binding sites have been found in the guinea pig trigeminal ganglion, but only at high radioligand concentration (WAEBER and MOSKOWITZ 1995a). Previous studies failed to detect [^{125}I]GTI 5-$HT_{1B/1D}$-binding sites in rat and human trigeminal ganglia (BRUINVELS et al. 1992, 1993); hence these [^{3}H]sumatriptan-binding sites might correspond to 5-HT_{1F} receptors. We have also detected low densities of [^{3}H]5-CT-binding sites in guinea pig trigeminal ganglion; the fact that labeling was displaced by 100 nM 8-OH-DPAT indicates that the ligand was probably binding to 5-HT_{1A} receptors. This is confirmed by the previous finding of intermediate densities of [3H]8-OH-DPAT of binding sites in the rabbit trigeminal ganglion (Waeber and Palacios, unpublished).

In contrast with the tissues mentioned above, where autoradiographic techniques detected few, if any, 5-HT-binding sites, the superficial laminae of

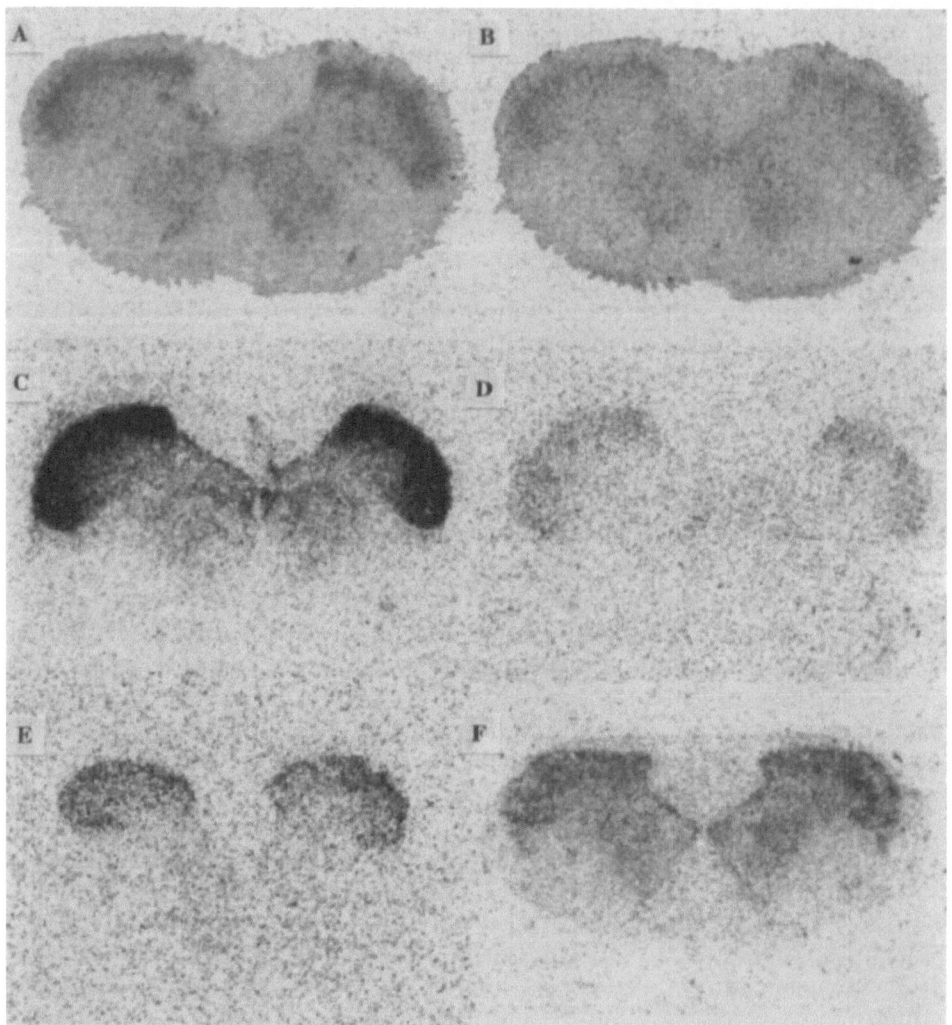

Fig. 1A–F. Distribution of binding sites labeled using ligands selective for different receptor subtypes in the guinea pig trigeminal nucleus. **A** [^{3}H]Sumatriptan binding in the absence of displacer (labels both 5-HT$_{1D}$ and 5-HT$_{1F}$ sites). **B** [^{3}H]Sumatriptan binding in the presence of 100 nM 5-CT (labels 5-HT$_{1F}$ sites). **C** [^{3}H]5-CT binding in the absence of displacer (labels 5-HT$_{1A}$, 5-HT$_{1D}$ and 5-HT$_7$ sites). **D** [^{3}H]5-CT binding in the presence of 100 nM dihydroergotamine (labels 5-HT$_7$ sites). **E** [^{3}H]8-OH-DPAT binding in the absence of displacer (labels 5-HT$_{1A}$ sites). **F** [^{3}H]L-694247 binding in the absence of displacer (labels 5-HT$_{1D}$ sites)

the trigeminal nucleus caudalis display intermediate to dense labeling with most of the 5-HT receptor ligands used (see Fig. 1). The selective 5-HT$_{1B/1D}$ ligand [^{125}I]GTI labels sites in the guinea pig medulla, while the selective 5-HT$_{1B/1D}$ ligand [^{3}H]L-694,247 labels sites in the guinea pig and human medulla.

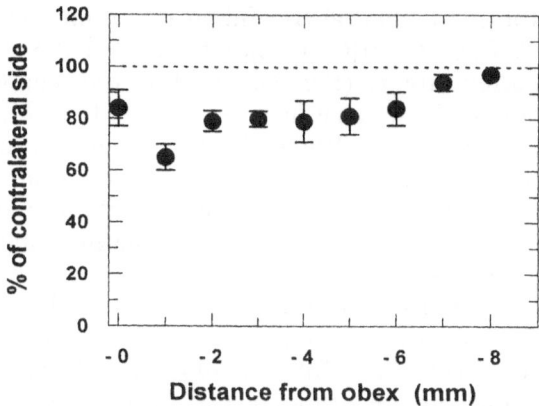

Fig. 2. Density   of   [^{125}I]serotonin-5-O-carboxymethyl-glycyl-iodo-tyrosinamide ([^{125}I]GTI)-binding sites in the superficial laminae (I and II) of the guinea pig trigeminal nucleus after unilateral trigeminal rhizotomy (see text for details)

[^{3}H]Sumatriptan labeling in the guinea pig trigeminal nucleus does not seem to be displaced by 100 nM 5-CT (WAEBER and MOSKOWITZ 1995a), suggesting that the labeling corresponds mostly to 5-HT$_{1F}$ receptors under those conditions. Finally, [^{3}H]5-CT seems to label 5-HT$_{1A}$, 5-HT$_{1D}$ and 5-HT$_7$ sites in this area (WAEBER and MOSKOWITZ 1995b). It should also be mentioned that the spinal trigeminal nucleus and the substantia gelatinosa of the spinal cord contain the highest densities of 5-HT$_3$ receptors found in the brain of several species (WAEBER et al. 1990).

   Although 5-CT seems to block plasma extravasation via a novel receptor, autoradiographic studies with [^{3}H]5-CT did not find evidence for labeling of receptors other than 5-HT$_{1A}$, 5-HT$_{1B/1D}$ and 5-HT$_7$ in the brain parenchyma (WAEBER and MOSKOWITZ 1995b; To et al. 1995). Despite their high affinity for 5-CT (HOYER et al. 1994), 5-HT$_5$ sites do not seem to be labeled under the conditions of these studies. 5-HT$_{1A}$ receptors did, however, present some atypical properties, in particular in hippocampus, where they showed a very low affinity for antagonists such as methiothepine and spiperone (WAEBER and MOSKOWITZ 1995b), but there is no indication that these sites may be related to the receptor inhibiting extravasation.

   In vitro receptor autoradiography cannot distinguish whether the observed binding sites are located on primary afferent terminals of the trigeminal ganglion, on the cell bodies of second-order neurons (or interneurons) in the trigeminal nucleus or on the terminals of other neurons projecting to this area. Two approaches have been used to gain insight into the cellular location of receptors. The first technique involves selective lesioning of tracts or pathways, or the destruction of specific cell populations. The second method relies on the use of in situ hybridization histochemistry to examine the location of the cell bodies synthesizing a given receptor.

We have performed unilateral trigeminal rhizotomy on guinea pigs and found a 25%–30% loss of [^{125}I]GTI-binding sites on the side ipsilateral to the lesion 8 days after surgery (Fig. 2). The loss was significant between 0 and 2 mm posterior to obex, which is in agreement with the known distribution of trigeminal sensory afferents. While the lost receptors might correspond to those mediating sumatriptan's prejunctional inhibition of neuropeptide release, the location and functional significance of the remaining sites is not certain. They might correspond to the sites mediating inhibition of central trigeminal neurons by sumatriptan only after blood-brain barrier disruption (Kaube et al. 1993; Shepheard et al. 1995b). Although their role in migraine and neurogenic inflammation is unknown, it is worth mentioning that about 20%–30% of 5-HT$_{1A}$ and 50% of 5-HT$_3$ receptors in the substantia gelatinosa of the spinal cord are located on capsaicin-sensitive fibers (Daval et al. 1987; Hamon et al. 1989) and are likely to correspond to the 5-HT$_{1A}$ and 5-HT$_3$ sites labeled in the dorsal root ganglia (Waeber and Palacios, unpublished). The presynaptic location of 5-HT$_{1A}$ receptors in the lumbar spinal cord was, however, not confirmed in a more recent study (Croul et al. 1995). Since the spinal trigeminal nucleus is generally considered as the rostral extension of the spinal substantia gelatinosa, it is likely that some presynaptic 5-HT$_{1A}$ and 5-HT$_3$ receptors also exist in the trigeminal system. Indeed, the 5-HT$_3$ receptor antagonist ICS-205,939 (5-HT is excitatory at 5-HT$_3$ receptors) has been shown to reduce c-Fos expression in the trigeminal nucleus after noxious chemical stimulation of the rat nasal mucosa (Ebersberger et al. 1995).

As the sequence of the mRNA encoding most of currently known receptor subtypes has become available, it has been possible to use in situ hybridization histochemistry to localize the cells expressing specific receptors. Thus, Bruinvels et al. (1992, 1993) have shown that trigeminal ganglion neurons contain mRNA for 5-HT$_{1B}$ and 5-HT$_{1D}$ receptors, confirming the prejunctional location of at least some 5-HT$_{1B}$ sites. Adham et al. (1996) also demonstrated the presence of 5-HT$_{1F}$ receptor mRNA in the guinea pig trigeminal ganglion, suggesting that 5-HT$_{1F}$ receptors might also play a role in the prejunctional inhibition of neuropeptide release. Unfortunately, most of the other published in situ hybridization studies have concentrated on the forebrain and no other data are available yet on the trigeminal ganglion.

II. Detection of the mRNA for Different 5-HT Receptor Subtypes

Rebeck et al. (1994) used reverse transcriptase polymerase chain reaction (RT-PCR) to obtain molecular evidence for the expression of an mRNA species encoding the h5-HT$_{1D}$ receptor subtype in human trigeminal ganglia; no h5-HT$_{1B}$ mRNA was detected. In guinea pigs, a single amplification product was detected, which showed an 85% sequence homology to h5-HT$_{1D}$ and 71% to h5-HT$_{1B}$. Recently, these results were partially confirmed (Bouchelet et al. 1996) by the demonstration of similar amounts of both h5-HT$_{1D}$ and h5-HT$_{1B}$

mRNA in human trigeminal ganglia. 5-HT_{1F} receptor mRNA was also found in human trigeminal ganglia, indicating a presynaptic location of these receptors. The facts that 5-HT_{1D} and 5-HT_{1F} receptors possess similar affinities for sumatriptan and act via similar second messenger systems suggest that both receptors might mediate the antimigraine effects of sumatriptan.

Interestingly, BOUCHELET et al. (1996) found both h5-HT_{1B} and 5-HT_{1F} receptor mRNAs on cerebrovascular tissues, but only trace amounts of h5-HT_{1D} receptor mRNA. h5-HT_{1B} mRNA was also found in one coronary artery. In view of the reported cardiovascular side effects of sumatriptan (MACINTYRE et al. 1993), these data suggest that selective h5-HT_{1D} agonists (with no effect at h5-HT_{1B} receptors) might alleviate headache with greatly reduced side effects, provided efficacy in the neurogenic inflammation model is predictive for antimigraine efficacy (see below for alternative models and theories).

It was mentioned earlier that 5-CT-induced inhibition of plasma extravasation might be mediated by 5-HT_7 receptors. Recently, the mRNA encoding these receptors was found on the smooth muscle cells of a variety of blood vessels (ULLMER et al. 1995); functional studies showed that 5-HT_7 receptors mediate endothelium-independent relaxation of the jugular vein and possibly other vessels (LEUNG et al. 1996 and references therein). Since cerebral arteries reportedly dilate during migraine attacks (FRIBERG et al. 1991), the vasorelaxant properties of 5-HT_7 receptors might weaken their suitability as a useful antimigraine target.

FOZARD and KALKMAN (1994) recently contended that 5-HT_{2B} (or possibly 5-HT_{2C}) receptor activation might be a key factor in the initiation of migraine. Their hypothesis was based on the fact that meta-chlorophenyl-piperazine (mCPP, an agonist at these receptors) triggers migraine attacks in control patients and migraineurs, and that the human daily doses of a series of antimigraine prophylactic drugs (methysergide, pizotifen, Org GC 94, cyproheptadine, mianserin, amitriptyline, chlorpromazine and propanolol) are in close correlation with their antagonist potencies at 5-HT_{2B} receptors, while ketanserin and pindolol are inactive as antimigraine agents and weak 5-HT_{2B} antagonists. 5-HT_{2B} receptor mRNA is present on all blood vessels investigated (ULLMER et al. 1995) and 5-HT_{2B} receptor activation induces endothelium-dependent vasodilation, presumably as a consequence of nitric oxide release. This release might activate and sensitize trigeminovascular neurons and initiate the neurogenic inflammation process associated with migraine. The linkage between nitric oxide and migraine is likely to be important although poorly understood at the present time (OLESEN et al. 1994).

III. Antimigraine Drugs and Vasoconstriction

Both vascular (HUMPHREY and FENIUK 1991) and neuronal effects (see above) have been invoked to explain the antimigraine action of sumatriptan and ergot alkaloids. Stretching of the arterial wall produced by extracranial vasodilation was traditionally thought to be the source of pain in migraine (GRAHAM and

WOLFF 1938). During migraine attack, the middle cerebral artery was found to be dilated on the headache side in one study and the dilation was reversed by sumatriptan injection (FRIBERG et al. 1991). It has also been shown that sumatriptan increases blood velocity in the middle cerebral and internal carotid arteries, suggesting that constriction of the large basal intracranial arteries might mediate the antimigraine effect of sumatriptan (CAEKEBEKE et al. 1992). Possible effects on arteriovenous anastomoses have also been hypothesized (PERREN et al. 1989; DEN BOER et al. 1991).

The study of contraction and relaxation of arterial ring segments from cerebral vessels is commonly used to determine the direct effect of drugs on the cerebral vasculature. Since marked species differences are known to exist with regard to transmitters and receptors on cerebral blood vessels (HAMEL et al. 1985), studies should ideally be conducted using human tissues. It has also been recognized that vessels from different vascular beds react differently (VANHOUTTE 1978). Despite these limitations, it is interesting to note that sumatriptan was developed by screening compounds for a vasoconstrictor activity in the dog saphenous vein (HUMPHREY et al. 1988). A number of blood vessels have been found to contract in response to sumatriptan (human, dog and rabbit saphenous vein, human dural vessels, the basilar artery of most species, as well as the renal, umbilical, iliac and coronary artery; see SAXENA and TFELT-HANSEN 1993 for individual references). It is, however, unclear whether these effects are all mediated by $5-HT_{1D}$ receptors or whether mere vasoconstriction is therapeutically important in migraine.

Using pharmacological correlation and Northern blot analysis, HAMEL et al. (1993a,b) demonstrated that the $h5-HT_{1B}$ receptor subtype mediates vasoconstriction in human cerebral blood vessels. These results were recently confirmed using RT-PCR analysis (BOUCHELET et al. 1996). Whether antimigraine efficacy can be dissociated from vasoactive properties still needs a definitive clinical demonstration; this might be provided by the effectiveness of drugs such as CP-122,288 or $h5-HT_{1D}$-selective agonists in clinical trials. The former drugs have been shown to inhibit neurogenic inflammation at doses far below those required to detect vasoconstrictor activity (GUPTA et al. 1995; BEATTIE and CONNOR 1995). Considering the presence of the mRNA for nearly all 5-HT receptor subtypes on most blood vessels (ULLMER et al. 1995), the lack of $5-HT_{1D\alpha}$ mRNA on cerebral vessels points to this subtype as a rare opportunity to target antimigraine agents devoid of cardiovascular side effects.

D. Conclusion

Although the safety profile of sumatriptan has been considerably improved over that of ergot alkaloids, it still displays some cardiovascular side effects that might reflect its action at several 5-HT receptor subtypes. If the described models in rodents are indeed relevant and predictive for antimigraine efficacy, it is likely that decreasing the affinity of antimigraine drugs at $h5\text{-}HT_{1B}$ receptors, while conserving their affinity at $h5\text{-}HT_{1D}$ receptors, might result in active agents with greatly reduced side effects. The target of CP-122,288 and 5-CT-mediating inhibition of neurogenic inflammation might also be promising for this purpose, provided an homologous site exists in humans.

References

Adham N, Bard JA, Zgombick JM, Durkin MM, Weinshank RL, Branchek TA (1996) Cloning and characterization of a recombinant guinea pig $5\text{-}HT_{1F}$ receptor. Soc Neurosci Abstr 22:528.9

Anton F, Herdegen T, Peppel P, Leah JD (1991) C-fos-like immunoreactivity in rat brainstem neurons following noxious chemical stimulation of the nasal mucosa. Neuroscience 41:629–641

Battaglia G, Rustioni A (1988) Coexistence of glutamate and substance P in dorsal root ganglion cells of the rat and monkey. J Comp Neurol 277:302–312

Beattie DT, Connor HE (1995) The pre- and postjunctional activity of CP-122,288, a conformationally restricted analogue of sumatriptan. Eur J Pharmacol 276:271–276

Bouchelet I, Cohen Z, Case B, Séguéla P, Hamel E (1996) Differential expression of sumatriptan-sensitive 5-hydroxytryptamine receptors in human trigeminal ganglia and cerebral blood vessels. Mol Pharmacol 50:219–223

Bouhelal R, Smounya L, Bockaert J (1988) $5\text{-}HT_{1B}$ receptors are negatively coupled with adenylate cyclase in rat substantia nigra. Eur J Pharmacol 151:189–196

Bruinvels AT (1993) $5\text{-}HT_{1D}$ receptors reconsidered: radioligand binding assays, receptor autoradiography and in situ hybridization histochemistry in the mammalian nervous system. Thesis, University of Utrecht

Bruinvels AT, Landwehrmeyer B, Moskowitz MA, Hoyer D (1992) Evidence for 5-HT_{1B} messenger RNA in rat trigeminal ganglia. Eur J Pharmacol 227:357–359

Buzzi MG, Moskowitz MA (1990) The antimigraine drug, sumatriptan (GR43175), selectively blocks neurogenic plasma extravasation from blood vessels in dura mater. Br J Pharmacol 99:202–206

Buzzi MG, Moskowitz MA, Peroutka SJ, Byun B (1991a) Further characterization of the putative 5-HT receptor which mediates blockade of neurogenic plasma protein extravasation in rat dura mater. Br J Pharmacol 103:1421–1428

Buzzi MG, Carter WB, Shimizu T, Heath H III, Moskowitz MA (1991b) Dihydroergotamine and sumatriptan attenuate levels of CGRP in plasma in rat superior sagittal sinus during electrical stimulation of the trigeminal ganglion. Neuropharmacology 30:1193–1200

Caekebeke JFV, Ferrari MD, Zwetsloot CP, Jansen J, Saxena PR (1992) Antimigraine drug sumatriptan increases blood flow velocity in large cerebral arteries during migraine attacks. Neurology 42:1522–1526

Croul S, Sverstiuk A, Radzievsky A, Murray M (1995) Modulation of neurotransmitter receptors following unilateral L1-S2 deafferentation: NK1, NK3, NMDA, and $5\text{-}HT_{1A}$ receptor binding autoradiography. J Comp Neurol 361:633–644

Cutrer FM, Schoenfeld D, Limmroth V, Panahian N, Moskowitz MA (1995a) Suppression by the sumatriptan analogue, CP-122,288, of c-fos immunoreactivity in trigeminal nucleus caudalis induced by intracisternal capsaicin. Br J Pharmacol 114:987–992

Cutrer FM, Moussaoui S, Garret C, Moskowitz MA (1995b) The non-peptide neurokinin-1 antagonist, RPR 100893, decreases c-fos expression in trigeminal nucleus caudalis following noxious chemical meningeal stimulation. Neurosci 64:741–750

Dauphin F, Hamel E (1992) Identification of multiple muscarinic binding site subtypes in cat and human cerebral vasculature. J Pharmacol Exp Ther 260:660–667

Daval G, Vergé D, Basbaum A, Bourgoin S, Hamon M (1987) Autoradiographic evidence of serotonin binding sites on primary afferent fibers in the dorsal horn of the rat spinal cord. Neurosci Lett 83:71–76

Davis KD, Dostrovsky JO (1986) Activation of trigeminal brain-stem nociceptive neurons by dural artery stimulation. Pain 25:395–401

De Keyser J, Vauquelin G, De Backer JP, De Vos H, Wilczak N (1993) What intracranial tissues in humans contain sumatriptan-sensitive serotonin 5-HT$_1$-type receptors. Neurosci Lett 164:63–66

Den Boer MO, Villalón CM, Heiligers JPC, Humphrey PPA, Saxena PR (1991) Role of 5-HT$_1$-like receptors in the reduction of porcine cranial arteriovenous anastomotic shunting by sumatriptan. Br J Pharmacol 102:323–330

Dimitriadou V, Buzzi MG, Moskowitz MA, Theoharides TC (1991) Trigeminal sensory fiber stimulation induces morphological changes in rat dura mater mast cells. Neurosci 44:97–112

Dimitriadou V, Buzzi MG, Theoharides TC, Moskowitz MA (1992) Ultrastructural evidence for neurogenically mediated changes in blood vessels of the rat dura mater and tongue following antidromic trigeminal stimulation. Neuroscience 48:187–203

Dougherty PM, Willis WD (1991) Enhancement of spinothalamic neuron responses to chemical and mechanical stimuli following combined microiontophoretic application of N-methyl-D-aspartic acid and substance P. Pain 47:85–93

Dragunow M, Faull R (1989) The use of c-fos as a metabolic marker in neuronal pathway tracing. J Neurosci Meth 29:261–265

Ebersberger A, Anton F, Tölle TR, Zieglgänsberger W (1995) Morphine, 5-HT$_2$ and 5-HT$_3$ receptor antagonists reduce c-fos expression in the trigeminal nuclear complex following noxious chemical stimulation of the rat nasal mucosa. Brain Res 676:336–342

Eide PK, Hole K, Berge OG (1988) Mechanisms by which the putative serotonin receptor antagonist metitepin alters nociception in mice. J Neural Transm 73:31–41

Fozard JR, Kalkman HO (1994) 5-Hydroxytryptamine (5-HT) and the initiation of migraine:new perspectives. Naunyn Schmiedebergs Arch Pharmacol 350:225–229

Friberg L, Olesen J, Iversen HK, Sperling B (1991) Migraine pain associated with middle cerebral artery dilatation:Reversal by sumatriptan. Lancet 338:13–17

Goadsby PJ, Edvinsson L (1991) Sumatriptan reverses the changes in calcitonin gene-related peptide seen in the headache phase of migraine. Cephalalgia 11 (Suppl 11):3–4

Goadsby PJ, Gundlach AL (1991) Localization of [^{3}H]-dihydroergotamine binding sites in the cat central nervous system: relevance to migraine. Ann Neurol 29:91–94

Goadsby PJ, Edvinsson L (1994) Peripheral and central trigeminovascular activation in cat is blocked by the serotonin 5HT$_{1D}$ receptor agonist 311C90. Headache 34:394–399

Goadsby PJ, Edvinsson L, Ekman R (1988) Release of vasoactive peptides in the extracerebral circulation of humans and the cat during activation of the trigeminovascular system. Ann Neurol 23:193–196

Goadsby PJ, Edvinsson L, Ekman R (1990) Vasoactive peptide release in the extracerebral circulation of human during migraine headache. Ann Neurol 28:183–187

Graham JR, Wolff HG (1938) Mechanism of migraine headache and action of ergotamine tartrate. Arch Neurol Psychiatr 39:737–763

Gupta P, Brown D, Butler P, Ellis P, Grayson KL, Land GC, Macor JE, Robson SF, Wythes MJ, Shepperson NB (1995) The in vivo pharmacological profile of a 5-HT$_1$ receptor agonist, CP-122,288, a selective inhibitor of plasma extravasation. Br J Pharmacol 116:2385–2390

Hamel E, Edvinsson L. MacKenzie E (1985) Reactivity of various cerebral arteries to vasoactive substances in different species. J Cereb Blood flow Metab 5 (Suppl 1):S553–S554

Hamel E, Grégoire L, Lau B (1993a) 5-HT$_1$ receptor mediating contraction in bovine cerebral arteries: a model for human cerebrovascular '5-HT$_{1D\beta}$' receptors. Eur J Pharmacol 242:75–82

Hamel E, Fan E, Linville D, Ting V, Villemure JG, Chia LS (1993b) Expression of mRNA for the serotonin 5-hydroxytryptamine$_{1D\beta}$ receptor subtype in human and bovine cerebral arteries. Mol Pharmacol 44:242–246

Hamon M, Gallisot M, Ménard F, Gozlan H, Bourgoin S, Vergé D (1989) Biochemical and autoradiographic evidence of 5-HT$_3$ receptor binding sites on capsaicin-sensitive fibers in rat spinal cord. Eur J Pharmacol 164:315–322

Hartig PR, Hoyer D. Humphrey PPA, Martin GR (1996) Alignment of receptor nomenclature with the human genome: classification of 5-HT$_{1B}$ and 5-HT$_{1D}$ receptor subtypes. Trends Pharmacol Sci 17:103–105

Hoyer D, Clarke DE, Fozard JR, Hartig PR, Martin GR, Mylecharane EJ, Saxena PR, Humphrey PPA (1994) VII. International Union of Pharmacology classification of receptors for 5-hydroxytryptamine (serotonin). Pharmacol Rev 46:157–203

Hoskin KL, Kaube H, Goadsby PJ (1995) A comparison of the effects of dihydroergotamine and sumatriptan of c-fos expression in the trigeminal nucleus of the cat. Cephalalgia 15 (Suppl 14):P190

Huang Z. Byun B, Matsubara T, Moskowitz MA (1993) Time-dependent blockade of neurogenic plasma extravasation in dura mater by 5-HT$_{1B/D}$ agonists and endopeptidase 24.11. Br J Pharmacol 108:331–335

Humphrey PPA, Feniuk W (1991) Mode of action of the anti-migraine drug sumatriptan. Trends Pharmacol Sci 12:444–446

Humphrey PPA, Feniuk W, Perren MJ, Connor HE. Oxford AW, Coates IH, Butina D (1988) GR43175, a selective agonist for the 5-HT$_1$-like receptor in dog isolated saphenous vein. Br J Pharmacol 94:1123–1132

Hunt SP. Pini A. Evan G (1987) Induction of c-fos-like protein in spinal cord neurons following sensory stimulation. Nature 328:632–634

Johnson KW, Schaus JM, Cohen ML, Audia JE, Kaldor SW, flaugh ME, Krushinski JH, Shenk KW, Kiefer Jr AD, Nissen JS, Dressman BA, Zgombick JM, Branchek TA, Adham N, Phebus LA (1996) Inhibition of neurogenic protein extravasation in the dura via 5-HT$_{1F}$ receptor activation. Soc Neurosci Abstr 22:528.10

Juul R, Edvinsson L, Gisvold SE, Ekman R, Brubakk AO, Fredriksen TA (1990) Calcitonin-gene related peptide-LI in subarachnoid haemorrhage in man. Signs of activation of trigemino-cerebrovascular system? Br J Neurosurg 4:171–180

Kaube H, Hoskin KL, Goadsby PJ (1993) Inhibition by sumatriptan of central trigeminal neurones only after blood-brain barrier disruption. Br J Pharmacol 109:788–792

Kenakin T (1993) Pharmacologic analysis of drug-receptor interaction. 2nd edn. Raven, New York

Knyihar-Csillik E, Tajti J, Mohtasham S, Sari G, Vecsei L (1995) Electrical stimulation of the Gasserian ganglion induces structural alterations of calcitonin gene-related peptide-immunoreactive perivascular sensory nerve terminals in the rat cerebral dura mater: a possible model of migraine headache. Neurosci Lett 184:189–192

Lasbennes F, Verrecchia C, Philipson V, Seylaz J (1992) Muscarinic binding of pial vessels and arachnoid membrane. J Neurochem 58:2230–2235

Lee WS, Moskowitz MA (1993) Conformationally restricted sumatriptan analogues, CP-122,288 and CP-122,638, exhibit enhanced potency against neurogenic inflammation in dura mater. Brain Res 626:303–305

Lee Y, Kawai Y, Shiosaka S, Takami K, Kiyama H, Hillyard CJ, Girgis S, MacIntyre I, Emson PC, Tohyama M (1985) Coexistence of calcitonin gene-related peptide and substance P-like peptide in single cells of the trigeminal ganglion of the rat: immunohistochemical analysis. Brain Res 330:194–196

Lee WS, Mousaoui SM, Moskowitz MA (1994) Oral or parenteral non-peptide NK1 receptor antagonist RPR 100893 blocks neurogenic plasma extravasation within guinea-pig dura mater and conjunctiva. Br J Pharmacol 112:920–924

Lee WS, Limmroth V, Ayata C, Cutrer FM, Waeber C, Yu X, Moskowitz MA (1995) Peripheral GABA$_A$ receptor mediated effects of sodium valproate on dural plasma extravasation to substance P and trigeminal stimulation. Br J Pharmacol 116:1661–1667

Leung E, Walsh LKM, Pulido-Rios MT, Eglen RM (1996) Characterization of putative 5-HT$_7$ receptors mediating direct relaxation in Cynomolgus monkey isolated jugular vein. Br J Pharmacol 117:926–930

Linnik MD, Sakas DE, Uhl GR, Moskowitz MA (1989) Subarachnoid blood and headache: altered trigeminal tachykinin gene expression. Ann Neurol 25:179–184

MacIntyre PD, Bhargava B, Hogg KJ, Gemmill JD, Hillis WS (1993) Effect of subcutaneous sumatriptan, a selective 5-HT$_1$ agonist, on the systemic pulmonary and coronary circulation. Circulation 87:401–405

Macor JE, Burkhart JH, Heym JH, Ives JL, Lebel LA, Newman ME, Nielsen JA, Ryan K, Schulz DW, Torgersen LK, Koe BK (1990) 3-(1,2,5,6-Tetrahydropyrid-4-yl)pyrrolo[3, 2-b]pyrid-5-one: a potent and selective serotonin (5-HT$_{1B}$) agonist and rotationally restricted phenolic analogue of 5-methyl-3-(1,2,5,6-tetrahydropyrid-4-yl)indole. J Med Chem 33:2087–2093

Markowitz S, Saito K, Moskowitz MA (1987) Neurogenically mediated leakage of plasma protein occurs from blood vessels in dura mater but not brain. J Neuroscience 7:4129–4136

Matsubara T, Moskowitz MA, Byun B (1991) CP-93,129, a potent and selective 5-HT$_{1B}$ receptor agonist, blocks neurogenic plasma extravasation within rat but not guinea pig dura mater. Br J Pharmacol 104:3–4

Matsubara T, Moskowitz MA, Huang Z (1992) UK-14,304, R(−)-α-methyl-histamine and octreotide (SMS201-995) block plasma protein leakage within dura mater by prejunctional mechanisms. Eur J Pharmacol 224:145–150

McCulloch J, Uddman R, Kingman TA, Edvinsson L (1986) Calcitonin gene-related peptide: functional role in cerebrovascular regulation. Proc Natl Acad Sci USA 83:5731–5735

Messlinger KB, Pawlak M, Kurosawa M, Carmody JJ (1995) Calcitonin gene-related peptide, but not substance P, mediates the increased meningeal blood flow elicited by electrical stimulation of rat dura mater encephali. In: Olesen J, Moskowitz MA (eds) Experimental headache models. Lippincott-Raven, Philadelphia (Frontiers in headache research, vol 5), p 101

Middlemiss DN, Bremer ME, Smith SM (1988) A pharmacological analysis of the 5-HT receptor mediating inhibition of 5-HT release in the guinea-pig frontal cortex. Eur J Pharmacol 157:101–107

Miller KJ, King A, Demchyshyn L, Niznik H, Teitler M (1992) Agonist activity of sumatriptan and metergoline at the human 5-HT$_{1Db}$ receptor: further evidence for a role of the 5-HT$_{1D}$ receptor in the action of sumatriptan. Eur J Pharmacol 227:99–102

Miyamoto A, Sakota T, Nishio A (1994) Characterization of 5-hydroxytryptamine receptors on the isolated pig basilar artery by functional and radioligand binding study. Jpn J Pharmacol 65:265–273

Moret C, Briley M (1993) The unique effect of methiothepin on the terminal serotonin autoreceptor in the rat hypothalamus could be an example of inverse agonism. J Psychopharmacol 7:331–337

Moskowitz MA (1992) Neurogenic versus vascular mechanisms of sumatriptan and ergot alkaloids in migraine. Trends Pharmacol Sci 13:307–311

Moskowitz MA (1993) Neurogenic inflammation in the pathophysiology and treatment of migraine. Neurology 43 (Suppl 3):S16–S20

Moskowitz MA, Brody M, Liu-Chen L-Y (1983) In vitro release of immunoreactive substance P from putative afferent nerve endings in bovine pia arachnoid. Neuroscience 9:809–814

Moskowitz MA, Nozaki K, Kraig RP (1993) Neocortical spreading depression provokes the expression of c-fos protein-like immunoreactivity within trigeminal nucleus caudalis via trigeminovascular mechanisms. J Neurosci 13:1167–1177

Nakai K, Itakura T, Naka Y, Nakakita K, Kamei I, Imai H, Yokote H, Komai N (1986) The distribution of adrenergic receptors in cerebral blood vessels: an autoradiographic study. Brain Res 381:148–152

Nozaki K, Moskowitz MA, Boccalini P (1992a) CP-93,129, sumatriptan, dihydroergotamine block c-fos expression within rat trigeminal nucleus caudalis caused by chemical stimulation of the meninges. Br J Pharmacol 106:409–415

Nozaki K, Boccalini P, Moskowitz MA (1992b) Expression of c-fos-like immunoreactivity in brain stem after meningeal irritation by blood in subarachnoid space. Neurosci 49:669–680

Olesen J, Thomsen LL, Iversen H (1994) Nitric oxide is a key molecule in migraine and other vascular headaches. Trends Pharmacol Sci 15:149–153

Onodera H, Kogure K, Ono Y, Igarashi K, Kiyota K, Nagaoka A (1989) Protooncogene c-fos is transiently induced in the rat cerebral cortex after forebrain ischemia. Neurosci Lett 98:101–104

O'Shaughnessy CT, Connor HE (1994) Activation of sensory nerves in guinea-pig isolated basilar artery by nicotine: evidence for inhibition of trigeminal sensory neurotransmission by sumatriptan. Eur J Pharmacol 259:37–42

Ottoson A, Hill SJ, Edvinsson L (1990) Histamine receptors in brain vessels of guinea-pig: in vitro pharmacology and ligand binding. Acta Physiol Scand 140:135–141

Pauwels P, Palmier C (1994) Inhibition by 5-HT of forskolin-induced cAMP formation in the renal opossum epithelial cell line OK: mediation by a 5-HT_{1B}-like receptor and antagonism by methiothepin. Neuropharmacology 33:67–75

Pauwels PJ, Colpaert FC (1995) The 5-HT_{1D} receptor antagonist GR-127,935 is an agonist at cloned human 5-HT_{1Da} receptor sites. Neuropharmacol 34:235–237

Pelto-Huikko M, Dagerlind U, Ceccatelli S, Hoekfelt T (1991) The immediate-early genes c-fos and c-jun are differentially expressed in the rat adrenal gland after capsaicin treatment. Neurosci Lett 126:163–166

Peroutka SJ, Kuhar MJ (1984) Autoradiographic localization of 5-HT_1 receptors to human and canine basilar arteries. Brain Res 310:193–196

Perren MJ, Feniuk W, Humphrey PPA (1989) The selective closure of feline carotid arteriovenous anastomoses (AVAs) by GR43175. Cephalalgia 9 (Suppl 9):41–46

Phebus LA, Johnson KW, Audia JE, Cohen ML, Dressman BA, Fritz JE, Kaldor SW, Krushinski JH, Schenck KW, Zgombick JM, Branchek TA, Adham N, Schaus JM (1996) Characterization of LY334370, a potent and selective 5-HT_{1F} receptor agonist, in the neurogenic dural inflammation model of migraine pain. Soc Neurosci Abstr 22:528.11

Rebeck GW, Maynard KI, Hyman BT, Moskowitz MA (1994) Selective 5-HT_{1Da} serotonin receptor gene expression in trigeminal ganglia: implications for antimigraine drug development. Proc Natl Acad Sci USA 91:3666–3669

Saito K, Markowitz S, Moskowitz MA (1988) Ergot alkaloids block neurogenic extravasation in dura mater: proposed action in vascular headaches. Ann Neurol 24:732–737

Saria A, Gamse R, Petermann J, Fischer JA, Theodorsson-Norheim E, Lundberg JM (1986) Simultaneous release of several tachykinins and calcitonin gene-related peptide from rat spinal cord slices. Neurosci Lett 63:310–314

Saxena PR, Tfelt-Hansen P (1993) Sumatriptan. In: Tfelt-Hansen P, Welch KMA (eds) The headaches. Raven, New York, pp 329–341

Schlicker E, Fink K, Göthert M, Hoyer D, Molderings G, Roschke I, Schoeffter P (1989) The pharmacological properties of the presynaptic serotonin autoreceptor in the pig brain cortex conform to the 5-HT_{1D} receptor subtype. Naunyn Schmiedebergs Arch Pharmacol 340:45–51

Schoeffter P, Hoyer D (1989) 5-Hydroxytryptamine 5-HT$_{1B}$ and 5-HT$_{1D}$ receptors mediating inhibition of adenylate cyclase activity. Pharmacological comparison with special reference to the effects of yohimbine, rauwolscine and some beta-adrenoceptor antagonists. Naunyn Schmiedebergs Arch Pharmacol 340:285–292

Schoeffter P, Hoyer D (1990) 5-Hydroxytryptamine (5-HT)-induced endothelium-dependent relaxation of pig coronary arteries is mediated by 5-HT receptors similar to the 5-HT$_{1D}$ receptor subtype. J Pharmacol Exp Ther 252:387–395

Schoeffter P, Waeber C, Palacios JM, Hoyer D (1988) The 5-hydroxytryptamine 5-HT$_{1D}$ receptor subtype is negatively coupled to adenylate cyclase in calf substantia nigra. Naunyn Schmiedebergs Arch Pharmacol 337:602–608

Sheng M, Greenberg ME (1990) The regulation and function of c-fos and other immediate early genes in the nervous system. Neuron 4:477–485

Shepheard S, Williamson D, Cook D, Baker R, Street L, Matassa V, Beer M, Middlemiss, D, Iversen L, Hill R, Hargreaves R (1995a) In vivo pharmacology of a novel 5-HT$_{1D}$ receptor agonist, MK-462. Cephalalgia 15 (Suppl 14):205

Shepheard SL, Williamson DJ, Williams J, Hill RG, Hargreaves RJ (1995b) Comparison of the effects of sumatriptan and the NK$_1$ antagonist CP-99,994 on plasma extravasation in dura mater and c-fos mRNA expression in trigeminal nucleus caudalis of rats. Neuropharmacology 34:255–261

Skingle M, Beattie DT, Scopes DIC, Starkey SJ, Connor HE, Feniuk W, Tyers MB (1996) GR127935: a potent and selective 5-HT$_{1D}$ receptor antagonist. Behav Brain Res 73:157–161

Strassman AM, Vos BP (1993) Somatotopic and laminar organization of fos-like immunoreactivity in the medullary and upper cervical dorsal horn induced by noxious facial stimulation in the rat. J Comp Neurol 331:495–516

Strassman A, Mason P, Moskowitz M, Maciewicz R (1986) Response of brainstem trigeminal neurons to electrical stimulation of the dura. Brain Res 379:242–250

Sweet WH, Wepsic JG (1974) Controlled thermocoagulation of trigeminal ganglion and rootlets for differential destruction of pain fibers. 1. Trigeminal neuralgia. J Neurosurg 40:143–156

Thomas DR, Faruq SA, Brown AM (1995) Characterization of [^{35}S]GTPγS binding to CHO cell membranes expressing human 5-HT$_{1Da}$ receptors: evidence for negative efficacy of 5-HT receptor antagonists. Br J Pharmacol 114:153P

To ZP, Bonhaus DW, Eglen RM, Jakeman LB (1995) Characterization and distribution of putative 5-HT$_7$ receptors in guinea-pig brain. Br J Pharmacol 115:107–116

Uddman R, Edvinsson L (1989) Neuropeptides in the cerebral circulation. Cerebrovasc Brain Metab Rev 1:230–252

Ullmer C, Schmuck K, Kalkman HO, Lübbert H (1995) Expression of serotonin receptor mRNAs in blood vessels. FEBS Lett 370:215–221

Vanhoutte PM (1978) Heterogeneity in vascular smooth muscle. In: Vanhoutte PM (ed) Serotonin and the cardiovascular system. Raven, New York, pp 181–309

Waeber C, Moskowitz MA (1995a) [^{3}H]Sumatriptan labels both 5-HT$_{1D}$ and 5-HT$_{1F}$ receptor binding sites in the guinea pig brain: an autoradiographic study. Naunyn Schmiedebergs Arch Pharmacol 352:263–275

Waeber C, Moskowitz (1995b) Autoradiographic visualization of [^{3}H]5-carboxamidotryptamine binding sites in the guinea pig and rat brain. Eur J Pharmacol 283:31–46

Waeber C, Pinkus L, Palacios JM (1990) The (S)-isomer of [^{3}H]zacopride labels 5-HT$_3$ receptors with high affinity in rat brain. Eur J Pharmacol 181:283–287

Watson J, Burton M, Price GW, Jones BJ, Thomas D, Faruq A, Middlemiss DN (1995) GR127935 acts as a partial agonist at recombinant human 5-HT$_{1D\alpha}$ and 5-HT$_{1D\beta}$ receptors. Br J Pharmacol 114:362P

Yu X-J, Waeber C, Castanon N, Scearce K, Hen R, Macor JE, Chauveau J, Moskowitz MA (1996) 5-Carboxamidotryptamine, CP-122,288 and dihydroergotamine but not sumatriptan, CP-93,129 and serotonin-5-O-carboxymethyl-glycyl-tyrosinamide block dural plasma protein extravasation in knockout mice that lack 5-hydroxytryptamine$_{1B}$ receptors. Mol Pharmacol 49:761–765

Yu X-J, Cutrer FM, Moskowitz MA, Waeber C (1997) The 5-HT$_{1D}$ receptor antagonist GR-127,935 prevents inhibitory effects of sumatriptan but not CP-122,288 and 5-CT on neurogenic plasma extravasation within guinea pig dura mater. Neuropharmacology 36:83–91

Zagami AS, Goadsby PJ, Edvinsson L (1990) Stimulation of the superior sagittal sinus in the cat causes release of vasoactive peptides. Neuropeptides 16:69–75

CHAPTER 24

Modulation of Nociception by Descending Serotoninergic Projections

J. Sawynok

A. Introduction

5-Hydroxytryptamine (5-HT) plays a multifaceted role in the regulation of nociceptive transmission. Complexity arises from actions at multiple sites within the pain transmission system (periphery, spinal cord, supraspinal sites), actions at multiple 5-HT receptor subtypes with the potential for opposing or interactive influences, and interactions with a number of endogenous mediators at all levels. At sensory nerve terminals in the periphery, 5-HT plays a role in the inflammatory response and mediates pronociceptive effects by activation of 5-HT$_{1A}$, 5-HT$_2$ and 5-HT$_3$ receptors (Richardson et al. 1985; Meller et al. 1991; Taiwo and Levine 1992). Within the spinal cord, 5-HT can modulate sensory neurotransmission in a complex manner, as there is evidence for both pain facilitatory (5-HT$_{1A}$ and 5-HT$_2$ receptors) and pain inhibitory actions (5-HT$_{1B}$, 5-HT$_2$ and 5-HT$_3$ receptors) (Sect. E). The spinal 5-HT system constitutes the area of termination of a descending system that originates in the brainstem and projects to the dorsal horn of the spinal cord, and is a significant substrate for antinociception by systemically administered morphine as well as for analgesia mediated by stimulation of midbrain and brainstem sites (reviewed by Basbaum and Fields 1984; Besson and Chaouch 1987). At supraspinal sites, forebrain projections of 5-HT also can contribute to analgesia by morphine, as well as providing a supraspinal circuitry that modifies pain transmission in ascending pain projection pathways (reviewed by LeBars 1988; Wang and Nakai 1994). In the present chapter, the focus is on the spinal regulation of nociception by the various 5-HT receptors which subserve the descending serotoninergic projection from the brainstem, and on the interactions of 5-HT receptors with endogenous systems within the spinal cord.

B. Raphe-Spinal Serotoninergic Projections

5-HT immunoreactive neurons within the dorsal horn of the spinal cord originate primarily within the rostroventral medulla, which includes the nucleus raphe magnus and adjacent reticular formation, with additional inputs from the caudal pons (Bowker et al. 1982; Skagerberg and Björklund 1985; Kwiat and Basbaum 1992). Descending 5-HT fibres project to the dorsal horn primarily via the ipsilateral dorsolateral funiculus, while those in more ventral areas arise from ventral and ventrolateral funiculi (BASBAUM and FIELDS 1979; SKAGERBERG and BJÖRKLUND 1985; BULLITT and LIGHT 1989). Pontine and medullary regions innervate all levels of the spinal cord while midbrain regions project mainly to cervical and rostral thoracic levels (BOWKER et al. 1982).

The highest concentrations of 5-HT immunoreactive profiles are in laminae I and II (II outer; II inner contains few profiles), with laminae III, IV and V containing intermediate levels (RUDA et al. 1986; MARLIER et al. 1991a). Cellular and dendritic targets have been identified for 5-HT nerve terminals in the superficial dorsal horn, indicating the establishment of synaptic contacts (RUDA et al. 1982; LIGHT et al. 1983; GLAZER and BASBAUM 1984). However, other studies have emphasized that most serotoninergic axons in this area do not establish classical synapses and may provide a diffuse innervation (MAXWELL et al. 1983; MARLIER et al. 1991a; RIDET et al. 1993). This profile is similar to that seen in telencephalic regions (DESCARRIES et al. 1990). A diffuse innervation may mediate diverse effects of 5-HT by actions at various cellular locations within the dorsal horn, and allow for considerable neuroplasticity following sensory denervation (MARLIER et al. 1990).

While 5-HT neurons originate primarily from the rostroventral medulla, it is important to appreciate that non-5-HT neurons outnumber 5-HT neurons in this region (KWIAT and BASBAUM 1992; JONES and LIGHT 1992). Peptides such as substance P, somatostatin, enkephalin and thyrotropin-releasing hormone have been shown to be colocalized with 5-HT in the same neurons in the rostroventral medulla and spinal cord (JOHANSSON et al. 1981; BOWKER and ABBOTT 1990), and may modify spinal actions of 5-HT in both the dorsal and ventral spinal cord.

C. 5-HT Receptors in the Spinal Cord

Multiple 5-HT receptors have been characterized using traditional pharmacological classification techniques such as binding techniques and functional responses to selective agonists and antagonists (5-HT$_1$, 5-HT$_2$, 5-HT$_3$, 5-HT$_4$). More recently, molecular biology techniques have revealed the existence of several further 5-HT receptors for which there was little prior pharmacological characterization (5-HT$_{1E}$, 5-HT$_{1F}$, 5-HT$_5$, 5-HT$_6$, 5-HT$_7$). Within the spinal cord, 5-HT$_1$, 5-HT$_2$ and 5-HT$_3$ receptors have been identified using binding and autoradiographic techniques. Both the laminar and rostrocaudal distributions of receptors have been defined, and specific lesion methods have provided some information on the localization of these receptor populations.

Quantitative analysis of high-affinity [^{3}H]5-HT binding (5-HT$_1$ receptors) reveals the highest levels of binding in laminae I and II of the dorsal horn along the rostrocaudal axis (MARLIER et al. 1991b). 5-HT$_{1A}$ sites account for 20%–50% of total 5-HT$_1$ sites (HUANG and PEROUTKA 1987; ZEMLAN et al. 1990; MARLIER et al. 1991b). 5-HT$_{1A}$ receptors are more numerous at sacral and lumbar levels than at thoracic and cervical levels (MARLIER et al. 1991b; THOR et al. 1993). 5-HT$_{1B}$ sites are less numerous than 5-HT$_{1A}$ sites, exhibit fewer differences between the different laminae, and show no rostrocaudal gradient along the spinal cord (HUANG and PEROUTKA 1987; DAVAL et al. 1987; ZEMLAN et al. 1990; MARLIER et al. 1991b). An additional 5-HT$_1$ receptor which is selectively localized in the spinal cord and may account for a significant proportion of total 5-HT$_1$ receptors has been identified and designated as a 5-HT$_{1S}$ receptor (ZEMLAN et al. 1990; ZEMLAN and SCHWAB 1991). This also is concentrated in the superficial laminae of the dorsal horn, showing additional high levels in laminae IX and X (MURPHY and ZEMLAN 1992). In contrast to 5-HT$_1$ receptors, 5-HT$_{2A}$ receptors occur in much lower levels throughout the spinal cord, and are more diffusely distributed throughout the various layers, with some of the highest levels occurring in the ventral motor area (MARLIER et al. 1991b; THOR et al. 1993). 5-HT$_{2C}$ receptors (formerly 5-HT$_{1C}$ receptors) also are present in the spinal cord (MOLINEAUX et al. 1989), but levels are quite low (HUANG and PEROUTKA 1987; ZEMLAN et al. 1990; CESSELIN et al. 1994). 5-HT$_3$-binding sites in the dorsal spinal cord (GLAUM and ANDERSON 1988) are selectively concentrated in the superficial dorsal spinal cord (HAMON et al. 1989).

The specific location of 5-HT receptors has been further investigated using lesion studies. Dorsal rhizotomy and/or capsaicin treatment of neonates reduce levels of 5-HT$_1$, 5-HT$_{1A}$, 5-HT$_{1B}$ (by 20%–30%, DAVAL et al. 1987; CESSELIN et al. 1994) and 5-HT$_3$ receptors (by 50%–80%, HAMON et al. 1989; KIDD et al. 1993) in the dorsal spinal cord, indicating the presence of a significant population of receptors on sensory afferent nerve terminals. Lesions to descending 5-HT pathways by intrathecal (i.t.) 5,7-dihydroxytryptamine (5,7-DHT) produces no change or an increase in receptor density for 5-HT$_{1A}$ and 5-HT$_{1B}$ receptors (BROWN et al. 1989; CESSELIN et al. 1994), indicating a predominant postsynaptic localization of these receptors in relation to 5-HT pathways. It needs to be recognized that postsynaptic upregulation probably masks detection of the loss of a presynaptic population of 5-HT$_{1B}$ receptors which function as autoreceptors to inhibit 5-HT release (BROWN et al. 1988; MURPHY and ZEMLAN 1988; MATSUMOTO et al. 1992). Lesions to descending noradrenergic pathways by DSP4 (N-2-chloroethyl-N-ethyl-2-bromobenzylamine) also produce an increase in 5-HT$_{1A}$ and 5-HT$_{1B}$ receptor binding (CESSELIN et al. 1994), providing binding evidence for a functional interaction between spinal 5-HT and NA systems (Sect. F.I). 5-HT$_3$-binding sites are unaffected by 5,7-DHT or DSP4 (KIDD et al. 1993), suggesting they are not present on spinally projecting monoaminergic nerve terminals.

D. Tonic Regulation of Nociceptive Transmission

The extent to which tonic activity in raphe-spinal projections regulates ongoing nociceptive thresholds has been addressed using a number of approaches. (a) 5-HT uptake inhibitors. 5-HT is released tonically from the spinal cord, as inferred from the presence of basal 5-HT levels in spinal cord dialysate fluid (Sorkin et al. 1988). The spinal administration of fluoxetine, a selective 5-HT reuptake inhibitor, has been reported to have no effect in the tail flick (Hwang and Wilcox 1987; but see Xu et al. 1994) or hot plate tests (Sawynok and Reid 1994). In all cases, the doses of fluoxetine were sufficient to enhance the action of 5-HT administered spinally. These observations question a significant role of tonically released 5-HT in regulating nociceptive thresholds in thermal tests. (b) Intrathecal antagonists. The i.t. administration of the non-selective 5-HT antagonists methysergide and metergoline has been reported to produce hyperalgesia in the tail flick and hot plate thermal threshold tests (Proudfit and Hammond 1981; Berge et al. 1983; Sawynok and Dickson 1985), and these observations are consistent with tonic activity. Hyperalgesia induced by systemic administration of 5-HT antagonists has been attributed to changes in skin temperature (Tjølsen et al. 1989), but the contribution of temperature changes to determining nociceptive thresholds subsequently has been questioned (Lichtman et al. 1993). (c) Intrathecal neurotoxins. The i.t. administration of the serotoninergic neurotoxins 5,6- and 5,7-DHT produces hyperalgesia in the tail flick test 2–5 days following administration but this is no longer observed at 10–14 days (Berge et al. 1983; Fasmer et al. 1983). The detection of hyperalgesia is intensity dependent, as it is more readily observed when a milder stimulus intensity is used (Sawynok and Dickson 1985). Some studies suggest that hyperalgesia is an apparent effect due to changes in skin temperature rather than reflecting tonic activity (Tjølsen et al. 1988; Eide et al. 1988). In the formalin test, a reduction in formalin-induced behaviours is seen at 3–4 days following i.t. 5,7-DHT, but this is not apparent at later time intervals (Fasmer et al. 1985; Tjølsen et al. 1991). In this case, data suggest a tonic facilitatory rather than tonic suppressive role for the 5-HT. The loss of responses seen at the later time interval most likely occurs due to the development of receptor supersensitivity which masks an acute effect (Howe and Yaksh 1982; Sawynok and Reid 1994). Supersensitivity may occur due to upregulation of 5-HT receptors, the removal of reuptake systems, or an increased efficacy in coupling between the receptor and its effector systems. Within the spinal cord, supersensitivity to agonists selective for 5-HT$_1$ but not 5-HT$_2$ or 5-HT$_3$ receptor agonists has been demonstrated following i.t. 5,7-DHT (Sawynok and Reid 1994).

E. 5-HT Agonists and Spinal Antinociception

The pharmacology of spinal 5-HT systems has been examined most directly following direct spinal application of 5-HT and agonists selective for specific receptor subtypes. While effects of systemic administration have been attributed to spinal actions on the basis of some mimicry by spinal application, there is always the potential for the confounding of observations by peripherally or supraspinally mediated responses that could either enhance or oppose spinal actions. The spinal application approach is limited by the selectivity of 5-HT agonists and antagonists. While selective agents are available for a number of receptors, the degree of selectivity is often less than optimal, and highly selective agonists exhibiting 1000-fold selectivity are only available for 5-HT$_{1A}$ receptors. Ideally, effects of agonists attributed to a particular 5-HT receptor subtype should be confirmed by the use of selective antagonists. Even with a direct spinal application of selective agonists, observations may still be confounded by: (a) dose, as this has the potential to activate one receptor subtype at some doses, and others at higher doses; (b) type of nociceptive test (e.g. thermal threshold versus pressure versus chemical models), which may have differences in the way they are regulated; and (c) influences due to effects on temperature, blood flow or motor function. With this potential for variability, it is perhaps not surprising that there is some controversy as to effects of the various 5-HT agonists on nociceptive transmission within the spinal cord. Nevertheless, some consensus has emerged with respect to such actions.

I. 5-HT$_{1A}$ Receptors

The activation of spinal 5-HT$_{1A}$ receptors most likely facilitates noxious sensory transmission as there are multiple lines of evidence to support this notion. I.t. administration of 5-HT$_{1A}$ ligands [8-hydroxy-2-(di-n-propylamino)-tetralin (8-OH-DPAT), buspirone] produces a decrease in nociceptive thresholds in the tail flick test (SOLOMON and GEBHART 1988; CRISP et al. 1991; ALHAIDER and WILCOX 1993; ALI et al. 1994) and enhances the scratching response following i.t. injection of substance P or n-methyl-D-aspartate (NMDA) (WILCOX and ALHAIDER 1990). In electrophysiological studies, 5-HT$_{1A}$ agonists enhance spinal reflexes (MURPHY and ZEMLAN 1990; NAGANO et al. 1988; WALLIS and WU 1992) and the responsiveness of dorsal horn neurons to noxious stimuli (ALI et al. 1994). 5-HT$_1$ receptors (most likely 5-HT$_{1A}$ in view of the absence of pronociceptive effects for other subtypes) may also mediate descending facilitation following brainstem stimulation (REN et al. 1991; ZHOU and GEBHART 1991). Spinal facilitatory effects of 5-HT$_{1A}$ agonists have been proposed to result from an expansion of receptive field, promotion of nociceptive reflexes, and the facilitation of rostral transmission (ALI et al. 1994).

Systemic administration of 8-OH-DPAT elicits spontaneous tail flicks (MILLAN et al. 1991) and attenuates analgesia by morphine and other μ-opioids (MILLAN and COLPAERT 1991a,b). These actions were attributed to spinally located 5-HT$_{1A}$ receptors as spinal administration of appropriate agonists produced spontaneous tail flicks and blocked the action of systemically administered opioid agonists; both actions were reversed by 5-HT$_{1A}$ antagonists (BERVOETS et al. 1993; ALHAIDER et al. 1993).

The effects of 5-HT$_{1A}$ agonists on spinal sensory transmission are controversial, as there are a number of reports that contradict the above observations. In some cases, the reason for the controversy is clear, but in others this is less apparent. In the tail flick test, the spinal application of 5-HT$_{1A}$ agonists produces an increase in reaction latency which appears independent of changes in skin temperature (EIDE et al. 1990; EIDE and HOLE 1991; XU et al. 1994). However, changes induced are mild in extent and a bell-shaped curve is expressed (XU et al. 1994). In the hot plate test, 8-OH-DPAT produces an increase in reaction latency (CRISP et al. 1991; SAWYNOK and REID 1994), but the involvement of 5-HT_{1A} receptors was not confirmed using selective antagonists. 5-HT_{1A} agonists have analgesic properties in a colorectal distension model (DANZEBRINK and GEBHART 1991), but this is not blocked by a non-selective 5-HT antagonist (methysergide), and actions could be mediated by receptor systems other than 5-HT. 5-HT_{1A} agonists decrease the amount of biting following i.t. substance P or NMDA (ALHAIDER and WILCOX 1993; MJELLEM et al. 1993), but the relationship of this behaviour to sensory transmission has been questioned (FRENK et al. 1988).

Administered systemically, 5-HT_{1A} agonists produce antinociception in the writhing and hot plate (but not tail flick) tests, but these effects are blocked by α_2-adrenergic antagonists rather than 5-HT_{1A} antagonists (MILLAN and COLPAERT 1991c; MILLAN 1994). Many 5-HT_{1A} agonists have appreciable affinity for α_2-receptors (VAN WIJNGARARDEN et al. 1990), and antinociception appears to be mediated by this receptor system. It appears that when antinociceptive actions of 5-HT_{1A} agonists have been reported, 5-HT_{1A} receptors have not necessarily been definitively implicated in these actions.

II. 5-HT_{1B} Receptors

5-HT_{1B} receptors have consistently been implicated in mediating anti-nociception. Spinal administration of 5-HT_{1B} agonists produces antinociception in the tail flick and hot plate tests (EIDE et al. 1990; CRISP et al. 1991; ALHAIDER and WILCOX 1993; XU et al. 1994; SAWYNOK and REID 1994; ALI et al. 1994) and the substance P and excitatory amino acid behavioural models (EIDE 1992; ALHAIDER and WILCOX 1993). Electrophysiological studies have revealed an inhibition of spinal reflex responses following spinal 5-HT_{1B} receptor activation (ZEMLAN et al. 1988; MURPHY et al. 1992) and inhibition of neuronal activity elicited by noxious stimulation (EL YASSIR et al. 1988; ALI et al. 1994). There are some reports of the 5-HT_{1B} agonist RU 24969 producing hyperalgesia (SOLOMON and GEBHART 1988) or lack of effect in the tail flick test (SAWYNOK and REID 1994), but this agent has a high affinity for 5-HT_{1A} receptors as well, and potential antinociceptive effects could be opposed by a simultaneous activation of this receptor system.

III. 5-HT_2 Receptors

There is evidence for both pain facilitatory and pain inhibitory effects following activation of spinal 5-HT_2 receptors. Thus, higher doses of 5-HT and 5-HT_2 agonists [1-(2,5-dimethoxy-4-iodophenyl)-2-aminopropane (DOI), α-Me-5-HT] produce a caudally directed biting, scratching behaviour similar to that seen with substance P and excitatory amino acids (HYLDEN and WILCOX 1983; WILCOX and ALHAIDER 1990; EIDE and HOLE 1991; MJELLEM et al. 1993), and augment responses produced by NMDA (MJELLEM et al. 1993). These actions are blocked by 5-HT$_2$ antagonists such as ketanserin and ritanserin. This response may result from the spinal release of substance P by 5-HT$_2$ receptor stimulation, as it is blocked by a substance P antagonist (EIDE and HOLE 1991).

In other nociceptive paradigms, 5-HT$_2$ agonists produce antinociception in the tail flick and hot plate tests (SOLOMON and GEBHART 1988; CRISP et al. 1991; EIDE and HOLE 1991; SAWYNOK and REID 1992a, 1994), and the colorectal distension model (DANZEBRINK and GEBHART 1991). One mechanism impli-cated in such antinociceptive actions is activation of inhibitory interneurons (Sect. F.II).

IV. 5-HT$_3$ Receptors

Activation of spinal 5-HT$_3$ receptors has uniformly been reported to inhibit spinal nociceptive processing. Thus, the i.t. administration of 2-Me-5-HT and phenylbiguanidine produces antinociception in the tail flick and hot plate tests (GLAUM et al. 1988, 1990; ALHAIDER et al. 1991; CRISP et al. 1991; SAWYNOK and REID 1991, 1994), the colorectal distension model (DANZEBRINK and GEBHART 1991), and inhibits biting scratching behaviours elicited by i.t. substance P and NMDA (WILCOX and ALHAIDER 1990; ALHAIDER et al. 1991). Electrophysi-ological studies have also produced evidence for a 5-HT$_3$-mediated inhibition of neurons that respond to noxious stimulation (ALHAIDER et al. 1991). Both behavioural and electrophysiological effects of 5-HT$_3$ agonists are blocked by 5-HT$_3$ receptor antagonists, confirming the involvement of this receptor sub-type in such actions (ALHAIDER et al. 1991). Activation of inhibitory interneu-rons is implicated in antinociception by 5-HT$_3$ agonists (Sect. F.II).

V. 5-HT$_{1S}$ Receptors

A novel spinal cord receptor, the 5-HT$_{1S}$ receptor (Sect. C), is anatomic-ally located in an ideal area to regulate pain transmission. 5-Methoxydimethyltryptamine (5-MeODMT) exhibits a high affinity for this receptor, yet it is difficult to clearly implicate this receptor subtype in the actions of 5-MeODMT as this agent also exhibits a high affinity for 5-HT$_{1A}$ and 5-HT$_{1B}$ receptors (TRICKLEBANK et al. 1985). The spinal application of 5-MeODMT produces antinociception in the tail flick and hot plate tests (SAWYNOK and REID 1994) and substance P and NMDA assays (ALHAIDER et al. 1993), and this spinal action most likely accounts for systemically mediated antinociception (e.g. ARCHER et al. 1986a,b). Antinociception by i.t. 5-MeODMT is variously blocked by antagonists for 5-HT$_{1A}$, 5-HT$_{1B}$, 5-HT$_{1S}$, 5-HT$_3$ and GABA receptors, and it has been suggested that this agent acts by the release of endogenous 5-HT and subsequent activation of multiple receptor populations (ALHAIDER et al. 1993). However, i.t. 5,7-DHT, which substantially depletes 5-HT levels in the spinal cord, does not inhibit but rather enhances (likely due to receptor supersensitivity) antinociception by 5-MeODMT (SAWYNOK and REID 1994), suggesting 5-HT release may not be a major factor in such antinociception. I.t. 5-MeODMT enhances scratching behaviours by substance P and this is inhibited by 5-HT$_{1A}$ antagonists, which is consistent with some 5-HT$_{1A}$ receptor activation by this agent (ALHAIDER et al. 1993).

VI. Receptor Subtypes Activated by 5-HT

In view of the heterogeneity of responses produced by selective 5-HT agonists, it is perhaps not surprising that i.t. 5-HT can produce both pain facilitatory and pain inhibitory responses. The most prominent effect of 5-HT is antinociception following i.t. administration (e.g. YAKSH and WILSON 1979; SCHMAUSS et al. 1983), but pronociceptive effects are observed at certain doses (HYLDEN and WILCOX 1983) and following spinal transection (ADVOKAT 1993). The latter observation indicates that important aspects of the action of 5-HT may be mediated by interactions with other systems that are eliminated by spinalization. The antinociceptive actions of 5-HT can be blocked by antagonists for 5-HT$_1$, 5-HT$_2$ and 5-HT$_3$ receptors (SCHMAUSS et al. 1983; GLAUM and ANDERSON 1988; CRISP et al. 1991), indicating a potential multiplicity of actions for the endogenous compound. Electrophysiologically, 5-HT has diverse effects on dorsal horn neurons, and these actions are mediated by different receptor subtypes (TAN and MILETIC 1992; ALI et al. 1994). 5-HT administered exogenously has the potential to activate receptors that normally are innervated in a conventional synaptic manner as well as synapses that may receive a more diffuse input (Sect. B), and this may contribute to the diversity of receptor subtypes involved in 5-HT responses.

F. Interactions with Endogenous Systems

The inhibitory effects of 5-HT on pain transmission in the spinal cord may involve a direct inhibition of nociceptive projection neurons, inhibition of excitatory interneurons, excitation of inhibitory interneurons, and interactions with primary afferents (reviewed by BESSON and CHAOUCH 1987; FIELDS et al. 1991). Anatomically, 5-HT-containing fibres establish synaptic contacts with spinothalamic tract neurons in the superficial dorsal horn, but axoaxonic profiles are rarely seen (RUDA et al. 1982; LIGHT et al. 1983; HYLDEN et al. 1986; BASBAUM and RALSTON 1986). This suggests that 5-HT effects on presynaptic control mechanisms may not be a prominent feature of its action. A number of endogenous mediators of the spinal actions of 5-HT have been identified.

I. Noradrenaline

There is a significant amount of evidence supporting an interaction of 5-HT with descending noradrenergic projection pathways in the dorsal horn. Spinal antinociception by 5-HT is dependent on an intact noradrenaline (NA) system as depletion of NA by the i.t. administration of 6-hydroxydopamine (6-OHDA) or DSP4 markedly reduces antinociception by i.t. 5-HT (ARCHER et al. 1986b; SAWYNOK and REID 1992a) and systemic 5-MeODMT (ARCHER et al. 1986b; POST et al. 1986; MINOR et al. 1988). Conversely, inhibition of the reuptake of NA with desipramine potentiates antinociception by i.t. 5-HT (SAWYNOK and REID 1992b). Both electrophysiological effects of 5-HT (NAKAGAWA et al. 1990) and antinociceptive effects of i.t. 5-HT (SAWYNOK and REID 1992a) and systemic 5-MeODMT (ARCHER et al. 1986a) are reduced by α_2-adrenergic antagonists. Collectively, such observations suggest that 5-HT and 5-MeODMT facilitate NA release from the spinal cord. This has been directly demonstrated using spinal cord slice and synaptosomal preparations (MATSUMOTO et al. 1990; REIMANN and SCHNEIDER 1993). The positive 5-HT-NA interaction is mediated by a 5-HT_1 (but not 5-HT_{1B}) receptor (SAWYNOK and REID 1992a, 1995).

Interactions of 5-HT with NA systems have the potential to be confounded by the activation of multiple receptor subtypes. Thus, 5-HT can inhibit NA release from the spinal cord by activation of 5-HT_2 receptors (CELUCH et al. 1992), which potentially could oppose 5-HT_1 receptor-mediated augmentation of release. An additional complexity in interpreting 5-HT-NA interactions is the fact that blockade by α_2-adrenergic antagonists may not be sufficient evidence to implicate endogenous NA in a particular process. Thus, α_2-antagonists can inhibit the actions of 5-HT_2 and 5-HT_3 receptor agonists without endogenous NA necessarily being involved (SAWYNOK and REID 1995). This could reflect a direct adrenergic receptor interaction of certain 5-HT ligands (VAN WIJNGAARDEN et al. 1990).

II. GABA and Glycine

The superficial laminae of the dorsal horn contain large numbers of γ-aminobutyric acid (GABA) immunoreactive profiles (TODD and McKENZIE 1989) as well as appreciable levels of glycine receptors and uptake sites (RIBEIRO-DA-SILVA and COIMBRA 1980; ZARBIN et al. 1981). Spinal antinociception mediated by 5-HT_3 receptor activation has been attributed to release of GABA from inhibitory interneurons, as GABA antagonists ($\text{GABA}_A > \text{GABA}_B$) inhibit antinociception and inhibitory electrophysiological actions produced by 2-methyl-5-hydroxytryptamine (2-Me-5-HT) (ALHAIDER et al. 1991). Activation of 5-HT_2 receptors also appears to promote GABA and glycine release in the medullary dorsal horn (SUGIYAMA and HUANG 1995). Such release may result from a direct depolarization of the inhibitory interneuron by 5-HT$_2$ and 5-HT$_3$ agonists (TAN and MILETIC 1992).

III. Enkephalin

5-HT immunoreactive profiles have been shown in close proximity to enkephalin containing interneurons (MILETIC et al. 1984; GLAZER and BASBAUM 1984), and it has been proposed that an enkephalin link mediates some of the actions of 5-HT in the dorsal spinal cord (BASBAUM and FIELDS 1984). Some studies have indeed observed that i.t. naloxone inhibits spinal antinociception by 5-HT (KELLSTEIN et al. 1988; YANG et al. 1994b). It has been noted that the dose of naloxone required to produce this effect exceeds that required to block the action of morphine, and activation of spinal κ- rather than μ-opioid receptors may be involved (YANG et al. 1994b). Other studies did not observe inhibition of the action of 5-HT with naloxone (YAKSH and WILSON 1979; YANG et al. 1994a), but this may have been due to inadequate dose.

IV. Adenosine

Adenosine deaminase and adenosine uptake sites have been selectively localized in capsaicin-sensitive small-diameter afferent fibres in the dorsal horn of the spinal cord (GEIGER and NAGY 1985; NAGY and DADDONA 1985). Methylxanthine adenosine receptor antagonists inhibit antinociception by i.t. 5-HT (DELANDER and HOPKINS 1987; SAWYNOK and REID 1992a; but see YANG et al. 1994a), and 5-HT has been shown to release adenosine from spinal cord preparations (SWEENEY et al. 1988, 1990). This release is capsaicin-sensitive, suggesting release originates from small-diameter primary afferent fibres. Methylxanthines inhibit antinociception by a number of 5-HT$_1$ (but not 5-HT$_2$ or 5-HT$_3$) receptor ligands (SAWYNOK and REID 1995), but the particular receptor subtype involved remains to be further clarified.

V. Peptides

While 5-HT$_{1A}$ and 5-HT$_{1B}$ receptors have been localized on sensory afferents (Sect. C), agonists for these receptors have generally not been shown to inhibit the release of peptides contained in sensory afferents (substance P, calcitonin gene-related peptide) (BOURGOIN et al. 1993; CESSELIN et al. 1994). Presynaptic inhibition thus does not appear to be a prominent mechanism by which activation of these receptors produces spinal antinociception. 5-HT does, however, inhibit substance P and somatostatin release from the spinal cord and trigeminal nucleus in a methysergide-sensitive manner (YONEHARA et al. 1991; KURAISHI et al. 1991), suggesting a possible 5-HT$_2$ receptor mediation of such actions. In the ventral spinal cord, 5-HT$_2$ receptor activation augments substance P release (IVERFELDT et al. 1986), but such release originates from a different pool than from the dorsal spinal cord. 5-HT$_3$ receptor activation does not inhibit (CESSELIN et al. 1994) and may actually increase peptide release from the spinal cord (SARIA et al. 1991), but the physiological significance of this latter action is unclear.

G. Conclusion

Multiple 5-HT receptor subtypes exist in the spinal cord and can contribute to the regulation of nociceptive transmission, in some cases in a complex manner. While the specific mechanisms recruited by each receptor subtype in mediating their respective actions are incompletely resolved, some of the endogenous substrates with which they interact are beginning to be understood. The use of multiple experimental approaches is required to further define these actions. Such studies are aimed at providing information which contributes to the development of strategies that can optimize the relief of pain, as well as understanding how analgesic paradigms (either drug- or stimulation-induced) can activate brainstem-spinal circuitry to produce antinociception.

References

Advokat C (1993) Intrathecal coadministration of serotonin and morphine differentially modulates the tail-flick reflex of intact and spinal rats. Pharmacol Biochem Behav 45:871–879

Alhaider AA, Wilcox GL (1993) Differential roles of 5-hydroxytryptamine$_{1A}$ and 5-hydroxytryptamine$_{1B}$ receptor subtypes in modulating spinal nociceptive transmission in mice. J Pharmacol Exp Ther 265:378–385

Alhaider AA, Lei SZ, Wilcox GL (1991) Spinal 5-HT$_3$ receptor-mediated antinociception: possible release of GABA. J Neurosci 11:1881–1888

Alhaider AA, Hamon M, Wilcox GL (1993) Intrathecal 5-methoxy-N,N-dimethyltryptamine in mice modulates 5-HT$_1$ and 5-HT$_3$ receptors. Eur J Pharmacol 249:151–160

Ali Z, Wu G, Kozlov A, Barasi S (1994) The actions of 5-HT$_1$ agonists and antagonists on nociceptive processing in the rat spinal cord: results from behavioural and electrophysiological studies. Brain Res 661:83–90

Archer T, Danysz W, Jonsson G, Minor BG, Post C (1986a) 5-Methoxy-N,N-dimethyltryptamine-induced analgesia is blocked by α-adrenoceptor antagonists in rats. Br J Pharmacol 89:293–298

Archer T, Jonsson G, Minor BG, Post C (1986b) Noradrenergic-serotonergic interactions and nociception in the rat. Eur J Pharmacol 120:295–307

Basbaum AI, Fields HL (1979) The origin of descending pathways in the dorsolateral funiculus of the spinal cord of the cat and rat: further studies on the anatomy of pain modulation. J Comp Neurol 187:513–532

Basbaum AI, Fields HL (1984) Endogenous pain control systems: brainstem spinal pathways and endorphin circuitry. Annu Rev Neurosci 7:309–338

Basbaum AI, Ralston HJ III (1986) Bulbospinal projections in the primate: a light and electron microscopic study of a pain modulating system. J Comp Neurol 250:311–323

Berge O-G, Fasmer OB, Flatmark T, Hole K (1983) Time course of changes in nociception after 5,6-dihydroxytryptamine lesions of descending 5-HT pathways. Pharmacol Biochem Behav 18:637–643

Bervoets K, Rivet J-M, Millan MJ (1993) 5-HT$_{1A}$ receptors and the tail-flick response. IV. Spinally localized 5-HT$_{1A}$ receptors postsynaptic to serotoninergic neurones mediate spontaneous tail-flicks in the rat. J Pharmacol Exp Ther 264:95–104

Besson JM, Chaouch A (1987) Peripheral and spinal mechanisms of nociception. Physiol Rev 67:67–186

Bourgoin S, Pohl M, Mauborgne A, Benoliel JJ, Collin E, Hamon M, Cesselin F (1993) Monoaminergic control of the release of calcitonin gene-related peptide- and substance P-like materials from spinal cord slices. Neuropharmacology 32:633–640

Bowker RM, Abbott LC (1990) Quantitative re-evaluation of descending serotonergic and non-serotonergic projections from the medulla of the rodent: evidence for extensive co-existence of serotonin and peptides in the same spinally projecting neurons, but not from the nucleus raphe magnus. Brain Res 512:15–25

Bowker RM, Westlund KN, Sullivan MC, Coulter JD (1982) Organization of descending serotonergic projections to the spinal cord. In: Kuypers HGJM, Martin GF (eds) Progress in brain research, vol 57. Elsevier Biomedical, Amsterdam, pp 239–265

Brown L, Amedro J, Williams G, Smith D (1988) A pharmacological analysis of the rat spinal cord serotonin (5-HT) autoreceptor. Eur J Pharmacol 145:163–171

Brown LM, Smith DL, Williams GM, Smith DJ (1989) Alterations in serotonin binding sites after 5,7-dihydroxytryptamine treatment in the rat spinal cord. Neurosci Lett 102:103–107

Bullitt E, Light AR (1989) Intraspinal course of descending serotoninergic pathways innervating the rodent dorsal horn and lamina X. J Comp Neurol 286:231–242

Celuch SM, Ramirez AJ, Enero MA (1992) Activation of 5-HT$_2$ receptors inhibits the evoked release of [^{3}H]noradrenaline in the rat spinal cord. Gen Pharmac 23:1063–1065

Cesselin F, Laporte A-M, Miquel M-C, Bourgoin S, Hamon M (1994) Serotonergic mechanisms of pain control. In: Gebhart GF, Hammond DL, Jensen TS (eds) Proceedings of the 7th world congress on pain, progress in pain research and management, vol 2. IASP, Seattle, pp 669–695

Crisp T, Stafinsky JL, Spanos LJ, Uram M, Perni VC, Donepudi HB (1991) Analgesic effects of serotonin and receptor-selective serotonin agonists in the rat spinal cord. Gen Pharmacol 22:247–251

Danzebrink RM, Gebhart GF (1991) Evidence that spinal 5-HT$_1$, 5-HT$_2$ and 5-HT$_3$ receptor subtypes modulate responses to noxious colorectal distension in the rat. Brain Res 538:64–75

Daval G, Vergé D, Basbaum AI, Bourgoin S, Hamon M (1987) Autoradiographic evidence of serotonin$_1$ binding sites on primary afferent fibres in the dorsal horn of the rat spinal cord. Neurosci Lett 83:71–76

DeLander GE, Hopkins CJ (1987) Interdependence of spinal adenosinergic, serotonergic and noradrenergic systems mediating antinociception. Neuropharmacology 26:1791–1794

Descarries L, Audet MA, Doucet G, Garcia S, Oleskevich S, Séguéla P, Soghomonian JJ, Watkins KC (1990) Morphology of central serotonin neurons. Brief review of quantified aspects of their distribution and ultrastructural relationships. Ann N Y Acad Sci 600:81–92

Eide PK (1992) Stimulation of 5-HT$_1$ receptors in the spinal cord changes substance P-induced behaviour. Neuropharmacology 31:541–545

Eide PK, Hole K (1991) Different role of 5-HT$_{1A}$ and 5-HT$_2$ receptors in spinal cord in the control of nociceptive responsiveness. Neuropharmacology 7:727–731

Eide PK, Berge O-G, Tjølsen A, Hole K (1988) Apparent hyperalgesia in the mouse tail-flick test due to increased tail skin temperature after lesioning of serotonergic pathways. Acta Physiol Scand 134:413–420

Eide PK, Joly NM, Hole K (1990) The role of spinal cord 5-HT$_{1A}$ and 5-HT$_{1B}$ receptors in the modulation of a spinal nociceptive reflex. Brain Res 536:195–200

El-Yassir N, Fleetwood-Walker SM, Mitchell R (1988) Heterogeneous effects of serotonin in the dorsal horn of rat: the involvement of 5-HT$_1$ receptor subtypes. Brain Res 456:147–158

Fasmer OB, Berge O-G, Walther B, Hole K (1983) Changes in nociception after intrathecal administration of 5,6-dihydroxytryptamine in mice. Neuropharmacology 22:1197–1201

Fasmer OB, Berge O-G, Hole K (1985) Changes in nociception after lesions of descending serotonergic pathways induced with 5,6-dihydroxytryptamine. Neuropharmacology 24:729–734

Fields HL, Heinricher MM, Mason P (1991) Neurotransmitters in nociceptive modulatory circuits. Annu Rev Neurosci 14:219–245

Frenk H, Bossut D, Urca G, Mayer DJ (1988) Is substance P a primary afferent neurotransmitter for nociceptive input? I. Analysis of pain-related behaviors resulting from intrathecal administration of substance P and 6 excitatory compounds. Brain Res 455:223–231

Geiger JD, Nagy JI (1985) Localization of [³H]nitrobenzylthioinosine binding sites in rat spinal cord and primary afferent neurons. Brain Res 347:321–327

Glaum SR, Anderson EG (1988) Identification of 5-HT₃ binding sites in rat spinal cord synaptosomal membranes. Eur J Pharmacol 156:287–290

Glaum SR, Proudfit HK, Anderson EG (1988) Reversal of the antinociceptive effects of intrathecally administered serotonin in the rat by a selective 5-HT₃ receptor antagonist. Neurosci Lett 95:313–317

Glaum SR, Proudfit HK, Anderson EG (1990) 5-HT₃ receptors modulate spinal nociceptive reflexes. Brain Res 510:12–16

Glazer EJ, Basbaum AI (1984) Axons which take up [³H]serotonin are presynaptic to enkephalin immunoreactive neurons in cat dorsal horn. Brain Res 298:386–391

Hamon M, Gallissot MC, Menard F, Gozlan H, Bourgoin S, Vergé D (1989) 5-HT₃ receptor binding sites are on capsaicin-sensitive fibres in the rat spinal cord. Eur J Pharmacol 164:315–322

Howe JR, Yaksh TL (1982) Changes in sensitivity to intrathecal norepinephrine and serotonin after 6-hydroxydopamine (6-OHDA) and 5,6-dihydroxytryptamine (5,6-DHT) or repeated monoamine administration. J Pharmacol Exp Ther 220:311–321

Huang JC, Peroutka SJ (1987) Identification of 5-hydroxytryptamine₁ binding site subtypes in rat spinal cord. Brain Res 436:173–176

Hwang AS, Wilcox GL (1987) Analgesic properties of intrathecally administered heterocyclic antidepressants. Pain 28:343–355

Hylden JLK, Wilcox GL (1983) Intrathecal serotonin in mice: analgesia and inhibition of a spinal action of substance P. Life Sci 33:789–795

Hylden JLK, Hayashi H, Ruda MA, Dubner R (1986) Serotonin innervation of physiologically identified lamina I projection neurons. Brain Res 370:401–404

Iverfeldt K, Peterson LL, Brodin E, Ogren SO, Bartfai T (1986) Serotonin type-2 receptor mediated regulation of substance P release in the ventral spinal cord and the effects of chronic antidepressant treatment. Naunyn Schmiedebergs Arch Pharmacol 333:1–6

Johansson O, Hökfelt T, Pernow B, Jeffcoate SL, White N, Steinbusch HWM, Verhofstad AAJ, Emson PC, Spindel E (1981) Immunohistochemical support for three putative transmitters in one neuron: coexistence of 5-hydroxytryptamine, substance P- and thyrotropin releasing hormone-like immunoreactivity in medullary neurons projecting to the spinal cord. Neuroscience 6:1857–1881

Jones SL, Light AR (1992) Serotoninergic medullary raphe spinal projection to the lumbar spinal cord in the rat: a retrograde immunohistochemical study. J Comp Neurol 322:599–610

Kellstein DE, Malseed RT, Goldstein FJ (1988) Opioid-monoamine interactions in spinal antinociception: evidence for serotonin but not norepinephrine reciprocity. Pain 34:85–92

Kidd EJ, Laporte AM, Langlois X, Fattaccini C-M, Doyen C, Lombard MC, Gozlan H, Hamon M (1993) 5-HT₃ receptors in the rat central nervous system are mainly located on nerve fibres and terminals. Brain Res 612:289–298

Kuraishi Y, Minami M, Satoh M (1991) Serotonin, but neither noradrenaline nor GABA, inhibits capsaicin-evoked release of immunoreactive somatostatin from slices of rat spinal cord. Neurosci Res 9:238–245

Kwiat GC, Basbaum AI (1992) The origin of brainstem noradrenergic and serotonergic projections to the spinal cord dorsal horn in the rat. Somatosens Motor Res 9:157–173

LeBars DL (1988) Serotonin and pain. In: Osborne NN, Hamon M (eds) Neuronal serotonin. Wiley, Chichester, pp 171–229

Lichtman AH, Smith FL, Martin BR (1993) Evidence that the antinociceptive tail-flick response is produced independently from changes in either tail-skin temperature or core temperature. Pain 55:283–295

Light AR, Kavookjian AM, Petrusz P (1983) The ultrastructure and synaptic connections of serotonin-immunoreactive terminals in spinal laminae I and II. Somatosens Res 1:33–50

Marlier L, Rajaofetra N, Poulat P, Privat A (1990) Modification of serotonergic innervation of the rat spinal cord dorsal horn after neonatal capsaicin treatment. J Neurosci Res 25:112–118

Marlier L, Sandillon F, Poulat P, Rajaofetra N, Geffard M, Privat A (1991a) Serotonergic innervation of the dorsal horn of rat spinal cord: light and electron microscope immunocytochemical study. J Neurocytol 20:310–322

Marlier L, Teilhac J-R, Cerruti C, Privat A (1991b) Autoradiographic mapping of 5-HT$_1$, 5-HT$_{1A}$, 5-HT$_{1B}$ and 5-HT$_2$ receptors in the rat spinal cord. Brain Res 550:15–23

Matsumoto I, Combs MR, Brannan S, Jones DJ (1990) Autoreceptor- and heteroreceptor-mediated regulation of monoamine release in spinal cord synaptosomes. Ann N Y Acad Sci 604:609–611

Matsumoto I, Combs MR, Jones DJ (1992) Characterization of 5-hydroxytryptamine$_{1B}$ receptors in rat spinal cord via [^{125}I]iodocyanopindolol binding and inhibition of [^{3}H]-5-hydroxytryptamine release. J Pharmacol Exp Ther 260:614–626

Maxwell DJ, Leranth CS, Verhofstad AAJ (1983) Fine structure of serotonin-containing axons in the marginal zone of the rat spinal cord. Brain Res 266:253–259

Meller ST, Lewis SJ, Brody MJ, Gebhart GF (1991) The peripheral nociceptive actions of intravenously administered 5-HT in the rat requires dual activation of both 5-HT$_2$ and 5-HT$_3$ receptor subtypes. Brain Res 561:61–68

Miletic V, Hoffert MJ, Ruda MA, Dubner R, Shigenaga Y (1984) Serotonergic axonal contacts on identified cat spinal dorsal horn neurons and their correlation with nucleus raphe magnus stimulation. J Comp Neurol 228:129–141

Millan MJ (1994) Serotonin and pain: evidence that activation of 5-HT$_{1A}$ receptors does not elicit antinociception against noxious thermal, mechanical and chemical stimuli in mice. Pain 58:45–61

Millan MJ, Colpaert FC (1991a) 5-Hydroxytryptamine (HT)$_{1A}$ receptors and the tail-flick response. II. High efficacy 5-HT$_{1A}$ agonists attenuate morphine-induced antinociception in mice in a competitive-like manner. J Pharmacol Exp Ther 256:983–992

Millan MJ, Colpaert FC (1991b) 5-Hydroxytryptamine (HT)$_{1A}$ receptors and the tail-flick response. III. Structurally diverse 5-HT$_{1A}$ partial agonists attenuate mu- but not kappa-opioid antinociception in mice and rats. J Pharmacol Exp Ther 256:993–1001

Millan MJ, Colpaert FC (1991c) α_2 Receptors mediate the antinociceptive action of 8-OH-DPAT in the hot-plate test in mice. Brain Res 539:342–346

Minor BG, Persson M-J, Post C, Jonsson G, Archer T (1988) Intrathecal noradrenaline restores 5-methoxy-N,N-dimethyltryptamine induced antinociception abolished by intrathecal 6-hydroxydopamine. J Neural Trans 72:107–120

Millan MJ, Bervoets K, Colpaert FC (1991) 5-Hydroxytryptamine (5-HT)$_{1A}$ receptors and the tail-flick response. I. 8-Hydroxy-2-(di-n-propylamino) tetralin HBr-induced spontaneous tail-flicks in the rat as an in vivo model of 5-HT$_{1A}$ receptor-mediated activity. J Pharmacol Exp Ther 256:973–982

Mjellem N, Lund A, Hole K (1993) Different functions of spinal 5-HT$_{1A}$ and 5-HT$_2$ receptor subtypes in modulating behaviour induced by excitatory amino acid receptor agonists in mice. Brain Res 626:78–82

Molineaux SM, Jessell TM, Axel R, Julius D (1989) 5-HT$_{1c}$ receptor is a prominent serotonin receptor subtype in the central nervous system. Proc Natl Acad Sci USA 86:6793–6797

Murphy AZ, Murphy RM, Zemlan FP (1992) Role of spinal serotonin$_1$ receptor subtypes in thermally and mechanically elicited nociceptive reflexes. Psychopharmacology 108:123–130

Murphy RM, Zemlan FP (1988) Selective 5-HT$_{1B}$ agonists identify the 5-HT autoreceptor in lumbar spinal cord of rat. Neuropharmacology 27:37–42

Murphy RM, Zemlan FP (1990) Selective serotonin$_{1A/1B}$ agonists differentially affect spinal nociceptive reflexes. Neuropharmacology 29:463–468

Murphy RM, Zemlan FP (1992) Quantitative autoradiographic mapping of a novel serotonin receptor-5-HT$_{1S}$. Neuroreport 3:837–840

Nagano N, Ono H, Fukuda H (1988) Functional significance of subtypes of 5-HT receptors in the rat spinal reflex pathway. Gen Pharmacol 19:789–793

Nagy JI, Daddona PE (1985) Anatomical and cytochemical relationships of adenosine deaminase-containing primary afferent neurons in the rat. Neuroscience 15:799–813

Nakagawa I, Omote K, Kitahata LM, Collins JG, Murata K (1990) Serotonergic mediation of spinal analgesia and its interaction with noradrenergic systems. Anesthesiology 73:474–478

Post C, Minor BG, Davies M, Archer T (1986) Analgesia induced by 5-hydroxytryptamine receptor agonists is blocked or reversed by noradrenaline-depletion in rats. Brain Res 363:18–27

Proudfit HK, Hammond DL (1981) Alterations in nociceptive threshold and morphine-induced analgesia produced by intrathecally administered amine antagonists. Brain Res 218:393–399

Reimann W, Schneider F (1993) The serotonin receptor agonist 5-methoxy-N,N-dimethyltryptamine facilitates noradrenaline release from rat spinal cord slices and inhibits monoamine oxidase activity. Gen Pharmacol 24:449–453

Ren K, Randich A, Gebhart GF (1991) Spinal serotonergic and kappa opioid receptors mediate facilitation of the tail flick reflex produced by vagal afferent stimulation. Pain 45:321–329

Ribeiro-da-silva A, Coimbra A (1980) Neuronal uptake of ^{3}H-GABA and ^{3}H-glycine in laminae I-III (substantia gelatinosa rolandi) of the rat spinal cord. An autoradiographic study. Brain Res 188:449–464

Richardson BP, Engel G, Donatsch P, Stadler PA (1985) Identification of serotonin M-receptor subtypes and their specific blockade by a new class of drugs. Nature 316:126–131

Ridet Jl, Rajaofetra N, Teilhac JR, Geffard M, Privat A (1993) Evidence for nonsynaptic serotonergic and noradrenergic innervation of the rat dorsal horn and possible involvement of neuron-glia interactions. Neuroscience 52:143–157

Ruda MA, Coffield J, Steinbusch HWM (1982) Immunocytochemical analysis of serotonergic axons in laminae I and II of the lumbar spinal cord of the cat. J Neurosci 2:1660–1671

Ruda MA, Bennett GJ, Dubner R (1986) Neurochemistry and neural circuitry in the dorsal horn. In: Emson PC, Rossor M, Tohyama M (eds) Progress in brain research, vol 66, Elsevier, Amsterdam, pp 219–268

Saria A, Javorsky F, Humpel C, Gamse R (1991) Endogenous 5-hydroxytryptamine modulates the release of tachykinins and calcitonin gene-related peptide from the rat spinal cord via 5-HT$_3$ receptors. Ann N Y Acad Sci 632:464–465

Sawynok J, Dickson C (1985) Evidence for the involvement of descending noradrenergic pathways in the antinociceptive effect of baclofen. Brain Res 335:89–97

Sawynok J, Reid A (1991) Noradrenergic and purinergic involvement in spinal antinociception by 5-hydroxytryptamine and 2-methyl-5-hydroxytryptamine. Eur J Pharmacol 204:301–309

Sawynok J, Reid A (1992a) Noradrenergic mediation of spinal antinociception by 5-hydroxytryptamine: characterization of receptor subtypes. Eur J Pharmacol 223:49–56

Sawynok J, Reid A (1992b) Desipramine potentiates spinal antinociception by 5-hydroxytryptamine, morphine and adenosine. Pain 50:113–118

Sawynok J, Reid A (1994) Spinal supersensitivity to 5-HT$_1$, 5-HT$_2$ and 5-HT$_3$ receptor agonists following 5,7-dihydroxytryptamine. Eur J Pharmacol 264:249–257

Sawynok J, Reid A (1995) Interactions of descending serotonergic systems with other neurotransmitters in the modulation of nociception. Behav Brain Res (in press)

Schmauss C, Hammond DL, Ochi JW, Yaksh TL (1983) Pharmacological antagonism of the antinociceptive effects of serotonin in the rat spinal cord. Eur J Pharmacol 90:349–357

Skagerberg G, Björklund A (1985) Topographic principles in the spinal projections of serotonergic and non-serotonergic brainstem neurons in the rat. Neuroscience 15:445–480

Solomon RE, Gebhart GF (1988) Mechanisms of effects of intrathecal serotonin on nociception and blood pressure in rats. J Pharmacol Exp Ther 245:905–912

Sorkin LS, Steinman JL, Hughes MG, Willis WD, McAdoo DJ (1988) Microdialysis recovery of serotonin released in spinal cord dorsal horn. J Neurosci Methods 23:131–138

Sugiyama BH, Huang L-YM (1995) Activation of 5-HT$_2$ receptors potentiates the spontaneous inhibitory postsynaptic currents (sIPSCs) in trigeminal neurons. Soc Neurosci Abstr 21:1415

Sweeney MI, White TD, Sawynok J (1988) 5-Hydroxytryptamine releases adenosine from primary afferent terminals in the spinal cord. Brain Res 462:346–349

Sweeney MI, White TD, Sawynok J (1990) 5-Hydroxytryptamine releases adenosine and cyclic AMP from primary afferent terminals in the spinal cord in vivo. Brain Res 528:55–61

Taiwo YO, Levine JD (1992) Serotonin is a directly-acting hyperalgesic agent in the rat. Neuroscience 48:485–490

Tan H, Miletic V (1992) Diverse actions of 5-hydroxytryptamine on frog spinal dorsal horn neurons in vitro. Neuroscience 49:913–923

Thor KB, Nickolaus S, Helke CJ (1993) Autoradiographic localization of 5-hydroxytryptamine$_{1A}$, 5-hydroxytryptamine$_{1B}$ and 5-hydroxytryptamine$_{1C/2}$ binding sites in the rat spinal cord. Neuroscience 55:235–252

Tjølsen A, Berge O-G, Eide PK, Broch OJ, Hole K (1988) Apparent hyperalgesia after lesions of the descending serotonergic pathways is due to increased tail skin temperature. Pain 33:225–231

Tjølsen A, Lund A, Eide PK, Berge O-G, Hole K (1989) The apparent hyperalgesic effect of a serotonin antagonist in tail flick test is mainly due to increased tail skin temperature. Pharmacol Biochem Behav 32:601–605

Tjølsen A, Berge O-G, Hole K (1991) Lesions of bulbo-spinal serotonergic or noradrenergic pathways reduce nociception as measured by the formalin test. Acta Physiol Scand 142:229–236

Todd AJ, McKenzie J (1989) GABA-immunoreactive neurons in the dorsal horn of the rat spinal cord. Neuroscience 31:799–806

Tricklebank MD, Forler C, Middlemiss DN, Fozard JR (1985) Subtypes of the 5-HT receptor mediating the behavioural responses to 5-methoxy-N,N-dimethyltryptamine in the rat. Eur J Pharmacol 117:15–24

Van Wijngaarden I, Tulp MThM, Soudijn W (1990) The concept of selectivity in 5-HT receptor research. Eur J Pharmacol 188:301–312

Wallis DI, Wu J (1992) Fast and slow ipsilateral and contralateral spinal reflexes in the neonate rat are modulated by 5-HT. Gen Pharmacol 23:1035–1044

Wang Q-P, Nakai Y (1994) The dorsal raphe: an important nucleus in pain modulation. Brain Res Bull 34:575–585

Wilcox GL, Alhaider AA (1990) Nociceptive and antinociceptive action of serotonergic agonists administered intrathecally. In: Besson J-M (ed) Serotonin and pain. Elsevier Science, Amsterdam, pp 205–219

Xu W, Qui XC, Han JS (1994) Serotonin receptor subtypes in spinal antinociception in the rat. J Pharmacol Exp Ther 269:1182–1189

Yaksh TL, Wilson PR (1979) Spinal serotonin terminal system mediates antinociception. J Pharmacol Exp Ther 208:446–453

Yang S-W, Zhang Z-H, Chen J-Y, Xie Y-F, Qiao J-T, Dafny N (1994a) Morphine and norepinephrine-induced antinociception at the spinal level is mediated by adenosine. Neuroreport 5:1441–1444

Yang S-W, Zhang Z-H, Wang R, Xie Y-F, Qiao J-T, Dafny N (1994b) Norepinephrine and serotonin-induced antinociception are blocked by naloxone with different dosages. Brain Res Bull 35:113–117

Yonehara N, Shibutani T, Imai Y, Sawada T, Inoki R (1991) Serotonin inhibits release of substance P evoked by tooth pulp stimulation in trigeminal nucleus caudalis in rabbits. Neuropharmacology 30:5–13

Zarbin MA, Wamsley JK, Kuhar MJ (1981) Glycine receptor: light microscopic autoradiographic localization with [^{3}H]strychnine. J Neurosci 1:532–547

Zemlan FP, Schwab EF (1991) Characterization of a novel serotonin receptor subtype (5-HT$_{1S}$) in rat CNS: interaction with a GTP binding protein. J Neurochem 57:2092–2099

Zemlan FP, Behbehani MM, Murphy RM (1988) Serotonin receptor subtypes and the modulation of pain transmission. Prog Brain Res 77:349–355

Zemlan FP, Schwab EF, Murphy RM, Behbehani MM (1990) Identification of a novel 5-HT$_1$ binding site in rat spinal cord. Neurochem Int 16:507–513

Zhuo M, Gebhart GF (1991) Spinal serotonin receptors mediate descending facilitation of a nociceptive reflex from the nuclei reticularis gigantocellularis and gigantocellularis pars alpha in the rat. Brain Res 550:35–48

Molecular Biology of Monoamine Oxidase A and B: Their Role in the Degradation of Serotonin

J.C. Shih, J. Grimsby, and K. Chen

A. Introduction

Monoamine oxidase [MAO; amine:oxygen oxidoreductase (deaminating) (flavin-containing), EC 1.4.3.4] catalyzes the oxidative deamination of various biogenic amines in the CNS and peripheral tissues. MAO is integral to the outer mitochondrial membrane (Greenwalt and Schnaitman 1970) and is classified as type A and type B (Johnston 1968; Squires 1968). MAOA preferentially oxidizes serotonin (5-hydroxytryptamine, 5-HT) and norepinephrine (NE) and is selectively inhibited by low concentrations of the irreversible inhibitor clorgyline (Johnston 1968) and by the reversible inhibitors Ro41-1049, brofaromine and moclobemide (Da Prada et al. 1990). MAOB has a high affinity for β-phenylethylamine (PEA) and is selectively inhibited by deprenyl (Knoll and Magyar 1972), Ro19-6327 (Da Prada et al. 1990) and MDL-72145 (Bey et al. 1984). Tyramine (TA) and dopamine (DA) are common substrates for both enzymes.

MAO is widely distributed in mammalian and nonmammalian species (Denney and Denney 1985; Berry et al. 1994). Both forms are coexpressed in most tissues; however, placenta (Thorpe et al. 1987) and platelets (Donnelly and Murphy 1977) contain predominantly MAOA and MAOB, respectively. In brain, MAOA is expressed predominantly in catecholaminergic neurons, whereas MAOB is expressed in serotoninergic neurons, astrocytes and glia (Levitt et al. 1982; Thorpe et al. 1987).

B. Molecular Biology of MAO

I. MAOA and B cDNAs and Genomic Structure

After the classification of MAOA and MAOB it was unknown whether they were separate polypeptides or a single polypeptide with active sites or different lipid attachments. Bach et al. (1988) unequivocally proved that MAOA and MAOB are distinct proteins, encoded by separate genes. We isolated full-length MAOA and MAOB cDNAs by screening a human liver cDNA library with oligonucleotide probes derived from the sequences of MAOA and MAOB peptide fragments. Comparison of the deduced amino acid sequence shows that MAOA and MAOB share 70% sequence identity and have mo-

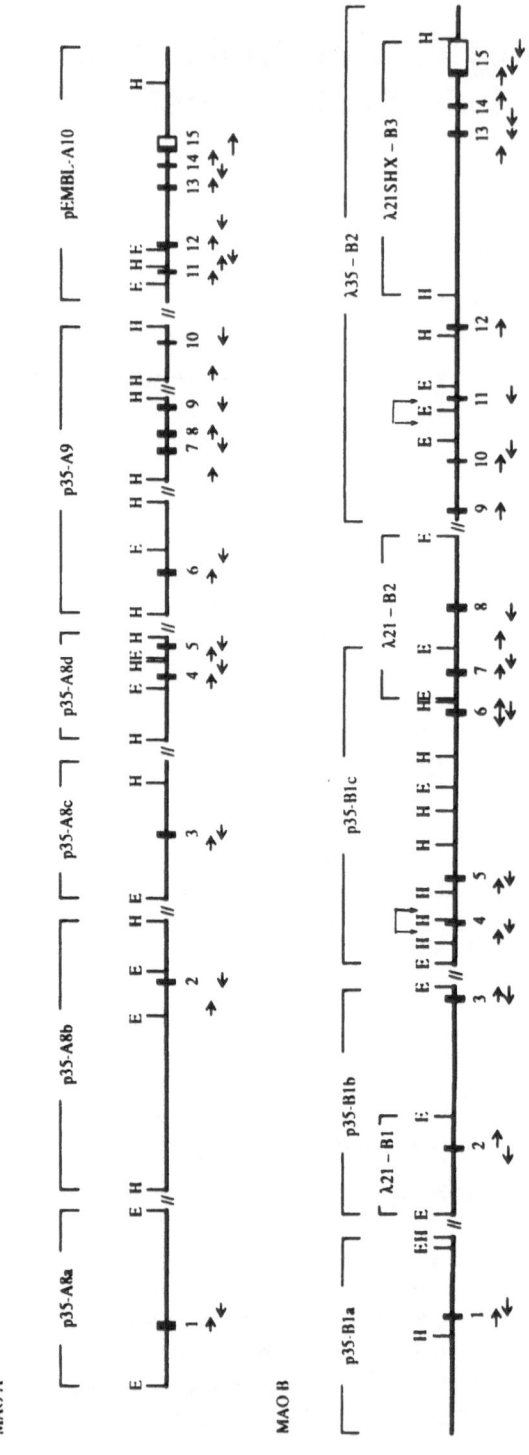

Scale: 1.6 kb/cm

Fig. 1. Genomic organization of the human MAOA and MAOB genes. *Filled bars* indicated coding regions and *unfilled bars* represent noncoding regions of the exons. The sequencing strategy is shown by *horizontal arrows*. *Double-headed arrows* represent equivocal assignment of an exon to either restriction fragment. *E, EcoRI; H, HindIII; //,* intron gap. (Data from Grimsby et al. 1991)

lecular weights of 59700 and 58800, respectively. Partial sequencing of a human placenta MAOA cDNA (Hsu et al. 1988) showed >99% homology to liver MAOA cDNA clones (Hsu et al. 1988) in 1.2 kb of the coding region and 150 nucleotides in the 3′-untranslated region. The deduced amino acid sequence for human liver MAOB (Bach et al. 1988) is identical to the sequence encoded by human platelet, frontal cortex (Chen et al. 1993) and retina MAOB cDNAs (Chen and Shih 1995). cDNA clones for trout liver MAO (Chen et al. 1994), bovine adrenal medulla MAOA (Powell et al. 1989), rat liver MAOA (Kuwahara et al. 1990), rat adrenal medulla MAOA (Kwan and Abell 1992) and rat liver MAOB (Ito et al. 1988) have been cloned thus far. Comparison of the amino acid sequence between human, bovine and rat MAOA shows ~88% sequence identity. Similarly, comparison of MAOB amino acid sequences from human and rat also show ~88% sequence identity.

The availability of MAO cDNA clones (review by Shih 1990) made possible the isolation and characterization of the human MAOA (Grimsby et al. 1991; Chen et al. 1991; Kwan et al. 1992) and MAOB (Grimsby et al. 1991; Kwan et al. 1992) gene structures. Both of the MAOA and MAOB genes are composed of 15 exons and span over 60 kb (Fig. 1). The exon-intron junctions for the MAO genes are at identical positions, which suggests the genes arose from duplication of an ancestor gene (Grimsby et al. 1991) more than 500 million years ago (Chen et al. 1991). The genes encoding MAOA and MAOB are closely linked on the X-chromosome at position p11.23–11.4 (Ozelius et al. 1988; Levy et al. 1989; Lan et al. 1989b). The core promoter elements for both MAOA (Zhu et al. 1992, 1994; Denney et al. 1994; Chen et al. 1995) and MAOB (Zhu et al. 1992) have been identified. Despite the coexpression of MAOA and MAOB in most tissues, the transcription factor-binding sequences are not conserved at corresponding positions, which is reflective of divergence after a duplication event. Loss of MAOA function in males results in mild retardation and prominent behavioral abnormalities (Brunner et al. 1993a,b). Both genes encoding MAOA and MAOB are deleted in some atypical Norrie disease patients (Lan et al. 1989b; Sims et al. 1989a).

1. Allelic Associations

Restriction fragment length polymorphisms (RFLPs) and dinucleotide repeats in the MAOA (Ozelius et al. 1988, 1989; Hotamisligil and Breakefield 1991; Black et al. 1991; Hinds et al. 1992) and MAOB genes (Konradi et al. 1992; Grimsby et al. 1992; Kurth et al. 1993) have been identified. Since abnormal MAO activity is associated with several neurological and psychiatric diseases such as "stimulus seeking," suicidal behavior, alcoholism, schizophrenia and affective disorders (Oreland et al. 1984), polymorphic markers are critical for identifying alleles associated with MAO activity or disease.

Many authors have used genetic markers to find an association or linkage between MAO and liability to a disorder. MAOA activity in human skin fibroblasts can vary by 50-fold among control individuals and evidence exists for a genetic determinant (Breakefield et al. 1980). This led to the question of whether the structural gene for MAOA determines enzymatic activity. Hotamisligil and Breakefield (1991) provided evidence for a strong correlation between allelic status and MAOA activity, indicating that noncoding, regulatory sequences in the MAOA gene, in part, control enzyme activity. MAOB activity can also vary among control populations (Murphy et al. 1976) and evidence exists for a genetic determinant (Rice et al. 1982). However, in contrast to the association of MAOA alleles and activity state, Girmen et al. (1992) found no correlation of platelet MAOB activity in 41 control males, which suggests that allelic status is not the primary determinant of enzymatic activity. This does not exclude the possibility that different MAOB alleles contribute to enzyme activity and play a role in disease.

Preliminary data by Vanyukov et al. (1995) showed an association between "long" dinucleotide repeats in the MAOA gene and increased risk of early onset alcoholism/substance abuse in males, but not females. This represents the first report showing an association between behavior and expansion of a dinucleotide $(CA)_n$ repeat. It is unknown whether the length of this repeat influences MAOA activity. However, the localization of both MAOA and MAOB dinucleotide repeats to intron 2 suggests a functional significance (Shih et al. 1993).

Brunner et al. (1993b) found linkage between MAOA and a kindred with affected males having borderline mental retardation and impulsive aggressive behavior. Subsequent studies by Brunner et al. (1993a) found that the afflicted individuals lack MAOA enzymatic activity due to a point mutation which introduces a premature stop codon in the structural gene for MAOA (see below for a discussion of altered biogenic amine amounts). Lim et al. (1995) hypothesized that the point mutation reported by Brunner et al. (1993a,b) or other MAOA mutations may be involved in bipolar affective disorder. They have reported evidence for an association of MAOA polymorphisms with bipolar disorder. However, these results were not replicated in a subsequent study (Craddock et al. 1995).

Allelic association studies have also been examined for the MAOB gene. Kurth et al. (1993), using single-stranded conformational polymorphism (SSCP) analysis of the MAOB gene, found an allele that is associated with a twofold relative risk for Parkinson's disease. In contrast, no association was found when using a MAOA polymorphism to determine allelic frequencies in the same population. However, Ho et al. (1994, 1995) found no association between MAOB alleles with regard to risk for Parkinson's disease.

Recently, Sobell et al. (1995) used a method called dideoxy fingerprinting (Sarkar et al. 1992) to scan for mutations primarily in the coding exons of the MAOB gene. They found no alterations of functional significance in 100 male schizophrenics.

The most consistently reproducible marker associated with alcoholism is lowered platelet MAOB activity when compared to controls (von Knorring et al. 1985; Devor et al. 1993a), and this does not reflect a decrease in enzyme molecules but possibly a variant form of MAO or endogenous inhibitor (DEVOR et al. 1993b). Genetic familial studies suggest that a single major locus controls MAOB activity and concentration, but only the enzymatic activity is significantly different between alcoholics and controls (Devor et al. 1993b).

2. Heterologous Expression

Heterologous expression of enzymatically active MAO is a valuable method for obtaining large homogeneous quantities of enzyme for structure and function studies. Expression of the MAOA and MAOB cDNAs in mammalian cells by LAN et al. (1989a) first demonstrated that a single polypeptide was enzymatically active and its substrate and inhibitor selectivities were similar to the endogenous enzymes (Fig. 2).

WEYLER et al. (1990) expressed catalytically active human MAOA in *Saccharomyces cerevisiae* and showed that it displayed properties such as human placental MAOA and contained a covalently bound cysteinylflavin. Since it is likely that yeast lacks a cysteinylflavinholoenzyme to catalyze FAD attachment, the reaction may proceed by an autoflavination mechanism (WEYLER et al. 1990). TAN et al. (1991) have compared the substrate specificities for various natural and synthetic substrates between human liver MAOA expressed in yeast and MAOA isolated from human placenta and found significant differences for bulky tetrahydropyridine derivatives. Comparison of the partial amino acid sequence of placental MAOA obtained from peptide sequencing with the deduced sequence of liver MAOA showed divergences in

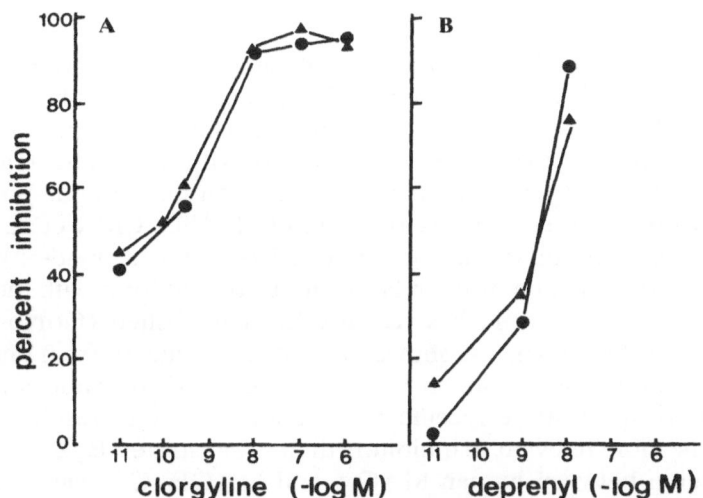

Fig. 2. Inhibition of MAOA and MAOB expressed in COS cells by clorgyline (*E*) and deprenyl (*J*). MAOA and MAOB activity was assayed using 5-HT and PEA, respectively, as substrates in the presence of various concentrations of inhibitors. (Data from LAN et al. 1989a)

five residues. However, because the human placenta cDNA shows >99% homology to the liver cDNA sequence (Hsu et al. 1988), it suggests that these sequence divergences are due to peptide sequencing errors.

a) Site-Directed Mutagenesis

Expression of site-directed mutants in COS cells has confirmed that covalent attachment of FAD to a cysteine (Cys) residue in MAOA, Cys-406 (Wu et al. 1993), and MAOB, Cys-397 (Wu et al. 1993; Gottowik et al. 1993), is critical for enzymatic activity. Furthermore, site-directed mutagenesis of a glutamic acid residue located in the ADP-binding β-α-β unit of human MAOB, a residue that plays a crucial role in binding to the 2'-hydroxyl group of ribose in the AMP moiety of FAD, drastically reduced catalytic activity (Kwan et al. 1995).

Since cysteine has been suggested to be an active-site residue in MAOB (Silverman and Zieske 1986), we systematically converted all cysteine residues in MAOA and MAOB to serine (Wu et al. 1993). In addition to the Cys residue that forms a covalent bond to FAD, we identified one Cys residue (374) in MAOA and two Cys residues (156 and 365) in MAOB that are critical for enzymatic activity. It is unknown whether these are active-site cysteine residues that are directly involved in catalytic reactions or are important for the conformation of active enzymes, for example, by forming disulphide bonds. Nevertheless, MAOA and MAOB have different active-site structures to accommodate broad and restrictive substrate specificities, respectively (Youdium 1978).

b) Chimeric MAO Forms

Expression of chimeric MAO forms in COS cells and yeast has been used by numerous laboratories to elucidate its substrate and inhibitor-binding region. Chimeric MAO forms created by exchanging an ADP-binding motif located in the NH$_2$-terminus of MAOA (residues 1–45) with the same region from MAOB (residues 1–36) showed no changes in substrate or inhibitor selectivity compared to parental enzymes (Gottowik et al. 1993; Chen et al. 1996). A detailed examination of 18 human chimeric MAO forms, made by progressively moving the junction of the NH$_2$-terminus of one form with the COOH-terminus of the other form, has recently been published (Gottowik et al. 1995). None of these chimeras showed a shift in specificity from one form to the other. However, one region in MAOB (residues 62–103) appears to play a role in conferring substrate specificity and another region (residues 146–220) appears to be more relevant to inhibitors than to substrates. By exchanging the internal region between human MAOA and MAOB, Grimsby et al. (1996) have localized a substrate- and inhibitor-binding site between MAOB residues 152–366. Taken together, it appears that MAOB residues 62–103 constitute part of a substrate-binding site that interacts with other determinants located between residues 152 and 366. Other studies using chimeric MAO

Table 1. Oxidation of 5-HT and PEA by rat brain MAOA and MAOB. Data are means ± SD

Inhibitor	Substrate	Form remaining	K_m (μM)	V_{max} (nmol/min/mg protein)
Deprenyl	5-HT[a]	MAOA	178 ± 2	0.73 ± 0.06
Clorgyline	5-HT[a]	MAOB	1170 ± 432	0.09 ± 0.02
None	5-HT[a]	MAOA + MAOB	162 ± 38	0.90 ± 0.01
Deprenyl	PEA[b]	MAOA	40.2 ± 7.6	1.09 ± 0.21
Clorgyline	PEA[b]	MAOB	12.2 ± 4.3	2.10 ± 0.13
None	PEA[b]	MAOA + MAOB	17.6 ± 2.1	3.31 ± 0.15

[a] Data obtained from FOWLER and TIPTON (1982).
[b] Data obtained from KINEMUCHI et al. (1980).

Table 2. MAO activity in transfected COS cells. Data are means ± SD. (Values obtained from LAN et al. 1989)

Enzyme	MAO activity (nmol/min/mg protein)	
	5-HT	PEA
MAOA	0.17 ± 0.06	ND
MAOB	ND	0.18 ± 0.13

ND, not detectable.

forms have concluded that residues 120–220 and 50–400 are responsible for the substrate specificity of rat MAOA and MAOB, respectively (TSUGENO et al. 1995).

C. Serotonin Oxidation by MAO

It is an oversimplification to suggest that 5-HT and PEA are substrates for only MAOA and MAOB, respectively. In fact, both of these substrates can be oxidized in vitro by MAOA and MAOB, albeit at reduced catalytic effectiveness compared to their preferred substrates (KINEMUCHI et al. 1980; FOWLER and TIPTON 1982). FOWLER and TIPTON (1982) treated rat brain homogenates with a concentration of clorgyline to selectively inhibit MAOA and found that the remaining MAOB was capable of deaminating 5-HT (Table 1). Similarly, KINEMUCHI et al. (1980) selectively inhibited MAOB with deprenyl and showed that the remaining MAOA oxidized PEA (Table 1).

Expression of human MAOA and MAOB cDNAs in mammalian COS cells by LAN et al. (1989a) showed that MAOA preferentially oxidizes 5-HT and MAOB preferentially oxidizes PEA (Table 2). However, the oxidation of PEA by MAOA and the oxidation of 5-HT by MAOB was not detected because of low enzyme expression levels.

Table 3. Kinetic constants for MAO expressed in yeast. Data are means ± SD. (Values obtained from GRIMSBY et al. 1996)

Enzyme	Substrate	K_m μM	K_{cat} (/s)	K_{cat}/K_m (/s/M)
MAOA	5-HT	66.1 ± 2.3	9.89 ± 1.17	1.5×10^5
MAOA	PEA	96.5 ± 8.4	2.07 ± 0.12	2.1×10^4
MAOB	5-HT	ND	ND	ND
MAOB	PEA	2.2 ± 0.06	1.77 ± 0.24	8.0×10^5

ND, 5-HT oxidation is not detectable using a substrate concentration of $100\,\mu M$.

Expression of MAOA and MAOB cDNAs in yeast (GRIMSBY et al. 1996) also demonstrated that MAOA can metabolize PEA at low substrate concentrations $(10\,\mu M)$ that are normally used to determine MAOB activity (Table 3). However, MAOB does not oxidize 5-HT using concentrations $(100\,\mu M)$ that are normally used to measure MAOA activity. This may reflect low expression levels of MAOB in the yeast expression system and/or low affinity of 5-HT for MAOB. TSUGENO et al. (1995) have also shown the lack of 5-HT oxidation by rat MAOB expressed in yeast.

I. Effect of MAO Deficiencies on 5-HT Levels

Altered 5-HT levels have been associated with various psychiatric disorders including depression, aggression, anxiety states, schizophrenia and alcoholism (review by VAN PRAAG et al. 1987). In order to understand the role of MAO in the oxidation of 5-HT and other biogenic amines, it is essential to determine the levels of these neurotransmitters after MAO inhibition. This provides insight into the in vivo role of MAOA and MAOB. The recent discovery of a MAOA deficiency in humans (BRUNNER et al. 1993a,b) and mice (CASES et al. 1995) and deletion of the MAO genes in atypical Norrie disease enables the functional significance of these enzymes towards the oxidation of 5-HT and other biogenic amines to be assessed without the concern of cross-inhibition as observed in some cases with MAO inhibitors.

1. MAOA-Deficient Humans and Mice

BRUNNER et al. (1993a,b) identified a Dutch kindred in which the affected males completely lack MAOA and display impulsive aggression and border-line mental retardation. They determined amine and amine metabolite excretion in three individuals lacking MAOA and showed that the concentrations of 5-HT, 3-methoxytyramine (3-MT, an O-methylated metabolite of DA), TA and normetanephrine (NMN, an O-methylated metabolite of NE) increased about fivefold compared to controls. In contrast, concentrations of the NE-deaminated metabolites vanillymandelic acid (VMA) and 3-methoxy-4-hydroxy-phenyl(vanil)glycol (MHPG) decreased by about tenfold. Furthermore, the concentrations of 5-hydroxyindole-3-acetic acid (5-HIAA, a deaminated metabolite of 5-HT) and vanilglycolic acid (HVA, a deaminated metabolite of DA) were marginally decreased. PEA excretion was not evaluated, but it is not expected to change based on its rapid inactivation by MAOB.

CASES et al. (1995) generated interferon-β transgenic mice and noticed X-linked recessive abnormal behavior of mouse pups in one of the transgenic lines. Based on the report by BRUNNER et al. (1993a), they surmised that these mice might be MAOA deficient and found that the transgene integrated into the MAOA gene, causing a deletion of exons 2 and 3.

In MAOA-deficient pup brains, 5-HT levels increased ninefold over control levels with a concomitant decrease in 5-HIAA. NE levels increased two-fold in pup and adult brains compared to wild-type mice. Surprisingly, with increasing age, the concentration of 5-HT increased and 5-HIAA levels decreased in the MAOA-deficient mice. By 210 days the 5-HT and 5-HIAA concentrations were normal. These age-related changes in the concentrations of 5-HT and 5-HIAA can be represented as a 5-HIAA/5-HT ratio, a value used to represent 5-HT turnover by MAO (Table 4). These ratios show a drastic difference in young MAOA-deficient mice compared to controls, but with increasing age the ratios approach normal values. MAOB activity shows an age-related increase in mice brain (SAURA et al. 1994), which suggests that 5-HT oxidation by MAOB is more significant in older mice (Table 4). CELADA and ARTIGAS (1993) compared the brain concentrations of 5-HT and 5-HIAA in rats after treatment with selective and nonselective MAO inhibitors. They observed that tranylcypromine, a nonselective MAO inhibitor, increased 5-HT and decreased 5-HIAA more than clorgyline. Furthermore, BEL and ARTIGAS (1995) used intracerebral microdialysis to measure 5-HT output in rats and showed that broforamine, a MAOA inhibitor, increased 5-HT levels by 200%, whereas a combination of broforamine and deprenyl (a MAOB-specific inhibitor) increased it to 500%. Similar results were obtained with

Table 4. Age-related changes in the 5-HIAA/5-HT ratios and MAOB activity for MAOA-deficient mice brains. The 5-HIAA/5-HT ratios were obtained from Cases et al. (1995). MAOB activity was determined using PEA as a substrate

Age (days)	5-HIAA/5-HT		MAOB activity (nmol/20 min/mg protein)
	MAOA deficient	Wild-type	
1	0.01	0.9	−
8	0.03	0.9	5.3
12	0.03	0.9	9.6
28	0.1	0.5	−
46	0.1	0.3	−
90	0.1	0.3	15.9
210	0.4	0.4	−
395	−	−	33.0

−, not determined.

clorgyline and deprenyl. Broforamine and clorgyline are both MAOA-specific inhibitors, except the former is also a 5-HT uptake blocker. These results suggest that when MAOA is inhibited the concentration of 5-HT increases to a level suitable for oxidation by MAOB.

The MAOB inhibitor deprenyl caused a near disappearance of 5-HIAA in MAOA-deficient mice but had no effect in control mice (CASES et al. 1995). These results demonstrate the important role of MAOB in 5-HT metabolism. However, the contribution of 5-HT oxidation by MAOB under normal conditions is unknown. 5-HT oxidation in cells coexpressing both forms of MAO would likely be mediated by MAOA because of its higher affinity for 5-HT. In contrast, the cellular localization of MAOB in serotoninergic neurons suggests that it plays an important role in regulating 5-HT levels in these neurons. These results demonstrate that 5-HT oxidation by MAOB is more pronounced with increasing age.

MAOA-deficient mice display signs of abnormal behavior that are strikingly different between pups and adults (CASES et al. 1995). The phenotypic traits of mice between the age of postnatal days 1 and 16 range from intense head nodding to frantic running. In contrast, adult males display offensive aggressive behavior. Treatment with the 5-HT synthesis inhibitor parachlorophenylalanine reversed these abnormal behavioral traits in MAOA-deficient pups, whereas the catecholamine synthesis inhibitor α-methylparatyrosine had no effect.

These results are consistent with earlier studies using MAO inhibitors. The effects of MAOA inhibitors on monoamine concentrations in rat brain show an increase in 5-HT, DA, NE, NMN and 3-MT whereas the concentrations of 5-HIAA, 3,4-dihydroxyphenylacetic acid (DOPAC), HVA and MHPG decrease (DA PRADA et al. 1990; COLZI et al. 1990; KUMAGAE et al. 1991).

2. Deletion of the MAOA and MAOB Genes in Norrie Disease

Loss of function of the Norrie disease (ND) gene results in congenital blindness (NORRIE 1927; WARBURG 1966), with half of the patients manifesting mental retardation and/or progressive sensorineural hearing loss (WARBURG 1975). GAL et al. (1985a,b) mapped the ND gene to Xp11.3–11.2 by linkage to locus *DXS7*. The deletion of *DXS7* (DE LA CHAPELLE et al. 1985; GAL et al. 1985a) and close proximity of the ND gene to the *MAOA/MAOB* locus prompted investigations into the status of the MAO genes. Hybridization of MAO cDNAs to DNA obtained from ND individuals having a deletion of the *DXS7* locus revealed the complete absence of MAO-hybridizing sequences (LAN et al. 1989b; SIMS et al. 1989a; review, CHEN et al. 1995). Atypical ND is associated with the loss of MAOA/MAOB function and is characterized by additional symptoms including somatic growth failure, severe mental retardation, autistic-like behavior, atonic seizures, autonomic dysfunction and altered sleep patterns (SIMS et al. 1989a,b; MURPHY et al. 1990; COLLINS et al. 1992).

Amine and amine metabolite amounts for Norrie disease patients with a deletion of the MAOA and MAOB have been determined. Marked elevations in the urinary excretion of PEA, 3-MT, NMN, *o*-tyramine, *m*-tyramine and *p*-tyramine (metabolites of tyramine) were observed, whereas the concentrations of VMA and MHPG decreased (MURPHY et al. 1990, 1991; COLLINS et al. 1992). Urinary excretion of 5-HT was not determined in these studies. Surprisingly, 5-HIAA, HVA and DOPAC were normal, which suggests that other metabolic pathways such as plasma amine oxidase and/or benzylamine oxidase or other atypical amine oxidases compensate for loss of the MAO genes (MURPHY et al. 1991).

D. Conclusions

Successful isolation of the MAOA and MAOB cDNA and genomic clones has provided a great deal of knowledge about the structure and function of these important enzymes. Site-directed mutagenesis and construction of chimeric MAO forms has made possible the localization of amino acid residues involved in FAD covalent attachment, putative active-site cysteine residues and substrate- and inhibitor-binding regions. Such advances are crucial to a better understanding of MAO function that may lend itself to the development of better MAO inhibitors. Heterologous expression of MAO in yeast or bacteria will produce milligram quantities of the enzymes which may be suitable for crystallization and atomic coordinate determinations. Alterations in the levels of biogenic amines and their metabolites in MAO-deficient humans and mice has clearly shown that 5-HT is predominately oxidized in vivo by MAOA, whereas the role of MAOB is important in the absence of MAOA. The ability of MAOB to metabolize 5-HT in MAOA-deficient mice is intriguing. What is MAOB's role in regulating 5-HT levels in normal individuals, especially in serotoninergic neurons where MAOB is the predominant MAO form? It is also interesting that MAOA inhibitors are effective antidepressants yet in MAOA-deficient humans and mice they produce aggressive behavior. This result suggests that MAO regulates biogenic amines during development and its absence may result in abnormal neuronal architecture. In conclusion, molecular biology has helped answer many questions that have been addressed over the 25 years of MAO research and has opened new questions for future investigations.

Acknowledgements. This work was supported by NIMH grants Nos. R37 MH 39085 (Merit award), R01 MH37020, Research Scientist Award K05 MH00796 and a Welin Professorship.

References

Bach AW, Lan NC, Johnson DL, Abell CW, Bembenek ME, Kwan S-W, Seeburg PH, Shih JC (1988) cDNA cloning of human liver monoamine oxidase A and B:

molecular basis of differences in enzymatic properties. Proc Natl Acad Sci USA 85:4934–4938

Bel N, Artigas F (1995) In vivo evidence for the reversible action of the monoamine oxidase inhibitor brofaromine on 5-hydroxytryptamine release in rat brain. Naunyn Schmiedebergs Arch Pharmacol 351:475–482

Berry MD, Juorio AV, Paterson IA (1994) The functional role of monoamine oxidase A and B in the mammalian central nervous system. Prog Neurobiol 42:375–391

Bey P, Fozard J, McDonald I, Palfreyman MG, Zreika M (1984) MDL 72145: a potent and selective inhibitor of MAO type B. Br J Pharmacol 81:50P

Black GCM, Chen ZY, Craig IW, Powell JF (1991) Dinucleotide repeat polymorphism at the MAOA locus. Nucleic Acids Res 19:689

Breakefield XO, Giller EL Jr, Nurnburger JI Jr, Castiglione CM, Buchsbaum MS, Gershon ES (1980) Monoamine oxidase type A in fibroblasts from patients with bipolar depressive illness. Psychiatry Res 2:307–314

Brunner HG, Nelen MR, Breakefield XO, Ropers HH, van Oost BA (1993a) Abnormal behavior associated with a point mutation in the structural gene for monoamine oxidase A. Science 262:578–580

Brunner HG, Nelen MR, van Zandvoort P, Abeling NGGM, van Gennip AH, Wolters EC, Kuiper MA, Ropers HH, van Oost BA (1993b) X-linked borderline mental retardation with prominent behavioral disturbance: phenotype, genetic localization, and evidence for disturbed monoamine metabolism. Am J Hum Genet 52:1032–1039

Cases O, Seif I, Grimsby J, Gaspar P, Chen K, Pournin S, Muller U, Aguet M, Babinet C, Shih JC, De Maeyer E (1995) Aggressive behavior and altered amounts of brain serotonin and norepinephrine in mice lacking MAOA. Science 268:1763–1766

Celada P, Artigas F (1993) Plasma 5-hydroxyindoleacetic acid as an indicator of monoamine oxidase-A inhibition in rat brain and peripheral tissues. J Neurochem 61:2191–2198

Chen K, Shih JC (1995) The amino acid sequences of human retina and liver monoamine oxidase (MAO) A are identical. Soc Neurosci Abstr 21:51

Chen K, Wu H-F, Shih JC (1993) The deduced amino acid sequences of human platelet and frontal cortex monoamine oxidase B are identical. J Neurochem 61:187–190

Chen K, Wu H-F, Grimsby J, Shih JC (1994) Cloning of a novel monoamine oxidase cDNA from trout liver. Mol Pharmacol 46:1226–1233

Chen K, Wu H-F, Shih JC (1996) The influence of C-terminus on MAOA and MAOB catalytic activity. J Neurochem 66:797–803

Chen Z-Y, Hotamisligil GS, Huang J-K, Wen L, Ezzeddine D, Aydin-Muderrisoglu N, Powell JF, Huang RH, Breakefield XO, Craig I, Hsu Y-PP (1991) Structure of the human gene for monoamine oxidase type A. Nucleic Acids Res 19:4537–4541

Chen Z-Y, Denney RM, Breakefield XO (1995) Norrie disease and MAO genes: nearest neighbors. Human Mol Genet 4:1729–1737

Collins FA, Murphy DL, Reiss AL, Sims KB, Lewis JG, Freund L, Karoum F, Zhu D, Maumenee IH, Antonarakis SE (1992) Clinical, biochemical, and neuropsychiatric evaluation of a patient with a contiguous gene syndrome due to a microdeletion Xp11.3 including the Norrie disease locus and monoamine oxidase (MAOA and MAOB) genes. Am J Med Genet 42:127–134

Colzi A, d'Agostino F, Kettler R, Borroni E, Da Prada M (1990) Effect of selective and reversible MAO inhibitors on dopamine outflow in rat striatum: a microdialysis study. J Neural Transm [Suppl] 32:79–84

Craddock N, Daniels J, Roberts E, Rees M, McGuffin P, Owen MJ (1995) No evidence for allelic association between bipolar disorder and monoamine oxidase A gene polymorphisms. Am J Hum Genet 60:322–324

Da Prada M, Kettler R, Keller HH, Cesura AM, Richards JG, Saura Marti J, Muggli-Maniglio D, Wyss P-C, Kyburz E, Imhoff R (1990) From moclobemide to Ro 19–6327 and Ro 41–1049: the development of a new class of reversible, selective MAO-A and MAO-B inhibitors. J Neural Transm [Suppl] 29:279–292 de la Chapelle A, Sankila E-M, Lindlof M, Aula P, Norio R (1985) Norrie disease caused by a gene deletion allowing carrier detection and prenatal diagnosis. Clin Genet 28:317–320

Denney RM, Denney CB (1985) An update on the identity crisis of monoamine oxidase: new and old evidence for the independence of MAO A and B. Pharmacol Ther 30:227–259

Denney RM, Sharma A, Dave SK, Waguespack A (1994) A new look at the promoter of the human monoamine oxidase A gene: mapping transcription initiation sites and capacity to drive luciferase expression. J Neurochem 63:843–856

Devor EJ, Cloninger CR, Hoffman PL, Tabakoff B (1993a) Association of monoamine oxidase (MAO) activity with alcoholism and alcoholic subtypes. Am J Hum Genet 48:209–213

Devor EJ, Cloninger CR, Kwan S-W, Abell CW (1993b) A genetic familial study of monoamine oxidase B activity and concentration in alcoholics. Alcohol Clin Exp Res 17:263–267

Donnelly CH, Murphy DL (1977) Substrate- and inhibitor-related characteristics of human platelet monoamine oxidase. Biochem Pharmacol 26:853–858

Fowler CJ, Tipton KF (1982) Deamination of 5-hydroxytryptamine by both forms of monoamine oxidase in the rat brain. J Neurochem 38:733–736

Gal A, Bleeker-Wagemakers LM, Wienker TF, Warburg M, Ropers HH (1985a) Localization of the gene for Norrie disease by linkage to the DXS7 locus. Cytogenet Cell Genet 40:633

Gal A, Stolzenberger C, Wienker TF, Wieacker P, Ropers HH, Friedrich U, Bleeker-Wagemakers LM, Pearson P, Warburg M (1985b) Norrie's disease: close linkage with genetic markers from the proximal short arm of the X-chromosome. Clin Genet 27:282–283

Girmen AS, Baenziger J, Hotamisligil GS, Konradi C, Shalish C, Sullivan JL, Breakefield XO (1992) Relationship between platelet monoamine oxidase B activity and alleles at the MAOB locus. J Neurochem 59:2063–2066

Gottowik J, Cesura AM, Malherbe P, Lang G, Da Prada M (1993) Characterization of wild-type and mutant forms of human monoamine oxidase A and B expressed in a mammalian cell line. FEBS Lett 317:152–156

Gottowik J, Malherbe P, Lang G, Da Prada M, Cesura AM (1995) Structure/function relationships of mitochondrial monoamine oxidase A and B chimeric forms. Eur J Biochem 230:934–942

Greenwalt JW, Schnaitman C (1970) An appraisal of the use of monoamine oxidase as an enzyme marker for the outer mitochondrial membrane. J Cell Biol 46:173–179

Grimsby J, Chen K, Wang L-J, Lan NC, Shih JC (1991) Human monoamine oxidase A and B genes exhibit identical exon-intron organization. Proc Natl Acad Sci USA 88:3637–3641

Grimsby J, Chen K, Devor EJ, Cloninger CR, Shih JC (1992) Dinucleotide repeat (TG)23 polymorphism in the MAOB gene. Nucleic Acids Res 20:924

Grimsby J, Zentner M, Shih JC (1996) Identification of a region important for human monoamine oxidase B substrate and inhibitor selectivity. Life Sci 58:777–787

Hinds HL, Hendriks RW, Craig IW, Chen ZY (1992) Characterization of a highly polymorphic region near the first exon of the human MAOA gene containing a GT dinucleotide and a novel VNTR motif. Genomics 13:896–897

Ho SL, Ramsden DB, Kapadi AL, Sturman SG, Williams AC (1994) Activity and polymorphism of monoamine oxidase-B gene in idiopathic Parkinson's disease. Biog Amines 10:579–586

Ho SL, Kapadi AL, Ramsden DB, Williams AC (1995) An allelic association study of monoamine oxidase B in Parkinson's disease. Ann Neurol 37:403–405

Hotamisligil GS, Breakefield XO (1991) Human monoamine oxidase A gene determines levels of enzyme activity. Am J Hum Genet 49:383–392

Hsu Y-PP, Weyler W, Chen S, Sims KB, Rinehart WB, Utterback MC, Powell JF, Breakefield XO (1988) Structural features of human monoamine oxidase A elucidated from cDNA and peptide sequences. J Neurochem 51:1321–1324

Ito A, Kuwahara T, Inadome S, Sagara Y (1988) Molecular cloning of a cDNA for rat liver monoamine oxidase B. Biochem Biophys Res Commun 157:970–976

Johnston JP (1968) Some observations upon a new inhibitor of monoamine oxidase in brain tissue. Biochem Pharmacol 17:1285–1297

Kinemuchi H, Wakui Y, Kamijo K (1980) Substrate selectivity of type A and type B monoamine oxidase in rat brain. J Neurochem 35:109–115

Knoll J, Magyar K (1972) Some puzzling pharmacological effect of monoamine oxidase inhibitors. Adv Biochem Psychopharmacol 5:393:408

Konradi C, Ozelius L, Breakefield XO (1992) Highly polymorphic (GT)n repeat sequence in intron II of the human MAO-B gene. Genomics 12:176–177

Kumagae Y, Matsui Y, Iwata (1991) Deamination of norepinephrine, dopamine, and serotonin by type A monoamine oxidase in discrete regions of the rat brain and inhibition by RS-8359. Jpn J Pharmacol 55:121–128

Kurth JH, Kurth MC, Poduslo SE, Schwankhaus JD (1993) Association of a monoamine oxidase B allele with Parkinson's disease. Ann Neurol 33:368–372

Kuwahara T, Takamoto S, Ito A (1990) Primary structure of rat monoamine oxidase A deduced from cDNA and its expression in rat tissues. Agric Biol Chem 54:253–257

Kwan S-W, Abell CW (1992a) cDNA cloning and sequencing of rat monoamine oxidase A: comparison with the human and bovine enzymes. Comp Biochem Physiol 102B:143–147

Kwan S-W, Bergeron JM, Abell CW (1992b) Molecular properties of monoamine oxidases A and B. Psychopharmacology 106:S1–S5

Kwan S-W, Lewis DA, Zhou BP, Abell CW (1995) Characterization of a dinucleotide-binding site in monoamine oxidase B by site-directed mutagenesis. Arch Biochem 316:385–391

Lan NC, Chen C, Shih JC (1989a) Expression of functional human monoamine oxidase A and B cDNAs in mammalian cells. J Neurochem 52:1652–1654

Lan NC, Heinzmann C, Gal A, Klisak I, Orth U, Lai E, Grimsby J, Sparkes RS, Mohandas T, Shih JC (1989b) Human monoamine oxidase A and B genes map to Xp11.23 and are deleted in a patient with Norrie disease. Genomics 4:552–559

Levitt P, Pintar JE, Breakefield XO (1982) Immunocytochemical demonstration of monoamine oxidase B in brain astrocytes and serotonergic neurons. Proc Natl Acad Sci USA 79:6385–6389

Levy ER, Powell JF, Buckle VJ, Hsu Y-PP, Breakefield XO, Craig IW (1989) Localization of human monoamine oxidase to Xp11.23–11.4 by in situ hybridization: implications for Norrie disease. Genomics 5:368–370

Lim LC, Powell J, Sham P, Castle D, Hunt N, Murray R, Gill M (1995) Evidence for a genetic association between alleles of monoamine oxidase A gene and bipolar affective disorder. Am J Hum Genet 60:325–331

Murphy DL, Wright C, Buchsbaum M, Nichols A, Costa JL, Wyatt RJ (1976) Platelet and plasma amine oxidase in activity in 680 normals; sex age differences and stability over time. Biochem Med 16:254–257

Murphy DL, Sims KB, Karoum F, de la Chapelle A, Norio R, Sankila E-M, Breakefield XO (1990) Marked amine and amine metabolite changes in Norrie disease patients with an X-chromosome deletion affecting monoamine oxidase. J Neurochem 54:242–247

Murphy DL, Sims KB, Karoum F, Garrick NA, de la Chapelle A, Sankila EM, Norio R, Breakefield XO (1991) Plasma amine oxidase activities in Norrie disease patients with an X-chromosomal deletion affecting monoamine oxidase. J Neural Trans [Gen Sect] 83:1–12

Norrie G (1927) Causes of blindness in children. Acta Ophthalmol (Copenh) 5:357–386

Oreland L, Knorring L, von Schalling D (1984) Connections between monoamine oxidase, temperament and disease. In: Patton SW, Mitchell J, Turner P (eds)

Proceedings of 11th international congress on pharmacology, vol 2. MacMillan, London, pp 193–202

Ozelius L, Hsu Y-PP, Bruns G, Powell JF, Chen S, Weyler W, Utterback M, Zucker D, Haines J, Trofatter JA, Conneally PM, Gusella JF, Breakefield XO (1988) Human monoamine oxidase gene (MAOA): chromosome position (Xp21–p11) and DNA polymorphism. Genomics 3:53–58

Ozelius L, Gusella JF, Breakefield XO (1989) MspI RFLP for human MAOA gene. Nucleic Acids Res 17:10516

Powell JF, Hsu Y-PP, Weyler W, Chen S, Salach J, Andrikopoulos K, Mallet J, Breakefield XO (1989) The primary structure of bovine monoamine oxidase type A. Biochem J 259:407–413

Rice J, McGuffin P, Shaskan E (1982) A comingling analysis of platelet monoamine oxidase activity. Psychiatry Res 7:325–335

Sarkar G, Yoon H-S, Sommer SS (1992) Dideoxy fingerprinting (ddF): a rapid and efficient screen for the presence of mutations. Genomics 13:441–443

Saura J, Richards JG, Mahy N (1994) Differential age-related changes of MAO-A and MAO-B in mouse brain and peripheral organs. Neurobiol Aging 15:399–408

Shih JC (1990) Molecular basis of human MAO A and B. Neuropsychopharmacology 4:1–7

Shih JC, Grimsby J, Chen K, Zhu Q-S (1993) Structure and promoter organization of the human monoamine oxidase A and B genes. J Psychiatry Neurosci 18:25–32

Silverman RB, Zieske PA (1986) Identification of the amino acid bound to the labile adduct formed during inactivation of monoamine oxidase by 1-phenylcyclopropylamine. Biochem Biophys Res Commun 135:154–159

Sims KB, de la Chapelle A, Norio R, Sankila E-M Hsu Y-PP, Rinehart WB, Corey TJ, Ozelius L, Powell JF, Bruns G, Gusella JF, Murphy DL, Breakefield XO (1989a) Monoamine oxidase deficiency in males with an X chromosome deletion. Neuron 2:1069–1076

Sims KB, Ozelius L, Corey T, Rinehart WB, Liberfarb R, Haines J, Chen WJ, Norio R, Sankila E, de la Chapelle A, Murphy DL, Gusella J, Breakefield XO (1989b) Norrie disease gene is distinct from the monoamine oxidase genes. Am J Hum Genet 45:424–434

Sobell JL, Lind TJ, Hebrink DD, Heston LL, Sommer SS (1995) Screening the monoamine oxidase B gene in 100 male schizophrenics: possible clustered mutations in African-Americans but lack of functionally significant sequence changes. Am J Hum Genet 57:A172

Squires RF (1968) Additional evidence for the existence of several forms of mitochondrial monoamine oxidase in the mouse. Biochem Pharmacol 17:1401–1409

Tan AK, Weyler W, Salach JI, Singer TP (1991) Differences in substrate specificities of monoamine oxidase A from human liver and placenta. Biochem Biophys Res Commun 181:1084–1088

Thorpe LW, Westlund KN, Kochersperger LM, Abell CW, Denney RM (1987) Immunocytochemical localization of monoamine oxidase A and B in human peripheral tissues and brain. J Histochem Cytochem 35:23–32

Tsugeno Y, Hirashiki I, Ogata F, Ito A (1995) Regions of the molecule responsible for substrate specificity of monoamine oxidase A and B: chimeric enzyme analysis. J Biochem 118:974–980

van Praag HM, Kahn RS, Asnis GM, Wetzler S, Brown SL, Bleich A, Korn ML (1987) Denosologization of biological psychiatry or the specificity of 5-HT disturbances in psychiatric disorders. J Affective Disord 13:1–8

Vanyukov MM, Moss HB, Yu LM, Tarter RE, Deka R (1995) Preliminary evidence for an association of a dinucleotide repeat polymorphism at the MAOA gene with early onset alcholism/substance abuse. Am J Med Genet 60:122–126

von Knorring A-L, Bohman M, von Knorring L, Oreland L (1985) Platelet MAO-activity as a biological marker in sub-groups of alcholism. Acta Psychiatr Scand 72:51–58

Warburg M (1966) Norrie's disease: a congenital progressive oculo-acoustico-cerebral degeneration. Acta Ophthalmol (Copenh) [Suppl] 89:1–47

Warburg M (1975) Norrie's disease – differential diagnosis and treatment. Acta Ophthalmol (Copenh) 53:217–236

Weyler W, Titlow CG, Salach JI (1990) Catalytically active monoamine oxidase type A from human liver expressed in Saccharomyces cerevisiae contains FAD. Biochem Biophys Res Commun 173:1205–1211

Wu H-F, Chen K, Shih JC (1993) Site-directed mutagenesis of monoamine oxidase A and B: role of cysteines. Mol Pharmacol 43:888–893

Youdium MBH (1978) The active centers of monoamine oxidase type A and B binding with ^{14}C-clorgyline and ^{14}C-deprenyl. J Neural Transm 43:199–205

Zhu QS, Grimsby J, Chen K, Shih JC (1992) Promoter organization and activity of human monoamine oxidase (MAO) A and B genes. J Neurosci 12:4437–4446

Zhu QS, Chen K, Shih JC (1994) Bidirectional promoter of human monoamine oxidase A (MAOA) controlled by transcription factor Sp1. J Neurosci 14:7393–7403

Molecular Biology, Pharmacology, and Genetics of the Serotonin Transporter: Psychobiological and Clinical Implications

K.-P. Lesch

A. Introduction

Serotonin (5-HT), which acts at multiple pre- and postsynaptic receptor sites, fulfills the criteria for a neurotransmitter and modulator of synaptic signal transduction in many functional systems of the brain. As a consequence of this enormous versatility, brainstem 5-HT systems affect many diverse functions, such as mood, cognition, appetite and satiation, sleep, motor behavior, neuroendocrine and circadian rhythms, social and reproductive behavior. By determining the magnitude and duration of postsynaptic receptor-mediated signaling, carrier-facilitated 5-HT transport into and release from the presynaptic neuron plays a key role in the spatiotemporal fine-tuning of 5-HT neurotransmission (Fig. 1). Recent advances have resulted from the molecular and functional characterization of the serotonin transporter (5-HTT), from pharmacological and neurochemical studies relating the actions of psychoactive drugs (e.g., tricyclics, selective 5-HT reuptake inhibitors, psychostimulants) to discrete effects on 5-HT uptake and release, and from the molecular dissection of the complex changes in functional expression as a consequence of altered gene transcription. On the basis of its molecular structure and anatomical distribution, substrate specificity, electrophysiological properties, and drug-binding profiles the 5-HTT has been placed in the extended gene family of Na^+/Cl^--dependent cell surface transport proteins.

The new insights into neurotransporter diversity provide the means for novel approaches of studying uptake processes at the molecular level. Current research strategies are focusing on functional mechanisms of 5-HT translocation and inhibitor binding, on molecular regulation of 5-HTT gene expression, and on posttranslational modification of the 5-HTT protein. Important information is also being derived from the analysis of gene organization and from modeling 5-HTT-dependent neuroplasticity or drug actions that target the 5-HTT in genetically engineered animals.

In the psychobiological context it is becoming increasingly evident that inadequate adaptive responses to environmental stressors in conjunction with predisposing genes and/or developmental vulnerability are likely to contribute to the etiopathogenesis of behavioral and psychiatric disorders.

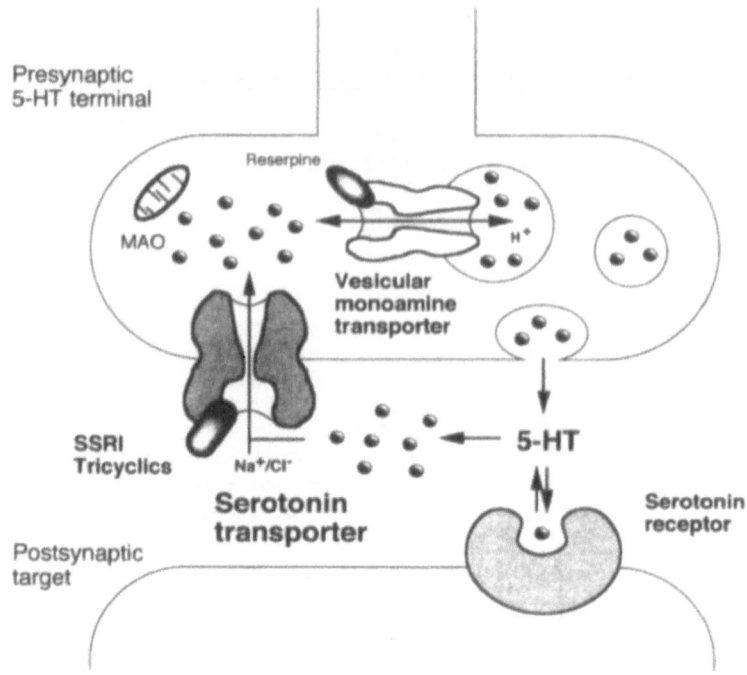

Fig. 1. Serotoninergic synapse. 5-HT is stored in vesicles within the presynaptic neuron and upon neuronal excitation is released into the synaptic cleft. It crosses the cleft and binds to postsynaptic (and presynaptic) 5-HT receptors. The action of 5-HT is terminated by reuptake back into the 5-HT neuron via the 5-HT transporter. The transport process can be blocked by reuptake inhibitors. Once inside the serotoninergic neuron, 5-HT can be enzymatically degraded by MAO or taken back into the storage vesicles via a vesicular monoamine transporter for reuse. The latter process can also be blocked by specific inhibitors, such as reserpine

While the early phases of development of the CNS are dominated by genetically determined local chemical signals between developing cells, it is now clear that at later stages of development and during adult life neural connectivity is affected by sensory inputs from the environment. A polymorphism in the regulatory region of the 5-HTT gene is associated with anxiety- and depression-related personality traits, and preliminary evidence suggests that it also affects the risk to develop affective disorders and alcohol dependence. To further validate the concept of the 5-HTT gene as a susceptibility locus for emotional instability transgenic strategies are gaining momentum. These strategies address the question of the extent to which targeted disruption of the 5-HTT gene affects the biochemistry, electro-physiology, and pharmacology of the 5-HT system and modulates neural development and synaptic plasticity. It may also provide a system to dissect successive events that lead to disease states and to test novel therapeutic concepts.

B. Molecular Biology

I. Gene Family of Na⁺/Cl⁻-Dependent Transporters

A repertoire of different cloning strategies have led to the molecular characterization of an extented gene family of Na^+/Cl^--dependent transporter proteins, including the transporters for 5-HT and catecholamines (norepinephrine, dopamine), γ-aminobutyric acid (GABA), glycine, proline, taurine, betaine, creatine, and several orphan carriers (for review see AMARA and KUHAR 1993; BLAKELY et al. 1994; UHL and JOHNSON 1994; and references therein). Moreover, several subtypes of the GABA transporters have been identified. On the basis of their remarkable amino acid identity of 69%–80% and their properties as targets of tri- and heterocyclic antidepressants, amphetamine, cocaine, and their analogs, the carriers for the biogenic amines 5-HT, dopamine, and norepinephrine constitute a distinct subfamily.

II. Molecular Structure

The cDNA encoding the 5-HTT has been characterized by a combination of expression cloning and homology screening and by PCR-assisted amplification of reverse transcribed mRNA with degenerate oligonucleotides (BLAKELY et al. 1991; HOFFMAN et al. 1991; MAYSER et al. 1991; LESCH et al. 1993a; RAMAMOORTHY et al. 1993). The sequence analysis of rat, mouse, and human cDNAs predicts a protein of 630 amino acids that lacks a signal peptide for promoting insertion into the membrane after translation. The 5-HTT cDNA from *Drosophila melanogaster* displays roughly 50% identity with the rodent and human 5-HTTs (COREY et al. 1994; DEMCHYSHYN et al. 1994). A single

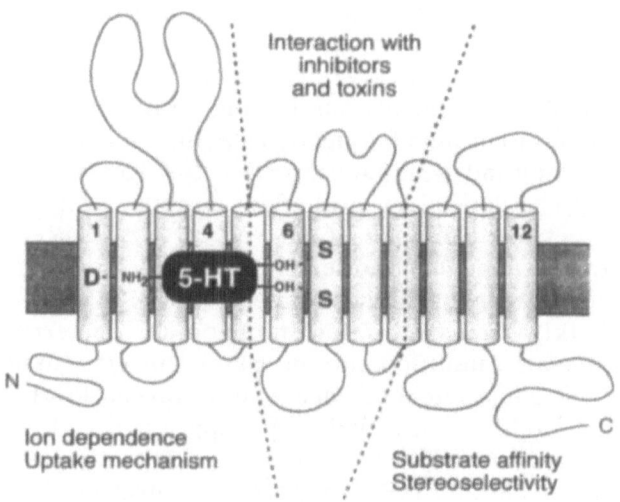

Fig. 2. Hypothetical functional domains of the 5-HT transporter

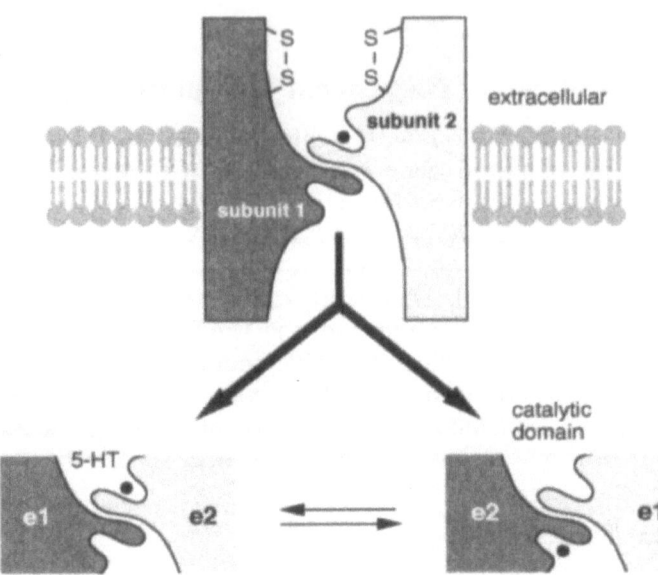

Fig. 3. Quarternary structure of the 5-HTT. The section is across two or four subunits that form the 5-HTT. The complex forms a cavity that extents across the bilayer, and the catalytic domain is at the center of the cavity. Each subunit contributes one transport site to the complex. Subunit 1 is shown in an e1 state while subunit 2 is in an e2 state binding extracellular 5-HT. When subunit 2 translocates bound 5-HT across the membrane, subunit 1 must undergo the antiparallel conformational change, and vice versa. The structure of each subunit is stabilized by extracellular, internal disulfide bridges

primary transcript is produced from mouse and rat cDNA while several mRNA species can be detected in humans. Although some variations exist, the structure of the 5-HTT protein is best described by a model with 12 transmembrane segments, intracellular amino and carboxyl termini, and an extended extracellular loop with several potential N-linked glycosylation sites between transmembrane segments 3 and 4 (Fig. 2). While it is most likely that this loop contributes to protein folding and stability, it may also participate in intracellular traficking and membrane insertion. N-Glycosylation of the 5-HTT protein is likely to be dependent on the biosynthetic constraints imposed by the 5-HTT expressing cell type (QIAN et al. 1995). Recent work has demonstrated that treatment of transiently expressed recombinant 5-HTT with sulfhydryl oxidizing agents generates dimeric (130–180kDa) and tetrameric adducts (220–270kDa), thus indicating an oligomeric quaternary structure of the 5-HTT protein after insertion into the membrane (JESS et al. 1996; Fig. 3).

The 5-HTT is subject to both acute and chronic levels of regulation. Posttranslational modification including phospporylation/dephospporylation through protein kinase C (PKC) and calmodulin-dependent protein kinase as well as via nitric oxide and cGMP represent central mechanisms of acute modulation of 5-HTT function (JAYANTHI et al. 1994; LAUNAY et al. 1994;

MILLER and HOFFMAN 1994). Long-term adaptation in 5-HTT function is likely to be more a consequence of regulation at the level of gene expression (see Sects. F.I–IV).

III. Electrogenicity

According to the classical model presented by NELSON and RUDNICK (1979), for each transport cycle, one Na^+ and one Cl^- are cotransported with each positively charged 5-HT molecule, while one K^+ is countertransported. 5-HT transport can also function in reverse when external K^+ concentration rises, and the cell is depolarized. This stoichiometry predicts that no net charge crosses the cell membrane during the transport cycle. However, recent studies using voltage clamp analysis indicate that the 5-HT uptake is potentially electrogenic, and that the 5-HTT (and several other structurally related transporters) displays complex ion channel-like properties.

Three distinct conducting states of the 5-HTT have been described: a steady-state transport-associated current (during the gating process 5–12 elementary charges cross the membrane for each 5-HT molecule transported, depending on the membrane potential), a hyperpolarization-dependent transient inward current, and a small leakage current (MAGER et al. 1994). Each of the three currents are blocked by 5-HT reuptake inhibitors (also see Sect. D). These results reveal that transport mechanisms may not be encompassed by classic carrier models and support an emerging view that transporter-mediated ionic currents contribute to signaling in the nervous system (SONDERS and AMARA 1996). Molecular strategies are likely to provide further insights into the regulation of 5-HT transport as a function of membrane potential, ion gradients, intracellular signal processing, and gene expression.

C. Developmental Biology and Anatomical Distribution

5-HT regulates morphogenetic activities such as cell proliferation, differentiation, and metamorphosis in early embryonic development. These phylogenetically old functions are reiterated in the developing mammalian brain. In addition to defined brain regions, the 5-HTT is expressed in platelets, lymphocytes, gut, lung, placenta, and several other organs. Studies in the developing rat brain indicate that the 5-HTT is present on embryonic day 15 (E15), even before the transformation of serotoninergic growth cones during cortical synaptogenesis (IVGY-MAY et al. 1994). While in adult life 5-HTT expression appears to be restricted to raphe neurons, it has been detected in the cingulate cortex and thalamus during postnatal development of the brain. It has been shown that the dense transient 5-HT innervation of the somatosensory, visual, and auditory cortices originates in the thalamus rather than in the midbrain raphe complex. 5-HT is detected in thalamocortical fibers and most 5-HT cortical labeling disappears after thalamic lesions. While thalamic glutamatergic neurons do not synthesize 5-HT, they take up exogenous 5-HT through transiently expressed 5-HTT located on thalamocortical axons and terminals. Intriguingly, internalized 5-HT might thus be stored and used as a "borrowed transmitter" for serotoninergic signaling or could exert an intraneuronal control on thalamic maturation (LEBRAND et al. 1996).

I. Autoradiography, Immunohistochemistry, and mRNA Expression

Quantitative autoradiographic analyses of radioligand binding (e.g., [³H]cyanoimipramine, [³H]paroxetine, [³H]citalopram, [¹²⁵I]RTI-55) in sections of the adult brain have demonstrated highest densities of 5-HTTs on serotoninergic cell bodies in the midbrain raphe complex and in its projection areas including cortical areas, entorhinal cortex, CA3 region of the hippocampus, amygdala, substantia nigra, caudate putamen, and hypothalamus (CORTES et al. 1988; HRDINA et al. 1989; HENSLER et al. 1994). Unlike the 5-HT$_{1A}$ receptor, the 5-HTT is relatively evenly distributed within the human dorsal raphe subnuclei (STOCKMEIER et al. 1996). In immunoblots of membranes prepared from rat midbrain and cortex, the 5-HTT migrates as a single 76-kDa polypeptide with a relative abundance consistent with the known distribution of 5-HT neurons and axonal projections (QIAN et al. 1995).

Immunocytochemical studies have revealed widespread and heterogeneous distribution of 5-HTT positive processes in the rat brain. High densities of immunoreactivity are detected within the caudate putamen, amygdaloid complex, substantia nigra, ventral pallidum, septal nuclei, interpeduncular nucleus, trigeminal motor nucleus, olfactory nuclei, and cortical areas (QIAN et al. 1995; SUR et al. 1996). High levels of expression are also found in the stratum oriens of area CA3 and to a lesser extent in the stratum oriens of CA1 and the stratum lacunosum moleculare of CA1 and CA3 regions of the hippocampus (SUR et al. 1996). Within the raphe nuclei a moderate to high incidence of labeled processes are observed, and immunopositive cell bodies are detected in the dorsal raphe nucleus. In addition, some immunoreactive fibers are present in the molecular and granular layers of the cerebellum or in the cochlear and olivary nuclei. Although 5-HTT protein expression is not detectable in glial cells, astrocytes may have a developmentally related potential to participate in the 5-HT clearance (DAVE and KIMELBERG 1994). mRNA encoding the 5-HTT is abundant in ascending neurons of the raphe nuclei, especially in the dorsal and median raphe (LESCH et al. 1993b; MCLAUGHLIN et al. 1996). 5-HTT mRNA is expressed at the highest levels in ventral and ventrolateral subregions of the dorsal raphe, with lower levels of expression in the median raphe, oral pontine reticular nuclei, and the supralemniscal cell groups.

II. In Vivo Imaging

Some ligands for the 5-HTT have been developed for use in imaging studies such as single-photon emission tomography (SPECT) and positron emission tomography (PET). [^{123}I]2β-Carbomethoxy-3β(4-iodophenyl)tropane (β-CIT; also designated RTI-55), a potent analog of cocaine with high binding affinity and selectivity for both the 5-HTT and the dopamine transporter, has been used to visualize the terminals of serotoninergic (and dopaminergic) neurons in nonhuman primate and human brain (STANLEY et al. 1982; STALEY et al. 1994; Fig. 4). SPECT experiments in nonhuman primates show that [^{123}I]β-CIT in vivo binding to dopamine transporters have a much slower washout than binding to 5-HTTs (LARUELLE et al. 1993). Another highly potent and 5-HT-selective tropane, 3β-[4(1-methylethenyl)phenyl]-2β-propanoyl-8-azabicyclo[3.2.1]octane, which has a K_i of 0.1 nM at 5-HTT and is 150 times more potent at 5-HTTs vs dopamine transporters and almost 1000 times more potent at 5-HTTs vs norepinephrine transporters, may be advantageous in the study of serotoninergic terminals with PET (DAVIES et al. 1996). Promising preliminary 5-HTT in vivo imaging results have also been obtained with the SPECT tracer [^{123}I]5-iodo-6-nitroquipazine (JAGUST et al. 1996).

D. Pharmacological Properties

I. Drug Interaction

The 5-HTT is regarded as an initial site of action of antidepressant drugs, of several presumably neurotoxic agents including the amphetamine derivatives 3,4-methylenedioxymethamphetamine (MDMA, "ecstasy"), nonstimulant fenfluramine, and the 1-methyl-4-phenylpyridinium ion (MPP$^+$), neurotoxic metabolite of 1-methyl-4-phenyl-1,2,3,6-tetrahydropyridine (MPTP), and its analogs. Tricyclics, such as imipramine or amitriptyline, and the selective 5-HT reuptake inhibitors (SSRIs) fluvoxamine, fluoxetine, sertraline, citalopram, and paroxetine are widely used in the treatment of depression, anxiety, obsessive-compulsive disorder, and eating disorders as well as substance abuse including alcoholism (LESCH and BENGEL 1995; and references therein; Fig. 4). While the therapeutic efficacy of SSRIs is associated with fewer serious side effects and overdose hazards than tricyclic antidepressants, their initial mechanism of action – inhibition of 5-HT reuptake – is not novel (MURPHY et al. 1995). Finding a conceptually reasonable middle course between 5-HT transport, 5-HT subsystem complexity, 5-HT receptor multiplicity, and related neurobiological mechanisms of action underlying the antidepressant effects of these drugs has been extraordinarily challenging; this is particularly the case when the focus of research is on the delayed onset of clinical efficacy reflecting neuroadaptational events during long-term administration (PINEYRO et al. 1994). Although chronic antidepressant treatment enhances the efficiency of serotoninergic signaling (BLIER et al. 1990), recent findings suggest that SSRIs are not effective following acute administration, because inhibition of 5-HT reuptake at the cell body or terminal of the serotoninergic neuron (with high densities of 5-HTT) is likely to result in stimulation of somatodendritic (5-

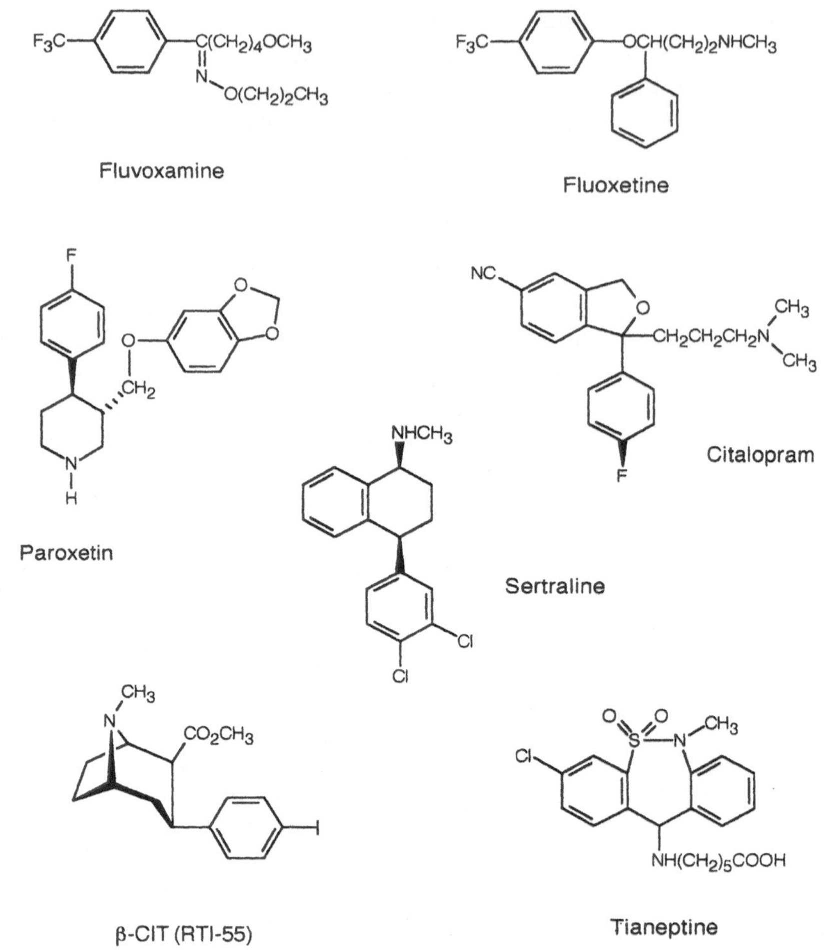

Fluvoxamine

Fluoxetine

Paroxetin

Citalopram

Sertraline

β-CIT (RTI-55)

Tianeptine

Fig. 4. Selective inhibitors and ligands of the 5-HT transporter

HT$_{1A}$) and terminal autoreceptors (5-HT$_{1B}$) which in turn impairs release of 5-HT. Concurrent antagonism of these autoreceptors during 5-HT reuptake blockade may have the potential to accelerate the antidepressant effect of 5-HTT inhibition and may be a practical concept for developing rapidly acting antidepressants (also see Sect. F.III and Fig. 10).

Tianeptine, an atypical tricyclic with a substituted dibenzothiadiazepine nucleus and an aminoheptanoic chain with a terminal acidic group, is a novel 5-HT uptake modulatory compound (for review see Murphy et al. 1995; Fig. 4). Tianeptine appears to enhance 5-HT reuptake in both animal models and humans. Although the mechanism of its 5-HT uptake enhancing action remains elusive, the preclinical and clinical profile of tianeptine is indicative of antidepressant activity.

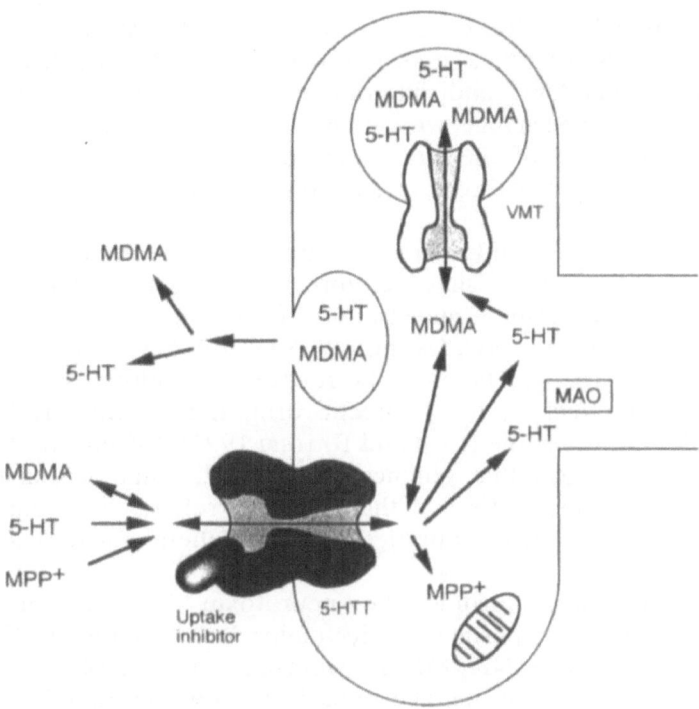

Fig. 5. The role of the 5-HTT in the mechanism of MDMA and MPP⁺ neurotoxicity. Amphetamines and their highly lipophilic, neurotoxic analogs, such as MDMA, are substrates of the 5-HTT, displace 5-HT from its storage vesicle, and ultimately induce 5-HT efflux by a carrier-dependent process. MDMA is therefore a potent releaser of 5-HT. Long-term administration results in toxic degeneration of serotoninergic terminals. MDMA-induced 5-HT release and neurotoxicity can be prevented by tri- and heterocyclic uptake inhibitors. MPP⁺, the neurotoxic metabolite of MPTP, also enters serotoninergic neurons via the 5-HTT. In pathological conditions a dysfunctional transport process may contribute to an increased susceptibility to exogenous MPP⁺-like neurotoxins. The vulnerability to MPP⁺ may be further aggravated by an impaired capacity of the vesicular monoamine transporter (*VMT*), which plays a central role in the sequestration of cytoplasmic toxins and thus in the limitation of mitochondrial damage

II. Toxins and Neurodegeneration

Of particular interest is the mechanism of toxin-induced modification of serotoninergic function and its impact on brain development, event-related synaptic plasticity, and CNS de-/regeneration (LESCH et al. 1996b; and references therein). While tri- and heterocyclic antidepressants inhibit 5-HT reuptake, several agents enhance serotoninergic neurotransmission by facilitating both exocytic and nonexocytic 5-HTT-mediated 5-HT release (Fig. 5). Amphetamines and their highly lipophilic, neurotoxic analogs, such as MDMA, are substrates of the 5-HTT, displace the biogenic amine from its storage vesicle, and ultimately induce 5-HT efflux by a carrier-dependent process and by stimulation of exocytosis (LEVI and RAITERI 1993; RUDNICK and WALL 1993; SCHULDINER et al. 1993). MDMA induces efflux in a stereo-specific, Na$^+$-dependent, and imipramine-sensitive manner characteristic of 5-HTT-mediated exchange, whereas MDMA-evoked vesicular 5-HT efflux is due to dissipation of the transmembrane pH difference generated by ATP hydrolysis and to direct interaction with the vesicular monoamine transporter.

Although the direct and indirect dopaminergic action of MDMA is likely to be related to the mechanism of reinforcement, it is a highly potent releaser of 5-HT and long-term administration results in toxic degeneration of serotoninergic terminals. MDMA-induced 5-HT release and neurotoxicity can be prevented by tri- and heterocyclic reuptake inhibitors. By analogy, the 5-HT releasing effect of the anorectic drug fenfluramine is blocked by clomipramine or fluoxetine (LEVI and RAITERI 1993). Recent studies in trans-fection models suggest that potency toward individual plasma membrane biogenic amine transporters and the ability to release accumulated amine substrates are independent properties of each amphetamine derivative (WALL et al. 1995).

5-HTT substrates include the neurotoxin MPTP, which induces Parkinson's syndrome, particularly its major brain metabolite MPP$^+$, and MPTP analogs such as NH$_2$-MPTP (BUCKMAN et al. 1988; ANDREWS and MURPHY 1993; Fig. 5). The uptake process is saturable, may be blocked by inhibitors of 5-HT uptake, and is accompanied by a decrease in intracellular ATP and an inhibition of mitochondrial state 3 respiration. These findings demonstrate that the dopamine uptake system is not the only means by which potent neurotoxin can be efficiently transported into cells.

III. Structure-Function Relationships

The cloning of multiple species homologs of the 5-HTT provides a unique opportunity for molecular comparisons to identify potential domains and residues involved in transport function. The relationship between the 5-HTT structure and the binding of substrates, ions, and antagonists has been studied in cell models transfected with species-specific 5-HTTs and by pharmacologi-cal comparisons of chimeric human and rodent 5-HTT (Fig. 2). The findings suggest that the substrate 5-HT as well as several structurally similar nontricyclic antidepressants and specific inhibitors of 5-HT transport interact largely with nonpolar and sterically confining environments, and that a par-ticular spatial coordination of the amino and phenyl groups (separated by an alkyl backbone) is important for transport interaction (CHANG et al. 1993). While this aminophenyl coordination motif seems to be a structural requisite for transport interaction, and is therefore likely to be part of the transport pharmacophore, additional phenyl rings present in some of the nontricyclic antidepressants may help to account for their relatively higher affinities in 5-HT transport interaction.

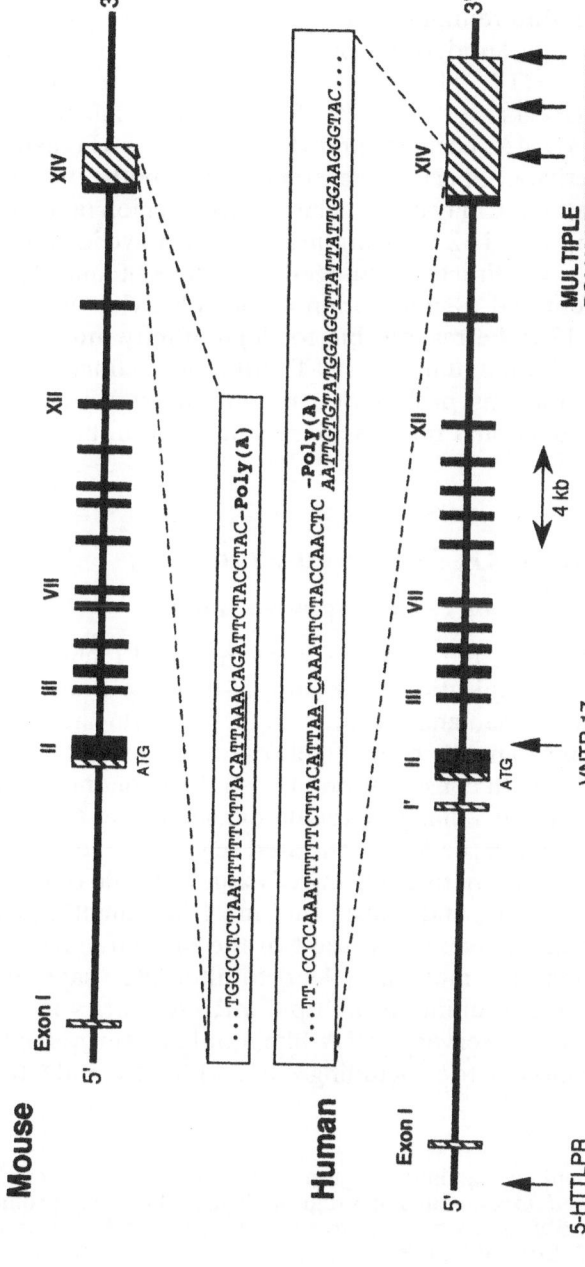

Fig. 6. Genomic organization of the murine and human 5-HTT. Positions of exons (I–XIV; *black boxes*, coding regions; *hatched boxes*, noncoding regions) and translation initiation site (*ATG*) are indicated. Exon 1' is alternatively spliced. Comparison of cDNA and genomic sequences indicate that the mouse signal for mRNA cleavage/polyadenylation is a natural variant which occurs in 12% of the known genes and exhibits 75%–80% of the activity of the classical cleavage/polyadenylation signal (AATAAA). The proximal polyadenylation signal-like motif in the human gene is considerably divergent, which is likely to lead to cell-specific alternate usage of additional polyadenylation sites and result in multiple mRNA species in humans

A possible molecular explanation for the distinct pharmacological actions of various antidepressants (and other ligands of the 5-HTT) has been provided by Schloss and Betz (1995) who reported the existence of two distinct binding sites with differential sensitivity for tricyclic and nontricyclic antidepressants. Most tricyclic antidepressants are significantly more potent at the human than the rat 5-HTT, whereas D-amphetamine is a more potent inhibitor of rat 5-HTT (Barker et al. 1994). Several other ligands such as fluoxetine, paroxetine, MDMA, cocaine, and the substrate 5-HT exhibit no significant species selectivity. Species preferences for both the tricyclic imipramine and D-amphetamine may be associated with a region in or near the transmembrane domains 11 and 12 as being involved in tricyclic antidepressant recognition. Using site-directed mutagenesis, Barker and Blakely (1995) have recently identified a single amino acid, phenylalanine 586, in transmembrane domain 12 to be responsible for high-affinity interactions of tricyclic antidepressants with the human 5-HTT. Although refined molecular tools are gaining momentum as probes of structure-activity relationships, structural domains with functional relevance are still far from being conclusively elucidated.

E. Gene Organization: Of Mice and Men

I. Chromosomal Mapping and Gene Organization

Progress is also being made in the elucidation of the 5-HTT chromosomal localization and genomic organization, including 5'-flanking transcriptional regulatory sequences. The human and murine 5-HTT genes (located on chromosome 17q12 and chromosome 11, respectively) are composed of 14 exons spanning approx. 35 kb and with conservation of both the exon/intron organization and to a lesser extent the 5'-flanking regulatory sequences (Lesch et al. 1994; Fig. 6) Length variation in the 5' untranslated region of midbrain raphe 5-HTT mRNA results from alternative splicing of exon 1'. While only a single mRNA species occurs in rats and mice, the most proximal signal for polyadenylation in the human gene is a variant of the rat and murine motif. This polyadenylation signal-like motif may lead to alternate usage of additional polyadenylation sites, resulting in multiple mRNA species in humans (Heils et al. 1995). A highly conserved TATA-like motif and several potential binding sites for transcription factors including AP1, AP2, SP1, and CRE-like

Fig. 7. Structure of the 5'-flanking regulatory region of the human 5-HTT gene. *Black boxes*, coding regions; *hatched boxes*, noncoding regions. The 5-HTT gene promotor is defined by a TATA-like motif, and several potential binding sites for transcription factors including AP1, AP2, SP1, and a CRE-like motif are present in the 5'-flanking region. The position (*shaded boxes*) and consensus sequence (*open boxes*) of the 5-HTTLPR are indicated. The potential of the 5-HTTLPR to form G4-DNA is also depicted (also see Sect. G.V)

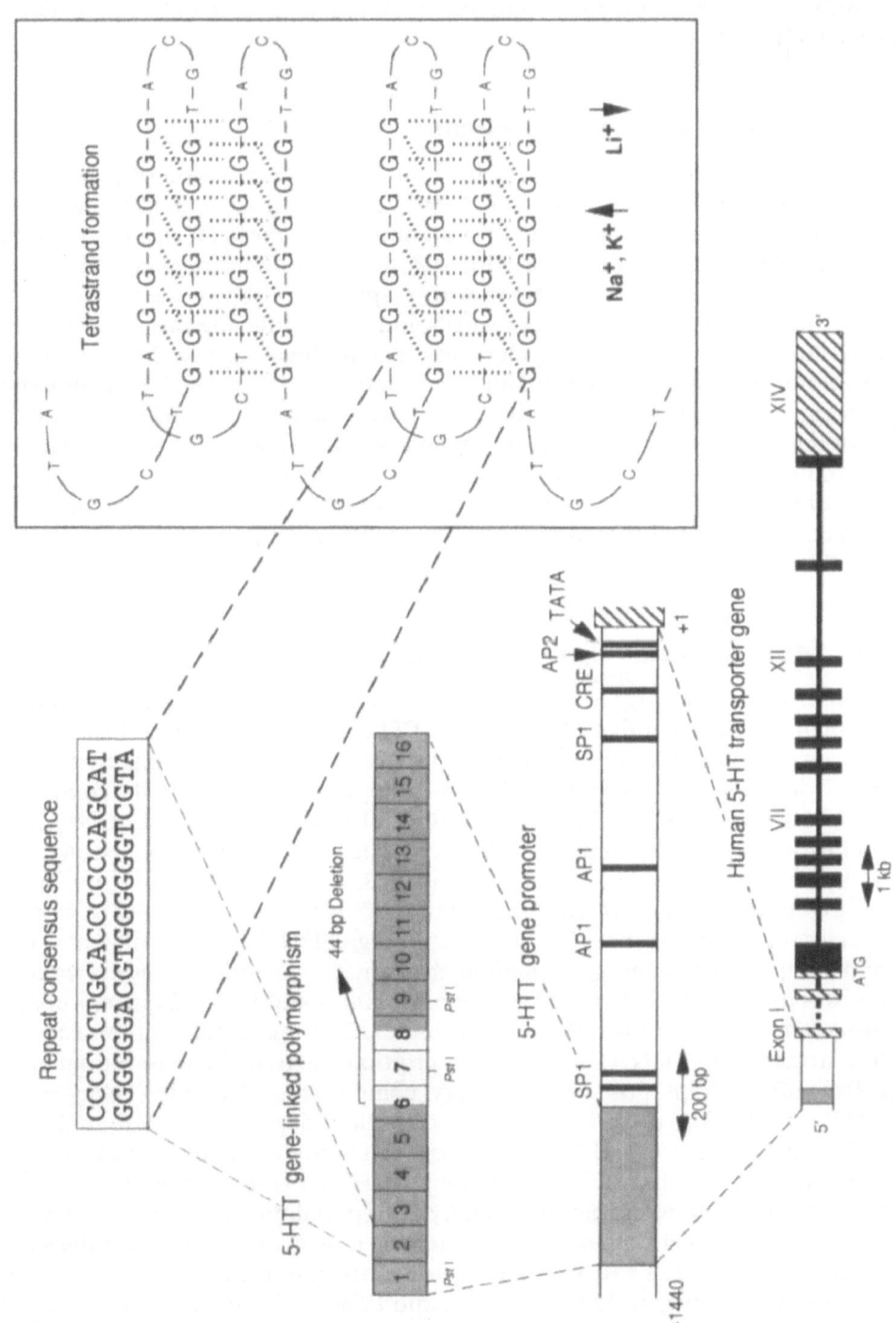

motifs are present in the 5'-flanking regulatory (promoter) region across species (Fig. 7).

II. Regulation of Gene Expression

The role of these transcription factor motifs in the regulation of human and murine 5-HTT gene expression has been studied by transfection studies using reporter gene fusion constructs of 5'-flanking sequences (Heils et al. 1995). Functional promoter mapping with deletional mutants revealed that core promoter sequences are contained within approx. 100 bp of the transcription start site, and that regulation of cAMP- and PKC-inducible promoter activity depends on multiple *cis*-acting elements. The findings suggest that the 5-HTT promoter is active in primary and continuous cell lines which constitutively express 5-HTT, that the information contained within approx. 2 kb of 5'-flanking sequence is sufficient to confer its cell-specific expression, that the promoter responds to cAMP/PKA and PKC induction, and that the expression of the 5-HTT gene is regulated by a combination of positive and negative *cis*-acting elements operating through a basal promoter unit defined by a TATA-like motif.

III. 5-HTT Linked Polymorphic Region

Transcriptional activity of the human 5-HTT gene promoter is modulated by a polymorphic repetitive element (5-HTTLPR). The 5-HTTLPR is located approximately 1 kb upstream of the 5-HTT gene transcription initiation site in a GC-rich region and is composed of 14–16 repeat elements with the consensus sequence CCCCCCTGCACCCCCCAGCAT (Fig. 7; Heils et al. 1996). The highly prevalent polymorphism, which consists of the presence or absence of a 44-bp segment involving repeat elements 6–8, is located close to a "hot spot" for deletion mutagenesis. Comparison of the 5-HTT gene promoter in various species revealed that the 5-HTTLPR is "foreign" DNA, possibly of viral origin and unique to humans and nonhuman primates including *Macaca mulatta* (rhesus monkeys; Bengel et al. 1997a,b; A. Heils and K.P. Lesch, in preparation). The 5-HTTTLPR may form a complex G-string-dependent tetrastrand-like structure (G4-DNA), silences transcriptional activity in nonserotoninergic cells, and contains positive regulatory components elements (Heils et al. 1995). In addition, the 5-HTTLPR confers allele-dependent differential transcriptional activity on 5-HTT gene promoter activity when fused to a luciferase reporter gene and transfected into human 5-HTT expressing cell lines (Heils et al. 1996). PCR-based genotype analysis in populations of varying ethnicity has revealed allele frequencies of 57% for the long allele and 43% for the short allele. The 5-HTTLPR genotypes are distributed according to Hardy-Weinberg equilibrium: 32% l/l, 49% l/s, and 19% s/s. A similar allele frequency in *Macaca mulatta* indicates evolutionary preservation of genotype distribution, which may reflect a gain of function with potential relevance to social interaction in human and nonhuman primate populations.

IV. Allelic Variation of Functional 5-HTT Expression

The effects of 5-HTTLPR variability on functional gene expression has been determined by studying the relationship between 5-HTTLPR genotype, 5-HTT gene transcription, and 5-HT uptake activity in human lymphoblastoid cell lines (LESCH et al. 1996a). Lymphoblasts constitutively express functional 5-HTT, display 5-HT/5-HT$_{1A}$ receptor-dependent proliferative responses, exhibit cAMP- and PKC-dependent 5-HTT gene regulation, and permit expression studies of the native 5-HTT gene in cell lines cultured from subjects with different 5-HTTLPR genotypes. The effect of the 5-HTTLPR on 5-HTT expression at the mRNA level has been assessed by competitive PCR and at the protein level by [^{125}I]RTI-55 binding and [^{3}H]5-HT uptake experiments. Cells homozygous for the l form of the 5-HTTLPR produce higher concentrations of 5-HTT mRNA than do cells containing one or two copies of the s form. Membrane preparations from l/l lymphoblasts bind more [^{125}I]RTI-55 than do s/s cells. Furthermore, the rate of imipramine-sensitive [^{3}H]5-HT uptake is about twice as high as in cells homozygous for the l form of the 5-HTTLPR as in cells carrying one or two endogenous copies of the s form of the promoter. The genotype-dependent differences in mRNA concentrations, [^{125}I]RTI-55 binding, and [^{3}H]5-HT uptake persists proportionally when 5-HTT gene transcription is induced with forskolin or phorbol myristate acetate.

The following section summarizes the converging lines of evidence that allelic variation in functional 5-HTT expression is likely to play a role in developmental neuroplasticity, thus determining the expression of complex traits and their associated behavior throughout adult life. Moreover, it is becoming increasingly evident that allelic variation of 5-HTT function, in conjunction with other predisposing genetic factors and with inadequate adaptive responses to environmental stressors, is likely to contribute to the etiopathogenesis of depression, anxiety disorders, alcohol abuse, and neurodegenerative processes.

F. Genetics: Listening to the Gene

I. Mutation Screening

5-HTT function appears to be dysregulated in a variety of complex behavioral traits and disorders such as depression, bipolar, anxiety, obsessive-compulsive, schizophrenic, and neurodegenerative disorders, autism, substance abuse, and eating disorders (BRILEY et al. 1980; PAUL et al. 1981; for review see ELLIS and SALMOND 1994; OWEN and NEMEROFF 1994). Moreover, 5-HT uptake capacity has been proposed to be a trait variable for affective disorder since it remains decreased after recovery. Twin studies suggest that 5-HT uptake is genetically controlled (Meltzer and Arora 1988), which is further supported by the fact that 5-HTT function is lower in first-degree relatives of patients with depression who themselves display decreased 5-HT uptake. The hypothesis that genetic control of and disease-related alteration in 5-HTT function is more likely to be related to differential regulation of 5-HTT expression than to amino acid substitutions is further supported by the fact that mutation screening of the gene's translated exons in several samples of patients with affective spectrum and obsessive-compulsive disorders detected only rare coding variants (Lesch et al. 1995; Altemus et al. 1996; Di Bella et al. 1996).

II. 5-HTT Linked Polymorphic Region and Personality Traits: A Susceptibility Locus?

Following the demonstration of allelic variation in functional expression, the 5-HTT has assumed importance as a piece in the mosaiclike texture of anxiety-related personality traits. Anxiety is a fundamental, enduring, and continuously distributed dimension of normal human personality, with a heritability of 40%–60% (Bouchard 1994; and references therein). The contribution of 5-HTTLPR variability to individual phenotypic differences in temperament was determined in a combined population and family genetic study on two independently collected groups of human subjects ($n = 505$) consisting of predominantly male siblings and other family members (Lesch et al. 1996a). Anxiety-related and other personality traits have been assessed by the NEO personality inventory (NEO-PI-R), a self-report inventory based on the five-factor model of personality. Both independently collected study populations displayed a significant association between 5-HTTLPR genotype and neuroticism (Fig. 8). Individuals with either one or two copies of the s form of the 5-HTTLPR (referred to together as group S) had higher

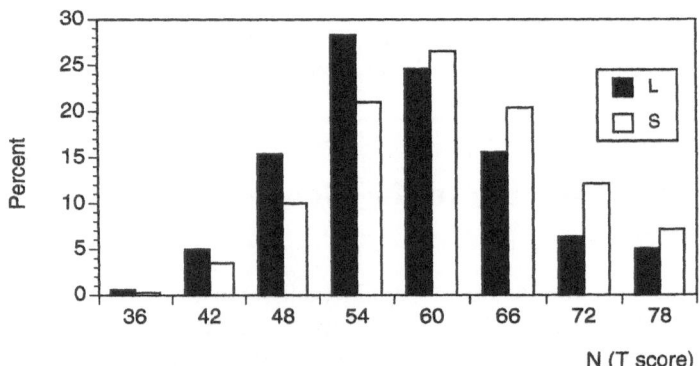

Fig. 8. Distributions of neuroticism (NEO-PI-R factor N) scores. *X-axis*, neuroticism scores separated into eight groups with the indicated median T scores; *Y-axis*, the distribution (percentage) of subjects from the L ($n = 163$) and S ($n = 342$) groups in each of the eight T score groups (Lesch et al. 1996)

neuroticism scores than individuals homozygous for the l variant of the 5-HTTLPR (group L).

The scores for the l/s and s/s genotypes were not significantly different, indicating that the polymorphism has more of a dominant/recessive than of a codominant/additive effect on neuroticism, just as it does on 5-HTT gene expression. The NEO-PI-R is based on a hierarchical model in which each personality factor is comprised of several related facets. Analysis of the six facets of neuroticism showed significant associations between 5-HTTLPR genotype and the facets of anxiety, angry hostility, depression and impulsiveness, but not self-consciousness or vulnerability. The association between 5-HTTLPR genotype and personality traits was also analyzed in 78 sib pairs who were discordant for the 5-HTTLPR. Despite the reduction in sample size, the difference between the L and S siblings was statistically significant even after conservatively correcting for the nonindependence of sib pairs from the same family. These family studies show that the observed associations between 5-HTTLPR genotype and anxiety-related traits are due to genetic transmission rather than population stratification.

These results show that the 5-HTTLPR affects a constellation of traits related to anxiety and depression. Allelic variation in functional 5-HTT expression contributes a modest but replicable 3%–4% of the total variance and 7%–9% of the genetic variance, based on estimates from twin studies using these and related measures which have consistently demonstrated that genetic factors contribute 40%–60% of the variance in neuroticism, harm avoidance, and other anxiety-related personality traits in large population samples (BOUCHARD 1994). The associations represent only a small portion of the genetic contribution to anxiety-related personality traits. If other genes are hypothesized to contribute similar gene-dosage effects to anxiety, approximately 10–15 genes might be predicted to be involved. Small, additive, or interactive contributions of this size have in fact been found in studies of other quantitative traits (LANDER and SCHORK 1994; and references therein). As other anxiety-related genes are identified, including perhaps some with effects that are larger than or interact with this polymorphism, it might become possible to use this information to enhance individualized pharmacological treatment of psychiatric disorders.

III. Affective Disorders

The impact of genetic and environmental factors on 5-HTT function and its consequences for synaptic plasticity of the developing and mature brain in psychiatric disorders and during therapeutic intervention is of particular interest. Genetic susceptibility factors contribute considerably to the etiology of affective illness, including bipolar affective disorder and unipolar depression, and heritability is estimated be up to 86% for bipolar disorder and 30%–37% for depression. Although the two disorders may share factors which determine depressive symptoms, with an associated high risk of suicide, genetic heterogeneity, and a substantial but varying environmental component, complicates identification of predisposing genes. Linkage between the 5-HTT locus on human chromosome 17q12 and bipolar affective disorder has not been detected although individual families with LOD scores of approximately 1 have been observed (Kelsoe et al. 1996). This is not necessarily unexpected since linkage analysis does not have the sensitivity to detect quantitative trait or vulnerability loci with modest effects and several other putative loci for bipolar affective disorder may obscure 5-HTTLPR effects.

Previous analysis of the 5-HTT gene for polymorphic variants revealed, in addition to some rare polymorphisms which are not associated with disease (Lesch et al. 1995; Altemus et al. 1996; Di Bella et al. 1996), a variable-number tandem repeats (VNTR-17) located in intron 2 with two common (with 10 or 12 repeats) and one rare allele (with 9 repeats; Lesch et al. 1994). This rare allele is reported to be associated with unipolar but not bipolar affective disorder (Ogilvie et al. 1996). While this finding has not been replicated (Stöber et al. 1996), analysis of a larger sample of subjects with affective disorders indicates a strong association between allele 12 and bipolar disorder (Collier et al. 1996a).

The effect of 5-HTTLPR-dependent variability in functional 5-HTT expression on susceptibility to affective illness has been assessed in a case-control association study in 454 patients with bipolar or unipolar affective disorder and 570 controls, derived from three different European centers (Collier et al. 1996b). In all three centers the frequency of the low activity allele (short 5-HTTLPR) was higher in patients than in controls. Although the differences for each individual center were not individually significant, a stratified analysis of all three samples gave a significant overall odds ratio of 1.23. The excess of the homozygous short 5-HTTLPR genotype among the patients was even greater with an odds ratio of 1.53 and a population-attributable risk of 29%. The allelic association between the intron 2 VNTR of the 5-HTT gene and affective disorders in these studies does not appear to be entirely a result of linkage disequilibrium between the two polymorphisms since only modest evidence for linkage disequilibrium between the intronic VNTR and 5-HTTLPR in patients with affective disorder and controls has been detected. These findings indicate that the low transporter-mediated 5-HT uptake function resulting from reduced activity of the 5-HTT gene's transcriptional apparatus increases susceptibility to both bipolar affective disorder and unipolar depression. Although the increase in risk for homozygotes for the low 5-HT uptake activity allele is relatively modest, the observation of a similar effect of this allele in both unipolar depression and bipolar disorder may indicate that affective spectrum disorders share a general genetic susceptibility locus for anxiety and depressive symptoms. This notion has been predicted by multiple studies reporting that individuals who score high in measures of neuroticism, harm avoidance, and other anxiety-related personality traits display an increased risk in developing affective illness and associated spectrum disorder such as alcoholism (Fig. 9).

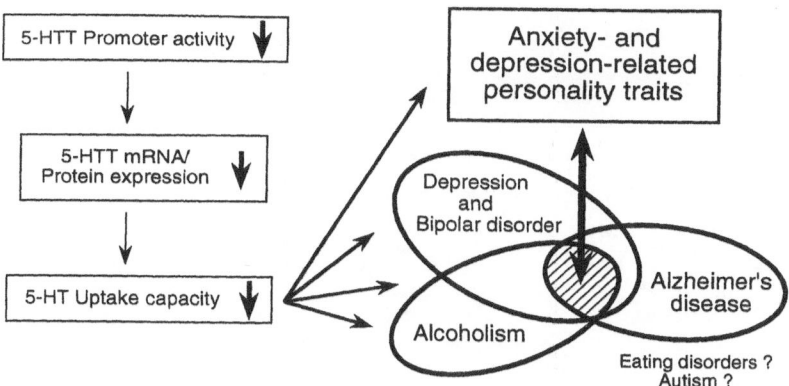

Fig. 9. Genetically driven low 5-HT transporter function as one of the determinants of anxiety- and depression-related personality traits and as a susceptibility factor for neuropsychiatric disorders

The findings that individuals with the short 5-HTTLPR allele and reduced 5-HTT function have greater anxiety-related personality characteristics and are at risk to develop affective illness would at first seem to conflict with the fact that tricyclic antidepressants and SSRIs such as fluoxetine, which competitively inhibit 5-HT uptake, are therapeutic agents in anxiety and depressive disorders. This apparent contradiction may be explained by regional variation and the complex autoregulatory processes which are operational in different brain areas (ROUTLEDGE and MIDDLEMISS 1996; Fig. 10). The genetically driven impaired ability of low 5-HTT function for a rapid 5-HT clearance following release into the synaptic cleft may elicit acute increases of 5-HT in the vicinity of serotoninergic cell bodies and dendrites in the raphe complex and may exert a somatodendritic 5-HT_{1A} receptor-mediated negative feedback that leads to an overall decrease in 5-HT neurotransmission in depressed patients with short 5-HTTLPR alleles. On the other hand, chronic antidepressant treatment induces adaptive changes in the $5\text{-HTT}/5\text{-HT}_{1A}$ and 5-HT_{1B} receptor-modulated negative feedback regulation that eventually leads to an overall enhancement of terminal 5-HT release (Fig. 10).

IV. Alcoholism

Alcohol dependence is an etiologically and clinically heterogeneous syndrome caused by a complex interaction of genetic and environmental factors. Although genetic liability to alcohol dependence varies widely among individuals, heritable factors account for about 50% of the total variance of alcohol dependence in men and about 25% in women. Therefore, differentiating psychobiological traits of addictive behavior is particularly important for dissection of the complex genetic susceptibility of alcohol dependence (LANDER and SCHORK 1994). Based on differences in genetic involvement, various

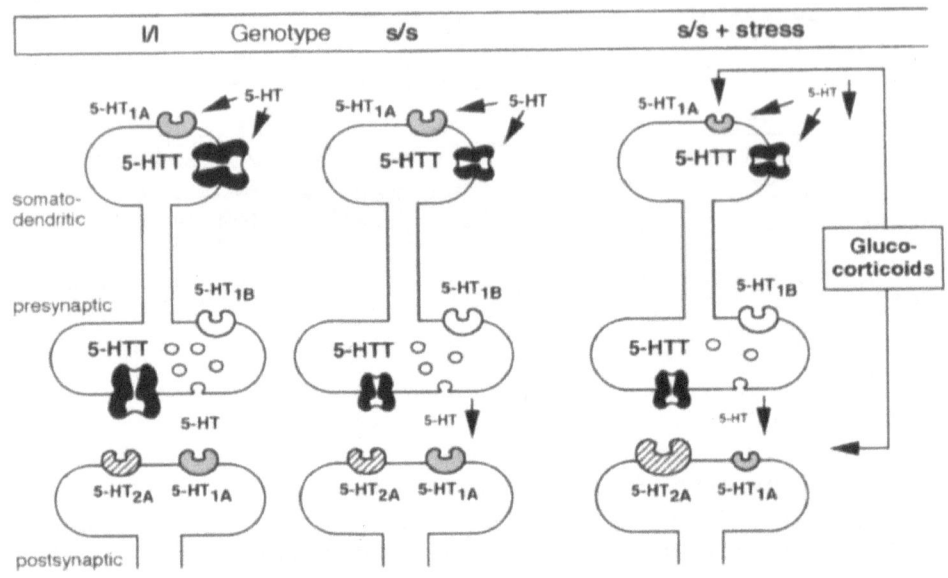

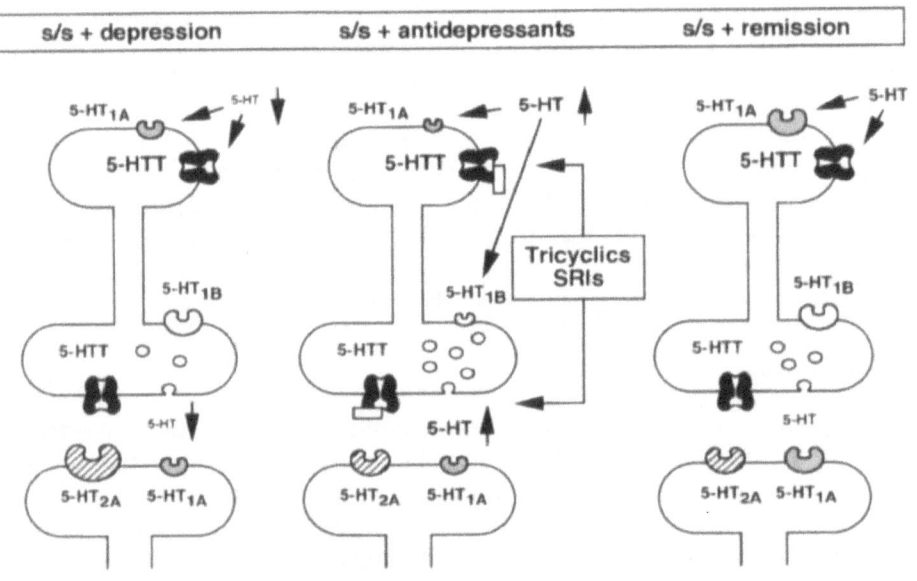

typologies of alcohol dependence have been proposed. The factors of impulsive-aggressive behavior and early onset define a subgroup of alcoholics (type 2) with strong genetic predisposition to alcohol dependence (CLONINGER 1987). A genetically based central serotoninergic deficit is thought to be involved in the developmental pathways of alcohol preference and dependence by modulating motivational behavior and neuroadaptive processes.

Studies in human alcoholics have demonstrated lower concentrations of 5-HT metabolites in cerebrospinal fluid and low platelet 5-HT content. 5-HT related impulsive, aggressive, and suicidal behavior has been linked to a primordial personality that is susceptible to alcoholism. Type 2 alcoholics are characterized by impulsive aggression (CLONINGER 1987), suggesting a link between alcohol-seeking behavior and low central 5-HT. Studies have assessed the role of allelic variation in functional 5-HTT expression in the genetic load for alcoholism and have found a trend toward greater expression of the homozygous genotype of the short 5-HTTLPR in patients with alcohol dependence ($n = 315$) than in controls ($n = 216$; SANDER et al. 1997). When the overall sample was subdivided into patients with type 2 alcoholism and patients with a history of withdrawal seizures or delirium, the gene effect reached significance, with odds ratios between 1.4 and 2.3. These modest but consistently significant genotypic associations emphasize the need for further studies to confirm that low 5-HTT function confers susceptibility to severe alcohol dependence (Fig. 9).

V. Neurodevelopmental and Neurodegenerative Disorders

Disturbances in functional 5-HTT expression are also likely to play a role in the pathophysiology of neurodevelopmental disorders such schizophrenia and autism (PIVEN et al. 1991; JOYCE et al. 1993), and there is a considerable body of evidence indicating an extensive loss of both cholinergic and 5-HT neurons

◄───────────────────────────────

Fig. 10. The impact of allelic variation of functional 5-HTT expression in 5-HT dependent, stress-precipitated depression and during long-term antidepressant drug treatment on adaptive autoregulation of 5-HT neurotransmission. Anxiety-related personality characteristics and the risk of developing affective illness of individuals with the short 5-HTTLPR allele and reduced 5-HTT function are a consequence of regional variation of 5-HT receptor subtype and 5-HTT distribution that underlie complex autoregulatory processes operational in different brain areas. The genetically driven impaired ability of low 5-HTT function for a rapid 5-HT clearance following release into the synaptic cleft may elicit acute increases of 5-HT in the vicinity of serotoninergic cell bodies and dendrites in the raphe complex and may exert a somatodendritic 5-HT$_{1A}$ receptor-mediated negative feedback that leads to an overall decrease in 5-HT neurotransmission in depressed patients with short 5-HTTLPR alleles. This mechanism may also explain the apparent contradiction that tricyclic antidepressants and SSRIs, which competitively inhibit 5-HT uptake, are therapeutic agents in anxiety and depressive disorders. Chronic antidepressant treatment induces adaptive changes in the 5-HTT/5-HT$_{1A}$ and 5-HT$_{1B}$ receptor-modulated negative feedback regulation that eventually leads to an overall enhancement of terminal 5-HT release in the brain of patients with Alzheimer's disease. Cholinergic-serotoninergic interactions based a 5-HT$_{1A}$ receptor-mediated modulation of septal cholinergic neurons have been suggested to play a central role in learning and memory and in the appearance of age-dependent or neurodegenerative cognitive deficits (Kia et al. 1996; and references therein). Unlike early-onset familial Alzheimer's disease with autosomal dominant mutations in the APP and Presenilin 1 and 2 genes, the late-onset disease (age at onset >60 years) is a common (up to 95% of all cases) complex disorder with a multifactorial and polygenic etiology. Alzheimer's disease associated reductions in the number of 5-HTTs have been reported in the dorsal raphe nucleus, hippocampus, entorhinal cortex, and platelets, the latter indicating that this effect is not an epiphenomenon of the disease process. In a population-based association study the frequency of the low-activity allele of the 5-HTTLPR was increased in patients with late-onset Alzheimer's disease ($n = 196$), and an excess of the homozygous genotype of the short 5-HTTLPR was found in comparison to controls ($n = 271$), with a significant odds ratio of 1.7 and a population-attributable risk of 33% (Li et al. 1997). The association was unrelated to age and to ε4 alleles of the APOE gene. These results indicate that central 5-HT function related to allelic variation in functional 5-HTT expression may be involved in the pathogenetic mechanism of cognitive deficits observed in both severe depression and senile dementias, and may be a risk factor for late-onset Alzheimer's disease (Fig. 9).

G. Talking Back to the Gene: Target or Tool?

I. Glucocorticoids

A variety of physiological factors regulate gene expression, and considerable progress has been made regarding the molecular mechanisms of regulation achieved by steroid hormones, thyroid hormone, retinoic acid, and vitamin D$_3$, all of which bind to receptors of the same gene family. Pre- and perinatal stress or exposure to excess glucocorticoids is known to alter central nervous system function and to result in lasting changes in reactions to stress. While glucocorticoids and serotoninergic neurons exert reciprocal regulatory actions, the contribution of 5-HTT is complex, and data on its regulation are still far from conclusive (Owen and Nemeroff 1994; Fumagalli et al. 1996; Slotkin et al. 1996a,b). It has previously been reported that glucocorticoids modulate 5-HT uptake and inhibitor binding (Tukiainen 1981; Arora and Meltzer 1986). Dexamethasone seems not to have a consistent effect on functional expression of the 5-HTT in the adult brain, although it decreases 5-HTT mRNA expression in both young and old rats, and there is now converging evidence that elevated glucocorticoids are unlikely to be responsible for reduced 5-HTTs in depression. Interestingly, dexamethasone given to pregnant rats on gestational days 17–19 produces a dose-dependent retardation of brainstem growth but induces a significant elevation of [^{3}H]paroxetine binding in offsprings after birth, with the peak effect at the intermediate dose that persists into young adulthood. Since this effect is not observed after postnatal treatment or by treatment in adulthood, it likely represents specific programming by glucocorticoids during the prenatal period (SLOTKIN et al. 1996a). Dysregulated 5-HTT expression may contribute to alterations of synaptic function that ultimately produce the psychobiological abnormalities seen after stress exposure. Sequence analysis of the 5'-flanking regulatory region of the 5-HTT gene revealed the presence of glucocorticoid responsive elements, and studies with the 5-HTTLPR-containing promoter indicate a pronounced allelic variability in glucocorticoid-dependent (but not in estrogen-dependent) induction of 5-HTT promoter activity (HEILS, MÖSSNER, and LESCH, unpublished).

II. 5-HT Reuptake Inhibitors

Several lines of evidence from kinetic studies of 5-HT transport and uptake inhibitor binding assays have suggested that chronic but not short-term treatment with mood-stabilizing drugs and psychomotor stimulants have a time-dependent regulatory effect on monoamine transporter systems. However, both approaches suffer from technical difficulties and the lack of selectivity for the 5-HTT, which has contributed to considerable controversy (PINEYRO et al. 1994). To test the hypothesis of antidepressant-induced modulation of the 5-HT uptake mechanism more directly and to determine whether pharmacological manipulation regulates the 5-HTT at the transcriptional level, 5-HTT mRNA expression was examined in the midbrain raphe complex by quantitative northern blot analysis (LESCH et al. 1993b). The effects of chronic treatment with a variety of monoamine uptake inhibiting antidepressant drugs including several tricyclics (imipramine, desipramine, and clomipramine), the SSRI fluoxetine, the monoamine oxidase (MAO) A inhibitor clorgyline, and several 5-HT receptor subtype agonists on 5-HTT mRNA expression were assessed. Long-term (21-day) administration of the nonselective 5-HT/norepinephrine uptake inhibitor imipramine robustly decreased 5-HTT mRNA. The highly selective 5-HT uptake inhibitor fluoxetine also significantly diminished 5-HTT mRNA expression. While a similar magnitude of reduction was found during chronic treatment with clomipramine, which preferentially inhibits the 5-HTT, and the relatively selective norepinephrine uptake inhibitor desipramine, it reached significance only at the trend level. The MAO-A inhibitor clorgyline, the nonselective 5-HT receptor agonist m-chlorophenylpiperazine, the selective 5-HT$_{1A}$ agonist 8-hydroxy-2-(di-n-propyl-amino)tetraline, and 5-HT$_{2A/C}$ agonist 1-(2,5-dimethoxy-4-iodophenyl)-2-aminopropane exhibited no effect. The findings provide direct evidence that 5-HTT mRNA transcription and/or stability is regulated by chronic treatment with selective and nonselective uptake inhibitors.

At least two mechanisms may be responsible for downregulation of 5-HTT expression. One possibility is that 5-HTT inhibitors acutely increase the availability of 5-HT in the synaptic cleft and produce, depending on the drug class, multiple neuroadaptational alterations involving receptors, regulatory proteins of signal transduction pathways, and transporter systems. In vivo electrophysiological studies in the rat have demonstrated that chronic administration of tricyclic antidepressants enhance synaptic transmission by increasing the sensitivity of postsynaptic 5-HT_{1A} receptors, while selective 5-HT transport blockers induce this effect by desensitizing terminal 5-HT_{1B} autoreceptors (Fig. 10). In analogy to agonist-induced receptor downregulation, increased 5-HT is likely to modify uptake and subsequent synaptic release of 5-HT by homologous regulation of its transporter protein involving both transcriptional and posttranscriptional components. An alternate possibility is that tricyclic and heterocyclic antidepressants exert a direct effect on the regulation of 5-HTT gene transcription.

A possible site of such direct action is the midbrain raphe complex, the region with the highest density of the 5-HTT. In agreement with this hypothesis it has recently been reported that mianserine directly modulates 5-HT_{2A} gene transcription (TOTH and SHENK 1994), and that tricyclic antidepressants including imipramine and desipramine alter type II glucocorticoid receptor mRNA levels in rat brain and increase the activity of the human glucocorticoid receptor gene promoter (PEIFFER et al. 1991; PEPIN et al. 1991). A recent study suggests a sequential decrease in 5-HTT and 5-HT_{1B} mRNA expression during chronic fluoxetine treatment. Fluoxetine reduced dorsal raphe 5-HT_{1B} mRNA levels in a time-dependent manner whereas 5-HTT mRNA was only transiently decreased within 21 days of treatment (NEUMAIER et al. 1996a). This reduction in 5-HT_{1B} mRNA was specific to dorsal raphe nucleus and was not found in several postsynaptic (nonserotoninergic) regions.

These results further support the notion that long-term antidepressant treatment increases 5-HT release from axonal terminals by downregulating the mRNA coding for presynaptic 5-HT_{1B} autoreceptors (NEUMAIER et al. 1996a,b). Interestingly, the 5-HTT mRNA decreasing effects of imipramine- and p-chlorophenylalanine-(PCPA-) induced 5-HT depletion of 5-HT appear to be additive in rat brain (LINNET et al. 1995). Regardless of mechanism, mRNA transcription, processing, transport and translation may be central in the regulation of 5-HTT and 5-HT_{1B} function and its modulation during long-term administration of drugs with therapeutic potential in affective disorders (Fig. 10).

III. Other Compounds and Environmental Factors

Several studies report a complex spatiotemporal modulation of functional 5-HTT mRNA expression following treatment with the irreversible tryptophan hydroxylase inhibitor PCPA (LINNET et al. 1995; YU et al. 1995; RATTRAY et al. 1996). PCPA first rapidly downregulates 5-HTT mRNA levels in 5-HT cell bodies within the ventromedial but not the dorsomedial portion of the dorsal raphe nucleus but then increases 5-HTT mRNA levels in both regions. While no change in inhibitor binding and synaptosomal 5-HT uptake in terminal regions is found after short-term treatment, 14 days of PCPA administration significantly reduces both these parameters by approximately 20% in the hippocampus and in cerebral cortex. The striatum shows a lower sensitivity to this effect, and no significant changes are observed in the levels of inhibitor binding to 5-HT cell bodies in the dorsal raphe nucleus (RATTRAY et al. 1996). These results show that acute depletion of 5-HT may induce an increase in the turnover of 5-HTT mRNA, that changes in 5-HTT mRNA are not temporally related to changes in 5-HTT protein levels, and that dorsal raphe 5-HT neurons are differentially regulated by drugs depending on their location.

The appetite-suppressing amphetamine analog fenfluramine, given in high doses, can produce long-lasting decreases in brain levels of 5-HT and 5-HTT protein, although no significant effect on 5-HTT mRNA expression in the dorsal raphe has been observed (SEMPLE-ROWLAND et al. 1996). On the other hand, 5,7-dihydroxytryptamine-(5,7-DHT-) induced 5-HT cell loss is accompanied by a progressive decline in both 5-HTT mRNA and protein. These findings indicate drug specificity of the loss of forebrain markers for the 5-HT system that reflects either the loss of fine-caliber 5-HT axon terminals or a decrease in the expression of these markers in the somata of these cells which are located in the dorsal raphe. In a in situ hybridization study fenfluramine produced a significant but transient downregulation of 5-HTT mRNA in cells which lie in the ventral portion of the dorsal raphe nucleus but not in the dorsal part of the dorsal raphe nucleus. This suggests that cells which lie in the ventral part of the dorsal raphe nucleus may be more sensitive to the effects of chronic fenfluramine administration, but that fenfluramine does not cause long-term changes in gene expression in 5-HT cell bodies (RATTRAY et al. 1994).

Since lymphoblasts constitutively express functional 5-HTT and display 5-HT/5-HT$_{1A}$ receptor-dependent proliferative responses, it can be assumed that 5-HTT modulated fine-tuning of 5-HT/5-HT$_{1A}$ receptor interaction is relevant to cellular immune responses. It is therefore not surprising that cytokines and other immunomodulators such as interleukin-1β (IL-1β) and tumor necrosis factor-α (as well as the previously discussed glucocorticoids) participate in the regulation of 5-HTT gene expression (RAMAMOORTHY et al. 1995; MÖSSNER et al. in preparation). Finally, lack of circadian rhythmicity of functional 5-HTT has been reported, but long-term changes in the density of cortical 5-HTT may be induced by lasting alterations of certain environmental variables such as starvation (ZHOU et al. 1996).

IV. Targeted Gene Disruption

Among the prime transgenic strategies directed at elucidating the role of the 5-HTT in normal behavior and in disease states is the generation of a mouse model with a targeted disruption of the 5-HTT gene. This approach addresses the question of the extent to which targeted disruption of the 5-HTT gene affects the biochemistry, electrophysiology, and pharmacology of the 5-HT system and modulates neural development and synaptic plasticity. For generation of 5-HTT-deficient mice a genomic segment containing exon 2 of the murine 5-HTT gene was deleted by homologous recombination (BENGEL et al. 1997a,b). This segment contains the start codon, a portion of the substrate binding domain, and several posttranslational modification site. Autoradiography of the brain 5-HTT protein using [^{125}I]RTI-55 binding confirmed an

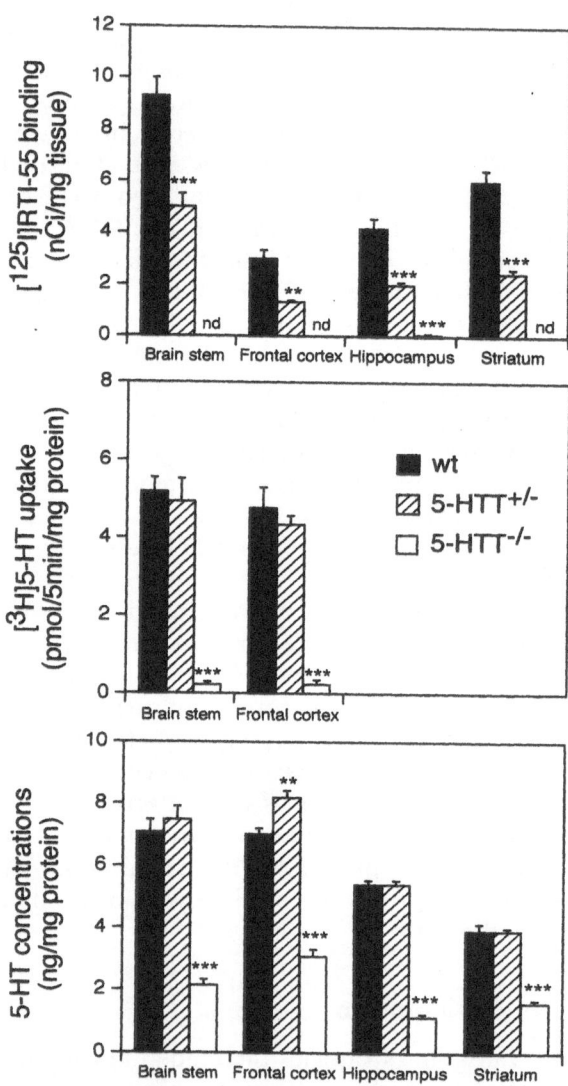

Fig. 11. [^{25}I]RTI-55 binding, [3H]5-HT uptake, and 5-HT concentrations in 5-HTT deficient mice (wild-type vs. 5-HTT$^{+/-}$ and 5-HTT$^{-/-}$; *$p < 0.05$, **$p < 0.01$, and ***$p < 0.001$, one-way ANOVA followed by t tests). For generation of 5-HTT-deficient mice a genomic segment containing exon 2 of the murine 5-HTT gene was deleted by homologous recombination (BENGEL et al. 1997b)

approx. 50% reduction in uptake site density in 5-HTT$^{+/-}$ mice, and a complete absence of binding across all regions in null mutants. 5-HTT-dependent function was further studied by examining [^{3}H]5-HT uptake kinetics in brain synaptomsomal preparations (Fig. 11).

The accumulation of [^{3}H]5-HT in both brainstem and frontal cortex in 5-HTT$^{-/-}$ mutant mice was approx. 97% less than in control mice. Together with the autoradiographic data, these results demonstrate a gene dose-dependent decrease in 5-HTT function. To evaluate the importance of the 5-HTT in regulating the serotoninergic system 5-HT concentrations in various regions of the brain were measured. In contrast to normal levels of 5-HT in 5-HTT$^{+/-}$ mice, 5-HTT$^{-/-}$ mutants showed marked reductions in 5-HT concentrations in brainstem, frontal cortex, hippocampus, and striatum. Possible behavioral consequences of the disruption of the 5-HTT were investigated in studies of MDMA-induced locomotor activity. Administration of 5mg/kg (+)MDMA to control mice resulted in a threefold increase in locomotor activity. In 5-HTT$^{+/-}$ mice (+)MDMA-induced hyperactivity was attenuated 50% and was nonexistent in 5-HTT$^{-/-}$ mutants. These results demonstrate the requirement for a functional 5-HTT in the locomoter for the stimulatory effects of (+)MDMA in mice (BENGEL et al. 1997b).

It is anticipated that 5-HTT-deficient mice will prove to be a useful model for continued study of the mechanism of action of drugs used in treating disorders thought to involve serotoninergic dysfunction. These mice may also represent a powerful tool for investigating selected drugs of abuse, including MDMA, which in addition to having psychedelic effects, may also produce 5-HT-related neurotoxicity, such as MDMA. Moreover, targeted disruption of the 5-HTT gene is likely to increase our understanding of how this gene is regulated, and what functions the expressed 5-HTT proteins have for development, physiology, metabolism, and behavior. It also may provide a system to dissect successive events that lead to disease states related to anxiety and depression and to test novel therapeutic concepts.

V. Promoter and Regulatory Elements of the 5-HTT Gene: Therapeutic Relevance?

Although still in its technical infancy, gene promoters and their associated *cis*- and *trans*-acting regulatory elements are potential targets for therapeutic intervention. One of the unique features of 5-HTT gene regulatory elements is to confer 5-HT cell-restricted actitvity such that a transferred gene is transcribed only in serotoninergic raphe neurons. Recent studies have identified and functionally mapped regulatory elements linked to 5-HTT gene promoter as molecular correlates of 5-HT cell-specific genes expression (HEILS et al. 1995). For confirmation of 5-HT cell-specific expression of a transferred gene the generation of transgenic mice is ultimately required, since the promoter/transgene construct is incorporated into the genome and thus present in all cell types. As a consequence, cell type-specific promoter activity has physiological meaning only when the introduced transgene is expressed according to a certain tissue-restricted pattern independent of its site of integration.

A variety of physiological factors regulate gene expression, and some of these could be applied to modulate the expression of promoters in gene therapy protocols. One of the first considerations is that, in the process of differentiation, the regulation of genes as a function of the stage of neural tissue maturation requires the use of developmentally regulated promoter and/or enhancer elements. Regulation of gene expression by steroid hormones is a potentially simple form of physiological influence that may be employed to modulate expression of transferred genes. The molecular definition of a hormonal response elements allows a choice of sequence elements which can

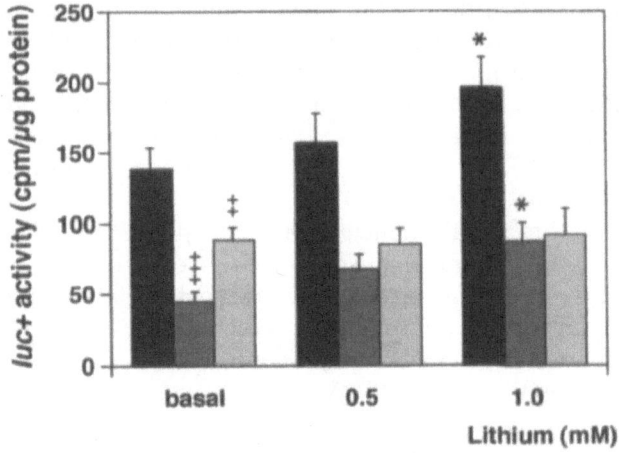

5-HTT promoter/luciferase reporter gene constructs

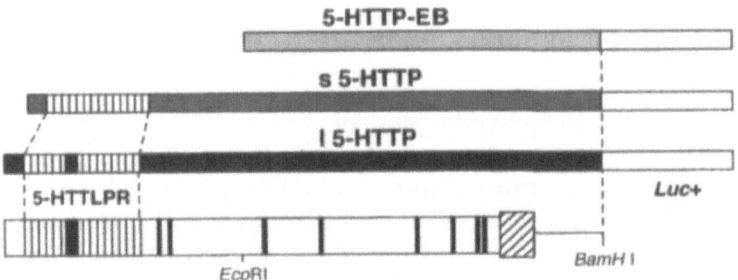

Fig. 12. Effect of lithium on the transcriptional activity of the long (*l*) and short (*s*) 5-HTT gene promoter (5-HTTP; l vs. s: *$p < 0.05$, ***$p < 0.01$; one-way ANOVA followed by *t* tests). The l and s variants of the 5-HTTP were ligated into a promoterless luciferase expression vector (l and s 5-HTTP). *5-HTTP-EB*, a deletional mutant of 5-HTTP lacking the 5-HTT linked polymorphic region (5-HTTLPR)

serve to regulate heterologous promoters. These elements would then be subject to modulation by natural or easily administered synthetic pharmacoactive molecules. Another possibility of influence is the regulation by ions. It has been shown, principally from studies on G-string-dependent tetrastrand DNA structure (G4-DNA), that ions such as Na^+ and K^+ facilitate, while lithium (Li^+) inhibits, the formation G4-DNA and thus have the potential to regulate the transcriptional activity of linked promoters. Preliminary results indicate that the G4-DNA structure formed by the 5-HTTLPR associated with 5-HTT gene promoter may be involved in Li^+- and other antibipolar drug-dependent modulation of 5-HT uptake function (HEILS et al. in preparation; Figs. 7, 12). The use of ions is therefore a feasible way of achieving promoter-based regulation of genes in cells where promoter-associated G4-DNA elements are found.

VI. Developing Drugs that Target 5-HTT Gene Expression

Given the fact that the 5-HTT continues to be an important target for large-scale drug production and development, novel therapeutic strategies targeting 5-HTT gene transcription need to be explored. Traditionally, the discovery of therapeutic agents has relied on random screening of large numbers of compounds for products that interfere with biological events associated with disease. Increasing knowledge of the molecular etiology of behavioral disorders, coupled with the identification of the key proteins and cellular pathways involved is now paving the way for rational drug design, targeting individual genes or their products.

The final stage of gene regulation involves proteins, generally termed nuclear transcription factors (NTFs), which activate or repress the transcription of specific genes, resulting in changes of the cellular phenotype. NTFs bind to specific motifs in target gene 5'-regulatory sequences resulting in temporal and/or tissue-specific alterations in transcription from that gene's promoter. Various stimuli lead to the activation or repression of different NTFs. Thus, NTFs play a key role in controlling differentiation and development, and this renders them tempting targets for therapeutic intervention for several reasons: (a) They are widely involved in human disease and may be activated in a tissue- or disease-specific manner. (b) Many mammalian NTFs have already been identified and characterized, making disease- and/or tissue-specific targeting more feasible. (c) NTFs are highly specific in their target recognition, and NTFs from different families interact with each other when bound to DNA at composite response elements (BUSTIN and McKAY 1994). This has two striking consequences: ubiquitous factors can affect cell specificity, and closely related factors from a given family can produce very different regulatory patterns.

Multiple lines of evidence indicate that NTFs are directly or indirectly targeted by many if not all psychoactive drugs. The mechanisms by which they exert therapeutic effects must still be studied in detail, but NTFs such as fos and zif268 are believed to mediate between receptor-activated second messenger systems and the transcriptional apparatus of genes involved in the complex functions of neuronal cells. Since preliminary results suggest that therapeutic drugs, hormones, and ions directly interfere with 5-HTT gene promoter activity, a future avenue of research will therefore be the investigation of indirect and direct effects of drugs known for their impact on the 5-HT system and of experimental agents on the transcriptional apparatus of the 5-HTT gene using in vitro systems and transgenic animals.

H. Future Directions

Novel strategies aimed at modifying 5-HT function at the level of gene transcription are likely to be derived from future studies. Identification of experimental drugs within the framework of these studies could set the stage for a rational approach to the design of drugs that act in a tissue- or disease-specific manner, target particular genes and proteins, and combine increased efficacy with reduced side effects. The enhanced understanding of promoters, their regulatory elements, and the protein factors involved in transcriptional regulation is beginning to find application in strategies of gene therapy. Because the structure and organization of promoter and other regulatory elements indicates a high degree of modularity, the design of an artificial transcriptional apparatus by combining core promoters and inducible and/or repressing elements that might allow modulation of gene expression in a therapeutic setting may now be an achievable goal. Developmental regulation and tissue-restricted expression in order to avoid unwanted side effects of the gene products and the modulation of expression by other physiological stimuli or by administered drugs could also be advantageous features of this approach.

In line with this thinking, technologies that allow the addition, alteration, or elimination of individual genes from the genome to create transgenic animal models for neuropsychiatric disorders are currently being developed and optimized. The 5-HTT-deficient mouse model will further increase our knowledge of how this gene is regulated, and what functions the expressed 5-HTT proteins have on development, physiology, metabolism, and behavior. There are also clear advantages of having a transgenic animal model for complex traits, for modeling 5-HTT-related behaviors, and for pathophysiological processes in affective disorders. Genetically engineered animals may provide the first unequivocal proof that a particular gene participates in the pathological changes that occur with psychiatric disorders. They can also provide a system for carefully dissecting the successive events that lead to the disease state and a custom-designed whole-animal system for testing potential therapies to treat and eventually cure psychiatric disorders.

Acknowledgement. The author is supported by the Hermann and Lilly Schilling Foundation.

References

Altemus M, Murphy DL, Greenberg B, Lesch KP (1996) Intact coding region of the serotonin transporter in obsessive-compulsive disorder. Am J Med Genet 67:409–411

Amara S, Kuhar M (1993) Neurotransmitter transporters: recent progress. Annu Rev Neurosci 16:73–93

Andrews AM, Murphy DL (1993) 2'-NH2-MPTP in Swiss Webster mice: evidence for long-term (6-month) depletion in cortical and hippocampal serotonin and norepinephrine, differential protection by selective uptake inhibitors or clorgyline and functional changes in central serotonin transmission. J Pharmacol Exp Therap 267:1432–1439

Arora RC, Meltzer HY (1986) Effect of adrenalectomy and corticosterone on [³H]imipramine binding in rat blood platelets and brain. Eur J Pharmacol 123:415–419

Barker EL, Blakely RD (1995) Norepinephrine and serotonin transporters: molecular targets of antidepressant drugs. In: Bloom FE, Kupfer DJ (eds) Psychopharmacology: the fourth generation of progress. Raven, New York, pp 321–333

Barker EL, Kimmel HL, Blakely RD (1994) Chimeric human and rat serotonin transporters reveal domains involved in recognition of transporter ligands. Mol Pharmacol 46:799–807

Bengel D, Heils A, Petri S, Teufel A, Seemann M, Riederer P, Murphy DL, Lesch KP (1997) Gene structure and 5'-flanking regulatory region of the murine serotonin transporter. Brain Res Mol Brain Res 44:286–293

Bengel D, Murphy DL, Feltner D, Seemann M, Heils A, Grinberg A, Westphal H, Lesch KP (1997) Targeted disruption of the murine serotonin transporter gene. Proc Natl Acad Sci USA (submitted)

Blakely RD, Berson HE, Fremeau RT, Caron MC, Peek MM, Prince HK, Bradley CC (1991) Cloning and expression of functional serotonin transporter from rat brain. Nature 354:66–70

Blakely R, De Felice LJ, Hartzell HC (1994) Molecular physiology of norepinephrine and serotonin transporters. J Exp Biol 196:263–281

Blier P, De Montigny C, Chaput Y (1990) A role for the serotonin system in the mechanisms of action of antidepressant treatments: preclinical evidence. J Clin Psychiatry 51S:14–20

Bouchard TJ Jr (1994) Genes, environment and personality. Science 264:1700–1701

Briley MS, Langer SZ, Raiman R, Sechter D, Zarifian E (1980) Tritiated imipramine binding sites are decreased in platelets of untreated depressed patients. Science 209:303–305

Buckman TD, Chang R, Sutphin MS, Eiduson S (1988) Interaction of 1-methyl-4-phenylpyridinium ion with human platelets. Biochem Biophys Res Commun 151:897–904

Bustin, SA, McKay IA (1994) Transcription factors: targets for new designer drugs. Br J Biomed Sci 51:157–157

Chang AS, Chang S, Starnes DM (1993) Structure-activity relationships of serotonin transport: relevance to nontricyclic antidepressant interactions. Eur J Pharmacol 247:239–248

Cloninger CR (1987) Neurogenetic adaptive mechanisms in alcoholism. Science 236:410–416

Collier DA, Arranz MJ, Sham P, Battersby S, Wallada H, Gill P, Aitchison KJ, Sodhi M, Li T, Roberts GW, Smith B, Morton J, Murray RM, Smith D, Kirov G (1996a) The serotonin transporter is a potential susceptibility factor for bipolar affective disorder. Neuroreport 7:1675–1679

Collier DA, Stöber G, Li T, Heils A, Catalano M, Di Bella D, Arranz MJ, Murray RM, Vallada HP, Bengel D, Müller CR, Roberts GW, Smeraldi E, Kirov G, Pak S, Lesch KP (1996b) A novel functional polymorphism within the promoter of the serotonin transporter gene: possible role in susceptibility to affective disorders. Mol Psychiatry 1:453–460

Corey JL, Quick MW, Davidson N, Lester HA, Guastella J (1994) A cocaine-sensitive Drosophila serotonin transporter: cloning, expression, and electrophysiological characterization. Proc Natl Acad Sci USA 91:1188–1192

Cortes R, Soriano E, Pazos A, Probst A, Palacios JM (1988) Autoradiography of antidepressant binding sites in the human brain: localization using [³H]imipramine and [³H]paroxetine. Neuroscience 27:473–496

Dave V, Kimelberg HK (1994) Na⁺-dependent, fluoxetine-sensitive serotonin uptake by astrocytes tissue-printed from rat cerebral cortex. J Neurosci 14:4972–4986

Davies, HML, Kuhn LA, Thornley C, Matasi JJ, Sexton T, Childers SR (1996) Synthesis of 3β-aryl-8-azabicyclo-(31)-octanes with high binding affinities and selectivities for the serotonin transporter site. J Med Chem 39:2554–2558

Demchyshyn LL, Pristupa B, Sugamori KS, Barker EL, Blakely RD, Wolfgang WJ, Forte MA, Niznik HB (1994) Cloning, expression, and localization of a chloride-facilitated, cocaine-sensitive serotonin transporter from Drosophila melanogaster. Proc Natl Acad Sci USA 91:5158–5162

Di Bella D, Catalano M, Balling U, Smeraldi E, Lesch KP (1996) Systematic screening for mutations in the coding region of the 5-HTT gene using PCR and DGGE. Am J Hum Genet 67:541–545

Ellis PM, Salmond C (1994) Imipramine binding in depression: a meta-analysis. Biol Psychiatry 36:292–299

Fumagalli F, Jones SR, Caron MG, Seidler FJ, Slotkin TA (1996) Expression of mRNA coding for the serotonin transporter in aged vs young rat brain: differential effects of glucocorticoids. Brain Res 719:225–228

Heils A, Teufel A, Petri S, Seemann M, Bengel D, Balling U, Riederer P, Lesch KP (1995) Functional promoter and polyadenylation site mapping of the human serotonin (5-HT) transporter gene. J Neural Transm 102:247–254

Heils A, Teufel A, Petri S, Stöber G, Riederer P, Bengel B, Lesch KP (1996) Allelic variation of human serotonin transporter gene expression. J Neurochem 6:2621–2624

Hensler JG, Ferry RC, Labow DM, Kovachich GB, Frazer A (1994) Quantitative autoradiography of the serotonin transporter to assess the distribution of serotonergic projections from the dorsal raphe nucleus. Synapse 17:1–15

Hoffman BJ, Mezey E, Brownstein M (1991) Cloning of a serotonin transporter affected by antidepressants. Science 254:579–580

Hrdina PD, Foy B, Hepner A, Summers RJ (1989) Antidepressant binding sites in brain: autoradiographic comparison of (3H)paroxetine and (3H)imipramine localization and relationship to serotonin transporter. J Pharmacol Exp Therap 252:410–417

Ivgy-May N, Tamir H, Gershon MD (1994) Synaptic properties of serotonergic growth cones in developing rat brain. J Neurosci 14:1011–1029

Jagust WJ, Eberling JL, Biegon A, Taylor SE, Van Brocklin HF, Jordan S, Handrahan SM, Roberts JA, Brennan KM, Mathis CA (1996) [¹²³I]5-Iodo-6-nitroquipazine: SPECT radiotracer to image the serotonin transporter. J Nuclear Med 37:1207–1214

Jayanthi LD, Ramamoorthy S, Mahesh VB, Leibach FH, Ganapathy V (1994) Calmodulin-dependent regulation of the catalytic function of the human serotonin transporter in placental choriocarcinoma cells. J Biol Chem 269:14424–14429

Jess U, Betz H, Schloss P (1996) The membrane bound rat serotonin transporter, SERT1, is an oligomeric protein. FEBS Lett 394:44–46

Joyce JN, Shane A, Lexow N, Winokur A, Casanova MF, Kleinman JE (1993) Serotonin uptake sites and serotonin receptors are altered in the limbic system of schizophrenics. Neuropsychopharmacology 8:315–336

Kelsoe JR, Remick RA, Sadovnick AD, Kristbjarnarson H, Flodman P, Spence MA, Morison M, Mroczkowskiparker Z, Bergesch P, Rapaport MH, Mirow AL, Blakely RD, Helgason T, Egeland J (1996) Genetic linkage study of bipolar disorder and the serotonin transporter. Am J Med Genet 67:215–227

Kia HK, Brisorgueil MJ, Daval G, Langlois X, Hamon M, Verge D (1996) Serotonin$_{1A}$ receptors are expressed by a subpopulation of cholinergic neurons in the rat medial septum and diagonal band of Broca-A double immunocytochemical study. Neurosci 74:143–154

Lander ES, Schork NJ (1994) Genetic dissection of complex traits. Science 265:2037–2048

Laruelle M, Abi-Dargham A, Casanova MF, Toti R, Weinberger DR, Kleinman JE (1993) Selective abnormalities of serotonergic receptors in schizophrenia: a postmortem study. Arch Gen Psychiatry 50:810–818

Launay JM, Bonduox D, Oset-Gasque MJ, Emami S, Mutel K, Haimart M, Gespach C (1994) Increases in human platelet serotonin uptake by atypical histamine receptors. Am J Physiol 266:526–536

Lebrand C, Cases O, Adelbrecht C, Doye A, Alvarez C, Elmestikawy S, Seif I, Gaspar P (1996) Transient uptake and storage of serotonin in developing thalamic neurons. Neuron 17:823–835

Lesch KP, Wolozin BL, Estler HC, Murphy DL, Riederer P (1993a) Isolation of a cDNA encoding the human brain serotonin transporter. J Neural Transm 91:67–73

Lesch KP, Aulakh CS, Wolozin BL, Tolliver TJ, Hill JL, Murphy DL (1993b) Regional brain expression of serotonin transporter mRNA and its regulation by reuptake inhibiting antidepressants. Mol Brain Res 17:31–35

Lesch KP, Balling U, Gross J, Strauss K, Wolozin BL, Murphy DL, Riederer P (1994) Organization of human serotonin transporter gene. J Neural Transm 95:157–164

Lesch KP, Bengel D (1995) Neurotransmitter reuptake mechanisms: targets for drugs to study and treat psychiatric, neurological and neurodegenerative disorders. CNS Drugs 4:302–322

Lesch KP, Franzek E, Gross J, Wolozin BL, Riederer P, Murphy DL (1995) Primary structure of the serotonin transporter in unipolar depression and bipolar disorder. Biol Psychiatry 37:215–223

Lesch KP, Bengel D, Heils A, Sabol SZ, Greenberg BD, Petri S, Benjamin J, Müller CR, Hamer DH, Murphy DL (1996a) Association of anxiety-related traits with a polymorphism in the serotonin transporter gene regulatory region. Science 274:1527–1531

Lesch KP, Heils A, Riederer P (1996b) The role of neurotransporters in excitotoxicity, neuronal cell death and other neurodegenerative processes. J Mol Med 74:365–378

Levi G, Raiteri M (1993) Carrier-mediated release of neurotransmitters. Trends Neurosci 16:415–419

Li T, Holmes C, Sham PC, Vallada H, Birkett J, Kirov G, Lesch KP, Powell J, Lovestone S, Collier D (1997) Allelic functional variation of serotonin transporter expression is a susceptibility factor for late onset Alzheimer's disease. Neuroreport 8:683–686

Linnet K, Koed K, Wiborg O, Gregersen N (1995) Serotonin depletion decreases serotonin transporter mRNA levels in rat brain. Brain Res 697:251–253

Mager S, Min C, Henry DJ, Chavkin C, Hoffman BJ, Davidson N, Lester HA (1994) Conducting states of a mammalian serotonin transporter. Neuron 12:845–859

Mayser W, Betz H, Schloss P (1991) Isolation of cDNA encoding a novel member of the neurotransmitter transporter gene family. FEBS Lett 295:203–206

McLaughlin DP, Little KY, Lopez JF, Watson SJ (1996) Expression of serotonin transporter mRNA in human brainstem raphe nuclei. Neuropsychopharmacology 15:523–529

Meltzer H, Arora RC (1988) Genetic control of serotonin uptake in blood platelets: a twin study. Psychiatry Res 24:263–269

Miller KJ, Hoffman BJ (1994) Adenosine A3 receptors regulate serotonin transport via nitric oxide and cGMP. J Biol Chem 269:27351–27356

Murphy DL, Mitchell PB, Potter WZ (1995) Novel pharmacological approaches to the treatment of depression. In: Bloom FE, Kupfer DJ (eds) Psychopharmacology: the fourth generation of progress. Raven, New York

Nelson PJ, Rudnick G (1979) Coupling between platelet 5-hydoxytryptamine and potassium transport. J Biol Chem 254:10084–10089

Neumaier JF, Root DC, Hamblin MW (1996a) Chronic fluoxetine reduces serotonin transporter mRNA and 5 HT1B mRNA in a secuential manner in the rat dorsal raphe nucleus. Neuropsychopharmacology 15:515–522

Neumaier JF, Szot P, Peskind ER, Dorsa DM, Hamblin MW (1996b) Serotonergic lesioning differentially affects presynaptic and postsynaptic 5-HT1B receptor mRNA levels in rat brain. Brain Res 722:50–58

Ogilvie AD, Battersby B, Bubb VJ, Fink G, Harmar AJ, Goodwin GM, Smith CAD (1996) Polymorphism in serotonin transporter gene associated with susceptibility to major depression. Lancet 347:731–733

Owen MJ, Nemeroff CB (1994) The role of serotonin in the pathophysiology of depression: focus on the serotonin transporter. Clin Chem 40:288–295

Paul S, Rehavi N, Skolnick P, Ballenger J, Goodwin F (1981) Depressed patients have decreased binding of tritiated imipramine to platelet serotonin transporter. Arch Gen Psychiatry 38:1315–1317

Peiffer A, Veilleux S, Barden N (1991) Antidepressant and other centrally acting drugs regulate glucocorticoid receptor messenger RNA levels in rat brain. Psychoneuroendocrinology 16:505–515

Pepin M, Govindan M, Barden N (1991) Increased glucocorticoid receptor gene promoter activity after antidepressant treatment. Mol Pharmacol 41:1016–1022

Pineyro G, Blier P, Dennis T, De Montigny C (1994) Desensitization of the neuronal 5-HT carrier following its long-term blockade. J Neurosci 14:3036–3047

Piven J, Tsai G, Nehme E, Coyle JT, Chase GA, Folstein SE (1991) Platelet serotonin, a possible marker for familial autism. J Autism Develop Disord 21:51–59

Qian Y, Melikian HE, Rye DB, Levey AI, Blakely RD (1995) Identification and characterization of antidepressant-sensitive serotonin transporter proteins using site-specific antibodies. J Neurosci 15:1261–1274

Ramamoorthy S, Bauman A, Moore K, Han H, Yang-Feng T, Chang A, Ganapathy V, Blakely R (1993) Antidepressant- and cocaine-sensitive human serotonin transporter: molecular cloning, expression, and chromosomal localization. Proc Natl Acad Sci USA 90:2542–2546

Ramamoorthy S, Ramamoorthy JD, Prasad PD, Bhat GK, Mahesh VB, Leibach FH, Ganapathy V (1995) Regulation of the human serotonin transporter by interleukin-1 beta. Biochem Biophys Res Commun 216:560–567

Rattray M, Wotherspoon G, Savery D, Baldessari S, Marden C, Priestley JV, Bendotti C (1994) Chronic D-fenfluramine decreases serotonin transporter messenger RNA expression in dorsal raphe nucleus. Eur J Pharmacol Mol Pharmacol 268:439–442

Rattray M, Baldessari S, Gobbi M, Mennini T, Saminin R, Bendotti C (1996) p-Chlorphenylalanine changes serotonin transporter mRNA levels and expression of the gene product. J Neurochem 67:463–472

Routledge C, Middlemiss DN (1996) The 5-HT hypothesis of depression revisited. Mol Psychiatry 1:437

Rudnick G, Wall SC (1993) Non-neurotoxic amphetamine derivatives release serotonin through serotonin transporters. Mol Pharmacol 43:271–276

Sander T, Harms H, Dufeu P, Kuhn S, Hoehe M, Lesch KP, Schmidt LG (1997) Serotonin transporter gene variants and type-2 alcohol dependence. Biol Psychiatry (in press)

Schloss P, Betz H (1995) Heterogeneity of antidepressant binding sites on the recombinant rat serotonin transporter SERT1. Biochemistry 34:12590–12595

Schuldiner S, Steiner-Mordoch S, Yelin R, Wall SC, Rudnick G (1993) Amphetamine derivatives interact with both plasma membrane and secretory vesicle biogenic amine transporters. Mol Pharmacol 44:1227–1231

Semple-Rowland SL, Mahatme A, Rowland NE (1996) Effects of dexfenfluramine or 5,7-dihydroxytryptamine on tryptophan hydroxylase and serotonin transporter mRNAs in rat dorsal raphe. Mol Brain Res 41:121–127

Slotkin TA, Barnes GA, McCook EC, Seidler FJ (1996a) Programming of brainstem serotonin transporter development by prenatal glucocorticoids. Dev Brain Res 93:155–161

Slotkin TA, McCook EC, Ritchie JC, Seidler FJ (1996b) Do glucocorticoids contribute to the abnormalities in serotonin transporter expression and function seen in depression: an animal model. Biol Psychiatry 40:576–584

Sonders MS, Amara SG (1996) Channels in transporters. Curr Opin Neurobiol 6:294–302

Staley JK, Basile M, Flynn DD, Mash DC (1994) Visualizing dopamine and serotonin transporters in the human brain with the potent cocaine analogue. J Neurochem 62:549–556

Stanley M, Virgilio J, Gershon S (1982) Tritiated imipramine binding sites are decreased in the frontal cortex of suicides. Science 216:1337–1339

Stöber G, Heils A, Lesch KP (1996) Serotonin transporter gene polymorphism and affective disorder. Lancet 347:1340–1341

Stockmeier CA, Shapiro LA, Haycock JW, Thompson PA, Lowy MT (1996) Quantitative subregional distribution of serotonin (1A) receptors and serotonin transporters in the human dorsal raphe. Brain Res 727:1–12

Sur C, Betz H, Schloss P (1996) Immunocytochemical detection of the serotonin transporter in rat brain. Neurosci 73:217–231

Toth M, Shenk T (1994) Antagonist-mediated down-regulation of 5-hydroxytryptamine type 2 receptor gene expression: modulation of transcription. Mol Pharmacol 45:1095–1100

Tukiainen E (1981) Effect of hypophysectomy and adrenalectomy on 5-hydroxytryptamine uptake by rat hypothalamic synaptosomes and blood platelets. Acta Pharmacol Toxicol 48:139–143

Uhl GR, Johnson PS (1994) Neurotransmitter transporters: three important gene families for neuronal function. J Exp Biol 196:229–236

Wall SC, Gu H, Rudnick G (1995) Biogenic amine flux mediated by cloned transporters stably expressed in cultured cell lines: amphetamine specificity for inhibition and efflux. Mol Pharmacol 47:544–550

Yu A, Yang J, Pawlyk AC, Tejani-Butt SM (1995) Acute depletion of serotonin down-regulates serotonin transporter mRNA in raphe neurons. Brain Res 688:209–212

Zhou D, Huether G, Wiltfang J, Hajak G, Rüther E (1996) Serotonin transporters in the rat frontal cortex: lack of circadian rhythmicity but down-regulation by food restriction. J Neurochem 67:656–661

CHAPTER 27

Animal Models of Integrated Serotoninergic Functions: Their Predictive Value for the Clinical Applicability of Drugs Interfering with Serotoninergic Transmission

P.E. KEANE and P. SOUBRIÉ

A. Introduction

Studies of serotonin in the brain began in the early 1950s, when TWAROG and PAGE (1953) used a sensitive bioassay to detect the compound's presence in cerebral tissue. Shortly afterwards, a possible central role for serotonin in mental illness was evoked by WOOLEY and SHAW (1954), based principally on the psychotomimetic activity of lysergic acid diethylamide (LSD), which had been found to antagonize the effects of serotonin on smooth muscle. Since this time, there has been an intense research effort towards an under-standing of this neurotransmitter, and it is now clear that serotoninergic systems are implicated to various extents in the expression of a wide range of CNS functions, including neuronal development (LAUDER 1983), thermoregu-lation (FELDBERG and MYERS 1964), pain (TENEN 1967), motor regulation (JACOBS 1991), sleep (JOUVET 1967), appetite (FERNSTROM and WURTMAN 1971), sexual behaviour (GORZALKA et al. 1990), aggression (SHEARD 1969; SOUBRIÉ 1986), anxiety (CHOPIN and BRILEY 1987) and mood (GOLDEN and GILMORE 1990).

The implication of serotonin in such a variety of functions is less surprising when one considers the overall structure of the serotoninergic systems in the brain. Anatomically, the serotoninergic systems are among the diffusely orga-nized projection systems of the brain, consisting of a morphologically diverse group of neurons whose cell bodies are located in the brainstem raphe nuclei and some regions of the reticular formation, and complex axonal systems which project to almost all regions of the CNS, but with particularly dense innervation of the cerebral cortex, limbic structures, basal ganglia, many parts of the brainstem and the grey matter of the spinal cord. Furthermore, a significant proportion of the neurons are branched, so that a single cell fre-quently projects to more than one brain area. The consequence of this struc-ture is that serotonin is released from nerve terminals in virtually all brain regions, a fact which might help to explain why this neurotransmitter is in-volved in functions of such widely different anatomical origins as emotion (principally associated with the limbic nuclei), feeding, sexual activity and thermoregulation (which are mainly hypothalamic in origin), or memory (generally linked to hippocampal activity). Another factor to be taken into account when considering the surprising diversity of the roles of serotoninergic systems in different CNS functions is the very large number of different 5-HT receptors (and receptor subtypes) which are now known to exist in the mammalian CNS (HOYER et al. 1994). Initial studies on the behavioural and other effects of serotonin used pharmacological tools [such as the tryptophan hydroxylase inhibitor p-chlorophenylalanine or the serotonin precursor 5-hydroxytryptophan (5-HTP)] which modified the activity of all serotoninergic pathways at the same time, and could not distinguish between the effects of different segments (either anatomical or functional) of brain serotoninergic systems in the modulation of CNS functions. More recently, studies using selective lesions, or relatively selective agonists and/or antagonists of different receptors and subtypes, have led to a better understanding of the different roles that serotonin plays in the regulation of behaviours.

Not surprisingly, in view of this diversity of actions, serotonin has also been implicated in a variety of pathological conditions, ranging from affective disorders to Alzheimer's disease. The validity of animal models for predicting the potential therapeutic effects of drugs is a central preoccupation in psychopharmacology (WILLNER 1991), and it is clear that many tests provide not only valid positive results, but also false positives and false negatives (see, e.g. PORSOLT 1991). The extreme diversity of the effects that have been induced by modifications of serotoninergic function raises the question of the extent of their predictive value for clinical applications. At present, clinical data exist relating to only a minority of the effects which have been associated with the serotoninergic system in animal studies, in particular depression, anxiety, psychosis, obsessive-compulsive disorders (OCDs) and the rather more recently addressed areas of sexual dysfunctions and eating disorders. A major object of the present article is to review the effects of serotoninergic compounds, with particular reference to compounds with selective affinity for one or other receptor subtype, in animal models of neuropsychiatric dysfunctions, and

Table 1. Principal animal models used for the evaluation of the CNS effects of drug interfering with serotoninergic transmission

Clinical indication	Model
Anxiety	**Conditioned avoidance response** • Conflict models • Conditioned suppression • Startle response **Unconditional response** • Staircase test • Light/dark boxes • Elevated plus maze • Social interaction • Novelty • Pentetrazol discrimination • Frustative non-reward

Table 1. *Continued*

Clinical indication	MODEL
Depression	**Stress models** • Learned helplessness • Forced swimming • Chronic unpredictable stress • Amphetamine withdrawal **Lesion models** • Olfactory bulbectomy **Pharmacological models** • Reserpine reversal • Yohimbine potentiation **Other models** • DRL 72 • Waiting behaviour
Schizophrenia	**Dopamine models** • Striatal (stereotypy) • Limbic (hyperactivity) • Cell body firing rates • Latent inhibition **Indoleamine models** • LSD antagonism • 5-HT syndrome antagonism
Aggressivity	**Offensive behaviour** • Resident-intruder • Hypothalamic stimulation **Defensive behaviour** • Foot-shock
Sexual dysfunction	**Female** • Lordosis **Male** • Mounting intromissions, ejaculations (latency and frequency)
Eating disorders	**Spontaneous food consumption** **Obesity models** • Lesion- or stimulation-induced • Genetic • Forced feeding **Anorexia models** • Hypothalamic lesions • Undernutrition
Obsessive-compulsive disorders	**Models selectively responsive to (chronic) serotonin uptake inhibition** • Marble burying • Schedule-induced polydipsia • Disruption of spontaneous alternation

compare them with the available results of clinical studies, in order to determine the degree of validity that may be attributed to these models. The principal models used for the evaluation of each of these areas are summarized in Table 1.

B. Anxiety

A possible role for serotonin in the regulation of the level of anxiety has been suspected for many years. A reduction of serotonin synthesis, or the administration of some non-selective serotonin antagonists, leads to a reduction in the suppression of behaviour by punishment, which has been attributed to an anxiolytic effect (GELLER and BLUM 1970; STEIN et al. 1973). More direct evidence has been forthcoming from studies of the effects of selective 5-HT$_{1A}$ receptor agonists on anxiety. Initial experiments with the first 5-HT$_{1A}$ receptor agonist shown to exhibit anxiolytic effects indicated that buspirone was only poorly active (TRABER and GLASER 1987) or even inactive (SIMIAND et al. 1993) in some anxiolytic models. Nevertheless, other 5-HT$_{1A}$ receptor agonists, including 8-hydroxy-N,N-dipropyl-2-amino tetralin (8-OH-DPAT) and SR 57746A, possess a larger spectrum of anxiolytic activity which is not dissimilar from that of the benzodiazepines (see, e.g. SIMIAND et al. 1993).

   5-HT$_{1A}$ receptors have two principal localizations, on the cell bodies of the 5-HT neurons themselves, in the midbrain raphe nuclei, where they function as autoreceptors to inhibit the activity of the 5-HT neurons (SPROUSE and AGHAJANIAN 1986, 1987; HAMON et al. 1988; HUTSON et al. 1989), and also on postsynaptic membranes in many forebrain areas, and notably in the hippocampus, where their activation reduces the firing rate of the postsynaptic cell bodies in this brain area (ANDRADE and NICOLL 1987; BLIER and DE MONTIGNY 1990). The anxiolytic effect of 5-HT$_{1A}$ receptor agonists may occur by the stimulation of the autoreceptors in the raphe nuclei, since intra-raphe injections of serotonin (THIÉBOT et al. 1984) and buspirone (COSTALL et al. 1988a) produce an anxiolytic activity in rodents. Recent studies have sought to determine whether other subtypes of serotonin receptor may also be involved in the regulation of anxiety. Overall, studies using 5-HT$_{1B/1D}$ and 5-HT$_{2A/2C}$ receptor ligands have produced little clear evidence of the implication of these receptors in anxiety (GRIEBEL 1995). An anxiolytic-like activity of various 5-HT$_3$ receptor antagonists has been described in animals (JONES et al. 1988; COSTALL et al. 1988b). In most of these studies, the antagonists were active over a very wide dose range, with the minimal active dose being frequently in the nanogram or even the picogram range. The activity was not accompanied by sedative or muscle-relaxant effects, and frequently disappeared at (relatively) higher doses. Despite considerable effort, other laboratories (e.g. FILE and JOHNSTON 1989) have been unable to reproduce these results.

C. Depression

Many procedures have been used to evaluate the antidepressant potential of compounds. Initially, the selection of these tests as possible models of antidepressant activity was based on their ability to detect an activity of the earliest known clinically effective antidepressants, i.e. the tricyclic antidepressants, and the monoamine oxidase inhibitors. Both of these classes of compound modify (among other things) serotoninergic neurotransmission, providing an initial indication for an involvement of serotonin in both animal models and clinical depression. This hypothesis has been amply confirmed by the marked antidepressant activity, observed both in the clinical setting and in laboratory tests, of a large number of selective serotonin reuptake inhibitors (SSRIs) including fluoxetine, fluvoxamine, citalopram, indalpine and paroxetine (PINDER and WIERINGA 1993).

Agonists or antagonists of a number of serotoninergic receptors have also been evaluated in animal models of depression. $5-HT_{1A}$ receptor agonists have been found to produce an antidepressant-like activity in a variety of animal models (CERVO and SAMANIN 1987; GIRAL et al. 1988; SIMIAND et al. 1993). A number of studies have been performed to determine the anatomical site underlying the antidepressant effects of $5-HT_{1A}$ receptor agonists. Some results suggest a presynaptic site for this effect, while others favour a postsynaptic site, as some data suggest that the effect of 8-OH-DPAT in the behavioural despair test is mediated by the stimulation of receptors in the raphe nuclei (CERVO and SAMANIN 1987; CERVO et al. 1988), while the drug's activity in the learned helplessness model of depression appears to be due to the stimulation of postsynaptic receptors in the septum (MARTIN et al. 1990). Several approaches have implicated $5-HT_{2A}$ receptors in depression. $5-HT_{2A}$ receptors have been found to be downregulated by chronic administration of antidepressant drugs to rats, and it was proposed that supersensitivity of $5-HT_{2A}$ receptors might play a role in the aetiology of depressive illness (PEROUTKA and SNYDER 1980). This has led to the evaluation of some $5-HT_{2A}$ receptor antagonists in antidepressant models, and several compounds have been shown to exhibit an activity in the learned helplessness and differential-reinforcement of low-rate (72 s) models (MAREK et al. 1989; RINALDI-CARMONA et al. 1992). Most $5-HT_{2A}$ receptor antagonists also bind with reasonable affinity to $5-HT_{2C}$ (formerly $5-HT_{1C}$) receptors, and this has raised the question of a possible implication of the latter in the antidepressant effects of the $5-HT_{2A}$ antagonists (JENCK et al. 1993). However, SR 46349B possesses high affinity for the $5-HT_{2A}$ receptor but not the $5-HT_{2C}$ receptor, and is active in the learned helplessness model of depression (RINALDI-CARMONA et al. 1992), reinforcing a probable role for $5-HT_{2A}$ receptors in this animal model of depression.

The effects of $5-HT_3$ receptors have also been evaluated in depression. In the learned helplessness model, MARTIN et al. (1992) found that zacopride, ondansetron and ICS 205–930 were all active at low doses in reversing escape deficits induced by inescapable foot-shocks in this model. However, a $5-HT_3$ receptor agonist has recently been shown to possess antidepressant-like activity in this and in other animal models, including the behavioural despair model (PONCELET et al. 1995), so that the role of $5-HT_3$ receptors in depression remains to be clarified.

D. Schizophrenia

Interest in serotonin as a prime factor in schizophrenia was initiated with the observation that LSD induced a profound psychotic-like state in normal subjects (GADDUM 1954; WOOLEY and SHAW 1954). A closer examination of the clinical state induced by LSD revealed many differences compared to the state of schizophrenia (see, e.g. SZARA 1967). This, together with the development of the dopamine hypothesis of schizophrenia, led to a gradual fading of interest in the role of serotonin in this disease state. Nevertheless, many studies continued to document schizophrenia-related serotoninergic abnormalities, including elevated levels of serotonin and its metabolite 5-hydroxyindoleacetic acid (5-HIAA) (CROW et al. 1979; FARLEY et al. 1980), and low 5-HT_{2A} receptor densities in prefrontal cortex (BENNETT et al. 1979; REYNOLDS et al. 1983).

A possible role of serotonin in schizophrenia again came to the fore with studies of the atypical antipsychotic clozapine, which has a superior efficacy/side effects profile and a higher affinity for several 5-HT receptors than for dopamine D_2 receptors. It was rapidly established that clozapine was an antagonist at 5-HT_{2A} receptors (FINK et al. 1984; FRIEDMAN et al. 1985; RASMUSSEN and AGHAJANIAN 1988) and a number of studies have attempted to characterize the differences between clozapine and haloperidol in animal models hoped to be predictive of their clinically different profiles. Most studies have shown that clozapine has a predominant effect on the mesolimbic dopaminergic system (generally believed to be implicated in the psychotic process), and little or no effect on the nigrostriatal system (which is usually considered to be the site at which the typical antipsychotics elicit their extrapyramidal side effects). Thus, clozapine preferentially blocked effects of dopamine injected into the limbic system rather than into the striatum (BORISON and DIAMOND 1983). The drug produced an inactivation of mesolimbic A10 (but not of nigrostriatal A9) dopaminergic cell bodies, while typical neuroleptics induced an identical inactivation of both A9 and A10 cells (CHIODO and BUNNEY 1983). Similar results showing a preferential effect of clozapine on limbic systems were obtained in other studies (CHIODO and BUNNEY 1985; HAND et al. 1987; BLAHA and LANE 1987; RIVEST et al. 1991).

Pinpointing the mechanisms underlying the limbic selectivity of clozapine has been the subject of much research, and is complicated by the fact that clozapine possesses high affinity for many other receptors apart from 5-HT_{2A}, including 5-HT_{2C}, 5-HT_3, 5-HT_6, 5-HT_7, D_4, m_1 and α_1 receptors (see MELTZER 1994), some of which have been proposed to be involved in the atypical profile of the compound (WANG et al. 1994; BALDESSARINI et al. 1992; KUOPPAMAKI et al. 1993). Some studies suggest that a strong 5-HT_{2A} affinity, linked to a weak D_2 affinity, is a major characteristic of atypical neuroleptics (MELTZER 1992), whereas other studies suggest that antagonism of the 5-HT_{2A} receptor, in the absence of any effect on D_2 receptors, may be sufficient to produce an atypical, limbic-selective profile (SORENSEN et al. 1993). Support for the latter concept is given by the encouraging results of some clinical trials with relatively selective $5-HT_{2A}$ antagonists such as ritanserin (GELDERS et al. 1985) and amperozide (AXELSSON et al. 1991).

There is some evidence that the $5-HT_3$ receptor may be implicated in psychotic processes. The hypothesis is based upon a series of observations indicating that $5-HT_3$ receptor antagonists can inhibit raised mesolimbic dopaminergic function. There are relatively few $5-HT_3$ receptors in the brain, but they are highly localized into a small number of areas, which include the amygdala and the limbic cortex (KILPATRICK et al. 1987). When amphetamine is injected into the rat nucleus accumbens, it elicits a hyperactivity which can be antagonized by typical and atypical neuroleptics, and by the $5-HT_3$ antagonist ondansetron (COSTALL et al. 1987). Similarly, the hyperactivity produced by chronic dopamine infusion into the nucleus accumbens is also abolished by neuroleptics and by $5-HT_3$ antagonists, the latter acting at doses in the $\mu g/kg$ range (COSTALL et al. 1987, 1990). The effects of dopamine infusions into the amygdala can also be prevented by $5-HT_3$ receptor antagonists (COSTALL et al. 1987). If, as mentioned above, some symptoms of schizophrenia are produced by limbic dopaminergic hyperactivity, the effects of the $5-HT_3$ antagonists should predict their efficacy as antipsychotics. A further indication of a possible relationship between $5-HT_3$ receptors and psychosis comes from studies by ASHBY et al. (1989), who have found that the suppression of cell firing produced by microiontophoretic application of the $5-HT_3$ receptor agonist 2-methyl-5-HT to medial prefrontal cortex cells was antagonized by $5-HT_3$ receptor antagonists, but also by the atypical neuroleptic clozapine. Other studies also suggest that clozapine can interact with $5-HT_3$ receptors (EDWARDS et al. 1991; VOLONTE et al. 1992).

E. Obsessive-Compulsive Disorders

Initial evidence in favour of a role for serotonin in obsessive-compulsive disorders stemmed from clinical studies of the effects of antidepressant drugs. Among the range of compounds tested, only antidepressants with a powerful effect on serotonin uptake have shown clear promise for the treatment of obsessive-compulsive disorders (OCDs) (GOODMAN et al. 1990; INSEL et al. 1990), whereas other antidepressants which in addition affected noradrenergic or other systems were found to be without effect (ANANTH et al. 1980; FOA et al. 1987).

Some animal models have been proposed to reflect the clinical condition of OCD, on the basis that they are reactive to compounds that relatively selectively inhibit serotoninergic uptake, while being unaffected by compounds that are devoid of this profile. These models include "marble burying" behaviour (BROEKKAMP et al. 1989) and schedule-induced polydipsia, in which food-deprived rats exposed to a procedure in which food is delivered intermittently will drink unusually large amounts of water (WOODS et al. 1993). In a third model proposed for OCDs (YADIN et al. 1991), the disruption of spontaneous alternation behaviour by 8-OH-DPAT was attenuated by chronic administration of fluoxetine. It is particularly interesting that chronic administration is required for the compounds to be effective in these models, in parallel with the clinical situation, in which several weeks' administration of selective serotonin reuptake inhibitors is necessary for a significant improvement of OCD (INSEL et al. 1990). At present, the efficacy of the SSRI is the most important evidence for a role of serotonin in OCD. Despite suggestions that downregulation of $5-HT_{2C}$ receptors may underly this efficacy (HOLLANDER et al. 1991; KENNETT et al. 1994), there is as yet insufficient evidence for this to be more than a working hypothesis (BARR et al. 1993). Similarly, the above-mentioned animal models still need to have their validity confirmed. It remains to be seen whether studies with selective 5-HT receptor subtype agonists or antagonists (or, indeed, totally unrelated compounds) in one or other of these models will lead to the development of more selective anti-OCD medications.

F. Aggressive Behaviour

Studies on aggressive behaviour in rats (mainly using the mouse-killing model) have shown that reducing serotoninergic transmission (e.g. by inhibition of tryptophan hydroxylase, or using chemical or electrolytic lesions) precipitates aggressive behaviour (WALDBILLIG 1979). More recent studies have shown that $5-HT_{1A}$ receptor agonists are effective in reducing aggressive behaviour in a variety of animal models, including maternal aggression (OLIVIER et al. 1989), shock-induced aggression (WHITE et al. 1991) and territorial aggression (SIMIAND et al. 1993). This is likely to be due to the activation of the $5-HT_{1A}$ autoreceptors in the raphe nuclei, as the injection of 8-OH-DPAT into this nucleus produces antiaggressive effects (Mos and OLIVIER 1991).

Some evidence also suggests that compounds with $5-HT_{1B}$ receptor agonist properties, such as RU 24969, eltoprazine and m-trifluoromethylphenylpiperazine (TFMPP), have an antiaggressive effect in resident-intruder and maternal aggression models in rodents (OLIVIER and Mos 1992). In contrast, $5-HT_{2A}$, $5-HT_{2C}$ and $5-HT_3$ receptor antagonists do not appear to specifically modify aggressive behaviour (OLIVIER and Mos 1986; WHITE et al. 1991). Studies with eltoprazine in rats lesioned with 5,7-dihydroxytryptamine (SIJBESMA et al. 1991) suggest a postsynaptic site of action for the antiaggressive effect. Since the $5-HT_{1B}$ receptor is absent in humans, where it is replaced by the $5-HT_{1D}$ receptor (HOYER et al. 1994), it is possible that $5-HT_{1D}$ receptor agonists possess antiaggressive properties in the clinic.

G. Sexual Dysfunction

Increased activity of the serotoninergic system as a whole has been proposed to inhibit sexual receptivity in the female rat (MEYERSON 1964), and also to inhibit sexual behaviour in the male rat (SHEARD 1969; SHILLITO 1969). A role for 5-HT_{1A} receptors in the regulation of lordosis is suggested by the observation that 8-OH-DPAT, buspirone, ipsapirone and gepirone inhibit lordosis in females primed with oestrogen and progesterone (AHLENIUS et al. 1986; MENDELSON and GORZALKA 1986). At lower doses, however, ipsapirone and gepirone facilitate lordosis (MENDELSON and GORZALKA 1986), an effect probably mediated by their stimulation of inhibitory somatodendritic 5-HT_{1A} autoreceptors in the raphe nuclei.

Although increasing overall serotoninergic release (by administering 5-hydroxytryptophan) inhibits lordosis in females, the antagonism of 5-HT_{2A} receptor function also inhibits lordosis (MENDELSON and GORZALKA 1986; GORZALKA et al. 1990). Studies with 5-HT_3 antagonists do not suggest a major role for this receptor in the regulation of female sexual activity (GORZALKA et al. 1990). Studies with ligands at 5-HT_2 receptors have produced conflicting evidence for a role of this receptor in the regulation of male sexual activity (MENDELSON and GORZALKA 1985; AHLENIUS et al. 1981).

In contrast to the effects observed in female rats, the stimulation of 5-HT_{1A} receptors tends to facilitate male copulatory activity, as indicated by a decreased number of mounts and intromissions that precede the ejaculation, as well as a decrease in the overall latency to ejaculation (AHLENIUS et al. 1981; AHLENIUS and LARSSON 1984; ARNONE et al. 1995). Similar results have been observed in rhesus monkeys (POMERANTZ et al. 1993). The fact that the sexual activity of male rats is also improved by treatment with p-chlorophenylalanine (AHLENIUS et al. 1971) and inhibited by 5-HTP (AHLENIUS et al. 1980) suggests that the mechanism of action of 5-HT_{1A} receptor agonists for the improvement of sexual performance is likely to be the activation of the somatodendritic receptors. Further support for this hypothesis comes from the observation that local injections of 8-OH-DPAT into the median, but not the dorsal raphe nucleus, or of 5-HT into both these nuclei, can produce a facilitation of sexual activity (HILLEGAART et al. 1989, 1991).

In contrast to these effects of 5-HT_{1A} agonists, male rat copulatory activity is inhibited by agents that stimulate 5-HT_{1B} and/or 5-HT_{2C} receptors (FERNANDEZ-GUASTI et al. 1989; MENDELSON and GORZALKA 1990). Conversely, penile erections in male rats (BERENDSEN et al. 1990) and monkeys (POMERANTZ et al. 1993) are elicited by agonists of 5-HT_{1B} and/or 5-HT_{2C} receptors, and inhibited by 5-HT_{1A} receptor agonists, further confirming the multifarious nature of serotoninergic control of sexual behaviour.

H. Eating Disorders

Much evidence has accumulated favouring an inhibitory role for serotonin on food intake (e.g. LEIBOWITZ and SHOR-POSNER 1986). Increasing the concentration of serotonin in the synaptic cleft using the serotonin releaser fenfluramine (BLUNDELL and LESHEM 1974) or by inhibition of uptake (GOUDIE et al. 1976) reduces food intake in rodents. Studies of the roles of different receptor subtypes on food intake shows that the situation is somewhat more complex, however. 8-OH-DPAT induces hyperphagia (Dourish et al. 1985; Bendotti and Samanin 1986), as do numerous other 5-HT$_{1A}$ receptor agonists (Dourish et al. 1986; Gilbert and Dourish 1987; Hutson et al. 1987). It has been proposed that this effect is produced by the stimulation of the 5-HT$_{1A}$ autoreceptors in the raphe nuclei, a proposal supported by the observation that depletion of serotonin by p-chlorophenylalanine abolishes the effect of the 5-HT$_{1A}$ agonists (Hutson et al. 1987; Dourish et al. 1986). Furthermore, the injection of 8-OH-DPAT into the raphe nuclei also increases feeding (Hutson et al. 1986). 5-HT$_{1B}$, 5-HT$_{2A}$ and 5-HT$_{2C}$ receptors also appear to play roles in food consumption. 1-(Metachlorophenyl)-piperazine (mCPP) and TFMPP, which have affinity for these receptors, cause hypophagia in normally fed and in food-deprived rats (Kennett et al. 1987; Samanin et al. 1979), probably by acting in the paraventricular nucleus of the hypothalamus (Hutson et al. 1988). Studies with antagonists suggest that mCPP and TFMPP produce hypophagia via both 5-HT$_{1B}$ and 5-HT$_{2C}$ receptors, whereas RU 24969 stimulates only the former receptor (Kennett and Curzon 1988). Fenfluramine may produce its anorexic effects by the (indirect) stimulation of 5-HT$_2$ receptors (Hewson et al. 1988), although this has been questioned (Neill and Cooper 1988).

I. Predictive Value of Animal Models for Clinical Applicability

The assessment of the value of any particular animal model can be approached in a variety of ways and this question has been recently discussed in detail (Willner 1991). One aspect is the extent to which animal models reflect the pathological processes themselves. This will vary from one model to the next, but it seems likely that the models in use at present reflect at best only a small part of the disease they are supposed to represent. This does not necessarily prevent them from being able to efficiently detect potential therapeutic agents for the diseases themselves, however, and the present discussion will attempt to determine the extent to which the various animal models can correctly predict the clinical effects of drugs acting via the modification of serotoninergic activity. This, in itself, is a rather complex problem, since a number of cases are known in which a test has been found to predict the potential therapeutic effect of many clinically active drugs, while being impervious to the effects of certain "families" of clinically active compounds. An example of this type of false-negative response is the generally noted lack of effect of SSRIs in the behavioural despair test of depression (Porsolt 1991), despite the test's ability to detect an activity of about 90% of known clinically active antidepressants (Borsini and Meli 1988). In view of this, we will approach the question of the predictive value of animal models in terms of therapeutic categories, rather than test by test.

Buspirone has been evaluated in clinical studies of anxiety, and found to be effective (GOLDBERG and FINNERTY 1979). It would appear clear therefore that the compound's activity in a variety of anxiolytic models, albeit weak and rather inconsistent, is predictive of the clinical efficacy in anxiety. Reasons why buspirone exhibits a limited anxiolytic profile in animals may include poor bioavailability, its affinity for dopaminergic D_2 receptors, and also its breakdown to 1-pyrimidinyl piperazine, a compound with non-negligible pharmacological actions of its own, including α_2 receptor and GABA-receptor antagonist activities, which might interfere with the expression of the effects of 5-HT_{1A} receptor stimulation. Another reason may be that the initially used anxiolytic tests were too closely oriented to detect compounds with a specific mechanism of action, i.e. the benzodiazepines. It remains to be seen whether activity in some anxiolytic models is predictive of a therapeutic effect of 5-HT_3 receptor antagonists, as little has been published on the profile of this category of drugs in psychatric illness (see BENTLEY and BARNES 1995).

The global antidepressant-like profile of SSRIs in animal models (despite their frequently documented inactivity in the behavioural despair test) appears clearly predictive of the well-documented clinical effects of this drug category (PINDER and WIERINGA 1993). The activity of 5-HT_{1A} receptor agonists in models of depression may also successfully predict an effect in the clinical situation, in view of the results of a number of clinical trials on these compounds (HELLER et al. 1990; RICKELS et al. 1991). The situation is less clear for 5-HT_{2A} antagonists, however, despite an isolated positive effect reported with ritanserin (PINDER and WIERINGA 1993). Similarly, the possible therapeutic relevance of 5-HT_3 receptor modulation must await the outcome of clinical trials.

Some clinical studies have given support to the concept that mixed $5\text{-HT}_{2A}/D_2$ antagonists, or even selective 5-HT_{2A} antagonists, may provide effective antipsychotic activity with little or no extrapyramidal side effects (MELTZER 1989; AXELSSON et al. 1991), suggesting that the "limbic-selective" profile of such compounds in animal studies may provide a valid basis for the selection of new antipsychotics. Again, however, this situation is less clear for 5-HT_3 antagonists, and will only be clarified with the publication of the results of clinical trials.

In the case of OCD, it is at present very difficult to approach the concept of predictability. The animal models that have been proposed for this indication have been selected because they appear to respond selectively (at least in chronic studies) to the only class of compounds (SSRIs) known to improve OCD at present. Although this parallel is compelling, it can hardly be considered proof of predictability. This will only come when a new (non-SSRI) substance found to be effective in these animal models is confirmed as being efficacious in clinical trials in OCD patients. Although aggression is a major problem in patients with psychiatric disorders, it does not constitute a single clinical syndrome, but instead is present in a range of diagnostic categories. In support of laboratory studies, preliminary clinical trials suggest that the 5-

$HT_{1A/1B}$ agonist eltoprazine can reduce the aggression of psychiatric patients, without modifying their level of social interest or exploration (RATEY and CHANDLER 1995).

The concept that an overall increase in serotonin levels in the synaptic cleft results in an inhibition of sexual activity appears to be validated by clinical experience indicating loss of sexual interest, anorgasmia or ejaculatory delay in patients treated with SSRI (KLINE 1989; ZAJECKA et al. 1991). The delayed ejaculation induced by sertraline has even been used with success as a treatment for premature ejaculation (MENDELS et al. 1995). In contrast to these effects of SSRI, relatively little has been published on the effects of 5-HT_{1A} receptor agonists on human sexual activity. One preliminary study has, however, indicated that buspirone normalizes decreased sexual function in patients with generalized anxiety disorder (OTHMER and OTHMER 1987), again in agreement with the results obtained in animal models.

The serotonin-releasing agent fenfluramine and the SSRI fluoxetine have been found to decrease weight in obese subjects (STEWART et al. 1986; FERGUSON and FEIGHNER 1987), and also to be effective for the treatment of bulimia nervosa (ROBINSON et al. 1985; FREEMAN and HAMPSON 1987), results which are consistent with their effects in animal models of food consumption. In contrast to animal data, however, the 5-HT_{1A} agonist ipsapirone has been found in one study to reduce bulimia nervosa (GERETSEGGER et al. 1995). One might speculate that this apparently conflicting observation is due to the preferential stimulation of postsynaptic 5-HT_{1A} receptors by ipsapirone in human brain, whereas the hyperphagic effect of 5-HT_{1A} receptor agonists in rodents has been attributed to the stimulation of the autoreceptor, which would ultimately reduce serotoninergic activity at the postsynaptic level.

Despite several discrepancies of this kind, the majority of available data appears to lend credence to the conclusion that the effects produced by modification of the serotoninergic neuronal system in rodents can in general be considered to be a reliable indication of what is likely to happen in the clinical situation. It is not certain whether this conclusion can be extended to the effects of 5-HT_3 receptor stimulation or antagonism, as practically nothing is known at present of the effects of such compounds in clinical trials of anxiety, depression or psychosis.

In view of the breadth of processes shown to be modulated by the manipulation of serotoninergic function, in both animals and humans, one final question may be addressed. Is there a basic, transnosographic dimension of CNS function that is controlled by serotoninergic activity which transcends individual pathologies and animal models? It has been proposed that serotoninergic processes may participate in the control of impulsive behaviour, or waiting capacity (SOUBRIÉ 1986). Since a number of pathological conditions, including depression, bulimia and OCD, involve dysfunctions of impulse control, this may be an underlying factor whose modulation could lead to a number of the clinically useful effects of serotoninergic agents.

Further efforts will be required, however, in order to extend this concept to other effects of serotonin.

References

Ahlenius S, Larsson K (1984) Failure to antagonize the 8-hydroxy-2-(di-n-propylamino) tetralin-induced facilitation of male rat sexual behavior by the administration of 5-HT receptor antagonists. Eur J Pharmacol 99:279–286

Ahlenius S, Eriksson H, Larsson K, Modigh K, Sodersten P (1971) Mating behavior in the male rat treated with p-chlorophenylalanine methyl ester alone and in combination with pargyline. Psychopharmacologia 20:383–388

Ahlenius S, Larsson K, Svensson L (1980) Further evidence for an inhibitory role of central 5-HT in male rat sexual behavior. Psychopharmacology 68:217–220

Ahlenius S, Larsson K, Svensson L, Hjorth S, Carlsson A, Lindberg P, Wikstrom H, Sanchez D, Arvidsson LE, Hacksell U, Nilsson JLG (1981) Effects of a new type of 5-HT receptor agonist on male rat sexual behavior. Pharmacol Biochem Behav 15:785–792

Ahlenius S, Fernandez-Guasti A, Hjorth S, Larsson K (1986) Suppression of lordosis behaviour by the putative 5-HT receptor agonist 8-OH-DPAT in the rat. Eur J Pharmacol 124:361–363

Ananth J, Pecknold JC, Van der Steen N, Engelsmann F (1980) Double blind comparative study of clomipramine and amitryptliline in obsessive neurosis. Prog Neuropsychopharmacol Biol Psychiatry 5:257–266

Andrade R, Nicoll RA (1987) Pharmacological distinct actions of serotonin on single pyramidal neurons of the rat hippocampus recorded in vitro. J Physiol (Lond) 394:99–124

Arnone M, Baroni M, Gai J, Guzzi U, Desclaux MF, Keane PE, Le Fur G, Soubrié P (1995) Effect of SR 59026A, a new 5-HT$_{1A}$ receptor agonist, on sexual activity in male rats. Behav Pharmacol 6:276–282

Ashby CR, Edwards E, Harkins KL, Wang RY (1989) Differential effect of typical and atypical antipsychotic drug on the suppressant action of 2-methyl-serotonin on medial prefrontal cortical cells: a microiontophoretic study. Eur J Pharmacol 166:583–584

Axelsson R, Nilsson A, Christensson E, Bjork R (1991) Effects of amperozide in schizophrenia: an open study of a potent 5-HT$_2$ receptor antagonist. Psychopharmacology 104:287–292

Baldessarini RJ, Huston-Lyons D, Campbell A, Marsh E, Cohen BM (1992) Do central antiadrenergic actions contribute to the atypical properties of clozapine? Br J Psychiatry 17 (Suppl):12–16

Barr LC, Goodman WK, Price LH (1993) The serotonin hypothesis of obsessive compulsive disorder. Int Clin Psychopharmacol 8 (Suppl 2):79–82

Bendotti C, Samanin R (1986) 8-Hydroxy-(di-N-propylamino) tetralin (8-OH-DPAT) elicits eating in free feeding rats by acting on central serotonin neurons. Eur J Pharmacol 121:147–154

Bennett JP, Enna SF, Bylund DB, Gillin JC, Wyatt RJ, Snyder SH (1979) Neurotransmitter receptors in frontal cortices of schizophrenics. Arch Gen Psychiatry 36:927–934

Bentley KR, Barnes NM (1995) Therapeutic potential of serotonin 5-HT$_3$ antagonists in neuropsychiatric disorders. CNS Drugs 3:363–392

Berendsen HG, Jenck F, Broekkamp CLE (1990) Involvement of 5-HT$_{1C}$ receptors in drug-induced penile erections in rats. Psychopharmacology 101:57–61

Blaha CD, Lane RF (1987) Chronic treatment with classical and atypical antipsychotic drug differentially decreases dopamine release in striatum and nucleus accumbens in vivo. Neurosci Lett 78:199–204

Blier P, De Montigny C (1990) Differential effect of gepirone on presynaptic and postsynaptic serotonin receptors: single-cell recording studies. J Clin Psychopharmacol 10:135–209

Blundell JE, Leshem MB (1974) Central action of anorexic agents: effects of amphetamine and fenfluramine in rats with lateral hypothalamic lesions. Eur J Pharmacol 28:81–88

Borison RL, Diamond BI (1983) Regional selectivity of neuroleptic drugs: an argument for site specificity. Brain Res Bull 11:215–218

Borsini F, Meli A (1988) Is the forced swimming test a suitable model for revealing antidepressant activity? Psychopharmacology 94:147–160

Broekkamp CLE, Berendsen HHE, Jenck F, Van Delft AML (1989) Animal models for anxiety and response to serotonergic drugs. Psychopathology 22 (Suppl 1):2–12

Cervo L, Samanin R (1987) Potential antidepressant properties of 8-OH-DPAT, a selective serotonin-1A receptor agonist. Eur J Pharmacol 144:223–229

Cervo L, Grignaschi G, Samanin R (1988) 8-OH-DPAT, a selective serotonin-1A receptor agonist, reduces the immobility of rats in the forced swimming test by acting on the nucleus raphe dorsalis. Eur J Pharmacol 158:53–59

Chiodo LA, Bunney BS (1983) Typical and atypical neuroleptics: differential effects of chronic administration on the activity of A9 and A10 midbrain dopaminergic neurons. J Neurosci 3:1607–1619

Chiodo LA, Bunney BS (1985) Possible mechanisms by which repeated clozapine administration differentially affects the activity of two subpopulations of midbrain dopamine neurons. J Neurosci 5:2539–2544

Chopin P, Briley M (1987) Animal models of anxiety: the effect of compounds that modify 5-HT neurotransmission. Trends Pharmacol Sci 8:383–388

Costall B, Domeney AM, Naylor RJ, Tyers MB (1987) Effects of the 5-HT$_3$ receptor antagonist GR38032F on raised dopaminergic activity in the mesolimbic system of the rat and marmoset brain. Br J Pharmacol 92:881–894

Costall B, Kelly ME, Naylor RJ, Onaivi ES (1988a) Actions of buspirone in a putative model of anxiety in the mouse. J Pharm Pharmacol 40:494–500

Costall B, Domeney AM, Gerrard PA, Kelly ME, Naylor RJ, Tyers MB (1988b) Zacopride: anxiolytic profile in rodent and primate models of anxiety. J Pharm Pharmacol 40:302–305

Costall B, Naylor RJ, Tyers MB (1990) The psychopharmacology of 5-HT$_3$ receptors. Pharmacol Ther 47:1816–202

Crow TJ, Baker H, Gross A, Josephy M, Lofthouse R, Longden A, Owen F, Riley G, Glover V, Killpack W (1979) Monamine mechanisms in chronic schizophrenia: post-mortem neurochemical findings. Br J Psychiatry 134:249–256

Dourish CT, Hutson PH, Curzon G (1985) Low doses of the putative serotonin agonist 8-hydroxy-2-(di-n-propylamino) tetralin (8-OH-DPAT) elicit feeding in the rat. Psychopharmacology 86:197–204

Dourish CT, Hutson PH, Curzon G (1986) Para-chlorophenylalanine prevents feeding induced by the serotonin agonist 8-hydroxy-2-(di-n-propylamino) tetralin (8-OH-DPAT). Psychopharmacology 89:467–471

Edwards E, Ashby CR Jr, Wang RY (1991) The effect of typical and atypical antipsychotic drugs on the stimulation of phosphoinositide hydrolysis produced by the 5-HT$_3$ receptor agonist 2-methyl-serotonin. Brain Res 545:276–278

Farley I, Shannak K, Hornykiewicz O (1980) Brain monoamine changes in chronic paranoid schizophrenia and their possible relation to increased dopamine receptor sensitivity. In: Pepeu G, Kuhar M, Enna S (eds) Receptors for neurotransmitters and peptide hormones. Raven Press, New York, pp 427–433

Feldberg W, Myers RD (1964) Effects on temperature of 5-hydroxytryptamine, adrenaline and noradrenaline injected into the cerebral ventricles or the hypothalamus of cats. J Physiol (Lond) 173:25

Ferguson JM, Feighner JP (1987) Fluoxetine-induced weight loss in overweight non-depressed humans. Int J Obes 11 (Suppl 3):163–170

Fernandez-Guasti A, Escalante A, Agmo A (1989) Inhibitory action of various 5-HT$_{1B}$ receptor agonists on rat masculine sexual behavior. Pharmacol Biochem Behav 34:811–816

Fernstrom JD, Wurtman RJ (1971) Brain serotonin content: increase following ingestion of carbohydrate diet. Science 174:1023–1025

File SE, Johnston AL (1989) Lack of effects of 5-HT$_3$ receptor antagonists in the social interaction and elevated plus-maze tests of anxiety in the rat. Psychopharmacology 99:248–251

Fink H, Morgenstern R, Oetssner W (1984) Clozapine – A serotonin antagonist? Pharmacol Biochem Behav 20:513–517

Foa EB, Steketee G, Kozak MJ, Dugger D (1987) Imipramine and placebo in the treatment of obsessive compulsives: their effect on depression and on obsessional symptoms. Psychopharmacol Bull 23:8–11

Freeman CP, Hampson M (1987) Fluoxetine as a treatment for bulimia nervosa. Int J Obes 11 (Suppl 3):171–177

Friedman RL, Sanders-Bush E, Barrett RL (1985) Clozapine blocks disruptive and discriminative stimulus effects of quipazine. Eur J Pharmacol 106:191–193

Gaddum JH (1954) Drugs antagonistic to 5-hydroxytryptamine. In: Wolstenholme GW (ed) Ciba Foundation Symposium on Hypertension. Little Brown and Company, Boston, pp 75–77

Gelders Y, Ceulemans DY, Hoppenbrouwers ML, Reyntjens A, Janssen P (1985) Ritanserin, a selective serotonin antagonist in chronic schizophrenia. In: IVth World Congress of Biological Psychiatry. Abstract 420.4, p 338

Geller I, Blum K (1970) The effect of 5-HTP on para-chlorophenylalanine (p-CPA) attenuation of conflict behaviour. Eur J Pharmacol 9:319–324

Geretsegger C, Greimel KV, Roed IS, Hesselink JMK (1995) Ipsapirone in the treatment of bulimia nervosa – an open pilot study. Int J Eating Disorders 17:359–363

Gilbert F, Dourish CT (1987) Effects of the novel anxiolytics gepirone, buspirone and ipsapirone on free feeding and on feeding induced by 8-OH-DPAT. Psychopharmacology 93:349–352

Giral P, Martin P, Soubrié P, Simon P (1988) Reversal of helpless behaviour in rats by putative 5-HT$_{1A}$ agonists. Biol Psychiatry 23:237–242

Goldberg HL, Finnerty RJ (1979) The comparative efficacy of buspirone and diazepam in the treatment of anxiety. Am J Psychiatry 136:1184–1187

Golden RN, Gilmore J (1990) Serotonin and mood disorder. Psychatr Ann 20:580–586

Goodman WK, Price LH, Delgado DL, Palumbo J, Krystal JH, Nagy LM, Rasmussen SA, Heninger GR, Charney DS (1990) Specificity of serotonin reuptake inhibitors in the treatment of obsessive-compulsive disorder. Arch Gen Psychiatry 47:577–585

Gorzalka BB, Mendelson SD, Watson NV (1990) Serotonin receptor subtypes and sexual behavior. Ann NY Acad Sci 600:435–446

Goudie AJ, Thomton EW, Wheeler TJ (1976) Effects of Lilly 110140, a specific inhibitor of 5-hydroxytryptamine uptake, on food intake and on 5-hydroxytryptophan-induced anorexia. Evidence for serotonergic inhibition of feeding. J Pharm Pharmacol 28:318–320

Griebel G (1995) 5-Hydroxytryptamine-interacting drugs in animal models of anxiety disorders: more than 30 years of research. Pharmacol Ther 65:319–395

Hamon M, Fattacini CM, Adrien J, Gallissot MC, Martin P, Gozlan H (1988) Alterations of central serotonin and dopamine turnover in rats treated with ipsapirone and other 5-hydroxytryptamine$_{1A}$ agonists with potential anxiolytic properties. J Pharmacol Exp Ther 246:745–752

Hand TH, Hu XT, Wang RY (1987) Differential effects of acute clozapine and haloperidol on the activity of ventral tegmental (A10) and nigrostriatal (A9) dopamine neurons. Brain Res 415:259–269

Heller AH, Beneke M, Kuemmel B, Spencer D, Kurtz NM (1990) Ipsapirone: evidence for efficacy in depression. Psychopharmacol Bull 26:219–222

Hewson G, Leighton GE, Hill RG, Hughes J (1988) Ketanserin antagonises the anorectic effect of dl-fenfluramine in the rat. Eur J Pharmacol 145:227–230

Hillegaart V, Ahlenius S, Larsson K (1989) Effects of local application of 5-HT into the median and dorsal raphe nuclei on male rat sexual and motor behaviour. Behav Brain Res 33:279–286

Hillegaart V, Ahlenius S, Larsson K (1991) Region-selective inhibition of male rat sexual behaviour and motor performance by localised forebrain 5-HT injections: a comparison with effects produced by 8-OH-DPAT. Behav Brain Res 42:169–180

Hollander E, De Carin C, Gully R, Nitescu A, Suckow RF, Gorman JM, Klein DF, Liebowitz MR (1991) Effects of chronic fluoxetine treatment on behavioural and neuroendocrine responses to methachlorophenylpiperazine. Psychiatr Res 36:1–17

Hoyer D, Clarke DE, Fozard JR, Hartig PR, Martin GR, Mylecharane EJ, Saxena PR, Humphrey PPA (1994) International Union of Pharmacology Classification of Receptors for 5-hydroxytryptamine (serotonin). Pharmacol Rev 46:157–204

Hutson PH, Dourish CT, Curzon G (1986) Neurochemical and behavioural evidence for mediation of the hyperphagic action of 8-OH-DPAT by 5-HT cell body autoreceptors. Eur J Pharmacol 129:347–352

Hutson PH, Donohoe TP, Curzon G (1987) Neurochemical and behavioural evidence for an agonist action of 1-[2-(4-aminophenyl)ethyl]-4-(3-trifluoromethylphenyl)-piperazine (LY 165163) at central 5-HT receptors. Eur J Pharmacol 138:215–223

Hutson PH, Donohoe TP, Curzon G (1988) Infusion of the 5-hydroxytryptamine agonists RU 24969 and TFMPP into the paraventricular nucleus of the hypothalamus causes hypophagia. Psychopharmacology 95:550–552

Hutson PH, Sarna GS, O'Connell MT, Curzon G (1989) Hippocampal 5-HT synthesis and release in vivo is decreased by infusion of 8-OH-DPAT into the nucleus raphe dorsalis. Neurosci Lett 100:276–280

Insel TR, Zohar J, Berkelfat C, Murphy D (1990) Serotonin in obsessions, compulsions and the control of aggressive impulses. Ann NY Acad Sci 600:547–586

Jacobs BL (1991) Serotonin and behaviour: emphasis on motor control. J Clin Psychiatry 52 (Suppl):17–23

Jenck F, Moreau JL, Mutel V, Martin JR, Haefely WE (1993) Evidence for a role of 5-HT$_{1c}$ receptors in the antiserotonergic properties of some antidepressant drugs. Eur J Pharmacol 231:223–229

Jones BJ, Costall B, Domeney AM, Kelly ME, Naylor RJ, Oakley NR, Tyers MB (1988) The potential anxiolytic activity of GR 38032F, a 5-HT$_3$ receptor antagonist. Br J Pharmacol 93:985–993

Jouvet M (1967) Neurophysiology of the states of sleep. Physiol Rev 47:117–177

Kennett GA, Curzon G (1988) Evidence that mCPP may have behavioural effects mediated by 5-HT$_{1C}$ receptors. Br J Pharmacol 94:137–147

Kennett GA, Dourish CT, Curzon G (1987) 5-HT$_{1B}$ agonists induce anorexia at a postsynaptic site. Eur J Pharmacol (Netherlands) 141:429–435

Kennett GA, Lightlower S, De Biasi V, Stevens NC, Wood MD, Tulloch IF, Blackburn TP (1994) Effect of chronic administration of selective 5-hydroxytryptamine and noradrenaline uptake inhibitors on a putative index of 5-HT$_{2C/2B}$ receptor function. Neuropharmacology 33:1581–1588

Kilpatrick GJ, Jones BJ, Tyers MB (1987) Identification and distribution of 5-HT$_3$ receptors in rat brain using radioligand binding. Nature 330:746–748

Kline MD (1989) Fluoxetine and anorgasmia. Am J Psychiatry 146:804–805

Kuoppamaki M, Seppala T, Syvalahti E, Hietala J (1993) Chronic clozapine treatment decreases 5-hydroxytryptamine$_{1C}$ receptor density in the rat choroid plexus: comparison with haloperidol. J Pharmacol Exp Ther 264:1262–1267

Lauder JM (1983) Hormonal and humoral influences on brain development. Psychoneuroendocrinology 8:121–155

Leibowitz SF, Shor-Posner G (1986) Brain serotonin and eating behavior. Appetite 7 (Suppl):1–14

Marek G, Li A, Seiden L (1989) Selective 5-hydroxytryptamine$_2$ antagonists have antidepressant-like effects on differential-reinforcement-of-low-rate 72-second schedule. J Pharmacol Exp Ther 250:52–59

Martin P, Beninger RJ, Hamon M, Puech AJ (1990) Antidepressant-like action of 8-OH-DPAT, a 5-HT$_{1A}$ agonist, in the learned helplessness paradigm: evidence for a postsynaptic mechanism. Behav Brain Res 38:135–140

Martin P, Gozlan H, Puech AJ. (1992) 5-HT$_3$ receptor antagonists reverse helpless behaviour in rats. Eur J Pharmacol 212:73–78

Meltzer HY (1989) Clinical studies on the mechanism of action of clozapine: the dopamine-serotonin hypothesis of schizophrenia. Psychopharmacology 99:S18–S27

Meltzer HY (1992) The importance of serotonin-dopamine interactions in the action of clozapine. Br J Psychatry 17 (Suppl):22–29

Meltzer HY (1994) An overview of the mechanism of action of clozapine. J Clin Psychiatry 55 (Suppl B):47–52

Mendels J, Camera A, Sikes C (1995) Sertraline treatment for premature ejaculation. J Clin Psychopharmacol 15:341–346

Mendelson SD, Gorzalka BB (1985) Serotonin antagonist piremperone inhibits sexual behavior in the male rat: attenuation by quipazine. Pharmacol Biochem Behav 22:565–571

Mendelson SD, Gorzalka BB (1986) Methysergide inhibits and facilitates lordosis behaviour in the female rat in a time-dependent manner. Neuropharmacol 25:749–755

Mendelson SD, Gorzalka BB (1990) Sex differences in the effects of 1-(m-trifluoromethylphenyl) piperazine and 1-(m-chlorophenyl) piperazine on copulatory behavior in the rat. Neuropharmacol 29:783–786

Meyerson BJ (1964) Estrus behaviour in spayed rats after estrogen or progesterone treatment in combination with reserpine or tetrabenazine. Psychopharmacologia 6:210–218

Mos J, Olivier B (1991) On the site of action of eltoprazine, TFMPP and 8-OH-DPAT in reducing aggression in rats. Biol Psychiatry 29:209S

Neill JC, Cooper SJ (1988) Evidence that d-fenfluramine anorexia is mediated by 5-HT$_1$ receptors. Psychopharmacology 97:213–218

Olivier B, Mos J (1992) Rodent models of aggressive behaviour and serotonergic drugs. Prog Neuropsychopharmacol Biol Psychiatry 16:847–870

Olivier B, Mos J (1986) Serenics and aggression. Stress Med 2:197–209

Olivier B, Mos J, Van der Heyden JAM, Hertoz J (1989) Serotonergic modulation of social interactions in isolated male mice. Psychopharmacology 97:154–156

Othmer E, Othmer SC (1987) Effect of buspirone on sexual dysfunction in patients with generalised anxiety disorder. J Clin Psychiatry 48:201–203

Peroutka SJ, Snyder SH (1980) Long-term antidepressant treatment decreases spiroperidol-labelled serotonin receptor binding. Science 210:956–962

Pinder RM, Wieringa JH (1993) Third generation antidepressants. Med Res Rev 13:259–325

Pomerantz SM, Hepner BC, Wertz JM (1993) 5-HT$_{1A}$ and 5-HT$_{1C/1D}$ receptor agonists produce reciprocal effects on male sexual behavior of rhesus monkeys. Eur J Pharmacol 243:227–234

Poncelet M, Pério A, Simiand J, Gout G, Soubrié P, Le Fur G (1995) Antidepressant-like effects of SR 57227A, a 5-HT$_3$ receptor agonist, in rodents. J Neural Transm (in press)

Porsolt RD (1991) Pharmacological models of depression. In: Olivier B, Mos J, Slangen JL (eds) Animal models in psychopharmacology. Birkhaüser-Verlag, Basel, pp 137–160

Rasmussen K, Aghajanian GK (1988) Potency of antipsychotics in reversing the effects of a hallucinogenic drug on locus coeruleus neurons correlates with compound binding affinity. Neuropsychopharmacology 1:101–107

Ratey JJ, Chandler HK (1995) Serenics. Therapeutic potential in aggression. CNS Drugs 4:256–260

Reynolds GP, Rossor MN, Iversen LL (1983) Preliminary studies of human cortical 5-HT$_2$ receptors and their involvement in schizophrenia and neuroleptic drug action. J Neural Transm 18 (Suppl):273–277

Rickels K, Amsterdam JD, Clary C, Puzzuoli G, Schweitzer E (1991) Buspirone in major depression: a controlled study. J Clin Psychiatry 52:34–38

Rinaldi-Carmona M, Congy C, Santucci V, Simiand J, Gautrel B, Neliat G, Labeeuw B, Le Fur G, Soubrié P, Brelière JC (1992) Biochemical and pharmacological properties of SR 46349B, a new potent and selective 5-hydroxytriptamine$_2$ receptor antagonist. J Pharmacol Exp Ther 262:759–768

Rivest R, Jolicoeur FB, Marsden CA (1991) Use of amfonelic acid to discriminate between classical and atypical neuroleptics and neurotensin: an in vivo voltammetric study. Brain Res 544:86–93

Robinson PH, Checkley SA, Russel GFM (1985) Suppression of eating by fenfluramine in patients with bulimia nervosa. Br J Psychiatry 146:169–176

Samanin R, Mennini T, Ferraris A, Bendotti C, Borsini F, Garaltini S (1979) m-Chlorophenylpiperazine: a central serotonin agonist causing powerful anorexia in rats. Naunyn Schmiedebergs Arch Pharmacol 308:159–163

Sheard MH (1969) The effect of pCPA on behavior in rats: relation to brain serotonin and 5-HIAA. Brain Res 15:524–528

Shillito EE (1969) The effect of p-chlorophenylalanine on social interactions of male rats. Br J Pharmacol 36:193–194

Sijbesma H, Schipper J, De Kloet ER, Mos J, Van Aken H, Olivier B (1991) Postsynaptic 5-HT$_1$ receptors and offensive aggression in rats: a combined behavioural and autoradiographic study with eltoprazin. Pharmacol Biochem Behav 38:447–458

Simiand J, Keane PE, Barnouin MC, Keane M, Soubrié P, Le Fur G (1993) Neuropsychopharmacological profile in rodents of SR 57746A, a new, potent 5-HT$_{1A}$ receptor agonist. Fund Clin Pharmacol 7:413–427

Sorensen SM, Kehne JH, Fadayel GM, Humphreys TM, Ketteler HJ, Sullivan CK, Taylor VL, Schmidt CJ (1993) Characterization of the 5-HT$_2$ receptor antagonist MDL 100,907 as a putative atypical antipsychotic: behavioral, electrophysiological and neurochemical studies. J Pharmacol Exp Ther 266:684–691

Soubrié P (1986) Reconciling the role of central serotonin neurons in human and animal behavior. Behav Brain Res 9:319–364

Sprouse JS, Aghajanian GK (1986) (–)-Propranolol blocks the inhibition of serotonergic dorsal raphe firing by 5-HT$_{1A}$ selective agonists. Eur J Pharmacol 128:295–298

Sprouse JS, Aghajanian GK (1987) Electrophysiological responses of serotonergic dorsal raphe neurons to 5-HT$_{1A}$ and 5-HT$_{1B}$ agonists. Synapse 1:3–9

Stein DJ, Wise CD, Berger BD (1973) Anti-anxiety action of benzodiazepines: decrease in activity of serotonin neurons in the punishment system. In: Costa E, Greengard P (eds) The benzodiazepines. Raven Press, New York, pp 299–326

Stewart IC, Dopald P, Munro JF (1986) Neuropharmacological treatment of obesity. In: Ferrari E, Brambilla F (eds) Disorders of eating behaviour: a psychoneuroendocrine approach. Pergamon Press, Oxford, pp 295–303

Szara S (1967) Hallucinogenic amines and schizophrenia. In: Himwich HE, Kety SS, Snythies JR (eds) Amines and schizophrenia. Pergamon Press, New York, pp 181–197

Tenen SS (1967) The effects of p-chlorophenylalanine, a serotonin depletor, on avoidance acquisition, pain sensitivity and related behaviour in the rat. Psychopharmacology 10:204–219

Thiébot MH, Soubrié P, Hamon M, Simon P (1984) Evidence against the involvement of serotonergic neurons in the anti-punishment activity of diazepam in the rat. Psychopharmacology 182:355–359

Traber J, Glaser T (1987) 5-HT$_{1A}$ receptor-related anxiolytics. Trends Pharmacol Sci 8:432–437

Twarog BM, Page JH (1953) Serotonin content of some mammalian tissues and urine and a method of its determination. Am J Physiol 175:157–161

Volonte M, Ceci A, Borsini F (1992) Effect of haloperidol and clozapine on (+) SKF 10047-induced dopamine release: role of 5-HT$_3$ receptors. Eur J Pharmacol 213:163–164

Waldbillig RJ (1979) The role of the dorsal and median raphe in the inhibition of muricide. Brain Res 160:341–346

Wang RY, Ashby CR, Edwards E, Zhang JY (1994) The role of 5-HT$_3$-like receptors in the action of clozapine. J Clin Psychiatry 55 (Suppl B):23–26

White SM, Kucharik RF, Moyer JA (1991) Effects of serotonergic agents on isolation-induced aggression. Pharmacol Biochem Behav 39:729–736

Willner P (1991) Behavioural models in psychopharmacology. Cambridge University Press, Cambridge

Woods A, Smith C, Szewczak M, Dunn RW, Comfeldt M, Borbett R (1993) Selective serotonin re-uptake inhibitors decrease schedule-induced polydipsia in rats: a potential model for obsessive compulsive disorder. Psychopharmacology 112:195–198

Wooley DM, Shaw E (1954) A biochemical and pharmacological suggestion about certain mental disorders. Proc Natl Acad Sci 40:228–231

Yadin E, Friedman E, Wagner HB (1991) Spontaneous alternation behavior: an animal model for obsessive-compulsive disorder? Pharmacol Biochem Behav 40:311–315

Zajecka J, Fawcett F, Schaff M, Jeffries H, Guy C (1991) The role of serotonin in sexual dysfunction: fluoxetine associated orgasm dysfunction. J Clin Psychiatry 52:66–68

CHAPTER 28

Current Psychiatric Uses of Drugs Acting on the Serotonin System

P. Blier and C. de Montigny

A. Introduction

The pharmacotherapy of many psychiatric disorders is carried out with medications that act either directly on elements regulating the function of the serotonin (5-HT) system or indirectly *via* other neurotransmitters impinging on 5-HT neurons. Some 40 years ago, tricyclic antidepressant (TCA) drugs and monoamine oxidase inhibitors (MAOI) were introduced for the treatment of major depression. These major therapeutic breakthroughs were due to serendipity and the neurochemical effects underlying their antidepressant effect remained unknown for several years. In 1964, it was postulated that TCA drugs acted by blocking the reuptake of noradrenaline (NE) and/or 5-HT (Glowinski and Axelrod 1964), and by then the capacity of MAOIs to inhibit the oxidative catabolism of these two neurotransmitters had been documented. In the following decade, several reports indicated that these drugs were also effective in the treatment of psychiatric disorders other than major depression. Later, selective 5-HT reuptake inhibitors (SSRIs) were synthesized and subsequently demonstrated in the 1980s to be effective in treating depression and some anxiety disorders. These discoveries established that the 5-HT system plays a pivotal role in the therapeutic effects of several drugs in psychiatry.

Despite considerable advances in knowledge concerning mechanisms of neurotransmission since the introduction of antidepressant drugs, it remained difficult to explain why such medications were effective in depressive and anxiety disorders because their therapeutic effects were delayed by at least a week and their acute biochemical actions, as well as side effects, occurred within hours. Numerous laboratories focused their efforts on the adaptive properties of the 5-HT system in different brain regions, while others focused on the identification of different 5-HT receptor subtypes (see Fig. 1). These research endeavours provided at least partial answers to the question formulated above. The present chapter will concentrate on the role of the 5-HT system in the treatment of four classes of psychiatric disorders, namely depressive disorders, obsessive compulsive disorders, anxiety disorders, and schizophrenia. Very little will be mentioned with regard to the pathogenesis of these conditions, as the etiology and the therapeutics of psychiatric disorders are two distinct issues. For instance, it is not because SSRIs are effective in several

disorders that 5-HT necessarily plays an important role in their pathophysiology. The model of Parkinson's disease, being attributable to a decrease of
dopamine in the nigrostriatal pathway and treated by giving a dopamine

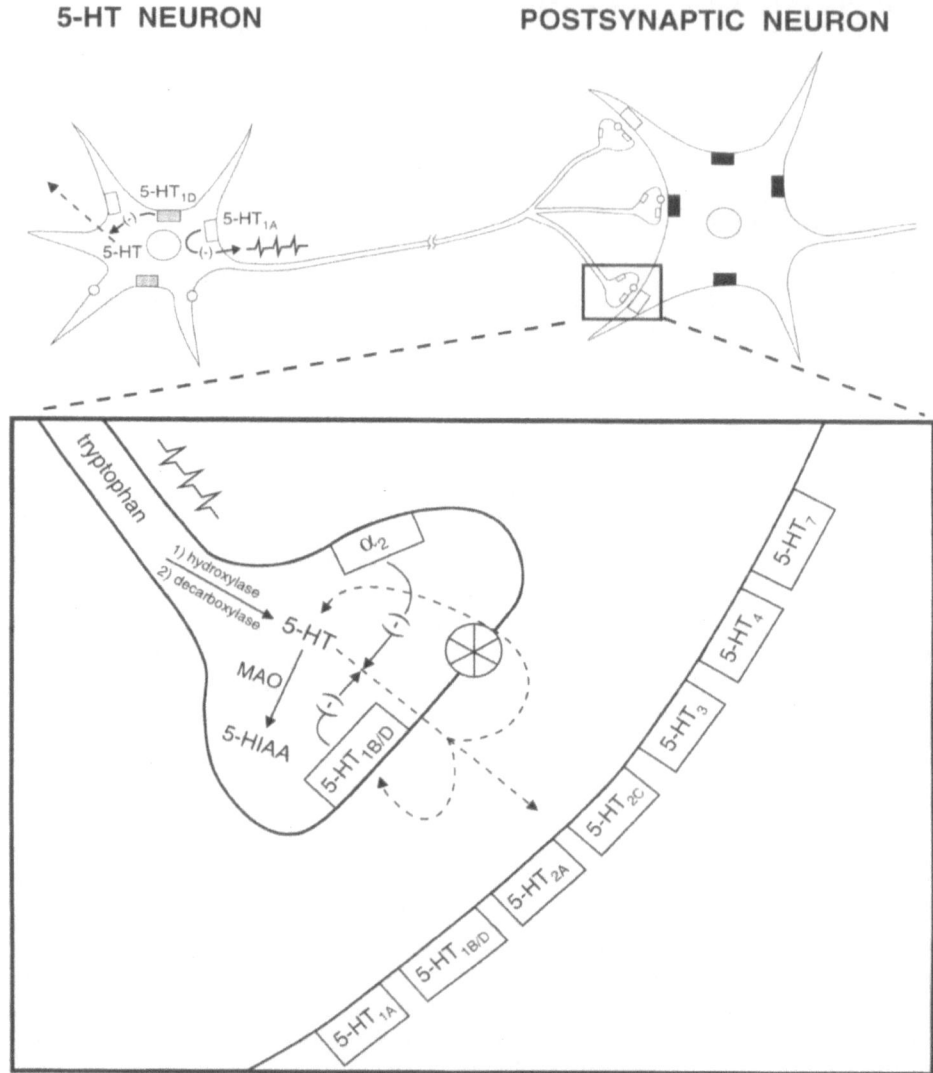

Fig. 1. Presynaptic and postsynaptic factors regulating the effectiveness of serotonin
(*5-HT*) neurotransmission. Only the subtypes of 5-HT receptors for which an electrophysiological response has been identified in unitary recordings are depicted. 5-HT$_{1A}$
receptors on the cell body of 5-HT-containing neurons mediate an inhibitory effect on
firing activity by the opening of potassium channels. 5-HT$_{1D}$ receptors on the cell body
and 5-HT$_{1B/D}$ receptors on the terminals exert an inhibitory action on 5-HT release. The
hub on the 5-HT terminal and on the 5-HT neuron represent the high-affinity reuptake
carrier. Different symbols have been used to depict the postsynaptic 5-HT receptors to
indicate that they have different pharmacological properties

precursor or a dopamine receptor agonist, is more the exception than the rule. For each diagnostic, a brief clinical description will be presented, based on the Diagnostic and Statistical Manual IV of the AMERICAN PSYCHIATRIC ASSOCIATION (1994), and their approximate prevalence given in order to give the readership an idea of the importance of the different disorders. The putative mechanisms of action of the different drugs used in these disorders will then be described.

B. Depression

Major depression is characterized by a depressed mood or a loss of interest or pleasure in nearly all activities for a period of at least 2 weeks. The individual must also present at least four of the following symptoms: changes in appetite or weight, sleep, and psychomotor activity (*i.e.*, either an increase or a decrease), fatigue or decreased energy, feeling of worthlessness or guilt, decreased ability to think, concentrate, or make decisions, and recurrent thoughts of death or suicidal ideation. There is an atypical form of depression in which appetite and sleep are increased. Although this presentation is much more rare than the typical one, it is interesting from a conceptual point of view as patients with this form of depression respond well to MAOIs but less so to the TCA imipramine (LIEBOWITZ et al. 1988). A less severe form of depression is now recognized and called dysthymia. In this case, the diagnosis requires that fewer symptoms be present in addition to the depressed mood and that the latter not necessarily be present every day, but for a 2-year duration. The incidence of major depression in the general population is around 5% with a female-to-male ratio of affected individuals of about 2 to 1.

I. Mechanisms of Action of Antidepressant Treatments

Extensive electrophysiological investigations carried out in our laboratory have documented that several types of antidepressant treatments enhance 5-HT neurotransmission in the rat hippocampus (see BLIER and DE MONTIGNY 1994). This net effect that is common to the major types of antidepressant treatments is, however, mediated via different mechanisms (see Fig. 1 and Table 1). All TCA drugs, independently of their capacity to inhibit the reuptake of 5-HT and/or NE, progressively enhance the responsiveness of postsynaptic 5-HT_{1A} receptors in the hippocampus with a time course that is congruent with the delayed onset of action of these drugs in major depression (DE MONTIGNY and AGHAJANIAN 1978). It has also been demonstrated by our group and other laboratories that this enhanced responsiveness to 5-HT also occurs in other, but not all, brain regions and that the sensitivity of 5-HT receptor subtypes other than that of the 5-HT_{1A} receptors is also altered. For instance, in the facial motor nucleus, the receptors mediating the effect of 5-HT are of the 5-HT_2 subtype, and they are sensitised following repeated TCA

Table 1. Effects of long-term administration of antidepressant treatments of 5-HT neurotransmission[a]

	Somatodendritic autoreceptor responsiveness[b]	Terminal autoreceptor responsiveness[c]	Postsynaptic responsiveness[b]	Net effect[d]
Tricyclic antidepressants	0	0	↑	↑′
Electroconvulsive shocks	0	0	↑	↑
Monoamide oxidase inhibitors	↓	0	0/↓	↑
Selective 5-HT reuptake inhibitors	↓	↓	0	↑
5-HT$_{1A}$ agonists	↓	0	↑	↑e

[a] Rats were treated for at least 14 days.
[b] Assessed by microiontophoresis or systemic injection of 5-HT receptor agonists.
[c] Assessed by comparing the effect of agonists or antagonists in control and treated rats.
[d] Determined from the firing activity of the presynaptic neurons and the effect of stimulating 5-HT fibers.
[e] Effect obtained by an enhanced tonic activation of postsynaptic 5-HT$_{1A}$ receptors resulting from a normal amount of 5-HT (normalized 5-HT neuronal firing) and the presence of the exogenous 5-HT$_{1A}$ agonist.
↑, increased; ↓, decreased; 0, no change.

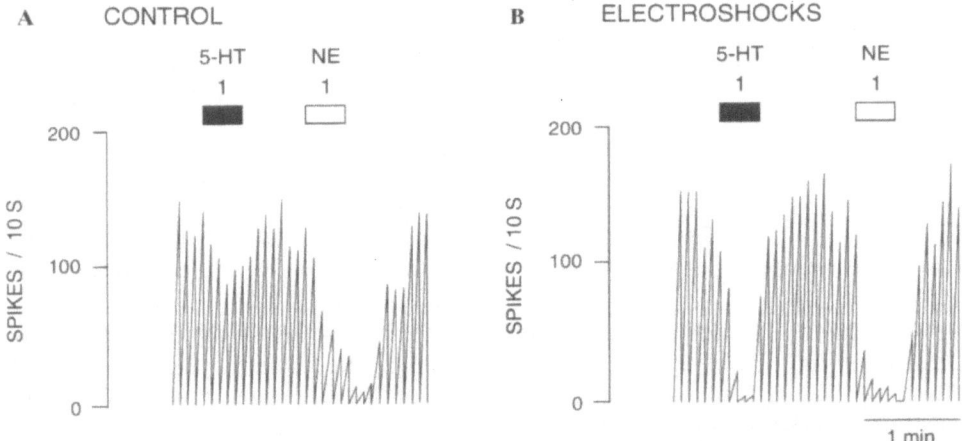

Fig. 2A,B. Integrated firing rate histograms of two CA$_3$ dorsal hippocampus pyramidal neurons showing their response to microiontophoretic applications of serotonin (*5-HT*) and noradrenaline (*NE*) in a control rat (**A**) and in a rat treated with six electroconvulsive shocks over a 2-week period (**B**). The shocks were delivered under halothane anesthesia and the recordings were carried out under chloral hydrate anesthesia. The *bars* above the traces indicate the duration of the application (50s) for which the ejection current is given in nanoamperes. The time base applies to both traces. Note that only the responsiveness to 5-HT is increased

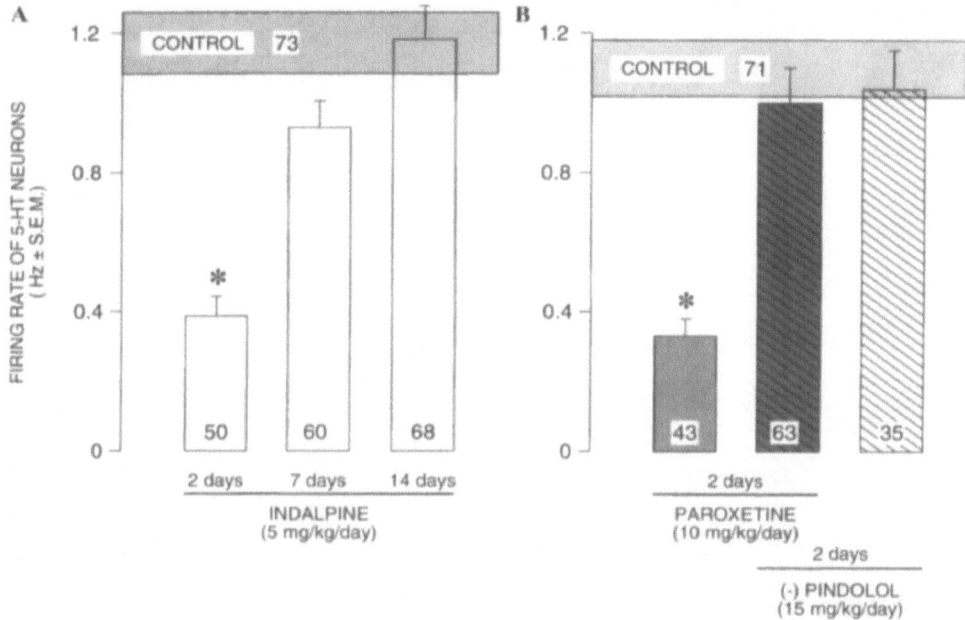

Fig. 3A,B. Effects of the administration of selective serotonin (5-HT) reuptake inhibitors and of the 5-HT$_{1A}$ antagonist (–)pindolol on the spontaneous firing activity of dorsal raphe 5-HT neurons recorded in chloral hydrate anesthetized rats. **A** Indalpine was given intraperitoneally on a daily basis and the recordings were carried out 24 h after the last dose. **B** Paroxetine and (–)pindolol were administered using osmotic minipumps implanted subcutaneously, and the recordings were carried out with the minipumps in place. The *numbers* at the bottom of each column and in the shaded areas represent the number of 5-HT neurons recorded in each group. *$p < 0.001$ with respect to the control group which is depicted by the corresponding *horizontal shaded areas*

drug administration (MENKES et al. 1980). Repeated, but not single, electroconvulsive shock administration also induces this sensitization to 5-HT in the dorsal hippocampus (DE MONTIGNY 1984; Fig. 2). This is consistent with the clinical effectiveness of repeated sessions.

MAOIs, SSRIs, and 5-HT$_{1A}$ agonists all induce an initial attenuation of the firing activity of 5-HT neurons upon treatment initiation because they increase the degree of activation of the somatodendritic 5-HT$_{1A}$ autoreceptors which control the firing rate of 5-HT neurons (BLIER and DE MONTIGNY 1983, 1985a, 1987; Fig. 3). This is followed by a gradual recovery to normal firing activity of 5-HT neurons when the treatment is continued for 2–3 weeks due to a desensitization of the 5-HT$_{1A}$ autoreceptors (Fig. 4). At this point in time, MAOIs enhance 5-HT transmission by increasing the amount of 5-HT released per action potential as a result of a greater concentration of 5-HT in the terminals (BLIER et al. 1986). SSRIs produce the same effect, not by augmenting the releasable pool of 5-HT like MAOIs do, but rather by desensitizing

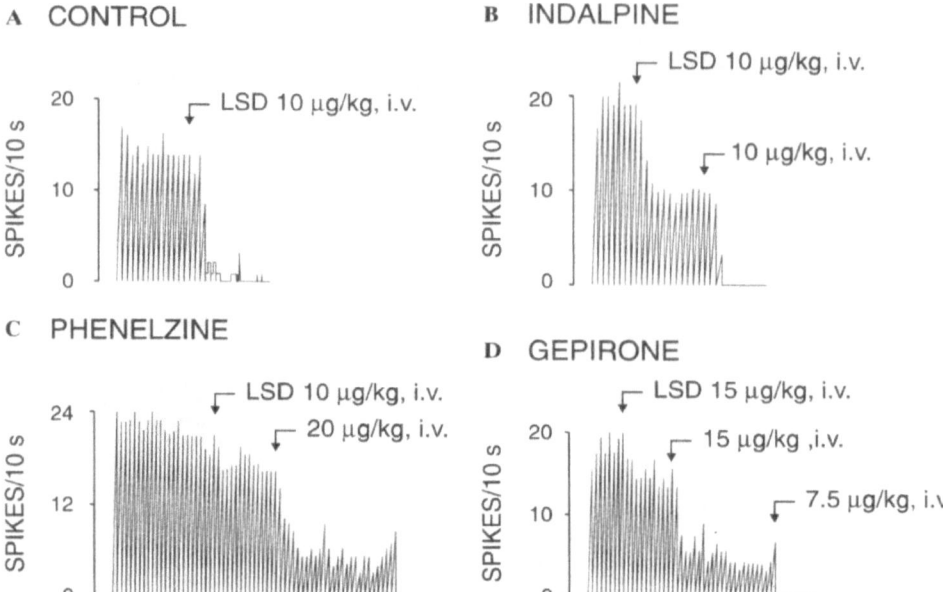

Fig. 4A–D. Integrated firing rate histograms of serotonin (5-HT) neurons recorded from the dorsal raphe nucleus showing the inhibitory effect of the 5-HT autoreceptor agonist lysergic acid diethylamide (*LSD*) in a control rat (**A**) and in rats that received long-term treatments with the selective 5-HT reuptake inhibitor indalpine (5 mg/kg per day, intraperitoneally × 14 days) (**B**), the monoamine oxidase (MAO) inhibitors phenelzine (15 mg/kg per day, intraperitoneally × 21 days) (**C**), or the 5-HT$_{1A}$ agonist gepirone (15 mg/kg per day, subcutaneously using an osmotic minipump) (**D**). Each peak represents the number of action potentials recorded per 10-s interval

the terminal 5-HT autoreceptor which exerts a negative influence on the amount of 5-HT that is released per impulse (CHAPUT et al. 1986). 5-HT$_{1A}$ agonists produce an enhanced tonic activation of postsynaptic 5-HT$_{1A}$ receptors, as a result of a normalized firing activity of 5-HT neurons (and of 5-HT release as well) in the presence of the exogenous 5-HT$_{1A}$ agonist acting on normosensitive postsynaptic 5-HT$_{1A}$ receptors (BLIER and DE MONTIGNY 1987).

II. An Accelerating Strategy: 5-HT$_{1A}$ Autoreceptor Antagonism

As previously mentioned, all antidepressant drugs require at least 1–2 weeks of sustained administration before exerting a clinically significant anti-depressant effect and another 4–6 weeks to produce their maximal therapeutic benefit. A novel strategy has recently been devised to accelerate the antide-pressant response (DE MONTIGNY et al. 1993). It consists in combining an SSRI with the 5-HT$_{1A}$/β-adrenoceptor antagonist pindolol (ARTIGAS et al. 1994; BLIER and BERGERON 1995; Fig. 5a). This approach is based on the capacity of pindolol to block the 5-HT$_{1A}$ autoreceptor on the cell body of 5-HT neurons preventing the initial decrease of the firing activity of these neurons at the beginning of the SSRI treatment, as described in the preceding section (Fig. 3). It is important to emphasize that pindolol does not block certain postsynaptic

A **PAROXETINE PLUS PINDOLOL**

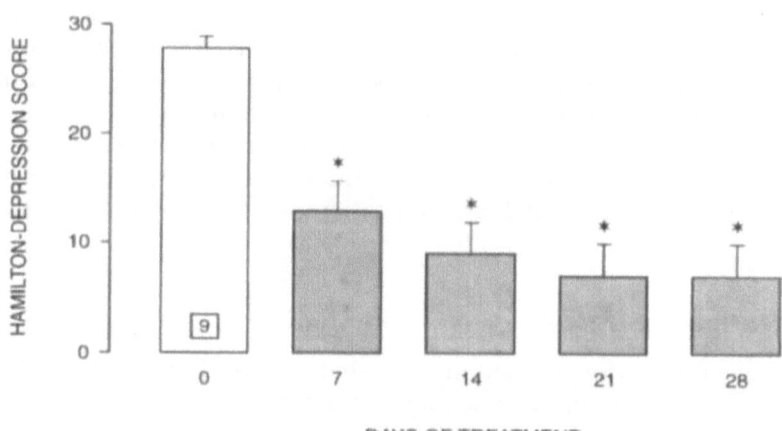

DAYS OF TREATMENT

B **TRIMIPRAMINE PLUS PINDOLOL**

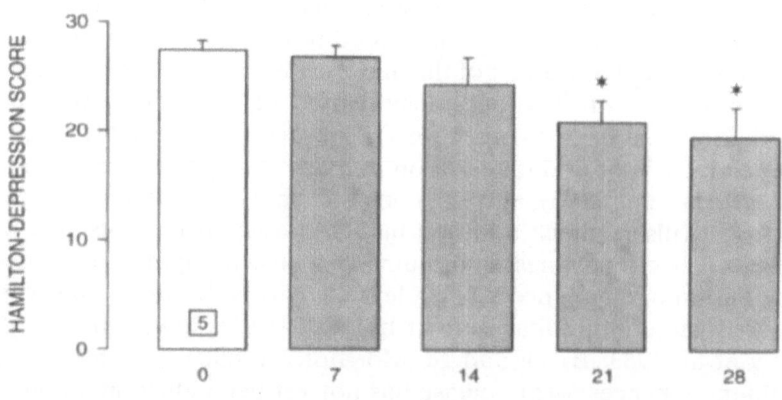

DAYS OF TREATMENT

Fig. 5A,B. Hamilton-Depression scores on the 21-item scale depicting the intensity of depressive symptomatology in depressed unipolar patients. Paroxetine was given at a 20 mg/day dose throughout the 28 days (**A**), whereas, the tricyclic agent trimipramine was administered at 75 mg at bedtime for the first week and at 150 mg at bedtime for the subsequent 3 weeks (**B**). Pindolol was given in both groups at a dose of 2.5 mg three times daily throughout the trials. The degree of improvement obtained in the trimipramine plus pindolol is generally that obtained with an antidepressant drug used alone. *$p < 0.05$ when compared to the corresponding baseline values (day 0)

5-HT$_{1A}$ receptors (ROMERO et al. 1996). Otherwise, the simultaneous blockade of both pre- and postsynaptic 5-HT$_{1A}$ receptors would prevent a net increase in 5-HT neurotransmission via postsynaptic 5-HT$_{1A}$ receptors, which are present in a particular high density in the limbic system. Since 5-HT release in most postsynaptic regions is highly dependent on electrical impulse flow in 5-HT axons, the prevention of the initial decrease of 5-HT neuronal firing activity by pindolol physiologically mimics the desensitization of the 5-HT$_{1A}$ autoreceptor which has been documented to occur after a 2-week SSRI treatment. In support of this hypothesis, pindolol fails to accelerate the onset of action of those TCAs which do not block 5-HT reuptake and, consequently, which do not initially inhibit the firing rate of 5-HT neurons (BLIER et al. 1997; Fig. 5b). Two placebo-controlled trials indicate that pindolol combination with an SSRI from the beginning of the treatment does in fact accelerate the therapeutic effect of SSRIs in depressed patients (ARTIGAS et al. 1996; ISAAC et al. 1996).

III. Potentiation Strategies

A major caveat of antidepressant drugs is that one third of patients will fail to improve even when given an adequate dose for a sufficient time. Unfortunately, extensive research has thus far not been successful in identifying predictive factors for the different classes of agents. In addition, many patients who do respond favorably fail to achieve remission. Substituting antidepressant drugs is not a time-efficient strategy as a washout is sometimes necessary, and the same delays mentioned above are necessary before obtaining a response. Adding a second agent, which need not be antidepressant *per se*, to potentiate the therapeutic effect of the first drug is now considered much more time- and cost-efficient. The best documented of these strategies so far is lithium addition (see DE MONTIGNY 1994). There are controlled trials supporting the effectiveness of lithium addition in TCA-resistant patients (HENINGER et al. 1983; JOFFE et al. 1993), and even one that shows equal efficacy of lithium and electroconvulsive shock addition in TCA-resistant patients (DINAN and BARRY 1989). The effectiveness of lithium addition is thought to result from the capacity of lithium to enhance 5-HT release in certain brain regions when it is administered acutely or subacutely at blood levels as low as 0.4 mEq/L (DE MONTIGNY et al. 1983; BLIER and DE MONTIGNY 1985b). The mechanism by which lithium enhances 5-HT release has not yet been elucidated, but it is at least clear that it is different from that of SSRIs or MAOIs. Furthermore, lithium does not modify the responsiveness of postsynaptic 5-HT receptors that TCA drugs sensitize (BLIER and DE MONTIGNY 1985b). It is, therefore, not surprising that lithium has been reported to be effective with TCA drugs, SSRIs, and MAOIs. Although the reported rate of response to this strategy shows considerable variation, it is estimated to be effective in about 50% of the cases. Considering that the rate of response to the three main classes of antidepressant drugs is about 60%, the use of lithium addition in patients not

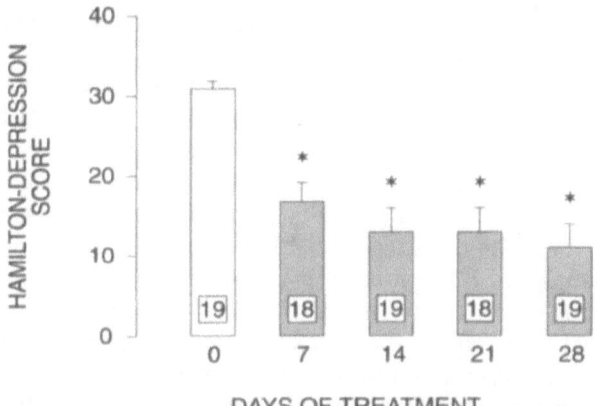

Fig. 6. Intensity of the depressive syndrome, using the 21-item Hamilton-Depression scale, in drug-resistant patients who received pindolol 2.5 mg three times daily without altering their antidepressant drug regimen. Eight patients were on paroxetine (20–40 mg/day), five were on sertraline (50–100 mg/day), three were on fluoxetine (20–40 mg/day), and two were on moclobemide (900 mg/day). The unequal numbers at the bottom of the columns are because one patient could not be assessed at days 7 and 21. None of the five patients on sertraline responded to pindolol addition. $*p < 0.01$ when compared with the corresponding day 0 value using one-way analysis of variance

responding to one of the above-mentioned drugs thus increases the success rate of these three types of drug treatments to about 80% (see Bauman et al. 1996).

Pindolol addition is not only an accelerating strategy, but it has also been reported to potentiate the antidepressant effect of SSRIs and MAOIs (Artigas et al. 1994; Blier and Bergeron 1995; Artigas et al. 1996; Fig. 6). In a double-blind trial carried out in nonresistant patients, the response rate in the SSRI plus placebo group was 58% whereas in the SSRI plus pindolol group it was 75% (Artigas et al. 1996). In another controlled trial using mainly resistant patients, the response rate was 73% in the trazodone (a weak 5-HT reuptake blocker) plus pindolol arm and 20% in the trazodone plus placebo arm (Maes et al. 1996). One may speculate that depressed patients who are rapidly and markedly improved by the addition of the 5-HT$_{1A}$ antagonist pindolol may have failed to respond to their antidepressant drug regimen because their 5-HT$_{1A}$ autoreceptors did not desensitize, thus precluding a recovery of the firing activity of 5-HT neurons.

The addition of the 5-HT$_{1A}$ agonist buspirone to an SSRI regimen is a clinically useful strategy in resistant patients. Although controlled studies are still lacking for this approach, its effectiveness is probably similar to that of lithium or pindolol addition (Jacobsen 1991; Joffe and Schuller 1993). When SSRI-resistant patients respond to buspirone addition, it is noteworthy that they do so much more rapidly than when buspirone is used alone in untreated depressed patients (Robinson et al. 1990). The beneficial effect of buspirone in SSRI-resistant patients is likely to result from a direct activation of postsynaptic 5-HT_{1A} receptors, with a dampening of its suppressant effect on 5-HT neuron firing activity by the prior desensitization of 5-HT_{1A} autoreceptors by the SSRI. Verification of this hypothesis, as well as that for that for the beneficial effect of pindolol in this same group of patients, awaits the development of a probe for the human 5-HT_{1A} autoreceptor.

The combination of an SSRI with a TCA blocking NE reuptake has been reported to exert a more rapid and robust antidepressant effect than with an antidepressant drug alone (Nelson et al. 1991). This strategy may be more effective because the SSRI gradually increases the function of 5-HT neurons and the TCA drug enhances postsynaptic 5-HT responsiveness (see Table 1), or the TCA drug progressively increases the function of NE neurons as a result of NE reuptake blockade. It is noteworthy that the putatively greater efficacy of the latter approach finds apparent support in the results of controlled clinical trials with the new antidepressant drug venlafaxine. This drug is an inhibitor of both the reuptake of 5-HT and of NE. Unlike the SSRIs, which present a relatively flat dose-response in the treatment of depression, the effectiveness of venlafaxine appears to be dose-dependent. This is based on three lines of evidence. First, venlafaxine seems to present a dose-response relationship under placebo-controlled conditions in the range of 75–375 mg per day (Guelfi et al. 1995). Second, 200 mg venlafaxine given daily has been shown to be superior to 40 mg a day of the SSRI fluoxetine by week four of a 6-week trial in hospitalized patients with major depression and melancholia (Clerc et al. 1994). Finally, the use of venlafaxine in treatment-resistant depression has been shown to be effective by two groups of investigators (Nierenberg et al. 1994; de Montigny et al. 1995). Furthermore, in the dose range of 300–375 mg per day (achieved within one week), venlafaxine induced a significantly greater treatment response after 1 and 2 weeks than was observed in the placebo group (see Montgomery 1995). These observations suggest that venlafaxine may be endowed with a more rapid onset of action as well as a greater efficacy.

C. Obsessive Compulsive Disorder

Obsessive-compulsive disorder (OCD) is characterized by recurrent and persistent thoughts, impulses, or images (obsessions) and/or repetitive behaviors or mental acts (compulsions) which are aimed at preventing or reducing distress. The obsessions are intrusive and inappropriate, and not merely worries about real-life problems. Patients actively attempt to ignore or suppress their obsessions, even though they recognize that these are the product of their own mind. The compulsions are performed in response to an obsession or in accordance with extremely rigid rules. Finally, an essential diagnostic feature is that those symptoms take up more than 1 hour per day, or significantly interfere with the individual's normal routine, occupational functioning, or usual social activities or relationships. OCD is presently classified with anxiety disorders in the American classification of psychiatric disorders (DSM-IV). Many experts in the field, however, share the opinion that it should stand on its own in a different category, as recognized in the International Classification of Diseases (ICD-10), perhaps with several of the so-called OCD-related disorders such as anorexia nervosa, bulimia, body dysmorphic disorder, and compulsive gambling. Large community studies carried out in the 1980s have documented OCD rates ranging from 1.6% to 3.0%. These studies, therefore, make OCD the fourth most common psychiatric disorder following substance abuse, affective disorder, and phobias. Notably, OCD is more prevalent than schizophrenia.

While there is little evidence for an important role of the 5-HT system in the etiopathology of OCD, there is conducive evidence for a major role of this monoaminergic system in the pharmacotherapy of this disorder. First and foremost, among all antidepressant treatments including electroconvulsive shocks, only the drugs which are potent 5-HT reuptake blockers exert a clear therapeutic effect in OCD patients (JENIKE 1993; Table 2). The first placebo-controlled trials of the TCA drug chlorimipramine, the only drug in this class with potent 5-HT reuptake blocking property, clearly demonstrated the effectiveness of this pharmacotherapy (MONTGOMERY 1980). Subsequently, elegant crossover studies with chlorimipramine and the selective NE reuptake blocker desipramine (which is also a TCA), have clearly shown it to be the capacity of the former drug to block the 5-HT reuptake process that is responsible for the anti-OCD effect. The advent of the newer nontricyclic 5-HT reuptake blockers truly established the pivotal role of the 5-HT system in the beneficial effect of certain antidepressant drugs. Since most of these drugs (fluoxetine, fluvoxamine, sertraline, and paroxetine) are from different chemical families and the only common neurobiological property they share is their capacity to potently block the 5-HT reuptake process, this leaves little doubt that the anti-OCD effect of these drugs is mediated via the 5-HT system. The second line of evidence for an important role of the 5-HT system in the anti-OCD effect of

Table 2. Effectiveness of antidepressant treatments in major depression and obsessive compulsive disorder

	Depression	Obsessive compulsive disorder
Tricyclic antidepressant drugs	+++	0[a]
Selective 5-HT reuptake inhibitors	+++	++
Monoamine oxidase inhibitors	+++	+
5-HT$_{1A}$ agonists	++	0
Electroconvulsive shocks	+++	0

+++, produces a remission in at least half the patients; ++, produces a partial response in about half of patients with obsessive compulsive disorder or a remission in a significant percentage of depressed patients; +, may be useful in some cases; 0, no therapeutic benefit.
[a] With the exception of chlorimipramine which is a drug with a potent 5-HT reuptake blocking property.

selected antidepressant drugs is the capacity of the nonselective 5-HT antagonist metergoline to increase anxiety and OCD symptoms in drug-remitted patients (BENKELFAT et al. 1989; GREENBERG et al. 1994).

The peculiar characteristic of SSRIs on the 5-HT system, as depicted in Table 1, is their capacity to desensitize terminal 5-HT autoreceptors. These results were, however, obtained in the rat hippocampus and subsequently in the rat hypothalamus after 2- to 3-week treatments (CHAPUT et al. 1986; MORET and BRILEY 1990). It is, however, hazardous to extrapolate such results to the treatment of OCD with SSRI for two reasons. First, the anti-OCD effect of SSRI generally requires a much longer delay before being maximal than the antidepressant effect (MONTGOMERY 1994). Second, positron emission tomography scanning studies in humans have clearly delineated the orbitofrontal cortex–head of the caudate nucleus–thalamus neuronal circuitry as being involved in mediating OCD symptomatology (for a review, see INSEL 1992), and not the hippocampus or the hypothalamus.

In order to overcome these limitations, the following experiments were undertaken. 5-HT release and terminal 5-HT autoreceptor sensitivity were assessed in guinea pigs, since their terminal 5-HT autoreceptors exhibit the same pharmacological properties as in humans (they are "h5-HT$_{1B\text{-LIKE}}$"; see new 5-HT subtype nomenclature of HARTIG et al. 1996). The treatment period was extended to 8 weeks in an attempt to approach the time necessary to obtain a near maximal anti-OCD effect. Finally, 5-HT release and terminal autoreceptor responsiveness was examined in the frontal cortex and in its orbitofrontal subdivision, as well as in the head of the caudate nucleus (EL MANSARI et al. 1995).

The SSRIs were administered using osmotic minipumps, implanted subcutaneously, so as to obtain a constant infusion of the drugs and mimic as much as possible the plasma levels achieved in humans. Indeed, SSRIs are metabolized much more rapidly by rodents than by humans. Consequently, administering such agents even on a twice daily basis to laboratory animals does not produce the sustained plasma levels obtained in humans with oral administration, thus making it to difficult to extrapolate the animal data to human therapeutics.

Fig. 7A,B. Effects of sustained administration of the selective serotonin (5-HT) reuptake inhibitor (SSRI) paroxetine (10 mg/kg per day, subcutaneously using osmotic minipumps) on (**A**) the electrically evoked release of [^{3}H]5-HT in guinea pig brain slices prepared 2 days after removing the minipumps, and (**B**) on the inhibitory effect of the 5-HT autoreceptor agonist 5-methoxytryptamine. The *numbers* at the bottom of each column indicate the number of slices prepared from controls and treated guinea pigs processed in parallel on the same day. The *open circles* represent the results obtained in the controls and the *filled circles* those obtained in the slices prepared from treated animals. In **B**, the agonist was introduced 20 min before the second stimulation period, and the overflow in the latter trial was always compared to that obtained in the first stimulation period in the absence of any drug for the same slices. *$p < 0.05$ when compared to the control value using the two-tailed Student's-t-test; *N.S.* not statistically significant ($p > 0.05$)

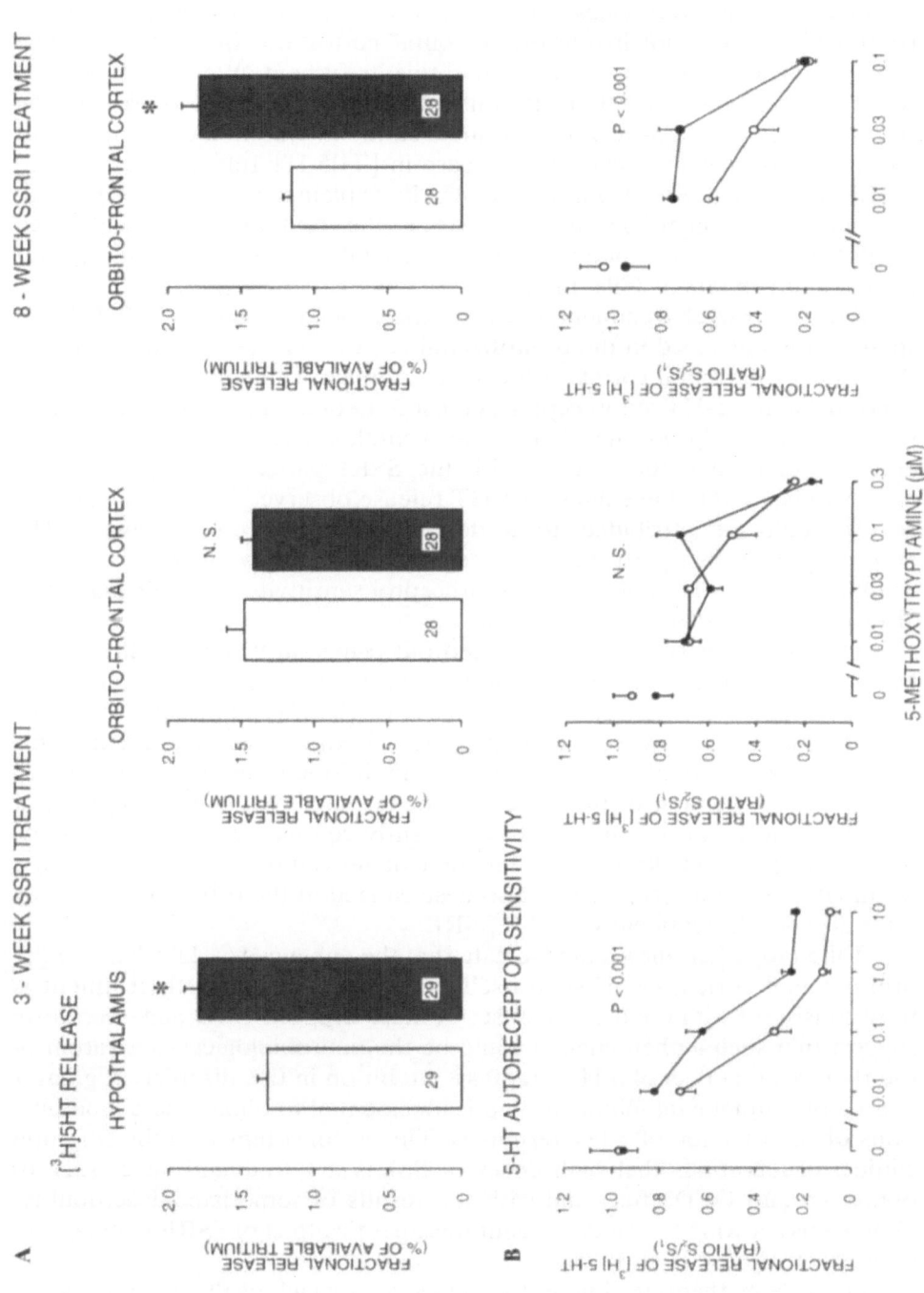

In brain slices preloaded with [^{3}H]5-HT and then continuously superfused, the electrically evoked release of [^{3}H]5-HT was significantly enhanced in the frontal cortex, but not in the orbitofrontal cortex nor in the head of the caudate nucleus after a 3-week paroxetine treatment. We had previously reported on this small but significant enhancement of 5-HT release in the frontal cortex after a 3-week paroxetine treatment. Unlike the hypothalamus and the hippocampus, where the increase in [^{3}H]5-HT release observed following a 3-week paroxetine treatment can be explained in part by a desensitization of the terminal 5-HT autoreceptor, in the frontal cortex the enhanced [^{3}H]5-HT release is attributable to a desensitization of the 5-HT transporter (Blier and Bouchard 1994; Fig. 7).

After an 8-week treatment, however, the evoked release of [^{3}H]5-HT was significantly enhanced in the orbitofrontal cortex to a greater extent than in the rest of the frontal cortex of the same animals. The concentration–effect curve using the 5-HT autoreceptor agonist 5-methoxytryptamine was shifted to the right in orbitofrontal, but not in frontal, cortex slices prepared from guinea pigs treated for 8 weeks with the SSRI paroxetine (Fig. 7). These results indicate that the enhanced 5-HT release observed in the orbitofrontal cortex could be attributed to a desensitization of the terminal 5-HT autoreceptor in that particular brain region. Interestingly, in the head of the caudate nucleus 5-HT release and autoreceptor sensitivity were still unaltered, after an 8-week paroxetine treatment.

The same experiments were carried out using another SSRI, fluoxetine, but at a lower dose regimen. Both in the orbitofrontal cortex and in the head of the caudate nucleus, the evoked release of [^{3}H]5-HT and terminal 5-HT autoreceptor sensitivity were unaltered after 3- and 8-week treatments. This suggests that a marked degree of reuptake inhibition is important in order to induce an attenuation of the sensitivity of terminal 5-HT autoreceptors, at least in some brain regions. We have recently completed analogous experiments using a 10mg/kg per day regimen of fluoxetine and found that the terminal 5-HT autoreceptor was also desensitized in the orbitofrontal cortex after an 8-week treatment with this SSRI.

Taken together, these data indicate that the enhanced 5-HT release in the orbitofrontal cortex, manifesting itself after a prolonged SSRI treatment is fully consistent with the delayed onset of these drugs in OCD and, therefore, suggest that such a phenomenon could be the neurobiological substratum of the therapeutic effect of 5-HT reuptake inhibition in this disorder. A greater degree of reuptake inhibition also appears essential to obtain these modifications of the function of 5-HT terminals. This is consistent with the common clinical observations that high doses of SSRIs are frequently necessary to obtain an anti-OCD effect and with the results of some fixed-dose double-blind trials showing a dose-dependent therapeutic effect of SSRIs (Tollefson et al. 1994; Greist et al. 1995).

On a more theoretical note, the changes – or lack of them – induced by SSRI in different brain regions clearly point to differential properties of 5-HT terminals projecting to the different regions (PIÑEYRO et al. 1994; EL MANSARI et al. 1995). This was also documented in our studies assessing the function of the 5-HT transporter after the above-mentioned SSRI treatment for 3 and 8 weeks. We observed that, after a 3- and an 8-week treatment with paroxetine, the 5-HT transporter was desensitized in the frontal cortex but not in the orbitofrontal cortex, as indicated by the loss of the capacity of paroxetine, when introduced in the superfusion medium, to enhance the electrically evoked overflow of [^{3}H]5-HT in slices prepared from the former but not latter brain region. We had previously documented that this attenuated function of 5-HT transporters is attributable to a decreased number of such sites in the rat hippocampus and frontal cortex (PIÑEYRO et al. 1994). These regional differences raise the question as to why the terminal 5-HT autoreceptors and the 5-HT transporters exhibit such differential changes during sustained and prolonged administration of SSRIs. The mechanisms underlying the differential desensitization of terminal 5-HT autoreceptors are presently under study in our laboratory.

These studies have identified the terminal 5-HT autoreceptor as a putative target for the anti-OCD effect of SSRI. This implies that the use of a 5-HT$_{1D}$ antagonist selective for the terminal 5-HT autoreceptor should produce the same functional changes as SSRIs do, but much more rapidly. Obtaining a more robust anti-OCD effect would also be possible with such an agent used as an adjuvant. This is indeed eagerly awaited, as SSRIs in the vast majority of patients produce only a partial remission, notwithstanding the important proportion of patients not responding at all to SSRIs and those who cannot tolerate their side effects. Unlike in resistant depression, the augmentation strategies described in the preceding section are successful in only a very low number of OCD patients (see BLIER and BERGERON 1996). An antagonist of the terminal 5-HT autoreceptor should, however, also be effective in major depression.

D. Anxiety Disorders

Several diagnostic entities have been lumped together in the group of anxiety disorders mainly because they share a common major symptom, anxiety. This does not imply that they are necessarily disorders with a similar pathogenesis, nor that drugs effective to treat these disorders act on the same brain structures or the same receptor subtypes. Only three disorders will be discussed, as extensive data, fundamental and/or clinical, have not yet been generated for all of them.

I. Panic Disorder

The essential feature of panic disorder is the presence of recurrent, unexpected panic attacks followed by at least 1 month of persistent concern about having another attack, worry about the possible implications or consequences of the panic attacks, or a significant behavioral change related to the attacks. An unexpected, spontaneous, and uncued panic attack is one that is not associated with a situational trigger, and at least two such attacks are required for the diagnosis. These attacks cause extreme anxiety and may mimic a life-threatening illness. The patients may also feel as if they are going crazy or losing control. The sensation is often described as losing control of a motor vehicle entering a curve at high speed. The lifetime prevalence of panic disorder is estimated to be between 1.5% and 3.5% of the general population, based on epidemiological studies carried out worldwide.

The first antidepressant drugs reported to be effective in this disorder were the TCA imipramine and an MAOI (KLEIN 1964; WEST and DALLY 1959). Following the introduction of the SSRIs, it was observed that the latter drugs were also effective in the treatment of panic disorder. It is striking to observe clinically that at the beginning of the treatment, whether a TCA or a SSRI is used at the usual starting dose for depression, an exacerbation of the symptoms occurs, usually an increase in the number of spontaneous attacks. This problem rarely arises when treating depression with the same drugs. Consequently, the starting dose is decreased by at least half to avoid this exacerbation and then progressively increased to the high therapeutic range for depression. This difference between depressed patients and those suffering from panic disorder clearly delineates a hypersensitivity to the serotoninergic and/or noradrenergic properties of the drugs. Subsequently, the therapeutic effects of the drugs occur gradually at about the same rate as in the treatment of major depression. Typically, after about 6 weeks of treatment spontaneous panic attacks do not occur anymore. This observation suggests that the same adaptive changes that mediate the antidepressant response may account for the antipanic action of these drugs. However, not all drugs effective in the treatment of depression are also useful in panic disorder. For example, $5\text{-}HT_{1A}$ agonists have thus far not been found of therapeutic value in panic disorder. The latter clinical observation indicates that postsynaptic 5-HT receptors other than those of the $5\text{-}HT_{1A}$ subtype are responsible for the antipanic effect of SSRIs, TCAs, and MAOIs (Table 3). Obviously, the changes responsible for the antipanic effect may also occur in brain areas other than those responsible for the antidepressant action. For instance, at a structural level, positron

Table 3. Effectiveness of antidepressant treatments in three anxiety disorders

	Generalized anxiety	Panic	Social phobia
Tricyclic antidepressant drugs	+++	+++	0
Selective 5-HT reuptake inhibitors	+++	+++	+++
Monoamine oxidase inhibitors	+++	+++	+++
$5\text{-}HT_{1A}$ agonists	+++	0	+
Electroconvulsive shocks	0	N.D.	N.D.

+++, produces a remission in at least half the patients; +, may be useful in some cases; 0, no therapeutic benefit; N.D., not determined.

emission tomography scan as well as structural studies have consistently identified abnormalities in the parahippocampal region in panic disorder patients (see BAXTER 1995). At a neurochemical level, it is interesting to note that the shortest active form of cholecystokinin, the four amino acid-long peptide, when injected intravenously in human subjects immediately triggers a classical panic attack (DE MONTIGNY 1989; BRADWEJN et al. 1991). Yet, cholecystokinin terminals in the brain are endowed with 5-HT$_3$ receptors and a subpopulation of 5-HT$_3$ receptors modulating the release of 5-HT was reported to be desensitized following long-term administration of an SSRI (PAUDICE and RAITERI 1991; BLIER and BOUCHARD 1994). Interestingly, the panicogenic response to cholecystokinin is markedly attenuated following long-term treatment with the TCA imipramine or the SSRI fluvoxamine (BRADWEJN and KOSZYCKI 1994; VAN MEGEN et al. 1996).

II. Generalized Anxiety Disorder

Generalized anxiety disorder is characterized by excessive anxiety and worry that occurs most days, for at least 6 months, about a number of events and activities. The worry is difficult to control by the individual. The anxiety and worry must be accompanied by at least three of the following symptoms: restlessness, being easily fatigued, difficulty concentrating, irritability, muscle tension, or perturbed sleep. The lifetime prevalence of this disorder is about 5%. In practice, however, it is relatively rare to see patients with a pure form of generalized anxiety disorder. There is commonly an associated disorder, such as panic disorder.

Several antidepressant drugs are equally beneficial in generalized anxiety disorder as in panic disorder. Indeed, SSRIs and TCA drugs are as effective as benzodiazepines in this disorder (KHAN et al. 1986; LAWS et al. 1990). However, unlike in panic disorder, 5-HT$_{1A}$ agonists have been shown to be as effective as benzodiazepines in the generalized anxiety disorder (PECKNOLD 1994). The prototypical 5-HT$_{1A}$ agent buspirone was in fact shown to be an effective anxiolytic agent before 5-HT$_{1A}$ binding sites were identified (GOLDBERG and FINNERTY 1979). The full therapeutic effect of SSRIs, TCAs, and of 5-HT$_{1A}$ agonists requires sustained administration and their full therapeutic response may not be expected before 4–6 weeks, as is the case for their use in depression (ROBINSON et al. 1990). On the one hand, benzodiazepines relieve patients quite rapidly, most likely due to their anxiolytic as well as to their additional and immediate sedative and myorelaxant properties. On the other hand, antidepressant drugs rapidly produce undesirable side effects. However, the adaptative changes on the serotoninergic and/or noradrenergic system they produce render their discontinuation much less problematic, as is the case following their use in depression, than that of benzodiazepines which often produce rebound phenomena. For these reasons, antidepressant drugs are considered first-line treatment for the long-term treatment of the generalized anxiety disorder.

III. Social Phobia

The main feature of social phobia is a marked and persistent fear of social or performance situations in which embarrassment may occur. Exposure to the social or performance situation almost invariably provokes immediate anxiety. Most often the situation is avoided, although it is sometimes endured with marked distress. As for most related conditions, the diagnosis is appropriate only if the avoidance, fear, or anxious anticipation of encountering the social or performance situation interferes significantly with the person's daily routine, occupational functioning, social life, or if the person is markedly distressed about the phobia. The lifetime prevalence of social phobia ranges from about 3% to 13%. Obviously, the prevalence may vary according to the threshold used to make the diagnosis.

The validity of this diagnosis may be challenged given the major influence that social and psychological factors may exert in the development of this condition. However, it is interesting to note that SSRIs and MAOIs have been demonstrated in placebo-controlled studies to be effective in the treatment of this disorder (VAN VLIET et al. 1994; KATZELNICK et al. 1995). Given that the two latter families of agents do not exert an immediate anxiolytic effect, as is the case with benzodiazepines, it may be argued that they gradually compensate or rectify a biological anomaly that nevertheless may have developed secondary to aversive conditioning. It is also important to mention that TCA drugs are not useful in this disorder (Table 3).

Taken together, the therapeutic observations summarized in this section on anxiety disorders clearly indicate that the neurobiological substratum for the beneficial effects of TCA agents, SSRIs, 5-HT$_{1A}$ agonists, and MAOIs varies from one disorder to another. In the case of OCD and panic disorder, one may assume that modifications in the function of the 5-HT system in specific brain structures implicated in mediating these disorders explain at least in part the therapeutic effects of the above-mentioned drugs. However, it is presently not possible to make this kind of inference with respect to generalized anxiety disorder and social phobia, as structural and positron emission tomography scan studies have not yet identified specific brain regions involved in these two conditions. Nevertheless, the clinical observations that, for instance, 5-HT$_{1A}$ agonists are useful in generalized anxiety disorder but not in panic disorder and that the TCA imipramine is beneficial in panic disorder but not in OCD clearly indicate that different 5-HT receptor subtypes are implicated in these therapeutic actions.

E. Schizophrenia

Schizophrenia is a chronic psychosis characterized by a mixture of positive and negative signs and symptoms that have been present for a significant portion of time for at least one month period, with some signs of the disorder persisting for more than 6 months. Positive symptoms include delusions, hallucinations, and disorganized speech and behavior. Delusions are false beliefs involving a misinterpretation of perceptions or experiences, whereas hallucinations are false perceptions in the absence of cues. Negative symptoms include affective flattening and decreased fluency and productivity not only of thought and speech, but also in the initiation of goal-directed behaviors. These signs and symptoms must be of sufficient intensity to cause marked social or occupational dysfunction. The lifetime prevalence of schizophrenia is about 1%.

Until recently, the cornerstone of the treatment of schizophrenia was the blockade of dopaminergic receptors. Classical or typical antipsychotic medications, which are potent dopaminergic D_2 receptor blockers, are very effective in attenuating the positive symptoms but have generally little impact on the negative symptoms. Clozapine is considered as atypical antipsychotic because it does not produce extrapyramidal side effects (*i.e.*, rigidity, bradykinesia, tremors) like all classical antipsychotics; it also has been demonstrated to be effective in a significant proportion of resistant cases, and it has a beneficial effect on the negative symptoms (MELTZER 1995). The latter effect is presently thought to result at least in part from the activity of clozapine at one or more subtypes of 5-HT receptors. Indeed, clozapine has an affinity that may range from moderate to high for 5-HT_{1A}, 5-HT_{2A}, 5-HT_{2C}, 5-HT_3, 5-HT_6 and 5-HT_7 binding sites. Until recently, major research endeavors have been centered on 5-HT_2 receptors. Several new antipsychotic drugs were developed with both dopaminergic D_2 and 5-HT_{2A} antagonistic properties. Risperidone is the prototypical agent of this new class of antipsychotics. However, there is no general consensus on the superiority of this drug over classical antipsychotics, either with regard to its efficacy in resistant cases or the lack of production of extrapyramidal side effects at higher doses within its therapeutic range. Olanzapine is a newly commercialized $D_2/D_4/5\text{-HT}_2$ antagonist with documented superiority over classical antipsychotics. However, there are no clozapine-controlled studies examining the effectiveness of this new class of agents. It is noteworthy that a selective 5-HT_{2A} antagonist (MDL 100,907) has been reported to produce a depolarization block of dopamine neurons (a complete suppression of firing resulting from the increase of the resting membrane potential) of the ventral tegmental area (A10) but not of those of the substantia nigra (A9; SORENSEN et al. 1993). The latter property is generally thought to be related to the ability to produce an antipsychotic effect without inducing extrapyramidal side effects (FREEMAN and BUNNEY 1987). This is because the atypical antipsychotic drug clozapine inactivates A10 but not A9 dopamine neurons and does not produce extrapyramidal side effects, whereas all the classical antipsychotic drugs inactivate both populations of dopamine neurons and generally require concomitant anti-parkinsonian medication. It will, therefore, be of interest to determine whether MDL 100,907 is endowed with antipsychotic properties with regard to the contribution of the antagonism of 5-HT_{2A} receptors to the antipsychotic effect. A contribution of 5-HT receptors in the attenuation of negative symptoms of schizophrenia could not be ruled out even if MDL 100 907 were devoid of antipsychotic effect. Indeed, recent reports indicate that negative symptoms may be attenuated by using antidepressant drugs, in particular SSRIs (SIRIS et al. 1991; SILVER and NASSAR 1992; GOFF et al. 1995).

F. Conclusion

A better understanding of the role of the 5-HT system in the therapeutic effects of several psychotropic medications has led to major improvements in the treatment of psychiatric disorders. The overall impact has already been important considering the combined prevalence of at least 10% for the five types of disorders reviewed in this chapter, taking into account comorbidity. There are now medications with much more tolerable side effect profiles than those of the TCA drugs and irreversible MAOIs that are useful in several disorders. The onset of the antidepressant action of SSRIs has been considerably shortened for a significant percentage of depressed patients, and the use of potentiating strategies in treatment-resistant patients has remarkably improved treatment outcome. There remains, however, a great deal of work to be done. For instance, identifying predictive factors for the response to the different classes of agents acting on specific 5-HT receptor subtypes would help shorten the time to response with the drugs that are already available. A better knowledge of the pharmacological profile of the 5-HT receptor subtypes present in the brain structures playing an important role in mediating symptoms of particular diseases should lead to the development of superior pharmacological agents. Ongoing research in the field of the pharmacology of 5-HT should thus help decrease morbidity and the suicide rates related to psychiatric disorders.

References

American Psychiatric Association (1994) Diagnostic and Statistical Manual of Mental Disorders, Fourth Edition. American Psychiatric Association, Washington, DC

Artigas F, Perez V, Alvarez E (1994) Pindolol induces a rapid improvement of depressed patients with serotonin reuptake inhibitors. Arch Gen Psychiatry 51:248–251

Artigas F, Romero L, de Montigny C, Blier P (1996) Accelerated effect of selected antidepressant drugs in combination with 5-HT$_{1A}$ antagonists. Trends Neurosci 19:378–383

Baumann P, Nil R, Souche A, Montaldi S, Baettig D, Lambert S, Uehlinger C, Kasas A, Amey M, Jonzier-Perey M (1996) A double-blind, placebo-controlled study of citalopram with and without lithium in the treatment of therapy-resistant depressive patients: a clinical, pharmacokinetic, and pharmacogenetic investigation. J Clin Psychopharmacol 16:307–314

Baxter Jr, LR (1995) Neuroimaging studies of human anxiety disorders. In: Bloom FE, Kupfer D (eds) Psychopharmacology: the fourth generation of progress. Raven, New York, pp 1287–1300

Benkelfat C, Murphy DL, Zohar J, Hill JL, Grover G, Insel TR (1989) Clomipramine in obsessive-compulsive disorder: further evidence for a serotonergic mechanism of action. Arch Gen Psychiatry 46:23–28

Blier P, Bergeron R (1995) Effectiveness of pindolol with selected antidepressant drugs in the treatment of major depression. J Clin Psychopharmacol 15:217–222

Blier P, Bergeron R (1996) Sequential administration of augmentation strategies in treatment-resistant obsessive-compulsive disorder: preliminary findings. Int Clin Psychopharmacol 11:37–44.

Blier P, Bouchard C (1994) Modulation of 5-HT release in the guinea-pig brain following long-term administration of antidepressant drugs. Br J Pharmacol 113:485–495

Blier P, de Montigny C (1994) Current advances and trends in the treatment of depression. Trends Pharmacological Sci 15:220–226

Blier P, de Montigny C (1983) Electrophysiological studies on the effect of repeated zimelidine administration on serotonergic neurotransmission in the rat. J Neurosci 3:1270–1278

Blier P, de Montigny C (1985a) Serotonergic but not noradrenergic neurons in rat CNS adapt to long-term treatment with monoamine oxidase inhibitors. Neuroscience 16:949–955

Blier P, de Montigny C (1985b) Short-term lithium administration enhances serotonergic neurotransmission: electrophysiological evidence in the rat CNS. Eur J Pharmacol 113:69–77

Blier P, de Montigny C (1987) Modifications of 5-HT neuron properties by sustained administration of the 5-HT$_{1A}$ agonist gepirone: electrophysiological studies in the rat brain. Synapse 1:470–480

Blier P, de Montigny C, Azzaro AJ (1986) Modification of serotonergic and noradrenergic neurotransmission by repeated administration of monoamine oxidase inhibitors: Electrophysiological studies in the rat CNS. J Pharmacol Exp Ther 227:987–994

Blier P, Bergeron R, de Montigny C (1997) Selective activation of postsynaptic 5-HT$_{1A}$ receptors produces a rapid antidepressant response. Neuropsychopharmacology 16:333–338

Bradwejn J, Koszycki D, Shriqui C (1991) Enhanced sensitivity to cholecystokinin tetrapeptide in panic disorder. Arch Gen Psychiatry 48:603–610

Bradwejn J, Koszycki D (1994) Imipramine antagonism of the panicogenic effects of cholecystokinin tetrapeptide in panic disorder patients. Am J Psychiatry 151:261–263

Chaput Y, de Montigny C, Blier P (1986) Effects of a selective 5-HT reuptake blocker, citalopram, on the sensitivity of 5-HT autoreceptors: electrophysiological studies in the rat. Naunyn-Schmiedeberg's Arch Pharmacol 333:342–348

Clerc GE, Ruimy P, Verdeay-Pailles J, on behalf of the Venlafaxine French Inpatient Study Group (1994) A double-blind comparison of venlafaxine and fluoxetine in patients hospitalized for major depression and melancholia. Int Clin Psychopharmacol 9:138–143

de Montigny C (1984) Electroconvulsive shock treatments enhance responsiveness of forebrain neurons to serotonin. J Pharmacol Exp Ther 228:230–234

de Montigny C (1989) Cholecystokinin tetrapeptide induces panic-like attacks in healthy volunteers: preliminary findings. Arch Gen Psychiatry 46:511–517

de Montigny C (1994) Lithium addition in treatment-resistant depression. Int Clin Psychopharmacol 9 [Suppl 2]:31–35

de Montigny C, Aghajanian GK (1978) Tricyclic antidepressants: long-term treatment increases responsivity of rat forebrain neurons to serotonin. Science 202:1303–1306

de Montigny C, Cournoyer G, Morissette R, Langlois R, Caillé G (1983) Lithium carbonate addition in tricyclic antidepressant-resistant unipolar depression. Arch Gen Psychiatry 40:1327–1334 de Montigny C, Chaput Y, Blier P, (1993) Classical and novel targets for antidepressant drugs. Int Acad Biomed Drug Res 5:8–17

de Montigny C, Debonnel G, Bergeron R, St-André I, Blier P (1995) Venlafaxine in treatment-resistant depression: an open-label multicenter study. Am Col Neuropharmacol 34:158

Dinan TG, Barry S (1989) A comparison of electroconvulsive therapy with a combined lithium and tricyclic combination among depressed tricyclic nonresponders. Acta Psychiatr Scand 80:97–100

El Mansari M, Bouchard C, Blier P (1995) Alteration of serotonin release by selective serotonin reuptake inhibitors in the guinea pig orbitofrontal cortex: relevance to the treatment of obsessive-compulsive disorder. Neuropsychopharmacology 13:117–127

Freeman AS, Bunney BS (1987) Chronic neuroleptic effects on dopamine neuron activity: a model for predicting therapeutic efficacy and side effects. In: Dahl SG, Gream LF, Paul SM, Potter WZ (eds) Clinical pharmacology in psychiatry. Selectivity in psychotropic drug action – promises or problems? Springer, Berlin Heidelberg New York (Psychopharmacology Series, vol 3)

Glowinski J, Axelrod J (1964) Inhibition of uptake of tritiated noradrenaline in the intact rat brain by imipramine and structurally related compounds. Nature 204:1318–1319

Goff DC, Midha KK, Sarid-Segal O, Hubbard JW, Amico E (1995) A placebo-controlled trial of fluoxetine added to neuroleptic in patients with schizophrenia. Psychopharmacology 117:417–423

Goldberg HL, Finnerty RJ (1979) The comparative efficacy of buspirone and diazepam in the treatment of anxiety. Am J Psychiatry 136:1184–1187

Greenberg B, Benjamin J, Murphy D (1994) Metergoline biphasically enhances, then antagonizes, fluoxetine's therapeutic effect in obsessive-compulsive disorder. Neuropsychopharmacology 10(suppl 3):256S

Greist J, Chouinard G, DuBoff, E Halaris A, Won Kim S, Koran, L, Liebowitz M, Lydiard RB, Rasmussen S, White K, Sikes C (1995) Double-blind parallel comparison of three dosages of sertraline and placebo in outpatients with obsessive-compulsive disorder. Arch Gen Psychiatry 52:289–295

Guelfi JD, White C, Hackett D, Guichoux JY, Magni G (1995) Effectiveness of venlafaxine in patients hospitalized for major depression an melancholia. J Clin Psych 56:450–458

Hartig PR, Hoyer D, Humphrey PPA, Martin GR (1996) Alignment of receptor nomenclature with the human genome: classification of 5-HT$_{1B}$ and 5-HT$_{1D}$ receptor subtypes. Trends Pharmacol Sci 17:103–105

Heninger GR, Charney DS, Sternberg DE (1983) Lithium carbonate augmentation of antidepressant treatment. Arch Gen Psychiatry 40:1335–1342

Insel TR (1992) Toward a neuroanatomy of obsessive-compulsive disorder. Arch Gen Psychiatry 49:739–744

Jacobsen FM (1991) Possible augmentation of antidepressant response by buspirone. J Clin Psychiatry 52:217–220

Jenike MA (1993) Obsessive-compulsive disorder: efficacy of specific treatment as assessed by controlled trials. Psychopharmacol Bull 29:487–499

Joffe RT, Schuller DR (1993) An open study of buspirone augmentation of serotonin reuptake inhibitors in refractory depression. J Clin Psychiatry 54:269–271

Joffe RT, Singer W, Levitt AJ, MacDonald C (1993) A placebo-controlled comparison of lithium and triiodothyronine augmentation of tricyclic antidepressants in unipolar refractory depression. Arch Gen Psychiatry 50:387–393

Katzelnick DJ, Kobak KA, Greist JH, Jefferson JW, Mantle JM, Serlin RC (1995) Sertraline in social phobia: a double-blind, placebo-controlled crossover study. Am J Psychiatry 152:1368–1371

Khan RJ, McNair DM, Lipman RS, Covi L, Rickels K, Downing R, Fisher S, Frankenthaler LM (1986). Imipramine and chlordiazepoxide in depressive and anxiety disorders. II Efficacy in anxious outpatients. Arch Gen Psych 43:79–85

Klein DF (1964) Delineation of two drug-response anxiety syndromes. Psychopharmacologia 5:397–408

Laws D, Ashford JJ, Anstee JA (1990) A multicentre double-blind comparative trial of fluvoxamine versus lorazepam in mixed anxiety and depression treated in general practice. Acta Psychiatr Scand 81:185–189

Liebowitz MR, Quitkin FM, Steward JW, McGrath PJ, Harrison WM, Markowitz JS, Rabkin JG, Tricamo E, Goetz DM Klein DF (1988) Antidepressant specificity in atypical depression. Arch Gen Psychiatry 45:129–137

Maes M, Vandoolaeghe E, Desnyder R (1996) Efficacy of treatment with trazodone in combination with pindolol or fluoxetine in major depression. J Affect Dis 41:201–210

Meltzer HY (1995) Atypical antipsychotic drugs. In: Bloom FE, Kupfer DJ (eds) Psychopharmacology: the fourth generation of progress. Raven, New York, pp 1277–1286

Menkes DB, Aghajanian GK, McCall RB (1980) Chronic antidepressant treatment enhances alpha-adrenergic and serotonergic responses in the facial nucleus. Life Sci 27:45–55

Montgomery SA (1980) Clomipramine in obsessional neurosis: a placebo controlled trial. Pharm Med 1:189–192

Montgomery SA (1994) Pharmacological treatment of obsessive-compulsive disorder. In: Hollander E, Zohar J, Marazziti D, Olivier B (eds) Current insights in obsessive disorder. Wiley, New York, pp 215–226

Montgomery SA (1995) Rapid onset of action of venlafaxine. Int Clin Psychopharmacol 10 [Suppl 2]:21–27

Moret C, Briley M (1990) Serotonin autoreceptor subsensitivity and antidepressant activity. Eur J Pharmacol 180:351–356

Nelson JC, Mazure CM, Bowers MB, Jaltow PI (1991) A preliminary, open study of the combinatfion of fluoxetine and desipramine for rapid treatment of major depression. Arch Gen Psychiatry 48:303–307

Nierenberg AA, Feighner JP, Rudolph R, Cole JO, Sullivan J (1994) Venlafaxine for treatment-resistant unipolar depression. J Clin Psychopharmacol 14:419–423

Paudice P, Raiteri M (1991) Cholecystokinin release mediated by 5-HT$_3$ receptors in rat cerebral cortex and nucleus accumbens. Br J Pharmacol 103:1790–1794

Pecknold JC (1994) Serotonin 5-HT$_{1A}$ agonists. A comparative review. CNS Drugs 2(3):234–251

Piñeyro G, Blier P, Dennis T, de Montigny C (1994) Desensitization of the neuronal 5-HT carrier following its long-term blockade. J Neurosci 14:3036–3047

Robinson DS, Rickels K, Feighner J, Fabre LF, Gammans RE, Shrotriya RC, Alms CR, Andary JJ, Messina ME (1990) Clinical effects of the 5-HT$_{1A}$ partial agonists in depression: a composite analysis of buspirone in the treatment of depression. J Clin Psychopharmacol 10:67S–76S

Romero L, Bel N, Artigas F, de Montigny C, Blier P (1996) Effect of pindolol at pre- and postsynaptic 5-HT$_{1A}$ receptors: in vivo microdialysis and electrophysiological studies in the rat brain. Neuropsychopharmacology 15:349–360

Silver H, Nassar A (1992) Fluvoxamine improves negative symptoms in treated chronic schizophrenia: an add-on double-blind, placebo-controlled study. Biol Psychiatry 31:698–704

Siris SG, Bermanzohn PC, Gonzalez A, Mason SE, White CV, Shuwall MA (1991) The use of antidepressants for negative symptoms in a subset of schizophrenic patients. Psychopharmacol Bull 27:331–335

Sorensen SM, Kehne JH, Fadayel GM, (1993) Characterization of the 5-HT$_2$ receptor antagonist MDL 100907 as a putative atypical antipsychotic: behavioral, electrophysiological and neurochemical studies. J Pharmacol Exp Ther 266:684–691

Tollefson GD, Ramprey Am, Potvin JH, Henike MA, Rush AJ, Dominquez RA, Koran LM, Shear MK, Goodman W, Genduso LA (1994) A multicenter investigation of fixed-dose fluoxetine in the treatment of obsessive-compulsive disorder. Arch Gen Psychiatry 51:559–567

Tome MB, Isaac MT, Harte R, Holland C (1997) Paroxetine and Pindolol: A randomized trial of serotonergic autoreceptor blockade in the reduction of antidepressant latency. Int Clin Psychopharmacol (in press)

van Megen HJGM, Westenberg HGM, den Boer JA, Kahn RS (1996) Cholecystokinin in anxiety. Eur Neuropsychopharmacol 6:263–280

van Vliet IM, den Boer JA, Westenberg HGM (1994) Psychopharmacological treatment of social phobia: A double-blind, placebo-controlled study with fluvoxamine. Psychopharmacology 115:128–134

West ED, Dally PJ (1959) Effects of iproniazid in depressive syndromes. Br Med J 5163:1491–1494

Subject Index